Advanced Theoretical and Numerical Electromagnetics

The ACES Series on Computational and Numerical Modelling in Electrical Engineering

Andrew F. Peterson, PhD – Series Editor
The volumes in this series will encompass the development and application of numerical techniques to electrical and electronic systems, including the modelling of electromagnetic phenomena over all frequency ranges and closely related techniques for acoustic and optical analysis. The scope includes the use of computation for engineering design and optimization, as well as the application of commercial modelling tools to practical problems. The series will include titles for senior undergraduate and postgraduate education, research monographs for reference, and practitioner guides and handbooks.

Titles in the Series
K. Warnick, **"Numerical Methods for Engineering,"** 2010.
W. Yu, X. Yang and W. Li, **"VALU, AVX and GPU Acceleration Techniques for Parallel FDTD Methods,"** 2014.
A.Z. Elsherbeni, P. Nayeri and C.J. Reddy, **"Antenna Analysis and Design Using FEKO Electromagnetic Simulation Software,"** 2014.
A.Z. Elsherbeni and V. Demir, **"The Finite-Difference Time-Domain Method in Electromagnetics with MATLAB® Simulations, 2nd Edition,"** 2015.
M. Bakr, A.Z. Elsherbeni and V. Demir, **"Adjoint Sensitivity Analysis of High Frequency Structures with MATLAB®,"** 2017.
O. Ergul, **"New Trends in Computational Electromagnetics,"** 2019.
D. Werner, **"Nanoantennas and Plasmonics: Modelling, design and fabrication,"** 2020.
K. Kobayashi and P.D. Smith, **"Advances in Mathematical Methods for Electromagnetics,"** 2020.
S. Roy, **"Uncertainty Quantification of Electromagnetic Devices, Circuits, and Systems,"** 2021.

Advanced Theoretical and Numerical Electromagnetics

Volume 1: Static, stationary and time-varying fields

Vito Lancellotti

The Institution of Engineering and Technology

Published by SciTech Publishing, an imprint of The Institution of Engineering and Technology, London, United Kingdom

The Institution of Engineering and Technology is registered as a Charity in England & Wales (no. 211014) and Scotland (no. SC038698).

First published 2021

The Institution of Engineering and Technology
Michael Faraday House
Six Hills Way, Stevenage
Herts, SG1 2AY, United Kingdom

www.theiet.org

British Library Cataloguing in Publication Data
A catalogue record for this product is available from the British Library

ISBN 978-1-83953-564-2 (Volume 1 hardback)
ISBN 978-1-83953-565-9 (Volume 1 PDF)
ISBN 978-1-83953-568-0 (Volume 2 hardback)
ISBN 978-1-83953-569-7 (Volume 2 PDF)
ISBN 978-1-83953-570-3 (2 Volume set hardback)

Typeset in India by MPS Limited
Printed in the UK by CPI Group (UK) Ltd, Croydon

To the memory of my parents,
Loreto and Angelina,
with deep gratitude

Contents

List of figures

List of tables

List of examples

About the author

Vito Lancellotti received the laurea degree (M.Sc.) with honors in Electrical Engineering and the Ph.D. degree in Electronics and Communications from Politecnico di Torino (Italy) in 1995 and 1999. In early 1999, he joined Telecom Italia Lab in Torino as a Senior Researcher and was involved in projects concerning TCP/IP and ATM networks. In June 2000, as a Senior Researcher, he joined the Milan-based subsidiary of Corning (now Avanex), where he worked on the design of broadband electro-optic lithium-niobate modulators and optical waveguides. From 2002 to 2008, he served as a Research Fellow and lecturer in the Department of Electrical Engineering of the Politecnico di Torino, where he contributed to developing the TOPICA code, a tool devised for the analysis and design of plasma-facing antennas utilized in magnetically controlled nuclear fusion. In 2005, he was appointed Visiting Scientist at the Massachusetts Institute of Technology (Cambridge, MA) and, in 2007, conducted research at the Max-Planck-Institut für Plasmaphysik (Garching, Germany). From April 2008 to August 2017, he was with the Faculty of Electrical Engineering of the Technical University of Eindhoven (Netherlands), where he served as a Senior Researcher and as an Assistant Professor. His specialties and interests include theoretical, applied and computational electromagnetics; antennas; plasma sources and antennas; plasma-facing antennas for nuclear fusion; biological effects of electromagnetic fields; metallic and dielectric waveguides; spectral methods and the Wiener-Hopf technique.

Foreword

The ACES/IET book series on Computational Electromagnetics and Engineering fosters the development and application of numerical techniques to electrical systems, including the modeling of electromagnetic phenomena over all frequency ranges and closely-related techniques for acoustic and optical analysis. While focusing on research monographs promoting computational practice for engineering design and optimization, the series also includes titles for undergraduate and graduate education.

Modern curricula in electromagnetic (EM) fields, especially at the graduate level, require a front-end course that efficiently reviews undergraduate materials and fills in gaps, since most students in undergraduate electrical engineering programs today are only required to take one course on fundamentals. The course must then progress to cover the advanced concepts needed to introduce the students to classical methods of solving boundary value problems, approximate asymptotic techniques, and modern numerical techniques. Along the way, students need to learn about material interactions and metamaterials, and a wide range of applications including microwave devices, antennas, EM propagation and scattering problems, modern communications, and bio-electromagnetics.

The present textbook, designed to be the principal source for a one- or two-course sequence on graduate-level EM, does an excellent job of meeting the needs of such a curriculum. This book begins by reviewing the fundamentals of static fields and EM waves, then progresses through more advanced concepts for time-varying fields, including scalar and dyadic Green's functions and source-field relations, surface and volume equivalence principles, and the spectral representation of EM fields. Dispersive materials are discussed at length. The text concludes by addressing integral equations and their numerical solution using the method of moments. Of particular interest is the inclusion of mathematical details seldom found in engineering texts. Appendices review topics such as complex variables, functional analysis, and dyads, all of which are needed for a solid foundation in EM theory.

This book should prove useful for both students and professionals with interest in a wide variety of EM applications. Dr. Lancellotti's approach of combining fundamental EM theory and computational techniques make this an efficient, self-contained resource.

Series Editor Andrew F. Peterson
School of ECE, Georgia Institute of Technology
Atlanta, GA USA
August 3, 2021

Preface

It appears to me, therefore, that the study of electromagnetism in all its extent has now become of the first importance as means of promoting the progress of science.
— James C. Maxwell

In the early spring of 2008, I gladly found myself joining the Electromagnetics Group of the Faculty of Electrical Engineering of the Technical University of Eindhoven, The Netherlands. From the very beginning, in addition to conducting research on integral equations and domain decomposition methods, I was involved in the lectures of an elective graduate course that, at the time, was known as 'Mathematics for Electromagnetism' and had been initiated, a few years earlier, by Prof. Antonius G. Tijhuis, head of the Electromagnetics Group. Around 2011 I had the pleasure and the honor to take over the course and was asked to redesign it in the context of a general overhauling that the electrical-engineering study program was concurrently undergoing.

By and large, this book is an outgrowth of the lectures I delivered from 2011 till 2017 for the said course which, at some point along the way, was renamed 'Advanced Electromagnetics and Moments Methods' for two reasons. First and foremost, in the meantime, integral equations and the Method of Moments had found their place in the course program, and this feature had to be properly advertised to prospective students. Secondly, experience indicated that hinting at mathematics inexplicably seemed to scare away the less-bold engineering students who, as a consequence, would not register for the course. That might have been so, though one thing is certain, i.e., the organization of the book, for the most part, closely mirrors the logical order with which the topics were presented to the class because I built the material starting from the slides I had prepared and used for the lectures over the years. Nevertheless, being reasonably unconstrained by criteria of time or space, the material collected in the book exceeds by far the contents of the original set of lectures.

As was expected of the brave students who dared to attend the course and ultimately enjoyed it — this I can say on the grounds of the positive anonymous feedback I received — in order to benefit from this book the Reader should have an undergraduate knowledge of classical electromagnetism and should know calculus reasonably well. Starting with the Lorentz force, the first part of the book covers classical electromagnetic theory, namely, Maxwell's equations, the limits of electrostatic and magneto-stationary fields, electromagnetic theorems, wave propagation, time-varying fields and potentials, integral representations and equivalence principles, fields in cavities and waveguides, and dispersion in material media. The second part, computational electromagnetics, deals with surface and volume integral equations and hybrid formulations in the frequency domain, as well as the solution of those equations through the Method of Moments.

I shall be the first to admit that I got carried away and perhaps included more mathematical details than would befit an electrical engineering book, advanced as it may be. And yet, it seems to me that the final result is absolutely in keeping with the original name and intent of the course which inspired me. Indeed, in an effort to make the book as self-contained as possible, detailed proof of most formulas and theorems is given, either in the scope of the main text or in dedicated appendices.

This applies, e.g., to the convergence of singular integrals that involve Green functions as well as to the asymptotic behavior of integrals at infinity.

The smoothness properties of the electrostatic volume, double- and single-layer potentials (a topic which, owing to time limitations was not part of the set of lectures for the course) are examined extensively in Sections 2.8-2.10, with due regard to the delicate matter of interchanging integrals and derivatives, and the check *a posteriori* that the integral representation does solve the Poisson equation. These results are also relevant to the smoothness of the integral representation of the stationary magnetic potential (Section 5.1) and the electrodynamic potentials (Section 9.1) since all of them exhibit the same singular character as the electrostatic potential. Still, I reckon that the Reader may safely skip Sections 2.8-2.10 and the related Appendices F and G without compromising the full understanding of the remaining topics.

Since Dirac delta distributions have the unquestionable merit of making it amazingly simple to include singular sources in the equations (which is why I used them a lot during the lectures), I covered the topic in Appendix C. However, in the text, I favored the classical approach — based on the isolation of the singular point with a sphere — and employed Dirac distributions sparingly, and only when avoiding them would have made the exposition too cumbersome. For instance, the classical approach is used for the derivation of integral formulas for the electrostatic potential in Section 2.7, the magnetic vector potential in Section 5.1, and the electrodynamic potentials in Section 9.1. By contrast, a not-so-trivial problem, in which surface delta distributions ultimately make their appearance, consists of computing the electromagnetic field produced by a point charge that moves with a uniform velocity larger than the speed of light in the underlying medium (Section 9.2). The free-space time-dependent three-dimensional Green function provides another typical example because it is a singular outward-going spherical wave and, as such, can only be expressed with the aid of a surface delta (Section 8.7).

On a related score, the potentials generated by surface charges and currents are often obtained by expressing such sources with the aid of suitable surface delta distributions and by invoking the properties thereof to transform volume integrals into surface ones. Contrariwise, in the text, I described the direct derivation based on homogeneous Poisson or Helmholtz equations supplemented with the relevant matching conditions, which naturally allow bringing the surface sources into the picture (Sections 3.1, 5.2, and 9.3).

The students of 'Advanced Electromagnetics and Moments Methods' would see dyads and dyadics for the first time at the beginning of the course, when I introduced the linear constitutive relationships in anisotropic media. On that occasion, I usually managed to deflect questions and put off the topic by discounting dyadic permittivity, permeability, and conductivity as nothing more than matrices. At a later stage, I explained the need for entities more general than vectors before I discussed the multipole expansion of the electrostatic scalar potential since the quadrupole moment is a dyadic. The short *intermezzo* also laid the groundwork for the dyadic Green functions, which are singular dyadic fields. Yet, the interested Reader might want to browse Appendix E upfront because I expanded the early discussion of the energy balance in Section 1.10 by deriving, in Sections 1.11 and 1.12, the conservation laws for linear and angular momentum, both of which involve dyadic fields.

At some time or the other, the course program also covered the topics of wave propagation in one-dimensional periodic structures and the spectral representation of fields and sources in situations where the geometry is invariant in one spatial direction. These subjects are dealt with in Sections 11.3 and 11.4, where proofs of the Floquet theorem and of the image principle for time-harmonic fields are given. Then again, since the modal expansion in cavities and hollow-pipe cylindrical waveguides is a spectral representation as well, I decided to include these topics in Sections 11.1 and 11.2 for the sake of completeness. In particular, the waveguide problem is reduced to that of finding waves along fictitious transmission lines by means of the approach introduced by N. Marcuvitz and J. Schwinger.

The specific choice of subjects for Chapters 13-15 (namely, integral equations and Method of Moments) was motivated, for the most part, by the research I conducted at the Technical University of Eindhoven. In particular, since I worked on domain decomposition methods for surface integral equations, it seemed almost natural to expose the students to that topic, which enabled them to undertake and carry out graduation projects in computational electromagnetics. Thus, integral equations for radiation and scattering problems are discussed at length, whereas the classical line integral equations for wire antennas are left out. Volume integral equations and hybrid techniques are also presented, as these subjects are in keeping with the idea of mathematically separating a problem into parts.

A popular tenet has it that explaining a topic to someone is a good, if not the best, way of learning and understanding it. Not only is this statement true, in my opinion, but it is also, in essence, the very reason why I decided to write this book. I was also spurred by the moderately selfish desire of producing a self-contained reference I can turn to for help whenever my memory fails me. In the process, I learned a great deal, because I was forced to ask myself questions and to come up with satisfactory answers, but more importantly, I had a lot of fun.

I wonder what would be of electromagnetics in one hundred years from now since, as I put these thoughts in writing, Maxwell's theory is already about one hundred and fifty years old and one of the most complete, at that. While it is indisputable that one need not master electromagnetics — or quantum field theory, for that matter — to use a mobile phone or a tv set, still, someone must have the knowledge to build those devices in the first place, right?

VITO LANCELLOTTI
Torino, Italy
November 25, 2021

Acknowledgements

A curious fact about this work is that, back in the spring of 2017, when I started putting the material together, I did not dream it would develop into being a 'real' book. However, now that *Advanced Theoretical and Numerical Electromagnetics* has indeed been published (and in two volumes, at that) it is my distinct pleasure to give well-deserved credit to all the people who, in different moments, either deliberately or unwittingly, played a role in making this book see the light of day.

Dr. Nicki Dennis, Commissioning Editor for Radar, Electromagnetics, and Signal Processing with IET Press, reacted enthusiastically to my book proposal, which I submitted for consideration in February 2021. She boldly believed in the potential of this project and patiently assisted me in the early stages of the publication process.

Mrs. Olivia Wilkins, Assistant Editor with IET Press, handled the manuscript submission and helped me stay on schedule by checking on my progress regularly, though unobtrusively. She addressed all of my concerns, especially about the nifty typesetting of the long and complicated equations found in this book.

Mrs. Nikki Tarplett, Senior Production Controller with IET Press, was my production contact within the IET. In particular, she made sure that the typesetting process ran smoothly.

Mr. Srinivasan N., Project Manager with MPS Ltd., India, was in charge of the final typesetting of the book. He handled all of my requests for amendments and changes in the proofs with considerable patience and professionalism.

The Reviewers of my book proposal devoted precious time to assessing this project. They all expressed favorable opinions on the usefulness of the subject, the organization of the book, and its potential audience. What is more, they were kind enough to vouch for my expertise in electromagnetics. One of the Reviewers even quipped he would look forward to reading the completed book, so I humbly hope this scientific work will live up to his optimistic expectations.

Prof. Andrew F. Peterson of Georgia Institute of Technology in Atlanta, GA, author of the book *Computational Methods for Electromagnetics* (IEEE Press, 1998) and Series Editor with IET Press, did me the honor of reviewing the manuscript and wrote an invaluable foreword which makes a convincing case for this work. His support, appreciation, and endorsement mean a lot to me and are gratefully acknowledged.

Prof. Antonius G. Tijhuis, Emeritus, literally provided the opportunity for this book when, in 2011, he entrusted me with the responsibility for the lectures of the graduate course 'Mathematics for Electromagnetism' at the Technical University of Eindhoven, NL, while he good-naturedly offered to serve as my teaching assistant. The slides of the 2016-2017 edition of said course are the skeleton on which this resource got fleshed out, as it were, in a three-year-long effort.

Prof. Renato Orta, Emeritus, was an excellent teacher of mine and also served as thesis adviser for both my graduation project and my doctoral dissertation. Under his authoritative guidance and thanks to his well-prepared lectures for the course 'Optical Circuits and Components' (which I attended in 1992), I developed a thorough and deep knowledge of waveguides, periodic structures, and spectral methods, all of which are described in Chapter 11.

Prof. Vito Daniele, Emeritus, an expert in the Wiener-Hopf method, was another great teacher of mine who, during his graduate course 'Advances in Electromagnetic Waves', used to joke that: 'Electromagnetics is a hard subject.' And yet, his inspiring lessons, which were jam-packed with mathematical derivations, were crystal clear (I still have and treasure the notes I took) and taught me how to apply the theory of complex functions to electromagnetics. I utilized these invaluable concepts, for example, when discussing the 2-D Green function (Section 7.6) and the Kramers-Krönig relations (Section 12.2).

Dr. Theodore Anderson, Haleakala Research and Development Inc., Brookfield, MA, author of the book *Plasma Antennas* (Artech House, 2011), kindly granted me permission to include pictures of his gaseous plasma antennas in Figure 13.23.

Dr. Davide Melazzi conducted part of his doctoral research on plasma engines for spacecraft propulsion as a guest Ph. D. student at the Technical University of Eindhoven, NL, in 2011. Our pleasant and productive cooperation led to the development of ADAMANT, a FORTRAN code for the numerical analysis of coupled plasma and antennas with integral equations and the Method of Moments. The mathematical and computational aspects of that electromagnetic problem inspired the contents of Sections 13.5.1 and 15.5. In particular, Dr. Melazzi contributed the colorful artist's impression of the plasma thruster reproduced, with permission, in Figure 13.21.

Mr. Anuar D. J. Fernandez-Olvera was one of the most brilliant students of my course 'Advanced Electromagnetics and Moments Methods'. In 2015, he conducted his graduation project, which entailed, for his part, the coding of a hybrid surface-integral-equation and wave-equation approach for the analysis of plasma antennas. The theoretical investigations stimulated by that activity constitute the subject of Sections 13.5.2, 15.6, and 15.7.

Last but not least, I would like to pay tribute to my beloved late parents, Loreto and Angelina. They helped me become the person and the scientist I am today by never failing to sustain me with unconditional love, constant encouragement, and unshakable practical and emotional support. Unfortunately, though, my father, an honest and hard-working man, never had a chance to know of my moving to Eindhoven and of my writing this book because he passed away in 2002. My elderly mother, strong as a lioness, passed away suddenly and unexpectedly, due to the complications from a severe COVID-19 infection, in early April of 2021, just a few weeks after I had excitedly broken the news to her that IET Press would publish my work. Therefore, this book is for my father and my mother: I wish they could see it, but, most of all, I hope that, wherever they are now, they may be proud.

Vito Lancellotti
Torino, Italy
November 25, 2021

Chapter 1
Fundamental notions and theorems

1.1 The electromagnetic field

Electric charge (physical dimension: coulomb, C) is as fundamental a property of ponderable matter as *mass* is [1, 2]. Moreover, electric charge comes in two types, i.e., it can be either *positive* or *negative*. As is probably known to the Reader, charges of opposite sign attract each other, whereas charges with like sign repel each other. However, not all elementary particles are charged, that is, possess a certain amount of charge, a case in point being the *neutron*. Charged particles are *discrete* physical entities which carry a well-defined amount of electric charge, the latter being an integer multiple of the charge of the *proton* $q_e > 0$ or the *electron* $-q_e < 0$, where $q_e = 1.6021 \cdot 10^{-19}$ C. Still, the charge carried by subatomic particles, such as *quarks*, is a multiple of $\pm q_e/3$.

The classical theory of electromagnetism [1, 3–18] is essentially concerned with the study of the interaction between charged particles and the *electromagnetic fields and waves* which they generate. The electromagnetic field comprises four three-dimensional *vector fields* which are function of position ($\mathbf{r} \in \mathbb{R}^3$) and time ($t \in \mathbb{R}$), namely, two entities of *intensity* and two entities of *quantity*, according to the classification proposed by A. Sommerfeld [12]. More specifically, the entities of intensity are:

- $\mathcal{E}(\mathbf{r}, t)$, the *electric field strength* or more simply the electric field, which carries physical dimension of volt per meter (V/m)
- $\mathcal{B}(\mathbf{r}, t)$, the *magnetic flux density* or magnetic induction with physical dimension of weber per square meter (Wb/m^2) or tesla (T)

whereas the entities of quantity are:

- $\mathcal{D}(\mathbf{r}, t)$, the *electric displacement field* or electric flux density or electric induction with physical dimension of coulomb per square meter (C/m^2)
- $\mathcal{H}(\mathbf{r}, t)$, the *magnetic field strength* or more simply the magnetic field, which carries physical dimension of ampere per meter (A/m).

Loosely speaking, $\mathcal{E}$ and $\mathcal{B}$ answer the question 'How strong is the field?', whereas $\mathcal{D}$ and $\mathcal{H}$ answer the question 'How much charge is there?'. The electric field and the magnetic induction are fundamental entities, and in fact they are sufficient to describe the electromagnetic phenomena in vacuum (free space). Conversely, the displacement vector and the magnetic field are secondary entities, as it were, related to the electric field and the magnetic induction. $\mathcal{D}$ and $\mathcal{H}$ play an important role in the analysis of electromagnetic phenomena in the presence of material media which admit a macroscopic description in terms of some global parameters (see Section 1.6 and Chapter 12).

The electromagnetic field is produced by *source fields* or *sources* for short, namely,

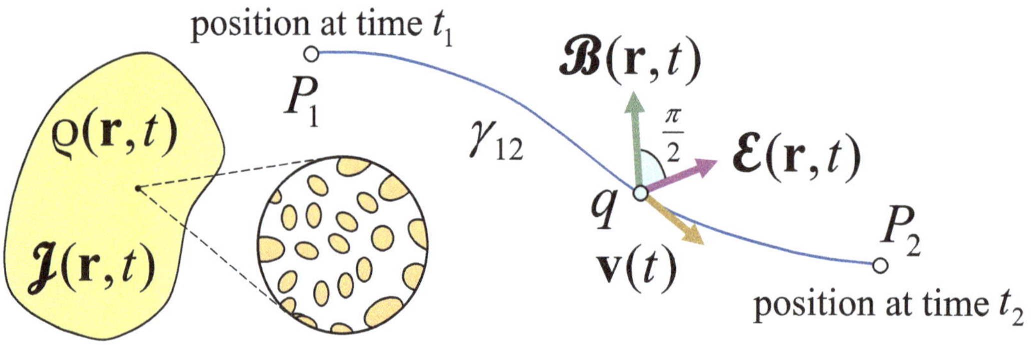

Figure 1.1 A region of space containing sources ϱ and $\mathcal{J}$, and a test point charge q which, through the Lorentz force (1.4), *'feels' the electromagnetic field produced by the sources.*

- $\varrho(\mathbf{r}, t)$, the impressed (volume) *electric charge density*, which carries physical dimension of coulomb per cubic meter (C/m^3)
- $\mathcal{J}(\mathbf{r}, t)$, the impressed (volume) *electric current density* with physical dimension of ampere per square meter (A/m^2)

and, as we shall see in Section 1.2.1, ϱ and $\mathcal{J}$ are directly related to $\mathcal{D}$ and $\mathcal{H}$, a link which supports the appellation of *entities of quantity*. Since electric charge exists in *lumps*, as recalled at the beginning, when we introduce ϱ and $\mathcal{J}$ we make the assumption that the charged particles involved are so numerous as to lend themselves to a collective *macroscopic* description. In other words, the whereabouts of each single particle is glossed over, and we actually think of the total charge as being *spread continuously* — though not necessarily *uniformly* — throughout a given region of space [7, Chapter 1], [15, Chapter 1], [18, Chapter 1], as is pictorially suggested in the left part of Figure 1.1. Accordingly, the *formal* definition of volume charge density reads

$$\varrho(\mathbf{r}, t) := \lim_{\Delta V \to 0^+} \frac{\Delta Q(t)}{\Delta V} \tag{1.1}$$

where $\Delta Q(t)$ is the amount of charge contained in a small spatial region with volume ΔV at time t.

The volume current density $\mathcal{J}(\mathbf{r}, t)$ tells us how much charge flows through a given point in space $\mathbf{r}$ at a specific time t as well as in which direction the charge is moving. To arrive at a formal definition of $\mathcal{J}$ we suppose that at time t a certain amount of charge Q moving with velocity $\mathbf{v} = v\hat{\mathbf{v}}$ crosses a small planar (mathematical) surface S oriented perpendicularly to $\hat{\mathbf{v}}$. At a later time $t' := t + \Delta t$ the same charge will have travelled a distance $v\Delta t$ and swept a small right cylinder of volume

$$\Delta V = v\, \Delta t\, \Delta A \tag{1.2}$$

with ΔA denoting the area of S. The total charge that during the time interval Δt has kept crossing S and piling up in the cylinder is $\Delta Q(t')$, thus the current density reads

$$\mathcal{J}(\mathbf{r}, t) = \mathcal{J}(\mathbf{r}, t)\hat{\mathbf{v}} := \lim_{\Delta A \to 0^+} \lim_{\Delta t \to 0^+} \frac{\Delta Q(t')}{\Delta A\, \Delta t}\hat{\mathbf{v}} = \lim_{\Delta V \to 0^+} \frac{\Delta Q(t')}{\Delta V}\mathbf{v} = \varrho(\mathbf{r}, t)\mathbf{v} \tag{1.3}$$

on account of (1.1) [18, Chapter 1].

As regards the physical dimensions of electromagnetic entities and sources we adopt the choices of the *Système International d'Unités* (SI), which was formerly known as Giorgi system or MKSA system (short for *meter*, *kilogram*, *second* and *ampere*, which are the fundamental units of length, mass, time and current) though other coherent sets of units are possible. A clear and comparative account on the topic can be found in the books by J. D. Jackson [7, Appendix] and K. Milton [19, Appendix].

The electromagnetic field is more than just a mathematical tool for describing the interaction between charged particles. Rather, being endowed with energy, momentum and even angular momentum [16, 17] — which are properties typically associated with ponderable matter — the electromagnetic field must be regarded as a physical reality in its own right. In fact, the one way we have to detect or measure the electromagnetic field consists of observing the effect that the electric field $\mathcal{E}$ and the magnetic intensity $\mathcal{B}$ do have on the motion of charged particles. This phenomenon is governed by the *Lorentz force* (physical dimension: newton, N) as [20, Section 4.3]

$$\mathcal{F}_{em}(\mathbf{r},t) := \underbrace{q\mathcal{E}(\mathbf{r},t)}_{\text{electric part}} + \underbrace{q\mathbf{v}(t)\times\mathcal{B}(\mathbf{r},t)}_{\text{magnetic part}} \tag{1.4}$$

where $\mathbf{v}(t)$ is the instantaneous velocity of the particle with charge q (see the right part of Figure 1.1). The collective force acting on a small finite amount of charge ΔQ confined in a region with volume ΔV reads

$$\Delta\mathcal{F}_{em}(\mathbf{r},t) := \Delta Q(t)\mathcal{E}(\mathbf{r},t) + \Delta Q(t)\mathbf{v}(t)\times\mathcal{B}(\mathbf{r},t) \tag{1.5}$$

and thus dividing through by ΔV and taking the limit $\Delta V \to 0^+$ yields the volume *density of Lorentz force* (physical dimension: newton per cubic meter, N/m^3) [20, pp. 106 and ff.]

$$\mathbf{f}_{em}(\mathbf{r},t) := \varrho(\mathbf{r},t)\mathcal{E}(\mathbf{r},t) + \mathcal{J}(\mathbf{r},t)\times\mathcal{B}(\mathbf{r},t) \tag{1.6}$$

having used (1.1) and (1.3).

The Lorentz force provides us with the important connection between electromagnetic entities and mechanical ones. However, while the electric part of $\mathcal{F}_{em}$ is aligned with $\mathcal{E}$, the magnetic part of the Lorentz force is always orthogonal to the magnetic induction and the instantaneous velocity of the charged particle. Indeed, the mechanical work [21, Chapters 13-14] done by $\mathcal{F}_{em}$ in the process of altering or determining the path γ_{12} followed by the particle as it moves from point P_1 to point P_2 reads

$$\begin{aligned} W_{12} &:= \int_{\gamma_{12}} \mathrm{d}s\,\hat{\mathbf{s}}(\mathbf{r})\cdot\mathcal{F}_{em}(\mathbf{r},t) = \int_{t_1}^{t_2} \mathrm{d}t\,\frac{\mathrm{d}s}{\mathrm{d}t}\hat{\mathbf{s}}(\mathbf{r})\cdot\mathcal{F}_{em}(\mathbf{r},t) \\ &= q\int_{t_1}^{t_2} \mathrm{d}t\,\mathbf{v}(t)\cdot[\mathcal{E}(\mathbf{r},t)+\mathbf{v}(t)\times\mathcal{B}(\mathbf{r},t)] = q\int_{t_1}^{t_2}\mathrm{d}t\,\mathbf{v}(t)\cdot\mathcal{E}(\mathbf{r},t) \end{aligned} \tag{1.7}$$

whence on account of the vector identities (H.13) and (H.7) we see that the magnetic field does *no work* on electric charges.

Then again, since electric charges are responsible for the generation of electromagnetic fields, should we not consider also the entities $\mathcal{E}$ and $\mathcal{B}$ produced by q in turn? Well, in principle $\mathcal{E}$ and $\mathcal{B}$ due to q should be included in the picture, so to speak, and would complicate the analysis by far [17, Section 11.2.2], [2, Chapter 9], [22, 23]. Therefore, it is customary to introduce an ideal type of charged particle, called a *test* point charge, which is endowed with two important features:

(1) a *test* point charge possesses *no* spatial extension,
(2) a *test* point charge produces *no* electromagnetic field.

These assumptions mean, in practice, that a test point charge affects neither the sources ϱ and $\mathcal{J}$ nor itself.

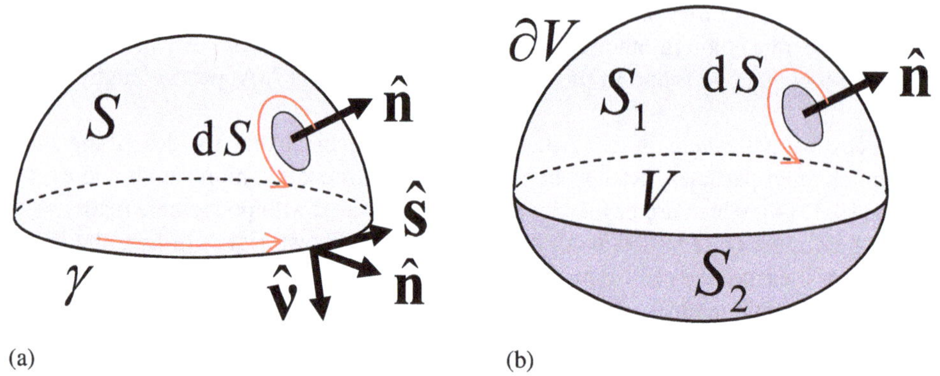

Figure 1.2 *For stating the Maxwell equations in integral form: (a) an open smooth surface* S *with smooth boundary* $\gamma := \partial S$ *and (b) a bounded domain* V *with smooth boundary* $\partial V := S_1 \cup S_2$.

1.2 The Maxwell equations

The *Maxwell equations* [7,8,10,15,18,24,25] constitute the mathematical formulation of the electromagnetic phenomena. Said equations are termed 'classic' in that the interactions between charged particles are not analyzed in the framework of quantum mechanics. On the other hand, the Maxwell equations are fully *relativistic* [7, Chapter 11], [24, Chapter 5], [26, Section 8.7], [27, Chapter 9], [28], i.e., not only do they fit perfectly within A. Einstein's theory of special relativity, but magnetic phenomena are also found to be a consequence of the *relative motion* of charged particles and observers (see Sections 1.9 and 9.2) [17, Section 12.3.1]. Interestingly, the set of four equations we nowadays attribute to J. C. Maxwell were actually derived from Maxwell's theory [3,4] independently by O. Heaviside and H. Hertz more than twenty years later (see [29,30] for a historical perspective).

1.2.1 Integral or global form

We begin by discussing the *integral* or *global* form of the four Maxwell equations [26, Chapter 3]. The electromagnetic entities involved in the equations are those detected by an observer at rest in the 'laboratory' *reference frame*. With the latter expression we indicate the combination of a *system of coordinates* plus a means to measure *distances* and *times*.

In the following, $S \subset \mathbb{R}^3$ denotes an *open* smooth surface whose boundary $\gamma := \partial S$ is a smooth closed curve (Figure 1.2a), and $V \subset \mathbb{R}^3$ represents a bounded domain with smooth boundary ∂V (Figure 1.2b) [31]. The unit vector $\hat{\mathbf{n}}$ perpendicular to S and the unit vector $\hat{\mathbf{s}}$ tangent to γ are oriented with respect to each other according to the *right-handed screw rule*. In words, a right-handed screw aligned with $\hat{\mathbf{n}}(\mathbf{r})$ and oriented in such a way that the tip points in the direction specified by $\hat{\mathbf{n}}$, advances while rotating in the direction prescribed by $\hat{\mathbf{s}}$. Besides, an ideal observer who stands on γ and walks in the direction indicated by $\hat{\mathbf{s}}$ sees the surface S at her left. Occasionally, we shall also need the unit vector $\hat{\boldsymbol{\nu}} := \hat{\mathbf{s}} \times \hat{\mathbf{n}}$, which is perpendicular to the line γ and tangential to S thereon. With our choices $\hat{\mathbf{s}}$, $\hat{\mathbf{n}}$ and $\hat{\boldsymbol{\nu}}$ constitute a *right-handed orthogonal triple* of unit vectors (also see Figure H.1). Finally, the unit normal $\hat{\mathbf{n}}$ on ∂V is positively oriented outward V.

The **Faraday law** (1825) — based on Faraday's experimentation with electric currents, magnets and circuits in relative motion — reads

$$\oint_\gamma \mathrm{d}s\, \hat{\mathbf{s}}(\mathbf{r}) \cdot \mathcal{E}(\mathbf{r},t) = -\int_S \mathrm{d}S\, \hat{\mathbf{n}}(\mathbf{r}) \cdot \frac{\partial}{\partial t}\mathcal{B}(\mathbf{r},t), \qquad t \in \mathbb{R} \tag{1.8}$$

which in words is formulated by stating that the circulation (Appendix A.4) of the electric field along a closed curve γ equals the negative of the flux (Appendix A.3) of the time rate of variation of the magnetic induction through the surface S bounded by γ. We notice that (1.8) holds true for any choice of S and, accordingly, γ, that is, the Faraday law merely establishes the equality of the quantities in the two sides of the equation, but the circulation of $\mathcal{E}(\mathbf{r},t)$ and the flux of the time rate of variation of $\mathcal{B}(\mathbf{r},t)$ for a given electromagnetic field depend on the shape and spatial position of S and γ.

Furthermore, if the surface S is fixed nor does it change shape in the laboratory frame, we may swap integration and time-differentiation [see Appendix A.6 and (A.65)] in the right-hand side of (1.8) and arrive at the alternative form

$$\oint_\gamma \mathrm{d}s\, \hat{\mathbf{s}}(\mathbf{r}) \cdot \mathcal{E}(\mathbf{r},t) = -\frac{\mathrm{d}}{\mathrm{d}t}\int_S \mathrm{d}S\, \hat{\mathbf{n}}(\mathbf{r}) \cdot \mathcal{B}(\mathbf{r},t) := -\frac{\mathrm{d}}{\mathrm{d}t}\Psi_B(t), \qquad t \in \mathbb{R} \tag{1.9}$$

under the assumption that both $\mathcal{B}(\mathbf{r},t)$ and $\partial\mathcal{B}/\partial t$ are continuous for $(\mathbf{r},t) \in S \times \mathbb{R}$ (cf. [32, Theorem 9.42]). In the form (1.9), the Faraday law states that the circulation of the electric field equals the negative of the time rate of variation of the magnetic flux $\Psi_B(t)$ (physical dimension: weber, Wb). In this regard, also see the discussion in Section 1.3.

The **magnetic Gauss law** reads

$$\Psi_B(t) := \int_{\partial V} \mathrm{d}S\, \hat{\mathbf{n}}(\mathbf{r}) \cdot \mathcal{B}(\mathbf{r},t) = 0, \qquad t \in \mathbb{R} \tag{1.10}$$

that is, the magnetic flux through a closed (Gaussian) surface — boundary of a finite domain — vanishes. Sometimes (1.10) is interpreted as the mathematical statement of the *non-existence* of magnetic charges (Section 6.5). Alternatively, we can state that the magnetic flux through any two open surfaces S_1 and S_2 with the same boundary γ takes on the same value. The result follows from the application of (1.10) to the surface $\partial V := S_1 \cup S_2$ (Figure 1.2b), viz.,

$$\begin{aligned} 0 &= \int_{\partial V} \mathrm{d}S\, \hat{\mathbf{n}}(\mathbf{r}) \cdot \mathcal{B}(\mathbf{r},t) \\ &= \int_{S_1} \mathrm{d}S\, \hat{\mathbf{n}}_1(\mathbf{r}) \cdot \mathcal{B}_1(\mathbf{r},t) - \int_{S_2} \mathrm{d}S\, \hat{\mathbf{n}}_2(\mathbf{r}) \cdot \mathcal{B}_2(\mathbf{r},t), \qquad t \in \mathbb{R} \end{aligned} \tag{1.11}$$

with due regard to the definition of the local normals, i.e., $\hat{\mathbf{n}}(\mathbf{r}) = \hat{\mathbf{n}}_1(\mathbf{r})$ for $\mathbf{r} \in S_1$ and $\hat{\mathbf{n}}(\mathbf{r}) = -\hat{\mathbf{n}}_2(\mathbf{r})$ for $\mathbf{r} \in S_2$. As a consequence, from (1.8) we conclude that the circulation of the electric field is the same for all the open surfaces S with the same boundary γ.

The **Ampère-Maxwell law** or **Ampère theorem** reads

$$\oint_\gamma \mathrm{d}s\, \hat{\mathbf{s}}(\mathbf{r}) \cdot \mathcal{H}(\mathbf{r},t) = \int_S \mathrm{d}S\, \hat{\mathbf{n}}(\mathbf{r}) \cdot \mathcal{J}_{\mathrm{tot}}(\mathbf{r},t) \tag{1.12}$$

$$:= \int_S \mathrm{d}S\, \hat{\mathbf{n}}(\mathbf{r}) \cdot \left[\frac{\partial}{\partial t}\mathcal{D}(\mathbf{r}, t) + \mathcal{J}(\mathbf{r}, t)\right], \qquad t \in \mathbb{R} \tag{1.13}$$

which in words can be formulated by saying that the circulation of the magnetic field along a closed curve γ equals the flux of the *total electric current density* $\mathcal{J}_{\mathrm{tot}}(\mathbf{r}, t)$ through S bounded by γ. Here, the total electric current density [33, Chapter 1] consists of two terms: *a*) the electric current density flowing through S and *b*) the time rate of variation of the displacement vector. Therefore, a magnetic field is invariably generated by an electric current density (either steady or time-varying) but also a time-varying electric displacement vector. Furthermore, if the surface S is fixed nor does it change shape in the laboratory frame, we may write [see Appendix A.6 and (A.65)]

$$\oint_\gamma \mathrm{d}s\, \hat{\mathbf{s}}(\mathbf{r}) \cdot \mathcal{H}(\mathbf{r}, t) = \frac{\mathrm{d}}{\mathrm{d}t}\int_S \mathrm{d}S\, \hat{\mathbf{n}}(\mathbf{r}) \cdot \mathcal{D}(\mathbf{r}, t) + \int_S \mathrm{d}S\, \hat{\mathbf{n}}(\mathbf{r}) \cdot \mathcal{J}(\mathbf{r}, t) \tag{1.14}$$

$$:= \frac{\mathrm{d}}{\mathrm{d}t}\Psi_D(t) + I(t), \qquad t \in \mathbb{R} \tag{1.15}$$

so long as both $\mathcal{D}(\mathbf{r}, t)$ and $\partial\mathcal{D}/\partial t$ are continuous for $(\mathbf{r}, t) \in S \times \mathbb{R}$ (cf. [32, Theorem 9.42]). We then state that the circulation of the magnetic field equals the time rate of variation of the electric flux $\Psi_D(t)$ (physical dimension: C) plus the current $I(t)$ (physical dimension: A).

The (electric) **Gauss law** — derived by Cavendish in 1772 — reads

$$\Psi_D(t) := \int_{\partial V} \mathrm{d}S\, \hat{\mathbf{n}}(\mathbf{r}) \cdot \mathcal{D}(\mathbf{r}, t) = \int_V \mathrm{d}V\, \varrho(\mathbf{r}, t) := Q(t), \qquad t \in \mathbb{R} \tag{1.16}$$

i.e., the electric flux through a closed surface (boundary of a bounded domain) equals the total electric charge enclosed by the surface. As a result, the electric flux through any two open surfaces S_1 and S_2 with the same boundary γ may be different, if a net non-zero charge is enclosed by $\partial V := S_1 \cup S_2$ (Figure 1.2b). Thus, unlike the time rate of variation of $\Psi_B(t)$ in (1.8), the corresponding contribution in the Ampère-Maxwell law is not necessarily the same for any two surfaces S_1 and S_2 with the same boundary γ, and this happens when S_1 cannot be deformed continuously into S_2 without crossing a region which contains some electric charge.

Maxwell's equations (1.8), (1.10), (1.13) and (1.16) are complemented with the **law of conservation of charge** in integral form, which reads

$$0 = \int_{\partial V} \mathrm{d}S\, \hat{\mathbf{n}}(\mathbf{r}) \cdot \mathcal{J}(\mathbf{r}, t) + \int_V \mathrm{d}V\, \frac{\partial}{\partial t}\varrho(\mathbf{r}, t) := I(t) + \frac{\mathrm{d}}{\mathrm{d}t}Q(t), \qquad t \in \mathbb{R} \tag{1.17}$$

and it states that the total charge in a bounded domain V is either constant — and thus conserved — or if it changes, the time rate of variation of the charge is balanced by a current flowing through the boundary ∂V. For instance, if the total charge is decreasing, a positive current $I(t)$ exists that accounts for charges leaving V. If the surface ∂V is *fixed* and *does not change* in time, the conservation of charge follows from (1.15) applied to the closed surface $S := \partial V$ and the Gauss law (1.16). If the total charge is constant in V, then (1.17) requires $\hat{\mathbf{n}}(\mathbf{r}) \cdot \mathcal{J}(\mathbf{r}, t) = 0$ for $(\mathbf{r}, t) \in \partial V \times \mathbb{R}$.

A more general form of (1.17) valid in the presence of surface charges is derived in Section 1.7 on page 35.

1.2.2 Differential or local or point form

The Maxwell equations (1.8), (1.10), (1.13) and (1.16) remain valid even if the open surface S or the boundary ∂V are moving or changing with time, though computing the integrals may not be straightforward. If a reference frame can be found in which S and ∂V are at rest and the shapes thereof do not depend on time, then we can determine an equivalent *differential* or *local* form of the Maxwell equations [24, Chapter III], [26, Cahpter 3]. The latter are usually more suitable for the solution of practical electromagnetic problems both analytically and through numerical techniques.

In order to obtain the **Faraday law** in differential form we start with (1.8) and invoke the Stokes theorem (A.55) on the left-hand side. This step is permissible as long as $\mathcal{E}(\mathbf{r}, t)$ is of class $\mathcal{C}^1(S)^3 \cap \mathcal{C}(\overline{S})^3$ (we refer the Reader to Appendix D.1 for the meaning of this notation). In symbols, we have

$$\int_S \mathrm{d}S\, \hat{\mathbf{n}}(\mathbf{r}) \cdot \left[\nabla \times \mathcal{E}(\mathbf{r}, t) + \frac{\partial}{\partial t}\mathcal{B}(\mathbf{r}, t)\right] = 0, \qquad t \in \mathbb{R} \tag{1.18}$$

whence by invoking the mean value theorem [34] we find

$$\hat{\mathbf{n}}(\mathbf{r}_0) \cdot \left[\nabla \times \mathcal{E}(\mathbf{r}, t)|_{\mathbf{r}=\mathbf{r}_0} + \frac{\partial}{\partial t}\mathcal{B}(\mathbf{r}_0, t)\right] = 0 \tag{1.19}$$

where $\mathbf{r}_0 \in S$ is a suitable point. Since the surface S is arbitrary, $\hat{\mathbf{n}}(\mathbf{r}_0)$ is non-null and in general not perpendicular to the vector field within brackets in the equation above, we conclude that

$$\nabla \times \mathcal{E}(\mathbf{r}, t) = -\frac{\partial}{\partial t}\mathcal{B}(\mathbf{r}, t) \tag{1.20}$$

for any $(\mathbf{r}, t) \in \mathbb{R}^3 \times \mathbb{R}$. In view of the assumptions we have made to arrive at this result, it should be evident that (1.20) is less general than its global counterpart (1.8). Besides, (1.20) may be interpreted by saying that a time-varying magnetic induction field generates an electric field. Conversely, it is seen from (1.20) that a constant though non-uniform magnetic induction $\mathcal{B}(\mathbf{r})$ is not enough to induce an electric field.

In order to derive the differential form of the **magnetic Gauss law** we start with (1.10) and apply the Gauss theorem (A.53), under the hypothesis that $\mathcal{B}(\mathbf{r}, t)$ is of class $\mathcal{C}^1(V)^3 \cap \mathcal{C}(\overline{V})^3$ (Appendix D.1). In symbols, this reads

$$\int_V \mathrm{d}V\, \nabla \cdot \mathcal{B}(\mathbf{r}, t) = 0, \qquad t \in \mathbb{R} \tag{1.21}$$

whence by invoking the mean value theorem [34] we find

$$\nabla \cdot \mathcal{B}(\mathbf{r}_0, t) = 0 \tag{1.22}$$

where $\mathbf{r}_0 \in V$ is a suitable point. From the arbitrariness of V and hence of $\mathbf{r}_0$ it follows

$$\nabla \cdot \mathcal{B}(\mathbf{r}, t) = 0 \tag{1.23}$$

for any $(\mathbf{r}, t) \in \mathbb{R}^3 \times \mathbb{R}$. A vector field whose divergence vanishes is said to be *solenoidal* or *divergence-free* (see Section 8.1).

The derivation of the **Ampère-Maxwell law** in differential form proceeds mostly along the same lines followed to obtain (1.20). However, care must be exercised in applying the Stokes theorem to the left-hand side of (1.13) because the magnetic field $\mathcal{H}(\mathbf{r}, t)$ may not be of class $\mathcal{C}^1(S)^3 \cap \mathcal{C}(\overline{S})^3$ (Appendix D.1), if the current density $\mathcal{J}(\mathbf{r}, t)$, $\mathbf{r} \in V_J$, crosses a surface $S_J := S \cap V_J$, i.e., $\mathcal{J}(\mathbf{r}, t) = 0$ for $\mathbf{r} \in S \setminus S_J$.

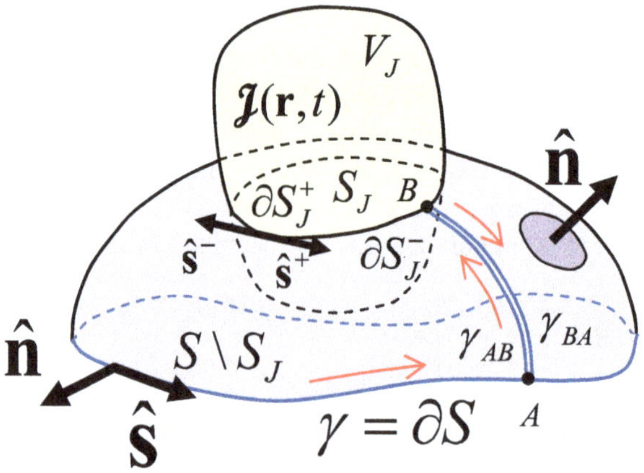

Figure 1.3 For the derivation of the Ampère-Maxwell law in local form when the surface S intersects a region V_J where a volume current density $\mathcal{J}(\mathbf{r}, t)$ flows.

Indeed, since we do not know whether or not $\mathcal{H}(\mathbf{r}, t)$ suffers jumps across the line ∂S_J and yet (1.13) holds separately on S_J and $S \setminus S_J$, we may write instead

$$\oint_{\partial S_J^+} \mathrm{d}s\, \hat{\mathbf{s}}^+(\mathbf{r}) \cdot \mathcal{H}^+(\mathbf{r}, t) = \int_{S_J} \mathrm{d}S\, \hat{\mathbf{n}}(\mathbf{r}) \cdot \left[\frac{\partial}{\partial t}\mathcal{D}(\mathbf{r}, t) + \mathcal{J}(\mathbf{r}, t)\right] \tag{1.24}$$

$$\oint_{\gamma} \mathrm{d}s\, \hat{\mathbf{s}}(\mathbf{r}) \cdot \mathcal{H}(\mathbf{r}, t) + \oint_{\partial S_J^-} \mathrm{d}s\, \hat{\mathbf{s}}^-(\mathbf{r}) \cdot \mathcal{H}^-(\mathbf{r}) = \int_{S\setminus S_J} \mathrm{d}S\, \hat{\mathbf{n}}(\mathbf{r}) \cdot \frac{\partial}{\partial t}\mathcal{D}(\mathbf{r}, t) \tag{1.25}$$

where $\mathcal{H}^+$, $\mathcal{H}^-$ indicate the values of the magnetic field on either side of the line ∂S_J (Figure 1.3). The orientations of S_J, $S \setminus S_J$ and S are consistent, meaning that the unit normal $\hat{\mathbf{n}}(\mathbf{r})$ is continuous across ∂S_J. As a result, the orientations of ∂S_J^+ and γ are the same.

The expression in the left member of (1.25) is justified by observing that the boundary of $S \setminus S_J$ is comprised of two *disjoint* closed loops, i.e., γ and ∂S_J^-. A *single*, closed loop may be recovered by performing a 'cut' in the surface $S \setminus S_J$ along an arbitrary smooth path which connects the points A on γ and B on ∂S_J^-, as is suggested in Figure 1.3. If we introduce the lines γ_{AB} and γ_{BA} — which are conceptually *distinct* but in practice *coincide* — the boundary of the *split* surface $S \setminus S_J$ becomes the piecewise-smooth loop $\Gamma := \gamma \cup \gamma_{AB} \cup \partial S_J^- \cup \gamma_{BA}$. Moreover, if we choose a consistent orientation for Γ in accordance with the right-handed screw rule (cf. Figure 1.2a), then (1.13) may be applied. Still, since the contributions of the line integrals of $\mathcal{H}(\mathbf{r}, t)$ along γ_{AB} and γ_{BA} have equal magnitude and opposite sign, they cancel out, and in the end we find (1.25). The procedure just described is sometimes referred to as the *cancellation principle*.

By summing (1.24) and (1.25) side by side and performing trivial manipulations on the line integrals as well as the flux ones we get

$$\oint_{\partial S_J^+} \mathrm{d}s\, \hat{\mathbf{s}}^+(\mathbf{r}) \cdot \mathcal{H}^+(\mathbf{r}, t) + \oint_{\gamma} \mathrm{d}s\, \hat{\mathbf{s}}(\mathbf{r}) \cdot \mathcal{H}(\mathbf{r}, t) + \oint_{\partial S_J^-} \mathrm{d}s\, \hat{\mathbf{s}}^-(\mathbf{r}) \cdot \mathcal{H}^-(\mathbf{r})$$

$$= \oint_{\gamma} \mathrm{d}s\, \hat{\mathbf{s}}(\mathbf{r}) \cdot \mathcal{H}(\mathbf{r}, t) + \oint_{\partial S_J} \mathrm{d}s\, \hat{\mathbf{s}}(\mathbf{r}) \cdot [\mathcal{H}^+(\mathbf{r}, t) - \mathcal{H}^-(\mathbf{r}, t)]$$

$$= \int_{S} \mathrm{d}S\, \hat{\mathbf{n}}(\mathbf{r}) \cdot \left[\frac{\partial}{\partial t}\mathcal{D}(\mathbf{r}, t) + \mathcal{J}(\mathbf{r}, t)\right] \tag{1.26}$$

having taken into account in the last step that $\mathcal{J}(\mathbf{r}, t)$ vanishes for $\mathbf{r} \in S \setminus \overline{S}_J$ by hypothesis. This equation differs from (1.13), which is also valid, for the presence of an extra line integral along ∂S_J. Of course, the two ways of applying the Ampère-Maxwell law to the physical configuration of Figure 1.3 ought to be equivalent for the sake of consistency. Only if we *postulate* the condition

$$\hat{\mathbf{s}}(\mathbf{r}) \cdot \mathcal{H}^+(\mathbf{r}, t) = \hat{\mathbf{s}}(\mathbf{r}) \cdot \mathcal{H}^-(\mathbf{r}, t), \qquad \mathbf{r} \in \partial S_J, \qquad t \in \mathbb{R} \tag{1.27}$$

that is, the continuity of $\hat{\mathbf{s}} \cdot \mathcal{H}$ across ∂S_J, does the contribution of the circulation along ∂S_J vanish in general, and hence equivalence with (1.13) is guaranteed.

Next, we invoke the Stokes theorem (A.55) on the line integrals in the left-hand sides of (1.24) and (1.25) and combine like flux integrals to arrive at

$$\int_{S_J} \mathrm{d}S\, \hat{\mathbf{n}}(\mathbf{r}) \cdot \left[\nabla \times \mathcal{H}(\mathbf{r}, t) - \frac{\partial}{\partial t}\mathcal{D}(\mathbf{r}, t) - \mathcal{J}(\mathbf{r}, t)\right] = 0 \tag{1.28}$$

$$\int_{S \setminus S_J} \mathrm{d}S\, \hat{\mathbf{n}}(\mathbf{r}) \cdot \left[\nabla \times \mathcal{H}(\mathbf{r}, t) - \frac{\partial}{\partial t}\mathcal{D}(\mathbf{r}, t)\right] = 0 \tag{1.29}$$

provided $\mathcal{H}(\mathbf{r}, t)$ is continuously differentiable separately in S_J and $S \setminus \overline{S}_J$. Since the integrands in (1.28) and (1.29) are also continuous, thanks to the mean value theorem [34] we find

$$\hat{\mathbf{n}}(\mathbf{r}_1) \cdot \left[\nabla \times \mathcal{H}(\mathbf{r}_1, t) - \frac{\partial}{\partial t}\mathcal{D}(\mathbf{r}_1, t) - \mathcal{J}(\mathbf{r}_1, t)\right] = 0 \tag{1.30}$$

$$\hat{\mathbf{n}}(\mathbf{r}_2) \cdot \left[\mathcal{H}(\mathbf{r}_2, t) - \frac{\partial}{\partial t}\mathcal{D}(\mathbf{r}_2, t)\right] = 0 \tag{1.31}$$

where $\mathbf{r}_1 \in S_J$ and $\mathbf{r}_2 \in S \setminus S_J$ are two suitable points. In view of the arbitrariness of the surface S, which we may select at will so as to intersect V_J across ever different surfaces S_J (Figure 1.3), we conclude that

$$\nabla \times \mathcal{H}(\mathbf{r}, t) = \frac{\partial}{\partial t}\mathcal{D}(\mathbf{r}, t) + \mathcal{J}(\mathbf{r}, t), \qquad \mathbf{r} \in V_J \tag{1.32}$$

$$\nabla \times \mathcal{H}(\mathbf{r}, t) = \frac{\partial}{\partial t}\mathcal{D}(\mathbf{r}, t), \qquad \mathbf{r} \in \mathbb{R}^3 \setminus \overline{V}_J \tag{1.33}$$

for any instant of time. We may combine these two equations into a single one, namely,

$$\nabla \times \mathcal{H}(\mathbf{r}, t) = \frac{\partial}{\partial t}\mathcal{D}(\mathbf{r}, t) + \mathcal{J}(\mathbf{r}, t) \tag{1.34}$$

valid for any $(\mathbf{r}, t) \in \mathbb{R}^3 \times \mathbb{R}$, under the hypothesis that $\hat{\mathbf{s}}(\mathbf{r}) \cdot \mathcal{H}(\mathbf{r}, t)$ is continuous across the boundary of $\mathcal{J}(\mathbf{r}, t)$.

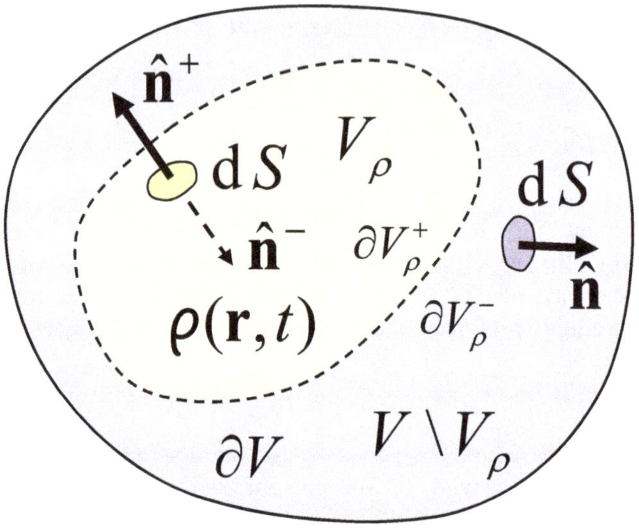

Figure 1.4 For the derivation of the electric Gauss law in local form when the region V encloses an electric charge density $\varrho(\mathbf{r}, t)$ in the sub-domain $V_\varrho \subset V$.

As a by product, if in (1.26) we invoke the Stokes theorem on the circulation along ∂S_J^+ as well as the pair of line integrals along γ and ∂S_J^-, by virtue of (1.27) we obtain

$$\oint_\gamma \mathrm{d}s\, \hat{\mathbf{s}}(\mathbf{r}) \cdot \mathcal{H}(\mathbf{r}, t) = \int_S \mathrm{d}S\, \hat{\mathbf{n}}(\mathbf{r}) \cdot \nabla \times \mathcal{H}(\mathbf{r}, t), \qquad t \in \mathbb{R} \tag{1.35}$$

which we interpret by saying that the Stokes theorem (A.55) is valid even though the field $\mathcal{H}(\mathbf{r}, t)$ is continuously differentiable separately on $S \setminus \overline{S}_J$ and S_J as long as the tangential component of $\mathcal{H}(\mathbf{r}, t)$ is continuous across the boundary ∂S_J. On the contrary, we see that the Stokes theorem may not be applied directly over S if (1.27) is false.

In like manner, for the derivation of the local **electric Gauss law** from (1.16) we are not allowed to apply the Gauss theorem directly to the flux integral, because the displacement vector $\mathcal{D}(\mathbf{r}, t)$ may not be of class $\mathcal{C}^1(V)^3 \cap \mathcal{C}(\overline{V})^3$ (Appendix D.1), if the charge density fills a region $V_\varrho \subset V$ only, i.e., $\varrho(\mathbf{r}, t) = 0$ for $\mathbf{r} \in V \setminus V_\varrho$, as is sketched in Figure 1.4.

Indeed, since we do not know beforehand whether or not $\mathcal{D}(\mathbf{r}, t)$ suffers jumps across ∂V_ϱ, we apply the Gauss law separately to the regions V_ϱ and $V \setminus V_\varrho$, viz.,

$$\int_{\partial V_\varrho^+} \mathrm{d}S\, \hat{\mathbf{n}}^+(\mathbf{r}) \cdot \mathcal{D}^+(\mathbf{r}, t) = \int_{V_\varrho} \mathrm{d}V\, \varrho(\mathbf{r}, t) \tag{1.36}$$

$$\int_{\partial V} \mathrm{d}S\, \hat{\mathbf{n}}(\mathbf{r}) \cdot \mathcal{D}(\mathbf{r}, t) + \int_{\partial V_\varrho^-} \mathrm{d}S\, \hat{\mathbf{n}}^-(\mathbf{r}) \cdot \mathcal{D}^-(\mathbf{r}, t) = 0 \tag{1.37}$$

where $\mathcal{D}^+, \mathcal{D}^-$ indicate the values of the displacement vector on either side of ∂V_ϱ (Figure 1.4). The orientations of ∂V_ϱ^+ and ∂V are consistent, meaning that the unit normals $\hat{\mathbf{n}}(\mathbf{r})$ point outwards V_ϱ and V, respectively.

The form of the left member of (1.37) is justified by noticing that the boundary of $V \setminus V_\varrho$ is made up of two *disjoint* closed surfaces, i.e., ∂V and ∂V_ϱ^-. In this case, though, even 'cutting' the region

$V \setminus V_\varrho$ with a suitable surface does not help us construct a single closed surface. For instance, if we take the intersection of $V \setminus V_\varrho$ with a plane S_P which passes also through V_ϱ, we end up with two domains V_1 and V_2 whose boundaries share the common surface $S_P \cap (V \setminus V_\varrho)$. Now, when we apply the Gauss law to V_1 and V_2 — which are devoid of charges — we obtain, among other terms, two flux integrals over either side of $S_P \cap (V \setminus V_\varrho)$. When the resulting relationships are summed side by side, the contributions of the two integrals over $S_P \cap (V \setminus V_\varrho)$ cancel one another, because they are equal in magnitude and opposite in sign. The remaining equation is precisely (1.37).

By summing (1.36) and (1.37) side by side and carrying out few simple manipulations we obtain

$$\begin{aligned}
\int_{\partial V} \mathrm{d}V\, \hat{\mathbf{n}}(\mathbf{r}) \cdot \mathfrak{D}(\mathbf{r},t) + \int_{\partial V_\varrho^-} \mathrm{d}V\, \hat{\mathbf{n}}^-(\mathbf{r}) \cdot \mathfrak{D}^-(\mathbf{r},t) + \int_{\partial V_\varrho^+} \mathrm{d}V\, \hat{\mathbf{n}}^+(\mathbf{r}) \cdot \mathfrak{D}^+(\mathbf{r},t) = \\
= \int_{\partial V} \mathrm{d}V\, \hat{\mathbf{n}}(\mathbf{r}) \cdot \mathfrak{D}(\mathbf{r},t) + \int_{\partial V_\varrho} \mathrm{d}V\, \hat{\mathbf{n}}(\mathbf{r}) \cdot [\mathfrak{D}^+(\mathbf{r},t) - \mathfrak{D}^-(\mathbf{r},t)] \\
= \int_{V} \mathrm{d}V\, \varrho(\mathbf{r},t) \qquad (1.38)
\end{aligned}$$

where in the last step we have extended the domain integral from V_ϱ to V, because $\varrho(\mathbf{r},t)$ vanishes for $\mathbf{r} \in V \setminus V_\varrho$. Evidently, this equation differs from (1.16), which also holds true, for the presence of an extra flux integral across ∂V_ϱ. But then, the two ways of applying the global Gauss law to the physical configuration of Figure 1.4 ought to be equivalent. Only if we *postulate* the condition

$$\hat{\mathbf{n}}(\mathbf{r}) \cdot \mathfrak{D}^+(\mathbf{r},t) = \hat{\mathbf{n}}(\mathbf{r}) \cdot \mathfrak{D}^-(\mathbf{r},t), \qquad \mathbf{r} \in \partial V_\varrho, \qquad t \in \mathbb{R} \qquad (1.39)$$

i.e., the continuity of $\hat{\mathbf{n}} \cdot \mathfrak{D}$ across ∂V_ϱ, does the flux through ∂V_ϱ vanish, and equivalence with (1.16) is achieved.

Next, we apply the divergence theorem (A.53) separately to the flux integrals in (1.36) and (1.37) to get

$$\int_{V_\varrho} \mathrm{d}V\, [\nabla \cdot \mathfrak{D}(\mathbf{r},t) - \varrho(\mathbf{r},t)] = 0 \qquad \int_{V \setminus V_\varrho} \mathrm{d}V\, \nabla \cdot \mathfrak{D}(\mathbf{r},t) = 0 \qquad (1.40)$$

provided $\mathfrak{D}(\mathbf{r},t)$ is continuously differentiable in V_ϱ and $V \setminus \overline{V}_\varrho$. Furthermore, since both integrands in (1.40) are continuous, by virtue of the mean value theorem [34] we obtain

$$\nabla \cdot \mathfrak{D}(\mathbf{r}_1,t) - \varrho(\mathbf{r}_1,t) = 0 \qquad \nabla \cdot \mathfrak{D}(\mathbf{r}_2,t) = 0 \qquad (1.41)$$

where $\mathbf{r}_1 \in V_\varrho$ and $\mathbf{r}_2 \in V \setminus \overline{V}_\varrho$ are two suitable points. Since we may select the volume V arbitrarily and also in such a way that the boundary ∂V intersects the charge region V_ϱ, we conclude that

$$\nabla \cdot \mathfrak{D}(\mathbf{r},t) = \varrho(\mathbf{r},t), \qquad \mathbf{r} \in V_\varrho \qquad (1.42)$$

$$\nabla \cdot \mathfrak{D}(\mathbf{r},t) = 0, \qquad \mathbf{r} \in V \setminus \overline{V}_\varrho \qquad (1.43)$$

for any instant of time. Lastly, by combining these two relationships, we obtain a single equation

$$\nabla \cdot \mathfrak{D}(\mathbf{r},t) = \varrho(\mathbf{r},t) \qquad (1.44)$$

for any $(\mathbf{r},t) \in \mathbb{R}^3 \times \mathbb{R}$, with $\hat{\mathbf{n}}(\mathbf{r}) \cdot \mathfrak{D}(\mathbf{r},t)$ continuous across ∂V_ϱ.

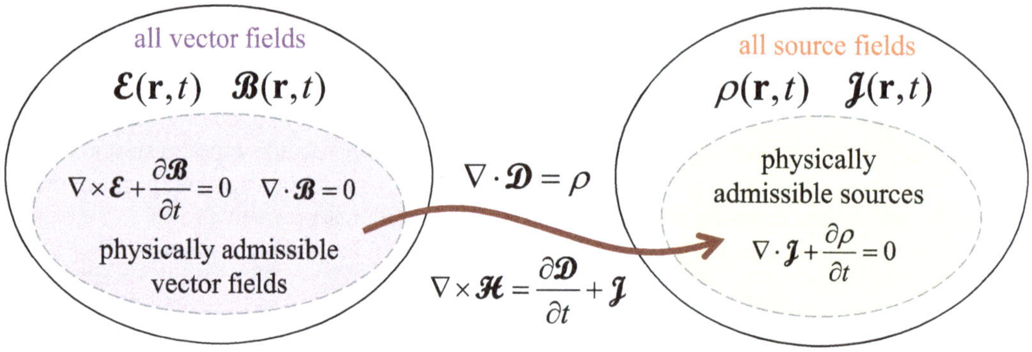

Figure 1.5 The role played by the Maxwell equations and the continuity equation.

As a byproduct, if in (1.38) we apply the Gauss theorem to the pair of flux integrals through ∂V and ∂V_ϱ^- as well as to the flux across ∂V_ϱ^+, in light of (1.39) we obtain

$$\int_{\partial V} \mathrm{d}S\, \hat{\mathbf{n}}(\mathbf{r}) \cdot \mathcal{D}(\mathbf{r},t) = \int_V \mathrm{d}V\, \nabla \cdot \mathcal{D}(\mathbf{r},t), \qquad t \in \mathbb{R} \tag{1.45}$$

which we interpret by saying that the Gauss theorem (A.53) is valid even if the field $\mathcal{D}(\mathbf{r},t)$ is continuously differentiable separately in the sub-domains $V \setminus \overline{V}_\varrho$ and V_ϱ and the normal component of $\mathcal{D}(\mathbf{r},t)$ is continuous across ∂V_ϱ. Conversely, (1.38) shows that the Gauss theorem may not be applied directly to the region V bounded by ∂V if (1.39) is false.

Finally, the **law of conservation of charge** in differential form follows from (1.17) by invoking the Gauss theorem on the flux integral. If the current density $\mathcal{J}(\mathbf{r},t)$ crosses a surface $S_J \subset \partial V$ only, then we can divide the domain V into two parts $V_J \cap V$ and $V \setminus (V_J \cap V)$ such that $\mathcal{J}(\mathbf{r},t) \neq 0$ for $\mathbf{r} \in V_J \cap V$. Then, we need to apply the Gauss theorem only to the region $V_J \cap V$. In any case, the mean value theorem and the arbitrariness of the domain V yield the result

$$\nabla \cdot \mathcal{J}(\mathbf{r},t) + \frac{\partial}{\partial t}\varrho(\mathbf{r},t) = 0 \tag{1.46}$$

for any $(\mathbf{r},t) \in V_J \times \mathbb{R}$. Condition (1.46) is also referred to as the *continuity equation*. A special form of continuity for surface charges and currents is derived in Section 1.7 on page 35.

The set of local Maxwell's equations plus the continuity equation must be supplemented with suitable *initial conditions* (i.e., the value of the fields at $t = t_0$ throughout the domain of interest) as well as *boundary conditions* (i.e., the expected behavior of fields at the boundary of the solution domain). Various types of boundary conditions for fields and sources are discussed in Section 1.7, whereas the role of initial and boundary conditions in the solution of the local Maxwell equations is explained in Sections 6.3 and 6.4.

We notice that, as anticipated above, the Ampère-Maxwell law and the Gauss law provide a relationship between the entities of quantity ($\mathcal{D}$ and $\mathcal{H}$) and the sources (ϱ and $\mathcal{J}$). Conversely, the Faraday law and the magnetic Gauss law involve entities of intensity ($\mathcal{E}$ and $\mathcal{B}$). As is visually suggested in Figure 1.5, (1.20) and (1.23) can be considered constraints of sorts imposed on the electromagnetic field and, in effect, are sometimes referred to as the *Bianchi identities* due to the resemblance they bear to the law obeyed by the four-dimensional Riemann tensor of curvature in general relativity [35, Chapter 15].

It is stated oft-times that Maxwell's equations are not independent of each other because, at a cursory glance, the two Gauss laws would appear to follow from the Faraday law and the Ampère-Maxwell law [6,8,9,13,14,36–42]. However, this statement is incorrect as it stands, and the Maxwell equations are all necessary. Then again, if (1.20), (1.23), (1.34) and (1.44) were not independent, why would we bother stating all of them?

First of all, let us check whether or not (1.23) can indeed be derived from (1.20). The first logical step consists of taking the divergence of both sides of (1.20), viz.,

$$0 = \nabla \cdot \nabla \times \mathcal{E}(\mathbf{r}, t) = -\frac{\partial}{\partial t}\nabla \cdot \mathcal{B}(\mathbf{r}, t) \tag{1.47}$$

where the rightmost equality follows from (A.39), and we have swapped the order of temporal and spatial derivatives on account of the Schwarz theorem [32, pp. 235–236] if $\mathcal{B}$ is twice differentiable. At this point we can only conclude that

$$\nabla \cdot \mathcal{B}(\mathbf{r}, t) = f(\mathbf{r}) \tag{1.48}$$

where $f(\mathbf{r})$ is some yet unspecified scalar field independent of time. Clearly, we need some additional hypothesis for us to claim that $f(\mathbf{r}) = 0$. If we *postulate* that the electromagnetic field, and hence $\mathcal{B}(\mathbf{r}, t)$, was everywhere zero at some time t_0, then (1.48) implies

$$f(\mathbf{r}) = \nabla \cdot \mathcal{B}(\mathbf{r}, t_0) = 0 \tag{1.49}$$

so that the scalar field $f(\mathbf{r})$ vanishes identically, and (1.23) is recovered. In conclusion, (1.23) does not merely follow from the Faraday law. In view of (1.49), the local magnetic Gauss law (1.23) is also referred to as the initial condition for the Faraday law (1.20).

Likewise, let us check whether or not (1.44) can be derived from (1.34). Again, we begin by taking the divergence of both sides

$$0 = \nabla \cdot \nabla \times \mathcal{H}(\mathbf{r}, t) = \nabla \cdot \frac{\partial}{\partial t}\mathcal{D}(\mathbf{r}, t) + \nabla \cdot \mathcal{J}(\mathbf{r}, t) = \frac{\partial}{\partial t}[\nabla \cdot \mathcal{D}(\mathbf{r}, t) - \varrho(\mathbf{r}, t)] \tag{1.50}$$

where we have swapped the order of temporal and spatial derivatives on account of the Schwarz theorem [32, pp. 235–236] if $\mathcal{D}$ is twice differentiable, and we have applied the continuity equation (1.46). We can now conclude that

$$\nabla \cdot \mathcal{D}(\mathbf{r}, t) - \varrho(\mathbf{r}, t) = g(\mathbf{r}) \tag{1.51}$$

where $g(\mathbf{r})$ is some unspecified scalar field independent of time. If we *postulate*, as before, that the sources and the electromagnetic field, and in particular $\mathcal{D}(\mathbf{r}, t)$, were everywhere zero at some time t_0, then (1.51) yields

$$g(\mathbf{r}) = \nabla \cdot \mathcal{D}(\mathbf{r}, t_0) - \varrho(\mathbf{r}, t_0) = 0 \tag{1.52}$$

so that the scalar field $g(\mathbf{r})$ vanishes identically and (1.44) is obtained. What is more, this assumption is supported by Coulomb's (1785) or Cavendish's experiments (1773) [30]. In summary, (1.44) does not simply follow from (1.34). Moreover, on account of (1.52), the Gauss law (1.44) can be thought of as an initial condition for the local Ampère-Maxwell law (1.34).

Although the postulates introduced above may seem *ad hoc*, experimental evidence supports both electric and magnetic Gauss laws, so that it seems physically sound to assume $f(\mathbf{r}) = 0 = g(\mathbf{r})$. Furthermore, by comparing (1.23) and (1.44), we are led to interpret (1.23) as a statement on the non-existence of magnetic charges (see Section 6.5).

1.3 The Faraday law for slowly moving conductors

In Sections 1.2.1 and 1.2.2 we discussed the Maxwell equations valid in a reference frame in which surfaces and volumes are at rest. According to Einstein's principle of relativity [43], the laws of physics take on the same functional *form* in all admissible reference frames. It follows that Maxwell equations in either global or local form look formally the same in all inertial frames of reference, i.e., frames which move with uniform relative velocity with respect to one another [44]. Nevertheless, the electromagnetic entities experienced by two observers in relative motion may actually be quite different.

The rules for relating the entities in two frames in relative uniform motion are provided by the *Lorentz transformations* of coordinates that were introduced in 1904 [1, 7, 15, 17, 24, 44–46]. These transformations allow describing the increase of mass, the shortening of length, and the time dilation for bodies which move at speeds close to the velocity of light. However, if the relative velocity of the two frames is far smaller than the speed of light (in vacuum), we can obtain a convenient form of the Faraday law in the case where the surface S is moving or changing shape with respect to the laboratory frame.

We suppose that an open smooth surface S with boundary γ drifts or deforms with velocity $\mathbf{v}(\mathbf{r}, t)$ with respect to the laboratory frame. Besides, we denote with $\mathcal{E}$ and $\mathcal{B}$ the entities of intensity as detected by an observer at rest in the lab, and with $\mathcal{E}'$ and $\mathcal{B}'$ the corresponding entities in the frame where the surface is instantaneously at rest. Under the hypothesis that $|\mathbf{v}| = v \ll c_0$, where

$$c_0 = 2.99792456 \cdot 10^8 \text{ m/s} \tag{1.53}$$

is the speed of light in vacuum, one can prove the integral identity (Appendix A.6)

$$\begin{aligned}\frac{\mathrm{d}}{\mathrm{d}t}\int_{S(t)} \mathrm{d}S\, \hat{\mathbf{n}}(\mathbf{r}) \cdot \mathcal{B}(\mathbf{r}, t) &= \int_{S(t)} \mathrm{d}S\, \hat{\mathbf{n}}(\mathbf{r}) \cdot \frac{\partial}{\partial t}\mathcal{B}(\mathbf{r}, t) \\ &+ \int_{S(t)} \mathrm{d}S\, \hat{\mathbf{n}}(\mathbf{r}) \cdot \mathbf{v}\, \nabla \cdot \mathcal{B}(\mathbf{r}, t) - \int_{S(t)} \mathrm{d}S\, \hat{\mathbf{n}}(\mathbf{r}) \cdot \nabla \times [\mathbf{v} \times \mathcal{B}(\mathbf{r}, t)]\end{aligned} \tag{1.54}$$

which is known as the *Helmholtz transport theorem* [10, 15, 43, 47–50]. Even though S does not deform — whereby $\hat{\mathbf{n}}(\mathbf{r})$ is constant in time — the integration limits of the flux integrals in the left-hand side depend on time, as $S(t)$ changes position with velocity $\mathbf{v}$. That is why we may not simply move the time derivative inside the flux integral. Conversely, if the velocity $\mathbf{v}$ is zero, we recover the simpler rule for interchanging derivative and integral that we already employed in (1.9).

Next, by invoking the magnetic Gauss law (1.23) and applying the Stokes theorem, we arrive at

$$\frac{\mathrm{d}}{\mathrm{d}t}\int_{S(t)} \mathrm{d}S\, \hat{\mathbf{n}}(\mathbf{r}) \cdot \mathcal{B}(\mathbf{r}, t) = \int_{S(t)} \mathrm{d}S\, \hat{\mathbf{n}}(\mathbf{r}) \cdot \frac{\partial}{\partial t}\mathcal{B}(\mathbf{r}, t) - \oint_{\gamma(t)} \mathrm{d}s\, \hat{\mathbf{s}}(\mathbf{r}) \cdot [\mathbf{v} \times \mathcal{B}(\mathbf{r}, t)] \tag{1.55}$$

which we may use in combination with (1.8), valid in the laboratory frame, to eliminate the flux of the time rate of change of $\mathcal{B}(\mathbf{r}, t)$. By pairing the circulation in the right member of (1.55) with the similar one in (1.8), we find

$$\oint_{\gamma(t)} \mathrm{d}s\, \hat{\mathbf{s}}(\mathbf{r}) \cdot [\mathcal{E}(\mathbf{r}, t) + \mathbf{v} \times \mathcal{B}(\mathbf{r}, t)] = -\frac{\mathrm{d}}{\mathrm{d}t}\int_{S(t)} \mathrm{d}S\, \hat{\mathbf{n}}(\mathbf{r}) \cdot \mathcal{B}(\mathbf{r}, t) \tag{1.56}$$

which sometimes goes by the name of **kinematic form of the Faraday law** [2, Section 2.2], [51, Sections 8.4-8.6], [44, Section 11.2], [15, 52, 53]. We now discuss the subtle implications of this deceptively simple result.

Since the time derivative in the right-hand side is taken *after* the integral has been computed, (1.56) looks *formally* the same as (1.9) which we stipulated with the proviso that the surface was not moving or changing shape. Therefore, we may interpret (1.56) as the global Faraday law stated in the reference frame where $S(t)$ is instantaneously at rest. More importantly, we conclude that

$$\mathcal{E}'(\mathbf{r}, t) = \mathcal{E}(\mathbf{r}, t) + \mathbf{v} \times \mathcal{B}(\mathbf{r}, t) \tag{1.57}$$

$$\mathcal{B}'(\mathbf{r}, t) = \mathcal{B}(\mathbf{r}, t) \tag{1.58}$$

are the relationships between the entities of intensity in the rest frame of $S(t)$ and those in the lab. However, because (1.57) and (1.58) have been derived by assuming that $|\mathbf{v}| = v \ll c_0$, these transformation rules are correct only *to the first order* in v/c_0 — as a matter of fact they would be exact if the speed of light were infinite [7, Chapter 11]. Moreover, an observer moving with the surface would refer the electromagnetic field to local space and time coordinates $(\mathbf{r}', t')$ which are related to $(\mathbf{r}, t)$ by the Galilean transformation [45, Section 13.2]

$$\mathbf{r}' = \mathbf{r} - \mathbf{v}t \tag{1.59}$$

$$t' = t \tag{1.60}$$

again under the assumption that $|\mathbf{v}| = v \ll c_0$. Therefore, $\mathcal{E}'(\mathbf{r}, t)$ and $\mathcal{B}'(\mathbf{r}, t)$ are an approximation of the entities detected by an observer who moves with $S(t)$, though measured at points in space and time as seen by the observer in the laboratory frame. Notice that this is not to say that the Faraday law is approximate; on the contrary, it is the way we compute $\mathcal{E}'$ and $\mathcal{B}'$ in the rest frame of $S(t)$ which fails to hold true if $v \approx c_0$.

The circulation of $\mathcal{E}'$ is called the *electromotive force* and indicated with the symbol $\mathcal{E}$ (physical dimension: volt, V) [17, Section 7.1], [52, Section 4.10]. Thereby, we may write (1.56) succinctly as

$$\mathcal{E}(t) = -\frac{\mathrm{d}}{\mathrm{d}t}\Psi_B(t) \tag{1.61}$$

with $\Psi_B(t)$ denoting the flux of $\mathcal{B}$ or $\mathcal{B}'$ through S; some Authors call (1.61) the *flux rule*. The contribution to $\mathcal{E}$ purely due to the motion of the circuit, namely, the line integral of $\mathbf{v} \times \mathcal{B}(\mathbf{r}, t)$, is called the *induced* or *motional* or *flux-cutting electromotive force* or also *electromotance* with a somewhat old-fashioned nomenclature [45, Chapter 23]. Unfortunately, electromotive force is quite a misnomer because $\mathcal{E}$ represents a voltage difference rather than a mechanical or electrical force.

Since we may choose the curve $\gamma(t)$ so that it coincides with the thin conducting wire constituting an electric circuit, the kinetic Faraday law comes in handy for the analysis of circuits which drift or have moving parts in the presence of a magnetic induction field (see Examples 1.1 and 1.2). Needless to say, the approach works fine so long as the velocities involved are small, or else one should use the full-fledged set of Maxwell equations for moving bodies [12, 44].

Example 1.1 (Electromotive force in a planar circuit with variable shape)

To illustrate a typical application of (1.56) and (1.61) we consider a circuit comprised of a resistor with resistance R connected to a sliding metallic strip by means of pieces of thin wire and two sliding contacts C_1 and C_2, as is depicted in Figure 1.6. The strip has height h and is gently dragged by some mechanical force, which, by balancing frictional forces, keeps the strip moving at a uniform velocity $\mathbf{v} = v\hat{\mathbf{x}}$. The whole setup is immersed in a uniform static magnetic induction field $\mathcal{B} = -\mathcal{B}_0\hat{\mathbf{z}}$

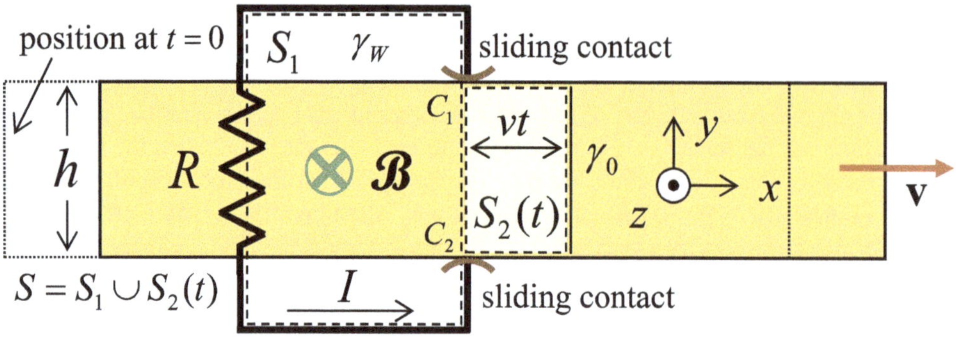

Figure 1.6 Sliding metallic strip: the static magnetic induction field in the laboratory frame induces a non-zero electric field in a planar circuit with variable shape.

perpendicular to the strip and the plane xOy which contains the circuit. Besides, the electric field $\mathcal{E}$ is zero in the laboratory frame by assumption.

We wish to determine the current I flowing in the resistor.

At time $t = t' = 0$, an observer at rest in the reference frame of the circuit notices that the latter coincides with the boundary ∂S_1 of the rectangle S_1. As the time goes by and the strip slides along the positive x-direction, the circuit changes shape and coincides with the boundary of the surface $S(t) = S_1 \cup S_2(t)$, where S_2 is the surface that has been swept out by the strip at time $t = t' > 0$. We emphasize that, while S is changing shape, the unit normal $\hat{\mathbf{n}} = \hat{\mathbf{z}}$ remains constant.

It is straightforward to compute the electromotive force with the flux rule (1.61). Indeed, the magnetic flux through $S(t)$ reads

$$\Psi_B(t) = \int_{S(t)} \mathrm{d}S\, \hat{\mathbf{n}}(\mathbf{r}) \cdot \mathcal{B}(\mathbf{r}, t) = \hat{\mathbf{z}} \cdot (A_1 + vth)(-\mathcal{B}_0 \hat{\mathbf{z}}) \tag{1.62}$$

where A_1 denotes the area of S_1, and we find

$$\mathcal{E} = vh\mathcal{B}_0 \tag{1.63}$$

in accordance with (1.61). But then, the kinematic form of the Faraday law must yield the same result. To apply (1.56), we notice that the only part of the boundary $\gamma(t) = \partial S(t)$ which contributes to the circulation is the straight line γ_0, because it moves with velocity $\mathbf{v}$ perpendicular to the unit tangent vector $\hat{\mathbf{s}} = \hat{\mathbf{y}}$. The other parts are either fixed or they change in such a way that $\hat{\mathbf{s}} \cdot \mathbf{v} \times \mathcal{B} = 0$. Therefore, we have

$$\mathcal{E} = \oint_{\gamma(t)} \mathrm{d}s\, \hat{\mathbf{s}}(\mathbf{r}) \cdot [\underbrace{\mathcal{E}(\mathbf{r}, t)}_{=0} + \mathbf{v} \times \mathcal{B}(\mathbf{r}, t)] = \int_0^h \mathrm{d}y\, \hat{\mathbf{y}} \cdot v\hat{\mathbf{x}} \times (-\mathcal{B}_0 \hat{\mathbf{z}}) = vh\mathcal{B}_0 \tag{1.64}$$

in perfect agreement with the previous finding, as expected.

The drifting strip behaves as a battery of strength $vh\mathcal{B}_0$ connected to the resistor through the sliding contacts. To determine how the poles of the equivalent battery are oriented, we resort to the **Lenz rule** [7,53], [14, Section 1.3.B]. The latter, formulated by the Russian physicist Heinrich Lenz in 1834, states that the current induced in the circuit must flow in such a way as to *oppose the magnetic field* which generated it in the first place. In Chapter 4 we shall see that the magnetic field lines encircle the currents and are positively oriented according to the right-handed screw rule. Therefore, if we assume that the current I flows in a counterclockwise fashion (around $\hat{\mathbf{z}}$), it produces on $S(t)$ a magnetic field oriented in the positive z-direction, which opposes $\mathcal{B}$ as prescribed by the Lenz rule.

However, this explanation may sound weak in that the arrangement of the wires and the resistor is somewhat arbitrary. Indeed, what if we took the surface S_1 at a right angle with the strip, say, in the yOz plane? The direction of the flow of I in the wire would then not be so obvious. We solve the issue by observing that the free electrons in the strip are set in motion by $\mathcal{E}'(\mathbf{r}, t) = v\mathcal{B}_0\hat{\mathbf{y}}$ in accordance with (1.57). Since the electron charge is negative — whereby the electrons actually flow along the negative y-direction — we presume that the electric field $\mathcal{E}'(\mathbf{r}, t)$ gives rise to a current density $\mathcal{J}(\mathbf{r}, t) = \mathcal{J}_x\hat{\mathbf{x}} + \mathcal{J}_y\hat{\mathbf{y}}$ in the strip that is positively oriented from C_2 to C_1. The y-component of $\mathcal{J}$ must vanish on the edges $x = 0$ and $x = h$ [see (1.195) in Section 1.7] for points away from the sliding contacts. Regardless, the current density in turn experiences the magnetic part of the Lorentz force density (1.6) [20, pp. 106 and ff.]

$$\mathbf{f}_{em}(\mathbf{r}, t) = \mathcal{J}(\mathbf{r}, t) \times \mathcal{B} = \mathcal{J}_x(\mathbf{r}, t)\mathcal{B}_0\hat{\mathbf{y}} - \mathcal{J}_y(\mathbf{r}, t)\mathcal{B}_0\hat{\mathbf{x}} \tag{1.65}$$

which thus has a component oriented opposite to the velocity of the strip, since $\mathcal{J}_y > 0$. We see that our choice is correct, because the Lorentz force tends to slow down the strip, thus hindering the motion that causes the current $\mathcal{J}(\mathbf{r}, t)$, as is dictated by the Lenz rule. Hence, we conclude that the positive pole of the equivalent battery is connected to the sliding contact C_1. Parenthetically, we must amend our initial statement on the task of the dragging force, which, we now know, must balance frictional forces as well as the Lorentz force acting on $\mathcal{J}(\mathbf{r}, t)$.

Finally, to determine I we break up the boundary $\gamma(t)$ into two lines, i.e., γ_W along the wire through the resistor and $\gamma_S(t)$ across the strip

$$\oint_{\gamma(t)} \mathrm{d}s\,\hat{\mathbf{s}} \cdot \mathcal{E}'(\mathbf{r}, t) = \int_{\gamma_W} \mathrm{d}s\,\hat{\mathbf{s}} \cdot \mathcal{E}'(\mathbf{r}, t) + \int_{\gamma_S(t)} \mathrm{d}s\,\hat{\mathbf{s}} \cdot \mathcal{E}'(\mathbf{r}, t) = RI + R_S I \tag{1.66}$$

where R_S denote the equivalent resistance of the strip. Application of (1.56) now yields

$$I = \frac{vh\mathcal{B}_0}{R + R_S}. \tag{1.67}$$

(End of Example 1.1)

Example 1.2 (Electromotive force generated by a Faraday disk)

The next application example — due to Faraday himself [53] — is a more practical implementation that extends the idea of the sliding strip circuit of Figure 1.6. To make the problem slightly less trivial we consider a capacitor with capacitance C and a resistor with resistance R connected to a rotating metallic disk by means of pieces of thin wire and sliding contacts C_1 and C_2, as is sketched in Figure 1.7. The contact C_1 brushes the axle, whereas C_2 touches the boundary of the disk. The latter has radius a and at time $t = t' = 0$ it starts getting gently pushed by some mechanical torque which, by balancing the frictional forces, keeps the disk rotating at constant angular velocity ω. If we also accounted for the finite radius $b < a$ of the axle, the rotating part of the circuit would be the annular region which extends from b up to a. We set $b = 0$, as this geometrical detail is not essential for the discussion.

The whole circuit is immersed in a static magnetic induction field $\mathcal{B} = -\mathcal{B}_0\hat{\mathbf{z}}$ perpendicular to the xOy plane, which contains the disk and the circuit, and the electric field is zero in the laboratory

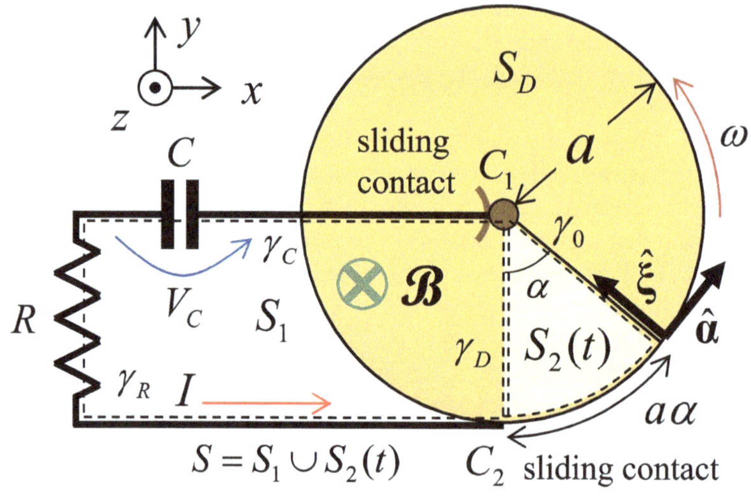

Figure 1.7 Faraday disk: the static magnetic induction field in the laboratory frame induces a non-zero electric field in a planar circuit with variable shape.

frame. Since the rotation of the metallic disk in the presence of the magnetic induction causes a voltage drop to appear between the two sliding contacts C_1 and C_2, this device — which can produce high voltages and low currents — is also known as a *homopolar generator* [45, 54], [55, Section 14.4.1].

We wish to determine the current $I(t)$ flowing in the circuit for $t = t' > 0$ by further assuming that the capacitor is initially uncharged.

At time $t = t' = 0$ an observer at rest in the reference frame of the circuit realizes that the latter coincides with the boundary ∂S_1 of the rectangle S_1. As soon as the torque kicks in, the disk starts rotating in a counterclockwise direction around the positive z-axis. Consequently, the circuit changes its shape and coincides with the boundary $\gamma(t) := \partial S(t)$ of the surface $S(t) = S_1 \cup S_2(t)$, where $S_2(t)$ is the angular sector that has been swept out by the disk at time $t = t' > 0$. Again, we notice that, although the circuit is changing shape, still the normal $\hat{\mathbf{n}} = \hat{\mathbf{z}}$ remains constant.

Also in this case it is relatively easier to compute the electromotive force with the aid of the magnetic flux linked by the circuit through $S(t)$. In symbols, this reads

$$\Psi_B(t) = \int_{S(t)} \mathrm{d}S\, \hat{\mathbf{n}} \cdot \mathcal{B} = \hat{\mathbf{z}} \cdot \left[A_1 + \frac{1}{2}\alpha(t)a^2 \right](-\mathcal{B}_0\hat{\mathbf{z}}) \tag{1.68}$$

where A_1 denotes the area of S_1 and $\alpha(t)$ the angle of the angular sector S_2. We find

$$\mathcal{E} = \frac{1}{2}\frac{\mathrm{d}\alpha}{\mathrm{d}t}a^2\mathcal{B}_0 = \frac{1}{2}\omega a^2\mathcal{B}_0 \tag{1.69}$$

from (1.61) and the very definition of angular velocity. On the other hand, to compute the circulation of $\mathcal{E}'$ we observe that the only part of $\gamma(t)$ which gives a net contribution is the radial line γ_0, whose points move with different velocities which depend on the distance from the axle but are always perpendicular to the unit tangent vector $\hat{\xi}$ and the magnetic field. The other parts of $\gamma(t)$ are either fixed or they change in such a way that $\hat{\mathbf{s}} \cdot \mathbf{v} \times \mathcal{B} = 0$, where $\hat{\mathbf{s}}$ is the unit vector tangent to $\gamma(t)$. Therefore, the electric field in the reference frame of the circuit at points on γ_0 reads (Figure 1.7)

$$\mathcal{E}'(s) = \mathbf{v} \times \mathcal{B} = \omega(a - \xi)\,\hat{\alpha} \times (-\mathcal{B}_0\hat{\mathbf{z}}) = \omega(a - \xi)\mathcal{B}_0\hat{\xi} \tag{1.70}$$

where $\xi \in [0, a]$ is a radial coordinate measured from the boundary of the disk towards the axle, $\hat{\boldsymbol{\xi}}$ is the corresponding unit vector tangent to γ_0, and $\hat{\boldsymbol{\alpha}} = \hat{\boldsymbol{\xi}} \times \hat{\mathbf{z}}$ is the unit vector tangent to the disk. All in all, we have

$$\mathcal{E} = \oint_{\gamma(t)} \mathrm{d}s\, \hat{\mathbf{s}}(\mathbf{r}) \cdot [\underbrace{\mathcal{E}(\mathbf{r}, t)}_{=0} + \mathbf{v} \times \mathcal{B}(\mathbf{r}, t)] = \int_0^a \mathrm{d}\xi\, \omega(a - \xi)\mathcal{B}_0 = \frac{1}{2}\omega a^2 \mathcal{B}_0 \tag{1.71}$$

as expected, though computing the circulation usually requires a little more ingenuity.

The electromotive force represents an equivalent battery whose positive (resp. negative) pole is connected to the sliding contact C_1 (resp. C_2) so that the current $I(t)$ flows in a counterclockwise fashion around the z-axis from the axle through the circuit and back to the disk. This orientation produces a magnetic field which on $S(t)$ is oriented along $\hat{\mathbf{z}}$ and thus it opposes $\mathcal{B}$ as required by the Lenz rule [7,53]. But then, as in Example 1.1, the arrangement of the wire should not play a critical role. A more cogent explanation hinges on the observation that the electric field (1.70) generates a current $\mathcal{J}(\mathbf{r}, t) = \mathcal{J}_\xi \hat{\boldsymbol{\xi}} + \mathcal{J}_\alpha \hat{\boldsymbol{\alpha}}$ in the disk. The ξ-component of $\mathcal{J}$ must vanish for $\xi = 0$ and $\alpha \neq 0$, i.e., on the boundary of the disk except at the location of the sliding contact C_2 [see (1.195) in Section 1.7]. Regardless, we surmise that $\mathcal{J}(\mathbf{r}, t)$ flows from the boundary of the disk toward the axle and observe that the volume density of Lorentz force (1.6) [20, pp. 106 and ff.] reads

$$\mathbf{f}_{em}(\mathbf{r}, t) = \mathcal{J}(\mathbf{r}, t) \times \mathcal{B} = \mathcal{J}_\alpha(\mathbf{r}, t)\mathcal{B}_0 \hat{\boldsymbol{\xi}} - \mathcal{J}_\xi(\mathbf{r}, t)\mathcal{B}_0 \hat{\boldsymbol{\alpha}} \tag{1.72}$$

and has a component oriented opposite to the velocity of the disk, since $\mathcal{J}_\xi > 0$. Hence, we see that our choice is correct, because the Lorentz force tries to slow down the disk, thus opposing the motion which causes $\mathcal{J}(\mathbf{r}, t)$ as demanded by the Lenz rule. Also in this device, to keep the disk going, the external torque must overcome the frictional forces and the braking action of the Lorentz force.

To determine the current $I(t)$ we split the boundary $\gamma(t)$ into three lines, namely, γ_C across the capacitor, γ_R across the resistor, and $\gamma_D(t)$ through the disk. In symbols, we have

$$\begin{aligned}\oint_{\gamma(t)} \mathrm{d}s\, \hat{\mathbf{s}} \cdot \mathcal{E}'(\mathbf{r}, t) &= \int_{\gamma_C} \mathrm{d}s\, \hat{\mathbf{s}} \cdot \mathcal{E}'(\mathbf{r}, t) + \int_{\gamma_R} \mathrm{d}s\, \hat{\mathbf{s}} \cdot \mathcal{E}'(\mathbf{r}, t) + \int_{\gamma_D(t)} \mathrm{d}s\, \hat{\mathbf{s}} \cdot \mathcal{E}'(\mathbf{r}, t) \\ &= V_C(t) + RI(t) + R_D I(t)\end{aligned} \tag{1.73}$$

where V_C denotes the voltage drop across the capacitor and R_D the equivalent resistance of the disk. By enforcing (1.56) explicitly, we obtain the circuit equation (cf. [44, Eqn. (5.86)])

$$V_C(t) + (R + R_D)C\frac{\mathrm{d}V_C}{\mathrm{d}t} = \frac{1}{2}\omega a^2 \mathcal{B}_0, \qquad t \geqslant 0 \tag{1.74}$$

which is solved by

$$V_C(t) = \frac{1}{2}\omega a^2 \mathcal{B}_0 \left[1 - \exp\left(-\frac{t}{(R + R_D)C}\right)\right], \qquad t \geqslant 0. \tag{1.75}$$

Finally, the current in the circuit is given by

$$I(t) = C\frac{\mathrm{d}V_C}{\mathrm{d}t} = \frac{\omega a^2 \mathcal{B}_0}{2(R + R_D)} \exp\left(-\frac{t}{(R + R_D)C}\right), \qquad t \geqslant 0 \tag{1.76}$$

and is discontinuous for $t = 0$ — it jumps from 0 to $\omega a^2 \mathcal{B}_0/[2(R+R_D)]$ — because we have assumed that the disk reaches its final angular velocity ω instantaneously, and this causes the electromotive force to jump as well at $t = 0$.

(End of Example 1.2)

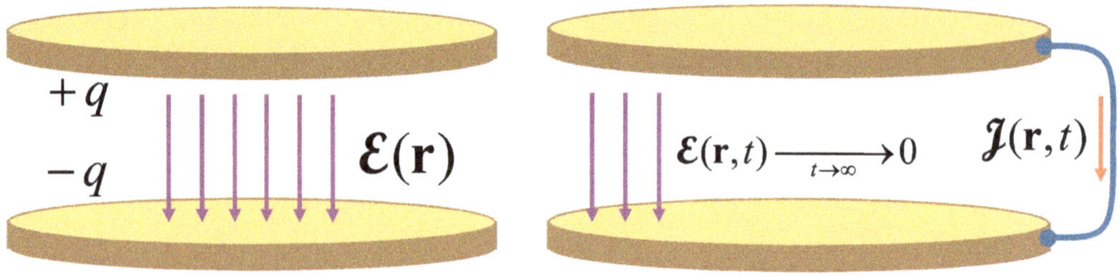

(a) (b)

Figure 1.8 For showing that the Ampère law (1.78) fails for time-varying currents: (a) charged parallel-plate capacitor in equilibrium; (b) discharge of the capacitor through a thin conducting wire.

1.4 Displacement current

A simple dimensional analysis of (1.34) shows that $\mathcal{J}(\mathbf{r}, t)$ and the time-derivative of the displacement vector field necessarily carry the same physical dimension. For this reason, $\partial\mathcal{D}/\partial t$ is called *displacement current* (density) [2, Section 5.1], [11, Section 10.3], [27, Section 3.A.2], [24, Section II.23], [56, Section9.2]. At the time of Maxwell's work on electromagnetism, it was already known that an electric current generates a magnetic field (based on Oersted's 1819 experiment with a current-carrying wire and a compass needle [4, page 128], [11, Section 1.9.1]), and (1.13) and (1.34) were written just as

$$\oint_{\gamma} \mathrm{d}s\, \hat{\mathbf{s}}(\mathbf{r}) \cdot \mathcal{H}(\mathbf{r}, t) = \int_{S} \mathrm{d}S\, \hat{\mathbf{n}}(\mathbf{r}) \cdot \mathcal{J}(\mathbf{r}, t) \tag{1.77}$$

$$\nabla \times \mathcal{H}(\mathbf{r}, t) = \mathcal{J}(\mathbf{r}, t) \tag{1.78}$$

which together were termed the **Ampère law**. Unfortunately, (1.77) and (1.78) are correct only for truly steady currents and stationary magnetic fields (Chapter 4), they remain approximately valid if the variation of the current with time is relatively slow (Section 9.8), but they may predict contradictory results if $\mathcal{J}(\mathbf{r}, t)$ has a general time dependence.

To get a feeling of what the problem is, we consider the charged parallel-plate capacitor sketched in Figure 1.8a [52, Section 7.9], [24, Section 23], [18, Section 3.1 and Fig. 1.2]. In principle, a static electric field exists everywhere in space, but in practice it is substantially non-zero only in the region between the plates, and the system is stable. Next, we suppose to connect the two plates with a thin conducting wire (Figure 1.8b): the charges on the plates have now a means (the wire) to redistribute themselves, in a manner of speaking, so that the system reaches a new final stable configuration. In a very short time the negative charges (electrons) from the lower plate flow along the wire, reach and neutralize the excess positive charges (ions) on the upper plate. By convention, the flow of negative charges along the wire is described as a time-varying current density $\mathcal{J}(\mathbf{r}, t)$ from the positively charged upper plate. In the process, both the current density and the electric field between the plates rapidly decrease with time until they vanish.

We would like to apply the global form (1.77) to the system formed by the capacitor and the wire. To this purpose, we are at liberty to pick up any smooth open surface S, but only two choices are relevant:

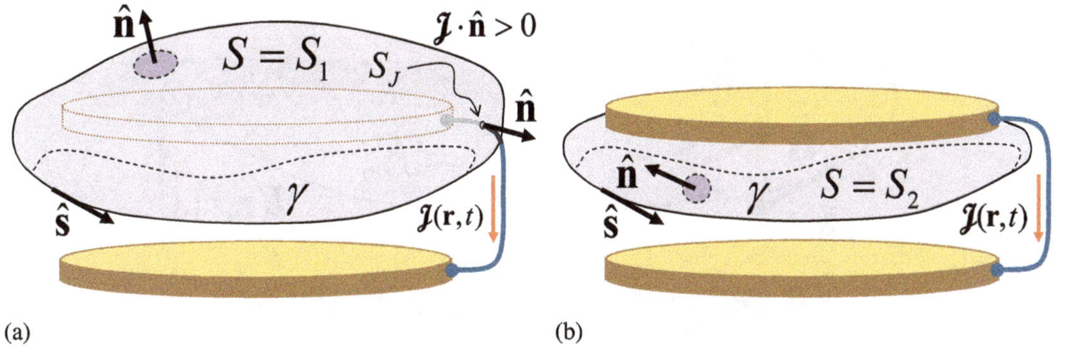

Figure 1.9 For showing that the Ampère law (1.78) fails for time-varying currents: (a) the surface S_1 intersects the wire; (b) the surface S_2 lies between the plates.

(1) the boundary γ sits between the plates and does not encircle the wire, and the surface $S = S_1$ is 'punctured' by the wire (Figure 1.9a);
(2) the boundary γ is taken the same as before, but $S = S_2$ is entirely contained between the plates so that it does not intersect the wire (Figure 1.9b).

In the first scenario, we have

$$\oint_\gamma \mathrm{d}s\, \hat{\mathbf{s}}(\mathbf{r}) \cdot \mathcal{H}(\mathbf{r}, t) = \int_{S_J} \mathrm{d}S\, \hat{\mathbf{n}}(\mathbf{r}) \cdot \mathcal{J}(\mathbf{r}, t) = I(t) > 0 \tag{1.79}$$

where the flux integral has been restricted to $S_J \subset S_1$, i.e., the part of S_1 over which $\mathcal{J}(\mathbf{r}, t)$ is actually non-zero. In the second case, we find

$$\oint_\gamma \mathrm{d}s\, \hat{\mathbf{s}}(\mathbf{r}) \cdot \mathcal{H}(\mathbf{r}, t) = 0 \tag{1.80}$$

because nowhere on S_2 is $\mathcal{J}(\mathbf{r}, t)$ non-null. These two findings are patently at odds with each other, whereas (1.77), to be truly useful, ought to predict the same result regardless of the specific choice made for S. In fact, the root cause of the discrepancy is that (1.78) together with (A.39) implies

$$\nabla \cdot \mathcal{J}(\mathbf{r}, t) = \nabla \cdot \nabla \times \mathcal{H}(\mathbf{r}, t) = 0 \tag{1.81}$$

but this equation cannot be true — the divergence of $\mathcal{J}(\mathbf{r}, t)$ cannot vanish — inasmuch as $\mathcal{J}(\mathbf{r}, t)$ must also satisfy the continuity equation (1.46) in the region occupied by the wire.

It took the genius of Maxwell to fix (1.78) in 1865 by noticing that a divergence-free current density is obtained if we make use of the Gauss law (1.44) to eliminate the charge density from the continuity equation [3, 5]. To elucidate:

$$0 = \nabla \cdot \mathcal{J}(\mathbf{r}, t) + \frac{\partial}{\partial t}\varrho(\mathbf{r}, t) = \nabla \cdot \left[\mathcal{J}(\mathbf{r}, t) + \frac{\partial}{\partial t}\mathcal{D}(\mathbf{r}, t)\right] = \nabla \cdot \mathcal{J}_{\text{tot}}(\mathbf{r}, t) \tag{1.82}$$

having swapped time and space derivatives of $\mathcal{D}(\mathbf{r}, t)$ if the latter is twice differentiable for $(\mathbf{r}, t) \in \mathbb{R}^3 \times \mathbb{R}$. Apparently, (1.82) is compatible with $\nabla \cdot \nabla \times \mathcal{H} = 0$, as implied by the Ampère-Maxwell law (1.34).

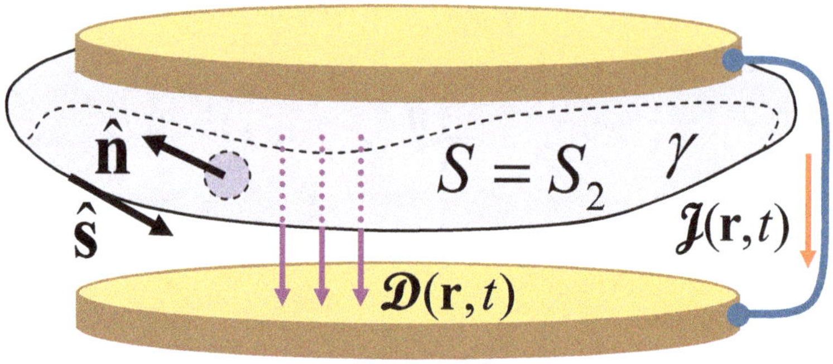

Figure 1.10 The Ampère-Maxwell law (1.13) *applied to the capacitor-wire system of Figure 1.8b.*

This prompts us to examine the capacitor-wire system again, but this time armed with the full-fledged global Ampère-Maxwell law (Figure 1.9a). For the first choice of S we have

$$\oint_\gamma \mathrm{d}s\, \hat{\mathbf{s}}(\mathbf{r}) \cdot \mathcal{H}(\mathbf{r}, t) = \int_{S_J} \mathrm{d}S\, \hat{\mathbf{n}}(\mathbf{r}) \cdot \mathcal{J}(\mathbf{r}, t) > 0 \tag{1.83}$$

where the contribution of the displacement current has been neglected on the grounds that the displacement vector is essentially confined between the plates of the capacitor, and $\hat{\mathbf{n}}(\mathbf{r}) \cdot \mathcal{J}(\mathbf{r}, t) > 0$ for $\mathbf{r} \in S_J$, because the current density flows from the upper plate to the lower plate, as argued above.

For the second choice (Figure 1.10) we obtain

$$\oint_\gamma \mathrm{d}s\, \hat{\mathbf{s}}(\mathbf{r}) \cdot \mathcal{H}(\mathbf{r}, t) = \int_{S_2} \mathrm{d}S\, \hat{\mathbf{n}}(\mathbf{r}) \cdot \frac{\partial}{\partial t}\mathcal{D}(\mathbf{r}, t) > 0 \tag{1.84}$$

where $\mathcal{J}(\mathbf{r}, t)$ contributes naught, since it does not puncture S_2. Also, to show that the right-hand side is positive, we observe that the displacement vector (like the electric field) is directed from the positively charged plate towards the negatively charged one. Therefore, $\hat{\mathbf{n}}(\mathbf{r}) \cdot \mathcal{D}(\mathbf{r}, t) < 0$ for $\mathbf{r} \in S$ with our choices of S and the normal thereon, and

$$\hat{\mathbf{n}}(\mathbf{r}) \cdot \frac{\partial}{\partial t}\mathcal{D}(\mathbf{r}, t) = \frac{\partial}{\partial t}[\hat{\mathbf{n}}(\mathbf{r}) \cdot \mathcal{D}(\mathbf{r}, t)] > 0 \tag{1.85}$$

since $\hat{\mathbf{n}}(\mathbf{r}) \cdot \mathcal{D}(\mathbf{r}, t) \to 0^-$ as $t \to +\infty$. Since the magnetic field generated during the capacitor discharge is one and only, the circulation of $\mathcal{H}(\mathbf{r}, t)$ along γ — which is the boundary of both S_1 and S_2 — is unique, and the left-hand sides of (1.83) and (1.84) coincide. Had we not introduced a suitable non-zero *displacement current density* between the plates of the capacitor, then (1.83) and (1.84) would have been inconsistent. In fact, by comparing (1.83) and (1.84) we find

$$I(t) := \int_{S_J} \mathrm{d}S\, \hat{\mathbf{n}}(\mathbf{r}) \cdot \mathcal{J}(\mathbf{r}, t) = \int_{S_2} \mathrm{d}S\, \hat{\mathbf{n}}(\mathbf{r}) \cdot \frac{\partial}{\partial t}\mathcal{D}(\mathbf{r}, t) \tag{1.86}$$

i.e., the electric current $I(t)$ that flows along the wire (more precisely, through S_J) equals the displacement current between the capacitor plates.

1.5 Time-harmonic fields and sources

The electromagnetic entities $\mathcal{E}$, $\mathcal{B}$, $\mathcal{D}$ and $\mathcal{H}$ introduced in Section 1.1 may exhibit an arbitrary dependence on time, and the temporal evolution of the electromagnetic field is determined by that of the sources $\varrho(\mathbf{r}, t)$ and $\mathcal{J}(\mathbf{r}, t)$.

In many practical engineering applications, e.g., the design of antenna systems, it is sufficient and convenient to consider sources and fields which are *periodic functions* of time, because this assumption greatly simplifies the solution of Maxwell's equations. A periodic time-dependence is, of course, a mathematical abstraction, in that a periodic phenomenon should have started at an infinitely remote point in time ($t = -\infty$) and should keep on repeating unchanged forever ($t = +\infty$) .

Still, we can think of a periodic electromagnetic field as the limiting *sinusoidal state* [18, Chapter 8] attained (after a transient has passed) by a time-varying field which was 'turned on' at some time t_0. The shortest time interval T (physical dimension: second, s) that must elapse in order for sources and electromagnetic entities to repeat themselves is called the *period*. The number of *repetitions* or *cycles* that occur in a unit of time is provided by the reciprocal of the period and referred to as the *frequency* (physical dimension: hertz, 1/s = Hz). A third parameter related to T and f is the *circular* or *angular frequency*, defined as

$$\omega := 2\pi f = \frac{2\pi}{T} \tag{1.87}$$

with physical dimension: radian per second, rad/s. If sources and electromagnetic entities are characterized by a single angular frequency ω, we can suppose that the spatial variation is separated from the temporal dependence in the form [53, Chapter 19], [26, Appendix A.5]

$$\mathcal{E}(\mathbf{r}, t) = \mathbf{E}_0(\mathbf{r}) \cos(\omega t + \phi_E) = \mathrm{Re}\{\mathbf{E}_0(\mathbf{r}) \mathrm{e}^{\mathrm{j}\phi_E} \mathrm{e}^{\mathrm{j}\omega t}\} = \mathrm{Re}\{\mathbf{E}(\mathbf{r}) \mathrm{e}^{\mathrm{j}\omega t}\} \tag{1.88}$$

where[1]

- $\mathbf{E}_0(\mathbf{r}) \in \mathbb{R}^3$ (physical dimension: V/m) is the *spatial* part of the field;
- $\cos(\omega t + \phi_E) \in [-1, 1]$ is the *temporal*, periodic part of the field, and $\phi_E \in \mathbb{R}$ is the phase;
- $\mathbf{E}(\mathbf{r}) := \mathbf{E}_0(\mathbf{r}) \exp(\mathrm{j}\,\phi_E)$ is a three-dimensional complex vector field which is called the *phasor* of the electric field $\mathcal{E}(\mathbf{r}, t)$;
- the magnitude $|\mathbf{E}_0(\mathbf{r})|$ represents the *peak value* of $|\mathcal{E}(\mathbf{r}, t)|$ that is attained whenever $\omega t + \phi_E = 2\pi n$, $n \in \mathbb{Z}$.

Perfectly analogous expressions hold true for the other three entities and the sources $\mathcal{J}(\mathbf{r}, t)$ and $\varrho(\mathbf{r}, t)$. Needless to say, the phasor $\rho(\mathbf{r})$ of the charge density is a complex *scalar* field. When representation (1.88) is adopted, we say that fields and sources are **time-harmonic** [15,33,52,58,59, 61] or **monochromatic** [14,20]. The latter designation refers to the fact that *light* of a given color may be identified as an electromagnetic disturbance of a single well-defined frequency.

For the sake of completeness, we mention that phasors might be introduced alternatively by assuming the Ansatz

$$\mathcal{E}(\mathbf{r}, t) = \mathbf{E}_0(\mathbf{r}) \sin(\omega t + \phi_E) = \mathrm{Im}\{\mathbf{E}_0(\mathbf{r}) \mathrm{e}^{\mathrm{j}\phi_E} \mathrm{e}^{\mathrm{j}\omega t}\} = \mathrm{Im}\{\mathbf{E}(\mathbf{r}) \mathrm{e}^{\mathrm{j}\omega t}\} \tag{1.89}$$

but rarely is this choice used. Furthermore, some Authors (e.g., [33]) prefer to include a factor $\sqrt{2}$ whereby definition (1.88) is replaced by

$$\mathcal{E}(\mathbf{r}, t) = \sqrt{2}\mathbf{E}_0(\mathbf{r}) \cos(\omega t + \phi_E) = \sqrt{2}\mathrm{Re}\{\mathbf{E}_0(\mathbf{r}) \mathrm{e}^{\mathrm{j}\phi_E} \mathrm{e}^{\mathrm{j}\omega t}\} = \sqrt{2}\mathrm{Re}\{\mathbf{E}(\mathbf{r}) \mathrm{e}^{\mathrm{j}\omega t}\} \tag{1.90}$$

[1] In electrical engineering (e.g. [37,57–59]) the imaginary unit is customarily indicated with the symbol j defined in (B.3), and the time dependence of periodic fields and sources takes the form $\exp(\mathrm{j}\,\omega t)$. By contrast, in the mathematical and physical literature (e.g. [7,20,60]) the imaginary unit is generally called $\mathrm{i} \equiv -\mathrm{j}$ whereby the time-varying part of (1.88) reads $\exp(-\mathrm{i}\omega t)$.

Table 1.1 Correspondence between time-harmonic fields and phasors

Field	Phasor	Field	Phasor
$\mathbf{E}_0(\mathbf{r})\cos(\omega t + \phi_E)$	$\mathbf{E}_0(\mathbf{r})\exp(\mathrm{j}\,\phi_E)$	$\frac{\partial}{\partial t}\mathbf{E}_0(\mathbf{r})\cos(\omega t + \phi_E)$	$\mathrm{j}\,\omega\mathbf{E}_0(\mathbf{r})\exp(\mathrm{j}\,\phi_E)$
$\mathbf{E}_0(\mathbf{r})\sin(\omega t + \phi_E)$	$-\mathrm{j}\,\mathbf{E}_0(\mathbf{r})\exp(\mathrm{j}\,\phi_E)$	$\frac{\partial}{\partial t}\mathbf{E}_0(\mathbf{r})\sin(\omega t + \phi_E)$	$\omega\mathbf{E}_0(\mathbf{r})\exp(\mathrm{j}\,\phi_E)$
$\mathbf{E}_0(\mathbf{r})\cos(\omega t)$	$\mathbf{E}_0(\mathbf{r})$	$\frac{\partial^n}{\partial t^n}\mathbf{E}_0(\mathbf{r})\cos(\omega t + \phi_E)$	$(\mathrm{j}\,\omega)^n\mathbf{E}_0(\mathbf{r})\exp(\mathrm{j}\,\phi_E)$
$\mathbf{E}_0(\mathbf{r})\sin(\omega t)$	$-\mathrm{j}\,\mathbf{E}_0(\mathbf{r})$	$\frac{\partial^n}{\partial t^n}\mathbf{E}_0(\mathbf{r})\sin(\omega t + \phi_E)$	$-\mathrm{j}(\mathrm{j}\,\omega)^n\mathbf{E}_0(\mathbf{r})\exp(\mathrm{j}\,\phi_E)$

in which case $|\mathbf{E}(\mathbf{r})| = |\mathbf{E}_0(\mathbf{r})|$ represents the *root-mean-square value* of $|\mathcal{E}(\mathbf{r}, t)|$, viz.,

$$\left(\frac{1}{T}\int_0^T \mathrm{d}t\,|\mathcal{E}(\mathbf{r}, t)|^2\right)^{1/2} = \left\{\frac{2}{T}\int_0^T \mathrm{d}t\,|\mathbf{E}_0(\mathbf{r})|^2[\cos(\omega t + \phi_E)]^2\right\}^{1/2} = |\mathbf{E}_0(\mathbf{r})| \tag{1.91}$$

whereas the peak value is given by $\sqrt{2}|\mathbf{E}_0(\mathbf{r})|$. Anyway, we shall adhere to the convention of (1.88) throughout.

Since the association between a phasor and its temporal counterpart is unique, we can equivalently work with phasors, solve Maxwell's equations, and at last reinstate the time-dependence, namely,

$$\mathcal{E}(\mathbf{r}, t) = \mathrm{Re}\{\mathbf{E}(\mathbf{r})e^{\mathrm{j}\,\omega t}\} = \mathrm{Re}\{[\mathbf{E}'(\mathbf{r}) + \mathrm{j}\,\mathbf{E}''(\mathbf{r})]e^{\mathrm{j}\,\omega t}\} = \mathbf{E}'(\mathbf{r})\cos(\omega t) - \mathbf{E}''(\mathbf{r})\sin(\omega t) \tag{1.92}$$

where $\mathbf{E}'(\mathbf{r})$ and $\mathbf{E}''(\mathbf{r})$ are the real and the imaginary parts of the phasor $\mathbf{E}(\mathbf{r})$.

The advantage of manipulating phasors such as (1.88) becomes apparent when we need to take the time derivatives of $\mathcal{E}(\mathbf{r}, t)$. To the first order, we have

$$\begin{aligned}\frac{\partial}{\partial t}\mathcal{E}(\mathbf{r}, t) &= \frac{\partial}{\partial t}\mathrm{Re}\{\mathbf{E}(\mathbf{r})e^{\mathrm{j}\,\omega t}\} = \frac{1}{2}\frac{\partial}{\partial t}\mathbf{E}(\mathbf{r})e^{\mathrm{j}\,\omega t} + \frac{1}{2}\frac{\partial}{\partial t}\mathbf{E}^*(\mathbf{r})e^{-\mathrm{j}\,\omega t} \\ &= \frac{1}{2}\left\{\mathrm{j}\,\omega\mathbf{E}(\mathbf{r})e^{\mathrm{j}\,\omega t} + [\mathrm{j}\,\omega\mathbf{E}(\mathbf{r})e^{\mathrm{j}\,\omega t}]^*\right\} = \mathrm{Re}\left\{\mathrm{j}\,\omega\mathbf{E}(\mathbf{r})e^{\mathrm{j}\,\omega t}\right\}\end{aligned} \tag{1.93}$$

on account of (B.6) applied to the Cartesian components of $\mathbf{E}(\mathbf{r})$ and the linearity of the differential operator. In light of the very definition (1.88), (1.93) provides us with a practical rule: the phasor of $\partial\mathcal{E}/\partial t$ is obtained by multiplying the phasor of $\mathcal{E}(\mathbf{r}, t)$ by $\mathrm{j}\,\omega$. Higher-order derivatives with respect to time are computed in like manner. For instance, to the order $n \in \mathbb{N}$ we have

$$\begin{aligned}\frac{\partial^n}{\partial t^n}\mathcal{E}(\mathbf{r}, t) &= \frac{\partial^n}{\partial t^n}\mathrm{Re}\{\mathbf{E}(\mathbf{r})e^{\mathrm{j}\,\omega t}\} = \frac{1}{2}\frac{\partial^n}{\partial t^n}\mathbf{E}(\mathbf{r})e^{\mathrm{j}\,\omega t} + \frac{1}{2}\frac{\partial^n}{\partial t^n}\mathbf{E}^*(\mathbf{r})e^{-\mathrm{j}\,\omega t} \\ &= \frac{1}{2}\left\{(\mathrm{j}\,\omega)^n\mathbf{E}(\mathbf{r})e^{\mathrm{j}\,\omega t} + \left[(\mathrm{j}\,\omega)^n\mathbf{E}(\mathbf{r})e^{\mathrm{j}\,\omega t}\right]^*\right\} = \mathrm{Re}\left\{(\mathrm{j}\,\omega)^n\mathbf{E}(\mathbf{r})e^{\mathrm{j}\,\omega t}\right\}\end{aligned} \tag{1.94}$$

which in practice means the phasor of the nth time-derivative of $\mathcal{E}(\mathbf{r}, t)$ is obtained by multiplying $\mathbf{E}(\mathbf{r})$ by $(\mathrm{j}\,\omega)^n$. A list of forward and backward transformation rules between time-harmonic fields and phasors thereof is given in Table 1.1.

Thanks to these preliminary considerations we can specialize the Maxwell equations to the case of time-harmonic fields and sources. We focus on the local form of the laws, though similar steps can be taken to turn the global forms into their time-harmonic counterparts. To be specific, we consider the Ampère-Maxwell law (1.34) and, in accordance with (1.88) and (1.93), introduce the phasors of $\mathcal{H}(\mathbf{r},t)$, $\partial\mathcal{D}(\mathbf{r},t)/\partial t$ and $\mathcal{J}(\mathbf{r},t)$, viz.,

$$\nabla\times\mathrm{Re}\{\mathbf{H}(\mathbf{r})e^{j\omega t}\} = \mathrm{Re}\{j\,\omega\mathbf{D}(\mathbf{r})e^{j\omega t}\} + \mathrm{Re}\{\mathbf{J}(\mathbf{r})e^{j\omega t}\} \tag{1.95}$$

$$\mathrm{Re}\{\nabla\times[\mathbf{H}(\mathbf{r})e^{j\omega t}]\} = \mathrm{Re}\{[j\,\omega\mathbf{D}(\mathbf{r}) + \mathbf{J}(\mathbf{r})]e^{j\omega t}\} \tag{1.96}$$

$$e^{j\omega t}\nabla\times\mathbf{H}(\mathbf{r}) = [j\,\omega\mathbf{D}(\mathbf{r}) + \mathbf{J}(\mathbf{r})]e^{j\omega t} \tag{1.97}$$

where we have invoked, in succession, (B.6), the linearity of the curl operator, (H.50) with $\exp(j\,\omega t)$ being independent of $\mathbf{r}$, and lastly the fact that the real parts of two complex-valued functions coincide if the functions coincide. Since the complex exponential is never zero for $\omega t \in \mathbb{C}$, (1.97) is satisfied if we enforce the equality of the vector coefficients in both sides. This procedure yields the differential form of the **Ampère-Maxwell law** for time-harmonic fields and sources

$$\nabla\times\mathbf{H}(\mathbf{r}) = j\,\omega\mathbf{D}(\mathbf{r}) + \mathbf{J}(\mathbf{r}) \tag{1.98}$$

which amounts, in practice, to hiding and implying the time-dependence $\exp(j\,\omega t)$.

Similar steps lead to the differential form of the remaining Maxwell equations and of the conservation of charge for time-harmonic fields and sources. We quote the result for future reference:

$$\nabla\times\mathbf{E}(\mathbf{r}) = -j\,\omega\mathbf{B}(\mathbf{r}) \tag{1.99}$$

$$\nabla\cdot\mathbf{D}(\mathbf{r}) = \rho(\mathbf{r}) \tag{1.100}$$

$$\nabla\cdot\mathbf{B}(\mathbf{r}) = 0 \tag{1.101}$$

$$\nabla\cdot\mathbf{J}(\mathbf{r}) = -j\,\omega\rho(\mathbf{r}) \tag{1.102}$$

where $\mathbf{B}(\mathbf{r}) := \mathbf{B}_0(\mathbf{r})\exp(j\,\phi_B)$ and so forth.

Equations (1.98)-(1.102) provide the mathematical formulation of Maxwell's theory of electromagnetism in the frequency domain, i.e., the domain of the variable ω. In this setup, the Maxwell equations relate sources and fields which are 'frozen' in space, so to speak. Furthermore, (1.98) and (1.99) state that the time-harmonic magnetic entities *lag behind* the electric ones by a quarter of a period $T = 1/f = 2\pi/\omega$. For instance, using the Faraday law yields

$$\begin{aligned}\mathrm{Re}\{\nabla\times\mathbf{E}_0(\mathbf{r})e^{j\phi_E}e^{j\omega t}\} &= -\mathrm{Re}\{j\,\omega\mathbf{B}_0(\mathbf{r})e^{j\phi_B}e^{j\omega t}\}\\ &= \mathrm{Re}\left\{\omega\mathbf{B}_0(\mathbf{r})\exp\left[j\left(\omega t + \phi_B - \frac{\pi}{2}\right)\right]\right\}\end{aligned} \tag{1.103}$$

$$\nabla\times\mathbf{E}_0(\mathbf{r})\cos(\omega t + \phi_E) = \omega\mathbf{B}_0(\mathbf{r})\cos\left(\omega t + \phi_B - \frac{\pi}{2}\right) \tag{1.104}$$

and in order for the vector fields on either side of this equation to exhibit the same time-dependence we must have

$$\phi_B = \phi_E + \frac{\pi}{2} = \phi_E + \frac{\omega T}{4} \tag{1.105}$$

which means precisely that the magnetic induction — proportional to $\cos(\omega t + \phi_B)$ — lags behind the electric field. We may also say that $\mathcal{E}$ and $\mathcal{B}$ are in *phase quadrature*. This situation is illustrated in Figure 1.11 for $\omega t \in [-2\pi, 2\pi]$. Similar observations apply to $\mathcal{J}$ and ϱ by virtue of the conservation of charge (1.102).

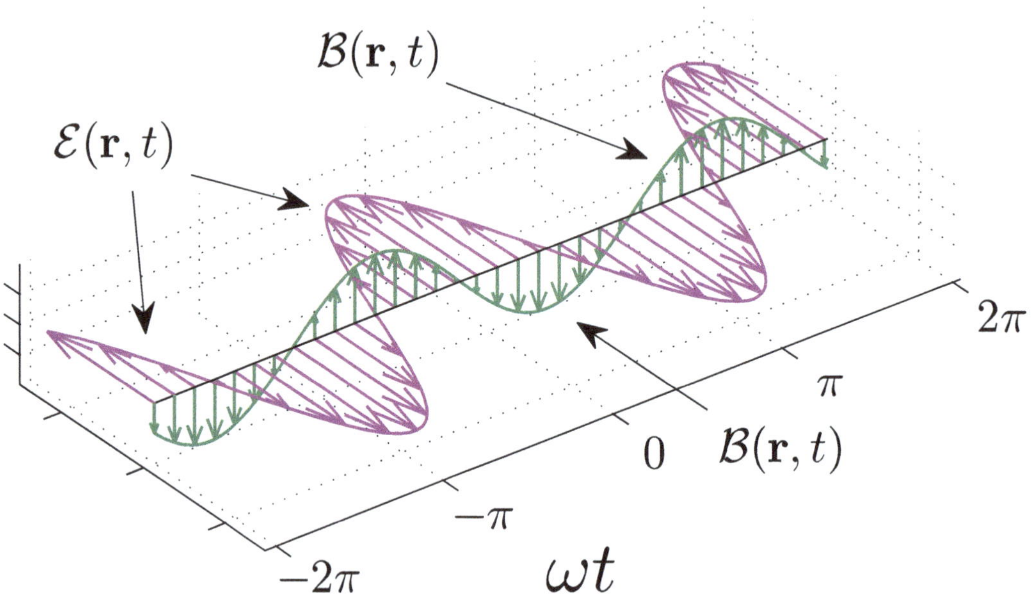

Figure 1.11 Temporal evolution of time-harmonic electric field and magnetic induction in a point in space: the magnetic vector lags behind the electric one by a quarter of a period.

We conclude by pointing out that in the time-harmonic regime it is correct to state that the two Gauss laws follow from (1.98) and (1.99) by taking the divergence of both sides and invoking the continuity equation (1.102). Indeed, direct calculation shows

$$0 = \nabla \cdot \nabla \times \mathbf{H}(\mathbf{r}) = \mathrm{j}\,\omega \nabla \cdot \mathbf{D}(\mathbf{r}) + \nabla \cdot \mathbf{J}(\mathbf{r}) = \mathrm{j}\,\omega[\nabla \cdot \mathbf{D}(\mathbf{r}) - \rho(\mathbf{r})] \tag{1.106}$$

$$0 = \nabla \cdot \nabla \times \mathbf{E}(\mathbf{r}) = -\mathrm{j}\,\omega \nabla \cdot \mathbf{B}(\mathbf{r}) \tag{1.107}$$

whence we obtain (1.100) and (1.101) since $\omega \neq 0$. In fact, there is no ambiguity in the process for no arbitrary constant need be set by resorting to additional hypotheses. But then, one can argue that assuming the fields and sources are periodic functions of time may precisely be regarded as the further condition employed to link the Gauss laws to (1.98) and (1.99).

1.6 Constitutive relationships

The Maxwell equations as stated in (1.20), (1.23), (1.34) and (1.44) are formulated in a coordinate-free notation, i.e., they do not rely on a specific system of coordinates and associated basis vectors (Appendices A.1 and A.2) to represent the differential operators and the electromagnetic entities and sources.

For the sake of argument we choose Cartesian coordinates and observe that (1.20) and (1.34) each amount to three coupled scalar equations, since $\mathcal{E}$, $\mathcal{B}$, $\mathcal{D}$, $\mathcal{H}$ and $\mathcal{J}$ are three-dimensional vectors. A quick check reveals that we have at our disposal a grand total of eight equations to be solved for twelve scalar fields, namely, the Cartesian components of $\mathcal{E}$, $\mathcal{B}$, $\mathcal{D}$, $\mathcal{H}$. Thus, even when supplemented with initial and boundary conditions, the system of equations (1.20), (1.23), (1.34) and (1.44) is under-determined and, as such, cannot provide a unique solution, that is, a set of formulas which relate $\mathcal{E}$, $\mathcal{B}$, $\mathcal{D}$, $\mathcal{H}$ univocally to $\mathcal{J}$ and ϱ.

Consequently, we need to reduce the number of unknowns by introducing some relation among the entities of the electromagnetic field. Since $\mathcal{E}$ and $\mathcal{B}$ are fundamental entities (Section 1.1), it seems logical to assume the following Ansatz (see Sections 3.7 and 5.6)

$$\mathcal{D} = g_D(\mathcal{E}, \mathcal{B}) \tag{1.108}$$

$$\mathcal{H} = g_H(\mathcal{E}, \mathcal{B}) \tag{1.109}$$

where g_D and g_H are quite arbitrary functions or, even more generally, operators acting on the components of $\mathcal{E}$ and $\mathcal{B}$. However, it has become customary to consider $\mathcal{D}$ and $\mathcal{B}$ (the flux densities) as secondary entities related to $\mathcal{E}$ and $\mathcal{H}$, namely,

$$\mathcal{D} = f_D(\mathcal{E}, \mathcal{H}) \tag{1.110}$$

$$\mathcal{B} = f_B(\mathcal{E}, \mathcal{H}) \tag{1.111}$$

where again f_D and f_B are arbitrary, possibly non-linear functions or operators. Equations (1.110) and (1.111) are called the **constitutive relationships** [10, Section 8.1], [11, Chapter 15] or sometimes **laws of behavior** (e.g., [62, pp. 71 and ff.]). The explicit form of f_D and f_B depends on the microscopic properties of the medium in which the electromagnetic field is generated by the sources: we postpone a detailed discussion until Chapter 12, and here we only recall a few preliminary concepts.

Before we proceed, though, we must ascertain whether or not the supplementary conditions (1.110) and (1.111) ensure the solvability of Maxwell's equations. If we use the constitutive relationships to eliminate $\mathcal{D}$ and $\mathcal{B}$ from (1.20), (1.23), (1.34) and (1.44), we obtain a differential system with just six unknowns (the scalar components of $\mathcal{E}$ and $\mathcal{H}$) whereas the total number of equations is unaltered and still equal to eight. Thus, it would appear that now the system in question is overspecified, inasmuch as we have, perhaps, two equations too many. As a matter of fact, a detailed analysis shows that Maxwell's equations together with the constitutive relationships are just the right mix of equations and unknowns necessary to obtain a unique solution so long as suitable boundary and initial conditions are specified. We shall elaborate on the topic in Chapters 6 and 8.

In **free space** or vacuum the constitutive relations take on a particularly simple form

$$\mathcal{D}(\mathbf{r}, t) := \varepsilon_0 \mathcal{E}(\mathbf{r}, t) \tag{1.112}$$

$$\mathcal{B}(\mathbf{r}, t) := \mu_0 \mathcal{H}(\mathbf{r}, t) \tag{1.113}$$

where ε_0 and μ_0 are the *constitutive parameters*. More precisely,

- ε_0 is a constant called the (electric) *permittivity of vacuum* and carries physical dimension farad per meter (F/m);
- μ_0 is a constant called the (magnetic) *permeability of vacuum* and has dimension henry per meter (H/m).

Not only are the values assigned to ε_0 and μ_0 a matter of choice and convenience [8, Section 1.8], they even depend on the specific system of units adopted for the electromagnetic entities. With the choices of the SI for $\mathcal{H}$ and $\mathcal{B}$ (Section 1.1) the permeability is set as

$$\mu_0 = 4\pi \cdot 10^{-7} \text{ H/m} \tag{1.114}$$

(parenthetically, the factor 4π is ubiquitous in electromagnetics) and the permittivity is determined so as to satisfy the condition

$$\frac{1}{\sqrt{\varepsilon_0 \mu_0}} = c_0 \tag{1.115}$$

whence

$$\varepsilon_0 \approx 8.854 \cdot 10^{-12} \text{ F/m} \tag{1.116}$$

where c_0 is the speed of light in vacuum [see (1.53)]. The rationale for this choice will become clear in Section 1.8 with the derivation of the wave equations from (1.20) and (1.34).

Owing to the simplicity of (1.112) and (1.113), it is desirable to retain the same functional form for **material media**, namely,

$$\mathcal{D}(\mathbf{r}, t) := \varepsilon(\mathbf{r})\mathcal{E}(\mathbf{r}, t) \tag{1.117}$$

$$\mathcal{B}(\mathbf{r}, t) := \mu(\mathbf{r})\mathcal{H}(\mathbf{r}, t) \tag{1.118}$$

where $\varepsilon(\mathbf{r})$ and $\mu(\mathbf{r})$ are scalar fields that take on values other than those of ε_0 and μ_0 in general. Media which admit constitutive relationships such as (1.117) and (1.118) are termed **isotropic**, because $\mathcal{D}$ and $\mathcal{B}$ are parallel to $\mathcal{E}$ and $\mathcal{H}$, respectively. It follows that vacuum is isotropic, too.

A medium for which $\varepsilon(\mathbf{r})$ is substantially different than ε_0, while $\mu(\mathbf{r}) \approx \mu_0$, is said **dielectric**. Conversely, a medium for which $\mu(\mathbf{r})$ departs substantially from μ_0, while $\varepsilon(\mathbf{r}) \approx \varepsilon_0$, is said to be **magnetic**. Furthermore, if $\varepsilon(\mathbf{r})$ and $\mu(\mathbf{r})$ are constant, the medium is said **homogeneous**, otherwise **inhomogeneous**. Parenthetically, we mention that (1.112) and (1.113) hold for a stationary observer in an empty space where gravitational effects are absent or negligible. If gravity is included, then (1.112) and (1.113) must be traded for (1.117) and (1.118), as if even vacuum were, in fact, an isotropic material medium [44, Sections 9.8, 11.13, and 11.14].

In **conductors** the free electrons are set in motion by an applied electric field, and this gives rise to the so-called conduction current density $\mathcal{J}_c(\mathbf{r}, t)$. Specifically, in the presence of a conduction current density the Ampère-Maxwell law in differential form reads

$$\nabla \times \mathcal{H}(\mathbf{r}, t) = \frac{\partial}{\partial t}\mathcal{D}(\mathbf{r}, t) + \mathcal{J}_c(\mathbf{r}, t) + \mathcal{J}(\mathbf{r}, t) \tag{1.119}$$

where $\mathcal{J}_c(\mathbf{r}, t)$ is not an independent source but rather an additional unknown. It should be realized that from a physical viewpoint $\mathcal{J}(\mathbf{r}, t)$ is a conduction current as well. However, in (1.119) we assume $\mathcal{J}(\mathbf{r}, t)$ to be a known, independent source term, whereas $\mathcal{J}_c(\mathbf{r}, t)$ is a secondary contribution which depends on the field. Thus, in order to ensure solvability (see the discussion above) we need to complement (1.117) and (1.118) with a similar relation for $\mathcal{J}_c$, viz.,

$$\mathcal{J}_c(\mathbf{r}, t) := \sigma(\mathbf{r})\mathcal{E}(\mathbf{r}, t) \tag{1.120}$$

which is known as the **Ohm law**. The scalar field $\sigma(\mathbf{r})$ is called the *conductivity* of the medium and carries the physical dimension of siemens per meter (S/m) or $1/(\Omega\text{m})$. The simple proportionality relation embodied in (1.120) can hold true only if the free distance between the charges is 'small' as compared to the local spatial variation of the electric field, or else more complicated, non-local relationships must be adopted [63, p. 193]. Conducting media which obey (1.120) are termed isotropic, and conductors for which $\sigma(\mathbf{r})$ is a constant are said homogeneous.

In a homogeneous conductor, endowed with constant parameters ε and σ, all charges *decay* with time, i.e., permanent volume charges are physically impossible in the absence of *generators* which can sustain an electromagnetic field (see Section 4.7.3). Indeed, by combining the continuity equation (1.46), the constitutive relationships (1.120) and (1.117), we find

$$0 = \frac{\partial \varrho(\mathbf{r}, t)}{\partial t} + \nabla \cdot \mathcal{J}_c(\mathbf{r}, t) = \frac{\partial \varrho(\mathbf{r}, t)}{\partial t} + \sigma \nabla \cdot \mathcal{E}(\mathbf{r}, t) = \frac{\partial \varrho(\mathbf{r}, t)}{\partial t} + \frac{\sigma}{\varepsilon}\nabla \cdot \mathcal{D}(\mathbf{r}, t) \tag{1.121}$$

which becomes

$$\frac{\partial}{\partial t}\varrho(\mathbf{r},t)+\frac{\sigma}{\varepsilon}\varrho(\mathbf{r},t)=0, \qquad t\in[t_0,+\infty) \tag{1.122}$$

by virtue of the Gauss law (1.44). The solution to this differential equation reads

$$\varrho(\mathbf{r},t)=\varrho(\mathbf{r},t_0)\mathrm{e}^{-t\sigma/\varepsilon}, \qquad t\in[t_0,+\infty) \tag{1.123}$$

where $\varrho(\mathbf{r},t_0)$ is the initial charge density in the conductor. Therefore, if $\sigma>0$, as is true for real conductors, (1.123) shows that the charge ultimately decays to zero as $t\to+\infty$, and the shorter the relaxation time constant ε/σ, the faster the decay rate. Since the charge is conserved in a closed system, (1.123) implies that for a finite-sized conductor the charge moves towards the surface until an equilibrium state is reached (see Chapter 2). Furthermore, after the transient has finished or at any rate in steady conditions (1.123) implies

$$\sigma\varrho(\mathbf{r})=0 \tag{1.124}$$

because $\partial\varrho/\partial t=0$. Under the same hypotheses, there is no conduction current, and hence $\mathcal{E}(\mathbf{r})=0$ in the conductor.[2] By contrast, we observe that in a homogeneous dielectric medium $\sigma=0$, the relaxation time is infinite, and hence charges are permanent (see Section 3.7).

The special case $\sigma(\mathbf{r})\to+\infty$ is termed Perfect Electric Conductor (PEC). Although no such medium exists in nature, still most good conductors exhibit a very large conductivity (Table 7.2, Section 12.3.1 and Figure 12.3), which makes the PEC assumption a very accurate one in practical applications. The electric field in a PEC vanishes identically, because

$$\mathcal{E}(\mathbf{r},t)=\lim_{\sigma\to+\infty}\frac{\mathcal{J}_c(\mathbf{r},t)}{\sigma(\mathbf{r})}=0 \tag{1.125}$$

if the conduction current is finite. Then, the displacement vector $\mathcal{D}(\mathbf{r},t)$ vanishes in view of (1.117), and hence the charge density $\varrho(\mathbf{r},t)$ vanishes as well because of the Gauss law (1.44). Finally, the Faraday law (1.20) yields $\partial\mathcal{B}/\partial t=0$, which requires that the magnetic induction be constant in time. However, following the same reasoning that led us to the formulation of (1.49), i.e., if we postulate that the electromagnetic field was zero at some point in time, we conclude that $\mathcal{B}(\mathbf{r})=0$ in a PEC. Then, as $\mathcal{D}(\mathbf{r},t)=0=\mathcal{H}(\mathbf{r},t)$, the Ampère-Maxwell law states that $\mathcal{J}_c(\mathbf{r},t)=0$, that is, no conduction current is possible inside a PEC.

Anisotropic media and **conductors** admit linear constitutive relationships and a form of the Ohm law which are more general than the simple proportionality expressed by (1.117), (1.118) and (1.120). This happens because $\mathcal{D}$, $\mathcal{J}_c$ and $\mathcal{B}$ are not everywhere parallel to $\mathcal{E}$ and $\mathcal{H}$, respectively, in the medium, and (1.117), (1.118) and (1.120) are modified as follows:

$$\mathcal{D}(\mathbf{r},t):=\overline{\varepsilon}(\mathbf{r})\cdot\mathcal{E}(\mathbf{r},t) \tag{1.126}$$

$$\mathcal{B}(\mathbf{r},t):=\overline{\mu}(\mathbf{r})\cdot\mathcal{H}(\mathbf{r},t) \tag{1.127}$$

$$\mathcal{J}_c(\mathbf{r},t):=\overline{\sigma}(\mathbf{r})\cdot\mathcal{E}(\mathbf{r},t) \tag{1.128}$$

where $\overline{\varepsilon}(\mathbf{r})$, $\overline{\mu}(\mathbf{r})$ and $\overline{\sigma}(\mathbf{r})$ denote dyadic fields [48, 61] and are called the dyadic permittivity, the dyadic permeability and the dyadic conductivity. The Reader is referred to Appendix E for an overview of dyadic calculus.

[2]This is a classic result. On the contrary, a quantum-mechanical description of the conductors shows that the electric field is not zero.

We emphasize that, simple and convenient as they are, the constitutive relationships (1.117), (1.118), (1.120), (1.126), (1.127) and (1.128) are only approximate. Indeed, they imply that a change in $\mathcal{E}$ and $\mathcal{H}$ is *instantaneously* mirrored by $\mathcal{D}$, $\mathcal{B}$ and $\mathcal{J}_c$, i.e., the 'disturbance' or the information propagates with *infinite* velocity within the medium — which violates the postulate of the speed of light being the maximum velocity attainable. In Chapter 12 we shall derive constitutive relationships which, though based on a classical view of the atomic and molecular structure of matter, are physically more sound.

Sometimes it may be convenient to express the constitutive parameters of one medium in terms of those pertinent to another medium which is taken as reference. Since vacuum is the simplest medium we can think of, we define the constitutive parameters relative to those of free space, namely,

$$\varepsilon_{\mathrm{r}}(\mathbf{r}) := \frac{\varepsilon(\mathbf{r})}{\varepsilon_0} \qquad \overline{\varepsilon}_{\mathrm{r}}(\mathbf{r}) := \frac{\overline{\varepsilon}(\mathbf{r})}{\varepsilon_0} \tag{1.129}$$

$$\mu_{\mathrm{r}}(\mathbf{r}) := \frac{\mu(\mathbf{r})}{\mu_0} \qquad \overline{\mu}_{\mathrm{r}}(\mathbf{r}) := \frac{\overline{\mu}(\mathbf{r})}{\mu_0} \tag{1.130}$$

which are dimensionless and called the *relative permittivity* and *relative permeability*, respectively.

1.7 Boundary conditions for fields and currents

The Maxwell equations in differential form (1.20), (1.23), (1.34) and (1.44) are certainly valid in free space as well as in an unbounded homogeneous isotropic medium. On the contrary, there is no guarantee *a priori* that the local form can be stated and applied if the constitutive parameters of the background medium exhibit jump discontinuities. This happens, for example, in the presence of a penetrable, conducting object which occupies a finite region of space V.

On the whole, the global form of the Maxwell equations holds true, as long as all the integrands are finite or possess integrable singularities. However, since we invoked both the Gauss theorem (A.53) and the Stokes theorem (A.55) to derive the local counterpart, the latter applies, strictly speaking, only under the hypotheses for which said theorems are valid.

So, how do we settle the question? Well, in the presence of a penetrable object, we may derive the local form of Maxwell's equations separately in V (i.e., within the object) and in $\mathbb{R}^3 \setminus \overline{V}$, the region complementary to the closure of V. We deliberately leave out the boundary ∂V, because that is precisely where the field entities and the sources may not be continuous and suffer jumps. We may then go on to solve the equations separately in V and $\mathbb{R}^3 \setminus \overline{V}$ and thereafter try and 'stitch' the two parts together on the boundary ∂V consistently.

When we talk of *boundary* or *continuity conditions*, we refer to the set of rules we need to 'stitch' the solutions of Maxwell's differential equations consistently across the interface between two regions endowed with different constitutive parameters. Other common names are *jump*, *matching*, *transmission* or *linking conditions* and to derive them two strategies can be devised:

(a) One applies the global form of Maxwell's equations by choosing suitable vanishingly small regions (open surfaces and volumes) and, in the limit, determines the jump conditions [7, 8, 10, 11, 11, 14, 17, 39, 64].

(b) One manipulates the global form of Maxwell's equations (always valid) and determines the additional hypotheses required for the application of the Gauss theorem and the Stokes theorem [15], [62, pp. 68–69].

The first approach effectively leads to the matching conditions for the fields on either side of a material discontinuity, though it does not provide information on the appropriate local form of

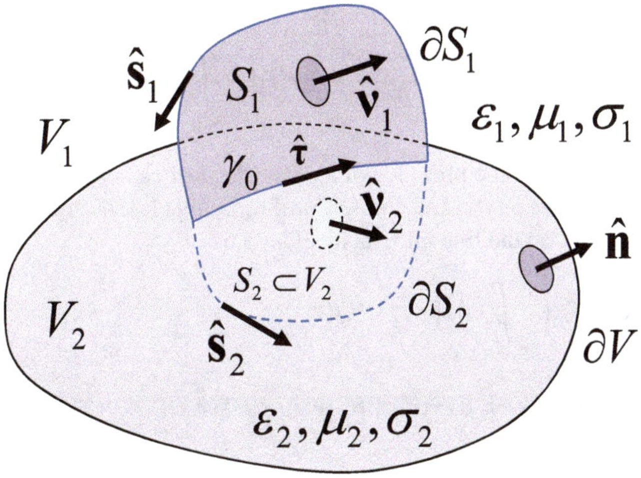

Figure 1.12 A material body confined in the region $V_2 \subset \mathbb{R}^3$ and the geometrical quantities for determining the jump conditions of $\mathcal{H}$ and $\mathcal{E}$ across ∂V.

the Maxwell equations. Conversely, the second strategy is more complete in that it gives the jump conditions as well as the differential Maxwell equations in each region where the material properties are smooth. To elaborate, we consider a bounded region $V_2 \subset \mathbb{R}^3$ with smooth boundary ∂V which constitutes the interface between two material media (Figure 1.12). We denote the unbounded complementary region $\mathbb{R}^3 \setminus \overline{V}_2$ with V_1 and we take the unit normal $\hat{\mathbf{n}}$ to ∂V directed outward V_2 (inward V_1).

For the jump conditions of *electric* and *magnetic* fields we take an open connected surface S which intersects ∂V along a line γ_0 (Figure 1.12). In this way, S ends up divided into two surfaces $S_1 \subset V_1$ and $S_2 \subset V_2$. We denote with ∂S_l, $l = 1, 2$, the part of the boundary of S contained in V_l, and hence we have $\partial S := \partial S_1 \cup \partial S_2$; the path ∂S is sometimes called an *amperian loop* [17]. Besides, we call $\hat{\boldsymbol{\nu}}_l$ the unit normal to S_l, $\hat{\mathbf{s}}_l$ the unit vector tangential to the loop $\partial S_l \cup \gamma_0$; $\hat{\boldsymbol{\nu}}_l$ and $\hat{\mathbf{s}}_l$ are oriented according to the right-handed screw rule (cf. Figure 1.2a). We also have the relations:

$$\hat{\boldsymbol{\nu}}(\mathbf{r}) \equiv \hat{\boldsymbol{\nu}}_l(\mathbf{r}), \qquad \mathbf{r} \in S_l \tag{1.131}$$

$$\hat{\mathbf{s}}(\mathbf{r}) \equiv \hat{\mathbf{s}}_l(\mathbf{r}), \qquad \mathbf{r} \in \partial S_l \tag{1.132}$$

$$\hat{\boldsymbol{\tau}}(\mathbf{r}) \equiv \hat{\mathbf{n}}(\mathbf{r}) \times \hat{\boldsymbol{\nu}}(\mathbf{r}) \equiv \hat{\mathbf{s}}_1(\mathbf{r}) \equiv -\hat{\mathbf{s}}_2(\mathbf{r}), \qquad \mathbf{r} \in \gamma_0 \tag{1.133}$$

where $\hat{\boldsymbol{\nu}}$ is the unit normal to $S := S_1 \cup S_2$, $\hat{\mathbf{s}}$ is the unit tangent to ∂S, and $\hat{\boldsymbol{\tau}}$ is the unit tangent to γ_0.

The trouble is that we are not allowed to apply the Stokes theorem (A.55) to the circulation of $\mathcal{H}(\mathbf{r}, t)$ along the loop ∂S, because the magnetic field may not be of class $\mathcal{C}^1(S)^3 \cap \mathcal{C}(\overline{S})^3$ (Appendix D.1) owing to the possible discontinuities caused by the material interface ∂V. But then, the global Ampère-Maxwell law (1.13) is invariably valid when stated over S_1, S_2 and S, namely,

$$\oint_{\partial S_1 \cup \gamma_0} \mathrm{d}s\, \hat{\mathbf{s}}_1(\mathbf{r}) \cdot \mathcal{H}_1(\mathbf{r}, t) = \int_{S_1} \mathrm{d}S\, \hat{\boldsymbol{\nu}}_1(\mathbf{r}) \cdot \left[\frac{\partial \mathcal{D}_1(\mathbf{r}, t)}{\partial t} + \mathcal{J}_1(\mathbf{r}, t) \right] \tag{1.134}$$

$$\oint_{\partial S_2 \cup \gamma_0} \mathrm{d}s\, \hat{\mathbf{s}}_2(\mathbf{r}) \cdot \mathcal{H}_2(\mathbf{r}, t) = \int_{S_2} \mathrm{d}S\, \hat{\boldsymbol{\nu}}_2(\mathbf{r}) \cdot \left[\frac{\partial \mathcal{D}_2(\mathbf{r}, t)}{\partial t} + \mathcal{J}_2(\mathbf{r}, t) \right] \tag{1.135}$$

and

$$\oint_{\partial S} \mathrm{d}s\, \hat{\mathbf{s}}(\mathbf{r}) \cdot \mathcal{H}(\mathbf{r},t) = \int_S \mathrm{d}S\, \hat{\nu}(\mathbf{r}) \cdot \left[\frac{\partial \mathcal{D}(\mathbf{r},t)}{\partial t} + \mathcal{J}(\mathbf{r},t) \right] + \int_{\gamma_0} \mathrm{d}S\, \hat{\nu}(\mathbf{r}) \cdot \mathcal{J}_S(\mathbf{r},t) \tag{1.136}$$

where we have also allowed for the presence of an electric *surface current density* $\mathcal{J}_S(\mathbf{r},t)$ (physical dimension: A/m) which flows on ∂V and crosses γ_0. Summing (1.134) and (1.135) side by side and carrying out a little algebra on the line integrals yields

$$\begin{aligned} &\oint_{\partial S_1 \cup \gamma_0} \mathrm{d}s\, \hat{\mathbf{s}}_1(\mathbf{r}) \cdot \mathcal{H}_1(\mathbf{r},t) + \oint_{\partial S_2 \cup \gamma_0} \mathrm{d}s\, \hat{\mathbf{s}}_2(\mathbf{r}) \cdot \mathcal{H}_2(\mathbf{r},t) = \\ &\quad = \oint_{\partial S} \mathrm{d}s\, \hat{\mathbf{s}}(\mathbf{r}) \cdot \mathcal{H}(\mathbf{r},t) + \int_{\gamma_0} \mathrm{d}s\, \hat{\boldsymbol{\tau}}(\mathbf{r}) \cdot [\mathcal{H}_1(\mathbf{r},t) - \mathcal{H}_2(\mathbf{r},t)] \\ &\quad = \int_S \mathrm{d}S\, \hat{\nu}(\mathbf{r}) \cdot \left[\frac{\partial \mathcal{D}(\mathbf{r},t)}{\partial t} + \mathcal{J}(\mathbf{r},t) \right] \end{aligned} \tag{1.137}$$

whence we obtain

$$\begin{aligned} \oint_{\partial S} \mathrm{d}s\, \hat{\mathbf{s}}(\mathbf{r}) \cdot \mathcal{H}(\mathbf{r},t) = \int_S \mathrm{d}S\, \hat{\nu}(\mathbf{r}) \cdot \left[\frac{\partial \mathcal{D}(\mathbf{r},t)}{\partial t} + \mathcal{J}(\mathbf{r},t) \right] \\ - \int_{\gamma_0} \mathrm{d}s\, \hat{\boldsymbol{\tau}}(\mathbf{r}) \cdot [\mathcal{H}_1(\mathbf{r},t) - \mathcal{H}_2(\mathbf{r},t)] \end{aligned} \tag{1.138}$$

which should be compared with (1.136). In fact, applying (1.13) either separately over S_1 and S_2 or directly over S ought to lead to the same global equations in order for the theory to be consistent. Apparently, the results are different, unless we *postulate* the condition

$$\int_{\gamma_0} \mathrm{d}s\, \{\hat{\nu}(\mathbf{r}) \cdot \mathcal{J}_S(\mathbf{r},t) + \hat{\boldsymbol{\tau}}(\mathbf{r}) \cdot [\mathcal{H}_1(\mathbf{r},t) - \mathcal{H}_2(\mathbf{r},t)]\} = 0 \tag{1.139}$$

and, if we assume that the scalar fields $\hat{\nu}(\mathbf{r}) \cdot \mathcal{J}_S(\mathbf{r},t)$ and $\hat{\boldsymbol{\tau}}(\mathbf{r}) \cdot \mathcal{H}_l(\mathbf{r},t)$ are continuous functions of $\mathbf{r} \in \gamma_0$, the mean value theorem [34] applied to the last line integral provides

$$\hat{\boldsymbol{\tau}}(\mathbf{r}_0) \cdot [\mathcal{H}_1(\mathbf{r}_0,t) - \mathcal{H}_2(\mathbf{r}_0,t)] + \hat{\nu}(\mathbf{r}_0) \cdot \mathcal{J}_S(\mathbf{r}_0,t) = 0 \tag{1.140}$$

where $\mathbf{r}_0 \in \gamma_0$ is a suitable point. Equivalently, we can write

$$\hat{\nu}(\mathbf{r}_0) \cdot \{\hat{\mathbf{n}}(\mathbf{r}_0) \times [\mathcal{H}_1(\mathbf{r}_0,t) - \mathcal{H}_2(\mathbf{r}_0,t)] - \mathcal{J}_S(\mathbf{r}_0,t)\} = 0 \tag{1.141}$$

in light of (1.133). However, the surface S is arbitrary and can be chosen so as to intersect ∂V along any possible curve $\gamma_0 \subset \partial V$, and the unit vector $\hat{\nu}(\mathbf{r}_0)$ is well-defined and not necessarily perpendicular to the vector field within brackets in the above equation. Hence, it follows that

$$\hat{\mathbf{n}}(\mathbf{r}) \times [\mathcal{H}_1(\mathbf{r},t) - \mathcal{H}_2(\mathbf{r},t)] = \mathcal{J}_S(\mathbf{r},t), \qquad \mathbf{r} \in \partial V \tag{1.142}$$

i.e., the (rotated) tangential component of the magnetic field is either continuous across the material interface ∂V — if $\mathcal{J}_S(\mathbf{r},t) = 0$ — or suffers a jump equal to surface current density on ∂V.

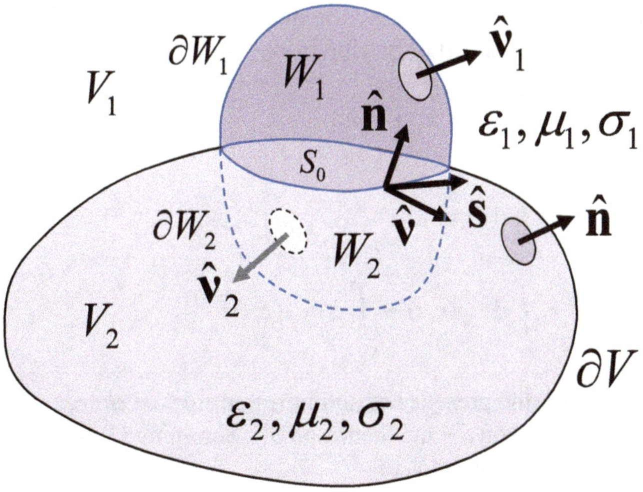

Figure 1.13 A material body confined in the region $V_2 \subset \mathbb{R}^3$ and the geometrical quantities for determining the jump conditions of $\mathfrak{D}$, $\mathfrak{B}$ and $\mathfrak{J}$ across ∂V.

Finally, the Stokes theorem may be invoked separately in the left-hand sides of (1.134) and (1.135) in that the fields $\mathcal{H}_l(\mathbf{r}, t)$, $l = 1, 2$, are of class $\mathcal{C}^1(S_l)^3 \cap \mathcal{C}(\overline{S}_l)^3$ by hypothesis. The mean value theorem applied to the resulting flux integrals over S_1 and S_2 along with the arbitrariness of S yields the local form of the Ampère-Maxwell law in each region where the constitutive parameters are smooth, viz.,

$$\nabla \times \mathcal{H}_l(\mathbf{r}, t) = \frac{\partial}{\partial t}\mathfrak{D}_l(\mathbf{r}, t) + \mathfrak{J}_l(\mathbf{r}, t), \qquad \mathbf{r} \in V_l \tag{1.143}$$

where the current $\mathfrak{J}_l$ denotes the part of $\mathfrak{J}$ flowing within V_l.

The jump conditions for the electric field are determined by applying the same procedure outlined above to the global Faraday law (1.8). With self-evident notation they read

$$[\mathcal{E}_1(\mathbf{r}, t) - \mathcal{E}_2(\mathbf{r}, t)] \times \hat{\mathbf{n}}(\mathbf{r}) = 0, \qquad \mathbf{r} \in \partial V \tag{1.144}$$

i.e., the (rotated) tangential component of the electric field is continuous across the material interface ∂V. Also, similarly to (1.143) the Faraday law holds true independently in V_1 and V_2.

As regards the matching conditions of *displacement* vector, *magnetic* induction and *current* density, we consider a bounded connected volume W which intersects V along a surface S_0 (Figure 1.13). Hence, W is divided into two parts $W_1 \subset V_1$ and $W_2 \subset V_2$. We call ∂W_l, $l = 1, 2$, the part of the boundary of W contained in V_l, whereby we have $\partial W := \partial W_1 \cup \partial W_2$. The boundaries of the open surfaces ∂W_l are two loops which coincide with the boundary ∂S_0. Additionally, we let $\hat{\boldsymbol{\nu}}_l$ the outward unit normal on ∂W_l. We also have the vector identities:

$$\hat{\boldsymbol{\nu}}(\mathbf{r}) \equiv \hat{\boldsymbol{\nu}}_l(\mathbf{r}), \qquad \mathbf{r} \in \partial W_l, \qquad l = 1, 2 \tag{1.145}$$

$$\hat{\mathbf{n}}(\mathbf{r}) \equiv -\hat{\boldsymbol{\nu}}_1(\mathbf{r}) \equiv \hat{\boldsymbol{\nu}}_2(\mathbf{r}), \qquad \mathbf{r} \in S_0 \tag{1.146}$$

$$\hat{\mathbf{s}}(\mathbf{r}) \equiv \hat{\mathbf{n}}(\mathbf{r}) \times \hat{\boldsymbol{\nu}}(\mathbf{r}), \qquad \mathbf{r} \in \partial S_0 \tag{1.147}$$

where $\hat{\boldsymbol{\nu}}$ is the unit normal on ∂W and $\hat{\mathbf{s}}$ the unit tangent to ∂S_0.

In this case the problem is that we may not apply the Gauss theorem (A.53) directly to the flux integral of $\mathfrak{D}(\mathbf{r}, t)$ through ∂W, since $\mathfrak{D}(\mathbf{r}, t)$ may not be of class $\mathcal{C}^1(W)^3 \cap \mathcal{C}(\overline{W})^3$ (Appendix D.1) in view of the possible discontinuities caused by the material interface ∂V. We proceed by observing that the global Gauss law (1.16) is certainly valid in W_1, W_2 and W, viz.,

$$\int_{\partial W_1 \cup S_0} \mathrm{d}S\ \hat{\boldsymbol{\nu}}_1(\mathbf{r}) \cdot \boldsymbol{\mathcal{D}}_1(\mathbf{r},t) = \int_{W_1} \mathrm{d}V\, \varrho_1(\mathbf{r},t) \tag{1.148}$$

$$\int_{\partial W_2 \cup S_0} \mathrm{d}S\ \hat{\boldsymbol{\nu}}_2(\mathbf{r}) \cdot \boldsymbol{\mathcal{D}}_2(\mathbf{r},t) = \int_{W_2} \mathrm{d}V\, \varrho_2(\mathbf{r},t) \tag{1.149}$$

$$\int_{\partial W} \mathrm{d}S\ \hat{\boldsymbol{\nu}}(\mathbf{r}) \cdot \boldsymbol{\mathcal{D}}(\mathbf{r},t) = \int_{W} \mathrm{d}V\, \varrho(\mathbf{r},t) + \int_{S_0} \mathrm{d}S\ \varrho_S(\mathbf{r},t) \tag{1.150}$$

where we have accounted for the presence of an impressed *surface charge density* $\varrho_S(\mathbf{r},t)$ (physical dimension: C/m^2) concentrated on ∂V and hence on S_0. Summing (1.148) and (1.149) side by side and combining the surface integrals leads to

$$\begin{aligned}\int_{W} \mathrm{d}V\, \varrho(\mathbf{r},t) &= \int_{\partial W_1 \cup S_0} \mathrm{d}S\ \hat{\boldsymbol{\nu}}_1(\mathbf{r}) \cdot \boldsymbol{\mathcal{D}}_1(\mathbf{r},t) + \int_{\partial W_2 \cup S_0} \mathrm{d}S\ \hat{\boldsymbol{\nu}}_2(\mathbf{r}) \cdot \boldsymbol{\mathcal{D}}_2(\mathbf{r},t) \\ &= \int_{\partial W} \mathrm{d}S\ \hat{\boldsymbol{\nu}}(\mathbf{r}) \cdot \boldsymbol{\mathcal{D}}(\mathbf{r},t) + \int_{S_0} \mathrm{d}S\ \hat{\mathbf{n}}(\mathbf{r}) \cdot [\boldsymbol{\mathcal{D}}_2(\mathbf{r},t) - \boldsymbol{\mathcal{D}}_1(\mathbf{r},t)]\end{aligned} \tag{1.151}$$

whence we get

$$\int_{\partial W} \mathrm{d}S\ \hat{\boldsymbol{\nu}}(\mathbf{r}) \cdot \boldsymbol{\mathcal{D}}(\mathbf{r},t) = \int_{W} \mathrm{d}V\, \varrho(\mathbf{r},t) - \int_{S_0} \mathrm{d}S\ \hat{\mathbf{n}}(\mathbf{r}) \cdot [\boldsymbol{\mathcal{D}}_2(\mathbf{r},t) - \boldsymbol{\mathcal{D}}_1(\mathbf{r},t)] \tag{1.152}$$

which should be identical with (1.150) for consistency. To accomplish this result we must *postulate* the condition

$$\int_{S_0} \mathrm{d}S\ \{\hat{\mathbf{n}}(\mathbf{r}) \cdot [\boldsymbol{\mathcal{D}}_2(\mathbf{r},t) - \boldsymbol{\mathcal{D}}_1(\mathbf{r},t)] + \varrho_S(\mathbf{r},t)\} = 0 \tag{1.153}$$

and, if we assume that the scalar fields $\varrho_S(\mathbf{r},t)$ and $\hat{\mathbf{n}}(\mathbf{r}) \cdot \boldsymbol{\mathcal{D}}_l(\mathbf{r},t)$ are continuous functions of $\mathbf{r} \in \partial V$, the mean value theorem [34] applied to the last integral yields

$$\hat{\mathbf{n}}(\mathbf{r}_0) \cdot [\boldsymbol{\mathcal{D}}_1(\mathbf{r}_0,t) - \boldsymbol{\mathcal{D}}_2(\mathbf{r}_0,t)] = \varrho_S(\mathbf{r}_0,t) \tag{1.154}$$

with $\mathbf{r}_0 \in S_0$ a suitable point. Since the domain W is arbitrary, it follows

$$\hat{\mathbf{n}}(\mathbf{r}) \cdot [\boldsymbol{\mathcal{D}}_1(\mathbf{r},t) - \boldsymbol{\mathcal{D}}_2(\mathbf{r},t)] = \varrho_S(\mathbf{r},t), \qquad \mathbf{r} \in \partial V \tag{1.155}$$

i.e., the normal component of the displacement vector is either continuous across ∂V — if $\varrho_S(\mathbf{r},t) = 0$ — or suffers a jump equal to the surface charge density on ∂V.

Finally, the Gauss theorem may be invoked separately in the left members of (1.148) and (1.149) in that $\boldsymbol{\mathcal{D}}_l(\mathbf{r},t)$ is of class $\mathcal{C}^1(W_l)^3 \cap \mathcal{C}(\overline{W_l})^3$ by assumption. The mean value theorem applied to the resulting domain integrals over W_1 and W_2 along with the arbitrariness of W gives the local form of the electric Gauss law in each region where the material parameters are smooth, viz.,

$$\nabla \cdot \boldsymbol{\mathcal{D}}_l(\mathbf{r},t) = \varrho_l(\mathbf{r},t), \qquad \mathbf{r} \in V_l, \qquad l = 1,2 \tag{1.156}$$

where ϱ_l denotes the part of ϱ contained in W_l. It is worth mentioning that the jump of $\hat{\mathbf{n}} \cdot \mathfrak{D}$ across ∂V still follows from (1.155) even if the permittivity is continuous through ∂V (see Figure 1.13).

The manipulation of the flux integral of $\mathfrak{B}(\mathbf{r}, t)$ over ∂W is similar and easier — so long as we do not allow for fictitious magnetic charges for the sake of completeness (cf. Section 6.6). The resulting boundary conditions read

$$\hat{\mathbf{n}}(\mathbf{r}) \cdot [\mathfrak{B}_1(\mathbf{r}, t) - \mathfrak{B}_2(\mathbf{r}, t)] = 0, \qquad \mathbf{r} \in \partial V \tag{1.157}$$

i.e., the normal component of the magnetic induction field is continuous across ∂V. The magnetic Gauss law in local form remains valid separately in W_l, $l = 1, 2$.

Lastly, we notice that the validity of the continuity equation in differential form (1.46) in the presence of material discontinuities is also predicated on the possibility of applying the Gauss theorem. To gain insight into the jump conditions of $\mathfrak{J}(\mathbf{r}, t)$ across ∂V, we consider the same geometrical setup of Figure 1.13 and begin by observing that the global form of the conservation of charge (1.17) holds true in W_1 and W_2, namely,

$$\int_{\partial W_1 \cup S_0} \mathrm{d}S \, \hat{\boldsymbol{\nu}}_1(\mathbf{r}) \cdot \mathfrak{J}_1(\mathbf{r}, t) + \int_{W_1} \mathrm{d}V \, \frac{\partial \varrho_1}{\partial t} = 0 \tag{1.158}$$

$$\int_{\partial W_2 \cup S_0} \mathrm{d}S \, \hat{\boldsymbol{\nu}}_2(\mathbf{r}) \cdot \mathfrak{J}_2(\mathbf{r}, t) + \int_{W_2} \mathrm{d}V \, \frac{\partial \varrho_2}{\partial t} = 0 \tag{1.159}$$

and by summing these equations side by side we get

$$\begin{aligned}
&\int_{\partial W_1 \cup S_0} \mathrm{d}S \, \hat{\boldsymbol{\nu}}_1(\mathbf{r}) \cdot \mathfrak{J}_1(\mathbf{r}, t) + \int_{\partial W_2 \cup S_0} \mathrm{d}S \, \hat{\boldsymbol{\nu}}_2(\mathbf{r}) \cdot \mathfrak{J}_2(\mathbf{r}, t) + \int_{W_1} \mathrm{d}V \, \frac{\partial \varrho_1}{\partial t} + \int_{W_2} \mathrm{d}V \, \frac{\partial \varrho_2}{\partial t} \\
&\quad = \int_{\partial W} \mathrm{d}S \, \hat{\boldsymbol{\nu}}(\mathbf{r}) \cdot \mathfrak{J}(\mathbf{r}, t) + \int_{W} \mathrm{d}V \, \frac{\partial \varrho}{\partial t} + \int_{S_0} \mathrm{d}S \, \hat{\mathbf{n}}(\mathbf{r}) \cdot [\mathfrak{J}_2(\mathbf{r}, t) - \mathfrak{J}_1(\mathbf{r}, t)] = 0
\end{aligned} \tag{1.160}$$

whereby we see that this result differs from the global conservation of charge (1.17) applied to W owing to the presence of the additional flux integral over S_0. Admittedly, we may not just dismiss this term by setting it to zero and hence enforce the continuity of the normal component of $\mathfrak{J}(\mathbf{r}, t)$ through ∂V, inasmuch as there may be surface charges and currents confined on ∂V.

Therefore, we investigate the problem by applying the Ampère-Maxwell law (1.13) to the *closed* surfaces $\partial W_1 \cup S_0$ and $\partial W_2 \cup S_0$, viz.,

$$\begin{aligned}
0 &= \int_{\partial W_1 \cup S_0} \mathrm{d}S \, \hat{\boldsymbol{\nu}}_1(\mathbf{r}) \cdot \left[\frac{\partial \mathfrak{D}_1}{\partial t} + \mathfrak{J}_1(\mathbf{r}, t) \right] \\
&= \int_{\partial W_1} \mathrm{d}S \, \hat{\boldsymbol{\nu}}_1(\mathbf{r}) \cdot \left[\frac{\partial \mathfrak{D}_1}{\partial t} + \mathfrak{J}_1(\mathbf{r}, t) \right] - \int_{S_0} \mathrm{d}S \, \hat{\mathbf{n}}(\mathbf{r}) \cdot \left[\frac{\partial \mathfrak{D}_1}{\partial t} + \mathfrak{J}_1(\mathbf{r}, t) \right] \\
&= \oint_{\partial S_0} \mathrm{d}s \, \hat{\mathbf{s}}(\mathbf{r}) \cdot \mathfrak{H}_1(\mathbf{r}, t) - \int_{S_0} \mathrm{d}S \, \hat{\mathbf{n}}(\mathbf{r}) \cdot \left[\frac{\partial \mathfrak{D}_1}{\partial t} + \mathfrak{J}_1(\mathbf{r}, t) \right]
\end{aligned} \tag{1.161}$$

$$0 = \int_{\partial W_2 \cup S_0} \mathrm{d}S \, \hat{\boldsymbol{\nu}}_2(\mathbf{r}) \cdot \left[\frac{\partial \mathfrak{D}_2}{\partial t} + \mathfrak{J}_2(\mathbf{r}, t) \right]$$

$$
\begin{aligned}
&= \int_{\partial W_2} \mathrm{d}S\, \hat{\boldsymbol{\nu}}_2(\mathbf{r}) \cdot \left[\frac{\partial \mathfrak{D}_2}{\partial t} + \mathfrak{J}_2(\mathbf{r}, t)\right] + \int_{S_0} \mathrm{d}S\, \hat{\mathbf{n}}(\mathbf{r}) \cdot \left[\frac{\partial \mathfrak{D}_2}{\partial t} + \mathfrak{J}_2(\mathbf{r}, t)\right] \\
&= -\oint_{\partial S_0} \mathrm{d}s\, \hat{\mathbf{s}}(\mathbf{r}) \cdot \mathfrak{H}_2(\mathbf{r}, t) + \int_{S_0} \mathrm{d}S\, \hat{\mathbf{n}}(\mathbf{r}) \cdot \left[\frac{\partial \mathfrak{D}_2}{\partial t} + \mathfrak{J}_2(\mathbf{r}, t)\right]
\end{aligned} \tag{1.162}
$$

on account of (1.146). The flux integrals through $\partial W_1 \cup S_0$ and $\partial W_2 \cup S_0$ vanish because the 'boundary' of a closed surface degenerates into a point (cf. Figures 1.2a and 1.2b). Alternatively, we reach the same conclusion by applying the Ampère-Maxwell law to the open surfaces ∂W_1, ∂W_2 and S_0 which are all bounded by the same loop ∂S_0. Next, summing (1.161) and (1.162) side by side yields

$$
\oint_{\partial S_0} \mathrm{d}s\, \hat{\mathbf{s}}(\mathbf{r}) \cdot [\mathfrak{H}_1(\mathbf{r}, t) - \mathfrak{H}_2(\mathbf{r}, t)] + \int_{S_0} \mathrm{d}S\, \hat{\mathbf{n}}(\mathbf{r}) \cdot \left[\frac{\partial \mathfrak{D}_2}{\partial t} - \frac{\partial \mathfrak{D}_1}{\partial t} + \mathfrak{J}_2(\mathbf{r}, t) - \mathfrak{J}_1(\mathbf{r}, t)\right] = 0 \tag{1.163}
$$

which we can further cast as

$$
\begin{aligned}
&-\oint_{\partial S_0} \mathrm{d}s\, \hat{\boldsymbol{\nu}}(\mathbf{r}) \cdot \hat{\mathbf{n}}(\mathbf{r}) \times [\mathfrak{H}_1(\mathbf{r}, t) - \mathfrak{H}_2(\mathbf{r}, t)] \\
&\qquad + \int_{S_0} \mathrm{d}S \left\{\frac{\partial}{\partial t} \hat{\mathbf{n}}(\mathbf{r}) \cdot [\mathfrak{D}_2(\mathbf{r}, t) - \mathfrak{D}_1(\mathbf{r}, t)] + \hat{\mathbf{n}}(\mathbf{r}) \cdot [\mathfrak{J}_2(\mathbf{r}, t) - \mathfrak{J}_1(\mathbf{r}, t)]\right\} \\
&= -\oint_{\partial S_0} \mathrm{d}s\, \hat{\boldsymbol{\nu}}(\mathbf{r}) \cdot \mathfrak{J}_S(\mathbf{r}, t) - \int_{S_0} \mathrm{d}S \left\{\frac{\partial \varrho_S}{\partial t} + \hat{\mathbf{n}}(\mathbf{r}) \cdot [\mathfrak{J}_1(\mathbf{r}, t) - \mathfrak{J}_2(\mathbf{r}, t)]\right\} \\
&= -\int_{S_0} \mathrm{d}S \left[\nabla_s \cdot \mathfrak{J}_S(\mathbf{r}, t) + \frac{\partial \varrho_S}{\partial t} + \hat{\mathbf{n}}(\mathbf{r}) \cdot \mathfrak{J}_1(\mathbf{r}, t) - \hat{\mathbf{n}}(\mathbf{r}) \cdot \mathfrak{J}_2(\mathbf{r}, t)\right] = 0
\end{aligned} \tag{1.164}
$$

by virtue of (1.147), the surface Gauss theorem (A.59) on S_0 under the assumption that $\mathfrak{J}_S(\mathbf{r}, t)$ is of class $\mathcal{C}^1(S_0)^3 \cap \mathcal{C}(\overline{S}_0)^3$, and the jump conditions (1.142) and (1.155). If, in addition, the scalar fields $\varrho_S(\mathbf{r}, t)$ and $\hat{\mathbf{n}} \cdot \mathfrak{J}_l(\mathbf{r}, t)$, $l = 1, 2$, are continuous functions of $\mathbf{r} \in \partial V$, by applying the mean value theorem to (1.164) and invoking the arbitrariness of W we obtain [15, Section 2.8.2], [44, p. 163]

$$
\nabla_s \cdot \mathfrak{J}_S(\mathbf{r}, t) + \frac{\partial}{\partial t} \varrho_S(\mathbf{r}, t) = \hat{\mathbf{n}}(\mathbf{r}) \cdot [\mathfrak{J}_2(\mathbf{r}, t) - \mathfrak{J}_1(\mathbf{r}, t)], \qquad \mathbf{r} \in \partial V \tag{1.165}
$$

which can be regarded as the local form of the **continuity equation** for surface charges and currents but also as the jump condition for $\mathfrak{J}(\mathbf{r}, t)$. We remark that $\hat{\mathbf{n}}(\mathbf{r}) \cdot \mathfrak{J}_1(\mathbf{r}, t)$ must be evaluated in the limit as $\mathbf{r}$ approaches ∂V from the positive side (V_1) whereas $\hat{\mathbf{n}}(\mathbf{r}) \cdot \mathfrak{J}_2(\mathbf{r}, t)$ is the limit for $\mathbf{r}$ reaching ∂V from within V_2. Therefore, (1.165) states that the normal component of the current density is either continuous across ∂V or suffers a jump if a charge density exists on ∂V. Conversely, from the viewpoint of the charges localized on the material interface ∂V, the right-hand side of (1.165) acts as a *source* or a *sink* of charges, that is, charges may move onto ∂V from V_l or leave ∂V and drift into V_l. Last but not least, either (1.164) or (1.165) may be employed to eliminate the normal components of $\mathfrak{J}(\mathbf{r}, t)$ on either side of S_0 from (1.160) to arrive at

$$
\int_{\partial W} \mathrm{d}S\, \hat{\boldsymbol{\nu}}(\mathbf{r}) \cdot \mathfrak{J}(\mathbf{r}, t) + \oint_{\partial S_0} \mathrm{d}s\, \hat{\boldsymbol{\nu}}(\mathbf{r}) \cdot \mathfrak{J}_S(\mathbf{r}, t) + \int_W \mathrm{d}V \frac{\partial \varrho}{\partial t} + \int_{S_0} \mathrm{d}S \frac{\partial \varrho_S}{\partial t} = 0 \tag{1.166}
$$

which extends (1.17) in the presence of surface charges and currents localized on $S_0 \subset W$. The validity of the local form of the continuity equation (1.46) in V_1 and V_2 — where the material properties are continuous and no surface charges exist — follows from (1.158) and (1.159).

With *steady* or *stationary state* we mean that all transients have finished and fields and sources do not depend on time any longer, though currents are still possible, as discussed in Chapter 4. Thus, in steady conditions we have $\partial \varrho_S/\partial t = 0$ and (1.165) becomes [13], [15, Section 3.2.2]

$$\nabla_s \cdot \mathcal{J}_S(\mathbf{r}) + \hat{\mathbf{n}} \cdot [\mathcal{J}_1(\mathbf{r}) - \mathcal{J}_2(\mathbf{r})] = 0, \qquad \mathbf{r} \in \partial V \tag{1.167}$$

i.e., the jump is only due to the possible presence of a stationary surface current.

As is apparent from the previous discussion, for the determination of the jump conditions it is sufficient to assume that the regions V_1 and V_2 are filled with different material media, because the actual constitutive relationships have not yet been used explicitly. Thus, we now examine a few special cases, namely, a PEC object in a lossless medium, a homogeneous conductor in a lossless medium in steady and static conditions, and two homogeneous conductors.

PEC body in a lossless medium

If $\sigma_1 = 0$ and $\sigma_2 \to \infty$, we know from the discussion in Section 1.6 on page 29 that $\mathcal{E}_2(\mathbf{r}, t) = 0 = \mathcal{H}_2(\mathbf{r}, t)$ for $\mathbf{r} \in V_2$. Therefore, the jump conditions (1.142), (1.144), (1.155) and (1.157) become

$$\hat{\mathbf{n}}(\mathbf{r}) \times \mathcal{H}_1(\mathbf{r}, t) = \mathcal{J}_S(\mathbf{r}, t), \qquad \mathbf{r} \in \partial V \tag{1.168}$$

$$\mathcal{E}_1(\mathbf{r}, t) \times \hat{\mathbf{n}}(\mathbf{r}) = 0, \qquad \mathbf{r} \in \partial V \tag{1.169}$$

$$\hat{\mathbf{n}}(\mathbf{r}) \cdot \mathcal{D}_1(\mathbf{r}, t) = \varrho_S(\mathbf{r}, t), \qquad \mathbf{r} \in \partial V \tag{1.170}$$

$$\hat{\mathbf{n}}(\mathbf{r}) \cdot \mathcal{B}_1(\mathbf{r}, t) = 0, \qquad \mathbf{r} \in \partial V \tag{1.171}$$

whereas (1.165) yields

$$\nabla_s \cdot \mathcal{J}_S(\mathbf{r}, t) + \frac{\partial \varrho_S(\mathbf{r}, t)}{\partial t} = 0, \qquad \mathbf{r} \in \partial V \tag{1.172}$$

a continuity equations for surface charges and currents over ∂V, because $\mathcal{J}_2(\mathbf{r}, t)$ vanishes identically inside a PEC, and $\mathcal{J}_1(\mathbf{r}, t)$ is null by hypothesis. Indeed, we recall that in a penetrable lossless medium the charges are permanent and no conduction currents are possible (see Section 1.6, page 29). We notice that (1.172) is also a necessary consequence of (1.168), (1.170) and the Ampère-Maxwell law (1.34). This can be verified by taking the surface divergence of both sides in (1.168) and applying (A.60). Using (1.34) with $\mathcal{J}(\mathbf{r}, t) = 0$ for $\mathbf{r} \in \partial V$ to eliminate the magnetic field and combining with (1.170) yields the result.

Conducting body in a lossless medium

We suppose $\sigma_1 = 0$ and $\sigma_2 > 0$ and that V_2 is a normal domain as in the geometry of Figure 1.13. In the *steady* state the electric field is curl-free everywhere, and for the solvability of the Ampère law

$$\nabla \times \mathcal{H}_2(\mathbf{r}) = \mathcal{J}_2(\mathbf{r}) = \sigma_2 \mathcal{E}_2(\mathbf{r}), \qquad \mathbf{r} \in V_2 \tag{1.173}$$

the conduction current density $\mathcal{J}_2(\mathbf{r})$ must be a solenoidal vector field. In the absence of a surface current on ∂V, we have

$$\mathcal{J}_2(\mathbf{r}) \cdot \hat{\mathbf{n}}(\mathbf{r}) = 0, \qquad \mathbf{r} \in \partial V \tag{1.174}$$

by virtue of (1.165) with $\partial \varrho_S/\partial t = 0 = \hat{\mathbf{n}} \cdot \mathcal{J}_1(\mathbf{r})$. Further, if the conductor is homogeneous, $\mathcal{J}_2(\mathbf{r})$ is also curl-free and hence it may be derived from an auxiliary scalar potential $\Upsilon(\mathbf{r})$ as (cf. Section 2.2)

$$\mathcal{J}_2(\mathbf{r}) := \nabla \Upsilon(\mathbf{r}), \qquad \mathbf{r} \in V_2 \tag{1.175}$$

whereby we have

$$\int_{V_2} dV\, |\mathcal{J}_2(\mathbf{r})|^2 = \int_{V_2} dV\, \mathcal{J}_2(\mathbf{r}) \cdot \nabla \Upsilon(\mathbf{r}) = \int_{V_2} dV\, \nabla \cdot [\mathcal{J}_2(\mathbf{r})\Upsilon(\mathbf{r})]$$
$$= \int_{\partial V_2} dS\, \hat{\mathbf{n}}(\mathbf{r}) \cdot \mathcal{J}_2(\mathbf{r})\Upsilon(\mathbf{r}) = 0 \tag{1.176}$$

where we have used (H.51) and the vanishing of the divergence of $\mathcal{J}_2(\mathbf{r})$, and applied the Gauss theorem (A.53) with the unit normal pointing outwards V_2. The final step is a consequence of the boundary condition (1.174). Since $|\mathcal{J}_2(\mathbf{r})|$ is a non-negative quantity, it follows that

$$\mathcal{J}_2(\mathbf{r}) = 0, \qquad \mathbf{r} \in V_2 \tag{1.177}$$

whence we also derive (cf. Section 4.7.3)

$$\mathcal{E}_2(\mathbf{r}) = 0, \qquad \mathbf{r} \in V_2 \tag{1.178}$$
$$\mathcal{D}_2(\mathbf{r}) = 0, \qquad \mathbf{r} \in V_2 \tag{1.179}$$

on account of (1.120), (1.117), and finally

$$\mathcal{E}_2(\mathbf{r}) \times \hat{\mathbf{n}}(\mathbf{r}) = 0 = \mathcal{E}_1(\mathbf{r}) \times \hat{\mathbf{n}}(\mathbf{r}), \qquad \mathbf{r} \in \partial V \tag{1.180}$$
$$\hat{\mathbf{n}}(\mathbf{r}) \cdot \mathcal{D}_2(\mathbf{r}) = 0, \qquad \mathbf{r} \in \partial V^- \tag{1.181}$$
$$\hat{\mathbf{n}}(\mathbf{r}) \cdot \mathcal{D}_1(\mathbf{r}) = \varrho_S(\mathbf{r}), \qquad \mathbf{r} \in \partial V^+ \tag{1.182}$$

from (1.144) and (1.155).

As regards the magnetic entities, we have

$$\hat{\mathbf{n}}(\mathbf{r}) \times \mathcal{H}_1(\mathbf{r}) = \hat{\mathbf{n}}(\mathbf{r}) \times \mathcal{H}_2(\mathbf{r}), \qquad \mathbf{r} \in \partial V \tag{1.183}$$
$$\hat{\mathbf{n}}(\mathbf{r}) \cdot \mathcal{B}_1(\mathbf{r}) = \hat{\mathbf{n}}(\mathbf{r}) \cdot \mathcal{B}_2(\mathbf{r}), \qquad \mathbf{r} \in \partial V \tag{1.184}$$

from (1.157) and (1.142).

In the *static* case (Chapters 2 and 3) we assume that all charges are at rest, and hence $\mathcal{J}_2(\mathbf{r}) = 0 = \mathcal{J}_S(\mathbf{r})$. This condition implies immediately (1.178) and (1.179). Then, it follows that the matching conditions (1.144) and (1.155) take the same form as (1.180)-(1.182). Finally, (1.183) and (1.184) do not apply because the magnetic field — which is a consequence of charge motion — is identically zero in the static regime.

Conducting body in a conducting medium

When both σ_1 and σ_2 are finite and the surface charge density ϱ_S is time-independent, a surface current density $\mathcal{J}_S(\mathbf{r})$ on ∂V does not appear, in that the free charges tend to drift away from the interface and migrate into the regions V_1 and V_2. Hence, the boundary condition (1.165) yields

$$0 = \hat{\mathbf{n}} \cdot \sigma_1 \mathcal{E}_1(\mathbf{r}, t) - \hat{\mathbf{n}} \cdot \sigma_2 \mathcal{E}_1(\mathbf{r}, t) = \hat{\mathbf{n}} \cdot \frac{\sigma_1}{\varepsilon_1} \mathcal{D}_1(\mathbf{r}, t) - \hat{\mathbf{n}} \cdot \frac{\sigma_2}{\varepsilon_2} \mathcal{D}_2(\mathbf{r}, t) \tag{1.185}$$

for $\mathbf{r} \in \partial V$. It follows that, unless the condition

$$\frac{\sigma_1}{\sigma_2} = \frac{\varepsilon_1}{\varepsilon_2} \tag{1.186}$$

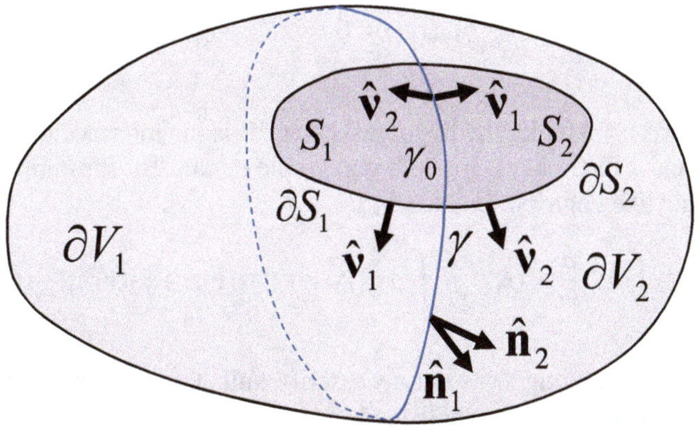

Figure 1.14 A piecewise-smooth material interface $\partial V := \partial V_1 \cup \partial V_2$ and the geometrical quantities for determining the jump conditions of $\mathfrak{J}_S$ across the line $\gamma \subset \partial V$.

holds true, the normal component of the displacement vector is discontinuous, and a surface charge ϱ_S exists on ∂V in accordance with (1.155).

The local surface continuity equation (1.165) is important because, as expressed by (1.168) and (1.170), surface densities of electric charge and current are routinely encountered at the interface ∂V between a penetrable medium and a good or perfect conductor. Nonetheless, while the validity of (1.165) rests on the smoothness of the source fields on ∂V, there are essentially two reasons why $\mathfrak{J}_S(\mathbf{r}, t)$ and $\varrho_S(\mathbf{r}, t)$ may not be continuous, let alone differentiable on ∂V, namely,

(a) the conducting material interface is piecewise-smooth, whereby the unit normal $\hat{\mathbf{n}}(\mathbf{r})$ on ∂V is not uniquely defined along edges, corners and tips,
(b) *line* densities of charge $\varrho_L(\mathbf{r}, t)$ (physical dimension: C/m) exist over ∂V,

and both situations demand we revise (1.165). For instance, the boundary ∂V of a PEC body may exhibit sharp edges *per se*, or a piecewise-smooth interface may result from the approximation of an otherwise smooth boundary by means of a triangular tessellation (see Section 14.1 and Figure 14.2). All in all, it is necessary to determine the relevant jump conditions for $\mathfrak{J}_S(\mathbf{r}, t)$ over ∂V.

For the sake of simplicity we suppose that a surface current $\mathfrak{J}_S(\mathbf{r}, t)$ flows over the closed surface ∂V comprised of two open parts ∂V_1 and ∂V_2 which are joined along the common boundary γ, as is suggested in Figure 1.14. We choose an auxiliary surface $S_0 \subset \partial V$ which is divided into two parts S_1 and S_2 by the line γ. We indicate with ∂S_l, $l = 1, 2$, the part of the boundary of S_0 contained in ∂V_l, whereby we have $\partial S_0 := \partial S_1 \cup \partial S_2$. Besides, we indicate with γ_0 the part of γ contained in S_0. We call $\hat{\nu}_1$ the unit vector normal to $\partial S_1 \cup \gamma_0$, tangential to ∂V_1 and oriented positively outward S_1. A similar definition holds for $\hat{\nu}_2$. We also let

$$\hat{\nu}(\mathbf{r}) \equiv \hat{\nu}_l(\mathbf{r}), \qquad \mathbf{r} \in \partial S_l, \qquad l = 1, 2 \tag{1.187}$$

and observe that in general $\hat{\nu}_1(\mathbf{r}) + \hat{\nu}_2(\mathbf{r}) \neq 0$ for $\mathbf{r} \in \gamma_0$, if ∂V is not smooth along γ. Finally, we suppose that a line density of charge $\varrho_L(\mathbf{r}, t)$ is localized along γ.

Next, we apply (1.164) separately on S_1 and S_2 to get

$$\oint_{\partial S_1 \cup \gamma_0} \mathrm{d}s\, \hat{\nu}_1(\mathbf{r}) \cdot \mathfrak{J}_{S1}(\mathbf{r}, t) + \int_{S_1} \mathrm{d}S\, \frac{\partial}{\partial t} \varrho_{S1}(\mathbf{r}, t) = 0 \tag{1.188}$$

$$\oint_{\partial S_2 \cup \gamma_0} \mathrm{d}s\, \hat{\nu}_2(\mathbf{r}) \cdot \mathcal{J}_{S2}(\mathbf{r}, t) + \int_{S_2} \mathrm{d}S\, \frac{\partial}{\partial t} \varrho_{S2}(\mathbf{r}, t) = 0 \tag{1.189}$$

where we have set $\mathcal{J}_l(\mathbf{r}, t) = 0$ under the hypothesis that ∂V is the interface between either a lossless penetrable medium and a PEC or two lossless penetrable media. By summing these equations side by side and combining like contributions we find

$$\int_{\partial S_0} \mathrm{d}s\, \hat{\nu}(\mathbf{r}) \cdot \mathcal{J}_S(\mathbf{r}) + \int_{S_0} \mathrm{d}S\, \frac{\partial}{\partial t} \varrho_S(\mathbf{r}, t) + \int_{\gamma_0} \mathrm{d}s\, [\hat{\nu}_1(\mathbf{r}) \cdot \mathcal{J}_{S1}(\mathbf{r}, t) + \hat{\nu}_2(\mathbf{r}) \cdot \mathcal{J}_{S2}(\mathbf{r}, t)] = 0 \tag{1.190}$$

where the extra line integral along γ_0 is not necessarily null. On the other hand, the application of (1.164) directly to the surface S_0 as a whole leads us to

$$\oint_{\partial S_0} \mathrm{d}s\, \hat{\nu}(\mathbf{r}) \cdot \mathcal{J}_S(\mathbf{r}) + \int_{S_0} \mathrm{d}S\, \frac{\partial}{\partial t} \varrho_S(\mathbf{r}, t) + \int_{\gamma_0} \mathrm{d}s\, \frac{\partial}{\partial t} \varrho_L(\mathbf{r}, t) = 0 \tag{1.191}$$

having added the contribution of the surmised line charge density localized along $\gamma_0 \subset \gamma$. Now, the two ways of applying the surface continuity equation as expressed by (1.190) and (1.191) ought to lead to the same conclusions for consistency. Evidently, this goal is achieved if we demand that

$$\int_{\gamma_0} \mathrm{d}s \left[\hat{\nu}_1(\mathbf{r}) \cdot \mathcal{J}_{S1}(\mathbf{r}, t) + \hat{\nu}_2(\mathbf{r}) \cdot \mathcal{J}_{S2}(\mathbf{r}, t) - \frac{\partial}{\partial t} \varrho_L(\mathbf{r}, t) \right] = 0. \tag{1.192}$$

Under the additional assumption that $\hat{\nu}_1(\mathbf{r}) \cdot \mathcal{J}_{S1}$, $\hat{\nu}_2(\mathbf{r}) \cdot \mathcal{J}_{S2}$ and $\partial \varrho_L / \partial t$ are continuous along γ and hence γ_0, we may invoke the mean value theorem [34] to arrive at

$$\hat{\nu}_1(\mathbf{r}_0) \cdot \mathcal{J}_{S1}(\mathbf{r}_0, t) + \hat{\nu}_2(\mathbf{r}_0) \cdot \mathcal{J}_{S2}(\mathbf{r}_0, t) - \frac{\partial}{\partial t} \varrho_L(\mathbf{r}_0, t) = 0 \tag{1.193}$$

where $\mathbf{r}_0 \in \gamma_0$ is a suitable point. Since the surface $S_0 \subset \partial V$ is arbitrary we conclude that

$$\hat{\nu}_1(\mathbf{r}) \cdot \mathcal{J}_{S1}(\mathbf{r}, t) + \hat{\nu}_2(\mathbf{r}) \cdot \mathcal{J}_{S2}(\mathbf{r}, t) = \frac{\partial}{\partial t} \varrho_L(\mathbf{r}, t), \qquad \mathbf{r} \in \gamma \subset \partial V \tag{1.194}$$

constitutes the *jump condition for surface currents*. The quantity $\hat{\nu}_1(\mathbf{r}) \cdot \mathcal{J}_{S1}(\mathbf{r}, t)$ must be evaluated in the limit as $\mathbf{r}$ tends to γ from ∂V_1, whereas the dual observation holds for $\hat{\nu}_2(\mathbf{r}) \cdot \mathcal{J}_{S2}(\mathbf{r}, t)$. In practice, (1.194) states that the component of $\mathcal{J}_S(\mathbf{r}, t)$ normal to γ is either continuous — if $\varrho_L(\mathbf{r}, t) = 0$ or is *constant* in time — or suffers a jump equal to the time rate of variation of the line charge flowing into or away from γ. The local form of the surface conservation of charge on ∂V_1 and ∂V_2 follows from (1.188) and (1.189) as usual by invoking the surface Gauss theorem and the mean value theorem.

A special instance of (1.194) obtains when the line γ constitutes the boundary of an open surface (Figure 1.2a). This happens when ∂V is the surface of an infinitely thin PEC conductor (see Section 13.2.1) immersed in a penetrable lossless medium. In which case $\mathcal{J}_{S2}(\mathbf{r}, t) = 0$ by definition and (1.194) passes over into

$$\hat{\nu}_1(\mathbf{r}) \cdot \mathcal{J}_{S1}(\mathbf{r}, t) = 0, \qquad \mathbf{r} \in \gamma \subset \partial V \tag{1.195}$$

since the presence of a line density of charge on γ does not make physical sense.

All the matching conditions we have derived thus far are valid for general time-varying fields and sources. In the case of time-harmonic dependence we obtain the relevant jump conditions by systematically substituting every entity and source with the associated phasor, in accordance with the transformation rules summarized in Table 1.1. We list the relevant formulas for future reference

$$\hat{\mathbf{n}}(\mathbf{r}) \times [\mathbf{H}_1(\mathbf{r}) - \mathbf{H}_2(\mathbf{r})] = \mathbf{J}_S(\mathbf{r}), \qquad \mathbf{r} \in \partial V \tag{1.196}$$

$$[\mathbf{E}_1(\mathbf{r}) - \mathbf{E}_2(\mathbf{r})] \times \hat{\mathbf{n}}(\mathbf{r}) = 0, \qquad \mathbf{r} \in \partial V \tag{1.197}$$

$$\hat{\mathbf{n}}(\mathbf{r}) \cdot [\mathbf{D}_1(\mathbf{r}) - \mathbf{D}_2(\mathbf{r})] = \rho_S(\mathbf{r}), \qquad \mathbf{r} \in \partial V \tag{1.198}$$

$$\hat{\mathbf{n}}(\mathbf{r}) \cdot [\mathbf{B}_1(\mathbf{r}) - \mathbf{B}_2(\mathbf{r})] = 0, \qquad \mathbf{r} \in \partial V \tag{1.199}$$

$$\nabla_s \cdot \mathbf{J}_S(\mathbf{r}) + \mathrm{j}\,\omega\rho_S(\mathbf{r}) = \hat{\mathbf{n}}(\mathbf{r}) \cdot [\mathbf{J}_2(\mathbf{r}) - \mathbf{J}_1(\mathbf{r})], \qquad \mathbf{r} \in \partial V \tag{1.200}$$

$$\hat{\nu}_1(\mathbf{r}) \cdot \mathbf{J}_{S1}(\mathbf{r}) + \hat{\nu}_2(\mathbf{r}) \cdot \mathbf{J}_{S2}(\mathbf{r}) = \mathrm{j}\,\omega\rho_L(\mathbf{r}), \qquad \mathbf{r} \in \gamma \subset \partial V \tag{1.201}$$

where the material interface ∂V and the line of discontinuity $\gamma \subset \partial V$ are pictorially represented in Figures 1.12, 1.13 and 1.14.

1.8 Wave equations

All of us are familiar with the ripples which are produced on the surface of an otherwise unperturbed pond when a pebble is thrown in the water. We call (surface) *waves* the circular pattern of alternating *crests* and *troughs* which appear to move away from the point where the pebble plunges and sinks. We are also easily convinced that such waves consist of the displacement of water molecules from some equilibrium position. This is equivalent to saying that there cannot exist a wave without a supporting medium, the water in this case.

The most important prediction of Maxwell's equations in local form is the existence of *electromagnetic waves*, which in free space travel at the speed of light (1.53). Electromagnetic waves were experimentally demonstrated by H. Hertz in Karlsruhe (1887) [65], [11, Section 1.12] about twenty years after Maxwell published his groundbreaking paper in 1865 [5]. And yet, what exactly is an electromagnetic wave?

As we know from Section 1.1, in Maxwell's theory electromagnetic phenomena are described by means of the four vector fields $\mathcal{E}$, $\mathcal{B}$, $\mathcal{D}$ and $\mathcal{H}$ which permeate the whole space at each given moment in time. The very form of (1.20), (1.23), (1.34) and (1.44) suggests that any variation of the sources $\mathcal{J}(\mathbf{r}, t)$ and $\varrho(\mathbf{r}, t)$ with time will be reflected, sooner or later, in the entities $\mathcal{E}(\mathbf{r}, t)$, $\mathcal{B}(\mathbf{r}, t)$ and so forth. Thus, we are led to consider electromagnetic waves as the propagation of 'ripples' or 'disturbances' in the fabric of an otherwise unperturbed electromagnetic field. We attempt to visualize this concept with the aid of Figures 1.15a and 1.15b.

To begin with, we suppose that a solitary fixed positive charge exists in free space. We shall see in Chapter 2 that, in this case, of the four electromagnetic entities only $\mathcal{E}(\mathbf{r})$ and $\mathcal{D}(\mathbf{r})$ are actually non-zero, are radially directed with respect to the position of the charge, and do not change with time. The situation is depicted in Figure 1.15a by means of *streamlines* [56, Section 2.6], [53, Section 4.1], i.e., lines which at every point in space are tangential to the field entity of concern, here the vector $\mathcal{E}(\mathbf{r})$ or, equivalently, $\mathcal{D}(\mathbf{r})$ by virtue of (1.112).

Next, we assume that somehow we manage to give the charge a gentle kick, so to speak, and nudge it slightly from its previous position. What happens to the electromagnetic field, then? Well, we might say that the charge 'tries' to carry the electric field along with it while it is being pushed. As a result, in the neighborhood of the charge the electric field is still static and looks pretty much the same as it did before the kick, except that $\mathcal{E}(\mathbf{r})$ has been rigidly shifted along. Farther away from the charge, though, the electric field is not aware yet, in a manner of speaking, that the charge has been nudged, so $\mathcal{E}(\mathbf{r})$ has still to be updated and looks *exactly* like it did before the kick. Therefore, in an intermediate region of space the electric streamlines must sharply warp in order to seamlessly

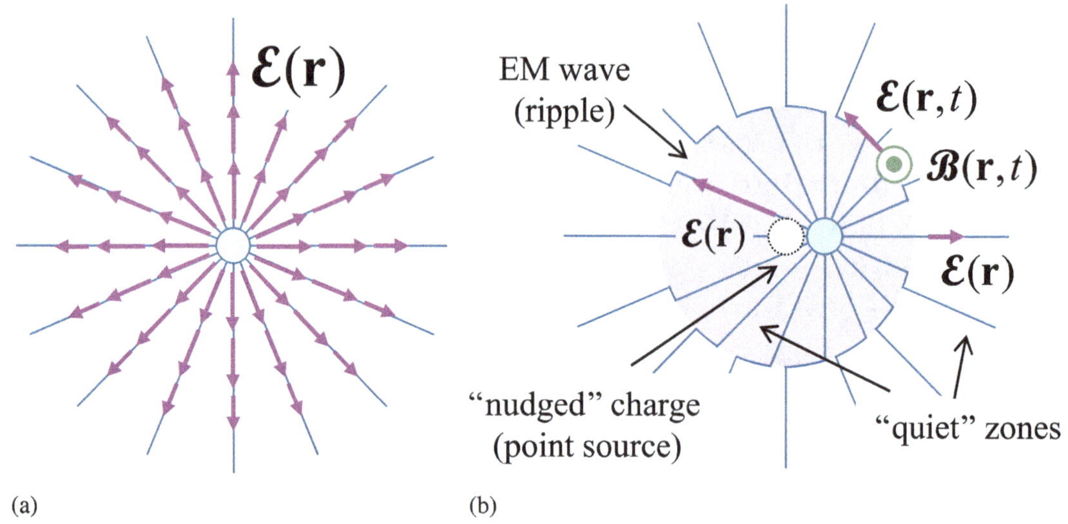

Figure 1.15 Accelerated charge (○) causing a ripple in the fabric of the electric field (→): (a) initial position of the charge at rest and static electric field, (b) final position and snapshot of the ripple propagating in the field.

match the new form of $\mathcal{E}(\mathbf{r})$ around the charge with the old form of $\mathcal{E}(\mathbf{r})$ farther away. The situation is exemplified in Figure 1.15b where a few bent streamlines are drawn.

Such abrupt variation of the streamlines describes a sudden change in the electric field. Besides, a magnetic induction appears as well, because a time-varying electric displacement generates a magnetic field in accordance with the Ampère-Maxwell law (1.13). We observe that the largest distortion occurs at right angles with respect to the direction of motion imparted to the charge with the kick, whereas the streamlines are not affected at all along the very direction of motion. The perturbed region — where the streamlines bend — constitute a *ripple* in the otherwise static electric field. As time goes by, the ripple moves ever farther away from the charge at the speed of light (1.53). The motion of the ripple is precisely the electromagnetic wave which has been generated by the sudden movement of the charge.

It is important to emphasize that, unlike the propagation of water waves on the surface of a pond, electromagnetic waves do not need a material medium to exist. Stated another way, the electromagnetic field *is* the 'medium' which sustains the waves, as is suggested by Figure 1.15b. What is more, Maxwell's equations predict that an electromagnetic wave can exists in source-free regions — where $\varrho(\mathbf{r},t) = 0 = \mathcal{J}(\mathbf{r},t)$ — essentially because a time-varying magnetic induction $\mathcal{B}(\mathbf{r},t)$ produces an electric field, according to (1.8) and, as already recalled, a time-varying displacement vector $\mathcal{D}(\mathbf{r},t)$ causes a magnetic field.

For the sake of completeness and historical rigor we must mention that Maxwell's original viewpoint was quite different [4]. Precisely guided by a mechanical view of the electromagnetic phenomena, he thought that electromagnetic waves should be disturbances in the *luminiferous ether*, an elastic, diaphanous medium which permeated everything. The sole purpose of the ether was, in effect, to sustain electromagnetic waves. Unfortunately or luckily, any experimental attempt at determining the relative velocity of the Earth and the ether — most notably the experiments of Fizeau (1851) [44, Section 4.6] and of Michelson and Morley (1886) [7, Chapter 11 and references therein], [21, Section 15.3], [11, Section 1.13], [44, Section 1.2] — was unsuccessful, and A. Einstein eventually discarded the idea of an elastic medium as the natural environment for the occurrence of electromagnetic phenomena.

We postpone an extensive discussion of electromagnetic waves until Chapter 7. Meanwhile, we focus on the manipulation of (1.20), (1.23), (1.34) and (1.44) in order to derive the so-called *wave equations* for $\mathcal{E}(\mathbf{r}, t)$ and $\mathcal{H}(\mathbf{r}, t)$ [66, Chapter 3], [67, Section 1.3]. We shall do this both in time domain and for a time-harmonic dependence. To proceed, we assume that $\mathcal{E}(\mathbf{r}, t)$ and $\mathcal{H}(\mathbf{r}, t)$ are at least twice continuously differentiable for $(\mathbf{r}, t) \in \mathbb{R}^3 \times \mathbb{R}$.

1.8.1 Time domain

We begin with the case of an unbounded homogeneous isotropic medium for which the constitutive relationships (1.112) and (1.113) hold true with $\varepsilon(\mathbf{r}) = \varepsilon$ and $\mu(\mathbf{r}) = \mu$. We take the curl of both sides of (1.20), viz.,

$$\nabla \times \nabla \times \mathcal{E}(\mathbf{r}, t) = -\nabla \times \frac{\partial}{\partial t} [\mu \mathcal{H}(\mathbf{r}, t)] = -\mu \frac{\partial}{\partial t} \nabla \times \mathcal{H}(\mathbf{r}, t) \tag{1.202}$$

where we have taken into account that the permeability is a constant, and we have swapped the order of space and time derivatives thanks to the Schwarz theorem [32, pp. 235–236]. We use the Ampère-Maxwell law (1.34) to eliminate $\nabla \times \mathcal{H}$ from the rightmost hand side,

$$\nabla \times \nabla \times \mathcal{E}(\mathbf{r}, t) = -\varepsilon\mu \frac{\partial^2}{\partial t^2} \mathcal{E}(\mathbf{r}, t) - \mu \frac{\partial}{\partial t} \mathcal{J}(\mathbf{r}, t) \tag{1.203}$$

where we have used the fact that the permittivity is a constant, too. The last equation involves the electric field only, as desired, but per se it is not equivalent to the original set of Maxwell's equations, unless it is paired with the Gauss law or the continuity equation. Therefore, an equivalent set of equations for the electric field reads

$$\nabla \times \nabla \times \mathcal{E}(\mathbf{r}, t) + \varepsilon\mu \frac{\partial^2}{\partial t^2} \mathcal{E}(\mathbf{r}, t) = -\mu \frac{\partial}{\partial t} \mathcal{J}(\mathbf{r}, t) \tag{1.204}$$

$$\nabla \cdot \mathcal{E}(\mathbf{r}, t) = \frac{1}{\varepsilon} \varrho(\mathbf{r}, t) \tag{1.205}$$

whereas the magnetic field can, in principle, be computed through the Faraday law, once $\mathcal{E}(\mathbf{r}, t)$ is known. We refer to (1.204) as the wave equation for $\mathcal{E}(\mathbf{r}, t)$ in a homogeneous isotropic medium, though we need one more step to justify this name. If we make use of (A.47) to transform the double curl operator in (1.204), we find

$$\nabla\nabla \cdot \mathcal{E}(\mathbf{r}, t) - \nabla^2 \mathcal{E}(\mathbf{r}, t) + \varepsilon\mu \frac{\partial^2}{\partial t^2} \mathcal{E}(\mathbf{r}, t) = -\mu \frac{\partial}{\partial t} \mathcal{J}(\mathbf{r}, t) \tag{1.206}$$

and, by invoking the Gauss law (1.205) to express $\nabla \cdot \mathcal{E}(\mathbf{r}, t)$ we finally arrive at

$$\nabla^2 \mathcal{E}(\mathbf{r}, t) - \varepsilon\mu \frac{\partial^2}{\partial t^2} \mathcal{E}(\mathbf{r}, t) = \frac{1}{\varepsilon} \nabla \varrho(\mathbf{r}, t) + \mu \frac{\partial}{\partial t} \mathcal{J}(\mathbf{r}, t) \tag{1.207}$$

which closely resembles the classic D'Alembert equation (1758) for elastic waves. In fact, the left-hand side can be written succinctly by introducing the *wave* or *D'Alembert operator* [45, Section 16.3], [24, Section 33], [11, Section 11.7]

$$\Box^2 := \nabla^2 - \frac{1}{c^2} \frac{\partial^2}{\partial t^2} \tag{1.208}$$

where

$$c := \frac{1}{\sqrt{\varepsilon\mu}} = \frac{c_0}{\sqrt{\varepsilon_r \mu_r}} \tag{1.209}$$

is the speed of the electromagnetic waves in the medium of concern. When the latter is free space, then (1.209) coincides with (1.115). In the mathematical literature (e.g., [68, Section XIV.3.1]) and in the special theory of relativity the D'Alembert operator is defined as the negative of the right member in (1.208).

In Cartesian coordinates, (1.207) separates into three uncoupled equations for the Cartesian components of $\mathcal{E}(\mathbf{r}, t)$, viz.

$$\frac{\partial^2 \mathcal{E}_\alpha}{\partial x^2} + \frac{\partial^2 \mathcal{E}_\alpha}{\partial y^2} + \frac{\partial^2 \mathcal{E}_\alpha}{\partial z^2} - \frac{1}{c^2}\frac{\partial^2 \mathcal{E}_\alpha}{\partial t^2} = \frac{1}{\varepsilon}\frac{\partial \varrho}{\partial \alpha} + \mu \frac{\partial \mathcal{J}_\alpha}{\partial t} \tag{1.210}$$

where the subscript α stands for x, y and z. We observe that (1.207) is equivalent to the original Maxwell equations, because the Gauss law has already been employed in the derivation and, indeed, the right-hand side involves all the sources. In this regard, ϱ must be at least continuous for $\mathbf{r} \in \mathbb{R}^3$, whereas $\mathcal{J}$ must be at least continuous for $t \in \mathbb{R}$, since we need the gradient and the time derivative, respectively.

To obtain the analogous result for the magnetic field, we take the curl of both sides of (1.34), i.e.,

$$\nabla \times \nabla \times \mathcal{H}(\mathbf{r}, t) = \nabla \times \frac{\partial}{\partial t}[\varepsilon \mathcal{E}(\mathbf{r}, t)] + \nabla \times \mathcal{J}(\mathbf{r}, t) = \varepsilon \frac{\partial}{\partial t} \nabla \times \mathcal{E}(\mathbf{r}, t) + \nabla \times \mathcal{J}(\mathbf{r}, t) \tag{1.211}$$

where we have recalled that the permittivity is a constant, and we have swapped the order of space and time derivative in the right-hand side. Next, we apply the Faraday law (1.20) to eliminate $\nabla \times \mathcal{E}$, namely,

$$\nabla \times \nabla \times \mathcal{H}(\mathbf{r}, t) = -\varepsilon\mu \frac{\partial^2}{\partial t^2}\mathcal{H}(\mathbf{r}, t) + \nabla \times \mathcal{J}(\mathbf{r}, t) \tag{1.212}$$

which is not fully equivalent to the original set of Maxwell's equations, unless it is paired with the magnetic Gauss law. In the end, an equivalent set of equations for the magnetic field reads:

$$\nabla \times \nabla \times \mathcal{H}(\mathbf{r}, t) + \varepsilon\mu \frac{\partial^2}{\partial t^2}\mathcal{H}(\mathbf{r}, t) = \nabla \times \mathcal{J}(\mathbf{r}, t) \tag{1.213}$$

$$\nabla \cdot \mathcal{H}(\mathbf{r}, t) = 0 \tag{1.214}$$

whereas the electric field can, in theory, be computed through the Ampère-Maxwell law (1.34). We refer to (1.213) as the wave equation for the magnetic field in an isotropic homogeneous medium. By applying (A.47) and the magnetic Gauss law (1.214) we can cast (1.213) into

$$\nabla^2 \mathcal{H}(\mathbf{r}, t) - \varepsilon\mu \frac{\partial^2}{\partial t^2}\mathcal{H}(\mathbf{r}, t) = -\nabla \times \mathcal{J}(\mathbf{r}, t) \tag{1.215}$$

which is the counterpart of (1.207) for $\mathcal{H}$ and, in Cartesian coordinates, takes on a form similar to (1.210).

If we allow for conduction losses in the underlying medium, then we need to extend the Ampère-Maxwell law with a conduction current $\mathcal{J}_c(\mathbf{r}, t)$ as in (1.119). The modified vector wave equations for lossy media may be obtained by means of the same steps followed before. More quickly, in light of the Ohm law (1.120), in (1.204) and (1.213) we may replace $\mathcal{J}$ with $\mathcal{J} + \sigma\mathcal{E}$ and carry out a few manipulations to arrive at

$$\nabla \times \nabla \times \mathcal{E}(\mathbf{r}, t) + \varepsilon\mu \frac{\partial^2}{\partial t^2}\mathcal{E}(\mathbf{r}, t) + \sigma\mu \frac{\partial}{\partial t}\mathcal{E}(\mathbf{r}, t) = -\mu \frac{\partial}{\partial t}\mathcal{J}(\mathbf{r}, t) \tag{1.216}$$

$$\nabla \times \nabla \times \mathcal{H}(\mathbf{r}, t) + \varepsilon\mu \frac{\partial^2}{\partial t^2}\mathcal{H}(\mathbf{r}, t) + \sigma\mu \frac{\partial}{\partial t}\mathcal{H}(\mathbf{r}, t) = \nabla \times \mathcal{J}(\mathbf{r}, t) \tag{1.217}$$

which we must complement with the Gauss laws (1.44) and (1.23), respectively.

We continue our derivation of wave equations by examining the case of an unbounded inhomogeneous isotropic medium, for which (1.117) and (1.118) hold true. The procedure is quite similar to the previous one, thus we outline the main steps.

Under the additional hypothesis $\mu(\mathbf{r}) \neq 0$ for $\mathbf{r} \in \mathbb{R}^3$, we divide (1.20) through by $\mu(\mathbf{r})$ and take the curl of both sides:

$$\nabla \times \frac{\nabla \times \mathcal{E}(\mathbf{r}, t)}{\mu(\mathbf{r})} = -\nabla \times \frac{\partial}{\partial t}\mathcal{H}(\mathbf{r}, t) = -\frac{\partial}{\partial t}\nabla \times \mathcal{H}(\mathbf{r}, t) \tag{1.218}$$

and then we use (1.34) to eliminate $\nabla \times \mathcal{H}$, viz.

$$\nabla \times \frac{\nabla \times \mathcal{E}(\mathbf{r}, t)}{\mu(\mathbf{r})} = -\varepsilon(\mathbf{r}) \frac{\partial^2}{\partial t^2}\mathcal{E}(\mathbf{r}, t) - \frac{\partial}{\partial t}\mathcal{J}(\mathbf{r}, t) \tag{1.219}$$

where we have noticed that $\varepsilon(\mathbf{r})$ does not depend on time. Finally, an equivalent set of equations for the electric field in an unbounded inhomogeneous isotropic medium is

$$\nabla \times \frac{\nabla \times \mathcal{E}(\mathbf{r}, t)}{\mu(\mathbf{r})} + \varepsilon(\mathbf{r}) \frac{\partial^2}{\partial t^2}\mathcal{E}(\mathbf{r}, t) = -\frac{\partial}{\partial t}\mathcal{J}(\mathbf{r}, t) \tag{1.220}$$

$$\nabla \cdot [\varepsilon(\mathbf{r})\mathcal{E}(\mathbf{r}, t)] = \varrho(\mathbf{r}, t) \tag{1.221}$$

and we refer to (1.220) as the wave equation for $\mathcal{E}$.

By assuming now $\varepsilon(\mathbf{r}) \neq 0$ for $\mathbf{r} \in \mathbb{R}^3$, we divide (1.34) through and take the curl of both sides

$$\nabla \times \frac{\nabla \times \mathcal{H}(\mathbf{r}, t)}{\varepsilon(\mathbf{r})} = \nabla \times \frac{\partial}{\partial t}\mathcal{E}(\mathbf{r}, t) + \nabla \times \frac{\mathcal{J}(\mathbf{r}, t)}{\varepsilon(\mathbf{r})} = \frac{\partial}{\partial t}\nabla \times \mathcal{E}(\mathbf{r}, t) + \nabla \times \frac{\mathcal{J}(\mathbf{r}, t)}{\varepsilon(\mathbf{r})} \tag{1.222}$$

and we use (1.20) to eliminate $\nabla \times \mathcal{E}$, viz.

$$\nabla \times \frac{\nabla \times \mathcal{H}(\mathbf{r}, t)}{\varepsilon(\mathbf{r})} = -\mu(\mathbf{r}) \frac{\partial^2}{\partial t^2}\mathcal{H}(\mathbf{r}, t) + \nabla \times \frac{\mathcal{J}(\mathbf{r}, t)}{\varepsilon(\mathbf{r})} \tag{1.223}$$

where we have noticed that $\mu(\mathbf{r})$ does not depend on time. Lastly, an equivalent set of equations for the magnetic field in an unbounded inhomogeneous isotropic medium reads

$$\nabla \times \frac{\nabla \times \mathcal{H}(\mathbf{r}, t)}{\varepsilon(\mathbf{r})} + \mu(\mathbf{r}) \frac{\partial^2}{\partial t^2}\mathcal{H}(\mathbf{r}, t) = \nabla \times \frac{\mathcal{J}(\mathbf{r}, t)}{\varepsilon(\mathbf{r})} \tag{1.224}$$

$$\nabla \cdot [\mu(\mathbf{r})\,\mathcal{H}(\mathbf{r}, t)] = 0 \tag{1.225}$$

and we refer to (1.224) as the wave equation for $\mathcal{H}$.

We finalize our discussion on wave equations in the time domain by extending the previous results to the more general situation of inhomogeneous anisotropic media for which (1.126) and (1.127) apply.

If we suppose that $\det\overline{\mu}(\mathbf{r}) \neq 0$ for $\mathbf{r} \in \mathbb{R}^3$, i.e., the dyadic permeability is invertible, we can dot-multiply (1.20) through by $[\overline{\mu}(\mathbf{r})]^{-1}$, and then take the curl of both sides, namely,

$$\nabla \times \left\{[\overline{\mu}(\mathbf{r})]^{-1} \cdot \nabla \times \mathcal{E}(\mathbf{r}, t)\right\} = -\nabla \times \frac{\partial}{\partial t}\mathcal{H}(\mathbf{r}, t) = -\frac{\partial}{\partial t}\nabla \times \mathcal{H}(\mathbf{r}, t) \tag{1.226}$$

and we use (1.34) to eliminate $\nabla \times \mathcal{H}$, viz.

$$\nabla \times \left\{[\overline{\mu}(\mathbf{r})]^{-1} \cdot \nabla \times \mathcal{E}(\mathbf{r}, t)\right\} = -\overline{\varepsilon}(\mathbf{r})\frac{\partial^2}{\partial t^2}\mathcal{E}(\mathbf{r}, t) - \frac{\partial}{\partial t}\mathcal{J}(\mathbf{r}, t) \tag{1.227}$$

where we have noticed that $\overline{\varepsilon}(\mathbf{r})$ is constant with time. An equivalent set of equations for the electric field in an unbounded inhomogeneous anisotropic medium reads

$$\nabla \times \left\{[\overline{\mu}(\mathbf{r})]^{-1} \cdot \nabla \times \mathcal{E}(\mathbf{r}, t)\right\} + \overline{\varepsilon}(\mathbf{r}) \cdot \frac{\partial^2}{\partial t^2}\mathcal{E}(\mathbf{r}, t) = -\frac{\partial}{\partial t}\mathcal{J}(\mathbf{r}, t) \tag{1.228}$$

$$\nabla \cdot [\overline{\varepsilon}(\mathbf{r}) \cdot \mathcal{E}(\mathbf{r}, t)] = \varrho(\mathbf{r}, t) \tag{1.229}$$

and we refer to (1.228) as the wave equation for $\mathcal{E}$.

Similarly, by further assuming that $\det\overline{\varepsilon}(\mathbf{r}) \neq 0$ for $\mathbf{r} \in \mathbb{R}^3$, we can dot-multiply (1.34) through by $[\overline{\varepsilon}(\mathbf{r})]^{-1}$ and take the curl of both sides, viz.,

$$\begin{aligned}\nabla \times \left\{[\overline{\varepsilon}(\mathbf{r})]^{-1} \cdot \nabla \times \mathcal{H}(\mathbf{r}, t)\right\} &= \nabla \times \frac{\partial}{\partial t}\mathcal{E}(\mathbf{r}, t) + \nabla \times \left\{[\overline{\varepsilon}(\mathbf{r})]^{-1} \cdot \mathcal{J}(\mathbf{r}, t)\right\} \\ &= \frac{\partial}{\partial t}\nabla \times \mathcal{E}(\mathbf{r}, t) + \nabla \times \left\{[\overline{\varepsilon}(\mathbf{r})]^{-1} \cdot \mathcal{J}(\mathbf{r}, t)\right\}\end{aligned} \tag{1.230}$$

and we make use of (1.20) to eliminate $\nabla \times \mathcal{E}$, namely,

$$\nabla \times \left\{[\overline{\varepsilon}(\mathbf{r})]^{-1} \cdot \nabla \times \mathcal{H}(\mathbf{r}, t)\right\} = -\overline{\mu}(\mathbf{r})\frac{\partial^2}{\partial t^2}\mathcal{H}(\mathbf{r}, t) + \nabla \times \left\{[\overline{\varepsilon}(\mathbf{r})]^{-1} \cdot \mathcal{J}(\mathbf{r}, t)\right\} \tag{1.231}$$

where we have observed that $\overline{\mu}(\mathbf{r})$ is a function of position only. Finally, an equivalent set of equations for the magnetic field in an unbounded inhomogeneous anisotropic medium reads

$$\nabla \times \left\{[\overline{\varepsilon}(\mathbf{r})]^{-1} \cdot \nabla \times \mathcal{H}(\mathbf{r}, t)\right\} + \overline{\mu}(\mathbf{r}) \cdot \frac{\partial^2}{\partial t^2}\mathcal{H}(\mathbf{r}, t) = \nabla \times \left\{[\overline{\varepsilon}(\mathbf{r})]^{-1} \cdot \mathcal{J}(\mathbf{r}, t)\right\} \tag{1.232}$$

$$\nabla \cdot [\overline{\mu}(\mathbf{r}) \cdot \mathcal{H}(\mathbf{r}, t)] = 0 \tag{1.233}$$

and we call (1.232) the wave equation for $\mathcal{H}$.

We conclude by observing that the obvious advantage of working with wave equations (as compared to the full set of Maxwell's equations in local form) consists of a further reduction of the unknown entities to just one (either $\mathcal{E}$ or $\mathcal{H}$). However, the wave equations are second order with respect to space and time coordinates, so (1.207), (1.215) and the like are not necessarily easier to solve in practical situations.

1.8.2 Frequency domain

The wave equations derived in the previous section take on a special form when time-harmonic fields are considered. In order to obtain the desired expressions we may start with the time-harmonic Maxwell equations (1.98), (1.99), (1.100) and (1.101), and follow the same steps described above.

However, a faster approach consists of transforming the time-domain wave equations directly by introducing the phasors of the corresponding entities and sources. The latter strategy works fine because the constitutive parameters defined in Section 1.6 are constant with time.

The practical rule for determining the phasor of the second-order time-derivative of a vector field is provided by (1.94) with $n = 2$. Keeping this in mind, we can simply write down the time-harmonic wave equations by inspecting their time-domain counterparts, namely,

$$\nabla \times \nabla \times \mathbf{E}(\mathbf{r}) - \omega^2 \left(\varepsilon - \mathrm{j}\frac{\sigma}{\omega}\right)\mu \mathbf{E}(\mathbf{r}) = -\mathrm{j}\,\omega\mu \mathbf{J}(\mathbf{r}) \tag{1.234}$$

$$\nabla \cdot \mathbf{E}(\mathbf{r}) = \frac{1}{\varepsilon}\rho(\mathbf{r}) \tag{1.235}$$

$$\nabla \times \nabla \times \mathbf{H}(\mathbf{r}) - \omega^2 \left(\varepsilon - \mathrm{j}\frac{\sigma}{\omega}\right)\mu \mathbf{H}(\mathbf{r}) = \nabla \times \mathbf{J}(\mathbf{r}) \tag{1.236}$$

$$\nabla \cdot \mathbf{H}(\mathbf{r}) = 0 \tag{1.237}$$

$$\nabla^2 \mathbf{E}(\mathbf{r}) + \omega^2 \varepsilon\mu \mathbf{E}(\mathbf{r}) = \frac{1}{\varepsilon}\nabla\rho(\mathbf{r}) + \mathrm{j}\,\omega\mu \mathbf{J}(\mathbf{r}) \tag{1.238}$$

$$\nabla^2 \mathbf{H}(\mathbf{r}) + \omega^2 \varepsilon\mu \mathbf{H}(\mathbf{r}) = -\nabla \times \mathbf{J}(\mathbf{r}) \tag{1.239}$$

$$\nabla \times \frac{\nabla \times \mathbf{E}(\mathbf{r})}{\mu(\mathbf{r})} - \mathrm{j}\,\omega^2 \varepsilon(\mathbf{r})\mathbf{E}(\mathbf{r}) = -\mathrm{j}\,\omega \mathbf{J}(\mathbf{r}) \tag{1.240}$$

$$\nabla \cdot [\varepsilon(\mathbf{r})\mathbf{E}(\mathbf{r})] = \rho(\mathbf{r}) \tag{1.241}$$

$$\nabla \times \frac{\nabla \times \mathbf{H}(\mathbf{r})}{\varepsilon(\mathbf{r})} - \omega^2 \mu(\mathbf{r})\mathbf{H}(\mathbf{r}) = \nabla \times \frac{\mathbf{J}(\mathbf{r})}{\varepsilon(\mathbf{r})} \tag{1.242}$$

$$\nabla \cdot [\mu(\mathbf{r})\mathbf{H}(\mathbf{r})] = 0 \tag{1.243}$$

$$\nabla \times \left\{[\overline{\mu}(\mathbf{r})]^{-1} \cdot \nabla \times \mathbf{E}(\mathbf{r})\right\} - \omega^2 \overline{\varepsilon}(\mathbf{r}) \cdot \mathbf{E}(\mathbf{r}) = -\mathrm{j}\,\omega \mathbf{J}(\mathbf{r}) \tag{1.244}$$

$$\nabla \cdot [\overline{\varepsilon}(\mathbf{r}) \cdot \mathbf{E}(\mathbf{r})] = \rho(\mathbf{r}) \tag{1.245}$$

$$\nabla \times \left\{[\overline{\varepsilon}(\mathbf{r})]^{-1} \cdot \nabla \times \mathbf{H}(\mathbf{r})\right\} - \omega^2 \overline{\mu}(\mathbf{r}) \cdot \mathbf{H}(\mathbf{r}, t) = \nabla \times \left\{[\overline{\varepsilon}(\mathbf{r})]^{-1} \cdot \mathbf{J}(\mathbf{r})\right\} \tag{1.246}$$

$$\nabla \cdot [\overline{\mu}(\mathbf{r}) \cdot \mathbf{H}(\mathbf{r})] = 0 \tag{1.247}$$

subject to the same hypotheses on the constitutive parameters.

In a homogeneous isotropic medium it is customary to introduce the quantity

$$k := \omega\sqrt{\varepsilon\mu} = \frac{\omega}{c} \tag{1.248}$$

which is called the *wavenumber* and carries the physical dimension of the inverse of a distance (1/m). If the medium is lossy, we define the *complex wavenumber* as

$$k := \omega\sqrt{\left(\varepsilon - \mathrm{j}\frac{\sigma}{\omega}\right)\mu} \tag{1.249}$$

on account of (1.234) and (1.236). Moreover, (1.238) and (1.239) can be written succinctly by defining the *Helmholtz operator* [2, Section 2.5], [10, Section 7.7], [26, Section 8.5], [36]

$$\nabla^2 + k^2 \tag{1.250}$$

which, in light of (1.94) for $n = 2$, can be regarded as the time-harmonic counterpart of the D'Alembert operator. For this reason, (1.238) and (1.239) are also called the (scalar) Helmholtz equations for $\mathbf{E}(\mathbf{r})$ and $\mathbf{H}(\mathbf{r})$, respectively.

We recall that the time-dependence in the form $\exp(\mathrm{j}\,\omega t)$ is implied in (1.238)-(1.247). Therefore, these equations govern the propagation of electromagnetic waves which are periodic in time. In fact, wave-like solutions in the form $\mathbf{E}(\mathbf{r})\exp(\mathrm{j}\,\omega t)$ are sometimes called *stationary waves* [62, Section 1.4.6].

1.9 Electromagnetic radiation

In Section 1.8 we have seen that Maxwell's equations can be cast into a form which more than suggests the possibility of electromagnetic waves. The sources of the electromagnetic field are charges and currents, but only time-varying currents can, in principle, generate electromagnetic waves. We say 'in principle' because, as we shall see in Section 7.8, highly symmetric configurations of charges, albeit *accelerated*, do not produce electromagnetic waves. We call *electromagnetic radiation* the net transfer or propagation of electromagnetic *energy*, *(linear) momentum* and *angular momentum* away from the sources by means of electromagnetic waves [2, Chapter 6]. On the contrary, charges at rest and steady (i.e., time-independent) currents do not radiate.

With the help of a fundamental classic result [17, Chapter 10], [69, Section 3.1], [20, Chapter 10], [2, Chapter 8], [10, Chapter 17], [1, Chapter 21], [24, Section 44], [55, Chapter 23], [7, 70, 71], we can gain insight into the structure of the electromagnetic field produced by a point charge q which moves with velocity $\mathbf{v}(t)$ along some trajectory $\mathbf{w}(t)$ in free space (Figure 1.16). At a given time t an observer at a location specified by $\mathbf{r}$ uses a probe (say, an antenna) to detect the electric field produced by the charge. Since the speed of light c_0 is finite, it takes a finite time in order for whatever disturbance of the electromagnetic field (caused by the moving charge) to reach the observer. Therefore, when he performs the experiment at $(\mathbf{r}, t)$ what he actually detects is the electromagnetic field that was produced by the charge when it occupied a different position $\mathbf{w}(t_r)$ at an earlier time t_r. Meanwhile, the disturbance has travelled for a time $t - t_r$ with velocity c_0 and has covered a distance R given by

$$R = |\mathbf{R}| = |\mathbf{r} - \mathbf{w}(t_r)| = c_0(t - t_r) \tag{1.251}$$

which defines t_r implicitly in terms of $(\mathbf{r}, t)$. For its meaning, t_r is referred to as the *retarded time* and ultimately depends on the form of the vector field $\mathbf{w}(t_r)$.

Lengthy and far-from-trivial calculations [2, Chapter 8], [7, Chapter 14], [24, Chapter 4], [72, Chapter 8], [17, 69] allow obtaining the following formulas for the electromagnetic field of q, viz.,

$$\mathcal{E}(\mathbf{r}, t) = \underbrace{\frac{q}{4\pi\varepsilon_0}\frac{R(c_0^2 - v^2)\mathbf{u}}{(\mathbf{R}\cdot\mathbf{u})^3}}_{\text{Coulomb field}} + \underbrace{\frac{q}{4\pi\varepsilon_0}\frac{R\mathbf{R}\times(\mathbf{u}\times\mathbf{a})}{(\mathbf{R}\cdot\mathbf{u})^3}}_{\text{acceleration field}} \tag{1.252}$$

$$\mathcal{B}(\mathbf{r}, t) = \frac{1}{c_0}\hat{\mathbf{R}}\times\mathcal{E}(\mathbf{r}, t) \tag{1.253}$$

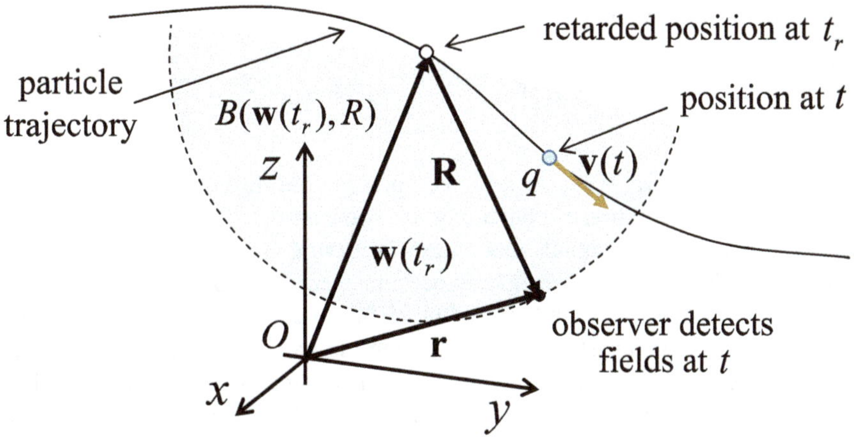

Figure 1.16 Relative positions of a moving point charge (∘) and an observer at rest. The ball $B(\mathbf{w}(t_r), R)$ has been drawn (−−) only partially.

where $\hat{\mathbf{R}} := \mathbf{R}/R$ denotes the unit vector in the direction of $\mathbf{R}$, $\mathbf{u}$ is the vector field

$$\mathbf{u}(\mathbf{r}, t_r) := c_0\hat{\mathbf{R}} - \mathbf{v}(t_r) \tag{1.254}$$

and

$$\mathbf{a}(t_r) := \left.\frac{\mathrm{d}}{\mathrm{d}t}\mathbf{v}(t)\right|_{t=t_r} \tag{1.255}$$

is the instantaneous acceleration of the charge. In (1.252) $\mathbf{R}$, $\mathbf{v}$, $\mathbf{u}$, and $\mathbf{a}$ are all evaluated at the retarded time t_r, which has to be computed through (1.251) for each pair $(\mathbf{r}, t)$; in (1.253) the dependence on t_r is implicit in $\hat{\mathbf{R}}$.

First of all, we observe that the electromagnetic field comes naturally decomposed into two contributions, namely, the Coulomb (or velocity or attached) field $\mathcal{E}_C(\mathbf{r}, t)$ and the acceleration (or radiation) field $\mathcal{E}_A(\mathbf{r}, t)$. The former depends only on the velocity of q both explicitly (v^2) and through $\mathbf{u}(\mathbf{r}, t_r)$, whereas the latter is essentially due to the charge acceleration. Hence, if the particle moves with uniform velocity along a straight line, then $\mathbf{a}(t_r) = 0$ at all times, and the acceleration field is absent (Section 9.2). Furthermore, for a particle at rest (1.252) correctly yields the known result (see Chapter 2), whereas (1.253) predicts a zero magnetic induction, thus confirming that magnetism is truly a consequence of the *relative* motion of charged particles and observers, as was argued in Section 1.1.

Secondly, we take a closer look at the dependence of the two contributions in (1.252) on the distance R between the observer and the retarded position of the charge. If we assume that the particle trajectory is contained in a bounded volume V — so that $|\mathbf{w}(t_r)|$ is bounded, though the trajectory does not have to be a closed curve — we have the estimates

$$|\mathcal{E}_C(\mathbf{r}, t)| \leqslant \frac{E_C}{R^2} \qquad\qquad |\mathcal{E}_A(\mathbf{r}, t)| \leqslant \frac{E_A}{R} \tag{1.256}$$

where E_C and E_A are the maximum values of the magnitudes of the vector fields that remain in (1.252) after singling out R. From (1.256) we conclude that the Coulomb field falls off with an inverse-square law, whereas the acceleration field decays with the inverse of the distance from the source of the disturbance. This different behavior is precisely what makes radiation possible.

Indeed, we shall show in Section 1.10.1 that the flux integral

$$P_F(t) := \int_{\partial V} \mathrm{d}S\ \hat{\mathbf{n}}(\mathbf{r}) \cdot \mathcal{E}(\mathbf{r},t) \times \mathcal{H}(\mathbf{r},t) \tag{1.257}$$

may be interpreted as the *rate of energy flow* (i.e., power) [69] through the closed surface $S = \partial V$ which is the boundary of a region V containing the sources; $\hat{\mathbf{n}}$ is the unit outward normal on ∂V. While a positive value of $P_F(t)$ signals that energy is leaving V, not all of the energy that crosses ∂V constitutes *radiation*, though. We may expect, in fact, that Coulomb and acceleration fields contribute differently to the right-hand side of (1.257). To clarify this point, we apply (1.257) to the case of a moving charged particle by considering the ball $B(\mathbf{w}(t_r), R)$ for which, after having specified $(\mathbf{r}, t)$, the retarded time t_r, the radius R, and the retarded position $\mathbf{w}(t_r)$ are all provided by the solution of (1.251). Physically, the sphere ∂B (see Figure 1.16) represents the locus of the points reached at time t by the disturbance (if any) that was produced by the particle at time $t_r = t - R/c_0$. In fact, if $\mathbf{a}(t_r) = 0$, then only the Coulomb field is non-null on ∂B at time t. Now, by inserting (1.253) into (1.257) with $\partial V = \partial B$ and $\hat{\mathbf{n}}(\mathbf{r}) = \hat{\mathbf{R}}$, and invoking (1.256), we have

$$\begin{aligned} P_F(t) &= \frac{1}{\mu_0} \int_{\partial B} \mathrm{d}S\ \hat{\mathbf{R}} \times \mathcal{E}(\mathbf{r},t) \cdot \mathcal{B}(\mathbf{r},t) = \frac{1}{\mu_0 c_0} \int_{\partial B} \mathrm{d}S\ \left|\hat{\mathbf{R}} \times \mathcal{E}(\mathbf{r},t)\right|^2 \\ &\leqslant \frac{1}{\mu_0 c_0} \int_{\partial B} \mathrm{d}S\ |\mathcal{E}(\mathbf{r},t)|^2 \leqslant \frac{1}{\mu_0 c_0} \int_{\partial B} \mathrm{d}S\ (|\mathcal{E}_C(\mathbf{r},t)| + |\mathcal{E}_A(\mathbf{r},t)|)^2 \\ &\leqslant \frac{1}{\mu_0 c_0} \int_{4\pi} \mathrm{d}\Omega\, R^2 \left(\frac{E_C^2}{R^4} + 2\frac{E_C E_A}{R^3} + \frac{E_A^2}{R^2} \right) \\ &= \frac{4\pi}{\mu_0 c_0} \left(\frac{E_C^2}{R^2} + 2\frac{E_C E_A}{R} + E_A^2 \right) \end{aligned} \tag{1.258}$$

where the symbol $\int_{4\pi} \mathrm{d}\Omega$ indicates integration over the full solid angle. Since in the limit as $R \to +\infty$ the first two terms in the last estimate vanish, we may conclude that only an accelerated particle generates an electromagnetic wave which carries non-zero power to infinity. However, (1.258) is unsatisfactory because it may give the wrong impression that the radiated power at infinity is non-zero. While it is true that the electromagnetic waves emitted by the particle carry power away towards infinity, still the velocity of propagation is finite, it takes an infinite amount of time for the power to reach infinity, so at any given time the radiated power at infinity is zero!

We wish to quantify this statement by evaluating (1.257). Since the calculation of the exact radiated power through ∂B for an arbitrary motion is quite complicated, we consider the case in which the charge q is instantaneously at rest in $\mathbf{w}(t_r)$ [17, Section 11.2]; this means that $\mathbf{v}(t_r) = 0$ and, as a result, $\mathbf{u}(\mathbf{r}, t_r) = c_0 \hat{\mathbf{R}}$. The Coulomb field is not zero, granted, but in consideration of (1.252) and the very definition of the vector $\mathbf{u}(\mathbf{r}, t_r)$, we see that $\mathcal{E}_C(\mathbf{r}, t)$ is aligned with $\hat{\mathbf{R}}$ and thus contributes naught to the rate of energy flow (1.257), namely,

$$P_F(t) := \frac{1}{\mu_0 c_0} \int_{\partial B} \mathrm{d}S\ \left|\hat{\mathbf{R}} \times \mathcal{E}_A(\mathbf{r},t)\right|^2 = \frac{1}{\mu_0 c_0} \int_{\partial B} \mathrm{d}S\ \frac{q^2 \left|\mathbf{a}(t_r) \times \hat{\mathbf{R}}\right|^2}{\left(4\pi\varepsilon_0 R c_0^2\right)^2} = \mu_0 \frac{q^2 a^2}{6\pi c_0} \tag{1.259}$$

where $a = |\mathbf{a}(t_r)|$, and the integration has been carried out in local polar spherical coordinates centered in $\mathbf{w}(t_r)$ with the polar axis aligned with the acceleration $\mathbf{a}(t_r)$. This result is referred to as the

Larmor formula [7,17,20], and yields the radiated power associated with the electromagnetic wave passing through ∂B at time $t = t_r + R/c_0$. Besides, since $P_F(t)$ is independent of the distance R, another observer located at a point $\mathbf{r}_1$, with $R_1 = |\mathbf{r}_1 - \mathbf{w}(t_r)| > R$, at the time $t_1 = t_r + R_1/c$ detects the same radiated power $P_F(t_1) = P_F(t)$.

We can repeat this reasoning for increasingly larger spheres and then conclude that a constant power is actually *travelling* towards infinity. Nonetheless, we can see that the radiated power *at* infinity at any given time t is *zero* by solving (1.251) for the retarded time, viz., $t_r = t - R/c_0$. For $R \to +\infty$ we need the acceleration of the particle for infinitely remote retarded times. However, if the particle has not been in accelerated motion forever, then $\mathbf{a}(t_r) \to 0$ for $t_r \to -\infty$, and hence $P_F(t)$ is null.

By the same line of thought we are led to state that the electromagnetic field *at* infinity at any given time is — at most — static or stationary. Again, this follows from the observation that the acceleration field must perforce be null for $t_r = t - R/c_0 \to -\infty$, unless the particle motion is truly *periodic* (an ideal situation we envisioned in Section 1.5). Therefore, it seems logical to assume the following boundary conditions for the fields at infinity [64]

$$\mathcal{E}(\mathbf{r}, t) = \mathcal{O}\left(\frac{1}{|\mathbf{r}|^2}\right), \qquad |\mathbf{r}| \to +\infty \tag{1.260}$$

$$\mathcal{H}(\mathbf{r}, t) = \mathcal{O}\left(\frac{1}{|\mathbf{r}|^2}\right), \qquad |\mathbf{r}| \to +\infty \tag{1.261}$$

uniformly with respect to space and time. This means, for instance, that there exist two positive constants C_E and b_E *independent* of time such that

$$|\mathcal{E}(\mathbf{r}, t)| \leqslant \frac{C_E}{|\mathbf{r}|^2} \qquad \text{for} \qquad |\mathbf{r}| \geqslant b_E \tag{1.262}$$

and similar estimates hold for the other electromagnetic entities.

1.10 Conservation of electromagnetic energy

We continue our discussion on radiation by determining a balance equation between electromagnetic entities and sources, on the one hand, and power and energy, on the other. In the process we shall justify formula (1.257) for the rate of energy flow. The Reader is supposed to be familiar with the definition of electric and magnetic energy [7,8,15–17].

1.10.1 Poynting theorem in the time domain

We begin our derivation by considering a region of free space V in which a distribution of charges of finite extent is at rest (Figure 1.17 and Section 2.1). The electric energy W_e of the configuration is [17, Section 2.4.3], [20, Chapter 2], [24, Chapter IV], [2, Section 6.2], [55, Section 15.4], [73, Section 3.3.1], [7,8]

$$W_e := \frac{1}{2}\int_V \mathrm{d}V\, \mathcal{E}(\mathbf{r}) \cdot \mathcal{D}(\mathbf{r}) = \frac{\varepsilon_0}{2}\int_V \mathrm{d}V\, \mathcal{E}(\mathbf{r}) \cdot \mathcal{E}(\mathbf{r}) \tag{1.263}$$

and carries the physical dimension of joule (J) or volt times coulomb (V C). It is worthwhile noticing that W_e involves the product of an entity of intensity ($\mathcal{E}$) and one of quantity ($\mathcal{D}$). We wish to

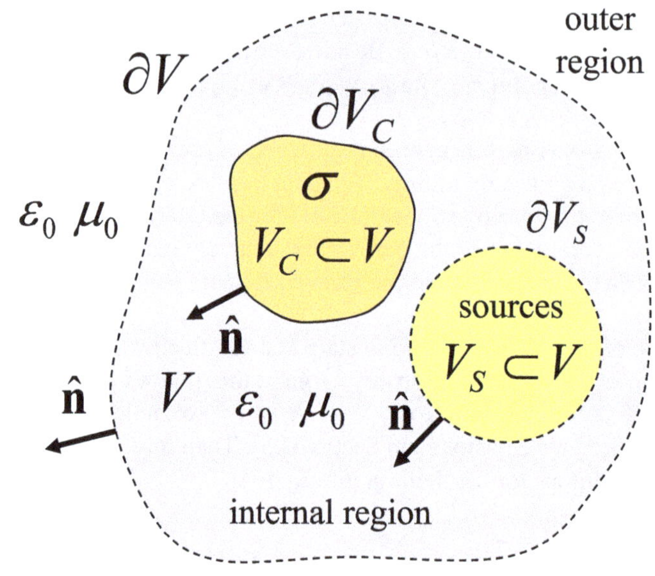

Figure 1.17 A bounded domain V in which sources and a conducting medium reside.

determine the change that W_e undergoes in response to small and slow variations of the charge distribution. For example, we can think of bringing in a small amount of charge from an infinitely remote distance; the procedure must be so slow as to prevent the generation of time-varying fields or electromagnetic waves in accordance to (1.252), because (1.263) holds true if the charges are at rest. Bearing this in mind, by denoting the variations of the energy and of the electric field with δW_e and $\delta\boldsymbol{\mathcal{E}}(\mathbf{r})$, we have

$$\begin{aligned} W_e + \delta W_e + o(\delta W_e) &= \frac{\varepsilon_0}{2}\int_V dV\,[\boldsymbol{\mathcal{E}}(\mathbf{r}) + \delta\boldsymbol{\mathcal{E}}(\mathbf{r})]\cdot[\boldsymbol{\mathcal{E}}(\mathbf{r}) + \delta\boldsymbol{\mathcal{E}}(\mathbf{r})] \\ &= \frac{\varepsilon_0}{2}\int_V dV\,[\boldsymbol{\mathcal{E}}(\mathbf{r})\cdot\boldsymbol{\mathcal{E}}(\mathbf{r}) + 2\delta\boldsymbol{\mathcal{E}}(\mathbf{r})\cdot\boldsymbol{\mathcal{E}}(\mathbf{r}) + |\delta\boldsymbol{\mathcal{E}}(\mathbf{r})|^2] \\ &= W_e + \varepsilon_0\int_V dV\,\delta\boldsymbol{\mathcal{E}}(\mathbf{r})\cdot\boldsymbol{\mathcal{E}}(\mathbf{r}) + \frac{\varepsilon_0}{2}\int_V dV\,|\delta\boldsymbol{\mathcal{E}}(\mathbf{r})|^2 \end{aligned} \tag{1.264}$$

under the hypothesis that $|\delta\boldsymbol{\mathcal{E}}| \ll |\boldsymbol{\mathcal{E}}|$. In conclusion, the variation of the electric energy reads

$$\delta W_e = \int_V dV\,\boldsymbol{\mathcal{E}}(\mathbf{r})\cdot\delta\boldsymbol{\mathcal{D}}(\mathbf{r}) \tag{1.265}$$

on account of the constitutive relationship (1.112).

Next, we consider a region of free space V in which steady currents exist (Section 4.1). The magnetic energy W_h of the system is [7], [24, Chapter IV], [20, Chapter 4]

$$W_h := \frac{1}{2}\int_V dV\,\boldsymbol{\mathcal{H}}(\mathbf{r})\cdot\boldsymbol{\mathcal{B}}(\mathbf{r}) = \frac{1}{2\mu_0}\int_V dV\,\boldsymbol{\mathcal{B}}(\mathbf{r})\cdot\boldsymbol{\mathcal{B}}(\mathbf{r}) \tag{1.266}$$

and carries the physical dimension of joule (J) or ampere times weber (A Wb). Similarly to W_e, W_h too is constructed from the product of an entity of intensity ($\mathcal{B}$) and one of quantity ($\mathcal{H}$). We wish to determine the changes of W_h that are caused by a small and slow variation of the current distribution. In this case, we can think of bringing in a closed loop of steady current from an infinitely remote distance. Again, the whole process must be so slow as to prevent the acceleration of the moving charges because (1.266) holds true for steady currents. We call δW_h and $\delta\mathcal{B}(\mathbf{r})$ the variations of the energy and the magnetic induction and, through the same steps that have led to (1.265), we find

$$\delta W_h = \int_V \mathrm{d}V\, \delta\mathcal{B}(\mathbf{r}) \cdot \mathcal{H}(\mathbf{r}) \tag{1.267}$$

under the hypothesis $|\delta\mathcal{B}| \ll |\mathcal{B}|$ and in light of (1.113).

Although (1.265) and (1.267) have been obtained for charges at rest and steady currents, it is *postulated* that the formulas remain valid even for a general time dependence of fields and sources. Then, we can employ (1.265) and (1.267) to write the time rates of variation of $W_e(t)$ and $W_h(t)$ and try to relate them to the sources of the electromagnetic field. If we assume the system of charges and currents was in equilibrium at time $t = t_0$ and let $\delta t = t - t_0$, we find

$$\left.\frac{\mathrm{d}W_e}{\mathrm{d}t}\right|_{t_0} = \lim_{\delta t\to 0}\frac{\delta W_e}{\delta t} = \int_V \mathrm{d}V\, \mathcal{E}(\mathbf{r}, t_0)\cdot \lim_{\delta t\to 0}\frac{\delta\mathcal{D}}{\delta t} = \int_V \mathrm{d}V\, \mathcal{E}(\mathbf{r}, t_0)\cdot \left.\frac{\partial\mathcal{D}}{\partial t}\right|_{t_0} \tag{1.268}$$

$$\left.\frac{\mathrm{d}W_h}{\mathrm{d}t}\right|_{t_0} = \lim_{\delta t\to 0}\frac{\delta W_h}{\delta t} = \int_V \mathrm{d}V\, \mathcal{H}(\mathbf{r}, t_0)\cdot \lim_{\delta t\to 0}\frac{\delta\mathcal{B}}{\delta t} = \int_V \mathrm{d}V\, \mathcal{H}(\mathbf{r}, t_0)\cdot \left.\frac{\partial\mathcal{B}}{\partial t}\right|_{t_0}. \tag{1.269}$$

Since t_0 is arbitrary for our purposes, we can consider any other time t and write the variation of the total energy of the system of charges and currents in the region V as

$$\frac{\mathrm{d}}{\mathrm{d}t}[W_e(t) + W_h(t)] = \int_V \mathrm{d}V \left[\mathcal{E}(\mathbf{r}, t)\cdot\frac{\partial\mathcal{D}(\mathbf{r}, t)}{\partial t} + \mathcal{H}(\mathbf{r}, t)\cdot\frac{\partial\mathcal{B}(\mathbf{r}, t)}{\partial t}\right] \tag{1.270}$$

which constitutes an intermediate result for the derivation of the balance equation.

To gain more generality we allow for the presence of a conducting medium with isotropic conductivity $\sigma(\mathbf{r})$ in the region $V_C \subset V$ (Figure 1.17). Besides, the sources are supposed to be confined in an smaller volume $V_S \subset V$, $V_S \cap V_C = \emptyset$. In order to relate the time rate of variation of the energy in (1.270) to the sources, from the local Ampère-Maxwell and Faraday laws we derive

$$\frac{\partial\mathcal{D}(\mathbf{r}, t)}{\partial t} = \nabla\times\mathcal{H}(\mathbf{r}, t) - \sigma(\mathbf{r})\mathcal{E}(\mathbf{r}, t) - \mathcal{J}(\mathbf{r}, t) \tag{1.271}$$

$$\frac{\partial\mathcal{B}(\mathbf{r}, t)}{\partial t} = -\nabla\times\mathcal{E}(\mathbf{r}, t) \tag{1.272}$$

where we have made use of (1.120) to express the conduction current flowing within V_C. By substituting the temporal derivatives into the integrand of the domain integral in the right-hand side of (1.270) we obtain

$$\begin{aligned}\mathcal{E}(\mathbf{r}, t)\cdot\frac{\partial\mathcal{D}(\mathbf{r}, t)}{\partial t} + \mathcal{H}(\mathbf{r}, t)\cdot\frac{\partial\mathcal{B}(\mathbf{r}, t)}{\partial t} &= \\ = \mathcal{E}(\mathbf{r}, t)\cdot\nabla\times\mathcal{H}(\mathbf{r}, t) - \mathcal{H}(\mathbf{r}, t)\cdot\nabla\times\mathcal{E}(\mathbf{r}, t) &- \sigma(\mathbf{r})|\mathcal{E}(\mathbf{r}, t)|^2 - \mathcal{E}(\mathbf{r}, t)\cdot\mathcal{J}(\mathbf{r}, t) \\ = -\nabla\cdot[\mathcal{E}(\mathbf{r}, t)\times\mathcal{H}(\mathbf{r}, t)] - \sigma(\mathbf{r})|\mathcal{E}(\mathbf{r}, t)|^2 &- \mathcal{E}(\mathbf{r}, t)\cdot\mathcal{J}(\mathbf{r}, t)\end{aligned} \tag{1.273}$$

which allows (1.270) to be rewritten as follows

$$\frac{\mathrm{d}}{\mathrm{d}t}[W_e(t) + W_h(t)] + \int_V \mathrm{d}V\, \nabla \cdot [\mathcal{E}(\mathbf{r}, t) \times \mathcal{H}(\mathbf{r}, t)] + \int_{V_C} \mathrm{d}V\, \sigma(\mathbf{r})|\mathcal{E}(\mathbf{r}, t)|^2 = -\int_{V_S} \mathrm{d}V\, \mathcal{E}(\mathbf{r}, t) \cdot \mathcal{J}(\mathbf{r}, t) \quad (1.274)$$

where we have restricted the last two volume integrals to the spatial regions where $\sigma(\mathbf{r})$ and $\mathcal{J}(\mathbf{r})$ are non-zero.

Finally, we would like to apply the Gauss theorem (A.53) to the integral of $\nabla \cdot [\mathcal{E}(\mathbf{r}, t) \times \mathcal{H}(\mathbf{r}, t)]$. To this purpose, we split V into three parts, i.e., V_C, V_S and $V \backslash (V_C \cup V_S)$, because $\mathcal{E}(\mathbf{r}, t) \times \mathcal{H}(\mathbf{r}, t)$ may not be differentiable everywhere in V due the presence of the conducting medium and the sources. By proceeding as in the derivation of (1.34) we find that the identity

$$\int_V \mathrm{d}V\, \nabla \cdot [\mathcal{E}(\mathbf{r}, t) \times \mathcal{H}(\mathbf{r}, t)] = \int_{\partial V} \mathrm{d}S\, \hat{\mathbf{n}}(\mathbf{r}) \cdot [\mathcal{E}(\mathbf{r}, t) \times \mathcal{H}(\mathbf{r}, t)] \quad (1.275)$$

holds true if one postulates the continuity of $\hat{\mathbf{n}}(\mathbf{r}) \cdot [\mathcal{E}(\mathbf{r}, t) \times \mathcal{H}(\mathbf{r}, t)]$ across ∂V_C and ∂V_S. However, this requirement is certainly fulfilled in light of the continuity of $\hat{\mathbf{n}}(\mathbf{r}) \times \mathcal{E}(\mathbf{r}, t)$ and $\hat{\mathbf{n}}(\mathbf{r}) \times \mathcal{H}(\mathbf{r}, t)$ across ∂V_C and ∂V_S, which was discussed in Section 1.7.

Therefore, we can cast (1.274) into its final form

$$\frac{\mathrm{d}}{\mathrm{d}t}[W_e(t) + W_h(t)] + \int_{\partial V} \mathrm{d}S\, \hat{\mathbf{n}}(\mathbf{r}) \cdot [\mathcal{E}(\mathbf{r}, t) \times \mathcal{H}(\mathbf{r}, t)] + \int_{V_C} \mathrm{d}V\, \sigma(\mathbf{r})|\mathcal{E}(\mathbf{r}, t)|^2 = -\int_{V_S} \mathrm{d}V\, \mathcal{E}(\mathbf{r}, t) \cdot \mathcal{J}(\mathbf{r}, t) \quad (1.276)$$

a relation which was first derived by John H. Poynting in 1884 and is rightfully known as the *Poynting theorem* in global (integral) form [2, 7, 8, 11, 15, 24, 27, 52, 74], [75, Section VIII.4]. Equation (1.276) constitutes the power balance for the electromagnetic field in the region V. In addition to the time rate of variation of the electromagnetic energy derived in (1.270) we identify the following three contributions:

$$P_F(t) := \int_{\partial V} \mathrm{d}S\, \hat{\mathbf{n}}(\mathbf{r}) \cdot [\mathcal{E}(\mathbf{r}, t) \times \mathcal{H}(\mathbf{r}, t)] \quad (1.277)$$

$$P_C(t) := \int_{V_C} \mathrm{d}V\, \sigma(\mathbf{r})|\mathcal{E}(\mathbf{r}, t)|^2 \quad (1.278)$$

$$P_S(t) := -\int_{V_S} \mathrm{d}V\, \mathcal{E}(\mathbf{r}, t) \cdot \mathcal{J}(\mathbf{r}, t) \quad (1.279)$$

where

- $P_F(t)$ is the same quantity already introduced in (1.257) and is interpreted as the *rate of energy flow* (instantaneous power) through ∂V. We observe that $P_F(t)$ may also be negative, if other sources are located outside V, and this simply means that a net amount of energy is actually *entering* V.

- $P_C(t)$ is interpreted as the instantaneous power *dissipated* into the conducting medium or, stated in another way, the rate at which electromagnetic energy is converted into heat and 'lost'. Evidently, $P_C(t)$ is a non-negative quantity, provided $\sigma(\mathbf{r}) \geqslant 0$ for $\mathbf{r} \in V_C$.
- $P_S(t)$ is interpreted as the instantaneous power *generated* by the sources or, in other words, the rate at which energy is delivered by the sources to the electromagnetic field.

In words, the Poynting theorem (1.276) states that the variation of the total energy in the electromagnetic field in the region V is due to the sources, the conduction losses and the flow through ∂V. In view of the very definitions (1.277)-(1.279) the integrands therein can be interpreted as surface or volume *densities* of instantaneous power. In particular, the vector field

$$\mathcal{S}(\mathbf{r},t) := \mathcal{E}(\mathbf{r},t) \times \mathcal{H}(\mathbf{r},t) \tag{1.280}$$

is called the (instantaneous) Poynting vector. While $\mathcal{S}$ has indeed the physical dimension of a power per unit area (W/m^2), still the surface power density flowing through the boundary ∂V is given by

$$\frac{\mathrm{d}P_F}{\mathrm{d}S} := \hat{\mathbf{n}}(\mathbf{r}) \cdot \mathcal{S}(\mathbf{r},t) \tag{1.281}$$

i.e., the component of the Poynting vector that is perpendicular to ∂V. Furthermore, (1.280) indicates that a flow of energy may only take place if electric and magnetic fields are present — which happens invariably for time-varying fields, because either entity generates the other one, as is prescribed by the Maxwell equations.

The Poynting theorem is valid even for an unbounded volume V or for $V \equiv \mathbb{R}^3$ [8]. To show that this is true, we start with (1.276) applied to a ball $V = B(0,b)$ and take the limit for $b \to +\infty$. In symbols, we have

$$\begin{aligned}\lim_{b\to+\infty} \frac{1}{2}\frac{\mathrm{d}}{\mathrm{d}t} \int_{B(0,b)} \mathrm{d}V \left[\varepsilon_0 |\mathcal{E}(\mathbf{r},t)|^2 + \mu_0 |\mathcal{H}(\mathbf{r},t)|^2\right] + \int_{V_C} \mathrm{d}V\, \sigma(\mathbf{r}) |\mathcal{E}(\mathbf{r},t)|^2 \\ + \lim_{b\to+\infty} \int_{\partial B} \mathrm{d}S\, \hat{\mathbf{n}}(\mathbf{r}) \cdot [\mathcal{E}(\mathbf{r},t) \times \mathcal{H}(\mathbf{r},t)] = -\int_{V_S} \mathrm{d}V\, \mathcal{E}(\mathbf{r},t) \cdot \mathcal{J}(\mathbf{r},t)\end{aligned} \tag{1.282}$$

and we need to show that the electromagnetic energy is finite, whereas the flux integral vanishes. For the electric energy, we pick up a ball $B(0,b_E)$ with $b_E < b$ where the radius b_E is the same constant used in (1.262) and estimate

$$\begin{aligned}\int_{B(0,b)} \mathrm{d}V\, |\mathcal{E}(\mathbf{r},t)|^2 &= \int_{B(0,b_E)} \mathrm{d}V\, |\mathcal{E}(\mathbf{r},t)|^2 + \int_{B(0,b)\setminus B(0,b_E)} \mathrm{d}V\, |\mathcal{E}(\mathbf{r},t)|^2 \\ &\leqslant \int_{B(0,b_E)} \mathrm{d}V\, |\mathcal{E}(\mathbf{r},t)|^2 + \int_{4\pi} \mathrm{d}\Omega \int_{b_E}^{b} \mathrm{d}r\, r^2 \frac{C_E^2}{r^4} \\ &= \int_{B(0,b_E)} \mathrm{d}V\, |\mathcal{E}(\mathbf{r},t)|^2 + 4\pi C_E^2 \left(\frac{1}{b_E} - \frac{1}{b}\right) \\ &\leqslant \int_{B(0,b_E)} \mathrm{d}V\, |\mathcal{E}(\mathbf{r},t)|^2 + 4\pi \frac{C_E^2}{b_E}\end{aligned} \tag{1.283}$$

and since the rightmost-hand side is independent of b, the electric energy remains finite as $b \to +\infty$. A perfectly similar conclusion holds for the magnetic energy. For the flux integral over ∂B we choose $b > \max\{b_E, b_H\}$, where the constants b_E, b_H are involved in the boundary conditions (1.260) and (1.261), and with the aid of (D.152) and (D.151) we estimate

$$\left| \int_{\partial B} dS\, \hat{\mathbf{n}}(\mathbf{r}) \cdot [\mathcal{E}(\mathbf{r}, t) \times \mathcal{H}(\mathbf{r}, t)] \right| \leqslant \int_{\partial B} dS\, |\mathcal{E}(\mathbf{r}, t)||\mathcal{H}(\mathbf{r}, t)| \leqslant \int_{4\pi} d\Omega\, b^2 \frac{C_E C_H}{b^4}$$
$$= 4\pi \frac{C_E C_H}{b^2} \xrightarrow[b \to +\infty]{} 0 \tag{1.284}$$

as anticipated. We shall use this result in Section 6.3 for the discussion on the uniqueness of the solutions to Maxwell's equations.

Next, we derive the local form of (1.276), namely, a balance equation which involves *densities* of power and energy. To proceed we re-write (1.274) as

$$\int_V dV \left[\mathcal{E}(\mathbf{r}, t) \cdot \frac{\partial \mathcal{D}(\mathbf{r}, t)}{\partial t} + \mathcal{H}(\mathbf{r}, t) \cdot \frac{\partial \mathcal{B}(\mathbf{r}, t)}{\partial t} \right] + \int_V dV\, \nabla \cdot \mathcal{S}(\mathbf{r}, t)$$
$$+ \int_V dV\, \sigma(\mathbf{r}) |\mathcal{E}(\mathbf{r}, t)|^2 + \int_V dV\, \mathcal{E}(\mathbf{r}, t) \cdot \mathcal{J}(\mathbf{r}, t) = 0 \tag{1.285}$$

on account of (1.270), (1.275), (1.280), and the fact that $\sigma(\mathbf{r}) = 0$ for $\mathbf{r} \notin \overline{V}_C$ and $\mathcal{J}(\mathbf{r}, t) = 0$ for $\mathbf{r} \notin \overline{V}_S$. By combining the terms in the left-hand sides of (1.285) into a single domain integral over V and then by invoking the mean value theorem [34] we obtain

$$\mathcal{E}(\mathbf{r}_0, t) \cdot \frac{\partial \mathcal{D}(\mathbf{r}_0, t)}{\partial t} + \mathcal{H}(\mathbf{r}_0, t) \cdot \frac{\partial \mathcal{B}(\mathbf{r}_0, t)}{\partial t} + \nabla \cdot \mathcal{S}(\mathbf{r}, t)|_{\mathbf{r}_0}$$
$$+ \sigma(\mathbf{r}_0) |\mathcal{E}(\mathbf{r}_0, t)|^2 + \mathcal{E}(\mathbf{r}_0, t) \cdot \mathcal{J}(\mathbf{r}_0, t) = 0 \tag{1.286}$$

where $\mathbf{r}_0 \in V$ is a suitable point. Since the volume V is arbitrary, we conclude that the following balance equation holds true for $(\mathbf{r}, t) \in \mathbb{R}^3 \times \mathbb{R}^+$

$$\mathcal{E}(\mathbf{r}, t) \cdot \frac{\partial \mathcal{D}(\mathbf{r}, t)}{\partial t} + \mathcal{H}(\mathbf{r}, t) \cdot \frac{\partial \mathcal{B}(\mathbf{r}, t)}{\partial t} + \nabla \cdot \mathcal{S}(\mathbf{r}, t) + \sigma(\mathbf{r}) |\mathcal{E}(\mathbf{r}, t)|^2 = -\mathcal{E}(\mathbf{r}, t) \cdot \mathcal{J}(\mathbf{r}, t) \tag{1.287}$$

which is called the *Poynting theorem* in local form. The first two terms represent the time rate of variation of the volume density of electromagnetic energy $w := w_e + w_h$, viz.,

$$\frac{\partial w}{\partial t} = \frac{\partial}{\partial t}(w_e + w_h) = \mathcal{E}(\mathbf{r}, t) \cdot \frac{\partial \mathcal{D}(\mathbf{r}, t)}{\partial t} + \mathcal{H}(\mathbf{r}, t) \cdot \frac{\partial \mathcal{B}(\mathbf{r}, t)}{\partial t} \tag{1.288}$$

by virtue of (1.270). Likewise, we can define the quantities

$$p_F(\mathbf{r}, t) := \nabla \cdot \mathcal{S}(\mathbf{r}, t) \tag{1.289}$$
$$p_C(\mathbf{r}, t) := \mathcal{E}(\mathbf{r}, t) \cdot \mathcal{J}_C(\mathbf{r}, t) = \sigma(\mathbf{r}) |\mathcal{E}(\mathbf{r}, t)|^2 \tag{1.290}$$
$$p_S(\mathbf{r}, t) := -\mathcal{E}(\mathbf{r}, t) \cdot \mathcal{J}(\mathbf{r}, t) \tag{1.291}$$

where

- $p_F(\mathbf{r}, t)$ is the rate of flow of the volume density of energy,
- $p_C(\mathbf{r}, t)$ is the volume density of power dissipated in the conductors,
- $p_S(\mathbf{r}, t)$ is volume density of power delivered by the sources.

With these positions, the Poynting theorem reads (cf. [16])

$$\frac{\partial}{\partial t}(w_e + w_h) + p_F + p_C = p_S \tag{1.292}$$

which in view of (1.289) is in the form of a continuity (conservation) equation for power densities [cf. (1.46)].

The Poynting theorem (1.276) and its local counterpart (1.287) are valid also in the case where the medium filling the region $V \setminus (V_C \cup V_S)$ is a dielectric or magnetic material with piecewise continuous constitutive parameters. In this regard, we can anticipate that the following matching condition for the Poynting vector

$$\hat{\mathbf{n}}(\mathbf{r}) \cdot [\mathcal{S}_1(\mathbf{r}, t) - \mathcal{S}_2(\mathbf{r}, t)] = 0 \tag{1.293}$$

holds true at the smooth interface between two material media (see Figure 1.12) by virtue of (1.142) and (1.144). The continuity condition of $\mathcal{S}(\mathbf{r}, t)$ across the interface between vacuum and a conductor is the same as (1.293) and has already been stated and used in the derivation of (1.276).

To determine the Poynting theorem in the presence of material boundaries, we modify a bit the situation depicted in Figure 1.17 by supposing that a penetrable body occupies a bounded connected volume $V_2 \subset V \setminus (V_C \cup V_S)$ and, just for ease of reference, we define $V_1 := V \setminus \overline{V}_2$. We may certainly apply (1.276) separately to V_1 and V_2, viz.,

$$\frac{\mathrm{d}}{\mathrm{d}t}[W_{e1}(t) + W_{h1}(t)] + \int_{\partial V} \mathrm{d}S\, \hat{\mathbf{n}}(\mathbf{r}) \cdot \mathcal{S}(\mathbf{r}, t) + \int_{\partial V_2} \mathrm{d}S\, \hat{\boldsymbol{\nu}}_1(\mathbf{r}) \cdot \mathcal{S}_1(\mathbf{r}, t) + \int_{V_C} \mathrm{d}V\, \sigma(\mathbf{r}) |\mathcal{E}(\mathbf{r}, t)|^2 = -\int_{V_S} \mathrm{d}V\, \mathcal{E}(\mathbf{r}, t) \cdot \mathcal{J}(\mathbf{r}, t) \tag{1.294}$$

$$\frac{\mathrm{d}}{\mathrm{d}t}[W_{e2}(t) + W_{h2}(t)] + \int_{\partial V_2} \mathrm{d}S\, \hat{\boldsymbol{\nu}}_2(\mathbf{r}) \cdot \mathcal{S}_2(\mathbf{r}, t) = 0 \tag{1.295}$$

where $\hat{\boldsymbol{\nu}}_1(\mathbf{r})$ and $\hat{\boldsymbol{\nu}}_2(\mathbf{r})$ are defined on ∂V_2 and point inwards V_2 and V_1, respectively. Since the electromagnetic energy is an additive set function, we may sum the equations above side by side to obtain

$$\frac{\mathrm{d}}{\mathrm{d}t}[W_e(t) + W_h(t)] + \int_{\partial V} \mathrm{d}S\, \hat{\mathbf{n}}(\mathbf{r}) \cdot \mathcal{S}(\mathbf{r}, t) + \int_{\partial V_2} \mathrm{d}S\, \hat{\boldsymbol{\nu}}(\mathbf{r}) \cdot [\mathcal{S}_2(\mathbf{r}, t) - \mathcal{S}_1(\mathbf{r}, t)] + \int_{V_C} \mathrm{d}V\, \sigma(\mathbf{r}) |\mathcal{E}(\mathbf{r}, t)|^2 = -\int_{V_S} \mathrm{d}V\, \mathcal{E}(\mathbf{r}, t) \cdot \mathcal{J}(\mathbf{r}, t) \tag{1.296}$$

where $W_e + W_h$ is the total energy stored in V, and $\hat{\boldsymbol{\nu}}(\mathbf{r}) = \hat{\boldsymbol{\nu}}_2(\mathbf{r}) = -\hat{\boldsymbol{\nu}}_1(\mathbf{r})$. The flux integral over ∂V_2 vanishes in consideration of (1.293), and we recover (1.276) applied to the whole region V.

Although the volume power densities and the time variation of the energy density are well-defined quantities, one may question the very meaning of a certain amount of power associated with a point or a finite region of space [8, 76, 77], [78, pp. 266–268], [1, Section 27-4]. Worse still, the notion of surface power density flowing through a point is even more troublesome because the commonly accepted definition of Poynting vector (1.280) is not the only possible one. Indeed, it is apparent that relation (1.287) is not affected if we trade the Poynting vector for the field

$$\mathcal{S}'(\mathbf{r}, t) = \mathcal{S}(\mathbf{r}, t) + \nabla \times \mathbf{f}(\mathbf{r}, t) \tag{1.297}$$

where $\mathbf{f}(\mathbf{r}, t)$ is any vector field twice differentiable with respect to $\mathbf{r}$, so that (A.39) can be invoked. The root cause for this troublesome arbitrariness lies in the fact that (1.287) sets a condition only for the divergence of $\mathcal{S}(\mathbf{r}, t)$. We shall prove in Section 8.1 that a vector field is fully determined if *divergence and curl* thereof are assigned. Then again, the power flow computed through (1.277) is well-defined and unique, because any contribution of the type $\hat{\mathbf{n}}(\mathbf{r}) \cdot \nabla \times \mathbf{f}(\mathbf{r}, t)$ — if present — integrates to zero over ∂V, as can be proved by applying the Gauss theorem. For this reason and in light of (1.276), (1.280) seems the most logical and natural choice.

It is worth mentioning that in some situations the scalar field $\hat{\mathbf{n}} \cdot \mathcal{S}$ in (1.281) does not represent a surface density of power flow. Consider, for instance, a charged parallel-plate capacitor placed between the poles of a horseshoe permanent magnet. The arrangement is such that, between the plates of the capacitor, the magnetic induction field $\mathcal{B}(\mathbf{r}) = \mu_0 \mathcal{H}(\mathbf{r})$ and the electric field $\mathcal{E}(\mathbf{r})$ produced by the fixed charges are orthogonal. While this assumption is fairly true only away from the edges of the plates and the magnet poles, the deviation from the ideal configuration is not critical for the discussion. Then, in accordance with (1.280) the Poynting vector between the plates can be substantially non-zero, and, more importantly, $\hat{\mathbf{n}} \cdot \mathcal{S}$ may not be null on the boundary ∂V of an arbitrary spatial region V within the capacitor. Yet, the system is in equilibrium, and hence there should be no net power flow through ∂V and certainly no radiation at all. But then, since the total electromagnetic energy is constant in V and neither conductors nor currents are present, (1.287) yields

$$\nabla \cdot \mathcal{S}(\mathbf{r}) = 0, \qquad \mathbf{r} \in V \tag{1.298}$$

that is, the Poynting vector must be solenoidal. Further, the Poynting theorem in global form (1.276) states that the total power flux $P_F(t)$ through ∂V vanishes identically, as we expect of this system, despite $\mathrm{d}P_F/\mathrm{d}S$ being possibly non-zero on ∂V.

Other examples that deal with similar situations are discussed in [50, p. 146 and p. 194].

1.10.2 Poynting theorem in the frequency domain

Since time-harmonic fields are periodic functions of time, they admit a special form of the Poynting theorem [39, Chapter 1], [33, Chapter 1], [74, Section 5.2], [75, Section VIII.5]. To obtain the relevant balance equation, we take the time average of both sides of (1.276) and (1.287) over a period $T := 2\pi/\omega$. Therefore, we first derive a few formulas for expressing the time average of time-harmonic quantities and combinations thereof in terms of the associated phasors (see Section 1.5).

To gain more generality, in the following $f(\mathbf{r}, t) \in \mathbb{R}$, $\mathcal{F}(\mathbf{r}, t) \in \mathbb{R}^3$ and $\mathcal{G}(\mathbf{r}, t) \in \mathbb{R}^3$ denote time-harmonic scalar and vector fields, respectively.

(a) Time average of a scalar field $f(\mathbf{r}, t)$

$$\langle f(\mathbf{r}, t)\rangle := \frac{1}{T}\int_{t_0}^{t_0+T} \mathrm{d}t\, f(\mathbf{r}, t) = \frac{1}{T}\int_{t_0}^{t_0+T} \mathrm{d}t\, \mathrm{Re}\{F(\mathbf{r})\mathrm{e}^{\mathrm{j}\omega t}\}$$

$$= \mathrm{Re}\left\{\frac{F(\mathbf{r})}{T}\int_{t_0}^{t_0+T} \mathrm{d}t\, \mathrm{e}^{\mathrm{j}\,\omega t}\right\} = 0 \tag{1.299}$$

(b) Time average of the square of a scalar field $[f(\mathbf{r},t)]^2$

$$\begin{aligned}
\left\langle [f(\mathbf{r},t)]^2 \right\rangle &:= \frac{1}{T}\int_{t_0}^{t_0+T} \mathrm{d}t\, [f(\mathbf{r},t)]^2 = \frac{1}{T}\int_{t_0}^{t_0+T} \mathrm{d}t \left[\mathrm{Re}\left\{F(\mathbf{r})\mathrm{e}^{\mathrm{j}\,\omega t}\right\}\right]^2 \\
&= \frac{1}{T}\int_{t_0}^{t_0+T} \mathrm{d}t \left[\frac{F(\mathbf{r})\mathrm{e}^{\mathrm{j}\,\omega t} + F^*(\mathbf{r})\mathrm{e}^{-\mathrm{j}\,\omega t}}{2}\right]^2 \\
&= \frac{1}{4T}\int_{t_0}^{t_0+T} \mathrm{d}t \left\{[F(\mathbf{r})]^2\mathrm{e}^{2\mathrm{j}\,\omega t} + [F^*(\mathbf{r})]^2\mathrm{e}^{-2\mathrm{j}\,\omega t} + 2|F(\mathbf{r})|^2\right\} \\
&= \frac{1}{2}|F(\mathbf{r})|^2
\end{aligned} \tag{1.300}$$

(c) Time average of the scalar product of two vector fields $\mathfrak{F}(\mathbf{r},t)\cdot\mathfrak{G}(\mathbf{r},t)$ [69, Section 1.28]

$$\begin{aligned}
\langle \mathfrak{F}\cdot\mathfrak{G}\rangle &:= \frac{1}{T}\int_{t_0}^{t_0+T} \mathrm{d}t\, \mathfrak{F}(\mathbf{r},t)\cdot\mathfrak{G}(\mathbf{r},t) = \frac{1}{T}\int_{t_0}^{t_0+T} \mathrm{d}t\, \mathrm{Re}\{\mathbf{F}(\mathbf{r})\mathrm{e}^{\mathrm{j}\,\omega t}\}\cdot\mathrm{Re}\{\mathbf{G}(\mathbf{r})\mathrm{e}^{\mathrm{j}\,\omega t}\} \\
&= \frac{1}{T}\int_{t_0}^{t_0+T} \mathrm{d}t\, \frac{\mathbf{F}(\mathbf{r})\mathrm{e}^{\mathrm{j}\,\omega t} + \mathbf{F}^*(\mathbf{r})\mathrm{e}^{-\mathrm{j}\,\omega t}}{2}\cdot\frac{\mathbf{G}(\mathbf{r})\mathrm{e}^{\mathrm{j}\,\omega t} + \mathbf{G}^*(\mathbf{r})\mathrm{e}^{-\mathrm{j}\,\omega t}}{2} \\
&= \frac{1}{4T}\int_{t_0}^{t_0+T} \mathrm{d}t\, [\mathbf{F}(\mathbf{r})\cdot\mathbf{G}^*(\mathbf{r}) + \mathbf{F}^*(\mathbf{r})\cdot\mathbf{G}(\mathbf{r})] \\
&\quad + \frac{1}{4T}\int_{t_0}^{t_0+T} \mathrm{d}t \left[\mathbf{F}(\mathbf{r})\cdot\mathbf{G}(\mathbf{r})\mathrm{e}^{2\mathrm{j}\,\omega t} + \mathbf{F}^*(\mathbf{r})\cdot\mathbf{G}^*(\mathbf{r})\mathrm{e}^{-2\mathrm{j}\,\omega t}\right] \\
&= \frac{1}{2}\mathrm{Re}\{\mathbf{F}(\mathbf{r})\cdot\mathbf{G}^*(\mathbf{r})\}
\end{aligned} \tag{1.301}$$

(d) Time average of the cross-product of two vector fields $\mathfrak{F}(\mathbf{r},t)\times\mathfrak{G}(\mathbf{r},t)$

$$\langle \mathfrak{F}\times\mathfrak{G}\rangle := \frac{1}{T}\int_{t_0}^{t_0+T} \mathrm{d}t\, \mathfrak{F}(\mathbf{r},t)\times\mathfrak{G}(\mathbf{r},t) = \frac{1}{2}\mathrm{Re}\{\mathbf{F}(\mathbf{r})\times\mathbf{G}^*(\mathbf{r})\} \tag{1.302}$$

The square root of $\langle [f(\mathbf{r},t)]^2\rangle$ is the *root-mean-square value* of $f(\mathbf{r},t)$ and, in accordance with (1.300), it equals $|F(\mathbf{r})|/\sqrt{2}$. As a special case, for $\mathfrak{F}(\mathbf{r},t) = \mathfrak{G}(\mathbf{r},t)$ (1.301) yields the average of $|\mathfrak{F}(\mathbf{r},t)|^2$. Furthermore, it follows that the *root-mean-square value* of $\mathfrak{F}(\mathbf{r},t)$ is $|\mathbf{F}(\mathbf{r})|/\sqrt{2}$.

We now apply the previous results to the contributions involved in the Poynting theorem (1.287). We begin with (1.277) and compute the average rate of energy flow

$$\langle P_F \rangle := \frac{1}{T} \int_{t_0}^{t_0+T} \mathrm{d}t \int_{\partial V} \mathrm{d}S\, \hat{\mathbf{n}}(\mathbf{r}) \cdot \boldsymbol{\mathcal{E}}(\mathbf{r}, t) \times \boldsymbol{\mathcal{H}}(\mathbf{r}, t) = \frac{1}{2} \mathrm{Re} \int_{\partial V} \mathrm{d}S\, \hat{\mathbf{n}}(\hat{\mathbf{r}}) \cdot \mathbf{E}(\mathbf{r}) \times \mathbf{H}^*(\mathbf{r}) \tag{1.303}$$

where t_0 is an arbitrary instant of time. We have swapped the order of spatial and temporal integration and applied (1.302). The vector field

$$\mathbf{S}(\mathbf{r}) := \frac{1}{2} \mathbf{E}(\mathbf{r}) \times \mathbf{H}^*(\mathbf{r}) \tag{1.304}$$

is called the *complex Poynting vector* [33,69]. It is important to realize that $\mathbf{S}(\mathbf{r})$ is not the phasor of $\boldsymbol{\mathcal{S}}(\mathbf{r}, t)$ in the sense of definition (1.88), whereas $\mathrm{Re}\{\mathbf{S}(\mathbf{r})\}$ represents the time average of $\boldsymbol{\mathcal{S}}(\mathbf{r}, t)$, viz.,

$$\langle \boldsymbol{\mathcal{S}} \rangle := \frac{1}{2} \mathrm{Re}\{\mathbf{E}(\mathbf{r}) \times \mathbf{H}^*(\mathbf{r})\} \tag{1.305}$$

by virtue of (1.302) applied to (1.280). Besides, on account of (1.281) and (1.304) the average surface power density flowing through ∂V reads

$$\left\langle \frac{\mathrm{d}P_F}{\mathrm{d}S} \right\rangle := \mathrm{Re}\{\hat{\mathbf{n}}(\mathbf{r}) \cdot \mathbf{S}(\mathbf{r})\} \tag{1.306}$$

and the volume density of average power flow is

$$\langle p_F \rangle := \mathrm{Re}\{\nabla \cdot \mathbf{S}(\mathbf{r})\} \tag{1.307}$$

in consideration of (1.289).

Secondly, we examine the average power dissipated in the conductors occupying the region $V_C \subset V$, i.e.,

$$\langle P_C \rangle := \frac{1}{T} \int_{t_0}^{t_0+T} \mathrm{d}t \int_{V_C} \mathrm{d}V\, \sigma(\mathbf{r}) |\boldsymbol{\mathcal{E}}(\mathbf{r}, t)|^2 = \frac{1}{2} \mathrm{Re} \int_{V_C} \mathrm{d}V\, \sigma(\mathbf{r}) |\mathbf{E}(\mathbf{r})|^2 \tag{1.308}$$

having swapped space and time integration and invoked (1.301).

The time average of the power delivered by the sources follows in like manner, viz.,

$$\langle P_S \rangle := -\frac{1}{T} \int_{t_0}^{t_0+T} \mathrm{d}t \int_{V_S} \mathrm{d}V\, \boldsymbol{\mathcal{E}}(\mathbf{r}, t) \cdot \boldsymbol{\mathcal{J}}(\mathbf{r}, t) = -\frac{1}{2} \mathrm{Re} \int_{V_S} \mathrm{d}V\, \mathbf{E}(\mathbf{r}) \cdot \mathbf{J}^*(\mathbf{r}) \tag{1.309}$$

on account of (1.279) and (1.301).

Next, we determine the average of the electric and magnetic energies by starting off with (1.263) and (1.266) extended to time-varying fields and, in particular, time-harmonic dependence. For the sake of argument, we also assume that the background medium in the volume $V \setminus V_C$ is free space, and that the conductor filling V_C exhibits the same permittivity and permeability as free space. Then, we have

$$\langle W_e \rangle := \frac{1}{T} \int_{t_0}^{t_0+T} \mathrm{d}t \frac{1}{2} \int_{V} \mathrm{d}V\, \boldsymbol{\mathcal{E}}(\mathbf{r}, t) \cdot \boldsymbol{\mathcal{D}}(\mathbf{r}, t)$$

$$= \frac{1}{4}\mathrm{Re}\int_V \mathrm{d}V\,\mathbf{E}(\mathbf{r})\cdot\mathbf{D}^*(\mathbf{r}) = \frac{\varepsilon_0}{4}\int_V \mathrm{d}V\,|\mathbf{E}(\mathbf{r})|^2 \geqslant 0 \tag{1.310}$$

$$\langle W_h\rangle := \frac{1}{T}\int_{t_0}^{t_0+T}\mathrm{d}t\,\frac{1}{2}\int_V \mathrm{d}V\,\mathcal{B}(\mathbf{r},t)\cdot\mathcal{H}(\mathbf{r},t)$$

$$= \frac{1}{4}\mathrm{Re}\int_V \mathrm{d}V\,\mathbf{B}(\mathbf{r})\cdot\mathbf{H}^*(\mathbf{r}) = \frac{\mu_0}{4}\int_V \mathrm{d}V\,|\mathbf{H}(\mathbf{r})|^2 \geqslant 0 \tag{1.311}$$

thanks to (1.301) and the constitutive relationships (1.112) and (1.113). Therefore, the average electromagnetic energy in V is non-negative.

Likewise, we can obtain the average of the time rate of variation of the electric and magnetic energies, namely,

$$\left\langle\frac{\mathrm{d}W_e}{\mathrm{d}t}\right\rangle := \frac{1}{T}\int_{t_0}^{t_0+T}\mathrm{d}t\int_V \mathrm{d}V\,\mathcal{E}(\mathbf{r},t)\cdot\frac{\partial\mathcal{D}(\mathbf{r},t)}{\partial t}$$

$$= \frac{1}{2}\mathrm{Re}\int_V \mathrm{d}V\,\mathbf{E}(\mathbf{r})\cdot[-\mathrm{j}\,\omega\mathbf{D}^*(\mathbf{r})] = -2\,\mathrm{Re}\{\mathrm{j}\,\omega\,\langle W_e\rangle\} = 0 \tag{1.312}$$

$$\left\langle\frac{\mathrm{d}W_h}{\mathrm{d}t}\right\rangle := \frac{1}{T}\int_{t_0}^{t_0+T}\mathrm{d}t\int_V \mathrm{d}V\,\frac{\partial\mathcal{B}(\mathbf{r},t)}{\partial t}\cdot\mathcal{H}(\mathbf{r},t)$$

$$= \frac{1}{2}\mathrm{Re}\int_V \mathrm{d}V\,\mathrm{j}\,\omega\mathbf{B}(\mathbf{r})\cdot\mathbf{H}^*(\mathbf{r}) = 2\,\mathrm{Re}\{\mathrm{j}\,\omega\,\langle W_h\rangle\} = 0 \tag{1.313}$$

where we have formally used (1.93). The average time rate of variation of the electromagnetic energy is zero because $\langle W_e\rangle$ and $\langle W_h\rangle$ are real positive quantities.

With these ingredients we can simply write down the time-averaged counterpart of (1.276) [15, 33]

$$\mathrm{Re}\int_{\partial V}\mathrm{d}S\,\hat{\mathbf{n}}(\mathbf{r})\cdot\mathbf{S}(\mathbf{r}) + \frac{1}{2}\int_{V_C}\mathrm{d}V\,\sigma(\mathbf{r})|\mathbf{E}(\mathbf{r})|^2 = -\frac{1}{2}\mathrm{Re}\int_{V_S}\mathrm{d}V\,\mathbf{E}(\mathbf{r})\cdot\mathbf{J}^*(\mathbf{r}) \tag{1.314}$$

or equivalently

$$\langle P_F\rangle + \langle P_C\rangle = \langle P_S\rangle \tag{1.315}$$

in light of (1.303), (1.308) and (1.309). The latter balance can also be given in local form in terms of power densities, namely,

$$\mathrm{Re}\{\nabla\cdot\mathbf{S}(\mathbf{r})\} + \frac{1}{2}\sigma(\mathbf{r})|\mathbf{E}(\mathbf{r})|^2 = -\frac{1}{2}\mathrm{Re}\{\mathbf{E}(\mathbf{r})\cdot\mathbf{J}^*(\mathbf{r})\} \tag{1.316}$$

for $\mathbf{r}\in\mathbb{R}^3$ since the volume V is arbitrary.

The extension of theorem (1.314) to the whole space leads to a significantly different result than the one we found for fields with general time dependence, essentially because the boundary conditions at infinity (1.260) and (1.261) *do not apply* to time-harmonic fields! Indeed, those relations were assumed on the grounds that at any finite time t the fields are at most static or stationary at infinity, since it takes an infinite amount of time for the acceleration field to reach infinity.

However, if the motion of the charged particle is *periodic* (and thus confined to a bounded region of space) the situation changes completely, inasmuch as the acceleration $\mathbf{a}(t_r)$ is a periodic function of time as well. Therefore, for any time t and an arbitrary large distance $R = |\mathbf{r} - \mathbf{w}(t_r)|$ from any retarded position of the particle $\mathbf{w}(t_r)$, the acceleration $\mathbf{a}(t_r)$ at the retarded time $t_r = t - R/c_0$ — no matter how remote t_r is — may be non-zero. It follows that time-harmonic electromagnetic fields may fall off *at most* as $1/R$ and, consequently, the complex Poynting vector (1.304) decays as $1/R^2$. Such asymptotic behavior implies that the flux integral of $\mathbf{S}$ may not vanish at infinity in general.

More specifically, let us consider a ball $B(0, b)$ with radius b large enough so that $V \subset B(0, b)$ with V being once again the bounded region of Figure 1.17. By applying the complex Poynting theorem to the *source-free lossless* region $B(0, b) \setminus V$ we find

$$\mathrm{Re} \int_{\partial B} \mathrm{d}S\, \hat{\mathbf{n}}(\mathbf{r}) \cdot \mathbf{S}(\mathbf{r}) - \mathrm{Re} \int_{\partial V} \mathrm{d}S\, \hat{\mathbf{n}}(\mathbf{r}) \cdot \mathbf{S}(\mathbf{r}) = 0 \tag{1.317}$$

with the unit normal $\hat{\mathbf{n}}(\mathbf{r})$ positively oriented outwards $B(0, b)$ and V on ∂B and ∂V, respectively. Now, since the second contribution in the left-hand side of (1.317) is the same quantity as in (1.314) applied to V, we see that the average power flow through ∂B coincides with the flow through ∂V. From this finding we conclude that:

(1) So long as we enclose all the relevant time-harmonic sources in a volume V, for the calculation of $\langle P_F \rangle$ we are entitled to choose any shape for V and the boundary thereof.

(2) More importantly, by taking the limit of the left-hand side of (1.317) as $b \to +\infty$, we see that the average power flow at infinity, unlike the instantaneous counterpart examined in (1.284), may be finite and non-null.

(3) Since the latter result is only due to the acceleration fields, $\langle P_F \rangle$ — even computed through ∂V at finite — represents the average flow of *radiated* power. In other words, the Coulomb fields do not contribute, on average, to the flow of energy through ∂V.

Actually, (1.317) may be used to show that time-harmonic fields must fall off *at least* as $1/R$ in order for the flux integral over the sphere at infinity to be finite. It is easy to check that any slower decay rate, say, $1/R^{\nu}$ with $\nu < 1$, gives rise to diverging radiated power at infinity — which is nonphysical. Finally, we anticipate that since the average power flow is non-null at infinity, it is more difficult to ensure uniqueness of solutions to the time-harmonic Maxwell equations in the whole space (see Section 6.4).

It is possible to prove the Poynting theorem for both general time-varying fields and in the time-harmonic regime by manipulating the local form of the Faraday law and the Ampère-Maxwell law in a purely mathematical fashion [33,52,58]. While there is no questioning the validity of the resulting balance equations, the procedure may sound artificial and unsatisfactory, inasmuch as it is not based on physical (energy) arguments. In the case of the time-harmonic fields and sources, though, the approach has the undisputable merit of producing a *complex-valued* relationship which as such is more general than (1.316).

To elaborate, we consider the time-harmonic Maxwell equations for the same configuration of sources and matter of Figure 1.17. We dot-multiply (1.99) by $\mathbf{H}^*(\mathbf{r})$ and the complex conjugate of (1.98) — augmented with the conduction current $\mathbf{J}_c(\mathbf{r}) := \sigma(\mathbf{r})\mathbf{E}(\mathbf{r})$ — by $\mathbf{E}(\mathbf{r})$, namely,

$$\mathbf{H}^*(\mathbf{r}) \cdot \nabla \times \mathbf{E}(\mathbf{r}) = -\mathrm{j}\,\omega \mathbf{H}^*(\mathbf{r}) \cdot \mathbf{B}(\mathbf{r}) \tag{1.318}$$

$$\mathbf{E}(\mathbf{r}) \cdot \nabla \times \mathbf{H}^*(\mathbf{r}) = -\mathrm{j}\,\omega \mathbf{E}(\mathbf{r}) \cdot \mathbf{D}^*(\mathbf{r}) + \mathbf{E}(\mathbf{r}) \cdot \sigma(\mathbf{r})\mathbf{E}^*(\mathbf{r}) + \mathbf{E}(\mathbf{r}) \cdot \mathbf{J}^*(\mathbf{r}) \tag{1.319}$$

with $\sigma(\mathbf{r}) = 0$ for $\mathbf{r} \notin V_C$ and $\mathbf{J}(\mathbf{r}) = 0$ for $\mathbf{r} \notin V_S$. Secondly, we exploit the linearity of the equations and we subtract the second from the first to get

$$\nabla \cdot [\mathbf{E}(\mathbf{r}) \times \mathbf{H}^*(\mathbf{r})] + \mathrm{j}\,\omega\,[\mathbf{H}^*(\mathbf{r}) \cdot \mathbf{B}(\mathbf{r}) - \mathbf{E}(\mathbf{r}) \cdot \mathbf{D}^*(\mathbf{r})] \\ + \sigma(\mathbf{r})|\mathbf{E}(\mathbf{r})|^2 = -\mathbf{E}(\mathbf{r}) \cdot \mathbf{J}^*(\mathbf{r}), \qquad \mathbf{r} \in \mathbb{R}^3 \quad (1.320)$$

having used the differential identity (H.49) in reverse gear. Finally, we divide through by two and write

$$\nabla \cdot \left[\frac{1}{2}\mathbf{E}(\mathbf{r}) \times \mathbf{H}^*(\mathbf{r})\right] + 2\,\mathrm{j}\,\omega\left[\frac{\mu_0}{4}|\mathbf{H}(\mathbf{r})|^2 - \frac{\varepsilon_0}{4}|\mathbf{E}(\mathbf{r})|^2\right] \\ + \frac{1}{2}\sigma(\mathbf{r})|\mathbf{E}(\mathbf{r})|^2 = -\frac{1}{2}\mathbf{E}(\mathbf{r}) \cdot \mathbf{J}^*(\mathbf{r}), \qquad \mathbf{r} \in \mathbb{R}^3 \quad (1.321)$$

on account of (1.112) and (1.113). We recognize the first term as the divergence of the complex Poynting vector $\mathbf{S}(\mathbf{r})$ introduced in (1.304). Moreover, in light of (1.310) and (1.311) we may define the quantities

$$\langle w_h(\mathbf{r})\rangle := \frac{\mu_0}{4}|\mathbf{H}(\mathbf{r})|^2 \geqslant 0 \qquad \langle w_e(\mathbf{r})\rangle := \frac{\varepsilon_0}{4}|\mathbf{E}(\mathbf{r})|^2 \geqslant 0 \quad (1.322)$$

and interpret them as the time-averaged volume densities of magnetic and electric energy stored in the field. With these positions (1.321) becomes

$$\nabla \cdot \mathbf{S}(\mathbf{r}) + 2\,\mathrm{j}\,\omega[\langle w_h(\mathbf{r})\rangle - \langle w_e(\mathbf{r})\rangle] + \frac{1}{2}\sigma(\mathbf{r})|\mathbf{E}(\mathbf{r})|^2 = -\frac{1}{2}\mathbf{E}(\mathbf{r}) \cdot \mathbf{J}^*(\mathbf{r}), \qquad \mathbf{r} \in \mathbb{R}^3 \quad (1.323)$$

which is the sought balance equation between complex vector fields.

Now, taking the real parts of both sides clearly yields (1.316), the balance equation between average densities of power. However, as anticipated, (1.323) also implies the equality of the imaginary parts of both sides, namely,

$$\mathrm{Im}\{\nabla \cdot \mathbf{S}(\mathbf{r})\} + 2\omega[\langle w_h(\mathbf{r})\rangle - \langle w_e(\mathbf{r})\rangle] = -\mathrm{Im}\left\{\frac{1}{2}\mathbf{E}(\mathbf{r}) \cdot \mathbf{J}^*(\mathbf{r})\right\}, \qquad \mathbf{r} \in \mathbb{R}^3 \quad (1.324)$$

under the assumption that $\sigma(\mathbf{r}) \in \mathbb{R}$. The interpretation of (1.324) — which does not follow by averaging the time-domain Poynting theorem (1.287) — is not obvious except for supposedly being a balance of volume densities of *reactive* power. Proceeding backwards, we may integrate (1.324) over the finite region of space V (Figure 1.17) to get the global form

$$\mathrm{Im}\int_{\partial V} \mathrm{d}S\, \hat{\mathbf{n}}(\mathbf{r}) \cdot \mathbf{S}(\mathbf{r}) + 2\omega(\langle W_h\rangle - \langle W_e\rangle) = -\mathrm{Im}\frac{1}{2}\int_{V_S} \mathrm{d}V\, \mathbf{E}(\mathbf{r}) \cdot \mathbf{J}^*(\mathbf{r}) \quad (1.325)$$

where we have invoked (1.310) and (1.311), and applied the Gauss theorem to the first integral. The procedure formally entails separating the integration into parts over V_C, V_S and $V \setminus (V_C \cup V_S)$ and enforcing the time-harmonic counterpart of the matching condition (1.293), whereby the flux integrals on either sides of ∂V_C and ∂V_S cancel each other out.

A special instance of (1.316) and (1.325) occurs when the region V is the inside of a *lossless cavity* (i.e., with $\sigma(\mathbf{r}) = 0$ for $\mathbf{r} \in V_C \subset V$) with PEC walls. In this configuration we have

$$\hat{\mathbf{n}}(\mathbf{r}) \cdot \mathbf{S}(\mathbf{r}) = \frac{1}{2}[\hat{\mathbf{n}}(\mathbf{r}) \times \mathbf{E}(\mathbf{r})] \cdot \mathbf{H}^*(\mathbf{r}) = 0, \qquad \mathbf{r} \in \partial V \quad (1.326)$$

in view of (1.304), the vector identity (H.13), and the boundary condition (1.169). This result combined with (1.314) means that no energy can flow in and out V, as is logical. The source-free time-harmonic Maxwell equations supplemented with PEC boundary conditions (1.169) admit an infinite, countable set of nonzero solutions $\mathbf{E}_\nu(\mathbf{r})$, $\mathbf{H}_\nu(\mathbf{r})$, with $\nu := (m, n, p) \in \mathbb{N}^3$ being a triple of indices, that are associated with particular values of the angular frequency $\omega = \omega_\nu > 0$ called *eigenvalues* [10, 33, 52] (see Sections 6.4.1 and 11.1). Therefore, we may apply (1.316) and (1.325) to $\mathbf{E}_\nu(\mathbf{r})$, $\mathbf{H}_\nu(\mathbf{r})$. Since by hypothesis $\mathbf{J}(\mathbf{r}) = 0 = \sigma(\mathbf{r})$ for $\mathbf{r} \in V$, then (1.316) becomes a trivial identity between null quantities. On the contrary, (1.325) demands

$$2\omega_\nu[\langle W_h(\omega_\nu)\rangle - \langle W_e(\omega_\nu)\rangle] = 0, \qquad \nu \in \mathbb{N}^3 \tag{1.327}$$

which is satisfied only if the average electric and magnetic energies of the fields $\mathbf{E}_\nu(\mathbf{r})$, $\mathbf{H}_\nu(\mathbf{r})$ associated with the each eigenvalue ω_ν are equal to one another. This occurrence is referred to as a *resonance* of the cavity.

1.11 Conservation of electromagnetic momentum

At the beginning of Section 1.9 we mentioned that electromagnetic waves carry energy but also momentum away from the sources. For this reason, it is meaningful to seek a conservation law [7, 8, 11, 15, 16, 20, 24, 43, 46, 74, 78], [55, Section 15.5], [73, Section 3.3.2] which involves the Lorentz force density $\mathbf{f}_{em}(\mathbf{r}, t)$ (1.6) and the density of linear momentum yet to be defined in the context of electrodynamics.

Unlike the electromagnetic entities which appear in the Lorentz force (1.4) for a *test* point charge, in (1.6) $\mathcal{E}(\mathbf{r}, t)$ and $\mathcal{B}(\mathbf{r}, t)$ indicate the fields *produced* by the sources $\varrho(\mathbf{r}, t)$ and $\mathcal{J}(\mathbf{r}, t)$. As a result, since sources and fields are related by the Maxwell equations (also see the diagram of Figure 1.5), the latter may be used to eliminate $\varrho(\mathbf{r}, t)$ and $\mathcal{J}(\mathbf{r}, t)$ from (1.6) and obtain an expression that involves only the fields.

To this purpose we suppose that $\varrho(\mathbf{r}, t)$ and $\mathcal{J}(\mathbf{r}, t)$ are confined to a bounded domain $V_J \subset V \subseteq \mathbb{R}^3$, wherefore $\mathbf{f}_{em}(\mathbf{r}, t)$ vanishes outside V_J. Then, by invoking the Gauss law (1.44) and the Ampère-Maxwell law (1.34) we have

$$\mathcal{E}(\mathbf{r}, t)\nabla \cdot \mathcal{D}(\mathbf{r}, t) - \mathcal{B}(\mathbf{r}, t) \times \left[\nabla \times \mathcal{H}(\mathbf{r}, t) - \frac{\partial \mathcal{D}}{\partial t}\right] = \mathbf{f}_{em}(\mathbf{r}, t), \qquad (\mathbf{r}, t) \in V \times \mathbb{R}^+ \tag{1.328}$$

whereas the following identity

$$-\mathcal{H}(\mathbf{r}, t)\nabla \cdot \mathcal{B}(\mathbf{r}, t) + \mathcal{D}(\mathbf{r}, t) \times \left[\nabla \times \mathcal{E}(\mathbf{r}, t) + \frac{\partial \mathcal{B}}{\partial t}\right] = 0, \qquad (\mathbf{r}, t) \in V \times \mathbb{R}^+ \tag{1.329}$$

holds trivially on account of the magnetic Gauss law (1.23) and the Faraday law (1.20). One should check that the vector field in the left member of (1.329) carries the physical dimension of a volume density of force. Consequently, we may subtract (1.328) from (1.329) side by side and combine the terms with temporal derivatives to obtain

$$\mathcal{D}(\mathbf{r}, t) \times [\nabla \times \mathcal{E}(\mathbf{r}, t)] - \mathcal{E}(\mathbf{r}, t)\nabla \cdot \mathcal{D}(\mathbf{r}, t) + \mathcal{B}(\mathbf{r}, t) \times [\nabla \times \mathcal{H}(\mathbf{r}, t)] - \mathcal{H}(\mathbf{r}, t)\nabla \cdot \mathcal{B}(\mathbf{r}, t)$$
$$+ \frac{\partial}{\partial t}[\mathcal{D}(\mathbf{r}, t) \times \mathcal{B}(\mathbf{r}, t)] = -\mathbf{f}_{em}(\mathbf{r}, t), \qquad (\mathbf{r}, t) \in V \times \mathbb{R}^+ \tag{1.330}$$

where the vector field

$$\mathbf{g}_{em} := \mathcal{D}(\mathbf{r}, t) \times \mathcal{B}(\mathbf{r}, t), \qquad (\mathbf{r}, t) \in V \times \mathbb{R}^+ \tag{1.331}$$

is customarily interpreted as the volume density of *linear electromagnetic momentum* (physical dimension: C Wb/m^4 = kg/(s m^2)) and is also known as the *Abraham density*. Still, for the debate that concerns this very definition in ponderable media the Reader may refer to the analysis presented in [79]. Further, in an isotropic homogeneous medium, such as free space, we may relate the density of momentum (1.331) to the Poynting vector (1.280), namely,

$$\mathbf{g}_{em}(\mathbf{r},t) = \varepsilon\mu\, \mathcal{E}(\mathbf{r},t) \times \mathcal{H}(\mathbf{r},t) = \frac{1}{c^2}\mathcal{S}(\mathbf{r},t) \tag{1.332}$$

thanks to the constitutive relationships (1.117) and (1.118) and definition (1.209).

In order for (1.330) to have the structure of a balance equation similar to (1.46) or (1.287) we must be able to turn the sum of the first four terms into the divergence of some *dyadic field* (Appendix E.2). Said transformation can indeed be accomplished when the medium that fills V is homogeneous and, in particular, either *isotropic*, or *anisotropic* with symmetric $\overline{\boldsymbol{\varepsilon}}$ and $\overline{\boldsymbol{\mu}}$, or even *bi-anisotropic* (a case we did not cover in Section 1.6) provided the pertinent constitutive matrix is symmetric [15], [14, Sections 1.5, 3.2]. For general constitutive relationships, though, (1.330) is the best we can hope for.

By assuming that the medium of concern is endowed with scalar constitutive parameters ε and μ, with the aid of the differential identity (H.53) we can write

$$\nabla(\mathcal{D}\cdot\mathcal{E}) = \varepsilon\nabla(\mathcal{E}\cdot\mathcal{E}) = 2\mathcal{D}\times(\nabla\times\mathcal{E}) + 2\mathcal{D}\cdot\nabla\mathcal{E} \tag{1.333}$$

$$\nabla(\mathcal{B}\cdot\mathcal{H}) = \mu\nabla(\mathcal{H}\cdot\mathcal{H}) = 2\mathcal{B}\times(\nabla\times\mathcal{H}) + 2\mathcal{B}\cdot\nabla\mathcal{H} \tag{1.334}$$

on the ground of (1.117) and (1.118). Next, we use these expressions to cast the first four terms of (1.330) as

$$\begin{aligned}\mathcal{D}\times(\nabla\times\mathcal{E}) - \mathcal{E}\nabla\cdot\mathcal{D} &= \frac{1}{2}\nabla(\mathcal{D}\cdot\mathcal{E}) - (\mathcal{D}\cdot\nabla\mathcal{E} + \mathcal{E}\nabla\cdot\mathcal{D})\\ &= \nabla\cdot\left(\frac{1}{2}\mathcal{D}\cdot\mathcal{E}\overline{\mathcal{I}} - \mathcal{D}\mathcal{E}\right)\end{aligned} \tag{1.335}$$

$$\begin{aligned}\mathcal{B}\times(\nabla\times\mathcal{H}) - \mathcal{H}\nabla\cdot\mathcal{B} &= \frac{1}{2}\nabla(\mathcal{B}\cdot\mathcal{H}) - (\mathcal{B}\cdot\nabla\mathcal{H} + \mathcal{H}\nabla\cdot\mathcal{B})\\ &= \nabla\cdot\left(\frac{1}{2}\mathcal{B}\cdot\mathcal{H}\overline{\mathcal{I}} - \mathcal{B}\mathcal{H}\right)\end{aligned} \tag{1.336}$$

having used the differential identities (H.65) and (H.71).

If the medium in V is anisotropic, (1.335) and (1.336) can be derived at the cost of some algebra. For ease of manipulation we expand vectors and differential operators in Cartesian coordinates and adopt the shorthand notation $\mathcal{D}_\alpha$ and $\mathcal{E}_\alpha$ with $\alpha \in \{x, y, z\}$ to indicate the components of $\mathcal{D}$ and $\mathcal{E}$. Further, we introduce the symbol ∂_α to signify differentiation with respect to the variable $\alpha \in \{x, y, z\}$. Then, we begin with the manipulation of the x-component of the electric part of (1.330), namely,

$$\begin{aligned}\hat{\mathbf{x}}\cdot\mathcal{D}\times(\nabla\times\mathcal{E}) - \mathcal{E}_x\nabla\cdot\mathcal{D} &= \mathcal{D}_y(\partial_x\mathcal{E}_y - \partial_y\mathcal{E}_x) - \mathcal{D}_z(\partial_z\mathcal{E}_x - \partial_x\mathcal{E}_z) - \mathcal{E}_x\nabla\cdot\mathcal{D}\\ &= \sum_{\alpha\in\{x,y,z\}}\mathcal{D}_\alpha\partial_x\mathcal{E}_\alpha - \sum_{\alpha\in\{x,y,z\}}\mathcal{D}_\alpha\partial_\alpha\mathcal{E}_x - \mathcal{E}_x\nabla\cdot\mathcal{D}\\ &= \mathcal{D}\cdot\partial_x\mathcal{E} - (\mathcal{D}\cdot\nabla\mathcal{E}_x + \mathcal{E}_x\nabla\cdot\mathcal{D})\\ &= \mathcal{D}\cdot\partial_x\mathcal{E} - \nabla\cdot(\mathcal{D}\mathcal{E}\cdot\hat{\mathbf{x}})\end{aligned} \tag{1.337}$$

having added and subtracted the term $\mathcal{D}_x\partial_x\mathcal{E}_x$ in the second step and used (H.51). Secondly, we exploit (1.126) and the stipulated symmetry of the dyadic $\overline{\varepsilon}$ (cf. Section 6.8)

$$\mathcal{D}\cdot\partial_x\mathcal{E} = \frac{1}{2}\mathcal{D}\cdot\partial_x\mathcal{E} + \frac{1}{2}\mathcal{E}\cdot\overline{\varepsilon}^T\cdot\partial_x\mathcal{E} = \frac{1}{2}\mathcal{D}\cdot\partial_x\mathcal{E} + \frac{1}{2}\mathcal{E}\cdot\partial_x(\overline{\varepsilon}\cdot\mathcal{E}) = \frac{1}{2}\partial_x(\mathcal{D}\cdot\mathcal{E}) \tag{1.338}$$

whereby we can write

$$\hat{\mathbf{x}}\hat{\mathbf{x}}\cdot\mathcal{D}\times(\nabla\times\mathcal{E}) - \hat{\mathbf{x}}\mathcal{E}_x\nabla\cdot\mathcal{D} = \frac{1}{2}\hat{\mathbf{x}}\,\partial_x(\mathcal{D}\cdot\mathcal{E}) - \nabla\cdot(\mathcal{D}\mathcal{E}\cdot\hat{\mathbf{x}}\hat{\mathbf{x}}) \tag{1.339}$$

on account of the differential identity (H.71) with $\mathcal{A} := \mathcal{D}\mathcal{E}\cdot\hat{\mathbf{x}}$ and $\mathcal{B} := \hat{\mathbf{x}}$. Similar expressions hold that involve $\hat{\mathbf{y}}$, $\mathcal{E}_y$, ∂_y and $\hat{\mathbf{z}}$, $\mathcal{E}_z$, ∂_z. By summing the three of them side by side we finally obtain

$$\begin{aligned}(\hat{\mathbf{x}}\hat{\mathbf{x}} + \hat{\mathbf{y}}\hat{\mathbf{y}} + \hat{\mathbf{z}}\hat{\mathbf{z}})\cdot\mathcal{D}\times(\nabla\times\mathcal{E}) - \mathcal{E}\nabla\cdot\mathcal{D} &= \\ &= \frac{1}{2}\nabla(\mathcal{D}\cdot\mathcal{E}) - \nabla\cdot[\mathcal{D}\mathcal{E}\cdot(\hat{\mathbf{x}}\hat{\mathbf{x}} + \hat{\mathbf{y}}\hat{\mathbf{y}} + \hat{\mathbf{z}}\hat{\mathbf{z}})] \\ &= \nabla\cdot\left(\frac{1}{2}\mathcal{D}\cdot\mathcal{E}\overline{\mathcal{I}} - \mathcal{D}\mathcal{E}\right)\end{aligned} \tag{1.340}$$

in light of (E.29) and (H.65). The magnetic part of (1.330) can be transformed in like manner if $\overline{\mu}$ is symmetric, and this ends the proof.

The dyadic fields (physical dimension: N/m^2 or J/m^3)

$$\overline{\mathcal{T}}_e(\mathbf{r},t) := \frac{1}{2}\mathcal{D}\cdot\mathcal{E}\overline{\mathcal{I}} - \mathcal{D}\mathcal{E}, \qquad (\mathbf{r},t)\in V\times\mathbb{R}^+ \tag{1.341}$$

$$\overline{\mathcal{T}}_h(\mathbf{r},t) := \frac{1}{2}\mathcal{B}\cdot\mathcal{H}\overline{\mathcal{I}} - \mathcal{B}\mathcal{H}, \qquad (\mathbf{r},t)\in V\times\mathbb{R}^+ \tag{1.342}$$

are called the *electric and magnetic stress tensors* or dyadics, and the dyadic field

$$\begin{aligned}\overline{\mathcal{T}}_{em}(\mathbf{r},t) &:= \overline{\mathcal{T}}_e(\mathbf{r},t) + \overline{\mathcal{T}}_h(\mathbf{r},t) \\ &= \frac{1}{2}(\mathcal{D}\cdot\mathcal{E} + \mathcal{B}\cdot\mathcal{H})\overline{\mathcal{I}} - \mathcal{D}\mathcal{E} - \mathcal{B}\mathcal{H}, \qquad (\mathbf{r},t)\in V\times\mathbb{R}^+\end{aligned} \tag{1.343}$$

is known as the Maxwell *electromagnetic stress tensor* or dyadic [2, 11, 15, 16, 78], [14, Chapter 1], [4, Chapter XI]. Some Authors [20, Section 6.2], [46, Section 1.2], [7,43,44,74] define the stress tensor as the negative of the one in (1.343). The stress tensor is patently symmetric if the underlying medium is isotropic, whereas this property is lost for anisotropic media even though $\overline{\varepsilon}$ and $\overline{\mu}$ are symmetric dyadics. The inherent separation of $\overline{\mathcal{T}}_{em}(\mathbf{r},t)$ into an electric and a magnetic part makes it possible to investigate the mechanical effects of electric and magnetic fields separately. It is also worth noticing that $\overline{\mathcal{T}}_{em}(\mathbf{r},t)$ is independent of the particular *orientation* of the electromagnetic field at a given location $\mathbf{r}$, for the right-hand side of (1.343) is unaffected by the substitutions $\mathcal{E} \Longrightarrow -\mathcal{E}$ and $\mathcal{H} \Longrightarrow -\mathcal{H}$ in combination with either (1.117), (1.118) or (1.126), (1.127).

Now, by virtue of (1.331), (1.335), (1.336) and (1.343) we can write (1.330) succinctly as

$$\nabla\cdot\overline{\mathcal{T}}_{em}(\mathbf{r},t) + \frac{\partial}{\partial t}\mathbf{g}_{em}(\mathbf{r},t) = -\mathbf{f}_{em}(\mathbf{r},t), \qquad (\mathbf{r},t)\in V\times\mathbb{R}^+ \tag{1.344}$$

which constitutes the desired local form of balance equation for the density of electromagnetic momentum. From (1.344) we infer that the momentum density is not conserved at points where charges and currents are present, inasmuch as the electromagnetic field may transfer momentum to the charges through the Lorentz force.

By integrating (1.344) over the region V we obtain the global form of the conservation law, viz.,

$$\int_V \mathrm{d}V\, \nabla \cdot \overline{\mathcal{T}}_{em}(\mathbf{r}, t) + \int_V \mathrm{d}V\, \frac{\partial}{\partial t}\mathbf{g}_{em}(\mathbf{r}, t) = -\mathcal{F}_{em}(t), \qquad t \in \mathbb{R}^+ \tag{1.345}$$

where

$$\mathcal{F}_{em}(t) := \int_{V_J} \mathrm{d}V\, \mathbf{f}_{em}(\mathbf{r}, t), \qquad t \in \mathbb{R}^+ \tag{1.346}$$

is the total Lorentz force acting on the sources within V_J. The nine components of $\overline{\mathcal{T}}_{em}(\mathbf{r}, t)$ are of class $\mathcal{C}^1(V) \cap \mathcal{C}(\overline{V})$ if so are the components of the four electromagnetic entities. Thus, we may invoke the Gauss theorem for dyadic fields (H.105) and swap integration and differentiation with respect to time to arrive at

$$\int_{\partial V} \mathrm{d}S\, \hat{\mathbf{n}}(\mathbf{r}) \cdot \overline{\mathcal{T}}_{em}(\mathbf{r}, t) + \frac{\mathrm{d}}{\mathrm{d}t}\int_V \mathrm{d}V\, \mathbf{g}_{em}(\mathbf{r}, t) = -\mathcal{F}_{em}(t), \qquad t \in \mathbb{R}^+ \tag{1.347}$$

under the hypotheses that the domain V and its boundary neither move nor change shape with time and the components of $\mathbf{g}_{em}(\mathbf{r}, t)$ and $\partial \mathbf{g}_{em}/\partial t$ are continuous for $(\mathbf{r}, t) \in V \times \mathbb{R}^+$ (cf. [32, Theorem 9.42]). The structure of (1.347) mirrors that of the Poynting theorem (1.276), and the interpretation is as follows.

- The quantity

$$\mathcal{G}_{em}(t) := \int_V \mathrm{d}V\, \mathbf{g}_{em}(\mathbf{r}, t) \tag{1.348}$$

  is interpreted as the instantaneous total linear *electromagnetic momentum* associated with the field and stored in the domain V. Odd as it sounds, momentum is stored in the field *even in stationary conditions* (see Section 4.1) pretty much in the same way as energy is stored under the same circumstances. We shall elaborate on the meaning of *potential momentum* available for charge motion in Section 5.3.
- The flux of $\overline{\mathcal{T}}_{em}$ may be regarded as the *rate of momentum flow* through the boundary ∂V [16, Section 3.2], [20, Section 6.2], though not all of this flow represents radiation. For instance, consider the special case of a point charge in free space and the fields thereof given by (1.252) and (1.253). An estimate similar to (1.258) carried out on the flux of the Maxwell stress tensor allows showing that only the acceleration field may contribute a non-zero momentum flow towards infinity. Alternatively, $\hat{\mathbf{n}}(\mathbf{r}) \cdot \overline{\mathcal{T}}_{em}(\mathbf{r}, t)$ for $\mathbf{r} \in \partial V$ represents a *force per unit area*, that is, a stress on the surface ∂V. These two viewpoints — rate of flow of momentum density or force density — are consistent inasmuch as any force may, in fact, be considered the time rate of variation of a momentum according to the Newton law of motion [21, Chapter 9].

We notice that, since (1.344) is perfectly similar to the fundamental equation which governs the propagation of small-amplitude elastic waves in solids [80, Chapter 10], [81], the interpretation of $\mathcal{T}_{em}$ and $\mathbf{g}_{em}$ is fully justified.

To extend (1.347) to all space $V \equiv \mathbb{R}^3$, we first apply the conservation of momentum to a ball $V := B(0,b)$ and secondly take the limit for $b \to +\infty$. In symbols, we have

$$\lim_{b\to+\infty} \int_{\partial B} \mathrm{d}S\, \hat{\mathbf{n}}(\mathbf{r}) \cdot \overline{\mathcal{T}}_{em}(\mathbf{r},t) + \lim_{b\to+\infty} \frac{\mathrm{d}}{\mathrm{d}t} \int_{B(0,b)} \mathrm{d}V\, \mathbf{g}_{em}(\mathbf{r},t) = -\mathcal{F}_{em}(t), \qquad t \in \mathbb{R}^+ \tag{1.349}$$

where we need to show that the rate of momentum flow vanishes at infinity whereas the total momentum remains finite. For the flux integral over ∂B we pick up a radius $b > \max\{b_E, b_H\}$, where the constants b_E, b_H enter the asymptotic boundary conditions (1.260) and (1.261), and thanks to (D.152), (D.151) we estimate

$$\begin{aligned} \left| \int_{\partial B} \mathrm{d}S\, \hat{\mathbf{n}}(\mathbf{r}) \cdot \overline{\mathcal{T}}_{em}(\mathbf{r},t) \right| &\leqslant \int_{\partial B} \mathrm{d}S\, \left| \hat{\mathbf{n}}(\mathbf{r}) \cdot \overline{\mathcal{T}}_{em}(\mathbf{r},t) \right| \\ &\leqslant \frac{3}{2} \int_{\partial B} \mathrm{d}S\, (|\mathcal{D}(\mathbf{r},t)||\mathcal{E}(\mathbf{r},t)| + |\mathcal{B}(\mathbf{r},t)||\mathcal{H}(\mathbf{r},t)|) \\ &\leqslant \frac{3}{2} \int_{4\pi} \mathrm{d}\Omega\, b^2 \left(\frac{C_D}{b^2}\frac{C_E}{b^2} + \frac{C_B}{b^2}\frac{C_H}{b^2} \right) \xrightarrow[b\to+\infty]{} 0 \end{aligned} \tag{1.350}$$

in light of either (1.117), (1.118) or (1.126), (1.127). For the integral of $\mathbf{g}_{em}$ we consider another ball $B(0,b_1)$ with $\max\{b_E, b_H\} < b_1 < b$ and estimate

$$\begin{aligned} \left| \int_{B(0,b)} \mathrm{d}V\, \mathbf{g}_{em}(\mathbf{r},t) \right| &= \left| \int_{B(0,b_1)} \mathrm{d}V\, \mathbf{g}_{em}(\mathbf{r},t) + \int_{B(0,b)\setminus B(0,b_1)} \mathrm{d}V\, \mathbf{g}_{em}(\mathbf{r},t) \right| \\ &\leqslant \int_{B(0,b_1)} \mathrm{d}V\, |\mathbf{g}_{em}(\mathbf{r},t)| + \int_{B(0,b)\setminus B(0,b_1)} \mathrm{d}V\, |\mathcal{D}(\mathbf{r},t)||\mathcal{B}(\mathbf{r},t)| \\ &\leqslant \int_{B(0,b_1)} \mathrm{d}V\, |\mathbf{g}_{em}(\mathbf{r},t)| + \int_{4\pi} \mathrm{d}\Omega \int_{b_1}^{b} \mathrm{d}r\, r^2 \frac{C_D C_B}{r^4} \\ &= \int_{B(0,b_1)} \mathrm{d}V\, |\mathbf{g}_{em}(\mathbf{r},t)| + 4\pi C_D C_B \left(\frac{1}{b_1} - \frac{1}{b} \right) \\ &\leqslant \int_{B(0,b_1)} \mathrm{d}V\, |\mathbf{g}_{em}(\mathbf{r},t)| + 4\pi \frac{C_D C_B}{b_1} \end{aligned} \tag{1.351}$$

and since the rightmost-hand side is independent of b, the momentum remains finite as $b \to +\infty$. Alternatively, if we argue that — the speed of light being finite — the field is electrostatic in character outside a sufficiently large ball $B(0,b)$, then $\mathbf{g}_{em}(\mathbf{r},t)$ vanishes for $\mathbf{r} \in \mathbb{R}^3 \setminus B[0,b]$, whereby the total momentum $\mathcal{G}_{em}(t)$ is perforce finite. At any rate, (1.347) reads

$$\frac{\mathrm{d}}{\mathrm{d}t}\mathcal{G}_{em}(t) = -\mathcal{F}_{em}(t) \tag{1.352}$$

when it is applied to the whole space. Since the Lorentz force describes the way the electromagnetic field affects the motion of charged particles (which have mass), in accordance with the Newton law of action and reaction [21, Chapter 10] it also holds

$$\frac{\mathrm{d}}{\mathrm{d}t}\mathcal{G}_{mech}(t) = \mathcal{F}_{em}(t) \tag{1.353}$$

where $\mathcal{G}_{mech}(\mathbf{r}, t)$ denotes the total *mechanical momentum* of the particles. Combination of (1.352) and (1.353) yields

$$\frac{\mathrm{d}}{\mathrm{d}t}\left[\mathcal{G}_{em}(t) + \mathcal{G}_{mech}(t)\right] = 0, \qquad t \in \mathbb{R}^+ \tag{1.354}$$

which means that for a *closed* system of charged particles the total momentum is conserved only if the electromagnetic field is endowed with its own momentum [20, 43, 44]. By contrast, from (1.347) and (1.353) we see that, in general, the total momentum is not conserved in a finite region of space $V \subset \mathbb{R}^3$, viz.,

$$\frac{\mathrm{d}}{\mathrm{d}t}\left[\mathcal{G}_{em}(t) + \mathcal{G}_{mech}(t)\right] = -\int_{\partial V} \mathrm{d}S\, \hat{\mathbf{n}}(\mathbf{r}) \cdot \overline{\mathcal{T}}_{em}(\mathbf{r}, t), \qquad t \in \mathbb{R}^+ \tag{1.355}$$

for electromagnetic momentum may flow in and out of V through the boundary ∂V, unless the flux of $\overline{\mathcal{T}}_{em}$ is null.

We continue by examining the properties of the stress tensor (1.343). The trace of $\overline{\mathcal{T}}_{em}(\mathbf{r}, t)$ reads [see Appendix E and (E.78)]

$$\begin{aligned}
\mathrm{Tr}\left\{\overline{\mathcal{T}}_{em}(\mathbf{r}, t)\right\} &= \frac{1}{2}(\mathcal{D}\cdot\mathcal{E} + \mathcal{B}\cdot\mathcal{H})\overline{\mathcal{I}} : \overline{\mathcal{I}} - \mathcal{D}\mathcal{E} : \overline{\mathcal{I}} - \mathcal{B}\mathcal{H} : \overline{\mathcal{I}} \\
&= \frac{3}{2}(\mathcal{D}\cdot\mathcal{E} + \mathcal{B}\cdot\mathcal{H}) - (\mathcal{D}\cdot\mathcal{E} + \mathcal{B}\cdot\mathcal{H}) \\
&= \frac{1}{2}(\mathcal{D}\cdot\mathcal{E} + \mathcal{B}\cdot\mathcal{H}) = w_e + w_h
\end{aligned} \tag{1.356}$$

that is, the volume density of electromagnetic energy in accordance with (1.263) and (1.266). Since in an isotropic medium the stress dyadic is symmetric, its eigenvalues are *real* and its eigenvectors form an *orthogonal* triplet. In this regard, it is immediately proved that the density of linear momentum (1.332) is an eigenvector of $\overline{\mathcal{T}}_{em}$ with eigenvalue the energy density $w := w_e + w_h$, namely,

$$\overline{\mathcal{T}}_{em} \cdot \mathbf{g}_{em} = w\, \mathbf{g}_{em} - \varepsilon\mathcal{E}\mathcal{E} \cdot \mathbf{g}_{em} - \mu\mathcal{H}\mathcal{H} \cdot \mathbf{g}_{em} = w\, \mathbf{g}_{em} \tag{1.357}$$

because the vector $\mathbf{g}_{em}$ is everywhere perpendicular to $\mathcal{E}$ and $\mathcal{H}$ by construction. For this reason, we may also expect the remaining eigenvectors of $\overline{\mathcal{T}}_{em}$ to be, in general, a linear combination of electric and magnetic field, say, $a\mathcal{E}/\sqrt{Z} + b\mathcal{H}\sqrt{Z}$, where a and b are real constants yet to be found and the scalar

$$Z := \sqrt{\frac{\mu}{\varepsilon}} = c\mu = \frac{1}{c\varepsilon} \tag{1.358}$$

which carries physical dimensions of ohms (Ω), is called the *intrinsic impedance* of the background medium. For instance, substituting (1.114) and (1.116) into (1.358) provides $Z = Z_0 \approx 376\ \Omega$ for free space. To proceed in our search for the remaining eigenvalues of $\overline{\mathcal{T}}_{em}$ we require

$$\overline{\mathcal{T}}_{em} \cdot \left(a\mathcal{E}/\sqrt{Z} + b\mathcal{H}\sqrt{Z}\right) = \nu\left(a\mathcal{E}/\sqrt{Z} + b\mathcal{H}\sqrt{Z}\right), \qquad \nu \in \mathbb{R} \tag{1.359}$$

which, after a little algebra, yields

$$\left[(w_h - w_e)a/\sqrt{Z} - \varepsilon\mathcal{E}\cdot\mathcal{H}b\sqrt{Z}\right]\mathcal{E} + \left[(w_e - w_h)b\sqrt{Z} - \mu\mathcal{E}\cdot\mathcal{H}a/\sqrt{Z}\right]\mathcal{H} = \nu(a\mathcal{E}/\sqrt{Z} + b\mathcal{H}\sqrt{Z}) \quad (1.360)$$

and this homogeneous vector equation may be separated into two scalar ones, since $\mathcal{E}$ and $\mathcal{H}$ are linearly independent, though not necessarily orthogonal in regions containing sources. Formally, we may dot-multiply both sides with the vectors $\mathcal{H}\times(\mathcal{E}\times\mathcal{H})$ and $\mathcal{E}\times(\mathcal{H}\times\mathcal{E})$ — which are perpendicular to $\mathcal{H}$ and $\mathcal{E}$, respectively — to obtain the homogeneous algebraic system

$$\begin{pmatrix} w_h - w_e - \nu & -Z\varepsilon\mathcal{E}\cdot\mathcal{H} \\ -\mu\mathcal{E}\cdot\mathcal{H}/Z & w_e - w_h - \nu \end{pmatrix}\begin{pmatrix} a \\ b \end{pmatrix} = \begin{pmatrix} 0 \\ 0 \end{pmatrix} \quad (1.361)$$

which admits two non-trivial solutions provided the determinant of the matrix vanishes [82–84]. In symbols, this happens if

$$\nu^2 = (w_e - w_h)^2 + \varepsilon\mu(\mathcal{E}\cdot\mathcal{H})^2 = w^2 - c^2|\mathbf{g}_{em}|^2 \quad (1.362)$$

and, more precisely, if ν takes on the values

$$\nu_+ = \left(w^2 - c^2|\mathbf{g}_{em}|^2\right)^{1/2} \qquad \nu_- = -\left(w^2 - c^2|\mathbf{g}_{em}|^2\right)^{1/2} \quad (1.363)$$

on account of (1.209) and (1.332). Since the trace of a dyadic coincides with the sum of its eigenvalues (Appendix E.2) we could have predicted that two out of three eigenvalues have opposite signs, because we know from (1.356) that the trace of $\overline{\mathcal{T}}_{em}$ equals the energy density w and from (1.357) that w is also an eigenvalue. Inserting ν_+ and ν_- back into (1.361) and solving for a and b allows us to write the remaining eigenvectors of $\overline{\mathcal{T}}_{em}$ as

$$\mathbf{u}_+ = \varepsilon\sqrt{Z}\mathcal{E}\cdot\mathcal{H}\mathcal{E} - \sqrt{Z}\left[w_e - w_h + \left(w^2 - c^2|\mathbf{g}_{em}|^2\right)^{1/2}\right]\mathcal{H} \quad (1.364)$$

$$\mathbf{u}_- = \left[w_e - w_h + \left(w^2 - c^2|\mathbf{g}_{em}|^2\right)^{1/2}\right]\mathcal{E}/\sqrt{Z} + \mu\mathcal{E}\cdot\mathcal{H}\mathcal{H}/\sqrt{Z} \quad (1.365)$$

which can be further normalized so as to carry the physical dimension of a momentum density for consistency with (1.357). On a related score, we notice that (1.362) implies

$$c|\mathbf{g}_{em}(\mathbf{r},t)| \leqslant w(\mathbf{r},t), \qquad (\mathbf{r},t) \in V\times\mathbb{R}^+ \quad (1.366)$$

because the eigenvalues of a real symmetric dyadic are real.

In Appendix E.2 it is shown that the unit dyadic $\overline{\mathcal{I}}$ may be expanded in terms of dyads formed with suitable combinations of three linearly independent vectors. In particular, specializing the second part of (E.37) with $\mathbf{a} = \mathcal{E}\times\mathcal{H}$, $\mathbf{b} = \mathcal{E}$, $\mathbf{c} = \mathcal{H}$ and $J = |\mathcal{E}\times\mathcal{H}|^2$ yields the representation

$$\overline{\mathcal{I}} = \frac{\mathbf{g}_{em}\mathbf{g}_{em}}{|\mathbf{g}_{em}|^2} + \frac{|\mathcal{H}|^2\mathcal{E} - \mathcal{H}\cdot\mathcal{E}\mathcal{H}}{c^4|\mathbf{g}_{em}|^2}\mathcal{E} + \frac{|\mathcal{E}|^2\mathcal{H} - \mathcal{E}\cdot\mathcal{H}\mathcal{E}}{c^4|\mathbf{g}_{em}|^2}\mathcal{H} \quad (1.367)$$

on account of (1.332). With the aid of (1.367) and the definition of stress tensor (1.343) we can cast $\overline{\mathcal{T}}_{em}$ into the insightful form

$$\overline{\mathcal{T}}_{em} = w\frac{\mathbf{g}_{em}\mathbf{g}_{em}}{|\mathbf{g}_{em}|^2} + \frac{(w_h - w_e)|\mathcal{H}|^2 + \varepsilon(\mathcal{E}\cdot\mathcal{H})^2}{c^4|\mathbf{g}_{em}|^2}\mathcal{E}\mathcal{E}$$

$$- \frac{w\mathcal{E} \cdot \mathcal{H}}{c^4 |\mathbf{g}_{em}|^2} (\mathcal{E}\mathcal{H} + \mathcal{H}\mathcal{E}) + \frac{(w_e - w_h)|\mathcal{E}|^2 + \mu(\mathcal{E} \cdot \mathcal{H})^2}{c^4 |\mathbf{g}_{em}|^2} \mathcal{H}\mathcal{H} \quad (1.368)$$

which indicates that $\overline{\mathfrak{T}}_{em}$, in general, is not diagonal when expressed in the vector basis formed with $\mathcal{E}$, $\mathcal{H}$ and $\mathcal{E} \times \mathcal{H}$, unless the dot product $\mathcal{E} \cdot \mathcal{H}$ vanishes.

In spatial regions devoid of charges and currents — where $\mathcal{E}$ and $\mathcal{H}$ are orthogonal and $w_e = w_h$ — (1.368) reduces to

$$\overline{\mathfrak{T}}_{em}(\mathbf{r}, t) = w \frac{\mathbf{g}_{em}\mathbf{g}_{em}}{|\mathbf{g}_{em}|^2} = w \frac{\mathcal{E} \times \mathcal{H}\mathcal{E} \times \mathcal{H}}{|\mathcal{E}|^2 |\mathcal{H}|^2}, \qquad \mathbf{r} \notin V_J \quad (1.369)$$

whereas the special procedure devised above to determine the eigenvectors of $\overline{\mathfrak{T}}_{em}$ breaks down and (1.364) and (1.365) are meaningless. But then, under the stated hypothesis, starting with (1.343) or (1.368) it is trivially shown that

$$\overline{\mathfrak{T}}_{em} \cdot \mathcal{E} = 0 = \overline{\mathfrak{T}}_{em} \cdot \mathcal{H}, \qquad \mathbf{r} \notin V_J \quad (1.370)$$

whereby we conclude that the remaining eigenvectors of $\overline{\mathfrak{T}}_{em}$ are $\mathcal{E}$ and $\mathcal{H}$, and the latter are associated with two degenerate null eigenvalues.

To determine the matching conditions for the stress tensor (1.343) at the interface of two spatial regions endowed with different constitutive parameters we refer once again to the geometrical setup of Figure 1.13 under the condition, though, that the media in V_1 and $V_2 \subset V_1$ are lossless, i.e., we let $\sigma_1 = \sigma_2 = 0$. Sources may exist in both V_1 and V_2, and a surface density of force $\mathbf{f}_{S\,em}$ may even be localized right at the material boundary $\partial V := \partial V_2$. Proceeding as in Section 1.7 we apply (1.347) separately to the domains W_1, W_2 and $W := W_1 \cup W_2$. In symbols, we have

$$\int_{\partial W_1 \cup S_0} \mathrm{d}S\ \hat{\boldsymbol{\nu}}_1(\mathbf{r}) \cdot \overline{\mathfrak{T}}_{1em}(\mathbf{r}, t) + \frac{\mathrm{d}}{\mathrm{d}t} \int_{W_1} \mathrm{d}V\ \mathbf{g}_{1em}(\mathbf{r}, t) = -\mathfrak{F}_{1em}(t) \quad (1.371)$$

$$\int_{\partial W_2 \cup S_0} \mathrm{d}S\ \hat{\boldsymbol{\nu}}_2(\mathbf{r}) \cdot \overline{\mathfrak{T}}_{2em}(\mathbf{r}, t) + \frac{\mathrm{d}}{\mathrm{d}t} \int_{W_2} \mathrm{d}V\ \mathbf{g}_{2em}(\mathbf{r}, t) = -\mathfrak{F}_{2em}(t) \quad (1.372)$$

$$\int_{\partial W} \mathrm{d}S\ \hat{\boldsymbol{\nu}}(\mathbf{r}) \cdot \overline{\mathfrak{T}}_{em}(\mathbf{r}, t) + \frac{\mathrm{d}}{\mathrm{d}t} \int_{W} \mathrm{d}V\ \mathbf{g}_{em}(\mathbf{r}, t) = -\mathfrak{F}_{em}(t) - \int_{S_0} \mathrm{d}S\ \mathbf{f}_{S\,em}(\mathbf{r}, t) \quad (1.373)$$

where $\mathfrak{F}_{1em}$ and $\mathfrak{F}_{2em}$ indicate the part of the Lorentz force associated with fields and sources in region W_1 and W_2, respectively. By adding (1.371) and (1.372) side by side and recalling relation (1.146) between $\hat{\boldsymbol{\nu}}_1$, $\hat{\boldsymbol{\nu}}_2$ and $\hat{\mathbf{n}}$ on S_0, after few manipulations we find

$$\int_{\partial W} \mathrm{d}S\ \hat{\boldsymbol{\nu}}(\mathbf{r}) \cdot \overline{\mathfrak{T}}_{em}(\mathbf{r}, t) + \frac{\mathrm{d}}{\mathrm{d}t} \int_{W} \mathrm{d}V\ \mathbf{g}_{em}(\mathbf{r}, t)$$

$$= -\mathfrak{F}_{em}(t) - \int_{S_0} \mathrm{d}S\ \hat{\mathbf{n}}(\mathbf{r}) \cdot \left[\overline{\mathfrak{T}}_{2em}(\mathbf{r}, t) - \overline{\mathfrak{T}}_{1em}(\mathbf{r}, t) \right] \quad (1.374)$$

which should coincide with (1.373) for consistency. This occurs if we *postulate* the condition

$$\int_{S_0} \mathrm{d}S\ \left[\hat{\mathbf{n}}(\mathbf{r}) \cdot \overline{\mathfrak{T}}_{2em}(\mathbf{r}, t) - \hat{\mathbf{n}}(\mathbf{r}) \cdot \overline{\mathfrak{T}}_{1em}(\mathbf{r}, t) - \mathbf{f}_{S\,em}(\mathbf{r}, t) \right] = 0 \quad (1.375)$$

whence we get

$$\hat{\mathbf{n}}(\mathbf{r}_0)\cdot\overline{\mathcal{T}}_{2em}(\mathbf{r}_0,t)-\hat{\mathbf{n}}(\mathbf{r}_0)\cdot\overline{\mathcal{T}}_{1em}(\mathbf{r}_0,t)-\mathbf{f}_{S\,em}(\mathbf{r}_0,t)=0, \qquad \mathbf{r}_0\in S_0 \tag{1.376}$$

by virtue of the mean value theorem [34] as long as $\hat{\mathbf{n}}(\mathbf{r})\cdot\overline{\mathcal{T}}_{1em}(\mathbf{r},t)$ and $\hat{\mathbf{n}}(\mathbf{r})\cdot\overline{\mathcal{T}}_{2em}(\mathbf{r},t)$ are continuous vector functions over $S_0\subset\partial V$. Since $S_0\subset\partial V$ is arbitrary, we obtain the desired jump condition

$$\hat{\mathbf{n}}(\mathbf{r})\cdot\overline{\mathcal{T}}_{1em}(\mathbf{r},t)-\hat{\mathbf{n}}(\mathbf{r})\cdot\overline{\mathcal{T}}_{2em}(\mathbf{r},t)=-\mathbf{f}_{S\,em}(\mathbf{r},t), \qquad \mathbf{r}\in\partial V \tag{1.377}$$

i.e., the normal component of the stress dyadic remains continuous across the interface unless a surface density of force is present. In the special case where V_2 is filled with a PEC, (1.377) simplifies as

$$\hat{\mathbf{n}}(\mathbf{r})\cdot\overline{\mathcal{T}}_{1em}(\mathbf{r},t)=-\mathbf{f}_{S\,em}(\mathbf{r},t), \qquad \mathbf{r}\in\partial V \tag{1.378}$$

because the electromagnetic field is null in V_2 (see discussion on page 29). Moreover, if the medium in V_1 is isotropic, in view of (1.169) and (1.171) $\boldsymbol{\mathcal{E}}_1$ is perpendicular to ∂V whereas $\boldsymbol{\mathcal{H}}_1$ is tangential to the interface. Then, with reference to (1.343) we have

$$\begin{aligned}\mathbf{f}_{S\,em}(\mathbf{r},t)&=-\frac{\varepsilon}{2}|\boldsymbol{\mathcal{E}}_1|^2\hat{\mathbf{n}}-\frac{\mu}{2}|\boldsymbol{\mathcal{H}}_1|^2\hat{\mathbf{n}}+\hat{\mathbf{n}}\cdot\boldsymbol{\mathcal{E}}_1\boldsymbol{\mathcal{E}}_1\varepsilon+\hat{\mathbf{n}}\cdot\boldsymbol{\mathcal{H}}_1\boldsymbol{\mathcal{H}}_1\mu\\&=-\frac{\varepsilon}{2}\mathcal{E}_{1n}^2\hat{\mathbf{n}}-\frac{\mu}{2}\mathcal{H}_{1t}^2\hat{\mathbf{n}}+\varepsilon\mathcal{E}_{1n}^2\hat{\mathbf{n}}\\&=\frac{1}{2}\left(\varepsilon\mathcal{E}_{1n}^2-\mu\mathcal{H}_{1t}^2\right)\hat{\mathbf{n}}, \qquad \mathbf{r}\in\partial V\end{aligned} \tag{1.379}$$

where $\mathcal{E}_{1n}$ denotes the normal component of $\boldsymbol{\mathcal{E}}_1$ and $\mathcal{H}_{1t}$ the tangential part of $\boldsymbol{\mathcal{H}}_1$. The total force exerted by the field on a PEC body follows by integrating (1.379) over ∂V.

1.12 Conservation of electromagnetic angular momentum

Since we introduced the linear momentum $\mathbf{g}_{em}$ of an electromagnetic wave in (1.331), on the grounds of the analogous definition of mechanical angular momentum with respect to the origin [21, Chapter 18] it is natural to regard the vector

$$\mathbf{l}_{em}(\mathbf{r},t):=\mathbf{r}\times\mathbf{g}_{em}(\mathbf{r},t), \qquad (\mathbf{r},t)\in V\times\mathbb{R}^+ \tag{1.380}$$

as the volume density of *electromagnetic angular momentum* (physical dimension: C Wb/m^3 = J s/m^3 = kg/(s m)) carried by a wave [16,20,74,79], [55, Section 15.6], [73, Section 3.3.4]. In order to derive the pertinent conservation law, if one can be found at all, we start by cross-multiplying (1.344) with $\mathbf{r}$, namely,

$$\mathbf{r}\times\left(\nabla\cdot\overline{\mathcal{T}}_{em}\right)+\frac{\partial}{\partial t}(\mathbf{r}\times\mathbf{g}_{em})=-\mathbf{r}\times\mathbf{f}_{em}(\mathbf{r},t), \qquad (\mathbf{r},t)\in V\times\mathbb{R}^+ \tag{1.381}$$

with the understanding that the density of Lorentz force $\mathbf{f}_{em}(\mathbf{r},t)$ is non-zero only for $\mathbf{r}\in V_J\subset V\subseteq\mathbb{R}^3$. 'Moving' $\mathbf{r}$ past the time derivative is permitted because $\mathbf{r}$ denotes a fixed observation point which does not depend on time. The desired balance equation follows from (1.381) if somehow we manage to turn the first term in the left-hand side into the divergence of a suitable dyadic field.

To this purpose, we employ Cartesian coordinates to write the Maxwell stress dyadic as the sum of three dyads, as is done in the first part of (E.11). Then, based on (E.90) we observe

$$\mathbf{r}\times\left(\nabla\cdot\overline{\mathcal{T}}_{em}\right)\cdot\hat{\mathbf{x}}=\mathbf{r}\times\left[(\nabla\cdot\mathcal{T}_x')\hat{\mathbf{x}}+\left(\nabla\cdot\mathcal{T}_y'\right)\hat{\mathbf{y}}+\left(\nabla\cdot\mathcal{T}_z'\right)\hat{\mathbf{z}}\right]\cdot\hat{\mathbf{x}}$$

$$
\begin{aligned}
&= \mathbf{r}\cdot\hat{\mathbf{y}}\left(\nabla\cdot\mathcal{T}_z'\right) - \mathbf{r}\cdot\hat{\mathbf{z}}\left(\nabla\cdot\mathcal{T}_y'\right) = y\left(\nabla\cdot\mathcal{T}_z'\right) - z\left(\nabla\cdot\mathcal{T}_y'\right)\\
&= y\left(\frac{\partial\mathcal{T}_{xz}}{\partial x} + \frac{\partial\mathcal{T}_{yz}}{\partial y} + \frac{\partial\mathcal{T}_{zz}}{\partial z}\right) - z\left(\frac{\partial\mathcal{T}_{xy}}{\partial x} + \frac{\partial\mathcal{T}_{yy}}{\partial y} + \frac{\partial\mathcal{T}_{zy}}{\partial z}\right)\\
&= \frac{\partial}{\partial x}\left(y\mathcal{T}_{xz} - z\mathcal{T}_{xy}\right) + \frac{\partial}{\partial y}\left(y\mathcal{T}_{yz} - z\mathcal{T}_{yy}\right) + \frac{\partial}{\partial z}\left(y\mathcal{T}_{zz} - z\mathcal{T}_{zy}\right) - \mathcal{T}_{yz} + \mathcal{T}_{zy}\\
&= \nabla\cdot\mathcal{M}_x'
\end{aligned}
\tag{1.382}
$$

where we have used (H.13) and (A.29), and also exploited the symmetry of $\overline{\mathcal{T}}_{em}$. The remaining combination of partial derivatives constitutes the divergence of a vector $\mathcal{M}_x'$ which in turn may be part of a dyadic $\overline{\mathcal{M}}_{em}$ according to (E.11). Indeed, two more equations are obtained when we examine the y- and the z-component, namely,

$$
\begin{aligned}
\mathbf{r}\times\left(\nabla\cdot\overline{\mathcal{T}}_{em}\right)\cdot\hat{\mathbf{y}} &= \frac{\partial}{\partial x}(z\mathcal{T}_{xx} - x\mathcal{T}_{xz}) + \frac{\partial}{\partial y}\left(z\mathcal{T}_{yx} - x\mathcal{T}_{yz}\right) + \frac{\partial}{\partial z}(z\mathcal{T}_{zx} - x\mathcal{T}_{zz}) - \mathcal{T}_{zx} + \mathcal{T}_{xz}\\
&= \nabla\cdot\mathcal{M}_y'
\end{aligned}
\tag{1.383}
$$

$$
\begin{aligned}
\mathbf{r}\times\left(\nabla\cdot\overline{\mathcal{T}}_{em}\right)\cdot\hat{\mathbf{z}} &= \frac{\partial}{\partial x}\left(x\mathcal{T}_{xy} - y\mathcal{T}_{xx}\right) + \frac{\partial}{\partial y}\left(x\mathcal{T}_{yy} - y\mathcal{T}_{yx}\right) + \frac{\partial}{\partial z}\left(x\mathcal{T}_{zy} - y\mathcal{T}_{zx}\right) - \mathcal{T}_{xy} + \mathcal{T}_{yx}\\
&= \nabla\cdot\mathcal{M}_z'
\end{aligned}
\tag{1.384}
$$

whence we conclude that

$$
\mathbf{r}\times\left(\nabla\cdot\overline{\mathcal{T}}_{em}\right) = \nabla\cdot\mathcal{M}_x'\hat{\mathbf{x}} + \nabla\cdot\mathcal{M}_y'\hat{\mathbf{y}} + \nabla\cdot\mathcal{M}_z'\hat{\mathbf{z}} = \nabla\cdot\overline{\mathcal{M}}_{em}
\tag{1.385}
$$

on account of (E.90) applied backwards.

Further, from (1.382)-(1.384) we gather that $\overline{\mathcal{M}}_{em}$ (physical dimension: N/m or J/m^2) explicitly reads

$$
\begin{aligned}
\overline{\mathcal{M}}_{em} := {}& (y\mathcal{T}_{xz} - z\mathcal{T}_{xy})\hat{\mathbf{x}}\hat{\mathbf{x}} + (z\mathcal{T}_{xx} - x\mathcal{T}_{xz})\hat{\mathbf{x}}\hat{\mathbf{y}} + (x\mathcal{T}_{xy} - y\mathcal{T}_{xx})\hat{\mathbf{x}}\hat{\mathbf{z}}\\
&+ (y\mathcal{T}_{yz} - z\mathcal{T}_{yy})\hat{\mathbf{y}}\hat{\mathbf{x}} + (z\mathcal{T}_{yx} - x\mathcal{T}_{yz})\hat{\mathbf{y}}\hat{\mathbf{y}} + (x\mathcal{T}_{yy} - y\mathcal{T}_{yx})\hat{\mathbf{y}}\hat{\mathbf{z}}\\
&+ (y\mathcal{T}_{zz} - z\mathcal{T}_{zy})\hat{\mathbf{z}}\hat{\mathbf{x}} + (z\mathcal{T}_{zx} - x\mathcal{T}_{zz})\hat{\mathbf{z}}\hat{\mathbf{y}} + (x\mathcal{T}_{zy} - y\mathcal{T}_{zx})\hat{\mathbf{z}}\hat{\mathbf{z}}\\
= {}& -\hat{\mathbf{x}}\mathcal{T}_x\times\mathbf{r} - \hat{\mathbf{y}}\mathcal{T}_y\times\mathbf{r} - \hat{\mathbf{z}}\mathcal{T}_z\times\mathbf{r} = -\overline{\mathcal{T}}_{em}\times\mathbf{r}
\end{aligned}
\tag{1.386}
$$

whereby (1.381) becomes

$$
\nabla\cdot\overline{\mathcal{M}}_{em}(\mathbf{r},t) + \frac{\partial}{\partial t}\left[\mathbf{r}\times\mathbf{g}_{em}(\mathbf{r},t)\right] = -\mathbf{r}\times\mathbf{f}_{em}(\mathbf{r},t), \qquad (\mathbf{r},t)\in V\times\mathbb{R}^+
\tag{1.387}
$$

which is has the structure of a continuity equation plus a source term. Some Authors [20, Section 6.3], [74, Section 2.3.3] define the dyadic $\overline{\mathcal{M}}_{em}$ as the negative of the one in (1.386) to be consistent with a different definition of $\overline{\mathcal{T}}_{em}$. Since we had to rely on the symmetry of $\overline{\mathcal{T}}_{em}$ to arrive at (1.387), in this form such conservation law applies only to homogeneous isotropic media.

Integrating both sides of (1.387) over V provides the global form of the balance equation, viz.,

$$
\int_{\partial V} \mathrm{d}V\,\nabla\cdot\overline{\mathcal{M}}_{em}(\mathbf{r},t) + \int_V \mathrm{d}V\,\frac{\partial}{\partial t}\mathbf{r}\times\mathbf{g}_{em}(\mathbf{r},t) = -\mathcal{N}_{em}(t), \qquad t\in\mathbb{R}^+
\tag{1.388}
$$

where

$$\mathcal{N}_{em}(t) := \int_{V_J} \mathrm{d}V\, \mathbf{r} \times \mathbf{f}_{em}(\mathbf{r}, t), \qquad t \in \mathbb{R}^+ \tag{1.389}$$

represents the *total torque* (physical dimension: Nm) exerted by the electromagnetic field on the charges and the currents existing in the region $V_J \subset V$. The nine components of $\overline{\mathcal{M}}_{em}(\mathbf{r}, t)$ are of class $\mathcal{C}^1(V) \cap \mathcal{C}(\overline{V})$ if so are the components of the four electromagnetic entities in (1.343). Thus, we may invoke the Gauss theorem for dyadic fields (H.105) and swap integration and differentiation with respect to time to arrive at

$$\int_{\partial V} \mathrm{d}S\, \hat{\mathbf{n}}(\mathbf{r}) \cdot \overline{\mathcal{M}}_{em}(\mathbf{r}, t) + \frac{\mathrm{d}}{\mathrm{d}t} \int_{V} \mathrm{d}V\, \mathbf{r} \times \mathbf{g}_{em}(\mathbf{r}, t) = -\mathcal{N}_{em}(t), \qquad t \in \mathbb{R}^+ \tag{1.390}$$

under the hypotheses that the domain V and its boundary neither move nor change shape with time and the components of $\mathbf{g}_{em}(\mathbf{r}, t)$ and $\partial \mathbf{g}_{em}/\partial t$ are continuous for $(\mathbf{r}, t) \in V \times \mathbb{R}^+$ (cf. [32, Theorem 9.42]). The terms in (1.390) are interpreted as follows.

- The quantity

$$\mathcal{L}_{em}(t) := \int_{V} \mathrm{d}V\, \mathbf{r} \times \mathbf{g}_{em}(\mathbf{r}, t) \tag{1.391}$$

  is regarded as the instantaneous total *angular electromagnetic momentum* associated with the field and stored in the domain V.
- The flux of $\overline{\mathcal{M}}_{em}(\mathbf{r}, t)$ given by (1.386) may be interpreted as the *rate of angular momentum flow* through the boundary ∂V [16], [20, Section 6.3], though not all of this efflux represents radiation. Alternatively, $\hat{\mathbf{n}}(\mathbf{r}) \cdot \overline{\mathcal{M}}_{em}(\mathbf{r}, t)$ for $\mathbf{r} \in \partial V$ may be understood as a *torque per unit area*. These two viewpoints — rate of angular momentum flow or torque density — are consistent because we may, in fact, regard any torque as the time rate of variation of angular momentum according to the Newton law for rotating bodies [21, Chapter 18].

Extending (1.390) to the whole space requires we check the convergence of the integrals when the domain of interest becomes $\mathbb{R}^3$. The vanishing of the flux of $\overline{\mathcal{M}}_{em}$ can be proved as in (1.350) for the flux of the stress tensor by invoking the asymptotic boundary conditions (1.260) and (1.261). By contrast, the latter are not enough for the finiteness of the total angular momentum. But then, we just have to realize that the charges which produce the field have been set in motion at a given moment in time, whereby the field is still static in character outside a ball $B(0, b)$ with sufficiently large radius b (cf. Figure 1.15b). Then, since $\mathcal{H}(\mathbf{r}, t)$ vanishes identically outside $B(0, b)$ so does the angular momentum density (1.380), and the integral in (1.391) remains finite inasmuch as it must be extended at most to $B(0, b)$.

Further, we may argue that after an initial transient has finished and a steady state has been attained, $\mathcal{E}(\mathbf{r}, t)$ and $\mathcal{H}(\mathbf{r}, t)$ do not depend on time any more. In which case we can derive from (4.29) in Chapter 4 and (5.155) in Chapter 5 that the stationary magnetic field $\mathcal{H}(\mathbf{r})$ produced by a localized steady current $\mathcal{J}(\mathbf{r})$ actually falls off as the *cube* of the distance from the source. Clearly, this is a stronger asymptotic condition than (1.261) and makes the angular momentum density (1.380) decay as the *fourth power* of the distance from the source. As a consequence, if we choose two concentric balls $B(0, b_1)$ and $B(0, b)$ with $b_1 < b$, we can estimate

$$\left| \int_{B(0,b)} \mathrm{d}V\, \mathbf{r} \times \mathbf{g}_{em}(\mathbf{r}, t) \right| = \left| \int_{B(0,b_1)} \mathrm{d}V\, \mathbf{r} \times \mathbf{g}_{em}(\mathbf{r}, t) + \int_{B(0,b)\setminus B(0,b_1)} \mathrm{d}V\, \mathbf{r} \times \mathbf{g}_{em}(\mathbf{r}, t) \right|$$

$$
\begin{aligned}
&\leqslant \int\limits_{B(0,b_1)} \mathrm{d}V\, r|\mathbf{g}_{em}(\mathbf{r},t)| + \int\limits_{B(0,b)\setminus B(0,b_1)} \mathrm{d}V\, r|\mathcal{D}(\mathbf{r},t)||\mathcal{B}(\mathbf{r},t)| \\
&\leqslant \int\limits_{B(0,b_1)} \mathrm{d}V\, r|\mathbf{g}_{em}(\mathbf{r},t)| + \int\limits_{4\pi} \mathrm{d}\Omega \int\limits_{b_1}^{b} \mathrm{d}r\, r^2 \frac{C_D C_B}{r^4} \\
&= \int\limits_{B(0,b_1)} \mathrm{d}V\, r|\mathbf{g}_{em}(\mathbf{r},t)| + 4\pi C_D C_B \left(\frac{1}{b_1} - \frac{1}{b}\right) \\
&\leqslant \int\limits_{B(0,b_1)} \mathrm{d}V\, r|\mathbf{g}_{em}(\mathbf{r},t)| + 4\pi \frac{C_D C_B}{b_1}
\end{aligned}
\tag{1.392}
$$

by virtue of (1.260) and the asymptotic behavior of stationary magnetic fields. Since the last contribution in the rightmost-hand side is independent of b, it remains constant and finite as $b \to +\infty$. Thus, the total angular momentum remains finite in $\mathbb{R}^3$ even in the steady-state regime.

In particular, (1.390) reads

$$
\frac{\mathrm{d}}{\mathrm{d}t}\mathcal{L}_{em}(t) = -\mathcal{N}_{em}(t) \tag{1.393}
$$

when it is applied to the whole space. Since the motion of rotating bodies obeys the equation [21, Chapter 18]

$$
\frac{\mathrm{d}}{\mathrm{d}t}\mathcal{L}_{mech}(t) = \mathcal{N}_{em}(t) \tag{1.394}
$$

where $\mathcal{L}_{mech}(t)$ indicates the total angular momentum of the particles, combination with (1.393) yields

$$
\frac{\mathrm{d}}{\mathrm{d}t}\left[\mathcal{L}_{em}(t) + \mathcal{L}_{mech}(t)\right] = 0, \qquad t \in \mathbb{R}^+ \tag{1.395}
$$

which is analogous to (1.354). In words, for a closed system of charged particles the total angular momentum is conserved only if the electromagnetic field carries its own angular momentum. On the contrary, from (1.390) and (1.394) we see that, in general, the total momentum is not conserved in a finite region of space $V \subset \mathbb{R}^3$

$$
\frac{\mathrm{d}}{\mathrm{d}t}\left[\mathcal{L}_{em}(t) + \mathcal{L}_{mech}(t)\right] = -\int\limits_{\partial V} \mathrm{d}S\, \hat{\mathbf{n}}(\mathbf{r}) \cdot \overline{\mathcal{M}}_{em}(\mathbf{r},t), \qquad t \in \mathbb{R}^+ \tag{1.396}
$$

since electromagnetic angular momentum flows in and out of V through the boundary ∂V, unless the flux of $\overline{\mathcal{M}}_{em}$ is null.

References

[1] Feynmann RP, Leighton R, Sands M. The Feynmann lectures on physics. vol. 2. Reading, MA: Addison-Wesley; 1964.

[2] Anderson N. The Electromagnetic Field. New York, NY: Springer Science+Business Media; 1968.

[3] Maxwell JC. A Treatise on Electricity and Magnetism. vol. 1. Oxford, UK: Clarendon Press; 1873.

[4] Maxwell JC. A Treatise on Electricity and Magnetism. vol. 2. Oxford, UK: Clarendon Press; 1873.
[5] Maxwell JC. A dynamical theory of the electromagnetic field. Phil Trans Royal Society London. 1865 January;155:459–512.
[6] Kong JA. Maxwell Equations. Cambridge, MA: EMW Publishing; 2002.
[7] Jackson JD. Classical Electrodynamics. 3rd ed. Chichester, UK: Wiley; 1999.
[8] Stratton JA. Electromagnetic theory. London, UK: McGraw-Hill; 1941.
[9] Schelkunoff SA. Electromagnetic Fields. New York, NY: Blaisdell Pub. Co.; 1963.
[10] Van Bladel JG. Electromagnetic Fields. Piscataway, NJ: IEEE Press; 2007.
[11] Helrich CS. The classical theory of fields - Electromagnetism. Berlin Heidelberg: Springer-Verlag; 2012.
[12] Sommerfeld A. Electrodynamics. vol. 3 of Lectures on theoretical physics. New York, NY: Academic Press; 1952.
[13] Cheng DK. Field and Wave Electromagnetics. New York, NY: Addison-Wesley Pub. Co.; 1989.
[14] Kong JA. Electromagnetic Wave Theory. 2nd ed. New York, NY: Wiley; 1990.
[15] Rothwell EJ, Cloud MJ. Electromagnetics. London, UK: CRC Press; 2001.
[16] Schwinger J, De Raad LL, Milton KA, et al. Classical electrodynamics. Perseus Books; 1998.
[17] Griffith DJ. Introduction to electrodynamics. 3rd ed. Upper Saddle River, NJ: Prentice Hall; 1999.
[18] Fano RM, Chu LJ, Adler RB. Electromagnetic Fields, Energy, and Forces. New York, NY: Wiley; 1960.
[19] Milton KA, Schwinger J. Electromagnetic Radiation: Variational Methods, Waveguides and Accelerators. Berlin Heidelberg: Springer-Verlag; 2006.
[20] Konopinski EJ. Electromagnetic Fields and Relativistic Particles. New York, NY: McGraw Hill; 1981.
[21] Feynmann RP, Leighton R, Sands M. The Feynmann lectures on physics. vol. 1. Reading, MA: Addison-Wesley; 1964.
[22] Frisch M. Inconsistency in classical electrodynamics. Philosophy of Science. 2004 October;71:525–549.
[23] Vickers P. Frisch, Muller, and Belot on an Inconsistency in Classical Electrodynamics. British Journal for the Philosophy of Science. 2008;59:767–792.
[24] Biggs HF. The Electromagnetic Field. London, UK: Oxford University Press; 1934.
[25] Popović Z, Popović BD. Introductory Electromagnetics. Upper Saddle River, NJ: Prentice Hall; 2000.
[26] Schwab AJ. Field Theory Concepts. Berlin Heidelberg: Springer-Verlag; 1988.
[27] Tonnelat MA. The principles of electromagnetic theory and of relativity. Dordrecht-Holland, NL: D. Reidel Publishing Company; 1966.
[28] Post EJ. Formal structure of electromagnetics. Amsterdam, NL: North-Holland Publishing Company; 1962.
[29] Bucci OM. From Electromagnetism to the Electromagnetic Field: The Genesis of Maxwell's Equations. IEEE Antennas and Propagation Magazine. 2014 December;56(6):299–307.
[30] Elliott RS. Electromagnetics: History, Theory, and Applications. Piscataway, NJ: IEEE Press; 1993.
[31] Kellogg OD. Foundations of potential theory. Berlin Heidelberg: Springer-Verlag; 1929.
[32] Rudin W. Principles of mathematical analysis. 3rd ed. New York, NY: McGraw-Hill; 1976.
[33] Harrington RF. Time-harmonic Electromagnetic Fields. London, UK: McGraw-Hill; 1961.

[34] Makarov B, Podkorytov A. Real Analysis: Measures, Integrals and Applications. Universitext. London, UK: Springer; 2013.
[35] Misner CW, Thorne KS, Wheeler JA. Gravitation. San Francisco, CA: W. H. Freeman and Company; 1970.
[36] Chew WC. Waves and Fields in Inhomogenous Media. New York, NY: Van Nostrand Teinhold; 1990.
[37] Pozar D. Microwave Engineering. 4th ed. New York, NY: John Wiley & Sons, Inc.; 2012.
[38] Jin JM. The Finite Element Method in Electromagnetics. New York, NY: John Wiley & Sons, Inc.; 1993.
[39] Collin RE. Field Theory of Guided Waves. Piscataway, NJ: IEEE press; 1991.
[40] Taflove A. Computational Electrodynamics: The Finite-Difference Time-Domain Method. Boston, MA: Artech House; 1995.
[41] Cessenat M. Mathematical Methods in Electromagnetism: Linear theory and applications. River Edge, NJ: World Scientific; 1996.
[42] Someda CG. Electromagnetic Waves. New York, NY: Chapman & Hall; 1998.
[43] Pauli W. Theory of relativity. New York, NY: Dover Publications, Inc.; 1981.
[44] Van Bladel JG. Relativity and Engineering. Berlin Heidelberg : Springer Berlin Heidelberg; 1984.
[45] Lorrain P, Corson DR, Lorrain F. Electromagnetic Fields and Waves. 3rd ed. New York, NY: W. H. Freeman and Company; 1988.
[46] Davis JL. Wave Propagation in Electromagnetic Media. New York, NY: Springer-Verlag; 1990.
[47] Sommerfeld A. Mechanics of Deformable Bodies. vol. 2 of Lectures on theoretical physics. New York, NY: Academic Press; 1950.
[48] Tai CT. Generalized Vector and Dyadic Analysis. New York, NY: IEEE Press; 1997.
[49] Loomis LH, Sternberg S. Advanced Calculus. Reading, MA:Addison-Wesley; 1968.
[50] Abraham M, Becker R. The Classical Theory of Electricity and Magnetism. 2nd ed. London, UK: Blackie; 1932.
[51] Kraus JD, Carver KR. Electromagnetics. 2nd ed. Tokyo, Japan: McGraw-Hill; 1973.
[52] Guru BS, Hiziroglu HR. Electromagnetic field theory fundamentals. 2nd ed. New York, NY: Cambridge University Press; 2004.
[53] Holt CA. Introduction to Electromagnetic Fields and Waves. New York, NY: John Wiley & Sons, Inc.; 1963.
[54] Shen LC, Kong JA. Applied electromagnetism. Monterey, CA: Brooks/Cole; 1983.
[55] Zangwill A. Modern electrodynamics. Cambridge, UK: Cambrigde University Press; 2013.
[56] Hayt WH, Buck JA. Engineering Electromagnetics. 8th ed. New York, NY: McGraw-Hill; 2012. International edition.
[57] Balanis CA. Antenna Theory: Analysis and Design. 2nd ed. New York, NY: John Wiley & Sons, Inc.; 1997.
[58] Balanis CA. Advanced Engineering Electromagnetics. New York, NY: John Wiley & Sons, Inc.; 2005.
[59] Orfanidis SJ. Electromagnetic Waves and Antennas. www.ece.rutgers.edu/~orfanidi/ewa; 2004.
[60] Morse P, Feshbach H. Methods of Theoretical Physics. New York, NY: McGraw-Hill; 1953.
[61] Lindell IV. Methods for electromagnetic field analysis. Piscataway, NJ: IEEE Press; 1992.
[62] Dautray R, Lions JL. Mathematical Analysis and Numerical Methods for Science and Technology - Physical Origins and Classical Methods. vol. 1. Berlin Heidelberg: Springer-Verlag; 1990.

[63] Fournet G. Electromagnétisme à partir des équations locales. Paris, FR: Masson; 1979.

[64] Bassanini P, Elcrat A. Mathematical Theory of Electromagnetism. Creative Commons; 2009.

[65] Hertz HR. On the propagation velocity of electrodynamic effects. Annalen der Physik. 1888;270(7):551–560. In German: Über die Ausbreitungsgeschwindigkeit der electrodynamischen Wirkungen.

[66] Dobbs R. Electromagnetic Waves. London, UK: Routledge & Kegan Paul; 1985.

[67] Zhang K, Li D. Electromagnetic Theory for Microwaves and Optoelectronics. Berlin Heidelberg: Springer-Verlag; 1998.

[68] Dautray R, Lions JL. Mathematical Analysis and Numerical Methods for Science and Technology - Evolution Problems I. vol. 5. Berlin Heidelberg: Springer-Verlag; 1992.

[69] Jones DS. The Theory of Electromagnetism. Oxford, UK: Pergamon; 1964.

[70] Landau LD, Lifshitz EM. The classical theory of fields. 3rd ed. Oxford, UK: Pergamon Press; 1971.

[71] Rozzi T, Mongiardo M. Open electromagnetic waveguides. London, UK: Institution of Electrical Engineers; 1997.

[72] Thidé B. Electromagnetic Field Theory. New York, NY: Dover Publications, Inc.; 2011.

[73] Melia F. Electrodynamics. Chicago, IL: The University of Chicago Press; 2001.

[74] Oughstun KE. Electromagnetic and Optical Pulse Propagation. vol. 1. New York, NY: Springer Science+Business Media; 2006.

[75] Slater JC, Frank NH. Electromagnetism. New York, NY: McGraw-Hill; 1947.

[76] Eyges L. The Classical Electromagnetic Field. New York, NY: Dover Publications, Inc.; 1980.

[77] Penfield P, Haus H. Electrodynamics of Moving Media. Cambridge, MA: MIT Press; 1967.

[78] Mason M, Weaver W. The electromagnetic field. New York, NY: Dover Publications, Inc.; 1929.

[79] Pfeifer RNC, Nieminen TA, Heckenberg NR, et al. Two controversies in classical electromagnetism. In: Dholakia K, Spalding GC, editors. Optical Trapping and Optical Micromanipulation III. vol. 6326 of Proceedings of SPIE; 2006. p. 1–10.

[80] de Hoop AT. Handbook of radiation and scattering of waves. London, UK: Academic Press; 1995.

[81] Auld BA. Acoustic Fields and Waves in Solids. New York, NY: John Wiley & Sons, Inc.; 1973.

[82] Golub GH, van Loan CF. Matrix Computations. Baltimore, MD: Johns Hopkins University Press; 1996.

[83] Bau III D, Trefethen LN. Numerical linear algebra. Philadelphia, PA: Soci. Indus. Ap. Math.; 1997.

[84] Blyth TS, Robertson EF. Basic Linear Algebra. 2nd ed. Springer Undergraduate Mathematics Series. London, UK: Springer-Verlag; 2002.

Chapter 2
Static electric fields I

2.1 Laws of electrostatics

The laws of electrostatics describe the so-called *static limit* of Maxwell's equations. With static limit we mean that *all* the electromagnetic entities and the sources introduced in Chapter 1 do not depend on time. In particular, this implies that the charges are *fixed* in space, because currents are not permitted. However, does this hypothesis make any sense at all? Well, it does and is quite helpful so long as we are clear that electrostatics is yet another idealized model of the physical reality. Truly electrostatic fields are an abstraction pretty much in the same way as truly time-harmonic fields (Section 1.5) do not exist in actuality. To elucidate, we learned in Section 1.6 that permanent charges in a bounded conductor are not stable and that they drift towards the surface of the conductor whereon they reach a final stable configuration, at least until some other distant charge is brought closer to the conductor, that is. Barring any more external intervention, though, after the drifting (the transient) has terminated, the electromagnetic entities remain constant with time. Thereby, it seems reasonable that Maxwell's equations ought to simplify, essentially because the time rate of variation of the displacement vector, the magnetic intensity and the charge density vanish identically. But then, we also know from (1.253) that magnetism (the magnetic field) is a consequence of the relative motion of charges and observers. In the realm of electrostatics charges are fixed to any observer, and hence no magnetic field exists. Moreover, if the charges are fixed, no currents exist, and the continuity equations (1.17) or (1.46) are meaningless. Since time does not appear in the equations to be derived below, they are, in principle, valid from any infinitely remote moment in time and will remain valid forever. In fact, this is the idealization mentioned above. This point is more subtle than it may sound because the assumption that an electrostatic field has existed since forever clashes with the hypothesis we made at the end of Section 1.2.1 to set the functions $f(\mathbf{r})$ and $g(\mathbf{r})$ in (1.48) and (1.51). There we postulated that the electromagnetic field entities were indeed null at some point in time.

Keeping all of the previous observations in mind, we can state the laws of electrostatic by dropping the time dependence in the Maxwell equations and ridding ourselves of the magnetic entities holusbolus. For the integral form we obviously use the same surfaces and volumes defined at the beginning of Section 1.2.1 and in Figures 1.2a and 1.2b.

The **Gauss law** in global form (1.16) becomes

$$\Psi_D := \int_{\partial V} \mathrm{d}S\, \hat{\mathbf{n}}(\mathbf{r}) \cdot \mathcal{D}(\mathbf{r}) = \int_V \mathrm{d}V\, \varrho(\mathbf{r}) := Q \tag{2.1}$$

i.e., the electric flux through a closed surface equals the electric charge enclosed by the surface. The main difference with (1.16) is that the total charge in V is fixed and thus constant, hence the flux does not depend on time.

The only other equation we need in electrostatics follows from the **Faraday law** (1.8), which turns into

$$\oint_{\gamma} \mathrm{d}s\, \hat{\mathbf{s}}(\mathbf{r}) \cdot \mathcal{E}(\mathbf{r}) = 0 \tag{2.2}$$

i.e., the net circulation (Appendix A.4) of the electric field along any closed curve is zero.

In order to determine the local counterpart of (2.1) and (2.2) we may, in principle, repeat all the steps outlined in Section 1.2.2. More directly, we can just drop the time dependence from the Maxwell equations in differential form, with the proviso that the relevant vector and scalar fields satisfy the conditions for the applicability of the Gauss theorem and the Stokes theorem. In either way, by assuming the charge density $\varrho(\mathbf{r})$ is non-zero in a bounded volume V_ϱ, we arrive at the following equations [1, Chapter 1], [2, Chapter 18], [3, Chapter 1]

$$\nabla \cdot \mathcal{D}(\mathbf{r}) = \varrho(\mathbf{r}), \qquad \mathbf{r} \in \mathbb{R}^3 \tag{2.3}$$

$$\nabla \times \mathcal{E}(\mathbf{r}) = 0, \qquad \mathbf{r} \in \mathbb{R}^3 \tag{2.4}$$

where $\mathcal{E}(\mathbf{r})$ must be of class $\mathcal{C}^1(\mathbb{R}^3)$, $\mathcal{D}(\mathbf{r})$ must be of class and $\mathcal{C}^1(\mathbb{R}^3 \setminus \partial V_\varrho)$ with $\hat{\mathbf{n}} \cdot \mathcal{D}$ continuous across ∂V_ϱ. Thus, when turned into local form, the condition on the net-circulation requires that $\mathcal{E}(\mathbf{r})$ be curl-free. A vector field which satisfies (2.2) or (2.4) is also said *lamellar* or *conservative* (see Section 8.1).

In electrostatics the sources of the electric fields are just *fixed* charges, and the Gauss law relates an entity of quantity with the sources, in keeping with the general form of Maxwell's equations. Both (2.2) and (2.4) play the role of a condition imposed on the admissible vector fields (cf. Figure 1.5). In words, among all the possible vector fields, we look for those with zero circulation or zero curl. From another point of view, (2.3) and (2.4) can be interpreted as the problem of inverting the divergence operator subject to the condition that the solution be conservative.

Finally, (2.3) and (2.4) are equivalent to four scalar equations in six unknowns. When the constitutive relationship for the displacement vector is used, the total number of unknowns reduces to three, i.e., the scalar components of $\mathcal{E}(\mathbf{r})$ in the basis of choice. It looks like we have got one equation too many but, as was the case for the general Maxwell equations, the system comprised of (2.3) and (2.4) is not over-determined, as we shall discuss further in Chapter 8 with the aid of a general result concerning vector fields in a three-dimensional space.

Example 2.1 (Electrostatic field of a spherical uniform distribution of charge)

Suppose that an amount of charge is distributed uniformly within the ball $B(0, a)$ (never mind how this arrangement can be achieved in practice). The charge density reads

$$\varrho(\mathbf{r}) = \begin{cases} \varrho_0, & r = |\mathbf{r}| < a \\ 0, & r > a \end{cases} \tag{2.5}$$

and clearly is a radial function discontinuous across the sphere $\partial B := \{\mathbf{r} \in \mathbb{R}^3 : |\mathbf{r}| = a\}$. We wish to compute the electric field produced by $\varrho(\mathbf{r})$ in free space and for $r \geqslant 0$. This seems an innocent enough problem which we can address by means of either the Gauss law in global form or the system (2.3) and (2.4).

However, the first strategy is not viable in general, essentially because $\mathcal{E}(\mathbf{r})$ is a three-dimensional vector field, whereas (2.1) is just a single global condition. Hence, at best we can hope to fix one of the three scalar components of $\mathcal{E}(\mathbf{r})$. In the present problem this approach pays off, because symmetry prompts us to conclude that $\mathcal{E}(\mathbf{r})$ is a radially directed vector, whereby (2.1) can be used to determine the radial component of $\mathcal{E}(\mathbf{r})$ in spherical coordinates (Appendix A.1). We also need to check that the net-circulation of the solution found is zero for any closed curve. This task is accomplished by recalling that (2.1) and (2.3) are equivalent, so we can examine $\nabla \times \mathcal{E}$.

The strategy based on the local form of the electrostatic equations works in general, but the difficulty lies exactly in the fact that we have to solve both of them simultaneously (see the next section). In this case, though, on invoking spherical symmetry again, we can solve (2.3) for the radial component of $\mathcal{E}(\mathbf{r})$. To make sure that the solution thus found is an admissible field, we actually need to check that it is curl-free!

Getting on with the first strategy, we choose a convenient surface to apply the Gauss law: the sphere $S := \{\mathbf{r} \in \mathbb{R}^3 : |\mathbf{r}| = b > 0\}$ will do the trick. Since the field is radial and the charge exists in free space, we have

$$\mathcal{D}(\mathbf{r}) = \varepsilon_0 \mathcal{E}_r(r)\hat{\mathbf{r}} \tag{2.6}$$

in spherical coordinates and in consideration of (1.112). The flux of $\mathcal{D}(\mathbf{r})$ through the sphere reads

$$\int_S \mathrm{d}S \; \varepsilon_0 \hat{\mathbf{r}} \cdot \hat{\mathbf{r}} \mathcal{E}_r(r) = 4\pi\varepsilon_0 b^2 \mathcal{E}_r(b) \tag{2.7}$$

because $\mathcal{E}_r(r)$ is constant over S for symmetry reasons. Next, we compute the total charge in the ball $B(0, b)$, i.e.,

$$Q(b) = \int_{B(0,b)} \mathrm{d}V \varrho(\mathbf{r}) = \begin{cases} \dfrac{4}{3}\pi b^3 \varrho_0, & b < a \\ \dfrac{4}{3}\pi a^3 \varrho_0, & b > a \end{cases} \tag{2.8}$$

and this must be equal to the flux of $\mathcal{D}$, whereby we can determine $\mathcal{E}_r(b)$ as a function of the radius b. By putting everything together, and renaming b to r, we find

$$\mathcal{E}(\mathbf{r}) = \begin{cases} \dfrac{\varrho_0}{3\varepsilon_0}\mathbf{r}, & r < a \\ \dfrac{\varrho_0}{3\varepsilon_0}\dfrac{a^3}{r^3}\mathbf{r}, & r > a \end{cases} \tag{2.9}$$

where we notice that the normal component of $\mathcal{E}(\mathbf{r})$ is continuous across the sphere ∂B, that is, where the charge density has a jump. This is not always true, rather it is a consequence of the linear constitutive relationship (1.112) and the continuity of $\hat{\mathbf{r}} \cdot \mathcal{D}$. According to (2.9) the electric field vanishes at the center of the charge distribution, though this result may not come as a surprise what with the spherical symmetry of the source. An interesting sidelight is that for a spherical homogeneous distribution of mass with radius a the Newtonian theory of gravitation predicts a radial gravitational force which, aside from different multiplicative factors, behaves as the electric field given by (2.9).

Secondly, for the other strategy we write the divergence operator in spherical coordinates on account of (A.31) and, since $\mathcal{E}$ is radial, we have from (2.3)

$$\frac{1}{r^2}\frac{\mathrm{d}}{\mathrm{d}r}(r^2 \varepsilon_0 \mathcal{E}_r) = \begin{cases} \varrho_0, & r < a \\ 0, & r > a \end{cases} \tag{2.10}$$

which is solved by

$$\mathcal{E}_r(r) = \begin{cases} \dfrac{\varrho_0 r}{3\varepsilon_0}, & r < a \\ \dfrac{A}{r^2}, & r > a \end{cases} \tag{2.11}$$

where A is a real constant yet unspecified. We must choose A so as to guarantee the continuity of $\hat{\mathbf{n}} \cdot \mathcal{D}$ across the sphere ∂B, and in so doing we arrive at the same result given in (2.9). The Gauss law in integral form takes care of the continuity of $\hat{\mathbf{n}} \cdot \mathcal{D}$ implicitly, whereas with the local form we need to enforce such condition explicitly. We also notice that $\mathcal{D}(\mathbf{r})$ is continuously differentiable everywhere but on the sphere ∂B, and this happens because $\varrho(\mathbf{r})$ is only piecewise continuous.

Lastly, we check that (2.9) is actually a conservative field. Direct calculation of the curl in spherical coordinates by means of (A.34) shows that this is indeed the case. In general, a radial field is curl-free.

The electrostatic energy associated with the field (2.9) is computed through (1.263) by means of standard polar spherical coordinates (Appendix A.1), viz.,

$$\begin{aligned} W_e &= \frac{\varepsilon_0}{2} \int\limits_{\mathbb{R}^3} \mathrm{d}V \, |\mathcal{E}(\mathbf{r})|^2 = 2\pi\varepsilon_0 \int\limits_0^a \mathrm{d}r \left(\frac{\varrho_0}{3\varepsilon_0}\right)^2 r^4 + 2\pi\varepsilon_0 \int\limits_a^{+\infty} \mathrm{d}r \left(\frac{\varrho_0}{3\varepsilon_0}\right)^2 \frac{a^6}{r^2} \\ &= 2\pi \frac{\varrho_0^2}{9\varepsilon_0}\left(\frac{a^5}{5} + a^5\right) = \frac{4\pi\varrho_0^2}{3\varepsilon_0}\frac{a^5}{5} \end{aligned} \tag{2.12}$$

and can be interpreted as the work done to assemble the distribution of charge within $B(0, a)$.

(End of Example 2.1)

Example 2.2 (Electrostatic field of a point charge)

The electrostatic field generated by a point charge q located at $\mathbf{r} = 0$ in free space is radially directed and spherically symmetric, and thus can be determined by applying the global Gauss law over a sphere of radius a centered in the origin. From (1.16) and (1.112) we have

$$q = \varepsilon_0 \int\limits_S \mathrm{d}S \, \hat{\mathbf{r}} \cdot \mathcal{E}(a) = 4\pi\varepsilon_0 a^2 \mathcal{E}_r(a) \tag{2.13}$$

which can be solved for $\mathcal{E}_r(a)$. On renaming a to the radial coordinate r (since the radius of the sphere is arbitrary) we find the fundamental result

$$\mathcal{E}(\mathbf{r}) = \frac{q}{4\pi\varepsilon_0 r^2}\hat{\mathbf{r}}, \qquad \mathbf{r} \in \mathbb{R}^3 \setminus \{0\} \tag{2.14}$$

whereby we see that the electric field of a point charge decays with the square of the distance r from the source and is *singular* at the location of the charge. Besides, since the electric field is a radial vector, it is also conservative. This statement can be verified by taking the curl of $\mathcal{E}(\mathbf{r})$, e.g., in spherical coordinates with the aid of (A.34).

The electrostatic energy (1.263) in a bounded region containing a point charge is infinite. This is one of the few quirks of the very notion of point charge [4, 5], [6, Section 2.4.4].

(End of Example 2.2)

2.2 Scalar potential and the Poisson equation

We can convince ourselves that the system formed by (2.3) and (2.4) is not over-determined by devising a general solution strategy. We have already mentioned that (2.4) plays the role of a condition of sorts which restricts the set of admissible electric fields. Is there a way to make sure that we invariably pick up a conservative field, so that we do not have to bother with checking *a posteriori* that (2.4) is satisfied by our solution? The answer is affirmative.

Indeed, any vector field which is the gradient of a continuously differentiable scalar field possesses no curl, as is stated by (A.38) [7, Lemma 3, p. 214]. Therefore, if we make the Ansatz

$$\mathcal{E}(\mathbf{r}) = -\nabla\Phi(\mathbf{r}), \qquad \mathbf{r} \in \mathbb{R}^3 \tag{2.15}$$

then (2.4) is always true by virtue of (A.38). In electrostatics the function $\Phi(\mathbf{r})$ is called the *scalar potential* and carries the physical dimension of volt (V). We remark that the scalar potential is defined up to an arbitrary additive constant, as is evident from (2.15). And yet, when we determine the electric field through (2.15), the constant contributes naught, and $\mathcal{E}(\mathbf{r})$ is unambiguously defined. Moreover, we notice that (2.15) satisfies (2.2) as well, as is known from calculus. Indeed,

$$\oint_\gamma \mathrm{d}s\, \hat{\mathbf{s}}(\mathbf{r}) \cdot \nabla\Phi(\mathbf{r}) = 0 \tag{2.16}$$

means that the scalar potential takes on the same value after a full turn around any closed curve γ. This property is concisely stated by saying that $\Phi(\mathbf{r})$ is a *one-valued* scalar field.

Evidently, if we can compute $\Phi(\mathbf{r})$ somehow, we can retrieve the electric field by means of (2.15). Since $\mathcal{D}(\mathbf{r})$ and $\mathcal{E}(\mathbf{r})$ are related by a suitable constitutive relationship (Section 1.6), we should manage to transform (2.3) into an equation for the scalar potential. For instance, in free space we have

$$\nabla \cdot [\varepsilon_0 \nabla\Phi(\mathbf{r})] = \varepsilon_0 \nabla^2\Phi(\mathbf{r}) = -\varrho(\mathbf{r}), \qquad \mathbf{r} \in \mathbb{R}^3 \tag{2.17}$$

by virtue of (1.112) and (2.15); this second-order partial differential equation is named after **Poisson** [8, Chapter 2]. The continuity of $\hat{\mathbf{n}} \cdot \mathcal{D}(\mathbf{r})$ for $\mathbf{r} \in \partial V_\varrho$ translates into a condition for the scalar potential, namely, $\hat{\mathbf{n}} \cdot \nabla\Phi(\mathbf{r})$ must be continuous across ∂V_ϱ in light of (2.15) and (2.3). Thanks to (2.15) we have effectively transformed the system (2.3) and (2.4) into a single scalar equation in just one unknown scalar field, and this confirms that the system comprised of (2.3) and (2.4) is well specified.

We can shed light on the behavior of the electrostatic field and the scalar potential at infinity by examining the electric energy associated with a finite amount of charge with density $\varrho(\mathbf{r})$ for $\mathbf{r} \in V_\varrho$. In particular, it is fruitful to write the potential as

$$\Phi(\mathbf{r}) = \Phi_1(\mathbf{r}) + \Phi_\infty \tag{2.18}$$

so as to exhibit the arbitrary constant explicitly. We begin with the energy in a bounded volume $V \supseteq V_\varrho$ with unit normal $\hat{\mathbf{n}}$ to ∂V positively directed outward V. According to definition (1.263) we have

$$\begin{aligned} W_e &:= \frac{1}{2}\int_V \mathrm{d}V\, \mathcal{E}(\mathbf{r}) \cdot \mathcal{D}(\mathbf{r}) = -\frac{1}{2}\int_V \mathrm{d}V\, \nabla\Phi(\mathbf{r}) \cdot \mathcal{D}(\mathbf{r}) \\ &= \frac{1}{2}\int_V \mathrm{d}V\, \Phi(\mathbf{r})\nabla \cdot \mathcal{D}(\mathbf{r}) - \frac{1}{2}\int_V \mathrm{d}V\, \nabla \cdot [\Phi(\mathbf{r})\mathcal{D}(\mathbf{r})] \end{aligned}$$

$$= \frac{1}{2}\int_{V_\varrho} \mathrm{d}V\,[\Phi_1(\mathbf{r}) + \Phi_\infty]\varrho(\mathbf{r}) - \frac{1}{2}\int_{\partial V} \mathrm{d}S\,[\Phi_1(\mathbf{r}) + \Phi_\infty]\hat{\mathbf{n}}(\mathbf{r})\cdot\mathcal{D}(\mathbf{r})$$

$$= \frac{1}{2}\int_{V_\varrho} \mathrm{d}V\,\Phi_1(\mathbf{r})\varrho(\mathbf{r}) - \frac{1}{2}\int_{\partial V} \mathrm{d}S\,\Phi_1(\mathbf{r})\hat{\mathbf{n}}(\mathbf{r})\cdot\mathcal{D}(\mathbf{r})$$

$$+ \frac{\Phi_\infty}{2}\left[\int_{V_\varrho} \mathrm{d}V\,\varrho(\mathbf{r}) - \int_{\partial V} \mathrm{d}S\,\hat{\mathbf{n}}(\mathbf{r})\cdot\mathcal{D}(\mathbf{r})\right] \tag{2.19}$$

where we have used (2.3) and the fact that $\varrho(\mathbf{r})$ is confined to V_ϱ. The surface integral over ∂V follows from the application of the Gauss theorem separately to V_ϱ and $V \setminus \overline{V}_\varrho$ and by assuming the continuity of $\hat{\mathbf{n}}\cdot\mathcal{D}$ through ∂V_ϱ (cf. Section 1.2.2). We observe that the last contribution in (2.19) is zero owing to the Gauss law (1.16) applied to V, and thus the value of the arbitrary constant Φ_∞ is irrelevant. Besides, the electric energy in V is finite and it should remain so even when we consider the whole space because the charge is finite. Therefore, we trade V for the ball $B(0, a)$, and require the limit of the integral over the sphere ∂B to be finite as $a \to +\infty$. This is certainly true if $\Phi_1(\mathbf{r})$ and $\mathcal{D}(\mathbf{r})$ admit the following estimates

$$|\Phi_1(\mathbf{r})| \leqslant \frac{C_\Phi}{|\mathbf{r}|}, \qquad \text{for} \qquad |\mathbf{r}| \geqslant b_\Phi \tag{2.20}$$

$$|\mathcal{D}(\mathbf{r})| \leqslant \frac{C_D}{|\mathbf{r}|^2}, \qquad \text{for} \qquad |\mathbf{r}| \geqslant b_D \tag{2.21}$$

with C_Φ, b_Φ, C_D and b_D being suitable positive constants. Indeed, we have

$$\left|\int_{\partial B} \mathrm{d}S\,\Phi_1(\mathbf{r})\hat{\mathbf{n}}(\mathbf{r})\cdot\mathcal{D}(\mathbf{r})\right| \leqslant \int_{\partial B} \mathrm{d}S\,|\Phi_1(\mathbf{r})\hat{\mathbf{n}}(\mathbf{r})\cdot\mathcal{D}(\mathbf{r})| \leqslant \int_{\partial B} \mathrm{d}S\,|\Phi_1(\mathbf{r})||\mathcal{D}(\mathbf{r})|$$

$$\leqslant \int_{\partial B} \mathrm{d}S\,\frac{C_\Phi}{a}\frac{C_D}{a^2} = 4\pi\frac{C_\Phi C_D}{a} \tag{2.22}$$

which vanishes as $a \to +\infty$. In free space, the estimate (2.21) applies to $\mathcal{E}(\mathbf{r})$ as well, whereby physically admissible electrostatic fields decay with the inverse square of the distance away from the sources (cf. Examples 2.1 and 2.2). In summary, we see that the asymptotic behavior of the potential is

$$\Phi(\mathbf{r}) = \mathcal{O}\left(\frac{1}{|\mathbf{r}|}\right) + \Phi_\infty, \qquad |\mathbf{r}| \to +\infty \tag{2.23}$$

if the sources are of finite extent.

The splitting (2.19) of the electric energy W_e into a volume and a surface integral — both involving the scalar potential — also provides insight into the meaning of the normal component of the displacement vector. If we choose the volume V so as to leave all the charges in the complementary unbounded region $\mathbb{R}^3 \setminus \overline{V}$, the volume integral vanishes identically. The surface integral over ∂V then represents the electrostatic energy stored in V and due to the charges in $\mathbb{R}^3 \setminus \overline{V}$; the minus sign is just a consequence of the orientation of $\hat{\mathbf{n}}$ towards $\mathbb{R}^3 \setminus \overline{V}$, and W_e is positive. By comparing the volume integral and the surface integral in (2.19) we are led to interpret $-\hat{\mathbf{n}}(\mathbf{r})\cdot\mathcal{D}(\mathbf{r})$ as a layer of fictitious charges distributed over ∂V and responsible for producing the electrostatic field in V. This conclusion agrees with the boundary condition (1.182) as well as the integral representation of the scalar potential we shall derive in Section 2.7.

For the sake of completeness, we notice that the global form of the Gauss law in free space becomes

$$\varepsilon_0 \int_S \mathrm{d}S\, \hat{\mathbf{n}}(\mathbf{r}) \cdot \nabla\Phi(\mathbf{r}) = \varepsilon_0 \int_S \mathrm{d}S\, \frac{\partial \Phi}{\partial \hat{\mathbf{n}}} = -\int_V \mathrm{d}V \varrho(\mathbf{r}) \tag{2.24}$$

and similar expressions hold for isotropic and anisotropic media. If the total charge is finite, (2.24) leads to estimate (2.20) for $\Phi(\mathbf{r})$.

When other constitutive relationships are contemplated, the Poisson equation looks slightly different than (2.17). In an unbounded inhomogeneous isotropic medium we have

$$\nabla \cdot [\varepsilon(\mathbf{r})\nabla\Phi(\mathbf{r})] = -\varrho(\mathbf{r}), \qquad \mathbf{r} \in \mathbb{R}^3 \tag{2.25}$$

with $\varepsilon(\mathbf{r}) \in \mathcal{C}^1(\mathbb{R}^3)$, whereas in an unbounded inhomogeneous anisotropic medium the Poisson equation reads

$$\nabla \cdot [\overline{\varepsilon}(\mathbf{r}) \cdot \nabla\Phi(\mathbf{r})] = -\varrho(\mathbf{r}), \qquad \mathbf{r} \in \mathbb{R}^3 \tag{2.26}$$

where the entries of $\overline{\varepsilon}(\mathbf{r})$ must be continuously differentiable in $\mathbb{R}^3$.

Example 2.3 (Electrostatic scalar potential of a point charge)

The scalar potential generated by a point charge q located at $\mathbf{r} = 0$ follows by integrating (2.15) where the electric field is given by (2.14), namely,

$$\frac{q}{4\pi\varepsilon_0 r^2}\hat{\mathbf{r}} = -\hat{\mathbf{r}}\frac{\mathrm{d}\Phi}{\mathrm{d}r} \tag{2.27}$$

which yields

$$\Phi(r) = \frac{q}{4\pi\varepsilon_0 r} + \Phi_\infty \tag{2.28}$$

with the arbitrary constant Φ_∞ representing the value of the scalar potential at infinity. This result is in agreement with the expected asymptotic behavior (2.23).

Finally, if the charge is located in a point $\mathbf{r}'$, the electrostatic potential reads

$$\Phi(\mathbf{r}) = \frac{q}{4\pi\varepsilon_0|\mathbf{r} - \mathbf{r}'|} + \Phi_\infty \tag{2.29}$$

on the grounds that the result must depend only on the distance from the charge.

(End of Example 2.3)

Example 2.4 (Electrostatic scalar potential of a spherical uniform distribution of charge)

The charge in question is confined within the ball $B(0, a)$ and the density is defined by (2.5) in Example 2.1. To determine the scalar potential we may apply definition (2.15) with the electrostatic field given by (2.9) and the gradient expressed in polar spherical coordinates (Appendix A.1). As a result we obtain an *ordinary* first-order differential equation for $\Phi(\mathbf{r})$ which is readily solved. This strategy is not viable in general, if $\Phi(\mathbf{r})$ depends on two or three coordinates, and thus the Poisson equation must be solved (see Example 2.7 and Section 2.7).

In symbols, we have

$$-\frac{\mathrm{d}\Phi}{\mathrm{d}r} = \begin{cases} \dfrac{\varrho_0}{3\varepsilon_0} r, & r < a \\ \dfrac{\varrho_0}{3\varepsilon_0} \dfrac{a^3}{r^2}, & r > a \end{cases} \tag{2.30}$$

which upon integration yields

$$\Phi(\mathbf{r}) = \begin{cases} A - \dfrac{\varrho_0}{6\varepsilon_0} r^2, & r < a \\ \Phi_\infty + \dfrac{\varrho_0}{3\varepsilon_0} \dfrac{a^3}{r}, & r > a \end{cases} \tag{2.31}$$

with the constants A and Φ_∞ yet unspecified. We choose A so as to ensure the continuity of the potential through the sphere ∂B

$$A - \frac{\varrho_0}{6\varepsilon_0} a^2 = \Phi_\infty + \frac{\varrho_0}{3\varepsilon_0} a^2 \qquad \Longrightarrow \qquad A = \Phi_\infty + \frac{\varrho_0}{2\varepsilon_0} a^2. \tag{2.32}$$

Substituting into (2.31) yields the result

$$\Phi(\mathbf{r}) = \begin{cases} \dfrac{\varrho_0}{6\varepsilon_0}(3a^2 - r^2) + \Phi_\infty, & r < a \\ \dfrac{\varrho_0}{3\varepsilon_0} \dfrac{a^3}{r} + \Phi_\infty, & r > a. \end{cases} \tag{2.33}$$

Comparing the expression of $\Phi(\mathbf{r})$ for $r > a$ with (2.28), which gives the potential generated by a point charge in the origin, we may interpret (2.33) by saying that outside the ball $B(0, a)$ the potential is equivalently produced by a charge $Q = 4\pi\varrho_0 a^3/3$ placed in $r = 0$. Indeed, Q is the total charge present in $B(0, a)$.

The electrostatic energy associated with the charge Q can be equivalently computed with the aid of (2.19) applied to $V \equiv \mathbb{R}^3$, whereby the surface integral over ∂V vanishes, and we have

$$W_e = \frac{1}{2} \int_{B(0,a)} \mathrm{d}V \, \varrho(\mathbf{r})\Phi_1(\mathbf{r}) = 2\pi \int_0^a \mathrm{d}r \, r^2 \frac{\varrho_0^2}{6\varepsilon_0}(3a^2 - r^2) = \pi \frac{\varrho_0^2}{3\varepsilon_0}\left(a^5 - \frac{a^5}{5}\right) = \frac{4\pi\varrho_0^2}{3\varepsilon_0} \frac{a^5}{5} \tag{2.34}$$

which coincides with the previous result (2.12).

(End of Example 2.4)

Example 2.5 (Electrostatic scalar potential and electric field of an electrostatic dipole)

The second simplest configuration of sources we can think of is the combination of two point charges, which is referred to as an *electrostatic dipole* [9, Chapter 3], [10, Section 4.7]. The rationale for the name is that a single point charge was sometimes also called a 'pole'. In fact, it can be shown there is a unique correspondence between electrostatics in a two-dimensional space and the theory of complex functions of one variable (Appendix B), so that the 'poles' of a function are seen to occur at the locations of charges [11].

A dipole becomes especially meaningful from a theoretical viewpoint when the charges have opposite signs and are infinitely close to one another, in which case we speak of an elementary static dipole. Although this arrangement is evidently an abstraction, the corresponding electrostatic field and scalar potential play an important role in the solution of (2.3) for general charge densities (see Section 2.7).

Let us suppose that a negative point charge $-q$ ($q > 0$) is located at $\mathbf{r} = 0$ in free space, and that a second positive charge q is placed at $\mathbf{r} = \mathbf{u}$ with $\mathbf{u}$ an arbitrary constant vector. The scalar potential of the dipole follows from the superposition of the potentials generated by the two charges computed via (2.29), viz.,

$$\Phi(\mathbf{r};\mathbf{u}) = \frac{q}{4\pi\varepsilon_0}\left(\frac{1}{|\mathbf{r}-\mathbf{u}|} - \frac{1}{|\mathbf{r}|}\right), \qquad \mathbf{r} \in \mathbb{R}^3 \setminus \{0, \mathbf{u}\} \tag{2.35}$$

and we wish to examine this expression in the limit as $|\mathbf{u}| \to 0$. To this purpose we employ the binomial theorem [12, Formula 5.3], i.e.,

$$\begin{aligned} |\mathbf{r}-\mathbf{u}|^{-\frac{1}{2}} &= (r^2 - 2\mathbf{r}\cdot\mathbf{u} + u^2)^{-\frac{1}{2}} = \frac{1}{r}\left(1 - 2\hat{\mathbf{r}}\cdot\frac{\mathbf{u}}{r} + \frac{u^2}{r^2}\right)^{-\frac{1}{2}} \\ &= \frac{1}{r}\left(1 + \hat{\mathbf{r}}\cdot\frac{\mathbf{u}}{r}\right) + o\left(\frac{|\mathbf{u}|}{|\mathbf{r}|^2}\right), \qquad |\mathbf{u}| \ll |\mathbf{r}|. \end{aligned} \tag{2.36}$$

Inserting this expression into the potential yields

$$\Phi(\mathbf{r};\mathbf{u}) = \frac{q\mathbf{u}\cdot\hat{\mathbf{r}}}{4\pi\varepsilon_0 r^2} + o\left(\frac{|\mathbf{u}|}{|\mathbf{r}|^2}\right), \qquad |\mathbf{u}| \ll |\mathbf{r}| \tag{2.37}$$

thus we see that the leading term of this asymptotic expansion decays with the square of the distance away from the source. In essence, since the charges have opposite signs, the scalar potential they produce cancel each other to the first order of approximation. Moreover, the configuration is not spherically symmetric, rather, axially symmetric around the direction specified by the vector $\mathbf{u}$. The quantity

$$\mathbf{p} := q\mathbf{u} \tag{2.38}$$

is called the *moment* of the dipole and carries the physical dimensions of coulomb meter (Cm).

We may now take the limit for vanishing distance between the charges and thus determine the potential of an elementary static dipole. However, under the present assumptions the dipole moment would simply vanish thus yielding a null potential. Therefore, we make the additional hypothesis that in the limit as $|\mathbf{u}| \to 0$, $\mathbf{p}$ remains finite. This clearly implies that q must become infinite in the process. In the end, the scalar potential of an elementary dipole located at $\mathbf{r} = 0$ reads

$$\Phi(\mathbf{r}) = \frac{\mathbf{p}\cdot\hat{\mathbf{r}}}{4\pi\varepsilon_0 r^2} = \frac{\mathbf{p}}{\varepsilon_0}\cdot\nabla\frac{-1}{4\pi r} = -\frac{1}{\varepsilon_0}\nabla\cdot\frac{\mathbf{p}}{4\pi r}, \qquad \mathbf{r} \in \mathbb{R}^3 \setminus \{0\}. \tag{2.39}$$

Shown in Figure 2.1 for the case $\mathbf{p} = p\hat{\mathbf{z}}$ are a few *equipotential lines*, i.e., the loci of points in which the scalar potential takes on the same constant value.

We notice that the total charge of a dipole is zero, whereby we expect the Gauss law to predict a null flux of the displacement vector through the boundary of any region containing the dipole. Indeed, if we take the surface S to be the sphere of radius a centered in origin, we have

$$\hat{\mathbf{r}}\cdot\mathcal{D}(a\hat{\mathbf{r}}) = -\frac{\mathrm{d}}{\mathrm{d}r}\frac{\mathbf{p}\cdot\hat{\mathbf{r}}}{4\pi r^2}\bigg|_{r=a} = \frac{2\mathbf{p}\cdot\hat{\mathbf{r}}}{4\pi a^3} = \frac{|\mathbf{p}|\cos\theta}{2\pi a^3} \tag{2.40}$$

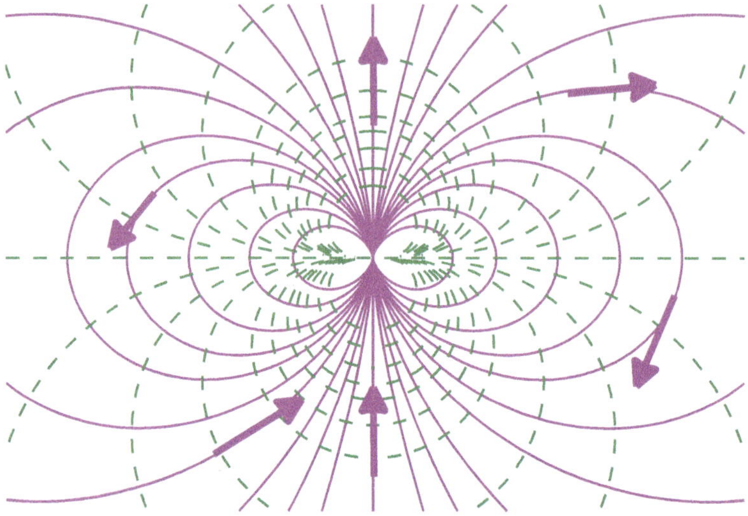

Figure 2.1 Lines of constant $\Phi(\mathbf{r})$ (−−) and streamlines of $\mathcal{E}(\mathbf{r})$ (−) in the xOz plane for an elementary electric dipole with moment $\mathbf{p} = p\hat{\mathbf{z}}$.

having taken the dipole moment oriented along the z-axis for simplicity. By applying (1.16) and integrating in polar spherical coordinates we find

$$\int_S \mathrm{d}S\, \hat{\mathbf{r}} \cdot \mathcal{D}(a\hat{\mathbf{r}}) = 2\pi \int_0^\pi \mathrm{d}\theta\, \sin\theta \frac{|\mathbf{p}|\cos\theta}{2\pi a} = 0 \tag{2.41}$$

as anticipated.

Finally, the potential of an elementary dipole located at a point $\mathbf{r}'$ reads

$$\Phi(\mathbf{r}) = \frac{\mathbf{p}\cdot(\mathbf{r}-\mathbf{r}')}{4\pi\varepsilon_0|\mathbf{r}-\mathbf{r}'|^3} = \frac{\mathbf{p}}{\varepsilon_0}\cdot\nabla'\frac{1}{4\pi|\mathbf{r}-\mathbf{r}'|}, \qquad \mathbf{r}\in\mathbb{R}^3\setminus\{\mathbf{r}'\} \tag{2.42}$$

which follows by replacing $\mathbf{r}$ with $\mathbf{r}-\mathbf{r}'$ in (2.39). The symbol ∇' indicates that the derivatives are taken with respect to the primed coordinates.

The electric field produced by an electric dipole with moment $\mathbf{p} = p\hat{\mathbf{z}}$ located in the origin can be computed through (2.15) applied to (2.39). Carrying out the derivatives in spherical coordinates (r, ϑ, φ) with the help of (A.28) yields

$$\mathcal{E}(\mathbf{r}) = \frac{p}{4\pi\varepsilon_0 r^3}(2\hat{\mathbf{r}}\cos\vartheta + \hat{\boldsymbol{\vartheta}}\sin\vartheta) = \frac{2\hat{\mathbf{r}}\hat{\mathbf{r}} - \hat{\boldsymbol{\vartheta}}\hat{\boldsymbol{\vartheta}}}{4\pi\varepsilon_0 r^3}\cdot\mathbf{p} = \frac{3\hat{\mathbf{r}}\hat{\mathbf{r}} - \overline{\mathcal{I}}}{4\pi\varepsilon_0 r^3}\cdot\mathbf{p} \tag{2.43}$$

for points $\mathbf{r}\in\mathbb{R}^3\setminus\{0\}$. The expression in the rightmost-hand side is valid in general for a moment $\mathbf{p}$ with arbitrary orientation. Streamlines of the electric field [10, Section 2.6], [2, Section 4.1], are superimposed to equipotential lines in Figure 2.1.

(End of Example 2.5)

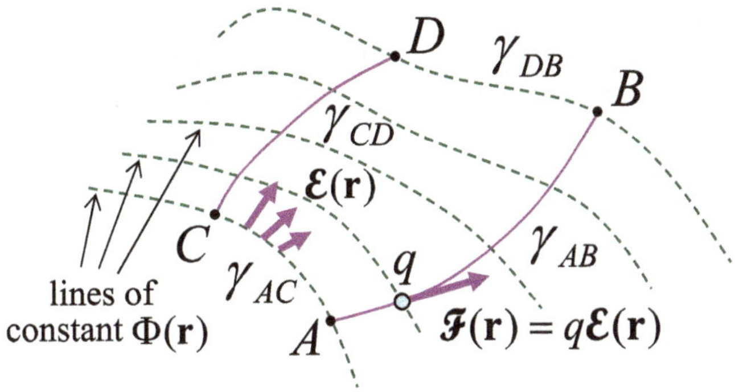

Figure 2.2 A test point charge immersed in an electrostatic field.

2.3 Physical meaning of the scalar potential

So far we have regarded the scalar potential as no more than an auxiliary field which facilitates the solution of (2.3) and (2.4). However, in view of the very definition (2.15) $\Phi(\mathbf{r})$ comes with a physical meaning of its own [13, Section 3.2.1], [14, Section 3.2], [6, Section 2.4.1], [15, Section 4.2], [16, Section 3.1]. To find out what that is, we consider a thought experiment in which a positive test point charge (see Section 1.1) finds itself in a region of space where an electrostatic field $\mathcal{E}(\mathbf{r})$ exists, as is exemplified in Figure 2.2. The field $\mathcal{E}(\mathbf{r})$ is produced by some distant charges which are inconsequential for the present discussion.

The dashed curves in Figure 2.2 represent *equipotential lines*. Since the electric field is defined as the negative of the gradient of $\Phi(\mathbf{r})$, it follows that $\mathcal{E}(\mathbf{r})$ is perpendicular everywhere to the equipotential lines. The *lines of force* or field lines or streamlines are the curves whose tangent vector is parallel to $\mathcal{E}(\mathbf{r})$ at each point. Therefore, lines of force and equipotential lines are everywhere perpendicular to each other [2, Section 18.2].

Next, we observe that the charge is subjected to the action of the electrostatic force $\mathcal{F}(\mathbf{r}) = q\mathcal{E}(\mathbf{r})$. The latter is a special instance of the Lorentz force (1.4) in the case where no magnetic fields are present. Furthermore, if the charge is positive, then $\mathcal{F}(\mathbf{r})$ has the same direction and orientation of the electric field. We can ask ourselves what is the work W_{AB} done by $\mathcal{F}(\mathbf{r})$ as it pushes the charge along the line of force γ_{AB}. This is a simple calculation [17, Chapters 13 and 14], namely,

$$W_{AB} := \int_{\gamma_{AB}} \mathrm{d}s\, \hat{\mathbf{s}}(\hat{\mathbf{r}}) \cdot \mathcal{F}(\mathbf{r}) = q \int_{\gamma_{AB}} \mathrm{d}s\, \hat{\mathbf{s}}(\hat{\mathbf{r}}) \cdot \mathcal{E}(\mathbf{r})$$

$$= -q \int_{\gamma_{AB}} \mathrm{d}s\, \hat{\mathbf{s}}(\hat{\mathbf{r}}) \cdot \nabla\Phi(\mathbf{r}) = q[\Phi(\mathbf{r}_A) - \Phi(\mathbf{r}_B)] > 0 \tag{2.44}$$

where $\mathbf{r}_A$ and $\mathbf{r}_B$ are the position vectors of the endpoints of γ_{AB}. From this result we draw two important conclusions.

First of all, we see that the scalar potential is somehow associated with the work done by the electrostatic force. However, not the scalar potential *per se* but rather the potential difference between two points is proportional to the work! This is a key aspect, because the scalar potential is defined within an arbitrary additive constant, as is evident from (2.15), but luckily when we compute W_{AB} the arbitrary constant has no effect. Secondly, since the work is positive and $q > 0$ by assumption, then the potential in B is smaller than the potential in A. This means the $\Phi(\mathbf{r})$ decreases in the direction of the electric field, an observation in agreement with (2.29) and (2.14). Notice that we would reach the same conclusions if the charge was negative, in which case the electrostatic force would be oriented oppositely with respect to $\mathcal{E}(\mathbf{r})$ and would drag the charge from B to A.

Finally, we remark that the result (2.44) does not change if the charge q is dragged along other field lines, so long as they begin and end on the same pair of equipotential surfaces, e.g., between the points C and D in Figure 2.2. This is a consequence of (2.2) applied to the closed curve $\gamma := \gamma_{AB} \cup \gamma_{DB} \cup \gamma_{CD} \cup \gamma_{AC}$, viz.,

$$\begin{aligned} 0 &= \oint_{\gamma} \mathrm{d}s\, \hat{\mathbf{s}}(\mathbf{r}) \cdot \mathcal{E}(\mathbf{r}) \\ &= \int_{\gamma_{AB}} \mathrm{d}s\, \hat{\mathbf{s}}(\mathbf{r}) \cdot \mathcal{E}(\mathbf{r}) - \int_{\gamma_{DB}} \mathrm{d}s\, \hat{\mathbf{s}}(\mathbf{r}) \cdot \mathcal{E}(\mathbf{r}) - \int_{\gamma_{CD}} \mathrm{d}s\, \hat{\mathbf{s}}(\mathbf{r}) \cdot \mathcal{E}(\mathbf{r}) - \int_{\gamma_{AC}} \mathrm{d}s\, \hat{\mathbf{s}}(\mathbf{r}) \cdot \mathcal{E}(\mathbf{r}) \end{aligned} \tag{2.45}$$

because the line integrals along γ_{DB} and γ_{AC} vanish since the electric field is perpendicular to the equipotential lines.

2.4 Boundary conditions for the scalar potential

In Section 1.7 we went to great lengths to obtain the differential form of Maxwell's equations in the presence of discontinuous constitutive parameters. In the process we came up with the set of conditions that the electromagnetic entities must satisfy at the (smooth) interface between regions in which the constitutive parameters are indeed differentiable. Under the same assumptions, we wish to derive the jump conditions for the electrostatic entities, namely, $\mathcal{D}$, $\mathcal{E}$ and Φ.

It is not necessary to start with (2.1) and (2.2) and repeat all the steps again because the matching conditions (1.155) and (1.144) apply for a general time-dependence and hence, in particular, remain valid for electrostatic fields at an interface separating two dielectric media. We list the results

$$\hat{\mathbf{n}}(\mathbf{r}) \cdot [\mathcal{D}_1(\mathbf{r}) - \mathcal{D}_2(\mathbf{r})] = \varrho_S(\mathbf{r}), \qquad \mathbf{r} \in \partial V \tag{2.46}$$

$$[\mathcal{E}_1(\mathbf{r}) - \mathcal{E}_2(\mathbf{r})] \times \hat{\mathbf{n}}(\mathbf{r}) = 0, \qquad \mathbf{r} \in \partial V \tag{2.47}$$

where ∂V is the interface between two regions V_1 and V_2 filled with different dielectric media as in Figure 1.13. The jump conditions for the scalar potential can be deduced by choosing a constitutive relationship and inserting (2.15) into (2.46) and (2.47).

If the dielectric media filling V_1 and V_2 are isotropic, (2.46) yields

$$\varrho_S(\mathbf{r}) = -\hat{\mathbf{n}}(\mathbf{r}) \cdot [\varepsilon_1(\mathbf{r})\nabla\Phi_1(\mathbf{r}) - \varepsilon_2(\mathbf{r})\nabla\Phi_2(\mathbf{r})] = \varepsilon_2(\mathbf{r})\frac{\partial\Phi_2}{\partial\hat{\mathbf{n}}} - \varepsilon_1(\mathbf{r})\frac{\partial\Phi_1}{\partial\hat{\mathbf{n}}}, \qquad \mathbf{r} \in \partial V \tag{2.48}$$

i.e., the normal derivative of the scalar potential undergoes a jump due to the discontinuity of the permittivity across ∂V and the possible presence of surface charges thereon.

If the dielectric media that fill V_1 and V_2 are anisotropic, (2.46) provides a more general matching condition

$$\varrho_S(\mathbf{r}) = \hat{\mathbf{n}}(\mathbf{r}) \cdot \left[\overline{\varepsilon}_2(\mathbf{r}) \cdot \nabla\Phi_2(\mathbf{r}) - \overline{\varepsilon}_1(\mathbf{r}) \cdot \nabla\Phi_1(\mathbf{r})\right], \qquad \mathbf{r} \in \partial V. \tag{2.49}$$

From the condition on the electric fields (2.47) we derive

$$\begin{aligned} 0 &= \hat{\mathbf{n}}(\mathbf{r}) \times \{[\mathcal{E}_1(\mathbf{r}) - \mathcal{E}_2(\mathbf{r})] \times \hat{\mathbf{n}}(\mathbf{r})\} \\ &= \hat{\mathbf{n}}(\mathbf{r}) \times \{[\nabla\Phi_2(\mathbf{r}) - \nabla\Phi_1(\mathbf{r})] \times \hat{\mathbf{n}}(\mathbf{r})\} \\ &= \nabla_s\Phi_2(\mathbf{r}) - \nabla_s\Phi_1(\mathbf{r}) = \nabla_s[\Phi_2(\mathbf{r}) - \Phi_1(\mathbf{r})], \qquad \mathbf{r} \in \partial V \end{aligned} \tag{2.50}$$

where $\nabla_s\{\bullet\}$ indicates the surface gradient on ∂V (cf. Appendix A.5). The last step is permitted in view of the linearity of the gradient operator and the assumption that $\Phi_1(\mathbf{r})$ and $\Phi_2(\mathbf{r})$ can be separately differentiated over ∂V. In fact, the normal derivatives of $\Phi_1(\mathbf{r})$ and $\Phi_2(\mathbf{r})$ on either side of ∂V are different according to (2.48) or (2.49). From (2.50) we conclude that the jump of $\Phi(\mathbf{r})$ across ∂V must be equal to some real constant, say,

$$\Phi_2(\mathbf{r}) - \Phi_1(\mathbf{r}) = M, \qquad \mathbf{r} \in \partial V \tag{2.51}$$

and it is a matter of convenience to set $M = 0$. However, in case more than one material interface is considered, then ∂V becomes surface-wise multiply connected (loosely speaking, comprised of different separated pieces). Suppose $\partial V := \cup_n \partial V_n$, then (2.50) holds separately on each boundary ∂V_n and

$$\Phi_{2n}(\mathbf{r}) - \Phi_{1n}(\mathbf{r}) = M_n, \qquad \mathbf{r} \in \partial V_n \tag{2.52}$$

i.e., the jump of $\Phi(\mathbf{r})$ may be different for each connected component ∂V_n of ∂V.

Next, we examine the situation in which the medium that fills V_2 is a conductor, whereas V_1 still contains a dielectric ($\sigma_1(\mathbf{r}) = 0$ for $\mathbf{r} \in V_1$). The boundary conditions for the field entities are given by (1.182) and (1.180). The corresponding jump conditions for the scalar potential read

$$-\varepsilon_1(\mathbf{r})\frac{\partial \Phi_1}{\partial \hat{\mathbf{n}}} = \varrho_S(\mathbf{r}), \qquad \mathbf{r} \in \partial V \tag{2.53}$$

$$-\hat{\mathbf{n}}(\mathbf{r}) \cdot \bar{\varepsilon}_1(\mathbf{r}) \cdot \nabla \Phi_1 = \varrho_S(\mathbf{r}), \qquad \mathbf{r} \in \partial V \tag{2.54}$$

for isotropic and anisotropic dielectric in V_1, respectively.

Since the electrostatic field is zero inside a conductor (cf. discussion on page 37), (2.15) says that

$$\Phi_2(\mathbf{r}) = \Phi_{02}, \qquad \mathbf{r} \in V_2 \tag{2.55}$$

i.e., the scalar potential is *constant* in a conductor. On the other hand condition (1.180) implies

$$0 = \hat{\mathbf{n}}(\mathbf{r}) \times [\mathcal{E}_1(\mathbf{r}) \times \hat{\mathbf{n}}(\mathbf{r})] = -\hat{\mathbf{n}}(\mathbf{r}) \times [\nabla \Phi_1(\mathbf{r}) \times \hat{\mathbf{n}}(\mathbf{r})] = -\nabla_s \Phi_1(\mathbf{r}), \qquad \mathbf{r} \in \partial V \tag{2.56}$$

whence we conclude that

$$\Phi_1(\mathbf{r}) = \Phi_{01}, \qquad \mathbf{r} \in \partial V \tag{2.57}$$

where Φ_{01} is yet another constant. The constants Φ_{01} and Φ_{02} can be different, and thus Φ may exhibit a jump across ∂V (cf. Section 2.9). If more than one conductor is present, the surface ∂V is surface-wise multiply connected. As a result, the jumps of the scalar potential can be different on each part ∂V_n of ∂V.

2.5 Uniqueness of the static solutions

2.5.1 Scalar potential

We wish to determine whether and under which conditions the solution to the Poisson equations (2.17), (2.25) and (2.26) may be unique. The result goes by *uniqueness theorem* for the electrostatic (scalar) potential and, when it applies, it guarantees that, regardless of the approach followed to

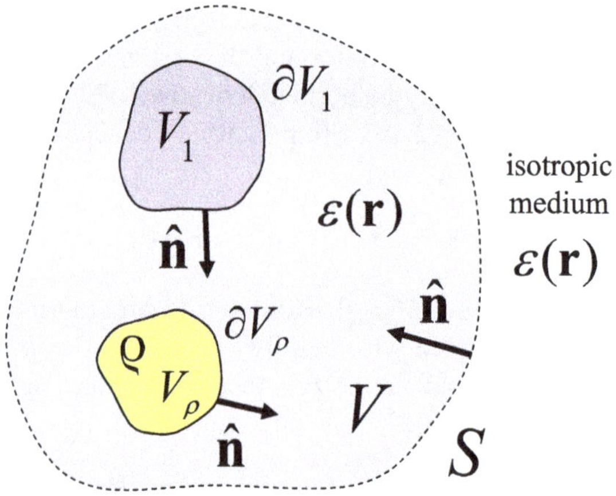

Figure 2.3 For proving uniqueness of solutions to the Poisson equation.

solve the Poisson equation, if we find a solution, that is the only possible one [1, 6]. Besides, since the electric field is related to the scalar potential through (2.15), the uniqueness of $\Phi(\mathbf{r})$ evidently implies that $\mathcal{E}(\mathbf{r})$ and $\mathcal{D}(\mathbf{r})$ are unique too. As is customary, we shall address the proof of uniqueness by contradiction, namely, we assume that two solutions to the Poisson equation exist and show that this hypothesis leads to contradictory conclusions.

We shall derive the result for a bounded region of space first. We consider a (connected) volume $V_1 \subset \mathbb{R}^3$ filled with a dielectric medium or a conductor immersed in an unbounded isotropic dielectric medium endowed with permittivity $\varepsilon(\mathbf{r})$ (see Figure 2.3). We call V the region of space surrounded by ∂V_1 and a smooth closed surface S, so that the boundary of V is $\partial V := S \cup \partial V_1$; the unit normal $\hat{\mathbf{n}}(\mathbf{r})$ for $\mathbf{r} \in \partial V$ points inwards V. Besides, electric charges with density $\varrho(\mathbf{r})$ are confined in a bounded volume $V_\varrho \subset V$.

We suppose that $\Phi_1(\mathbf{r})$ and $\Phi_2(\mathbf{r})$ are two solutions to (2.25) for $\mathbf{r} \in V$, i.e.,

$$\nabla \cdot [\varepsilon(\mathbf{r})\nabla\Phi_1(\mathbf{r})] = -\varrho(\mathbf{r}), \qquad \mathbf{r} \in V \tag{2.58}$$

$$\nabla \cdot [\varepsilon(\mathbf{r})\nabla\Phi_2(\mathbf{r})] = -\varrho(\mathbf{r}), \qquad \mathbf{r} \in V \tag{2.59}$$

and we define the difference $\Phi_0(\mathbf{r}) := \Phi_1(\mathbf{r}) - \Phi_2(\mathbf{r})$. Since the Poisson equation is linear, we obtain an equation for $\Phi_0(\mathbf{r})$ by subtracting (2.58) and (2.59) side by side, viz.,

$$\nabla \cdot [\varepsilon(\mathbf{r})\nabla\Phi_0(\mathbf{r})] = 0, \qquad \mathbf{r} \in V \tag{2.60}$$

and in the process we have lost the effect of the sources. Therefore, showing that $\Phi_1(\mathbf{r})$ and $\Phi_2(\mathbf{r})$ are actually the same scalar potential is equivalent to showing that the homogeneous equation (2.60) — also known as the **Laplace** equation [18, Chapter 9], [1, Chapters 2, 3] — admits only the trivial solution $\Phi_0(\mathbf{r}) = 0$ for $\mathbf{r} \in V$. To this purpose we exploit the differential identity (H.51)

$$\nabla \cdot [\Phi_0(\mathbf{r})\varepsilon(\mathbf{r})\nabla\Phi_0(\mathbf{r})] = \Phi_0(\mathbf{r})\nabla \cdot [\varepsilon(\mathbf{r})\nabla\Phi_0(\mathbf{r})] + \nabla\Phi_0(\mathbf{r}) \cdot [\varepsilon(\mathbf{r})\nabla\Phi_0(\mathbf{r})] = \varepsilon(\mathbf{r})|\nabla\Phi_0(\mathbf{r})|^2 \tag{2.61}$$

where the last passage is a consequence of (2.60). Next, we integrate both sides of this equation over V to get

$$\int_V \mathrm{d}V\, \nabla \cdot [\Phi_0(\mathbf{r})\varepsilon(\mathbf{r})\nabla\Phi_0(\mathbf{r})] = \int_V \mathrm{d}V\, \varepsilon(\mathbf{r})|\nabla\Phi_0(\mathbf{r})|^2 \tag{2.62}$$

and we would like to apply the Gauss theorem (A.53) on the first integral. By following the same steps outlined in Section 1.2.2 for the derivation of the local electric Gauss law, we find out that this is possible provided the vector field $\Phi_0(\mathbf{r})\varepsilon(\mathbf{r})\nabla\Phi_0(\mathbf{r})$ is continuously differentiable separately in $V \setminus \overline{V}_\varrho$ and V_ϱ and $\hat{\mathbf{n}}(\mathbf{r}) \cdot \Phi_0(\mathbf{r})\varepsilon(\mathbf{r})\nabla\Phi_0(\mathbf{r})$ is continuous across ∂V_ϱ. But this is certainly true if $\hat{\mathbf{n}}(\mathbf{r}) \cdot \mathcal{D}_0(\mathbf{r})$ is continuous ($\mathcal{D}_0(\mathbf{r})$ is the difference displacement) and $\Phi_0(\mathbf{r})$ is continuous. The former condition was already invoked in Section 1.2.2 to obtain the local form of the Gauss law, whereas the latter requirement is reasonable because ∂V_ϱ does not constitute a material interface. In conclusion, we may write

$$-\int_{\partial V} \mathrm{d}S\, \Phi_0(\mathbf{r})\varepsilon(\mathbf{r})\hat{\mathbf{n}}(\mathbf{r}) \cdot \nabla\Phi_0(\mathbf{r}) = \int_V \mathrm{d}V\, \varepsilon(\mathbf{r})|\nabla\Phi_0(\mathbf{r})|^2 \tag{2.63}$$

where the minus sign is a consequence of the orientation of the normal $\hat{\mathbf{n}}$. We observe that the surface integral vanishes identically if one of the following conditions is true

$$\Phi_0(\mathbf{r}) = 0, \qquad \mathbf{r} \in \partial V \tag{2.64}$$

$$\frac{\partial}{\partial \hat{\mathbf{n}}}\Phi_0(\mathbf{r}) = 0, \qquad \mathbf{r} \in \partial V \tag{2.65}$$

that is, either Φ_0 or its normal derivative must be zero on the boundary of the region of interest. Actually, we might as well assume the difference potential zero on ∂V_1 and the normal derivative zero on S, or the other way round. We postpone for a moment the interpretation and the feasibility of these requirements and proceed with the proof.

If the surface integral vanishes, (2.63) leads to the requirement

$$\int_V \mathrm{d}V\, \varepsilon(\mathbf{r})|\nabla\Phi_0(\mathbf{r})|^2 = 0 \tag{2.66}$$

which in view of (1.263) means that the electric energy stored in the volume V and associated with the difference field must be zero. In fact, we could have arrived at the same intermediate result by applying (2.19) to the difference potential $\Phi_0(\mathbf{r})$ in V. On a side note, since $\varepsilon(\mathbf{r}) > 0$, the volume integral in (2.66) can be interpreted as the squared norm of the vector field $\nabla\Phi_0(\mathbf{r})$ in the volume V. Therefore, (2.66) requires the norm of $\nabla\Phi_0(\mathbf{r})$ to be zero and hence this implies

$$\nabla\Phi_0(\mathbf{r}) = 0, \qquad \mathbf{r} \in V \tag{2.67}$$

by the very definition of norm. We already know that the latter is satisfied if $\Phi_0(\mathbf{r})$ is equal to some constant throughout V. What can we say about this constant? It depends on which condition we choose between (2.64) and (2.65).

If we assume that (2.64) holds, then $\Phi_0(\mathbf{r})$ is zero on ∂V and, since it is a constant in V, $\Phi_0(\mathbf{r})$ vanishes everywhere in V. Still, if the difference potential is identically null, then $\Phi_1(\mathbf{r}) = \Phi_2(\mathbf{r})$ for $\mathbf{r} \in V$ — contrary to our assumption — and the solution to the Poisson equation (2.25) is unique for $\mathbf{r} \in V$.

If we adopt (2.65) for the behavior of $\Phi_0(\mathbf{r})$ on the boundary, we can only conclude that $\Phi_0(\mathbf{r}) = \Phi_0$, a constant, for $\mathbf{r} \in V$. Therefore, (2.65) allows us to determine the scalar potential up to an arbitrary constant, i.e., $\Phi_1(\mathbf{r}) = \Phi_2(\mathbf{r}) + \Phi_0$ and hence, strictly speaking, the solution to (2.25) is not unique. This is not a critical issue, because on the one hand Φ_0 has not effect on the electric field — which is thus unique — and on the other the work done by electric forces is related to the potential difference rather than the potential (Section 2.3).

Lastly, we need to investigate whether or not (2.64) and (2.65) are reasonable at all. In light of the definition of the difference potential, (2.64) implies that the two speculated solutions $\Phi_1(\mathbf{r})$ and $\Phi_2(\mathbf{r})$ take on the same values for $\mathbf{r} \in \partial V$. In like manner, (2.65) requires that the normal derivative of $\Phi_1(\mathbf{r})$ and $\Phi_2(\mathbf{r})$ be assigned for $\mathbf{r} \in \partial V$. In summary, we see that uniqueness is achieved (possibly up to an arbitrary additive constant, that is) if the Poisson equation is supplemented with conditions on the potential or its normal derivative on the boundary of the domain of interest. In symbols, the problems

$$\begin{cases} \nabla \cdot [\varepsilon(\mathbf{r})\nabla\Phi(\mathbf{r})] = -\varrho(\mathbf{r}), & \mathbf{r} \in V \\ \Phi(\mathbf{r}) = f(\mathbf{r}), & \mathbf{r} \in \partial V \end{cases} \tag{2.68}$$

$$\begin{cases} \nabla \cdot [\varepsilon(\mathbf{r})\nabla\Phi(\mathbf{r})] = -\varrho(\mathbf{r}), & \mathbf{r} \in V \\ \dfrac{\partial}{\partial \hat{\mathbf{n}}}\Phi(\mathbf{r}) = g(\mathbf{r}), & \mathbf{r} \in \partial V \end{cases} \tag{2.69}$$

admit a unique solution. The requirement on the potential for $\mathbf{r} \in \partial V$ is called a *Dirichlet boundary condition*, whereas the constraint on the normal derivative is called a *Neumann boundary condition* [1, 11, 13, 19].

It is worthwhile mentioning that (2.69) may not have a solution at all if $\varrho(\mathbf{r})$ and $g(\mathbf{r})$ are arbitrarily chosen because, in effect, the two of them are not independent. If we integrate the Poisson equation over the bounded region of interest, apply the Gauss theorem (in Figure 2.3 the unit normal points inwards V) and invoke the boundary condition on ∂V, we find [19, Section 4.5], [20, Section 4.2.1]

$$\int_{V_\varrho} \mathrm{d}V\, \varrho(\mathbf{r}) = -\int_V \mathrm{d}V\, \nabla \cdot [\varepsilon(\mathbf{r})\nabla\Phi(\mathbf{r})] = \int_{\partial V} \mathrm{d}S\; \varepsilon(\mathbf{r})\frac{\partial}{\partial \hat{\mathbf{n}}}\Phi(\mathbf{r}) = \int_{\partial V} \mathrm{d}S\; \varepsilon(\mathbf{r})g(\mathbf{r}) \tag{2.70}$$

whence

$$\int_{\partial V} \mathrm{d}S\; \varepsilon(\mathbf{r})g(\mathbf{r}) = \int_{V_\varrho} \mathrm{d}V\, \varrho(\mathbf{r}) \tag{2.71}$$

which is a necessary condition for the solvability of (2.69). From a physical viewpoint (2.71) stems from the global Gauss law (1.16), that is, the flux of $\mathcal{D}(\mathbf{r})$ through ∂V must equal the charge enclosed in V.

We now extend the proof of uniqueness to the case where the region V is unbounded by letting the surface S recede to infinity. In fact, we can assume S to be a sphere of radius a and take the limit as $a \to +\infty$. Thus, the only step we need to revise is the assumption that the surface integral over S in (2.63) is zero *and* that the energy integral remains finite. But we know this to be true from (2.22) on account of the estimates (2.20) and (2.21) applied to the difference potential.

Thus, the solution to the Poisson equation is unique in an unbounded region, provided the potential decays at least as the inverse of the distance away from the sources. In symbols, the exterior problems

$$\begin{cases} \nabla \cdot [\varepsilon(\mathbf{r})\nabla\Phi(\mathbf{r})] = -\varrho(\mathbf{r}), & \mathbf{r} \in \mathbb{R}^3 \setminus \overline{V}_1 \\ \Phi(\mathbf{r}) = f_1(\mathbf{r}), & \mathbf{r} \in \partial V_1 \\ \Phi(\mathbf{r}) = \mathcal{O}\left(\dfrac{1}{|\mathbf{r}|}\right), & |\mathbf{r}| \to +\infty \end{cases} \tag{2.72}$$

and

$$\begin{cases} \nabla \cdot [\varepsilon(\mathbf{r})\nabla\Phi(\mathbf{r})] = -\varrho(\mathbf{r}), & \mathbf{r} \in \mathbb{R}^3 \setminus \overline{V}_1 \\ \dfrac{\partial}{\partial \hat{\mathbf{n}}}\Phi(\mathbf{r}) = g_1(\mathbf{r}), & \mathbf{r} \in \partial V_1 \\ \Phi(\mathbf{r}) = \mathcal{O}\left(\dfrac{1}{|\mathbf{r}|}\right), & |\mathbf{r}| \to +\infty \end{cases} \tag{2.73}$$

have a unique solution, though we need to verify the validity of the conditions on ∂V_1.

If the medium filling V_1 is a dielectric, then we know from (2.51) that the potential can be made continuous through ∂V_1; alternatively, (2.48) holds with $\varrho_S(\mathbf{r}) = 0$. In practice, we need to solve two separate problems, namely, one in V_1 and another in $\mathbb{R}^3 \setminus \overline{V}_1$ and then join the solutions on ∂V_1 by enforcing both (2.48) and (2.51).

If the medium filling V_1 is a conductor, we already know that the potential is constant in V_1 and on ∂V_1. The normal derivative of the potential is related to the charge on the surface of the conductor, as is prescribed by (2.53). Therefore, the problems

$$\begin{cases} \nabla \cdot [\varepsilon(\mathbf{r})\nabla\Phi(\mathbf{r})] = -\varrho(\mathbf{r}), & \mathbf{r} \in \mathbb{R}^3 \setminus \overline{V}_1 \\ \Phi(\mathbf{r}) = \Phi_{01}, & \mathbf{r} \in \partial V_1 \\ \Phi(\mathbf{r}) = \mathcal{O}\left(\dfrac{1}{|\mathbf{r}|}\right), & |\mathbf{r}| \to +\infty \end{cases} \tag{2.74}$$

$$\begin{cases} \nabla \cdot [\varepsilon(\mathbf{r})\nabla\Phi(\mathbf{r})] = -\varrho(\mathbf{r}), & \mathbf{r} \in \mathbb{R}^3 \setminus \overline{V}_1 \\ -\varepsilon(\mathbf{r})\dfrac{\partial}{\partial \hat{\mathbf{n}}}\Phi(\mathbf{r}) = \varrho_S(\mathbf{r}), & \mathbf{r} \in \partial V_1 \\ \Phi(\mathbf{r}) = \mathcal{O}\left(\dfrac{1}{|\mathbf{r}|}\right), & |\mathbf{r}| \to +\infty \end{cases} \tag{2.75}$$

have a unique solution.

There are no additional requirements concerning the two Neumann problems (2.73) and (2.75) inasmuch as the region of interest is unbounded. Indeed, integrating the Poisson equation in (2.73) over $B(0,a) \setminus \overline{V}_1$, applying the Gauss theorem, and letting the radius a grow infinitely large yields

$$\begin{aligned} \int\limits_{V_\varrho} \mathrm{d}V\, \varrho(\mathbf{r}) &= -\int\limits_{B(0,a)\setminus V_1} \mathrm{d}V\, \nabla \cdot [\varepsilon(\mathbf{r})\nabla\Phi(\mathbf{r})] = \int\limits_{\partial V_1} \mathrm{d}S\; \varepsilon(\mathbf{r})\frac{\partial}{\partial \hat{\mathbf{n}}}\Phi(\mathbf{r}) + \int\limits_{\partial B} \mathrm{d}S\; \varepsilon(\mathbf{r})\frac{\partial}{\partial \hat{\mathbf{n}}}\Phi(\mathbf{r}) \\ &\xrightarrow[a\to+\infty]{} \int\limits_{\partial V_1} \mathrm{d}S\; \varepsilon(\mathbf{r})g_1(\mathbf{r}) - Q_\infty \end{aligned} \tag{2.76}$$

whence

$$\int\limits_{V_\varrho} \mathrm{d}V\, \varrho(\mathbf{r}) = \int\limits_{\partial V_1} \mathrm{d}S\; \varepsilon(\mathbf{r})g_1(\mathbf{r}) - Q_\infty \tag{2.77}$$

where $-Q_\infty$ is the limiting constant value of the integral over ∂B. Since Q_∞ is not known a priori we may, in fact, assign $\varrho(\mathbf{r})$ and $g_1(\mathbf{r})$ at will, and (2.77) sets the value of Q_∞. Similar arguments hold for (2.75).

The extension of these results to the case of an anisotropic medium filling V or $\mathbb{R}^3 \setminus \overline{V}_1$ is not difficult and left to the Reader.

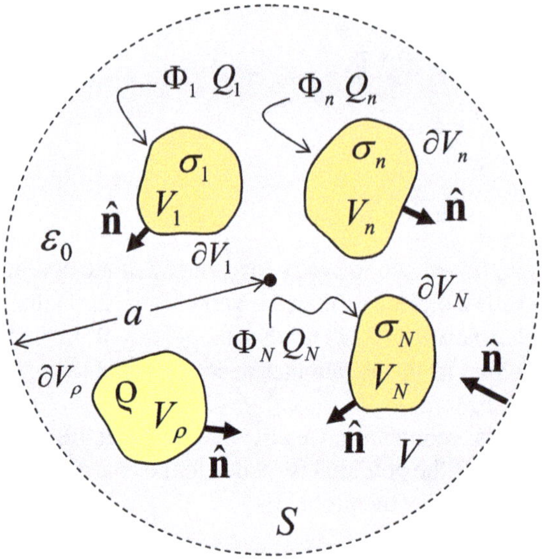

Figure 2.4 For proving uniqueness of solutions to the Poisson equation in the presence of N conducting bodies.

A classic problem in electrostatics consists of finding the potential and then the electric field due to a charge distribution $\varrho(\mathbf{r})$, $\mathbf{r} \in V_\varrho \subset \mathbb{R}^3$, in the presence of N conducting bodies [1, 19]. As is suggested in Figure 2.4, the conductors occupy the normal domains V_n, $n = 1, \ldots, N$, are endowed with conductivities σ_n, and are immersed in free space. We may wonder, then, under which conditions the solution to the relevant Poisson equation is unique.

We already know that $\Phi(\mathbf{r})$ is constant on the multiply-connected smooth boundary $\cup_n \partial V_n$, though on each surface ∂V_n the potential may have different values, say, Φ_n. Besides, the conductors may carry charges Q_n which are confined to the surfaces ∂V_n (see discussion on page 28). We define the difference potential $\Phi_0(\mathbf{r})$ as the solution to (2.60) in the unbounded region $V := \mathbb{R}^3 \setminus \cup_n \overline{V}_n$ and let S be a sphere of radius a large enough to enclose all the conductors and the charge region V_ϱ. Then, for the geometry under investigation (2.63) passes over into

$$\int_V \mathrm{d}V\, \varepsilon_0 |\nabla \Phi_0(\mathbf{r})|^2 = \sum_{n=1}^{N} \Phi_{0n} \int_{\partial V_n} \mathrm{d}S\, \hat{\mathbf{n}}(\mathbf{r}) \cdot \mathcal{D}_0(\mathbf{r}) = \sum_{n=1}^{N} \Phi_{0n} Q_{0n} \tag{2.78}$$

since the integral over S vanishes in the limit as $a \to +\infty$ by virtue of (2.20) and (2.21). In particular, Φ_{0n} and Q_{0n} indicate the constant difference potential and the constant difference charge on ∂V_n. Now, we can conclude that the rightmost-hand side vanishes — and hence uniqueness is guaranteed — if either one of the following boundary conditions is true:

(1) the potentials Φ_n, $n = 1, \ldots, N$, are assigned (*Dirichlet problem*), in which case the constants Φ_{0n} are all zero;
(2) the charges Q_n on ∂V_n, $n = 1, \ldots, N$, are assigned (*Neumann problem*), whereby $Q_{0n} = 0$.

We are already familiar with the first condition, but the second calls for some thinking. Indeed, since the constants Q_n do not enter (2.17) directly, how do we enforce the desired requirements on the charges? Well, the way out consists of obtaining a preliminary formal expression for $\Phi(\mathbf{r})$ that we may use to write down the surface charge densities on ∂V_n and, ultimately, the charges Q_n. To this purpose, we consider $N+1$ *auxiliary Dirichlet problems*, namely,

$$\begin{cases} \nabla^2\Psi(\mathbf{r}) = -\dfrac{\varrho(\mathbf{r})}{\varepsilon}, & \mathbf{r} \in V \\ \Psi(\mathbf{r}) = 0, & \mathbf{r} \in \bigcup\limits_{m=1}^{N} \partial V_m \\ \Psi(\mathbf{r}) = \mathcal{O}\left(\dfrac{1}{|\mathbf{r}|}\right), & |\mathbf{r}| \to +\infty \end{cases} \tag{2.79}$$

and

$$\begin{cases} \nabla^2\Upsilon_n(\mathbf{r}) = 0, & \mathbf{r} \in V \\ \Upsilon_n(\mathbf{r}) = \delta_{nm}, & \mathbf{r} \in \partial V_m \\ \Upsilon_n(\mathbf{r}) = \mathcal{O}\left(\dfrac{1}{|\mathbf{r}|}\right), & |\mathbf{r}| \to +\infty \end{cases} \qquad n, m \in \{1, \ldots, N\} \tag{2.80}$$

where

$$\delta_{nm} := \begin{cases} 1, & n = m \\ 0, & n \neq m \end{cases} \tag{2.81}$$

denotes a Kronecker delta. The solution of the first problem yields the electrostatic potential $\Psi(\mathbf{r})$ in V with *homogeneous* Dirichlet boundary conditions, whereas the N remaining ones [11] provide the *dimensionless* scalar fields $\Upsilon_n(\mathbf{r})$ which vanish on the boundary of all conductors except ∂V_n, whereon they equal unity. All in all, according to (2.78), $\Psi(\mathbf{r})$ and $\Upsilon_n(\mathbf{r})$ are *uniquely* determined, since one way or another the potential is specified on $\cup_n \partial V_n$ and at infinity.

Then, since the Laplace operator is linear (Appendix D.3), the solutions to the Poisson equations are *additive*, and hence the potential $\Phi(\mathbf{r})$ which solves (2.17) for $\mathbf{r} \in V$ may be written as

$$\Phi(\mathbf{r}) = \Psi(\mathbf{r}) + \sum_{n=1}^{N} \Phi_n \Upsilon_n(\mathbf{r}), \qquad \mathbf{r} \in V \tag{2.82}$$

where the constants Φ_n are the *unknown* potentials on ∂V_n. It is a straightforward matter to check that the right member of (2.82) equals Φ_n on the boundary of the nth conducting body. Therefore, if all of the Φ_n's were indeed assigned, (2.82) would yield the unique solution to the Dirichlet problem of the N conductors in Figure 2.4.

By contrast, since we do not know the values of the coefficients Φ_n as yet, we go on and invoke the jump condition (1.182) to get the N surface charge densities, viz.,

$$\varrho_{Sn}(\mathbf{r}) = -\varepsilon_0 \hat{\mathbf{n}}(\mathbf{r}) \cdot \nabla\Psi(\mathbf{r}) - \sum_{m=1}^{N} \Phi_m \varepsilon_0 \hat{\mathbf{n}}(\mathbf{r}) \cdot \nabla\Upsilon_m(\mathbf{r}), \qquad \mathbf{r} \in \partial V_n \tag{2.83}$$

on the grounds of (2.82), (2.15) and (1.117). Next, by integrating over each and every conducting surface ∂V_n we obtain

$$Q_n = -\int\limits_{\partial V_n} \mathrm{d}S\, \varepsilon_0 \hat{\mathbf{n}}(\mathbf{r}) \cdot \nabla\Psi(\mathbf{r}) - \sum_{m=1}^{N} \int\limits_{\partial V_n} \mathrm{d}S\, \varepsilon_0 \hat{\mathbf{n}}(\mathbf{r}) \cdot \nabla\Upsilon_m(\mathbf{r}) \Phi_m, \qquad n = 1, \ldots, N \tag{2.84}$$

which, the charges Q_n being assigned, form an algebraic system of rank N to be solved for the potentials Φ_n. Once the latter have been found, the solution to the Neumann problem follows from (2.82). Contrariwise, (2.84) provides the values of the charges in the Dirichlet problem of Figure 2.4. The constants

$$Q_{\varrho n} := -\varepsilon_0 \int_{\partial V_n} \mathrm{d}S \, \hat{\mathbf{n}}(\mathbf{r}) \cdot \nabla \Psi(\mathbf{r}), \qquad n = 1, \ldots, N \tag{2.85}$$

represent the charge induced on the surfaces ∂V_n by the nearby distribution $\varrho(\mathbf{r})$, when the N conductors are held at zero potential. The coefficients

$$C_{nm} := -\varepsilon_0 \int_{\partial V_n} \mathrm{d}S \, \hat{\mathbf{n}}(\mathbf{r}) \cdot \nabla \Upsilon_m(\mathbf{r}), \qquad n, m \in \{1, \ldots, N\} \tag{2.86}$$

form the entries of the $N \times N$ *capacitance matrix* $[C]$ (physical dimensions: farad, F) [21, Section 6.8], [6, Section 2.5.4], [22, Section D.5], [10, Chapter 6], [19, Chapter 4], [23, Chapter 8], [24, Chapter 5]. Correspondingly, the scalar fields $\Upsilon_n(\mathbf{r})$ are called *capacitary potentials* [25]. Interestingly, $[C]$ depends only on the shape and relative position of the conductors, aside from the permittivity of the background dielectric medium (see Example 3.3).

With these positions we can write (2.84) succinctly in matrix form as

$$[Q] = [Q_\varrho] + [C][\Phi] \tag{2.87}$$

where

$$[Q] := \begin{pmatrix} Q_1 \\ \vdots \\ Q_N \end{pmatrix}, \qquad [Q_\varrho] := \begin{pmatrix} Q_{\varrho 1} \\ \vdots \\ Q_{\varrho N} \end{pmatrix}, \qquad [\Phi] := \begin{pmatrix} \Phi_1 \\ \vdots \\ \Phi_N \end{pmatrix} \tag{2.88}$$

are N-dimensional column vectors. Solving (2.87) for $[\Phi]$ and substituting back into (2.82) yields the desired solution to the Neumann problem, viz.,

$$\Phi(\mathbf{r}) = \Psi(\mathbf{r}) + [\Upsilon]^T [C]^{-1} \left([Q] - [Q_\varrho] \right) \qquad \mathbf{r} \in V \tag{2.89}$$

where $[\Upsilon]$ is an abstract column vector which stores the capacitary potentials $\Upsilon_n(\mathbf{r})$ seriatim. We notice that, thanks to (2.87), assigning the charges Q_n is equivalent to prescribing the potentials Φ_n. Therefore, also in the Neumann problem of Figure 2.4 the potential $\Phi(\mathbf{r})$ is entirely determined and there is no room for an arbitrary additive constant, though this is not evident from (2.78). A closed-form instance of (2.82) or (2.89) exists, e.g., for the case of a point charge in the presence of a conducting sphere [1, Section 2.3]. For general geometries, though, we may resort to integral equations and numerical techniques to determine $\Psi(\mathbf{r})$ and the capacitary potentials, but we may also want to formulate the problem of Figure 2.4 directly in terms of integral equations without the intervention of auxiliary potentials [19].

The electrostatic energy stored in the space $V := \mathbb{R}^3 \setminus \cup_n \overline{V}_n$ around the conductors is the sum of two contributions, namely, the energy due to $\varrho(\mathbf{r})$ with the bodies held at zero potential plus the energy due to the charge present on the boundaries ∂V_n. Indeed, by starting with definition (1.263) and using (2.82) we have

$$W_e := \frac{\varepsilon_0}{2} \int_V \mathrm{d}V \, |\mathcal{E}(\mathbf{r})|^2 = \frac{\varepsilon_0}{2} \int_V \mathrm{d}V \left| \nabla \Psi(\mathbf{r}) + \sum_{n=1}^{N} \Phi_n \nabla \Upsilon_n(\mathbf{r}) \right|^2$$

$$
\begin{aligned}
&= \frac{\varepsilon_0}{2}\int_V \mathrm{d}V\,|\nabla\Psi(\mathbf{r})|^2 + \frac{\varepsilon_0}{2}\int_V \mathrm{d}V \left|\sum_{n=1}^{N}\Phi_n\nabla\Upsilon_n(\mathbf{r})\right|^2 + \varepsilon_0\sum_{n=1}^{N}\Phi_n\int_V \mathrm{d}V\,\nabla\Psi(\mathbf{r})\cdot\nabla\Upsilon_n(\mathbf{r}) \\
&= \frac{\varepsilon_0}{2}\int_V \mathrm{d}V\,|\nabla\Psi(\mathbf{r})|^2 + \frac{\varepsilon_0}{2}\int_V \mathrm{d}V \left|\sum_{n=1}^{N}\Phi_n\nabla\Upsilon_n(\mathbf{r})\right|^2 \\
&\quad + \varepsilon_0\sum_{n=1}^{N}\Phi_n\int_V \mathrm{d}V\,\nabla\cdot[\Psi(\mathbf{r})\nabla\Upsilon_n(\mathbf{r})] - \varepsilon_0\sum_{n=1}^{N}\Phi_n\int_V \mathrm{d}V\,\Psi(\mathbf{r})\nabla^2\Upsilon_n(\mathbf{r}) \\
&= \underbrace{\frac{\varepsilon_0}{2}\int_V \mathrm{d}V\,|\nabla\Psi(\mathbf{r})|^2}_{\text{energy due to } \varrho} + \underbrace{\frac{\varepsilon_0}{2}\int_V \mathrm{d}V \left|\sum_{n=1}^{N}\Phi_n\nabla\Upsilon_n(\mathbf{r})\right|^2}_{\text{energy due to } Q_n} \\
&\quad - \varepsilon_0\sum_{n=1}^{N}\Phi_n\int_{\cup_m\partial V_m} \mathrm{d}S\,\Psi(\mathbf{r})\hat{\mathbf{n}}(\mathbf{r})\cdot\nabla\Upsilon_n(\mathbf{r}) - \varepsilon_0\sum_{n=1}^{N}\Phi_n\int_V \mathrm{d}V\,\Psi(\mathbf{r})\nabla^2\Upsilon_n(\mathbf{r})
\end{aligned} \tag{2.90}
$$

having applied (H.51) and the Gauss theorem, and recalled that the relevant integral over the sphere S tends to zero in the limit as $a \to +\infty$. The last two contributions vanish identically because $\Psi(\mathbf{r}) = 0$ for $\mathbf{r} \in \cup_m\partial V_m$ and $\Upsilon_n(\mathbf{r})$ obeys the Laplace equation in V. The other terms can be further manipulated in order to exhibit $\varrho(\mathbf{r})$ and the capacitance matrix explicitly, viz.,

$$
\begin{aligned}
\frac{\varepsilon_0}{2}\int_V \mathrm{d}V\,|\nabla\Psi(\mathbf{r})|^2 &= \frac{\varepsilon_0}{2}\int_V \mathrm{d}V\,\nabla\cdot[\Psi(\mathbf{r})\nabla\Psi(\mathbf{r})] - \frac{\varepsilon_0}{2}\int_V \mathrm{d}V\,\Psi(\mathbf{r})\nabla^2\Psi(\mathbf{r}) \\
&= -\frac{\varepsilon_0}{2}\int_{\cup_n\partial V_n} \mathrm{d}S\,\Psi(\mathbf{r})\hat{\mathbf{n}}(\mathbf{r})\cdot\nabla\Psi(\mathbf{r}) - \frac{\varepsilon_0}{2}\int_V \mathrm{d}V\,\Psi(\mathbf{r})\nabla^2\Psi(\mathbf{r}) \\
&= \frac{1}{2}\int_V \mathrm{d}V\,\varrho(\mathbf{r})\Psi(\mathbf{r})
\end{aligned} \tag{2.91}
$$

and

$$
\begin{aligned}
\frac{\varepsilon_0}{2}\int_V \mathrm{d}V \left|\sum_{n=1}^{N}\Phi_n\nabla\Upsilon_n(\mathbf{r})\right|^2 &= \frac{\varepsilon_0}{2}\sum_{n=1}^{N}\sum_{m=1}^{N}\Phi_n\Phi_m\int_V \mathrm{d}V\,\nabla\Upsilon_n(\mathbf{r})\cdot\nabla\Upsilon_m(\mathbf{r}) \\
&= \frac{\varepsilon_0}{2}\sum_{n=1}^{N}\sum_{m=1}^{N}\Phi_n\Phi_m\int_V \mathrm{d}V\,\left\{\nabla\cdot[\Upsilon_n(\mathbf{r})\cdot\nabla\Upsilon_m(\mathbf{r})] - \Upsilon_n(\mathbf{r})\nabla^2\Upsilon_m(\mathbf{r})\right\} \\
&= \frac{1}{2}\sum_{n=1}^{N}\sum_{m=1}^{N}\Phi_n\Phi_m\left[-\varepsilon_0\int_{\partial V_n} \mathrm{d}S\,\hat{\mathbf{n}}(\mathbf{r})\cdot\nabla\Upsilon_m(\mathbf{r})\right] = \frac{1}{2}[\Phi]^T[C][\Phi]
\end{aligned} \tag{2.92}
$$

on account of (2.79), (2.80) and (2.86). With these results, the energy reads

$$
W_e = \frac{1}{2}\int_V \mathrm{d}V\,\varrho(\mathbf{r})\Psi(\mathbf{r}) + \frac{1}{2}[\Phi]^T[C][\Phi] \tag{2.93}
$$

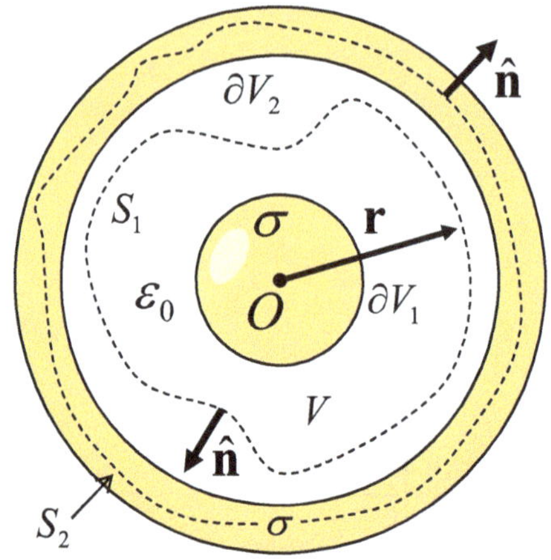

Figure 2.5 Meridian cross-section of an ideal spherical capacitor.

and this ends the proof of the initial statement. Furthermore, it can be shown that $[C]$ is symmetric (see Example 6.9) and from (2.92) we have

$$[\Phi]^T[C][\Phi] = \varepsilon_0 \int_V \mathrm{d}V \left| \sum_{n=1}^{N} \Phi_n \nabla \Upsilon_n(\mathbf{r}) \right|^2 > 0 \tag{2.94}$$

which says that the capacitance matrix is actually symmetric positive definite [26]. For a solitary conductor $[C]$ degenerates into the capacitance C_{11} [27, Section 6.3], and (2.94) states that the capacitance is positive (see Example 3.3). Moreover, in the absence of other charges ($\varrho(\mathbf{r}) = 0$ in Figure 2.4), we can formally solve (2.94) for C_{11}, namely,

$$C_{11} = \frac{\varepsilon_0}{\Phi_1^2} \int_V \mathrm{d}V\, |\Phi_1 \nabla \Upsilon_1(\mathbf{r})|^2 = \frac{2W_e}{\Phi_1^2} \tag{2.95}$$

whence we may also interpret the capacitance as *twice the energy* stored in the field around the conductor when the latter is held at unitary potential ($\Phi_1 = 1$ V).

Example 2.6 (Potential, field and capacitance matrix of a spherical capacitor)

An ideal spherical capacitor (or condenser) is a device comprised of two concentric conducting spheres with radii $r_1 > 0$ and $r_2 > r_1$, as is suggested in Figure 2.5. In practice, though, on the larger sphere a small opening must exist that provides access to the inner one in order to connect the capacitor to a battery with a wire [18, Example 4.3].

Since the spatial region V between the two spheres is bounded, the capacitary potentials $\Upsilon_n(\mathbf{r})$, $n = 1, 2$, solve the Dirichlet problems

$$\begin{cases} \nabla^2 \Upsilon_n(\mathbf{r}) = 0, & \mathbf{r} \in V \\ \Upsilon_n(\mathbf{r}) = \delta_{nm}, & \mathbf{r} \in \partial V_m \end{cases} \qquad n, m \in \{1, 2\} \tag{2.96}$$

where spherical symmetry prompts us to expand the Laplace operator in a system of polar spherical coordinates (Appendices A.1 and A.2) with origin in the center of the capacitor, and to seek solutions which depend only on the radial distance r (see Section 3.5.1 further on).

It is a simple matter to ascertain that the capacitary potentials are

$$\Upsilon_1(\mathbf{r}) = \frac{\frac{1}{r} - \frac{1}{r_2}}{\frac{1}{r_1} - \frac{1}{r_2}}, \qquad \Upsilon_2(\mathbf{r}) = \frac{\frac{1}{r_1} - \frac{1}{r}}{\frac{1}{r_1} - \frac{1}{r_2}}, \qquad r \in [r_1, r_2] \tag{2.97}$$

whence electrostatic potential and field in V read

$$\Phi(\mathbf{r}) = \Phi_1 \frac{\frac{1}{r} - \frac{1}{r_2}}{\frac{1}{r_1} - \frac{1}{r_2}} + \Phi_2 \frac{\frac{1}{r_1} - \frac{1}{r}}{\frac{1}{r_1} - \frac{1}{r_2}}, \qquad r \in [r_1, r_2] \tag{2.98}$$

$$\mathcal{E}(\mathbf{r}) = -\nabla\Phi(\mathbf{r}) = \frac{\Phi_1 - \Phi_2}{\frac{1}{r_1} - \frac{1}{r_2}} \frac{1}{r^2}, \qquad r \in [r_1, r_2] \tag{2.99}$$

where Φ_1 and Φ_2 are the constant potentials of inner and outer sphere.

The surface densities of charge on the spheres are computed by means of the jump condition (1.182), viz.,

$$\varrho_{S1} = -\varepsilon_0 \hat{\mathbf{r}} \cdot \nabla\Phi|_{r=r_1} = \varepsilon_0 \frac{\Phi_1 - \Phi_2}{\frac{1}{r_1} - \frac{1}{r_2}} \frac{1}{r_1^2} \qquad \varrho_{S2} = \varepsilon_0 \hat{\mathbf{r}} \cdot \nabla\Phi|_{r=r_2} = -\varepsilon_0 \frac{\Phi_1 - \Phi_2}{\frac{1}{r_1} - \frac{1}{r_2}} \frac{1}{r_2^2} \tag{2.100}$$

assuming that the medium 'filling' the region V is vacuum. Lastly, integrating the densities over ∂V_1 and ∂V_2 yields

$$Q_1 = 4\pi\varepsilon_0 \frac{\Phi_1 - \Phi_2}{\frac{1}{r_1} - \frac{1}{r_2}} \qquad Q_2 = -4\pi\varepsilon_0 \frac{\Phi_1 - \Phi_2}{\frac{1}{r_1} - \frac{1}{r_2}} \tag{2.101}$$

whence we obtain the 2×2 capacitance matrix of the system, i.e.,

$$[C] = \begin{pmatrix} C_{11} & C_{12} \\ C_{21} & C_{22} \end{pmatrix}, \qquad C_{11} = -C_{12} = -C_{21} = C_{22} = \frac{4\pi\varepsilon_0}{\frac{1}{r_1} - \frac{1}{r_2}} \tag{2.102}$$

with the aid of (2.87). Both the stored charges and the entries of $[C]$ grow larger as the conducting spheres are brought closer together. Besides, comparing (2.99) and (2.101) with (2.14) suggests that we interpret the electric field in V as being produced by an equivalent point charge Q_1 placed in the center of the capacitor.

We notice from (2.101) that the charges Q_1 and Q_2 are proportional to the potential difference and, in effect, are the negative of one another. The reason for this result lies in the fact that one conducting sphere encircles the other, and hence the region V is surface-wise multiply connected.[1] To elucidate, we consider a surface $S_1 \subset V$ which contains the inner sphere. Further, we suppose that the outer sphere ∂V_2 is the inner side of a conducting spherical shell which has some finite non-zero thickness. So, it is possible to pick up a surface S_2 that contains ∂V_2 and is entirely located within the shell (Figure 2.5). Invoking the electric Gauss law (2.1) on S_1 and S_2 gives

$$Q_1 = \varepsilon_0 \int_{S_1} dS\, \hat{\mathbf{n}}(\mathbf{r}) \cdot \mathcal{E}(\mathbf{r}), \qquad Q_1 + Q_2 = \varepsilon_0 \int_{S_2} dS\, \hat{\mathbf{n}}(\mathbf{r}) \cdot \mathcal{E}(\mathbf{r}) = 0 \tag{2.103}$$

because the electrostatic field is null within the conducting shell and hence on S_2.

[1] Indeed, this geometry is of interest for the Helmholtz decomposition of vector fields in bounded regions (Section 8.1.2) and the search for eigenfunctions in cavities (Section 11.1).

Next, by inspection of (2.102) we conclude that the capacitance matrix is singular and, in particular, has rank one. This property is a consequence of the vanishing of the total charge and as such can be proved without knowing the capacitance matrix explicitly. Indeed, from (2.103) and the very definition of $[C]$ we have

$$0 = Q_1 + Q_2 = (C_{11} + C_{21})\Phi_1 + (C_{12} + C_{22})\Phi_2 \tag{2.104}$$

which implies

$$C_{11} + C_{21} = 0 \qquad\qquad C_{12} + C_{22} = 0 \tag{2.105}$$

since Φ_1 and Φ_2 may be chosen at will. Thus, the rows of $[C]$ are linearly dependent, and the determinant thereof vanishes.

From the previous discussion we gather that, if the potentials are prescribed (Dirichlet problem) the charges can be found from (2.101), whereas the relevant Neumann problem cannot be solved univocally. Even if Q_1 and Q_2 are assigned in agreement with (2.103), still only the potential difference $\Phi_1 - \Phi_2$ can be obtained. This is equivalent to specifying the potential of one conductor with respect to the potential of the other one taken as a reference. Nevertheless, the electric field is unambiguously determined, as is attested by (2.99).

The proof that the total charge must vanish is independent of the shape of the conductors and can be extended to $N - 1$ conducting bodies all contained inside a closed surface S_N flush with yet another conductor. In which instance, (2.105) generalizes to N equations which give one row of $[C]$ as a linear combinations of the remaining ones. All in all, only $N - 1$ potentials can be determined with respect to the potential of one conductor taken as reference.

(End of Example 2.6)

2.5.2 *Electrostatic field*

The conditions derived in the previous section for the uniqueness of the scalar potential apply, in fact, to $\Phi(\mathbf{r})$ and the normal derivative thereof on the boundary of the region of concern. The requirement on the normal derivative of $\Phi(\mathbf{r})$ is readily translated into a condition for the normal component of $\mathcal{E}(\mathbf{r})$, though this conclusion is somewhat unsatisfactory. For instance, we do not know what role is played by the tangential component of $\mathcal{E}(\mathbf{r})$ on the boundary of the solution domain.

As a matter of fact, we can address the issue of uniqueness for the electrostatic equations (2.3) and (2.4) directly with the aid of both a scalar potential and, more generally, a vector potential. To this purpose, we consider different type of domains $V \subseteq \mathbb{R}^3$ in free space and determine the conditions for the solution to be unique. Since the proof by contradiction requires allowing for two distinct solutions $\mathcal{E}_1(\mathbf{r})$ and $\mathcal{E}_2(\mathbf{r})$, we define the difference electrostatic field $\mathcal{E}_0 := \mathcal{E}_1(\mathbf{r}) - \mathcal{E}_2(\mathbf{r})$ which satisfies the homogeneous set of electrostatic equations

$$\nabla \times \mathcal{E}_0(\mathbf{r}) = 0, \qquad \mathbf{r} \in V \tag{2.106}$$

$$\nabla \cdot \mathcal{E}_0(\mathbf{r}) = 0, \qquad \mathbf{r} \in V \tag{2.107}$$

and our task reduces to finding the conditions under which $\mathcal{E}_0(\mathbf{r}) = 0$ is the only admissible solution. The proofs for all the domains of interest are based on the zero-energy argument, i.e., we need to make sure that the electrostatic energy of $\mathcal{E}_0(\mathbf{r})$ in V is actually null.

We begin with a normal domain V (Figure 1.2b), in which case $\mathcal{E}_0(\mathbf{r})$ can be derived from *either* a scalar potential (see Section 2.2 and 2.5.1) *or* a vector potential. If we introduce a scalar potential $\Phi_0(\mathbf{r})$ as in (2.15), we may transform the electrostatic energy of the difference field as follows

$$\begin{aligned} W_e &:= \frac{\varepsilon_0}{2}\int_V \mathrm{d}V\,|\mathcal{E}_0(\mathbf{r})|^2 = -\frac{\varepsilon_0}{2}\int_V \mathrm{d}V\,\mathcal{E}_0(\mathbf{r})\cdot\nabla\Phi_0(\mathbf{r}) \\ &= -\frac{\varepsilon_0}{2}\int_V \mathrm{d}V\,\nabla\cdot[\mathcal{E}_0(\mathbf{r})\Phi_0(\mathbf{r})] = \frac{\varepsilon_0}{2}\int_{\partial V} \mathrm{d}S\,\hat{\mathbf{n}}(\mathbf{r})\cdot\mathcal{E}_0(\mathbf{r})\Phi_0(\mathbf{r}) \end{aligned} \tag{2.108}$$

having used (H.51), (2.107) and the Gauss theorem with the unit normal oriented inwards V. The flux integral in the rightmost-hand side vanishes if

$$\hat{\mathbf{n}}(\hat{\mathbf{r}})\cdot\mathcal{E}_0(\mathbf{r}) = 0, \qquad \mathbf{r}\in\partial V \tag{2.109}$$

which is guaranteed if $\mathcal{E}_1(\mathbf{r})$ and $\mathcal{E}_2(\mathbf{r})$ have been assigned the same normal component for $\mathbf{r}\in\partial V$. Then, it follows that the energy is zero and, since $\mathcal{E}_0(\mathbf{r})$ is continuous, $\mathcal{E}_0(\mathbf{r})$ must vanish identically in V as we wanted to prove. This conclusion was already known from the condition on the normal derivative of the scalar potential.

If, on the other hand, we determine $\mathcal{E}_0(\mathbf{r})$ from a vector potential, namely,

$$\mathcal{E}_0(\mathbf{r}) := \nabla\times\mathcal{F}_0(\mathbf{r}) \tag{2.110}$$

(2.107) is automatically satisfied.[2] The electrostatic energy can be cast as follows

$$\begin{aligned} W_e &:= \frac{\varepsilon_0}{2}\int_V \mathrm{d}V\,|\mathcal{E}_0(\mathbf{r})|^2 = \frac{\varepsilon_0}{2}\int_V \mathrm{d}V\,\mathcal{E}_0(\mathbf{r})\cdot\nabla\times\mathcal{F}_0(\mathbf{r}) \\ &= \frac{\varepsilon_0}{2}\int_V \mathrm{d}V\,\nabla\cdot[\mathcal{F}_0(\mathbf{r})\times\mathcal{E}_0(\mathbf{r})] = \frac{\varepsilon_0}{2}\int_{\partial V} \mathrm{d}S\,\hat{\mathbf{n}}(\mathbf{r})\times\mathcal{E}_0(\mathbf{r})\cdot\mathcal{F}_0(\mathbf{r}) \end{aligned} \tag{2.111}$$

where we have used (H.49), (2.106) and applied the Gauss theorem with the unit normal pointing inwards V. In this case the surface integral vanishes if

$$\hat{\mathbf{n}}(\hat{\mathbf{r}})\times\mathcal{E}_0(\mathbf{r}) = 0, \qquad \mathbf{r}\in\partial V \tag{2.112}$$

which is guaranteed if $\mathcal{E}_1(\mathbf{r})$ and $\mathcal{E}_2(\mathbf{r})$ have been assigned the same tangential component for $\mathbf{r}\in\partial V$. By the same arguments as before, the difference field is zero in V and the solution is unique. With this new finding we have clarified the role of the tangential component of the electric field — something which could not be inferred from the proofs of Section 2.5.1.

Next, we investigate the issue of uniqueness in the whole space, i.e., $V = \mathbb{R}^3$; this is a special case of the previous configuration and, to obtain the result, we start with a ball $B(0,a)$ and we let the radius a approach infinity. It is convenient to derive $\mathcal{E}_0(\mathbf{r})$ from a scalar potential, whereby the energy in the ball $B(0,a)$ reads

$$W_e := \frac{\varepsilon_0}{2}\int_{B(0,a)} \mathrm{d}V\,|\mathcal{E}_0(\mathbf{r})|^2 = \frac{\varepsilon_0}{2}\int_{\partial B} \mathrm{d}S\,\hat{\mathbf{n}}(\mathbf{r})\cdot\mathcal{E}_0(\mathbf{r})\Phi_0(\mathbf{r}) \leqslant \frac{\varepsilon_0}{2}\int_{\partial B} \mathrm{d}S\,|\mathcal{E}_0(\mathbf{r})||\Phi_0(\mathbf{r})| \tag{2.113}$$

[2] If the medium of concern were inhomogeneous, the definition would be $\varepsilon(\mathbf{r})\mathcal{E}_0(\mathbf{r}) := \nabla\times\mathcal{F}_0(\mathbf{r})$, and if it were anisotropic we would assume $\overline{\overline{\varepsilon}}(\mathbf{r})\cdot\mathcal{E}_0(\mathbf{r}) := \nabla\times\mathcal{F}_0(\mathbf{r})$.

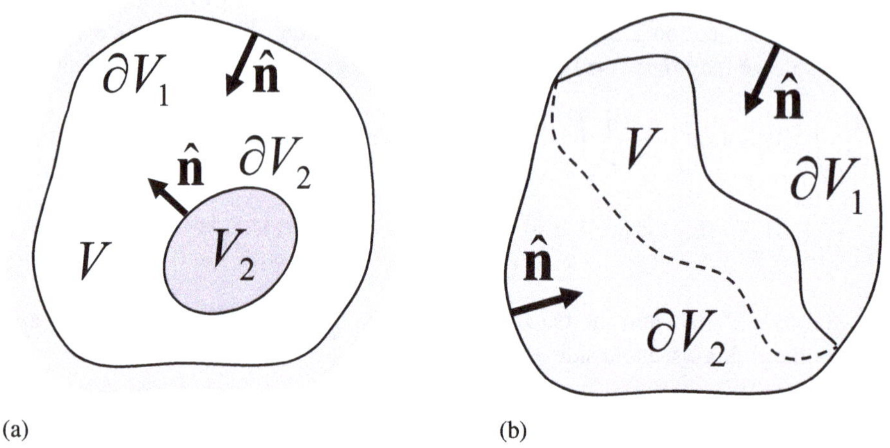

Figure 2.6 For proving the uniqueness of the solutions to the electrostatic equations: (a) a surface-wise multiply-connected domain V bounded by two disjoint closed surfaces ∂V_1 and ∂V_2; (b) a normal domain with the boundary conceptually divided into two parts ∂V_1 and ∂V_2.

and the last surface integral is null for $a \to +\infty$ if the scalar potential and the electric field satisfy the estimates

$$|\Phi_0(\mathbf{r})| \leqslant \frac{C_\Phi}{|\mathbf{r}|}, \qquad |\mathbf{r}| \to +\infty \tag{2.114}$$

$$|\mathcal{E}_0(\mathbf{r})| \leqslant \frac{C_E}{|\mathbf{r}|^2}, \qquad |\mathbf{r}| \to +\infty \tag{2.115}$$

that is, both the potential and, as a consequence, the field are regular at infinity. In conclusion, the solution to the electrostatic equations is unique if the sources are assigned and the field decays with the inverse square of the distance.

The proof of uniqueness in a surface-wise multiply-connected region V (Figure 2.6a) has to be modified a bit; a typical example is given by the space between two concentric spheres. In this case, it is not possible to introduce a single-valued vector potential throughout V [25, Chapter 1], and we need to resort to $\Phi_0(\mathbf{r})$. Let us call ∂V_1 and ∂V_2 the two closed *disjoint* surfaces which comprise the boundary of V. The electrostatic energy can be written as

$$W_e := \frac{\varepsilon_0}{2}\int_V \mathrm{d}V\, |\mathcal{E}_0(\mathbf{r})|^2 = \frac{\varepsilon_0}{2}\sum_{n=1}^{2}\int_{\partial V_n} \mathrm{d}S\, \hat{\mathbf{n}}(\mathbf{r}) \cdot \mathcal{E}_0(\mathbf{r})\Phi_0(\mathbf{r}) \tag{2.116}$$

in light of the Gauss theorem applied with the unit normals oriented inwards V. We see that the flux integrals vanish — and uniqueness is achieved — if the normal component of the difference field is null on $\partial V_1 \cup \partial V_2$. This is guaranteed if $\hat{\mathbf{n}}(\mathbf{r}) \cdot \mathcal{E}_1(\mathbf{r})$ and $\hat{\mathbf{n}}(\mathbf{r}) \cdot \mathcal{E}_2(\mathbf{r})$ have been assigned the same values on the boundary. We notice that $V = \mathbb{R}^3 \setminus V_2$ is a special case of the previous configuration, where ∂V_1 recedes to infinity. Uniqueness is achieved if $\hat{\mathbf{n}}(\mathbf{r}) \cdot \mathcal{E}_0(\mathbf{r}) = 0$ on ∂V_2 and (2.115) holds true.

Finally, we examine a situation of a normal domain V with the boundary a connected closed surface comprised of two parts, i.e., $\partial V = \partial V_1 \cup \partial V_2$ (Figure 2.6b). We can show that uniqueness is guaranteed if we assign the normal component of $\mathcal{E}(\mathbf{r})$ on ∂V_1 and the tangential component of $\mathcal{E}(\mathbf{r})$ on ∂V_2. To this purpose, we introduce a scalar potential $\Phi_0(\mathbf{r})$ and write the energy as follows

$$W_e := \frac{\varepsilon_0}{2} \int_V \mathrm{d}V \, |\mathcal{E}_0(\mathbf{r})|^2 = \frac{\varepsilon_0}{2} \sum_{n=1}^{2} \int_{\partial V_n} \mathrm{d}S \; \hat{\mathbf{n}}(\mathbf{r}) \cdot \mathcal{E}_0(\mathbf{r}) \Phi_0(\mathbf{r}) \tag{2.117}$$

with the usual positions. The flux integral on ∂V_1 vanishes because $\mathcal{E}_0(\mathbf{r}) \cdot \hat{\mathbf{n}}(\mathbf{r}) = 0$ for $\mathbf{r} \in \partial V_1$. For the flux integral over ∂V_2 we observe that

$$\hat{\mathbf{n}}(\mathbf{r}) \times \mathcal{E}_0(\mathbf{r}) = -\hat{\mathbf{n}}(\mathbf{r}) \times \nabla \Phi_0(\mathbf{r}) = 0, \qquad \mathbf{r} \in \partial V_2 \tag{2.118}$$

implies that $\Phi_0(\mathbf{r}) = C_\Phi$, a constant. Moreover, the Gauss theorem applied to (2.107) yields

$$0 = \int_V \mathrm{d}V \, \nabla \cdot \mathcal{E}_0(\mathbf{r}) = -\int_{\partial V_1} \mathrm{d}S \; \hat{\mathbf{n}}(\mathbf{r}) \cdot \mathcal{E}_0(\mathbf{r}) - \int_{\partial V_2} \mathrm{d}S \; \hat{\mathbf{n}}(\mathbf{r}) \cdot \mathcal{E}_0(\mathbf{r}) = -\int_{\partial V_2} \mathrm{d}S \; \hat{\mathbf{n}}(\mathbf{r}) \cdot \mathcal{E}_0(\mathbf{r}) \tag{2.119}$$

since $\mathcal{E}_0(\mathbf{r}) \cdot \hat{\mathbf{n}}(\mathbf{r}) = 0$ for $\mathbf{r} \in \partial V_1$. Thus, also the second integral in (2.117) vanishes

$$\frac{\varepsilon_0}{2} \int_{\partial V_2} \mathrm{d}S \; \hat{\mathbf{n}}(\mathbf{r}) \cdot \mathcal{E}_0(\mathbf{r}) \Phi_0(\mathbf{r}) = C_\Phi \frac{\varepsilon_0}{2} \int_{\partial V_2} \mathrm{d}S \; \hat{\mathbf{n}}(\mathbf{r}) \cdot \mathcal{E}_0(\mathbf{r}) = 0 \tag{2.120}$$

and this implies zero energy and, as a result, uniqueness as before.

Extension of these proofs to an inhomogeneous and possibly anisotropic dielectric medium is not difficult and left to the Reader.

2.6 The three-dimensional static Green function

The various instances of the Poisson equation discussed in Sections 2.2 and 2.5.1 can be solved in closed form only for a limited set of canonical geometries and boundary conditions. In general situations one resorts to numerical techniques to obtain an approximation to the solution. Among the cases that we do know how to deal with analytically are the so-called fundamental solutions of the Poisson equation.

In electrostatics the fundamental solutions are termed *static Green's functions*, after the English mathematician George Green (1793-1841) or alternatively *influence* or *source* functions and may be interpreted as special scalar potentials, namely, those produced by a point charge or a combination of point charges [1, 6, 11], [18, Chapter 8]. The Green functions — which are also known as *transfer functions* in system theory and as *impulse responses* in signal processing — play an important role in the construction of solutions to the Poisson equations when arbitrary sources and boundary conditions are assigned. Intuitively, we understand that the potential generated by a number of point charges is just the linear combination of the individual potentials in light of the linearity of the Poisson equation. In Section 2.7 we shall extend this idea rigorously to the case of a continuous distribution of charges with density $\varrho(\mathbf{r})$.

We recall from Section 2.1 that the local form of the Gauss law (2.3) was derived under the hypothesis of the charge density $\varrho(\mathbf{r})$ being a continuous scalar field. Here we are concerned with the density of a point charge Q. A point charge, by definition, has no spatial extent and is concentrated in just one point, hence in keeping with (1.1) the corresponding density should be zero everywhere and infinite at the location of the charge. Clearly, no ordinary function can represent such density and, even if that were the case, the density would be strongly discontinuous. The modern way out consists of introducing the notion of distributions and, in particular, of *Dirac delta* (Appendix C). The Poisson equation is then interpreted in the context of distributions. We follow the classic approach, though, because we think it provides more insight into the very formulation of the problem.

To obtain the differential Gauss law and the associated Poisson equation for a point charge located at $\mathbf{r} = 0$ we go back to the integral form (1.16) and we choose a bounded region $V \subset \mathbb{R}^3 \setminus \{0\}$ which *does not* contain the charge. Then, the displacement vector is of class $\mathcal{C}^1(V)^3 \cap \mathcal{C}(\overline{V})^3$ and we can apply the Gauss theorem (A.53) backwards. The mean value theorem [28] and the arbitrariness of V allows us to write

$$\nabla \cdot \mathcal{D}(\mathbf{r}) = 0, \qquad \mathbf{r} \in \mathbb{R}^3 \setminus \{0\}. \tag{2.121}$$

In the following we use (2.121) to determine two fundamental solutions, namely, the Green functions in a homogeneous unbounded medium with scalar and dyadic permittivity, respectively.

2.6.1 Unbounded homogeneous isotropic medium

The appropriate Poisson equation for an unbounded homogeneous isotropic medium follows from (2.121) in combination with $\mathcal{D}(\mathbf{r}) = \varepsilon\, \mathcal{E}(\mathbf{r})$, viz.,

$$\varepsilon \nabla^2 \Phi(\mathbf{r}) = 0, \qquad \mathbf{r} \in \mathbb{R}^3 \setminus \{0\} \tag{2.122}$$

where (2.15) has been used. The Green function is then defined through

$$\Phi(\mathbf{r}) = \frac{Q}{\varepsilon} G(\mathbf{r}) \tag{2.123}$$

with Q the point charge at $\mathbf{r} = 0$. In this case, the fundamental solution $G(\mathbf{r})$ is expected to be spherically symmetric because to an ideal observer sitting at $\mathbf{r} \neq 0$ the point charge looks just the same from any view angle. Therefore, $G(\mathbf{r})$ must satisfy the problem

$$\begin{cases} \nabla^2 G(|\mathbf{r}|) = 0, & \mathbf{r} \in \mathbb{R}^3 \setminus \{0\} \\ G(|\mathbf{r}|) = \mathcal{O}\left(\dfrac{1}{|\mathbf{r}|}\right), & |\mathbf{r}| \to +\infty \end{cases} \tag{2.124}$$

where we have explicitly highlighted the dependence on the distance $|\mathbf{r}|$ from the charge. It should be noted that we cannot conclude the solution to (2.124) is unique, since we may not invoke the arguments of Section 2.5.1. We arbitrarily decide to pick up the solution that obeys the additional condition

$$\int_S \mathrm{d}S\, \hat{\mathbf{n}} \cdot \nabla G(|\mathbf{r}|) = -1 \tag{2.125}$$

where S is any closed surface surrounding the point $\mathbf{r} = 0$. This constraint arises from the Gauss law (2.1) — which is always valid, even though $G(|\mathbf{r}|)$ is singular for $\mathbf{r} = 0$ — because we have

$$\mathcal{D}(\mathbf{r}) = -Q \nabla G(|\mathbf{r}|) \tag{2.126}$$

by virtue of (2.123).

The solution of (2.124) is more easily carried out in spherical coordinates, in that the Poisson equation becomes essentially one-dimensional. In light of (A.42) we then have

$$\frac{1}{r^2}\frac{\mathrm{d}}{\mathrm{d}r}\left(r^2\frac{\mathrm{d}G}{\mathrm{d}r}\right) = 0, \qquad r = |\mathbf{r}| \neq 0 \tag{2.127}$$

which is solved by

$$G(r) = \frac{A}{r} + B \tag{2.128}$$

where A and B are two real constants. Evidently, B must be set to zero on account of the desired behavior as $r \to +\infty$. The other constant must be chosen so as to satisfy the remaining condition (2.125). By letting S be a sphere of radius r_0 centered on the charge we find ($\hat{\mathbf{n}} \equiv \hat{\mathbf{r}}$)

$$1 = -\int_S \mathrm{d}S\, \hat{\mathbf{r}} \cdot \nabla G(r) = \int_S \mathrm{d}S\, \frac{A}{r_0^2} = 4\pi A \tag{2.129}$$

whereby $A = 1/(4\pi)$. Thus, the Green function that satisfies (2.124) and (2.125) reads

$$G(|\mathbf{r}|) = \frac{1}{4\pi|\mathbf{r}|}, \qquad \mathbf{r} \in \mathbb{R}^3 \tag{2.130}$$

and it has an integrable singularity at the location of the charge. Since this case is particularly simple, it could have been handled by just invoking the integral Gauss law and the spherical symmetry, as was done in Example 2.1.

We can extend the result to the case of a point charge located at $\mathbf{r}' \neq 0$ by denoting with $\mathbf{R} = \mathbf{r}-\mathbf{r}'$ the position vector with respect to the charge. Hence, the relevant Green function reads

$$G(\mathbf{r},\mathbf{r}') = \frac{1}{4\pi R} = \frac{1}{4\pi|\mathbf{R}|} = \frac{1}{4\pi|\mathbf{r}-\mathbf{r}'|}, \qquad \mathbf{r},\mathbf{r}' \in \mathbb{R}^3 \tag{2.131}$$

and it follows from (2.130) by observing that the desired fundamental solution must depend only on the distance R from the charge.

2.6.2 *Unbounded homogeneous anisotropic medium*

The relevant Poisson equation for a point charge in an unbounded homogeneous anisotropic medium reads

$$\nabla \cdot [\overline{\overline{\varepsilon}} \cdot \nabla\Phi(\mathbf{r})] = 0, \qquad \mathbf{r} \neq 0 \tag{2.132}$$

in light of (2.121) and (1.126). We define the Green function through

$$\Phi(\mathbf{r}) = \frac{Q}{\varepsilon_0} G(\mathbf{r}) \tag{2.133}$$

with Q being the point charge at $\mathbf{r} = 0$. Then, $G(\mathbf{r})$ must solve the problem

$$\begin{cases} \nabla \cdot [\overline{\overline{\varepsilon}}_\mathrm{r} \cdot \nabla G(\mathbf{r})] = 0, & \mathbf{r} \in \mathbb{R}^3 \setminus \{0\} \\ G(\mathbf{r}) = \mathcal{O}\left(\dfrac{1}{|\mathbf{r}|}\right), & |\mathbf{r}| \to +\infty \end{cases} \tag{2.134}$$

where $\overline{\varepsilon}_{\mathrm{r}} := \overline{\varepsilon}/\varepsilon_0$ is the dyadic permittivity relative to the permittivity of free space. In general, $G(\mathbf{r})$ depends on $\mathbf{r}$ rather than on $|\mathbf{r}|$, insofar as an observer 'sees' the charge through a different permittivity for each view angle. Nonetheless, if $\overline{\varepsilon}_{\mathrm{r}}$ is a *symmetric positive definite* dyadic, viz.,

$$\mathbf{u} \cdot \overline{\varepsilon}_{\mathrm{r}} \cdot \mathbf{u} > 0, \qquad \mathbf{u} \in \mathbb{R}^3 \setminus \{0\} \tag{2.135}$$

the eigenvalues of $\overline{\varepsilon}_{\mathrm{r}}$ are strictly positive. Therefore, without loss of generality we can always choose a system of coordinates in which the permittivity is *diagonal*, viz.,

$$\overline{\varepsilon}_{\mathrm{r}} = \varepsilon_{xx,\mathrm{r}}\hat{\mathbf{x}}\hat{\mathbf{x}} + \varepsilon_{yy,\mathrm{r}}\hat{\mathbf{y}}\hat{\mathbf{y}} + \varepsilon_{zz,\mathrm{r}}\hat{\mathbf{z}}\hat{\mathbf{z}} \tag{2.136}$$

in order to facilitate the solution of (2.134), which becomes

$$\varepsilon_{xx,\mathrm{r}}\frac{\partial^2 G}{\partial x^2} + \varepsilon_{yy,\mathrm{r}}\frac{\partial^2 G}{\partial y^2} + \varepsilon_{zz,\mathrm{r}}\frac{\partial^2 G}{\partial z^2} = 0, \qquad \mathbf{r} \in \mathbb{R}^3 \setminus \{0\} \tag{2.137}$$

subject to the condition that $G(\mathbf{r})$ vanish at infinity.

To this purpose, we make a change of independent variables, namely,

$$x = \sqrt{\varepsilon_{xx,\mathrm{r}}}\xi, \qquad y = \sqrt{\varepsilon_{yy,\mathrm{r}}}\eta, \qquad z = \sqrt{\varepsilon_{zz,\mathrm{r}}}\zeta \tag{2.138}$$

which is a special case of affine transformation of the space [29, Section 4.3]. Then, we define $\tilde{\mathbf{r}} = \xi\hat{\boldsymbol{\xi}} + \eta\hat{\boldsymbol{\eta}} + \zeta\hat{\boldsymbol{\zeta}}$, the position vector in the new system of Cartesian coordinates, and obtain

$$|\tilde{\mathbf{r}}|^2 = \xi^2 + \eta^2 + \zeta^2 = \frac{x^2}{\varepsilon_{xx,\mathrm{r}}} + \frac{y^2}{\varepsilon_{yy,\mathrm{r}}} + \frac{z^2}{\varepsilon_{zz,\mathrm{r}}} = \mathbf{r} \cdot \overline{\varepsilon}_{\mathrm{r}}^{-1} \cdot \mathbf{r} = \left|\overline{\varepsilon}_{\mathrm{r}}^{-\frac{1}{2}} \cdot \mathbf{r}\right|^2 \tag{2.139}$$

$$\frac{\partial G}{\partial x} = \frac{1}{\sqrt{\varepsilon_{xx,\mathrm{r}}}}\frac{\partial G}{\partial \xi}, \qquad \frac{\partial G}{\partial y} = \frac{1}{\sqrt{\varepsilon_{yy,\mathrm{r}}}}\frac{\partial G}{\partial \eta}, \qquad \frac{\partial G}{\partial z} = \frac{1}{\sqrt{\varepsilon_{zz,\mathrm{r}}}}\frac{\partial G}{\partial \zeta} \tag{2.140}$$

$$\frac{\partial^2 G}{\partial x^2} = \frac{1}{\varepsilon_{xx,\mathrm{r}}}\frac{\partial^2 G}{\partial \xi^2}, \qquad \frac{\partial^2 G}{\partial y^2} = \frac{1}{\varepsilon_{yy,\mathrm{r}}}\frac{\partial^2 G}{\partial \eta^2}, \qquad \frac{\partial^2 G}{\partial z^2} = \frac{1}{\varepsilon_{zz,\mathrm{r}}}\frac{\partial^2 G}{\partial \zeta^2}, \tag{2.141}$$

and finally

$$\begin{cases} \dfrac{\partial^2 G}{\partial \xi^2} + \dfrac{\partial^2 G}{\partial \eta^2} + \dfrac{\partial^2 G}{\partial \zeta^2} = 0, & \tilde{\mathbf{r}} \in \mathbb{R}^3 \setminus \{0\} \\ G(\tilde{\mathbf{r}}) = \mathcal{O}\left(\dfrac{1}{|\tilde{\mathbf{r}}|}\right), & |\tilde{\mathbf{r}}| \to +\infty \end{cases} \tag{2.142}$$

which, having formally the same structure as (2.124), is solved by

$$G(\xi, \eta, \zeta) = \frac{A}{|\tilde{\mathbf{r}}|} = \frac{A}{\sqrt{\xi^2 + \eta^2 + \zeta^2}} = \frac{A}{\left(\dfrac{x^2}{\varepsilon_{xx,\mathrm{r}}} + \dfrac{y^2}{\varepsilon_{yy,\mathrm{r}}} + \dfrac{z^2}{\varepsilon_{zz,\mathrm{r}}}\right)^{1/2}} \tag{2.143}$$

where we decide to determine the constant A so as to satisfy the global Gauss law (2.1). Since we have

$$\mathcal{D}(\mathbf{r}) = -Q\overline{\varepsilon}_{\mathrm{r}} \cdot \nabla G(\mathbf{r}) \tag{2.144}$$

this choice amounts to the normalization condition

$$1 = -\int_S \mathrm{d}S\, \hat{\mathbf{n}} \cdot \overline{\varepsilon}_\mathrm{r} \cdot \nabla G(|\mathbf{r}|) \tag{2.145}$$

and a suitable closed surface S for the calculation of the flux is the ellipsoid with axes $\sqrt{\varepsilon_{xx,\mathrm{r}}}$, $\sqrt{\varepsilon_{yy,\mathrm{r}}}$, $\sqrt{\varepsilon_{zz,\mathrm{r}}}$, and centered in the origin. In which case the integral can be computed in ellipsoidal coordinates [30, Table 1.10] not without some sweat.

A smarter way to obtain the result consists of observing that the very same ellipsoid is transformed into the unitary sphere in the space of the variables ξ, η, ζ, where we can equally well state the Gauss law. In this regard, we also need to determine what charge $\tilde{q}$ in the new coordinates corresponds to Q in the original Cartesian system. We use the definition of charge density (1.17) applied to the whole space

$$Q = \int_{\mathbb{R}^3} \mathrm{d}V\, \varrho(\mathbf{r}) = \sqrt{\varepsilon_{xx,\mathrm{r}}\varepsilon_{yy,\mathrm{r}}\varepsilon_{zz,\mathrm{r}}} \int_{-\infty}^{+\infty} \mathrm{d}\xi \int_{-\infty}^{+\infty} \mathrm{d}\eta \int_{-\infty}^{+\infty} \mathrm{d}\zeta\, \varrho(\tilde{\mathbf{r}}) = \sqrt{\det(\overline{\varepsilon}_\mathrm{r})}\, \tilde{q} \tag{2.146}$$

whence $\tilde{q} = Q/\sqrt{\det(\overline{\varepsilon}_\mathrm{r})}$. Finally, we state (2.1) in the space of the variables ξ, η, ζ, i.e.,

$$\frac{Q}{\sqrt{\det(\overline{\varepsilon}_\mathrm{r})}} = -\int_{\tilde{S}} \mathrm{d}\tilde{S}\, Q \left.\frac{\partial G}{\partial |\tilde{\mathbf{r}}|}\right|_{|\tilde{\mathbf{r}}|=1} = Q \int_{\tilde{S}} \mathrm{d}\tilde{S} \left.\frac{A}{|\tilde{\mathbf{r}}|^2}\right|_{|\tilde{\mathbf{r}}|=1} = 4\pi Q A \tag{2.147}$$

where $\tilde{S}$ is the unitary sphere centered in the origin. The latter equation provides the desired value of the constant A. By substituting into (2.143) and reverting to the original Cartesian coordinates we obtain the relevant Green function

$$G(\mathbf{r}) = \frac{1}{4\pi\sqrt{\det(\overline{\varepsilon}_\mathrm{r})\mathbf{r} \cdot \overline{\varepsilon}_\mathrm{r}^{-1} \cdot \mathbf{r}}} = \frac{1}{4\pi\sqrt{\det(\overline{\varepsilon}_\mathrm{r})}\left|\overline{\varepsilon}_\mathrm{r}^{-1/2} \cdot \mathbf{r}\right|}, \qquad \mathbf{r} \in \mathbb{R}^3. \tag{2.148}$$

We notice that (2.148) is a coordinate-free expression, and hence it is valid even if the permittivity dyadic is not diagonal, in spite of the way it was derived. Last but not least, (2.148) is consistent with the Green function in an isotropic medium (2.130), as can be checked by letting $\overline{\varepsilon}_\mathrm{r} = \varepsilon_\mathrm{r}\overline{\mathcal{I}}$ and $\varepsilon_\mathrm{r} = 1$.

2.7 Integral representation of the scalar potential

Our next goal is to find an explicit expression for the scalar potential $\Phi(\mathbf{r})$ in a given region of space in terms of the charge distribution and the boundary conditions. We can think of this procedure as of actually inverting the Poisson equation. Which is why the desired result for the potential is sometimes referred to as the *integral solution* to the Poisson equation. However, we shall see that this interpretation is deceptive owing to a crucial aspect [1].

For reasons that will become apparent at the end, we denote the position vector with $\mathbf{r}' = x'\hat{\mathbf{x}} + y'\hat{\mathbf{y}} + z'\hat{\mathbf{z}}$. Hence, the scalar potential and the charge density will be functions of $\mathbf{r}'$; the Laplace operator will be indicated with ∇'^2 to signify that the derivatives are computed with respect to the primed coordinates.

For the sake of argument, suppose we wish to solve the Poisson equation relevant to the geometry of Figure 2.7a. With $\Phi_1(\mathbf{r}')$ we denote the scalar potential produced in the volume V by the

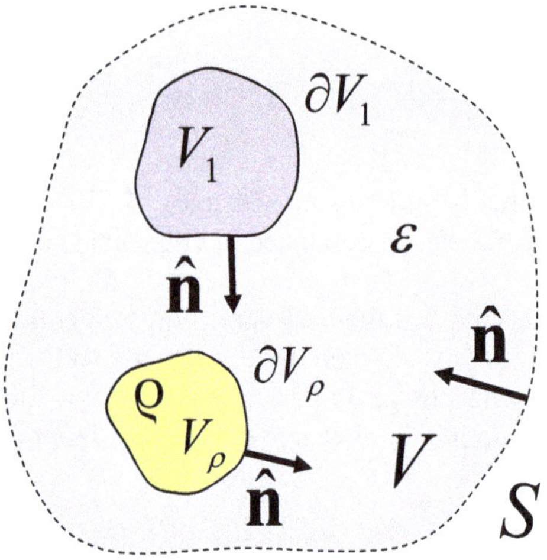

(a) *original problem*

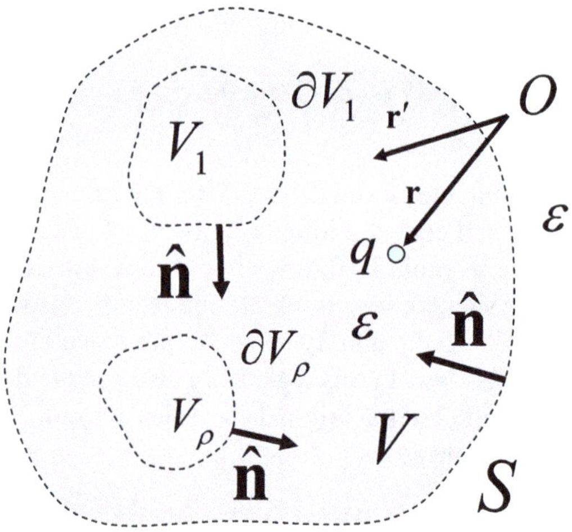

(b) *auxiliary problem*

Figure 2.7 For the derivation of the integral representation of the scalar potential.

charge distribution with density $\varrho(\mathbf{r}')$ for $\mathbf{r}' \in V_\varrho$ in the presence of a medium that occupies the domain V_1. The isotropic medium filling V is a homogeneous dielectric with permittivity ε. We know from Section 2.5.1 that we need boundary conditions on $\partial V = S \cup \partial V_1$ to ensure a unique solution. Therefore, in symbols the problem reads

$$\begin{cases} \varepsilon \nabla'^2 \Phi_1(\mathbf{r}') = -\varrho(\mathbf{r}'), & \mathbf{r}' \in V \\ \Phi_1(\mathbf{r}') = f(\mathbf{r}') \quad \text{or} \quad \dfrac{\partial}{\partial \hat{\mathbf{n}}} \Phi_1(\mathbf{r}') = g(\mathbf{r}'), & \mathbf{r}' \in \partial V = S \cup \partial V_1 \end{cases} \tag{2.149}$$

where f and g are regular functions on ∂V.

We also consider an auxiliary problem (Figure 2.7b) comprised of a point charge q located at $\mathbf{r} \in V \setminus \overline{V}_\varrho$ in an otherwise homogeneous medium with permittivity ε. We denote the potential produced by q with $\Phi_2(\mathbf{r}')$, and the corresponding problem reads

$$\begin{cases} \varepsilon \nabla'^2 \Phi_2(\mathbf{r}') = 0, & \mathbf{r}' \in \mathbb{R}^3 \setminus \{\mathbf{r}\} \qquad \mathbf{r} \in V \setminus \overline{V}_\varrho \\ \Phi_2(\mathbf{r}') = \mathcal{O}\left(\dfrac{1}{|\mathbf{r}'|}\right), & |\mathbf{r}'| \to +\infty \end{cases} \tag{2.150}$$

thus, apart from a normalization factor, $\Phi_2(\mathbf{r}')$ is the Green function (2.131). The auxiliary problem is quite discretionary, but we are prompted to choose one that we can solve analytically because the knowledge of $\Phi_2(\mathbf{r}')$ is a key ingredient of the integral representation.

We are now ready to begin the derivation proper. We multiply (2.149) by $\Phi_2(\mathbf{r}')$ and (2.150) by $\Phi_1(\mathbf{r}')$ to obtain

$$-\Phi_2(\mathbf{r}')\frac{\varrho(\mathbf{r}')}{\varepsilon} = \Phi_2(\mathbf{r}')\nabla'^2\Phi_1(\mathbf{r}') = \nabla' \cdot [\Phi_2(\mathbf{r}')\nabla'\Phi_1(\mathbf{r}')] - \nabla'\Phi_2(\mathbf{r}') \cdot \nabla'\Phi_1(\mathbf{r}') \tag{2.151}$$

$$0 = \Phi_1(\mathbf{r}')\nabla'^2\Phi_2(\mathbf{r}') = \nabla' \cdot [\Phi_1(\mathbf{r}')\nabla'\Phi_2(\mathbf{r}')] - \nabla'\Phi_1(\mathbf{r}') \cdot \nabla'\Phi_2(\mathbf{r}') \tag{2.152}$$

which are valid simultaneously for $\mathbf{r}' \in V \setminus \{\mathbf{r}\}$. In fact, since $\Phi_2(\mathbf{r}')$ is singular in $\mathbf{r}' = \mathbf{r}$, we exclude the point charge with a ball $B(\mathbf{r}, a)$, as is suggested in Figure 2.8; the radius a is chosen small enough

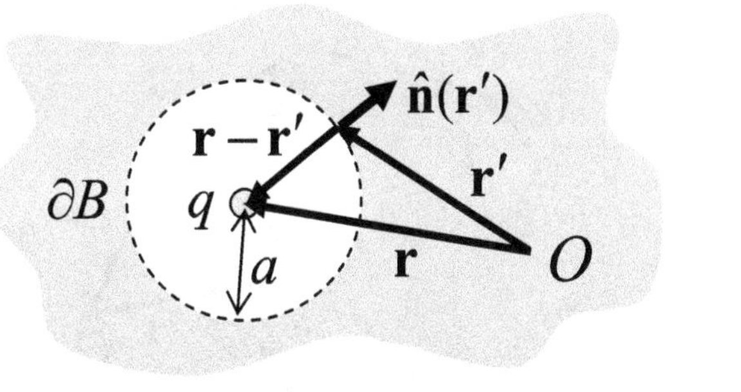

Figure 2.8 A ball $B(\mathbf{r}, a)$ for isolating the location of the charge where the Green function is singular.

so that $B[\mathbf{r}, a] \subset V$. Then, we subtract the equations above side by side and integrate over $V \setminus B(\mathbf{r}, a)$ with respect to $\mathbf{r}'$ to get

$$\int\limits_{V\setminus B(\mathbf{r},a)} \mathrm{d}V'\, \nabla' \cdot [\Phi_2(\mathbf{r}')\nabla'\Phi_1(\mathbf{r}') - \Phi_1(\mathbf{r}')\nabla'\Phi_2(\mathbf{r}')] = -\int\limits_{V_\varrho} \mathrm{d}V'\, \Phi_2(\mathbf{r}')\frac{\varrho(\mathbf{r}')}{\varepsilon} \tag{2.153}$$

where we have restricted the integration to the support of $\varrho(\mathbf{r}')$ in the right-hand side.

We may apply the Gauss theorem (A.53) to the volume integral in the left member, because the vector fields $\Phi_2(\mathbf{r}')\nabla'\Phi_1(\mathbf{r}')$ and $\Phi_1(\mathbf{r}')\nabla'\Phi_2(\mathbf{r}')$ are differentiable in $V \setminus B[\mathbf{r}, a]$ and continuous in $\overline{V} \setminus B(\mathbf{r}, a)$, and hence we find

$$\int\limits_{S\cup\partial V_1} \mathrm{d}S'\, \hat{\mathbf{n}}(\mathbf{r}') \cdot [\Phi_2(\mathbf{r}')\nabla'\Phi_1(\mathbf{r}') - \Phi_1(\mathbf{r}')\nabla'\Phi_2(\mathbf{r}')] + \int\limits_{\partial B} \mathrm{d}S'\, \Phi_2(\mathbf{r}')\hat{\mathbf{n}}(\mathbf{r}') \cdot \nabla'\Phi_1(\mathbf{r}')$$
$$- \int\limits_{\partial B} \mathrm{d}S'\, \Phi_1(\mathbf{r}')\hat{\mathbf{n}}(\mathbf{r}') \cdot \nabla'\Phi_2(\mathbf{r}') = \int\limits_{V_\varrho} \mathrm{d}V'\, \Phi_2(\mathbf{r}')\frac{\varrho(\mathbf{r}')}{\varepsilon} \tag{2.154}$$

where we have taken into account that the unit normal $\hat{\mathbf{n}}$ on ∂V and ∂B is oriented inward the domain $V \setminus B(\mathbf{r}, a)$. Lastly, we take the limit of both sides as $a \to 0$.

The flux integral over $S \cup \partial V_1$ as well as the volume integral do not depend on a and as such pose no difficulties. For the flux integrals over ∂B we observe that (see Figure 2.8)

$$\mathbf{r}' = \mathbf{r} + a\hat{\mathbf{n}}(\mathbf{r}'), \qquad \mathbf{r}' \in \partial B \tag{2.155}$$

$$\Phi_2(\mathbf{r}') = \frac{q}{4\pi\varepsilon|\mathbf{r} - \mathbf{r}'|} = \frac{q}{4\pi\varepsilon a}, \qquad \mathbf{r}' \in \partial B \tag{2.156}$$

$$\hat{\mathbf{n}}(\mathbf{r}') \cdot \nabla'\Phi_2(\mathbf{r}') = \hat{\mathbf{n}}(\mathbf{r}') \cdot \frac{q(\mathbf{r} - \mathbf{r}')}{4\pi\varepsilon|\mathbf{r} - \mathbf{r}'|^3} = -\frac{q}{4\pi\varepsilon a^2}, \qquad \mathbf{r}' \in \partial B \tag{2.157}$$

in view of (2.123) and (2.131). We have

$$\int\limits_{\partial B} \mathrm{d}S'\, \Phi_2(\mathbf{r}')\hat{\mathbf{n}}(\mathbf{r}') \cdot \nabla'\Phi_1(\mathbf{r}') = \frac{q}{4\pi\varepsilon a} \int\limits_{\partial B} \mathrm{d}S'\, \hat{\mathbf{n}}(\mathbf{r}') \cdot \nabla'\Phi_1(\mathbf{r}')$$
$$= a\frac{q}{\varepsilon}\hat{\mathbf{n}}(\mathbf{r}'_0) \cdot \nabla'\Phi_1(\mathbf{r}')|_{\mathbf{r}'_0} \xrightarrow[a\to 0]{} 0 \tag{2.158}$$

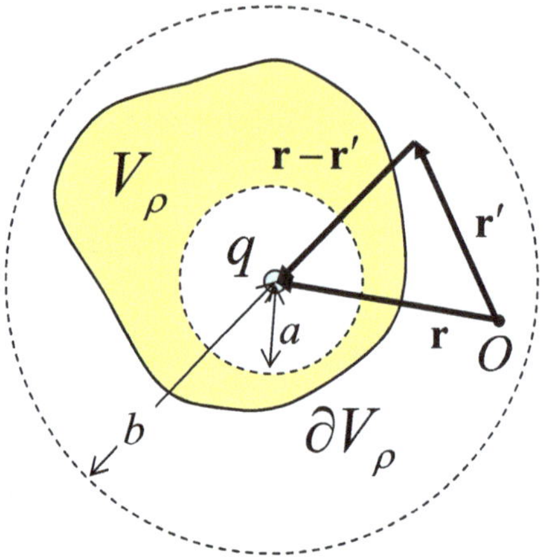

Figure 2.9 Two concentric balls $B(\mathbf{r}, a)$ and $B(\mathbf{r}, b)$ for showing that the volume integral in (2.160) is bounded for observation points $\mathbf{r}$ within the source region.

$$\int_{\partial B} dS'\, \Phi_1(\mathbf{r}')\hat{\mathbf{n}}(\mathbf{r}') \cdot \nabla'\Phi_2(\mathbf{r}') = -\frac{q}{4\pi\varepsilon a^2} \int_{\partial B} dS'\, \Phi_1(\mathbf{r} + a\hat{\mathbf{n}}(\mathbf{r}'))$$
$$= -\frac{q}{\varepsilon}\Phi_1(\mathbf{r} + a\hat{\mathbf{n}}(\mathbf{r}'_0)) \underset{a\to 0}{\longrightarrow} -\frac{q}{\varepsilon}\Phi_1(\mathbf{r}) \tag{2.159}$$

where the mean value theorem [28] has been invoked, with $\mathbf{r}'_0 \in \partial B$ being a suitable point, since the potentials and the normal derivatives thereof are regular on $\partial B \subset V$, and finally $\mathbf{r}'_0 \to \mathbf{r}$ as $a \to 0$ (Figure 2.8).

By putting everything together, dividing through by q/ε and dropping the subscript 1 for the potential of the first problem, we find

$$\Phi(\mathbf{r}) = \frac{1}{\varepsilon}\int_{V_\varrho} dV'\, \frac{\varrho(\mathbf{r}')}{4\pi R} + \int_{S\cup\partial V_1} dS'\, \Phi(\mathbf{r}')\frac{\partial}{\partial \hat{\mathbf{n}}'}\frac{1}{4\pi R} - \int_{S\cup\partial V_1} dS'\, \frac{1}{4\pi R}\frac{\partial \Phi}{\partial \hat{\mathbf{n}}'} \tag{2.160}$$

where $\hat{\mathbf{n}}' = \hat{\mathbf{n}}(\mathbf{r}')$ and $R = |\mathbf{r} - \mathbf{r}'|$. This is the desired integral representation of the scalar potential. Owing to the way (2.160) has been derived, it is valid for points $\mathbf{r} \in V \setminus \overline{V}_\varrho$, because the point charge for the auxiliary problem was set outside the source region V_ϱ. We can, however, extend the result to all points $\mathbf{r} \in V$.

To this purpose, we choose the auxiliary problem so that the point charge is located in the volume V_ϱ, and we set a so that $B[\mathbf{r}, a] \subset V_\varrho$ (Figure 2.9). We arrive at an expression similar to (2.154) in which the volume integral extends over $V_\varrho \setminus B(\mathbf{r}, a)$. Besides, $\Phi_1(\mathbf{r}')$ and $\hat{\mathbf{n}}(\mathbf{r}') \cdot \nabla'\Phi_1(\mathbf{r}')$ are regular in the region occupied by the sources — if $\varrho(\mathbf{r}')$ is at least continuous — whereas $\Phi_2(\mathbf{r}')$ and $\hat{\mathbf{n}}(\mathbf{r}') \cdot \nabla'\Phi_2(\mathbf{r}')$ are certainly regular outside $B(\mathbf{r}, a)$. Therefore, the integrals over ∂B yield the same results as before in the limit as $a \to 0$.

As regards the integral over $V_\varrho \setminus B(\mathbf{r}, a)$ we show that it exists and is bounded for any value of a and, in particular, for $a \to 0$. We choose a ball $B(\mathbf{r}, b)$ with b large enough so that $B(\mathbf{r}, a) \subset V_\varrho \subset B(\mathbf{r}, b)$ (Figure 2.9)

$$\left| \int_{V_\varrho\setminus B(\mathbf{r},a)} dV'\, \frac{\varrho(\mathbf{r}')}{4\pi R} \right| \leqslant \int_{V_\varrho\setminus B(\mathbf{r},a)} dV'\, \frac{|\varrho(\mathbf{r}')|}{4\pi R} \leqslant \int_{V_\varrho\setminus B(\mathbf{r},a)} dV'\, \frac{\|\varrho\|_\infty}{4\pi R} \leqslant \int_{B(\mathbf{r},b)\setminus B(\mathbf{r},a)} dV'\, \frac{\|\varrho\|_\infty}{4\pi R}$$

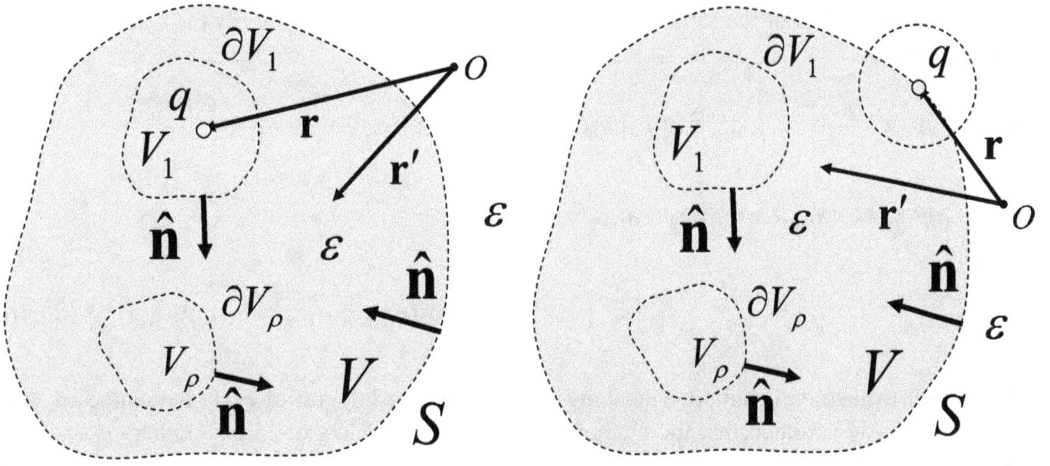

(a) *charge outside V* (b) *charge on the boundary ∂V*

Figure 2.10 Auxiliary problems for the derivation of the integral representation of the scalar potential.

$$= \|\varrho\|_\infty \int_a^b \mathrm{d}R\, R = \frac{1}{2}\|\varrho\|_\infty \left(b^2 - a^2\right) \leqslant \frac{1}{2}\|\varrho\|_\infty\, b^2 \tag{2.161}$$

where we have used the boundedness of $\varrho(\mathbf{r}')$, and we have integrated in a system of local polar spherical coordinates (R, α', β') centered in $\mathbf{r}$. With these results it remains proved that (2.160) holds for $\mathbf{r} \in V$, including the source region.

A key step in the derivation of (2.160) was the choice of a suitable auxiliary problem. In particular, we placed the point charge q inside V. Suppose now that we locate the point charge outside V, say, within the domain V_1 (Figure 2.10a) and follow the same steps which led us to (2.160). Since $\mathbf{r} \notin V$, (2.151) and (2.152) hold true for $\mathbf{r}' \in V$, and therefore we can integrate over V with no need to exclude the charge with a ball. Proceeding with the derivation, we arrive at

$$0 = \frac{1}{\varepsilon}\int_{V_\varrho} \mathrm{d}V' \frac{\varrho(\mathbf{r}')}{4\pi R} + \int_{S\cup\partial V_1} \mathrm{d}S'\, \Phi(\mathbf{r}')\frac{\partial}{\partial \hat{\mathbf{n}}'}\frac{1}{4\pi R} - \int_{S\cup\partial V_1} \mathrm{d}S'\, \frac{1}{4\pi R}\frac{\partial \Phi}{\partial \hat{\mathbf{n}}'} \tag{2.162}$$

for $\mathbf{r} \notin V$. We observe that the right-hand side of this last expression is nothing but the right member of (2.160) though evaluated for $\mathbf{r} \notin V$. Therefore, we conclude that the integral representation (2.160) still applies for observation points outside the domain of interest and, more specifically, it returns a null value for the scalar potential. It is important to recall that $\Phi(\mathbf{r})$ represents the solution to the Poisson problem (2.149), possibly extended to points outside V, and the potential may not be zero at all for $\mathbf{r} \notin V$. In other words, (2.160) provides the correct value of $\Phi(\mathbf{r})$ only for points within the domain of derivation. Then again, if (2.160) yields zero for $\mathbf{r} \notin V$, we conclude from (2.162) that the three integrals in the left-hand side must cancel each other. In practice, the effect of the sources located in $V_\varrho \subset V$ is exactly neutralized, as it were, by the contribution of the surface integrals.

Finally, we examine the case of a point charge q located on the boundary of V, say, $\mathbf{r} \in S$, as is shown in Figure 2.10b. We exclude the charge with a ball $B(\mathbf{r}, a)$ whereby the boundary ∂V remains divided into three parts, namely, ∂V_1, the open surface $S' := \{\mathbf{r}' \in S : |\mathbf{r}' - \mathbf{r}| \geqslant a\}$, and the open surface $S'' := \partial B \cap \overline{V}$. For the same reasons that led to (2.154), we may apply the Gauss theorem to (2.153) to obtain

$$\int_{V_\varrho} \mathrm{d}V'\, \Phi_2(\mathbf{r}')\frac{\varrho(\mathbf{r}')}{\varepsilon} = \int_{\partial V_1} \mathrm{d}S'\, \hat{\mathbf{n}}(\mathbf{r}')\cdot[\Phi_2(\mathbf{r}')\nabla'\Phi_1(\mathbf{r}') - \Phi_1(\mathbf{r}')\nabla'\Phi_2(\mathbf{r}')]$$
$$+ \int_{S'} \mathrm{d}S'\, \hat{\mathbf{n}}(\mathbf{r}')\cdot[\Phi_2(\mathbf{r}')\nabla'\Phi_1(\mathbf{r}') - \Phi_1(\mathbf{r}')\nabla'\Phi_2(\mathbf{r}')]$$
$$+ \int_{S''} \mathrm{d}S'\, \hat{\mathbf{n}}(\mathbf{r}')\cdot[\Phi_2(\mathbf{r}')\nabla'\Phi_1(\mathbf{r}') - \Phi_1(\mathbf{r}')\nabla'\Phi_2(\mathbf{r}')] \quad (2.163)$$

and we wish to take the limit for vanishing a. The volume integral of $\varrho(\mathbf{r}')$ is unaffected, since we have implicitly assumed that the charge density is strictly contained in V. Otherwise, we may proceed as in (2.161) for the case of q within V_ϱ. In like manner, the flux integral over ∂V_1 is not affected.

For the integral over S' we have

$$\lim_{a\to 0} \int_{S'} \mathrm{d}S'\, \hat{\mathbf{n}}(\mathbf{r}')\cdot[\Phi_2(\mathbf{r}')\nabla'\Phi_1(\mathbf{r}') - \Phi_1(\mathbf{r}')\nabla'\Phi_2(\mathbf{r}')] =$$
$$= \mathrm{PV}\int_S \mathrm{d}S'\, \hat{\mathbf{n}}(\mathbf{r}')\cdot[\Phi_2(\mathbf{r}')\nabla'\Phi_1(\mathbf{r}') - \Phi_1(\mathbf{r}')\nabla'\Phi_2(\mathbf{r}')]$$
$$= \mathrm{PV}\int_S \mathrm{d}S'\, \frac{q\hat{\mathbf{n}}(\mathbf{r}')\cdot\nabla'\Phi_1(\mathbf{r}')}{4\pi\varepsilon|\mathbf{r}-\mathbf{r}'|} - \mathrm{PV}\int_S \mathrm{d}S'\, \frac{q\hat{\mathbf{n}}(\mathbf{r}')\cdot(\mathbf{r}-\mathbf{r}')}{4\pi\varepsilon|\mathbf{r}-\mathbf{r}'|^3}\Phi_1(\mathbf{r}') \quad (2.164)$$

and observe that the strong singularity of the second integral in the rightmost member is only apparent, inasmuch as this contribution can be further estimated as

$$\left|\mathrm{PV}\int_S \mathrm{d}S'\, \frac{q\hat{\mathbf{n}}(\mathbf{r}')\cdot(\mathbf{r}-\mathbf{r}')}{4\pi\varepsilon|\mathbf{r}-\mathbf{r}'|^3}\Phi_1(\mathbf{r}')\right| \leqslant \mathrm{PV}\int_S \mathrm{d}S'\, \frac{|\hat{\mathbf{n}}(\mathbf{r}')\cdot(\mathbf{r}-\mathbf{r}')|}{4\pi\varepsilon|\mathbf{r}-\mathbf{r}'|^3}|q\Phi_1(\mathbf{r}')|$$
$$\leqslant \mathrm{PV}\int_S \mathrm{d}S'\, \frac{M|q\Phi_1(\mathbf{r}')|}{4\pi\varepsilon|\mathbf{r}-\mathbf{r}'|} \quad (2.165)$$

by virtue of (F.1) provided S is smooth at $\mathbf{r}$ (Appendix F.1). Thus, this term is dominated by an integral which has the same structure as the first contribution in the rightmost-hand side of (2.164). The proof that integrals of this type are finite — i.e., the singularity of $1/R$ can be integrated over a surface — is addressed in Section 2.10 and, in particular, with (2.256) and (2.257).

For the integral over S'' we recall (2.156) and (2.157), and apply the mean value theorem [28] to obtain

$$\int_{S''} \mathrm{d}S'\, \hat{\mathbf{n}}(\mathbf{r}')\cdot[\Phi_2(\mathbf{r}')\nabla'\Phi_1(\mathbf{r}') - \Phi_1(\mathbf{r}')\nabla'\Phi_2(\mathbf{r}')] =$$
$$= \frac{q}{4\pi\varepsilon a}\int_{S''} \mathrm{d}S'\, \hat{\mathbf{n}}(\mathbf{r}')\cdot\nabla'\Phi_1(\mathbf{r}') + \frac{q}{4\pi\varepsilon a^2}\int_{S''} \mathrm{d}S'\, \Phi_1(\mathbf{r}')$$
$$= \frac{q}{4\pi\varepsilon} a\Omega(a)\hat{\mathbf{n}}(\mathbf{r}_0')\cdot\nabla'\Phi_1(\mathbf{r}')|_{\mathbf{r}'=\mathbf{r}_0'} + \frac{q}{4\pi\varepsilon}\Omega(a)\Phi_1(\mathbf{r}+a\hat{\mathbf{n}}(\mathbf{r}_0')) \quad (2.166)$$

where $\mathbf{r}'_0 \in S''$ is a suitable point, and $\Omega(a)$ is the solid angle subtended by S'' with respect to $\mathbf{r}$. Since S is smooth by hypothesis, the solid angle approaches 2π as $a \to 0$, that is, S is approximated infinitely well by the tangent plane in $\mathbf{r}' = \mathbf{r}$, and S'' tends to the half-sphere centered in $\mathbf{r}$ and contained in V. This statement is proved in Appendix F.2. By collecting all these results, dropping the common factor q/ε, and renaming Φ_1 to Φ, we arrive at the formula

$$\frac{1}{2}\Phi(\mathbf{r}) = \frac{1}{\varepsilon}\int\limits_{V_\varrho} \mathrm{d}V' \frac{\varrho(\mathbf{r}')}{4\pi R} + \int\limits_{\partial V_1} \mathrm{d}S' \left[\Phi(\mathbf{r}')\frac{\partial}{\partial \hat{\mathbf{n}}'}\frac{1}{4\pi R} - \frac{1}{4\pi R}\frac{\partial \Phi}{\partial \hat{\mathbf{n}}'}\right] + \mathrm{PV}\int\limits_{S} \mathrm{d}S' \left[\Phi(\mathbf{r}')\frac{\partial}{\partial \hat{\mathbf{n}}'}\frac{1}{4\pi R} - \frac{1}{4\pi R}\frac{\partial \Phi}{\partial \hat{\mathbf{n}}'}\right] \tag{2.167}$$

for $\mathbf{r} \in \partial V$. In conclusion, if the observation point is on the boundary of the region of interest, the integral representation yields half the value of the scalar potential. Incidentally, since the point $\mathbf{r}$ is located outside the region bounded by $S' \cup S'' \cup \partial V_1$ (Figure 2.10b) we could have obtained (2.167) also by starting with (2.162) and then by taking the limit for vanishing a.

The integral representation (2.160) states that the scalar potential in a region of space V containing charges can be computed so long as we know the charge density and the values of $\Phi(\mathbf{r})$ and its normal derivative on the surfaces that bound the region of interest. Therefore, it would appear that we have an explicit formula for the general solution to the problem (2.149). However, we learned in Section 2.5.1 that the Poisson equation admits a unique solution if *either* the potential *or* its normal derivative is assigned on the boundary of the solution domain, whereas (2.160) involves both quantities! In fact, $\Phi(\mathbf{r})$ and $\hat{\mathbf{n}} \cdot \nabla\Phi(\mathbf{r})$ for $\mathbf{r} \in \partial V$ are not independent, and specifying them arbitrarily can lead to non-physical results [1]. That is why it is more appropriate to refer to (2.160) as an integral representation rather than the solution to the Poisson equation.

A closer look at (2.160) and (2.162) suggests that the surface integrals over $\partial V = S \cup \partial V_1$ play a role similar to the volume integral involving the charges. More precisely, the surface integrals represent the effect of sources other than $\varrho(\mathbf{r})$ contained in V. Where are these sources located? Well, they surely cannot be inside V because they are already accounted for by $\varrho(\mathbf{r})$. Therefore, we are led to conclude that the surface integrals describe the effects of charges in $\mathbb{R}^3 \setminus \overline{V}$ and this includes V_1 in our setup.

In truth, we can show that if no sources are located externally to V, then the surface integrals in (2.160) are actually null. (We shall reach analogous conclusions for the integral representations of the magnetic vector potential in Section 5.1 and of the electromagnetic fields in the time-harmonic regime in Section 10.2.) Indeed, since (2.160) is a general result, we can apply it to find an expression for $\Phi(\mathbf{r})$ in the domain V_1. If there are no sources in V_1, the volume integral vanishes and we arrive at

$$\Phi(\mathbf{r}) = -\int\limits_{\partial V_1} \mathrm{d}S'\, \Phi(\mathbf{r}')\frac{\partial}{\partial \hat{\mathbf{n}}'}\frac{1}{4\pi R} + \int\limits_{\partial V_1} \mathrm{d}S' \frac{1}{4\pi R}\frac{\partial \Phi}{\partial \hat{\mathbf{n}}'} \tag{2.168}$$

for points $\mathbf{r} \in V_1$. Notice that we have taken the unit normal directed outward V_1, hence the different signs with respect to (2.160). Still, we have shown with (2.162) that the integral representation provides a null potential for observation points outside the domain of derivation. Therefore, we conclude from (2.168) that the right-hand side vanishes for $\mathbf{r} \notin V_1$ or, specifically, for $\mathbf{r} \in V$. But these surface integrals are the same as those appearing in (2.160), and thus we have proved that they vanish when no sources are present in V_1. Likewise, we can show that the combination of integrals over S vanishes as well, if no sources are located in $\mathbb{R}^3 \setminus (V \cup V_1)$.

We can now extend (2.160) to an unbounded region V, as we did in Section 2.5.1 for the uniqueness theorem, by letting S recede to infinity, under the assumption that no charges are present other

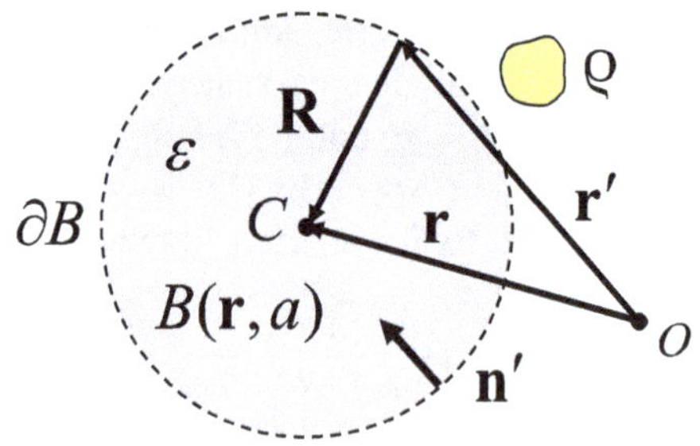

Figure 2.11 For proving the mean value theorem of electrostatics.

than those in V_ϱ. We choose $V := B(0,a) \setminus V_1$ and let a approach infinity. Then, the surface integral over S vanishes because both $\Phi(\mathbf{r})$ and the static Green function (2.131) satisfy estimates (2.21) and (2.20). Finally, if we take the medium in V_1 to be a dielectric with permittivity ε and also assume that no sources are located in V_1, the integral formula (2.160) becomes

$$\Phi(\mathbf{r}) = \int_{V_\varrho} \mathrm{d}V' \, \frac{\varrho(\mathbf{r}')/\varepsilon}{4\pi|\mathbf{r}-\mathbf{r}'|}, \qquad \mathbf{r} \in \mathbb{R}^3 \tag{2.169}$$

as we need not exclude the region V_1 at all. From a mathematical viewpoint the integral operator

$$\mathfrak{L}\{\bullet\} := \int_{V_\varrho} \mathrm{d}V' \, \frac{\{\bullet\}}{4\pi|\mathbf{r}-\mathbf{r}'|}, \qquad \mathbf{r} \in \mathbb{R}^3 \tag{2.170}$$

can be regarded as the inverse of $-\nabla^2\{\bullet\}$, i.e., the negative of the Laplace operator which appears in the Poisson equation (2.17).

As an application of the integral representation (2.160), we can prove the *mean value theorem* of electrostatics [1, 13] also known as the *Gauss theorem of arithmetic mean* [16, Section 8.6]. The latter states that the value of the potential at the center of a charge-free ball $B(\mathbf{r}, a)$ filled with a homogeneous isotropic dielectric medium equals the average of the values taken on by $\Phi(\mathbf{r}')$ over ∂B, the sphere of radius a. The relevant geometry is depicted in Figure 2.11.

Indeed, by invoking (2.160) with $V := B(\mathbf{r}, a)$ we have

$$\begin{aligned}\Phi(\mathbf{r}) &= \int_{\partial B} \mathrm{d}S' \, \Phi(\mathbf{r}') \frac{\partial}{\partial \hat{\mathbf{n}}'} \frac{1}{4\pi R} - \int_{\partial B} \mathrm{d}S' \, \frac{1}{4\pi R} \frac{\partial}{\partial \hat{\mathbf{n}}'} \Phi(\mathbf{r}') \\ &= \frac{1}{4\pi a^2} \int_{\partial B} \mathrm{d}S' \, \Phi(\mathbf{r}') + \frac{1}{4\pi a \varepsilon} \int_{\partial B} \mathrm{d}S' \, \hat{\mathbf{n}}(\mathbf{r}') \cdot \mathfrak{D}(\mathbf{r}') = \frac{1}{A_{\partial B}} \int_{\partial B} \mathrm{d}S' \, \Phi(\mathbf{r}')\end{aligned} \tag{2.171}$$

since $R = |\mathbf{r} - \mathbf{r}'| = a$, and the flux of $\mathfrak{D}(\mathbf{r})$ vanishes by virtue of the Gauss law (2.1) because the sources are located outside $B(\mathbf{r}, a)$. Alternatively, since the second integrand is in the form $\hat{\mathbf{n}}(\mathbf{r}') \cdot \nabla' \Phi$, one can apply the Gauss theorem (A.53) in reverse gear and then invoke the fact that $\Phi(\mathbf{r}')$ solves the Laplace equation within $B(\mathbf{r}, a)$.

A second mean value theorem can now be derived with the aid of (2.171) by integrating the potential over the entire ball $B(\mathbf{r}, a)$. By introducing a local system of polar spherical coordinates (τ', α', β') centered on $\mathbf{r}$, we have $\mathbf{r}' = \mathbf{r} + \boldsymbol{\tau}'$ and

$$\int_{B(\mathbf{r},a)} \mathrm{d}V' \, \Phi(\mathbf{r}') = \int_0^a \mathrm{d}\tau' \int_{|\mathbf{r}'-\mathbf{r}|=\tau'} \mathrm{d}S' \, \Phi(\mathbf{r}') = \int_0^a \mathrm{d}\tau' 4\pi\tau'^2 \Phi(\mathbf{r}) = \frac{4}{3}\pi a^3 \Phi(\mathbf{r}) \tag{2.172}$$

where the second step follows precisely from (2.171) applied over the ball $B(\mathbf{r}, \tau') \subseteq B(\mathbf{r}, a)$. Solving formally for $\Phi(\mathbf{r})$ yields the result

$$\Phi(\mathbf{r}) = \frac{3}{4\pi a^3} \int_{B(\mathbf{r},a)} \mathrm{d}V' \, \Phi(\mathbf{r}') = \frac{1}{V_B} \int_{B(\mathbf{r},a)} \mathrm{d}V' \, \Phi(\mathbf{r}') \tag{2.173}$$

that is, the average value of the potential over $B(\mathbf{r}, a)$ equals the value of Φ at the center of the ball, so long as the latter is devoid of charges.

Both (2.171) and (2.173) arise from the potential being a solution to the Laplace equation in a region V which contains the ball $B(\mathbf{r}, a)$, and as such these formulas can be extended to scalar fields defined in higher-dimensional spaces [8, Chapter 2], [20, Chapter 4].

Example 2.7 (Scalar potential of a spherical uniform distribution of charge (reprise))

Once again we consider the charge density $\varrho(\mathbf{r})$ given by (2.5) and use the integral solution (2.169) to compute $\Phi(\mathbf{r})$. Of course, we expect to find the same results as in Example 2.4. To calculate the volume integral we consider a polar spherical system of coordinates in which the polar angle α' is taken from the direction specified by the position vector $\mathbf{r}$. Thereby, the distance between source and observation points reads

$$R = |\mathbf{r} - \mathbf{r}'| = \left(r'^2 - 2rr' \cos\alpha' + r^2\right)^{1/2} \tag{2.174}$$

and we have

$$\begin{aligned} \Phi(r) &= \int_{B(0,a)} \mathrm{d}V' \frac{\varrho_0}{4\pi\varepsilon R} = \frac{\varrho_0}{4\pi\varepsilon} \int_0^a \mathrm{d}r' \int_0^\pi \mathrm{d}\alpha' \frac{2\pi r'^2 \sin\alpha'}{(r'^2 - 2rr' \cos\alpha' + r^2)^{1/2}} \\ &= \frac{\varrho_0}{2r\varepsilon} \int_0^a \mathrm{d}r' \, r' \left[\left(r'^2 - 2rr' \cos\alpha' + r^2\right)^{1/2}\right]_{\alpha'=0}^{\pi} = \frac{\varrho_0}{2r\varepsilon} \int_0^a \mathrm{d}r' \, r'(r' + r - |r - r'|) \end{aligned} \tag{2.175}$$

having observed that the problem is axial symmetric around $\mathbf{r}$. To calculate the integral over r' we distinguish two cases, namely, $r < a$ and $r > a$

$$\int_0^a \mathrm{d}r' \, r'(r' + r - |r - r'|) = \begin{cases} \displaystyle\int_0^r \mathrm{d}r' \, 2r'^2 + \int_r^a \mathrm{d}r' \, 2rr' = r\left(a^2 - \frac{r^2}{3}\right), & r < a \\ \displaystyle\int_0^a \mathrm{d}r' \, 2r'^2 = \frac{2}{3}a^3, & r > a. \end{cases} \tag{2.176}$$

By putting everything together we obtain

$$\Phi(r) = \begin{cases} \dfrac{\varrho_0}{6\varepsilon}\left(3a^2 - r^2\right), & r < a \\ \dfrac{\varrho_0 a^3}{3\varepsilon r}, & r > a \end{cases} \tag{2.177}$$

which is in agreement with the known result.

(End of Example 2.7)

An interesting situation is constituted by the presence of a conductor that entirely fills V_1. In this case the integral representation simplifies, because the normal derivative of $\Phi(\mathbf{r})$ is related to the surface charge density (cf. Section 2.4), whereas the scalar potential is constant on ∂V_1. For the flux integrals over ∂V_1 we have

$$-\int_{\partial V_1} dS' \frac{1}{4\pi R} \frac{\partial \Phi}{\partial \hat{\mathbf{n}}'} = \frac{1}{\varepsilon} \int_{\partial V_1} dS' \frac{\varrho_S(\mathbf{r}')}{4\pi R} \tag{2.178}$$

in light of (2.53), and also

$$\int_{\partial V_1} dS' \, \Phi(\mathbf{r}') \frac{\partial}{\partial \hat{\mathbf{n}}'} \frac{1}{4\pi R} = \Phi_{01} \int_{\partial V_1} dS' \, \hat{\mathbf{n}}(\mathbf{r}') \cdot \nabla' \frac{1}{4\pi R} = \Phi_{01} \int_{V_1} dV' \, \nabla'^2 \frac{1}{4\pi R} = 0 \tag{2.179}$$

where we have applied the Gauss theorem since $1/(4\pi R)$ is twice differentiable for $\mathbf{r} \notin \overline{V}_1$, $\mathbf{r}' \in \partial V_1$. Thus, (2.160) reads

$$\Phi(\mathbf{r}) = \frac{1}{\varepsilon} \int_{V_\varrho} dV' \frac{\varrho(\mathbf{r}')}{4\pi R} + \frac{1}{\varepsilon} \int_{\partial V_1} dS' \frac{\varrho_S(\mathbf{r}')}{4\pi R} + \int_S dS' \, \Phi(\mathbf{r}') \frac{\partial}{\partial \hat{\mathbf{n}}'} \frac{1}{4\pi R} - \int_S dS' \frac{1}{4\pi R} \frac{\partial \Phi}{\partial \hat{\mathbf{n}}'} \tag{2.180}$$

where the remaining flux integrals may be null if no sources are located outside $V \cup V_1$.

When the scalar potential vanishes on the boundary of the region of interest (this is the homogeneous instance of the Dirichlet problem) we can obtain a particular integral representation for $\Phi(\mathbf{r})$ by starting from the general result (2.160). For the sake of simplicity, in the geometry of Figure 2.7a we let the excluded region V_1 shrink to a point in that, if necessary, we can always think of V as being surface-wise multiply-connected in the end. Therefore, for any point $\mathbf{r} \in V$ we have

$$\Phi(\mathbf{r}) = \frac{1}{\varepsilon} \int_{V_\varrho} dV' \frac{\varrho(\mathbf{r}')}{4\pi R} - \int_S dS' \frac{1}{4\pi R} \frac{\partial \Phi}{\partial \hat{\mathbf{n}}'} = -\int_V dV' \frac{\nabla'^2 \Phi}{4\pi R} - \int_S dS' \frac{\hat{\mathbf{n}}' \cdot \nabla' \Phi}{4\pi R} \tag{2.181}$$

where we have invoked the Poisson equation (2.149) with $f(\mathbf{r}') = 0$ on S.

In order to transform the volume integral over V by means of the Gauss theorem (A.53) we isolate the singular point $\mathbf{r}$ with a small ball $B(\mathbf{r}, a) \subset V$ (cf. Figure 2.8) and define the surface-wise multiply-connected region $V_a := V \setminus B[\mathbf{r}, a]$. Besides, since $\nabla\Phi$ may not be continuously differentiable throughout V owing to the presence of discontinuities across ∂V_ϱ, in principle we should split the integration into two parts, namely, over V_ϱ and over $V \setminus \overline{V}_\varrho$. However, from the discussion of Section 1.2.2 (pages 10 and ff.) we know that the Gauss theorem can still be applied over V so long as $\hat{\mathbf{n}} \cdot \nabla\Phi$ is continuous through ∂V_ϱ. Thus, we make this assumption and explicitly deal only with the singularity of the Green function (2.131). Then, thanks to the differential identity (H.51) we have

$$\begin{aligned}
\Phi(\mathbf{r}) &= \int_{V_a} dV' \, \nabla'\Phi \cdot \nabla' \frac{1}{4\pi R} - \int_{V_a} dV' \, \nabla' \cdot \frac{\nabla'\Phi}{4\pi R} - \int_{B(\mathbf{r},a)} dV' \frac{\nabla'^2 \Phi}{4\pi R} - \int_S dS' \frac{\hat{\mathbf{n}}' \cdot \nabla' \Phi}{4\pi R} \\
&= \int_{V_a} dV' \frac{\nabla'\Phi \cdot \hat{\mathbf{R}}}{4\pi R^2} + \int_S dS' \frac{\hat{\mathbf{n}}' \cdot \nabla' \Phi}{4\pi R} + \int_{\partial B} dS' \frac{\hat{\mathbf{n}}' \cdot \nabla' \Phi}{4\pi R} - \int_{B(\mathbf{r},a)} dV' \frac{\nabla'^2 \Phi}{4\pi R} - \int_S dS' \frac{\hat{\mathbf{n}}' \cdot \nabla' \Phi}{4\pi R}
\end{aligned}$$

$$= \int_{V_a} dV' \, \nabla'\Phi \cdot \frac{\mathbf{r}-\mathbf{r}'}{4\pi R^3} + \frac{1}{4\pi a} \int_{\partial B} dS' \, \hat{\mathbf{n}}' \cdot \nabla'\Phi - \int_{B(\mathbf{r},a)} dV' \, \frac{\nabla'^2\Phi}{4\pi R}$$

$$= \int_{V_a} dV' \, \nabla'\Phi \cdot \frac{\mathbf{r}-\mathbf{r}'}{4\pi R^3} + \int_{B(\mathbf{r},a)} dV' \left(\frac{1}{a} - \frac{1}{R}\right) \frac{\nabla'^2\Phi}{4\pi} \tag{2.182}$$

having applied the Gauss theorem twice with the unit normals on S and ∂B positively oriented inwards V_a (cf. Figure 2.7a). We wish to show that, in the limit as $a \to 0$, the integral over V_a remains finite and the integral over $B(\mathbf{r}, a)$ vanishes.

Indeed, since $\nabla'\Phi$ is bounded over V, by virtue of the Cauchy-Schwarz inequality (D.151) we have

$$\left| \int_{V_a} dV' \, \frac{\nabla'\Phi \cdot \hat{\mathbf{R}}}{4\pi R^2} \right| \leqslant \int_{V_a} dV' \, \frac{|\nabla'\Phi \cdot \hat{\mathbf{R}}|}{4\pi R^2} \leqslant \int_{V_a} dV' \, \frac{\|\nabla'\Phi\|_\infty}{4\pi R^2} \leqslant \int_{B(\mathbf{r},b)\setminus B(\mathbf{r},a)} dV' \, \frac{\|\nabla'\Phi\|_\infty}{4\pi R^2}$$
$$= \left\|\nabla'\Phi\right\|_\infty (b-a) \leqslant \left\|\nabla'\Phi\right\|_\infty b \tag{2.183}$$

where $B(\mathbf{r}, b)$ is a larger ball which contains V (see Figure 2.9 for a similar geometrical setup). This string of inequalities indicates that the integral over V_a is bounded for any value of a and in particular for $a = 0$. Concerning the contribution from $B(\mathbf{r}, a)$ in (2.182) we first observe that if $B(\mathbf{r}, a) \cap V_\varrho = \emptyset$ (i.e., if the ball is located outside the source region) then the integral vanishes because so does $\nabla'^2\Phi$ by virtue of (2.149). On the other hand, if $B(\mathbf{r}, a) \cap V_\varrho \neq \emptyset$ we have

$$\left| \int_{B(\mathbf{r},a)} dV' \left(\frac{1}{a} - \frac{1}{R}\right) \frac{\nabla'^2\Phi}{4\pi} \right| = \left| \int_{B(\mathbf{r},a)\cap V_\varrho} dV' \left(\frac{1}{R} - \frac{1}{a}\right) \frac{\varrho(\mathbf{r}')}{4\pi\varepsilon} \right| \leqslant \int_{B(\mathbf{r},a)\cap V_\varrho} dV' \left(\frac{1}{R} - \frac{1}{a}\right) \frac{\|\varrho\|_\infty}{4\pi\varepsilon}$$
$$\leqslant \int_{B(\mathbf{r},a)} dV' \left(\frac{1}{R} - \frac{1}{a}\right) \frac{\|\varrho\|_\infty}{4\pi\varepsilon} = \frac{a^2 \, \|\varrho\|_\infty}{6\varepsilon} \xrightarrow[a\to 0]{} 0 \tag{2.184}$$

where we have integrated in a system of local polar spherical coordinates (R, α', β') centered in $\mathbf{r}$.

All in all, we have proved that for $a \to 0$ (2.182) yields

$$\Phi(\mathbf{r}) = \int_V dV' \, \nabla'\Phi \cdot \frac{\mathbf{r}-\mathbf{r}'}{4\pi R^3} = \int_V dV' \, \nabla'\Phi \cdot \nabla' \frac{1}{4\pi R}, \qquad \mathbf{r} \in V \tag{2.185}$$

which we can regard as an identity of sorts that holds in case the potential is null on the boundary of V. In fact, for observation points outside the region of interest the right-hand side of (2.185) returns zero. This property should be expected in light of the analogous behavior exhibited by (2.160) from which (2.185) has been derived. Still, we can check this claim directly by using (H.51) again and considering

$$\int_V dV' \, \nabla'\Phi \cdot \nabla' \frac{1}{4\pi R} = \int_V dV' \, \nabla' \cdot \left[\Phi(\mathbf{r}')\nabla' \frac{1}{4\pi R}\right] - \int_V dV' \, \Phi(\mathbf{r}')\nabla'^2 \frac{1}{4\pi R}$$
$$= -\int_S dS' \, \Phi(\mathbf{r}') \, \hat{\mathbf{n}}' \cdot \nabla' \frac{1}{4\pi R} = 0 \tag{2.186}$$

where the Gauss theorem has been applied since $\Phi(\mathbf{r}')$ is continuously differentiable in V and $\nabla'(1/R)$ is regular for $\mathbf{r}' \in V$ and $\mathbf{r} \notin \overline{V}$. The result follows because $\Phi(\mathbf{r}') = 0$ on S and $1/R$ solves the Laplace equation for $\mathbf{r}' \in V$ and $\mathbf{r} \notin \overline{V}$ (see Section 2.6.1).

The three types of integrals which combine in (2.160) or (2.180) to give the scalar potential are so important as to deserve names of their own, viz.,

- the *volume* potential

$$\mathcal{V}(\mathbf{r}) := \int_{V_\varrho} \mathrm{d}V' \frac{\varrho(\mathbf{r}')/\varepsilon}{4\pi|\mathbf{r}-\mathbf{r}'|} = \frac{1}{\varepsilon} G(\mathbf{r}) * \varrho(\mathbf{r}) \tag{2.187}$$

- the *double-layer* potential

$$\mathcal{W}(\mathbf{r}) := \int_{S} \mathrm{d}S' \, \Phi(\mathbf{r}') \frac{\partial}{\partial \hat{\mathbf{n}}'} \frac{1}{4\pi|\mathbf{r}-\mathbf{r}'|} = [\Phi(\mathbf{r})\hat{\mathbf{n}}(\mathbf{r})] \overset{*}{\cdot} \nabla G(\mathbf{r}) \tag{2.188}$$

- the *single-layer* potential

$$\mathcal{S}(\mathbf{r}) := -\int_{S} \mathrm{d}S' \frac{1}{4\pi|\mathbf{r}-\mathbf{r}'|} \frac{\partial \Phi}{\partial \hat{\mathbf{n}}'} = G(\mathbf{r}) * [-\hat{\mathbf{n}}(\mathbf{r}) \cdot \nabla \Phi(\mathbf{r})] \tag{2.189}$$

where '$*$' indicates the convolution product and '$\overset{*}{\cdot}$' denotes convolution and dot-product between vector operands. We devote the following sections to discussing the physical meaning and the mathematical properties of each type of potential.

2.8 Volume potential

In the context of (2.160) and (2.180) the volume potential $\mathcal{V}(\mathbf{r})$ represents the contribution of the charges located in the region of interest. Indeed, the integration over V_ϱ conforms with the intuitive idea that the potential due to a set of charges coincides with the sum of the individual contributions.

From a physical point of view, $\mathcal{V}(\mathbf{r})$ is the electrostatic potential generated by $\varrho(\mathbf{r})$ in the background unbounded homogeneous medium endowed with permittivity ε, that is, as if in the geometrical setup of Figure 2.7a the material medium in V_1 were *absent*. Therefore, $\mathcal{V}(\mathbf{r})$ is the unique solution to the Poisson equation

$$\begin{cases} \varepsilon \nabla^2 \Phi(\mathbf{r}) = -\varrho(\mathbf{r}), & \mathbf{r} \in \mathbb{R}^3 \\ \Phi(\mathbf{r}) = \mathcal{O}\left(\dfrac{1}{|\mathbf{r}|}\right), & r = |\mathbf{r}| \to +\infty \end{cases} \tag{2.190}$$

with $\varrho(\mathbf{r}) \neq 0$ in $V_\varrho \subset \mathbb{R}^3$. We can show that the boundary condition at infinity is actually satisfied by observing that [cf. (3.196) and (3.204) in Section 3.6]

$$\frac{1}{|\mathbf{r}-\mathbf{r}'|} = \frac{1}{r}\left(1 - 2\hat{\mathbf{r}} \cdot \frac{\mathbf{r}'}{r} + \frac{r'^2}{r^2}\right)^{-\frac{1}{2}} = \frac{1}{r}\left[1 + \hat{\mathbf{r}} \cdot \frac{\mathbf{r}'}{r} + \mathcal{O}\left(\frac{1}{r^2}\right)\right] = \frac{1}{r} + \mathcal{O}\left(\frac{1}{r^2}\right), \quad r \to +\infty \tag{2.191}$$

whence by inserting this asymptotic expansion into (2.187) we find

$$\int_{V_\varrho} \mathrm{d}V' \frac{\varrho(\mathbf{r}')/\varepsilon}{4\pi R} = \frac{1}{4\pi\varepsilon r} \int_{V_\varrho} \mathrm{d}V' \, \varrho(\mathbf{r}') + \mathcal{O}\left(\frac{1}{r^2}\right), \qquad r \to +\infty \tag{2.192}$$

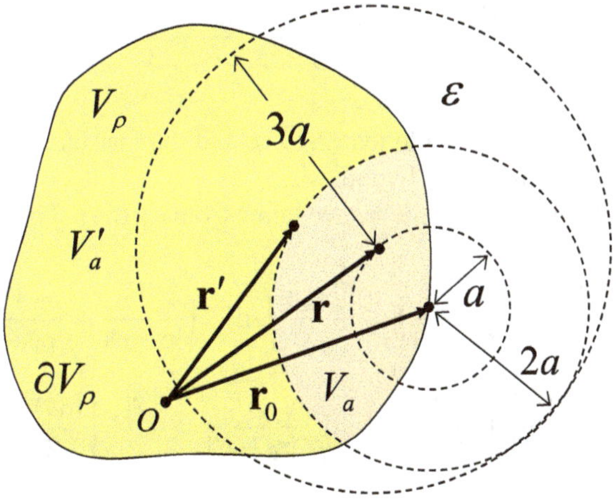

Figure 2.12 Geometrical construction for proving the continuity of the static volume potential.

which is in line with (2.23).

We wish to show that the static volume potential $\mathcal{V}(\mathbf{r})$ defined by (2.187) is a *continuous* function for $\mathbf{r} \in \mathbb{R}^3$ under the only assumption that the charge density $\varrho(\mathbf{r})$ is *bounded* in V_ϱ [20, pp. 199-200], [16,31]. Accordingly, given two points $\mathbf{r}, \mathbf{r}_0 \in \mathbb{R}^3$ and an arbitrary small number $\eta > 0$ there must exist a distance $d > 0$ such that

$$|\mathcal{V}(\mathbf{r}) - \mathcal{V}(\mathbf{r}_0)| < \eta \tag{2.193}$$

provided $|\mathbf{r} - \mathbf{r}_0| < d$. This is the usual definition or criterion for continuity in a three-dimensional space, but the difficulty here lies in the kernel of $\mathcal{V}(\mathbf{r})$ being singular for $\mathbf{r} \in V_\varrho$. To get around this hurdle we isolate the point $\mathbf{r}_0$ with the ball $B(\mathbf{r}_0, 2a)$ and define the subsets $V_a := V_\varrho \cap B(\mathbf{r}_0, 2a)$ and $V'_a := V_\varrho \setminus V_a$, as is suggested in Figure 2.12 in particular for $\mathbf{r}_0 \in \partial V_\varrho$. We split the integration into two parts, namely, over V'_a and over V_a

$$\mathcal{V}(\mathbf{r}) - \mathcal{V}(\mathbf{r}_0) = \frac{1}{4\pi\varepsilon} \int_{V'_a} \mathrm{d}V' \, \varrho(\mathbf{r}') \left(\frac{1}{|\mathbf{r} - \mathbf{r}'|} - \frac{1}{|\mathbf{r}_0 - \mathbf{r}'|} \right)$$
$$- \frac{1}{4\pi\varepsilon} \int_{V_a} \mathrm{d}V' \, \varrho(\mathbf{r}') \left(\frac{1}{|\mathbf{r} - \mathbf{r}'|} - \frac{1}{|\mathbf{r}_0 - \mathbf{r}'|} \right) \tag{2.194}$$

under the assumption that $\mathbf{r} \in B(\mathbf{r}_0, a)$, which is true so long as $d \leqslant a$. Notice that V_a may happen to be the empty set if $\mathbf{r}, \mathbf{r}_0 \notin V_\varrho$, in which instance the integral over V_a need not be defined at all.

First of all, we observe that the function to be integrated over V'_a is regular and continuous therein because

$$|\mathbf{r} - \mathbf{r}'| \geqslant a \qquad\qquad |\mathbf{r}_0 - \mathbf{r}'| \geqslant 2a \tag{2.195}$$

for $\mathbf{r} \in B(\mathbf{r}_0, a)$ (see Figure 2.12). Therefore, we can estimate

$$\frac{1}{4\pi\varepsilon} \left| \int_{V'_a} \mathrm{d}V' \varrho(\mathbf{r}') \left(\frac{1}{|\mathbf{r} - \mathbf{r}'|} - \frac{1}{|\mathbf{r}_0 - \mathbf{r}'|} \right) \right| \leqslant \frac{1}{4\pi\varepsilon} \int_{V'_a} \mathrm{d}V' \, |\varrho(\mathbf{r}')| \left| \frac{1}{|\mathbf{r} - \mathbf{r}'|} - \frac{1}{|\mathbf{r}_0 - \mathbf{r}'|} \right|$$

$$\leqslant \frac{\|\varrho\|_\infty}{4\pi\varepsilon}\int_{V_a'} \mathrm{d}V' \frac{|\mathbf{r}_0-\mathbf{r}|}{|\mathbf{r}-\mathbf{r}'||\mathbf{r}_0-\mathbf{r}'|} \leqslant \frac{\|\varrho\|_\infty}{4\pi\varepsilon}\int_{V_a'} \mathrm{d}V' \frac{d}{2a^2} < \frac{\eta}{3} \tag{2.196}$$

where we have invoked (H.20) and exploited (2.195), and the last inequality follows because it is possible to choose a suitable small d by hypothesis.

Secondly, the integral over V_a is finite, as we know from (2.161). Therefore, we estimate

$$\begin{aligned}\frac{1}{4\pi\varepsilon}\left|\int_{V_a} \mathrm{d}V'\,\varrho(\mathbf{r}')\left(\frac{1}{|\mathbf{r}-\mathbf{r}'|}-\frac{1}{|\mathbf{r}_0-\mathbf{r}'|}\right)\right| &\leqslant \frac{\|\varrho\|_\infty}{4\pi\varepsilon}\int_{V_a} \mathrm{d}V'\left(\frac{1}{|\mathbf{r}-\mathbf{r}'|}+\frac{1}{|\mathbf{r}_0-\mathbf{r}'|}\right)\\ \leqslant \frac{\|\varrho\|_\infty}{4\pi\varepsilon}\int_{B(\mathbf{r},3a)} \mathrm{d}V'\,\frac{1}{|\mathbf{r}-\mathbf{r}'|} + \frac{\|\varrho\|_\infty}{4\pi\varepsilon}\int_{B(\mathbf{r}_0,2a)} \mathrm{d}V'\,\frac{1}{|\mathbf{r}_0-\mathbf{r}'|} &= \frac{\|\varrho\|_\infty}{\varepsilon}\left[\frac{(3a)^2}{2}+\frac{(2a)^2}{2}\right] < \frac{2}{3}\eta\end{aligned} \tag{2.197}$$

where the second-to-last estimates follow by observing that $|\mathbf{r}-\mathbf{r}'| \leqslant |\mathbf{r}-\mathbf{r}_0| + |\mathbf{r}_0-\mathbf{r}'| \leqslant 3a$ and that $V_a \subseteq B(\mathbf{r}_0, 2a)$ (see Figure 2.12). Moreover, we can choose the radius a small enough for the last inequality to hold, which also makes $\mathbf{r}$ approach $\mathbf{r}_0$ since $|\mathbf{r}-\mathbf{r}_0| < d < a$.

By combining these results we find

$$\begin{aligned}|\mathcal{V}(\mathbf{r})-\mathcal{V}(\mathbf{r}_0)| \leqslant{}& \frac{1}{4\pi\varepsilon}\left|\int_{V_a'} \mathrm{d}V'\,\varrho(\mathbf{r}')\left(\frac{1}{|\mathbf{r}-\mathbf{r}'|}-\frac{1}{|\mathbf{r}_0-\mathbf{r}'|}\right)\right|\\ &+\frac{1}{4\pi\varepsilon}\left|\int_{V_a} \mathrm{d}V'\,\varrho(\mathbf{r}')\left(\frac{1}{|\mathbf{r}-\mathbf{r}'|}-\frac{1}{|\mathbf{r}_0-\mathbf{r}'|}\right)\right| < \frac{\eta}{3}+\frac{2}{3}\eta = \eta\end{aligned} \tag{2.198}$$

and thus we conclude that $\mathcal{V}(\mathbf{r})$ is continuous for $\mathbf{r} \in \mathbb{R}^3$.

We wish to show that the static volume potential is *continuously differentiable* for $\mathbf{r} \in \mathbb{R}^3$ if the charge density is at least *bounded* in V_ϱ. In particular, it holds true

$$\nabla\mathcal{V}(\mathbf{r}) = \frac{1}{4\pi\varepsilon}\int_{V_\varrho} \mathrm{d}V'\,\varrho(\mathbf{r}')\nabla\frac{1}{|\mathbf{r}-\mathbf{r}'|} = -\frac{1}{4\pi\varepsilon}\int_{V_\varrho} \mathrm{d}V'\,\varrho(\mathbf{r}')\frac{\mathbf{r}-\mathbf{r}'}{|\mathbf{r}-\mathbf{r}'|^3}, \qquad \mathbf{r} \in \mathbb{R}^3 \tag{2.199}$$

that is to say, it is permitted to *interchange* the order of differentiation and integration and to compute the gradient of the static Green function $G(\mathbf{r},\mathbf{r}')$ first, despite the singular nature of the integrand [16, 20, 31]. That the improper integral in the rightmost-hand side remains finite for $\mathbf{r} \in \mathbb{R}^3$ can be proved by isolating the singular point $\mathbf{r}$ by means of a ball $B(\mathbf{r}, a)$ and by integrating over $V_\varrho \setminus B(\mathbf{r}, a)$, as in Figures 2.8 and 2.9. Specifically, if the charge density $\varrho(\mathbf{r}')$ is bounded, we have the estimates

$$\begin{aligned}\left|\int_{V_\varrho\setminus B(\mathbf{r},a)} \mathrm{d}V'\,\varrho(\mathbf{r}')\nabla'\frac{1}{4\pi R}\right| &\leqslant \int_{V_\varrho\setminus B(\mathbf{r},a)} \mathrm{d}V'\,\frac{\|\varrho\|_\infty}{4\pi R^2} \leqslant \int_{B(\mathbf{r},b)\setminus B(\mathbf{r},a)} \mathrm{d}V'\,\frac{\|\varrho\|_\infty}{4\pi R^2}\\ &= \|\varrho\|_\infty \int_a^b \mathrm{d}R = \|\varrho\|_\infty\,(b-a) \leqslant \|\varrho\|_\infty\, b\end{aligned} \tag{2.200}$$

and this shows that the integral remains bounded for any value of a.

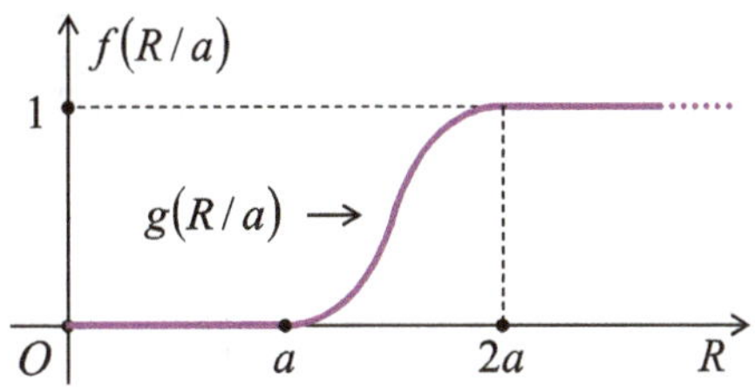

Figure 2.13 The graph of an instance of the radial three-dimensional step function (2.201).

We proceed by introducing the real-valued radial function $f(|\mathbf{r}-\mathbf{r}'|/a) = f(R/a) \in \mathcal{C}^1(\mathbb{R}^+)$ with $a > 0$, defined as (Figure 2.13)

$$f\left(\frac{R}{a}\right) = \begin{cases} 0, & 0 < R/a < 1 \\ g(R/a) \in [0,1], & 1 < R/a < 2 \\ 1, & R/a > 2 \end{cases} \tag{2.201}$$

where $g(R/a) \in \mathcal{C}^1([1,2])$ obeys the boundary conditions

$$g(1^+) = 0, \qquad g(2^-) = 1, \qquad g'(1^+) = g'(2^-) = 0 \tag{2.202}$$

in order to ensure the overall differentiability of $f(R/a)$, though the exact form of $g(R/a)$ is not important for the discussion. Since $f(R/a)$ vanishes in the ball $B(\mathbf{r}, a)$ and is equal to one outside $B(\mathbf{r}, 2a)$, it can be regarded as a three-dimensional step function of sorts with smooth transition from 0 to 1 [cf. (C.49)].

Then, we define the auxiliary potential

$$\Upsilon(\mathbf{r}; a) := \frac{1}{4\pi\varepsilon} \int_{V_\varrho} dV' \, \frac{\varrho(\mathbf{r}')}{|\mathbf{r}-\mathbf{r}'|} f\left(\frac{|\mathbf{r}-\mathbf{r}'|}{a}\right), \qquad a > 0 \tag{2.203}$$

and observe that, as a function of a, $\Upsilon(\mathbf{r}; a)$ tends to $\mathcal{V}(\mathbf{r})$ for $a \to 0^+$ thanks to the chosen form of $f(R/a)$. As a matter of fact, $\mathcal{V}(\mathbf{r})$ and $\Upsilon(\mathbf{r}; a)$ even coincide when $V_\varrho \cap B(\mathbf{r}, 2a) = \emptyset$. Otherwise, since $\varrho(\mathbf{r}')$ is bounded and $0 \leqslant 1 - f(R/a) \leqslant 1$ by construction, for $\mathbf{r} \in V_\varrho$ we have

$$|\, \mathcal{V}(\mathbf{r}) - \Upsilon(\mathbf{r}; a)| \leqslant \frac{\|\varrho\|_\infty}{4\pi\varepsilon} \int_{B(\mathbf{r},2a)} dV' \, \frac{1}{|\mathbf{r}-\mathbf{r}'|} = \frac{\|\varrho\|_\infty}{\varepsilon} \int_0^{2a} d\tau' \, \tau' = 2a^2 \frac{\|\varrho\|_\infty}{\varepsilon} \xrightarrow[a \to 0^+]{} 0 \tag{2.204}$$

where we have integrated in local polar spherical coordinates (τ', α', β') centered on $\mathbf{r}$. The convergence is uniform in $\mathbb{R}^3$ because the leftmost member of (2.204) is dominated by a vanishing constant which does not depend on the observation point $\mathbf{r}$.

Next, we show that $\nabla\Upsilon(\mathbf{r}; a)$ tends to the expression in the rightmost-hand side of (2.199) for $a \to 0^+$. This goal can be accomplished in a component-wise manner by expanding the gradient operator in Cartesian coordinates [cf. (A.26)]. Since the integrand in (2.203) is continuously differentiable for $(\mathbf{r}, \mathbf{r}') \in \mathbb{R}^3 \times V_\varrho$, we may bring the spatial derivatives inside the integral, and thus $\Upsilon(\mathbf{r}; a) \in \mathcal{C}^1(\mathbb{R}^3)$. For instance, we have

$$\left| \int_{V_\varrho} dV' \, \frac{\varrho(\mathbf{r}')}{4\pi\varepsilon} \frac{\partial}{\partial x} \frac{1}{|\mathbf{r}-\mathbf{r}'|} - \frac{\partial}{\partial x} \int_{V_\varrho} dV' \, \frac{\varrho(\mathbf{r}')}{4\pi\varepsilon |\mathbf{r}-\mathbf{r}'|} f\left(\frac{|\mathbf{r}-\mathbf{r}'|}{a}\right) \right| =$$

$$= \left| \int_{V_\varrho} \mathrm{d}V' \frac{\varrho(\mathbf{r}')}{4\pi\varepsilon} \frac{\partial}{\partial x} \frac{1}{|\mathbf{r}-\mathbf{r}'|} - \int_{V_\varrho} \mathrm{d}V' \frac{\varrho(\mathbf{r}')}{4\pi\varepsilon} \frac{\partial}{\partial x} \left[\frac{1}{|\mathbf{r}-\mathbf{r}'|} f\left(\frac{|\mathbf{r}-\mathbf{r}'|}{a} \right) \right] \right|$$

$$= \left| \int_{B(\mathbf{r},2a)} \mathrm{d}V' \frac{\varrho(\mathbf{r}')}{4\pi\varepsilon} \frac{\partial}{\partial x} \left\{ \frac{1}{|\mathbf{r}-\mathbf{r}'|} \left[1 - f\left(\frac{|\mathbf{r}-\mathbf{r}'|}{a} \right) \right] \right\} \right|$$

$$\leqslant \int_{B(\mathbf{r},2a)} \mathrm{d}V' \frac{\|\varrho\|_\infty}{4\pi\varepsilon} \left(\left| \frac{\partial}{\partial x} \frac{1}{|\mathbf{r}-\mathbf{r}'|} \right| + \frac{\|f'\|_\infty}{a|\mathbf{r}-\mathbf{r}'|} \right) \leqslant \frac{\|\varrho\|_\infty}{4\pi\varepsilon} \int_{B(\mathbf{r},2a)} \mathrm{d}V' \left(\frac{1}{R^2} + \frac{\|f'\|_\infty}{aR} \right)$$

$$= \frac{\|\varrho\|_\infty}{\varepsilon} \int_0^{2a} \mathrm{d}\tau' \left(1 + \frac{\tau'}{a} \|f'\|_\infty \right) = 2 \frac{\|\varrho\|_\infty}{\varepsilon} \left(1 + \|f'\|_\infty \right) a \xrightarrow[a \to 0^+]{} 0 \tag{2.205}$$

and the same conclusion holds for the derivatives with respect to y and z. The symbol $f'(\bullet)$ denotes the derivative of $f(\bullet)$ with respect to the argument. Also in this case the convergence is uniform in $\mathbb{R}^3$ because the leftmost member of (2.205) is dominated by a vanishing constant which is independent of the observation point $\mathbf{r}$. Finally, we notice that the proof holds trivially if $V_\varrho \cap B(\mathbf{r}, 2a) = \emptyset$ because then $\varrho(\mathbf{r}') = 0$ for $\mathbf{r}' \in B(\mathbf{r}, 2a)$.

Lastly, we need to show that the limit of $\nabla\Upsilon(\mathbf{r}; a)$ is precisely $\nabla\mathcal{V}(\mathbf{r})$. Although this claim sounds intuitively true in light of (2.204), still the issue is quite a subtle point, as it ultimately concerns the possibility of interchanging the order of two limits, and there is no guarantee, in general, that such operation is permitted.

For ease of manipulation we temporarily adopt the symbol $\Xi(\mathbf{r})$ to indicate the vector field in the rightmost member of (2.199). Then, we pick up two points $\mathbf{r}_1$ and $\mathbf{r}_2$ in $\mathbb{R}^3$ and consider the following function of the real dimensionless parameter $\tau \in [0, 1]$

$$\Psi((1-\tau)\mathbf{r}_1 + \tau\mathbf{r}_2) := \mathcal{V}(\mathbf{r}_1) + \int_0^\tau \mathrm{d}\tau' \, (\mathbf{r}_2 - \mathbf{r}_1) \cdot \Xi((1-\tau')\mathbf{r}_1 + \tau'\mathbf{r}_2) \tag{2.206}$$

which essentially arises from the line integral of the field $\Xi(\mathbf{r})$ from $\mathbf{r}_1$ to $\mathbf{r}$ along the straight segment that connects $\mathbf{r}_1$ to $\mathbf{r}_2$, as is suggested in Figure 2.14. Definition (2.206) makes sense because by means of (2.205) and analogous estimates we have proved that the Cartesian components of $\Xi(\mathbf{r})$ are the uniform limit of the components of the continuous field $\nabla\Upsilon(\mathbf{r}; a)$, and hence $\Xi(\mathbf{r})$ is continuous [32, Chapter 7] and as such it can be integrated. More importantly, the function $\Psi((1-\tau)\mathbf{r}_1 + \tau\mathbf{r}_2)$ is constructed so as to coincide with the potential $\mathcal{V}(\mathbf{r})$ when $\tau = 0$ or, equivalently, $\mathbf{r} = \mathbf{r}_1$. Likewise, the auxiliary potentials $\Upsilon(\mathbf{r}; a)$ can be expressed as

$$\Upsilon((1-\tau)\mathbf{r}_1 + \tau\mathbf{r}_2; a) = \Upsilon(\mathbf{r}_1; a) + \int_0^\tau \mathrm{d}\tau' \, (\mathbf{r}_2 - \mathbf{r}_1) \cdot \nabla\Upsilon((1-\tau')\mathbf{r}_1 + \tau'\mathbf{r}_2; a) \tag{2.207}$$

when restricted to points $\mathbf{r}(\tau) := (1-\tau)\mathbf{r}_1 + \tau\mathbf{r}_2$. By subtracting (2.206) and (2.207) side by side and taking the absolute values we get

$$|\Psi(\mathbf{r}(\tau)) - \Upsilon(\mathbf{r}(\tau); a)|$$

$$= \left| \mathcal{V}(\mathbf{r}_1) - \Upsilon(\mathbf{r}_1; a) + \int_0^\tau \mathrm{d}\tau' \, (\mathbf{r}_2 - \mathbf{r}_1) \cdot [\Xi((1-\tau')\mathbf{r}_1 + \tau'\mathbf{r}_2) - \nabla\Upsilon((1-\tau')\mathbf{r}_1 + \tau'\mathbf{r}_2; a)] \right|$$

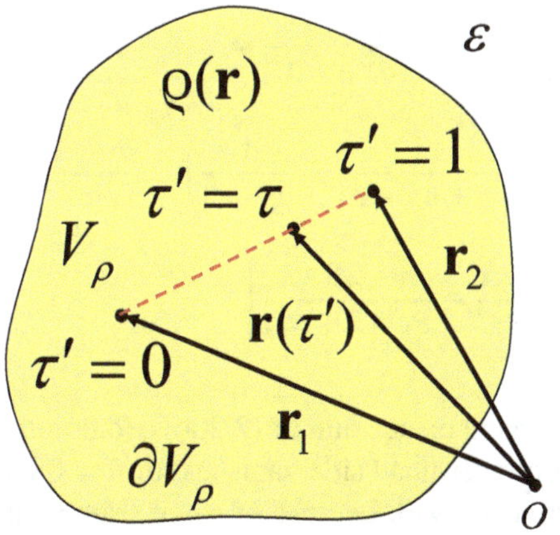

Figure 2.14 Geometrical construction for computing the gradient of the volume potential $\mathcal{V}(\mathbf{r})$*.*

$$\leqslant |\mathcal{V}(\mathbf{r}_1) - \Upsilon(\mathbf{r}_1; a)| + \int_0^1 \mathrm{d}\tau' \, |\mathbf{r}_2 - \mathbf{r}_1| \left| \Xi((1-\tau')\mathbf{r}_1 + \tau'\mathbf{r}_2) - \nabla\Upsilon((1-\tau')\mathbf{r}_1 + \tau'\mathbf{r}_2; a) \right|$$

$$\leqslant 2a^2 \frac{\|\varrho\|_\infty}{\varepsilon} + 6|\mathbf{r}_2 - \mathbf{r}_1| \frac{\|\varrho\|_\infty}{\varepsilon} \left(1 + \|f'\|_\infty\right) a \xrightarrow[a \to 0^+]{} 0 \tag{2.208}$$

where we have used the triangle inequality (H.19) and the Cauchy-Schwarz inequality (D.151), and applied (2.204) once (with $\mathbf{r} = \mathbf{r}_1$) and (2.205) three times before computing the integral.

This intermediate result says that $\Upsilon(\mathbf{r}; a)$ tends to $\Psi(\mathbf{r})$ for all observation points on the straight segment $\mathbf{r} = (1-\tau)\mathbf{r}_1 + \tau\mathbf{r}_2$. However, since $\mathbf{r}_1$ and $\mathbf{r}_2$ are arbitrary, we conclude that $\Upsilon(\mathbf{r}; a)$ tends to $\Psi(\mathbf{r})$ everywhere in $\mathbb{R}^3$. But then, since $\Upsilon(\mathbf{r}; a)$ tends also to $\mathcal{V}(\mathbf{r})$ in view of (2.204), $\Psi(\mathbf{r})$ must needs coincide with $\mathcal{V}(\mathbf{r})$, because this limit is unique. By taking the derivative with respect to τ of both sides of (2.206) we obtain

$$\frac{\mathrm{d}}{\mathrm{d}\tau} \Psi((1-\tau)\mathbf{r}_1 + \tau\mathbf{r}_2) = (\mathbf{r}_2 - \mathbf{r}_1) \cdot \nabla\Psi(\mathbf{r}) = (\mathbf{r}_2 - \mathbf{r}_1) \cdot \Xi(\mathbf{r}) \tag{2.209}$$

whence we conclude

$$-\frac{1}{4\pi\varepsilon} \int_{V_\varrho} \mathrm{d}V' \, \varrho(\mathbf{r}') \frac{\mathbf{r} - \mathbf{r}'}{|\mathbf{r} - \mathbf{r}'|^3} = \Xi(\mathbf{r}) = \nabla\Psi(\mathbf{r}), \qquad \mathbf{r} \in \mathbb{R}^3 \tag{2.210}$$

in view of the arbitrariness of $\mathbf{r}_1$ and $\mathbf{r}_2$ already argued above. Therefore, as we have shown that $\Psi(\mathbf{r})$ in (2.206) and $\mathcal{V}(\mathbf{r})$ in (2.187) are, in fact, the same potential, (2.210) proves (2.199) and, in combination with (2.204) and (2.205), also that

$$\nabla\left[\lim_{a \to 0^+} \Upsilon(\mathbf{r}; a)\right] = \nabla\mathcal{V}(\mathbf{r}) = \nabla\Psi(\mathbf{r}) = \Xi(\mathbf{r}) = \lim_{a \to 0^+} \nabla\Upsilon(\mathbf{r}; a), \qquad \mathbf{r} \in \mathbb{R}^3 \tag{2.211}$$

as anticipated.

For the sake of completeness, we outline an alternative proof based on the classic notion of derivative [16]. We let $\mathbf{R}_h = \mathbf{r} + h\hat{\mathbf{x}} - \mathbf{r}'$ and $\mathbf{R} = \mathbf{r} - \mathbf{r}'$ and observe

$$\frac{1}{h}\left(\frac{1}{|\mathbf{r} + h\hat{\mathbf{x}} - \mathbf{r}'|} - \frac{1}{|\mathbf{r} - \mathbf{r}'|}\right) = \frac{R^2 - R_h^2}{hR_hR(R + R_h)} = -\frac{h + 2\mathbf{R} \cdot \hat{\mathbf{x}}}{R_hR(R + R_h)} \tag{2.212}$$

$$\frac{\mathcal{V}(\mathbf{r}+h\hat{\mathbf{x}})-\mathcal{V}(\mathbf{r})}{h}-\frac{1}{4\pi\varepsilon}\int_{V_\varrho}\mathrm{d}V'\,\varrho(\mathbf{r}')\frac{\partial}{\partial x}\frac{1}{|\mathbf{r}-\mathbf{r}'|}=$$
$$=\frac{1}{4\pi\varepsilon}\int_{V_\varrho}\mathrm{d}V'\,\varrho(\mathbf{r}')\left[\frac{1}{h}\left(\frac{1}{|\mathbf{r}+h\hat{\mathbf{x}}-\mathbf{r}'|}-\frac{1}{|\mathbf{r}-\mathbf{r}'|}\right)-\frac{\partial}{\partial x}\frac{1}{|\mathbf{r}-\mathbf{r}'|}\right]$$
$$=\frac{1}{4\pi\varepsilon}\int_{V_\varrho}\mathrm{d}V'\,\varrho(\mathbf{r}')\left[\frac{\mathbf{R}\cdot\hat{\mathbf{x}}}{R^3}-\frac{h+2\mathbf{R}\cdot\hat{\mathbf{x}}}{R_hR(R+R_h)}\right] \tag{2.213}$$

where the last integral is convergent on account of (2.200) and because $|h+2\mathbf{R}\cdot\hat{\mathbf{x}}| \leqslant |x-x'|+|x+h-x'| \leqslant R+R_h$. Besides, the rightmost-hand side vanishes for $h=0$ and, if it represents a continuous function, then it tends to zero as $h\to 0$, i.e., $\mathbf{r}+h\hat{\mathbf{x}}\to\mathbf{r}$. Hence, the rest of the proof consists of showing that the rightmost member is a continuous function of $\mathbf{r}\in\mathbb{R}^3$. The latter statement may be confirmed by splitting the integration as in (2.194). Considering finite increments along $\hat{\mathbf{y}}$ and $\hat{\mathbf{z}}$ provides the proof for the other partial derivatives of $\mathcal{V}(\mathbf{r})$.

Thanks to (2.205), the Ansatz (2.15) and the integral solution (2.169) we have the expression [1]

$$\mathcal{E}(\mathbf{r})=-\nabla\frac{1}{4\pi\varepsilon}\int_{V_\varrho}\mathrm{d}V'\,\varrho(\mathbf{r}')\frac{1}{|\mathbf{r}-\mathbf{r}'|}=\frac{1}{4\pi\varepsilon}\int_{V_\varrho}\mathrm{d}V'\,\varrho(\mathbf{r}')\frac{\mathbf{r}-\mathbf{r}'}{|\mathbf{r}-\mathbf{r}'|^3} \tag{2.214}$$

for the electric field produced by $\varrho(\mathbf{r}')$ in the whole space.

Finally, checking *a posteriori* that (2.187) solves (2.190) for observation points *outside* the charge region V_ϱ is a relatively simple matter, since the static Green function (2.131) is regular for $\mathbf{r}\in\mathbb{R}^3\setminus\overline{V}_\varrho$ and $\mathbf{r}'\in V_\varrho$ and thus we may safely interchange the Laplace operator with the integral over V_ϱ. In symbols, we have

$$\nabla^2\mathcal{V}(\mathbf{r})=\nabla^2\int_{V_\varrho}\mathrm{d}V'\,\frac{\varrho(\mathbf{r}')/\varepsilon}{4\pi|\mathbf{r}-\mathbf{r}'|}=\frac{1}{\varepsilon}\int_{V_\varrho}\mathrm{d}V'\,\varrho(\mathbf{r}')\nabla^2\frac{1}{4\pi|\mathbf{r}-\mathbf{r}'|}=0 \tag{2.215}$$

where the result follows because the Green function in turn solves the Laplace equation for $\mathbf{r}\neq\mathbf{r}'$ [see (2.123)].

By contrast, the check is way more involved when $\mathbf{r}\in V_\varrho$, essentially because the second-order partial derivatives of the static Green function (2.131) exhibit a singularity of the type $1/|\mathbf{r}-\mathbf{r}'|^3$ which cannot be integrated over a three-dimensional domain. (In Section 9.4 we will face the same problem with time-harmonic fields.) Still, various proofs are possible depending on the degree of regularity one requires of the charge density. For instance, J. D. Jackson [1, Section 1.7] assumes that $\varrho(\mathbf{r})$ admits a Taylor expansion and, by considering the auxiliary potential function

$$\Upsilon(\mathbf{r};w):=\frac{1}{4\pi\varepsilon}\int_{V_\varrho}\mathrm{d}V'\,\frac{\varrho(\mathbf{r}')}{\left(|\mathbf{r}-\mathbf{r}'|^2+w^2\right)^{1/2}},\qquad \mathbf{r}\in\mathbb{R}^3 \tag{2.216}$$

where $w\geqslant 0$ is a parameter, he develops a limiting procedure for $w\to 0^+$. The denominator of the integrand in (2.216) may be interpreted as the distance between the source point $\mathbf{r}'\in V_\varrho$ and a *complex observation point*, say, $\mathbf{r}=(r+\mathrm{j}w)\hat{\mathbf{r}}\in\mathbb{C}^3$, whereby $\Upsilon(\mathbf{r};w)$ represents the potential produced by $\varrho(\mathbf{r}')$ at the complex location $(r+\mathrm{j}w)\hat{\mathbf{r}}$. The proof under the assumption that $\varrho(\mathbf{r})$ is continuously differentiable can be found in [20, Theorem 2.17]. Hereinbelow we show that (2.187) solves (2.190) under the milder hypothesis that the charge density $\varrho(\mathbf{r})$ is *Hölder continuous* in V_ϱ, namely,

$$\left|\varrho(\mathbf{r}) - \varrho(\mathbf{r}')\right| \leqslant M|\mathbf{r} - \mathbf{r}'|^\nu, \qquad \nu \in]0,1[\qquad \mathbf{r},\mathbf{r}' \in V_\varrho \tag{2.217}$$

where $M > 0$ is a suitable constant (Appendix F.4) [20, 31].

To this purpose, we consider the auxiliary vector field

$$\boldsymbol{\Xi}(\mathbf{r};a) := \frac{1}{\varepsilon}\int_{V_\varrho} \mathrm{d}V'\, f\left(\frac{|\mathbf{r}-\mathbf{r}'|}{a}\right)\varrho(\mathbf{r}')\nabla\frac{1}{4\pi|\mathbf{r}-\mathbf{r}'|}, \qquad \mathbf{r} \in \mathbb{R}^3 \tag{2.218}$$

where $f(R/a)$ is the same radial function introduced in (2.201) on page 123. The integral is well-defined despite the singular character of the gradient of the Green function, because $f(R/a)$ vanishes for $\mathbf{r}'$ inside the ball $B(\mathbf{r},a) \subset V_\varrho$ by construction (Figure 2.13). Apparently, $\boldsymbol{\Xi}(\mathbf{r};a)$ and $\nabla\mathcal{V}(\mathbf{r})$ coincide if $V_\varrho \cap B(\mathbf{r},2a) = \emptyset$ because then $f(R/a) = 1$ in V_ϱ thanks to definition (2.201). Otherwise, in the limit as $a \to 0^+$, $\boldsymbol{\Xi}(\mathbf{r};a)$ tends to $\nabla\mathcal{V}(\mathbf{r})$ given in the rightmost member of (2.199), since $\varrho(\mathbf{r}')$ is also bounded and $0 \leqslant 1 - f(R/a) \leqslant 1$ by construction, and hence for $\mathbf{r} \in V_\varrho$ we have

$$\left|\frac{\partial}{\partial x}\mathcal{V}(\mathbf{r}) - \Xi_x(\mathbf{r};a)\right| \leqslant \frac{\|\varrho\|_\infty}{4\pi\varepsilon}\int_{B(\mathbf{r},2a)} \mathrm{d}V'\,\frac{1}{|\mathbf{r}-\mathbf{r}'|^2} = \frac{\|\varrho\|_\infty}{\varepsilon}\int_0^{2a}\mathrm{d}\tau' = 2a\frac{\|\varrho\|_\infty}{\varepsilon} \xrightarrow[a\to 0^+]{} 0 \tag{2.219}$$

where we have integrated in local polar spherical coordinates (τ',α',β') centered on $\mathbf{r}$. The same estimate holds for the other Cartesian components of $\nabla\mathcal{V}(\mathbf{r})$ and $\boldsymbol{\Xi}(\mathbf{r};a)$. The convergence is uniform in V_ϱ because the leftmost member of (2.219) is dominated by a vanishing constant which does not depend on the observation point $\mathbf{r}$.

We set $\mathbf{r} \in V_\varrho$, suppose that $B(\mathbf{r},2a) \subset V_\varrho$ — which is not a limitation, as we shall let a approach zero anyway — and examine the derivatives of the Cartesian components of $\boldsymbol{\Xi}(\mathbf{r})$. In this regard, we may swap the order of differentiation and integration, e.g.,

$$\frac{\partial}{\partial x}\Xi_x(\mathbf{r};a) = \frac{1}{\varepsilon}\frac{\partial}{\partial x}\int_{V_\varrho}\mathrm{d}V'\,\varrho(\mathbf{r}')f\left(\frac{R}{a}\right)\frac{\partial}{\partial x}\frac{1}{4\pi R} = \frac{1}{\varepsilon}\int_{V_\varrho}\mathrm{d}V'\,\varrho(\mathbf{r}')\frac{\partial}{\partial x}\left[f\left(\frac{R}{a}\right)\frac{\partial}{\partial x}\frac{1}{4\pi R}\right] \tag{2.220}$$

since the integrand in (2.218) is continuously differentiable with respect to x, y, z for $(\mathbf{r},\mathbf{r}') \in \mathbb{R}^3 \times V_\varrho$. The remaining integral can be transformed into one over a larger domain D (with piecewise-smooth boundary) which strictly contains V_ϱ, as is pictured in Figure 2.15, provided we let

$$\tilde{\varrho}(\mathbf{r}) = \begin{cases}\varrho(\mathbf{r}), & \mathbf{r} \in V_\varrho \\ 0, & \mathbf{r} \in D \setminus \overline{V}_\varrho\end{cases} \tag{2.221}$$

i.e., so long as we extend the charge density to a function which is null for points $\mathbf{r} \in D \setminus \overline{V}_\varrho$. In principle $\tilde{\varrho}(\mathbf{r})$ may not be continuous across ∂V_ϱ, which is why we have moved the derivatives inside the integral prior to the extension. Then, we perform a little algebra

$$\begin{aligned}\frac{\partial}{\partial x}\Xi_x(\mathbf{r};a) &= \frac{1}{\varepsilon}\int_{V_\varrho}\mathrm{d}V'\,\varrho(\mathbf{r}')\frac{\partial}{\partial x}\left[f\left(\frac{R}{a}\right)\frac{\partial}{\partial x}\frac{1}{4\pi R}\right] = \frac{1}{\varepsilon}\int_D \mathrm{d}V'\,\tilde{\varrho}(\mathbf{r}')\frac{\partial}{\partial x}\left[f\left(\frac{R}{a}\right)\frac{\partial}{\partial x}\frac{1}{4\pi R}\right] \\ &= \frac{\varrho(\mathbf{r})}{\varepsilon}\int_D \mathrm{d}V'\,\frac{\partial}{\partial x'}\left[f\left(\frac{R}{a}\right)\frac{\partial}{\partial x'}\frac{1}{4\pi R}\right] + \frac{1}{\varepsilon}\int_D \mathrm{d}V'\,[\tilde{\varrho}(\mathbf{r}') - \varrho(\mathbf{r})]\frac{\partial}{\partial x}\left[f\left(\frac{R}{a}\right)\frac{\partial}{\partial x}\frac{1}{4\pi R}\right]\end{aligned}$$

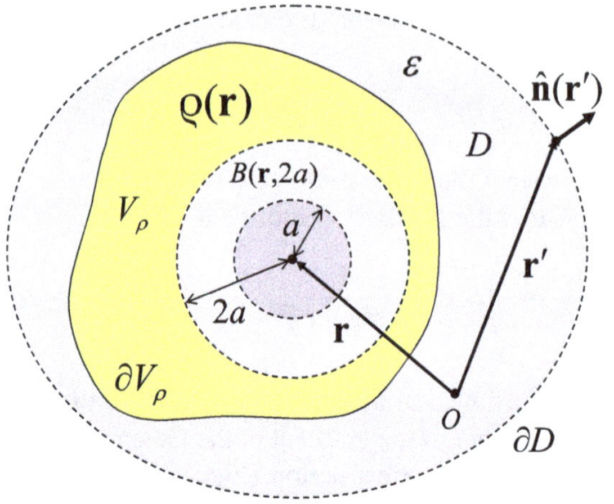

Figure 2.15 Geometrical construction for proving that the volume potential $\mathcal{V}(\mathbf{r})$ solves the Poisson equation for points **r** *in the source region V_ϱ.*

$$= \frac{\varrho(\mathbf{r})}{\varepsilon}\int_{\partial D} dS'\, n_x(\mathbf{r}')\frac{\partial}{\partial x'}\frac{1}{4\pi R}$$
$$+\frac{1}{\varepsilon}\int_D dV'\,[\tilde{\varrho}(\mathbf{r}')-\varrho(\mathbf{r})]\left[f\left(\frac{R}{a}\right)\frac{3(x-x')^2-R^2}{4\pi R^5}-f'\left(\frac{R}{a}\right)\frac{x-x'}{4\pi a R^3}\right] \tag{2.222}$$

where $n_x(\mathbf{r}')=\hat{\mathbf{x}}\cdot\hat{\mathbf{n}}(\mathbf{r}')$, we have invoked the Gauss lemma (H.88) with the unit normal pointing outwards D, and noticed that $f(R/a)=1$ for $\mathbf{r}'\in\partial D$ by construction. The symbol $f'(\bullet)$ denotes the derivative of $f(\bullet)$ with respect to the argument. The surface integral is independent of the radius a and is always well-defined because $\mathbf{r}'\in\partial D$ and $\mathbf{r}\in V_\varrho\subset D$ by hypothesis. Similarly, the domain integral is well-defined since the integrand is regular because $f(R/a)$ and its derivative vanish for $R<a$ (see Figure 2.13).

We wish to show that

$$\lim_{a\to 0^+}\frac{\partial}{\partial x}\Xi_x(\mathbf{r};a)=\frac{\varrho(\mathbf{r})}{\varepsilon}\int_{\partial D} dS'\, n_x(\mathbf{r}')\frac{\partial}{\partial x'}\frac{1}{4\pi R}+\frac{1}{\varepsilon}\int_D dV'\,[\tilde{\varrho}(\mathbf{r}')-\varrho(\mathbf{r})]\,\frac{3(x-x')^2-R^2}{4\pi R^5} \tag{2.223}$$

where the domain integral exists finite. Indeed, it can be split into the sum of two contributions, namely, one over $D\setminus V_\varrho$ and one over V_ϱ. When $\mathbf{r}'\in D\setminus V_\varrho$ the integrand is regular because $\mathbf{r}\in V_\varrho$, whereas for $\mathbf{r},\mathbf{r}'\in V_\varrho$, the Hölder condition (2.217) applies and we have

$$\left|\int_{V_\varrho} dV'\,[\varrho(\mathbf{r}')-\varrho(\mathbf{r})]\,\frac{3(x-x')^2-R^2}{4\pi R^5}\right|\leqslant\int_{V_\varrho} dV'\,\left|\varrho(\mathbf{r}')-\varrho(\mathbf{r})\right|\frac{\left|3(x-x')^2-R^2\right|}{4\pi R^5}$$
$$\leqslant\int_{V_\varrho} dV'\, MR^\nu\frac{4R^2}{4\pi R^5}\leqslant M\int_{B(\mathbf{r},b)} dV'\,\frac{R^{\nu-3}}{\pi}=4\int_0^b d\tau'\,\tau'^{\nu-1}=\frac{4b^\nu}{\nu} \tag{2.224}$$

where $B(\mathbf{r}, b)$ is a larger ball that contains V_ϱ. Therefore, we just have to estimate the difference between the right-hand sides of (2.222) and (2.223). In particular, this entails

$$\begin{aligned}
&\left| \int_D \mathrm{d}V' \, [\tilde{\varrho}(\mathbf{r}') - \varrho(\mathbf{r})] \left[1 - f\left(\frac{R}{a}\right)\right] \frac{3(x - x')^2 - R^2}{4\pi\varepsilon R^5} \right| = \\
&\quad = \left| \int_{B(\mathbf{r},2a)} \mathrm{d}V' \, [\varrho(\mathbf{r}') - \varrho(\mathbf{r})] \left[1 - f\left(\frac{R}{a}\right)\right] \frac{3(x - x')^2 - R^2}{4\pi\varepsilon R^5} \right| \\
&\quad \leqslant \int_{B(\mathbf{r},2a)} \mathrm{d}V' \, |\varrho(\mathbf{r}') - \varrho(\mathbf{r})| \frac{|3(x - x')^2 - R^2|}{4\pi\varepsilon R^5} \leqslant \frac{4M}{\varepsilon} \int_0^{2a} \mathrm{d}\tau' \, \tau'^{\nu - 1} \\
&\quad = \frac{4M}{\varepsilon} \frac{(2a)^\nu}{\nu} \xrightarrow[a \to 0^+]{} 0
\end{aligned} \tag{2.225}$$

and

$$\begin{aligned}
&\left| \int_D \mathrm{d}V' \, [\tilde{\varrho}(\mathbf{r}') - \varrho(\mathbf{r})] f'\left(\frac{R}{a}\right) \frac{x - x'}{4\pi\varepsilon a R^3} \right| = \left| \int_{B(\mathbf{r},2a)\setminus B(\mathbf{r},a)} \mathrm{d}V' \, [\varrho(\mathbf{r}') - \varrho(\mathbf{r})] f'\left(\frac{R}{a}\right) \frac{x - x'}{4\pi\varepsilon a R^3} \right| \\
&\quad \leqslant \int_{B(\mathbf{r},2a)\setminus B(\mathbf{r},a)} \mathrm{d}V' \, |\varrho(\mathbf{r}') - \varrho(\mathbf{r})| \, \|f'\|_\infty \frac{|x - x'|}{4\pi\varepsilon a R^3} \leqslant \int_{B(\mathbf{r},2a)\setminus B(\mathbf{r},a)} \mathrm{d}V' \, M R^\nu \, \|f'\|_\infty \frac{R}{4\pi\varepsilon a R^3} \\
&\quad = M \frac{\|f'\|_\infty}{\varepsilon a} \int_a^{2a} \mathrm{d}\tau' \, \tau'^{\nu} = M \frac{\|f'\|_\infty}{\varepsilon} \frac{2^{\nu+1} - 1}{\nu + 1} a^\nu \xrightarrow[a \to 0^+]{} 0
\end{aligned} \tag{2.226}$$

whereby (2.223) is proved. What is more, the same result holds for $\Xi_y(\mathbf{r}; a)$ and $\Xi_z(\mathbf{r}; a)$ with obvious modifications. The convergence in (2.223) is uniform for $\mathbf{r} \in V_\varrho$ in that the quantities in the leftmost members of (2.225) and (2.226) are dominated by vanishing constants which are independent of $\mathbf{r}$. It is also apparent that simple uniform continuity of $\varrho(\mathbf{r})$ — which in (2.217) obtains for $\nu = 0$ — is not enough for (2.225) and (2.226) to be true.

For ease of subsequent manipulations we call $g_x(\mathbf{r})$, $\mathbf{r} \in V_\varrho$, the function in the right-hand side of (2.223) and, by picking up two points $\mathbf{r}_1$ and $\mathbf{r}_2$ in V_ϱ such that $x_1 = \hat{\mathbf{x}} \cdot \mathbf{r}_1 < \hat{\mathbf{x}} \cdot \mathbf{r}_2 = x_2$, we define

$$\Psi_x(\mathbf{r}) = \frac{\partial}{\partial x} \mathcal{V}(\mathbf{r}) \bigg|_{x = x_1} + \int_{x_1}^{x} \mathrm{d}x' \, g_x(x', y, z), \qquad \mathbf{r} \in \{\mathbf{r} \in V_\varrho : x \in [x_1, x_2]\} \tag{2.227}$$

which amounts to an integration along $\hat{\mathbf{x}}$ between the planes $x' = x_1$ and $x' = x \leqslant x_2$. This definition makes sense because by means of (2.223), (2.225) and (2.226) we have proved that $g_x(\mathbf{r})$ is the uniform limit of a sequence of continuous functions. Thus, $g_x(\mathbf{r})$ is continuous in V_ϱ [32, Chapter 7] and as such can be integrated. Besides, the function $\Psi_x(\mathbf{r})$ equals the x-derivative of the volume potential (2.187) when $x = x_1$. Similarly, we can write

$$\Xi_x(\mathbf{r}; a) = \Xi_x(x_1, y, z; a) + \int_{x_1}^{x} \mathrm{d}x' \, \frac{\partial}{\partial x'} \Xi_x(x', y, z; a) \tag{2.228}$$

for points $\mathbf{r} \in V_\varrho$ between the planes $x' = x_1$ and $x' = x \leqslant x_2$. By subtracting (2.227) and (2.228) side by side and taking the magnitudes we get

$$\begin{aligned}|\Psi_x(\mathbf{r}) - \Xi_x(\mathbf{r}; a)| &= \left| \frac{\partial}{\partial x} \mathcal{V}(\mathbf{r}) \right|_{x=x_1} - \Xi_x(x_1, y, z; a) + \int_{x_1}^{x} \mathrm{d}x' \left[g_x(x', y, z) - \frac{\partial}{\partial x'} \Xi_x(x', y, z; a) \right] \Bigg| \\ &\leqslant \left| \frac{\partial}{\partial x} \mathcal{V}(\mathbf{r}) \right|_{x=x_1} - \Xi_x(x_1, y, z; a) \Bigg| + \int_{x_1}^{x_2} \mathrm{d}x' \left| g_x(x', y, z) - \frac{\partial}{\partial x'} \Xi_x(x', y, z; a) \right| \\ &\leqslant 2a \frac{\|\varrho\|_\infty}{\varepsilon} + \frac{M}{\varepsilon}(x_2 - x_1) \left(4\frac{2^\nu}{\nu} + \|f'\|_\infty \frac{2^{\nu+1} - 1}{\nu + 1} \right) a^\nu \xrightarrow[a \to 0^+]{} 0 \end{aligned} \tag{2.229}$$

by virtue of (2.219) applied with $\mathbf{r} = x_1\hat{\mathbf{x}} + y\hat{\mathbf{y}} + z\hat{\mathbf{z}}$ and (2.225) and (2.226).

This intermediate result says that the functions $\Xi_x(\mathbf{r}; a)$ tend to $\Psi_x(\mathbf{r})$ for points $\mathbf{r} \in V_\varrho$ between the planes $x = x_1$ and $x = x_2$. Then again, since $\mathbf{r}_1$ and $\mathbf{r}_2$ are arbitrary, we conclude that $\Xi_x(\mathbf{r}; a)$ tends to $\Psi_x(\mathbf{r})$ everywhere in V_ϱ. Since $\Xi_x(\mathbf{r}; a)$ tends also to the x-derivative of the volume potential in light of (2.219), and this limit is unique, we obtain that $\Psi_x(\mathbf{r})$ and $\partial\mathcal{V}/\partial x$ coincide for $\mathbf{r} \in V_\varrho$. By taking the derivative with respect to x of both sides of (2.227) we first get

$$\frac{\partial \Psi_x}{\partial x} = g_x(\mathbf{r}), \qquad \mathbf{r} \in \{\mathbf{r} \in V_\varrho : x \in [x_1, x_2]\} \tag{2.230}$$

whence we conclude that

$$\begin{aligned}\frac{\partial^2 \mathcal{V}}{\partial x^2} = \frac{\partial \Psi_x}{\partial x} &= \frac{\varrho(\mathbf{r})}{\varepsilon} \int_{\partial D} \mathrm{d}S' \, n_x(\mathbf{r}') \frac{\partial}{\partial x'} \frac{1}{4\pi R} \\ &+ \frac{1}{\varepsilon} \int_D \mathrm{d}V' \, [\tilde{\varrho}(\mathbf{r}') - \varrho(\mathbf{r})] \frac{3(x - x')^2 - R^2}{4\pi R^5}, \qquad \mathbf{r} \in V_\varrho \end{aligned} \tag{2.231}$$

in view of the arbitrariness of $\mathbf{r}_1$ and $\mathbf{r}_2$. Perfectly similar steps can be taken to prove the analogous of (2.231) for the second-order partial derivatives of the volume potential with respect to y and z.

Now, summing (2.231) and the relevant formulas for $\partial^2\mathcal{V}/\partial y^2$ and $\partial^2\mathcal{V}/\partial z^2$ sides by sides yields

$$\begin{aligned}\nabla^2 \mathcal{V}(\mathbf{r}) &= \frac{\varrho(\mathbf{r})}{\varepsilon} \int_{\partial D} \mathrm{d}S' \left[n_x(\mathbf{r}') \frac{\partial}{\partial x'} \frac{1}{4\pi R} + n_y(\mathbf{r}') \frac{\partial}{\partial y'} \frac{1}{4\pi R} + n_y(\mathbf{r}') \frac{\partial}{\partial y'} \frac{1}{4\pi R} \right] \\ &= \frac{\varrho(\mathbf{r})}{\varepsilon} \int_{\partial D} \mathrm{d}S' \, \hat{\mathbf{n}}(\mathbf{r}') \cdot \nabla' \frac{1}{4\pi R} = \frac{\varrho(\mathbf{r})}{\varepsilon} \int_{\partial D} \mathrm{d}S' \, \frac{\hat{\mathbf{n}}(\mathbf{r}') \cdot \hat{\mathbf{R}}}{4\pi R^2} \end{aligned} \tag{2.232}$$

inasmuch as the three domain integrals add up to zero. The remaining flux integral depends neither on the shape nor the size of ∂D. Indeed, we first notice that in a local system of polar spherical coordinates (τ', α', β') with center in $\mathbf{r}$ and unit radial vector $\hat{\boldsymbol{\tau}}' = -\hat{\mathbf{R}}$ we have [see (A.31)]

$$\nabla' \cdot \frac{\hat{\mathbf{R}}}{4\pi R^2} = 0, \qquad \mathbf{r}' \in B(\mathbf{r}, b) \setminus D, \qquad \mathbf{r} \in V_\varrho \subset D \subset B(\mathbf{r}, b) \tag{2.233}$$

where the ball $B(\mathbf{r}, b)$ is large enough to contain D. By integrating this identity we find

$$0 = \int_{B(\mathbf{r},b)\setminus D} \mathrm{d}V \, \nabla' \cdot \frac{\hat{\mathbf{R}}}{4\pi R^2} = -\int_{\partial D} \mathrm{d}S' \, \frac{\hat{\mathbf{n}}(\mathbf{r}') \cdot \hat{\mathbf{R}}}{4\pi R^2} - \int_{\partial B} \mathrm{d}S' \, \frac{\hat{\mathbf{R}} \cdot \hat{\mathbf{R}}}{4\pi b^2} \tag{2.234}$$

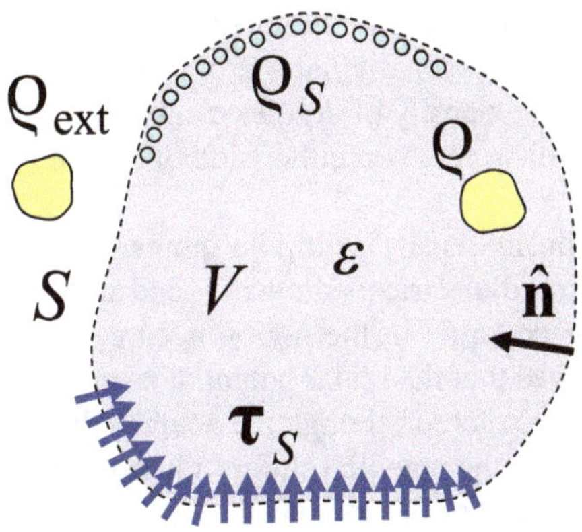

Figure 2.16 Physical meaning of electrostatic double-layer and single-layer potentials.

having applied the Gauss theorem with the unit normal positively oriented inwards $B(\mathbf{r}, b) \setminus D$ so as to be consistent with (2.232). Then, we have

$$\int_{\partial D} \mathrm{d}S' \, \frac{\hat{\mathbf{n}}(\mathbf{r}') \cdot \hat{\mathbf{R}}}{4\pi R^2} = -\int_{\partial B} \mathrm{d}S' \, \frac{1}{4\pi b^2} = -1 \tag{2.235}$$

independently of b. Lastly, inserting (2.235) back into (2.232) proves that $\mathcal{V}(\mathbf{r})$ given by (2.187) solves (2.190) when $\varrho(\mathbf{r})$ meets condition (2.217). As a byproduct, by virtue of (2.219), (2.231) and (2.223) we have also found

$$\begin{aligned} \nabla \cdot \left[\lim_{a \to 0^+} \boldsymbol{\Xi}(\mathbf{r}; a) \right] &= \nabla \cdot \nabla \mathcal{V}(\mathbf{r}) = \frac{\partial \Psi_x}{\partial x} + \frac{\partial \Psi_y}{\partial y} + \frac{\partial \Psi_z}{\partial z} \\ &= g_x(\mathbf{r}) + g_y(\mathbf{r}) + g_z(\mathbf{r}) = \lim_{a \to 0^+} \nabla \cdot \boldsymbol{\Xi}(\mathbf{r}; a), \qquad \mathbf{r} \in V_\varrho \end{aligned} \tag{2.236}$$

and hence $\nabla \cdot \boldsymbol{\Xi}(\mathbf{r}; a)$ approaches $\nabla^2 \mathcal{V}(\mathbf{r})$ in the source region. Alternative proofs which rely on assumption (2.217) are outlined in [20, Exercise 2.5] and [33, Lemma 10, pp. 41–45].

2.9 Double-layer potential

The double-layer potential $\mathcal{W}(\mathbf{r})$ depends on the value of $\Phi(\mathbf{r})$ on the boundary S of the region of interest V. We rewrite (2.188) in a slightly different form, namely,

$$\mathcal{W}(\mathbf{r}) := \frac{1}{\varepsilon} \int_S \mathrm{d}S' \, \varepsilon \Phi(\mathbf{r}') \hat{\mathbf{n}}(\mathbf{r}') \cdot \nabla' \frac{1}{4\pi |\mathbf{r} - \mathbf{r}'|}, \qquad \mathbf{r} \in \mathbb{R}^3 \tag{2.237}$$

in order to understand the physical meaning of this term. Comparing the integrand with the scalar potential of an elementary dipole — obtained in Example 2.5 and given by (2.42) — suggests that $\mathcal{W}(\mathbf{r})$ may be interpreted as the superposition of the effect of *equivalent* (fictitious) dipoles continuously distributed over S. Said dipoles are perpendicular to S, as is pictorially shown in Figure 2.16, and the quantity

$$\tau_S(\mathbf{r}) := \varepsilon \Phi(\mathbf{r}) \hat{\mathbf{n}}(\mathbf{r}), \qquad \mathbf{r} \in S \tag{2.238}$$

represents a *surface density* of electric dipole moments (physical dimension: C/m = Cm/m^2). But then, since an elementary static dipole originates from the pairing of two infinitely close point charges of opposite sign, a surface density of dipoles can be thought of as the pairing of two layers of charges with opposite signs distributed on either side of S, and this interpretation better justifies the name for $\mathcal{W}(\mathbf{r})$.

The special case of a uniform density of dipole moments $\boldsymbol{\tau}_S$ has physical relevance when S represents the interface between a dielectric medium in V and a conductor in $\mathbb{R}^3 \setminus \overline{V}$, in which instance $\mathcal{W}(\mathbf{r})$ contributes to the scalar potential in the region occupied by the dielectric [25]. Indeed, in Section 2.4, page 91, we observed that the scalar potential is constant both within a region occupied by a conductor and on the surface thereof, though the actual values of these two constants need not be the same. When that is the case, the potential suffers a jump across S. Therefore, the double-layer potential accounts for the effect of possible conductors flush with the surface S.

As regards the behavior of $\mathcal{W}(\mathbf{r})$ in an unbounded region for points far away from the sources, we consider the asymptotic expansion

$$\nabla' \frac{1}{4\pi R} = \frac{\mathbf{r} - \mathbf{r}'}{4\pi|\mathbf{r} - \mathbf{r}'|^3} = \frac{\hat{\mathbf{r}}}{4\pi r^2} + \mathcal{O}\left(\frac{1}{r^3}\right), \qquad r \to +\infty \tag{2.239}$$

based on (2.191). If S_τ is a closed smooth surface which is the support of a layer of dipoles with density $\boldsymbol{\tau}_S(\mathbf{r})$, we have

$$\frac{1}{\varepsilon} \int_{S_\tau} \mathrm{d}S'\, \boldsymbol{\tau}_S(\mathbf{r}') \cdot \nabla' \frac{1}{4\pi R} = \frac{\hat{\mathbf{r}}}{4\pi\varepsilon r^2} \cdot \int_{S_\tau} \mathrm{d}S'\, \boldsymbol{\tau}_S(\mathbf{r}') + \mathcal{O}\left(\frac{1}{r^3}\right), \qquad r \to +\infty \tag{2.240}$$

which is consistent with the potential (2.39) of an electrostatic dipole. More precisely, in light of (2.238) the surface integral of $\boldsymbol{\tau}_S(\mathbf{r})$ does represent the moment $\mathbf{p}_{eq}$ of an equivalent electrostatic dipole located in the origin (cf. Section 3.6).

Although the integrand of the double-layer potential is singular for points $\mathbf{r} \in S$, the limits for $\mathbf{r} \to S^\pm$ are finite, and the integral exists at least in the Cauchy principal value sense. To determine the limiting values of $\mathcal{W}(\mathbf{r})$ we exclude the observation point $\mathbf{r} \in S$ with a small ball $B(\mathbf{r}, a)$ and define the open surfaces $S' := \{\mathbf{r}' \in S : |\mathbf{r}' - \mathbf{r}| \geqslant a\}$, $S_1 := \partial B \cap (\mathbb{R}^3 \setminus \overline{V})$ and $S_2 := \partial B \cap \overline{V}$.

If we think of the observation point $\mathbf{r}$ as approaching S from the positive side (as is specified by the normal $\hat{\mathbf{n}}$) then we start with the double-layer potential over the surface $S' \cup S_1$ and take the limit as $a \to 0$ (Figure 2.17):

$$\begin{aligned}
\lim_{a\to 0} \mathcal{W}(\mathbf{r}; a) &= \lim_{a\to 0} \int_{S'} \mathrm{d}S'\, \Phi(\mathbf{r}') \frac{\partial}{\partial \hat{\mathbf{n}}'} \frac{1}{4\pi R} + \lim_{a\to 0} \int_{S_1} \mathrm{d}S'\, \Phi(\mathbf{r}') \frac{\partial}{\partial \hat{\mathbf{n}}'} \frac{1}{4\pi R} \\
&= \mathrm{PV} \int_S \mathrm{d}S'\, \Phi(\mathbf{r}') \frac{\partial}{\partial \hat{\mathbf{n}}'} \frac{1}{4\pi R} + \lim_{a\to 0} \frac{1}{4\pi a^2} \int_{S_1} \mathrm{d}S'\, \Phi(\mathbf{r} - a\hat{\mathbf{n}}(\mathbf{r}')) \\
&= \mathrm{PV} \int_S \mathrm{d}S'\, \Phi(\mathbf{r}') \hat{\mathbf{n}}(\mathbf{r}') \cdot \nabla' \frac{1}{4\pi R} + \lim_{a\to 0} \frac{\Omega(a)}{4\pi} \Phi(\mathbf{r} - a\hat{\mathbf{n}}(\mathbf{r}_0')) \\
&= \mathrm{PV} \int_S \mathrm{d}S'\, \Phi(\mathbf{r}') \frac{\hat{\mathbf{n}}(\mathbf{r}') \cdot (\mathbf{r} - \mathbf{r}')}{4\pi|\mathbf{r} - \mathbf{r}'|^3} + \frac{1}{2}\Phi(\mathbf{r})
\end{aligned} \tag{2.241}$$

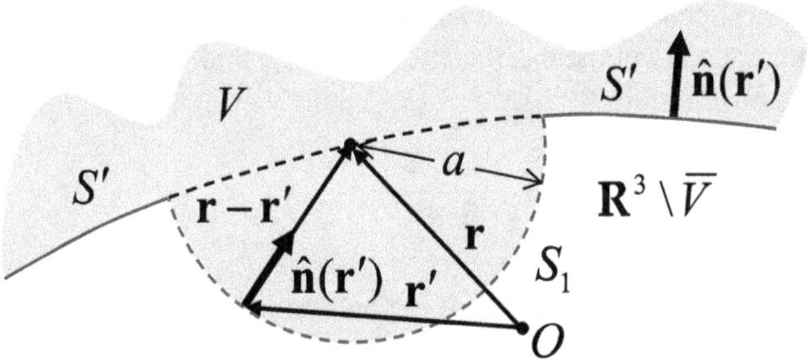

Figure 2.17 For computing the limiting values of $\mathcal{W}(\mathbf{r})$ *as* $\mathbf{r} \to S^+$.

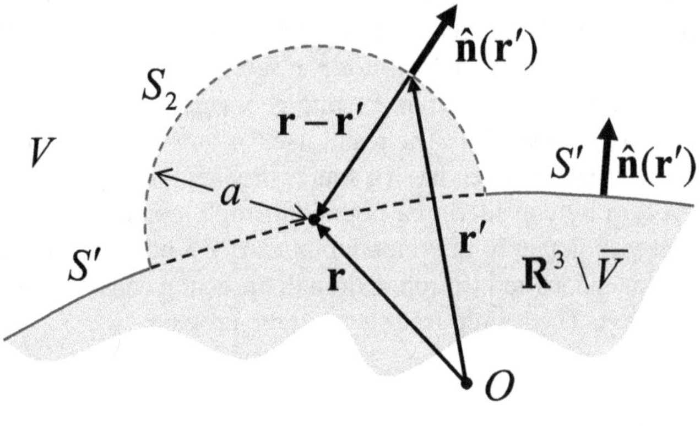

Figure 2.18 For computing the limiting values of $\mathcal{W}(\mathbf{r})$ *as* $\mathbf{r} \to S^-$.

where the last contribution follows from the mean value theorem [28] and the fact that the solid angle $\Omega(a)$ subtended by S_1 at $\mathbf{r}$ tends to 2π, if S is smooth (Appendix F.2). The integral over S is dominated by a single-layer potential of type (2.189) with 'surface density' $|\Phi(\mathbf{r}')|$, namely,

$$\left| \mathrm{PV} \int_S \mathrm{d}S' \, \Phi(\mathbf{r}') \frac{\hat{\mathbf{n}}(\mathbf{r}') \cdot (\mathbf{r} - \mathbf{r}')}{4\pi |\mathbf{r} - \mathbf{r}'|^3} \right| \leqslant \mathrm{PV} \int_S \mathrm{d}S' \, |\Phi(\mathbf{r}')| \frac{|\hat{\mathbf{n}}(\mathbf{r}') \cdot (\mathbf{r} - \mathbf{r}')|}{4\pi |\mathbf{r} - \mathbf{r}'|^3} \leqslant \int_S \mathrm{d}S' \, \frac{M|\Phi(\mathbf{r}')|}{4\pi |\mathbf{r} - \mathbf{r}'|} \tag{2.242}$$

because $\Phi(\mathbf{r}')$ is regular on S and condition (F.1) holds true for $\mathbf{r}, \mathbf{r}' \in S$, provided S is smooth. We shall prove in Section 2.10 and in particular with (2.257) that a single-layer potential is finite for $\mathbf{r} \in \mathbb{R}^3$.

Conversely, if the observation point $\mathbf{r}$ approaches S from the negative side (that is, from the complementary domain $\mathbb{R}^3 \setminus \overline{V}$) then we consider the surface $S' \cup S_2$ and take the limit as $a \to 0$ (Figure 2.18):

$$\begin{aligned}
\lim_{a \to 0} \mathcal{W}(\mathbf{r}; a) &= \lim_{a \to 0} \int_{S'} \mathrm{d}S' \, \Phi(\mathbf{r}') \frac{\partial}{\partial \hat{\mathbf{n}}'} \frac{1}{4\pi R} + \lim_{a \to 0} \int_{S_2} \mathrm{d}S' \, \Phi(\mathbf{r}') \frac{\partial}{\partial \hat{\mathbf{n}}'} \frac{1}{4\pi R} \\
&= \mathrm{PV} \int_S \mathrm{d}S' \, \Phi(\mathbf{r}') \frac{\partial}{\partial \hat{\mathbf{n}}'} \frac{1}{4\pi R} - \lim_{a \to 0} \frac{1}{4\pi a^2} \int_{S_2} \mathrm{d}S' \, \Phi(\mathbf{r} + a\hat{\mathbf{n}}(\mathbf{r}')) \\
&= \mathrm{PV} \int_S \mathrm{d}S' \, \Phi(\mathbf{r}') \hat{\mathbf{n}}(\mathbf{r}') \cdot \nabla' \frac{1}{4\pi R} - \frac{1}{2} \Phi(\mathbf{r})
\end{aligned} \tag{2.243}$$

where the surface integral can be estimated as in (2.242), and the minus sign in the last term is a consequence of the manifestly opposite orientations of $\hat{\mathbf{n}}(\mathbf{r}')$ and $\mathbf{r} - \mathbf{r}'$ on S_2.

In conclusion, we have found

$$\lim_{\mathbf{r}\to S^\pm} \mathcal{W}(\mathbf{r}) = \pm\frac{1}{2}\Phi(\mathbf{r}) + \mathrm{PV}\int_S \mathrm{d}S'\,\Phi(\mathbf{r}')\frac{\partial}{\partial \hat{\mathbf{n}}'}\frac{1}{4\pi R} \tag{2.244}$$

whence we compute the jump of the double-layer potential across S as

$$\lim_{\mathbf{r}\to S^+} \mathcal{W}(\mathbf{r}) - \lim_{\mathbf{r}\to S^-} \mathcal{W}(\mathbf{r}) = \Phi(\mathbf{r}). \tag{2.245}$$

Finally, we observe that $\mathcal{W}(\mathbf{r})$ can be defined for open surfaces and represents the potential due to a layer of elementary dipoles. Moreover, for an open flat surface, the principal value of the integral in (2.244) vanishes, because $\hat{\mathbf{n}}(\mathbf{r}') \cdot (\mathbf{r} - \mathbf{r}') = 0$ for $\mathbf{r}, \mathbf{r}' \in S$.

The double-layer potential also has an interesting geometrical interpretation which goes by the name of *Gauss solid angle formula* [25]. The result itself is an ingenious application of (2.160) for representing a unitary electrostatic potential in a source-free region of space.

To elucidate, we begin by considering a bounded simply connected volume $V \subset \mathbb{R}^3$ and we pick the unit vector $\hat{\mathbf{n}}$ perpendicular to the smooth boundary ∂V positively oriented outward V. The medium filling V is homogeneous and isotropic, though the actual constitutive parameters are inconsequential for the discussion. The source-free electrostatic problem

$$\begin{cases} \nabla^2\Phi(\mathbf{r}) = 0, & \mathbf{r} \in V \\ \Phi(\mathbf{r}) = 1, & \mathbf{r} \in \partial V \end{cases} \tag{2.246}$$

clearly admits the solution $\Phi(\mathbf{r}) = 1$ for $\mathbf{r} \in V$, which is unique according to the discussion of Section 2.5.1. Besides, since the electrostatic potential is constant in V, the normal derivative $\hat{\mathbf{n}} \cdot \nabla\Phi$ on ∂V vanishes. Notice that we do not assign the normal derivative on ∂V, rather we determine the value thereof *after* the Poisson equation has been solved with just $\Phi(\mathbf{r})$ specified on ∂V.

Next, we observe that we can apply the integral representation of Section 2.7 to express $\Phi(\mathbf{r})$ in terms of sources (but none are present in V by hypothesis) and boundary conditions. Thus, from (2.160), (2.162) and (2.167) we find that only the double-layer potential contributes, namely,

$$-\int_{\partial V} \mathrm{d}S'\,\Phi(\mathbf{r}')\hat{\mathbf{n}}(\mathbf{r}')\cdot\nabla'\frac{1}{4\pi R} = \begin{cases} \Phi(\mathbf{r}), & \mathbf{r} \in V \\ \frac{1}{2}\Phi(\mathbf{r}), & \mathbf{r} \in \partial V \\ 0, & \mathbf{r} \in \mathbb{R}^3 \setminus \overline{V} \end{cases} \tag{2.247}$$

where the leading minus sign in the left-hand side accounts for the opposite orientation of $\hat{\mathbf{n}}$ in the present geometrical setup. Finally, we set $\Phi(\mathbf{r}) = 1 = \Phi(\mathbf{r}')$ in the previous intermediate result and multiply through by 4π, viz.,

$$\varpi(\mathbf{r}) = \int_{\partial V} \mathrm{d}S'\,\frac{\hat{\mathbf{n}}(\mathbf{r}')\cdot(\mathbf{r}' - \mathbf{r})}{|\mathbf{r}' - \mathbf{r}|^3} = \begin{cases} 4\pi, & \mathbf{r} \in V \\ 2\pi, & \mathbf{r} \in \partial V \\ 0, & \mathbf{r} \in \mathbb{R}^3 \setminus \overline{V} \end{cases} \tag{2.248}$$

and this completes the derivation. Equation (2.248) provides the value of the solid angle $\varpi(\mathbf{r})$ subtended by the closed surface ∂V with respect to the observation point $\mathbf{r}$, and this property justifies

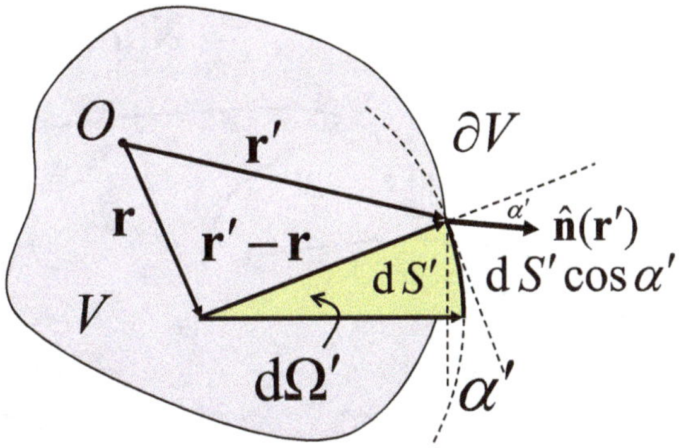

Figure 2.19 For defining the differential solid angle $\mathrm{d}\Omega'$ *with respect to* $\mathbf{r}$.

the name. To be specific, the differential solid angle is related to the area element of ∂V in (2.248) as follows (see Figure 2.19)

$$\mathrm{d}\Omega' = \frac{\cos\alpha'}{|\mathbf{r}' - \mathbf{r}|^2}\mathrm{d}S' = \frac{\hat{\mathbf{n}}(\mathbf{r}') \cdot (\mathbf{r}' - \mathbf{r})}{|\mathbf{r}' - \mathbf{r}|^3}\,\mathrm{d}S' \tag{2.249}$$

which means $|\mathbf{r}' - \mathbf{r}|^2\mathrm{d}\Omega'$ is the projection of $\mathrm{d}S'$ onto the sphere that has radius $|\mathbf{r}' - \mathbf{r}|$, passes through $\mathbf{r}' \in \partial V$ and is centered in $\mathbf{r}$. Further, a more general result is obtained in the scope of Section 5.4.

2.10 Single-layer potential

The single-layer potential $\mathcal{S}(\mathbf{r})$ in (2.189) depends on the normal derivative of $\Phi(\mathbf{r})$ over the boundary of the region of interest. We recall from the discussion in Section 2.4 that the jump of $\partial\Phi/\partial\hat{\mathbf{n}}$ across S is related to material discontinuities and the possible presence of surface charges on S. Also, comparison with the volume potential and examination of (2.180) strongly suggest that $\partial\Phi/\partial\hat{\mathbf{n}}$ plays a role analogous to a layer of surface density of charges on S, hence the name for $\mathcal{S}(\mathbf{r})$.

Indeed, in agreement with (1.155) and (1.182) the contribution of $\mathcal{S}(\mathbf{r})$ to the electrostatic potential can be construed as the effect of an *equivalent* fictitious layer of charges (see Figure 2.16) with surface density given by

$$\varrho_S(\mathbf{r}) = -\varepsilon\,\hat{\mathbf{n}}(\mathbf{r}) \cdot \nabla\Phi(\mathbf{r}) = \hat{\mathbf{n}}(\mathbf{r}) \cdot \mathcal{D}(\mathbf{r}), \qquad \mathbf{r} \in S \tag{2.250}$$

but it should be kept in mind that this insight does not necessarily imply the actual physical presence of charges on S. More importantly, in general situations, such as the problem of Figure 2.7a, $\hat{\mathbf{n}}(\mathbf{r}) \cdot \nabla\Phi(\mathbf{r})$ is actually unknown.

The similarity with the volume potential also applies to the asymptotic behavior of $\mathcal{S}(\mathbf{r})$ in an unbounded spatial region. If S_ϱ is a closed smooth surface which is the support of a surface layer of charges with density $\varrho_S(\mathbf{r})$, we have

$$\int_{S_\varrho} \mathrm{d}S'\,\frac{\varrho_S(\mathbf{r}')/\varepsilon}{4\pi R} = \frac{1}{4\pi\varepsilon r}\int_{S_\varrho} \mathrm{d}S'\,\varrho_S(\mathbf{r}') + \mathcal{O}\left(\frac{1}{r^2}\right), \qquad r \to +\infty \tag{2.251}$$

on account of (2.191). The surface integral is interpreted as an equivalent point charge Q_{eq} placed in the origin (cf. Section 3.6). Finally, the single-layer potential can be extended to a layer of charges distributed over an open surface (Figure 1.2a).

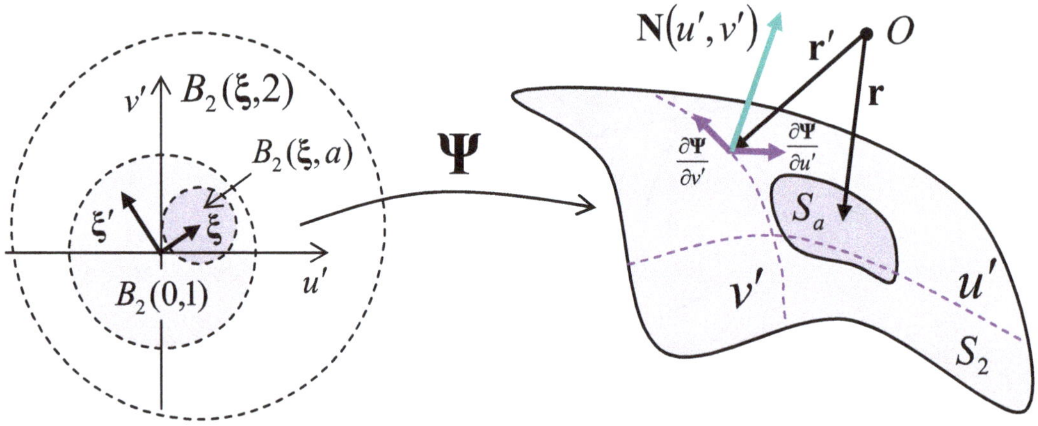

Figure 2.20 Geometrical construction for showing that the static single-layer potential exists for points $\mathbf{r} \in S$.

The scalar field $\mathcal{S}(\mathbf{r})$ exists also for observation points on the smooth surface S where the charges (either real or fictitious) reside. This property was anticipated in Section 2.7 for the proof of (2.167). Indeed, as complicated as the shape of S may happen to be, the relevant integral can always be split into two parts, namely, over $S_1 := S \setminus S_2$ and $S_2 \subset S$, where the latter is a 'small' and simple-shaped 'cutout' of S such that $\mathbf{r} \in S_2$. Since the integral over S_1 exists because the integrand is regular thereon, we need only examine the behavior of the integral over S_2.

We introduce a set of local dimensionless coordinates $\boldsymbol{\xi} := (u, v)$ which are related to the points $\mathbf{r} \in S_2$ through the bijective mapping

$$\boldsymbol{\Psi} : \mathbb{R}^2 \supset B_2(0,1) \longrightarrow S_2 \subset \mathbb{R}^3 \tag{2.252}$$

where $B_2(0,1)$ is the unit circle (Appendix D.1). Likewise, we let $\boldsymbol{\xi}' := (u', v') \in B_2(0,1)$ be the local coordinates which are mapped to $\mathbf{r}' \in S_2$ by $\boldsymbol{\Psi}$. This setup is graphically represented in Figure 2.20. Since $S_2 \subset S$ is smooth, the inverse map $\boldsymbol{\Psi}^{-1}(\mathbf{r})$ is differentiable at least once, and hence on account of (A.96) we have the estimate

$$|\boldsymbol{\xi} - \boldsymbol{\xi}'| = \left|\boldsymbol{\Psi}^{-1}(\mathbf{r}) - \boldsymbol{\Psi}^{-1}(\mathbf{r}')\right| \leqslant c_1|\mathbf{r} - \mathbf{r}'| = c_1|\boldsymbol{\Psi}(\boldsymbol{\xi}) - \boldsymbol{\Psi}(\boldsymbol{\xi}')| \tag{2.253}$$

where $c_1 > 0$ is a suitable constant. We further isolate the point $\boldsymbol{\xi}$ — which is mapped onto $\mathbf{r}$ — with another circle $B_2(\boldsymbol{\xi}, a)$ whose (dimensionless) radius a is chosen small enough so that $B_2(\boldsymbol{\xi}, a) \subset B_2(0,1)$. The circle $B_2(\boldsymbol{\xi}, a)$ is mapped onto the surface $S_a \subset S_2$ and thus it effectively isolates the point $\mathbf{r}$ (Figure 2.20). We are left with the task of showing that the integral over $S_2 \setminus S_a$ remains finite no matter how small a radius a we choose. To this purpose we observe that the vector

$$\mathbf{N}(u', v') := \frac{\partial \boldsymbol{\Psi}}{\partial u'} \times \frac{\partial \boldsymbol{\Psi}}{\partial v'}, \qquad (u', v') \in B_2(0,1) \tag{2.254}$$

is perpendicular to S_2 at $\mathbf{r}'$ in that the vectors

$$\frac{\partial}{\partial u'}\boldsymbol{\Psi}(\boldsymbol{\xi}'), \qquad \frac{\partial}{\partial v'}\boldsymbol{\Psi}(\boldsymbol{\xi}') \tag{2.255}$$

are tangential to S_2 in the same point. Therefore, we estimate

$$\left| \int_{S_2 \setminus S_a} \mathrm{d}S' \frac{\varrho_S(\mathbf{r}')}{4\pi\varepsilon|\mathbf{r}-\mathbf{r}'|} \right| = \left| \int_{B_2(0,1)\setminus B_2(\boldsymbol{\xi},a)} \mathrm{d}u'\mathrm{d}v' \frac{\varrho_S(\mathbf{r}')}{4\pi\varepsilon|\boldsymbol{\Psi}(\boldsymbol{\xi})-\boldsymbol{\Psi}(\boldsymbol{\xi}')|} \left| \frac{\partial \boldsymbol{\Psi}}{\partial u'} \times \frac{\partial \boldsymbol{\Psi}}{\partial v'} \right| \right|$$
$$\leqslant \int_{B_2(0,1)\setminus B_2(\boldsymbol{\xi},a)} \mathrm{d}u'\mathrm{d}v' \frac{1}{4\pi|\boldsymbol{\Psi}(\boldsymbol{\xi})-\boldsymbol{\Psi}(\boldsymbol{\xi}')|} \underbrace{\frac{\|\varrho_S\|_\infty}{\varepsilon} \left\| \frac{\partial \boldsymbol{\Psi}}{\partial u'} \right\|_\infty \left\| \frac{\partial \boldsymbol{\Psi}}{\partial v'} \right\|_\infty}_{c_2>0}$$
$$\leqslant \int_{B_2(0,1)\setminus B_2(\boldsymbol{\xi},a)} \mathrm{d}u'\mathrm{d}v' \frac{c_1 c_2}{4\pi|\boldsymbol{\xi}-\boldsymbol{\xi}'|} \leqslant \int_{B_2(\boldsymbol{\xi},2)\setminus B_2(\boldsymbol{\xi},a)} \mathrm{d}u'\mathrm{d}v' \frac{c_1 c_2}{4\pi|\boldsymbol{\xi}-\boldsymbol{\xi}'|} \tag{2.256}$$

having used (2.253) and the fact that $B_2(0,1) \subset B_2(\boldsymbol{\xi},2)$ because $1 + |\boldsymbol{\xi}| < 2$ by construction (Figure 2.20). The last integral in (2.256) can now be computed by using a local system of polar coordinates (τ', α') centered in $\boldsymbol{\xi}$, viz.,

$$\left| \int_{S_2 \setminus S_a} \mathrm{d}S' \frac{\varrho_S(\mathbf{r}')}{4\pi\varepsilon|\mathbf{r}-\mathbf{r}'|} \right| \leqslant \int_0^{2\pi} \mathrm{d}\alpha' \int_a^2 \mathrm{d}\tau' \tau' \frac{c_1 c_2}{4\pi\tau'} = \frac{c_1 c_2}{4\pi} 2\pi(2-a) \leqslant c_1 c_2 \tag{2.257}$$

and since the quantity in the rightmost-hand side remains finite for $a \to 0^+$, this analysis proves that the single-layer potential exists for $\mathbf{r} \in \mathbb{R}^3$ so long as the density $\varrho_S(\mathbf{r})$ is bounded.

We wish to show that the single-layer potential (2.189) is *uniformly Hölder continuous* for $\mathbf{r} \in \mathbb{R}^3$, i.e., for any two points $\mathbf{r}_1, \mathbf{r}_2 \in \mathbb{R}^3$ it holds [20, page 127]

$$|\mathcal{S}(\mathbf{r}_1) - \mathcal{S}(\mathbf{r}_2)| \leqslant M|\mathbf{r}_1 - \mathbf{r}_2|^\alpha, \qquad \alpha \in]0,1[\tag{2.258}$$

with $M > 0$ being a suitable constant, provided the surface charge density $\varrho_S(\mathbf{r})$ is at least continuous for $\mathbf{r} \in S$. Notice that Hölder continuity is stronger than simple continuity, which occurs for $\alpha = 0$. For the proof we resort to checking the three conditions (Hölder a)-(Hölder c) of the criterion discussed in Appendix F.4.

First of all, we observe that $\mathcal{S}(\mathbf{r})$ is *bounded* as has been proved by means of (2.256) and (2.257), and hence condition (Hölder a) is verified. Secondly, we write

$$|\mathcal{S}(\mathbf{r}_1) - \mathcal{S}(\mathbf{r}_2)| \leqslant \frac{\|\varrho_S\|_\infty}{4\pi\varepsilon} \int_S \mathrm{d}S' \left| \frac{1}{|\mathbf{r}_1 - \mathbf{r}'|} - \frac{1}{|\mathbf{r}_2 - \mathbf{r}'|} \right| \tag{2.259}$$

because the density is continuous and thus bounded by hypothesis. For the remaining integrand we consider the strip U_b of S [see (F.36)] and take the points $\mathbf{r}_1, \mathbf{r}_2$ in $\mathbb{R}^3 \setminus U_b$ where, by definition of strip, it holds $|\mathbf{r}_l - \mathbf{r}'| > b$, $l = 1, 2$, for $\mathbf{r}' \in S$. Then, we observe

$$\left| \frac{1}{|\mathbf{r}_1 - \mathbf{r}'|} - \frac{1}{|\mathbf{r}_2 - \mathbf{r}'|} \right| = \frac{\big||\mathbf{r}_1 - \mathbf{r}'| - |\mathbf{r}_2 - \mathbf{r}'|\big|}{|\mathbf{r}_1 - \mathbf{r}'||\mathbf{r}_2 - \mathbf{r}'|} \leqslant \frac{|\mathbf{r}_1 - \mathbf{r}_2|}{|\mathbf{r}_1 - \mathbf{r}'||\mathbf{r}_2 - \mathbf{r}'|} \leqslant \frac{1}{b^2}|\mathbf{r}_1 - \mathbf{r}_2| \tag{2.260}$$

by virtue of (H.20). It follows that condition (Hölder b) (i.e., Lipschitz continuity in $\mathbb{R}^3 \setminus U_b$) is also verified.

To show that $\mathcal{S}(\mathbf{r})$ is Hölder continuous in an open neighborhood H_a of S [see (F.15) and Figure F.2] we choose two points $\mathbf{r}_l \in H_a$, $l = 1, 2$, such that

$$\mathbf{r}_l = \mathbf{r}'_l + w_l \hat{\mathbf{n}}(\mathbf{r}'_l), \qquad \mathbf{r}'_l \in S, \qquad |w_l| < a \tag{2.261}$$

and we separate the surface S into two parts, namely,

$$S_{\mathbf{r}_1',b} := \{\mathbf{r}' \in S : |\mathbf{r}' - \mathbf{r}_1'| < b\} = S \cap B(\mathbf{r}_1', b) \qquad \text{with} \qquad b = 3|\mathbf{r}_1 - \mathbf{r}_2| \tag{2.262}$$

and $S \setminus S_{\mathbf{r}_1',b}$. We preliminary observe that for points $\mathbf{r}' \in S_{\mathbf{r}_1',b}$ we have

$$\begin{aligned} |\mathbf{r}' - \mathbf{r}_2'| &\leqslant |\mathbf{r}' - \mathbf{r}_1'| + |\mathbf{r}_1' - \mathbf{r}_2'| \leqslant |\mathbf{r}' - \mathbf{r}_1'| + 2|\mathbf{r}_1 - \mathbf{r}_2| \\ &\leqslant |\mathbf{r}' - \mathbf{r}_1'| + 3|\mathbf{r}_1 - \mathbf{r}_2| = |\mathbf{r}' - \mathbf{r}_1'| + b < 2b \end{aligned} \tag{2.263}$$

on account of estimate (F.31) and definition (2.262). If additionally we introduce the surface

$$S_{\mathbf{r}_2',2b} := \{\mathbf{r}' \in S : |\mathbf{r}' - \mathbf{r}_2'| < 2b\} = S \cap B(\mathbf{r}_2', 2b) \qquad \text{with} \qquad b = 3|\mathbf{r}_1 - \mathbf{r}_2| \tag{2.264}$$

then (2.263) implies that the points of $S_{\mathbf{r}_1',b}$ belong also to $S_{\mathbf{r}_2',2b}$, and hence $S_{\mathbf{r}_1',b}$ is contained in $S_{\mathbf{r}_2',2b}$. This finding will come in handy in a moment and again in Appendix G.

Now, the contribution of $S_{\mathbf{r}_1',b}$ — which is one part of the right-hand side of (2.259) — is estimated as follows

$$\begin{aligned} &\frac{\|\varrho_S\|_\infty}{4\pi\varepsilon} \int_{S_{\mathbf{r}_1',b}} dS' \left| \frac{1}{|\mathbf{r}_1 - \mathbf{r}'|} - \frac{1}{|\mathbf{r}_2 - \mathbf{r}'|} \right| \\ &\leqslant \frac{\|\varrho_S\|_\infty}{4\pi\varepsilon} \int_{S_{\mathbf{r}_1',b}} dS' \frac{1}{|\mathbf{r}_1 - \mathbf{r}'|} + \frac{\|\varrho_S\|_\infty}{4\pi\varepsilon} \int_{S_{\mathbf{r}_1',b}} dS' \frac{1}{|\mathbf{r}_2 - \mathbf{r}'|} \\ &\leqslant \frac{\|\varrho_S\|_\infty}{2\pi\varepsilon} \int_{S_{\mathbf{r}_1',b}} dS' \frac{1}{|\mathbf{r}_1' - \mathbf{r}'|} + \frac{\|\varrho_S\|_\infty}{2\pi\varepsilon} \int_{S_{\mathbf{r}_1',b}} dS' \frac{1}{|\mathbf{r}_2' - \mathbf{r}'|} \\ &\leqslant \frac{\|\varrho_S\|_\infty}{2\pi\varepsilon} \int_{S_{\mathbf{r}_1',b}} dS' \frac{1}{|\mathbf{r}_1' - \mathbf{r}'|} + \frac{\|\varrho_S\|_\infty}{2\pi\varepsilon} \int_{S_{\mathbf{r}_2',2b}} dS' \frac{1}{|\mathbf{r}_2' - \mathbf{r}'|} \end{aligned} \tag{2.265}$$

where we have also employed (F.26) since $\mathbf{r}_l \in H_a$. The integral over $S_{\mathbf{r}_1',b}$ is dominated by the integral over $S_{\mathbf{r}_2',2b}$ because the integrand is positive and $S_{\mathbf{r}_1',b} \subset S_{\mathbf{r}_2',2b}$, as we have shown with (2.263). Besides, the last two integrals are single-layer potentials with unitary densities and thus they are bounded in accordance with (2.257). In particular, by formally defining two bijective mappings from the circles $B_2(\boldsymbol{\xi}_1', b)$ and $B_2(\boldsymbol{\xi}_2', 2b)$ onto $S_{\mathbf{r}_1',b}$ and $S_{\mathbf{r}_2',2b}$, respectively, we have

$$\int_{S_{\mathbf{r}_1',b}} dS' \frac{1}{2\pi|\mathbf{r}_1' - \mathbf{r}'|} \leqslant \int_{B_2(\boldsymbol{\xi}_1',b)} du'dv' \frac{c_1}{2\pi|\boldsymbol{\xi}_1' - \boldsymbol{\xi}'|} = bc_1 \tag{2.266}$$

$$\int_{S_{\mathbf{r}_2',2b}} dS' \frac{1}{2\pi|\mathbf{r}_2' - \mathbf{r}'|} \leqslant \int_{B_2(\boldsymbol{\xi}_2',2b)} du'dv' \frac{c_2}{2\pi|\boldsymbol{\xi}_2' - \boldsymbol{\xi}'|} = 2bc_2 \tag{2.267}$$

where c_1 and c_2 are suitable constants, and we have integrated in polar coordinates centered in $\boldsymbol{\xi}_1'$ and $\boldsymbol{\xi}_2'$. Substituting (2.266) and (2.267) back into (2.265) yields

$$\frac{\|\varrho_S\|_\infty}{4\pi\varepsilon} \int_{S_{\mathbf{r}_1',b}} dS' \left| \frac{1}{|\mathbf{r}_1 - \mathbf{r}'|} - \frac{1}{|\mathbf{r}_2 - \mathbf{r}'|} \right| \leqslant \frac{\|\varrho_S\|_\infty}{\varepsilon} (c_1 + 2c_2) b$$

$$= \frac{\|\varrho_S\|_\infty}{\varepsilon}(c_1 + 2c_2)3|\mathbf{r}_1 - \mathbf{r}_2|^{1-\alpha}|\mathbf{r}_1 - \mathbf{r}_2|^\alpha \leqslant M|\mathbf{r}_1 - \mathbf{r}_2|^\alpha \tag{2.268}$$

where the constant M is independent of $\mathbf{r}_1$ and $\mathbf{r}_2$. This is possible because the distance $|\mathbf{r}_1 - \mathbf{r}_2|$ is dominated by the characteristic size of H_a (see Figure F.2).

To continue with the integral over $S \setminus S_{\mathbf{r}_1',b}$ — the other part of the right-hand side of (2.259) — we observe that for points $\mathbf{r}' \in S \setminus S_{\mathbf{r}_1',b}$ we have

$$3|\mathbf{r}_1 - \mathbf{r}_2| = b \leqslant |\mathbf{r}' - \mathbf{r}_1'| \leqslant 2|\mathbf{r}' - \mathbf{r}_1| \tag{2.269}$$

in light of the definition of $S_{\mathbf{r}_1',b}$ and estimate (F.26). Therefore, we have

$$|\mathbf{r}_1 - \mathbf{r}'| = |\mathbf{r}_1 - \mathbf{r}_2 + \mathbf{r}_2 - \mathbf{r}'| \leqslant |\mathbf{r}_1 - \mathbf{r}_2| + |\mathbf{r}' - \mathbf{r}_2| \tag{2.270}$$

$$|\mathbf{r}_2 - \mathbf{r}'| \geqslant |\mathbf{r}_1 - \mathbf{r}'| - |\mathbf{r}_1 - \mathbf{r}_2| \geqslant \left(1 - \frac{2}{3}\right)|\mathbf{r}_1 - \mathbf{r}'| = \frac{1}{3}|\mathbf{r}_1 - \mathbf{r}'| \tag{2.271}$$

by virtue of (2.269). Then, we find

$$\left|\frac{1}{|\mathbf{r}_1 - \mathbf{r}'|} - \frac{1}{|\mathbf{r}_2 - \mathbf{r}'|}\right| \leqslant \frac{|\mathbf{r}_1 - \mathbf{r}_2|}{|\mathbf{r}_1 - \mathbf{r}'||\mathbf{r}_2 - \mathbf{r}'|} \leqslant 3\frac{|\mathbf{r}_1 - \mathbf{r}_2|}{|\mathbf{r}_1 - \mathbf{r}'|^2} \leqslant 12\frac{|\mathbf{r}_1 - \mathbf{r}_2|}{|\mathbf{r}_1' - \mathbf{r}'|^2} \tag{2.272}$$

thanks to (2.271) used backwards and (F.26). Next, we consider the second part of the right member of (2.259), viz.,

$$\begin{aligned}
\frac{\|\varrho_S\|_\infty}{4\pi\varepsilon} \int\limits_{S\setminus S_{\mathbf{r}_1',b}} \mathrm{d}S' \left|\frac{1}{|\mathbf{r}_1 - \mathbf{r}'|} - \frac{1}{|\mathbf{r}_2 - \mathbf{r}'|}\right| &\leqslant \frac{\|\varrho_S\|_\infty}{4\pi\varepsilon} \int\limits_{S\setminus S_{\mathbf{r}_1',b}} \mathrm{d}S'\, 12\frac{|\mathbf{r}_1 - \mathbf{r}_2|}{|\mathbf{r}_1' - \mathbf{r}'|^2} \\
&= \frac{\|\varrho_S\|_\infty}{\pi\varepsilon}|\mathbf{r}_1 - \mathbf{r}_2|^\alpha \int\limits_{S\setminus S_{\mathbf{r}_1',b}} \mathrm{d}S' \left(\frac{b}{3}\right)^{1-\alpha} \frac{3}{|\mathbf{r}_1' - \mathbf{r}'|^2} \\
&\leqslant \frac{\|\varrho_S\|_\infty}{\pi\varepsilon} \int\limits_{S\setminus S_{\mathbf{r}_1',b}} \mathrm{d}S' \frac{3^\alpha}{|\mathbf{r}_1' - \mathbf{r}'|^{1+\alpha}}|\mathbf{r}_1 - \mathbf{r}_2|^\alpha = M|\mathbf{r}_1 - \mathbf{r}_2|^\alpha
\end{aligned} \tag{2.273}$$

because $b^{1-\alpha} \leqslant |\mathbf{r}' - \mathbf{r}_1'|^{1-\alpha}$ for $\mathbf{r}' \in S \setminus S_{\mathbf{r}_1',b}$ [see (2.269)]. The remaining integral is certainly bounded so long as $b > 0$, in that the integrand is not even singular for $\mathbf{r}' \notin S_{\mathbf{r}_1',b}$. To prove that the integral of concern is finite for any value of b we choose a surface S_2 such that $S_{\mathbf{r}_1',b} \subset S_2 \subset S$ and further split the calculation into two parts, namely, over $S_1 := S \setminus S_2$ and over $S_2 \setminus S_{\mathbf{r}_1',b}$. The contribution of S_1 is finite because it is independent of b and the integrand is regular for $\mathbf{r}' \notin S_{\mathbf{r}_1',b}$. For the integral over $S_2 \setminus S_{\mathbf{r}_1',b}$ we define a set of local coordinates $\boldsymbol{\xi}' = (u', v')$ and formally introduce a bijective mapping $\boldsymbol{\Psi}$ as in (2.252) from $B_2(0, 1)$ to S_2. For simplicity, this time we require that $\boldsymbol{\Psi}$ map the point $\boldsymbol{\xi}_1' = (0, 0)$ onto $\mathbf{r}_1'$ and the smaller circle $B_2(0, \nu)$, $\nu < 1$, onto the surface $S_{\mathbf{r}_1',b}$. Then, we consider

$$\begin{aligned}
\int\limits_{S_2\setminus S_{\mathbf{r}_1',b}} \mathrm{d}S' \frac{1}{|\mathbf{r}_1' - \mathbf{r}'|^{1+\alpha}} &= \int\limits_{B_2(0,1)\setminus B_2(0,\nu)} \mathrm{d}u'\mathrm{d}v' \frac{1}{|\boldsymbol{\Psi}(\boldsymbol{\xi}_1') - \boldsymbol{\Psi}(\boldsymbol{\xi}')|^{1+\alpha}} \left|\frac{\partial \boldsymbol{\Psi}}{\partial u'} \times \frac{\partial \boldsymbol{\Psi}}{\partial v'}\right| \\
&\leqslant \int\limits_{B_2(0,1)\setminus B_2(0,\nu)} \mathrm{d}u'\mathrm{d}v' \frac{c_1 c_2}{|\boldsymbol{\xi}_1' - \boldsymbol{\xi}'|^{1+\alpha}}
\end{aligned} \tag{2.274}$$

by virtue of (2.253). The last integral can be computed by using a local system of polar coordinates (τ', η') centered in ξ_1', viz.,

$$\int_{S_2 \setminus S_{\mathbf{r}_1', b}} \mathrm{d}S' \frac{1}{|\mathbf{r}_1' - \mathbf{r}'|^{1+\alpha}} \leqslant \int_0^{2\pi} \mathrm{d}\eta' \int_\nu^1 \mathrm{d}\tau' \, \tau' \frac{c_1 c_2}{\tau'^{1+\alpha}} = 2\pi c_1 c_2 \frac{1 - \nu^{1-\alpha}}{1 - \alpha} \leqslant \frac{2\pi c_1 c_2}{1 - \alpha} \tag{2.275}$$

where the last term remains finite for any $\nu = \nu(b)$ provided $\alpha < 1$. Thus, using (2.268) and (2.273) in (2.259) shows that condition (Hölder c) is verified for arbitrary points $\mathbf{r}_1, \mathbf{r}_2 \in H_a$.

In summary, since the three criteria (Hölder a)-(Hölder c) on page 1185 are fulfilled, the single-layer potential is Hölder continuous for $\mathbf{r} \in \mathbb{R}^3$.

We go on to show that the *gradient* of the single-layer potential $\mathcal{S}(\mathbf{r})$ can be *continuously extended* for observation points $\mathbf{r} \in S$ under the hypothesis that the charge density $\varrho_S(\mathbf{r}) \in \mathcal{C}^\alpha(S)$, i.e., it is Hölder continuous on S with $\alpha \in]0, 1[$. More precisely, the tangential component $\nabla_s \mathcal{S}(\mathbf{r})$ is continuous across S, whereas the normal components on either side of S read

$$\lim_{\mathbf{r} \to S^\pm} \frac{\partial}{\partial \hat{\mathbf{n}}} \mathcal{S}(\mathbf{r}) = \mp \frac{1}{2\varepsilon} \varrho_S(\mathbf{r}) + \mathrm{PV} \int_S \mathrm{d}S' \varrho_S(\mathbf{r}') \hat{\mathbf{n}}(\mathbf{r}) \cdot \nabla \frac{1}{4\pi\varepsilon |\mathbf{r} - \mathbf{r}'|} \tag{2.276}$$

where the integral exists as an improper one. Actually, the integral is bounded because — S being a $\mathcal{C}^2$-smooth surface — we can estimate the integrand as

$$\left| \hat{\mathbf{n}}(\mathbf{r}) \cdot \nabla \frac{1}{|\mathbf{r} - \mathbf{r}'|} \right| = \frac{|\hat{\mathbf{n}}(\mathbf{r}) \cdot (\mathbf{r} - \mathbf{r}')|}{|\mathbf{r} - \mathbf{r}'|^3} \leqslant \frac{M}{|\mathbf{r} - \mathbf{r}'|} \tag{2.277}$$

by virtue of (F.1) with an obvious exchange of the role of $\mathbf{r}'$ and $\mathbf{r}$. The remaining integral — an instance of single-layer potential — is dominated as in (2.257).

To begin with, we examine the case of a constant unitary density and indicate the pertinent potential with the symbol $\mathcal{S}_1(\mathbf{r})$. The conclusions we shall draw on $\nabla \mathcal{S}_1(\mathbf{r})$ will help us prove the result for a general $\varrho_S(\mathbf{r}')$. We pick up an arbitrary constant vector $\mathbf{u}$ and consider observation points $\mathbf{r}$ away from the surface S. Then, we have

$$\mathbf{u} \cdot \nabla \mathcal{S}_1(\mathbf{r}) = \mathbf{u} \cdot \nabla \int_S \mathrm{d}S' \frac{1}{4\pi\varepsilon R} = \int_S \mathrm{d}S' \mathbf{u} \cdot \nabla \frac{1}{4\pi\varepsilon R} = - \int_S \mathrm{d}S' \mathbf{u} \cdot \nabla' \frac{1}{4\pi\varepsilon R} \tag{2.278}$$

where the interchange of integration and derivatives is possible because the integrand is regular for $\mathbf{r} \in \mathbb{R}^3 \setminus S$. We separate both $\mathbf{u}$ and the gradient into two parts, namely, perpendicular and tangential to S, whereby we have

$$\begin{aligned} \mathbf{u} \cdot \nabla \mathcal{S}_1(\mathbf{r}) &= - \int_S \mathrm{d}S' \left[\mathbf{u} \cdot \hat{\mathbf{n}}(\mathbf{r}') \frac{\partial}{\partial \hat{\mathbf{n}}'} \frac{1}{4\pi\varepsilon R} + \mathbf{u}_t(\mathbf{r}') \cdot \nabla_s' \frac{1}{4\pi\varepsilon R} \right] \\ &= - \int_S \mathrm{d}S' \mathbf{u} \cdot \hat{\mathbf{n}}(\mathbf{r}') \frac{\partial}{\partial \hat{\mathbf{n}}'} \frac{1}{4\pi\varepsilon R} + \int_S \mathrm{d}S' \frac{\nabla_s' \cdot \mathbf{u}_t(\mathbf{r}')}{4\pi\varepsilon R} \end{aligned} \tag{2.279}$$

where $\nabla_s'\{\bullet\}$ and $\nabla_s' \cdot \{\bullet\}$ indicate the surface gradient (A.48) and divergence (A.49) with respect to the primed coordinates, and we have applied the differential identity (H.77) and the surface Gauss theorem (A.59) over the closed surface S. Although $\mathbf{u}$ is a constant vector, the tangential part $\mathbf{u}_t(\mathbf{r}')$, viz.,

$$\mathbf{u}_t(\mathbf{r}') := \hat{\mathbf{n}}(\mathbf{r}') \times [\mathbf{u} \times \hat{\mathbf{n}}(\mathbf{r}')], \qquad \mathbf{r}' \in S \tag{2.280}$$

is a surface vector field which depends on $\mathbf{r}'$ because $\hat{\mathbf{n}}(\mathbf{r}')$ changes over S. If we let $\mathbf{u} \equiv \hat{\mathbf{x}}$, we obtain

$$\hat{\mathbf{x}} \cdot \nabla \mathcal{S}_1(\mathbf{r}) = \frac{\partial \mathcal{S}_1}{\partial x} = -\int_S \mathrm{d}S'\, n_x(\mathbf{r}') \frac{\partial}{\partial \hat{\mathbf{n}}'} \frac{1}{4\pi\varepsilon R} + \int_S \mathrm{d}S'\, \frac{\nabla'_\mathrm{s} \cdot \hat{\mathbf{x}}_t(\mathbf{r}')}{4\pi\varepsilon R} \tag{2.281}$$

and analogous expressions hold by choosing $\mathbf{u} \equiv \hat{\mathbf{y}}$ and $\mathbf{u} \equiv \hat{\mathbf{z}}$. Again, notice that $\hat{\mathbf{x}}_t(\mathbf{r}')$ is tangential to S and hence not constant. We may combine the formal representation of the derivatives of $\mathcal{S}_1(\mathbf{r})$ with respect to the three Cartesian coordinates as in (A.26) in order to form the gradient, viz.,

$$\nabla \mathcal{S}_1(\mathbf{r}) = -\int_S \mathrm{d}S'\, \hat{\mathbf{n}}(\mathbf{r}') \frac{\partial}{\partial \hat{\mathbf{n}}'} \frac{1}{4\pi\varepsilon R} + \int_S \mathrm{d}S'\, \frac{\boldsymbol{\Xi}(\mathbf{r}')}{4\pi\varepsilon R}, \qquad \mathbf{r} \in \mathbb{R}^3 \setminus S \tag{2.282}$$

where

$$\boldsymbol{\Xi}(\mathbf{r}') = \hat{\mathbf{x}} \nabla'_\mathrm{s} \cdot \hat{\mathbf{x}}_t(\mathbf{r}') + \hat{\mathbf{y}} \nabla'_\mathrm{s} \cdot \hat{\mathbf{y}}_t(\mathbf{r}') + \hat{\mathbf{z}} \nabla'_\mathrm{s} \cdot \hat{\mathbf{z}}_t(\mathbf{r}'), \qquad \mathbf{r}' \in S \tag{2.283}$$

is a surface vector field on S. From (2.282) we see that $\nabla \mathcal{S}_1(\mathbf{r})$ is comprised of two terms which we recognize as a double-layer potential with vector density $\hat{\mathbf{n}}(\mathbf{r}')$ and a single-layer potential with vector density $\boldsymbol{\Xi}(\mathbf{r}')$. In Section 2.9 we proved that the double-layer potential remains bounded for $\mathbf{r} \to S^\pm$, although, as summarized in (2.244), the value of the limit depends on whether $\mathbf{r}$ approaches S from the positive side or the negative one. On the other hand, we have just showed above that the single-layer potential is finite and even Hölder continuous for $\mathbf{r} \in \mathbb{R}^3$.

Thus, by applying (2.244) to the Cartesian components of the right-hand side of (2.282) we arrive at

$$\lim_{\mathbf{r} \to S^\pm} \nabla \mathcal{S}_1(\mathbf{r}) = \mp \frac{1}{2\varepsilon} \hat{\mathbf{n}}(\mathbf{r}) - \mathrm{PV} \int_S \mathrm{d}S'\, \hat{\mathbf{n}}(\mathbf{r}') \frac{\partial}{\partial \hat{\mathbf{n}}'} \frac{1}{4\pi\varepsilon R} + \int_S \mathrm{d}S'\, \frac{\boldsymbol{\Xi}(\mathbf{r}')}{4\pi\varepsilon R} \tag{2.284}$$

whence we conclude that $\nabla \mathcal{S}_1(\mathbf{r})$ has a continuous extension for observation points which tend to the surface S from either side thereof. We remark that the two integrals above are bounded for $\mathbf{r} \in S$ because they are special instances of double-layer and single-layer potentials. In particular, we realize that the component of $\nabla \mathcal{S}_1(\mathbf{r})$ tangential to S is continuous because only the normal component is responsible for the jump through S. Indeed, the normal derivative reads

$$\lim_{\mathbf{r} \to S^\pm} \frac{\partial \mathcal{S}_1}{\partial \hat{\mathbf{n}}} = \mp \frac{1}{2\varepsilon} - \mathrm{PV} \int_S \mathrm{d}S'\, \hat{\mathbf{n}}(\mathbf{r}) \cdot \hat{\mathbf{n}}(\mathbf{r}') \frac{\partial}{\partial \hat{\mathbf{n}}'} \frac{1}{4\pi\varepsilon R} + \int_S \mathrm{d}S'\, \frac{\hat{\mathbf{n}}(\mathbf{r}) \cdot \boldsymbol{\Xi}(\mathbf{r}')}{4\pi\varepsilon R} \tag{2.285}$$

and, in fact, can be written in the alternative, compact form

$$\lim_{\mathbf{r} \to S^\pm} \frac{\partial \mathcal{S}_1}{\partial \hat{\mathbf{n}}} = \mp \frac{1}{2\varepsilon} + \mathrm{PV} \int_S \mathrm{d}S'\, \hat{\mathbf{n}}(\mathbf{r}) \cdot \nabla \frac{1}{4\pi\varepsilon |\mathbf{r} - \mathbf{r}'|} \tag{2.286}$$

in principle by combining the integrals into a single one by partly tracing back the steps that led us to (2.285) in the first place. However, we may not invoke the surface Gauss theorem in (2.285) right away inasmuch as the relevant integrand is now singular for $\mathbf{r}' = \mathbf{r}$. Therefore, we follow a limiting procedure which consists of smoothing the integrands with the aid of the function $f(R/a) \in \mathcal{C}^1(\mathbb{R}^+)$ introduced in (2.201) on page 123.

For ease of manipulation we let

$$\mathcal{U}(\mathbf{r}) := -\mathrm{PV}\int_S \mathrm{d}S'\,\hat{\mathbf{n}}(\mathbf{r})\cdot\hat{\mathbf{n}}(\mathbf{r}')\frac{\partial}{\partial\hat{\mathbf{n}}'}\frac{1}{4\pi\varepsilon R} + \int_S \mathrm{d}S'\,\frac{\hat{\mathbf{n}}(\mathbf{r})\cdot\Xi(\mathbf{r}')}{4\pi\varepsilon R} \tag{2.287}$$

and also define the auxiliary potential

$$\Upsilon(\mathbf{r};a) := -\mathrm{PV}\int_S \mathrm{d}S'\,\hat{\mathbf{n}}(\mathbf{r})\cdot\hat{\mathbf{n}}(\mathbf{r}')\frac{\partial}{\partial\hat{\mathbf{n}}'}\frac{f(R/a)}{4\pi\varepsilon R} + \int_S \mathrm{d}S'\,\hat{\mathbf{n}}(\mathbf{r})\cdot\Xi(\mathbf{r}')\frac{f(R/a)}{4\pi\varepsilon R} \tag{2.288}$$

where the length a is chosen small enough for the intersection $S \cap B(\mathbf{r}, 2a)$ to be non-empty. Both integrals in (2.288) are well-defined because $f(R/a)$ vanishes for $R \leqslant a$ (Figure 2.13).

First of all we show that $\Upsilon(\mathbf{r};a)$ reduces to $\mathcal{U}(\mathbf{r})$ as a approaches zero. By subtracting (2.287) and (2.288) side by side we get

$$\begin{aligned}
|\Upsilon(\mathbf{r};a) - \mathcal{U}(\mathbf{r})| &= \left|\mathrm{PV}\int_S \mathrm{d}S'\,\hat{\mathbf{n}}(\mathbf{r})\cdot\hat{\mathbf{n}}(\mathbf{r}')\frac{\partial}{\partial\hat{\mathbf{n}}'}\frac{1-f(R/a)}{4\pi\varepsilon R} - \int_S \mathrm{d}S'\,\hat{\mathbf{n}}(\mathbf{r})\cdot\Xi(\mathbf{r}')\frac{1-f(R/a)}{4\pi\varepsilon R}\right| \\
&\leqslant \mathrm{PV}\int_S \mathrm{d}S'\left|\left[1-f\left(\frac{R}{a}\right)\right]\frac{\hat{\mathbf{n}}(\mathbf{r}')\cdot\hat{\mathbf{R}}}{4\pi\varepsilon R^2} + f'\left(\frac{R}{a}\right)\frac{\hat{\mathbf{n}}(\mathbf{r}')\cdot\hat{\mathbf{R}}}{4\pi\varepsilon Ra}\right| + \int_S \mathrm{d}S'\left[1-f\left(\frac{R}{a}\right)\right]\frac{|\hat{\mathbf{n}}(\mathbf{r})\cdot\Xi(\mathbf{r}')|}{4\pi\varepsilon R} \\
&\leqslant \int_{S\cap B(\mathbf{r},2a)} \mathrm{d}S'\,\frac{M}{4\pi\varepsilon R} + \int_{S\cap(B(\mathbf{r},2a)\setminus B(\mathbf{r},a))} \mathrm{d}S'\,\frac{M\,\|f'\|_\infty}{2\pi\varepsilon R} + \int_{S\cap B(\mathbf{r},2a)} \mathrm{d}S'\,\frac{\|\Xi\|_\infty}{4\pi\varepsilon R}
\end{aligned} \tag{2.289}$$

where we have used $1 - f(R/a) \leqslant 1$, estimate (F.1), the boundedness of $\Xi(\mathbf{r}')$, and also noticed that $R \leqslant 2a$ for $\mathbf{r}' \in S \cap (B(\mathbf{r}, 2a) \setminus B(\mathbf{r}, a))$. Since the remaining integrals are instances of single-layer potentials — which we know to be invariably finite from (2.256) and (2.257) — they should vanish when evaluated on a surface that shrinks to a point. To be specific, we may introduce a set of local dimensionless coordinates $\boldsymbol{\xi}' := (u', v')$ which are related to the points $\mathbf{r}' \in S \cap B(\mathbf{r}, 2a)$ through the bijective mapping

$$\boldsymbol{\Psi} : \mathbb{R}^2 \supset B_2(0, 2b) \longrightarrow S \cap B(\mathbf{r}, 2a) \tag{2.290}$$

where the constant $b = b(a)$ tends to zero as $a \to 0^+$. For the sake of simplicity we may assume that $\boldsymbol{\Psi}$ maps the origin $\boldsymbol{\xi}' = 0$ onto $\mathbf{r}$. Thus, we have

$$\begin{aligned}
\int_{S\cap B(\mathbf{r},2a)} \mathrm{d}S'\,\frac{1}{4\pi|\mathbf{r}-\mathbf{r}'|} &= \int_{B_2(0,2b)} \mathrm{d}u'\mathrm{d}v'\,\frac{1}{4\pi|\boldsymbol{\Psi}(0)-\boldsymbol{\Psi}(\boldsymbol{\xi}')|}\left|\frac{\partial\boldsymbol{\Psi}}{\partial u'}\times\frac{\partial\boldsymbol{\Psi}}{\partial v'}\right| \\
&\leqslant \int_{B_2(0,2b)} \mathrm{d}u'\mathrm{d}v'\,\frac{c_2}{4\pi|\boldsymbol{\Psi}(0)-\boldsymbol{\Psi}(\boldsymbol{\xi}')|} \leqslant \int_{B_2(0,2b)} \mathrm{d}u'\mathrm{d}v'\,\frac{c_1c_2}{4\pi\xi'} \\
&= \frac{c_1c_2}{2}\int_0^{2b} \mathrm{d}\xi' = c_1c_2b(a) \underset{a\to 0^+}{\longrightarrow} 0
\end{aligned} \tag{2.291}$$

where we have invoked estimate (2.253) with $\boldsymbol{\xi} = 0$ and computed the last integral in polar coordinates (ξ', η'). This proves that the integrals over $S \cap B(\mathbf{r}, 2a)$ in the rightmost-hand side of (2.289) tend to zero. For the one over the annular region $S \cap (B(\mathbf{r}, 2a) \setminus B(\mathbf{r}, a))$ we consider

$$\boldsymbol{\Psi} : \mathbb{R}^2 \supset (B_2(0, 2b) \setminus B_2(0, b)) \longrightarrow S \cap (B(\mathbf{r}, 2a) \setminus B(\mathbf{r}, a)) \tag{2.292}$$

under the same hypotheses made above. Then, we have

$$\begin{aligned}\int\limits_{S\cap(B(\mathbf{r},2a)\setminus B(\mathbf{r},a))} \mathrm{d}S' \frac{1}{2\pi|\mathbf{r}-\mathbf{r}'|} &= \int\limits_{B_2(0,2b)\setminus B_2(0,b)} \mathrm{d}u'\mathrm{d}v' \frac{1}{2\pi|\mathbf{\Psi}(0)-\mathbf{\Psi}(\xi')|}\left|\frac{\partial\mathbf{\Psi}}{\partial u'}\times\frac{\partial\mathbf{\Psi}}{\partial v'}\right| \\ &\leqslant \int\limits_{B_2(0,2b)\setminus B_2(0,b)} \mathrm{d}u'\mathrm{d}v' \frac{c_2}{2\pi|\mathbf{\Psi}(0)-\mathbf{\Psi}(\xi')|} \leqslant \int\limits_{B_2(0,2b)\setminus B_2(0,b)} \mathrm{d}u'\mathrm{d}v' \frac{c_1c_2}{2\pi\xi'} \\ &= c_1c_2\int_b^{2b}\mathrm{d}\xi' = c_1c_2 b(a) \xrightarrow[a\to 0^+]{} 0\end{aligned} \tag{2.293}$$

again thanks to (2.253).

Secondly, in view of the definition of $\Xi(\mathbf{r}')$ and the regularity of the integrands in (2.288) we can transform $\Upsilon(\mathbf{r};a)$ as follows

$$\begin{aligned}\Upsilon(\mathbf{r};a) &= -\mathrm{PV}\int_S \mathrm{d}S' \left[n_x(\mathbf{r})\hat{\mathbf{x}} + n_y(\mathbf{r})\hat{\mathbf{y}} + n_z(\mathbf{r})\hat{\mathbf{z}}\right]\cdot\hat{\mathbf{n}}(\mathbf{r}')\frac{\partial}{\partial\hat{\mathbf{n}}'}\frac{f(R/a)}{4\pi\varepsilon R} \\ &\quad + \int_S \mathrm{d}S'\left[n_x(\mathbf{r})\nabla_{\mathrm{s}}'\cdot\hat{\mathbf{x}}_t(\mathbf{r}') + n_y(\mathbf{r})\nabla_{\mathrm{s}}'\cdot\hat{\mathbf{y}}_t(\mathbf{r}') + n_z(\mathbf{r})\nabla_{\mathrm{s}}'\cdot\hat{\mathbf{z}}_t(\mathbf{r}')\right]\frac{f(R/a)}{4\pi\varepsilon R} \\ &= -\mathrm{PV}\int_S \mathrm{d}S'\left[n_x(\mathbf{r})x_n(\mathbf{r}') + n_y(\mathbf{r})y_n(\mathbf{r}') + n_z(\mathbf{r})z_n(\mathbf{r}')\right]\frac{\partial}{\partial\hat{\mathbf{n}}'}\frac{f(R/a)}{4\pi\varepsilon R} \\ &\quad - \int_S \mathrm{d}S'\left[n_x(\mathbf{r})\hat{\mathbf{x}}_t(\mathbf{r}') + n_y(\mathbf{r})\hat{\mathbf{y}}_t(\mathbf{r}') + n_z(\mathbf{r})\hat{\mathbf{z}}_t(\mathbf{r}')\right]\cdot\nabla_{\mathrm{s}}'\frac{f(R/a)}{4\pi\varepsilon R} \\ &= -\mathrm{PV}\int_S \mathrm{d}S'\,\hat{\mathbf{n}}(\mathbf{r})\cdot\nabla'\frac{f(R/a)}{4\pi\varepsilon R} = \mathrm{PV}\int_S \mathrm{d}S'\,\hat{\mathbf{n}}(\mathbf{r})\cdot\nabla\frac{f(R/a)}{4\pi\varepsilon R}\end{aligned} \tag{2.294}$$

where $x_n(\mathbf{r}')$ denotes the normal component of $\hat{\mathbf{x}}$ at $\mathbf{r}'$, and so forth. In the second step we have used (H.77) and applied the surface Gauss theorem (A.59) three times over the closed surface S.

Finally, we need to show that the rightmost member of (2.294) indeed tends to the integral in (2.286). In symbols, we have

$$\begin{aligned}\left|\Upsilon(\mathbf{r};a) - \mathrm{PV}\int_S \mathrm{d}S'\,\hat{\mathbf{n}}(\mathbf{r})\cdot\nabla\frac{1}{4\pi\varepsilon R}\right| &= \left|\mathrm{PV}\int_S \mathrm{d}S'\,\hat{\mathbf{n}}(\mathbf{r})\cdot\nabla'\frac{1-f(R/a)}{4\pi\varepsilon R}\right| \\ &\leqslant \mathrm{PV}\int_S \mathrm{d}S'\left|\left[1-f\left(\frac{R}{a}\right)\right]\frac{\hat{\mathbf{n}}(\mathbf{r})\cdot\hat{\mathbf{R}}}{4\pi\varepsilon R^2} + f'\left(\frac{R}{a}\right)\frac{\hat{\mathbf{n}}(\mathbf{r})\cdot\hat{\mathbf{R}}}{4\pi\varepsilon Ra}\right| \\ &\leqslant \int\limits_{S\cap B(\mathbf{r},2a)} \mathrm{d}S'\frac{M}{4\pi\varepsilon R} + \int\limits_{S\cap(B(\mathbf{r},2a)\setminus B(\mathbf{r},a))} \mathrm{d}S'\frac{M\,\|f'\|_\infty}{2\pi\varepsilon R} \xrightarrow[a\to 0^+]{} 0\end{aligned} \tag{2.295}$$

and the result follows in light of (2.291) and (2.293). Since $\Upsilon(\mathbf{r};a)$ also tends to $\mathcal{U}(\mathbf{r})$ in (2.287), and the limit is unique, this proves the validity of (2.286).

Now we recall that we are interested in the gradient of the single-layer potential with arbitrary charge density $\varrho_S(\mathbf{r}')$. We pick up a point $\mathbf{r} = \mathbf{r}_0' + w\hat{\mathbf{n}}(\mathbf{r}_0') \in H_a \setminus S$ where $|w| < a$, $\mathbf{r}_0' \in S$, and H_a is an open neighborhood of S [see (F.15) and Figure F.2]. Then, since $1/R$ is regular for $\mathbf{r} \in H_a \setminus S$ we may write $\nabla \mathcal{S}(\mathbf{r})$ as

$$\begin{aligned}\nabla \mathcal{S}(\mathbf{r}) &= \int_S \mathrm{d}S' \, \varrho_S(\mathbf{r}') \nabla \frac{1}{4\pi\varepsilon R} \\ &= \int_S \mathrm{d}S' \left[\varrho_S(\mathbf{r}') - \varrho_S(\mathbf{r}'_0)\right] \nabla \frac{1}{4\pi\varepsilon R} + \varrho_S(\mathbf{r}'_0) \nabla \mathcal{S}_1(\mathbf{r}), \qquad \mathbf{r} \in H_a \setminus S \end{aligned} \tag{2.296}$$

where the Cartesian components of the first vector field in the rightmost-hand side can be reduced to scalar fields of type (G.1) with wavenumber $k = 0$, $\mathbf{V}(\mathbf{r}') \in \{\hat{\mathbf{x}}, \hat{\mathbf{y}}, \hat{\mathbf{z}}\}$ and $\psi(\mathbf{r}') = \varrho_S(\mathbf{r}')/\varepsilon$. Since the charge density is Hölder continuous on S by hypothesis, then the proof of Appendix G applies. Specifically, the integral exists also for $\mathbf{r} = \mathbf{r}'_0$ as an improper one, and the field is a Hölder continuous function of $\mathbf{r} \in H_a$ for any exponent $\beta < \alpha \in]0, 1[$. Combining this latest finding with the above conclusions on $\nabla \mathcal{S}_1(\mathbf{r})$ proves the initial claim, namely, that $\nabla \mathcal{S}(\mathbf{r})$ can be continuously extended for points $\mathbf{r} \to \mathbf{r}'_0$. Lastly, by letting $\mathbf{r} = \mathbf{r}'_0$, for the normal derivative we have

$$\begin{aligned}\lim_{\mathbf{r} \to S^\pm} \frac{\partial \mathcal{S}}{\partial \hat{\mathbf{n}}} &= \int_S \mathrm{d}S' \, [\varrho_S(\mathbf{r}') - \varrho_S(\mathbf{r})] \, \hat{\mathbf{n}}(\mathbf{r}) \cdot \nabla \frac{1}{4\pi\varepsilon R} \mp \frac{\varrho_S(\mathbf{r})}{2\varepsilon} + \varrho_S(\mathbf{r}) \mathrm{PV} \int_S \mathrm{d}S' \, \hat{\mathbf{n}}(\mathbf{r}) \cdot \nabla \frac{1}{4\pi\varepsilon R} \\ &= \mp \frac{\varrho_S(\mathbf{r})}{2\varepsilon} + \mathrm{PV} \int_S \mathrm{d}S' \, \varrho_S(\mathbf{r}') \hat{\mathbf{n}}(\mathbf{r}) \cdot \nabla \frac{1}{4\pi\varepsilon|\mathbf{r} - \mathbf{r}'|}\end{aligned} \tag{2.297}$$

thanks to (2.286), and this proves (2.276). The jump of the normal derivative of $\mathcal{S}(\mathbf{r})$ through S reads

$$\lim_{\mathbf{r} \to S^+} \frac{\partial \mathcal{S}}{\partial \hat{\mathbf{n}}} - \lim_{\mathbf{r} \to S^-} \frac{\partial \mathcal{S}}{\partial \hat{\mathbf{n}}} = -\frac{\varrho_S(\mathbf{r})}{\varepsilon} \tag{2.298}$$

which is consistent with the matching condition (2.46) in light of (1.117) and (2.15).

References

[1] Jackson JD. Classical Electrodynamics. 3rd ed. Chichester, UK: Wiley; 1999.

[2] Holt CA. Introduction to Electromagnetic Fields and Waves. New York, NY: John Wiley & Sons, Inc.; 1963.

[3] Tonnelat MA. The principles of electromagnetic theory and of relativity. Dordrecht-Holland, NL: D. Reidel Publishing Company; 1966.

[4] Frisch M. Inconsistency in classical electrodynamics. Philosophy of Science. 2004 October;71:525–549.

[5] Vickers P. Frisch, Muller, and Belot on an Inconsistency in Classical Electrodynamics. British Journal for the Philosophy of Science. 2008;59:767–792.

[6] Griffith DJ. Introduction to electrodynamics. 3rd ed. Upper Saddle River, NJ: Prentice Hall; 1999.

[7] Dautray R, Lions JL. Mathematical Analysis and Numerical Methods for Science and Technology - Spectral Theory and Applications. vol. 3. Berlin Heidelberg: Springer-Verlag; 1990.

[8] Dautray R, Lions JL. Mathematical Analysis and Numerical Methods for Science and Technology - Physical Origins and Classical Methods. vol. 1. Berlin Heidelberg: Springer-Verlag; 1990.

[9] Cottingham WN, Greenwood DA. Electricity and Magnetism. Cambridge, UK: Cambridge University Press; 1991.

[10] Hayt WH, Buck JA. Engineering Electromagnetics. 8th ed. New York, NY: McGraw-Hill; 2012. International edition.

[11] Morse P, Feshbach H. Methods of Theoretical Physics. New York, NY: McGraw-Hill; 1953.

[12] Dwight HB. Tables of Integrals and Other Mathematical Data. 4th ed. New York, NY: MacMillan Publishing Co., Inc.; 1961.

[13] Rothwell EJ, Cloud MJ. Electromagnetics. London, UK: CRC Press; 2001.

[14] Anderson N. The Electromagnetic Field. New York, NY: Springer Science+Business Media; 1968.

[15] Schwab AJ. Field Theory Concepts. Berlin Heidelberg: Springer-Verlag; 1988.

[16] Kellogg OD. Foundations of potential theory. Berlin Heidelberg: Springer-Verlag; 1929.

[17] Feynmann RP, Leighton R, Sands M. The Feynmann lectures on physics. vol. 1. Reading, MA: Addison-Wesley; 1964.

[18] Helrich CS. The classical theory of fields - Electromagnetism. Berlin Heidelberg: Springer-Verlag; 2012.

[19] Van Bladel JG. Electromagnetic Fields. Piscataway, NJ: IEEE Press; 2007.

[20] Bassanini P, Elcrat AR. Theory and applications of partial differential equations. Mathematical Concepts And Methods In Science And Engineering. New York, NY: Plenum Press; 1997.

[21] Feynmann RP, Leighton R, Sands M. The Feynmann lectures on physics. vol. 2. Reading, MA: Addison-Wesley; 1964.

[22] Konopinski EJ. Electromagnetic Fields and Relativistic Particles. New York, NY: McGraw Hill; 1981.

[23] Popović Z, Popović BD. Introductory Electromagnetics. Upper Saddle River, NJ: Prentice Hall; 2000.

[24] Zangwill A. Modern electrodynamics. Cambridge, UK: Cambrigde University Press; 2013.

[25] Bassanini P, Elcrat A. Mathematical Theory of Electromagnetism. Creative Commons; 2009.

[26] Golub GH, van Loan CF. Matrix Computations. Baltimore, MD: Johns Hopkins University Press; 1996.

[27] Lorrain P, Corson DR, Lorrain F. Electromagnetic Fields and Waves. 3rd ed. New York, NY: W. H. Freeman and Company; 1988.

[28] Makarov B, Podkorytov A. Real Analysis: Measures, Integrals and Applications. Universitext. London, UK: Springer; 2013.

[29] Lindell IV. Methods for electromagnetic field analysis. Piscataway, NJ: IEEE Press; 1992.

[30] Moon P, Spencer DE. Field Theory Handbook: Including Coordinate Systems, Differential Equations and Their Solutions. 2nd ed. Berlin Heidelberg: Springer-Verlag; 1971.

[31] Van Bladel JG. Singular Electromagnetic Fields and Sources. Piscataway, NJ: IEEE Press; 1991.

[32] Rudin W. Principles of mathematical analysis. 3rd ed. New York, NY: McGraw-Hill; 1976.

[33] Müller C. Foundations of the Mathematical Theory of Electromagnetic Waves. Berlin Heidelberg: Springer-Verlag; 1969.

Chapter 3
Static electric fields II

3.1 Scalar potential due to surface charges

We mentioned that the single- and double-layer potentials (2.189) and (2.188) can be interpreted as the contributions of surface distributions of charges and electric dipoles, respectively. Here we wish to extend this notion and address the question of finding the potential generated by surface densities of charge $\varrho_S(\mathbf{r})$ and dipoles $\boldsymbol{\tau}_S(\mathbf{r})$.

We begin with $\varrho_S(\mathbf{r})$ [1] and assume that the charges are immersed in a dielectric medium endowed with permittivity ε and confined to a smooth closed surface $S := \partial V$ which constitutes the boundary of a finite domain V (Figure 3.1). The relevant electrostatic potential $\Phi(\mathbf{r})$ is the unique solution to the Neumann problem

$$\begin{cases} \nabla^2\Phi(\mathbf{r}) = 0, & \mathbf{r} \in \mathbb{R}^3 \setminus S \\ \Phi(\mathbf{r})|_{\mathbf{r}\in S^-} = \Phi(\mathbf{r})|_{\mathbf{r}\in S^+}, & \mathbf{r} \in S \\ \varrho_S(\mathbf{r}) = \varepsilon \left.\dfrac{\partial\Phi}{\partial\hat{\mathbf{n}}}\right|_{\mathbf{r}\in S^-} - \varepsilon \left.\dfrac{\partial\Phi}{\partial\hat{\mathbf{n}}}\right|_{\mathbf{r}\in S^+}, & \mathbf{r} \in S \\ \Phi(\mathbf{r}) = \mathcal{O}\left(\dfrac{1}{|\mathbf{r}|}\right), & |\mathbf{r}| \to +\infty \end{cases} \tag{3.1}$$

where the unit vector $\hat{\mathbf{n}}(\mathbf{r})$ normal to S points towards the outside of V. The potential *per se* is continuous through S, but the normal component of $\nabla\Phi(\mathbf{r})$ suffers a jump across the surface S owing to the presence of the charges. The discontinuity is a consequence of the matching condition (2.46).

At first blush it may sound strange or even wrong that in (3.1) we specify constraints for both Φ and $\hat{\mathbf{n}} \cdot \nabla\Phi$ on the boundary of V, inasmuch as in Section 2.5.1 we showed that uniqueness is achieved by assigning either the potential or the normal derivative thereof. As a matter of fact, the conditions derived in Section 2.5.1 apply to a single Laplace equation defined in a spatial region where the potential and its derivatives are continuous, whereas strictly speaking problem (3.1) is comprised of two separate Laplace equations formulated in V and the complementary domain. A detailed analysis carried out along the same lines followed in Section 2.5.1 evidences that uniqueness correctly requires conditions for both Φ and $\hat{\mathbf{n}} \cdot \nabla\Phi$ across S. Intuitively, from the viewpoint of the problem in V we may enforce, say, the potential on S^-, but this in turn is not known until we solve the problem outside V. Conversely, from the standpoint of the complementary domain we then enforce $\hat{\mathbf{n}} \cdot \nabla\Phi$ on S^+, and this quantity too is unknown unless we find the potential in V. Therefore, the matching conditions (2.48) and (2.51) *couple* the two problems in V and $\mathbb{R}^3 \setminus \overline{V}$ and guarantee uniqueness.

To take advantage of the integral representation (2.160) we separate the problem into two parts, namely, finding the potential in $\mathbb{R}^3 \setminus \overline{V}$ and the one for $\mathbf{r} \in V$. Since the complementary domain $\mathbb{R}^3 \setminus \overline{V}$ is unbounded, we formally apply (2.160) to the region of space bounded by S and a ball $B(0, a)$ with

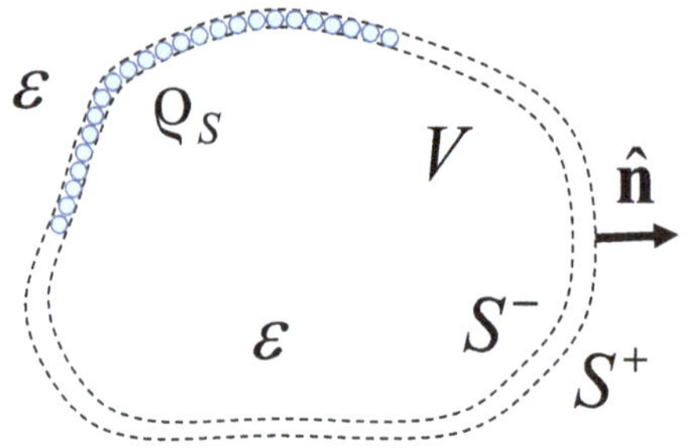

Figure 3.1 *For the derivation of the electrostatic potential generated by a layer of charges with density $\varrho_S(\mathbf{r})$ in a homogeneous isotropic dielectric medium. For visualization's sake the two sides of S (−−) are drawn slightly away from one another.*

the radius a large enough for V to be contained in $B(0, a)$. In the limit as the radius a approaches infinity the contribution of the surface integral over the sphere ∂B vanishes in that both $\Phi(\mathbf{r})$ and the static Green function (2.131) obey the estimate (2.20). Therefore, we obtain

$$\Phi(\mathbf{r}) = \int_S dS'\, \Phi(\mathbf{r}')|_{S^+} \frac{\partial}{\partial \hat{\mathbf{n}}'} \frac{1}{4\pi R} - \int_S dS' \frac{1}{4\pi R} \left.\frac{\partial \Phi}{\partial \hat{\mathbf{n}}'}\right|_{S^+}, \qquad \mathbf{r} \in \mathbb{R}^3 \setminus \overline{V} \tag{3.2}$$

where $R = |\mathbf{r} - \mathbf{r}'|$, and we have indicated that within the integrals the potential and the normal derivative thereof are evaluated on the positive side of S.

For the potential in V we use (2.160) again to arrive at

$$\Phi(\mathbf{r}) = -\int_S dS'\, \Phi(\mathbf{r}')|_{S^-} \frac{\partial}{\partial \hat{\mathbf{n}}'} \frac{1}{4\pi R} + \int_S dS' \frac{1}{4\pi R} \left.\frac{\partial \Phi}{\partial \hat{\mathbf{n}}'}\right|_{S^-}, \qquad \mathbf{r} \in V \tag{3.3}$$

where we have taken into account the fact that the unit normal is oriented towards the outside of V. Here, within the integrals the potential and the normal derivative thereof are evaluated on the negative side of S.

Next, we recall that (2.160) yields zero when the right-hand side is evaluated for observation points $\mathbf{r}$ set outside the region for which the representation was derived in the first place. As a result, (3.2) predicts a zero potential for $\mathbf{r} \in V$, whereas (3.3) returns zero for $\mathbf{r}$ in the complementary domain. Thus, if we sum (3.2) and (3.3) side by side we get

$$\Phi(\mathbf{r}) = \int_S dS' \left[\Phi(\mathbf{r}')|_{S^+} - \Phi(\mathbf{r}')|_{S^-}\right] \frac{\partial}{\partial \hat{\mathbf{n}}'} \frac{1}{4\pi R} + \int_S dS' \frac{1}{4\pi R} \left[\left.\frac{\partial \Phi}{\partial \hat{\mathbf{n}}'}\right|_{S^-} - \left.\frac{\partial \Phi}{\partial \hat{\mathbf{n}}'}\right|_{S^+} \right], \qquad \mathbf{r} \in \mathbb{R}^3 \setminus S \tag{3.4}$$

which is a single representation valid for every point $\mathbf{r} \in \mathbb{R}^3 \setminus S$. To finalize the derivation we recall that the potential is continuous across S whereas the jump of the normal derivatives is related to the surface charge density. In symbols, we find

$$\Phi(\mathbf{r}) = \frac{1}{\varepsilon} \int_S dS' \frac{\varrho_S(\mathbf{r}')}{4\pi R}, \qquad \mathbf{r} \in \mathbb{R}^3 \setminus S \tag{3.5}$$

which is the desired solution to (3.1). The integral exists also for observation points on the surface S (Section 2.10). We remark that forgetting to prescribe the continuity of Φ across S would leave an unknown contribution in (3.4) and formula (3.5) could not be stated.

Example 3.1 (Electrostatic potential and field of a uniform spherical layer of charges)

We consider an amount of charge Q, possibly fictitious, which is uniformly distributed over the boundary of a ball $B(0, a)$ in a homogeneous dielectric medium. This configuration is ideal inasmuch as on the one hand electric charges are discrete entities (cf. Figure 1.1) whereas on the other the distribution is unstable (see the Earnshaw theorem in Section 3.8). At any rate, with the surface density given by

$$\varrho_S(\mathbf{r}) := \varrho_{S0} = \frac{Q}{4\pi a^2}, \qquad \mathbf{r} = a\hat{\mathbf{r}} \tag{3.6}$$

we compute the electrostatic potential through (3.5).

For the calculation of the surface integral it is convenient to choose a system of spherical coordinates (r', α', β') with the polar axis aligned with the position vector $\mathbf{r}$ and the angle α' such that $\mathbf{r} \cdot \mathbf{r}' = rr' \cos\alpha'$. For reasons of symmetry the potential must depend only on the distance r from the center of $B(0, a)$. We have

$$\begin{aligned}\Phi(r) &= \int_{\partial B} \mathrm{d}S' \frac{\varrho_{S0}}{4\pi\varepsilon R} = \frac{\varrho_{S0}}{4\pi\varepsilon} \int_0^{\pi} \mathrm{d}\alpha' \frac{2\pi a^2 \sin\alpha'}{(a^2 - 2ra\cos\alpha' + r^2)^{1/2}} \\ &= \frac{a\varrho_{S0}}{2r\varepsilon} \left[\left(a^2 - 2ra\cos\alpha' + r^2\right)^{1/2} \right]_{\alpha'=0}^{\pi} = \frac{a\varrho_{S0}}{2r\varepsilon}(a + r - |r - a|) = \begin{cases} \dfrac{a\varrho_{S0}}{\varepsilon}, & r < a \\ \dfrac{a^2\varrho_{S0}}{r\varepsilon}, & r > a \end{cases}\end{aligned} \tag{3.7}$$

having observed that the problem is axially symmetric around $\mathbf{r}$.

By comparing $\Phi(r)$ for observation points outside the sphere with the potential (2.28) of a point charge in the origin we may interpret the result by stating that for $r > a$ the potential is equivalently produced by a point charge $4\pi a^2 \varrho_{S0} = Q$ placed at $r = 0$. Furthermore, we notice that $\Phi(r)$ is continuous across the charge layer.

On account of (2.15) the electric field produced by $\varrho_S(\mathbf{r})$ reads

$$\mathcal{E}(\mathbf{r}) = \begin{cases} 0, & r < a \\ \dfrac{a^2 \varrho_{S0}}{r^2 \varepsilon} \hat{\mathbf{r}}, & r > a \end{cases} \tag{3.8}$$

and, in particular, is radially directed and discontinuous across the charge layer. The jump of the normal component is in agreement with (2.46) and (2.298). The vanishing of the electric field *everywhere* inside the surface distribution of charge can be understood by noticing that in each point $\mathbf{r} \in B(0, a)$ the contributions to $\mathcal{E}(\mathbf{r})$ that arise from different parts of ∂B combine in such a way that the net result is zero.

(End of Example 3.1)

We continue with the potential produced by a layer of dipoles [1] confined to a smooth closed surface $S := \partial V$ which constitutes the boundary of a finite domain V (Figure 3.2). Since a dipole is conceptually comprised of two point charges separated by an infinitely short distance, the density

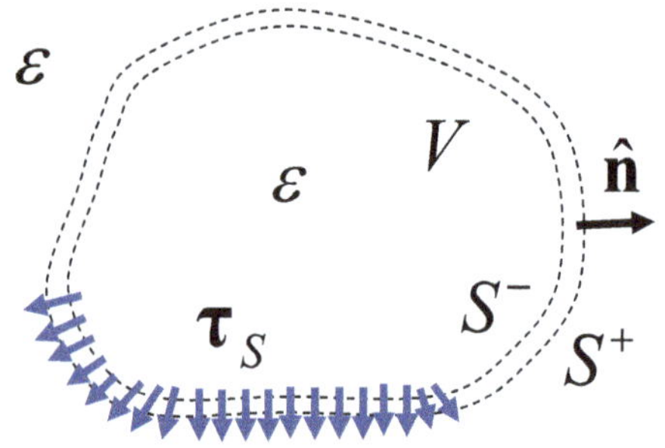

Figure 3.2 For the derivation of the electrostatic potential generated by a layer of dipoles with density $\boldsymbol{\tau}_S(\mathbf{r})$ in a homogeneous isotropic dielectric medium. For visualization's sake the two sides of S (−−) are drawn slightly away from one another.

$\boldsymbol{\tau}_S(\mathbf{r})$ may also be interpreted as the result of two layers of positive and negative charge densities set on either side of S. The relevant electrostatic potential is the unique solution to the Dirichlet problem

$$\begin{cases} \nabla^2\Phi(\mathbf{r}) = 0, & \mathbf{r} \in \mathbb{R}^3 \setminus S \\ \boldsymbol{\tau}_S(\mathbf{r}) = \varepsilon\hat{\mathbf{n}}(\mathbf{r})\,[\Phi(\mathbf{r})|_{\mathbf{r}\in S^+} - \Phi(\mathbf{r})|_{\mathbf{r}\in S^-}], & \mathbf{r} \in S \\ \left.\dfrac{\partial\Phi}{\partial\hat{\mathbf{n}}}\right|_{\mathbf{r}\in S^-} = \left.\dfrac{\partial\Phi}{\partial\hat{\mathbf{n}}}\right|_{\mathbf{r}\in S^+}, & \mathbf{r} \in S \\ \Phi(\mathbf{r}) = \mathcal{O}\left(\dfrac{1}{|\mathbf{r}|}\right), & |\mathbf{r}| \to +\infty \end{cases} \tag{3.9}$$

where $\Phi(\mathbf{r})$ suffers a jump across the surface S owing to the presence of the dipoles (see Section 2.9). By contrast, the normal derivative of the potential is continuous through S.

Proceeding as before we separate the problem into two parts, namely, for the outside and the inside of S. The pertinent potentials are given again by (3.2) and (3.3). The same remarks apply, and by combining (3.2) and (3.3) we obtain (3.4). This time, though, the normal derivative is continuous whereas the potential is not. Taking the second and the third line of (3.9) into account we get

$$\Phi(\mathbf{r}) = \int_S \mathrm{d}S'\,[\Phi(\mathbf{r}')|_{\mathbf{r}\in S^+} - \Phi(\mathbf{r}')|_{\mathbf{r}\in S^-}]\,\hat{\mathbf{n}}(\mathbf{r}')\cdot\nabla'\frac{1}{4\pi R} = \frac{1}{\varepsilon}\int_S \mathrm{d}S'\,\boldsymbol{\tau}_S(\mathbf{r}')\cdot\nabla'\frac{1}{4\pi R} \tag{3.10}$$

valid for points $\mathbf{r} \in \mathbb{R}^3 \setminus S$. The integral exists also for points on S but the values on either side of S are different. According to (2.244) we have

$$\lim_{\mathbf{r}\to S^\pm} \Phi(\mathbf{r}) = \frac{1}{\varepsilon}\mathrm{PV}\int_S \mathrm{d}S'\,\boldsymbol{\tau}_S(\mathbf{r}')\cdot\nabla'\frac{1}{4\pi R} \pm \frac{1}{2\varepsilon}\boldsymbol{\tau}_S(\mathbf{r})\cdot\hat{\mathbf{n}}(\mathbf{r}) \tag{3.11}$$

whence we recover the original jump

$$\varepsilon\,[\Phi(\mathbf{r})|_{\mathbf{r}\in S^+} - \Phi(\mathbf{r})|_{\mathbf{r}\in S^-}] = \boldsymbol{\tau}_S(\mathbf{r})\cdot\hat{\mathbf{n}}(\mathbf{r}) \tag{3.12}$$

exactly as assumed in (3.9).

In deriving (3.5) and (3.10) we considered the charges and the dipoles as being distributed over a closed surface S because we wished to employ the integral representation (2.160). However, our formulas remain valid even if the densities of concern are defined on a *open* smooth surface (Figure 1.2a). This statement follows simply by assuming that the boundary conditions in (3.1) and (3.9) hold true only for a part $S_o \subset S$ with S still representing the boundary of V. Therefore, in (3.5) and (3.10) we only have to compute the integrals over S_o in that across $S \setminus S_o$ the normal derivative of $\Phi(\mathbf{r})$ or the potential itself are continuous as well.

3.2 Integral representation of the electrostatic field

We may attempt to find an integral representation of the electrostatic field in a region of space V filled with an isotropic dielectric medium. With reference to Figure 2.7a we consider the electrostatic problem

$$\begin{cases} \varepsilon\nabla' \cdot \mathcal{E}_1(\mathbf{r}') = \varrho(\mathbf{r}'), & \mathbf{r}' \in V \\ \nabla' \times \mathcal{E}_1(\mathbf{r}') = 0, & \mathbf{r}' \in V \\ \hat{\mathbf{n}}(\mathbf{r}') \cdot \mathcal{E}_1(\mathbf{r}') = f(\mathbf{r}') \quad \text{or} \quad \hat{\mathbf{n}}(\mathbf{r}') \times \mathcal{E}_1(\mathbf{r}') = \mathbf{g}(\mathbf{r}'), & \mathbf{r}' \in S \cup \partial V_1 \end{cases} \tag{3.13}$$

which admits a unique solution. As an auxiliary problem we take the equations for the electrostatic field produced by an electric dipole of moment $\mathbf{p}$ in an unbounded homogeneous isotropic medium endowed with permittivity ε. The actual value of the moment is inessential, and we will remove $\mathbf{p}$ from the final result. To begin with we assume the dipole is located at a point $\mathbf{r} \in V \setminus V_\varrho$ (see Figure 2.7b)

$$\begin{cases} \varepsilon\nabla' \cdot \mathcal{E}_2(\mathbf{r}') = -\varepsilon\nabla'^2\Phi_2(\mathbf{r}') = 0, & \mathbf{r}' \in \mathbb{R}^3 \setminus \{\mathbf{r}\} \quad \mathbf{r} \notin V_\varrho \\ \nabla' \times \mathcal{E}_2(\mathbf{r}') = 0, & \mathbf{r}' \in \mathbb{R}^3 \\ \mathcal{E}_2(\mathbf{r}) = \mathcal{O}\left(\dfrac{1}{|\mathbf{r}'|^3}\right), & |\mathbf{r}'| \to +\infty \end{cases} \tag{3.14}$$

where the scalar potential $\Phi_2(\mathbf{r}')$ is given by (2.42) with ε in lieu of ε_0. Since $\mathcal{E}_2(\mathbf{r}')$ and $\Phi_2(\mathbf{r}')$ are singular at the location of the dipole, we exclude that point by means of a ball $B(\mathbf{r}, a) \subset V$. We multiply the Gauss law of problem (3.13) by $\Phi_2(\mathbf{r}')$ to obtain

$$\frac{\varrho(\mathbf{r}')}{\varepsilon}\Phi_2(\mathbf{r}') = \Phi_2(\mathbf{r}')\nabla' \cdot \mathcal{E}_1(\mathbf{r}') = \nabla' \cdot [\Phi_2(\mathbf{r}')\mathcal{E}_1(\mathbf{r}')] - \nabla'\Phi_2(\mathbf{r}') \cdot \mathcal{E}_1(\mathbf{r}') \tag{3.15}$$

which is valid for $\mathbf{r}' \in V \setminus B(\mathbf{r}, a)$. To cast the last term into the divergence of a vector field we observe

$$-\nabla'\Phi_2(\mathbf{r}') = -\mathbf{p} \cdot \nabla'\nabla'\frac{1}{4\pi\varepsilon R} = \nabla' \times \left(\mathbf{p} \times \nabla'\frac{1}{4\pi\varepsilon R}\right) \tag{3.16}$$

on account of (2.42) and the differential identity (H.54) applied to $\mathbf{p}$ and $\nabla'(1/R)$. Thanks to this result and in view of the curl-free nature of $\mathcal{E}_1(\mathbf{r}')$, we arrive at

$$\frac{\varrho(\mathbf{r}')}{\varepsilon}\mathbf{p} \cdot \nabla'\frac{1}{4\pi R} = \nabla' \cdot \left[\left(\mathbf{p} \cdot \nabla'\frac{1}{4\pi R}\right)\mathcal{E}_1(\mathbf{r}') + \left(\mathbf{p} \times \nabla'\frac{1}{4\pi R}\right) \times \mathcal{E}_1(\mathbf{r}')\right] \tag{3.17}$$

which we integrate with respect to $\mathbf{r}'$ over $V \setminus B(\mathbf{r}, a)$. We may apply the Gauss theorem because the vector field between brackets is continuously differentiable for $\mathbf{r}' \neq \mathbf{r}$. In this respect, recall that the unit normal points inwards $V \setminus B(\mathbf{r}, a)$. In symbols, we have

$$\frac{\mathbf{p}}{\varepsilon} \cdot \int_{V_\varrho} \mathrm{d}V'\, \varrho(\mathbf{r}')\nabla'\frac{1}{4\pi R}$$

$$= -\mathbf{p}\cdot\int\limits_{S\cup\partial V_1} \mathrm{d}S'\left\{\hat{\mathbf{n}}(\mathbf{r}')\cdot\mathcal{E}_1(\mathbf{r}')\nabla'\frac{1}{4\pi R}+\nabla'\frac{1}{4\pi R}\times[\mathcal{E}_1(\mathbf{r}')\times\hat{\mathbf{n}}(\mathbf{r}')]\right\}$$
$$-\mathbf{p}\cdot\int\limits_{\partial B} \mathrm{d}S'\left\{\hat{\mathbf{n}}(\mathbf{r}')\cdot\mathcal{E}_1(\mathbf{r}')\nabla'\frac{1}{4\pi R}+\nabla'\frac{1}{4\pi R}\times[\mathcal{E}_1(\mathbf{r}')\times\hat{\mathbf{n}}(\mathbf{r}')]\right\} \quad (3.18)$$

and we need to take the limit as $a \to 0$. The volume integral over the source and the surface integral over the $S \cup \partial V_1$ are unaffected in that we have placed the dipole outside V_ϱ and away from the boundary. As for the integral over the sphere ∂B (cf. Figure 2.8) we recall (2.155) and observe

$$\nabla'\frac{1}{4\pi R} = -\frac{\hat{\mathbf{n}}(\mathbf{r}')}{4\pi a^2}, \qquad \mathbf{r}' \in \partial B \quad (3.19)$$

$$-\mathbf{p}\cdot\int\limits_{\partial B} \mathrm{d}S'\left\{\hat{\mathbf{n}}(\mathbf{r}')\cdot\mathcal{E}_1(\mathbf{r}')\nabla'\frac{1}{4\pi R}+\nabla'\frac{1}{4\pi R}\times[\mathcal{E}_1(\mathbf{r}')\times\hat{\mathbf{n}}(\mathbf{r}')]\right\} =$$
$$= \frac{\mathbf{p}}{4\pi a^2}\cdot\int\limits_{\partial B}\mathrm{d}S'\{\hat{\mathbf{n}}(\mathbf{r}')\hat{\mathbf{n}}(\mathbf{r}')\cdot\mathcal{E}_1(\mathbf{r}')+\hat{\mathbf{n}}(\mathbf{r}')\times[\mathcal{E}_1(\mathbf{r}')\times\hat{\mathbf{n}}(\mathbf{r}')]\}$$
$$= \mathbf{p}\cdot\mathcal{E}_1(\mathbf{r}+a\hat{\mathbf{n}}(\mathbf{r}'_0)) \xrightarrow[a\to 0]{} \mathbf{p}\cdot\mathcal{E}_1(\mathbf{r}) \quad (3.20)$$

where we have used (H.14) and the mean value theorem with $\mathbf{r}'_0 \in \partial B$ a suitable point. By letting $\mathbf{p}$ equal to $\hat{\mathbf{x}}$, $\hat{\mathbf{y}}$ and $\hat{\mathbf{z}}$ in succession we obtain three scalar formulas for the Cartesian components of $\mathcal{E}_1(\mathbf{r})$. Therefore, by dropping the inconsequential subscript 1, we can write a single vector expression for the electrostatic field as

$$\mathcal{E}(\mathbf{r}) = \frac{1}{\varepsilon}\int\limits_{V_\varrho}\mathrm{d}V'\varrho(\mathbf{r}')\nabla'\frac{1}{4\pi R}+\int\limits_{S\cup\partial V_1}\mathrm{d}S'\left\{\hat{\mathbf{n}}(\mathbf{r}')\cdot\mathcal{E}(\mathbf{r}')\nabla'\frac{1}{4\pi R}+\nabla'\frac{1}{4\pi R}\times[\mathcal{E}(\mathbf{r}')\times\hat{\mathbf{n}}(\mathbf{r}')]\right\} \quad (3.21)$$

for observation points $\mathbf{r} \in V \setminus \overline{V}_\varrho$. The extension of (3.21) to points in the source region is provided by (2.200).

By placing the dipole in the complementary domain $\mathbb{R}^3 \setminus \overline{V}$ we may show that (3.21) returns zero. On the other hand, the surface integrals as they stand are not well defined for observation points on the boundary $S \cup \partial V_1$. Therefore, to determine the limit value of (3.21) as $\mathbf{r} \to S \cup \partial V_1$ we first transform the relevant integrand as follows

$$\underbrace{\hat{\mathbf{n}}'\times(\nabla'\times\mathcal{E})\frac{1}{R}}_{=0}+(\hat{\mathbf{n}}'\times\mathcal{E})\times\nabla'\frac{1}{R}+\hat{\mathbf{n}}'\cdot\mathcal{E}\nabla'\frac{1}{R} =$$
$$= \hat{\mathbf{n}}'\times\left(\nabla'\times\frac{\mathcal{E}}{R}\right)-\hat{\mathbf{n}}'\times\left(\nabla'\frac{1}{R}\times\mathcal{E}\right)+(\hat{\mathbf{n}}'\times\mathcal{E})\times\nabla'\frac{1}{R}+\hat{\mathbf{n}}'\cdot\mathcal{E}\nabla'\frac{1}{R}$$
$$= \hat{\mathbf{n}}'\times\left(\nabla'\times\frac{\mathcal{E}}{R}\right)-\hat{\mathbf{n}}'\cdot\mathcal{E}\nabla'\frac{1}{R}+\mathcal{E}\hat{\mathbf{n}}'\cdot\nabla'\frac{1}{R}+\mathcal{E}\nabla'\frac{1}{R}\cdot\hat{\mathbf{n}}'-\hat{\mathbf{n}}'\nabla'\frac{1}{R}\cdot\mathcal{E}+\hat{\mathbf{n}}'\cdot\mathcal{E}\nabla'\frac{1}{R}$$
$$= \hat{\mathbf{n}}'\times\left(\nabla'\times\frac{\mathcal{E}}{R}\right)-\hat{\mathbf{n}}'\nabla'\cdot\frac{\mathcal{E}}{R}+2\mathcal{E}\hat{\mathbf{n}}'\cdot\nabla'\frac{1}{R} \quad (3.22)$$

where we have used (H.14) and the fact that $\mathcal{E}(\mathbf{r})$ is curl-free everywhere as well as divergence-less in charge-free regions.[1] Next, we transform the first two terms in the rightmost-hand side above with identity (H.57) applied to $\mathcal{E}/R$

$$
\begin{aligned}
&\hat{\mathbf{n}}' \times \left(\nabla' \times \frac{\mathcal{E}}{R}\right) - \hat{\mathbf{n}}'\nabla' \cdot \frac{\mathcal{E}}{R} + 2\mathcal{E}\hat{\mathbf{n}}' \cdot \nabla' \frac{1}{R} = \\
&= (\hat{\mathbf{n}}' \times \nabla') \times \frac{\mathcal{E}}{R} - \hat{\mathbf{n}}' \cdot \nabla' \frac{\mathcal{E}}{R} + 2\mathcal{E}\hat{\mathbf{n}}' \cdot \nabla' \frac{1}{R} \\
&= (\hat{\mathbf{n}}' \times \nabla') \times \frac{\mathcal{E}}{R} - \left(\hat{\mathbf{n}}' \cdot \nabla' \frac{1}{R}\right)\mathcal{E} - \frac{\hat{\mathbf{n}}'}{R} \cdot \nabla'\mathcal{E} + 2\mathcal{E}\hat{\mathbf{n}}' \cdot \nabla' \frac{1}{R} \\
&= (\hat{\mathbf{n}}' \times \nabla') \times \frac{\mathcal{E}}{R} + \mathcal{E}\hat{\mathbf{n}}' \cdot \nabla' \frac{1}{R} - \frac{1}{R}\hat{\mathbf{n}}' \cdot \nabla'\mathcal{E}
\end{aligned}
\tag{3.23}
$$

where we have also employed identity (H.52). When integrated over S and ∂V_1, the first term in the rightmost member above contributes naught

$$
\int_{S \cup \partial V_1} \mathrm{d}S' \, [\hat{\mathbf{n}}(\mathbf{r}') \times \nabla'] \times \frac{\mathcal{E}(\mathbf{r}')}{R} = 0
\tag{3.24}
$$

in light of (H.94) [3, Eq. (A1.144)] applied on S and ∂V_1 which are closed surfaces by construction. Thanks to (3.23) we can cast (3.21) as

$$
\mathcal{E}(\mathbf{r}) = \frac{1}{\varepsilon}\int_{V_\varrho} \mathrm{d}V' \, \varrho(\mathbf{r}')\nabla' \frac{1}{4\pi R} + \int_{S \cup \partial V_1} \mathrm{d}S' \left[\mathcal{E}(\mathbf{r}')\hat{\mathbf{n}}'(\mathbf{r}') \cdot \nabla' \frac{1}{4\pi R} - \frac{1}{4\pi R}\hat{\mathbf{n}}'(\mathbf{r}') \cdot \nabla'\mathcal{E}(\mathbf{r}')\right]
\tag{3.25}
$$

where now we recognize the Cartesian components of the last two terms as special instances of double-layer and single-layer potentials (2.188) and (2.189), in which case we may apply the relevant results of Sections 2.9 and 2.10. More specifically, we observe that the single-layer potential is continuous across $S \cup \partial V_1$, whereas the double-layer potential is discontinuous, though finite across $S \cup \partial V_1$, with limiting values given by

$$
\lim_{\mathbf{r} \to (S \cup \partial V_1)^\pm} \int_{S \cup \partial V_1} \mathrm{d}S' \, \mathcal{E}(\mathbf{r}')\hat{\mathbf{n}}'(\mathbf{r}') \cdot \nabla' \frac{1}{4\pi R} = \pm\frac{1}{2}\mathcal{E}(\mathbf{r}) + \mathrm{PV}\int_{S \cup \partial V_1} \mathrm{d}S' \, \mathcal{E}(\mathbf{r}')\hat{\mathbf{n}}'(\mathbf{r}') \cdot \nabla' \frac{1}{4\pi R}
\tag{3.26}
$$

on account of (2.244). Now, using (3.26) in (3.25) shows that the integral representation returns $\mathcal{E}/2$ for $\mathbf{r} \in S \cup \partial V_1$.

Last but not least, we may conclude that the surface integrals represent the contribution of sources other than $\varrho(\mathbf{r}')$ and located outside V. In particular, if no sources are contemplated in $\mathbb{R}^3 \setminus (\overline{V} \cup V_1)$ we may let the surface S recede to infinity and show that the value of the corresponding integral is zero by virtue of estimate (2.21).

In principle, we can use (3.21) to compute the electrostatic field in points $\mathbf{r} \in V$, so long as the boundary values of $\mathcal{E}(\mathbf{r}')$ are known. However, as we know from Section 2.5.2, uniqueness of solutions for (3.13) is achieved by assigning either the tangential or the normal component of the electric field on $S \cup \partial V_1$, and hence the latter two may not be specified independently. Nonetheless, (3.21) comes in handy to formulate electrostatic problems by means of surface integral equations in which the unknown quantities are precisely either $\hat{\mathbf{n}}(\mathbf{r}') \cdot \mathcal{E}(\mathbf{r}')$ or $\mathcal{E}(\mathbf{r}') \times \hat{\mathbf{n}}(\mathbf{r}')$ or both for $\mathbf{r}' \in \partial V$.

[1] Also see (5.60) further on and the book by Stratton [2, Section 8.15] for a similar result concerning time-harmonic fields.

For instance, if V_1 is filled with a conductor, then $\mathcal{E}(\mathbf{r}') \times \hat{\mathbf{n}}(\mathbf{r}') = 0$ on ∂V_1, whereas — in view of (2.15) and (2.53) — $\varepsilon\mathcal{E}(\mathbf{r}') \cdot \hat{\mathbf{n}}(\mathbf{r}')$ represents the *unknown* surface charge induced on ∂V_1 by the charges in V_ϱ.

Finally, we discuss the physical meaning of $\hat{\mathbf{n}}(\mathbf{r}') \cdot \mathcal{E}(\mathbf{r}')$ or $\hat{\mathbf{n}}(\mathbf{r}') \times \mathcal{E}(\mathbf{r}')$. Upon division and multiplication by ε, the surface integral of the normal component of $\mathcal{E}(\mathbf{r}')$ may be cast in a form similar to the volume integral of $\varrho(\mathbf{r}')$. Thereby, we are led to interpret $\varepsilon\hat{\mathbf{n}}(\mathbf{r}') \cdot \mathcal{E}(\mathbf{r}')$ as the surface density of a layer of (equivalent) fictitious charges distributed on $S \cup \partial V_1$, though inside V.

The interpretation of the tangential component of $\mathcal{E}(\mathbf{r}') \times \hat{\mathbf{n}}(\mathbf{r}')$ as a steady surface *magnetic current* density flowing on the plus side of $S \cup \partial V_1$ is far from obvious at this stage. The conclusion is based on the apparent symmetry between (3.21) and the analogous integral representation of the stationary magnetic field (5.58) which we shall derive in Section 5.1.2. We mention in passing that a volume density of steady magnetic current would appear as a source term in the right-hand side of (2.4) and, thus, would make the electric field non-conservative in the very region occupied by the magnetic current.

3.3 Other Green functions for static problems

The integral representation of the scalar potential derived in Section 2.7 was based on the choice of a suitable auxiliary problem which we could solve to determine a Green function; in particular, we fixed the boundary condition at infinity. Thus, we may wonder whether other choices are possible that lead to other Green functions and, consequently, other integral representations.

3.3.1 The Dirichlet Green function

Following the previous remark and with reference to the geometry of Figure 2.7b, we consider a Green function with homogeneous Dirichlet boundary conditions on ∂V, i.e., the unique solution to the Poisson equation [4, Section 8.5]

$$\begin{cases} \nabla'^2 G_D(\mathbf{r}, \mathbf{r}') = 0, & \mathbf{r}' \in V \setminus \{\mathbf{r}\} \qquad \mathbf{r} \in V \\ G_D(\mathbf{r}, \mathbf{r}') = 0, & \mathbf{r}' \in \partial V := S \cup \partial V_1 \\ \displaystyle\int_{\partial B} \mathrm{d}S' \, \hat{\mathbf{n}}(\mathbf{r}') \cdot \nabla G_D(\mathbf{r}, \mathbf{r}') = -1 & \end{cases} \tag{3.27}$$

where $B(\mathbf{r}, a) \subset V$, $a > 0$, is a small ball centered on the excluded point $\mathbf{r}$, and the normalization condition of the type (2.125) is included to ensure uniqueness. If $G_D(\mathbf{r}, \mathbf{r}')$ can be found analytically, then the integral representation (2.160) simplifies inasmuch as the contribution of the single-layer potential is identically null. In symbols, we have

$$\Phi(\mathbf{r}) = \frac{1}{\varepsilon} \int_{V_\varrho} \mathrm{d}V \, \varrho(\mathbf{r}') G_D(\mathbf{r}, \mathbf{r}') + \int_{S \cup \partial V_1} \mathrm{d}S' \, \Phi(\mathbf{r}') \frac{\partial}{\partial \hat{\mathbf{n}}'} G_D(\mathbf{r}, \mathbf{r}'), \qquad \mathbf{r} \in \overline{V} \tag{3.28}$$

where the surface integrals are intended in the Cauchy principal-value sense for $\mathbf{r} \in \partial V := S \cup \partial V_1$. It is worthwhile noticing that the remaining surface integral vanishes if also the potential $\Phi(\mathbf{r})$ obeys homogeneous Dirichlet boundary conditions for $\mathbf{r} \in S \cup \partial V_1$.

At any rate, the problem with $G_D(\mathbf{r}, \mathbf{r}')$ lies in the difficulty of finding a closed-form solution to (3.27) in practical configurations. The Dirichlet Green function can only be computed for a limited number of canonical problems, e.g., for the inside and the outside of a sphere [1,2].

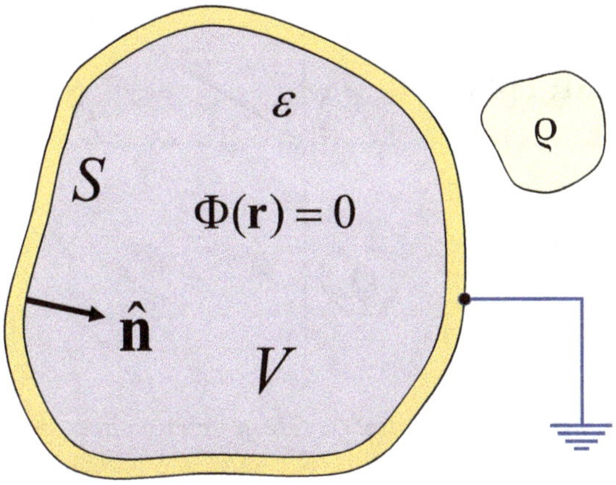

Figure 3.3 Electrostatic shielding achieved with a grounded conducting shell.

Nonetheless, $G_D(\mathbf{r}, \mathbf{r}')$ has theoretical relevance, and we can use it to prove that the electrostatic potential in a region V bounded by a conducting shell is not affected by the presence of external charges (see Figure 3.3). The key point is that the electrostatic potential on the surface $S := \partial V$ of the shell is a constant Φ_0 in general (Section 2.4). Additionally, we can nullify the potential of the shell by 'grounding' it, i.e., by connecting the shell to the Earth, which is at zero potential. The Earth acts as an infinitely large conductor of sorts, so whatever charges may have been present on the shell get dispersed in the ground, and the final stable configuration yields $\Phi(\mathbf{r}) = 0$ for $\mathbf{r} \in S$. Notice that since there are no charges within the shell and the potential is zero on the boundary, then the unique solution to the homogeneous Poisson equation is $\Phi(\mathbf{r}) = 0$ for $\mathbf{r} \in \overline{V}$.

What happens to $\Phi(\mathbf{r})$ if we bring some *external* charge closer to the shell? To answer this question we may employ the integral representation (3.28) for points $\mathbf{r} \in V$, viz.,

$$\Phi(\mathbf{r}) = \int_S \mathrm{d}S'\, \Phi(\mathbf{r}')\hat{\mathbf{n}}(\mathbf{r}') \cdot \nabla' G_D(\mathbf{r}, \mathbf{r}') = \Phi_0 \int_S \mathrm{d}S'\, \hat{\mathbf{n}}(\mathbf{r}') \cdot \nabla' G_D(\mathbf{r}, \mathbf{r}') = 0 \tag{3.29}$$

with no contribution from the volume potential because the charges, if present, are located outside V. The results follows by observing that Φ_0 is zero, since the shell is grounded. Thereby, $\Phi(\mathbf{r})$ remains zero for $\mathbf{r} \in V \cup S$ irrespective of the actual distribution of charges outside the shell. This phenomenon — known as electrostatic shielding — is exploited for the realization of the so-called *Faraday cages*, which are employed for the electrical insulation of devices or humans from hazardous high level of potential in the surroundings.

It is important to realize that the actual knowledge of the Dirichlet Green function is irrelevant for the proof outlined above; we have just assumed that $G_D(\mathbf{r}, \mathbf{r}')$ exists.

Example 3.2 (The Dirichlet Green function of a dielectric half space)

One of the few geometries for which the Dirichlet Green function can be determined in closed form (with elementary means, at that) is constituted by a half space filled with an isotropic homogeneous dielectric medium bounded by a conducting planar interface. For the sake of argument, suppose that in a system of Cartesian coordinates the interface S_0 is the plane $z = 0$ and that the half space of

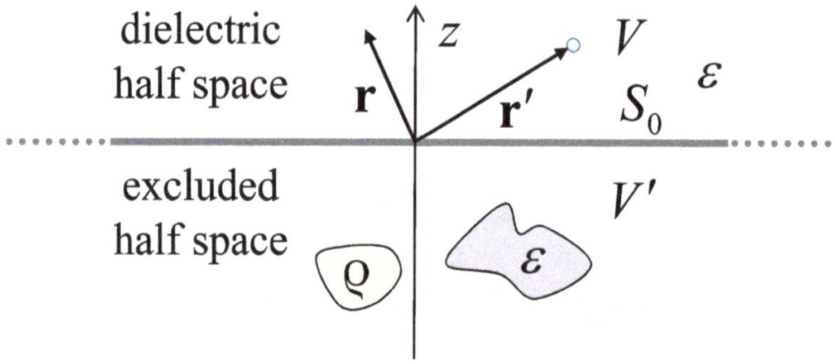

Figure 3.4 Geometry for the Dirichlet Green function in a dielectric half space.

interest is the unbounded region $V := \{\mathbf{r} \in \mathbb{R}^3 : z > 0\}$, as is illustrated in Figure 3.4. Thus, $G_D(\mathbf{r}, \mathbf{r}')$ is the unique solution to the problem

$$\begin{cases} \nabla^2 G_D(\mathbf{r}, \mathbf{r}') = 0, & \mathbf{r} \in V \setminus \{\mathbf{r}'\} \qquad \mathbf{r}' \in V \\ G_D(\mathbf{r}, \mathbf{r}') = 0, & \mathbf{r} \in S_0 := \{\mathbf{r} \in \mathbb{R}^3 : z = 0\} \\ G_D(\mathbf{r}, \mathbf{r}') = \mathcal{O}\left(\dfrac{1}{|\mathbf{r}|}\right), & |\mathbf{r}| \to +\infty \\ \displaystyle\int_{\partial B} \mathrm{d}S\, \hat{\mathbf{n}}(\mathbf{r}) \cdot \nabla G_D(\mathbf{r}, \mathbf{r}') = -1 & \end{cases} \tag{3.30}$$

where $B(\mathbf{r}', a) \subset V$ is a small ball centered on the excluded point $\mathbf{r}'$.

The physical meaning of $G_D(\mathbf{r}, \mathbf{r}')$ and the definition thereof imply that the conducting interface S_0 is grounded. As a result, the content of the complementary region $V' := \mathbb{R}^3 \setminus \overline{V}$ is utterly irrelevant, in that the potential within the dielectric is affected only by charges and matter placed in V owing to the shielding effect of the interface (see Figure 3.3).

From (2.130) we know the Green function for a homogeneous unbounded isotropic dielectric medium. Evidently, although (2.131) has the right asymptotic behavior, yet it does not solve (3.30) because it does not vanish on S_0. Nonetheless, since for $\mathbf{r} \to \mathbf{r}'$ we expect $G_D(\mathbf{r}, \mathbf{r}')$ to exhibit the same singular nature as that of $G(\mathbf{r}, \mathbf{r}')$ given by (2.131), we speculate that $G_D(\mathbf{r}, \mathbf{r}')$ ought to have the functional form

$$G_D(\mathbf{r}, \mathbf{r}') = \frac{1}{4\pi|\mathbf{r} - \mathbf{r}'|} + g(\mathbf{r}, \mathbf{r}'), \qquad \mathbf{r} \in V \setminus \{\mathbf{r}'\} \tag{3.31}$$

where $g(\mathbf{r}, \mathbf{r}')$ is a harmonic function for $\mathbf{r}, \mathbf{r}' \in V$, i.e., solves the Laplace equation (Section 3.5 and, e.g., [5, Section 8.7], [6, Section 2.2]).

What is the physical meaning of $g(\mathbf{r}, \mathbf{r}')$? Well, $G_D(\mathbf{r}, \mathbf{r}')$ is the normalized electrostatic potential of a point charge at $\mathbf{r}'$ in the presence of the grounded interface. Since $1/(4\pi|\mathbf{r} - \mathbf{r}'|)$ is the potential of the same charge in an unbounded medium, then $g(\mathbf{r}, \mathbf{r}')$ must account for the potential of some other charges located outside V. As a matter of fact, $g(\mathbf{r}, \mathbf{r}')$ is produced by a suitable distribution of surface charges induced on S_0 by the point charge at $\mathbf{r}'$. More importantly, $g(\mathbf{r}, \mathbf{r}')$ should make it possible to satisfy the boundary condition on ∂V. Last but not least, notice that with this choice the normalization condition is satisfied partly because the first term obeys (2.125) and partly because

$$\int_{\partial B} \mathrm{d}S\, \hat{\mathbf{n}}(\mathbf{r}) \cdot \nabla g(\mathbf{r}, \mathbf{r}') = \int_{B(\mathbf{r}', a)} \mathrm{d}V\, \nabla^2 g(\mathbf{r}, \mathbf{r}') = 0 \tag{3.32}$$

as $g(\mathbf{r}, \mathbf{r}')$ is harmonic throughout V by hypothesis.

In light of these observations, we *try* for $G_D(\mathbf{r}, \mathbf{r}')$ the functional form

$$G_D(\mathbf{r}, \mathbf{r}') = \frac{1}{4\pi|\mathbf{r} - (x'\hat{\mathbf{x}} + y'\hat{\mathbf{y}} + z'\hat{\mathbf{z}})|} + \frac{A}{4\pi|\mathbf{r} - (x'\hat{\mathbf{x}} + y'\hat{\mathbf{y}} - z'\hat{\mathbf{z}})|} \tag{3.33}$$

and we hope to determine the real constant A in such a way that $G_D(\mathbf{r}, \mathbf{r}')$ invariably vanishes for $\mathbf{r} \in S_0$; with this Ansatz the asymptotic behavior is satisfied. Indeed, for $z = 0$ the expression above leads to

$$G_D(\mathbf{r}, \mathbf{r}')|_{z=0} = \frac{1 + A}{4\pi|(x - x')^2 + (y - y')^2 + z'^2|^{1/2}} = 0 \tag{3.34}$$

and this equation holds true for any $\mathbf{r} \in S_0$, if $A = -1$. We observe that the vector $x'\hat{\mathbf{x}} + y'\hat{\mathbf{y}} - z'\hat{\mathbf{z}}$ is the specular counterpart of $\mathbf{r}'$ with respect to the xOy plane. With the aid of (E.29) it is easy to check that

$$x'\hat{\mathbf{x}} + y'\hat{\mathbf{y}} - z'\hat{\mathbf{z}} = \mathbf{r}' - 2z'\hat{\mathbf{z}} = (\overline{\mathfrak{I}} - 2\hat{\mathbf{z}}\hat{\mathbf{z}}) \cdot \mathbf{r}' = \overline{\mathfrak{U}} \cdot \mathbf{r}' \tag{3.35}$$

where $\overline{\mathfrak{U}} = \overline{\mathfrak{U}}^{-1}$ is a symmetric unitary dyadic. In linear algebra the matrix $[U]$ associated with $\overline{\mathfrak{U}}$ is called an elementary reflector and the multiplication of $[U]$ and a vector is referred to as a Householder transformation [7]. In words, $\overline{\mathfrak{U}}$ *reflects* a vector about the plane $z = 0$. With this definition the relevant Dirichlet Green function can be written as

$$G_D(\mathbf{r}, \mathbf{r}') = \frac{1}{4\pi|\mathbf{r} - \mathbf{r}'|} - \frac{1}{4\pi\left|\mathbf{r} - \overline{\mathfrak{U}} \cdot \mathbf{r}'\right|} \tag{3.36}$$

for $\mathbf{r}, \mathbf{r}' \in V$. Since (3.36) solves (3.30) it is also the right and unique solution, in spite of the contrived way we have followed to obtain the result. In view of the presence of the interface S_0, $G_D(\mathbf{r}, \mathbf{r}')$ depends on $\mathbf{r}$ and $\mathbf{r}'$ separately, rather than just on the distance $|\mathbf{r} - \mathbf{r}'|$, as is the case for the unbounded space solution (2.131). Comparison of the two contributions in the right-hand side of (3.36) suggests that the second term represents the potential of a *fictitious* point charge located at $\mathbf{r} = x'\hat{\mathbf{x}} + y'\hat{\mathbf{y}} - z'\hat{\mathbf{z}}$; such charge is the negative of the one placed in $\mathbf{r}'$. This observation is the basis for the so called *image principle* of electrostatics discussed in Section 3.9.

Finally, we notice that, as a function of $\mathbf{r}$ and $\mathbf{r}'$, $G_D(\mathbf{r}, \mathbf{r}')$ is defined in the whole space except for $\mathbf{r} = \mathbf{r}'$ and $\mathbf{r} = x'\hat{\mathbf{x}} + y'\hat{\mathbf{y}} - z'\hat{\mathbf{z}}$ where it is singular. However, the physical meaning of (3.36) as potential of a point charge in $\mathbf{r}' \in V$ in the presence of the conducting interface is valid *only* for $\mathbf{r} \in V$. In general, (3.36) does not yield the right potential in the complementary domain V' (Figure 3.4). For instance, if V' is filled with a conductor, the potential therein should be constant, whereas evidently the right-hand side of (3.36) is anything but constant for $\mathbf{r} \in V'$.

Since $G_D(\mathbf{r}, \mathbf{r}')$ is singular at the locations of both the physical and the fictitious charge, formula (3.28) does not yield a null potential when evaluated for $\mathbf{r} \notin \overline{V}$. Indeed, if we follow the procedure of Section 2.7 by using $G_D(\mathbf{r}, \mathbf{r}')$ given by (3.36), when we place the sampling charge in V', the fictitious one 'migrates' into V. The exclusion of the singularity at the location of the fictitious charge produces, in the limit, the negative of the value of the potential in that point. This glitch may be overcome if we agree to define the Dirichlet Green function as

$$G_D(\mathbf{r}, \mathbf{r}') := \begin{cases} \dfrac{1}{4\pi|\mathbf{r} - \mathbf{r}'|} - \dfrac{1}{4\pi\left|\mathbf{r} - \overline{\mathfrak{U}} \cdot \mathbf{r}'\right|}, & \mathbf{r} \in V \\ 0, & \mathbf{r} \in V' \end{cases} \tag{3.37}$$

with $\mathbf{r}' \in V$. This extended definition, though not necessary for the problem (3.30), ensures that the right-hand side of the integral representation (3.28) yields zero for $\mathbf{r} \notin \overline{V}$.

In accordance with (3.37) $G_D(\mathbf{r}, \mathbf{r}')$ is continuous across S_0 but the normal derivative thereof is not, and this correctly accounts for a single layer of charge induced on the interface S_0 by the point charge at $\mathbf{r}'$. We may compute the normalized charge on S_0 by observing that the normalized surface charge density is given by $-\hat{\mathbf{z}} \cdot \nabla G_D$ for $z = 0^+$ by virtue of (1.170) and (2.15). Therefore, we consider

$$-\left.\frac{\mathrm{d}G_D}{\mathrm{d}z}\right|_{z=0^+} = \frac{-2z'}{4\pi[(x-x')^2 + (y-y')^2 + z'^2]^{3/2}} \tag{3.38}$$

and integrate over S_0 with respect to (x, y). To this purpose, it is fruitful to choose a system of polar coordinates (τ, α) centered in $(x', y', 0)$, whereby $(x-x')^2 + (y-y')^2 = \tau^2$ and the induced charge follows from

$$-\int_{\partial V} \mathrm{d}S \left.\frac{\mathrm{d}G_D}{\mathrm{d}z}\right|_{z=0^+} = -\int_0^{+\infty} \mathrm{d}\tau \frac{4\pi z'\tau}{4\pi(\tau^2 + z'^2)^{3/2}} = \left[\frac{z'}{(\tau^2 + z'^2)^{1/2}}\right]_{\tau=0}^{+\infty} = -\frac{z'}{|z'|} = -1 \tag{3.39}$$

having observed that the integrand is independent of α. This result in tandem with the normalization condition in (3.30) shows that the total induced charge is the negative of the unitary charge placed at $\mathbf{r}'$.

(End of Example 3.2)

3.3.2 *The Neumann Green function*

By arguments similar to those employed to derive the Dirichlet Green function, we may be tempted to impose homogeneous Neumann boundary conditions for $\mathbf{r}' \in \partial V$, that is,

$$\hat{\mathbf{n}}(\mathbf{r}') \cdot \nabla' G_N(\mathbf{r}, \mathbf{r}') = 0, \qquad \mathbf{r}' \in \partial V := S \cup \partial V_1 \tag{3.40}$$

again with reference to the geometry of Figure 2.7b. Unfortunately, this requirement cannot be met for a bounded volume. In Section 2.6 we have defined the Green function as a normalized electrostatic potential generated at $\mathbf{r}'$ by a point charge located at $\mathbf{r}$. Therefore, the Gauss law applied to $G_N(\mathbf{r}, \mathbf{r}')$ yields

$$\int_{\partial V} \mathrm{d}S' \, \hat{\mathbf{n}}(\mathbf{r}') \cdot \nabla' G_N(\mathbf{r}, \mathbf{r}') = 1 \tag{3.41}$$

for a charge located at $\mathbf{r} \in V$. Except for a minus sign due to the unit normal pointing inward V (Figure 2.7b), this is the same condition (2.129) which we invoked to set the arbitrary constant of the Green function. We see that taking the normal derivative of $G_N(\mathbf{r}, \mathbf{r}')$ to be zero on ∂V causes the Gauss law to be violated.

Thus, the best we can do under the circumstances is set $\hat{\mathbf{n}}(\mathbf{r}') \cdot \nabla' G_N(\mathbf{r}, \mathbf{r}')$ equal to a constant adjusted so that (3.41) is fulfilled. The corresponding fundamental solution is called Neumann Green function [1, 8] and satisfies the problem

$$\begin{cases} \nabla'^2 G_N(\mathbf{r}, \mathbf{r}') = 0, & \mathbf{r}' \in V \setminus \{\mathbf{r}\} \qquad \mathbf{r} \in V \\ \hat{\mathbf{n}}(\mathbf{r}') \cdot \nabla' G_N(\mathbf{r}, \mathbf{r}') = \dfrac{1}{A_{\partial V}}, & \mathbf{r}' \in \partial V := S \cup \partial V_1 \end{cases} \tag{3.42}$$

where $A_{\partial V}$ denotes the area of the boundary ∂V. This assumption still works in the limit of an unbounded region — in which case $1/A_{\partial V} \to 0$.

With the Neumann Green function the integral representation applied to the configuration of Figure 2.7a reads

$$\Phi(\mathbf{r}) = \frac{1}{\varepsilon} \int_{V_\varrho} \mathrm{d}V \, \varrho(\mathbf{r}') G_N(\mathbf{r}, \mathbf{r}') + \frac{1}{A_{\partial V}} \int_{S \cup \partial V_1} \mathrm{d}S' \, \Phi(\mathbf{r}') - \int_{S \cup \partial V_1} \mathrm{d}S' \, G_N(\mathbf{r}, \mathbf{r}') \frac{\partial}{\partial \hat{\mathbf{n}}'} \Phi(\mathbf{r}') \tag{3.43}$$

for observation points $\mathbf{r} \in V$. Notice that we may not extend this formula as it stands to points $\mathbf{r} \in \mathbb{R}^3 \setminus \overline{V}$. Indeed, if the sampling charge is placed outside V (Figure 2.10a), the Gauss law dictates

$$\int_{\partial V} \mathrm{d}S' \, \hat{\mathbf{n}}(\mathbf{r}') \cdot \nabla' G_N(\mathbf{r}, \mathbf{r}') = 0, \qquad \mathbf{r} \notin V \tag{3.44}$$

which is not satisfied with the Neumann boundary condition imposed on $G_N(\mathbf{r}, \mathbf{r}')$.

If the region V is bounded by conductors, then $\Phi(\mathbf{r}') = \Phi_0$ for $\mathbf{r}' \in \partial V := S \cup \partial V_1$ and

$$\frac{1}{A_{\partial V}} \int_{S \cup \partial V_1} \mathrm{d}S' \, \Phi(\mathbf{r}') = \Phi_0 \tag{3.45}$$

otherwise the integral above provides the mean value of $\Phi(\mathbf{r}')$ on the boundary.

As was the case for $G_D(\mathbf{r}, \mathbf{r}')$, also the Neumann Green function can be computed analytically for a limited number of canonical problems.

3.4 Properties of the static Green functions

We have already pointed out (Section 2.6) that the static Green function in a homogeneous unbounded isotropic medium depends only on the relative distance between the position of the charge ($\mathbf{r}'$) and that of the observer ($\mathbf{r}$). It is also evident, by virtue of its very definition, that the Green function (2.131) satisfies the following identity

$$G(\mathbf{r}, \mathbf{r}') = G(\mathbf{r}', \mathbf{r}) \tag{3.46}$$

i.e., $G(\mathbf{r}, \mathbf{r}')$ is invariant (symmetric) under an exchange of observation and source points. This property is known as *reciprocity* and, in the present case, is a consequence of the isotropy of the underlying medium. From a mathematical viewpoint, reciprocity holds because the Laplace operator in (2.124) is *self-adjoint* (see discussion on page 385 further on and Appendix D.6) [9, Section 10.4]. The symmetry property in the form (3.46) applies to the Green function (2.148) in an unbounded homogeneous anisotropic medium as well, as it can be verified by inspection, because we assumed the permittivity dyadic to be symmetric.

We can show that reciprocity holds also for the Dirichlet Green function [10, Chapter 10]. For simplicity, we dispense with the small excluded region V_1 of Figure 2.7a and consider a volume V bounded by a smooth surface $S := \partial V$ with unit normal $\hat{\mathbf{n}}$ oriented outward V. We examine the equations satisfied by $G_D(\mathbf{r}, \mathbf{r}')$ corresponding to two point charges located in $\mathbf{r}_1 \in V$ and $\mathbf{r}_2 \in V$

$$\nabla'^2 G_D(\mathbf{r}_1, \mathbf{r}') = 0, \qquad \mathbf{r}' \in V \setminus \{\mathbf{r}_1\}, \qquad \mathbf{r}_1 \in V \tag{3.47}$$

$$\nabla'^2 G_D(\mathbf{r}_2, \mathbf{r}') = 0, \qquad \mathbf{r}' \in V \setminus \{\mathbf{r}_2\}, \qquad \mathbf{r}_2 \in V \tag{3.48}$$

with $\mathbf{r}_1 \neq \mathbf{r}_2$. We multiply the first equation by $G_D(\mathbf{r}_2, \mathbf{r}')$, the second by $G_D(\mathbf{r}_1, \mathbf{r}')$, and we subtract the resulting expressions side by side to get

$$\begin{aligned} 0 &= G_D(\mathbf{r}_2, \mathbf{r}')\nabla'^2 G_D(\mathbf{r}_1, \mathbf{r}') - G_D(\mathbf{r}_1, \mathbf{r}')\nabla'^2 G_D(\mathbf{r}_2, \mathbf{r}') \\ &= \nabla' \cdot [G_D(\mathbf{r}_2, \mathbf{r}')\nabla' G_D(\mathbf{r}_1, \mathbf{r}') - G_D(\mathbf{r}_1, \mathbf{r}')\nabla' G_D(\mathbf{r}_2, \mathbf{r}')] \end{aligned} \tag{3.49}$$

which holds true for $\mathbf{r}' \in V \setminus \{\mathbf{r}_1, \mathbf{r}_2\}$. Next, we exclude the source points with two balls $B_1(\mathbf{r}_1, a)$ and $B_2(\mathbf{r}_2, a)$, and integrate over the surface-wise multiply-connected region $V_a := V \setminus B(\mathbf{r}_1, a) \setminus B(\mathbf{r}_2, a)$, viz.,

$$\begin{aligned} &\int_{V_a} \mathrm{d}V' \, \nabla' \cdot [G_D(\mathbf{r}_2, \mathbf{r}')\nabla' G_D(\mathbf{r}_1, \mathbf{r}') - G_D(\mathbf{r}_1, \mathbf{r}')\nabla' G_D(\mathbf{r}_2, \mathbf{r}')] \\ &= \int_{\partial V} \mathrm{d}S' \, \hat{\mathbf{n}}(\mathbf{r}') \cdot [G_D(\mathbf{r}_2, \mathbf{r}')\nabla' G_D(\mathbf{r}_1, \mathbf{r}') - G_D(\mathbf{r}_1, \mathbf{r}')\nabla' G_D(\mathbf{r}_2, \mathbf{r}')] \\ &+ \int_{\partial B_1} \mathrm{d}S' \, \hat{\mathbf{n}}(\mathbf{r}') \cdot [G_D(\mathbf{r}_2, \mathbf{r}')\nabla' G_D(\mathbf{r}_1, \mathbf{r}') - G_D(\mathbf{r}_1, \mathbf{r}')\nabla' G_D(\mathbf{r}_2, \mathbf{r}')] \\ &\qquad + \int_{\partial B_2} \mathrm{d}S' \, \hat{\mathbf{n}}(\mathbf{r}') \cdot [G_D(\mathbf{r}_2, \mathbf{r}')\nabla' G_D(\mathbf{r}_1, \mathbf{r}') - G_D(\mathbf{r}_1, \mathbf{r}')\nabla' G_D(\mathbf{r}_2, \mathbf{r}')] \end{aligned} \tag{3.50}$$

where the Gauss theorem has been applied on the grounds that the Green function and its gradient are continuously differential for $\mathbf{r}' \in V_a$. We observe that the integral over ∂V is null owing to the Dirichlet boundary conditions imposed on $G_D(\mathbf{r}, \mathbf{r}')$. Thus, we are left with the identity

$$\begin{aligned} &\int_{\partial B_1} \mathrm{d}S' \, \hat{\mathbf{n}}(\mathbf{r}') \cdot [G_D(\mathbf{r}_2, \mathbf{r}')\nabla' G_D(\mathbf{r}_1, \mathbf{r}') - G_D(\mathbf{r}_1, \mathbf{r}')\nabla' G_D(\mathbf{r}_2, \mathbf{r}')] \\ &\qquad = \int_{\partial B_2} \mathrm{d}S' \, \hat{\mathbf{n}}(\mathbf{r}') \cdot [G_D(\mathbf{r}_1, \mathbf{r}')\nabla' G_D(\mathbf{r}_2, \mathbf{r}') - G_D(\mathbf{r}_2, \mathbf{r}')\nabla' G_D(\mathbf{r}_1, \mathbf{r}')] \end{aligned} \tag{3.51}$$

and we would like to compute integrals in the limit as $a \to 0$. To this purpose we observe that $G_D(\mathbf{r}_l, \mathbf{r}')$, $l = 1, 2$, are singular functions of $\mathbf{r}'$ at the respective locations of the charges. While the actual form of $G_D(\mathbf{r}_l, \mathbf{r}')$ may not be known, still the singular behavior must be the same as that of the Green function (2.131). Therefore, we may assume the Ansatz

$$G_D(\mathbf{r}, \mathbf{r}') = g(\mathbf{r}, \mathbf{r}') + \frac{1}{4\pi|\mathbf{r} - \mathbf{r}'|}, \qquad \mathbf{r}' \in V \setminus \{\mathbf{r}\}, \qquad \mathbf{r} \in V \tag{3.52}$$

where $g(\mathbf{r}, \mathbf{r}')$ solves the Laplace equation in V and is regular for $\mathbf{r}' = \mathbf{r}$. Since the contribution of the charge in $\mathbf{r}$ has been singled out, $g(\mathbf{r}, \mathbf{r}')$ physically represents the potential due to charges induced on the conducting boundary ∂V. These charges adjust themselves on ∂V so as to produce the right additional potential which allows satisfying the Dirichlet boundary condition.

In particular, on ∂B_l we have

$$G_D(\mathbf{r}_l, \mathbf{r}') = g(\mathbf{r}_l, \mathbf{r}') + \frac{1}{4\pi a}, \qquad \mathbf{r}' \in \partial B_l \tag{3.53}$$

$$\hat{\mathbf{n}}(\mathbf{r}') \cdot \nabla' G_D(\mathbf{r}_l, \mathbf{r}') = \frac{\partial}{\partial \hat{\mathbf{n}}'} g(\mathbf{r}_l, \mathbf{r}') + \frac{1}{4\pi a^2}, \qquad \mathbf{r}' \in \partial B_l \tag{3.54}$$

by means of which we may proceed to compute the integrals in (3.51). We determine four contributions

$$\begin{aligned}&\int_{\partial B_1} \mathrm{d}S'\, G_D(\mathbf{r}_2,\mathbf{r}')\hat{\mathbf{n}}(\mathbf{r}')\cdot\nabla' G_D(\mathbf{r}_1,\mathbf{r}') = \\ &\quad = \int_{\partial B_1} \mathrm{d}S'\, G_D(\mathbf{r}_2,\mathbf{r}_1 - a\hat{\mathbf{n}}(\mathbf{r}'))\left[\frac{\partial}{\partial\hat{\mathbf{n}}'}g(\mathbf{r}_1,\mathbf{r}') + \frac{1}{4\pi a^2}\right] \\ &\quad = 4\pi a^2 G_D(\mathbf{r}_2,\mathbf{r}_1 - a\hat{\mathbf{n}}(\mathbf{r}'_0))\left[\frac{\partial}{\partial\hat{\mathbf{n}}'}g(\mathbf{r}_1,\mathbf{r}')\bigg|_{\mathbf{r}'_0} + \frac{1}{4\pi a^2}\right] \xrightarrow[a\to 0]{} G_D(\mathbf{r}_2,\mathbf{r}_1)\end{aligned} \tag{3.55}$$

$$\begin{aligned}&\int_{\partial B_1} \mathrm{d}S'\, G_D(\mathbf{r}_1,\mathbf{r}')\hat{\mathbf{n}}(\mathbf{r}')\cdot\nabla' G_D(\mathbf{r}_2,\mathbf{r}') = \\ &\quad = \int_{\partial B_1} \mathrm{d}S' \left[g(\mathbf{r}_1,\mathbf{r}') + \frac{1}{4\pi a}\right]\hat{\mathbf{n}}(\mathbf{r}')\cdot\nabla' G_D(\mathbf{r}_2,\mathbf{r}') \\ &\quad = 4\pi a^2\left[g(\mathbf{r}_1,\mathbf{r}'_0) + \frac{1}{4\pi a}\right]\hat{\mathbf{n}}(\mathbf{r}'_0)\cdot\nabla' G_D(\mathbf{r}_2,\mathbf{r}')\Big|_{\mathbf{r}'_0} \xrightarrow[a\to 0]{} 0\end{aligned} \tag{3.56}$$

$$\begin{aligned}&\int_{\partial B_2} \mathrm{d}S'\, G_D(\mathbf{r}_1,\mathbf{r}')\hat{\mathbf{n}}(\mathbf{r}')\cdot\nabla' G_D(\mathbf{r}_2,\mathbf{r}') = \\ &\quad = \int_{\partial B_2} \mathrm{d}S'\, G_D(\mathbf{r}_1,\mathbf{r}_2 - a\hat{\mathbf{n}}(\mathbf{r}'))\left[\frac{\partial}{\partial\hat{\mathbf{n}}'}g(\mathbf{r}_2,\mathbf{r}') + \frac{1}{4\pi a^2}\right] \\ &\quad = 4\pi a^2 G_D(\mathbf{r}_1,\mathbf{r}_2 - a\hat{\mathbf{n}}(\mathbf{r}'_0))\left[\frac{\partial}{\partial\hat{\mathbf{n}}'}g(\mathbf{r}_2,\mathbf{r}')\bigg|_{\mathbf{r}'_0} + \frac{1}{4\pi a^2}\right] \xrightarrow[a\to 0]{} G_D(\mathbf{r}_1,\mathbf{r}_2)\end{aligned} \tag{3.57}$$

$$\begin{aligned}&\int_{\partial B_2} \mathrm{d}S'\, G_D(\mathbf{r}_2,\mathbf{r}')\hat{\mathbf{n}}(\mathbf{r}')\cdot\nabla' G_D(\mathbf{r}_1,\mathbf{r}') = \\ &\quad = \int_{\partial B_2} \mathrm{d}S' \left[g(\mathbf{r}_2,\mathbf{r}') + \frac{1}{4\pi a}\right]\hat{\mathbf{n}}(\mathbf{r}')\cdot\nabla' G_D(\mathbf{r}_1,\mathbf{r}') \\ &\quad = 4\pi a^2\left[g(\mathbf{r}_2,\mathbf{r}'_0) + \frac{1}{4\pi a}\right]\hat{\mathbf{n}}(\mathbf{r}'_0)\cdot\nabla' G_D(\mathbf{r}_1,\mathbf{r}')\Big|_{\mathbf{r}'_0} \xrightarrow[a\to 0]{} 0\end{aligned} \tag{3.58}$$

having repeatedly invoked the mean value theorem, because all the integrands involved are regular on the respective surfaces of integration. In conclusion, (3.51) yields

$$G_D(\mathbf{r}_2,\mathbf{r}_1) = G_D(\mathbf{r}_1,\mathbf{r}_2) \tag{3.59}$$

and since $\mathbf{r}_l$ are arbitrary points in V, this proves the symmetry of the Dirichlet Green function.

A similar result cannot be obtained for the Neumann Green function because the requirement on the normal derivative in (3.42) is not sufficient to make the integral over ∂V vanish identically. Thus, the symmetry relation

$$G_N(\mathbf{r}_2,\mathbf{r}_1) + \frac{1}{A_{\partial V}}\int_{\partial V} \mathrm{d}S'\, G_N(\mathbf{r}_2,\mathbf{r}') = G_N(\mathbf{r}_1,\mathbf{r}_2) + \frac{1}{A_{\partial V}}\int_{\partial V} \mathrm{d}S'\, G_N(\mathbf{r}_1,\mathbf{r}') \tag{3.60}$$

is the best we can hope for in general.

Another interesting property of the Dirichlet Green function is that $G_D(\mathbf{r}, \mathbf{r}')$ is always *positive*, if so is the unitary generating point charge. To prove the statement we consider a bounded region V with smooth boundary ∂V, whereby $G_D(\mathbf{r}, \mathbf{r}')$ is the unique solution to the problem

$$\begin{cases} \nabla^2 G_D(\mathbf{r}, \mathbf{r}') = 0, & \mathbf{r} \in V \setminus \{\mathbf{r}'\} \qquad \mathbf{r}' \in V \\ G_D(\mathbf{r}, \mathbf{r}') = 0, & \mathbf{r} \in \partial V \\ \displaystyle\int_{\partial V} \mathrm{d}V\, \hat{\mathbf{n}}(\mathbf{r}) \cdot \nabla G_D(\mathbf{r}, \mathbf{r}') = 1 & \end{cases} \tag{3.61}$$

where the unit normal on ∂V is oriented positively inwards V. Since we may choose the origin of the coordinates at will, to lighten the formulas we let $\mathbf{r}' = 0$ without loss of generality, so that we deal with the scalar field $G_D(\mathbf{r})$. We further exclude the origin, which is a singular point, by means of a small ball $B(0, a) \subset V$, $a > 0$, and define the surface-wise multiply-connected region $V_a := V \setminus B(0, a)$.

First and foremost, since it follows from (3.61) that $G_D(\mathbf{r})$ is harmonic in V_a, then $G_D(\mathbf{r})$ cannot have local maxima or minima in V_a [11, Section 8.6], [12]. Indeed, if pick up any point $\mathbf{r}_0 \in V_a$ and introduce another ball $B(\mathbf{r}_0, b) \subset V_a$ centered in $\mathbf{r}_0$, we have

$$G_D(\mathbf{r}_0) = \frac{3}{4\pi b^3} \int_{B(\mathbf{r}_0,b)} \mathrm{d}V\, G_D(\mathbf{r}) \qquad \Longrightarrow \qquad \int_{B(\mathbf{r}_0,b)} \mathrm{d}V\, [G_D(\mathbf{r}) - G_D(\mathbf{r}_0)] = 0 \tag{3.62}$$

by virtue of the (second) mean value theorem of electrostatic (2.173). Now, if $G_D(\mathbf{r}_0)$ *happened* to be a local *maximum*, then we could choose a conveniently small radius b so that $G_D(\mathbf{r}) - G_D(\mathbf{r}_0) < 0$ for any $\mathbf{r} \in B(\mathbf{r}_0, b) \setminus \{\mathbf{r}_0\}$. As a result, the rightmost integral in (3.62) would always be *negative*, because the integrand would vanish at most for $\mathbf{r} = \mathbf{r}_0$. Hence, since (3.62) would be obviously violated, we should conclude that $\mathbf{r}_0$ cannot be a local maximum. By means of a similar argument, we infer that $\mathbf{r}_0$ cannot be a local minimum either. Lastly, in light of the arbitrariness of $\mathbf{r}_0$ we draw the conclusion that $G_D(\mathbf{r})$ does not reach its extrema in the region V_a. Therefore, maxima and minima must necessarily reside on ∂V and the boundary of the excluded ball $B(0, a)$.

Accordingly, since $G_D(\mathbf{r}) = 0$ everywhere on ∂V, we have two possibilities, namely, $\mathbf{r} \in \partial V$ are either points of maximum or points of minimum for $G_D(\mathbf{r})$. If the latter circumstance is true, then $G_D(\mathbf{r}) > 0$ for $\mathbf{r} \in V_a$, as we claimed at the beginning. On the contrary, if the points on the boundary are maxima for $G_D(\mathbf{r})$, we can show that the normalization condition in (3.61) cannot be satisfied. To this purpose, we need to ascertain what the sign of $\hat{\mathbf{n}}(\mathbf{r}) \cdot \nabla G_D(\mathbf{r})$ is on the boundary ∂V.

Why, intuition suggests that if $G_D(\mathbf{r})$ is maximum (and null, at that) everywhere on ∂V, then it should decrease for points away from the boundary, and this behavior would make the normal derivative on ∂V negative or at least non-positive. To substantiate this hunch we pick up a point $\mathbf{r}_1 \in \partial V$ and define the auxiliary function of *one variable*

$$f(s) := G_D(\mathbf{r}(s)) = G_D(\mathbf{r}_1 + s\hat{\mathbf{n}}(\mathbf{r}_1)), \qquad s \in [0, h] \tag{3.63}$$

where h is a suitably small positive number such that $\mathbf{r}(h) = \mathbf{r}_1 + h\hat{\mathbf{n}}(\mathbf{r}_1) \in V_a$. If $f(s)$ *decreases* in the interval $[0, h]$, it follows that $f(s)$ is necessarily negative since $f(0) = G_D(\mathbf{r}_1) = 0$ by construction, and we have

$$\frac{f(s) - f(0)}{s} = \frac{f(s)}{s} \leqslant 0, \qquad s \in]0, h] \tag{3.64}$$

whereby in the limit as $s \to 0^+$ we get

$$\left.\frac{\mathrm{d}f}{\mathrm{d}s}\right|_{s=0^+} = \lim_{s\to 0^+} \frac{f(s)}{s} \leqslant 0 \tag{3.65}$$

whence we conclude that the (right) derivative of $f(s)$ in $s = 0^+$ is non-positive. Now, let us look at this intermediate result from the viewpoint of $G_D(\mathbf{r}(s))$, namely,

$$0 \geqslant \left.\frac{\mathrm{d}f}{\mathrm{d}s}\right|_{s=0^+} = \hat{\mathbf{n}}(\mathbf{r}_1) \cdot \nabla G_D(\mathbf{r})|_{\mathbf{r}=\mathbf{r}_1} \tag{3.66}$$

that is, the normal derivative of the Dirichlet Green function (towards the inside of V_a) is non-positive at $\mathbf{r}_1 \in \partial V$. But this point is completely arbitrary, and we can repeat the same steps for any position vector $\mathbf{r} \in \partial V$ in order to determine that $\hat{\mathbf{n}}(\mathbf{r}) \cdot \nabla G_D(\mathbf{r}) \leqslant 0$ on the boundary of V.

With the help of this latest finding, when we integrate the normal derivative over ∂V, we obtain

$$\int_{\partial V} \mathrm{d}S\, \hat{\mathbf{n}}(\mathbf{r}) \cdot \nabla G_D(\mathbf{r}) \leqslant 0 \tag{3.67}$$

but this inequality cannot be true because it patently violates the normalization condition included in (3.61). Therefore, we are only left with the other possibility, namely, that $G_D(\mathbf{r})$ reaches its minimum on ∂V and hence $G_D(\mathbf{r}) > 0$ for $\mathbf{r} \in V_a$.

It is not difficult to extend the proof for an arbitrary position of the source point $\mathbf{r}' \in V$. Needless to say, if we were to place a negative generating unitary charge at $\mathbf{r}'$, then we could prove that $G_D(\mathbf{r}, \mathbf{r}') < 0$ for $\mathbf{r} \in V \setminus \{\mathbf{r}'\}$.

3.5 Laplace equation and boundary value problems

A boundary value problem (BVP for short) in electrostatics consists of solving the Poisson or the Laplace equation in a region of space with assigned boundary conditions. Although the integral representations derived in Section 2.7 allow rephrasing a BVP as as integral equation [3, Section 4.2.2], the direct solution of the Laplace equation is a classical problem. However, analytical expressions can only be found when the region of interest is bounded by canonical shapes such as cubes, cylinders, cones, spheres or ellipsoids [13]. In this section we consider the construction of solutions to the three-dimensional Laplace equation

$$\nabla^2\Phi(\mathbf{r}) = 0, \qquad \mathbf{r} \in V \tag{3.68}$$

with the method of *separation of variables* [1, 2, 8, 13, 14], [15, Chapter 9], [16, Section 11.3], [4, Chapter 7], [17, Chapter 3], [18, Chapter 3]. Functions which solve (3.68) are termed *harmonic*. We present the details for the cases of polar spherical coordinates and circular cylindrical coordinates. Separation and solution in Cartesian coordinates is simpler and can be found, e.g., in [8, Appendix A.4], [15, Section 9.3], [17, Section 3.3], [18, Section 3.1], [19, Section 7.2].

3.5.1 *Polar spherical coordinates*

When the region of concern V is either the whole space or, if it exhibits spherical symmetry, it is bounded by surfaces which are (parts of) spheres, cones and planes, the Laplace operator is best expressed in a system of polar spherical coordinates (r, φ, ϑ) (Appendix A.1) because this choice facilitates the inclusion of the boundary conditions. Still, a closed-form solution may not be feasible anyway, in which case we may attempt to represent the potential as a series of known functions, say,

$$\Phi(\mathbf{r}) = \sum_{\nu} \Phi_\nu \Psi_\nu(r, \varphi, \vartheta), \qquad \mathbf{r} \in V \tag{3.69}$$

where the constant coefficients Φ_ν must be determined so as to satisfy the given boundary conditions on ∂V. We postpone a discussion on the nature of the index ν until further on; for the time being, the symbol ν simply signals that we *linearly* combine different, possibly infinitely numerous contributions.

A very convenient set of expansion functions $\{\Psi_\nu(\mathbf{r})\}$ is constituted by the fundamental solutions of (3.68) with the operator expressed in spherical coordinates r, φ and ϑ. Specifically, on account of (A.42) we require that $\Psi(\mathbf{r}) = \Psi_\nu(\mathbf{r})$ solve

$$\frac{1}{r^2}\frac{\partial}{\partial r}\left(r^2 \frac{\partial \Psi}{\partial r}\right) + \frac{1}{r^2 \sin\vartheta}\frac{\partial}{\partial \vartheta}\left(\sin\vartheta \frac{\partial \Psi}{\partial \vartheta}\right) + \frac{1}{r^2 \sin^2\vartheta}\frac{\partial^2 \Psi}{\partial \varphi^2} = 0 \tag{3.70}$$

in the region V. To determine $\Psi(\mathbf{r})$ we invoke the so-called *separation argument* or *separation of variables* [8, Appendix A.4], [13, Section IV], that is, we make the assumption that, although $\Psi(\mathbf{r})$ must depend on the three coordinates, it can be written as a product of three functions of just one variable

$$\Psi(\mathbf{r}) := \Xi(r)\Theta(\vartheta)F(\varphi) \tag{3.71}$$

whereby the dependence on r, ϑ and φ is effectively separated. This is the starting point which allows us to formulate (3.70) equivalently as three *uncoupled* equations.

To proceed we substitute the Ansatz (3.71) into (3.70), multiply through by $r^2 \sin^2\vartheta/\Psi$ and rearrange terms to get

$$\underbrace{\frac{\sin^2\vartheta}{\Xi}\frac{\mathrm{d}}{\mathrm{d}r}\left(r^2 \frac{\mathrm{d}\Xi}{\mathrm{d}r}\right) + \frac{\sin\vartheta}{\Theta}\frac{\mathrm{d}}{\mathrm{d}\vartheta}\left(\sin\vartheta \frac{\mathrm{d}\Theta}{\mathrm{d}\vartheta}\right)}_{\text{depends on } (r,\vartheta)} = \underbrace{-\frac{1}{F}\frac{\mathrm{d}^2 F}{\mathrm{d}\varphi^2}}_{\text{depends only on } \varphi} \tag{3.72}$$

where the left-hand side is equal to some function of two variables (r, ϑ) but the right-hand side is a function of φ only. Now, this equation is certainly satisfied if we set both sides equal to the *same* arbitrary constant m^2. The latter need not be the square of a whole number, though, if it is real, we shall show that m^2 is non-negative. Regardless, we obtain two separate equations

$$-\frac{1}{F}\frac{\mathrm{d}^2 F}{\mathrm{d}\varphi^2} = m^2 \tag{3.73}$$

$$\frac{\sin^2\vartheta}{\Xi}\frac{\mathrm{d}}{\mathrm{d}r}\left(r^2 \frac{\mathrm{d}\Xi}{\mathrm{d}r}\right) + \frac{\sin\vartheta}{\Theta}\frac{\mathrm{d}}{\mathrm{d}\vartheta}\left(\sin\vartheta \frac{\mathrm{d}\Theta}{\mathrm{d}\vartheta}\right) = m^2 \tag{3.74}$$

and we now try to split the second one in like manner. If we divide through by $\sin^2\theta$ and rearrange terms, we find

$$\underbrace{\frac{1}{\Xi}\frac{\mathrm{d}}{\mathrm{d}r}\left(r^2 \frac{\mathrm{d}\Xi}{\mathrm{d}r}\right)}_{\text{depends only on } r} = \underbrace{\frac{m^2}{\sin^2\vartheta} - \frac{1}{\Theta \sin\vartheta}\frac{\mathrm{d}}{\mathrm{d}\vartheta}\left(\sin\vartheta \frac{\mathrm{d}\Theta}{\mathrm{d}\vartheta}\right)}_{\text{depends only on } \vartheta} \tag{3.75}$$

and, by the same argument as that invoked above, the equation is satisfied if both sides are set equal to the *same* arbitrary constant. For reasons that will become apparent at the end of this discussion, it is customary to write the constant as $n(n+1)$, where n is not necessarily a whole number. Therefore, we obtain

$$\frac{1}{\Xi}\frac{\mathrm{d}}{\mathrm{d}r}\left(r^2\frac{\mathrm{d}\Xi}{\mathrm{d}r}\right) = n(n+1) \tag{3.76}$$

$$\frac{m^2}{\sin^2\vartheta} - \frac{1}{\Theta\sin\vartheta}\frac{\mathrm{d}}{\mathrm{d}\vartheta}\left(\sin\vartheta\frac{\mathrm{d}\Theta}{\mathrm{d}\vartheta}\right) = n(n+1) \tag{3.77}$$

whereby (3.70) is now fully separated into three equations. Let us write them again in a more polished way for further analysis:

$$\frac{\mathrm{d}^2F_m}{\mathrm{d}\varphi^2} + m^2F_m = 0 \tag{3.78}$$

$$\frac{1}{\sin\vartheta}\frac{\mathrm{d}}{\mathrm{d}\vartheta}\left(\sin\vartheta\frac{\mathrm{d}\Theta_{mn}}{\mathrm{d}\vartheta}\right) + \left[n(n+1) - \frac{m^2}{\sin^2\vartheta}\right]\Theta_{mn} = 0 \tag{3.79}$$

$$\frac{\mathrm{d}}{\mathrm{d}r}\left(r^2\frac{\mathrm{d}\Xi_n}{\mathrm{d}r}\right) - n(n+1)\Xi_n = 0 \tag{3.80}$$

having highlighted the dependence of the functions on the parameters m and n.

In summary, we have obtained three relatively simpler *ordinary* differential equations which depend, though, on the separation constants m and n. Any product of the type (3.71) that is formed with solutions to (3.78)-(3.79) is a legitimate candidate function for the set $\{\Psi_\nu\}$. Therefore, we see that the index ν actually stands for the pair (m, n) of separation constants. However, the nature of the numbers m and n as well as the ranges of values they can take is determined by the very shape of the region V. More precisely, the boundary conditions of $\Psi(\mathbf{r})$ on ∂V pass over into requirements for the three functions $\Xi(r)$, $\Theta(\vartheta)$ and $F(\varphi)$; said requirements in turn pose constraints on the admissible values for m and n. In fact, when supplemented with boundary conditions (3.78)-(3.79) become one-dimensional *eigenvalue* problems, that is, they admit non-trivial solutions (called *eigenfunctions*) only for selected values of m and n, which we call eigenvalues (Appendix D.7). We discuss the details for each spherical coordinate by considering a few relevant sets of boundary conditions.

We begin with (3.78), which is the familiar *harmonic equation* but also a special instance of the more general one-dimensional Stürm-Liouville problem [8, Appendix A.4], [20, Section 3.2], [21, Section 6.3]. If the volume V is rotationally symmetric around the z-axis, the boundary conditions read

$$F_m(0) = F_m(2\pi) \qquad \text{and} \qquad \left.\frac{\mathrm{d}F_m}{\mathrm{d}\varphi}\right|_{\varphi=0} = \left.\frac{\mathrm{d}F_m}{\mathrm{d}\varphi}\right|_{\varphi=2\pi} \tag{3.81}$$

i.e., $F_m(\varphi)$ must be a periodic function. Furthermore, if the volume V is limited along φ by two half-planes defined by $\varphi = \varphi_1$ and $\varphi = \varphi_2$ with $0 \leqslant \varphi_1 < \varphi_2 \leqslant 2\pi$, we may assume either homogeneous Dirichlet boundary conditions

$$F_m(\varphi_1) = F_m(\varphi_2) = 0 \tag{3.82}$$

or homogeneous Neumann boundary conditions

$$\left.\frac{\mathrm{d}F_m}{\mathrm{d}\varphi}\right|_{\varphi_1} = \left.\frac{\mathrm{d}F_m}{\mathrm{d}\varphi}\right|_{\varphi_2} = 0 \tag{3.83}$$

or even a combination thereof.

We may now show that, as anticipated, the eigenvalues m^2 are non-negative. To this purpose, we multiply (3.78) through by $F_m^*(\varphi)$ and integrate from φ_1 to φ_2 to obtain

$$0 = \int_{\varphi_1}^{\varphi_2} \mathrm{d}\varphi F_m^* \frac{\mathrm{d}^2 F_m}{\mathrm{d}\varphi^2} + m^2 \int_{\varphi_1}^{\varphi_2} \mathrm{d}\varphi F_m^* F_m = \left[F_m^* \frac{\mathrm{d}F_m}{\mathrm{d}\varphi}\right]_{\varphi_1}^{\varphi_2} - \int_{\varphi_1}^{\varphi_2} \mathrm{d}\varphi \left|\frac{\mathrm{d}F_m}{\mathrm{d}\varphi}\right|^2 + m^2 \int_{\varphi_1}^{\varphi_2} \mathrm{d}\varphi |F_m|^2 \tag{3.84}$$

having integrated by parts [22]. For any choice of boundary conditions listed above the difference of the products $F_m^* \mathrm{d}F_m/\mathrm{d}\varphi$ evaluated at the endpoints of the interval $I := [\varphi_1, \varphi_2]$ invariably vanishes, so we are left with

$$m^2 = \frac{\displaystyle\int_{\varphi_1}^{\varphi_2} \mathrm{d}\varphi \left|\frac{\mathrm{d}F_m}{\mathrm{d}\varphi}\right|^2}{\displaystyle\int_{\varphi_1}^{\varphi_2} \mathrm{d}\varphi\, |F_m|^2} = \frac{\left\|\dfrac{\mathrm{d}F_m}{\mathrm{d}\varphi}\right\|^2_{\mathcal{L}^2(I)}}{\|F_m\|^2_{\mathcal{L}^2(I)}} \tag{3.85}$$

where the ratio in the right-hand side — called the *Rayleigh quotient* — is evidently a non-negative quantity. Notice that the occurrence of a function $F_m(\varphi)$ null everywhere for $\varphi \in I$ — whereby $\|F_m\|_{\mathcal{L}^2(I)} = 0$ — is precluded by the very definition of eigenfunction.

Any two eigenfunctions associated with different eigenvalues m and m' are *orthogonal* in the sense that their inner product over I is null. To prove this, we consider the identity

$$F_{m'}^* \frac{\mathrm{d}^2 F_m}{\mathrm{d}\varphi^2} - F_m \frac{\mathrm{d}^2 F_{m'}^*}{\mathrm{d}\varphi^2} = \frac{\mathrm{d}}{\mathrm{d}\varphi}\left(F_{m'}^* \frac{\mathrm{d}F_m}{\mathrm{d}\varphi} - F_m \frac{\mathrm{d}F_{m'}^*}{\mathrm{d}\varphi}\right) = -(m^2 - m'^2) F_m F_{m'}^* \tag{3.86}$$

which integrated over I yields

$$0 = (m^2 - m'^2) \int_{\varphi_1}^{\varphi_2} \mathrm{d}\varphi F_m F_{m'}^* = (m^2 - m'^2)(F_m, F_{m'})_I \tag{3.87}$$

by virtue of any type of boundary conditions considered above. Since $m \neq m'$ by hypothesis, then the inner product must vanish.

As is well known, any function of the form

$$F_m(\varphi) = c_{m1} \mathrm{e}^{-\mathrm{j} m\varphi} + c_{m2} \mathrm{e}^{\mathrm{j} m\varphi} \tag{3.88}$$

solves (3.78). Nevertheless, m and one of the constants c_{m1}, c_{m2} are determined by the boundary conditions, whereas the remaining constant may be chosen so that the eigenfunction has unitary norm. In particular, the calculations yield

$$F_m(\varphi) = c_m \mathrm{e}^{-\mathrm{j} m\varphi}, \qquad \varphi \in [0, 2\pi], \qquad m \in \mathbb{Z} \tag{3.89}$$

for periodic conditions,

$$F_m(\varphi) = c_m \sin[m(\varphi - \varphi_1)], \qquad \varphi \in [\varphi_1, \varphi_2], \qquad m = \frac{p\pi}{\varphi_2 - \varphi_1}, \qquad p \in \mathbb{N} \setminus \{0\} \tag{3.90}$$

for homogeneous Dirichlet boundary conditions, and finally

$$F_m(\varphi) = c_m \cos[m(\varphi - \varphi_1)], \qquad \varphi \in [\varphi_1, \varphi_2], \qquad m = \frac{p\pi}{\varphi_2 - \varphi_1}, \qquad p \in \mathbb{N} \tag{3.91}$$

for homogeneous Neumann boundary conditions. We remark that $m = 0$ is not an eigenvalue for (3.90) because otherwise the corresponding eigenfunction would be null everywhere. Also notice that for a general choice of endpoints of the interval I the eigenvalues m in (3.90) and (3.91) are not whole numbers.

In the special case $m = 0$ (3.78) has the more general solution

$$F_0(\varphi) = b_0\varphi + c_0, \qquad \varphi \in \mathbb{R} \tag{3.92}$$

which in combination with $\Xi(r)$ and $\Theta(\vartheta)$ gives rise to a *many-valued* scalar field $\Psi(\mathbf{r})$. This happens because φ represents the azimuth angle in the ordinary three-dimensional space and triples such as $(r, \varphi_0 + 2\pi l, \vartheta)$ with $l \in \mathbb{Z}$, all identify the very same point in which $\Psi(\mathbf{r})$ takes on infinitely many different values (see Section 4.4 and Example 4.4).

Next, we consider (3.79) which is called the *associated Legendre equation* [1,2,14], [8, Appendix A.4] and is a special case of the Stürm-Liouville problem as well. If the volume V is limited along ϑ by two conical surfaces defined by $\vartheta = \vartheta_1$ and $\vartheta = \vartheta_2$ with $0 < \vartheta_1 < \vartheta_2 < \pi$, we may choose homogeneous Dirichlet or Neumann boundary conditions

$$\Theta_{mn}(\vartheta_1) = \Theta_{mn}(\vartheta_2) = 0 \tag{3.93}$$

$$\left.\frac{\mathrm{d}\Theta_{mn}}{\mathrm{d}\vartheta}\right|_{\vartheta_1} = \left.\frac{\mathrm{d}\Theta_{mn}}{\mathrm{d}\vartheta}\right|_{\vartheta_2} = 0 \tag{3.94}$$

or a combination thereof. Although (3.79) can be solved independently of (3.78), we have to keep in mind that the values of m are already set by the boundary conditions along φ. Therefore, (3.79) is an eigenvalue problem for n only with m fixed.

We may show that the quantity $n(n+1)$ is non-negative by multiplying through (3.79) with $\Theta^*_{mn}(\vartheta)\sin\vartheta$ and integrating over the interval $I := [\vartheta_1, \vartheta_2]$, viz.,

$$\begin{aligned} 0 &= \int_{\vartheta_1}^{\vartheta_2} \mathrm{d}\vartheta\, \Theta^*_{mn} \frac{\mathrm{d}}{\mathrm{d}\vartheta}\left(\sin\vartheta \frac{\mathrm{d}\Theta_{mn}}{\mathrm{d}\vartheta}\right) + \int_{\vartheta_1}^{\vartheta_2} \mathrm{d}\vartheta\, \Theta^*_{mn}\Theta_{mn}\left[n(n+1)\sin\vartheta - \frac{m^2}{\sin\vartheta}\right] \\ &= -\int_{\vartheta_1}^{\vartheta_2} \mathrm{d}\vartheta\, \sin\vartheta \left|\frac{\mathrm{d}\Theta_{mn}}{\mathrm{d}\vartheta}\right|^2 + n(n+1)\int_{\vartheta_1}^{\vartheta_2} \mathrm{d}\vartheta\, \sin\vartheta |\Theta_{mn}|^2 - m^2 \int_{\vartheta_1}^{\vartheta_2} \mathrm{d}\vartheta \frac{|\Theta_{mn}|^2}{\sin\vartheta} \end{aligned} \tag{3.95}$$

having integrated by parts [22] and applied the boundary conditions. We arrive at

$$n(n+1) = \frac{\displaystyle\int_{\vartheta_1}^{\vartheta_2} \mathrm{d}\vartheta\, \sin\vartheta \left|\frac{\mathrm{d}\Theta_{mn}}{\mathrm{d}\vartheta}\right|^2 + m^2 \int_{\vartheta_1}^{\vartheta_2} \mathrm{d}\vartheta \frac{|\Theta_{mn}|^2}{\sin\vartheta}}{\displaystyle\int_{\vartheta_1}^{\vartheta_2} \mathrm{d}\vartheta\, \sin\vartheta |\Theta_{mn}|^2} \tag{3.96}$$

where the ratio in the right-hand side is non-negative. Besides, we may prove that any two eigenfunctions associated with distinct eigenvalues n and n' are orthogonal with respect to the inner product over I. We consider the identity

$$\begin{aligned}
&\frac{\mathrm{d}}{\mathrm{d}\vartheta}\left(\Theta^*_{mn'}\sin\vartheta\frac{\mathrm{d}\Theta_{mn}}{\mathrm{d}\vartheta}\right)-\frac{\mathrm{d}}{\mathrm{d}\vartheta}\left(\Theta_{mn}\sin\vartheta\frac{\mathrm{d}\Theta^*_{mn'}}{\mathrm{d}\vartheta}\right)=\\
&=\Theta^*_{mn'}\frac{\mathrm{d}}{\mathrm{d}\vartheta}\left(\sin\vartheta\frac{\mathrm{d}\Theta_{mn}}{\mathrm{d}\vartheta}\right)-\Theta_{mn}\frac{\mathrm{d}}{\mathrm{d}\vartheta}\left(\sin\vartheta\frac{\mathrm{d}\Theta^*_{mn'}}{\mathrm{d}\vartheta}\right)\\
&=-\left[n(n+1)-n'(n'+1)\right]\Theta_{mn}\Theta^*_{mn'}\sin\vartheta
\end{aligned} \tag{3.97}$$

which integrated over I provides

$$\begin{aligned}
0&=\left[n(n+1)-n'(n'+1)\right]\int_{\vartheta_1}^{\vartheta_2}\mathrm{d}\vartheta\,\Theta_{mn}\Theta^*_{mn'}\sin\vartheta\\
&=\left[n(n+1)-n'(n'+1)\right](\Theta_{mn},\Theta_{mn'})_I
\end{aligned} \tag{3.98}$$

on account of either one of the homogeneous boundary conditions stated above. So long as $n \neq n'$, the inner product must vanish.

The general solutions to (3.79) are called the *associated Legendre functions* of the first and the second kind and are indicated with the symbols $P_n^m(\cos\vartheta)$ and $Q_n^m(\cos\vartheta)$, respectively [1, Chapter 3], [23, Chapter 20], [24, Chapter 8]. We recall that $Q_n^m(\cos\vartheta)$ exhibits a logarithmic singularity for $\vartheta \in \{0, \pi\}$ (see Appendix H.5), whereas $P_n^m(\cos\vartheta)$ is regular for $\vartheta \in \{0, \pi\}$ only if $n \in \mathbb{N}$; moreover, $P_n^m(\cos\vartheta) = 0$ for $n < |m|$. Therefore, any combination of two associated Legendre functions in the form

$$\Theta_{mn}(\vartheta) = c_1 P_n^m(\cos\vartheta) + c_2 Q_n^m(\cos\vartheta), \qquad |m| \leqslant n \tag{3.99}$$

solves (3.79). However, n and either c_1 or c_2 are determined by the boundary conditions along ϑ, although it may not be possible, in general, to determine the eigenvalues in closed form. The remaining constant may be chosen so that the eigenfunctions have unitary norm. For homogeneous Dirichlet boundary conditions with $0 < \vartheta_1 < \vartheta_2 < \pi$, the eigenvalues n satisfy the transcendental equation

$$P_n^m(\cos\vartheta_1)Q_n^m(\cos\vartheta_2) - Q_n^m(\cos\vartheta_1)P_n^m(\cos\vartheta_2) = 0 \tag{3.100}$$

and a similar one holds for the homogeneous Neumann boundary conditions.

A relatively simpler situation arises when the region V is the whole space or it includes (parts of) the z-axis, in that the usual boundary conditions in $\vartheta_1 = 0$ and $\vartheta_2 = \pi$ do not apply. Then again, if the function $\Phi(\mathbf{r})$ we wish to represent with the set $\{\Psi_\nu\}$ is regular for $\theta \in \{0, \pi\}$ we have to exclude the function $Q_n^m(\cos\vartheta)$ owing to the singular behavior thereof along the z-axis. This requires setting $c_2 = 0$, and the solution $\Theta_{mn}(\vartheta)$ will have the form

$$\Theta_{mn}(\vartheta) = c_{mn} P_n^m(\cos\vartheta), \qquad \vartheta \in [0, \pi] \qquad n \in \mathbb{N}, \qquad n \geqslant |m| \tag{3.101}$$

with $m \in \mathbb{Z}$ in accordance with (3.89). Under the same assumptions, it is customary to pair the Legendre functions of the first kind with the solutions to (3.78) and to introduce the functions

$$Y_{mn}(\vartheta, \varphi) := \sqrt{\frac{2n+1}{4\pi}\frac{(n-m)!}{(n+m)!}}\, P_n^m(\cos\vartheta)\mathrm{e}^{-\mathrm{j}m\varphi}, \qquad n \in \mathbb{N}, \qquad n \geqslant |m| \tag{3.102}$$

which are called *spherical harmonics* [25, 26] (or zonal or tesseral harmonics [2]). They form a complete orthonormal set for $\vartheta \in [0, \pi]$ and $\varphi \in [0, 2\pi]$, that is,

$$\int_0^{2\pi}\int_0^{\pi} \mathrm{d}\varphi\, \mathrm{d}\vartheta \sin\vartheta Y_{mn}(\vartheta,\varphi) Y^*_{m'n'}(\vartheta,\varphi) = \delta_{n'n}\delta_{m'm} \tag{3.103}$$

where the symbols

$$\delta_{n'n} := \begin{cases} 1 & n' = n \\ 0 & n' \neq n \end{cases} \qquad \delta_{m'm} := \begin{cases} 1 & m' = m \\ 0 & m' \neq m \end{cases} \tag{3.104}$$

denote two Kronecker deltas.

Lastly, we examine (3.80), which can be solved by elementary means. It is important to notice that, if the parameter n has been chosen to meet the boundary conditions in ϑ_1 and ϑ_2, (3.80) is not an eigenvalue problem per se. If the region V is bounded by two spheres $r = r_1$ and $r = r_2$ with $0 < r_1 < r_2 < +\infty$, we may choose homogeneous Diriclet or Neumann boundary conditions

$$\Xi_n(r_1) = \Xi_n(r_2) = 0 \tag{3.105}$$

$$\left.\frac{\mathrm{d}\Xi_n}{\mathrm{d}r}\right|_{r_1} = \left.\frac{\mathrm{d}\Xi_n}{\mathrm{d}r}\right|_{r_2} = 0 \tag{3.106}$$

or a combination thereof. On the whole, it is easily checked that any function of the form

$$\Xi_n(r) = c_{n1} r^n + \frac{c_{n2}}{r^{n+1}}, \qquad n \in \mathbb{N} \tag{3.107}$$

constitutes a solution to (3.80). When the parameter n is given, the two arbitrary constants c_{n1}, c_{n2} are determined by the boundary conditions along r. Again, special situations arise if $r_1 = 0$ and $V = B(0, r_2)$ or $r_2 \to +\infty$ and $V = \mathbb{R}^3 \setminus B(0, r_1)$. In the former case, we need to exclude contributions of the type $1/r^{n+1}$ — which are singular for $r = 0$ — if the function $\Phi(\mathbf{r})$ we want to represent is regular in the origin. In the latter case, if the function $\Phi(\mathbf{r})$ is regular (bounded) at infinity, then positive powers of r are disallowed in that they diverge as $r \to +\infty$.

From the discussion above it is apparent that we run into trouble if we try to satisfy homogeneous boundary conditions along all of the three spherical coordinates. More precisely, if we set m and n by enforcing conditions along φ and ϑ, the application of either (3.105) or (3.106) produces $c_{n1} = c_{n2} = 0$, whereby $\Xi_n(r) = 0$ and the corresponding function $\Psi_{mn}(\mathbf{r})$ vanishes identically. A moment's thought tells us that this disappointing result is correct. Why? The reason is simple: the Laplace equation (3.68) in V subject to homogeneous boundary conditions on ∂V is only solved by $\Phi(\mathbf{r}) = 0$. Indeed, this is precisely the argument we exploited in Section 2.5.1 to prove the uniqueness of the solutions to the Poisson equation.

We are now ready to write the expansion functions in practical cases. For instance, if $V = B(0, r_2)$ from the previous discussion we conclude that

$$\Psi_{mn}(\mathbf{r}) := r^n P_n^m(\cos\theta) \mathrm{e}^{-\mathrm{j}m\varphi}, \qquad n \in \mathbb{N}, \qquad |m| \leqslant n \tag{3.108}$$

and for $V = \mathbb{R}^3 \setminus B(0, r_1)$ we have

$$\Psi_{mn}(\mathbf{r}) := \frac{1}{r^{n+1}} P_n^m(\cos\theta) \mathrm{e}^{-\mathrm{j}m\varphi}, \qquad n \in \mathbb{N}, \qquad |m| \leqslant n. \tag{3.109}$$

Notice that the singular solution for $n = 0 = m$, $r > 0$, is essentially the static Green function (2.130).

Example 3.3 (Conducting sphere in a uniform electrostatic field)

A relatively simple electrostatic BVP consists of finding the electric potential and field due to the presence of a conducting sphere immersed in an otherwise uniform electrostatic field $\mathcal{E}^{i}(\mathbf{r}) = \mathcal{E}_0\hat{\mathbf{z}}$ [1, Section 2.5], [27, Section 9.C]. The latter does not satisfy the required boundary conditions (1.182) and (1.180) on the surface of the sphere. As a result, the charges redistribute themselves on the surface of the sphere so as to generate an appropriate secondary field $\mathcal{E}^{s}(\mathbf{r})$ which combines with $\mathcal{E}^{i}(\mathbf{r})$ so as to ensure the vanishing of the tangential component of $\mathcal{E}(\mathbf{r})$ (Figure 3.5).

Without loss of generality we suppose that the conducting sphere occupies the inside of the ball $B(0, a)$. Then, the relevant electrostatic potential is the unique solution to the following Dirichlet problem

$$\begin{cases} \nabla^2\Phi(\mathbf{r}) = 0, & \mathbf{r} \in \mathbb{R}^3 \setminus B(0, a) \\ \Phi(\mathbf{r}) = \Phi_0, & \mathbf{r} \in \partial B \\ \Phi(\mathbf{r}) = \mathcal{O}\left(\dfrac{1}{|\mathbf{r}|}\right), & |\mathbf{r}| \to +\infty \end{cases} \tag{3.110}$$

where Φ_0 is the known potential of the sphere. The solution is best accomplished by introducing polar spherical coordinates centered in $\mathbf{r} = 0$ with the polar axis parallel to the impressed electric field. In light of the remarks above we write the potential as

$$\Phi(\mathbf{r}) = \Phi^{i}(\mathbf{r}) + \Phi^{s}(\mathbf{r}), \qquad \mathbf{r} \in \mathbb{R}^3 \setminus B(0, a) \tag{3.111}$$

where

- $\Phi^{i}(\mathbf{r})$ is the 'incident' potential corresponding to the impressed electric field, which in turn is due to infinitely extended and infinitely remote sources;
- $\Phi^{s}(\mathbf{r})$ is the 'secondary' potential associated with the electric field $\mathcal{E}^{s}(\mathbf{r})$ produced by the yet unknown surface charge density on ∂B.

We do not concern ourselves with the potential inside the sphere because we already know that $\Phi(\mathbf{r})$ is constant and equal to Φ_0 for $\mathbf{r} \in B(0, a)$, since the sphere is made of a conducting medium.

To find the potential $\Phi^{i}(\mathbf{r})$ we write (2.15) explicitly in Cartesian coordinates

$$\mathcal{E}^{i}(\mathbf{r}) = \mathcal{E}_0\hat{\mathbf{z}} = -\hat{\mathbf{x}}\frac{\partial}{\partial x}\Phi^{i}(\mathbf{r}) - \hat{\mathbf{y}}\frac{\partial}{\partial y}\Phi^{i}(\mathbf{r}) - \hat{\mathbf{z}}\frac{\partial}{\partial z}\Phi^{i}(\mathbf{r}) \tag{3.112}$$

and observe that $\Phi^{i}(\mathbf{r})$ must be a function of z only inasmuch as the $\mathcal{E}^{i}(\mathbf{r})$ does not have components along x and y. Integrating with respect to z yields

$$\Phi^{i}(z) = -\mathcal{E}_0 z = -\mathcal{E}_0 r\cos\vartheta = -\mathcal{E}_0 r P_1(\cos\vartheta) \tag{3.113}$$

on account of (A.17) and (H.113). Besides, we have set to zero the arbitrary constant implied in the definition of the scalar potential $\Phi^{i}(\mathbf{r})$.

In view of the axial symmetry of the problem — i.e., since there is no dependence on the azimuth angle φ — we write the secondary potential as the linear superposition of Legendre polynomials, viz.,

$$\Phi^{s} = \sum_{n=0}^{+\infty} \frac{A_n}{r^{n+1}} P_n(\cos\theta), \qquad r \geqslant a \tag{3.114}$$

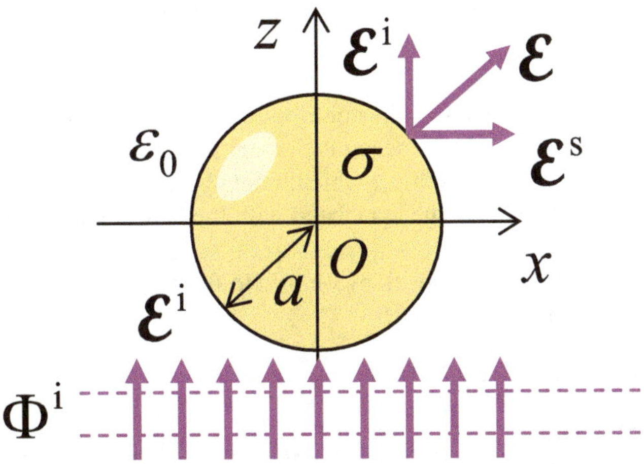

Figure 3.5 Conducting sphere immersed in an impressed uniform electrostatic field $\mathcal{E}^i$ (→). The dashed lines are the traces of the planes (perpendicular to $\mathcal{E}^i$) whereon the electrostatic potential Φ^i is constant.

where we have also ruled out the occurrence of positive powers of the radial distance r inasmuch as the secondary potential, being produced by a localized set of charges, must be regular at infinity.

While both Φ^i and the chosen expression (3.114) solve the Laplace equation separately for $r \geqslant a$, we have to determine the unknown expansion coefficients A_n so that for $\mathbf{r} \in \partial B$ the following is true

$$\sum_{n=0}^{+\infty} \frac{A_n}{a^{n+1}} P_n(\cos\theta) - \mathcal{E}_0 a P_1(\cos\vartheta) = \Phi_0 \tag{3.115}$$

for any value of ϑ. By pairing like terms proportional to polynomials of the same degree we find

$$\frac{A_0}{a} P_0(\cos\vartheta) = \Phi_0 P_0(\cos\vartheta) \qquad n = 0 \tag{3.116}$$

$$\left(\frac{A_1}{a^2} - \mathcal{E}_0 a\right) P_1(\cos\vartheta) = 0 \qquad n = 1 \tag{3.117}$$

$$\frac{A_n}{a^{n+1}} P_n(\cos\vartheta) = 0 \qquad n \geqslant 2 \tag{3.118}$$

and since these conditions must hold for any angle ϑ we obtain

$$A_0 = a\Phi_0, \qquad A_1 = a^3\mathcal{E}_0, \qquad A_n = 0 \qquad \text{for} \qquad n \geqslant 2. \tag{3.119}$$

Substituting these results back into (3.114) yields

$$\Phi^s(\mathbf{r}) = \frac{a}{r}\Phi_0 + \frac{a^3}{r^2}\mathcal{E}_0 \cos\vartheta = \underbrace{4\pi\varepsilon_0 a\Phi_0}_{\text{charge}} \frac{1}{4\pi\varepsilon_0 r} + \underbrace{4\pi\varepsilon_0 a^3 \mathcal{E}_0 \hat{\mathbf{z}}}_{\text{dipole moment}} \cdot \frac{\hat{\mathbf{r}}}{4\pi\varepsilon_0 r^2}, \qquad r \geqslant a \tag{3.120}$$

which, by comparison with (2.28) and (2.39), we may interpret by saying that the secondary potential is comprised of two terms, namely,

(1) the potential produced by an *equivalent charge*

$$q_{eq} = 4\pi\varepsilon_0 a\Phi_0 \tag{3.121}$$

(2) the potential produced by an *equivalent elementary electric dipole* with moment

$$\mathbf{p}_{eq} = 4\pi\varepsilon_0 a^3 \mathcal{E}_0 \hat{\mathbf{z}} \tag{3.122}$$

both located in the point $\mathbf{r} = 0$. Actually, the contribution of the equivalent charge is not caused by the impressed electric field, but rather it accounts for the potential produced by the excess charge already present on the surface of the sphere, which, being held at the non-zero potential Φ_0, is, in fact, charged. Conversely, the net charge associated with the equivalent dipole contribution should be zero.

To check the validity of these suppositions we turn to the boundary condition (1.182) and compute the surface charge density spread on ∂B

$$\begin{aligned} \varrho_S(\mathbf{r}) &= \varepsilon_0 \hat{\mathbf{r}} \cdot \mathcal{E}(\mathbf{r})|_{r=a} = -\varepsilon_0 \left.\frac{\mathrm{d}\Phi}{\mathrm{d}r}\right|_{r=a} \\ &= \varepsilon_0 \mathcal{E}_0 \cos\vartheta + \varepsilon_0 \frac{a}{r^2}\Phi_0\bigg|_{r=a} + \varepsilon_0 \frac{2a^3}{r^3}\mathcal{E}_0 \cos\vartheta\bigg|_{r=a} = \varepsilon_0 \frac{\Phi_0}{a} + 3\varepsilon_0 \mathcal{E}_0 \cos\vartheta \end{aligned} \tag{3.123}$$

which, when integrated over ∂B, yields

$$Q = \int_{\partial B} \mathrm{d}S\, \varrho_S(\mathbf{r}) = 4\pi\varepsilon_0 a \Phi_0 = q_{eq} \tag{3.124}$$

that is, the charge distributed over the surface of the conducting sphere coincides with the equivalent charge in the origin. Thus, the net surface charge induced by the impressed electric field is zero, as expected.

For a one-conductor system which is held at potential Φ_0 and carries a charge Q the quantity

$$C := \frac{Q}{\Phi_0} \tag{3.125}$$

is called the *capacitance* (physical dimensions: farad, F) of the conducting body [28, Section 6.8], [29, Section 2.5.4], [30, Section D.5], [31, Chapter 6]. The capacitance 'measures' the ability of the system to store electric charge. From (3.124) we see that

$$C = 4\pi\varepsilon_0 a \tag{3.126}$$

is the capacitance of the conducting sphere [15, Example 4.4]. Notice that C depends solely on the radius of the sphere, i.e., a geometrical quantity, and the permittivity of the surrounding medium.

It is interesting to compute the force — if any exists — exerted by the incident electric field on the sphere. To this purpose we recall from Section 1.11 that the normal component of the Maxwell stress tensor $\overline{\mathcal{T}}_{em}$ represents the rate of momentum flow through a surface or the force per unit area on the same surface. On the outer boundary of the sphere the total field is $\mathcal{E}(\mathbf{r}) = \mathcal{E}\hat{\mathbf{r}}$ by virtue of (1.180) (Figure 3.5) and hence the surface density of force reads

$$\mathbf{f}_{S\,em}(\mathbf{r}) = -\hat{\mathbf{r}} \cdot \overline{\mathcal{T}}_{em}(\mathbf{r}) = -\left(\frac{\varepsilon_0}{2}\mathcal{E}^2\hat{\mathbf{r}} - \varepsilon_0 \mathcal{E}^2 \hat{\mathbf{r}}\right) = \frac{\varepsilon_0}{2}\mathcal{E}^2\hat{\mathbf{r}} = \frac{[\varrho_S(\mathbf{r})]^2}{2\varepsilon_0}\hat{\mathbf{r}}, \qquad r = a \tag{3.127}$$

on account of (1.343) and the jump condition (1.182). By integrating this density in polar spherical coordinates over the boundary of the sphere we obtain the force, i.e.,

$$\mathcal{F}_{em} = \frac{\varepsilon_0}{2}\int_{\partial B} \mathrm{d}S \left(\frac{\Phi_0}{a} + 3\mathcal{E}_0 \cos\vartheta\right)^2 \hat{\mathbf{r}} = 4\pi\varepsilon_0 a \Phi_0 \mathcal{E}_0 \hat{\mathbf{z}} = q_{eq}\mathcal{E}^{\mathrm{i}}(0) \tag{3.128}$$

whence we gather that

(1) the force has the same direction as the incident electric field;
(2) the incident electric field acts only on the equivalent charge q_{eq} which accounts for the charge *already* distributed on the surface of the sphere;
(3) the net force on the part of the surface charge which is induced by the incident electric field is null essentially because this term — proportional with $\cos\vartheta$ — has odd symmetry with respect to the plane $z = 0$.

(End of Example 3.3)

Example 3.4 (Dielectric sphere in a uniform electrostatic field)

A slightly more general BVP consists of computing the electrostatic potential and the electric field due to a dielectric sphere endowed with uniform scalar permittivity ε_2 and immersed in an otherwise uniform unbounded isotropic medium endowed with permittivity ε_1 (Figure 3.6) [1, Section 4.4], [8, Section 3.2.10], [27, Section 9.D], [18, Section IV.6]. The sphere is exposed to an impressed electrostatic field $\mathcal{E}^{\mathrm{i}}(\mathbf{r}) = \mathcal{E}_0\hat{\mathbf{z}}$ which, evidently, does not obey the matching conditions (2.46) and (2.47) at the interface between the two media. Therefore, the bound charges within the dielectric arrange themselves in order to generate a secondary field $\mathcal{E}^{\mathrm{s}}(\mathbf{r})$ which in combination with $\mathcal{E}^{\mathrm{i}}(\mathbf{r}) = \mathcal{E}_0\hat{\mathbf{z}}$ makes (2.46) and (2.47) satisfied.

We assume that the dielectric sphere occupies the ball $B(0, a)$ and introduce the total scalar potential $\Phi(\mathbf{r})$ which is the unique solution to the homogeneous problem

$$\begin{cases} \nabla^2\Phi(\mathbf{r}) = 0, & \mathbf{r} \in \mathbb{R}^3 \setminus \partial B \\ \Phi_+(\mathbf{r}) = \Phi_-(\mathbf{r}), & \mathbf{r} \in \partial B \\ \varepsilon_1 \dfrac{\partial\Phi_+}{\partial\hat{\mathbf{n}}} = \varepsilon_2 \dfrac{\partial\Phi_-}{\partial\hat{\mathbf{n}}}, & \mathbf{r} \in \partial B \\ \Phi(\mathbf{r}) = \mathcal{O}\left(\dfrac{1}{|\mathbf{r}|}\right), & |\mathbf{r}| \to +\infty \end{cases} \tag{3.129}$$

where Φ_+ and Φ_- denote the potential on the positive and negative side of ∂B, with the unit normal on ∂B pointing into the unbounded background medium. Concerning uniqueness and the number of constraints enforced across ∂B we refer to the remark following (3.1) on page 147.

Since the problem exhibits azimuthal symmetry we look for a solution in the form

$$\Phi(\mathbf{r}) = \begin{cases} \displaystyle\sum_{n=0}^{+\infty} \frac{A_n}{r^{n+1}} P_n(\cos\vartheta) + \Phi^{\mathrm{i}}(\mathbf{r}), & r > a \\ \displaystyle\sum_{n=0}^{+\infty} B_n r^n P_n(\cos\vartheta), & r < a \end{cases} \tag{3.130}$$

where $\Phi^{\mathrm{i}}(\mathbf{r})$ is the potential associated with the impressed electric field produced by infinitely extended and infinitely far away charges. The explicit expression of $\Phi^{\mathrm{i}}(\mathbf{r})$ is given by (3.113) and will not be derived again here. We have excluded negative powers of r for $r < a$ because the potential must be finite at the center of the sphere. Similarly, positive powers of r are precluded for $r > a$ inasmuch as the contribution to the potential due to the sphere — which has finite size — must vanish at infinity.

We have to determine the expansion coefficients A_n and B_n so that for $r = a$ the matching conditions for the potential and the normal derivative thereof are true. In symbols, this means

$$\sum_{n=0}^{+\infty} \frac{A_n}{a^{n+1}} P_n(\cos\vartheta) - \mathcal{E}_0 a P_1(\cos\vartheta) = \sum_{n=0}^{+\infty} B_n a^n P_n(\cos\vartheta) \tag{3.131}$$

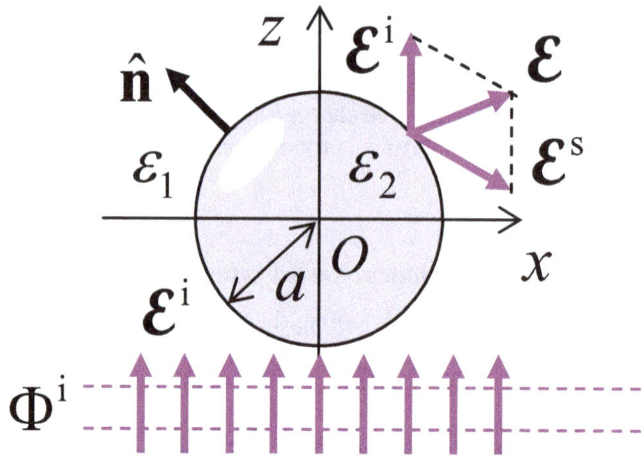

Figure 3.6 Dielectric sphere immersed in an impressed uniform electrostatic field $\mathcal{E}^i$ (→). The dashed lines are the traces of the planes (perpendicular to $\mathcal{E}^i$) whereon the electrostatic potential Φ^{i} is constant.

$$-\sum_{n=0}^{+\infty}\varepsilon_1\frac{n+1}{a^{n+2}}A_nP_n(\cos\vartheta)-\varepsilon_1\mathcal{E}_0P_1(\cos\vartheta)=\sum_{n=0}^{+\infty}\varepsilon_2na^{n-1}B_nP_n(\cos\vartheta) \tag{3.132}$$

for any value of the polar angle ϑ. By pairing terms proportional to polynomials with like powers we get

$$\left(\frac{A_n}{a^{n+1}}-B_na^n\right)P_n(\cos\vartheta)=0 \tag{3.133}$$

$$\left(\varepsilon_1\frac{n+1}{a^{n+2}}A_n+\varepsilon_2na^{n-1}B_n\right)P_n(\cos\vartheta)=0 \tag{3.134}$$

for $n \neq 1$ and

$$\left(\frac{A_1}{a^2}-\mathcal{E}_0a-B_1a\right)P_1(\cos\vartheta)=0 \tag{3.135}$$

$$\left(\varepsilon_1\frac{2}{a^3}A_1+\varepsilon_1\mathcal{E}_0+\varepsilon_2B_1\right)P_1(\cos\vartheta)=0 \tag{3.136}$$

for $n = 1$. The first set of equations yields a homogeneous linear system which admits only the trivial solution $A_n = 0 = B_n$, $n \neq 1$. For $n = 1$ from the second set we have

$$\frac{A_1}{a^2}-\mathcal{E}_0a=B_1a \qquad\qquad 2\varepsilon_1\frac{A_1}{a^3}+\varepsilon_1\mathcal{E}_0=-\varepsilon_2B_1 \tag{3.137}$$

whose solution is

$$A_1=\frac{\varepsilon_2-\varepsilon_1}{\varepsilon_2+2\varepsilon_1}a^3\mathcal{E}_0, \qquad\qquad B_1=-\frac{3\varepsilon_1}{\varepsilon_2+2\varepsilon_1}\mathcal{E}_0 \tag{3.138}$$

whereby the secondary potential reads

$$\Phi^{\mathrm{s}}(\mathbf{r})=\begin{cases}\dfrac{\varepsilon_2-\varepsilon_1}{\varepsilon_2+2\varepsilon_1}\dfrac{a^3}{r^2}\mathcal{E}_0\cos\vartheta, & r>a\\ -\dfrac{3\varepsilon_1}{\varepsilon_2+2\varepsilon_1}\mathcal{E}_0r\cos\vartheta, & r<a.\end{cases} \tag{3.139}$$

We may cast the expression of $\Phi^s(\mathbf{r})$ for $\mathbf{r} \in \mathbb{R}^3 \setminus B[0, a]$ as

$$\Phi^s(\mathbf{r}) = \underbrace{4\pi\varepsilon_1 a^3 \frac{\varepsilon_2 - \varepsilon_1}{\varepsilon_2 + 2\varepsilon_1}\mathcal{E}_0\hat{\mathbf{z}}}_{\text{dipole moment}} \cdot \frac{\hat{\mathbf{r}}}{4\pi\varepsilon_1 r^2} = \frac{p_{eq}\cos\vartheta}{4\pi\varepsilon_1 r^2}, \qquad r > a \tag{3.140}$$

which by comparison with (2.39) shows that the effect of the sphere is equivalent to the potential generated by an elementary dipole which has moment

$$\mathbf{p}_{eq} = 4\pi\varepsilon_1 a^3 \frac{\varepsilon_2 - \varepsilon_1}{\varepsilon_2 + 2\varepsilon_1}\mathcal{E}_0\hat{\mathbf{z}} \tag{3.141}$$

and is located in the origin. This dipole is oriented along $\hat{\mathbf{z}}$ if the permittivity of the sphere is larger than the permittivity of the background medium. The orientation of $\mathbf{p}_{eq}$ is reversed if $\varepsilon_2 < \varepsilon_1$, as would be the case, e.g., for a hollow spherical cavity bored in a dielectric medium.

The secondary and also total potential within the sphere reads

$$\Phi(\mathbf{r}) = \Phi^s(\mathbf{r}) = -\frac{3\varepsilon_1}{\varepsilon_2 + 2\varepsilon_1}\mathcal{E}_0 r\cos\vartheta \tag{3.142}$$

whence the electric field for $r < a$ is found to be

$$\boldsymbol{\mathcal{E}}^s(\mathbf{r}) = -\nabla\Phi^s(\mathbf{r}) = \frac{3\varepsilon_1}{\varepsilon_2 + 2\varepsilon_1}\mathcal{E}_0\nabla z = \frac{3\varepsilon_1}{\varepsilon_2 + 2\varepsilon_1}\mathcal{E}_0\hat{\mathbf{z}} \tag{3.143}$$

having made use of (A.17) and computed the gradient in Cartesian coordinates. If the permittivity of the sphere (ε_2) is larger than that of the background medium (ε_1) we have

$$3\varepsilon_1 = \varepsilon_1 + 2\varepsilon_1 < \varepsilon_2 + 2\varepsilon_1 \tag{3.144}$$

whence we find

$$\frac{3\varepsilon_1}{\varepsilon_2 + 2\varepsilon_1} < 1 \tag{3.145}$$

whereby we see that the electric field within the sphere is *less strong* than the impressed field; the converse is true if $\varepsilon_1 > \varepsilon_2$. To explain this phenomenon we need to understand what happens to the bound charges within a dielectric medium when they are exposed to an electric field. This topic is addressed in Section 3.7 in general terms and in Example 3.6 for a dielectric sphere.

(End of Example 3.4)

3.5.2 *Circular cylindrical coordinates*

It is convenient to represent the Laplace operator in a system of circular cylindrical coordinates (ρ, φ, z) (Appendix A.1), if the region of interest V is either the whole space or it possesses cylindrical symmetry and is bounded by surfaces which are (parts of) circular cylinders, half-planes and planes. Thanks to this choice the boundary conditions are more easily enforced. When a closed-form solution to (3.68) is not feasible we may try and express the potential as a series of known functions, viz.,

$$\Phi(\mathbf{r}) = \sum_\nu \Phi_\nu \Psi_\nu(\rho, \varphi, z), \qquad \mathbf{r} \in V \tag{3.146}$$

where the constant coefficients Φ_ν must be determined so as to satisfy the boundary conditions required on ∂V. As was the case for the expansion (3.69), the symbol ν signifies that we linearly combine different, possibly infinitely many terms.

In particular, as a set of expansion functions $\{\Psi_\nu(\mathbf{r})\}$ we may adopt the fundamental solutions of (3.68) with the operator expressed in circular cylindrical coordinates ρ, φ and z, namely,

$$\frac{1}{\rho}\frac{\partial}{\partial \rho}\left(\rho \frac{\partial \Psi}{\partial \rho}\right) + \frac{1}{\rho^2}\frac{\partial^2 \Psi}{\partial \rho^2} + \frac{\partial^2 \Psi}{\partial z^2} = 0 \tag{3.147}$$

to be solved in the region V.

To proceed we apply the *separation argument* or *separation of variables* [1, Chapter 3], [8, Appendix A.4]. Accordingly, we look for solutions in the form of a product of three functions of one variable

$$\Psi(\mathbf{r}) := B(\rho)F(\varphi)Z(z) \tag{3.148}$$

whereby the dependence on ρ, φ and z is effectively separated. With this position we can split (3.147) into three *uncoupled* equations. First of all, we substitute (3.148) into (3.147), divide through by Ψ and rearrange terms to get

$$\underbrace{\frac{1}{\rho B}\frac{\mathrm{d}}{\mathrm{d}\rho}\left(\rho \frac{\mathrm{d}B}{\mathrm{d}\rho}\right) + \frac{1}{\rho^2 F}\frac{\mathrm{d}^2 F}{\mathrm{d}\varphi^2}}_{\text{depends on } (\rho,\varphi)} = \underbrace{-\frac{1}{Z}\frac{\mathrm{d}^2 Z}{\mathrm{d}z^2}}_{\text{depends on } z} \tag{3.149}$$

where the quantity in the left-hand side is equal to some function of two variables (ρ, φ) but the right-hand side is a function of z only. Evidently, this equation is true if we set both sides equal to the *same* arbitrary constant $-k_z^2$ which we do not demand to be real at all. More importantly, there are no limitations on the values $-k_z^2$ may take on, unless the volume V is bounded in the z direction, in which case $-k_z^2$ will be restricted to a discrete set of *eigenvalues*. Notice that for the correct dimensional balance of (3.149), k_z carries the physical dimension of the inverse of a length, similarly to a wavenumber, because the coordinates ρ and z represent distances.

We obtain two separate equations

$$\frac{1}{\rho B}\frac{\mathrm{d}}{\mathrm{d}\rho}\left(\rho \frac{\mathrm{d}B}{\mathrm{d}\rho}\right) + \frac{1}{\rho^2 F}\frac{\mathrm{d}^2 F}{\mathrm{d}\varphi^2} = -k_z^2 \tag{3.150}$$

$$\frac{1}{Z}\frac{\mathrm{d}^2 Z}{\mathrm{d}z^2} = k_z^2 \tag{3.151}$$

and we attempt to split the first one likewise. If we multiply through by ρ^2 and rearrange terms, we have

$$\underbrace{\frac{\rho}{B}\frac{\mathrm{d}}{\mathrm{d}\rho}\left(\rho \frac{\mathrm{d}B}{\mathrm{d}\rho}\right) + \rho^2 k_z^2}_{\text{depends on } \rho} = \underbrace{-\frac{1}{F}\frac{\mathrm{d}^2 F}{\mathrm{d}\varphi^2}}_{\text{depends on } \varphi} \tag{3.152}$$

and this equation is true if we set both sides equal to the *same* arbitrary constant m^2, where m is not necessarily a whole number. Since the range of the variable φ is intrinsically limited, m may take on discrete, though infinite values. Thus, we obtain

$$\frac{\rho}{B}\frac{\mathrm{d}}{\mathrm{d}\rho}\left(\rho \frac{\mathrm{d}B}{\mathrm{d}\rho}\right) + \rho^2 k_z^2 = m^2 \tag{3.153}$$

$$-\frac{1}{F}\frac{\mathrm{d}^2F}{\mathrm{d}\varphi^2} = m^2 \tag{3.154}$$

whereby (3.147) is fully separated into three independent equations. We write them down neatly for the subsequent solution

$$\frac{\mathrm{d}^2}{\mathrm{d}z^2}Z(k_z) - k_z^2 Z(k_z) = 0 \tag{3.155}$$

$$\frac{\mathrm{d}^2F_m}{\mathrm{d}\varphi^2} + m^2F_m = 0 \tag{3.156}$$

$$\frac{1}{\rho}\frac{\mathrm{d}}{\mathrm{d}\rho}\left[\rho\frac{\mathrm{d}}{\mathrm{d}\rho}B_m(k_z)\right] + \left(k_z^2 - \frac{m^2}{\rho^2}\right)B_m(k_z) = 0 \tag{3.157}$$

where we have emphasized the dependence on the separation constants k_z and m.

We have come up with three simpler ordinary differential equations. Any product of the form (3.148) which is constructed with solutions to (3.155)-(3.157) may be taken as a member of the set $\{\Psi_\nu\}$. At this stage we may state that the index ν is a shorthand for the pair (k_z, m) of separation constants. The specific values of (k_z, m) and the range in which they are to be found depend on the shape of the region V as well as the boundary conditions imposed on the potential for $\mathbf{r} \in \partial V$. Indeed, the boundary conditions translate into requirements on the solutions $B_m(k_z)$, F_m and $Z(k_z)$ and these, in turn, limit the sets of admissible values for k_z and m. When complemented with boundary conditions, (3.155)-(3.157) become one-dimensional *eigenvalue problems* which, as a result, have non-trivial solutions (*eigenfunctions*) only for selected values of k_z and m (Appendix D.7). We go on to examine each equation in details.

We start off with (3.155), which constitutes a special case of the Stürm-Liouville problem [8, Appendix A.4], [20, Section 3.2], [21, Section 6.3]. If the volume V is bounded along z by two planes defined by $z = z_1$ and $z = z_2$ with $z_1 < z_2$, we may assume periodicity, i.e.,

$$Z(z_1; k_z) = Z(z_2; k_z), \qquad \left.\frac{\mathrm{d}Z(z;k_z)}{\mathrm{d}z}\right|_{z_1} = \left.\frac{\mathrm{d}Z(z;k_z)}{\mathrm{d}z}\right|_{z_2} \tag{3.158}$$

or homogeneous Dirichlet conditions

$$Z(z_1; k_z) = Z(z_2; k_z) = 0 \tag{3.159}$$

or homogeneous Neumann conditions

$$\left.\frac{\mathrm{d}Z(z;k_z)}{\mathrm{d}z}\right|_{z_1} = \left.\frac{\mathrm{d}Z(z;k_z)}{\mathrm{d}z}\right|_{z_2} = 0 \tag{3.160}$$

or a suitable combination of the last two. Under these assumptions we may show that k_z^2 must be *negative*. To proceed we multiply (3.155) through by $Z^*(z; k_z)$ and integrate from z_1 to z_2 to arrive at

$$\begin{aligned} 0 &= \int_{z_1}^{z_2} \mathrm{d}z\, Z^*(z;k_z)\frac{\mathrm{d}^2Z(z;k_z)}{\mathrm{d}z^2} - k_z^2\int_{z_1}^{z_2}\mathrm{d}z\, Z^*(z;k_z)Z(z;k_z) \\ &= \left[Z^*(z;k_z)\frac{\mathrm{d}Z(z;k_z)}{\mathrm{d}z}\right]_{z_1}^{z_2} - \int_{z_1}^{z_2}\mathrm{d}z\left|\frac{\mathrm{d}Z(z;k_z)}{\mathrm{d}z}\right|^2 - k_z^2\int_{z_1}^{z_2}\mathrm{d}z\,|Z(z;k_z)|^2 \end{aligned} \tag{3.161}$$

having integrated by parts [22]. For any choice of boundary conditions mentioned above, the finite increment of the quantity $Z^*(z;k_z)\mathrm{d}Z(z;k_z)/\mathrm{d}z$ over the interval $I := [z_1, z_2]$ vanishes, and we end up with

$$-k_z^2 = \frac{\displaystyle\int_{z_1}^{z_2} \mathrm{d}z \left|\frac{\mathrm{d}Z(z;k_z)}{\mathrm{d}z}\right|^2}{\displaystyle\int_{z_1}^{z_2} \mathrm{d}z \, |Z(z;k_z)|^2} = \frac{\left\|\dfrac{\mathrm{d}Z(z;k_z)}{\mathrm{d}z}\right\|^2_{\mathcal{L}^2(I)}}{\|Z(z;k_z)\|^2_{\mathcal{L}^2(I)}} \tag{3.162}$$

where the *Rayleigh quotient* in the right-hand side is evidently a non-negative quantity, and this proves that $k_z^2 \leqslant 0$. The possibility of a function $Z(z;k_z)$ null everywhere — in which case $\|Z(z;k_z)\|_{\mathcal{L}^2(I)} = 0$ — is forbidden for it contradicts the definition of eigenfunction. Notice that the condition $k_z^2 \leqslant 0$ implies that k_z is imaginary.

Any two eigenfunctions associated with two distinct eigenvalues k_z and k_z' are *orthogonal* in the sense that the inner product over I vanishes. Indeed, by integrating from z_1 to z_2 the identity

$$\begin{aligned}\frac{\mathrm{d}}{\mathrm{d}z}\left[Z^*(z;k_z')\frac{\mathrm{d}Z(z;k_z)}{\mathrm{d}z} - Z(z;k_z)\frac{\mathrm{d}Z^*(z;k_z')}{\mathrm{d}z}\right] &= \\ = Z^*(z;k_z')\frac{\mathrm{d}^2Z(z;k_z)}{\mathrm{d}z^2} - Z(z;k_z)\frac{\mathrm{d}^2Z^*(z;k_z')}{\mathrm{d}z^2} \\ = \left[k_z^2 - (k_z'^*)^2\right] Z(z;k_z)Z^*(z;k_z')\end{aligned} \tag{3.163}$$

we find

$$0 = \left[k_z^2 - (k_z'^*)^2\right] \int_{z_1}^{z_2} \mathrm{d}z \, Z(z;k_z)Z^*(z;k_z') = \left[k_z^2 - (k_z'^*)^2\right] \left(Z(k_z), Z(k_z')\right)_I \tag{3.164}$$

on account of any type of boundary conditions discussed above. The inner product must vanish, since $k_z \neq k_z'$ by assumption.

Any function of the form

$$Z(z;k_z) = c_1(k_z)\mathrm{e}^{k_z z} + c_2(k_z)\mathrm{e}^{-k_z z} \tag{3.165}$$

with $c_1(k_z)$ and $c_2(k_z)$ being two constants, solves (3.155). In general, the separation constant k_z may take on any value if the region V is unbounded. Otherwise, k_z and either one of $c_1(k_z)$ and $c_2(k_z)$ are determined by the boundary conditions. The remaining constant may be computed so as to ensure that the eigenfunction has unitary norm. More specifically, we have

$$Z(z;k_z) = c(k_z)\mathrm{e}^{k_z z}, \qquad z \in [z_1, z_2], \qquad k_z = -\frac{2\pi \mathrm{j} p}{z_2 - z_1}, \qquad p \in \mathbb{N} \tag{3.166}$$

for periodic boundary conditions,

$$Z(z;k_z) = c(k_z)\sinh[k_z(z - z_1)], \qquad z \in [z_1, z_2], \qquad k_z = \frac{-\mathrm{j} p\pi}{z_2 - z_1}, \qquad p \in \mathbb{N} \setminus \{0\} \tag{3.167}$$

for homogeneous Dirichlet boundary conditions and

$$Z(z;k_z) = c(k_z)\cosh[k_z(\varphi - \varphi_1)], \qquad z \in [z_1, z_2], \qquad k_z = \frac{-\mathrm{j} p\pi}{z_2 - z_1}, \qquad p \in \mathbb{N} \tag{3.168}$$

for homogeneous Neumann boundary conditions. We must discard $k_z = 0$ for the problem with homogeneous Dirichlet boundary conditions, because otherwise the corresponding eigenfunction would be null everywhere. Nonetheless, the special case $k_z = 0$ of (3.155) has the more general solution

$$Z(z;0) = b_0 z + c_0, \qquad z \in [z_1, z_2] \tag{3.169}$$

which is acceptable if the required boundary conditions lead to non-zero values for at least either one of the constants b_0 and c_0. For instance, inhomogeneous Dirichlet conditions give

$$Z(z;0) = \frac{(z - z_1)Z_2 - (z - z_2)Z_1}{z_2 - z_1}, \qquad z \in [z_1, z_2] \tag{3.170}$$

where Z_1 and Z_2 are the expected values of $Z(z;0)$ in z_1 and z_2.

The second equation of concern (3.156) is the usual harmonic equation and also the same as (3.78), which we extensively investigated in Section 3.5.1. Therefore, we do not repeat the analysis here, though we remark that the boundary conditions along φ determine the discrete values m^2 may take on.

Lastly, we examine (3.157) which is referred to as the *Bessel equation* [23–25,32,33] and is also a special instance of the Stürm-Liouville problem [8, Appendix A.4], [20, Section 3.2], [21, Section 6.3]. If the volume V is limited along the radial direction by two cylindrical surfaces defined by $\rho = \rho_1$ and $\rho = \rho_2$ with $0 < \rho_1 < \rho_2 < +\infty$, we may impose homogeneous Dirichlet or Neumann conditions

$$B_m(\rho_1; k_z) = B_m(\rho_2; k_z) = 0 \tag{3.171}$$

$$\rho_1 \frac{\mathrm{d}B_m(\rho; k_z)}{\mathrm{d}\rho}\bigg|_{\rho_1} = \rho_2 \frac{\mathrm{d}B_m(\rho; k_z)}{\mathrm{d}\rho}\bigg|_{\rho_2} = 0 \tag{3.172}$$

or a combination of the two. Although we may solve (3.157) independently of (3.155) and (3.156), the latter two already set the permissible values of k_z and m, respectively. Therefore, (3.157) may not be an eigenvalue problem per se. In fact, the situation is no different than the one described at the end of Section 3.5.1, namely, if we try to enforce homogeneous boundary conditions along the three coordinates, we end up with a solution which is zero everywhere in V. As already argued, this result is correct.

On the other hand, we may show that the eigenvalue k_z^2 is real and positive if either (3.171) or (3.172) is viable. We multiply (3.157) through by $\rho B_m^*(\rho; k_z)$ and integrate over the interval $I := [\rho_1, \rho_2]$, viz.,

$$\begin{aligned} 0 &= \int_{\rho_1}^{\rho_2} \mathrm{d}\rho\, B_m^*(\rho; k_z) \frac{\mathrm{d}}{\mathrm{d}\rho}\left[\rho \frac{\mathrm{d}B_m(\rho; k_z)}{\mathrm{d}\rho}\right] + \int_{\rho_1}^{\rho_2} \mathrm{d}\rho \left(k_z^2 \rho - \frac{m^2}{\rho}\right) |B_m(\rho; k_z)|^2 \\ &= -\int_{\rho_1}^{\rho_2} \mathrm{d}\rho\, \rho \left|\frac{\mathrm{d}B_m(\rho; k_z)}{\mathrm{d}\rho}\right|^2 + k_z^2 \int_{\rho_1}^{\rho_2} \mathrm{d}\rho\, \rho\, |B_m(\rho; k_z)|^2 - m^2 \int_{\rho_1}^{\rho_2} \mathrm{d}\rho \frac{|B_m(\rho; k_z)|^2}{\rho} \end{aligned} \tag{3.173}$$

where we have integrated by parts [22] and applied the boundary conditions (3.171) or (3.172). Solving with respect to k_z^2 yields

$$k_z^2 = \frac{\displaystyle\int_{\rho_1}^{\rho_2} \mathrm{d}\rho\,\rho \left|\frac{\mathrm{d}B_m(\rho;k_z)}{\mathrm{d}\rho}\right|^2 + m^2 \int_{\rho_1}^{\rho_2} \mathrm{d}\rho\,\frac{|B_m(\rho;k_z)|^2}{\rho}}{\displaystyle\int_{\rho_1}^{\rho_2} \mathrm{d}\rho\,\rho\,|B_m(\rho;k_z)|^2} \tag{3.174}$$

and this proves our assertion, for the right-hand side is a non-negative quantity. We notice in passing that this conclusion is just the opposite of that found while studying (3.155), which was likewise based on homogeneous conditions. There is nothing wrong, and this contradiction signals again that the Laplace equation is solved by $\Phi(\mathbf{r}) = 0$ if the potential is to meet homogeneous conditions all over the boundary of the region of interest.

Next, we show that two eigenfunctions associated with different eigenvalues k_z and k_z' and the same m are orthogonal over the interval I. We begin with the identity

$$\begin{aligned}
&\frac{\mathrm{d}}{\mathrm{d}\rho}\left[\rho B_m^*(\rho;k_z')\frac{\mathrm{d}B_m(\rho;k_z)}{\mathrm{d}\rho}\right] - \frac{\mathrm{d}}{\mathrm{d}\rho}\left[\rho B_m(\rho;k_z)\frac{\mathrm{d}B_m^*(\rho;k_z')}{\mathrm{d}\rho}\right] = \\
&= B_m^*(\rho;k_z')\frac{\mathrm{d}}{\mathrm{d}\rho}\left[\rho\frac{\mathrm{d}B_m(\rho;k_z)}{\mathrm{d}\rho}\right] - B_m(\rho;k_z)\frac{\mathrm{d}}{\mathrm{d}\rho}\left[\rho\frac{\mathrm{d}B_m^*(\rho;k_z')}{\mathrm{d}\rho}\right] \\
&= -(k_z^2 - k_z'^2)\rho B_m(\rho;k_z)B_m^*(\rho;k_z')
\end{aligned} \tag{3.175}$$

which integrated from ρ_1 to ρ_2 yields

$$0 = (k_z^2 - k_z'^2)\int_{\rho_1}^{\rho_2} \mathrm{d}\rho\,\rho B_m(\rho;k_z)B_m^*(\rho;k_z') = (k_z^2 - k_z'^2)\left(B_m(\rho;k_z), B_m^*(\rho;k_z')\right)_I \tag{3.176}$$

on account of either one of (3.171) and (3.172). Since k_z and k_z' differ by assumption, then the inner product must vanish.

To obtain the general solutions to (3.157) we cast it into the standard form [32]

$$\frac{1}{\xi}\frac{\mathrm{d}}{\mathrm{d}\xi}\left(\xi\frac{\mathrm{d}B_m}{\mathrm{d}\xi}\right) + \left(1 - \frac{m^2}{\xi^2}\right)B_m(\xi) = 0 \tag{3.177}$$

by letting $\xi = k_z\rho$ be the independent variable. If m is not an integer, (3.177) has two linearly independent solutions indicated with $J_m(\xi)$ and $J_{-m}(\xi)$, called *Bessel functions of the first kind* and order m [24, Chapter 9]. However, when m is an integer, $J_m(\xi)$ and $J_{-m}(\xi)$ become proportional to one another, and hence a second solution is customarily written as

$$Y_m(\xi) = \lim_{\alpha\to m} \frac{J_\alpha(\xi)\cos(\pi\alpha) - J_{-\alpha}(\xi)}{\sin(\pi\alpha)}, \qquad m \in \mathbb{Z} \tag{3.178}$$

which is referred to as *Neumann* or *Weber function* or also *Bessel function of the second kind* and order m [1]. We recall that $Y_m(\xi)$ is singular for $\xi = 0$, whereas $J_m(\xi)$ is regular for $\xi \geqslant 0$. The Bessel functions of order $m = 0$ are plotted in Figure 3.7a.

In light of the position $\xi = k_z\rho$ any expression of the form

$$B_m(\rho;k_z) = c_{m1}(k_z)J_m(k_z\rho) + c_{m2}(k_z)Y_m(k_z\rho) \tag{3.179}$$

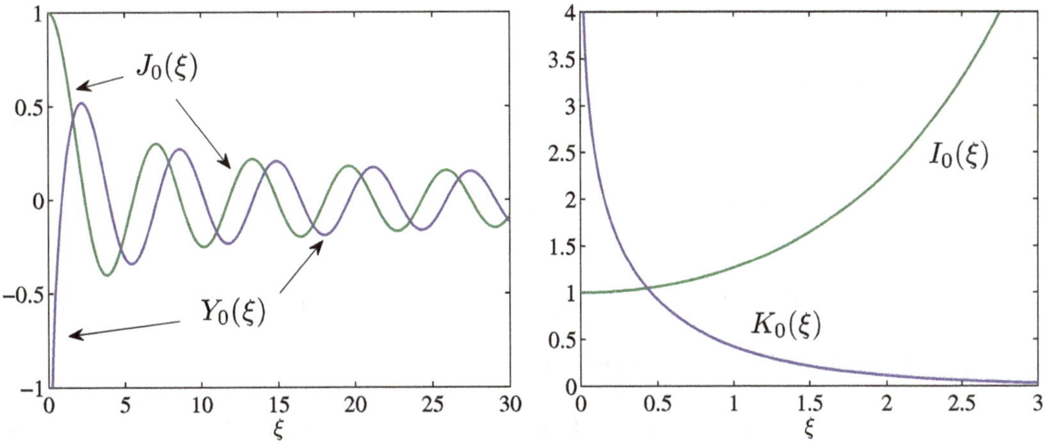

(a) $J_0(\xi)$ *and* $Y_0(\xi)$ (b) $I_0(\xi)$ *and* $K_0(\xi)$

Figure 3.7 Graphical representation of Bessel functions for real values of the argument.

solves (3.157). However, the eigenvalue k_z and either one of the constants $c_{m1}(k_z)$ and $c_{m2}(k_z)$ are constrained by the boundary conditions along the radial coordinate, but it is not possible to determine the eigenvalues analytically, not least because $J_m(\xi)$ and $Y_m(\xi)$ are available as series. For example, for Dirichlet boundary conditions the eigenvalues k_z follow from the transcendental equation

$$J_m(k_z\rho_1)Y_m(k_z\rho_2) - Y_m(k_z\rho_1)J_m(k_z\rho_2) = 0 \tag{3.180}$$

and a similar relation apply for Neumann conditions. We may choose the remaining constant in such a way that the eigenfunctions have unitary norm.

The usual boundary conditions do not apply if the region V is the whole space ($\rho_2 \to +\infty$) or it encompasses (parts of) the z-axis ($\rho_1 = 0$). If the solution we wish to represent with the set $\{\Psi_\nu\}$ must be regular for $\rho = 0$, then we cannot use the function $Y_m(k_z\rho)$ because of the singularity in the origin. Accordingly, we set $c_{m2}(k_z) = 0$ and the solution to (3.157) takes on the form

$$B_m(\rho; k_z) = c_m(k_z)J_m(k_z\rho), \qquad \rho \in [0, +\infty[\tag{3.181}$$

with m specified by the conditions along φ. Since both $J_m(k_z\rho)$ and $Y_m(k_z\rho)$ fall off as the inverse of $\sqrt{\rho}$ for $\rho \to +\infty$, the Bessel functions are intrinsically regular at infinity (see Appendix H.6).

If the separation constant k_z happens to be imaginary because of the boundary conditions along z, the argument of the Bessel functions becomes imaginary, too. Thus, it is customary to write the solutions along ρ by introducing alternative functions with real argument as follows

$$B_m(\rho; k_z) = c_{m1}(k_z)I_m(jk_z\rho) + c_{m2}(k_z)K_m(jk_z\rho) \tag{3.182}$$

where I_m and K_m are called *modified Bessel functions* of the first and second kind, respectively, and given in (H.140) and (H.141) [24, 32]. The defining equation follows from (3.157) by letting $\xi = \mathrm{j}\, k_z\rho \in \mathbb{R}$. The functions of order $m = 0$ are shown in Figure 3.7b.

The special case $k_z = 0$ turns (3.157) into [cf. (3.80)]

$$\frac{1}{\rho}\frac{\mathrm{d}}{\mathrm{d}\rho}\left(\rho\frac{\mathrm{d}}{\mathrm{d}\rho}B_m\right) - \frac{m^2}{\rho^2}B_m = 0 \tag{3.183}$$

which is solved by

$$B_m(\rho;0) = \begin{cases} c_{01}\log\rho + c_{02} & m = 0 \\ c_{m1}\rho^m + \dfrac{c_{m2}}{\rho^m} & m \neq 0. \end{cases} \tag{3.184}$$

Expansion functions well-suited to represent the potential outside an infinitely long circular cylinder of radius a are

$$\Psi_m(\mathbf{r};k_z) := \mathrm{e}^{k_z z - \mathrm{j}\, m\varphi} K_m(\mathrm{j}\, k_z\rho), \qquad k_z = -\mathrm{j}\,|k_z|, \qquad m \in \mathbb{N}\setminus\{0\} \tag{3.185}$$

whereas for the inside of the same cylinder we can use

$$\Psi_m(\mathbf{r};k_z) := \mathrm{e}^{k_z z - \mathrm{j}\, m\varphi} I_m(\mathrm{j}\, k_z\rho), \qquad k_z = -\mathrm{j}\,|k_z|, \qquad m \in \mathbb{N}\setminus\{0\}. \tag{3.186}$$

3.6 Multipole expansion of the scalar potential

In Section 2.2 we speculated the asymptotic behavior of the potential $\Phi(\mathbf{r})$ in the form (2.23) by observing that the electrostatic energy W_e stored in the whole space must be finite, if the field is produced by a charge density $\varrho(\mathbf{r})$ confined in a bounded volume $V_\varrho \subset \mathbb{R}^3$. We also found out that (2.23) is satisfied by the potential of a point charge (2.29) and, more importantly, by the volume potential $\mathcal{V}(\mathbf{r})$ in (2.192). Furthermore, examination of the potential (2.39) produced by an elementary static dipole shows that $\Phi(\mathbf{r})$ decays with the inverse square of the distance away from the source, and so does the double-layer potential $\mathcal{W}(\mathbf{r})$ in (2.239).

On the whole, aside from the immaterial constant Φ_∞ — which we may set to zero — it would appear that the functional form of $\Phi(\mathbf{r})$ for $|\mathbf{r}| \to +\infty$ cannot contains terms other than inverse powers of $|\mathbf{r}|$. As a matter of fact, positive powers of $|\mathbf{r}|$ for $|\mathbf{r}| \to +\infty$ may be justified only if we admit the presence of an infinite amount of charge in an unbounded region of space (see Examples 3.3 and 3.4).

Therefore, our next goal is to derive an approximate expression of $\Phi(\mathbf{r})$ valid for $|\mathbf{r}| \to +\infty$ given a general, though finite, distribution of charges with density $\varrho(\mathbf{r})$ for $\mathbf{r} \in V_\varrho \subset \mathbb{R}^3$. The result is called the *multipole expansion* of $\Phi(\mathbf{r})$ [1–3, 8, 14, 15, 29, 34, 35], [4, Chapter 4] and essentially follows from the expansion of the Green function (2.131). The rationale for this name is that $\Phi(\mathbf{r})$ can be construed as the superposition of potentials generated by groups of charges (or *poles*) and, in particular, by a *single pole* to the first order, a *dipole* to the second and a *quadrupole* to the third order. A quadrupole, as the name suggests, is a group of four charges, two positive and two negative; alternatively, an elementary quadrupole can also be seen as a pair of infinitely close parallel dipoles with opposite orientations. The next moment in line is an *octupole* which is a pair of infinitely close quadrupoles.

3.6.1 *Taylor series of the Green function*

It is relatively easy to determine the first few terms of the multiple expansion by computing the truncated Taylor series of the Green function around the source point $\mathbf{r}' = 0$. Unfortunately, the derivation of the higher order terms becomes increasingly involved and cumbersome. A better and systematic approach consists of writing the Green function as superposition of fundamental solutions to the Laplace equation in spherical polar coordinates (see Section 3.5.1). Either way, by inserting the expansion into the scalar potential (2.169) we arrive at the desired asymptotic behavior of $\Phi(\mathbf{r})$.

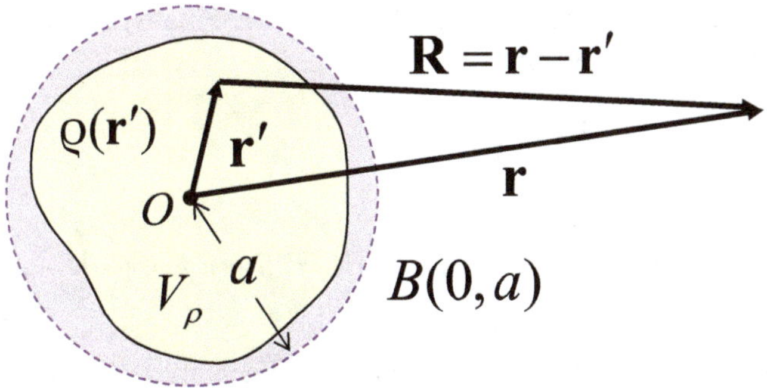

Figure 3.8 *The asymptotic expansion* (3.187) *holds true for points outside a ball* $B(0, a) \supset V_\varrho$.

To proceed with the first strategy, we consider the ball $B(a, 0)$ with the radius a chosen large enough so that $V_\varrho \subset B(a, 0)$, as is suggested in Figure 3.8. We observe that $1/|\mathbf{r} - \mathbf{r}'|$ — as a function of $\mathbf{r}'$ — can be written as

$$\frac{1}{|\mathbf{r}-\mathbf{r}'|} = \frac{1}{|\mathbf{r}|} + \nabla' \frac{1}{|\mathbf{r}-\mathbf{r}'|}\bigg|_{\mathbf{r}'=0} \cdot \mathbf{r}' + \frac{1}{2}\mathbf{r}' \cdot \nabla'\nabla' \frac{1}{|\mathbf{r}-\mathbf{r}'|}\bigg|_{\mathbf{r}'=0} \cdot \mathbf{r}' + o\left(|\mathbf{r}'|^2\right), \qquad |\mathbf{r}'| \to 0 \tag{3.187}$$

and this holds true for observation points $\mathbf{r} \notin B(a, 0)$. When considered as a function of $\mathbf{r}$, the expansion above becomes increasingly accurate as $|\mathbf{r}| \to +\infty$.

For ease of manipulation and to lighten the notation we let $\mathbf{R} = \mathbf{r} - \mathbf{r}'$, $|\mathbf{r}| = r$, $|\mathbf{r}'| = r'$, and compute the following derivatives with respect to $\mathbf{r}'$

$$\nabla'|\mathbf{r}-\mathbf{r}'| = -\frac{\mathbf{r}-\mathbf{r}'}{|\mathbf{r}-\mathbf{r}'|} = -\hat{\mathbf{R}} \tag{3.188}$$

$$\nabla'\frac{1}{|\mathbf{r}-\mathbf{r}'|} = \frac{\mathbf{r}-\mathbf{r}'}{|\mathbf{r}-\mathbf{r}'|^3} = \frac{\mathbf{R}}{R^3} \tag{3.189}$$

$$\nabla'\frac{1}{|\mathbf{r}-\mathbf{r}'|^3} = \nabla'\frac{1}{R^3} = \frac{3}{R^4}\hat{\mathbf{R}} = \frac{3}{R^5}\mathbf{R} \tag{3.190}$$

$$\nabla'(\mathbf{r}-\mathbf{r}') = -\hat{\mathbf{x}}\hat{\mathbf{x}} - \hat{\mathbf{y}}\hat{\mathbf{y}} - \hat{\mathbf{z}}\hat{\mathbf{z}} = -\overline{\mathfrak{I}} \tag{3.191}$$

$$\nabla'\nabla'\frac{1}{|\mathbf{r}-\mathbf{r}'|} = \nabla'\frac{\mathbf{R}}{R^3} = \left(\nabla'\frac{1}{R^3}\right)\mathbf{R} + \frac{1}{R^3}\nabla'\mathbf{R} = \frac{3\mathbf{R}\mathbf{R} - R^2\overline{\mathfrak{I}}}{R^5} \tag{3.192}$$

whereby we find

$$\nabla'\frac{1}{|\mathbf{r}-\mathbf{r}'|}\bigg|_{\mathbf{r}'=0} \cdot \mathbf{r}' = \frac{\hat{\mathbf{r}}}{r^2} \cdot \mathbf{r}' \tag{3.193}$$

$$\begin{aligned}\mathbf{r}' \cdot \nabla'\nabla'\frac{1}{|\mathbf{r}-\mathbf{r}'|}\bigg|_{\mathbf{r}'=0} \cdot \mathbf{r}' &= \mathbf{r}' \cdot \frac{3\,\hat{\mathbf{r}}\hat{\mathbf{r}} - \overline{\mathfrak{I}}}{r^3} \cdot \mathbf{r}' = \frac{3\,(\mathbf{r}' \cdot \hat{\mathbf{r}})(\hat{\mathbf{r}} \cdot \mathbf{r}') - r'^2\hat{\mathbf{r}} \cdot \hat{\mathbf{r}}}{r^3} \\ &= \hat{\mathbf{r}} \cdot \frac{3\,\mathbf{r}'\mathbf{r}' - r'^2\overline{\mathfrak{I}}}{r^3} \cdot \hat{\mathbf{r}}.\end{aligned} \tag{3.194}$$

With these results the asymptotic expansion (3.187) reads

$$\frac{1}{|\mathbf{r}-\mathbf{r}'|} = \frac{1}{r} + \frac{\hat{\mathbf{r}}}{r^2} \cdot \mathbf{r}' + \hat{\mathbf{r}} \cdot \frac{3\,\mathbf{r}'\mathbf{r}' - r'^2\overline{\mathfrak{I}}}{2r^3} \cdot \hat{\mathbf{r}} + o\left(r'^2\right), \qquad r' \to 0 \tag{3.195}$$

which already shows the expected behavior as a function of r. Finally, we insert (3.195) into (2.169) and we rearrange the contributions to get

$$\Phi(\mathbf{r}) = \frac{1}{4\pi\varepsilon r}\int_{V_\varrho} \mathrm{d}V'\varrho(\mathbf{r}') + \frac{\hat{\mathbf{r}}}{4\pi\varepsilon r^2}\cdot\int_{V_\varrho} \mathrm{d}V'\mathbf{r}'\varrho(\mathbf{r}') + \frac{1}{2}\frac{\hat{\mathbf{r}}}{4\pi\varepsilon r^3}\cdot\int_{V_\varrho} \mathrm{d}V'(3\,\mathbf{r}'\mathbf{r}' - r'^2\overline{\mathfrak{I}})\varrho(\mathbf{r}')\cdot\hat{\mathbf{r}} + o\left(\frac{1}{r^3}\right), \quad r\to+\infty \tag{3.196}$$

which is the desired multipole expansion of the potential. Each integral in this expansion is referred to as a *moment*, and the first three of them are

- the total charge (a scalar) in V_ϱ

$$Q := \int_{V_\varrho} \mathrm{d}V'\varrho(\mathbf{r}') \tag{3.197}$$

- the dipole moment (a vector) associated with the charge in V_ϱ

$$\mathbf{p} := \int_{V_\varrho} \mathrm{d}V'\mathbf{r}'\varrho(\mathbf{r}') \tag{3.198}$$

- the quadrupole moment (a dyadic, physical dimension Cm^2) associated with the charge in V_ϱ

$$\overline{\mathfrak{Q}} := \int_{V_\varrho} \mathrm{d}V'(3\,\mathbf{r}'\mathbf{r}' - r'^2\overline{\mathfrak{I}})\varrho(\mathbf{r}'). \tag{3.199}$$

The total charge is evidently invariant under translations and rotations of the coordinates. The dipole moment is invariant under translations, say, $\mathbf{r}' = \mathbf{r}'' + \mathbf{u}$, if the total charge is null

$$\mathbf{p}_{O'} := \int_{V_\varrho} \mathrm{d}V''\,(\mathbf{r}'' + \mathbf{u})\varrho(\mathbf{r}'' + \mathbf{u}) = \mathbf{p}_{O''} + Q\mathbf{u} \tag{3.200}$$

where $\mathbf{p}_{O'}$ and $\mathbf{p}_{O''}$ are the dipole moments computed with respect to the primed and double-primed systems of coordinates. The quadrupole moment is invariant under translations, if the total charge and the dipole moment are both null

$$\begin{aligned}\overline{\mathfrak{Q}}_{O'} &:= \int_{V_\varrho} \mathrm{d}V''\,[3(\mathbf{r}'' + \mathbf{u})(\mathbf{r}'' + \mathbf{u}) - |\mathbf{r}'' + \mathbf{u}|^2\overline{\mathfrak{I}}]\varrho(\mathbf{r}'' + \mathbf{u}) \\ &= \overline{\mathfrak{Q}}_{O''} + 3\mathbf{p}_{O''}\mathbf{u} + 3\mathbf{u}\mathbf{p}_{O''} - 2\mathbf{p}_{O''}\cdot\mathbf{u}\overline{\mathfrak{I}} + Q(3\mathbf{u}\mathbf{u} - u^2\overline{\mathfrak{I}})\end{aligned} \tag{3.201}$$

where $\overline{\mathfrak{Q}}_{O'}$ and $\overline{\mathfrak{Q}}_{O''}$ denote the quadrupole moments with respect to the primed and double-primed systems of coordinates. Furthermore, $\overline{\mathfrak{Q}}$ is a symmetric traceless dyadic (Appendix E). Symmetry is apparent from definition (3.199) because the integrand is a diagonal dyadic in polar spherical coordinates [see (E.31)]. As regards the trace we may examine the diagonal elements in a system of Cartesian coordinates

$$\mathrm{Tr}\left\{\overline{\mathfrak{Q}}\right\} := Q_{xx} + Q_{yy} + Q_{zz} = \int_{V_\varrho} \mathrm{d}V'\,(3x'^2 - r'^2)\varrho(\mathbf{r}')$$

$$+\int_{V_\varrho} \mathrm{d}V'\,(3y'^2 - r'^2)\varrho(\mathbf{r}') + \int_{V_\varrho} \mathrm{d}V'\,(3z'^2 - r'^2)\varrho(\mathbf{r}') = 0 \quad (3.202)$$

or, more generally, form the double dot product with the identity dyadic, i.e.,

$$\mathrm{Tr}\left\{\overline{\mathfrak{Q}}\right\} = \overline{\mathfrak{I}} : \overline{\mathfrak{Q}} = \int_{V_\varrho} \mathrm{d}V'\,\overline{\mathfrak{I}} : (3\mathbf{r}'\mathbf{r}' - r'^2\overline{\mathfrak{I}})\varrho(\mathbf{r}') = \int_{V_\varrho} \mathrm{d}V'\,(3r'^2 - 3r'^2)\varrho(\mathbf{r}') = 0 \qquad (3.203)$$

on account of (E.78) and (E.80).

With these definitions the first three terms of the multipole expansion of $\Phi(\mathbf{r})$ may be succinctly written as

$$\Phi(\mathbf{r}) = \frac{Q}{4\pi\varepsilon r} + \frac{\hat{\mathbf{r}}\cdot\mathbf{p}}{4\pi\varepsilon r^2} + \frac{1}{2}\frac{\hat{\mathbf{r}}\cdot\overline{\mathfrak{Q}}\cdot\hat{\mathbf{r}}}{4\pi\varepsilon r^3} + o\left(\frac{1}{r^3}\right), \qquad r \to +\infty. \qquad (3.204)$$

In words, the electrostatic potential of any distribution of charges can be thought of as being generated by an equivalent point charge Q in the origin, plus an equivalent dipole of moment $\mathbf{p}$, plus an equivalent quadrupole of moment $\overline{\mathfrak{Q}}$, and so forth with higher order moments.

One or more moments may happen to be null in configurations with a high degree of symmetry. For instance, in the case of a spherical uniform distribution of charge we see from Examples 2.4 and 2.7 that only the first moment is non-zero. Conversely, for an electrostatic dipole the first moment is null, and the dominant term is the dipole moment.

Although we have obtained (3.204) for a finite amount of charge described by a density function $\varrho(\mathbf{r})$, the same formula holds true for a discrete distribution of N point charges q_n located at points $\mathbf{r}_n$ with $n = 1, \ldots, N$. Starting with the definition of the relevant scalar potential in an unbounded homogeneous isotropic medium

$$\Phi(\mathbf{r}) = \sum_{n=1}^{N} \frac{q_n}{4\pi\varepsilon|\mathbf{r} - \mathbf{r}_n|}, \qquad \mathbf{r} \in \mathbb{R}^3 \setminus \{\mathbf{r}_1, \ldots, \mathbf{r}_N\} \qquad (3.205)$$

and by expanding the functions $1/|\mathbf{r} - \mathbf{r}_n|$ as in (3.195) with $\mathbf{r}' = \mathbf{r}_n$, we obtain

$$Q := \sum_{n=0}^{N} q_n, \qquad \mathbf{p} := \sum_{n=0}^{N} \mathbf{r}_n q_n, \qquad \overline{\mathfrak{Q}} := \sum_{n=0}^{N} (3\mathbf{r}_n\mathbf{r}_n - r_n^2\overline{\mathfrak{I}})\, q_n \qquad (3.206)$$

as the discrete counterparts of the first three moments. Of course, (3.204) now holds true outside a ball $B(0, a)$ with $a > \max_n |\mathbf{r}_n|$. The invariance properties discussed above apply to the discrete counterparts of the moments as well.

Example 3.5 (Multipole expansion of a quadrupole distribution of charges)

The simplest configuration of charges which gives rise to a quadrupole moment as the dominant contribution in the multipole expansion is constituted by a set of four point charges q_n, $n = 1, 2, 3, 4$, arranged at the vertices of a parallelogram. Since the total charge must be zero, two charges must be positive whereas the other two must be negative. Moreover, since the dipole moment must vanish as well, then positive and negative charges must be placed alternately along the boundary of the parallelogram, as is pictorially shown in Figure 3.9.

The calculation of the potential may still be carried out by combining the contributions from the charges and expanding the functions $1/|\mathbf{r} - \mathbf{r}_n|$, as we did for the elementary dipole in Example 2.5.

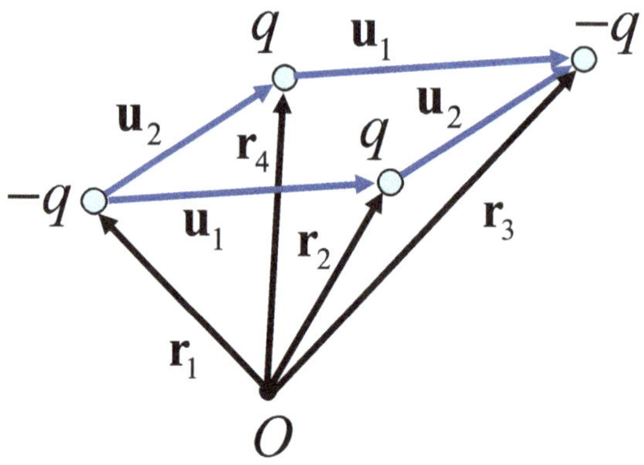

Figure 3.9 Geometrical setup for the calculation of the electric quadrupole moment.

However, we already have the necessary formulas for the moments at our disposal. For instance, if we choose $-q_1 = -q_3 = q_2 = q_4 = q$ and define the position vectors

$$\mathbf{r}_2 = \mathbf{r}_1 + \mathbf{u}_1, \qquad \mathbf{r}_3 = \mathbf{r}_1 + \mathbf{u}_1 + \mathbf{u}_2, \qquad \mathbf{r}_4 = \mathbf{r}_1 + \mathbf{u}_2 \tag{3.207}$$

we readily find

$$Q := \sum_{n=1}^{4} q_n = 0, \qquad \mathbf{p} := \sum_{n=1}^{4} \mathbf{r}_n q_n = 0, \qquad \overline{\mathcal{Q}} := q(2\mathbf{u}_1 \cdot \mathbf{u}_2 \overline{\mathcal{I}} - 3\mathbf{u}_1\mathbf{u}_2 - 3\mathbf{u}_2\mathbf{u}_1) \tag{3.208}$$

by applying (3.206). The quadrupole moment thus obtained is in coordinate-free form, so we need to introduce a system of coordinates to express the vectors and the identity dyadic for actual calculations. As expected, $\overline{\mathcal{Q}}$ is a symmetric dyadic, depends only on the relative position of the charges, and is independent of $\mathbf{r}_1$ or the position vector of any other charge, for that matter.

As an example, we consider the quadrupole moment of four charges $-q_1 = -q_3 = q_2 = q_4 = q$ arranged at the vertices of a square defined by $\mathbf{u}_1 = d(\hat{\mathbf{x}} + \hat{\mathbf{z}})$ and $\mathbf{u}_2 = d(\hat{\mathbf{z}} - \hat{\mathbf{x}})$. This information is sufficient to determine the associated quadrupole moment, which reads

$$\overline{\mathcal{Q}} = 6qd^2(\hat{\mathbf{x}}\hat{\mathbf{x}} - \hat{\mathbf{z}}\hat{\mathbf{z}}) \tag{3.209}$$

but to make a plot of the potential, we assume that the center of the square coincides with the origin of two superimposed systems of Cartesian and spherical coordinates. The charges are then located along the x- and z-axis with $\mathbf{r}_1 = -\hat{\mathbf{z}}d/\sqrt{2}$ and so forth.

Figure 3.10 shows a few equipotential lines in the half-planes $\varphi \in \{0, \pi/4, \pi/2\}$ as a function of z/d and of the normalized distance from the z-axis. The lines represent the loci of points for which the normalized dimensionless potential

$$\frac{4\pi\varepsilon\Phi(\mathbf{r})}{6q}d = \frac{1}{2}\left(\frac{d}{r}\right)^3 (\sin^2\vartheta\cos^2\varphi - \cos^2\vartheta) \tag{3.210}$$

is a constant and equal to the values indicated in the plots. We see that the potential is maximum along the direction determined by the pair of positive charges (x-axis) and minimum along the direction specified by the pair of negative charges (z-axis), as is evidenced by the leftmost plot of Figure 3.10. Besides, the potential is null in the direction (y-axis) perpendicular to the plane that contains the charges (rightmost plot of Figure 3.10).

(End of Example 3.5)

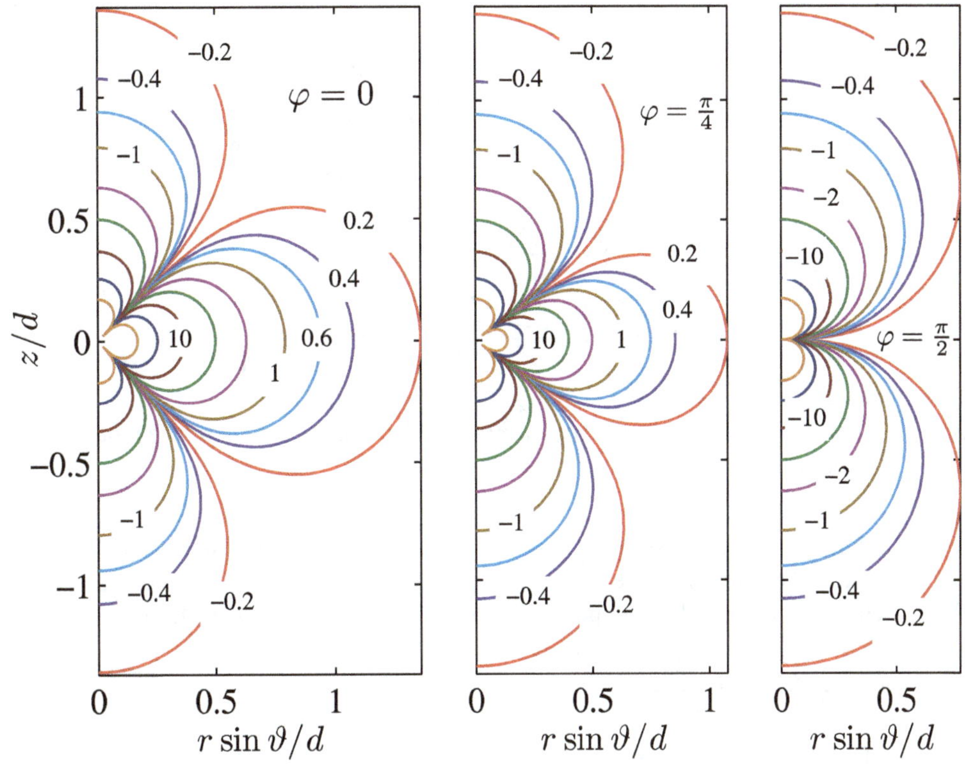

Figure 3.10 Equipotential lines of the normalized potential generated by an electric quadrupole with moment $\overline{\mathbf{Q}} = 6qd^2(\hat{\mathbf{x}}\hat{\mathbf{x}} - \hat{\mathbf{z}}\hat{\mathbf{z}})$.

3.6.2 Spherical harmonics

The alternative strategy for generating the multipole expansion of $\Phi(\mathbf{r})$ relies on the observation that $1/R$ — considered as a function of $\mathbf{r}$ with $\mathbf{r}'$ the fixed position of a point charge — can be expanded in a series of fundamental solutions of the Laplace equation in spherical coordinates (Section 3.5.1 and [1–3, 8, 14]) inasmuch as $1/R$ itself solves a problem of the type (2.124).

To find the relevant expansion, we consider an auxiliary system of spherical coordinates (r, α, β) in which the polar axis is aligned with $\mathbf{r}'$, and α and β denote the polar and the azimuth angles, respectively, as illustrated in Figure 3.11. In this setup, as the angle between $\mathbf{r}$ and $\mathbf{r}'$ is α, we notice that

$$R = |\mathbf{r} - \mathbf{r}'| = \left(r^2 - 2rr' \cos\alpha + r'^2\right)^{1/2} \tag{3.211}$$

is a function of r and α only, as r' is fixed. Consequently, $1/R$ admits a series expansion of the form (3.69) with the index $\nu = (n, m) = (n, 0)$

$$\frac{1}{|\mathbf{r} - \mathbf{r}'|} = \sum_{n=0}^{+\infty} \begin{cases} A_n r^n P_n(\cos\alpha), & r < r' \\ \dfrac{B_n}{r^{n+1}} P_n(\cos\alpha), & r > r' \end{cases} \tag{3.212}$$

because there is no dependence on the azimuth angle β, i.e., the function R is axially symmetric around $\mathbf{r}'$. It is necessary to treat the cases $r < r'$ and $r > r'$ separately in order to choose the

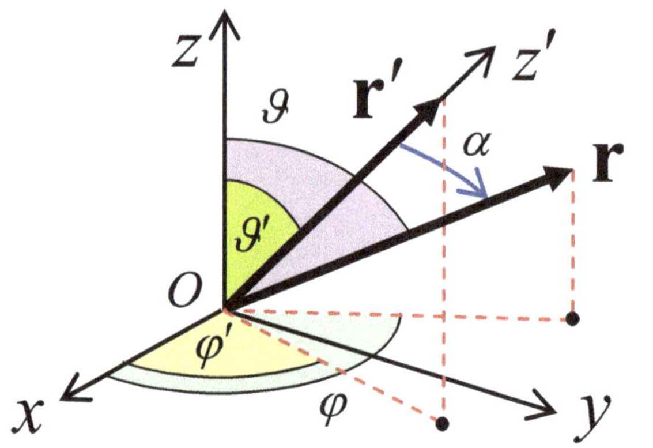

Figure 3.11 Three systems of spherical coordinates with the same origin for the proof of the addition theorem for spherical harmonics.

correct radial functions which ensure regularity in the origin $r = 0$ and at infinity. To determine the expansion coefficients A_n and B_n we examine the special arrangement of $\mathbf{r}$ aligned with $\mathbf{r}'$ for which $\alpha = 0$. (The choice of $\mathbf{r}$ parallel and opposite to $\mathbf{r}'$ works fine, too.) Then, we have

$$\frac{1}{|\mathbf{r}-\mathbf{r}'|} = \begin{cases} \dfrac{1}{r'(1-r/r')} = \displaystyle\sum_{n=0}^{+\infty} \frac{r^n}{r'^{n+1}}, & r < r' \\ \dfrac{1}{r(1-r'/r)} = \displaystyle\sum_{n=0}^{+\infty} \frac{r'^n}{r^{n+1}}, & r > r' \end{cases} \tag{3.213}$$

by invoking the geometrical series in powers of r/r' and r'/r [36, Formula 9.04]. Comparison with (3.212) for $\alpha = 0$ yields the desired coefficients because $P_n(1) = 1$ according to (H.135). If we define the two lengths

$$r_< := \min\{r, r'\}, \qquad r_> := \max\{r, r'\} \tag{3.214}$$

we may write (3.212) succinctly as

$$\frac{1}{|\mathbf{r}-\mathbf{r}'|} = \sum_{n=0}^{+\infty} \frac{r_<^n}{r_>^{n+1}} P_n(\cos\alpha). \tag{3.215}$$

Next, we would like to transform (3.215) into an expression that involves the spherical coordinates of $\mathbf{r}$ and $\mathbf{r}'$ (Figure 3.11). This must be possible, in that the angle α is related to (ϑ, φ) and (ϑ', φ') via

$$\cos\alpha = \cos\vartheta\cos\vartheta' + \sin\vartheta\sin\vartheta'\cos(\varphi - \varphi') \tag{3.216}$$

a formula which can be proved by computing $\mathbf{r}\cdot\mathbf{r}'$ in Cartesian coordinates and transforming to (ϑ, φ) and (ϑ', φ') with the aid of (A.17). So, what we need is a representation of the Legendre polynomial $P_n(\cos\alpha)$ as a sum of spherical harmonics of the type (3.102).

To obtain this result we begin by recalling that a generic function $g(\alpha, \beta)$ can be written as

$$g(\alpha, \beta) = \sum_{n=0}^{+\infty} \sum_{m=-n}^{n} g_{mn} P_n^m(\cos\alpha) \mathrm{e}^{-\mathrm{j}m\beta} \tag{3.217}$$

with expansion coefficients given by

$$g_{mn} = \frac{2n+1}{4\pi}\frac{(n-m)!}{(n+m)!}\int_{4\pi} \mathrm{d}\Omega\, g(\alpha,\beta)P_n^m(\cos\alpha)\mathrm{e}^{\mathrm{j}m\beta} \tag{3.218}$$

where the symbol $\int_{4\pi}\mathrm{d}\Omega$ signifies integration over the full solid angle. The series (3.217) is absolutely convergent, if $g(\alpha,\beta)$ is at least twice continuously differentiable for $(\alpha,\beta) \in [0,\pi]\times[0,2\pi]$. In fact, (3.217) constitutes a special instance of (3.69) with $r = 1$ and expansion functions of the type (3.108) or (3.109). Besides, for $\alpha = 0$ the β coordinate becomes *ignorable*, and hence $g(0,\beta)$ must not depend on β. Indeed, upon setting $\alpha = 0$ the series passes over into

$$g(0,\beta) = \sum_{n=0}^{+\infty} g_{0n} \tag{3.219}$$

with non-zero coefficients

$$g_{0n} = \frac{2n+1}{4\pi}\int_{4\pi} \mathrm{d}\Omega'\, g(\alpha',\beta')P_n(\cos\alpha') \tag{3.220}$$

in light of (H.135). In the interest of clarity we have renamed the *dummy* variables of integration as α' and β'.

We consider the functions $F_k(\alpha,\beta)$ defined as

$$F_k(\alpha,\beta) := \sum_{l=-k}^{k} f_{lk}P_k^l(\cos\alpha)\mathrm{e}^{-\mathrm{j}l\beta}, \qquad k \in \mathbb{N} \tag{3.221}$$

where f_{lk} are arbitrary coefficients. By applying (3.220) with $g(\alpha,\beta) = F_k(\alpha,\beta)$ we find

$$\begin{aligned} g_{0n} &= \frac{2n+1}{4\pi}\int_{4\pi} \mathrm{d}\Omega'\, F_k(\alpha',\beta')P_n(\cos\alpha') \\ &= \frac{2n+1}{4\pi}\sum_{l=-k}^{k} f_{lk}\int_{4\pi} \mathrm{d}\Omega'\, P_k^l(\cos\alpha')P_n(\cos\alpha')\mathrm{e}^{-\mathrm{j}l\beta'} = \begin{cases} 0, & l \neq 0, \quad n = 0,1,\ldots \\ 0, & l = 0, \quad n \neq k \\ f_{lk}, & l = 0, \quad n = k \end{cases} \end{aligned} \tag{3.222}$$

that is, only one term of the series (3.219) is non-null. If we rename k to n in (3.221), we have just proved the identity

$$\int_{4\pi} \mathrm{d}\Omega'\, F_n(\alpha',\beta')P_n(\cos\alpha') = \frac{4\pi}{2n+1}F_n(\alpha,\beta)|_{\alpha=0}, \qquad n \in \mathbb{N} \tag{3.223}$$

in accordance with (3.219) and (3.220). We shall use this formula in a moment.

Now, we notice that

$$\nabla^2 \frac{P_n(\cos\alpha)}{r^{n+1}} = 0, \qquad r > r' \tag{3.224}$$

$$\nabla^2[r^n P_n(\cos\alpha)] = 0, \qquad r < r' \tag{3.225}$$

with the Laplace operator expressed in polar spherical coordinates (r, α, β). However, in view of (3.216) $P_n(\cos\alpha)$ is also a function of ϑ and φ, and we may equivalently expand ∇^2 in terms of derivatives with respect to (r, ϑ, φ). Since equations of the type

$$\nabla^2 \frac{f(\vartheta,\varphi)}{r^{n+1}} = 0, \qquad r > r' \tag{3.226}$$

$$\nabla^2 [r^n f(\vartheta,\varphi)] = 0, \qquad r < r' \tag{3.227}$$

are satisfied when $f(\vartheta,\varphi)$ is any one of the $2n+1$ spherical harmonics $Y_{mn}(\vartheta,\varphi)$ with $|m| \leqslant n$, on the grounds of (3.78)-(3.80), it follows that $P_n(\cos\alpha)$ can be written as

$$P_n(\cos\alpha) = \sum_{m=-n}^{n} c_m P_n^m(\cos\vartheta) e^{-j m\varphi}, \qquad n \in \mathbb{N}. \tag{3.228}$$

Indeed, the case $P_0(\cos\alpha) = 1$ is trivial, and $P_1(\cos\alpha)$ is given explicitly by (3.216).

To compute the coefficients c_m in general we multiply through by $P_n^l(\cos\vartheta)\exp(j\, l\varphi)$ and integrate over the complete solid angle, viz.,

$$\int_{4\pi} d\Omega' \underbrace{P_n^l(\cos\vartheta) e^{j\, l\varphi}}_{=F_n(\alpha',\beta')} P_n(\cos\alpha') = \frac{4\pi}{2n+1} \frac{(n+l)!}{(n-l)!} c_l \tag{3.229}$$

on account of the orthogonality relation (H.128). We have highlighted the fact that $P_n^l(\cos\vartheta)\exp(j\, l\varphi)$ is also a function of α and β (see Figure 3.11) though the explicit dependence on these spherical coordinates is not needed. Therefore, the remaining integral can be carried out with respect to ϑ and φ, but also over α and β, which is more convenient. Indeed, since we have

$$\nabla^2 \left[\frac{P_n^l(\cos\vartheta)}{r^{n+1}} e^{j\, l\varphi} \right] = 0, \qquad r > r' \tag{3.230}$$

$$\nabla^2 \left[r^n P_n^l(\cos\vartheta) e^{j\, l\varphi} \right] = 0, \qquad r < r' \tag{3.231}$$

and ∇^2 may be expanded in polar spherical coordinates (r, α, β), it follows that $P_n^l(\cos\vartheta)\exp(j\, l\varphi)$ can be written as linear combination of the $2n+1$ spherical harmonics $Y_{ln}(\alpha,\beta)$ with $|l| \leqslant n$. In other words, $P_n^l(\cos\vartheta)\exp(j\, l\varphi)$ is a function of the type (3.221) with $k = n$, and hence we may invoke formula (3.223) to evaluate the integral in (3.229), namely,

$$\int_{4\pi} d\Omega' F_n(\alpha',\beta') P_n(\cos\alpha') = \frac{4\pi}{2n+1} P_n^l(\cos\vartheta) e^{j\, l\varphi}\Big|_{\alpha=0} = \frac{4\pi}{2n+1} P_n^l(\cos\vartheta') e^{j\, l\varphi'} \tag{3.232}$$

where we have noticed that for $\alpha = 0$, $\mathbf{r}$ and $\mathbf{r}'$ coincide and $\vartheta = \vartheta'$ and $\varphi = \varphi'$ (Figure 3.11). Inserting this result back into (3.229) and renaming l to m yields

$$c_m = \frac{(n-m)!}{(n+m)!} P_n^m(\cos\vartheta') e^{j m\varphi'}. \tag{3.233}$$

Finally, substituting this expression into (3.228) gives

$$P_n(\cos\alpha) = \sum_{m=-n}^{n} \frac{(n-m)!}{(n+m)!} P_n^m(\cos\vartheta) P_n^m(\cos\vartheta') e^{j m(\varphi'-\varphi)}$$

$$= \frac{4\pi}{2n+1} \sum_{m=-n}^{n} Y_{mn}(\vartheta, \varphi) Y^{*}_{mn}(\vartheta', \varphi') \tag{3.234}$$

the desired expansion of the Legendre polynomial. This result goes under the name of *addition theorem* for spherical harmonics [1, Sections 3.3, 3.6], [2, Section 7.5], [37, Theorem 2.8 p. 26], [14]. Inserting (3.234) into (3.215) yields

$$\frac{1}{|\mathbf{r} - \mathbf{r}'|} = \sum_{n=0}^{+\infty} \frac{4\pi}{2n+1} \frac{r_<^n}{r_>^{n+1}} \sum_{m=-n}^{n} Y_{mn}(\vartheta, \varphi) Y^{*}_{mn}(\vartheta', \varphi') \tag{3.235}$$

for $\mathbf{r} \neq \mathbf{r}'$. This formula provides $1/R$ in a completely factorized form (all the six variables are separated in the sense of Section 3.5.1) which comes in handy for the evaluation of integrals having $1/R$ as as a kernel, a case in point being the scalar potential $\Phi(\mathbf{r})$. The price to be paid is an infinite summation, so the approach is really convenient if just the first few terms of the series are significant.

The general multipole expansion of the potential follows by substituting (3.235) into the expression of the scalar potential (2.169) for $\mathbf{r} \in \mathbb{R}^3 \setminus B(0, a)$ (see Figure 3.8)

$$\begin{aligned} \Phi(\mathbf{r}) &= \frac{1}{\varepsilon} \int_{V_\varrho} \mathrm{d}V' \sum_{n=0}^{+\infty} \sum_{m=-n}^{n} \frac{\varrho(\mathbf{r}')}{2n+1} \frac{r'^n}{r^{n+1}} Y_{mn}(\vartheta, \varphi) Y^{*}_{mn}(\vartheta', \varphi') \\ &= \frac{1}{\varepsilon} \sum_{n=0}^{+\infty} \frac{1}{2n+1} \frac{1}{r^{n+1}} \sum_{m=-n}^{n} Y_{mn}(\vartheta, \varphi) \int_{V_\varrho} \mathrm{d}V' \varrho(\mathbf{r}') r'^n Y^{*}_{mn}(\vartheta', \varphi') \end{aligned} \tag{3.236}$$

where the interchange of integration and summation is permitted because the series converges for $\mathbf{r}' \in V_\varrho$ and $\mathbf{r} \notin B(0, a)$. The coefficients

$$q_{mn} = \int_{V_\varrho} \mathrm{d}V' \varrho(\mathbf{r}') r'^n Y^{*}_{mn}(\vartheta', \varphi'), \qquad n \in \mathbb{N}, \qquad |m| \leqslant n \tag{3.237}$$

are the multipole moments. It is easily verified that q_{00} is proportional to the total charge Q.

3.7 Polarization vector

In Section 1.6 we introduced a simple functional form for relating the displacement vector $\mathfrak{D}(\mathbf{r}, t)$ to the electric field $\mathcal{E}(\mathbf{r}, t)$ in a material medium. We wish to justify this assumption for the static case, while deferring a more rigorous derivation of the dielectric permittivity to Chapter 12 [4, Chapter 6].

For the present purposes, it is sufficient to think of a dielectric body as a collection of atoms or molecules which are held together by mutual electric forces. In a dielectric medium the charges are permanent and in equilibrium, i.e., bound to their atoms and molecules, and no net electric field exists (Figure 3.12a). However, if we expose the body to an external electric field, e.g., by placing it between the plates of a charged planar capacitor, the charges in the dielectric, though bound, may get displaced a bit from their equilibrium positions.

Since these modifications may occur non-uniformly in all the atoms or molecules involved, a single atom or molecule in the body experiences a net non-zero electric field which encompasses two contributions, namely, the external field and the additional field due to unbalanced charges located in the immediate surroundings of the atom in question (Figure 3.12b). We refer to this field as to the *local electric field* (also see Section 12.3.2 and Figure 12.5). The latter has the effect of nudging

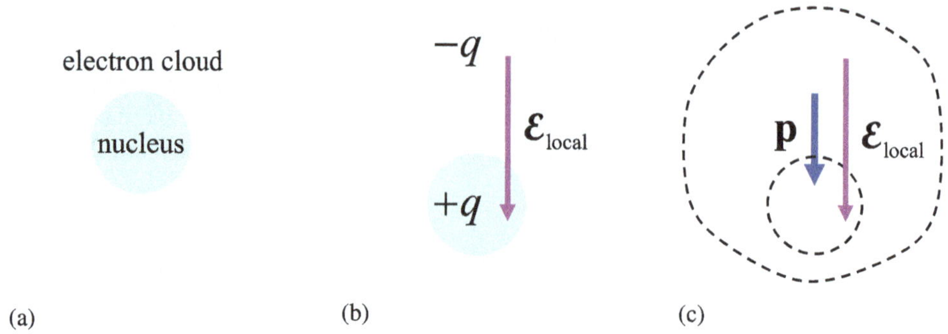

Figure 3.12 For defining the polarization vector: (a) an atom in equilibrium; (b) an atom subject to an electric field; (c) the polarized atom modelled with an equivalent static electric dipole of moment **p**.

the electrons and the protons of the atom or molecule in opposite directions, and this displacement results in an excess of positive and negative charge.

In Example 2.5 we introduced the notion of electric dipole as a pair of opposite point charges and the idea of elementary electrostatic dipole as the limit of two opposite charges placed an infinitesimal distance apart. Given that the total charge of the atom or molecule is zero (it was so in equilibrium), to a first approximation the scalar potential due to the unbalanced charges in the atom may be expressed as the contribution of a dipole moment (see Section 3.6). This is equivalent to modelling the atom or molecule as an equivalent elementary electrostatic dipole, as is pictorially suggested in Figure 3.12c, and to neglecting the effect of higher-order moments.

Since a dielectric body exposed to an electric field contains a huge number of such dipoles, it is convenient to consider the body as a continuous volumetric distribution of dipole moments in a bounded region V_D. We make this idea explicit by defining the **polarization vector** $\mathcal{P}(\mathbf{r})$ (physical dimension: $\mathrm{Cm/m^3 = C/m^2}$) as

$$\mathcal{P}(\mathbf{r}) := \lim_{\Delta V \to 0} \frac{\Delta \mathbf{p}(\mathbf{r})}{\Delta V} = \frac{\mathrm{d}\mathbf{p}(\mathbf{r})}{\mathrm{d}V}, \qquad \mathbf{r} \in V_D \tag{3.238}$$

i.e., a volume density of dipole moments. A dielectric body which exhibits a non-zero polarization vector when subject to an external electric field is said to be *polarized*. We emphasize that $\mathcal{P}(\mathbf{r})$ is a secondary entity of quantity because it depends on the applied electric field, pretty much in the same way as the conduction current $\mathcal{J}_c(\mathbf{r})$ in a conductor. Therefore, the polarization disappears after the external electric field has been switched off.

Building on formula (2.42) which gives the scalar potential produced by an elementary electrostatic dipole in free space, we may assume, as is reasonable, that the scalar potential produced by a distribution of charges $\varrho(\mathbf{r})$, $\mathbf{r} \in V_\varrho$, $V_\varrho \cap V_D = \emptyset$ in the presence of a dielectric medium takes on the form

$$\Phi(\mathbf{r}) = \frac{1}{\varepsilon_0} \int_{V_\varrho} \mathrm{d}V' \, \frac{\varrho(\mathbf{r}')}{4\pi R} + \frac{1}{\varepsilon_0} \int_{V_D} \mathrm{d}V' \, \mathcal{P}(\mathbf{r}') \cdot \nabla' \frac{1}{4\pi R}, \qquad \mathbf{r} \in \mathbb{R}^3 \tag{3.239}$$

that is, the effect of $\varrho(\mathbf{r})$ is augmented with a term which accounts for the state of polarization of the dielectric medium in V_D. We remark that the charges and the equivalent elementary static dipoles exist in free space.

Next, we wish to show that the contribution of the polarization vector may be interpreted as the effect of polarization charges distributed within V_D and over the boundary ∂V_D, under the assumption that $\mathcal{P}(\mathbf{r})$ is continuously differentiable. As is customary, we distinguish two cases, namely, observation points outside V_D and inside V_D. In the first case, we cast the volume integral over V_D into the form

$$\frac{1}{\varepsilon_0}\int_{V_D} dV'\,\mathcal{P}(\mathbf{r}')\cdot\nabla'\frac{1}{4\pi R} = -\frac{1}{\varepsilon_0}\int_{V_D} dV'\,\frac{\nabla'\cdot\mathcal{P}(\mathbf{r}')}{4\pi R} + \frac{1}{\varepsilon_0}\int_{V_D} dV'\,\nabla'\cdot\frac{\mathcal{P}(\mathbf{r}')}{4\pi R} \tag{3.240}$$

and we may apply the Gauss theorem to the last integral because $\mathcal{P}(\mathbf{r}')/R$ is of class $\mathcal{C}^1(V_D)^3\cap\mathcal{C}(\overline{V}_D)^3$. In the second situation, we isolate the singularity of the Green function with a ball $B(\mathbf{r},a)\subset V_D$ and split the integral into two parts, viz., over $V_a := V_D\setminus B(\mathbf{r},a)$ and $B(\mathbf{r},a)$

$$\begin{aligned}
&\frac{1}{\varepsilon_0}\int_{V_D} dV'\,\mathcal{P}(\mathbf{r}')\cdot\nabla'\frac{1}{4\pi R} = \\
&= \frac{1}{\varepsilon_0}\int_{V_a} dV'\,\mathcal{P}(\mathbf{r}')\cdot\nabla'\frac{1}{4\pi R} + \frac{1}{\varepsilon_0}\int_{B(\mathbf{r},a)} dV'\,\mathcal{P}(\mathbf{r}')\cdot\nabla'\frac{1}{4\pi R} \\
&= -\frac{1}{\varepsilon_0}\int_{V_a} dV'\,\frac{\nabla'\cdot\mathcal{P}(\mathbf{r}')}{4\pi R} + \frac{1}{\varepsilon_0}\int_{V_a} dV'\,\nabla'\cdot\frac{\mathcal{P}(\mathbf{r}')}{4\pi R} + \frac{1}{\varepsilon_0}\int_{B(\mathbf{r},a)} dV'\,\mathcal{P}(\mathbf{r}')\cdot\nabla'\frac{1}{4\pi R} \\
&= -\frac{1}{\varepsilon_0}\int_{V_a} dV'\,\frac{\nabla'\cdot\mathcal{P}(\mathbf{r}')}{4\pi R} + \frac{1}{\varepsilon_0}\int_{\partial V_D} dS'\,\frac{\hat{\mathbf{n}}(\mathbf{r}')\cdot\mathcal{P}(\mathbf{r}')}{4\pi R} \\
&\quad + \frac{1}{\varepsilon_0}\int_{\partial B} dS'\,\frac{\hat{\mathbf{n}}(\mathbf{r}')\cdot\mathcal{P}(\mathbf{r}')}{4\pi R} + \frac{1}{\varepsilon_0}\int_{B(\mathbf{r},a)} dV'\,\mathcal{P}(\mathbf{r}')\cdot\nabla'\frac{1}{4\pi R}
\end{aligned} \tag{3.241}$$

where the Gauss theorem has been applied to the integral over V_a because $\mathcal{P}(\mathbf{r}')/R$ is of class $\mathcal{C}^1(V_a)^3\cap\mathcal{C}(\overline{V}_a)^3$. We now examine the four remaining integrals in the limit as $a\to 0$. For the integral over V_a we pick up a ball $B(\mathbf{r},b)$ with radius b large enough so that $V_a\subset B(\mathbf{r},b)$ and estimate

$$\begin{aligned}
\left|\int_{V_a} dV'\,\frac{\nabla'\cdot\mathcal{P}(\mathbf{r}')}{4\pi R}\right| &\leqslant \int_{V_a} dV'\,\frac{\|\nabla'\cdot\mathcal{P}(\mathbf{r}')\|_\infty}{4\pi R} \leqslant \int_{B(\mathbf{r},b)\setminus B(\mathbf{r},a)} dV'\,\frac{\|\nabla'\cdot\mathcal{P}(\mathbf{r}')\|_\infty}{4\pi R} \\
&= \frac{1}{2}(b^2-a^2)\|\nabla'\cdot\mathcal{P}(\mathbf{r}')\|_\infty \leqslant \frac{1}{2}b^2\|\nabla'\cdot\mathcal{P}(\mathbf{r}')\|_\infty
\end{aligned} \tag{3.242}$$

so this contribution is bounded for any value of a. For the integral over ∂B we have

$$\left|\int_{\partial B} dS'\,\frac{\hat{\mathbf{n}}(\mathbf{r}')\cdot\mathcal{P}(\mathbf{r}')}{4\pi R}\right| \leqslant \|\mathcal{P}\|_\infty\int_{\partial B} dS'\,\frac{1}{4\pi a} = \|\mathcal{P}\|_\infty a \xrightarrow[a\to 0]{} 0 \tag{3.243}$$

and finally

$$\left|\int_{B(\mathbf{r},a)} dV'\,\mathcal{P}(\mathbf{r}')\cdot\nabla'\frac{1}{4\pi R}\right| \leqslant \|\mathcal{P}\|_\infty\int_{B(\mathbf{r},a)} dV'\,\frac{1}{4\pi R^2} = \|\mathcal{P}\|_\infty a \xrightarrow[a\to 0]{} 0 \tag{3.244}$$

whereas the integral over ∂V_D does not depend on a. Putting everything together we have proved the identity

$$\frac{1}{\varepsilon_0}\int_{V_D} \mathrm{d}V'\, \mathcal{P}(\mathbf{r}')\cdot\nabla'\frac{1}{4\pi R} = \frac{1}{\varepsilon_0}\int_{\partial V_D} \mathrm{d}S'\,\frac{\hat{\mathbf{n}}(\mathbf{r}')\cdot\mathcal{P}(\mathbf{r}')}{4\pi R} - \frac{1}{\varepsilon_0}\int_{V_D} \mathrm{d}V'\,\frac{\nabla'\cdot\mathcal{P}(\mathbf{r}')}{4\pi R} \tag{3.245}$$

for observation points $\mathbf{r} \in \mathbb{R}^3$. By comparing this result with the potential (2.169) generated by a volume density of charges in a homogeneous unbounded isotropic medium, we conclude that the effect of the dielectric may be construed as a distribution of volume and surface equivalent polarization charges with densities

$$\varrho_{VP}(\mathbf{r}) := -\nabla\cdot\mathcal{P}(\mathbf{r}), \qquad \mathbf{r}\in V_D \tag{3.246}$$

$$\varrho_{SP}(\mathbf{r}) := \hat{\mathbf{n}}(\mathbf{r})\cdot\mathcal{P}(\mathbf{r}), \qquad \mathbf{r}\in \partial V_D \tag{3.247}$$

with the unit normal pointing outward V_D; this interpretation goes by the name of Poisson's theorem. Notice that if $\mathcal{P}(\mathbf{r})$ is constant, the volume charge density is zero. On the other hand, the unfailing onset of a polarization charge on the surface of the body can be understood by observing that the equivalent dipoles along the boundary are only partially balanced by dipoles in the bulk of the body.

Lastly, we investigate the Gauss law in the presence of a dielectric body. The latter is now modelled as a collection of polarization charges in free space with densities given by (3.246) and (3.247). We consider a region of free space V bounded by the smooth surface ∂V which contains free charges in V_ϱ and polarization charges in $\overline{V}_D := V_D \cup \partial V_D$ (Figure 3.13). The global Gauss law reads

$$\varepsilon_0\int_{\partial V} \mathrm{d}S\,\hat{\mathbf{n}}(\mathbf{r})\cdot\mathcal{E}(\mathbf{r}) = \int_{V_\varrho}\mathrm{d}V\varrho(\mathbf{r}) + \underbrace{\int_{V_D}\mathrm{d}V\varrho_{VP}(\mathbf{r}) + \int_{\partial V_D}\mathrm{d}S\varrho_{SP}(\mathbf{r})}_{=Q_{VP}+Q_{SP}} = \int_{V_\varrho}\mathrm{d}V\varrho(\mathbf{r}) \tag{3.248}$$

because the total polarization charge in $\overline{V}_D$ is null! This conclusion also follows from (3.246) and the divergence theorem applied to the domain integral over V_D, which is then turned into the negative of the integral of ϱ_{SP} over ∂V_D by virtue of (3.247). Finally, notice that even in the presence of the dielectric body, the displacement vector $\mathcal{D}(\mathbf{r})$ over ∂V is defined through the constitutive relation (1.112), because the boundary ∂V precisely resides in free space.

We would like to obtain the local form of the Gauss law, but we may not invoke (A.53) in the leftmost-hand side of (3.248) inasmuch as the electric field may not be of class $\mathcal{C}^1(V)^3 \cap \mathcal{C}(\overline{V})^3$. Therefore, with reference to Figure 3.13 we introduce a bounded connected volume W which intersects V_D along a surface S_0. As a result, W is divided into two sub-domains $W_1 := W \cap (\mathbb{R}^3 \setminus \overline{V}_D)$ and $W_2 := W \cap V_D$. We indicate with ∂W_1 and ∂W_2 the parts of the boundary of W contained in $\mathbb{R}^3 \setminus \overline{V}_D$ and V_D, respectively, whereby we have $\partial W := \partial W_1 \cup \partial W_2$. The boundaries of the open surfaces ∂W_1 and ∂W_2 are two loops which coincide with the boundary ∂S_0. The geometrical relationships (1.145)-(1.147) also hold true. To gain more generality we suppose that the charge region V_ϱ is contained in W_1.

We state the global Gauss law in free space separately in W_1, W_2 and W, viz.,

$$\int_{\partial W_1\cup S_0} \mathrm{d}S\,\hat{\mathbf{v}}_1(\mathbf{r})\cdot\mathcal{E}(\mathbf{r})\varepsilon_0 = \int_{V_\varrho}\mathrm{d}V\,\varrho(\mathbf{r}) \tag{3.249}$$

$$\int_{\partial W_2\cup S_0} \mathrm{d}S\,\hat{\mathbf{v}}_2(\mathbf{r})\cdot\mathcal{E}(\mathbf{r})\varepsilon_0 = -\int_{W_2}\mathrm{d}V\,\nabla\cdot\mathcal{P}(\mathbf{r}) \tag{3.250}$$

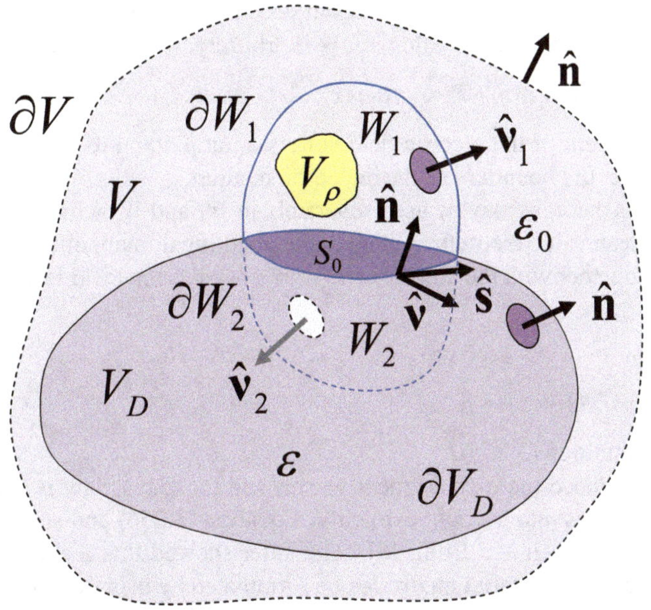

Figure 3.13 For determining the local Gauss law in the presence of a dielectric body.

$$\int_{\partial W} \mathrm{d}S\, \hat{\nu}(\mathbf{r}) \cdot \mathcal{E}(\mathbf{r})\varepsilon_0 = \int_{V_\varrho} \mathrm{d}V\, \varrho(\mathbf{r}) - \int_{W_2} \mathrm{d}V\, \nabla \cdot \mathcal{P}(\mathbf{r}) + \int_{S_0} \mathrm{d}S\, \hat{\mathbf{n}}(\mathbf{r}) \cdot \mathcal{P}(\mathbf{r}) \tag{3.251}$$

where we have included the contribution of the surface polarization charges on ∂V_D and hence S_0. We mention in passing that the issue of the continuity across V_ϱ was addressed in Section 1.2.2, and thus it does not concern us any more.

By summing (3.249) and (3.250) side by side and combining the flux integrals we get

$$\begin{aligned}
&\int_{\partial W_1 \cup S_0} \mathrm{d}S\, \hat{\nu}_1(\mathbf{r}) \cdot \mathcal{E}(\mathbf{r})\varepsilon_0 + \int_{\partial W_2 \cup S_0} \mathrm{d}S\, \hat{\nu}_2(\mathbf{r}) \cdot \mathcal{E}(\mathbf{r})\varepsilon_0 = \\
&\quad = \int_{\partial W} \mathrm{d}S\, \hat{\nu}(\mathbf{r}) \cdot \mathcal{E}(\mathbf{r})\varepsilon_0 + \int_{S_0} \mathrm{d}S\, \hat{\mathbf{n}}(\mathbf{r}) \cdot [\mathcal{E}^-(\mathbf{r}) - \mathcal{E}^+(\mathbf{r})]\varepsilon_0 \\
&\quad = \int_{V_\varrho} \mathrm{d}V\, \varrho(\mathbf{r}) - \int_{W_2} \mathrm{d}V\, \nabla \cdot \mathcal{P}(\mathbf{r})
\end{aligned} \tag{3.252}$$

where $\mathcal{E}^+(\mathbf{r})$ and $\mathcal{E}^-(\mathbf{r})$ indicate the values of the electric field on either side of ∂V_D. This relationship ought to be identical with (3.251) for consistency. We see that this indeed the case if we enforce the condition

$$\int_{S_0} \mathrm{d}S\, \hat{\mathbf{n}}(\mathbf{r}) \cdot \{[\mathcal{E}^-(\mathbf{r}) - \mathcal{E}^+(\mathbf{r})]\varepsilon_0 + \mathcal{P}(\mathbf{r})\} = 0 \tag{3.253}$$

and, if we assume that the vector fields $\mathcal{E}^-(\mathbf{r})$, $\mathcal{E}^+(\mathbf{r})$ and $\mathcal{P}(\mathbf{r})$ are continuous for $\mathbf{r} \in \partial V_D$, then the mean value theorem [5] applied to the integral above yields

$$\hat{\mathbf{n}}(\mathbf{r}_0) \cdot \{[\mathcal{E}^-(\mathbf{r}_0) - \mathcal{E}^+(\mathbf{r}_0)]\varepsilon_0 + \mathcal{P}(\mathbf{r}_0)\} = 0 \tag{3.254}$$

with $\mathbf{r}_0 \in S_0$ a suitable point. Notice that this result does not depend on whether or not W_1 contains true electric charges. Besides, since the domain W is arbitrary, we conclude

$$\hat{\mathbf{n}}(\mathbf{r}) \cdot [\mathcal{E}^+(\mathbf{r}) - \mathcal{E}^-(\mathbf{r})]\varepsilon_0 = \hat{\mathbf{n}}(\mathbf{r}) \cdot \mathcal{P}(\mathbf{r}) = \varrho_{SP}(\mathbf{r}), \qquad \mathbf{r} \in \partial V_D \tag{3.255}$$

i.e., the normal component of the electric field suffers a jump equal to the amount of polarization charges distributed over the boundary of the dielectric medium.

Finally, the Gauss theorem may be used separately in W_1 and W_2 where $\varepsilon_0 \mathcal{E}(\mathbf{r})$ is continuously differentiable. The mean value theorem applied to the resulting domain integrals over W_1 and W_2 in (3.249) and (3.250) together with the arbitrariness of W provides the local form of the electric Gauss law in each point, viz.,

$$\varepsilon_0 \nabla \cdot \mathcal{E}(\mathbf{r}) = \begin{cases} \varrho(\mathbf{r}), & \mathbf{r} \in \mathbb{R}^3 \setminus V_D \\ -\nabla \cdot \mathcal{P}(\mathbf{r}), & \mathbf{r} \in V_D \end{cases} \tag{3.256}$$

subject to the jump condition (3.255).

We have not introduced the displacement vector, and the Gauss law is actually stated in free space by taking all charges into account explicitly. Equation (3.256) and similar ones for the other laws are named after Maxwell and Boffi [8]. The latter showed that a consistent set of Maxwell equations in the presence of material media can be formulated by using $\mathcal{E}$, $\mathcal{B}$ as entities of intensity and $\mathcal{P}$ and a corresponding magnetization vector $\mathcal{M}$ (see Section 5.6) as entities of quantity instead of $\mathcal{D}$ and $\mathcal{H}$.

Nonetheless, we can cast (3.256) into the familiar form of the Gauss law involving the displacement vector. Indeed, we need to pair the polarization vector and the electric field and define

$$\mathcal{D}(\mathbf{r}) := \begin{cases} \varepsilon_0 \mathcal{E}(\mathbf{r}), & \mathbf{r} \in \mathbb{R}^3 \setminus V_D \\ \varepsilon_0 \mathcal{E}(\mathbf{r}) + \mathcal{P}(\mathbf{r}), & \mathbf{r} \in V_D \end{cases} \tag{3.257}$$

so that (3.256) reads

$$\nabla \cdot \mathcal{D}(\mathbf{r}) = \varrho(\mathbf{r}), \qquad \mathbf{r} \in \mathbb{R}^3 \tag{3.258}$$

and (3.255) becomes

$$\hat{\mathbf{n}}(\mathbf{r}) \cdot \underbrace{[\varepsilon_0 \mathcal{E}^-(\mathbf{r}) + \mathcal{P}(\mathbf{r})]}_{\mathcal{D}^-(\mathbf{r})} = \hat{\mathbf{n}}(\mathbf{r}) \cdot \underbrace{\varepsilon_0 \mathcal{E}^+(\mathbf{r})}_{\mathcal{D}^+(\mathbf{r})}, \qquad \mathbf{r} \in \partial V_D \tag{3.259}$$

which is a statement of the continuity of the normal component of $\mathcal{D}$ already found in Section 1.7.

The polarization vector may depend in some complicated manner on the external electric field. In linear isotropic media we may assume a simple proportionality relation, say,

$$\mathcal{P}(\mathbf{r}) := \varepsilon_0 \chi_e(\mathbf{r}) \mathcal{E}(\mathbf{r}), \qquad \mathbf{r} \in V_D \tag{3.260}$$

where the dimensionless scalar field $\chi_e(\mathbf{r})$ is called the dielectric susceptibility (cf. Section 12.3.2). Thus, from (3.257) we find

$$\mathcal{D}(\mathbf{r}) = \varepsilon_0 [1 + \chi_e(\mathbf{r})] \mathcal{E}(\mathbf{r}) = \varepsilon_0 \varepsilon_r \mathcal{E}(\mathbf{r}) = \varepsilon(\mathbf{r}) \mathcal{E}(\mathbf{r}), \qquad \mathbf{r} \in V_D \tag{3.261}$$

with ε the dielectric permittivity. In like manner in linear anisotropic media we may introduce a dyadic susceptibility and define

$$\overline{\varepsilon}(\mathbf{r}) = \varepsilon_0 \left[\overline{\mathcal{I}} + \overline{\chi}_e(\mathbf{r}) \right], \qquad \mathbf{r} \in V_D \tag{3.262}$$

as dielectric dyadic permittivity.

We observe that in the Maxwell-Boffi description of electromagnetism dielectric media are substituted by a suitable vector field which describes the distribution of polarization charges in free space. Therefore (3.256) is also referred to as the *microscopic* form of the Gauss law. By contrast, with the Gauss law (1.44) we assign the dielectric body an effective *macroscopic* parameter (the permittivity) and we overlook, as it were, the actual appearance of polarization charges.

Example 3.6 (Dielectric sphere in a uniform electrostatic field (reprise))

We briefly revisit the problem of a dielectric sphere exposed to a uniform electric field (Figure 3.6) from the viewpoint of the onset of polarization charges within the region $B(0, a)$. In accordance with definitions (3.257) and (3.261) we compute the polarization vector as

$$\mathcal{P}(\mathbf{r}) = \mathcal{D}(\mathbf{r}) - \varepsilon_1 \mathcal{E}(\mathbf{r}) = (\varepsilon_2 - \varepsilon_1)\mathcal{E}^{\mathrm{s}}(\mathbf{r}) = 3\varepsilon_1 \frac{\varepsilon_2 - \varepsilon_1}{\varepsilon_2 + 2\varepsilon_1} \mathcal{E}_0 \hat{\mathbf{z}}, \qquad r < a \tag{3.263}$$

where we have recalled that $\mathcal{E}(\mathbf{r}) = \mathcal{E}^{\mathrm{s}}(\mathbf{r})$ is given by (3.143) for $r < a$. Consequently, the densities of polarization charge within the sphere and on the surface thereof read

$$\varrho_{VP}(\mathbf{r}) = -\nabla \cdot \mathcal{P}(\mathbf{r}) = -(\varepsilon_2 - \varepsilon_1)\nabla \cdot \mathcal{E}^{\mathrm{s}}(\mathbf{r}) = 0, \qquad r < a \tag{3.264}$$

$$\varrho_{SP} = \hat{\mathbf{n}}(\mathbf{r}) \cdot \mathcal{P}(\mathbf{r}) = 3\varepsilon_1 \frac{\varepsilon_2 - \varepsilon_1}{\varepsilon_2 + 2\varepsilon_1} \mathcal{E}_0 \cos\vartheta, \qquad r = a \tag{3.265}$$

by virtue of (3.246) and (3.247). The total polarization charge is found by integrating ϱ_{SP} over the sphere and is null.

Since the polarization vector physically represents a volume density of elementary electric dipoles induced within the sphere by the external impressed field, we can obtain the equivalent dipole moment of the sphere by integrating $\mathcal{P}(\mathbf{r})$ over $B(0, a)$, namely,

$$\mathbf{p}_{eq} := \int_{B(0,a)} \mathrm{d}V\, \mathcal{P}(\mathbf{r}) = \int_{B(0,a)} \mathrm{d}V\, 3\varepsilon_1 \frac{\varepsilon_2 - \varepsilon_1}{\varepsilon_2 + 2\varepsilon_1} \mathcal{E}_0 \hat{\mathbf{z}} = 4\pi\varepsilon_1 a^3 \frac{\varepsilon_2 - \varepsilon_1}{\varepsilon_2 + 2\varepsilon_1} \mathcal{E}_0 \hat{\mathbf{z}} \tag{3.266}$$

which is in agreement with the previous result (3.141).

When we adopt the microscopic viewpoint embodied in the Maxwell-Boffi equations, we are led to ascribe the electric field within the sphere to the combination of the impressed one — which exists in an unbounded medium equipped with permittivity ε_1 — and the electric field produced by the polarization charges induced over the surface ∂B. Indeed, if we write $\mathcal{E}^{\mathrm{s}}(\mathbf{r})$ given by (3.143) by singling out $\mathcal{E}^{\mathrm{i}}(\mathbf{r})$ we get

$$\mathcal{E}^{\mathrm{s}}(\mathbf{r}) = \underbrace{\mathcal{E}_0 \hat{\mathbf{z}}}_{\text{impressed}} \underbrace{- \frac{\varepsilon_2 - \varepsilon_1}{\varepsilon_2 + 2\varepsilon_1} \mathcal{E}_0 \hat{\mathbf{z}}}_{\text{polarization}} = \mathcal{E}_0 \hat{\mathbf{z}} - \frac{1}{3\varepsilon_1} \mathcal{P}(\mathbf{r}) \tag{3.267}$$

whereby we see that the contribution of the polarization charges opposes the external field if $\varepsilon_1 < \varepsilon_2$ and, as a result, the field within the sphere is *less strong* than the impressed one; the converse is true if $\varepsilon_1 > \varepsilon_2$.

(End of Example 3.6)

3.8 The Kelvin and Earnshaw theorems

We discuss two theorems which concern the properties of electric fields produced by *fixed* distributions of charges. The first one was obtained by Lord Kelvin and is also known as Thomson

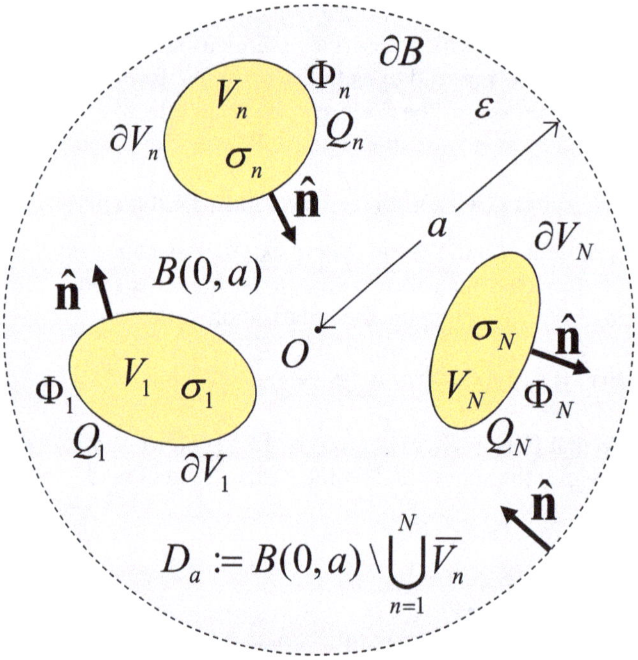

Figure 3.14 A system of N charged conductors for proving the Kelvin theorem.

theorem [38, p. 12], [2, pp. 114-116], [8, 39]. The second one was derived by S. Earnshaw in 1842 [2, pp. 116-117], [39, 40], that is, when the *luminiferous ether* (see Section 1.8) was still reckoned to play a role in the physics of electric and magnetic phenomena.

With reference to Figure 3.14 we suppose that N metallic objects with conductivities σ_n, $n = 1, \ldots, N$, occupy the volumes V_n, are held at potentials Φ_n, and are immersed in an unbounded homogeneous isotropic dielectric medium endowed with permittivity ε. We indicate the total charge on the surface ∂V_n of the nth conductor with Q_n and let $\mathcal{E}_0(\mathbf{r})$ denote the field produced by $\sum_n Q_n$ for $\mathbf{r} \in V := \mathbb{R}^3 \setminus \cup_n \overline{V}_n$. The displacement vector $\mathcal{D}_0(\mathbf{r}) := \varepsilon \mathcal{E}_0(\mathbf{r})$ and the electric field $\mathcal{E}_0(\mathbf{r}) = -\nabla \Phi_0(\mathbf{r})$ satisfy the Gauss law (2.3) and the constraint (2.4), respectively.

The Kelvin theorem says that, if the potentials Φ_n — and hence the charges Q_n — are kept constant, then the *charge densities* on the boundaries ∂V_n arrange themselves in such a way that the electrostatic energy (1.263) stored in the field $\mathcal{E}_0(\mathbf{r})$ is a *minimum*. To prove the statement we start with the surface-wise multiply-connected domain $D_a := B(0, a) \setminus \cup_n \overline{V}_n$, where $B(0, a)$ is a ball of radius a large enough to contain all the conducting bodies, and we let a grow indefinitely so that D_a passes over into V.

While the energy associated with the finite region D_a strictly reads

$$W_e(\mathcal{E}_0; a) := \frac{1}{2} \int_{D_a} dV \, \mathcal{E}_0(\mathbf{r}) \cdot \mathcal{D}_0(\mathbf{r}) = \frac{\varepsilon}{2} \int_{D_a} dV \, |\mathcal{E}_0(\mathbf{r})|^2 \tag{3.268}$$

from a mathematical viewpoint we may regard $W_e(\mathcal{E}; a)$ as an operator or a *functional* which takes arbitrary square-integrable vector fields $\mathcal{E}(\mathbf{r})$ and returns a positive real number (Appendix D.3). For our purposes we restrict the set of acceptable fields (the domain of the operator) by demanding that $\mathcal{E}(\mathbf{r})$ be in the form

$$\mathcal{E}(\mathbf{r}) := \mathcal{E}_0(\mathbf{r}) + \mathcal{E}'(\mathbf{r}) = -\nabla\Phi_0(\mathbf{r}) - \nabla\Phi'(\mathbf{r}), \qquad \mathbf{r} \in D_a \tag{3.269}$$

and obey

$$\varepsilon \int_{\partial V_n} \mathrm{d}S\, \hat{\mathbf{n}}(\mathbf{r}) \cdot [\mathcal{E}_0(\mathbf{r}) + \mathcal{E}'(\mathbf{r})] = Q_n, \qquad n = 1, \ldots, N \tag{3.270}$$

$$\Phi(\mathbf{r}) := \Phi_0(\mathbf{r}) + \Phi'(\mathbf{r}) = \Phi_n, \qquad n = 1, \ldots, N, \qquad \mathbf{r} \in \partial V_n \tag{3.271}$$

$$\Phi(\mathbf{r}) = \mathcal{O}\left(\frac{1}{|\mathbf{r}|}\right), \qquad |\mathbf{r}| \to +\infty \tag{3.272}$$

i.e., $\mathcal{E}(\mathbf{r})$ and $\Phi(\mathbf{r})$ satisfy the same boundary conditions as $\mathcal{E}_0(\mathbf{r})$ and $\Phi_0(\mathbf{r})$. From the left-hand side of (3.270) we gather that $\varepsilon\hat{\mathbf{n}}(\mathbf{r}) \cdot \mathcal{E}'(\mathbf{r})$ constitutes a perturbation of the 'true' charge density on the boundary of the nth conductor, but the total charge associated with $\varepsilon\hat{\mathbf{n}}(\mathbf{r}) \cdot \mathcal{E}'(\mathbf{r})$ is actually zero. Moreover, $\Phi(\mathbf{r})$ cannot solve the homogeneous Dirichlet problem for $\mathbf{r} \in D_a$ because, in accordance with the findings of Section 2.5.1, $\Phi_0(\mathbf{r})$ is the unique solution to the problem of Figure 3.14.

Under these circumstances, we have to show that, no matter how we choose $\mathcal{E}(\mathbf{r})$, the energy-like quantity $W_e(\mathcal{E}; a)$ is always larger than $W_e(\mathcal{E}_0; a)$ for $a \to +\infty$. Stated another way, the functional $W_e(\mathcal{E}; a)$ is *minimized* when $\mathcal{E}(\mathbf{r})$ *coincides* with the solution to (2.3) and (2.4), with either the Φ_n's or the Q_n's being assigned.

For the proof we examine

$$\begin{aligned}
W_e(\mathcal{E}; a) &:= \frac{\varepsilon}{2} \int_{D_a} \mathrm{d}V\, \mathcal{E}(\mathbf{r}) \cdot \mathcal{E}(\mathbf{r}) \\
&= \frac{\varepsilon}{2} \int_{D_a} \mathrm{d}V\, |\mathcal{E}_0(\mathbf{r})|^2 + \varepsilon \int_{D_a} \mathrm{d}V\, \mathcal{E}_0(\mathbf{r}) \cdot \mathcal{E}'(\mathbf{r}) + \frac{\varepsilon}{2} \int_{D_a} \mathrm{d}V\, |\mathcal{E}'(\mathbf{r})|^2 \\
&= W_e(\mathcal{E}_0; a) + \varepsilon \int_{D_a} \mathrm{d}V\, \mathcal{E}_0(\mathbf{r}) \cdot \mathcal{E}'(\mathbf{r}) + W_e(\mathcal{E}'; a)
\end{aligned} \tag{3.273}$$

where the energy of $\mathcal{E}_0(\mathbf{r})$ and the positive quantity $W_e(\mathcal{E}'; a)$ remain finite in the limit as $a \to +\infty$, inasmuch all fields satisfy the asymptotic condition (2.21) [see (2.19) and the subsequent discussion]. In particular, since the quantity $W_e(\mathcal{E}'; a)$ is non-negative, we obtain the estimate

$$W_e(\mathcal{E}) \geqslant W_e(\mathcal{E}_0) + \lim_{a \to +\infty} \varepsilon \int_{D_a} \mathrm{d}V\, \mathcal{E}_0(\mathbf{r}) \cdot \mathcal{E}'(\mathbf{r}) \tag{3.274}$$

whereby the Kelvin theorem follows if we can conclude that the last integral is invariably null under the stated assumptions and in the limit as $a \to +\infty$.

To this purpose, we employ the scalar potentials $\Phi_0(\mathbf{r})$ and $\Phi'(\mathbf{r})$

$$\begin{aligned}
\varepsilon \int_{D_a} \mathrm{d}V\, \mathcal{E}_0(\mathbf{r}) \cdot \mathcal{E}'(\mathbf{r}) &= \varepsilon \int_{D_a} \mathrm{d}V\, \nabla\Phi_0(\mathbf{r}) \cdot \nabla\Phi'(\mathbf{r}) \\
&= \varepsilon \int_{D_a} \mathrm{d}V\, \nabla \cdot [\Phi'(\mathbf{r})\nabla\Phi_0(\mathbf{r})] - \varepsilon \int_{D_a} \mathrm{d}V\, \Phi'(\mathbf{r})\nabla^2\Phi_0(\mathbf{r})
\end{aligned}$$

$$= -\sum_{n=1}^{N} \varepsilon \int_{\partial V_n} \mathrm{d}S\, \Phi'(\mathbf{r})\hat{\mathbf{n}}(\mathbf{r}) \cdot \nabla\Phi_0(\mathbf{r}) - \varepsilon \int_{\partial B} \mathrm{d}S\, \Phi'(\mathbf{r})\hat{\mathbf{n}}(\mathbf{r}) \cdot \nabla\Phi_0(\mathbf{r})$$
$$- \varepsilon \int_{D_a} \mathrm{d}V\, \Phi'(\mathbf{r})\nabla^2\Phi_0(\mathbf{r}) \tag{3.275}$$

having made use of (H.51) and the Gauss theorem with the unit normals on ∂V_n and ∂B pointing inwards D_a (Figure 3.14). The flux integrals over ∂V_n vanish because $\Phi'(\mathbf{r}) = 0$ on the conducting surfaces in view of (3.271). The integral over ∂B vanishes for $a \to +\infty$ in that both $\Phi_0(\mathbf{r})$ and $\Phi'(\mathbf{r})$ obey the asymptotic behavior (2.20). Finally, the remaining domain integral over D_a is identically null because $\Phi_0(\mathbf{r})$ satisfies the Laplace equation (3.68) by hypothesis. Therefore, the rightmost-hand side of (3.275) vanishes as $a \to +\infty$ and we are left with

$$W_e(\mathcal{E}) \geqslant W_e(\mathcal{E}_0) \tag{3.276}$$

and this concludes the proof.

Loosely speaking, the Earnshaw theorem states that a scalar field $\Upsilon(\mathbf{r})$ obeying the Laplace equation cannot have any local maxima or minima, whence it follows that points for which $\nabla\Upsilon(\mathbf{r}) = 0$ (i.e., stationary points) are *saddle points* of $\Upsilon(\mathbf{r})$. More precisely, if $\Upsilon(\mathbf{r})$ solves (3.68) in a bounded region of space $V \subset \mathbb{R}^3$, then the theorem forbids the occurrence of local maxima or minima for points strictly inside V, though $\Upsilon(\mathbf{r})$ may reach maxima or minima on the boundary ∂V. A perfectly analogous property exists for the real and the imaginary parts of a complex holomorphic function $f(z)$, $z := x + \mathrm{j}y \in U \subset \mathbb{C}$, because $\mathrm{Re}\{f(z)\}$ and $\mathrm{Im}\{f(z)\}$ — regarded as functions of x and y — solve the two-dimensional Laplace equation (see Appendix B.1). In electrostatics, where we are interested in $\Upsilon(\mathbf{r}) \equiv \Phi(\mathbf{r})$, this result implies that a system of charges or charged conducting bodies cannot maintain a stable fixed configuration under the sole action of static electric forces. Hereinbelow we present three proofs of the theorem.

Maxwell's argument in favor of the instability of a set of charges [41, §116] is essentially based on an unreconcilable contradiction which stems from the application of the Gauss law (2.1). To elucidate, we suppose that (3.68) holds true in a finite region $V \subset \mathbb{R}^3$ — which is thus devoid of charges — and also that a positive *test point charge* q placed in $\mathbf{r}_0 \in V$ does happen to be in stable equilibrium. Then, Maxwell reasoned, the electric field $\mathcal{E}(\mathbf{r})$ in V is produced by charges located *outside* V, and the Coulomb force $q\mathcal{E}(\mathbf{r}_0)$ acting on the charge should be *zero*, viz.,

$$0 = q\mathcal{E}(\mathbf{r}_0) = -q\nabla\Phi(\mathbf{r})|_{\mathbf{r}=\mathbf{r}_0}, \qquad \mathbf{r}_0 \in V \tag{3.277}$$

which makes $\mathbf{r}_0$ a stationary point for the potential $\Phi(\mathbf{r})$. Now, if we 'nudge' q a little away from the presumed point of equilibrium $\mathbf{r}_0$, the non-zero force acting on q should read $q\mathcal{E}(\mathbf{r}_0 + \Delta\mathbf{r})$ where $\Delta\mathbf{r}$ is the 'small' displacement undergone by the charge or, equivalently, the position of the charge with respect to $\mathbf{r}_0$. What can be said of the force acting on q in $\mathbf{r} := \mathbf{r}_0 + \Delta\mathbf{r}$?

Well, if $\mathbf{r}_0$ really *were* a point of stable equilibrium, the force *would* be *restoring* in character, i.e., one that tries to pull q back to $\mathbf{r}_0$ along $-\Delta\mathbf{r}$. As a result, we would have

$$q\mathcal{E}(\mathbf{r}_0 + \Delta\mathbf{r}) \cdot \Delta\mathbf{r} = -q|\mathcal{E}(\mathbf{r}_0 + \Delta\mathbf{r})||\Delta\mathbf{r}| < 0, \qquad \mathbf{r}_0 \in V \tag{3.278}$$

because the force is aligned with the displacement $\Delta\mathbf{r}$ and directed toward $\mathbf{r}_0$. Lastly, we invoke the Gauss law (2.1) with the closed smooth surface ∂V being the boundary of the ball $B(\mathbf{r}_0, |\Delta\mathbf{r}|) \subset V$, namely,

$$\varepsilon \int_{\partial B} \mathrm{d}S\, \hat{\mathbf{n}}(\mathbf{r}) \cdot \mathcal{E}(\mathbf{r}) = \varepsilon \int_{\partial B} \mathrm{d}S\, \frac{\Delta\mathbf{r}}{|\Delta\mathbf{r}|} \cdot \mathcal{E}(\mathbf{r}_0 + \Delta\mathbf{r}) < 0 \tag{3.279}$$

where the rightmost inequality is a direct consequence of (3.278). Now, if the Gauss law is to be trusted, then (3.279) implies that a certain amount of *negative* charge must be present within $B(\mathbf{r}_0, |\Delta\mathbf{r}|)$ and hence inside V, an occurrence which violates the initial hypothesis that the region V is charge-free. By the way, notice that q itself does not count inasmuch as it is a *test* charge (see Section 1.1). At any rate, having incurred into a glaring contradiction, we are compelled to dismiss $\mathbf{r}_0$ as a point of stable equilibrium in V. The very same conclusion is reached when we repeat this thought experiment by starting off with a negative charge.

For the second proof — under the same assumptions made above — we consider the asymptotic expansion (or truncated Taylor series) of the real-valued scalar field $\Phi(\mathbf{r}_0 + \Delta\mathbf{r})$ in a neighborhood $B(\mathbf{r}_0, a) \subset V$ of $\mathbf{r}_0 \in V$, viz.,

$$\Phi(\mathbf{r}_0 + \Delta\mathbf{r}) = \Phi(\mathbf{r}_0) + \frac{1}{2}\Delta\mathbf{r} \cdot \nabla\nabla\Phi(\mathbf{r})|_{\mathbf{r}=\mathbf{r}_0} \cdot \Delta\mathbf{r} + o\left(|\Delta\mathbf{r}|^2\right), \qquad \Delta\mathbf{r} \to 0 \tag{3.280}$$

where $\nabla\Phi(\mathbf{r})|_{\mathbf{r}=\mathbf{r}_0} = 0$ because *supposedly* $\mathbf{r}_0$ is a stationary point. The matrix associated with the dyadic $\nabla\nabla\Phi(\mathbf{r})$ is called the Hessian matrix. It is well known from calculus that the nature of $\mathbf{r}_0$ depends on the sign of the quadratic form

$$F(\Delta\mathbf{r}) := \Delta\mathbf{r} \cdot \nabla\nabla\Phi(\mathbf{r})|_{\mathbf{r}=\mathbf{r}_0} \cdot \Delta\mathbf{r}, \qquad |\Delta\mathbf{r}| \leqslant a \tag{3.281}$$

or more precisely

$$\begin{array}{lll} F(\Delta\mathbf{r}) > 0 & \Longrightarrow & \mathbf{r}_0 \text{ is a local minimum} \\ F(\Delta\mathbf{r}) < 0 & \Longrightarrow & \mathbf{r}_0 \text{ is a local maximum} \\ F(\Delta\mathbf{r}) \lesseqgtr 0 & \Longrightarrow & \mathbf{r}_0 \text{ is a saddle point} \\ F(\Delta\mathbf{r}) \lesseqqgtr 0 & \Longrightarrow & \text{higher-order derivatives needed to decide} \end{array} \qquad |\Delta\mathbf{r}| \leqslant a \tag{3.282}$$

and these conditions in turn may be checked by examining the eigenvalues λ_n, $n = 1, 2, 3$, of the dyadic $\nabla\nabla\Phi(\mathbf{r})|_{\mathbf{r}=\mathbf{r}_0}$. The latter is certainly real and — its nine Cartesian components being all the second derivatives of $\Phi(\mathbf{r})$ — also symmetric (Appendix E.2) by virtue of the Schwarz theorem [42, pp. 235–236]. Consequently, the eigenvectors $\mathbf{u}_n$ of $\nabla\nabla\Phi(\mathbf{r})|_{\mathbf{r}=\mathbf{r}_0}$ are *orthogonal* to one another and may even be chosen to be real with unitary magnitude. As a result, we may write any vector $\Delta\mathbf{r}$, $|\Delta\mathbf{r}| \leqslant a$, as a linear combination of $\{\mathbf{u}_n\}_{n=1}^3$, say,

$$\Delta\mathbf{r} = \sum_{n=1}^{3} \mathbf{u}_n c_n \qquad \text{with} \qquad c_n = \mathbf{u}_n \cdot \Delta\mathbf{r} \tag{3.283}$$

having assumed that $|\mathbf{u}_n| = 1$. With these positions we can cast the quadratic form $F(\Delta\mathbf{r})$ into an alternative format which involves the expansion coefficients c_n and the eigenvalues λ_n, namely,

$$F(\Delta\mathbf{r}) = \sum_{m=1}^{3} c_m \mathbf{u}_m \cdot \nabla\nabla\Phi(\mathbf{r})|_{\mathbf{r}=\mathbf{r}_0} \cdot \sum_{n=1}^{3} \mathbf{u}_n c_n = \sum_{m=1}^{3} c_m \mathbf{u}_m \cdot \sum_{n=1}^{3} \lambda_n \mathbf{u}_n c_n$$

$$= \lambda_1 c_1^2 + \lambda_2 c_2^2 + \lambda_3 c_3^2 \tag{3.284}$$

and hence we just have to examine the signs of λ_n, since $c_n^2 \geqslant 0$. Furthermore, by using, e.g., Cartesian coordinates we can show that

$$\mathrm{Tr}\left\{\nabla\nabla\Phi(\mathbf{r})|_{\mathbf{r}=\mathbf{r}_0}\right\} := \bar{\mathfrak{I}} : \nabla\nabla\Phi(\mathbf{r})|_{\mathbf{r}=\mathbf{r}_0} = \nabla^2\Phi(\mathbf{r})|_{\mathbf{r}_0} \tag{3.285}$$

and, since $\Phi(\mathbf{r})$ obeys the Laplace equation in $B(\mathbf{r}_0, a) \subset V$, on account of the proof (E.89), we get

$$\lambda_1 + \lambda_2 + \lambda_3 = \nabla^2\Phi(\mathbf{r})|_{\mathbf{r}_0} = 0 \tag{3.286}$$

that is, a direct and quite useful link between the Laplace equation and the eigenvalues of the dyadic $\nabla\nabla\Phi(\mathbf{r})|_{\mathbf{r}=\mathbf{r}_0}$. In particular, for this last constraint to hold true, the three *real* eigenvalues λ_n must necessarily have *different signs*.

In summary, from (3.284) it follows that — barring the occurrence of one or more null eigenvalues — the quadratic form $F(\Delta\mathbf{r})$ may take on both positive and negative values for $\mathbf{r} = \mathbf{r}_0 + \Delta\mathbf{r} \in B(\mathbf{r}_0, a)$, and the stationary point $\mathbf{r}_0$ is neither a maximum nor a minimum but at most a saddle for $\Phi(\mathbf{r})$. So, a test charge q placed in $\mathbf{r}_0$ is not in stable equilibrium — even though the Coulomb force $-q\nabla\Phi(\mathbf{r})|_{\mathbf{r}=\mathbf{r}_0}$ is indeed null — and any 'nudge', however small, may set the charge on a course which leads it away from the 'saddle'.

The third proof we consider [39, 43] is perhaps the most rigorous as it builds on the Kelvin theorem (3.276) and the fact, already mentioned, that scalar fields which obey the Laplace equation in a region $V \subset \mathbb{R}^3$ cannot have local maxima or minima for points $\mathbf{r} \in V$ [12], [11, Section 8.6].

We suppose that N isolated metallic objects with conductivities σ_n, $n = 1, \ldots, N$, occupy the volumes V_n and are immersed in free space, as is pictorially suggested in Figure 3.15. We indicate the surface charge density and the total charge on the surface ∂V_n of the nth conductor with $\varrho_{Sn}(\mathbf{r})$ and Q_n, respectively. Besides, we assume that the first conductor is free to move, whereas the remaining $N-1$ conducting bodies are held fixed in their positions. This hypothesis does not restrict the discussion at all, inasmuch as we can decide to label the bodies as we please. Last but not least, we presume that the set of conductors is indeed in *stable mechanical equilibrium*. In essence, the proof consists of showing that such assumption leads to a glaring contradiction.

We set the origin O of the main system of Cartesian coordinates (x, y, z) somewhere inside the region V_1 and suppose to displace the first conducting body along a vector $\mathbf{u}$. In order for the resulting configuration to make physical sense the shifted volume V_1' must not intersect any of the other $N-1$ regions V_m, $m = 2, \ldots, N$. We indicate the point corresponding to the position vector $\mathbf{u}$ with O_1 and also introduce a second system of Cartesian coordinates, say, (x', y', z'), with origin in O_1, so that the relation between old and new coordinates reads $\mathbf{r} = \mathbf{r}' + \mathbf{u}$ (Figure 3.15).

The rigid translation of V_1 into V_1' does not alter any of the total charges Q_n, because the latter are assigned by hypothesis. Nevertheless, for general choices of $\mathbf{u}$ the new surface densities of charge, say, $\varrho_{Sn}'(\mathbf{r})$, are most certainly different than the original ones, since the charges will redistribute themselves on the conducting surfaces so that even in the new configuration the electric energy is again a minimum, as the Kelvin theorem (3.276) dictates. Indeed, the charges Q_n being all known, the relevant potential formally solves a Neumann problem in the unbounded charge-free domain $V := (\mathbb{R}^3 \backslash \overline{V}_1')\backslash \cup_m \overline{V}_m$ [see Section 2.5.1 and in particular (2.87) and (2.89)]. For the proof of concern, though, in the new arrangement of conductors we examine the *fictitious* electric field $\boldsymbol{\mathcal{E}}'(\mathbf{r})$ generated by the *very same* charge densities $\varrho_{Sn}(\mathbf{r})$ that existed on $\cup_n \partial V_n$ prior to the translation of V_1.

We consider the finite surface-wise multiply-connected domain $D_a := (B(0, a) \setminus \overline{V}_1') \setminus \cup_m \overline{V}_m$, where the ball $B(0, a)$ is centered in O and has a radius a large enough to contain all the conducting bodies before and after the translation of V_1. We shall let a grow indefinitely so that D_a passes over

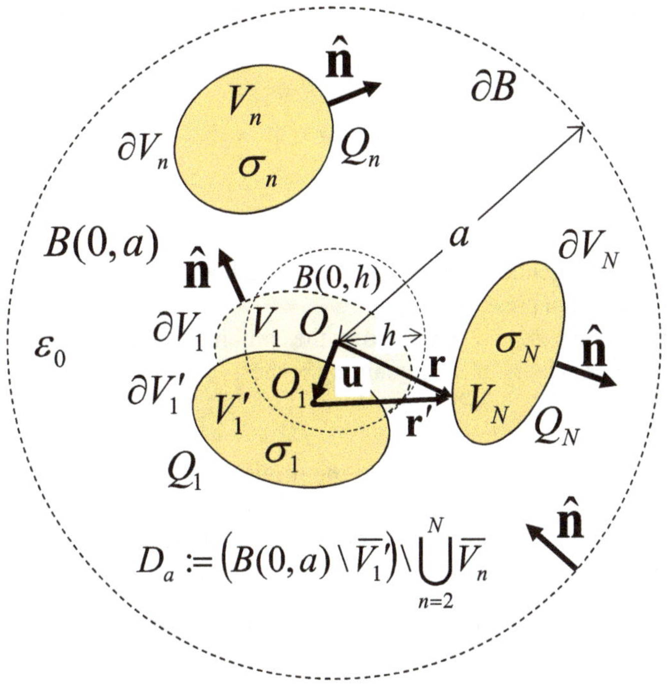

Figure 3.15 A system of N charged conductors for deriving the Earnshaw theorem.

into the unbounded region V. To lighten the notation a bit we define S_{N-1} as the multiply-connected surface $\cup_m \partial V_m$ formed with the union of the boundaries of all the fixed bodies. Thus, the boundary ∂D_a is the surface $\partial B \cup \partial V_1' \cup S_{N-1}$.

If we tailor the general expression of the electric energy (1.263) to the fictitious field $\mathcal{E}'(\mathbf{r})$ in D_a, we end up with a positive scalar function $W_e'(\mathbf{u};a)$ which depends on the translation vector $\mathbf{u}$ and, for the moment, also on the radius a. Strictly speaking, though, $W_e'(\mathbf{u};a)$ does not represent an energy inasmuch as $\mathcal{E}'(\mathbf{r})$ is a non-physical electric field. Regardless, our next goal is to show that $W_e'(\mathbf{u};a)$, as a function of $\mathbf{u}$, *solves* a Laplace equation.

To this purpose, since the laws of electrostatics (2.3) and (2.4) apply to $\mathcal{E}'(\mathbf{r})$ as well, we can introduce the associated fictitious potential $\Phi'(\mathbf{r})$ in accordance with (2.15). Further, we want to express $W_e'(\mathbf{u};a)$ in terms of $\Phi'(\mathbf{r})$ and charge densities, because the functions $\varrho_{Sn}(\mathbf{r})$ are assigned, rather than computed. As a result, the fictitious potential $\Phi'(\mathbf{r})$ is *not* necessarily constant on the surface of the conductors. In symbols, we have

$$\begin{aligned} W_e'(\mathbf{u};a) &:= \frac{1}{2}\int_{D_a} \mathrm{d}V\, \mathcal{E}'(\mathbf{r})\cdot\mathcal{D}'(\mathbf{r}) = -\frac{1}{2}\int_{D_a}\mathrm{d}V\,\nabla\cdot[\Phi'(\mathbf{r})\mathcal{D}'(\mathbf{r})] = \frac{1}{2}\int_{\partial D_a}\mathrm{d}S\,\hat{\mathbf{n}}(\mathbf{r})\cdot\mathcal{D}'(\mathbf{r})\Phi'(\mathbf{r}) \\ &= \frac{1}{2}\int_{\partial V_1'}\mathrm{d}S\,\varrho_{S1}(\mathbf{r})\Phi'(\mathbf{r}) + \sum_{m=2}^{N}\frac{1}{2}\int_{\partial V_m}\mathrm{d}S\,\varrho_{Sm}(\mathbf{r})\Phi'(\mathbf{r}) + \frac{1}{2}\int_{\partial B}\mathrm{d}S\,\hat{\mathbf{n}}(\mathbf{r})\cdot\mathcal{D}'(\mathbf{r})\Phi'(\mathbf{r}) \end{aligned} \tag{3.287}$$

having used the differential identity (H.51) under the assumption that $\mathcal{E}'(\mathbf{r})$ is divergence-free in D_a, invoked the Gauss theorem with the unit normal pointing inwards D_a, and recalled the boundary condition (1.182). The integral over ∂B is of no consequence as, for $a \to +\infty$, it tends to zero in view of the asymptotic behaviors (2.20) and (2.21).

Next, thanks to the linearity of (2.3), (2.4) and (2.15) we may split $\Phi'(\mathbf{r})$ into the sum of two contributions, namely,

$$\Phi'(\mathbf{r}) = \Phi'_1(\mathbf{r}) + \Phi'_{N-1}(\mathbf{r}), \qquad \mathbf{r} \in V := \left(\mathbb{R}^3 \setminus \overline{V'_1}\right) \setminus \bigcup_{m=2}^{N} \overline{V}_m \tag{3.288}$$

where

- $\Phi'_1(\mathbf{r})$ represents the potential due to $\varrho_{S1}(\mathbf{r}')$ on the boundary $\partial V'_1$ of the conductor that has undergone the translation prescribed by $\mathbf{u}$.
- $\Phi'_{N-1}(\mathbf{r})$ constitutes the potential generated by the combined effect of the charge densities $\varrho_{Sm}(\mathbf{r})$, $m = 2, \ldots, N$, on the surfaces of the remaining bodies.

It is important to notice that

(1) the potential $\Phi'_1(\mathbf{r}')$ — when referred to the origin O_1 — does not depend on $\mathbf{u}$, though the value of $\Phi'_1(\mathbf{r}') = \Phi'_1(\mathbf{r} - \mathbf{u})$ for $\mathbf{r}' \in S_{N-1}$ is affected by the choice of $\mathbf{u}$;
(2) the potential $\Phi'_{N-1}(\mathbf{r})$ does not depend on $\mathbf{u}$ either, but the value of $\Phi'_{N-1}(\mathbf{r}) = \Phi'_{N-1}(\mathbf{r}' + \mathbf{u})$ on $\partial V'_1$ does change with $\mathbf{u}$.

By keeping the above remarks in mind and inserting the Ansatz (3.288) in (3.287) we can write $W'_e(\mathbf{u}) := W'_e(\mathbf{u}; \infty)$ as

$$\begin{aligned} W'_e(\mathbf{u}) = \frac{1}{2}\int_{\partial V'_1} \mathrm{d}S'\, \varrho_{S1}(\mathbf{r}')\Phi'_1(\mathbf{r}') + \frac{1}{2}\int_{\partial V'_1} \mathrm{d}S'\, \varrho_{S1}(\mathbf{r}')\Phi'_{N-1}(\mathbf{r}' + \mathbf{u}) \\ + \sum_{m=2}^{N} \frac{1}{2}\int_{\partial V_m} \mathrm{d}S\, \varrho_{Sm}(\mathbf{r})\Phi'_1(\mathbf{r} - \mathbf{u}) + \sum_{m=2}^{N} \frac{1}{2}\int_{\partial V_m} \mathrm{d}S\, \varrho_{Sm}(\mathbf{r})\Phi'_{N-1}(\mathbf{r}) \end{aligned} \tag{3.289}$$

where we notice that the functional dependence on $\mathbf{u}$ manifests itself only through the potentials $\Phi'_1(\mathbf{r} - \mathbf{u})$ and $\Phi'_{N-1}(\mathbf{r}' + \mathbf{u})$. Since the latter are *harmonic functions* of the respective arguments [5, Section 8.7], [6, Section 2.2], we claim that $W'_e(\mathbf{u})$ solves the Laplace equation

$$\frac{\partial^2}{\partial u_x^2} W'_e(\mathbf{u}) + \frac{\partial^2}{\partial u_y^2} W'_e(\mathbf{u}) + \frac{\partial^2}{\partial u_z^2} W'_e(\mathbf{u}) = 0 \tag{3.290}$$

at least for points $\mathbf{u} = u_x\hat{\mathbf{x}} + u_y\hat{\mathbf{y}} + u_z\hat{\mathbf{z}}$ in a 'small' ball $B(0, h)$ centered in O (Figure 3.15).

To prove this statement we need to take the derivatives of $W'_e(\mathbf{u})$ with respect to u_x, u_y and u_z. The first and the fourth summand in the right member of (3.289) are independent of $\mathbf{u}$. The function $\varrho_{S1}(\mathbf{r}')\Phi'_{N-1}(\mathbf{r}' + \mathbf{u})$ and the derivatives thereof of any order with respect to u_x, u_y and u_z are continuous for $(\mathbf{r}', \mathbf{u}) \in \partial V'_1 \times B(0, h)$ essentially because the potential $\Phi'_{N-1}(\mathbf{r}' + \mathbf{u})$ is evaluated on $\partial V'_1$ and away from S_{N-1}. In like manner, the functions $\varrho_{Sm}(\mathbf{r})\Phi'_1(\mathbf{r} - \mathbf{u})$ and the derivatives thereof of any order with respect to u_x, u_y and u_z are continuous for $(\mathbf{r}, \mathbf{u}) \in \partial V_m \times B(0, h)$ essentially because the potential $\Phi'_1(\mathbf{r} - \mathbf{u})$ is evaluated on ∂V_m and away from $\partial V'_1$. Long story short, we are allowed to swap integration and differentiation. For instance, we have

$$\frac{\partial^2}{\partial u_x^2}\int_{\partial V'_1} \mathrm{d}S'\, \varrho_{S1}(\mathbf{r}')\Phi'_{N-1}(\mathbf{r}' + \mathbf{u}) = \int_{\partial V'_1} \mathrm{d}S'\, \varrho_{S1}(\mathbf{r}')\frac{\partial^2}{\partial u_x^2}\Phi'_{N-1}(\mathbf{r}' + \mathbf{u})$$

$$= \int_{\partial V_1'} dS' \varrho_{S1}(\mathbf{r}') \frac{\partial^2}{\partial x^2} \Phi_{N-1}'(\mathbf{r}) \tag{3.291}$$

and similarly for the derivatives with respect to u_y and u_z. The same steps may be applied to the third term in the right-hand side of (3.289). Summing the six partial results side by side shows that $W_e'(\mathbf{u})$ indeed solves the Laplace equation (3.290) in $B(0,h)$ because so do $\Phi_{N-1}'(\mathbf{r})$ and $\Phi_1'(\mathbf{r}')$ in the unbounded region V.

One more key observation is that $W_e'(0)$ coincides with the energy of the original configuration of conductors sketched in Figure 3.15. The assertion follows from the very definition (1.263) applied to the *true field* $\mathcal{E}(\mathbf{r})$ and the fact that, when $\mathbf{u} = 0$, $\Phi'(\mathbf{r})$ reduces to the *true potential* $\Phi(\mathbf{r})$. Besides, $W_e'(0)$ ought to be a minimum for $W_e'(\mathbf{u})$ by virtue of the Kelvin theorem (3.276), because we have assumed the original setup to be in static equilibrium. And yet, as $W_e'(\mathbf{u})$ is a solution to the Laplace equation (3.290) in $B(0,h)$, it *cannot* have a minimum for $\mathbf{u} = 0$ [12], [11, Section 8.6].

On the contrary, suppose that by happenstance $W_e'(\mathbf{u}) > W_e'(0)$ actually holds for any vector $\mathbf{u} \in B(0,h)$, $\mathbf{u} \neq 0$. Then, from the second mean value theorem of electrostatics (2.173) — which holds on the grounds that $W_e'(\mathbf{u})$ solves (3.290) — we get

$$\frac{4\pi}{3} h^3 W_e'(0) = \int_{B(0,h)} dV\, W_e'(\mathbf{u}) \qquad \Longrightarrow \qquad \int_{B(0,h)} dV\, [W_e'(\mathbf{u}) - W_e'(0)] = 0 \tag{3.292}$$

an identity which cannot be true, since the integrand is always non-negative owing to our hypothesis. It follows that $\mathbf{u} = 0$ is not a local minimum for $W_e'(\mathbf{u})$, though, by the same token, $\mathbf{u} = 0$ cannot be a local maximum either. Therefore, $W_e'(\mathbf{u})$ may at best have a saddle for $\mathbf{u} = 0$, and hence the condition

$$W_e'(\mathbf{u}) < W_e'(0), \qquad \mathbf{u} \in B(0,h) \tag{3.293}$$

must hold for some suitable translation vector $\mathbf{u}$. Then again, since the Kelvin theorem says that the energy of a system of charges in static equilibrium is a minimum, the *true energy* $W_e(\mathbf{u})$ of the new distribution of N conductors may not exceed $W_e'(\mathbf{u})$, the 'energy' of the *fictitious* field. Thus, we obtain the chain of inequalities

$$W_e(\mathbf{u}) < W_e'(\mathbf{u}) < W_e'(0), \qquad \mathbf{u} \in B(0,h) \tag{3.294}$$

whence, in particular, we conclude

$$W_e(\mathbf{u}) < W_e'(0), \qquad \mathbf{u} \in B(0,h) \tag{3.295}$$

that is to say, the energy of the configuration with the shifted conductor happens to be *smaller* than the energy of the original arrangement of conducting bodies. This final finding patently clashes with the initial hypothesis that the original configuration is mechanically stable, which consequently is proven fallacious. Clearly, one can repeat this reasoning all over again by starting with the new configuration, and so on so forth, never to find a stable arrangement, and this proves the Earnshaw theorem.

The historical importance of Earnshaw's theorem lies in the dire consequences that the result bears on the soundness of the atomic models and, ultimately, the very stability of matter as we know it [44]. Specifically, since at atomic level the interactions between electrons and protons are of electrical nature, the planetary model of the atom proposed by Rutherford (1911) cannot be stable.

At a closer look one realizes that the Earnshaw theorem applies to electrostatic fields, because the Coulomb force between two electric charges obeys an *inverse-square law*, that is, the force as well as the static field fall off with the inverse square of the distance from the source (see Examples 2.1 and 2.2). Likewise, in the context of Newton's theory of gravitation [45, Chapter 7], [11, Chapter 1] also the gravitational force is governed by an inverse-square law, and thus, mutatis mutandis, the Earnshaw theorem prevents a system of N material bodies with masses M_n, $n = 1, \ldots, N$, from being in stable equilibrium under the sole action of gravity.

On the other hand, the Earnshaw theorem is not necessarily true for magnetic forces and related stationary fields, as stable *magnetic levitation* can indeed be achieved. This happens partly because the sources of magnetic fields are, in effect, charges in motion rather than fixed ones, and partly because the corresponding magnetic potential $\Psi(\mathbf{r})$ (Section 4.4) can have local minima in the presence of diamagnetic materials (Section 5.6) [46]. This does not mean that the Earnshaw theorem is violated, but rather that it does not apply when the underlying hypotheses are not met in the first place.

3.9 Image principle in electrostatics

Broadly speaking, the image principle [1, 8], [11, Section 9.1], [4, Section 8.3] may be regarded as a tool for solving the electrostatic equations (2.3) and (2.4) in a dielectric medium bounded by conducting interfaces. Specifically, instead of tackling the original problem directly, one tries to cast it into an equivalent form which, hopefully, is easier to solve either analytically or with the aid of some numerical technique. In this regard, the image principle is a special case of the general surface equivalence principles that we shall address in Section 10.4.

In the equivalent problem the conductors are removed, as it were, and the effect thereof is *exactly* accounted for by *image charges* placed in suitable points within the regions that are occupied by the conductors in the original problem. In this way, all charges — physical and image ones — exist in an unbounded dielectric medium, and one may just apply (2.169) if the medium is homogeneous or (2.160) if the region of interest contains dielectric bodies which we exclude.

In electrostatics the procedure we have just outlined is *always possible theoretically*, no matter how complicated the shape of the conductors. However, in practice the locations and the values of the image charges can be found in closed form only when the conducting interfaces possess canonical shapes such as, e.g., planes, cylinders or spheres. In the following, we describe the image principle in details for the case of an infinite planar conducting interface. We show that the solution obtained by means of the equivalent problem is the correct one for the original one in two different ways.

To begin with, suppose we wish to determine the electrostatic field generated by a point charge q located at $\mathbf{r}'$ inside a homogeneous dielectric half space $V := \{\mathbf{r} \in \mathbb{R}^3 : z > 0\}$ in the presence of a conducting half space $V' := \mathbb{R}^3 \setminus \overline{V}$, as is sketched in Figure 3.16a; we refer to the domain V as the physical space. By virtue of (2.14) we conclude that

$$\mathcal{E}(\mathbf{r}; q) = \frac{q(\mathbf{r} - \mathbf{r}')}{4\pi\varepsilon|\mathbf{r} - \mathbf{r}'|^3}, \qquad \mathbf{r} \in \mathbb{R}^3, \qquad \mathbf{r}' \in V \tag{3.296}$$

is the field produced by q in an unbounded homogeneous dielectric medium. Evidently, this field alone satisfies neither the boundary condition $\mathcal{E}(\mathbf{r}) \times \hat{\mathbf{z}} = 0$ for points $\mathbf{r} \in S_0 := \{\mathbf{r} \in \mathbb{R}^3 : z = 0\}$ nor the requirement that $\mathcal{E}(\mathbf{r})$ be zero for $\mathbf{r} \in V'$. Hence, $\mathcal{E}(\mathbf{r}; q)$ cannot be the solution in the presence of the conducting half space. In fact, the field lines of $\mathcal{E}(\mathbf{r}; q)$ are straight, diverge from the location of q, and form an angle with the interface S_0, whereas the electric field we are looking for should be perpendicular to S_0 in accordance with (1.180). So, can we find a combination of charges in an *unbounded* homogeneous medium that produce an electric field with $\mathcal{E}(\mathbf{r}) \times \hat{\mathbf{z}} = 0$ for $\mathbf{r} \in S_0$?

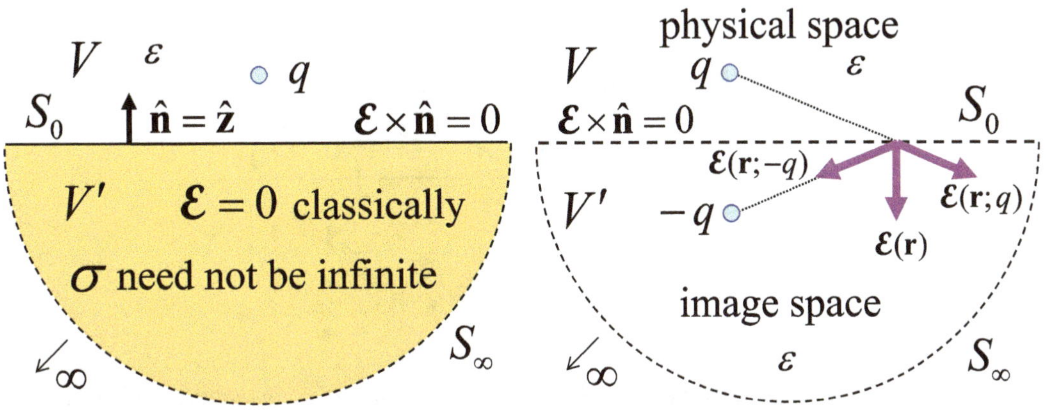

(a) *original problem* (b) *equivalent problem*

Figure 3.16 The image principle applied to the calculation of the electrostatic field generated by a charge (○) in the presence of a conducting half space.

Well, intuition and simple geometrical considerations suggest that we may fulfill this requirement if we combine $\mathcal{E}(\mathbf{r}; q)$ with the field $\mathcal{E}(\mathbf{r}; -q)$ produced by a charge $-q$ symmetrically placed across S_0 from q, as is illustrated in Figure 3.16b. We call the complementary domain V' the *image space* and the charge $-q$ the image charge. We choose the negative of the original charge so that the corresponding field lines are directed *towards* the charge. It is a straightforward matter to check that $\mathcal{E}(\mathbf{r}) = \mathcal{E}(\mathbf{r}; q) + \mathcal{E}_2(\mathbf{r}; -q)$ is perpendicular everywhere to S_0.

Does this nice result solve our original problem, though? Yes and no! For one thing, the field produced by q and $-q$ is *not null* in the image space, whereas the problem of q in the presence of the conductor demands for $\mathcal{E}(\mathbf{r})$ to be zero for $\mathbf{r} \in V'$. But then, $\mathcal{E}(\mathbf{r})$ does satisfy the prescribed boundary conditions on S_0 by construction. What is more, $\mathcal{E}(\mathbf{r})$ is also regular for $|\mathbf{r}| \to +\infty$, $\mathbf{r} \in V$ (and $\mathbf{r} \in V'$). Since both the original problem and the equivalent one obey the same conditions on the boundary of the domain, the two solutions must coincide in V according to the uniqueness results of Section 2.5.2.

In summary, the application of the image principle consists of three steps.

(1) Replace the conducting half space in V' with a dielectric medium endowed with the same permittivity as that of the medium filling V.
(2) Introduce a fictitious image charge in the image space V'.
(3) Solve the equivalent problem in the whole space and use the result in the physical space V.

We remark that by applying the image principle we have found a solution which is perfectly equivalent to the original one in the physical space, though not correct in the image space. This does not mean that the equivalent solution is anywhere wrong, rather, that it does not represent the solution to the original problem in the image space.

We now follow an alternative approach based on the Dirichlet Green function of Example 3.2 and the integral representation of the scalar potential to prove the image principle again in the situation of Figure 3.16a. To make the discussion a little more general we consider a charge density $\varrho(\mathbf{r})$ for $\mathbf{r} \in V_\varrho \subset V$ as well as other dielectric bodies confined within a bounded region V_1 (Figure 3.17). As a result, the physical space is 'limited' by $S_0 \cup \partial V_1$ and the half sphere at infinity. We take the unit normal on ∂V positively directed inwards V.

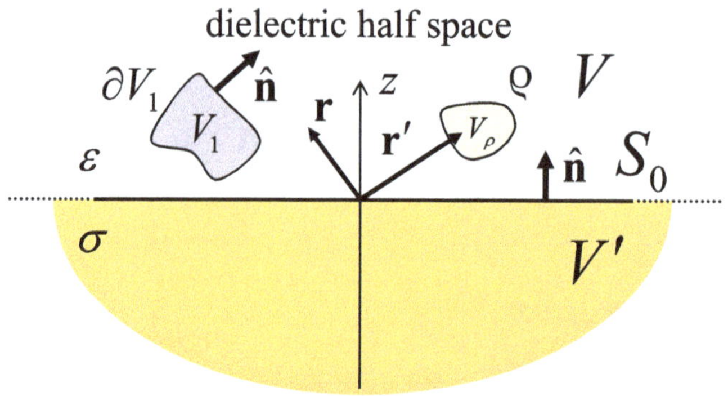

Figure 3.17 Charges and dielectric bodies in the presence of a conducting half space.

Then, the differential problem

$$\begin{cases} \varepsilon \nabla^2 \Phi(\mathbf{r}) = -\varrho(\mathbf{r}) & \mathbf{r} \in V \\ \Phi(\mathbf{r}) = 0 & \mathbf{r} \in S_0 \\ \Phi(\mathbf{r}) = \mathcal{O}\left(\dfrac{1}{|\mathbf{r}|}\right) & |\mathbf{r}| \to +\infty \\ \Phi(\mathbf{r}) = f(\mathbf{r}) \quad \text{or} \quad \dfrac{\partial \Phi}{\partial \hat{\mathbf{n}}} = g(\mathbf{r}) & \mathbf{r} \in \partial V_1 \end{cases} \tag{3.297}$$

admits a unique solution. Besides, from Section 2.7 we know that $\Phi(\mathbf{r})$ for $\mathbf{r} \in V$ can be written as a combination of suitable volume, single-layer and double-layer potentials, viz.,

$$\Phi(\mathbf{r}) = \frac{1}{\varepsilon} \int_{V_\varrho} \mathrm{d}V' G(\mathbf{r}, \mathbf{r}')\varrho(\mathbf{r}') + \int_{\partial V_1} \mathrm{d}S' \left[\Phi(\mathbf{r}') \frac{\partial G}{\partial \hat{\mathbf{n}}'} - G(\mathbf{r}, \mathbf{r}') \frac{\partial \Phi}{\partial \hat{\mathbf{n}}'}\right] - \int_{S_0} \mathrm{d}S' G(\mathbf{r}, \mathbf{r}') \frac{\partial \Phi}{\partial \hat{\mathbf{n}}'} \tag{3.298}$$

where $G(\mathbf{r}, \mathbf{r}')$ is the Green function (2.131) relevant to an unbounded homogeneous isotropic medium. There is no contribution from a double-layer potential over S_0, because $\Phi(\mathbf{r})$ vanishes thereon by hypothesis. The integral over S_0 represents the potential due to the single layer of charges induced on the conducting interface.

Then again, we are entitled to use any Green function to write down the integral representation of $\Phi(\mathbf{r})$. If we choose the Dirichlet Green function $G_D(\mathbf{r}, \mathbf{r}')$ given by (3.37), for points $\mathbf{r} \in V$ we find

$$\Phi(\mathbf{r}) = \frac{1}{\varepsilon} \int_{V_\varrho} \mathrm{d}V' G_D(\mathbf{r}, \mathbf{r}')\varrho(\mathbf{r}') + \int_{\partial V_1} \mathrm{d}S' \left[\Phi(\mathbf{r}') \frac{\partial G_D}{\partial \hat{\mathbf{n}}'} - G_D(\mathbf{r}, \mathbf{r}') \frac{\partial \Phi}{\partial \hat{\mathbf{n}}'}\right] \tag{3.299}$$

where even the contribution from the single-layer potential on S_0 vanishes in that thereon $G_D(\mathbf{r}, \mathbf{r}') = 0$ by construction. Notice that right-hand sides of (3.298) and (3.299), though formally different, must return the same result for $\mathbf{r} \in V$, inasmuch as the solution is unique in V. To elaborate, we split the Dirichlet Green function as

$$G_D(\mathbf{r}, \mathbf{r}') = G(\mathbf{r}, \mathbf{r}') + G_I(\mathbf{r}, \mathbf{r}') \tag{3.300}$$

where $G(\mathbf{r}, \mathbf{r}')$ and $G_I(\mathbf{r}, \mathbf{r}')$ can be identified by comparison with the explicit expression (3.36). In particular, the first contribution is the same as (2.131), whereas the index I in the other term stands for *image* — not surprisingly — and the reason for this will become clear in a moment. We rewrite (3.299) by separating the terms that involve $G(\mathbf{r}, \mathbf{r}')$ and $G_I(\mathbf{r}, \mathbf{r}')$, namely,

$$\Phi(\mathbf{r}) = \frac{1}{\varepsilon} \int_{V_\varrho} \mathrm{d}V' G(\mathbf{r}, \mathbf{r}')\varrho(\mathbf{r}') + \int_{\partial V_1} \mathrm{d}S' \left[\Phi(\mathbf{r}') \frac{\partial G}{\partial \hat{\mathbf{n}}'} - G(\mathbf{r}, \mathbf{r}') \frac{\partial \Phi}{\partial \hat{\mathbf{n}}'} \right]$$
$$+ \frac{1}{\varepsilon} \int_{V_\varrho} \mathrm{d}V' G_I(\mathbf{r}, \mathbf{r}')\varrho(\mathbf{r}') + \int_{\partial V_1} \mathrm{d}S' \left[\Phi(\mathbf{r}') \frac{\partial G_I}{\partial \hat{\mathbf{n}}'} - G_I(\mathbf{r}, \mathbf{r}') \frac{\partial \Phi}{\partial \hat{\mathbf{n}}'} \right] \quad (3.301)$$

and we compare this last form with (3.298). It becomes apparent that the single-layer contribution from S_0 must coincide with the integrals involving $G_I(\mathbf{r}, \mathbf{r}')$ for $\mathbf{r} \in V$. More importantly, the latter can be interpreted as the potentials produced by fictitious charges and matter symmetrically placed across S_0 in the image space V'. To convince ourselves that this is indeed true we need to cast the integrals in a form which involves the usual Green function (2.131), at least formally.

We begin with the volume integral and define the coordinates

$$z_1 := \inf_{\mathbf{r}' \in V_\varrho} z', \qquad z_2 := \sup_{\mathbf{r}' \in V_\varrho} z' \quad (3.302)$$

so that, in words, the region V_ϱ is confined between the planes $z' = z_l$, $l = 1, 2$, and any other plane $z' = z_0$ with $z_1 < z_0 < z_2$ intersects V_ϱ along the planar surface $S(z')$. With these definitions and in light of (3.37) we may write the integral as

$$\begin{aligned}
\frac{1}{\varepsilon} \int_{V_\varrho} \mathrm{d}V'\, G_I(\mathbf{r}, \mathbf{r}')\varrho(\mathbf{r}') &= \frac{1}{\varepsilon} \int_{V_\varrho} \mathrm{d}V' \frac{-\varrho(\mathbf{r}')}{4\pi \left| \mathbf{r} - \overline{\mathcal{U}} \cdot \mathbf{r}' \right|} \\
&= \frac{1}{\varepsilon} \int_{z_1}^{z_2} \mathrm{d}z' \int_{S(z')} \mathrm{d}S' \frac{-\varrho(\mathbf{r}')}{4\pi |\mathbf{r} - (x'\hat{\mathbf{x}} + y'\hat{\mathbf{y}} - z'\hat{\mathbf{z}})|} \\
&= \frac{1}{\varepsilon} \int_{-z_2}^{-z_1} \mathrm{d}\zeta' \int_{S(-\zeta')} \mathrm{d}S' \frac{-\varrho\overbrace{(x', y', -\zeta')}^{\overline{\mathcal{U}} \cdot \boldsymbol{\tau}'}}{4\pi |\mathbf{r} - \underbrace{(x'\hat{\mathbf{x}} + y'\hat{\mathbf{y}} + \zeta'\hat{\mathbf{z}})}_{\boldsymbol{\tau}'}|} \\
&= -\frac{1}{\varepsilon} \int_{V_{\varrho I}} \mathrm{d}V'_\tau\, \varrho\left(\overline{\mathcal{U}} \cdot \boldsymbol{\tau}'\right) G(\mathbf{r}, \boldsymbol{\tau}')
\end{aligned} \quad (3.303)$$

having changed the dummy integration variable z' into $-\zeta'$ and introduced the position vector $\boldsymbol{\tau}'$, as indicated. The dyadic $\overline{\mathcal{U}}$ was defined in (3.35), the differential $\mathrm{d}V'_\tau$ signifies integration with respect to $\boldsymbol{\tau}'$, and $V_{\varrho I}$ is the specular image of V_ϱ by construction. In the last step we have recognized that, as expected, the kernel is the three-dimensional static Green function (2.131) except for a different name of the source point.

We proceed with the potential of the single layer over ∂V_1 and define

$$z_1 := \inf_{\mathbf{r}' \in \partial V_1} z', \qquad z_2 := \sup_{\mathbf{r}' \in \partial V_1} z' \quad (3.304)$$

so that the closed surface ∂V_1 is confined between the planes $z' = z_l$, $l = 1, 2$, and any other plane $z' = z_0$ with $z_1 < z_0 < z_2$ intersects ∂V_1 along the planar closed curve $\gamma(z')$. We manipulate the integral as follows

$$
\begin{aligned}
-\int_{\partial V_1} \mathrm{d}S'\, G_I(\mathbf{r}, \mathbf{r}') \frac{\partial \Phi}{\partial \hat{\mathbf{n}}'} &= \int_{\partial V_1} \mathrm{d}S' \frac{\hat{\mathbf{n}}(\mathbf{r}') \cdot \nabla' \Phi(\mathbf{r}')}{4\pi \left|\mathbf{r} - \overline{\mathcal{U}} \cdot \mathbf{r}'\right|} \\
&= \int_{z_1}^{z_2} \mathrm{d}z' \oint_{\gamma(z')} \mathrm{d}s' \frac{\hat{\mathbf{n}}(\mathbf{r}') \cdot \nabla' \Phi(\mathbf{r}')}{4\pi |\mathbf{r} - (x'\hat{\mathbf{x}} + y'\hat{\mathbf{y}} - z'\hat{\mathbf{z}})|} \\
&= \int_{-z_2}^{-z_1} \mathrm{d}\zeta' \oint_{\gamma(-\zeta')} \mathrm{d}s' \frac{\hat{\mathbf{n}}\left(\overline{\mathcal{U}} \cdot \boldsymbol{\tau}'\right) \cdot \overline{\mathcal{U}} \cdot \nabla'_{\tau} \Phi\left(\overline{\mathcal{U}} \cdot \boldsymbol{\tau}'\right)}{4\pi |\mathbf{r} - \boldsymbol{\tau}'|} \\
&= \frac{1}{\varepsilon} \int_{\partial V_{1I}} \mathrm{d}S'_{\tau}\, G(\mathbf{r}, \boldsymbol{\tau}') \left[\varepsilon \hat{\mathbf{n}}\left(\overline{\mathcal{U}} \cdot \boldsymbol{\tau}'\right) \cdot \overline{\mathcal{U}} \cdot \nabla'_{\tau} \Phi\left(\overline{\mathcal{U}} \cdot \boldsymbol{\tau}'\right)\right]
\end{aligned}
\tag{3.305}
$$

having changed z' into $-\zeta'$ and introduced $\boldsymbol{\tau}'$ as before. The differential $\mathrm{d}S'_{\tau}$ signifies integration with respect to $\boldsymbol{\tau}'$, and ∂V_{1I} is the specular image of ∂V_1 by construction. The change of dummy variable has an effect on the calculation of the gradient, viz.,

$$
\nabla' \Phi(\mathbf{r}') = \hat{\mathbf{x}} \frac{\partial \Phi}{\partial x'} + \hat{\mathbf{y}} \frac{\partial \Phi}{\partial y'} + \hat{\mathbf{z}} \frac{\partial \Phi}{\partial z'} = \hat{\mathbf{x}} \frac{\partial \Phi}{\partial x'} + \hat{\mathbf{y}} \frac{\partial \Phi}{\partial y'} - \hat{\mathbf{z}} \frac{\partial \Phi}{\partial \zeta'} = \overline{\mathcal{U}} \cdot \nabla'_{\tau} \Phi(\overline{\mathcal{U}} \cdot \boldsymbol{\tau}')
\tag{3.306}
$$

where ∇'_{τ} denotes differentiation with respect to $\boldsymbol{\tau}'$. The vector $\hat{\mathbf{n}}(\overline{\mathcal{U}} \cdot \boldsymbol{\tau}') \cdot \overline{\mathcal{U}}$ represents the unit normal to ∂V_{1I} and is the specular image of $\hat{\mathbf{n}}(\mathbf{r}')$ on ∂V_1.

Finally, we modify the contribution of the double layer

$$
\begin{aligned}
\int_{\partial V_1} \mathrm{d}S'\, \Phi(\mathbf{r}') \frac{\partial G_I}{\partial \hat{\mathbf{n}}'} &= \int_{\partial V_1} \mathrm{d}S'\, \Phi(\mathbf{r}') \hat{\mathbf{n}}(\mathbf{r}') \cdot \nabla' \frac{-1}{4\pi \left|\mathbf{r} - \overline{\mathcal{U}} \cdot \mathbf{r}'\right|} \\
&= \int_{z_1}^{z_2} \mathrm{d}z' \oint_{\gamma(z')} \mathrm{d}s'\, \Phi(\mathbf{r}') \hat{\mathbf{n}}(\mathbf{r}') \cdot \nabla' \frac{-1}{4\pi |\mathbf{r} - (x'\hat{\mathbf{x}} + y'\hat{\mathbf{y}} - z'\hat{\mathbf{z}})|} \\
&= \int_{-z_2}^{-z_1} \mathrm{d}\zeta' \oint_{\gamma(-\zeta')} \mathrm{d}s'\, \Phi\left(\overline{\mathcal{U}} \cdot \boldsymbol{\tau}'\right) \hat{\mathbf{n}}\left(\overline{\mathcal{U}} \cdot \boldsymbol{\tau}'\right) \cdot \overline{\mathcal{U}} \cdot \nabla'_{\tau} \frac{-1}{4\pi |\mathbf{r} - \boldsymbol{\tau}'|} \\
&= \frac{1}{\varepsilon} \int_{\partial V_{1I}} \mathrm{d}S'_{\tau} \left[-\varepsilon \Phi\left(\overline{\mathcal{U}} \cdot \boldsymbol{\tau}'\right) \hat{\mathbf{n}}\left(\overline{\mathcal{U}} \cdot \boldsymbol{\tau}'\right) \cdot \overline{\mathcal{U}}\right] \cdot \nabla'_{\tau} G(\mathbf{r}, \boldsymbol{\tau}')
\end{aligned}
\tag{3.307}
$$

with the same steps employed above for the single layer.

Now we may interpret these contributions by comparing them with the corresponding integrals in (3.298) and we might as well rename the dummy variable $\boldsymbol{\tau}'$ to $\mathbf{r}'$.

- The integral over $V_{\varrho I}$ represents the potential of a charge density $-\varrho(x', y', -\zeta')$ in an unbounded homogeneous isotropic medium; owing to the minus sign and the dependence on $-\zeta'$, this charge is the negative of $\varrho(\mathbf{r}')$ and is virtually located in the image space V' (Figure 3.18).
- The integral of $\varepsilon \hat{\mathbf{n}}(\overline{\mathcal{U}} \cdot \boldsymbol{\tau}') \cdot \overline{\mathcal{U}} \cdot \nabla'_{\tau} \Phi(\overline{\mathcal{U}} \cdot \boldsymbol{\tau}')$ is the potential of a single layer of charges on ∂V_{1I} in an unbounded homogeneous isotropic medium; due to the dependence on $-\zeta'$, this surface

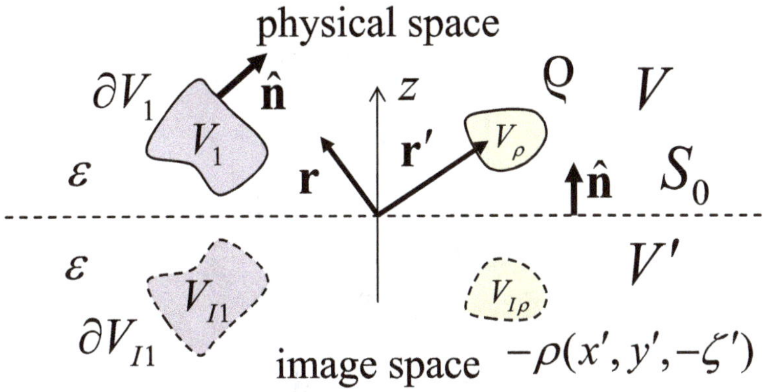

Figure 3.18 Equivalent problem for the potential obtained with the image principle.

charge density is virtually located in V'. Since the electric field is the negative of $\nabla'_\tau \Phi(\overline{\mathcal{U}} \cdot \boldsymbol{\tau}')$, the charge is the negative of the single layer on ∂V_1.

- The integral of $-\varepsilon\Phi(\overline{\mathcal{U}} \cdot \boldsymbol{\tau}')\hat{\mathbf{n}}(\overline{\mathcal{U}} \cdot \boldsymbol{\tau}') \cdot \overline{\mathcal{U}}$ constitutes the potential of a double layer of charges or a single layer of electric dipoles on ∂V_{I1} in an unbounded homogeneous isotropic medium; the dependence on $-\zeta'$ implies that the dipoles are virtually located in V'. Since $-\hat{\mathbf{n}}(\overline{\mathcal{U}} \cdot \boldsymbol{\tau}') \cdot \overline{\mathcal{U}}$ is the unit normal vector pointing *inwards* ∂V_{I1}, we see that the imaging process preserves the orientation of the z-component of an electric dipole moment and inverts the orientation of the component parallel to S_0. This conclusion can also be reached by considering the image of two charges with opposites sign in Figure 3.16b.

All in all, this discussion provides a more sound proof of the image principle in the case where the conducting interface is a plane. In the process we have shown that not only charges but also entire regions of space may be imaged, since ∂V_{1I} is indeed the image of the boundary ∂V_1. Furthermore, not only is our approach predicated on the availability of the relevant Dirichlet Green function, it shows that applying the image principle is actually equivalent to using $G_D(\mathbf{r}, \mathbf{r}')$, and this corroborates the statement that the image principle in electrostatics may always be applied, since the Dirichlet Green function may always be conceptually defined.

3.10 Singular electric fields

The singular behavior of the electric field encountered so far is strictly related to the notion of point charge. Besides, in the derivation of the boundary conditions and of the integral representations for $\mathcal{E}(\mathbf{r})$ and $\Phi(\mathbf{r})$ we have repeatedly assumed the smoothness of the interface and of the boundary of the region of interest. Smoothness is essentially required for the applicability of the Gauss and the Stokes theorems, since one needs to define local normals and tangents univocally (Appendix F).

Although objects with truly sharp corners or tips are an abstraction, in many practical applications — e.g., propagation of waves in a urban environment in the presence of high-rise buildings — one is interested in computing electromagnetic fields in the proximity of material boundaries which possess a very steep curvature. In which case, it is convenient to make the assumption of sharp corners or tips to facilitate the mathematical description of the problem. This simplification, though, comes at a price, namely, the electromagnetic field becomes *singular* (infinite) at the corner or tip in question, and the order of infinity depends on the local geometry of the boundary [47]. We may anticipate, as is logical, that the sharper a corner or tip is, the more singular the field is. Nonetheless,

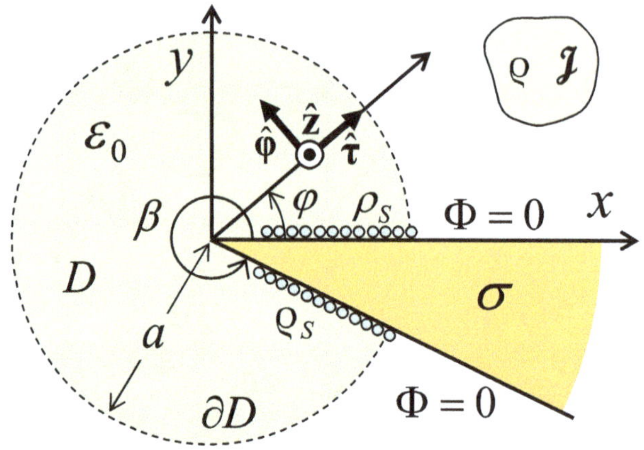

Figure 3.19 Geometry for studying potential, electric field and induced charges near the corner of a grounded conducting wedge.

we shall see that the electromagnetic energy in a region which includes the sharp boundary remains finite.

While the geometries which may give rise to singular fields are countless, it turns out that we need to investigate just a few cases involving canonical shapes, namely, the sharp edge of a wedge and the tip of a cone or a pyramid. The rationale is that the singular behavior is, in fact, confined to the immediate surroundings of edges or tips, so that, for the analysis, we may *locally* approximate the body in question as a wedge or a cone or a pyramid [47]. In particular, in the case of a wedge, the mathematical treatment further simplifies, since we may cleverly choose a system of circular cylindrical coordinates in which the z-axis is aligned with the corner. As a result, the problem becomes two-dimensional, as the geometry is independent of z.

More importantly, even though it appears that we need the full-fledged system of Maxwell equations for the study of the singularities of general time-dependent fields, it turns out that for our investigations we may actually apply electrostatics or the laws of stationary fields (see Chapter 4), and the reason for this simplification is as follows. First of all, whether the object is metallic or dielectric, the true sources of the field will induce conduction or polarization charges, respectively, on the boundary and, in particular, near the edges or tips thereof. These charges, in turn, act as sources of secondary fields which are indeed singular. Secondly, we know from (1.252) that the electric field produced by a moving charge amounts to two contributions, namely, the acceleration field and the Coulomb field. While the former is dominant far away from the sources but may even be absent, the latter is substantial in the neighborhood of the sources and is invariably present! Therefore, we see that the secondary fields close to the sources induced on the boundary of a metallic or dielectric body are essentially static or steady (cf. Example 9.3). Consequently, if we wish to study the singular character of the electric field around a sharp corner or tip, we may employ the laws of electrostatics.

The analysis of the singularities at the tip of a cone or a pyramid is more involved (see [47]), so we content ourselves with studying the behavior of the fields around the sharp edge of a metallic wedge in free space [3, Section 5.2], [38, Section 1.5]. For general time-dependence we assume the wedge to be a PEC, but the conductivity σ may be finite for truly electrostatic fields. We introduce a system of circular cylindrical coordinates (τ, φ, z) in which the z-axis coincides with the corner of the wedge. We abandon the more common notation ρ (Appendix A.1) for the radial coordinate in favor of τ, so as to avoid confusion with the charge density. One side of the wedge lies in the half plane $\varphi = 0$, the other one in the half plane $\varphi = \beta$ with $\beta \in]0, 2\pi]$, so that the aperture of the wedge is given by $2\pi - \beta$. The limiting case $\beta = 2\pi$ corresponds to an infinitely thin conducting half plane. We let $V := \{\mathbf{r} \in \mathbb{R}^3 : 0 < \varphi < \beta\}$ be the unbounded domain complementary to the region occupied by the wedge.

We consider a 'small' cylindrical region $D := \{\mathbf{r} \in \mathbb{R}^3 : 0 \leqslant \tau < a, 0 < \varphi < \beta, 0 < z < h\}$ containing part of the edge and no sources (Figure 3.19). We choose the radius a and the height h in such a way that the time rate of variation of $\mathfrak{B}(\mathbf{r}, t)$ and $\mathfrak{D}(\mathbf{r}, t)$ is negligible within D. Therefore, the electric field close to the corner of the wedge satisfies (2.3) and (2.4) subject to the jump conditions (1.169) or (1.180) on $\partial D \cap \partial V$. As a result, we may introduce the scalar potential $\Phi(\mathbf{r})$ in accordance with (2.15) and obtain the Laplace equation

$$\begin{cases} \nabla^2\Phi(\mathbf{r}) := \dfrac{1}{\tau}\dfrac{\partial}{\partial \tau}\left(\tau\dfrac{\partial \Phi}{\partial \tau}\right) + \dfrac{1}{\tau^2}\dfrac{\partial^2 \Phi}{\partial \varphi^2} + \dfrac{\partial^2 \Phi}{\partial z^2} = 0, & \mathbf{r} \in D \\ \Phi(\mathbf{r}) := \Phi(\tau, \varphi, z) = 0, & \mathbf{r} \in \partial D \cap \partial V \end{cases} \tag{3.308}$$

where the Dirichlet boundary condition for $\mathbf{r} \in \partial D \cap \partial V$ follows by assuming that the wedge is grounded. Then again, (3.308) cannot be solved *univocally* because we have not included proper matching conditions on $\partial D \cap V$. This lack of determinacy is tolerable for our purposes, since we are actually interested in the functional dependence of $\Phi(\mathbf{r})$ for small values of the radial coordinate.

We try a representation of $\Phi(\mathbf{r})$ as a linear combination of elementary solutions of the Laplace equation in circular cylindrical coordinates. We follow the general procedure of Section 3.5.2 and notice that from the viewpoint of an observer sitting very close to the corner of the wedge, the domain D appears practically *unbounded* along the z-direction. In which case the appropriate functional dependence with z is provided by (3.165) with a *real* separation constant k_z. The behavior along the azimuthal direction ought to incorporate the desired boundary condition on the sides of the wedge, which in accordance with (3.90) requires

$$F_{\nu_n}(\varphi) = c_{\nu_n} \sin(\nu_n \varphi), \qquad \varphi \in [0, \beta], \qquad \nu_n = \frac{n\pi}{\beta}, \qquad n = 1, 2, \dots \tag{3.309}$$

having traded the index m for ν_n, since the separation constant is not a whole number in general. The homogeneous Dirichlet condition imposed on $\Phi(\mathbf{r})$ for $\mathbf{r} \in \partial D \cap \partial V$ restricts ν_n to a set of discrete *eigenvalues*, as indicated. The index n cannot be zero because the corresponding function vanishes identically. Lastly, as we have already constrained k_z and ν_n, we are led to choose the dependence on τ in the form

$$B_{\nu_n}(\tau; k_z) = J_{\nu_n}(k_z \tau) \tag{3.310}$$

having noticed that the Bessel functions of the second kind $Y_{\nu_n}(k_z\tau)$ must be discarded if the scalar potential is to remain finite at the very corner of the wedge.

By putting everything together as in (3.148) we get

$$\begin{aligned} &\Psi_{\nu_n}(\mathbf{r}; k_z) := a_n(z; k_z) \sin(\nu_n \varphi) J_{\nu_n}(k_z \tau), \qquad \mathbf{r} \in D \\ &\qquad k_z \in \mathbb{R}, \qquad \nu_n := \frac{n\pi}{\beta}, \qquad n = 1, 2, \dots \end{aligned} \tag{3.311}$$

where $a_n(z; k_z)$ includes multiplicative constant factors and the dependence on z, because this is irrelevant for the singular behavior of the potential. In fact, since we are interested in a representation of $\Phi(\mathbf{r})$ for points $\mathbf{r} \in D$ and the radius a is 'small', we may as well use the small-argument approximation (H.152) of the Bessel functions and write

$$\Psi_{\nu_n}(\mathbf{r}) := A_n(z) \sin(\nu_n \varphi) \tau^{\nu_n}, \qquad \mathbf{r} \in D \tag{3.312}$$

with all the immaterial multiplicative constants included in $A_n(z)$.

Having thus determined the appropriate eigenfunctions for (3.308), we may write the potential as [8, Section 3.2.9], [1, Section 2.11], [47]

$$\Phi(\mathbf{r}) = \sum_{n=1}^{+\infty} A_n(z) \sin(\nu_n \varphi)\tau^{\nu_n}, \qquad \mathbf{r} \in D \tag{3.313}$$

where $A_n(z)$ denote unknown coefficients. This is the best we can do, unless we complement (3.308) with matching conditions for $\mathbf{r} \in \partial D \cap V$. Nonetheless, $\Phi(\mathbf{r})$ satisfies the matching conditions on the surface of the wedge by construction.

The electric field follows from (2.15), viz.,

$$\begin{aligned}\mathcal{E}(\mathbf{r}) &= -\nabla \sum_{n=1}^{+\infty} A_n(z) \sin(\nu_n \varphi)\tau^{\nu_n} \\ &= -\sum_{n=1}^{+\infty} \left\{ A_n(z)\nu_n \left[\hat{\boldsymbol{\tau}} \sin(\nu_n\varphi) + \hat{\boldsymbol{\varphi}} \cos(\nu_n\varphi)\right] \tau^{\nu_n - 1} + \hat{\mathbf{z}} \frac{\mathrm{d}A_n}{\mathrm{d}z} \tau^{\nu_n} \sin(\nu_n\varphi) \right\}\end{aligned} \tag{3.314}$$

for $\mathbf{r} \in D$. The surface charge density induced on $\partial D \cap \partial V$ may be computed from the jump condition (1.170) or (2.46)

$$\varrho_S(\tau, z) = \sum_{n=1}^{+\infty} \begin{cases} -\varepsilon_0 A_n(z) \nu_n \tau^{\nu_n - 1} & \varphi = 0 \\ \varepsilon_0 (-1)^n A_n(z) \nu_n \tau^{\nu_n - 1} & \varphi = \beta \end{cases} \qquad 0 \leqslant \tau \leqslant a. \tag{3.315}$$

From (3.314) we observe that the z-component of the electric field (i.e., the one parallel to the corner) tends to zero as $\tau \to 0^+$, because the exponent of τ is always positive. By contrast, the transverse component of $\mathcal{E}(\mathbf{r})$ (i.e., the one perpendicular to the edge) may become singular, if the exponent $\nu_n - 1$ of τ is negative, and this may happen only for $n = 1$ and $\beta > \pi$ (Figure 3.20a). Besides, from (3.315) we see that the surface charge density exhibits a similar singular behavior. Since the order of infinity is $1 - \pi/\beta$, the most singular case occurs for $\beta = 2\pi$, which corresponds to the half-plane (Figure 3.20b). For $\beta = \pi$ the wedge degenerates into a conducting half space and — not surprisingly — both field and surface charge are regular. This is also true for $0 < \beta < \pi$, that is, a reentrant wedge.

In spite of the possible singular character of the electric field at the corner of the wedge, the electric energy stored in the field remains finite. In this regard, we may formally compute the energy in the small cylindrical region D as follows

$$\begin{aligned}W_e(D) &:= \frac{\varepsilon_0}{2} \int_D \mathrm{d}V \, |\mathcal{E}(\mathbf{r})|^2 \\ &= \frac{\varepsilon_0}{2} \int_0^h \mathrm{d}z \int_0^\beta \mathrm{d}\varphi \int_0^a \mathrm{d}\tau \sum_{n=1}^{+\infty} \left\{ [A_n(z)]^2 \nu_n^2 \tau^{2\nu_n - 1} + \left(\frac{\mathrm{d}A_n}{\mathrm{d}z}\right)^2 \tau^{2\nu_n + 1} \sin^2(\nu_n\varphi) \right\} \\ &= \frac{\varepsilon_0}{4} \sum_{n=1}^{+\infty} \beta a^{2\nu_n} \left\{ \nu_n \int_0^h \mathrm{d}z \, [A_n(z)]^2 + \frac{a^2}{2\nu_n + 2} \int_0^h \mathrm{d}z \left(\frac{\mathrm{d}A_n}{\mathrm{d}z}\right)^2 \right\}\end{aligned} \tag{3.316}$$

where the orthogonality (3.87) of the eigenfunctions along φ has been exploited in computing the integral of the product of two series given by (3.314). It can be ascertained by inspection that all the terms in the series above are finite for any value of the angle $\beta > 0$.

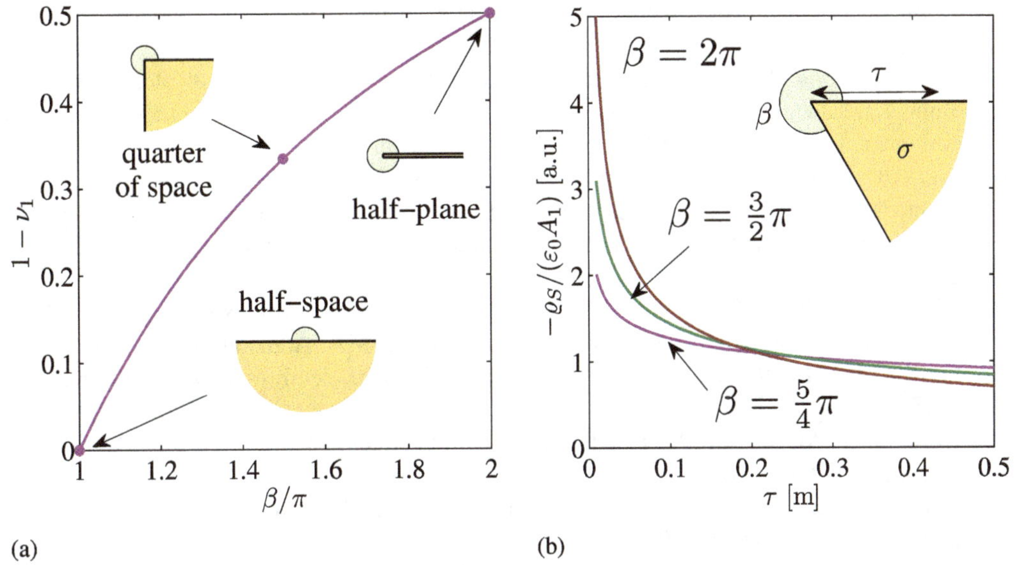

Figure 3.20 *Singular behavior near the corner of a conducting wedge: (a) singularity exponent $1-\nu_1$ versus the angle β; (b) dominant contribution to the surface charge density as a function of the distance from the corner.*

References

[1] Jackson JD. Classical Electrodynamics. 3rd ed. Chichester, UK: Wiley; 1999.

[2] Stratton JA. Electromagnetic theory. London, UK: McGraw-Hill; 1941.

[3] Van Bladel JG. Electromagnetic Fields. Piscataway, NJ: IEEE Press; 2007.

[4] Zangwill A. Modern electrodynamics. Cambridge, UK: Cambrigde University Press; 2013.

[5] Makarov B, Podkorytov A. Real Analysis: Measures, Integrals and Applications. Universitext. London, UK: Springer; 2013.

[6] Dautray R, Lions JL. Mathematical Analysis and Numerical Methods for Science and Technology - Physical Origins and Classical Methods. vol. 1. Berlin Heidelberg: Springer-Verlag; 1990.

[7] Bau III D, Trefethen LN. Numerical linear algebra. Philadelphia, PA: Soci. Indus. Ap. Math.; 1997.

[8] Rothwell EJ, Cloud MJ. Electromagnetics. London, UK: CRC Press; 2001.

[9] Muscat J. Functional Analysis. London, UK: Springer; 2014.

[10] Myint-U T. Partial Differential Equations of Mathematical Physics. 2nd ed. New York, NY: North-Holland; 1980.

[11] Kellogg OD. Foundations of potential theory. Berlin Heidelberg: Springer-Verlag; 1929.

[12] Bassanini P, Elcrat AR. Theory and applications of partial differential equations. Mathematical Concepts And Methods In Science And Engineering. New York, NY: Plenum Press; 1997.

[13] Moon P, Spencer DE. Field Theory Handbook: Including Coordinate Systems, Differential Equations and Their Solutions. 2nd ed. Berlin Heidelberg: Springer-Verlag; 1971.

[14] Harrington RF. Time-harmonic Electromagnetic Fields. London, UK: McGraw-Hill; 1961.

[15] Helrich CS. The classical theory of fields - Electromagnetism. Berlin Heidelberg: Springer-Verlag; 2012.

[16] Shen LC, Kong JA. Applied electromagnetism. Monterey, CA: Brooks/Cole; 1983.

[17] Binns KJ, Lawrenson PJ, Trowbridge CW. The analytical and numerical solution of electric and magnetic fields. Chichester: John Wiley & Sons, Inc.; 1992.

[18] Slater JC, Frank NH. Electromagnetism. New York, NY: McGraw-Hill; 1947.

[19] Kraus JD, Carver KR. Electromagnetics. 2nd ed. Tokyo, Japan: McGraw-Hill; 1973.

[20] Felsen LB, Marcuvitz N. Radiation and scattering of waves. Piscataway, NJ: IEEE Press; 2001.

[21] Morse P, Feshbach H. Methods of Theoretical Physics. New York, NY: McGraw-Hill; 1953.

[22] Apostol TM. Mathematical analysis. 2nd ed. Reading, MA: Addison-Wesley; 1974.

[23] Schwinger J, De Raad LL, Milton KA, et al. Classical electrodynamics. Perseus Books; 1998.

[24] Abramowitz M, Stegun IA. Handbook of mathematical functions. New York, NY: Dover Publications, Inc.; 1965.

[25] Müller C. Foundations of the Mathematical Theory of Electromagnetic Waves. Berlin Heidelberg: Springer-Verlag; 1969.

[26] Milton KA, Schwinger J. Electromagnetic Radiation: Variational Methods, Waveguides and Accelerators. Berlin Heidelberg: Springer-Verlag; 2006.

[27] Sommerfeld A. Electrodynamics. vol. 3 of Lectures on theoretical physics. New York, NY: Academic Press; 1952.

[28] Feynmann RP, Leighton R, Sands M. The Feynmann lectures on physics. vol. 2. Reading, MA: Addison-Wesley; 1964.

[29] Griffith DJ. Introduction to electrodynamics. 3rd ed. Upper Saddle River, NJ: Prentice Hall; 1999.

[30] Konopinski EJ. Electromagnetic Fields and Relativistic Particles. New York, NY: McGraw Hill; 1981.

[31] Hayt WH, Buck JA. Engineering Electromagnetics. 8th ed. New York, NY: McGraw-Hill; 2012. International edition.

[32] Watson GN. A Treatise on the Theory of Bessel Functions. 2nd ed. Cambridge Mathematical Library. Cambridge, UK: Cambridge University Press; 1995.

[33] Sommerfeld A. Partial Differential Equations in Physics. vol. 1 of Lectures on theoretical physics. New York, NY: Academic Press; 1949.

[34] Raab RE, Lange OLD. Multipole Theory in Electromagnetism. Oxford, UK: Clarendon Press; 2005.

[35] Landau LD, Lifshitz EM. The classical theory of fields. 3rd ed. Oxford, UK: Pergamon Press; 1971.

[36] Dwight HB. Tables of Integrals and Other Mathematical Data. 4th ed. New York, NY: MacMillan Publishing Co., Inc.; 1961.

[37] Colton DL, Kress R. Inverse Acoustic and Electromagnetic Scattering Theory. 3rd ed. New York, NY: Springer; 2013.

[38] Collin RE. Field Theory of Guided Waves. Piscataway, NJ: IEEE press; 1991.

[39] Bassanini P, Elcrat A. Mathematical Theory of Electromagnetism. Creative Commons; 2009.

[40] Earnshaw S. On the Nature of the Molecular Forces which Regulate the Constitution of the Luminiferous Ether. Trans Cambridge Phil Soc. 1842;7:97–112.

[41] Maxwell JC. A Treatise on Electricity and Magnetism. vol. 1. Oxford, UK: Clarendon Press; 1873.

[42] Rudin W. Principles of mathematical analysis. 3rd ed. New York, NY: McGraw-Hill; 1976.

[43] Graffi D. Teoria matematica dell'elettromagnetismo. Bologna, IT: Patron; 1972.
[44] Jones W. Earnshaw's theorem and the stability of matter. European Journal of Physics. 1980;1:85–88. Printed in Northern Ireland.
[45] Feynmann RP, Leighton R, Sands M. The Feynmann lectures on physics. vol. 1. Reading, MA: Addison-Wesley; 1964.
[46] Cazacu E, Maricaru M, Stanciulescu A, et al. Escaping from Earnshaw theorem. In: Rev. Roum. Sci. Techn.– Électrotechn. et Énerg.. vol. 49; 2004. p. 1–6.
[47] Van Bladel JG. Singular Electromagnetic Fields and Sources. Piscataway, NJ: IEEE Press; 1991.

Chapter 4
Stationary magnetic fields I

4.1 Stationary limit of Maxwell's equations

In Chapters 2 and 3 we considered the laws of electromagnetism in the very special case where all the charges of concern are fixed in a region of space. After examining the subtle limitations of such idea, we found out that magnetic phenomena are precluded, and that we can make do with just two field entities, viz., $\mathcal{E}(\mathbf{r})$ and either $\mathcal{D}(\mathbf{r})$ or $\mathcal{P}(\mathbf{r})$ in the presence of dielectric media.

Next in line in order of complexity comes the situation in which we allow for the occurrence of stationary (steady) currents. This assumption means that we let electric charges roam freely somehow, as long as the total charge in any region of space remains constant in time. Accordingly, the **conservation of charge** (1.17) dictates

$$\int_{\partial V} \mathrm{d}S\ \hat{\mathbf{n}}(\mathbf{r}) \cdot \mathcal{J}(\mathbf{r}) = 0 \tag{4.1}$$

where ∂V is a closed smooth surface bounding a volume V.

There are basically two ways for a current density to satisfy (4.1), namely, either $\mathcal{J}(\mathbf{r})$ is entirely confined in V, whence $\hat{\mathbf{n}}(\mathbf{r}) \cdot \mathcal{J}(\mathbf{r}) = 0$ everywhere on $S := \partial V$, or the current density must flow back and forth through S in such a way that the *net* amount of charge entering or leaving V is null at any time. We may wonder whether this hypothesis is physically sound. Well, the idea of stationary currents is fraught with practical difficulties no less than the notion of fixed charges is. To understand why, we examine what options we have at our disposal to realize a steady current.

In order for the condition $\hat{\mathbf{n}}(\mathbf{r}) \cdot \mathcal{J}(\mathbf{r}) = 0$ to be true for a current confined in the bounded volume V, we may surmise that $\mathcal{J}(\mathbf{r})$ swirls in a smaller region $V_J \subset V$ or that it flows along a path (an 'inner tube' of sorts) which, no matter how convoluted, is entirely contained in V. A simple case is pictorially represented in Figure 4.1a. If the tube in question is finite, then it needs to be closed, or else, if it were to begin and end somewhere within V, charges would keep accumulating at one end and leaving the other. In which instance, the conservation law (4.1) applied to a bounded volume encompassing only one end of the tube would be violated, because the net charge crossing the boundary of that region would not be zero. It may also happen, in principle at least, that the current tube is not closed but rather winds endlessly in V with no beginning nor end and densely fills V [1, p. 71].

The second configuration mentioned above (Figure 4.1b) just means that the path followed by $\mathcal{J}(\mathbf{r})$ intersects the boundary of V in at least two surfaces $S_l \subset \partial V$, $l = 1, 2$, in which case condition (4.1) is equivalent to

$$I := \int_{S_1} \mathrm{d}S\ \hat{\mathbf{n}}_1(\mathbf{r}) \cdot \mathcal{J}(\mathbf{r}) = \int_{S_2} \mathrm{d}S\ \hat{\mathbf{n}}_2(\mathbf{r}) \cdot \mathcal{J}(\mathbf{r}) \tag{4.2}$$

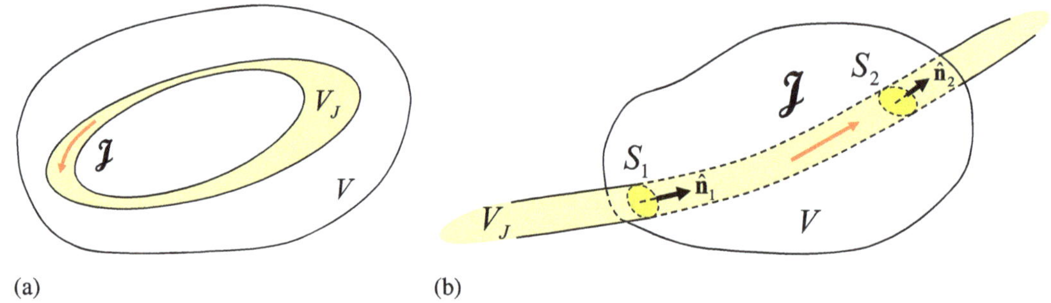

Figure 4.1 Possible ways of realizing a stationary current $\mathfrak{J}(\mathbf{r})$: (a) a finite-sized closed tube and (b) an infinitely extended tube.

i.e., the flux of $\mathfrak{J}(\mathbf{r})$ across any open surface which intersects a tube of current flow is constant. Notice that if the tube happens to intersect the boundary ∂V in more than two surfaces, relations more general than (4.2) may hold. Obviously, the tube still needs to be closed somewhere outside V, though in an unbounded region (say, the whole space) we can think of $\mathfrak{J}(\mathbf{r})$ flowing along an infinitely extended open path: we usually say that the path is 'closed at infinity'. This arrangement works nicely, because the endpoints are infinitely far off and charge depletion or accumulation therein is prevented, as it would take an infinitely long time to occur. Then again, since it is not possible to realize an infinitely extended tube of current in practice, such option remains no more than a theoretical solution.

So, $\mathfrak{J}(\mathbf{r})$ must follow a closed finite path to be stationary. For instance, this might be a conducting wire fashioned in the shape of a circular loop. In which case the charges (i.e., electrons) which constitute $\mathfrak{J}(\mathbf{r})$ are constrained to move endlessly in circles along the wire. Now, even though the electrons were moving according to a uniform circular motion — that is, with $|\mathbf{v}|$ constant — still the velocity $\mathbf{v}$ would change direction at any point of the wire, and this variation technically would result in an acceleration. Besides, the electrons 'bump' into the ions forming the conductor as well, and these collisions too cause the velocity to change unpredictably along the loop. In this regard, we already know from (1.252) and (1.253) that accelerated charges produce electromagnetic waves. In so doing, the charges spend part of their kinetic energy, which is transformed into electromagnetic energy carried away by the waves in the form of radiation, and hence the velocity must diminish, since the energy is conserved. We have thus reached a conclusion in apparent contradiction with the very hypothesis of charges moving with constant velocity, which in turn is instrumental to the notion of stationary current. It would look as if we cannot obtain a steady current in reality, not even along a finite-size closed path. By the way, an infinitely long straight wire would partly do the trick, because the velocity would not undergo any change of direction, but we would be back to facing the impracticality of such configuration.

We need to stick with closed tubes of finite extent and, additionally, make the working assumption that in the stationary regime the charges move in such a way that the net collective radiation from all of them is zero [2]. Nonetheless, collisions with the ions of the conductor lattice and acceleration due to bends in the path occur all the same and cause the charges to lose energy and momentum. That is why we need a battery, i.e., a source of energy, that keeps the charges going at a macroscopic velocity constant with time (cf. Section 4.7.3).

Furthermore, in the realm of steady currents the electromagnetic entities do not depend on time either, they have been constant since forever and will never change, which is of course an abstraction. Therefore, we are led to interpret the stationary regime as the limiting configuration attained by charges, currents and fields after all the transients have finished. One more problem with the steady state is that it implies that the fields have been non-null forever. Again, this condition does not conform with the idea that the fields were indeed null at some remote moment in time, a notion we invoked to set the arbitrary constants in (1.48) and (1.51).

Keeping all these difficulties in mind, we may go on to drop the time dependence from the global form of Maxwell equations. As was the case for the static regime, the electric **Gauss law** (1.16) and the **Faraday law** (1.8) become

$$\Psi_D := \int_{\partial V} \mathrm{d}S\, \hat{\mathbf{n}}(\mathbf{r}) \cdot \mathcal{D}(\mathbf{r}) = \int_V \mathrm{d}V\, \varrho(\mathbf{r}) := Q \tag{4.3}$$

$$\oint_\gamma \mathrm{d}s\, \hat{\mathbf{s}}(\mathbf{r}) \cdot \mathcal{E}(\mathbf{r}) = 0 \tag{4.4}$$

where we may interpret the second equation as the requirement that admissible electric fields produced by steady currents be conservative. The **Ampère-Maxwell** law (1.13) and the **magnetic Gauss law** (1.10) reduce to

$$\oint_\gamma \mathrm{d}s\, \hat{\mathbf{s}}(\mathbf{r}) \cdot \mathcal{H}(\mathbf{r}) = \int_S \mathrm{d}S\, \hat{\mathbf{n}}(\mathbf{r}) \cdot \mathcal{J}(\mathbf{r}) := I \tag{4.5}$$

$$\Psi_B := \int_{\partial V} \mathrm{d}S\, \hat{\mathbf{n}}(\mathbf{r}) \cdot \mathcal{B}(\mathbf{r}) = 0 \tag{4.6}$$

where (4.5), which was already well established before Maxwell's introduction of the displacement current, is referred to simply as the **Ampère law** in global form or sometimes the **Ampère circuital law**. We notice in passing that, when (4.5) is applied to a closed surface $S = \partial V$, it implies the charge conservation (4.1).

To derive the local form of Maxwell's equations for steady currents we apply the Gauss theorem to (4.3) and (4.6) and the Stokes theorem to (4.4) and (4.5). By assuming that $\varrho(\mathbf{r})$ is confined in a volume $V_\varrho \subset V$ and $\mathcal{J}(\mathbf{r})$ crosses a multiply-connected surface $S_J \subset \partial V$, we have

$$\nabla \cdot \mathcal{D}(\mathbf{r}) = \varrho(\mathbf{r}), \qquad \mathbf{r} \in \mathbb{R}^3 \tag{4.7}$$

$$\nabla \times \mathcal{E}(\mathbf{r}) = 0, \qquad \mathbf{r} \in \mathbb{R}^3 \tag{4.8}$$

and

$$\nabla \times \mathcal{H}(\mathbf{r}) = \mathcal{J}(\mathbf{r}), \qquad \mathbf{r} \in \mathbb{R}^3 \tag{4.9}$$

$$\nabla \cdot \mathcal{B}(\mathbf{r}) = 0, \qquad \mathbf{r} \in \mathbb{R}^3 \tag{4.10}$$

under the hypotheses that the field entities are continuously differentiable everywhere, $\hat{\mathbf{n}} \cdot \mathcal{D}$ is continuous through the boundary ∂V_ϱ, and $\hat{\mathbf{s}} \cdot \mathcal{H}$ is continuous through the line ∂S_J (see Section 1.2.2). In the presence of material bodies (dielectric, magnetic and conducting) differentiability is ensured everywhere except at the interfaces between regions with different constitutive parameters, and the boundary conditions derived in Section 1.7 apply.

Finally, the local form of the continuity equation follows by applying the Gauss theorem to (4.1)

$$\nabla \cdot \mathcal{J}(\mathbf{r}) = 0, \qquad \mathbf{r} \in V_J \tag{4.11}$$

by assuming that the current density is continuously differentiable. Thus, we see that a steady current is necessarily solenoidal. This conclusion can also be reached by taking the divergence of both sides in (4.9). Thereby, the requirement that $\mathcal{J}(\mathbf{r})$ be divergence-free is a necessary condition for the solvability of (4.9) [3, Section 6.1].

We see immediately that the most important consequence of having assumed steady currents only is the *decoupling* of electrical and magnetic entities. Indeed, the first two laws listed above are the same as those encountered in electrostatics, whereas the second two are needed to describe stationary magnetic phenomena, which are totally absent if the charges are fixed. It is important to realize that, although Maxwell equations decouple into two independent sets, this separation does not imply that a stationary current produces only magnetic fields, as one might mistakenly infer from (4.5) and (4.6). Rather, charges do produce electric fields which obey (4.3) and (4.4). The main difference with the laws of electrostatics is that, since the charges are moving, $\mathcal{E}(\mathbf{r})$ and $\mathcal{D}(\mathbf{r})$ depend on the velocity and the acceleration thereof, as we gather from (1.252). Yet, as we have discussed above, we assume that the charges move collectively in such a way that the macroscopic velocity is constant in time and no net radiation occurs. When dealing with steady currents, though, we are most often interested in the magnetic effects, and therefore we work with (4.9), (4.10) and (4.11) only.

The significance of the electric equations was already addressed in Section 2.1. The magnetic equations — in combination with, e.g., the constitutive relation (1.118) for an isotropic stationary medium — require that among all the possible magnetic fields we look for those with zero flux or zero divergence generated by solenoidal currents. Alternatively, (4.9) may be construed as the problem of inverting the curl operator subject to the condition that the solution be solenoidal. When complemented with a general constitutive relationship between $\mathcal{B}$ and $\mathcal{H}$, (4.9) and (4.10) become a set of four scalar equations in three unknowns. Although it appears that we have one equation too many, in Section 4.2 (and even more generally in Section 8.1) we shall show that the system is not over-determined.

Example 4.1 (Magnetic field of a circular cylindrical uniform stationary current density)

We consider a uniform stationary current density $\mathcal{J}(\mathbf{r})$ flowing in the inside of a circular cylinder of radius a. For our calculations, it is convenient to adopt a system of circular cylindrical coordinates (τ, φ, z) with the z-axis coincident with the axis of the cylinder. We choose the symbol τ for the radial coordinate in lieu of the standard ρ (Appendix A.1) so as to avoid confusion with the charge density. Hence, the current density reads

$$\mathcal{J}(\mathbf{r}) = \begin{cases} \mathcal{J}_0 \hat{\mathbf{z}}, & \tau = |\boldsymbol{\tau}| < a \\ 0, & \tau > a \end{cases} \tag{4.12}$$

i.e., $\mathcal{J}(\mathbf{r})$ is a vector field which depends on the polar coordinate only and is discontinuous across the cylinder $S := \{\mathbf{r} \in \mathbb{R}^3 : \tau = a\}$. Such current is an abstraction, of course, and generates a magnetic field with infinite energy. This is not surprising, as (4.12) implies the endless flow of an infinite amount of charge. Regardless, we wish to determine the magnetic field produced by $\mathcal{J}(\mathbf{r})$ in free space for $\tau \geqslant 0$. The problem can be tackled by means of the Ampère law in global form (4.5) or the system (4.9) and (4.10).

The first strategy is not viable in general, because $\mathcal{H}(\mathbf{r})$ is a three-dimensional vector field, and (4.5) represents just one scalar equation. However, as we shall see in a moment, the symmetry of the configuration allows us to conclude that the magnetic field possesses only one non-zero component which we can indeed retrieve with a clever usage of (4.5). We may then verify that the flux of $\mathcal{B}(\mathbf{r})$ is indeed zero across any unbounded surface which surrounds the cylinder. Alternatively, and more quickly, we may check whether $\mathcal{B}(\mathbf{r})$ is solenoidal, as (4.6) and (4.10) are equivalent.

The approach based on the local form of the stationary equations always works, though, as was the case for electrostatics, we have to solve both of them simultaneously (see Section 4.2). What is

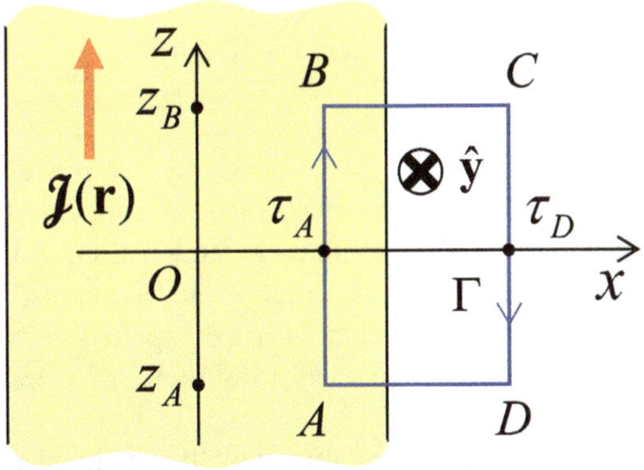

Figure 4.2 *Rectangular path* Γ *in the xOz plane for the calculation of* $\mathcal{H}_z(\tau)$ *produced by the current density* (4.12).

more, it is not immediately clear which component of $\mathcal{H}(\mathbf{r})$ is non-zero, and to draw our conclusions beforehand we need to tinker a bit with the Ampère law (4.5).

First of all, since the current density does not depend on the z coordinate, neither does the magnetic field; we say that the problem is invariant with respect to z. Secondly, since the current looks the same to an observer sitting in any point in space, we conclude that the magnetic field must not depend on the azimuthal coordinate φ; we say that the problem is invariant with respect to φ, too. All in all, $\mathcal{H}(\mathbf{r})$ must be a function of the distance τ from the z-axis.

In view of the invariance along z and φ, it is convenient to apply the global Ampère law (4.5) along the circular loop $\gamma := \{\mathbf{r} \in \mathbb{R}^3 : \tau = b,\ z = 0\}$, in which case the circulation of the magnetic field allows us to determine the polar component $\mathcal{H}_\varphi$. However, what can we say about the other two components $\mathcal{H}_\tau$ and $\mathcal{H}_z$?

Let us suppose that $\mathcal{J}(\mathbf{r})$ given by (4.12) generates a radial component $\mathcal{H}_\tau(\tau)$ that is non-null and positive. Now, since the Ampère law is a linear relationship, changing the sign of the current density (i.e., substituting $-\mathcal{J}_0$ for $\mathcal{J}_0$) makes the radial component $\mathcal{H}_\tau(\tau)$ flip and become negative. Next, what happens if we *conceptually execute* a one-hundred-eighty-degree turn of the current tube rigidly around, say, the x-axis? Well, this mechanical rotation of sorts causes the current flow to change orientation, namely, if $\mathcal{J}_0 > 0$ after the rotation the charges advance from plus infinity towards minus infinity, the negative z-direction. As a matter of fact, the rotation being purely mechanical, it has no effect whatsoever on the orientation of the radial component $\mathcal{H}_\tau(\tau)$, which, as a result, is still positively directed away from the z-axis. In summary, we have carried out two legitimate thought experiments whose outcomes are in evident contradiction. The only way to resolve the paradox is to conclude that the radial component $\mathcal{H}_\tau(\tau)$ produced by $\mathcal{J}(\mathbf{r})$ is necessarily null for any value of the radial coordinate.

We go on to investigate the occurrence of a non-zero z-component of the magnetic field. Experimental evidence suggests that $\mathcal{H}_z(\tau)$ ought to be zero, in that magnetic fields are always perpendicular to the current which generates them. Nonetheless, we can back up this expectation by applying (4.5) to a rectangular path Γ in the xOz plane, as is sketched in Figure 4.2. Even if Γ runs partly within the current tube, the total current encircled by Γ is invariably null because $\mathcal{J}(\mathbf{r}) \cdot \hat{\mathbf{y}} = 0$. Thus, we split up the integration into four contributions along the sides of the rectangle Γ, viz.,

$$0 = \oint_\Gamma \mathrm{d}s\, \hat{\mathbf{s}}(\mathbf{r}) \cdot \mathcal{H}(\mathbf{r}) = \int_{\Gamma_{AB}} \mathrm{d}s\, \hat{\mathbf{s}}(\mathbf{r}) \cdot \mathcal{H}(\mathbf{r}) + \int_{\Gamma_{CD}} \mathrm{d}s\, \hat{\mathbf{s}}(\mathbf{r}) \cdot \mathcal{H}(\mathbf{r})$$

$$= \int_{z_A}^{z_B} \mathrm{d}z\, [\mathcal{H}_z(\tau_A) - \mathcal{H}_z(\tau_D)] = (z_B - z_A)[\mathcal{H}_z(\tau_A) - \mathcal{H}_z(\tau_D)] \tag{4.13}$$

where we have taken into account that $\mathcal{H}_\tau(\tau) = 0$. Since we can move Γ around in the xOz plane and even change its size at will, the equation above lets us conclude that $\mathcal{H}_z(\tau)$ is constant and independent of τ. Still, even though $\mathcal{J}(\mathbf{r})$ is infinitely extended along z, the current tube has a finite cross section, and hence any component of the magnetic field $\mathcal{H}(\mathbf{r})$ must decay away from the z-axis with some negative power of τ. We see, then, that we are compelled to set $\mathcal{H}_z(\tau) = 0$, which ensures the z-component is zero at infinity or, more precisely, infinitely far away from the current along the radial direction $\hat{\boldsymbol{\tau}}$.

In the end, we are left with only the polar component $\mathcal{H}_\varphi$, which we can retrieve from (4.5)

$$\oint_\gamma \mathrm{d}s\, \hat{\boldsymbol{\varphi}} \cdot \hat{\boldsymbol{\varphi}}\, \mathcal{H}_\varphi(\tau) = 2\pi b\, \mathcal{H}_\varphi(\tau) \tag{4.14}$$

because $\mathcal{H}_\varphi(\tau)$ is constant along the circular loop γ. Next, we compute the total current flowing through the circle $C := \{\mathbf{r} \in \mathbb{R}^3 : \tau \leqslant b,\ z = 0\}$ bounded by $\gamma := \partial C$, i.e.,

$$I(b) = \int_C \mathrm{d}S\, \hat{\mathbf{z}} \cdot \mathcal{J}(\mathbf{r}) = \begin{cases} \mathcal{J}_0 \pi b^2, & b < a \\ \mathcal{J}_0 \pi a^2, & b > a \end{cases} \tag{4.15}$$

and this must be equal to the circulation of $\mathcal{H}(\mathbf{r})$, whereby we determine $\mathcal{H}_\varphi(b)$ as a function of the radius b. By putting everything together and renaming b to τ we find

$$\mathcal{H}(\mathbf{r}) = \begin{cases} \dfrac{\mathcal{J}_0}{2} \tau \hat{\boldsymbol{\varphi}}, & \tau < a \\ \dfrac{\mathcal{J}_0 a^2}{2\tau} \hat{\boldsymbol{\varphi}}, & \tau > a \end{cases} \tag{4.16}$$

which is a vector field continuous through the cylinder S where the current density suffers a jump.

The solution strategy based on the differential form of the equations for stationary currents and definition (A.33) leads to

$$\frac{1}{\tau} \frac{\mathrm{d}}{\mathrm{d}\tau} (\tau \mathcal{H}_\varphi) = \begin{cases} \mathcal{J}_0, & \tau < a \\ 0, & \tau > a \end{cases} \tag{4.17}$$

which is solved by

$$\mathcal{H}_\varphi(\mathbf{r}) = \begin{cases} \dfrac{\mathcal{J}_0}{2} \tau, & \tau < a \\ \dfrac{C_\varphi}{\tau}, & \tau > a \end{cases} \tag{4.18}$$

where C_φ is a constant we must determine so as to ensure the continuity of $\mathcal{H}_\varphi$ through the cylinder S. This procedure yields the same result as in (4.16). Lastly, direct calculation shows of the divergence with (A.30) that $\mathcal{B}(\mathbf{r}) = \mu_0 \mathcal{H}(\mathbf{r})$ is solenoidal.

The typical behavior of $\mathcal{H}_\varphi(\tau)$ is plotted in Figure 4.3 as a function of the normalized distance from the axis of the cylinder and for a few values of $\mathcal{J}_0$. Remarkably, it takes a comparatively large

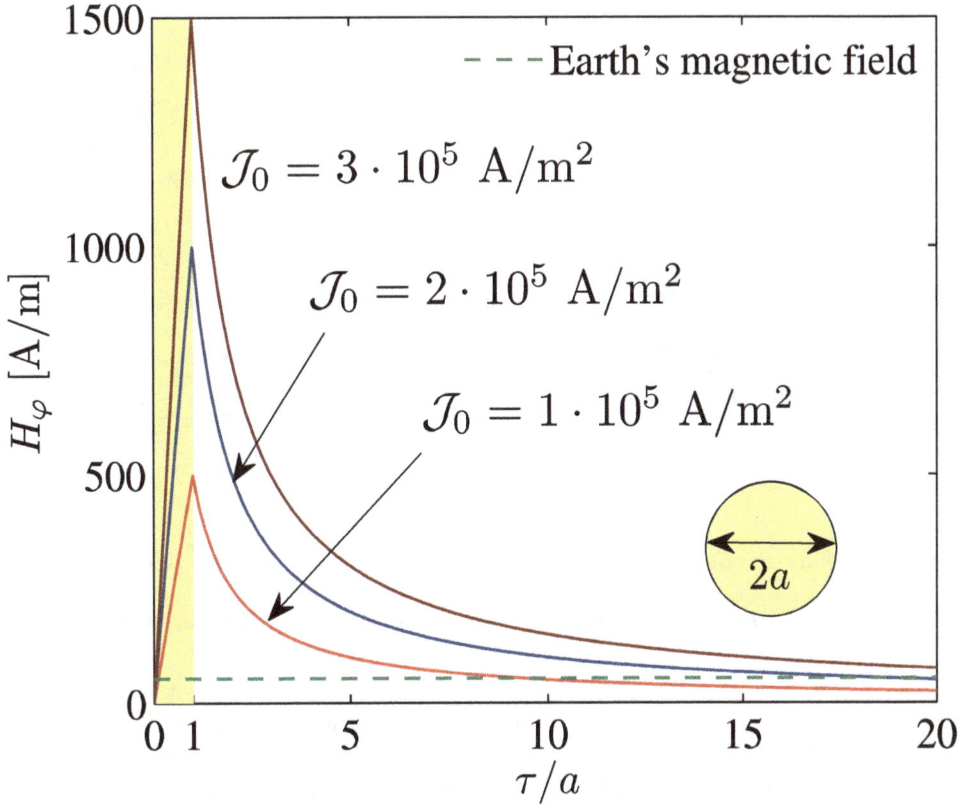

Figure 4.3 *The magnetic field component $\mathcal{H}_\varphi(\tau)$ generated by an infinite straight uniform current density of circular cross section according to* (4.16). *The Earth's geomagnetic field (−−) is shown for reference.*

current density to obtain values of the magnetic field that are substantially larger than the natural geomagnetic field of the Earth ($|\mathfrak{B}(\mathbf{r})| \approx 67\ \mu$T or $|\mathcal{H}(\mathbf{r})| \approx 53$ A/m at the poles [4]).

In summary, the Ampère law in global form handles the continuity of $\hat{\mathfrak{s}} \cdot \mathcal{H}$ implicitly, whereas the matching condition has to be enforced when working with the differential equations. The magnetic field thus found is continuously differentiable everywhere except on the cylinder, and this happens because $\mathfrak{J}(\mathbf{r})$ is piecewise continuous.

(End of Example 4.1)

Example 4.2 (Magnetic field of an infinite solenoid)

A solenoid is a simple magnetic device realized by winding a thin conducting wire in the shape of a coil, as is suggested in Figure 4.4a. In this configuration the *pitch* is the angle formed by the vector tangential to the wire and any plane perpendicular to the longitudinal symmetry axis. To get an idea of the magnetic field generated by a solenoid we consider an idealized model consisting of a tightly wound and infinitely long solenoid in which a steady line current I flows (Figure 4.4b). This problem, too, can be solved with the aid of the Ampère law in global form (4.5) — which is a single scalar equation — so long as we can conclude beforehand that the magnetic field possesses only one non-zero component in a suitable system of coordinates.

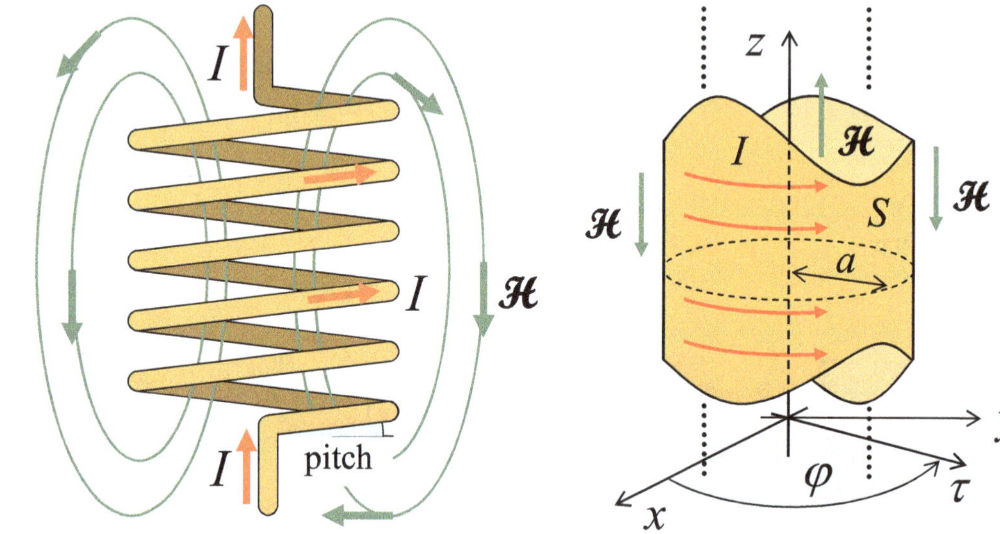

(a) *practical finite-size device* (b) *idealized infinite model*

Figure 4.4 For the calculation of the magnetic field generated by a solenoid.

In light of the chosen geometry it is convenient to consider circular cylindrical coordinates (τ, φ, z) and examine the components $\mathcal{H}_\tau$, $\mathcal{H}_\varphi$ and $\mathcal{H}_z$. The solenoid is aligned with the z-axis and has radius a. Besides, for the sake of simplicity we assume that the pitch of the winding is zero, which allows modelling the solenoid as a continuous distribution of coaxial circular coils with uniform linear density $\mathcal{N}$ (physical dimension: 1/m) formally defined as

$$\mathcal{N} := \lim_{\Delta z \to 0} \frac{\Delta N}{\Delta z} \tag{4.19}$$

where ΔN is the number of coils located between the planes $z = z_0$ and $z = z_0 + \Delta z$. This configuration in turn amounts to a surface current density

$$\mathcal{J}_S(\mathbf{r}) := \mathcal{N} I \hat{\boldsymbol{\varphi}} \tag{4.20}$$

localized on the circular cylinder $S := \{\mathbf{r} \in \mathbb{R}^3 : \tau = a\}$ (Figure 4.4b). By the way, it is straightforward to show that $\mathcal{J}_S(\mathbf{r})$ given by (4.20) satisfies the steady-state counterpart of the continuity equation (1.172), inasmuch as $\mathcal{N}I$ is independent of the azimuth angle.

First of all we notice that the azimuthal symmetry of the solenoid demands that $\boldsymbol{\mathcal{H}}$ be independent of φ. Likewise, $\boldsymbol{\mathcal{H}}$ must not depend on z for the geometry is invariant along the z-direction. Secondly, we try and rule out the appearance of a non-zero radial component by reasoning as follows. If we inverted the orientation of $\mathcal{J}_S(\mathbf{r})$ by changing the sign of I, then $\mathcal{H}_\tau(\tau)$ ought to change sign as well for the Ampère law is a linear relationship. On the other hand, if we rotated the whole solenoid rigidly by 180° around the x-axis, the current $\mathcal{J}_S(\mathbf{r})$ would similarly change orientation, whereas the process would have no effect on $\mathcal{H}_\tau(\tau)$. Yet, the two configurations are identical in all respects, and since the τ-component of $\boldsymbol{\mathcal{H}}(\tau)$ may not have two opposite orientations for the same value of the radial coordinate, we are compelled to conclude that $\mathcal{H}_\tau(\tau) = 0$ everywhere.

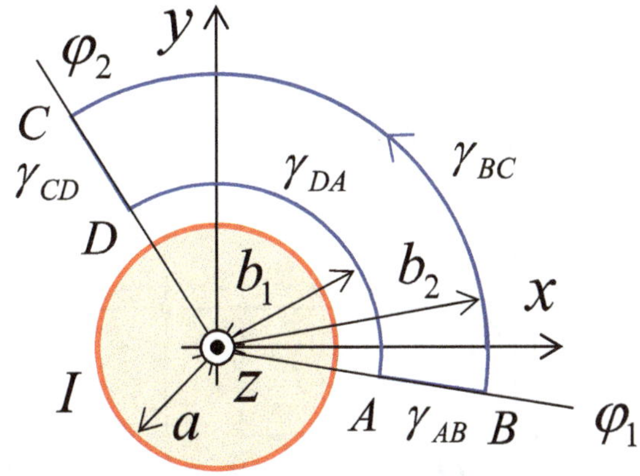

Figure 4.5 The boundary of an annular sector in a plane perpendicular to the solenoid of Figure 4.4b for showing that $\mathcal{H}_\varphi(\tau) = 0$.

Armed with these findings, we go on to examine the φ-component by applying the Ampère law (4.5) on the boundary γ of an annular sector perpendicular to the z-axis, as shown in Figure 4.5. In symbols, we have

$$\begin{aligned} 0 &= \oint_\gamma \mathrm{d}s\,\hat{\mathbf{s}}(\mathbf{r}) \cdot \mathcal{H}(\mathbf{r}) = \int_{\varphi_1}^{\varphi_2} \mathrm{d}\varphi\, b_2 \mathcal{H}_\varphi(b_2) - \int_{\varphi_1}^{\varphi_2} \mathrm{d}\varphi\, b_1 \mathcal{H}_\varphi(b_1) \\ &= (\varphi_2 - \varphi_1)[b_2 \mathcal{H}_\varphi(b_2) - b_1 \mathcal{H}_\varphi(b_1)] \end{aligned} \tag{4.21}$$

having used the fact that $\mathcal{H}_\tau(\tau) = 0$ and that $\mathcal{H}_\varphi(\tau)$ is perpendicular to the edges γ_{AB} and γ_{CD} of the contour. This relation implies that the product $\tau\mathcal{H}_\varphi(\tau)$ is a constant or, equivalently, that

$$\mathcal{H}_\varphi(\tau) = \frac{C_\tau}{\tau} \tag{4.22}$$

for $\tau \geqslant 0$, since we may also choose a path γ entirely contained within the solenoid or even one which crosses the current — the flux of $\mathcal{J}_S(\mathbf{r})$ would be zero anyway because $\hat{\mathbf{z}} \cdot \mathcal{J}_S(\mathbf{r}) = 0$. However, there is no reason why the magnetic field should be singular for $\tau = 0$, and hence we are forced to take $C_\tau = 0$ and conclude that $\mathcal{H}_\varphi(\tau)$ is identically null. This result is consistent with the experimental evidence that magnetic fields are perpendicular to the direction of the generating current, i.e., $\hat{\boldsymbol{\varphi}}$ in the present example.

All in all, we are left with the calculation of $\mathcal{H}_z(\tau)$. To this purpose we apply the Ampère law (4.5) along three different rectangular paths in the plane xOz (Figure 4.6). We choose the first path entirely outside the solenoid and compute

$$\begin{aligned} 0 &= \oint_\gamma \mathrm{d}s\,\hat{\mathbf{s}}(\mathbf{r}) \cdot \mathcal{H}(\mathbf{r}) = \int_{\gamma_{AB}} \mathrm{d}s\,\hat{\mathbf{s}}(\mathbf{r}) \cdot \mathcal{H}(\mathbf{r}) + \int_{\gamma_{CD}} \mathrm{d}s\,\hat{\mathbf{s}}(\mathbf{r}) \cdot \mathcal{H}(\mathbf{r}) \\ &= \int_{z_A}^{z_B} \mathrm{d}z\,[\mathcal{H}_z(\tau_A) - \mathcal{H}_z(\tau_D)] = (z_B - z_A)[\mathcal{H}_z(\tau_A) - \mathcal{H}_z(\tau_D)] \end{aligned} \tag{4.23}$$

whereby we conclude

$$\mathcal{H}_z(\tau) = C_z \tag{4.24}$$

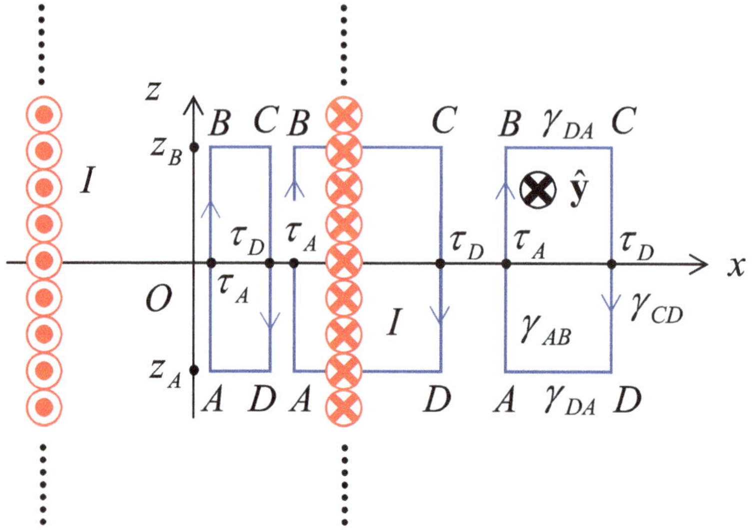

Figure 4.6 Longitudinal cut of the solenoid of Figure 4.4b and three rectangular paths in the plane xOz for the calculation of $\mathcal{H}_z(\tau)$ produced by the current I.

for $\tau > a$. But then, although the current is infinite in the z-direction, still it has finite size in the xOy plane, and thus we expect the magnetic field to fall off with the radial distance τ and to vanish at infinity. The only way to meet this requirement is by setting $C_z = 0$ in (4.27). In conclusion, the magnetic field is identically zero outside the solenoid.

For the second path we pick up a rectangle which lies entirely within the solenoid and evaluate

$$\begin{aligned} 0 &= \oint_{\gamma} \mathrm{d}s\,\hat{\mathbf{s}}(\mathbf{r}) \cdot \boldsymbol{\mathcal{H}}(\mathbf{r}) = \int_{\gamma_{AB}} \mathrm{d}s\,\hat{\mathbf{s}}(\mathbf{r}) \cdot \boldsymbol{\mathcal{H}}(\mathbf{r}) + \int_{\gamma_{CD}} \mathrm{d}s\,\hat{\mathbf{s}}(\mathbf{r}) \cdot \boldsymbol{\mathcal{H}}(\mathbf{r}) \\ &= \int_{z_A}^{z_B} \mathrm{d}z\,[\mathcal{H}_z(\tau_A) - \mathcal{H}_z(\tau_D)] = (z_B - z_A)[\mathcal{H}_z(\tau_A) - \mathcal{H}_z(\tau_D)] \end{aligned} \tag{4.25}$$

which implies that $\mathcal{H}_z(\tau)$ is equal to a constant — yet to be determined — for $\tau < a$.

Lastly, we choose a rectangular contour which lies partly within and partly outside the solenoid so as to encircle the current $(z_B - z_A)\mathcal{N}I$. The Ampère law (4.5) then yields

$$\begin{aligned} (z_B - z_A)\mathcal{N}I &= \oint_{\gamma} \mathrm{d}s\,\hat{\mathbf{s}}(\mathbf{r}) \cdot \boldsymbol{\mathcal{H}}(\mathbf{r}) = \int_{\gamma_{AB}} \mathrm{d}s\,\hat{\mathbf{s}}(\mathbf{r}) \cdot \boldsymbol{\mathcal{H}}(\mathbf{r}) \\ &= \int_{z_A}^{z_B} \mathrm{d}z\,\mathcal{H}_z(\tau_A) = (z_B - z_A)\mathcal{H}_z(\tau_A) \end{aligned} \tag{4.26}$$

whence we find

$$\mathcal{H}_z(\tau) = \mathcal{N}I \tag{4.27}$$

for $\tau < a$ since τ_A is arbitrary. We notice that with (4.20), (4.27) and $\mathcal{H}_z(\tau) = 0$ for $\tau > a$ the jump condition (1.142) is satisfied.

In conclusion, we have just proved that

$$\mathcal{H}(\mathbf{r}) = \begin{cases} \mathcal{N}I\hat{\mathbf{z}}, & \tau < a \\ 0, & \tau > a \end{cases} \tag{4.28}$$

that is the magnetic field inside an infinitely long solenoid is uniform and directed along the axis thereof. Although a practical solenoid must have a finite size, it is nonetheless a device able to generate very strong and approximately uniform magnetic fields in its interior.

(End of Example 4.2)

4.2 Vector potential and the vector Poisson equation

Despite the striking similarity between the magnetic equations (plus continuity equation) and the electric equations, solving (4.9), (4.10) and (4.11) is more difficult because the source term, $\mathcal{J}(\mathbf{r})$, is a vector and the Ampère law amounts to three scalar equations in a suitable systems of coordinates. Then again, we have already noticed that the magnetic Gauss law plays the role of a constraint of sorts to be imposed on the physically acceptable magnetic induction fields.

Similarly to what we did for the electrostatic case, we can make sure *a priori* that we pick a divergence-free $\mathcal{B}(\mathbf{r})$ by requiring it to be the curl of some — yet unknown — vector field, namely, [5, Sections 5.3, 5.4], [6, Lemma 3, p. 214]

$$\mathcal{B}(\mathbf{r}) = \nabla \times \mathcal{A}(\mathbf{r}), \qquad \mathbf{r} \in \mathbb{R}^3 \tag{4.29}$$

whereby (4.10) is invariably satisfied in view of (A.39), provided $\mathcal{A}(\mathbf{r}) \in \mathcal{C}^2(\mathbb{R}^3)^3$. In the context of steady currents and stationary magnetic fields, $\mathcal{A}(\mathbf{r})$ is called the *magnetic vector potential* and carries the physical dimension of Wb/m or Tm. Furthermore, thanks to the Ansatz (4.29) the global magnetic Gauss law is satisfied as well

$$\int_{\partial V} \mathrm{d}S\, \hat{\mathbf{n}}(\mathbf{r}) \cdot \nabla \times \mathcal{A}(\mathbf{r}) = \int_{V} \mathrm{d}V\, \nabla \cdot \nabla \times \mathcal{A}(\mathbf{r}) = 0 \tag{4.30}$$

in light of the divergence theorem and (A.39) again. On the other hand, the flux of $\mathcal{B}(\mathbf{r})$ through a smooth open surface S with smooth boundary $\gamma := \partial S$ (Figure 1.2a) may be related to the circulation of $\mathcal{A}(\mathbf{r})$ as

$$\Psi_B := \int_{S} \mathrm{d}S\, \hat{\mathbf{n}}(\mathbf{r}) \cdot \mathcal{B}(\mathbf{r}) = \oint_{\gamma} \mathrm{d}s\, \hat{\mathbf{s}}(\mathbf{r}) \cdot \mathcal{A}(\mathbf{r}) \tag{4.31}$$

by virtue of (4.29) and the Stokes theorem, which may be applied for $\mathcal{A}(\mathbf{r})$ is of class $\mathcal{C}^1(S)^3 \cap \mathcal{C}(\overline{S})^3$ by assumption. Comparing (4.31) with the Ampère law (4.5) suggests we intepret either the magnetic flux or $\mathcal{B}(\mathbf{r})$ as being the source of $\mathcal{A}(\mathbf{r})$.

Implicit in the definition of $\mathcal{A}(\mathbf{r})$ is a certain amount of arbitrariness, as was the case with the scalar potential in (2.15). For instance, adding any constant to $\mathcal{A}(\mathbf{r})$ does not alter the values of the magnetic induction field. More generally, we may combine $\mathcal{A}(\mathbf{r})$ with any conservative field, viz.,

$$\mathcal{A}(\mathbf{r}) = \mathcal{A}_1(\mathbf{r}) + \nabla\chi(\mathbf{r}) \tag{4.32}$$

because the curl of the field $\nabla\chi(\mathbf{r})$ is zero. The scalar field $\chi(\mathbf{r})$ has no effect even on the right-hand side of (4.31) for the circulation of $\nabla\chi(\mathbf{r})$ vanishes. We shall see in Section 5.1 that we may exploit this freedom by choosing a convenient scalar field $\chi(\mathbf{r})$ so as to simplify the task of solving the stationary magnetic equations.

Broadly speaking, the possibility of determining $\mathcal{B}(\mathbf{r})$ from a vector potential stems from the non-existence of isolated magnetic charges (poles), whereby we do not have a source term (i.e., a magnetic charge density) in the right-hand side of (4.10). Still, we know from experience that *magnetic dipoles* do exist, e.g., in the form of permanent magnets. Since the occurrence of isolated magnetic dipoles requires introducing a source term in (4.10), it turns out that we may define a vector potential $\mathcal{A}(\mathbf{r})$ only for points away from the sources, because $\mathcal{B}(\mathbf{r})$ is not solenoidal at the very location of the magnetic dipoles. We postpone this topic until Section 4.5.

Moving on with the case of steady currents for the time being, since $\mathcal{B}(\mathbf{r})$ and $\mathcal{H}(\mathbf{r})$ are related by a constitutive relationship (Section 1.6) we should manage to turn the Ampère law into an equation for the vector potential. In free space we have

$$\nabla \times [\nabla \times \mathcal{A}(\mathbf{r})] = \mu_0 \mathcal{J}(\mathbf{r}), \qquad \mathbf{r} \in \mathbb{R}^3 \tag{4.33}$$

on account of (1.113) and (4.29). This equation is sometimes called the **vector Poisson** equation because its role is analogous to that of (2.17) for electrostatics. However, (4.33) amounts to three second-order scalar equations for the components of $\mathcal{A}(\mathbf{r})$. The continuity of $\hat{\mathbf{s}} \cdot \mathcal{H}(\mathbf{r})$ for $\mathbf{r} \in \partial V_J$ translates into a condition for the vector potential, i.e., $\hat{\mathbf{s}} \cdot \nabla \times \mathcal{A}(\mathbf{r})$ must be continuous across ∂V_J owing to (4.29) and (1.113). Although we have managed to transform (4.9) and (4.10) into three scalar equations in three unknowns — with $\mathcal{J}(\mathbf{r})$ subject to (4.11) — it is far from obvious that (4.33) can be inverted to yield the vector potential and hence $\mathcal{B}(\mathbf{r})$ (see Section 5.1).

We can investigate the behavior of the stationary magnetic field and the vector potential at infinity by considering the magnetic energy [2,7,8] associated with a steady current density $\mathcal{J}(\mathbf{r})$ for $\mathbf{r} \in V_J$. As usual, we start with the energy stored in a bounded volume $V \supseteq V_J$ with unit normal $\hat{\mathbf{n}}$ on ∂V pointing outward V. In symbols, we have

$$\begin{aligned}
W_h &:= \frac{1}{2}\int_V \mathrm{d}V\, \mathcal{H}(\mathbf{r}) \cdot \mathcal{B}(\mathbf{r}) = \frac{1}{2}\int_V \mathrm{d}V\, \mathcal{H}(\mathbf{r}) \cdot \nabla \times \mathcal{A}(\mathbf{r}) = \\
&= \frac{1}{2}\int_V \mathrm{d}V\, \mathcal{A}(\mathbf{r}) \cdot \nabla \times \mathcal{H}(\mathbf{r}) - \frac{1}{2}\int_V \mathrm{d}V\, \nabla \cdot [\mathcal{H}(\mathbf{r}) \times \mathcal{A}(\mathbf{r})] \\
&= \frac{1}{2}\int_{V_J} \mathrm{d}V\, [\mathcal{A}_1(\mathbf{r}) + \nabla\chi(\mathbf{r})] \cdot \mathcal{J}(\mathbf{r}) - \frac{1}{2}\int_{\partial V} \mathrm{d}S\, \hat{\mathbf{n}}(\mathbf{r}) \times \mathcal{H}(\mathbf{r}) \cdot [\mathcal{A}_1(\mathbf{r}) + \nabla\chi(\mathbf{r})] \\
&= \frac{1}{2}\int_{V_J} \mathrm{d}V\, \mathcal{A}_1(\mathbf{r}) \cdot \mathcal{J}(\mathbf{r}) - \frac{1}{2}\int_{\partial V} \mathrm{d}S\, \mathcal{A}_1(\mathbf{r}) \cdot \hat{\mathbf{n}}(\mathbf{r}) \times \mathcal{H}(\mathbf{r}) \\
&\quad + \frac{1}{2}\int_{V_J} \mathrm{d}V\, \mathcal{J}(\mathbf{r}) \cdot \nabla\chi(\mathbf{r}) - \frac{1}{2}\int_{\partial V} \mathrm{d}S\, \hat{\mathbf{n}}(\mathbf{r}) \times \mathcal{H}(\mathbf{r}) \cdot \nabla\chi(\mathbf{r})
\end{aligned} \tag{4.34}$$

where we have used the splitting (4.32) and the fact the $\mathcal{J}(\mathbf{r})$ is confined to V_J. The surface integral over ∂V follows by applying the Gauss theorem separately to V_J and $V \setminus \overline{V_J}$ under the hypothesis that $\hat{\mathbf{n}} \times \mathcal{H}$ is continuous through ∂V_J (cf. Section 1.2.2). The last two contributions should cancel one another, because the energy ought not to depend on the particular choice of a vector potential. Indeed, if we dot-multiply (4.9) with $\nabla\chi$ and invoke (H.49) we find

$$\nabla\chi(\mathbf{r}) \cdot \mathcal{J}(\mathbf{r}) = \nabla\chi(\mathbf{r}) \cdot \nabla \times \mathcal{H}(\mathbf{r}) = \nabla \cdot [\mathcal{H}(\mathbf{r}) \times \nabla\chi(\mathbf{r})] \tag{4.35}$$

and by integrating both sides over V and applying the Gauss theorem we obtain the result anticipated above.

The magnetic energy in V is finite and it should remain so even when we extend the boundary to infinity. To this purpose, we consider the ball $B(0, a)$, and demand that limit of the integral over ∂B be finite as $a \to +\infty$. This condition is satisfied if $\mathcal{A}_1(\mathbf{r})$ and $\mathcal{H}(\mathbf{r})$ admit the estimates

$$|\mathcal{A}_1(\mathbf{r})| \leqslant \frac{C_A}{|\mathbf{r}|}, \qquad \text{for} \qquad |\mathbf{r}| \geqslant b_A \tag{4.36}$$

$$|\mathcal{H}(\mathbf{r})| \leqslant \frac{C_H}{|\mathbf{r}|^2}, \qquad \text{for} \qquad |\mathbf{r}| \geqslant b_H \tag{4.37}$$

with C_A, b_A, C_H and b_H being suitable positive constants. Indeed, it follows

$$\begin{aligned} \left| \int_{\partial B} \mathrm{d}S\, \hat{\mathbf{n}}(\mathbf{r}) \times \mathcal{H}(\mathbf{r}) \cdot \mathcal{A}_1(\mathbf{r}) \right| &\leqslant \int_{\partial B} \mathrm{d}S\, |\hat{\mathbf{n}}(\mathbf{r}) \times \mathcal{H}(\mathbf{r}) \cdot \mathcal{A}_1(\mathbf{r})| \\ &\leqslant \int_{\partial B} \mathrm{d}S\, |\mathcal{H}(\mathbf{r})||\mathcal{A}_1(\mathbf{r})| \leqslant \int_{\partial B} \mathrm{d}S\, \frac{C_A}{a}\frac{C_H}{a^2} = 4\pi \frac{C_A C_H}{a} \end{aligned} \tag{4.38}$$

and the last quantity vanishes as $a \to +\infty$. We shall see in Section 5.5 that the vector potential and the magnetic field fall off even more rapidly than what we surmise here, essentially because *magnetic charges* do not exist.

An estimate of the type (4.37) applies to the magnetic induction as well since

$$|\mathcal{B}(\mathbf{r})| = \left|\overline{\overline{\mu}}(\mathbf{r}) \cdot \mathcal{H}(\mathbf{r})\right| \leqslant \left\|\overline{\overline{\mu}}(\mathbf{r})\right\| |\mathcal{H}(\mathbf{r})| \leqslant |\mathcal{H}(\mathbf{r})| \sup_{\mathbf{r}\in\mathbb{R}^3} \left\|\overline{\overline{\mu}}(\mathbf{r})\right\| \tag{4.39}$$

by virtue of (1.127) and (E.76). We conclude that physically admissible stationary magnetic fields in the whole space obey an inverse square law, and the asymptotic behavior of the vector potential is

$$\mathcal{A}(\mathbf{r}) = \mathcal{O}\left(\frac{1}{|\mathbf{r}|}\right) + \nabla\chi, \qquad |\mathbf{r}| \to +\infty \tag{4.40}$$

if the sources are of finite extent.

The splitting (4.34) of W_h into a volume plus a surface integral — both involving the vector potential — lends itself to an interesting interpretation. If we choose the volume V so as to leave all the stationary currents in the complementary unbounded domain $\mathbb{R}^3 \setminus \overline{V}$, the volume integral in the last part of (4.34) vanishes identically. The remaining surface integral over ∂V then represents the magnetic energy stored in V due to the currents placed in $\mathbb{R}^3 \setminus \overline{V}$. The minus sign should not worry us, as it is just a consequence of the orientation of $\hat{\mathbf{n}}$ towards $\mathbb{R}^3 \setminus \overline{V}$, and W_h is positive nonetheless. However, comparison of the surface and the volume integrals in (4.34) leads us to interpret $-\hat{\mathbf{n}}\times\mathcal{H}(\mathbf{r})$ as a fictitious steady surface current density flowing over ∂V and producing the stationary magnetic field inside V. This finding is perfectly in line with the boundary conditions (1.168) and the integral representation of the vector potential derived in Section 5.1.

For the sake of completeness, we observe that the integral form of the Ampère law in free space reads

$$\mu_0 \int_{S_J} \mathrm{d}S\, \hat{\mathbf{n}}(\mathbf{r}) \cdot \mathcal{J}(\mathbf{r}) = \oint_{\gamma} \mathrm{d}s\, \hat{\mathbf{s}}(\mathbf{r}) \cdot \nabla \times \mathcal{A}(\mathbf{r}) = \int_{S} \mathrm{d}S\, \hat{\mathbf{n}}(\mathbf{r}) \cdot \nabla \times [\nabla \times \mathcal{A}(\mathbf{r})] \tag{4.41}$$

with $S_J \subseteq S$. The Stokes theorem may be applied separately to $S \setminus \overline{S}_J$ and S_J where $\nabla \times \mathcal{A}(\mathbf{r})$ is continuously differentiable, and the result follows by postulating the continuity of $\hat{\mathbf{s}}(\mathbf{r}) \cdot [\nabla \times \mathcal{A}(\mathbf{r})]$ for $\mathbf{r} \in \partial S_J$. Similar expressions apply for isotropic and anisotropic magnetic media.

At a first glance it would appear that (4.41) does not hold true if we let the loop γ recede to infinity because the flux of $\mathcal{J}(\mathbf{r})$ through S_J is seemingly unaffected by the limiting process, whereas the circulation of $\nabla \times \mathcal{A}(\mathbf{r})$ vanishes in view of (4.29), (1.113) and estimate (4.37). (This can be checked by taking γ to be a circumference of radius a and by estimating the integral for $a \to +\infty$.) We resolve this contradiction by observing that, contrary to the previous erroneous conclusion, the flux of $\mathcal{J}(\mathbf{r})$ through S_J is, in fact, affected in the limiting process: it is not possible to extend γ to infinity without crossing $\mathcal{J}(\mathbf{r})$ at least in one more surface, because $\mathcal{J}(\mathbf{r})$, being finite and steady, flows along a closed tube! Thereby, S_J becomes a multiply-connected surface, the flux integral comprises more than one separate contributions which cancel each other out, thus ensuring the validity of (4.41) under the estimate (4.36).

Conversely, if the current density is unbounded, as in Figure 4.1b, indeed γ can be chosen so that S_J is simply connected, and when we let γ recede to infinity the circulation must not vanish because the flux of $\mathcal{J}(\mathbf{r})$ is not zero. This means that the asymptotic behavior of the vector potential is not given by (4.36) but rather by

$$|\mathcal{A}_1(\mathbf{r})| \leqslant C_A \log |\mathbf{r}|, \qquad |\mathbf{r}| \to +\infty \tag{4.42}$$

with C_A is a suitable positive constant. Accordingly, an infinitely extended current density produces a stationary magnetic field which decays as the inverse of the distance from the source. This behavior is in agreement with the finding (4.16) of Example 4.1 for a infinitely long straight current density.

When we choose other constitutive relationships, the vector Poisson equation looks different that (4.33). In an unbounded inhomogeneous isotropic medium we have

$$\nabla \times \left[\frac{1}{\mu(\mathbf{r})} \nabla \times \mathcal{A}(\mathbf{r}) \right] = \mathcal{J}(\mathbf{r}), \qquad \mathbf{r} \in \mathbb{R}^3 \tag{4.43}$$

with $1/\mu(\mathbf{r}) \in \mathcal{C}^1(\mathbb{R}^3)$, whereas in an unbounded inhomogeneous anisotropic magnetic medium the vector Poisson equation reads

$$\nabla \times \left\{ [\overline{\mu}(\mathbf{r})]^{-1} \cdot \nabla \times \mathcal{A}(\mathbf{r}) \right\} = \mathcal{J}(\mathbf{r}), \qquad \mathbf{r} \in \mathbb{R}^3, \tag{4.44}$$

where the entries of the inverse dyadic permeability $[\overline{\mu}(\mathbf{r})]^{-1}$ must be continuously differentiable in $\mathbb{R}^3$.

Example 4.3 (Vector potential of an infinite solenoid)

We consider again the solenoid of Figure 4.4b and compute the relevant vector potential $\mathcal{A}(\mathbf{r})$. The result will come in handy in Section 5.3 while investigating the physical meaning of $\mathcal{A}(\mathbf{r})$, but it is also an interesting example of a situation in which the vector potential is non-zero in a region of space where, conversely, the magnetic field vanishes.

Since we already know the magnetic field from Example 4.2, we may determine $\mathcal{A}(\mathbf{r})$ by means of the global relation (4.31) instead of attempting the solution of (4.33) which is fraught with pitfalls. The procedure parallels the calculation of $\mathcal{H}(\mathbf{r})$ produced by a cylindrical steady current $\mathcal{J}(\mathbf{r})$ in Example 4.1 for the formal structure of (4.31) is the same as that of the Ampère law (4.5).

In a system of circular cylindrical coordinates (τ, φ, z) with the z-axis aligned with the axis of the solenoid (Figure 4.4b) the magnetic vector potential must be a function of τ only for the geometry and the magnetic intensity are invariant along φ and z. We may argue that the radial component $\mathcal{A}_\tau(\tau)$ is null everywhere by means of the same reasoning followed in Example 4.2 to show that $\mathcal{H}_\tau(\tau) = 0$. The longitudinal component $\mathcal{A}_z(\tau)$ may be proven to be constant by calculating the circulation of $\boldsymbol{\mathcal{A}}(\tau)$ along the rectangular contours of Figure 4.6 for the flux Ψ_B vanishes in all three cases and $\mathcal{B}_\tau(\tau) = \mu_0\mathcal{H}_\tau(\tau) = 0$ from (4.28). However, the magnetic intensity is bounded in any plane perpendicular to $\hat{\mathbf{z}}$ and, in light of the formal equality of (4.9) and (4.29), we expect the vector potential to fall off with the distance τ from the 'source' (i.e., the magnetic intensity). We may fulfill this condition only by setting $\mathcal{A}_z(\tau) = 0$.

Thus, we are left with the azimuthal component $\mathcal{A}_\varphi(\tau)$ which we may determine by applying (4.31) on the circle $C := \{\mathbf{r} \in \mathbb{R}^3 : \tau \leqslant b,\ z = 0\}$. For the circulation of $\boldsymbol{\mathcal{A}}(\tau)$ along $\gamma := \partial C$ we have

$$\oint_\gamma \mathrm{d}s\, \hat{\mathbf{s}}(\mathbf{r}) \cdot \boldsymbol{\mathcal{A}}(\mathbf{r}) = \int_0^{2\pi} \mathrm{d}\varphi\, b\, \hat{\boldsymbol{\varphi}} \cdot \hat{\boldsymbol{\varphi}} \mathcal{A}_\varphi(b) = 2\pi b \mathcal{A}_\varphi(b) \tag{4.45}$$

for $\mathcal{A}_\varphi(\tau)$ is constant on γ. Next, we compute the flux of $\boldsymbol{\mathcal{B}}(\mathbf{r})$ through C

$$\Psi_B(b) = \int_C \mathrm{d}S\, \hat{\mathbf{z}} \cdot \boldsymbol{\mathcal{B}}(\mathbf{r}) = \begin{cases} \mu_0 \mathcal{N} I \pi b^2, & b < a \\ \mu_0 \mathcal{N} I \pi a^2, & b > a \end{cases} \tag{4.46}$$

having used (4.28). Application of (4.31) yields $\mathcal{A}_\varphi(b)$ as a function of the radius b. By putting everything together and renaming b to τ we obtain

$$\boldsymbol{\mathcal{A}}(\tau) = \begin{cases} \dfrac{\mu_0 \mathcal{N} I}{2} \tau \hat{\boldsymbol{\varphi}}, & \tau < a \\ \dfrac{\mu_0 \mathcal{N} I a^2}{2\tau} \hat{\boldsymbol{\varphi}}, & \tau > a \end{cases} \tag{4.47}$$

which indicates that $\boldsymbol{\mathcal{A}}(\tau)$ is continuous across the circular cylinder $S := \{\mathbf{r} \in \mathbb{R}^3 : \tau = a\}$ where the magnetic intensity $\boldsymbol{\mathcal{B}}(\mathbf{r})$ suffers a jump.

(End of Example 4.3)

4.3 Boundary conditions for the vector potential

The matching conditions for the stationary magnetic entities simply follow from (1.142), (1.157), (1.168) and (1.171) by dropping the time dependence. We list the formulas for reference

$$\hat{\mathbf{n}}(\mathbf{r}) \times [\boldsymbol{\mathcal{H}}_1(\mathbf{r}) - \boldsymbol{\mathcal{H}}_2(\mathbf{r})] = \boldsymbol{\mathcal{J}}_S(\mathbf{r}), \qquad \mathbf{r} \in \partial V \tag{4.48}$$

$$\hat{\mathbf{n}}(\mathbf{r}) \cdot [\boldsymbol{\mathcal{B}}_1(\mathbf{r}) - \boldsymbol{\mathcal{B}}_2(\mathbf{r})] = 0, \qquad \mathbf{r} \in \partial V \tag{4.49}$$

$$\hat{\mathbf{n}}(\mathbf{r}) \times \boldsymbol{\mathcal{H}}_1(\mathbf{r}) = \boldsymbol{\mathcal{J}}_S(\mathbf{r}), \qquad \mathbf{r} \in \partial V \tag{4.50}$$

$$\hat{\mathbf{n}}(\mathbf{r}) \cdot \boldsymbol{\mathcal{B}}_1(\mathbf{r}) = 0, \qquad \mathbf{r} \in \partial V \tag{4.51}$$

where ∂V is the material interface between two media as in Figures 1.12 and 1.13. Here we wish to derive the jump conditions for the vector potential across the smooth interface ∂V between two magnetic media.

Since $\mathcal{A}(\mathbf{r})$ may not be continuously differentiable in the presence of a jump of the magnetic permeability, the best we can do is apply the Ansatz (4.29) separately in each medium. With reference to the geometry of Figure 1.12 we indicate the magnetic vector potentials in the regions V_1 and V_2 with $\mathcal{A}_1(\mathbf{r})$ and $\mathcal{A}_2(\mathbf{r})$. Then, from the steady-state condition (4.49) we find

$$\begin{aligned} 0 &= \hat{\mathbf{n}}(\mathbf{r}) \cdot [\mathcal{B}_1(\mathbf{r}) - \mathcal{B}_2(\mathbf{r})] = \hat{\mathbf{n}}(\mathbf{r}) \cdot [\nabla \times \mathcal{A}_1(\mathbf{r}) - \nabla \times \mathcal{A}_2(\mathbf{r})] \\ &= \nabla_s \cdot [\hat{\mathbf{n}}(\mathbf{r}) \times \mathcal{A}_2(\mathbf{r})] - \nabla_s \cdot [\hat{\mathbf{n}}(\mathbf{r}) \times \mathcal{A}_1(\mathbf{r})] \\ &= \nabla_s \cdot \{\hat{\mathbf{n}}(\mathbf{r}) \times [\mathcal{A}_2(\mathbf{r}) - \mathcal{A}_1(\mathbf{r})]\}, \qquad \mathbf{r} \in \partial V \end{aligned} \tag{4.52}$$

where we have used the differential identity (A.60), the linearity of the surface divergence $\nabla_s \cdot \{\bullet\}$, and the differentiability of the tangential vector field $\hat{\mathbf{n}}(\mathbf{r}) \times \mathcal{A}(\mathbf{r})$ over either side of ∂V. Indeed, we do not know anything yet about the smoothness of the normal component of $\mathcal{A}(\mathbf{r})$. From (4.52) we conclude that the jump of $\hat{\mathbf{n}}(\mathbf{r}) \times \mathcal{A}(\mathbf{r})$ across ∂V must be equal to a solenoidal surface vector field

$$\hat{\mathbf{n}}(\mathbf{r}) \times [\mathcal{A}_1(\mathbf{r}) - \mathcal{A}_2(\mathbf{r})] = \mathcal{V}_t(\mathbf{r}), \qquad \mathbf{r} \in \partial V \tag{4.53}$$

and for convenience we may set $\mathcal{V}_t(\mathbf{r}) = 0$. Therefore, we obtain the matching condition

$$\hat{\mathbf{n}}(\mathbf{r}) \times \mathcal{A}_1(\mathbf{r}) = \hat{\mathbf{n}}(\mathbf{r}) \times \mathcal{A}_2(\mathbf{r}), \qquad \mathbf{r} \in \partial V \tag{4.54}$$

that is, the (rotated) tangential component of the vector potential is continuous across the material interface ∂V [8, Section 26.2].

An alternative way of obtaining (4.54) consists of applying the global magnetic Gauss law (4.6) to the arbitrary surface $\partial W := \partial W_1 \cup \partial W_2$ in the geometrical setup of Figure 1.13. In symbols, we have

$$\begin{aligned} 0 &= \int_{\partial W} \mathrm{d}S\, \hat{\boldsymbol{\nu}}(\mathbf{r}) \cdot \mathcal{B}(\mathbf{r}) = \int_{\partial W_1} \mathrm{d}S\, \hat{\boldsymbol{\nu}}_1(\mathbf{r}) \cdot \mathcal{B}_1(\mathbf{r}) + \int_{\partial W_2} \mathrm{d}S\, \hat{\boldsymbol{\nu}}_2(\mathbf{r}) \cdot \mathcal{B}_2(\mathbf{r}) \\ &= \int_{\partial W_1} \mathrm{d}S\, \hat{\boldsymbol{\nu}}_1(\mathbf{r}) \cdot \nabla \times \mathcal{A}_1(\mathbf{r}) + \int_{\partial W_2} \mathrm{d}S\, \hat{\boldsymbol{\nu}}_2(\mathbf{r}) \cdot \nabla \times \mathcal{A}_2(\mathbf{r}) \\ &= \oint_{\partial S_0} \mathrm{d}s\, \hat{\mathbf{s}}(\mathbf{r}) \cdot [\mathcal{A}_1(\mathbf{r}) - \mathcal{A}_2(\mathbf{r})] \end{aligned} \tag{4.55}$$

where we have used (4.29), which is valid separately in W_1 and W_2, and also applied the Stokes theorem to the fluxes through the open surfaces ∂W_1 and ∂W_2 with due regard to the orientation of the unit normals $\hat{\boldsymbol{\nu}}_1(\mathbf{r})$ and $\hat{\boldsymbol{\nu}}_2(\mathbf{r})$. If the scalar field $\hat{\mathbf{s}}(\mathbf{r}) \cdot [\mathcal{A}_1(\mathbf{r}) - \mathcal{A}_2(\mathbf{r})]$ is continuous for $\mathbf{r} \in \partial S_0$, the mean value theorem [9] leads to

$$\hat{\mathbf{s}}(\mathbf{r}_0) \cdot [\mathcal{A}_1(\mathbf{r}_0) - \mathcal{A}_2(\mathbf{r}_0)] = 0, \qquad \mathbf{r}_0 \in \partial S_0 \tag{4.56}$$

which is equivalent to (4.54) on the grounds of (1.147) and the arbitrariness of S_0.

Thanks to (4.54) we may show that the global relation (4.31) remains valid even in the presence of material interfaces between media endowed with different magnetic permeability. By computing the flux of $\mathcal{B}_1(\mathbf{r})$ and $\mathcal{B}_2(\mathbf{r})$ through the auxiliary surfaces S_1 and S_2 defined in Figure 1.12 we find

$$\begin{aligned} \int_{S_1} \mathrm{d}S\, \hat{\boldsymbol{\nu}}_1(\mathbf{r}) \cdot \mathcal{B}_1(\mathbf{r}) &= \int_{S_1} \mathrm{d}S\, \hat{\boldsymbol{\nu}}_1(\mathbf{r}) \cdot \nabla \times \mathcal{A}_1(\mathbf{r}) = \oint_{\partial S_1 \cup \gamma_0} \mathrm{d}s\, \hat{\mathbf{s}}_1(\mathbf{r}) \cdot \mathcal{A}_1(\mathbf{r}) \\ \int_{S_2} \mathrm{d}S\, \hat{\boldsymbol{\nu}}_2(\mathbf{r}) \cdot \mathcal{B}_2(\mathbf{r}) &= \int_{S_2} \mathrm{d}S\, \hat{\boldsymbol{\nu}}_2(\mathbf{r}) \cdot \nabla \times \mathcal{A}_2(\mathbf{r}) = \oint_{\partial S_2 \cup \gamma_0} \mathrm{d}s\, \hat{\mathbf{s}}_2(\mathbf{r}) \cdot \mathcal{A}_2(\mathbf{r}) \end{aligned} \tag{4.57}$$

where we have used the Stokes theorem separately on S_1 and S_2. By summing the above relations side by side and rearranging the line integrals we have

$$\int_{S_0} \mathrm{d}S\, \hat{\boldsymbol{\nu}}(\mathbf{r}) \cdot \mathcal{B}(\mathbf{r}) = \int_{S_1} \mathrm{d}S\, \hat{\boldsymbol{\nu}}_1(\mathbf{r}) \cdot \mathcal{B}_1(\mathbf{r}) + \int_{S_2} \mathrm{d}S\, \hat{\boldsymbol{\nu}}_2(\mathbf{r}) \cdot \mathcal{B}_2(\mathbf{r})$$
$$= \oint_{\partial S_1 \cup \gamma_0} \mathrm{d}s\, \hat{\mathbf{s}}_1(\mathbf{r}) \cdot \mathcal{A}_1(\mathbf{r}) + \oint_{\partial S_2 \cup \gamma_0} \mathrm{d}s\, \hat{\mathbf{s}}_2(\mathbf{r}) \cdot \mathcal{A}_2(\mathbf{r})$$
$$= \oint_{\partial S_0} \mathrm{d}s\, \hat{\mathbf{s}}(\mathbf{r}) \cdot \mathcal{A}(\mathbf{r}) + \int_{\gamma_0} \mathrm{d}s\, \hat{\boldsymbol{\tau}}(\mathbf{r}) \cdot [\mathcal{A}_1(\mathbf{r}) - \mathcal{A}_2(\mathbf{r})] \tag{4.58}$$

where the integral along γ_0 vanishes by virtue of (1.133), (H.13), and (4.54). Thus, thanks to the arbitrariness of S_0 we have extended (4.31) to the cases where the surface S happens to cross material interfaces, even though the local form (4.29) is only valid individually in each region where $\mathcal{A}(\mathbf{r})$ is continuously differentiable.

Definition (4.29) contains too little information, as it were, for the vector potential to be unambiguously identified. Indeed, in Section 8.1 we shall prove the *Helmholtz theorem* [3], which states that a vector field is univocally determined if both the curl *and* the divergence thereof are specified. Accordingly, in Section 4.7 we shall find that *uniqueness* is achieved if the divergence of $\mathcal{A}(\mathbf{r})$ is assigned as well, e.g., in the form

$$\nabla \cdot \mathcal{A}(\mathbf{r}) = f(\mathbf{r}) \tag{4.59}$$

where $f(\mathbf{r})$ is a regular scalar field. This is always possible because, as we argued in Section 4.2, (4.29) defines $\mathcal{A}(\mathbf{r})$ within an additive *curl-free* field $\nabla\chi(\mathbf{r})$ which, albeit not necessary, can be determined once $\nabla\cdot\mathcal{A}(\mathbf{r})$ has been prescribed. On the other hand, in the presence of material interfaces, as is suggested in Figure 1.13, we must enforce condition (4.59) separately in each region where $\mathcal{A}(\mathbf{r})$ is continuously differentiable. Now, can we use (4.59) to determine appropriate matching conditions for the component of the magnetic vector potential perpendicular to the material interface ∂V? After all, this was the starting point for arriving at (1.155) and (1.157).

To address this question, we consider the geometrical setup of Figure 1.13 and, based on two instances of (4.59) in the regions V_1 and V_2, we write down the following integral identities for fields and potentials in the auxiliary domains W_1 and W_2

$$\int_{W_1} \mathrm{d}V\, f_1(\mathbf{r}) = \int_{W_1} \mathrm{d}V\, \nabla \cdot \mathcal{A}_1(\mathbf{r}) = \int_{\partial W_1 \cup S_0} \mathrm{d}S\, \hat{\boldsymbol{\nu}}_1(\mathbf{r}) \cdot \mathcal{A}_1(\mathbf{r}) \tag{4.60}$$

$$\int_{W_2} \mathrm{d}V\, f_2(\mathbf{r}) = \int_{W_2} \mathrm{d}V\, \nabla \cdot \mathcal{A}_2(\mathbf{r}) = \int_{\partial W_2 \cup S_0} \mathrm{d}S\, \hat{\boldsymbol{\nu}}_2(\mathbf{r}) \cdot \mathcal{A}_2(\mathbf{r}) \tag{4.61}$$

where the Gauss theorem has been applied because $\mathcal{A}_1(\mathbf{r})$ and $\mathcal{A}_2(\mathbf{r})$ are continuously differentiable in W_1 and W_2. By summing (4.60) and (4.61) side by side and combining like integrals we obtain

$$\int_{W} \mathrm{d}V\, f(\mathbf{r}) = \int_{\partial W_1 \cup S_0} \mathrm{d}S\, \hat{\boldsymbol{\nu}}_1(\mathbf{r}) \cdot \mathcal{A}_1(\mathbf{r}) + \int_{\partial W_2 \cup S_0} \mathrm{d}S\, \hat{\boldsymbol{\nu}}_2(\mathbf{r}) \cdot \mathcal{A}_2(\mathbf{r})$$
$$= \int_{\partial W} \mathrm{d}V\, \hat{\boldsymbol{\nu}}(\mathbf{r}) \cdot \mathcal{A}(\mathbf{r}) + \int_{S_0} \mathrm{d}S\, \hat{\mathbf{n}}(\mathbf{r}) \cdot [\mathcal{A}_2(\mathbf{r}) - \mathcal{A}_1(\mathbf{r})] \tag{4.62}$$

where $f(\mathbf{r}) := f_l(\mathbf{r})$, $\mathbf{r} \in V_l$, $l = 1, 2$, is a scalar field, possibly piecewise continuous. This last condition is formally no different than (1.152), which, when compared to (1.150), led us to (1.155). However, in the present case there is *a priori* no reason why the relevant global law

$$\int_{\partial W} \mathrm{d}V \, \hat{\boldsymbol{\nu}}(\mathbf{r}) \cdot \mathcal{A}(\mathbf{r}) = \int_{W} \mathrm{d}V \, f(\mathbf{r}) \tag{4.63}$$

should hold true — which would demand the vanishing of the integral over S_0 in (4.62). As a matter of fact, we have begun by imposing the local law (4.59) in V_1 and V_2, rather than (4.63). Nonetheless, it is desirable to *enforce* the continutiy of the normal component of $\mathcal{A}(\mathbf{r})$, viz.,

$$\hat{\mathbf{n}}(\mathbf{r}) \cdot \mathcal{A}_1(\mathbf{r}) = \hat{\mathbf{n}}(\mathbf{r}) \cdot \mathcal{A}_2(\mathbf{r}), \qquad \mathbf{r} \in \partial V \tag{4.64}$$

whereby the flux integral over S_0 in (4.62) vanishes and (4.63) is satisfied. Notice that (4.64) is independent of the values taken on by $f(\mathbf{r})$ in V_1 and V_2.

Still, are we sure that (4.64) can be achieved in practice? Once again, we are helped by the intrinsic indeterminacy of $\mathcal{A}(\mathbf{r})$ which is specified except for an additive conservative field $\nabla\chi$. Indeed, even imposing (4.59) is not enough to determine the arbitrary field $\chi(\mathbf{r})$ (see Section 4.7.1). As it turns out, we can exploit the extra degrees of freedom to achieve (4.63) and (4.64). If in accordance with (4.32) we let

$$\mathcal{A}_l(\mathbf{r}) = \mathcal{A}'_l(\mathbf{r}) + \nabla\chi_l(\mathbf{r}), \qquad \mathbf{r} \in V_l, \qquad l = 1, 2 \tag{4.65}$$

the integral over S_0 becomes

$$\int_{S_0} \mathrm{d}S \, \hat{\mathbf{n}}(\mathbf{r}) \cdot [\mathcal{A}_2(\mathbf{r}) - \mathcal{A}_1(\mathbf{r})] = \int_{S_0} \mathrm{d}S \left[\hat{\mathbf{n}}(\mathbf{r}) \cdot \mathcal{A}'_2(\mathbf{r}) + \frac{\partial \chi_1}{\partial \hat{\mathbf{n}}} - \hat{\mathbf{n}}(\mathbf{r}) \cdot \mathcal{A}'_1(\mathbf{r}) - \frac{\partial \chi_2}{\partial \hat{\mathbf{n}}} \right] \tag{4.66}$$

whence we see that we can always choose the normal derivatives of $\chi_1(\mathbf{r})$ and $\chi_2(\mathbf{r})$ in order for the integrand in the right member to vanish.

Finally, we observe that the jump condition for $\mathcal{H}(\mathbf{r})$ (4.48) can be rephrased in terms of the vector potential thanks to (4.29) and, e.g., the constitutive relationship (1.118)

$$\hat{\mathbf{n}}(\mathbf{r}) \times \frac{\nabla \times \mathcal{A}_1(\mathbf{r})}{\mu_1(\mathbf{r})} = \hat{\mathbf{n}}(\mathbf{r}) \times \frac{\nabla \times \mathcal{A}_2(\mathbf{r})}{\mu_2(\mathbf{r})}, \qquad \mathbf{r} \in \partial V \tag{4.67}$$

in the absence of an electric surface current density on ∂V.

4.4 Magnetic scalar potential

In spatial regions devoid of electric currents (and hence, of charges) the magnetic field strength $\mathcal{H}(\mathbf{r})$ is curl-free or conservative according to (4.9), in addition to being divergence-free by virtue of (1.118) and (4.10). Thus, in analogy to (2.15), it is tempting to introduce a magnetic scalar potential $\Psi(\mathbf{r})$ (physical dimension: A) and let [10, Section 3.3], [8, Chapter 29], [5, Sections 5.1,5.2]

$$\mathcal{H}(\mathbf{r}) = -\nabla\Psi(\mathbf{r}), \qquad \mathbf{r} \in \mathbb{R}^3 \setminus \overline{V}_J \tag{4.68}$$

where V_J is the domain occupied by the steady current current $\mathcal{J}(\mathbf{r})$. However, this assumption is fraught with difficulties inasmuch as, unlike the electrostatic potential $\Phi(\mathbf{r})$, $\Psi(\mathbf{r})$ is a *many-valued* scalar field. The reason for this behavior becomes apparent if we invoke the global Ampère law (4.5).

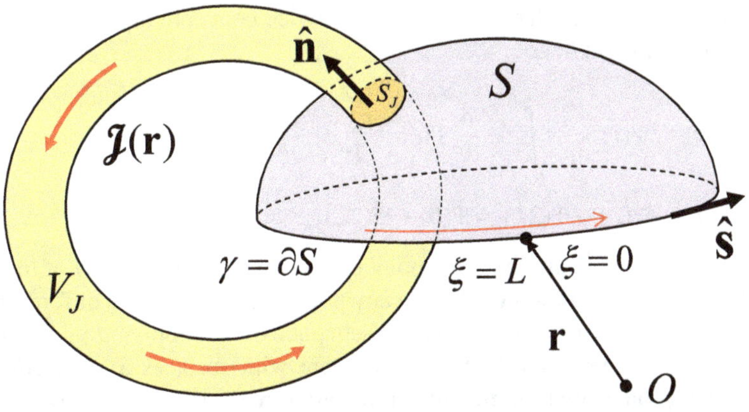

Figure 4.7 An open smooth surface S that intersects a region V_J occupied by the current density $\mathcal{J}(\mathbf{r})$ for showing that the magnetic scalar potential $\Psi(\mathbf{r})$ is many-valued.

We choose the surface S and the boundary $\gamma := \partial S$ so that S intersects the region of current density in $S_J := S \cap V_J$, and the closed path γ necessarily encircles the current $\mathcal{J}(\mathbf{r})$, as is pictorially suggested in Figure 4.7. To compute the circulation along γ we introduce the bijective mapping

$$\boldsymbol{\Upsilon} : \mathbb{R} \supset [0, L] \longrightarrow \gamma \subset \mathbb{R}^3 \tag{4.69}$$

whereby $\mathbf{r} = \boldsymbol{\Upsilon}(\xi)$, with $\xi \in [0, L]$, is a point on γ. The vector field

$$\boldsymbol{\Upsilon}'(\xi) = \frac{\mathrm{d}\boldsymbol{\Upsilon}}{\mathrm{d}\xi} \tag{4.70}$$

is tangential to γ at $\mathbf{r} = \boldsymbol{\Upsilon}(\xi)$. Now, combining (4.5) and (4.68) yields

$$\begin{aligned} I &:= \int_{S_J} \mathrm{d}S\, \hat{\mathbf{n}}(\mathbf{r}) \cdot \mathcal{J}(\mathbf{r}) = -\oint_{\gamma} \mathrm{d}s\, \hat{\mathbf{s}}(\mathbf{r}) \cdot \nabla\Psi(\mathbf{r}) = -\int_0^L \mathrm{d}\xi\, \frac{\mathrm{d}\boldsymbol{\Upsilon}}{\mathrm{d}\xi} \cdot \nabla\Psi[\boldsymbol{\Upsilon}(\xi)] \\ &= -\int_0^L \mathrm{d}\xi \left(\frac{\partial\Psi}{\partial x}\frac{\mathrm{d}x}{\mathrm{d}\xi} + \frac{\partial\Psi}{\partial y}\frac{\mathrm{d}y}{\mathrm{d}\xi} + \frac{\partial\Psi}{\partial z}\frac{\mathrm{d}z}{\mathrm{d}\xi} \right) = -\int_0^L \mathrm{d}\xi\, \frac{\mathrm{d}}{\mathrm{d}\xi}\Psi(\xi) = \Psi(0) - \Psi(L) \end{aligned} \tag{4.71}$$

because the Cartesian components of $\boldsymbol{\Upsilon}(\xi)$ are the coordinates x, y and z. Since γ is closed, though, the point specified by the position vector $\mathbf{r} = \boldsymbol{\Upsilon}(0)$ coincides with the point identified by $\mathbf{r} = \boldsymbol{\Upsilon}(L)$, and hence (4.71) tells us that, if the current I does not vanish, the magnetic scalar potential takes on two values in the local origin on γ. Moreover, since we may choose the local origin on γ at will, (4.71) shows that the magnetic scalar potential may take on at least *two* different values in any point on γ. Last but not least, since even the surface S is somewhat arbitrary, (4.71) says that the magnetic scalar potential takes on *two* different values in any point $\mathbf{r} \in \mathbb{R}^3 \setminus \overline{V}_J$. The difference between the values in question is precisely the current I which is encircled *once* by the loop γ. But this is not the end of the story.

We can think of S as being formed by two identical overlapped sheets S_1 and S_2 in such a way that the boundary ∂S is actually comprised of two adjoining instances γ_1 and $\gamma_2 \equiv \gamma_1$ of the same loop. The parameterization of ∂S now reads

$$\boldsymbol{\Upsilon} : \mathbb{R} \supset [0, 2L] \longrightarrow (\gamma_1 \cup \gamma_2) \subset \mathbb{R}^3 \tag{4.72}$$

and by taking into account that $\mathcal{J}(\mathbf{r})$ cuts through S twice conceptually, whereby the flux of $\mathcal{J}(\mathbf{r})$ through S_J amounts to $2I$, we have

$$\begin{aligned} 2I &= -\oint_{\gamma_1 \cup \gamma_2} \mathrm{d}s\, \hat{\mathbf{s}}(\mathbf{r}) \cdot \nabla\Psi(\mathbf{r}) = -\int_0^L \mathrm{d}\xi \frac{\mathrm{d}}{\mathrm{d}\xi}\Psi(\xi) - \int_L^{2L} \mathrm{d}\xi \frac{\mathrm{d}}{\mathrm{d}\xi}\Psi(\xi) \\ &= \Psi(0) - \Psi(L) + \Psi(L) - \Psi(2L) = \Psi(0) - \Psi(2L) \end{aligned} \tag{4.73}$$

where $\mathbf{r} = \Upsilon(0)$ and $\mathbf{r} = \Upsilon(2L)$ are the same point on $\gamma_1 \cup \gamma_2$. Again, in view of the arbitrariness of S we see that Ψ takes on *two* different values in any point $\mathbf{r} \in \mathbb{R}^3 \setminus \overline{V}_J$. The difference between these values is now *twice* the current I, and this is a consequence of the loop $\gamma_1 \cup \gamma_2$ encircling $\mathcal{J}(\mathbf{r})$ *twice*.

Evidently, we may repeat this thought experiment with surfaces S comprised of ever more sheets and hence envision a situation in which the boundary ∂S is a closed loop that encircles $\mathcal{J}(\mathbf{r})$ N times. Essentially, this means that an ideal observer has to complete N full laps around $\mathcal{J}(\mathbf{r})$ in order to walk along the whole boundary ∂S. After each lap, she notices that the magnetic potential $\Psi[\Upsilon(\xi)] = \Psi(\mathbf{r})$ has *decreased* with respect to the previous value by an amount equal to I. The quantity $-I$ is sometimes referred to as the *period* of the many-valued function [1].

We observe that if the surface S of Figure 4.7 is chosen in such a way that $S_J = \emptyset$, i.e., neither does S intersect the current density nor does the loop ∂S encircle $\mathcal{J}(\mathbf{r})$, then the circulation of $\mathcal{H}(\mathbf{r})$ along ∂S vanishes in accordance with the Ampère law (4.5). Likewise, if the surface S intersects V_J through two surfaces, say S_{J1} and S_{J2}, the net value of the flux of $\mathcal{J}(\mathbf{r})$ vanishes because $\mathcal{J}(\mathbf{r})$ is solenoidal [cf. (4.1)]. The same conclusion also holds if S cuts V_J across any *even* number of distinct surfaces. Under these circumstances from (4.71) or (4.73) we see that the magnetic potential can be defined *locally* as a single-valued scalar field.

In general the line integral of $\mathcal{H}(\mathbf{r})$ along a line which joins two points A and B is not path-independent (cf. Section 2.3). Actually, if we can transform the chosen line γ_1 into a new one, say, γ_2 without being forced to cross a current-carrying region of space (Figure 4.8a) then on account of (4.68) we find

$$\int_{\gamma_1} \mathrm{d}s\, \hat{\mathbf{s}}(\mathbf{r}) \cdot \mathcal{H}(\mathbf{r}) = \int_{\gamma_2} \mathrm{d}s\, \hat{\mathbf{s}}(\mathbf{r}) \cdot \mathcal{H}(\mathbf{r}) = \Psi(\mathbf{r}_A) - \Psi(\mathbf{r}_B) \tag{4.74}$$

because the circulation along the loop $\gamma_1 \cup (-\gamma_2)$ is null, as argued above. So, for the purpose of computing the line integral of $\mathcal{H}(\mathbf{r})$ the paths γ_1 and γ_2 are equivalent. On the contrary, if in the process of warping γ_1 into yet another line γ_3 we must cross a region occupied by a current density $\mathcal{J}(\mathbf{r})$ (Figure 4.8b), we have

$$\int_{\gamma_1} \mathrm{d}s\, \hat{\mathbf{s}}(\mathbf{r}) \cdot \mathcal{H}(\mathbf{r}) = I + \int_{\gamma_3} \mathrm{d}s\, \hat{\mathbf{s}}(\mathbf{r}) \cdot \mathcal{H}(\mathbf{r}) = \Psi(\mathbf{r}_A) - \Psi(\mathbf{r}_B) \tag{4.75}$$

in view of (4.5) because now $\gamma_1 \cup (-\gamma_3)$ encircles $\mathcal{J}(\mathbf{r})$ once. If we assume that $\Psi(\mathbf{r}_B)$ is the final value of the potential at the end of both γ_1 and γ_3, and $\Psi(\mathbf{r}_A)$ is the initial value of the potential at the beginning of γ_1, then for the consistency of the Ampère law the initial value of the potential at the beginning of γ_3 must be $\Psi(\mathbf{r}_A) - I$.

4.5 Magnetic dipoles

In principle, we might think of a magnetic dipole as a pair of magnetic point charges separated by a distance d, pretty much as we did for the definition of electric dipole in Example 2.5. However, this

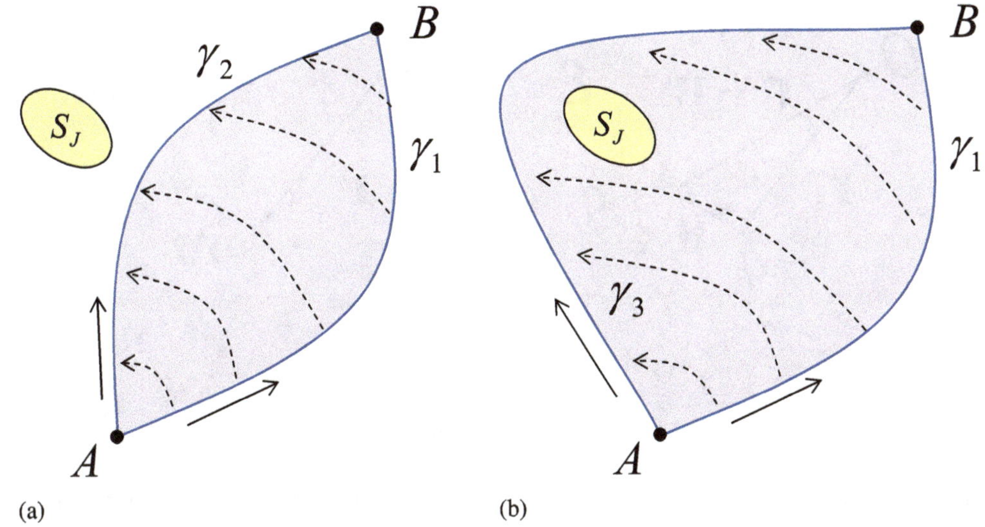

Figure 4.8 The line integral of a stationary magnetic field between two points A and B is path-dependent in general: (a) deformation of the line γ_1 into γ_2 without intersecting a current-carrying region of space; (b) deformation of the line γ_1 into γ_3 while crossing a current tube in S_J.

approach is unsatisfactory, in that the existence of isolated magnetic poles has never been proved. In fact, we shall see in Section 5.5 that any stationary current density $\mathcal{J}(\mathbf{r})$ confined in a bounded volume V_J behaves like a magnetic dipole when the magnetic field is observed at distances which are large as compared to the characteristic dimension of V_J.

On the other hand, practical examples of finite-size magnetic dipoles are constituted by permanent bar magnets made of iron or steel [11, Chapter 9]. Therefore, it is convenient to introduce the notion of a test point magnet, i.e., an elementary magnetic dipole of zero extension localized at a point in space. As was the case for an electric test point charge, we make the hypothesis that the test point magnet does not affect an external magnetic field. We wish to determine formulas for the potential energy of the magnetic dipole immersed in a magnetic induction field and for the torque produced on the dipole by such field [12, Section 12.4]. To get a feeling for the result, we start with the analogous formulas for an electrostatic dipole because, although electric test point charges are an abstraction as well, they are a model of true isolated charges.

Let us consider two point charges q and $-q$ ($q > 0$) located at points $\mathbf{r} + \mathbf{u}$ and $\mathbf{r}$ where $\mathbf{u}$ is a constant vector whose magnitude will be made to shrink to zero (Figure 4.9a). The combined potential energy of the charges reads (see Example 2.5)

$$
\begin{aligned}
W_e(\mathbf{r}; \mathbf{u}) &:= q\Phi(\mathbf{r} + \mathbf{u}) - q\Phi(\mathbf{r}) = q|\mathbf{u}|\frac{\Phi(\mathbf{r} + \mathbf{u}) - \Phi(\mathbf{r})}{|\mathbf{u}|} \\
&\xrightarrow[|\mathbf{u}|\to 0]{} p\frac{\partial}{\partial \hat{\mathbf{u}}}\Phi(\mathbf{r}) = p\hat{\mathbf{u}} \cdot \nabla\Phi(\mathbf{r}) = -\mathbf{p} \cdot \mathcal{E}(\mathbf{r})
\end{aligned}
\tag{4.76}
$$

where $\Phi(\mathbf{r})$ is the electrostatic potential associated with the *external* static electric field $\mathcal{E}(\mathbf{r})$ as in (2.15). In the limiting process we have assumed that, as the distance $|\mathbf{u}|$ of the charges tends to zero, the latter become infinite so as to yield a finite non-zero electric dipole moment $\mathbf{p}$.

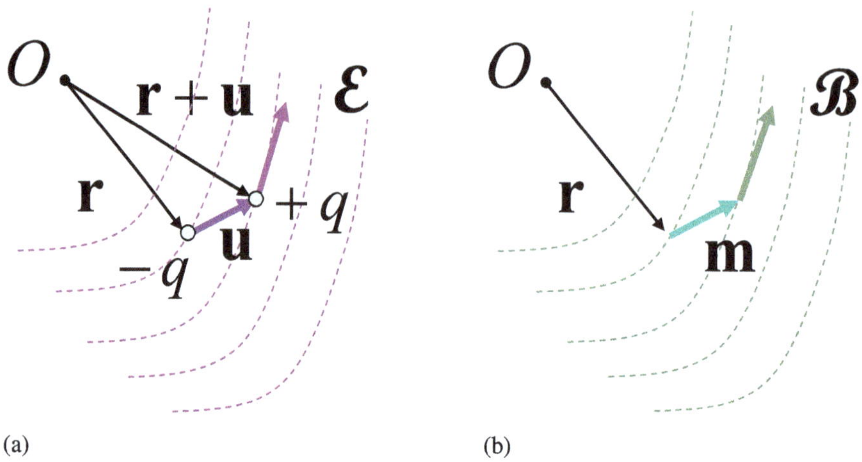

Figure 4.9 Elementary dipoles: (a) electric dipole in an electric field and (b) magnetic dipole in a magnetic induction field.

Similarly, if the two charges were somehow rigidly bound together at the end points of the vector **u** and could rotate freely around the point **r**, the external electric field would exert on q a torque — with respect to **r** — given by [13, Chapter 20]

$$\mathcal{N}_e(\mathbf{r};\mathbf{u}) := \mathbf{u} \times q\mathcal{E}(\mathbf{r}+\mathbf{u}) = q|\mathbf{u}|\hat{\mathbf{u}} \times \mathcal{E}(\mathbf{r}+\mathbf{u}) \xrightarrow[|\mathbf{u}|\to 0]{} \mathbf{p} \times \mathcal{E}(\mathbf{r}) \tag{4.77}$$

again with the same limiting procedure followed above.

Intuition suggests that analogous expressions ought to hold true for an elementary test magnetic dipole. Indeed, by recalling that the magnetic entity of intensity is the magnetic induction, we can surmise (Figure 4.9b) that the relevant potential energy and torque for an elementary magnetic dipole read

$$W_h(\mathbf{r}) := -\mathbf{m} \cdot \mathcal{B}(\mathbf{r}) \tag{4.78}$$

$$\mathcal{N}_h(\mathbf{r}) := \mathbf{m} \times \mathcal{B}(\mathbf{r}) \tag{4.79}$$

where **m** denotes the *magnetic dipole moment*. The dimensional analysis of either (4.78) or (4.79) indicates that **m** carries the physical dimension of a current times an area (Am^2). The formula for the torque is based on experimental evidence, e.g., the magnetized needle of a compass rotates in the magnetic field of the Earth. In fact, a magnetic dipole tends to align itself with respect to the magnetic field in such a way that the potential energy W_h is minimized. By denoting with α the angle between $\mathcal{B}(\mathbf{r})$ and **m** from (4.78) we have

$$\frac{\mathrm{d}W_h}{\mathrm{d}\alpha} = -\frac{\mathrm{d}}{\mathrm{d}\alpha}|\mathbf{m}||\mathcal{B}|\cos\alpha = |\mathbf{m}||\mathcal{B}|\sin\alpha \tag{4.80}$$

and the rightmost-hand side vanishes for $\alpha \in \{0, \pi\}$, i.e., when the dipole moment is parallel to the induction field.

Lastly, we observe that the conservative force associated with the potential energy (4.78) is computed as

$$\mathcal{F}(\mathbf{r}) := -\nabla W_h(\mathbf{r}) = \nabla[\mathbf{m} \cdot \mathcal{B}(\mathbf{r})] = \mathbf{m} \cdot \nabla\mathcal{B}(\mathbf{r}) = -\nabla \times [\mathbf{m} \times \mathcal{B}(\mathbf{r})] = -\nabla \times \mathcal{N}_h(\mathbf{r}) \tag{4.81}$$

by virtue of (H.53) and (H.54), since $\mathbf{m}$ is a constant vector and $\mathcal{B}(\mathbf{r})$ satisfies the magnetic Gauss law (1.23). Because the gradient of $\mathcal{B}(\mathbf{r})$ is involved, a magnetic dipole experiences a force only if the magnetic induction is not uniform.

Next, we examine the stationary magnetic equations for the field produced by an elementary magnetic dipole in free space. We are going to use the usual symbols for the magnetic entities, but now $\mathcal{B}(\mathbf{r})$ and $\mathcal{H}(\mathbf{r})$ represent the fields *due to the dipole*, which is not a test dipole any more. Again, analogy with the electrostatic case prompts us to write down the equations as follows

$$\nabla \times \mathcal{H}(\mathbf{r}) = 0, \qquad \mathbf{r} \in \mathbb{R}^3 \tag{4.82}$$

$$\nabla \cdot \mathcal{H}(\mathbf{r}) = 0, \qquad \mathbf{r} \in \mathbb{R}^3 \setminus \{0\} \tag{4.83}$$

for a dipole located in $\mathbf{r} = 0$. The magnetic field must be conservative, since it is not produced by a steady current but, at least conceptually, by a pair of magnetic charges, so this requirement is consistent with the similar conservative character of the electrostatic field. Moreover, the magnetic field must be solenoidal everywhere except perhaps at the very location of the dipole, where it may be singular. The global magnetic Gauss (4.6) law applied over a surface which includes the dipole must return a zero magnetic flux in agreement with the idea that the total magnetic charge is zero.

Since the magnetic field is curl-free everywhere (cf. Section 4.4) it is possible to introduce a single-valued magnetic scalar potential $\Psi(\mathbf{r})$ perfectly analogous to $\Phi(\mathbf{r})$ in (2.15), namely,

$$\mathcal{H}(\mathbf{r}) = -\nabla\Psi(\mathbf{r}) \tag{4.84}$$

and owing to the similarities considered so far between electric and magnetic dipoles we may surmise that

$$\Psi(\mathbf{r}) = \frac{\mathbf{m} \cdot \hat{\mathbf{r}}}{4\pi r^2} = -\mathbf{m} \cdot \nabla \frac{1}{4\pi r} = -\nabla \cdot \frac{\mathbf{m}}{4\pi r}, \qquad r = |\mathbf{r}| \neq 0 \tag{4.85}$$

is the potential produced by the elementary magnetic dipole. In fact, it is not difficult to verify that (4.85) satisfy the magnetic stationary equations.

By applying definition (4.84) along with (4.85), we can obtain an alternative representation for the magnetic field with the help of the differential identities (H.53) and (H.54), namely,

$$\mathcal{H}(\mathbf{r}) = \nabla\left(\mathbf{m} \cdot \nabla \frac{1}{4\pi r}\right) = \mathbf{m} \cdot \nabla\nabla \frac{1}{4\pi r} = -\nabla \times \left(\mathbf{m} \times \nabla \frac{1}{4\pi r}\right), \qquad r \neq 0 \tag{4.86}$$

having used the fact that $\mathbf{m}$ is a constant vector and $1/r$ is a harmonic function for $|\mathbf{r}| \neq 0$ [9, Section 8.7], [14, Section 2.2]. The rightmost-hand side is in the form of the curl of a vector field. Therefore, we have proved that the magnetic vector potential associated with an elementary magnetic dipole reads

$$\mathcal{A}(\mathbf{r}) = -\mu_0 \mathbf{m} \times \nabla \frac{1}{4\pi r} = \mu_0 \mathbf{m} \times \frac{\hat{\mathbf{r}}}{4\pi r^2}, \qquad \mathbf{r} \neq 0 \tag{4.87}$$

in light of (4.29) and the constitutive relationship (1.113) for free space. We emphasize that, since the magnetic field is not solenoidal at the location of the dipole, $\mathcal{H}(\mathbf{r})$ can be retrieved from the vector potential only for points $\mathbf{r} \neq 0$.

Finally, if the dipole is located at a point $\mathbf{r}'$ the vector potential becomes

$$\mathcal{A}(\mathbf{r}) = \mu_0 \mathbf{m} \times \nabla' \frac{1}{4\pi R}, \qquad \mathbf{r} \in \mathbb{R}^3 \setminus \{\mathbf{r}'\} \tag{4.88}$$

with $R = |\mathbf{R}| = |\mathbf{r} - \mathbf{r}'|$.

4.6 Energy and momentum in the stationary limit

For purely electrostatic fields the electric energy W_e is constant and no balance equation exists. On the other hand, in the stationary limit steady currents produce electric and magnetic fields which combine to give special instances of (1.276) and (1.347) where the total electromagnetic energy $W_e + W_h$ and the momentum $\mathcal{G}_{em}$ stored in the field remain constant.

With reference to the geometry of Figure 1.17 from (1.276) for the Poynting theorem we find immediately

$$\int_{\partial V} \mathrm{d}S\, \hat{\mathbf{n}}(\mathbf{r}) \cdot [\mathcal{E}(\mathbf{r}) \times \mathcal{H}(\mathbf{r})] + \int_{V_C} \mathrm{d}V\, \sigma(\mathbf{r}) |\mathcal{E}(\mathbf{r})|^2 = -\int_{V_S} \mathrm{d}V\, \mathcal{E}(\mathbf{r}) \cdot \mathcal{J}(\mathbf{r}) \tag{4.89}$$

because the electromagnetic entities and, in particular, the energy do not depend on time. The interpretation of the last two terms is the same as in the general time-dependent case. The first contribution, however, represents a *steady flow of electromagnetic energy* [15, p. 155] in and out of the domain V but none of it constitutes *radiation*, for the latter is associated with the generation and propagation of electromagnetic waves.

To extend the result to an unbounded domain V or the whole space $V \equiv \mathbb{R}^3$ we apply (4.89) to the ball $V := B(0, b)$ and take the limit as $b \to +\infty$. The flux integral is the only term affected in the limiting process, and hence we estimate

$$\left| \int_{\partial B} \mathrm{d}S\, \hat{\mathbf{n}}(\mathbf{r}) \cdot [\mathcal{E}(\mathbf{r}) \times \mathcal{H}(\mathbf{r})] \right| \leqslant \int_{\partial B} \mathrm{d}S\, |\mathcal{E}(\mathbf{r})||\mathcal{H}(\mathbf{r})| \leqslant \int_{\partial B} \mathrm{d}S\, \frac{C_E}{b^2} \frac{C_H}{b^2}$$
$$= 4\pi \frac{C_E C_H}{b^2} \xrightarrow[b \to +\infty]{} 0 \tag{4.90}$$

where $b > \max\{b_E, b_H\}$ with b_E and b_H the radii implied in the asymptotic behavior (1.260) and (1.261). Therefore, (4.89) passes over into

$$\int_{V_C} \mathrm{d}V\, \sigma(\mathbf{r}) |\mathcal{E}(\mathbf{r})|^2 = -\int_{V_S} \mathrm{d}V\, \mathcal{E}(\mathbf{r}) \cdot \mathcal{J}(\mathbf{r}) \tag{4.91}$$

which also states that in the absence of conducting media the steady power delivered by the sources is actually null.

For the balance of momentum and forces from (1.347) we find

$$\int_{\partial V} \mathrm{d}S\, \hat{\mathbf{n}}(\mathbf{r}) \cdot \overline{\mathcal{T}}_{em}(\mathbf{r}) = -\mathcal{F}_{em} \tag{4.92}$$

where the flux integral is a *steady flow of momentum* across the boundary ∂V but no radiation takes place. Alternatively, the normal component $\hat{\mathbf{n}}(\mathbf{r}) \cdot \overline{\mathcal{T}}_{em}(\mathbf{r})$ represents a surface density of force (a stress) on ∂V.

4.7 Uniqueness of the stationary solutions

4.7.1 *Vector potential*

We wish to determine the necessary conditions for the vector Poisson equations (4.33), (4.43) and (4.44) to admit just one solution. The result is the *uniqueness theorem* for the magnetic vector potential and, when it holds, it allows concluding that no matter how we determine the solution it also constitutes the only possible one. Furthermore, thanks to definition (4.29) and the constitutive relationship, the uniqueness of $\mathcal{A}(\mathbf{r})$, if any may be proved, implies that $\mathcal{B}(\mathbf{r})$ and $\mathcal{H}(\mathbf{r})$ are unique too. We try to draw our conclusions by contradiction, that is, we suppose that the solution is not unique and then show that such assumption leads to contradictory findings.

We begin with a bounded connected region of space containing stationary currents and a magnetic medium. Suppose that the latter occupies a connected volume $V_1 \subset \mathbb{R}^3$ immersed in an unbounded isotropic, possibly inhomogeneous, magnetic medium endowed with permeability $\mu(\mathbf{r})$, as is exemplified in Figure 4.10. We indicate with V the region of space bounded by ∂V_1 and a smooth closed surface S, and hence the boundary of V is $\partial V := S \cup \partial V_1$; we choose the unit normal $\hat{\mathbf{n}}(\mathbf{r})$ for $\mathbf{r} \in \partial V$ positively oriented towards the inside of V. Finally, the stationary current density $\mathcal{J}(\mathbf{r})$ flows in a closed tube $V_J \subset V$.

If the solution to (4.43) is not unique, then there may exist two distinct vector potentials $\mathcal{A}_1(\mathbf{r})$ and $\mathcal{A}_2(\mathbf{r})$, $\mathbf{r} \in V$, which satisfy (4.43), viz.,

$$\nabla \times \left[\frac{\nabla \times \mathcal{A}_1(\mathbf{r})}{\mu(\mathbf{r})} \right] = \mathcal{J}(\mathbf{r}), \qquad \mathbf{r} \in V \tag{4.93}$$

$$\nabla \times \left[\frac{\nabla \times \mathcal{A}_2(\mathbf{r})}{\mu(\mathbf{r})} \right] = \mathcal{J}(\mathbf{r}), \qquad \mathbf{r} \in V \tag{4.94}$$

and for ease of manipulation we introduce the difference potential $\mathcal{A}_0(\mathbf{r}) := \mathcal{A}_1(\mathbf{r}) - \mathcal{A}_2(\mathbf{r})$. Since the vector Poisson equation is linear, by subracting (4.93) and (4.94) side by side we arrive at

$$\nabla \times \left[\frac{\nabla \times \mathcal{A}_0(\mathbf{r})}{\mu(\mathbf{r})} \right] = 0, \qquad \mathbf{r} \in V \tag{4.95}$$

a homogeneous equation for $\mathcal{A}_0(\mathbf{r})$. Ideally, we would like to show that only the trivial solution $\mathcal{A}_0(\mathbf{r}) = 0$ is admissible for $\mathbf{r} \in V$, because this would ensure uniqueness of the vector potential. However, given that $\mathcal{A}(\mathbf{r})$ is specified up to an arbitrary conservative field, which has no effect on the fulfillment of (4.95), we may expect that the requirement $\mathcal{A}_0(\mathbf{r}) = 0$ cannot be guaranteed without additional conditions.

To see that this is indeed the case, based on (H.49) we consider the following differential identity

$$\begin{aligned}
\nabla \cdot & \left\{ \left[\frac{\nabla \times \mathcal{A}_0(\mathbf{r})}{\mu(\mathbf{r})} \right] \times \mathcal{A}_0(\mathbf{r}) \right\} = \\
&= \mathcal{A}_0(\mathbf{r}) \cdot \nabla \times \left[\frac{\nabla \times \mathcal{A}_0(\mathbf{r})}{\mu(\mathbf{r})} \right] - \nabla \times \mathcal{A}_0(\mathbf{r}) \cdot \frac{\nabla \times \mathcal{A}_0(\mathbf{r})}{\mu(\mathbf{r})} \\
&= - \frac{|\nabla \times \mathcal{A}_0(\mathbf{r})|^2}{\mu(\mathbf{r})}
\end{aligned} \tag{4.96}$$

where we have used (4.95) in the last passage. We go on and integrate both sides of this equation over V to obtain

$$\int_V \mathrm{d}V \, \nabla \cdot \left\{ \left[\frac{\nabla \times \mathcal{A}_0(\mathbf{r})}{\mu(\mathbf{r})} \right] \times \mathcal{A}_0(\mathbf{r}) \right\} = - \int_V \mathrm{d}V \frac{|\nabla \times \mathcal{A}_0(\mathbf{r})|^2}{\mu(\mathbf{r})} \tag{4.97}$$

and we would like to apply the divergence theorem to the first integral. Since a stationary current is present in $V_J \subset V$, we apply the Gauss theorem separately to $V \setminus \overline{V}_J$ and V_J where the vector

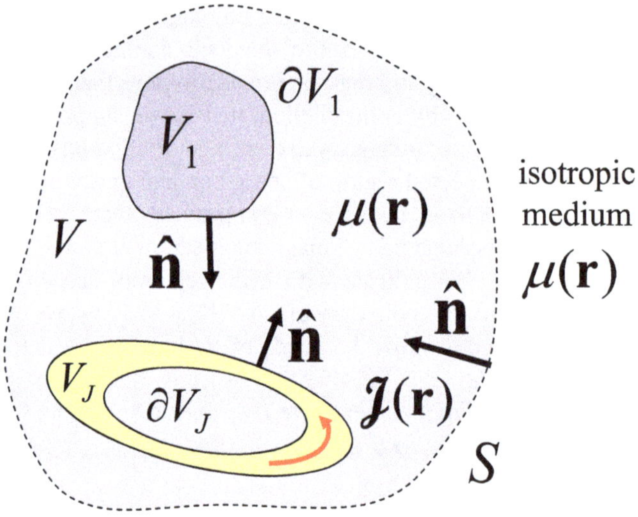

Figure 4.10 For proving uniqueness of the solutions to the vector Poisson equation.

field $[\nabla \times \mathcal{A}_0(\mathbf{r})/\mu(\mathbf{r})] \times \mathcal{A}_0(\mathbf{r})$ is continuously differentiable. The flux integrals on either side of ∂V_J cancel each other provided we assume that $\hat{\mathbf{n}}(\mathbf{r}) \times [\nabla \times \mathcal{A}_0(\mathbf{r})/\mu(\mathbf{r})] \cdot \mathcal{A}_0(\mathbf{r})$ is continuous across ∂V_J. This condition is satisfied if, for $\mathbf{r} \in \partial V_J$, $\hat{\mathbf{n}}(\mathbf{r}) \times \mathcal{H}_0(\mathbf{r})$ is continuous ($\mathcal{H}_0(\mathbf{r})$ is the difference magnetic field) and $\hat{\mathbf{n}}(\mathbf{r}) \times \mathcal{A}_0(\mathbf{r})$ is continuous. The former requirement was already invoked in Section 4.1 to obtain the differential form of the Ampère law, whereas the latter is reasonable if $\mathcal{J}(\mathbf{r}) = \varrho(\mathbf{r})\mathbf{v}(\mathbf{r})$ is a convection current flowing in a medium with permeability $\mu(\mathbf{r})$, as ∂V_J does not represent a material interface. In the end, we arrive at the relationship

$$\int_{\partial V} \mathrm{d}S\, \hat{\mathbf{n}}(\mathbf{r}) \times \left[\frac{\nabla \times \mathcal{A}_0(\mathbf{r})}{\mu(\mathbf{r})}\right] \cdot \mathcal{A}_0(\mathbf{r}) = \int_V \mathrm{d}V\, \frac{|\nabla \times \mathcal{A}_0(\mathbf{r})|^2}{\mu(\mathbf{r})} \tag{4.98}$$

where the minus sign has disappeared thanks to our choice of the orientation of the normal $\hat{\mathbf{n}}$ on ∂V (Figure 4.10). The surface integral vanishes identically if one of the following conditions is true

$$\hat{\mathbf{n}}(\mathbf{r}) \times \mathcal{A}_0(\mathbf{r}) = 0, \qquad \mathbf{r} \in \partial V \tag{4.99}$$

$$\hat{\mathbf{n}}(\mathbf{r}) \times [\nabla \times \mathcal{A}_0(\mathbf{r})] = 0, \qquad \mathbf{r} \in \partial V \tag{4.100}$$

that is, either the tangential components of the vector potential or the tangential curl of $\mathcal{A}_0$ must be null on the boundary of V. More generally, we can impose either condition on different parts of ∂V. We defer a discussion of the feasibility of these requirements and proceed with the proof. If the surface integral is null, (4.98) yields

$$\int_V \mathrm{d}V\, \frac{|\nabla \times \mathcal{A}_0(\mathbf{r})|^2}{\mu(\mathbf{r})} = 0 \tag{4.101}$$

which by virtue of (1.266) means that the magnetic energy stored in the region V associated with the difference stationary magnetic field must be zero. Besides, since $1/\mu(\mathbf{r}) > 0$ for ordinary magnetic media, the volume integral in (4.101) may be interpreted as the norm of the vector field $\nabla \times \mathcal{A}_0(\mathbf{r})$ in the volume V. Thus, the latter is null if, and only if, the vector field vanishes, and (4.101) implies

$$\nabla \times \mathcal{A}_0(\mathbf{r}) = 0, \qquad \mathbf{r} \in V \tag{4.102}$$

by definition of norm. Unfortunately, this equation for $\mathcal{A}_0(\mathbf{r})$ is not sufficient to conclude that the difference potential must be identically null throughout V. On the contrary, since any conservative field would do the trick, we can only state that

$$\mathcal{A}_1(\mathbf{r}) = \mathcal{A}_2(\mathbf{r}) + \nabla\chi(\mathbf{r}) \tag{4.103}$$

where $\chi(\mathbf{r})$ is an arbitrary scalar field twice differentiable for $\mathbf{r} \in V$. Therefore, the two conditions (4.99) and (4.100) allow us at best to determine the vector potential up to a conservative field $\nabla\chi$, which is in keeping with (4.32). And yet, the additive field $\nabla\chi$ bears no consequence on the energy stored in the field in V or the values of the magnetic induction field computed through (4.29).

Uniqueness for $\mathcal{A}(\mathbf{r})$ is achieved nonetheless, if we arbitrarily set $\chi(\mathbf{r})$, and this in turn may be accomplished by assigning the divergence of the vector potential, say,

$$\nabla \cdot \mathcal{A}_1(\mathbf{r}) = \nabla \cdot \mathcal{A}_2(\mathbf{r}) = f(\mathbf{r}), \qquad \mathbf{r} \in V \tag{4.104}$$

with $f(\mathbf{r})$ a scalar field regular for $\mathbf{r} \in V$. On account of (4.103), the condition on the divergence of the potential difference turns into

$$\nabla \cdot \mathcal{A}_0(\mathbf{r}) = \nabla^2\chi(\mathbf{r}) = 0, \qquad \mathbf{r} \in V \tag{4.105}$$

that is, $\chi(\mathbf{r})$ must be a *harmonic* function in V [9, Section 8.7], [14, Section 2.2]. Thus, we are left with the task of finding the conditions under which $\nabla\chi(\mathbf{r}) = 0$ for $\mathbf{r} \in V$. We observe

$$\int_V \mathrm{d}V\, |\nabla\chi(\mathbf{r})|^2 = \int_V \mathrm{d}V\, \nabla \cdot [\chi(\mathbf{r})\nabla\chi(\mathbf{r})] = -\int_{\partial V} \mathrm{d}S\, \chi(\mathbf{r})\, \hat{\mathbf{n}}(\mathbf{r}) \cdot \mathcal{A}_0(\mathbf{r}) \tag{4.106}$$

thanks to (4.105) and the Gauss theorem. The surface integral vanishes identically, if either one of these conditions holds

$$\chi(\mathbf{r}) = 0, \qquad \mathbf{r} \in \partial V \tag{4.107}$$

$$\frac{\partial}{\partial \hat{\mathbf{n}}}\chi(\mathbf{r}) = \hat{\mathbf{n}}(\mathbf{r}) \cdot \mathcal{A}_0(\mathbf{r}) = 0, \qquad \mathbf{r} \in \partial V \tag{4.108}$$

and, since they force the volume integral to be null too, we conclude that

$$\mathcal{A}_1(\mathbf{r}) - \mathcal{A}_2(\mathbf{r}) = \mathcal{A}_0(\mathbf{r}) = \nabla\chi(\mathbf{r}) = 0 \tag{4.109}$$

for $\mathbf{r} \in V$. Notice that the second condition above does not specify $\chi(\mathbf{r})$ univocally, but this is of no concern, because our goal was to show that the potential difference is zero throughout the region V, so long as we assign the divergence of $\mathcal{A}_0(\mathbf{r})$ and suitable conditions on ∂V.

We observe that (4.99) implies that $\hat{\mathbf{n}}(\mathbf{r}) \times \mathcal{A}_l(\mathbf{r})$, $l = 1, 2$, take on the same values for $\mathbf{r} \in \partial V$. Likewise, (4.108) translates into assigning the values of the normal component of $\mathcal{A}_l(\mathbf{r})$ on the boundary. Equation (4.100) requires assigning the values of the tangential curl for $\mathbf{r} \in \partial V$. Lastly, the condition on $\chi(\mathbf{r})$ (4.107) is quite strong, as it forces the auxiliary scalar field $\chi(\mathbf{r})$ to be zero throughout V. Indeed, $\nabla\chi(\mathbf{r}) = 0$ is solved by any constant χ_0, but since this constant must be zero on ∂V, then it is zero.

In summary, we have proved that the following systems of equations

$$\begin{cases} \nabla \times \left[\dfrac{\nabla \times \mathcal{A}(\mathbf{r})}{\mu(\mathbf{r})}\right] = \mathcal{J}(\mathbf{r}), & \mathbf{r} \in V \\ \nabla \cdot \mathcal{A}(\mathbf{r}) = f(\mathbf{r}) & \mathbf{r} \in V \\ \hat{\mathbf{n}}(\mathbf{r}) \times \mathcal{A}(\mathbf{r}) = \mathbf{g}(\mathbf{r}), & \mathbf{r} \in \partial V \\ \hat{\mathbf{n}}(\mathbf{r}) \cdot \mathcal{A}(\mathbf{r}) = h(\mathbf{r}), & \mathbf{r} \in \partial V \end{cases} \tag{4.110}$$

$$\begin{cases} \nabla \times \left[\dfrac{\nabla \times \mathcal{A}(\mathbf{r})}{\mu(\mathbf{r})} \right] = \mathcal{J}(\mathbf{r}), & \mathbf{r} \in V \\ \nabla \cdot \mathcal{A}(\mathbf{r}) = f(\mathbf{r}) & \mathbf{r} \in V \\ \hat{\mathbf{n}}(\mathbf{r}) \times [\nabla \times \mathcal{A}(\mathbf{r})] = \mathbf{g}(\mathbf{r}), & \mathbf{r} \in \partial V \\ \hat{\mathbf{n}}(\mathbf{r}) \cdot \mathcal{A}(\mathbf{r}) = h(\mathbf{r}), & \mathbf{r} \in \partial V \end{cases} \tag{4.111}$$

admit a unique solution.

It is worthwhile noticing that in both (4.110) and (4.111) the functions $f(\mathbf{r})$ and $h(\mathbf{r})$ are not independent of one another. Indeed, integrating the requirement imposed on the divergence of $\mathcal{A}(\mathbf{r})$ yields

$$\int_{\partial V} \mathrm{d}S\, \hat{\mathbf{n}}(\mathbf{r}) \cdot \mathcal{A}(\mathbf{r}) = \int_{\partial V} \mathrm{d}S\, h(\mathbf{r}) = -\int_{V} \mathrm{d}V\, f(\mathbf{r}) \tag{4.112}$$

having applied the Gauss theorem with the unit normal pointing inwards V. Likewise, in (4.111) the tangential component of $\nabla \times \mathcal{A}(\mathbf{r})$ may not be assigned at will on ∂V. The restriction follows by integrating the pertinent vector Poisson equation, viz.,

$$\int_{\partial V} \mathrm{d}S\, \hat{\mathbf{n}}(\mathbf{r}) \times \frac{\nabla \times \mathcal{A}(\mathbf{r})}{\mu(\mathbf{r})} = \int_{\partial V} \mathrm{d}S\, \frac{\mathbf{g}(\mathbf{r})}{\mu(\mathbf{r})} = -\int_{V_J} \mathrm{d}V\, \mathcal{J}(\mathbf{r}) = 0 \tag{4.113}$$

on account of the curl theorem (H.91) with the unit normal pointing inwards V. The vanishing of the volume integral of a steady current density is proved on page 297 in Section 5.5.

We may extend the proof of uniqueness to the case where the region V is unbounded. To this purpose we assume the surface S to be a sphere of radius a and take the limit as $a \to +\infty$. To draw our conclusion we need to review the hypothesis that the surface integral over S in (4.98) is null and that the energy integral stays finite. However, in Section 4.1 we have already verified that this is true, provided the vector potential and the magnetic field satisfy estimates (4.36) and (4.37).

Therefore, the vector Poisson equation admits a unique solution in an unbounded region of space, if the vector potential decays at least as the inverse of the distance away from a finite distribution of steady currents. In symbols, the problems

$$\begin{cases} \nabla \times \left[\dfrac{\nabla \times \mathcal{A}(\mathbf{r})}{\mu(\mathbf{r})} \right] = \mathcal{J}(\mathbf{r}), & \mathbf{r} \in \mathbb{R}^3 \setminus V_1 \\ \nabla \cdot \mathcal{A}(\mathbf{r}) = f(\mathbf{r}), & \mathbf{r} \in \mathbb{R}^3 \setminus V_1 \\ \hat{\mathbf{n}}(\mathbf{r}) \times \mathcal{A}(\mathbf{r}) = \mathbf{g}_1(\mathbf{r}), & \mathbf{r} \in \partial V_1 \\ \hat{\mathbf{n}}(\mathbf{r}) \cdot \mathcal{A}(\mathbf{r}) = h_1(\mathbf{r}), & \mathbf{r} \in \partial V_1 \\ \mathcal{A}(\mathbf{r}) = \mathcal{O}\left(\dfrac{1}{|\mathbf{r}|}\right), & |\mathbf{r}| \to +\infty \end{cases} \tag{4.114}$$

$$\begin{cases} \nabla \times \left[\dfrac{\nabla \times \mathcal{A}(\mathbf{r})}{\mu(\mathbf{r})} \right] = \mathcal{J}(\mathbf{r}), & \mathbf{r} \in \mathbb{R}^3 \setminus V_1 \\ \nabla \cdot \mathcal{A}(\mathbf{r}) = f(\mathbf{r}), & \mathbf{r} \in \mathbb{R}^3 \setminus V_1 \\ \hat{\mathbf{n}}(\mathbf{r}) \times [\nabla \times \mathcal{A}(\mathbf{r})] = \mathbf{g}_1(\mathbf{r}), & \mathbf{r} \in \partial V_1 \\ \hat{\mathbf{n}}(\mathbf{r}) \cdot \mathcal{A}(\mathbf{r}) = h_1(\mathbf{r}), & \mathbf{r} \in \partial V_1 \\ \mathcal{A}(\mathbf{r}) = \mathcal{O}\left(\dfrac{1}{|\mathbf{r}|} \right), & |\mathbf{r}| \to +\infty \end{cases} \tag{4.115}$$

admit a unique solution.

We need to check the validity of the conditions on ∂V_1. If the medium occupying the volume V_1 is a magnetic material, we know from (4.54), (4.64) and (4.67), that $\mathcal{A}(\mathbf{r})$ and the tangential curl thereof are continuous through ∂V_1. In practice, we need to solve two separate problems in V_1 and $\mathbb{R}^3 \setminus \overline{V}_1$ up to arbitrary constants which are then determined by enforcing (4.64) and either one of (4.54) or (4.67).

The extension of these results to the case of an anisotropic medium filling V or $\mathbb{R}^3 \setminus \overline{V}_1$ is not difficult and left to the Reader.

4.7.2 Magnetic entities in the presence of magnetic media

In the previous section we have derived necessary conditions for the uniqueness of the solution to the vector Poisson equation in terms of the boundary values of the vector potential and the tangential curl thereof. While the latter may be translated into an equivalent requirement for the tangential component of the magnetic induction $\mathcal{B}(\mathbf{r})$ in light of (4.29), it is not easy to draw conclusions for the normal component of $\mathcal{B}(\mathbf{r})$ from the knowledge gained so far.

Therefore, we consider the issue of uniqueness of the stationary magnetic equations (4.9) and (4.10). For the sake of argument, we focus on free space as a background medium and determine the necessary conditions for the solution to be unique in a region V. We suppose that two solutions $\mathcal{B}_1(\mathbf{r})$ and $\mathcal{B}_2(\mathbf{r})$ are possible and define the difference stationary magnetic field $\mathcal{B}_0(\mathbf{r}) := \mathcal{B}_1(\mathbf{r}) - \mathcal{B}_2(\mathbf{r})$ which satisfies the homogeneous equations

$$\nabla \times \mathcal{B}_0(\mathbf{r}) = 0, \qquad \mathbf{r} \in V \tag{4.116}$$

$$\nabla \cdot \mathcal{B}_0(\mathbf{r}) = 0, \qquad \mathbf{r} \in V \tag{4.117}$$

and our goal is to show that $\mathcal{B}_0(\mathbf{r}) = 0$ is the only admissible solution if suitable conditions are imposed on the values $\mathcal{B}_1(\mathbf{r})$ and $\mathcal{B}_2(\mathbf{r})$ assume on the boundary ∂V. We outline the proofs for the domains of interest, keeping in mind that most details are the same as those for the electrostatic equations in Section 2.5.2.

The simplest case is a normal domain (Figure 1.2b), in which case $\mathcal{B}_0(\mathbf{r})$ can be determined *either* from a vector potential (see Section 4.2 and 4.7.1) *or* a magnetic scalar potential (see Section 4.4). By introducing a vector potential $\mathcal{A}_0(\mathbf{r})$ as in (4.29), we have the following expression for the static magnetic energy in V

$$\begin{aligned} W_h &:= \frac{1}{2\mu_0} \int_V \mathrm{d}V \, |\mathcal{B}_0(\mathbf{r})|^2 = -\frac{1}{2\mu_0} \int_V \mathrm{d}V \, \nabla \cdot [\mathcal{B}_0(\mathbf{r}) \times \mathcal{A}_0(\mathbf{r})] \\ &= \frac{1}{2\mu_0} \int_{\partial V} \mathrm{d}S \, \hat{\mathbf{n}}(\mathbf{r}) \times \mathcal{B}_0(\mathbf{r}) \cdot \mathcal{A}_0(\mathbf{r}) \end{aligned} \tag{4.118}$$

having used (H.51), (4.116) and the Gauss theorem with the unit normal oriented inwards V. The flux integral in the rightmost-hand side vanishes if

$$\hat{\mathbf{n}}(\mathbf{r}) \times \mathcal{B}_0(\mathbf{r}) = 0, \qquad \mathbf{r} \in \partial V \tag{4.119}$$

which is granted if $\mathcal{B}_1(\mathbf{r})$ and $\mathcal{B}_2(\mathbf{r})$ are given the same tangential component for $\mathbf{r} \in \partial V$. Hence, if follows that the magnetic energy is zero and, since $\mathcal{B}_0(\mathbf{r})$ is continuous, $\mathcal{B}_0(\mathbf{r})$ must vanish identically in V, as we wanted to prove. This conclusion was already known from (4.100), i.e., the condition on the tangential curl of $\mathcal{A}(\mathbf{r})$.

Conversely, if we derive the magnetic induction from a magnetic scalar potential in the form

$$\mathcal{B}_0(\mathbf{r}) = -\mu_0 \nabla \Psi_0(\mathbf{r}) \tag{4.120}$$

(4.116) is invariably satisfied.[1] As regards the magnetic energy we have

$$\begin{aligned} W_h &:= \frac{1}{2\mu_0} \int_V \mathrm{d}V\, |\mathcal{B}_0(\mathbf{r})|^2 = -\frac{1}{2} \int_V \mathrm{d}V\, \nabla \cdot [\mathcal{B}_0(\mathbf{r})\Psi_0(\mathbf{r})] \\ &= \frac{1}{2} \int_{\partial V} \mathrm{d}S\, \hat{\mathbf{n}}(\mathbf{r}) \cdot \mathcal{B}_0(\mathbf{r})\Psi_0(\mathbf{r}) \end{aligned} \tag{4.121}$$

where we have used (H.51), (4.117) and applied the Gauss theorem with the unit normal pointing inwards V. Therefore, the surface integral is null if

$$\hat{\mathbf{n}}(\hat{\mathbf{r}}) \cdot \mathcal{B}_0(\mathbf{r}) = 0, \qquad \mathbf{r} \in \partial V \tag{4.122}$$

which happens if $\mathcal{B}_1(\mathbf{r})$ and $\mathcal{B}_2(\mathbf{r})$ have been assigned the same normal component for $\mathbf{r} \in \partial V$. By the same arguments as before, the difference magnetic induction is zero in V and the solution is unique. With this new result we have shed light on the role of the normal component of $\mathcal{B}(\mathbf{r})$ — which could not be simply guessed from the proofs of Section 4.7.1.

The proof of uniqueness in the whole space ($V \equiv \mathbb{R}^3$) is a special instance of the previous arguments. It is sufficient to consider a ball $B(0,a)$ and determine $\mathcal{B}_0(\mathbf{r})$ from a vector potential. The magnetic energy reads

$$\begin{aligned} W_h &:= \frac{1}{2\mu_0} \int_{B(0,a)} \mathrm{d}V\, |\mathcal{B}_0(\mathbf{r})|^2 = -\frac{1}{2\mu_0} \int_{B(0,a)} \mathrm{d}V\, \nabla \cdot [\mathcal{B}_0(\mathbf{r}) \times \mathcal{A}_0(\mathbf{r})] \\ &= \frac{1}{2\mu_0} \int_{\partial B} \mathrm{d}S\, \hat{\mathbf{n}}(\mathbf{r}) \times \mathcal{B}_0(\mathbf{r}) \cdot \mathcal{A}_0(\mathbf{r}) \end{aligned} \tag{4.123}$$

and the last surface integral is null for $a \to +\infty$ if the vector potential and the magnetic induction satisfy the estimates

$$|\mathcal{A}_0(\mathbf{r})| \leqslant \frac{C_A}{|\mathbf{r}|}, \qquad |\mathbf{r}| \to +\infty \tag{4.124}$$

$$|\mathcal{B}_0(\mathbf{r})| \leqslant \frac{C_B}{|\mathbf{r}|^2}, \qquad |\mathbf{r}| \to +\infty \tag{4.125}$$

[1]For an inhomogeneous medium with permeability $\mu(\mathbf{r})$ the definition would read $\mathcal{B}_0(\mathbf{r}) = -\mu(\mathbf{r})\nabla\Psi_0(\mathbf{r})$, and for an anisotropic magnetic medium we would set $\mathcal{B}_0(\mathbf{r}) = -\overline{\mu}(\mathbf{r}) \cdot \nabla\Psi_0(\mathbf{r})$.

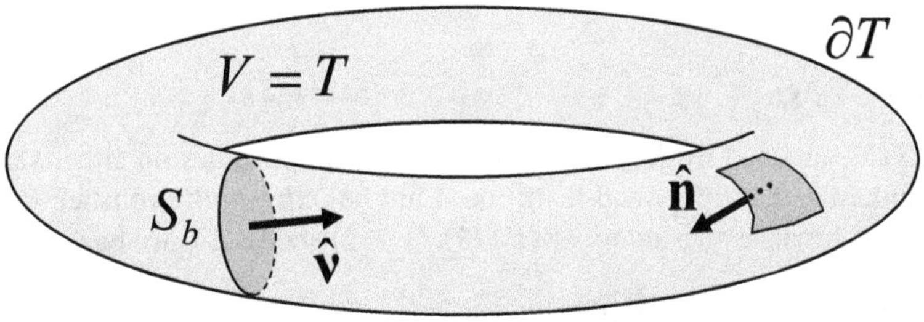

Figure 4.11 For proving uniqueness of solutions to the stationary magnetic equations: a torus (contour-wise multiply-connected domain) and a branch surface $S_b \subset T$.

that is, both the potential and the field are regular at infinity. In summary, if the current density is assigned and the field decays with the inverse square of the distance, the solution to the stationary magnetic equations is unique in $\mathbb{R}^3$.

Particularly meaningful for the stationary regime is the case of a contour-wise multiply-connected domain. A typical example is the inside of a torus T (Figure 4.11) though also the complementary unbounded domain $\mathbb{R}^3 \setminus \overline{T}$ is contour-wise multiply connected. In these cases there does not exist a single-valued magnetic scalar potential in V, and we might want to resort to a vector potential.

For $V = T$ the energy can be cast as in (4.118), whereby we conclude that uniqueness is achieved if $\hat{\mathbf{n}}(\mathbf{r}) \times \mathcal{B}(\mathbf{r})$ is assigned on ∂V. Similarly, for $V = \mathbb{R}^3 \setminus \overline{T}$, we start with the energy in a region comprised between a ball $B(0, a)$ and the torus T, viz.,

$$\begin{aligned} W_h &:= \frac{1}{2\mu_0} \int_{B(0,a)\setminus T} dV\, |\mathcal{B}_0(\mathbf{r})|^2 = -\frac{1}{2\mu_0} \int_{B(0,a)\setminus T} dV\, \nabla \cdot [\mathcal{B}_0(\mathbf{r}) \times \mathcal{A}_0(\mathbf{r})] \\ &= \frac{1}{2\mu_0} \int_{\partial B} dS\, \hat{\mathbf{n}}(\mathbf{r}) \times \mathcal{B}_0(\mathbf{r}) \cdot \mathcal{A}_0(\mathbf{r}) + \frac{1}{2\mu_0} \int_{\partial T} dS\, \hat{\mathbf{n}}(\mathbf{r}) \times \mathcal{B}_0(\mathbf{r}) \cdot \mathcal{A}_0(\mathbf{r}) \end{aligned} \tag{4.126}$$

with the unit normal pointing inwards $B(0, a) \setminus T$. The last surface integral is null if (4.119) holds, whereas the contribution on the sphere ∂B vanishes in the limit as $a \to +\infty$ if estimates (4.124) and (4.125) apply. Therefore, specifying the tangential component of $\mathcal{B}(\mathbf{r})$ and requiring regularity at infinity ensures uniqueness of (4.9) and (4.10) in this case.

By contrast, specifying the normal component of $\mathcal{B}(\mathbf{r})$ on the boundary of a contour-wise multiply-connected domain does not ensure uniqueness! This happens because the magnetic scalar potential $\Psi_0(\mathbf{r})$ associated with $\mathcal{B}_0(\mathbf{r})$ through (4.120) is many-valued, and we need to introduce a *branch surface* S_b, as is sketched in Figure 4.11, so as to turn $V = T$ into a simply connected region and apply the Gauss theorem. Specifically, we have

$$\begin{aligned} W_h &:= \frac{1}{2\mu_0} \int_{T} dV\, |\mathcal{B}_0(\mathbf{r})|^2 = -\frac{1}{2} \int_{T} dV\, \nabla \cdot [\mathcal{B}_0(\mathbf{r})\Psi_0(\mathbf{r})] \\ &= \frac{1}{2} \int_{\partial T} dS\, \hat{\mathbf{n}}(\mathbf{r}) \cdot \mathcal{B}_0(\mathbf{r})\Psi_0(\mathbf{r}) + \frac{1}{2} \int_{S_b} dS\, \hat{\mathbf{v}}(\mathbf{r}) \cdot \mathcal{B}_0(\mathbf{r}) \left[\Psi_0^+(\mathbf{r}) - \Psi_0^-(\mathbf{r})\right] \\ &= \frac{1}{2} \int_{\partial T} dS\, \hat{\mathbf{n}}(\mathbf{r}) \cdot \mathcal{B}_0(\mathbf{r})\Psi_0(\mathbf{r}) + \frac{1}{2}\Delta\Psi_b \int_{S_b} dS\, \hat{\mathbf{v}}(\mathbf{r}) \cdot \mathcal{B}_0(\mathbf{r}) \end{aligned} \tag{4.127}$$

where

$$\Delta\Psi_b := \Psi_0^+(\mathbf{r}) - \Psi_0^-(\mathbf{r}), \qquad \mathbf{r} \in S_b \tag{4.128}$$

is the *constant* value attained by the difference of the potential values on either side of the branch surface. We emphasize that $\Psi_0^+(\mathbf{r})$ and $\Psi_0^-(\mathbf{r})$ need not be *individually* constant on S_b, though the difference thereof is because on account of (1.113), (1.142) and (H.14) we have

$$\begin{aligned} 0 &= \hat{\boldsymbol{\nu}}(\mathbf{r}) \times \{\hat{\boldsymbol{\nu}}(\mathbf{r}) \times [\mathcal{B}_0^+(\mathbf{r}) - \mathcal{B}_0^-(\mathbf{r})]\} \\ &= \hat{\boldsymbol{\nu}}(\mathbf{r}) \times \left\{\left[\mu_0 \nabla\Psi_0^+(\mathbf{r}) - \mu_0 \nabla\Psi_0^-(\mathbf{r})\right] \times \hat{\boldsymbol{\nu}}(\mathbf{r})\right\} \\ &= \mu_0 \nabla_s \Psi_0^+(\mathbf{r}) - \mu_0 \nabla_s \Psi_0^-(\mathbf{r}) = \mu_0 \nabla_s \left[\Psi_0^+(\mathbf{r}) - \Psi_0^-(\mathbf{r})\right], \qquad \mathbf{r} \in S_b \end{aligned} \tag{4.129}$$

where $\nabla_s\{\bullet\}$ indicates the surface gradient on S_b (cf. Appendix A.5). Even though we require $\hat{\mathbf{n}}(\mathbf{r}) \cdot \mathcal{B}_0(\mathbf{r}) = 0$ on ∂T, the contribution of the remaining flux integral may not be zero in that $\mathcal{B}_1(\mathbf{r})$ and $\mathcal{B}_2(\mathbf{r})$ are not necessarily equal on S_b (as a matter of fact, this is precisely what we want to prove in the first place) and the scalar potential suffers a jump across the branch surface. So, we may not invoke the zero-energy argument, and the solution is not unique.

Stated another way, there may exist non-null *source-free* stationary magnetic fields inside a torus that exhibit zero normal component on the boundary. These solutions are called *Neumann vector fields* and can be derived from a harmonic many-valued scalar potential (cf. Sections 8.1.2 and 11.1.3) [1,3]. In practice, inside a ring-like or any contour-wise multiply-connected domain filled with a homogeneous isotropic magnetic medium the solution to (4.9) and (4.10) with $\hat{\mathbf{n}}(\mathbf{r}) \cdot \mathcal{B}(\mathbf{r}) = 0$ for $\mathbf{r} \in \partial V$ can be written as

$$\mathcal{B}(\mathbf{r}) = \mathcal{B}_J(\mathbf{r}) + \mathcal{B}_N(\mathbf{r}) = \mathcal{B}_J(\mathbf{r}) - \mu_0 \nabla\Psi(\mathbf{r}), \qquad \mathbf{r} \in V \tag{4.130}$$

where $\mathcal{B}_J(\mathbf{r})$ is a particular solution depending on $\mathcal{J}(\mathbf{r})$ and $\mathcal{B}_N(\mathbf{r})$ is a Neumann vector field. The integral of $\hat{\mathbf{n}}(\mathbf{r}) \times \mathcal{B}_N(\mathbf{r})$ for $\mathbf{r} \in \partial V$ (a quantity called the *vector circulation* in aerodynamics) is null in that

$$\int_{\partial V} \mathrm{d}S\, \hat{\mathbf{n}}(\mathbf{r}) \times \mathcal{B}_N(\mathbf{r}) = \int_V \mathrm{d}V\, \nabla \times \mathcal{B}_N(\mathbf{r}) = 0 \tag{4.131}$$

by virtue of the curl theorem (H.91). Nonetheless, the tangential component of $\mathcal{B}_N(\mathbf{r})$ cannot vanish on the boundary ∂V, or else the Neumann field would be identically null within V, as can be inferred from the proof given in (4.118).

Example 4.4 (The Neumann vector field inside a spherical-conical cavity)

We consider the stationary magnetic induction and the scalar potential in the source-free contour-wise multiply-connected bounded domain V defined as

$$V := \{\mathbf{r} \in \mathbb{R}^3 : 0 < r_1 < |\mathbf{r}| < r_2 < +\infty, 0 < \vartheta_1 < \vartheta < \vartheta_2 < \pi\} \tag{4.132}$$

that is, a ring-like region limited by two concentric spheres and a two-sheeted cone. Besides, we assume that the boundary ∂V is flush with a conductor and the medium inside V is vacuum. By expanding the Laplace operator in polar spherical coordinates (r, φ, ϑ) as in (A.42) and using the separation argument (Section 3.5.1) it is easily verified that the problem

$$\begin{cases} \nabla^2\Psi(\mathbf{r}) = 0, & \mathbf{r} \in V \\ \hat{\mathbf{n}}(\mathbf{r}) \cdot \nabla\Psi(\mathbf{r}) = 0, & \mathbf{r} \in \partial V \end{cases} \tag{4.133}$$

is solved by the family of scalar fields [see (3.92)]

$$\Psi(\mathbf{r}) = c_1\varphi + c_2, \qquad \varphi \in \mathbb{R} \tag{4.134}$$

with c_1 and c_2 arbitrary constants. Since $\Psi(r, 0, \vartheta) = c_2$ and $\Psi(r, 2\pi, \vartheta) = 2\pi c_1 + c_2$, then $\Psi(\mathbf{r})$ is many-valued with period $-2\pi c_1$.

As a check, we notice that

$$\hat{\mathbf{n}}(\mathbf{r}) \cdot \nabla\Psi(\mathbf{r}) = \hat{\mathbf{n}}(\mathbf{r}) \cdot \frac{c_1\hat{\boldsymbol{\varphi}}}{r\sin\vartheta} = \begin{cases} \hat{\mathbf{r}} \cdot \dfrac{c_1\hat{\boldsymbol{\varphi}}}{r_1\sin\vartheta} = 0, & r = r_1 \\ -\hat{\mathbf{r}} \cdot \dfrac{c_1\hat{\boldsymbol{\varphi}}}{r_2\sin\vartheta} = 0, & r = r_2 \\ \hat{\boldsymbol{\vartheta}} \cdot \dfrac{c_1\hat{\boldsymbol{\varphi}}}{r\sin\vartheta_1} = 0, & \vartheta = \vartheta_1 \\ -\hat{\boldsymbol{\vartheta}} \cdot \dfrac{c_1\hat{\boldsymbol{\varphi}}}{r\sin\vartheta_2} = 0, & \vartheta = \vartheta_2 \end{cases} \tag{4.135}$$

and $\Psi(\mathbf{r})$ is obviously harmonic. In keeping with (4.120) the magnetic induction reads

$$\mathcal{B}(\mathbf{r}) = -\frac{\mu_0 c_1}{r\sin\vartheta}\hat{\boldsymbol{\varphi}}, \qquad \mathbf{r} \in V \tag{4.136}$$

and is not affected by the additive constant c_2. The other one may be fixed by invoking the Ampère law (4.5)

$$\oint_\gamma \mathrm{d}s\, \hat{\mathbf{s}}(\mathbf{r}) \cdot \mathcal{B}(\mathbf{r}) = \mu_0 I, \qquad \gamma \subset V \tag{4.137}$$

where

- γ is a piecewise-smooth irreducible contour, i.e., a loop that is entirely contained in V *and* cannot be deformed into a point without crossing ∂V;
- I is the steady current flowing on ∂V and due to a surface current density $\mathcal{J}_S(\mathbf{r})$.

Since the result of the integration is independent of the specific shape of γ, we choose the circumference

$$\gamma := \{\mathbf{r} \in V : r = (r_1 + r_2)/2, \vartheta = (\vartheta_1 + \vartheta_2)/2, \varphi \in [0, 2\pi]\} \tag{4.138}$$

whereby $\hat{\mathbf{s}}(\mathbf{r}) = \hat{\boldsymbol{\varphi}}$, $\mathrm{d}s = r\sin\vartheta\mathrm{d}\varphi$ and

$$\oint_\gamma \mathrm{d}s\, \hat{\mathbf{s}}(\mathbf{r}) \cdot \mathcal{B}(\mathbf{r}) = -c_1\mu_0 \oint_\gamma \mathrm{d}s\, \frac{\hat{\boldsymbol{\varphi}} \cdot \hat{\boldsymbol{\varphi}}}{r\sin\vartheta} = -c_1\mu_0 \int_0^{2\pi} \mathrm{d}\varphi = -2\pi\mu_0 c_1 = \mu_0 I \tag{4.139}$$

whence

$$\mathcal{B}(\mathbf{r}) = \frac{\mu_0 I}{2\pi r\sin\vartheta}\hat{\boldsymbol{\varphi}}, \qquad \mathbf{r} \in V. \tag{4.140}$$

The steady surface current flowing on the inside of ∂V follows from condition (1.183) and the constitutive relation (1.113), viz.,

$$\mathcal{J}_S(\mathbf{r}) = \frac{1}{\mu_0}\hat{\mathbf{n}}(\mathbf{r}) \times \mathcal{B}(\mathbf{r}) = \begin{cases} -\dfrac{I\hat{\boldsymbol{\vartheta}}}{2\pi r_1 \sin\vartheta}, & r = r_1 \\ \dfrac{I\hat{\mathbf{r}}}{2\pi r \sin\vartheta_1}, & \vartheta = \vartheta_1 \\ \dfrac{I\hat{\boldsymbol{\vartheta}}}{2\pi r_2 \sin\vartheta}, & r = r_2 \\ -\dfrac{I\hat{\mathbf{r}}}{2\pi r \sin\vartheta_2}, & \vartheta = \vartheta_2 \end{cases} \qquad \varphi \in [0, 2\pi] \tag{4.141}$$

and hence the streamlines are closed loops in the meridian half-planes $\varphi = const.$

(End of Example 4.4)

4.7.3 *Magnetic entities in the presence of conductors*

Concerning uniqueness of solutions, quite an interesting situation arises when considering stationary fields in the presence of a conducting medium. For the sake of argument we suppose that a non-magnetic isotropic homogeneous conductor occupies a finite contour-wise multiply-connected region $V_C \subset \mathbb{R}^3$ and is exposed to the stationary electromagnetic field produced by a nearby distribution of steady current $\mathcal{J}(\mathbf{r})$ confined in V_J, as is suggested in Figure 4.12. We further assume that the underlying unbounded homogeneous medium is free space. The relevant steady-state Maxwell equations for points $\mathbf{r} \in \mathbb{R}^3$ read

$$\nabla \times \mathcal{H}(\mathbf{r}) = \mathcal{J}_c(\mathbf{r}) + \mathcal{J}(\mathbf{r}) \tag{4.142}$$

$$\mathcal{J}_c(\mathbf{r}) = \sigma \mathcal{E}(\mathbf{r}) \tag{4.143}$$

$$\mu_0 \nabla \cdot \mathcal{H}(\mathbf{r}) = 0 \tag{4.144}$$

$$\nabla \cdot \mathcal{E}(\mathbf{r}) = \varrho(\mathbf{r})/\varepsilon_0 \tag{4.145}$$

$$\nabla \times \mathcal{E}(\mathbf{r}) = 0 \tag{4.146}$$

where the conduction current $\mathcal{J}_c(\mathbf{r})$ — if it exists — flows within V_C by definition, and the electric charge density $\varrho(\mathbf{r})$ is also restricted to V_J. The electric field in the conductor is solenoidal because so is $\mathcal{J}_c(\mathbf{r})$, and this requirement is included in (4.145). Is the solution to (4.142)-(4.146) complemented with suitable boundary conditions unique?

To answer the question we make the assumption that two solutions are possible and examine the difference fields $\mathcal{H}_0(\mathbf{r}) = \mathcal{H}_1(\mathbf{r}) - \mathcal{H}_2(\mathbf{r})$ and $\mathcal{E}_0(\mathbf{r}) = \mathcal{E}_1(\mathbf{r}) - \mathcal{E}_2(\mathbf{r})$, which solve the source-free instance of (4.142)-(4.146). We also introduce a ball $B(0, a)$ with radius a large enough so that $B(0, a)$ encloses both V_J and V_C.

First of all, the stationary Poynting theorem (4.89) applied to the difference fields in $B(0, a)$ yields

$$\int_{\partial B} \mathrm{d}V\, \hat{\mathbf{n}}(\mathbf{r}) \cdot \mathcal{E}_0(\mathbf{r}) \times \mathcal{H}_0(\mathbf{r}) + \int_{V_C} \mathrm{d}V\, \sigma |\mathcal{E}_0(\mathbf{r})|^2 = 0 \tag{4.147}$$

and the flux of $\mathcal{E}_0(\mathbf{r}) \times \mathcal{H}_0(\mathbf{r})$ over ∂B vanishes as $a \to +\infty$ by virtue of the constitutive relationship (1.112) and the asymptotic conditions (2.21) and (4.37). Since the integrand of the domain integral over V_C is non-negative, we conclude that $\mathcal{E}_0(\mathbf{r})$ vanishes identically within the conductor.

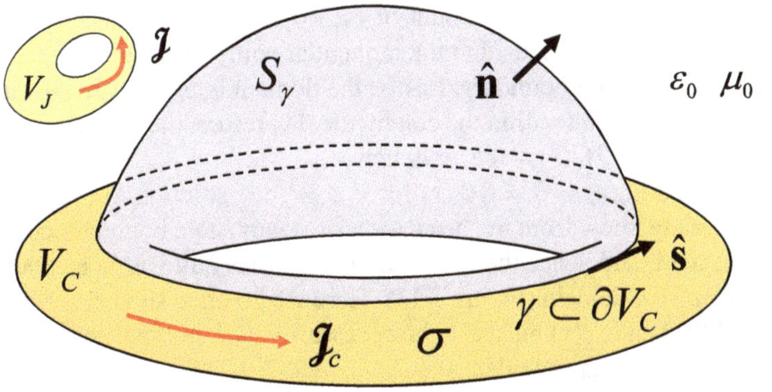

Figure 4.12 Ring-shaped isotropic conductor 'excited' by external sources.

Secondly, the electric energy in the contour-wise multiply-connected region $V_a := B(0,a) \setminus \overline{V}_C$ reads

$$W_e(a) := \frac{\varepsilon_0}{2}\int_{V_a} \mathrm{d}V\, |\mathcal{E}_0(\mathbf{r})|^2 = \frac{\varepsilon_0}{2}\int_{V_a} \mathrm{d}V\, \mathcal{E}_0(\mathbf{r})\cdot\nabla\times\mathcal{F}_0(\mathbf{r}) = \frac{\varepsilon_0}{2}\int_{V_a} \mathrm{d}V\, \nabla\cdot[\mathcal{F}_0(\mathbf{r})\times\mathcal{E}_0(\mathbf{r})]$$
$$= \frac{\varepsilon_0}{2}\int_{\partial B} \mathrm{d}S\, \hat{\mathbf{n}}(\mathbf{r})\cdot\mathcal{F}_0(\mathbf{r})\times\mathcal{E}_0(\mathbf{r}) - \frac{\varepsilon_0}{2}\int_{\partial V_C} \mathrm{d}S\, \hat{\mathbf{n}}(\mathbf{r})\times\mathcal{E}_0(\mathbf{r})\cdot\mathcal{F}_0(\mathbf{r}) \tag{4.148}$$

where $\mathcal{F}_0(\mathbf{r})$ is the vector potential associated with $\mathcal{E}_0(\mathbf{r})$ as in (2.110), and we have used the Gauss theorem with the unit normals oriented positively outwards V_a. The integral over the sphere ∂B vanishes in the limit as $a \to +\infty$ if $\mathcal{F}_0(\mathbf{r})$ falls off as the magnetic counterpart in (4.36) and $\varepsilon_0\mathcal{E}_0(\mathbf{r})$ obeys (2.21). The integral over ∂V_C is null because so is the tangential component of $\mathcal{E}_0(\mathbf{r})$. This happens because $\hat{\mathbf{n}}(\mathbf{r})\times\mathcal{E}_0(\mathbf{r})$ is continuous across ∂V_C, and we have established above that $\mathcal{E}_0(\mathbf{r}) = 0$ within the conductor. Thus, it follows that the electric energy is null for $\mathbf{r} \in \mathbb{R}^3 \setminus \overline{V}_C$ and, as a result, the difference electric field vanishes also outside the conductor.

Lastly, we examine the magnetic energy within $B(0,a)$, namely,

$$W_h(a) := \frac{\mu_0}{2}\int_{B(0,a)} \mathrm{d}V\, |\mathcal{H}_0(\mathbf{r})|^2 = \frac{1}{2}\int_{B(0,a)} \mathrm{d}V\, \mathcal{H}_0(\mathbf{r})\cdot\nabla\times\mathcal{A}_0(\mathbf{r})$$
$$= \frac{1}{2}\int_{V_a} \mathrm{d}V\, \nabla\cdot[\mathcal{A}_0(\mathbf{r})\times\mathcal{H}_0(\mathbf{r})]$$
$$+ \frac{1}{2}\int_{V_C} \mathrm{d}V\, \nabla\cdot[\mathcal{A}_0(\mathbf{r})\times\mathcal{H}_0(\mathbf{r})] + \frac{1}{2}\int_{V_C} \mathrm{d}V\, \mathcal{A}_0(\mathbf{r})\cdot\nabla\times\mathcal{H}_0(\mathbf{r})$$
$$= \frac{1}{2}\int_{\partial B} \mathrm{d}S\, \hat{\mathbf{n}}(\mathbf{r})\cdot\mathcal{A}_0(\mathbf{r})\times\mathcal{H}_0(\mathbf{r}) + \frac{1}{2}\int_{\partial V_C^+} \mathrm{d}S\, \hat{\mathbf{n}}(\mathbf{r})\cdot\mathcal{A}_0(\mathbf{r})\times\mathcal{H}_0(\mathbf{r})$$
$$- \frac{1}{2}\int_{\partial V_C^-} \mathrm{d}S\, \hat{\mathbf{n}}(\mathbf{r})\cdot\mathcal{A}_0(\mathbf{r})\times\mathcal{H}_0(\mathbf{r}) + \frac{1}{2}\int_{V_C} \mathrm{d}V\, \mathcal{A}_0(\mathbf{r})\cdot\sigma\mathcal{E}_0(\mathbf{r}) \tag{4.149}$$

where in accordance with (4.29) $\mathcal{A}_0(\mathbf{r})$ is the vector potential associated with $\mathcal{B}_0(\mathbf{r})$, and we have invoked the Gauss theorem separately in V_a and V_C. The flux integral over ∂B tends to zero as the radius a grows infinitely large on account of (4.36) and (4.37). The flux integrals over either side of ∂V_C cancel out on the grounds that the tangential components of both $\mathcal{H}_0(\mathbf{r})$ and $\mathcal{A}_0(\mathbf{r})$ are continuous across the conductor boundary. Lastly, the domain integral over V_C vanishes because we have just shown that $\mathcal{E}_0(\mathbf{r})$ is null within the conductor. Therefore, the magnetic energy is identically zero in the whole space, and $\mathcal{H}_0(\mathbf{r}) = 0$ everywhere.

In conclusion, since $\mathcal{E}_0(\mathbf{r}) = 0 = \mathcal{H}_0(\mathbf{r})$ for $\mathbf{r} \in \mathbb{R}^3$, the solution to (4.142)-(4.146) is unique. What is more, we already know from the discussion of steady-state boundary conditions on page 37 that the true electric field (and hence the conduction current) is null within a conductor whose shape can be reduced to a *normal* domain (Figure 1.13). Here we go on to show that within a contour-wise multiply-connected $\mathcal{E}(\mathbf{r})$ and $\mathcal{J}_c(\mathbf{r})$ must also be necessarily *null*, lest the electric field outside V_C fail to obey (4.146). Indeed, we can state that

(1) $\mathcal{J}_c(\mathbf{r})$ needs to be a solenoidal field for the solvability of (4.142), in that the left-hand side thereof is divergence-free on account of (A.39);
(2) $\mathcal{J}_c(\mathbf{r})$ is curl-free by virtue of the Ohm law (4.143) and (4.146) because the conductor is homogeneous by hypothesis;
(3) $\mathcal{J}_c(\mathbf{r})$ also obeys the condition $\hat{\mathbf{n}}(\mathbf{r}) \cdot \mathcal{J}_c(\mathbf{r}) = 0$ on ∂V_C in light of (1.174) inasmuch the surrounding medium is free space.

Thus, it follows that both $\mathcal{J}_c(\mathbf{r})$ and $\mathcal{E}(\mathbf{r})$ may be *Neumann vector fields* within V_C (cf. Example 4.4). Accordingly, by assuming that $\mathcal{J}_c(\mathbf{r})$ is derived from a many-valued harmonic scalar potential $\Upsilon_c(\mathbf{r})$ with period p (cf. Section 4.4), we have

$$-p = \oint_{\gamma} \mathrm{d}s\, \hat{\mathbf{s}}(\mathbf{r}) \cdot \mathcal{J}_c(\mathbf{r}) = \sigma \oint_{\gamma} \mathrm{d}s\, \hat{\mathbf{s}}(\mathbf{r}) \cdot \mathcal{E}(\mathbf{r}) \tag{4.150}$$

where $\gamma \subset \overline{V}_C$ is a piecewise-smooth irreducible contour. Evidently, the circulation of $\mathcal{E}(\mathbf{r})$ within $\overline{V}_C$ does not vanish. Now, if we choose $\gamma \subset \partial V_C$ as the boundary of a smooth open surface S_γ entirely located outside V_C (Figure 4.12) we may invoke the Stokes theorem (A.55) backwards and obtain

$$-p = \sigma \oint_{\gamma} \mathrm{d}s\, \hat{\mathbf{s}}(\mathbf{r}) \cdot \mathcal{E}(\mathbf{r}) = \sigma \int_{S_\gamma} \mathrm{d}S\, \hat{\mathbf{n}}(\mathbf{r}) \cdot \nabla \times \mathcal{E}(\mathbf{r}) \tag{4.151}$$

whence we see that the electric field cannot be conservative in $\mathbb{R}^3 \setminus \overline{V}_C$ so long as $p \neq 0$, and consequently (4.146) is violated! This contradiction is resolved by requiring that $\mathcal{E}(\mathbf{r})$ and hence $\mathcal{J}_c(\mathbf{r})$ be null within the conductor, as anticipated. Then again, since by virtue of (4.146) the electric field is curl-free everywhere, it can be derived from a single-valued scalar electrostatic potential $\Phi(\mathbf{r})$ as in (2.15). Therefore, for $\mathbf{r} \in V_C$ we have

$$\begin{aligned}\int_{V_C} \mathrm{d}V\, |\mathcal{J}_c(\mathbf{r})|^2 &= -\int_{V_C} \mathrm{d}V\, \mathcal{J}_c(\mathbf{r}) \cdot \sigma \nabla \Phi(\mathbf{r}) = -\sigma \int_{V_C} \mathrm{d}V\, \nabla \cdot [\Phi(\mathbf{r}) \mathcal{J}_c(\mathbf{r})] \\ &= -\sigma \int_{\partial V_C} \mathrm{d}S\, \underbrace{\hat{\mathbf{n}}(\mathbf{r}) \cdot \mathcal{J}_c(\mathbf{r})}_{=0} \Phi(\mathbf{r}) = 0\end{aligned} \tag{4.152}$$

and this is sufficient to conclude that $\mathcal{J}_c(\mathbf{r})$ and $\mathcal{E}(\mathbf{r})$ vanish inside the conductor. Put another way, an external *steady* current cannot generate a conduction current within a nearby homogeneous conductor not even when the latter has a ring-like shape [1, Section 3.1], [16, Section 10.1].

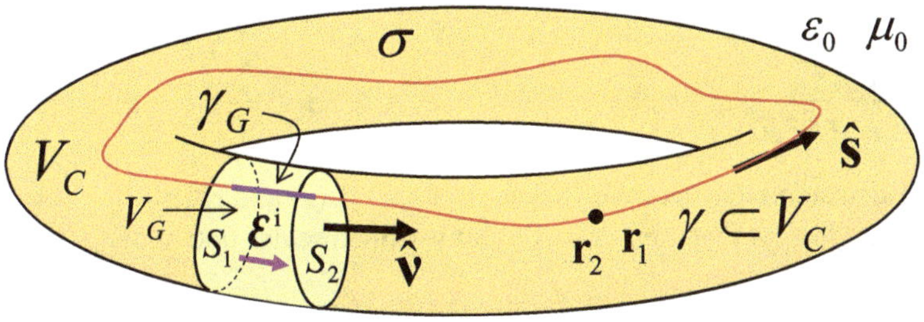

Figure 4.13 Ring-shaped homogeneous isotropic conductor excited by a localized impressed electric field (source of electromotive force).

By contrast, steady currents can be produced and maintained in ring-shaped conductors (e.g., closed electrical circuits comprised of thin wires) if the steady conduction losses are counteracted by a source of electromotive force (cf. Section 1.3) such as a battery [17, Section 1.8], [18, Section 4.10]. Accordingly, we must modify the Ohm law (4.143) by separating the electric field into *impressed* and *secondary* contributions, namely,

$$\mathcal{J}_c(\mathbf{r}) = \sigma\mathcal{E}^{\mathrm{i}}(\mathbf{r}) + \sigma\mathcal{E}^{\mathrm{s}}(\mathbf{r}), \qquad \mathbf{r} \in V_C \tag{4.153}$$

where $\mathcal{J}_c(\mathbf{r})$ is still a Neumann vector field, and $\mathcal{E}^{\mathrm{i}}(\mathbf{r})$ is non-null only in the region $V_G \subset V_C$ occupied by the generator. As is suggested in Figure 4.13, the region V_G is a thin 'slice' of V_C limited by the boundary ∂V_C and two cross-sections S_1 and S_2 of V_C. The secondary field $\mathcal{E}^{\mathrm{s}}(\mathbf{r})$ is conservative and can be derived from a single-valued scalar potential $\Phi(\mathbf{r})$ which becomes discontinuous if we let V_G shrink down to just a cross section S_G of V_C. For the sake of simplicity, we are also assuming that the conductivity within V_G is the same as that in the conductor. Although this might not be true in practice, it does not affect the key points of the discussion.

By choosing an irreducible contour $\gamma \subset \overline{V}_C$ with endpoints $\mathbf{r}_1 \notin V_G$ and $\mathbf{r}_2 \equiv \mathbf{r}_1$ and indicating with $\gamma_G := \gamma \cap V_G$ the part of γ that runs through V_G (Figure 4.13), we have

$$\begin{aligned}
-p &= \oint_\gamma \mathrm{d}s\, \hat{\mathbf{s}}(\mathbf{r}) \cdot \mathcal{J}_c(\mathbf{r}) = \sigma \oint_\gamma \mathrm{d}s\, \hat{\mathbf{s}}(\mathbf{r}) \cdot \left[\mathcal{E}^{\mathrm{i}}(\mathbf{r}) + \mathcal{E}^{\mathrm{s}}(\mathbf{r})\right] \\
&= \sigma \int_{\gamma_G} \mathrm{d}s\, \hat{\mathbf{s}}(\mathbf{r}) \cdot \mathcal{E}^{\mathrm{i}}(\mathbf{r}) + \sigma \oint_\gamma \mathrm{d}s\, \hat{\mathbf{s}}(\mathbf{r}) \cdot \mathcal{E}^{\mathrm{s}}(\mathbf{r}) \\
&= \sigma\mathcal{E} + \sigma[\Phi(\mathbf{r}_1) - \Phi(\mathbf{r}_2)] = \sigma\mathcal{E}
\end{aligned} \tag{4.154}$$

where $\mathcal{E}$ is the electromotive force or the strength of the battery. The circulation of $\mathcal{E}^{\mathrm{s}}(\mathbf{r})$ is null, since $\Phi(\mathbf{r}_2) = \Phi(\mathbf{r}_1)$, and this ensures that the electric field is curl-free outside the conductor on account of (4.151). Clearly, by assigning $\mathcal{E}$ we set the period of the many-valued harmonic scalar potential $\Upsilon_c(\mathbf{r})$, and this determines the current $\mathcal{J}_c(\mathbf{r})$ univocally. The following example will help clarify these claims.

Example 4.5 (Steady conduction current inside a ring-shaped conductor)

We consider an isotropic homogeneous conductor which occupies the spherical-conical region $V = V_C$ defined by (4.132) in Example 4.4. An impressed electric field $\mathcal{E}^{\mathrm{i}}(\mathbf{r})$ exists in a small domain $V_G \subset V_C$

$$V_G := \{\mathbf{r} \in V_C : \varphi \in [\varphi_1, \varphi_2] \subset [0, 2\pi]\} \tag{4.155}$$

and has the form

$$\mathcal{E}^{\mathrm{i}}(\mathbf{r}) = \frac{\mathcal{E}\hat{\boldsymbol{\varphi}}}{(\varphi_2 - \varphi_1) r \sin\vartheta}, \qquad \mathbf{r} \in V_G \tag{4.156}$$

where $\mathcal{E}$ is the electromotive force. We wish to compute $\mathcal{J}_c(\mathbf{r})$ and $\mathcal{E}^{\mathrm{s}}(\mathbf{r})$.

Since we already know that the family of many-valued scalar potentials

$$\Upsilon_c(\mathbf{r}) = c_1\varphi + c_2, \qquad \varphi \in \mathbb{R}, \qquad c_1, c_2 \in \mathbb{R} \tag{4.157}$$

gives rise to Neumann vector fields within V_C, we write the density of conduction current as

$$\mathcal{J}_c(\mathbf{r}) = -\nabla\Upsilon_c(\mathbf{r}) = -\frac{c_1}{r\sin\vartheta}\hat{\boldsymbol{\varphi}}, \qquad \mathbf{r} \in V_C \tag{4.158}$$

where the constant c_1 is determined in accordance with (4.154). By choosing the irreducible contour γ as the circumference defined in (4.138), whereby $\hat{\mathbf{s}}(\mathbf{r}) = \hat{\boldsymbol{\varphi}}$, $\mathrm{d}s = r\sin\vartheta\mathrm{d}\varphi$, we have

$$\sigma\mathcal{E} = \oint_\gamma \mathrm{d}s\, \hat{\mathbf{s}}(\mathbf{r}) \cdot \mathcal{J}_c(\mathbf{r}) = -2\pi c_1 \tag{4.159}$$

and by solving for c_1 we find

$$\mathcal{J}_c(\mathbf{r}) = \frac{\sigma\mathcal{E}}{2\pi r\sin\vartheta}\hat{\boldsymbol{\varphi}}, \qquad \mathbf{r} \in V_C. \tag{4.160}$$

The secondary electric field is obtained from (4.153) and reads

$$\begin{aligned}\mathcal{E}^{\mathrm{s}}(\mathbf{r}) &= \mathcal{J}_c(\mathbf{r})/\sigma - \mathcal{E}^{\mathrm{i}}(\mathbf{r}) \\ &= \begin{cases} \dfrac{\mathcal{E}\hat{\boldsymbol{\varphi}}}{2\pi r\sin\vartheta}\left(1 - \dfrac{2\pi}{\varphi_2 - \varphi_1}\right), & \varphi \in [\varphi_1, \varphi_2] \\ \dfrac{\mathcal{E}\hat{\boldsymbol{\varphi}}}{2\pi r\sin\vartheta}, & \varphi \in [0, \varphi_1] \cup [\varphi_2, 2\pi] \end{cases} \qquad \mathbf{r} \in V_C\end{aligned} \tag{4.161}$$

so it is a piecewise-constant function of φ. In particular, $\mathcal{E}^{\mathrm{s}}(\mathbf{r})$ is discontinuous across either side of the source region V_G and, since $\varphi_2 - \varphi_1 < 2\pi$, it opposes the impressed field for $\mathbf{r} \in V_G$. This can be explained by noticing that $\mathcal{E}^{\mathrm{i}}(\mathbf{r})$ moves the free electrons in the conductor from the section $\varphi = \varphi_2$ to the section $\varphi = \varphi_1 < \varphi_2$ of V_C. Consequently, excesses of positive and negative charges are produced for $\varphi = \varphi_2$ and $\varphi = \varphi_1$, respectively. Thus, the streamlines of $\mathcal{E}^{\mathrm{s}}(\mathbf{r})$ originate in $\varphi = \varphi_2$ and end onto $\varphi = \varphi_1$ by running through both V_G and $V_C \setminus V_G$. Further, application of (A.34) shows that $\mathcal{E}^{\mathrm{s}}(\mathbf{r})$ is curl-free, and hence (4.146) is fulfilled in V_C. In order to check that the circulation of $\mathcal{E}^{\mathrm{s}}(\mathbf{r})$ is null we use the circumference γ again, viz.,

$$\oint_\gamma \mathrm{d}s\, \hat{\mathbf{s}}(\mathbf{r}) \cdot \mathcal{E}^{\mathrm{s}}(\mathbf{r}) = \int_0^{\varphi_1} \mathrm{d}\varphi \frac{\mathcal{E}}{2\pi} + \int_{\varphi_1}^{\varphi_2} \mathrm{d}\varphi \frac{\mathcal{E}}{2\pi}\left(1 - \frac{2\pi}{\varphi_2 - \varphi_1}\right) + \int_{\varphi_2}^{2\pi} \mathrm{d}\varphi \frac{\mathcal{E}}{2\pi} = 0 \tag{4.162}$$

but since the result above is independent of the choice of γ, we can pick up a contour on ∂V_C, invoke the Stokes theorem (cf. Figure 4.12), and conclude that the electric field is curl-free also outside the conductor.

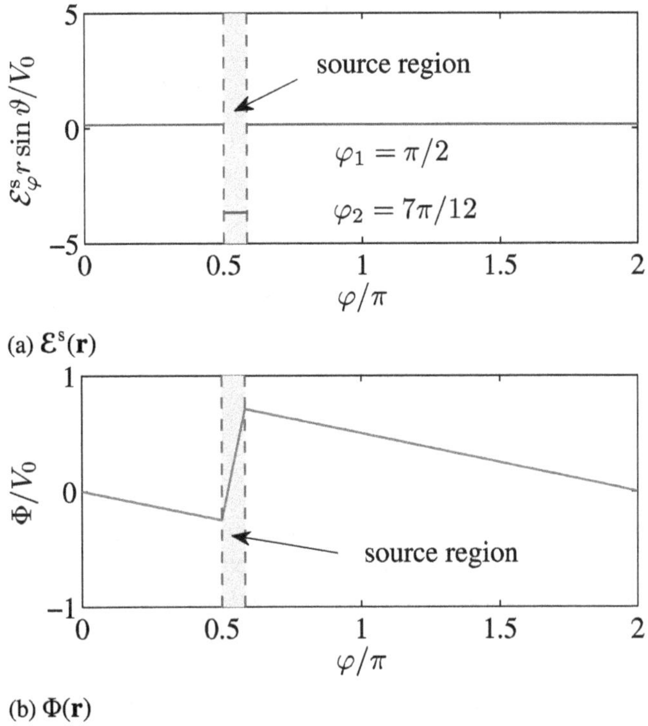

(a) $\mathcal{E}^s(\mathbf{r})$

(b) $\Phi(\mathbf{r})$

Figure 4.14 Secondary field and associated potential in a ring-shaped conductor.

Since $\mathcal{E}^s(\mathbf{r})$ has only the φ-component, the generating scalar potential $\Phi(\mathbf{r})$ of the secondary field is easily computed by requiring

$$\hat{\boldsymbol{\varphi}} \cdot \mathcal{E}^s(\mathbf{r}) = -\hat{\boldsymbol{\varphi}} \cdot \nabla\Phi(\mathbf{r}) \tag{4.163}$$

and expressing the gradient in polar spherical coordinates with (A.28). After a little algebra we arrive at

$$\Phi(\mathbf{r}) = \begin{cases} -\dfrac{\mathcal{E}}{2\pi}\varphi, & \varphi \in [0, \varphi_1] \\ \dfrac{\mathcal{E}}{2\pi}\left(2\pi\dfrac{\varphi - \varphi_1}{\varphi_2 - \varphi_1} - \varphi\right), & \varphi \in [\varphi_1, \varphi_2] \\ \dfrac{\mathcal{E}}{2\pi}(2\pi - \varphi), & \varphi \in [\varphi_2, 2\pi] \end{cases} \qquad \mathbf{r} \in V_C \tag{4.164}$$

whereby we see that $\Phi(\mathbf{r})$ depends on φ only and in particular $\Phi(0) = \Phi(2\pi) = 0$. The normalized secondary electric field and generating potential are drawn in Figures 4.14a and 4.14b for fixed values of r and ϑ as a function of the azimuthal angle for $\varphi_1 = \pi/2$, $\varphi_2 = 7\pi/12$, and $\mathcal{E} = V_0$.

In the limit as $\varphi_2 \to \varphi_1$ the source region V_G degenerates into a cross section S_G of V_C. The impressed electric field (4.156) becomes null everywhere in V_C but on S_G whereon it grows infinitely large. Thus, $\mathcal{E}^i(\mathbf{r})$ is best interpreted as a Dirac delta distribution by requiring (see Appendix C)

$$\lim_{\varphi_2 \to \varphi_1} \oint_\gamma ds\, \hat{\mathbf{s}}(\mathbf{r}) \cdot \mathcal{E}^i(\mathbf{r}) = \mathcal{E} \tag{4.165}$$

which in shorthand notation reads

$$\mathcal{E}^{\mathrm{i}}(\mathbf{r}) = \mathcal{E}\frac{\delta(\varphi - \varphi_1)}{r\sin\vartheta}\hat{\boldsymbol{\varphi}}, \qquad \mathbf{r} \in V_C. \tag{4.166}$$

By virtue of (4.153) the secondary field in the conductor has a meaning as a distribution as well, viz.,

$$\mathcal{E}^{\mathrm{s}}(\mathbf{r}) = \frac{1 - 2\pi\delta(\varphi - \varphi_1)}{2\pi r\sin\vartheta}\mathcal{E}\hat{\boldsymbol{\varphi}}, \qquad \mathbf{r} \in V_C \tag{4.167}$$

whereas the current $\mathcal{J}_c(\mathbf{r})$ is an ordinary vector field also because it is well-defined for $\varphi_2 \to \varphi_1$. Finally, the generating potential $\Phi(\mathbf{r})$ becomes

$$\Phi(\mathbf{r}) = \begin{cases} -\dfrac{\mathcal{E}}{2\pi}\varphi, & \varphi \in [0, \varphi_1] \\ \dfrac{\mathcal{E}}{2\pi}(2\pi - \varphi), & \varphi \in [\varphi_1, 2\pi] \end{cases} \qquad \mathbf{r} \in V_C \tag{4.168}$$

and suffers a jump equal to $\mathcal{E}$ across S_G. This is not surprising in that for $\varphi_2 \to \varphi_1$ the layers of positive and negative charges in $\varphi = \varphi_2$ and $\varphi = \varphi_1$ are brought infinitely close to one another and end up forming a double layer or a surface distribution of electric dipoles with density

$$\boldsymbol{\tau}_S(\mathbf{r}) = \varepsilon_0\mathcal{E}\hat{\boldsymbol{\varphi}}, \qquad \mathbf{r} \in S_G \tag{4.169}$$

which we know is responsible for the discontinuity in the electric scalar potential (see Sections 2.9 and 3.1). As a result, an ideal battery whose longitudinal dimension is negligible with respect to the characteristic size of the conductor may be modelled as a surface density of electric dipoles.

(End of Example 4.5)

At the beginning of this discussion we mentioned that the purpose of the impressed electric field in the system of Figure 4.13 is to counteract the power loss in the conductor and ultimately to keep the current density $\mathcal{J}_c(\mathbf{r})$ flowing. In this regard, it may come as a surprise that the steady power supplied by the generator is actually delivered to the free charges in the conductor by the electromagnetic field existing in the *complementary region* $\mathbb{R}^3 \setminus \overline{V}_C$ (cf. [19, Section 17] for the case of an infinite straight wire). To prove this statement we apply the steady-state Poynting theorem (4.89) to the source region V_G, namely,

$$\begin{aligned} -\int_{V_G} \mathrm{d}V\, \underbrace{\sigma\mathcal{E}^{\mathrm{i}}(\mathbf{r})}_{=\mathcal{J}^{\mathrm{i}}(\mathbf{r})}\cdot\mathcal{E}^{\mathrm{s}}(\mathbf{r}) &= \sigma\int_{V_G} \mathrm{d}V\,|\mathcal{E}^{\mathrm{s}}(\mathbf{r})|^2 + \int_{\partial V_G} \mathrm{d}S\,\hat{\mathbf{n}}(\mathbf{r})\cdot[\mathcal{E}^{\mathrm{s}}(\mathbf{r})\times\mathcal{H}^{\mathrm{s}}(\mathbf{r})] \\ &= \sigma\int_{V_G} \mathrm{d}V\,|\mathcal{E}^{\mathrm{s}}(\mathbf{r})|^2 + \int_{\partial V_G\cap\partial V_C} \mathrm{d}S\,\hat{\mathbf{n}}(\mathbf{r})\cdot[\mathcal{E}^{\mathrm{s}}(\mathbf{r})\times\mathcal{H}^{\mathrm{s}}(\mathbf{r})] \\ &\quad + \sum_{l=1}^{2}\int_{S_l} \mathrm{d}S\,\hat{\mathbf{n}}(\mathbf{r})\cdot[\mathcal{E}^{\mathrm{s}}(\mathbf{r})\times\mathcal{H}^{\mathrm{s}}(\mathbf{r})] \end{aligned} \tag{4.170}$$

but the integrals over the sections S_1 and S_2 vanish identically because thereon we have $\hat{\mathbf{n}} \times \mathcal{E}^{\mathrm{s}} = 0$. The power flux through $\partial V_G \cap \partial V_C$ is positively directed outwards V_G since $\mathcal{E}^{\mathrm{s}}$ opposes the impressed field and $\mathcal{J}_c$, whereas the streamlines of $\mathcal{H}^{\mathrm{s}}$ are closed loops around $\mathcal{J}_c$. All in all, this gives

$$\hat{\mathbf{n}}(\mathbf{r})\cdot[\mathcal{E}^{\mathrm{s}}(\mathbf{r})\times\mathcal{H}^{\mathrm{s}}(\mathbf{r})] > 0, \qquad \mathbf{r} \in \partial V_G \cap \partial V_C \tag{4.171}$$

and hence a steady efflux of power (not radiation, though) into the medium surrounding the conductor exists. This power seeps back into the conductor (away from the slice generator, that is) in order to sustain the current flow. Indeed, if we apply (4.89) again to $W := V_C \setminus \overline{V}_G$ bounded by ∂V_C and the cross sections S_2 and S_1 we get

$$0 = \sigma \int_W dV\, |\mathcal{E}^s(\mathbf{r})|^2 + \int_{\partial W} dS\, \hat{\mathbf{n}}(\mathbf{r}) \cdot [\mathcal{E}^s(\mathbf{r}) \times \mathcal{H}^s(\mathbf{r})]$$

$$= \sigma \int_W dV\, |\mathcal{E}^s(\mathbf{r})|^2 + \int_{\partial W \cap \partial V_C} dS\, \hat{\mathbf{n}}(\mathbf{r}) \cdot [\mathcal{E}^s(\mathbf{r}) \times \mathcal{H}^s(\mathbf{r})]$$

$$+ \sum_{l=1}^{2} \int_{S_l} dS\, \hat{\mathbf{n}}(\mathbf{r}) \cdot [\mathcal{E}^s(\mathbf{r}) \times \mathcal{H}^s(\mathbf{r})] \tag{4.172}$$

where the integrals over S_1 and S_2 are null for the same reason mentioned above, and hence no power flows directly into the conductor from the generator. Since the power dissipated in W is positive, we obtain

$$\int_{\partial W \cap \partial V_C} dS\, \hat{\mathbf{n}}(\mathbf{r}) \cdot [\mathcal{E}^s(\mathbf{r}) \times \mathcal{H}^s(\mathbf{r})] = -\sigma \int_W dV\, |\mathcal{E}^s(\mathbf{r})|^2 < 0 \tag{4.173}$$

and this is enough to conclude that a power efflux takes place from the outer medium into W across the conductor boundary, as anticipated. In effect, since $\mathcal{E}^s$ is tangential to $\partial W \cap \partial V_C$ and aligned with the current, and $\mathcal{H}^s$ has streamlines oriented as before around V_G, we can convince ourselves that the Poynting vector on $\partial W \cap \partial V_C$ is directed into W.

In important practical situations V_C is the region occupied by a thin conducting wire connected to a battery. In which case, we may safely assume that $\mathcal{J}_c(\mathbf{r})$ is uniform and write

$$\mathcal{J}_c(\mathbf{r}) = \frac{I}{A}\hat{\mathbf{s}}(\mathbf{r}), \qquad \mathbf{r} \in V_C \setminus V_G \tag{4.174}$$

where I is the constant current flowing in the circuit, and A is the area of the wire cross section. Besides, since the current density is solenoidal throughout V_C, given two cross sections $S_C \subset V_C \setminus V_G$ and $S_G \subset V_G$ we have

$$I = \int_{S_C} dS\, \hat{\boldsymbol{\nu}}(\mathbf{r}) \cdot \mathcal{J}_c(\mathbf{r}) = \int_{S_G} dS\, \hat{\boldsymbol{\nu}}(\mathbf{r}) \cdot \mathcal{J}_c(\mathbf{r}) \tag{4.175}$$

whence we conclude that also in the battery $\mathcal{J}_c(\mathbf{r})$ is necessarily proportional to I. Therefore, we can define the quantity

$$R_G := \frac{1}{\sigma I} \int_{\gamma_G} ds\, \hat{\mathbf{s}}(\mathbf{r}) \cdot \mathcal{J}_c(\mathbf{r}) \tag{4.176}$$

as the internal resistance (physical dimension: Ω) of the generator. Then, from (4.154) we get

$$\mathcal{E} = \frac{1}{\sigma} \oint_{\gamma} ds\, \hat{\mathbf{s}}(\mathbf{r}) \cdot \mathcal{J}_c(\mathbf{r}) = \frac{1}{\sigma} \int_{\gamma_G} ds\, \hat{\mathbf{s}}(\mathbf{r}) \cdot \mathcal{J}_c(\mathbf{r}) + \frac{1}{\sigma} \int_{\gamma_{21}} ds\, \hat{\mathbf{s}}(\mathbf{r}) \cdot \mathcal{J}_c(\mathbf{r})$$

$$= R_G I + \frac{l_{21}}{\sigma A} I = R_G I + RI \tag{4.177}$$

where γ_{21} is the part of γ that runs through the wire from S_2 back to S_1, l_{21} is the length of γ_{21}, and R is the resistance of the wire. Since the impressed field is confined to V_G, we can obtain an alternative expression for the integral along γ_{21}, namely,

$$RI = \frac{1}{\sigma} \int_{\gamma_{21}} \mathrm{d}s\, \hat{\mathbf{s}}(\mathbf{r}) \cdot \mathcal{J}_c(\mathbf{r}) = \int_{\gamma_{21}} \mathrm{d}s\, \hat{\mathbf{s}}(\mathbf{r}) \cdot \mathcal{E}^s(\mathbf{r}) = -\int_{\gamma_{21}} \mathrm{d}s\, \hat{\mathbf{s}}(\mathbf{r}) \cdot \nabla\Phi(\mathbf{r}) = \Phi_2 - \Phi_1 = V \tag{4.178}$$

where Φ_1 and Φ_2 are the values of the potential $\Phi(\mathbf{r})$ in the sections S_1 and S_2, i.e., on either side of the battery (see Figure 4.12). Therefore, V represents the *voltage drop* along the wire and (4.177) may be written as

$$V = RI = \mathcal{E} - R_G I \tag{4.179}$$

which we recognize as the Kirchhoff voltage law (1845) applied along the circuit. When the electromotive force is assigned and the resistances are known, the current I can be computed.

References

[1] Bassanini P, Elcrat A. Mathematical Theory of Electromagnetism. Creative Commons; 2009.
[2] Jackson JD. Classical Electrodynamics. 3rd ed. Chichester, UK: Wiley; 1999.
[3] Van Bladel JG. Electromagnetic Fields. Piscataway, NJ: IEEE Press; 2007.
[4] Mild KH, Greenebaum B. Environmental and Occupationally Encountered Electromagnetic Fields. In: Greenebaum B, Barnes FS, editors. Bioengineering and Biophysical Aspects of Electromagnetic Fields. Boca Raton, FL: CRC press; 2007. p. 1–34.
[5] Schwab AJ. Field Theory Concepts. Berlin Heidelberg: Springer-Verlag; 1988.
[6] Dautray R, Lions JL. Mathematical Analysis and Numerical Methods for Science and Technology - Spectral Theory and Applications. vol. 3. Berlin Heidelberg: Springer-Verlag; 1990.
[7] Griffith DJ. Introduction to electrodynamics. 3rd ed. Upper Saddle River, NJ: Prentice Hall; 1999.
[8] Schwinger J, De Raad LL, Milton KA, et al. Classical electrodynamics. Perseus Books; 1998.
[9] Makarov B, Podkorytov A. Real Analysis: Measures, Integrals and Applications. Universitext. London, UK: Springer; 2013.
[10] Rothwell EJ, Cloud MJ. Electromagnetics. London, UK: CRC Press; 2001.
[11] Saslow WM. Electricity, Magnetism, and Light. New York, NY: Academic Press; 2002.
[12] Zangwill A. Modern electrodynamics. Cambridge, UK: Cambrigde University Press; 2013.
[13] Feynmann RP, Leighton R, Sands M. The Feynmann lectures on physics. vol. 1. Reading, MA: Addison-Wesley; 1964.
[14] Dautray R, Lions JL. Mathematical Analysis and Numerical Methods for Science and Technology - Physical Origins and Classical Methods. vol. 1. Berlin Heidelberg: Springer-Verlag; 1990.
[15] Konopinski EJ. Electromagnetic Fields and Relativistic Particles. New York, NY: McGraw Hill; 1981.

[16] Van Bladel JG. Relativity and Engineering. Berlin Heidelberg : Springer Berlin Heidelberg; 1984.

[17] Helrich CS. The classical theory of fields - Electromagnetism. Berlin Heidelberg: Springer-Verlag; 2012.

[18] Guru BS, Hiziroglu HR. Electromagnetic field theory fundamentals. 2nd ed. New York, NY: Cambridge University Press; 2004.

[19] Sommerfeld A. Electrodynamics. vol. 3 of Lectures on theoretical physics. New York, NY: Academic Press; 1952.

Chapter 5
Stationary magnetic fields II

5.1 Integral representations

Similarly to expression (2.160) obtained in Section 2.7 for the electrostatic potential, we wish to derive integral representations for the vector potential and the stationary magnetic entities in a given region of space. Also in this case, the results will be a *formal* solution to the vector Poisson equation or to the stationary magnetic equations (4.9) and (4.10) subject to (4.11). In the following we denote the position vector with $\mathbf{r}' = x'\hat{\mathbf{x}} + y'\hat{\mathbf{y}} + z'\hat{\mathbf{z}}$ and, to specify derivatives with respect to the primed coordinates, the differential operators will be indicated with $\nabla'\{\bullet\}$, $\nabla' \cdot \{\bullet\}$, $\nabla' \times \{\bullet\}$ and $\nabla'^2\{\bullet\}$ as needed.

5.1.1 Vector potential in an isotropic medium

We address the problem of formally inverting the vector Poisson equation in a homogeneous magnetic medium endowed with magnetic permeability μ [1,2]; the result for free space then follows as a special case. Suppose we wish to find the vector potential $\mathcal{A}_1(\mathbf{r}')$ produced in a bounded volume V by the stationary current density $\mathcal{J}(\mathbf{r}')$ flowing in a tube $V_J \subset V$ in the presence of an another medium which occupies a smaller region V_1, as is exemplified in Figure 5.1a. In Section 4.7 we have found out that specifying the boundary conditions on $\partial V := S \cup \partial V_1$ and the divergence of the vector potential is sufficient to determine $\mathcal{A}_1(\mathbf{r}')$ univocally. But then, since the vector Poisson equation involves the curl of $\mathcal{A}_1(\mathbf{r}')$, we are at liberty to choose the arbitrary scalar field $\chi(\mathbf{r}')$ so as to fix the divergence of $\mathcal{A}_1(\mathbf{r}')$ in a convenient way. When we require the divergence to vanish, we say that $\mathcal{A}_1(\mathbf{r}')$ is determined in the *Coulomb gauge*. Other names for this choice are *radiation* or *transverse gauge* [3, 4] and the rationale for these appellations will be clarified in Sections 8.2 and 9.6.

With these positions the problem

$$\begin{cases} \nabla' \times [\nabla' \times \mathcal{A}_1(\mathbf{r}')] = \mu\mathcal{J}(\mathbf{r}'), & \mathbf{r}' \in V \\ \nabla' \cdot \mathcal{A}_1(\mathbf{r}') = 0, & \mathbf{r}' \in V \\ \hat{\mathbf{n}}(\mathbf{r}')\times\mathcal{A}_1(\mathbf{r}') = \mathbf{f}(\mathbf{r}') \quad \text{or} \quad \hat{\mathbf{n}}(\mathbf{r}')\times[\nabla'\times\mathcal{A}_1(\mathbf{r}')] = \mathbf{g}(\mathbf{r}'), & \mathbf{r}' \in \partial V \\ \hat{\mathbf{n}}(\mathbf{r}') \cdot \mathcal{A}_1(\mathbf{r}') = h(\mathbf{r}'), & \mathbf{r}' \in \partial V \end{cases} \tag{5.1}$$

admits a unique solution, with $\mathbf{f}$, $\mathbf{g}$ and h known regular fields on ∂V.

Next, we consider an auxiliary, possibly simpler problem for which the solution is available. Comparison with the strategy followed to derive (2.160) suggests that we should pick up an elementary stationary current density located at a point $\mathbf{r} \in V \setminus V_J$ in a homogeneous medium with permeability μ, hence the resulting vector potential would play the role of fundamental solution

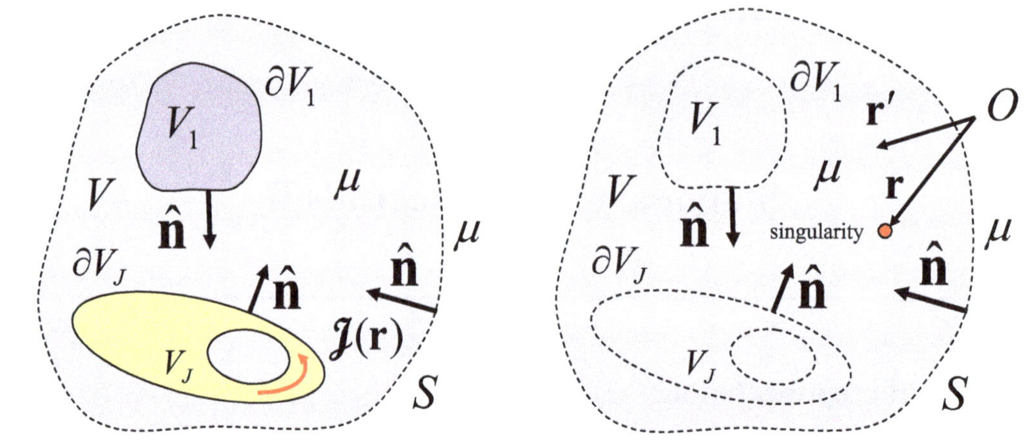

(a) *original problem* (b) *auxiliary problem*

Figure 5.1 For the derivation of the integral representation of the vector potential.

for the vector Poisson equation. However, owing to the vectorial nature of the entities and of the equations involved, it is not easy to figure out what that current density may look like.

Thus, we proceed backwards by starting with a known vector potential, i.e.,

$$\mathcal{A}_2(\mathbf{r}') = \frac{\mu\mathbf{u}}{4\pi|\mathbf{r}-\mathbf{r}'|}, \qquad \mathbf{r}' \in \mathbb{R}^3 \setminus \{\mathbf{r}\}, \qquad \mathbf{r} \in V \setminus V_J \tag{5.2}$$

with $\mathbf{u}$ being an arbitrary constant vector which we shall get rid of at the end (cf. [5, Appendix 1.1]). This vector field is singular for $\mathbf{r}' = \mathbf{r}$ (Figure 5.1b), so it stands as a good candidate for a Green function of sorts. To find the generating current density $\mathcal{J}_2(\mathbf{r}')$, we substitute $\mathcal{A}_2(\mathbf{r}')$ into the vector Poisson equation, viz.,

$$\begin{aligned}\nabla' \times [\nabla' \times \mathcal{A}_2(\mathbf{r}')] &= \nabla'\nabla'\frac{1}{4\pi R}\cdot\mu\mathbf{u} - \mu\mathbf{u}\nabla'^2\frac{1}{4\pi R}\\ &= \frac{3\mathbf{R}\mathbf{R} - R^2\overline{\mathcal{I}}}{4\pi R^5}\cdot\mu\mathbf{u} = \mu\mathcal{J}_2(\mathbf{r}'), \qquad \mathbf{r}' \neq \mathbf{r}\end{aligned} \tag{5.3}$$

where $\mathbf{R} = \mathbf{r} - \mathbf{r}'$ and $R = |\mathbf{R}|$. Besides, we have used (3.192) and the fact that $1/R$ is a harmonic function for $\mathbf{r}' \neq \mathbf{r}$ [6, Section 8.7], [7, Section 2.2]. This current is evidently singular for $\mathbf{r}' = \mathbf{r}$, though it is not therein concentrated, but rather it permeates the whole space. Nonetheless, $\mathcal{J}_2(\mathbf{r}')$ is solenoidal because

$$\nabla'\cdot\mathcal{J}_2(\mathbf{r}') = \nabla'\cdot\nabla'\nabla'\cdot\frac{\mathbf{u}}{4\pi R} - \nabla'\cdot\nabla'^2\frac{\mathbf{u}}{4\pi R} = \nabla'\cdot\nabla'^2\frac{\mathbf{u}}{4\pi R} - \nabla'\cdot\nabla'^2\frac{\mathbf{u}}{4\pi R} = 0 \tag{5.4}$$

everywhere except for $\mathbf{r}' = \mathbf{r}$. Yet, we can extend the above identity for $\mathbf{r}' = \mathbf{r}$ by taking 0 to be the limit of the undetermined form as $\mathbf{r}' \to \mathbf{r}$, whereby $\mathcal{J}_2(\mathbf{r})$ becomes solenoidal for $\mathbf{r}' \in \mathbb{R}^3$, as expected of a stationary current.

Regardless, the actual expression of $\mathcal{J}_2(\mathbf{r}')$ is not important for our purposes. In fact, we can state the auxiliary problem formally as

$$\begin{cases} \nabla' \times [\nabla' \times \mathcal{A}_2(\mathbf{r}')] = \nabla'\nabla' \cdot \mathcal{A}_2(\mathbf{r}'), & \mathbf{r}' \in \mathbb{R}^3 \setminus \{\mathbf{r}\} \\ \nabla' \cdot \mathcal{A}_2(\mathbf{r}') = \mu \mathbf{u} \cdot \nabla' \dfrac{1}{4\pi|\mathbf{r} - \mathbf{r}'|}, & \mathbf{r}' \in \mathbb{R}^3 \setminus \{\mathbf{r}\} \\ \mathcal{A}_2(\mathbf{r}') = \mathcal{O}\left(\dfrac{1}{|\mathbf{r}'|}\right), & |\mathbf{r}'| \to +\infty \end{cases} \tag{5.5}$$

which has unique solution given by (5.2) and known by construction, for that matter. Notice that it is essential that the divergence of $\mathcal{A}_2(\mathbf{r}')$ be non-null, or else the vector Poisson equation would be homogeneous!

We dot-multiply the first of (5.1) by $\mathcal{A}_2(\mathbf{r}')$ and the first of (5.5) by $\mathcal{A}_1(\mathbf{r}')$ to obtain

$$\begin{aligned} \mu \mathcal{A}_2(\mathbf{r}') \cdot \mathcal{J}(\mathbf{r}') &= \mathcal{A}_2(\mathbf{r}') \cdot \nabla' \times [\nabla' \times \mathcal{A}_1(\mathbf{r}')] \\ &= \nabla' \cdot \{[\nabla' \times \mathcal{A}_1(\mathbf{r}')] \times \mathcal{A}_2(\mathbf{r}')\} + \nabla' \times \mathcal{A}_2(\mathbf{r}') \cdot \nabla' \times \mathcal{A}_1(\mathbf{r}') \end{aligned} \tag{5.6}$$

$$\begin{aligned} \mathcal{A}_1(\mathbf{r}') \cdot \nabla'\nabla' \cdot \mathcal{A}_2(\mathbf{r}') &= \nabla' \cdot [\mathcal{A}_1(\mathbf{r}')\nabla' \cdot \mathcal{A}_2(\mathbf{r}')] \\ &= \nabla' \cdot \{[\nabla' \times \mathcal{A}_2(\mathbf{r}')] \times \mathcal{A}_1(\mathbf{r}')\} + \nabla' \times \mathcal{A}_1(\mathbf{r}') \cdot \nabla' \times \mathcal{A}_2(\mathbf{r}') \end{aligned} \tag{5.7}$$

which hold simultaneously for $\mathbf{r}' \in V \setminus \{\mathbf{r}\}$ where $\mathcal{A}_2(\mathbf{r}')$ is not singular. In the second equation we have exploited the Coulomb gauge imposed on $\mathcal{A}_1(\mathbf{r}')$. We exclude the singular point $\mathbf{r}$ with a ball $B(\mathbf{r}, a)$ whose radius a is taken small enough so that $B[\mathbf{r}, a] \subset V$. We go on to subtract the equations above side by side and integrate over $V \setminus B(\mathbf{r}, a)$

$$\begin{aligned} &\int\limits_{V\setminus B(\mathbf{r},a)} \mathrm{d}V'\, \nabla' \cdot \{\mathcal{A}_1(\mathbf{r}')\nabla' \cdot \mathcal{A}_2(\mathbf{r}') - [\nabla' \times \mathcal{A}_2(\mathbf{r}')] \times \mathcal{A}_1(\mathbf{r}')\} \\ &\qquad + \int\limits_{V\setminus B(\mathbf{r},a)} \mathrm{d}V'\, \nabla' \cdot \{[\nabla' \times \mathcal{A}_1(\mathbf{r}')] \times \mathcal{A}_2(\mathbf{r}')\} = \mu \int\limits_{V_J} \mathrm{d}V'\, \mathcal{A}_2(\mathbf{r}') \cdot \mathcal{J}(\mathbf{r}') \end{aligned} \tag{5.8}$$

where the integration in the right-hand side has been restricted to the support of $\mathcal{J}(\mathbf{r}')$.

We may apply the Gauss theorem to the volume integrals because the vector fields obtained from the combination of potentials and derivatives thereof are continuously differentiable in $V \setminus B(\mathbf{r}, a)$. Thus, we find

$$\begin{aligned} \mu \int\limits_{V_J} \mathrm{d}V'\, \mathcal{A}_2(\mathbf{r}') \cdot \mathcal{J}(\mathbf{r}') &= \int\limits_{S\cup\partial V_1} \mathrm{d}S'\, \hat{\mathbf{n}}(\mathbf{r}') \cdot \{[\nabla' \times \mathcal{A}_2(\mathbf{r}')] \times \mathcal{A}_1(\mathbf{r}') - \mathcal{A}_1(\mathbf{r}')\nabla' \cdot \mathcal{A}_2(\mathbf{r}')\} \\ &- \int\limits_{S\cup\partial V_1} \mathrm{d}S'\, \hat{\mathbf{n}}(\mathbf{r}') \cdot [\nabla' \times \mathcal{A}_1(\mathbf{r}')] \times \mathcal{A}_2(\mathbf{r}') \\ &+ \int\limits_{\partial B} \mathrm{d}S'\, \hat{\mathbf{n}}(\mathbf{r}') \cdot \{[\nabla' \times \mathcal{A}_2(\mathbf{r}')] \times \mathcal{A}_1(\mathbf{r}') - \mathcal{A}_1(\mathbf{r}')\nabla' \cdot \mathcal{A}_2(\mathbf{r}')\} \\ &- \int\limits_{\partial B} \mathrm{d}S'\, \hat{\mathbf{n}}(\mathbf{r}') \cdot [\nabla' \times \mathcal{A}_1(\mathbf{r}')] \times \mathcal{A}_2(\mathbf{r}') \end{aligned} \tag{5.9}$$

where we have taken into account that the unit normal $\hat{\mathbf{n}}$ on ∂V and ∂B is positively oriented inward $V \setminus B(\mathbf{r}, a)$. (The geometrical setup is similar to that of Figure 2.8 on page 111.) To take the limit of both sides as $a \to 0$ we examine the various integrals separately.

The flux integrals over $S \cup \partial V_1$ as well as the volume integral of the current density are independent of a, thus we just have to substitute the known expression for $\mathcal{A}_2(\mathbf{r}')$ and factor out the constant vector $\mathbf{u}$. In symbols, this gives

$$\int\limits_{S\cup\partial V_1} \mathrm{d}S'\, \hat{\mathbf{n}}(\mathbf{r}') \cdot [\nabla' \times \mathcal{A}_2(\mathbf{r}')] \times \mathcal{A}_1(\mathbf{r}') = \mu \int\limits_{S\cup\partial V_1} \mathrm{d}S'\, \hat{\mathbf{n}}(\mathbf{r}') \cdot \left[\nabla' \frac{1}{4\pi R} \times \mathbf{u}\right] \times \mathcal{A}_1(\mathbf{r}')$$
$$= \mu\mathbf{u} \cdot \int\limits_{S\cup\partial V_1} \mathrm{d}S'\, \nabla' \frac{1}{4\pi R} \times [\hat{\mathbf{n}}(\mathbf{r}') \times \mathcal{A}_1(\mathbf{r}')] \tag{5.10}$$

$$\int\limits_{S\cup\partial V_1} \mathrm{d}S'\, \hat{\mathbf{n}}(\mathbf{r}') \cdot \mathcal{A}_1(\mathbf{r}')\nabla' \cdot \mathcal{A}_2(\mathbf{r}') = \mu\mathbf{u} \cdot \int\limits_{S\cup\partial V_1} \mathrm{d}S'\, \hat{\mathbf{n}}(\mathbf{r}') \cdot \mathcal{A}_1(\mathbf{r}')\nabla' \frac{1}{4\pi R} \tag{5.11}$$

$$\int\limits_{S\cup\partial V_1} \mathrm{d}S'\, \hat{\mathbf{n}}(\mathbf{r}') \cdot [\nabla' \times \mathcal{A}_1(\mathbf{r}')] \times \mathcal{A}_2(\mathbf{r}') = \mu \int\limits_{S\cup\partial V_1} \mathrm{d}S'\, \hat{\mathbf{n}}(\mathbf{r}') \cdot [\nabla' \times \mathcal{A}_1(\mathbf{r}')] \times \frac{\mathbf{u}}{4\pi R}$$
$$= \mu\mathbf{u} \cdot \int\limits_{S\cup\partial V_1} \mathrm{d}S'\, \hat{\mathbf{n}}(\mathbf{r}') \times [\nabla' \times \mathcal{A}_1(\mathbf{r}')] \frac{1}{4\pi R} \tag{5.12}$$

$$\mu \int\limits_{V_J} \mathrm{d}V'\, \mathcal{A}_2(\mathbf{r}') \cdot \mathcal{J}(\mathbf{r}') = \mu^2\mathbf{u} \cdot \int\limits_{V_J} \mathrm{d}V'\, \frac{\mathcal{J}(\mathbf{r}')}{4\pi R}. \tag{5.13}$$

Concerning the flux integrals over the sphere ∂B we observe that for $\mathbf{r}' \in \partial B$ we have

$$\mathbf{r}' = \mathbf{r} + a\hat{\mathbf{n}}(\mathbf{r}'), \qquad \mathcal{A}_2(\mathbf{r}') = \frac{\mu\mathbf{u}}{4\pi a} \tag{5.14}$$

$$\nabla' \cdot \mathcal{A}_2(\mathbf{r}') = -\hat{\mathbf{n}}(\mathbf{r}') \cdot \frac{\mu\mathbf{u}}{4\pi a^2}, \qquad \nabla' \times \mathcal{A}_2(\mathbf{r}') = -\hat{\mathbf{n}}(\mathbf{r}') \times \frac{\mu\mathbf{u}}{4\pi a^2} \tag{5.15}$$

in view of definition (5.2). We consider the following two contributions

$$\int\limits_{\partial B} \mathrm{d}S'\, \hat{\mathbf{n}}(\mathbf{r}') \cdot \{[\nabla' \times \mathcal{A}_2(\mathbf{r}')] \times \mathcal{A}_1(\mathbf{r}') - \mathcal{A}_1(\mathbf{r}')\nabla' \cdot \mathcal{A}_2(\mathbf{r}')\}$$
$$= -\frac{\mu}{4\pi a^2} \int\limits_{\partial B} \mathrm{d}S'\, \{\hat{\mathbf{n}}(\mathbf{r}') \times [\hat{\mathbf{n}}(\mathbf{r}') \times \mathbf{u}] \cdot \mathcal{A}_1(\mathbf{r}') - \hat{\mathbf{n}}(\mathbf{r}') \cdot \mathcal{A}_1(\mathbf{r}')\, \mathbf{u} \cdot \hat{\mathbf{n}}(\mathbf{r}')\}$$
$$= \frac{\mu}{4\pi a^2} \int\limits_{\partial B} \mathrm{d}S'\, \mathbf{u} \cdot \mathcal{A}_1(\mathbf{r}') = \mu\mathbf{u} \cdot \mathcal{A}_1(\mathbf{r} + a\hat{\mathbf{n}}(\mathbf{r}'_0)) \xrightarrow[a\to 0]{} \mu\mathbf{u} \cdot \mathcal{A}_1(\mathbf{r}) \tag{5.16}$$

$$\int\limits_{\partial B} \mathrm{d}S'\, \hat{\mathbf{n}}(\mathbf{r}') \cdot [\nabla' \times \mathcal{A}_1(\mathbf{r}')] \times \mathcal{A}_2(\mathbf{r}') = \frac{\mu}{4\pi a} \int\limits_{\partial B} \mathrm{d}S'\, \hat{\mathbf{n}}(\mathbf{r}') \cdot [\nabla' \times \mathcal{A}_1(\mathbf{r}')] \times \mathbf{u}$$
$$= a\mu\hat{\mathbf{n}}(\mathbf{r}'_0) \cdot [\nabla' \times \mathcal{A}_1(\mathbf{r}')]|_{\mathbf{r}'_0} \times \mathbf{u} \xrightarrow[a\to 0]{} 0 \tag{5.17}$$

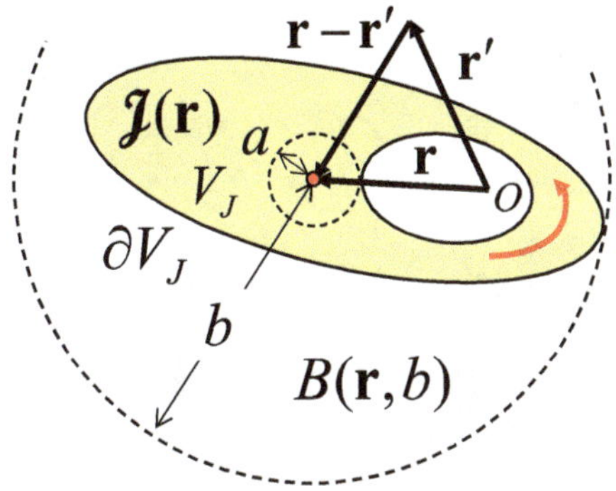

Figure 5.2 Geometrical construction for showing that the volume integral in (5.18) *is bounded for observation points within the current region. The ball* $B(\mathbf{r}, b)$ *has been drawn* (−−) *only partially.*

where the mean value theorem [6] has been applied, since $\mathcal{A}_1(\mathbf{r}')$ and its derivatives are continuous functions on $\partial B \subset V$, $\mathbf{r}'_0 \in \partial B$ is a suitable point, and $\mathbf{r}'_0 \to \mathbf{r}$ as $a \to 0$.

We collect these intermediate results, divide through by μ and drop the inessential subscript 1. All the contributions are in the form of a dot product between the constant vector $\mathbf{u}$ and some vector field which results from the calculation of a surface or a volume integral. By setting $\mathbf{u}$ equal to $\hat{\mathbf{x}}$, $\hat{\mathbf{y}}$, and $\hat{\mathbf{z}}$ in succession, we obtain three scalar relations for the Cartesian components of $\mathcal{A}(\mathbf{r})$, which we may combine to yield the vector relationship

$$\mathcal{A}(\mathbf{r}) = \mu \int\limits_{V_J} \mathrm{d}V' \frac{\mathcal{J}(\mathbf{r}')}{4\pi R} + \int\limits_{S \cup \partial V_1} \mathrm{d}S' \, \hat{\mathbf{n}}(\mathbf{r}') \times [\nabla' \times \mathcal{A}(\mathbf{r}')] \frac{1}{4\pi R}$$
$$+ \int\limits_{S \cup \partial V_1} \mathrm{d}S' \, [\hat{\mathbf{n}}(\mathbf{r}') \times \mathcal{A}(\mathbf{r}')] \times \nabla' \frac{1}{4\pi R} + \int\limits_{S \cup \partial V_1} \mathrm{d}S' \, \hat{\mathbf{n}}(\mathbf{r}') \cdot \mathcal{A}(\mathbf{r}') \nabla' \frac{1}{4\pi R} \tag{5.18}$$

which constitutes the desired result [1, 2]. Since we have placed the singular point $\mathbf{r}$ in $V \setminus V_J$, for the time being the representation is valid only outside the current tube V_J.

We can extend (5.18) to all points $\mathbf{r} \in V$ by choosing the auxiliary problem with $\mathbf{r} \in V_J$ and isolating the singular point with a ball $B(\mathbf{r}, a) \subset V_J$. Proceeding as before, we obtain an expression similar to (5.9) in which the volume integral is over $V_J \setminus B(\mathbf{r}, a)$. Since $\mathcal{A}_1(\mathbf{r}')$ and the derivatives thereof are regular also in the source region if $\mathcal{J}(\mathbf{r}')$ is at least continuous, the flux integrals over ∂B yield the same results as before in the limit as $a \to 0$. Lastly, the integral over $V_J \setminus B(\mathbf{r}, a)$ exists and is bounded for any value of a. We choose a second ball $B(\mathbf{r}, b)$ with the radius $b > a$ and large enough so that $B(\mathbf{r}, a) \subset V_J \subset B(\mathbf{r}, b)$ (Figure 5.2) and we consider the estimates

$$\left| \int\limits_{V_J \setminus B(\mathbf{r},a)} \mathrm{d}V' \frac{\mathcal{J}(\mathbf{r}')}{4\pi R} \right| \leqslant \int\limits_{V_J \setminus B(\mathbf{r},a)} \mathrm{d}V' \frac{|\mathcal{J}(\mathbf{r}')|}{4\pi R} \leqslant \int\limits_{V_J \setminus B(\mathbf{r},a)} \mathrm{d}V' \frac{\|\mathcal{J}\|_\infty}{4\pi R} \leqslant \int\limits_{B(\mathbf{r},b) \setminus B(\mathbf{r},a)} \mathrm{d}V' \frac{\|\mathcal{J}\|_\infty}{4\pi R}$$
$$= \|\mathcal{J}\|_\infty \int_a^b \mathrm{d}R \, R = \frac{1}{2} \|\mathcal{J}\|_\infty (b^2 - a^2) \leqslant \frac{1}{2} \|\mathcal{J}\|_\infty b^2 \tag{5.19}$$

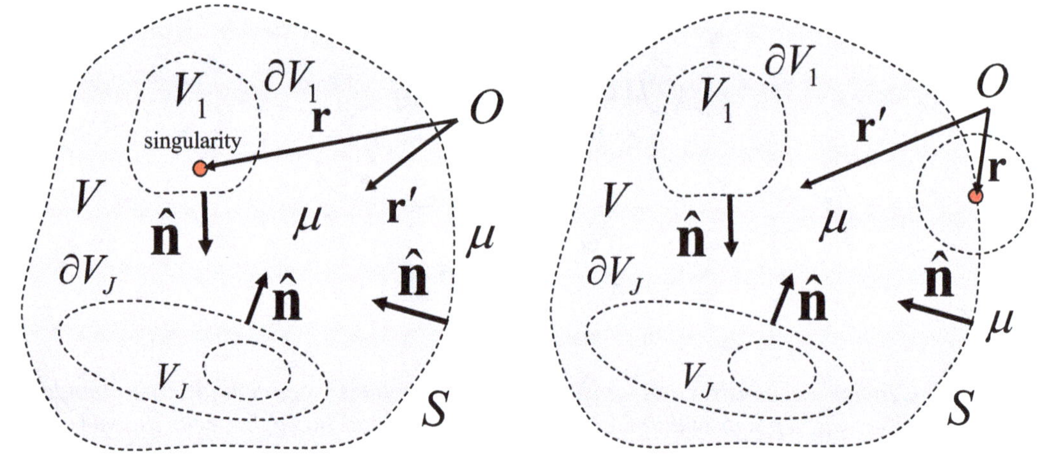

(a) *singularity outside V* (b) *singularity on the boundary ∂V*

Figure 5.3 Auxiliary problems for the derivation of the integral representation of the vector potential.

where we have used the continuity of $\mathcal{J}(\mathbf{r}')$ and integrated in local spherical coordinates (R, α', β') centered in $\mathbf{r}$. It thus remains proved that (5.18) holds true for $\mathbf{r} \in V$, current region included.

If we choose the auxiliary vector potential $\mathcal{A}_2(\mathbf{r}')$ — and hence the current density $\mathcal{J}_2(\mathbf{r}')$ — with the singular point $\mathbf{r}$ in the complementary domain $\mathbb{R}^3 \setminus \overline{V}$ (see Figure 5.3a) and repeat the steps followed thus far, we arrive at

$$0 = \mu \int_{V_J} \mathrm{d}V' \frac{\mathcal{J}(\mathbf{r}')}{4\pi R} + \int_{S \cup \partial V_1} \mathrm{d}S' \, \hat{\mathbf{n}}(\mathbf{r}') \times [\nabla' \times \mathcal{A}(\mathbf{r}')] \frac{1}{4\pi R} + \int_{S \cup \partial V_1} \mathrm{d}S' \, [\hat{\mathbf{n}}(\mathbf{r}') \times \mathcal{A}(\mathbf{r}')] \times \nabla' \frac{1}{4\pi R} + \int_{S \cup \partial V_1} \mathrm{d}S' \, \hat{\mathbf{n}}(\mathbf{r}') \cdot \mathcal{A}(\mathbf{r}') \nabla' \frac{1}{4\pi R} \tag{5.20}$$

for $\mathbf{r} \notin V$. As the right-hand side of this equation is formally the same as in (5.18), we conclude that the latter may be used for observation points located outside the domain of interest where it predicts a null value for the vector potential. As was the case for the scalar potential $\Phi(\mathbf{r})$, $\mathcal{A}(\mathbf{r})$ constitutes the unique solution to the problem (4.93), possibly extended to the points outside V, and this solution is not necessarily null for $\mathbf{r} \in V$. Therefore, (5.18) yields the actual values of $\mathcal{A}(\mathbf{r})$ only at points within the volume for which it has been derived. In this regard, (5.20) implies that all the integrals in the right-hand side must cancel each other. This means that the potential generated by the stationary current $\mathcal{J}(\mathbf{r})$ in $V_J \subset V$ is neutralized by the contribution of the three surface integrals over ∂V. Notice that this conclusion applies also for $\mathbf{r} \in V_1$, because V_1 is not part of the domain V.

Lastly, we discuss the case of the auxiliary potential with the singularity occurring on the boundary ∂V, e.g., $\mathbf{r} \in S$, as is depicted in Figure 5.3b. By introducing a ball $B(\mathbf{r}, a)$ we divide the boundary ∂V into three parts, namely, ∂V_1, the open surface $S' := \{\mathbf{r}' \in S : |\mathbf{r}' - \mathbf{r}| \geqslant a\}$, and the open surface $S'' := \partial B \cap \overline{V}$. Under the assumptions that led to (5.9) we may apply the divergence theorem to (5.8) to obtain

$$\mu \int_{V_J} dV' \mathcal{A}_2(\mathbf{r}') \cdot \mathcal{J}(\mathbf{r}') = \int_{\partial V_1} dS' \,\hat{\mathbf{n}}(\mathbf{r}') \cdot \{[\nabla' \times \mathcal{A}_2(\mathbf{r}')] \times \mathcal{A}_1(\mathbf{r}') - \mathcal{A}_1(\mathbf{r}')\nabla' \cdot \mathcal{A}_2(\mathbf{r}')\}$$

$$- \int_{\partial V_1} dS' \,\hat{\mathbf{n}}(\mathbf{r}') \cdot [\nabla' \times \mathcal{A}_1(\mathbf{r}')] \times \mathcal{A}_2(\mathbf{r}')$$

$$+ \int_{S'} dS' \,\hat{\mathbf{n}}(\mathbf{r}') \cdot \{[\nabla' \times \mathcal{A}_2(\mathbf{r}')] \times \mathcal{A}_1(\mathbf{r}') - \mathcal{A}_1(\mathbf{r}')\nabla' \cdot \mathcal{A}_2(\mathbf{r}')\}$$

$$- \int_{S'} dS' \,\hat{\mathbf{n}}(\mathbf{r}') \cdot [\nabla' \times \mathcal{A}_1(\mathbf{r}')] \times \mathcal{A}_2(\mathbf{r}')$$

$$+ \int_{S''} dS' \,\hat{\mathbf{n}}(\mathbf{r}') \cdot \{[\nabla' \times \mathcal{A}_2(\mathbf{r}')] \times \mathcal{A}_1(\mathbf{r}') - \mathcal{A}_1(\mathbf{r}')\nabla' \cdot \mathcal{A}_2(\mathbf{r}')\}$$

$$- \int_{S''} dS' \,\hat{\mathbf{n}}(\mathbf{r}') \cdot [\nabla' \times \mathcal{A}_1(\mathbf{r}')] \times \mathcal{A}_2(\mathbf{r}') \quad (5.21)$$

and we need to take the limit as $a \to 0$. Having assumed the current density strictly confined in $V_J \subset V$, the volume integral of $\mathcal{J}(\mathbf{r}')$ is not affected by the limiting procedure; otherwise, we may follow the strategy adopted for the singularity within V_J. In like manner the integrals over ∂V_1 are not affected.

The improper integrals over S' diverge for $a \to 0$ because some integrands contain terms proportional to $1/|\mathbf{r}-\mathbf{r}'|^2$ which cannot be integrated over a surface. Since the root cause of the problem is the singular behavior of the derivatives of the auxiliary potential $\mathcal{A}_2(\mathbf{r})$, we perform a little algebra before taking the limit for vanishing radius a.[1] To speed up the derivation we refer to (5.10)-(5.12) and write

$$\int_{S'} dS' \,\hat{\mathbf{n}}(\mathbf{r}') \cdot \{[\nabla' \times \mathcal{A}_2(\mathbf{r}')] \times \mathcal{A}_1(\mathbf{r}') - \mathcal{A}_1(\mathbf{r}')\nabla' \cdot \mathcal{A}_2(\mathbf{r}')\}$$

$$- \int_{S'} dS' \,\hat{\mathbf{n}}(\mathbf{r}') \cdot [\nabla' \times \mathcal{A}_1(\mathbf{r}')] \times \mathcal{A}_2(\mathbf{r}')$$

$$= \mu\mathbf{u} \cdot \int_{S'} dS' \,\nabla' \frac{1}{4\pi R} \times [\hat{\mathbf{n}}(\mathbf{r}') \times \mathcal{A}_1(\mathbf{r}')] - \mu\mathbf{u} \cdot \int_{S'} dS' \,\hat{\mathbf{n}}(\mathbf{r}') \cdot \mathcal{A}_1(\mathbf{r}')\nabla' \frac{1}{4\pi R}$$

$$- \mu\mathbf{u} \cdot \int_{S'} dS' \,\hat{\mathbf{n}}(\mathbf{r}') \times [\nabla' \times \mathcal{A}_1(\mathbf{r}')] \frac{1}{4\pi R}$$

$$= -\mu\mathbf{u} \cdot \nabla \times \int_{S'} dS' \frac{1}{4\pi R} [\hat{\mathbf{n}}(\mathbf{r}') \times \mathcal{A}_1(\mathbf{r}')] + \mu\mathbf{u} \cdot \nabla \int_{S'} dS' \,\hat{\mathbf{n}}(\mathbf{r}') \cdot \mathcal{A}_1(\mathbf{r}') \frac{1}{4\pi R}$$

$$- \mu\mathbf{u} \cdot \int_{S'} dS' \,\hat{\mathbf{n}}(\mathbf{r}') \times [\nabla' \times \mathcal{A}_1(\mathbf{r}')] \frac{1}{4\pi R} \quad (5.22)$$

[1]It is worthwhile mentioning that if the integrands on S' and S'' are first cast as in (5.43) further on, by virtue of (H.94) we can transform the troublesome terms into two line integrals over $\partial S' \equiv \partial S''$ that cancel one another out inasmuch as $\mathcal{A}_1(\mathbf{r})$ is continuous on $S' \cup S''$ and hence across $\partial S' \equiv \partial S''$.

where we have invoked the symmetry of the Green function (2.131) with respect to $\mathbf{r}$ and $\mathbf{r}'$, and interchanged the order of integration and differentiation since the integrands are regular for $(\mathbf{r}, \mathbf{r}') \in \partial V \times S'$. The limit for $a \to 0$ may now be taken as all the improper integrals exist and are finite.

As regards the integrals over S'' we may apply the mean value theorem [6] and obtain

$$\int_{S''} \mathrm{d}S'\, \hat{\mathbf{n}}(\mathbf{r}') \cdot [\nabla' \times \mathcal{A}_1(\mathbf{r}')] \times \mathcal{A}_2(\mathbf{r}') = \frac{\mu \mathbf{u}}{4\pi a} \cdot \int_{S''} \mathrm{d}S'\, \hat{\mathbf{n}}(\mathbf{r}') \times [\nabla' \times \mathcal{A}_1(\mathbf{r}')]$$

$$= a\mu \frac{\Omega(a)}{4\pi} \mathbf{u} \cdot \hat{\mathbf{n}}(\mathbf{r}_0') \times [\nabla' \times \mathcal{A}_1(\mathbf{r}')]\Big|_{\mathbf{r}'=\mathbf{r}_0'} \tag{5.23}$$

$$\int_{S''} \mathrm{d}S'\, \hat{\mathbf{n}}(\mathbf{r}') \cdot \{[\nabla' \times \mathcal{A}_2(\mathbf{r}')] \times \mathcal{A}_1(\mathbf{r}') - \mathcal{A}_1(\mathbf{r}')\nabla' \cdot \mathcal{A}_2(\mathbf{r}')\}$$

$$= \frac{\mu \mathbf{u}}{4\pi a^2} \cdot \int_{S''} \mathrm{d}S'\, \{\hat{\mathbf{n}}(\mathbf{r}') \times [\mathcal{A}_1(\mathbf{r}') \times \hat{\mathbf{n}}(\mathbf{r}')] + \hat{\mathbf{n}}(\mathbf{r}')\, \hat{\mathbf{n}}(\mathbf{r}') \cdot \mathcal{A}_1(\mathbf{r}')\}$$

$$= \mu \frac{\Omega(a)}{4\pi} \mathbf{u} \cdot \mathcal{A}_1(\mathbf{r} + a\hat{\mathbf{n}}(\mathbf{r}_0')) \tag{5.24}$$

where $\mathbf{r}_0' \in S''$ is a suitable point, and $\Omega(a)$ is the solid angle subtended by S'' with respect to $\mathbf{r} \in S$. The solid angle $\Omega(a)$ approaches 2π as $a \to 0$, as the surface S is smooth by hypothesis [see (F.8)]. This means that S is approximated infinitely well by the tangent plane in $\mathbf{r}' = \mathbf{r}$ and S'' tends to the half-sphere centered in $\mathbf{r}$ and contained in V.

So, by taking the limit, dividing through by μ, dropping the subscript 1, and invoking the arbitrariness of $\mathbf{u}$ we obtain

$$\frac{1}{2}\mathcal{A}(\mathbf{r}) = \mu \int_{V_J} \mathrm{d}V' \frac{\mathcal{J}(\mathbf{r}')}{4\pi R} + \int_{\partial V_1} \mathrm{d}S'\, \hat{\mathbf{n}}(\mathbf{r}') \times [\nabla' \times \mathcal{A}(\mathbf{r}')] \frac{1}{4\pi R}$$
$$+ \int_{\partial V_1} \mathrm{d}S'\, [\hat{\mathbf{n}}(\mathbf{r}') \times \mathcal{A}(\mathbf{r}')] \times \nabla' \frac{1}{4\pi R} + \int_{\partial V_1} \mathrm{d}S'\, \hat{\mathbf{n}}(\mathbf{r}') \cdot \mathcal{A}(\mathbf{r}') \nabla' \frac{1}{4\pi R}$$
$$+ \int_{S} \mathrm{d}S'\, \hat{\mathbf{n}}(\mathbf{r}') \times [\nabla' \times \mathcal{A}(\mathbf{r}')] \frac{1}{4\pi R} + \nabla \times \int_{S} \mathrm{d}S'\, [\hat{\mathbf{n}}(\mathbf{r}') \times \mathcal{A}(\mathbf{r}')] \frac{1}{4\pi R}$$
$$- \nabla \int_{S} \mathrm{d}S'\, \hat{\mathbf{n}}(\mathbf{r}') \cdot \mathcal{A}(\mathbf{r}') \frac{1}{4\pi R} \tag{5.25}$$

for points $\mathbf{r} \in S$. We need to modify the integrals over ∂V_1 in like fashion to allow for points $\mathbf{r} \in \partial V_1$. In summary, if the observation point is on the boundary of the domain of interest, the integral representation returns half the value of the vector potential.

The integral formulas (5.18) and (5.25) allow computing $\mathcal{A}(\mathbf{r})$ in a closed bounded domain provided we know the current density, the values of $\mathcal{A}(\mathbf{r})$ and of its tangential curl on the boundary. Furthermore, the undetermined scalar field χ in (4.32) was implicity specified when we selected the Coulomb gauge for $\mathcal{A}(\mathbf{r})$, so this condition is included in (5.18) and (5.25). We learned in Section 4.7 that the vector potential is uniquely determined if a gauge is selected, the normal component of $\mathcal{A}(\mathbf{r})$ is specified on the boundary of the solution domain and, lastly, *either* the tangential component *or* the tangential curl is imposed on the boundary. However, (5.18) and (5.25) require knowing both quantities on ∂V. It is expected that we may not specify $\hat{\mathbf{n}}(\mathbf{r}) \times \mathcal{A}(\mathbf{r})$ and $\hat{\mathbf{n}}(\mathbf{r}) \times [\nabla \times \mathcal{A}(\mathbf{r})]$ for $\mathbf{r} \in \partial V$ independently, because that may give rise to non-physical results. For this reason, (5.18) and (5.25) are integral representations of the vector potential rather than the solution to the Poisson equation.

In particular, the surface integrals may be construed as vector potentials produced by suitable layers of *equivalent sources* located on the boundary $S \cup \partial V_1$, though *inside* V. For instance, the integral involving the tangential curl of $\mathcal{A}(\mathbf{r}')$ possesses the same kernel as the volume integral of $\mathcal{J}(\mathbf{r}')$. Thus, we are led to interpret this contribution as the vector potential of a *fictitious* steady electric surface current density given by

$$\mathcal{J}_S(\mathbf{r}') := \frac{1}{\mu}\hat{\mathbf{n}}(\mathbf{r}') \times [\nabla' \times \mathcal{A}(\mathbf{r}')] = \hat{\mathbf{n}}(\mathbf{r}') \times \mathcal{H}(\mathbf{r}'), \qquad \mathbf{r}' \in S \cup \partial V_1 \tag{5.26}$$

in agreement with the jump condition (1.142). Comparing the surface integral of the tangential component of $\mathcal{A}(\mathbf{r}')$ with (4.88) suggests that we consider this contribution as the vector potential produced by a fictitious *layer of magnetic dipoles* (physical dimension: $\mathrm{Am}^2/\mathrm{m}^2 = \mathrm{A}$) with surface density

$$\mathcal{M}_S(\mathbf{r}') := \frac{1}{\mu}\hat{\mathbf{n}}(\mathbf{r}') \times \mathcal{A}(\mathbf{r}'), \qquad \mathbf{r}' \in S \cup \partial V_1 \tag{5.27}$$

and the dipoles are oriented parallel to the boundary. Finally, the meaning of the integral of $\hat{\mathbf{n}}(\mathbf{r}') \cdot \mathcal{A}(\mathbf{r}')$ is similar, though it is not easy to associate the normal component of the vector potential with a layer of magnetic dipoles and charges [1].

As was the case for the representation of the electrostatic potential (2.160), we expect the surface integrals in (5.18) to bring in the effect of sources other than $\mathcal{J}(\mathbf{r})$ in V_J (Figure 5.1a) as the latter is already explicitly accounted for. Furthermore, if no other sources are contemplated, then the surface integrals must vanish. Indeed, if we apply (5.18) to V_1, which we assume to be devoid of current and dipoles, we find

$$0 = \int_{\partial V_1} \mathrm{d}S'\, \hat{\mathbf{n}}(\mathbf{r}') \times [\nabla' \times \mathcal{A}(\mathbf{r}')]\frac{1}{4\pi R} + \int_{\partial V_1} \mathrm{d}S'\, [\hat{\mathbf{n}}(\mathbf{r}') \times \mathcal{A}(\mathbf{r}')] \times \nabla' \frac{1}{4\pi R} + \int_{\partial V_1} \mathrm{d}S'\, \hat{\mathbf{n}}(\mathbf{r}') \cdot \mathcal{A}(\mathbf{r}')\nabla' \frac{1}{4\pi R} \tag{5.28}$$

for observation points $\mathbf{r} \in V$. Since the right-hand side comprises the same surface integrals as in (5.18), we conclude that they are null if no currents are present in V_1. Similarly, we can show that the surface integrals over S are zero, as long as there are no other currents in $\mathbb{R}^3 \setminus (\overline{V} \cup \overline{V}_1)$.

Next, we extend (5.18) to an unbounded region of space filled with a homogeneous magnetic medium under the hypothesis that there are no other steady currents besides $\mathcal{J}(\mathbf{r})$ in V_J. In particular, we choose $V = B(0, a) \setminus V_1$ and take the limit as $a \to +\infty$. We observe that $\mathcal{A}(\mathbf{r})$ satisfies the asymptotic condition (4.36), whereby $\hat{\mathbf{n}} \times [\nabla \times \mathcal{A}(\mathbf{r})]$ decays with the inverse square of the distance away from the currents. With these estimates the surface integrals over ∂B may be proven to tend to zero as $a \to +\infty$. Additionally, if we assume that the medium filling V_1 has the same permeability μ and that no currents are placed in V_1, we do not have to exclude that region at all, and the integral representation reduces to

$$\mathcal{A}(\mathbf{r}) = \mu \int_{V_J} \mathrm{d}V' \frac{\mathcal{J}(\mathbf{r}')}{4\pi R} \tag{5.29}$$

for points $\mathbf{r} \in \mathbb{R}^3$.

We check that the integral representation (5.18) yields a vector potential which satisfies the Coulomb gauge by computing the divergence of each integral so as to show that they are solenoidal.

We start with the divergence of the volume integral of $\mathfrak{J}(\mathbf{r}')$ and observe that a key step is the interchange of integration and differentiation. The Cartesian components of said integral are of the same type as the volume potential discussed in Section 2.8 where, among other properties, we proved the validity of (2.199). By means of the same technique and (2.205) it remains proved that

$$\nabla \cdot \mu \int_{V_J} \mathrm{d}V' \frac{\mathfrak{J}(\mathbf{r}')}{4\pi R} = \mu \int_{V_J} \mathrm{d}V' \mathfrak{J}(\mathbf{r}') \cdot \nabla \frac{1}{4\pi R}, \qquad \mathbf{r} \in \mathbb{R}^3 \tag{5.30}$$

so long as $\mathfrak{J}(\mathbf{r}')$ is at least bounded for $\mathbf{r}' \in V_J$. Then, we distinguish two cases, namely, observation point $\mathbf{r} \in \mathbb{R}^3 \setminus \overline{V}_J$ and $\mathbf{r} \in V_J$. In the first case we have

$$\begin{aligned} \mu \int_{V_J} \mathrm{d}V' \mathfrak{J}(\mathbf{r}') \cdot \nabla \frac{1}{4\pi R} &= -\mu \int_{V_J} \mathrm{d}V' \mathfrak{J}(\mathbf{r}') \cdot \nabla' \frac{1}{4\pi R} = \mu \int_{V_J} \mathrm{d}V' \left[\frac{\nabla' \cdot \mathfrak{J}(\mathbf{r}')}{4\pi R} - \nabla' \cdot \frac{\mathfrak{J}(\mathbf{r}')}{4\pi R} \right] \\ &= -\mu \int_{\partial V_J} \mathrm{d}S' \frac{\hat{\mathbf{n}}(\mathbf{r}') \cdot \mathfrak{J}(\mathbf{r}')}{4\pi R} = 0 \end{aligned} \tag{5.31}$$

and the result follows because $\mathfrak{J}(\mathbf{r}')$ is solenoidal and confined within V_J. The Gauss theorem has been applied on the grounds that $\mathfrak{J}(\mathbf{r}')/R$ is regular for observation points $\mathbf{r} \in \mathbb{R}^3 \setminus \overline{V}_J$.

Since for $\mathbf{r} \in V_J$ the kernel is singular, we isolate the point $\mathbf{r}$ with a ball $B(\mathbf{r}, a) \subset V_J$ and split the integral into two parts, i.e., over $V_a := V_J \setminus B[\mathbf{r}, a]$ and over the ball (see Figure 2.8 for a similar setup)

$$\begin{aligned} \mu \int_{V_J} \mathrm{d}V' \mathfrak{J}(\mathbf{r}') \cdot \nabla \frac{1}{4\pi R} &= -\mu \int_{V_J} \mathrm{d}V' \mathfrak{J}(\mathbf{r}') \cdot \nabla' \frac{1}{4\pi R} \\ &= \mu \int_{V_a} \mathrm{d}V' \left[\frac{\nabla' \cdot \mathfrak{J}(\mathbf{r}')}{4\pi R} - \nabla' \cdot \frac{\mathfrak{J}(\mathbf{r}')}{4\pi R} \right] - \mu \int_{B(\mathbf{r},a)} \mathrm{d}V' \mathfrak{J}(\mathbf{r}') \cdot \frac{\hat{\mathbf{R}}}{4\pi R^2} \\ &= -\mu \int_{\partial B} \mathrm{d}S' \frac{\hat{\mathbf{n}}(\mathbf{r}') \cdot \mathfrak{J}(\mathbf{r}')}{4\pi R} - \mu \int_{B(\mathbf{r},a)} \mathrm{d}V' \mathfrak{J}(\mathbf{r}') \cdot \frac{\hat{\mathbf{R}}}{4\pi R^2} \end{aligned} \tag{5.32}$$

and we are left with the task of showing that the last two integrals vanish for $a \to 0^+$. We estimate

$$\left| \int_{\partial B} \mathrm{d}S' \frac{\hat{\mathbf{n}}(\mathbf{r}') \cdot \mathfrak{J}(\mathbf{r}')}{4\pi R} \right| \leqslant \int_{\partial B} \mathrm{d}S' \frac{\|\mathfrak{J}\|_\infty}{4\pi a} = a \|\mathfrak{J}\|_\infty \xrightarrow[a \to 0^+]{} 0 \tag{5.33}$$

and

$$\left| \int_{B(\mathbf{r},a)} \mathrm{d}V' \mathfrak{J}(\mathbf{r}') \cdot \frac{\hat{\mathbf{R}}}{4\pi R^2} \right| \leqslant \int_{B(\mathbf{r},a)} \mathrm{d}V' \frac{\|\mathfrak{J}\|_\infty}{4\pi R^2} = \|\mathfrak{J}\|_\infty \int_0^a \mathrm{d}R = a \|\mathfrak{J}\|_\infty \xrightarrow[a \to 0^+]{} 0 \tag{5.34}$$

whereby it remains proved that the volume integral in (5.18) or (5.29) is solenoidal.

As regards the surface integrals, we observe that interchanging derivatives and integral is possible as the integrands are continuously differentiable for observation points $\mathbf{r} \in V$ and source points

$\mathbf{r}' \in S \cup \partial V_1$; in particular, the kernel $1/R$ is not singular. For the surface integral of $\hat{\mathbf{n}}(\mathbf{r}') \times \mathcal{A}(\mathbf{r}')$ we have

$$\begin{aligned}\nabla \cdot \int_{S\cup\partial V_1} \mathrm{d}S' \, [\hat{\mathbf{n}}(\mathbf{r}') \times \mathcal{A}(\mathbf{r}')] \times \nabla' \frac{1}{4\pi R} &= \int_{S\cup\partial V_1} \mathrm{d}S' \, \nabla \cdot \left\{ [\hat{\mathbf{n}}(\mathbf{r}') \times \mathcal{A}(\mathbf{r}')] \times \nabla' \frac{1}{4\pi R} \right\} \\ &= \int_{S\cup\partial V_1} \mathrm{d}S' \, [\hat{\mathbf{n}}(\mathbf{r}') \times \mathcal{A}(\mathbf{r}')] \cdot \nabla' \times \nabla' \frac{1}{4\pi R} = 0 \end{aligned} \tag{5.35}$$

in light of (H.49) and (A.38). The same result holds for the integral of $\hat{\mathbf{n}}(\mathbf{r}') \cdot \mathcal{A}(\mathbf{r}')$ since $1/R$ is a harmonic function for $\mathbf{r}' \neq \mathbf{r}$.

For the surface integral of the tangential curl of $\mathcal{A}(\mathbf{r}')$ we have

$$\nabla \cdot \int_{S\cup\partial V_1} \mathrm{d}S' \, \frac{\hat{\mathbf{n}}(\mathbf{r}') \times [\nabla' \times \mathcal{A}(\mathbf{r}')]}{4\pi R} = -\int_{S\cup\partial V_1} \mathrm{d}S' \, \hat{\mathbf{n}}(\mathbf{r}') \times \mathcal{B}(\mathbf{r}') \cdot \nabla'_{\mathrm{s}} \frac{1}{4\pi R} \tag{5.36}$$

on account of (4.29), (H.51) and the fact that $\hat{\mathbf{n}}(\mathbf{r}') \times \mathcal{B}(\mathbf{r}')$ is a tangential vector field defined on $S \cup \partial V_1$. Next, in light of (H.77) we write

$$\begin{aligned} -\int_{S\cup\partial V_1} \mathrm{d}S' \, \hat{\mathbf{n}}(\mathbf{r}') \times \mathcal{B}(\mathbf{r}') \cdot \nabla'_{\mathrm{s}} \frac{1}{4\pi R} &= \int_{S\cup\partial V_1} \mathrm{d}S' \, \frac{\nabla'_{\mathrm{s}} \cdot [\hat{\mathbf{n}}(\mathbf{r}') \times \mathcal{B}(\mathbf{r}')]}{4\pi R} - \int_{S\cup\partial V_1} \mathrm{d}S' \, \nabla'_{\mathrm{s}} \cdot \frac{\hat{\mathbf{n}}(\mathbf{r}') \times \mathcal{B}(\mathbf{r}')}{4\pi R} \\ &= -\int_{S\cup\partial V_1} \mathrm{d}S' \, \frac{\hat{\mathbf{n}}(\mathbf{r}') \cdot \nabla' \times \mathcal{B}(\mathbf{r}')}{4\pi R} - 0 = 0 \end{aligned} \tag{5.37}$$

where we have used (H.82) and applied the surface Gauss theorem (A.59) to the closed surfaces S and ∂V_1. The remaining integral is zero because $\mathcal{B}(\mathbf{r}')$ is conservative on $S \cup \partial V_1$.

All in all, these partial results allow us to conclude that the divergence of the expression in the right-hand side of (5.18) is zero for $\mathbf{r} \in V$.

Thanks to (5.18) we can obtain a mean value theorem for the vector potential (see Section 2.7 for the analogous result in electrostatics). We suppose all the steady current densities are located outside a ball $B(\mathbf{r}, a)$ and apply (5.18)

$$\begin{aligned} \mathcal{A}(\mathbf{r}) &= \int_{\partial B} \mathrm{d}S' \, \hat{\mathbf{n}}(\mathbf{r}') \times [\nabla' \times \mathcal{A}(\mathbf{r}')] \frac{1}{4\pi R} \\ &\quad + \int_{\partial B} \mathrm{d}S' \, [\hat{\mathbf{n}}(\mathbf{r}') \times \mathcal{A}(\mathbf{r}')] \times \nabla' \frac{1}{4\pi R} + \int_{\partial B} \mathrm{d}S' \, \hat{\mathbf{n}}(\mathbf{r}') \cdot \mathcal{A}(\mathbf{r}') \nabla' \frac{1}{4\pi R} \\ &= \frac{1}{4\pi a} \int_{\partial B} \mathrm{d}S' \, \hat{\mathbf{n}}(\mathbf{r}') \times \mathcal{B}(\mathbf{r}') + \frac{1}{4\pi a^2} \int_{\partial B} \mathrm{d}S' \, \{ [\hat{\mathbf{n}}(\mathbf{r}') \times \mathcal{A}(\mathbf{r}')] \times \hat{\mathbf{n}}(\mathbf{r}') + \hat{\mathbf{n}}(\mathbf{r}') \cdot \mathcal{A}(\mathbf{r}') \hat{\mathbf{n}}(\mathbf{r}') \} \\ &= \frac{1}{4\pi a} \int_{B(\mathbf{r},a)} \mathrm{d}V' \, \nabla' \times \mathcal{B}(\mathbf{r}') + \frac{1}{4\pi a^2} \int_{\partial B} \mathrm{d}S' \, \mathcal{A}(\mathbf{r}') \end{aligned} \tag{5.38}$$

having invoked (4.29) and used (H.14), (H.91). The result follows by observing that the stationary magnetic induction field is curl-free in a region devoid of steady currents.

An alternative and somewhat faster way of finding an integral representation of the vector potential that solves problem (5.1) relies on the reduction of the vector Poisson equation to the familiar, more amenable scalar Poisson equation we encountered in electrostatics (Section 2.7). This step is based on the differential identity (H.59) and works better when we enforce the Coulomb gauge on $\mathcal{A}(\mathbf{r})$, viz.,

$$\nabla' \times [\nabla' \times \mathcal{A}(\mathbf{r}')] = \nabla'\nabla' \cdot \mathcal{A}(\mathbf{r}') - \nabla'^2\mathcal{A}(\mathbf{r}') = -\nabla'^2\mathcal{A}(\mathbf{r}') = \mu\mathcal{J}(\mathbf{r}') \tag{5.39}$$

which we want to solve subject to the same boundary conditions listed in (5.1).

Next, we would like to separate the vector equation above into three independent scalar equations for the components of $\mathcal{A}(\mathbf{r})$. While the splitting may, in principle, be achieved in any system of coordinates, the resulting equations are *decoupled* only for the Cartesian coordinates, because the fundamental vectors $\hat{\mathbf{x}}$, $\hat{\mathbf{y}}$ and $\hat{\mathbf{z}}$ do not depend on (x', y', z') themselves and, as a result, the Laplace operator applies only to $\mathcal{A}_x$, $\mathcal{A}_y$ and $\mathcal{A}_z$. In summary, we may cast the original problem into the following one

$$\begin{cases} \nabla'^2\mathcal{A}_\alpha = -\mu\mathcal{J}_\alpha(\mathbf{r}'), & \alpha \in \{x, y, z\} \qquad\qquad \mathbf{r}' \in V, \\ \hat{\mathbf{n}}(\mathbf{r}')\times\mathcal{A}(\mathbf{r}') = \mathbf{f}(\mathbf{r}') & \text{or} \quad \hat{\mathbf{n}}(\mathbf{r}')\times[\nabla'\times\mathcal{A}(\mathbf{r}')] = \mathbf{g}(\mathbf{r}'), \quad \mathbf{r}' \in \partial V \\ \hat{\mathbf{n}}(\mathbf{r}') \cdot \mathcal{A}(\mathbf{r}') = h(\mathbf{r}'), & \qquad\qquad\qquad\qquad\qquad\qquad \mathbf{r}' \in \partial V \end{cases} \tag{5.40}$$

with the Coulomb gauge implied. Since each Cartesian component of $\mathcal{A}(\mathbf{r}')$ now satisfies a scalar Poisson equation, we may just as well exploit the integral representation (2.160) three times with $\mathcal{A}_x$, $\mathcal{A}_y$ and $\mathcal{A}_z$ in lieu of Φ and μJ_x, μJ_y and μJ_z instead of ϱ/ε. Notice, though, that the boundary conditions in (5.40) still involve a mix of the three Cartesian components of $\mathcal{A}(\mathbf{r}')$. Finally, we combine the resulting expressions to obtain a single vector formula for $\mathcal{A}(\mathbf{r})$; again, this step is permissible, because $\hat{\mathbf{x}}$, $\hat{\mathbf{y}}$ and $\hat{\mathbf{z}}$ are constant vectors. In symbols, we have

$$\mathcal{A}(\mathbf{r}) = \mu \int_{V_J} \mathrm{d}V' \frac{\mathcal{J}(\mathbf{r}')}{4\pi R} + \int_{S\cup\partial V_1} \mathrm{d}S'\, \mathcal{A}(\mathbf{r}')\,\hat{\mathbf{n}}(\mathbf{r}') \cdot \nabla' \frac{1}{4\pi R} - \int_{S\cup\partial V_1} \mathrm{d}S' \frac{\hat{\mathbf{n}}(\mathbf{r}')}{4\pi R} \cdot \nabla'\mathcal{A}(\mathbf{r}') \tag{5.41}$$

for points $\mathbf{r} \in V$. The main drawback of this integral representation — which is fully equivalent to (5.18) — is that it does not explicitly involve the values of the vector potential and of the tangential curl thereof on $S \cup \partial V_1$. Besides, it is less obvious that $\mathcal{A}(\mathbf{r})$ as given by (5.41) satisfies the Coulomb gauge.

As quick as it was to derive (5.41) based on our knowledge of electrostatics, we need quite some algebra to cast the surface integrals back into the form they have in (5.18) or the other way round. We follow the latter strategy and, to keep the derivation lucid, we omit showing the dependence on $\mathbf{r}$ and $\mathbf{r}'$ and we write $\hat{\mathbf{n}}'$ as a shorthand for $\hat{\mathbf{n}}(\mathbf{r}')$. Starting with the integrands of the surface integrals in (5.18), we have

$$\begin{aligned} &\hat{\mathbf{n}}' \times (\nabla' \times \mathcal{A})\frac{1}{R} + (\hat{\mathbf{n}}' \times \mathcal{A}) \times \nabla'\frac{1}{R} + \hat{\mathbf{n}}' \cdot \mathcal{A}\nabla'\frac{1}{R} = \\ &\quad = \hat{\mathbf{n}}' \times \left(\nabla' \times \frac{\mathcal{A}}{R}\right) - \hat{\mathbf{n}}' \times \left(\nabla'\frac{1}{R} \times \mathcal{A}\right) + (\hat{\mathbf{n}}' \times \mathcal{A}) \times \nabla'\frac{1}{R} + \hat{\mathbf{n}}' \cdot \mathcal{A}\nabla'\frac{1}{R} \\ &\quad = \hat{\mathbf{n}}' \times \left(\nabla' \times \frac{\mathcal{A}}{R}\right) - \hat{\mathbf{n}}' \cdot \mathcal{A}\nabla'\frac{1}{R} + \mathcal{A}\hat{\mathbf{n}}' \cdot \nabla'\frac{1}{R} + \mathcal{A}\nabla'\frac{1}{R} \cdot \hat{\mathbf{n}}' - \hat{\mathbf{n}}'\nabla'\frac{1}{R} \cdot \mathcal{A} + \hat{\mathbf{n}}' \cdot \mathcal{A}\nabla'\frac{1}{R} \\ &\quad = \hat{\mathbf{n}}' \times \left(\nabla' \times \frac{\mathcal{A}}{R}\right) - \hat{\mathbf{n}}'\nabla' \cdot \frac{\mathcal{A}}{R} + 2\mathcal{A}\hat{\mathbf{n}}' \cdot \nabla'\frac{1}{R} \end{aligned} \tag{5.42}$$

where we have used (H.50), (H.14) and the fact that $\mathcal{A}(\mathbf{r}')$ satisfies the Coulomb gauge.[2]

[2]See the book by Stratton [1, Section 8.15] for a similar result concerning time-harmonic fields.

Secondly, we transform the first two terms in the rightmost-hand side above with the identity (H.57) applied to $\mathcal{A}/R$

$$
\begin{aligned}
\hat{\mathbf{n}}' \times \left(\nabla' \times \frac{\mathcal{A}}{R}\right) - \hat{\mathbf{n}}'\nabla' \cdot \frac{\mathcal{A}}{R} + 2\mathcal{A}\hat{\mathbf{n}}' \cdot \nabla' \frac{1}{R} = \\
&= (\hat{\mathbf{n}}' \times \nabla') \times \frac{\mathcal{A}}{R} - \hat{\mathbf{n}}' \cdot \nabla' \frac{\mathcal{A}}{R} + 2\mathcal{A}\hat{\mathbf{n}}' \cdot \nabla' \frac{1}{R} \\
&= (\hat{\mathbf{n}}' \times \nabla') \times \frac{\mathcal{A}}{R} - \left(\hat{\mathbf{n}}' \cdot \nabla' \frac{1}{R}\right)\mathcal{A} - \frac{\hat{\mathbf{n}}'}{R} \cdot \nabla'\mathcal{A} + 2\mathcal{A}\hat{\mathbf{n}}' \cdot \nabla' \frac{1}{R} \\
&= (\hat{\mathbf{n}}' \times \nabla') \times \frac{\mathcal{A}}{R} + \mathcal{A}\hat{\mathbf{n}}' \cdot \nabla' \frac{1}{R} - \frac{1}{R}\hat{\mathbf{n}}' \cdot \nabla'\mathcal{A}
\end{aligned}
\tag{5.43}
$$

where we have also employed the identity (H.52). The surface integral of the first term above vanishes

$$
\int\limits_{S \cup \partial V_1} \mathrm{d}S' \; [\hat{\mathbf{n}}(\mathbf{r}') \times \nabla'] \times \frac{\mathcal{A}(\mathbf{r}')}{4\pi R} = 0 \tag{5.44}
$$

thanks to (H.94) [8, Eq. (A1.144)] applied on S and ∂V_1, which are closed surfaces. Finally, the last two terms in (5.43) are the same as in the surface integrals in (5.41) aside from the immaterial multiplicative factor $1/(4\pi)$. Thus, with these calculations the equivalence between (5.41) and (5.18) is established. Moreover, the surface integrals in (5.41) remain finite even for $\mathbf{r} \in \partial V$ essentially because the singularity of $\hat{\mathbf{n}}(\mathbf{r}') \cdot \nabla'(1/R)$ is integrable in light of (F.1) [cf. (2.256) and (2.257)].

5.1.2 *Magnetic induction and magnetic field*

In principle, an integral representation of $\mathcal{B}(\mathbf{r})$ in a homogeneous magnetic medium follows from (5.18) and (5.25) by taking the curl of both sides on account of (4.29). However, the calculations are far from trivial, as it is not evident how the vector potential on the boundary ∂V is related to the corresponding values of $\mathcal{B}(\mathbf{r})$ thereon. A better and faster approach consists of solving the magnetic stationary equations directly in combination with an auxiliary problem, as we did in Section 5.1.1.

Suppose we wish to find an explicit expression for the magnetic induction field which solves the problem (see Figure 5.1a)

$$
\begin{cases}
\nabla' \times \mathcal{B}_1(\mathbf{r}') = \nabla' \times [\nabla' \times \mathcal{A}_1(\mathbf{r}')] = \mu\mathcal{J}(\mathbf{r}'), & \mathbf{r}' \in V \\
\nabla' \cdot \mathcal{B}_1(\mathbf{r}') = 0 = \nabla' \cdot \mathcal{A}_1(\mathbf{r}'), & \mathbf{r}' \in V \\
\hat{\mathbf{n}}(\mathbf{r}') \cdot \mathcal{B}_1(\mathbf{r}') = f(\mathbf{r}') \quad \text{or} \quad \hat{\mathbf{n}}(\mathbf{r}') \times \mathcal{B}_1(\mathbf{r}') = \mathbf{g}(\mathbf{r}'), & \mathbf{r}' \in \partial V
\end{cases}
\tag{5.45}
$$

with $f(\mathbf{r}')$ and $\mathbf{g}(\mathbf{r}')$ regular scalar and vector fields on $\partial V = S \cup \partial V_1$. We have indicated the vector potential $\mathcal{A}_1(\mathbf{r})$ and the choice of the Coulomb gauge.

For the auxiliary problem we consider the magnetic induction generated by an elementary magnetic dipole of moment $\mathbf{m}$ located at the observation point $\mathbf{r} \in V \setminus V_J$ in a homogeneous isotropic magnetic medium with permeability μ. The corresponding vector and scalar magnetic potentials (see Section 4.5) read

$$
\mathcal{A}_2(\mathbf{r}') = \mu\mathbf{m} \times \nabla' \frac{1}{4\pi R} \tag{5.46}
$$

$$
\Psi_2(\mathbf{r}') = \mu\mathbf{m} \cdot \nabla' \frac{1}{4\pi R} \tag{5.47}
$$

with $R = |\mathbf{r} - \mathbf{r}'|$. Finally, the stationary magnetic equations for the dipole are

$$\begin{cases} \nabla' \times \mathcal{B}_2(\mathbf{r}') = \nabla' \times [\nabla' \times \mathcal{A}_2(\mathbf{r}')] = 0, & \mathbf{r}' \in \mathbb{R}^3 \\ \nabla' \cdot \mathcal{B}_2(\mathbf{r}') = -\nabla'^2 \Psi_2(\mathbf{r}') = 0, & \mathbf{r}' \in \mathbb{R}^3 \setminus \{\mathbf{r}\} \qquad \mathbf{r} \in V \setminus V_J \\ \mathcal{B}_2(\mathbf{r}') = \mathcal{O}\left(\dfrac{1}{|\mathbf{r}'|^2}\right), & |\mathbf{r}'| \to +\infty \end{cases} \tag{5.48}$$

that is, $\mathcal{B}_2(\mathbf{r}')$ is conservative in the whole space and solenoidal everywhere except at the location of the dipole, where it is singular. There is no use in combining the vector Poisson equations as in Section 5.1.1 because the source term for the auxiliary problem actually appears as a singular point for the magnetic scalar potential. Thus, we isolate the dipole with a ball $B(\mathbf{r}, a)$ and observe that

$$\nabla' \times \mathcal{A}_2(\mathbf{r}') = -\nabla' \Psi_2(\mathbf{r}'), \qquad \mathbf{r}' \in \mathbb{R}^3 \setminus B(\mathbf{r}, a) \tag{5.49}$$

based on the findings of Section 4.5. We proceed by dot-multiplying the first of (5.45) with $\mathcal{A}_2(\mathbf{r}')$

$$\begin{aligned} \mu \mathcal{J}_1(\mathbf{r}') \cdot \mathcal{A}_2(\mathbf{r}') &= \mathcal{A}_2(\mathbf{r}') \cdot \nabla' \times \mathcal{B}_1(\mathbf{r}') \\ &= \nabla' \cdot [\mathcal{B}_1(\mathbf{r}') \times \mathcal{A}_2(\mathbf{r}')] + \nabla' \times \mathcal{A}_2(\mathbf{r}') \cdot \mathcal{B}_1(\mathbf{r}') \\ &= \nabla' \cdot [\mathcal{B}_1(\mathbf{r}') \times \mathcal{A}_2(\mathbf{r}')] - \mathcal{B}_1(\mathbf{r}') \cdot \nabla' \Psi_2(\mathbf{r}') \\ &= -\nabla' \cdot [\mathcal{A}_2(\mathbf{r}') \times \mathcal{B}_1(\mathbf{r}') + \mathcal{B}_1(\mathbf{r}') \Psi_2(\mathbf{r}')] \end{aligned} \tag{5.50}$$

which holds for $\mathbf{r}' \in V \setminus B(\mathbf{r}, a)$ on account of (5.49). Then, we integrate this expression over $V \setminus B(\mathbf{r}, a)$

$$\mu \int_{V_J} \mathrm{d}V' \, \mathcal{J}_1(\mathbf{r}') \cdot \mathcal{A}_2(\mathbf{r}') = -\int_{V \setminus B(\mathbf{r},a)} \mathrm{d}V' \, \nabla' \cdot [\mathcal{A}_2(\mathbf{r}') \times \mathcal{B}_1(\mathbf{r}') + \Psi_2(\mathbf{r}') \mathcal{B}_1(\mathbf{r}')] \tag{5.51}$$

where the domain integral in the left-hand side has been restricted to the flux tube of $\mathcal{J}_1(\mathbf{r}')$. We apply the Gauss theorem in the right-hand side (the unit normal points inwards $V \setminus B(\mathbf{r}, a)$)

$$\begin{aligned} \mu \int_{V_J} \mathrm{d}V' \, \mathcal{J}_1(\mathbf{r}') \cdot \mathcal{A}_2(\mathbf{r}') = &\int_{S \cup \partial V_1} \mathrm{d}S' \, \hat{\mathbf{n}}(\mathbf{r}') \cdot [\mathcal{A}_2(\mathbf{r}') \times \mathcal{B}_1(\mathbf{r}') + \Psi_2(\mathbf{r}') \mathcal{B}_1(\mathbf{r}')] \\ &+ \int_{\partial B} \mathrm{d}S' \, \hat{\mathbf{n}}(\mathbf{r}') \cdot [\mathcal{A}_2(\mathbf{r}') \times \mathcal{B}_1(\mathbf{r}') + \Psi_2(\mathbf{r}') \mathcal{B}_1(\mathbf{r}')] \end{aligned} \tag{5.52}$$

because $\mathcal{A}_2(\mathbf{r}')$, $\Psi_2(\mathbf{r}')$ and $\mathcal{B}_1(\mathbf{r}')$ are continuously differentiable in $V \setminus B(\mathbf{r}, a)$. To finalize the result, we need to insert the explicit expressions of the potentials of the dipole, take the limit as $a \to 0$, and get rid of the inessential dipole moment $\mathbf{m}$. For the integrals over V_J and $S \cup \partial V_1$ we have

$$\begin{aligned} \mu \int_{V_J} \mathrm{d}V' \, \mathcal{A}_2(\mathbf{r}') \cdot \mathcal{J}_1(\mathbf{r}') &= \mu^2 \int_{V_J} \mathrm{d}V' \left(\mathbf{m} \times \nabla' \frac{1}{4\pi R} \right) \cdot \mathcal{J}_1(\mathbf{r}') \\ &= \mu^2 \mathbf{m} \cdot \int_{V_J} \mathrm{d}V' \, \nabla' \frac{1}{4\pi R} \times \mathcal{J}_1(\mathbf{r}') \end{aligned} \tag{5.53}$$

and

$$\int_{S \cup \partial V_1} \mathrm{d}S' \, \hat{\mathbf{n}}(\mathbf{r}') \cdot [\mathcal{A}_2(\mathbf{r}') \times \mathcal{B}_1(\mathbf{r}') + \Psi_2(\mathbf{r}') \mathcal{B}_1(\mathbf{r}')] =$$

$$= \mu \int_{S\cup\partial V_1} dS' \left(\mathbf{m}\times\nabla'\frac{1}{4\pi R}\right)\cdot \mathcal{B}_1(\mathbf{r}')\times\hat{\mathbf{n}}(\mathbf{r}') + \mu \int_{S\cup\partial V_1} dS' \left(\mathbf{m}\cdot\nabla'\frac{1}{4\pi R}\right)\hat{\mathbf{n}}(\mathbf{r}')\cdot\mathcal{B}_1(\mathbf{r}')$$

$$= \mu\mathbf{m}\cdot \int_{S\cup\partial V_1} dS' \, [\hat{\mathbf{n}}(\mathbf{r}')\times\mathcal{B}_1(\mathbf{r}')]\times\nabla'\frac{1}{4\pi R} + \mu\mathbf{m}\cdot\int_{S\cup\partial V_1} dS' \, \hat{\mathbf{n}}(\mathbf{r}')\cdot\mathcal{B}_1(\mathbf{r}')\nabla'\frac{1}{4\pi R}. \tag{5.54}$$

As regards the flux integral over ∂B, for $\mathbf{r}' \in \partial B$ we have the relations

$$\mathbf{r}' = \mathbf{r} + a\hat{\mathbf{n}}(\mathbf{r}'), \qquad \Psi_2(\mathbf{r}') = -\mu\mathbf{m}\cdot\frac{\hat{\mathbf{n}}(\mathbf{r}')}{4\pi a^2}, \qquad \mathcal{A}_2(\mathbf{r}') = -\mu\mathbf{m}\times\frac{\hat{\mathbf{n}}(\mathbf{r}')}{4\pi a^2} \tag{5.55}$$

whereby it follows

$$\int_{\partial B} dS' \, \hat{\mathbf{n}}(\mathbf{r}')\cdot[\mathcal{A}_2(\mathbf{r}')\times\mathcal{B}_1(\mathbf{r}') + \Psi_2(\mathbf{r}')\mathcal{B}_1(\mathbf{r}')] =$$

$$= -\frac{\mu}{4\pi a^2}\int_{\partial B} dS \, [\mathbf{m}\times\hat{\mathbf{n}}(\mathbf{r}')\cdot\mathcal{B}_1(\mathbf{r}')\times\hat{\mathbf{n}}(\mathbf{r}') + \mathbf{m}\cdot\hat{\mathbf{n}}(\mathbf{r}')\hat{\mathbf{n}}(\mathbf{r}')\cdot\mathcal{B}_1(\mathbf{r}')]$$

$$= -\frac{\mu\mathbf{m}}{4\pi a^2}\cdot\int_{\partial B} dS \, \{\hat{\mathbf{n}}(\mathbf{r}')\times[\mathcal{B}_1(\mathbf{r}')\times\hat{\mathbf{n}}(\mathbf{r}')] + \hat{\mathbf{n}}(\mathbf{r}')\hat{\mathbf{n}}(\mathbf{r}')\cdot\mathcal{B}_1(\mathbf{r}')\}$$

$$= -\frac{\mu\mathbf{m}}{4\pi a^2}\cdot\int_{\partial B} dS \, \mathcal{B}_1(\mathbf{r}') = -\mu\mathbf{m}\cdot\mathcal{B}_1(\mathbf{r}+a\hat{\mathbf{n}}(\mathbf{r}_0')) \underset{a\to 0}{\longrightarrow} -\mu\mathbf{m}\cdot\mathcal{B}_1(\mathbf{r}) \tag{5.56}$$

by virtue of the mean value theorem [6].

We put all these intermediate results together, divide through by μ and drop the subscript 1. Since the dipole moment $\mathbf{m}$ is arbitrary, we can take it oriented along $\hat{\mathbf{x}}$, $\hat{\mathbf{y}}$ and $\hat{\mathbf{z}}$ in succession to obtain three scalar relations for $\mathcal{B}_x(\mathbf{r})$, $\mathcal{B}_y(\mathbf{r})$ and $\mathcal{B}_z(\mathbf{r})$. In the end, we may write the result as a single vector expression for $\mathcal{B}(\mathbf{r})$ [1], namely,

$$\mathcal{B}(\mathbf{r}) = \mu\int_{V_J} dV' \, \mathcal{J}(\mathbf{r}')\times\nabla'\frac{1}{4\pi R} + \int_{S\cup\partial V_1} dS' \, \hat{\mathbf{n}}(\mathbf{r}')\cdot\mathcal{B}(\mathbf{r}')\nabla'\frac{1}{4\pi R}$$
$$+ \int_{S\cup\partial V_1} dS' \, [\hat{\mathbf{n}}(\mathbf{r}')\times\mathcal{B}(\mathbf{r}')]\times\nabla'\frac{1}{4\pi R} \tag{5.57}$$

which is the desired integral representation of the magnetic induction valid for points $\mathbf{r} \in V \setminus V_J$. In light of the constitutive relationship (1.118), we also have the integral representation of the magnetic field

$$\mathcal{H}(\mathbf{r}) = \int_{V_J} dV' \, \mathcal{J}(\mathbf{r}')\times\nabla'\frac{1}{4\pi R}$$
$$+ \int_{S\cup\partial V_1} dS' \left\{\frac{1}{\mu}\hat{\mathbf{n}}(\mathbf{r}')\cdot\mathcal{B}(\mathbf{r}')\nabla'\frac{1}{4\pi R} + [\hat{\mathbf{n}}(\mathbf{r}')\times\mathcal{H}(\mathbf{r}')]\times\nabla'\frac{1}{4\pi R}\right\} \tag{5.58}$$

again for $\mathbf{r} \in V \setminus V_J$. The extension to points $\mathbf{r} \in V_J$ is analogous to that of (5.18). In particular, for the integral over $V_J \setminus B(\mathbf{r}, a)$, if the current density $\mathcal{J}(\mathbf{r}')$ is continuous, we have the estimates

$$\left| \int_{V_J \setminus B(\mathbf{r},a)} dV' \, \nabla' \frac{1}{4\pi R} \times \mathcal{J}(\mathbf{r}') \right| \leqslant \int_{V_J \setminus B(\mathbf{r},a)} dV' \frac{|\mathcal{J}(\mathbf{r}')|}{4\pi R^2} \leqslant \int_{V_J \setminus B(\mathbf{r},a)} dV' \frac{\|\mathcal{J}\|_\infty}{4\pi R^2} \leqslant \int_{B(\mathbf{r},b) \setminus B(\mathbf{r},a)} dV' \frac{\|\mathcal{J}\|_\infty}{4\pi R^2}$$
$$= \|\mathcal{J}\|_\infty \int_a^b dR = \|\mathcal{J}\|_\infty (b - a) \leqslant \|\mathcal{J}\|_\infty b \tag{5.59}$$

which proves that the integral remains bounded for any value of a.

By means of arguments similar to those laid down in Section 4.7.1 it is not difficult to show that (5.57) and (5.58) return zero for points outside V. Provided the derivatives of the kernel are brought outside the surface integrals first, (5.57) and (5.58) yield half the value of $\mathcal{B}(\mathbf{r})$ and $\mathcal{H}(\mathbf{r})$, respectively, for points $\mathbf{r} \in S \cup \partial V_1$. Alternatively, we can transform the integrands as in (5.43)

$$\underbrace{\hat{\mathbf{n}}' \times (\nabla' \times \mathcal{B}) \frac{1}{R}}_{=0} + (\hat{\mathbf{n}}' \times \mathcal{B}) \times \nabla' \frac{1}{R} + \hat{\mathbf{n}}' \cdot \mathcal{B} \nabla' \frac{1}{R}$$
$$= (\hat{\mathbf{n}}' \times \nabla') \times \frac{\mathcal{B}}{R} + \mathcal{B} \hat{\mathbf{n}}' \cdot \nabla' \frac{1}{R} - \frac{\hat{\mathbf{n}}'}{R} \cdot \nabla' \mathcal{B} \tag{5.60}$$

keeping into account that $\nabla' \times \mathcal{B} = 0$ for $\mathbf{r}' \in S \cup \partial V_1$ by hypothesis. Then, the first contribution in the right-hand side integrates to zero by virtue of (H.94) as S and ∂V_1 are closed, and the remaining terms are integrable over a surface. Furthermore, we can convince ourselves that the surface integrals represent the effect of steady current densities and magnetic media located in the complementary domain and thus, if none are present outside V, the surface integrals must vanish. Finally, in the absence of other sources, the boundary S can be brought to infinity in light of estimate (4.37) and, if the medium that fills V_1 is characterized by the same permeability μ as the material in $\mathbb{R}^3 \setminus \overline{V}_1$, the contribution from ∂V_1 vanishes as well. All in all, we have

$$\mathcal{B}(\mathbf{r}) = \mu \int_{V_J} dV' \, \mathcal{J}(\mathbf{r}') \times \nabla' \frac{1}{4\pi R}, \qquad \mathbf{r} \in \mathbb{R}^3 \tag{5.61}$$

which is often referred to as the **law of Biot and Savart** (1820) [4,9,10], [11, Section 10.2], although strictly speaking they obtained the result for straight currents.

Expressions (5.57) and (5.58) provide the magnetic induction and the magnetic field, respectively, in terms of the current density in the region of interest and the boundary values of the magnetic entities. It is important to realize, though, that both the normal and the tangential components of $\mathcal{B}(\mathbf{r})$ or $\mathcal{H}(\mathbf{r})$ are involved, whereas we know from Section 4.7.2 that to ensure uniqueness only $\hat{\mathbf{n}}(\mathbf{r}) \cdot \mathcal{B}(\mathbf{r})$ or $\hat{\mathbf{n}}(\mathbf{r}) \times \mathcal{B}(\mathbf{r})$ needs to be prescribed on the boundary of the domain.

All the integrals (volume and surface as well) in (5.57) and (5.58) involve the same kernel which we recognize as the gradient of the fundamental solution to the three-dimensional scalar Poisson equation (2.124). Therefore, we are led to interpret $\hat{\mathbf{n}}(\mathbf{r}') \times \mathcal{H}(\mathbf{r}')$ for $\mathbf{r}' \in S \cup \partial V_1$ as an equivalent (fictitious) electric surface current density. Similarly, we may argue that $\hat{\mathbf{n}}(\mathbf{r}') \cdot \mathcal{B}(\mathbf{r}')$ for $\mathbf{r}' \in S \cup \partial V_1$ plays the role of an equivalent (really fictitious) surface density of *magnetic charges*.

As an application of the integral representation of $\mathcal{B}(\mathbf{r})$ we can obtain a mean value theorem for the value of the magnetic induction at the center of a ball $B(\mathbf{r}, a)$. We assume, as is customary, that all the steady currents are located outside the ball and apply (5.57), viz.,

$$\begin{aligned}\mathcal{B}(\mathbf{r}) &= \int_{\partial B} \mathrm{d}S'\, \hat{\mathbf{n}}(\mathbf{r}') \cdot \mathcal{B}(\mathbf{r}')\nabla' \frac{1}{4\pi R} + \int_{\partial B} \mathrm{d}S'\, [\hat{\mathbf{n}}(\mathbf{r}') \times \mathcal{B}(\mathbf{r}')] \times \nabla' \frac{1}{4\pi R} \\ &= \frac{1}{4\pi a^2} \int_{\partial B} \mathrm{d}S'\, \{\hat{\mathbf{n}}(\mathbf{r}') \cdot \mathcal{B}(\mathbf{r}')\hat{\mathbf{n}}(\mathbf{r}') + [\hat{\mathbf{n}}(\mathbf{r}') \times \mathcal{B}(\mathbf{r}')] \times \hat{\mathbf{n}}(\mathbf{r}')\} \\ &= \frac{1}{4\pi a^2} \int_{\partial B} \mathrm{d}S'\, \mathcal{B}(\mathbf{r}')\end{aligned} \tag{5.62}$$

by virtue of (H.14). The value of $\mathcal{B}(\mathbf{r})$ at the center of the ball is the mean of the values taken on the sphere ∂B. Of course, a perfectly analogous result holds true for the magnetic field $\mathcal{H}(\mathbf{r})$.

5.1.3 Vector potential and magnetic entities in an anisotropic medium

The results of Section 5.1.1 can be extended to the case of a homogeneous magnetic medium endowed with *symmetric* dyadic permeability $\overline{\mu}$. For the sake of simplicity we restrict the analysis to the magnetic potential produced in the whole space by a finite steady current $\mathcal{J}(\mathbf{r})$ with $\mathbf{r} \in V_J$. The following problem

$$\begin{cases}\nabla \times \left[\overline{\mu}^{-1} \cdot \mathcal{B}(\mathbf{r})\right] = \mathcal{J}(\mathbf{r}), & \mathbf{r} \in \mathbb{R}^3 \\ \nabla \cdot \mathcal{B}(\mathbf{r}) = 0, & \mathbf{r} \in \mathbb{R}^3 \\ \mathcal{B}(\mathbf{r}) = \mathcal{O}\left(\dfrac{1}{|\mathbf{r}|^2}\right), & |\mathbf{r}| \to +\infty\end{cases} \tag{5.63}$$

admits a unique solution, in accordance with the findings of Section 4.7.2. By introducing the magnetic vector potential as in (4.29) we transform the Ampère law into the vector Poisson equation

$$\begin{cases}\nabla \times \left[\overline{\mu}^{-1} \cdot \nabla \times \mathcal{A}(\mathbf{r})\right] = \mathcal{J}(\mathbf{r}), & \mathbf{r} \in \mathbb{R}^3 \\ \mathcal{A}(\mathbf{r}) = \mathcal{O}\left(\dfrac{1}{|\mathbf{r}|}\right), & |\mathbf{r}| \to +\infty\end{cases} \tag{5.64}$$

though this problem is not fully equivalent to (5.63), because at this stage $\mathcal{A}(\mathbf{r})$ is determined up to an additional conservative field $\nabla\chi$ as in (4.32). We can exploit this degree of freedom to facilitate the solution of (5.64) but we first need a little algebra to figure out a convenient condition to impose on $\mathcal{A}(\mathbf{r})$. Ideally, this would be a more general Coulomb gauge.

We assume that $\overline{\mu}$ is a *symmetric positive definite dyadic*, namely,

$$\mathbf{u} \cdot \overline{\mu} \cdot \mathbf{u} > 0, \qquad \mathbf{u} \in \mathbb{R}^3 \setminus \{0\} \tag{5.65}$$

so that the eigenvalues of $\overline{\mu}$ are strictly positive. Therefore, without loss of generality we can always choose a system of coordinates in which the permeability is *diagonal*, viz.,

$$\overline{\mu} = \mu_{xx}\hat{\mathbf{x}}\hat{\mathbf{x}} + \mu_{yy}\hat{\mathbf{y}}\hat{\mathbf{y}} + \mu_{zz}\hat{\mathbf{z}}\hat{\mathbf{z}} \tag{5.66}$$

in order to facilitate the solution of (5.64).

We begin by expanding the differential operators with the aid of (A.32)

$$\overline{\mu}^{-1} \cdot \nabla \times \mathcal{A}(\mathbf{r}) = \frac{\hat{\mathbf{x}}}{\mu_{xx}}\left(\frac{\partial \mathcal{A}_z}{\partial y} - \frac{\partial \mathcal{A}_y}{\partial z}\right) + \frac{\hat{\mathbf{y}}}{\mu_{yy}}\left(\frac{\partial \mathcal{A}_x}{\partial z} - \frac{\partial \mathcal{A}_z}{\partial x}\right) + \frac{\hat{\mathbf{z}}}{\mu_{zz}}\left(\frac{\partial \mathcal{A}_y}{\partial x} - \frac{\partial \mathcal{A}_x}{\partial y}\right) \tag{5.67}$$

$$
\begin{aligned}
\mathcal{J}(\mathbf{r}) &= \nabla\times\left[\overline{\mu}^{-1}\cdot\nabla\times\mathcal{A}(\mathbf{r})\right] \\
&= -\frac{\hat{\mathbf{x}}}{\mu_{yy}\mu_{zz}}\left[\mu_{yy}\frac{\partial^2\mathcal{A}_x}{\partial^2 y}+\mu_{zz}\frac{\partial^2\mathcal{A}_x}{\partial^2 z}-\frac{\partial}{\partial x}\left(\mu_{yy}\frac{\partial\mathcal{A}_y}{\partial y}+\mu_{zz}\frac{\partial\mathcal{A}_z}{\partial z}\right)\right] \\
&\quad -\frac{\hat{\mathbf{y}}}{\mu_{xx}\mu_{zz}}\left[\mu_{zz}\frac{\partial^2\mathcal{A}_y}{\partial^2 z}+\mu_{xx}\frac{\partial^2\mathcal{A}_y}{\partial^2 x}-\frac{\partial}{\partial y}\left(\mu_{zz}\frac{\partial\mathcal{A}_z}{\partial z}+\mu_{xx}\frac{\partial\mathcal{A}_x}{\partial x}\right)\right] \\
&\quad -\frac{\hat{\mathbf{z}}}{\mu_{xx}\mu_{yy}}\left[\mu_{xx}\frac{\partial^2\mathcal{A}_z}{\partial^2 x}+\mu_{yy}\frac{\partial^2\mathcal{A}_z}{\partial^2 y}-\frac{\partial}{\partial z}\left(\mu_{xx}\frac{\partial\mathcal{A}_x}{\partial x}+\mu_{yy}\frac{\partial\mathcal{A}_y}{\partial y}\right)\right].
\end{aligned}
\tag{5.68}
$$

We provisionally separate the latter equation into three scalar identities, viz.,

$$
\begin{aligned}
-\mu_{yy}\mu_{zz}\mathcal{J}_x(\mathbf{r}) &= -\frac{\det\overline{\mu}}{\mu_{xx}}\mathcal{J}_x(\mathbf{r}) = -\hat{\mathbf{x}}\cdot\overline{\mu}^{-1}\cdot\mathcal{J}(\mathbf{r})\det\overline{\mu} \\
&= \mu_{xx}\frac{\partial^2\mathcal{A}_x}{\partial^2 x}+\mu_{yy}\frac{\partial^2\mathcal{A}_x}{\partial^2 y}+\mu_{zz}\frac{\partial^2\mathcal{A}_x}{\partial^2 z}-\frac{\partial}{\partial x}\left(\mu_{xx}\frac{\partial\mathcal{A}_x}{\partial x}+\mu_{yy}\frac{\partial\mathcal{A}_y}{\partial y}+\mu_{zz}\frac{\partial\mathcal{A}_z}{\partial z}\right) \\
&= \nabla\cdot[\overline{\mu}\cdot\nabla\mathcal{A}_x(\mathbf{r})]-\hat{\mathbf{x}}\cdot\nabla\nabla\cdot[\overline{\mu}\cdot\mathcal{A}(\mathbf{r})]
\end{aligned}
\tag{5.69}
$$

$$
\begin{aligned}
-\mu_{zz}\mu_{xx}\mathcal{J}_y(\mathbf{r}) &= -\frac{\det\overline{\mu}}{\mu_{yy}}\mathcal{J}_y(\mathbf{r}) = -\hat{\mathbf{y}}\cdot\overline{\mu}^{-1}\cdot\mathcal{J}(\mathbf{r})\det\overline{\mu} \\
&= \mu_{xx}\frac{\partial^2\mathcal{A}_y}{\partial^2 x}+\mu_{yy}\frac{\partial^2\mathcal{A}_y}{\partial^2 y}+\mu_{zz}\frac{\partial^2\mathcal{A}_y}{\partial^2 z}-\frac{\partial}{\partial y}\left(\mu_{xx}\frac{\partial\mathcal{A}_x}{\partial x}+\mu_{yy}\frac{\partial\mathcal{A}_y}{\partial y}+\mu_{zz}\frac{\partial\mathcal{A}_z}{\partial z}\right) \\
&= \nabla\cdot\left[\overline{\mu}\cdot\nabla\mathcal{A}_y(\mathbf{r})\right]-\hat{\mathbf{y}}\cdot\nabla\nabla\cdot[\overline{\mu}\cdot\mathcal{A}(\mathbf{r})]
\end{aligned}
\tag{5.70}
$$

$$
\begin{aligned}
-\mu_{xx}\mu_{yy}\mathcal{J}_z(\mathbf{r}) &= -\frac{\det\overline{\mu}}{\mu_{zz}}\mathcal{J}_z(\mathbf{r}) = -\hat{\mathbf{z}}\cdot\overline{\mu}^{-1}\cdot\mathcal{J}(\mathbf{r})\det\overline{\mu} \\
&= \mu_{xx}\frac{\partial^2\mathcal{A}_z}{\partial^2 x}+\mu_{yy}\frac{\partial^2\mathcal{A}_z}{\partial^2 y}+\mu_{zz}\frac{\partial^2\mathcal{A}_z}{\partial^2 z}-\frac{\partial}{\partial z}\left(\mu_{xx}\frac{\partial\mathcal{A}_x}{\partial x}+\mu_{yy}\frac{\partial\mathcal{A}_y}{\partial y}+\mu_{zz}\frac{\partial\mathcal{A}_z}{\partial z}\right) \\
&= \nabla\cdot[\overline{\mu}\cdot\nabla\mathcal{A}_z(\mathbf{r})]-\hat{\mathbf{z}}\cdot\nabla\nabla\cdot[\overline{\mu}\cdot\mathcal{A}(\mathbf{r})]
\end{aligned}
\tag{5.71}
$$

and by putting everything together we arrive at a single vector equation again

$$
\nabla\cdot[\overline{\mu}\cdot\nabla\mathcal{A}(\mathbf{r})]-\nabla\nabla\cdot[\overline{\mu}\cdot\mathcal{A}(\mathbf{r})] = -\overline{\mu}^{-1}\cdot\mathcal{J}(\mathbf{r})\det\overline{\mu}, \qquad \mathbf{r}\in\mathbb{R}^3
\tag{5.72}
$$

which generalizes (5.39). Although we shall return to scalar equations for the Cartesian components of $\mathcal{A}(\mathbf{r})$, the form (5.72) of the vector Poisson equation suggests that we set

$$
\nabla\cdot[\overline{\mu}\cdot\mathcal{A}(\mathbf{r})] = 0, \qquad \mathbf{r}\in\mathbb{R}^3
\tag{5.73}
$$

which constitutes the desired extension of the Coulomb gauge. In this way, on the one hand the vector potential is uniquely determined, whereas, on the other, the solution of (5.72) can be tackled with the same procedure followed in Section 2.6.2 for the three-dimensional Green function in an anisotropic dielectric medium. In particular, (5.63) is now fully equivalent to the three problems

$$
\begin{cases}
\mu_{xx}\dfrac{\partial^2\mathcal{A}_\alpha}{\partial^2 x}+\mu_{yy}\dfrac{\partial^2\mathcal{A}_\alpha}{\partial^2 y}+\mu_{zz}\dfrac{\partial^2\mathcal{A}_\alpha}{\partial^2 z} = -\dfrac{\det\overline{\mu}}{\mu_{\alpha\alpha}}\mathcal{J}_\alpha(\mathbf{r}), & \mathbf{r}\in\mathbb{R}^3 \\
\mathcal{A}_\alpha(\mathbf{r}) = \mathcal{O}\left(\dfrac{1}{|\mathbf{r}|}\right), \qquad \alpha\in\{x,y,z\} & |\mathbf{r}|\to+\infty
\end{cases}
\tag{5.74}
$$

with the gauge (5.73) implied. To carry out the solution we introduce new independent variables

$$x = \sqrt{\mu_{xx}}\xi, \qquad y = \sqrt{\mu_{yy}}\eta, \qquad z = \sqrt{\mu_{zz}}\zeta \tag{5.75}$$

which are a special case of affine transformation of the space [12, Section 4.3]. Then, we define $\tilde{\mathbf{r}} = \xi\hat{\boldsymbol{\xi}} + \eta\hat{\boldsymbol{\eta}} + \zeta\hat{\boldsymbol{\zeta}}$, the position vector in the new system of coordinates, and obtain

$$\frac{\partial \mathcal{A}_\alpha}{\partial x} = \frac{1}{\sqrt{\mu_{xx}}}\frac{\partial \mathcal{A}_\alpha}{\partial \xi}, \qquad \frac{\partial \mathcal{A}_\alpha}{\partial y} = \frac{1}{\sqrt{\mu_{yy}}}\frac{\partial \mathcal{A}_\alpha}{\partial \eta}, \qquad \frac{\partial \mathcal{A}_\alpha}{\partial z} = \frac{1}{\sqrt{\mu_{zz}}}\frac{\partial \mathcal{A}_\alpha}{\partial \zeta} \tag{5.76}$$

$$\frac{\partial^2 \mathcal{A}_\alpha}{\partial x^2} = \frac{1}{\mu_{xx}}\frac{\partial^2 \mathcal{A}_\alpha}{\partial \xi^2}, \qquad \frac{\partial^2 \mathcal{A}_\alpha}{\partial y^2} = \frac{1}{\mu_{yy}}\frac{\partial^2 \mathcal{A}_\alpha}{\partial \eta^2}, \qquad \frac{\partial^2 \mathcal{A}_\alpha}{\partial z^2} = \frac{1}{\mu_{zz}}\frac{\partial^2 \mathcal{A}_\alpha}{\partial \zeta^2} \tag{5.77}$$

whereby (5.74) passes over into

$$\begin{cases} \dfrac{\partial^2 \mathcal{A}_\alpha}{\partial \xi^2} + \dfrac{\partial^2 \mathcal{A}_\alpha}{\partial \eta^2} + \dfrac{\partial^2 \mathcal{A}_\alpha}{\partial \zeta^2} = -\dfrac{\det\overline{\boldsymbol{\mu}}}{\mu_{\alpha\alpha}}\mathcal{J}_\alpha(\tilde{\mathbf{r}}), & \tilde{\mathbf{r}} \in \mathbb{R}^3 \\ \mathcal{A}_\alpha(\tilde{\mathbf{r}}) = \mathcal{O}\left(\dfrac{1}{|\tilde{\mathbf{r}}|}\right), \qquad \alpha \in \{x, y, z\} & |\tilde{\mathbf{r}}| \to +\infty. \end{cases} \tag{5.78}$$

These equations involve the Laplace operator in the Cartesian coordinates (ξ, η, ζ) and are formally identical with the scalar Poisson equation (2.190). Therefore, (5.78) is solved by three volume potentials of the type (2.187) with the density $(\det\overline{\boldsymbol{\mu}})\mathcal{J}_\alpha/\mu_{\alpha\alpha}$ in lieu of ϱ/ε, viz.,

$$\mathcal{A}_\alpha(\tilde{\mathbf{r}}) = \det\overline{\boldsymbol{\mu}} \int_{\tilde{V}_J} \mathrm{d}\tilde{V}' \, \frac{\mathcal{J}_\alpha(\tilde{\mathbf{r}}')/\mu_{\alpha\alpha}}{4\pi|\tilde{\mathbf{r}} - \tilde{\mathbf{r}}'|}, \qquad \alpha \in \{x, y, z\} \tag{5.79}$$

with the source point defined by $\tilde{\mathbf{r}}' = \xi'\hat{\boldsymbol{\xi}} + \eta'\hat{\boldsymbol{\eta}} + \zeta'\hat{\boldsymbol{\zeta}}$, where ξ', η' and ζ' are given by expressions analogous to (5.75). If we revert back to the original system of Cartesian coordinates for source and observation points by means of (5.75) and also combine the three scalar relationships into a single vector expression, we find

$$\mathcal{A}(\mathbf{r}) = \sqrt{\det\overline{\boldsymbol{\mu}}} \int_{V_J} \mathrm{d}V' \, \frac{\overline{\boldsymbol{\mu}}^{-1} \cdot \mathcal{J}(\mathbf{r}')}{4\pi\left(\mathbf{R} \cdot \overline{\boldsymbol{\mu}}^{-1} \cdot \mathbf{R}\right)^{1/2}}, \qquad \mathbf{r} \in \mathbb{R}^3 \tag{5.80}$$

with $\mathbf{R} = \mathbf{r} - \mathbf{r}'$. We notice that (5.80) is a coordinate-free representation for the vector potential and as such holds true even though $\overline{\boldsymbol{\mu}}$ is not diagonal. Moreover, (5.80) correctly reduces to (5.29) when $\overline{\boldsymbol{\mu}} = \mu\overline{\mathcal{I}}$.

We can make statements about the properties of the solution (5.80) by examining the Cartesian components of $\mathcal{A}(\mathbf{r})$ given by the equivalent expressions (5.79). Since the latter are in the form of a volume potential as in (2.187), we know from Section 2.8 that $\mathcal{A}_\alpha(\tilde{\mathbf{r}})$, $\alpha \in \{x, y, z\}$, are continuous functions for $\tilde{\mathbf{r}} \in \mathbb{R}^3$, provided $\mathcal{J}(\mathbf{r}') = \mathcal{J}(\tilde{\mathbf{r}}')$ is bounded. Since the change of variables (5.75) has no singular points, it only alters the shape of V_J but does not affect the properties of the integrand, and hence $\mathcal{A}(\mathbf{r})$ is continuous for $\mathbf{r} \in \mathbb{R}^3$.

As regards the derivatives with respect to the observation point $\mathbf{r}$ we have, e.g.,

$$\frac{\partial \mathcal{A}_\alpha}{\partial x} = \frac{1}{\sqrt{\mu_{xx}}}\frac{\partial \mathcal{A}_\alpha}{\partial \xi} = \frac{\det\overline{\boldsymbol{\mu}}}{\sqrt{\mu_{xx}}}\frac{\partial}{\partial \xi}\int_{\tilde{V}_J} \mathrm{d}\tilde{V}' \, \frac{\mathcal{J}_\alpha(\tilde{\mathbf{r}}')/\mu_{\alpha\alpha}}{4\pi|\tilde{\mathbf{r}} - \tilde{\mathbf{r}}'|}$$

$$= \det\overline{\boldsymbol{\mu}} \int_{\tilde{V}_J} \mathrm{d}\tilde{V}' \frac{\mathcal{J}_\alpha(\tilde{\mathbf{r}}')}{\mu_{\alpha\alpha}} \frac{1}{\sqrt{\mu_{xx}}} \frac{\partial}{\partial \xi} \frac{1}{4\pi|\tilde{\mathbf{r}} - \tilde{\mathbf{r}}'|}$$

$$= \sqrt{\det\overline{\boldsymbol{\mu}}} \int_{V_J} \mathrm{d}V' \frac{\mathcal{J}_\alpha(\mathbf{r}')}{\mu_{\alpha\alpha}} \frac{\partial}{\partial x} \frac{1}{4\pi\left(\mathbf{R}\cdot\overline{\boldsymbol{\mu}}^{-1}\cdot\mathbf{R}\right)^{1/2}}, \qquad \mathbf{r}\in\mathbb{R}^3 \tag{5.81}$$

on account of (5.76) and (5.79). The proof that the derivative with respect to ξ can be moved inside the integral is the same as that given in Section 2.8 for the volume potential, whereas the fact that resulting integral remains finite in the space of the variable $\tilde{\mathbf{r}}$ for $\tilde{\mathbf{r}}' \in \tilde{V}_J$ can be proved as in Section 3.2 for the electrostatic field. We only demand that the current density $\mathcal{J}(\tilde{\mathbf{r}}')$ be bounded in $\tilde{V}_J$. Reverting back to the original coordinates concludes the argument. Analogous steps are required for the other two spatial variables.

Thanks to these preliminary results, we can go on to show that $\mathcal{A}(\mathbf{r})$ satisfies the Coulomb gauge (5.73). The first step consists of moving the derivatives inside the integral, i.e.,

$$\nabla\cdot[\overline{\boldsymbol{\mu}}\cdot\mathcal{A}(\mathbf{r})] = \nabla\cdot\int_{V_J} \mathrm{d}V' \frac{\sqrt{\det\overline{\boldsymbol{\mu}}}\,\mathcal{J}(\mathbf{r}')}{4\pi\left(\mathbf{R}\cdot\overline{\boldsymbol{\mu}}^{-1}\cdot\mathbf{R}\right)^{1/2}}$$

$$= \int_{V_J} \mathrm{d}V'\,\nabla\frac{\sqrt{\det\overline{\boldsymbol{\mu}}}}{4\pi\left(\mathbf{R}\cdot\overline{\boldsymbol{\mu}}^{-1}\cdot\mathbf{R}\right)^{1/2}}\cdot\mathcal{J}(\mathbf{r}'), \qquad \mathbf{r}\in\mathbb{R}^3 \tag{5.82}$$

which is now rigorously justified in light of (5.81) and similar formulas. Secondly, we distinguish the cases of observation point $\mathbf{r}$ away from V_J or within the source region. For $\mathbf{r} \in \mathbb{R}^3 \setminus \overline{V}_J$ we manipulate the integrand directly with the aid of (H.51), namely,

$$\nabla\cdot\int_{V_J} \mathrm{d}V' \frac{\sqrt{\det\overline{\boldsymbol{\mu}}}\,\mathcal{J}(\mathbf{r}')}{4\pi\left(\mathbf{R}\cdot\overline{\boldsymbol{\mu}}^{-1}\cdot\mathbf{R}\right)^{1/2}} = -\int_{V_J} \mathrm{d}V'\,\nabla'\frac{\sqrt{\det\overline{\boldsymbol{\mu}}}}{4\pi\left(\mathbf{R}\cdot\overline{\boldsymbol{\mu}}^{-1}\cdot\mathbf{R}\right)^{1/2}}\cdot\mathcal{J}(\mathbf{r}')$$

$$= \int_{V_J} \mathrm{d}V' \frac{\sqrt{\det\overline{\boldsymbol{\mu}}}\,\nabla'\cdot\mathcal{J}(\mathbf{r}')}{4\pi\left(\mathbf{R}\cdot\overline{\boldsymbol{\mu}}^{-1}\cdot\mathbf{R}\right)^{1/2}} - \int_{V_J} \mathrm{d}V'\,\nabla'\cdot\frac{\sqrt{\det\overline{\boldsymbol{\mu}}}\,\mathcal{J}(\mathbf{r}')}{4\pi\left(\mathbf{R}\cdot\overline{\boldsymbol{\mu}}^{-1}\cdot\mathbf{R}\right)^{1/2}}$$

$$= -\int_{\partial V_J} \mathrm{d}S' \frac{\sqrt{\det\overline{\boldsymbol{\mu}}}\,\hat{\mathbf{n}}(\mathbf{r}')\cdot\mathcal{J}(\mathbf{r}')}{4\pi\left(\mathbf{R}\cdot\overline{\boldsymbol{\mu}}^{-1}\cdot\mathbf{R}\right)^{1/2}} = 0 \tag{5.83}$$

where the result follows because $\mathcal{J}(\mathbf{r}')$ is solenoidal and, being confined to V_J, it does not flow through ∂V_J. The Gauss theorem has been applied because the vector field $\mathcal{J}(\mathbf{r}')/(\mathbf{R}\cdot\overline{\boldsymbol{\mu}}^{-1}\cdot\mathbf{R})^{1/2}$ is of class $\mathcal{C}^1(V_J)^3 \cap \mathcal{C}(\overline{V}_J)^3$ so long as $\mathbf{r} \in \mathbb{R}^3 \setminus \overline{V}_J$.

For points $\mathbf{r} \in V_J$ the kernel is singular and thus we exclude $\mathbf{r}$ with a ball $B(\mathbf{r}, a) \subset V_J$. To compute the gradient of the kernel we observe

$$\hat{\mathbf{x}}\frac{\partial}{\partial x}\frac{1}{\left(\mathbf{R}\cdot\overline{\boldsymbol{\mu}}^{-1}\cdot\mathbf{R}\right)^{1/2}} = -\frac{\hat{\mathbf{x}}(x-x')/\mu_{xx}}{\left(\mathbf{R}\cdot\overline{\boldsymbol{\mu}}^{-1}\cdot\mathbf{R}\right)^{3/2}} = -\frac{\hat{\mathbf{x}}\hat{\mathbf{x}}}{\mu_{xx}}\cdot\frac{\mathbf{R}}{\left(\mathbf{R}\cdot\overline{\boldsymbol{\mu}}^{-1}\cdot\mathbf{R}\right)^{3/2}} \tag{5.84}$$

and analogous formulas hold for the derivatives with respect to y and z. Therefore, we arrive at

$$\nabla\frac{1}{\left(\mathbf{R}\cdot\overline{\boldsymbol{\mu}}^{-1}\cdot\mathbf{R}\right)^{1/2}} = -\frac{\overline{\boldsymbol{\mu}}^{-1}\cdot\mathbf{R}}{\left(\mathbf{R}\cdot\overline{\boldsymbol{\mu}}^{-1}\cdot\mathbf{R}\right)^{3/2}} \tag{5.85}$$

by combining the results for the three Cartesian components. We split the integration into integrals over $V_a := V_J \setminus B(\mathbf{r}, a)$ and the ball (see Figure 2.8 for a similar setup)

$$\begin{aligned}
&\int_{V_J} \mathrm{d}V' \, \nabla \frac{\sqrt{\det \overline{\mu}}}{4\pi \left(\mathbf{R} \cdot \overline{\mu}^{-1} \cdot \mathbf{R}\right)^{1/2}} \cdot \mathfrak{J}(\mathbf{r}') = \\
&= -\int_{V_a} \mathrm{d}V' \, \nabla' \frac{\sqrt{\det \overline{\mu}}}{4\pi \left(\mathbf{R} \cdot \overline{\mu}^{-1} \cdot \mathbf{R}\right)^{1/2}} \cdot \mathfrak{J}(\mathbf{r}') + \int_{B(\mathbf{r},a)} \mathrm{d}V' \, \nabla \frac{\sqrt{\det \overline{\mu}}}{4\pi \left(\mathbf{R} \cdot \overline{\mu}^{-1} \cdot \mathbf{R}\right)^{1/2}} \cdot \mathfrak{J}(\mathbf{r}') \\
&= \int_{V_a} \mathrm{d}V' \, \frac{\sqrt{\det \overline{\mu}} \, \nabla' \cdot \mathfrak{J}(\mathbf{r}')}{4\pi \left(\mathbf{R} \cdot \overline{\mu}^{-1} \cdot \mathbf{R}\right)^{1/2}} - \int_{V_a} \mathrm{d}V' \, \nabla' \cdot \frac{\sqrt{\det \overline{\mu}} \, \mathfrak{J}(\mathbf{r}')}{4\pi \left(\mathbf{R} \cdot \overline{\mu}^{-1} \cdot \mathbf{R}\right)^{1/2}} \\
&\quad - \int_{B(\mathbf{r},a)} \mathrm{d}V' \, \mathfrak{J}(\mathbf{r}') \cdot \frac{\sqrt{\det \overline{\mu}} \, \overline{\mu}^{-1} \cdot \mathbf{R}}{4\pi \left(\mathbf{R} \cdot \overline{\mu}^{-1} \cdot \mathbf{R}\right)^{3/2}} \\
&= -\int_{\partial B} \mathrm{d}S' \, \frac{\sqrt{\det \overline{\mu}} \, \hat{\mathbf{R}} \cdot \mathfrak{J}(\mathbf{r}')}{4\pi \left(\mathbf{R} \cdot \overline{\mu}^{-1} \cdot \mathbf{R}\right)^{1/2}} - \int_{B(\mathbf{r},a)} \mathrm{d}V' \, \mathfrak{J}(\mathbf{r}') \cdot \frac{\sqrt{\det \overline{\mu}} \, \overline{\mu}^{-1} \cdot \mathbf{R}}{4\pi \left(\mathbf{R} \cdot \overline{\mu}^{-1} \cdot \mathbf{R}\right)^{3/2}}
\end{aligned} \tag{5.86}$$

and we are left with the task of showing that the last two integrals vanish for $a \to 0^+$. By introducing a local system of spherical coordinates (R, α', β') centered in $\mathbf{r}$ we have

$$\mathbf{R} \cdot \overline{\mu}^{-1} \cdot \mathbf{R} = R^2 \hat{\mathbf{R}} \cdot \overline{\mu}^{-1} \cdot \hat{\mathbf{R}} \tag{5.87}$$

where $\hat{\mathbf{R}}$ points towards $\mathbf{r}$ and the radial vector is $-\hat{\mathbf{R}}$. When $\mathbf{r}' \in \partial B$, $R = a$ and we estimate

$$\begin{aligned}
\left| \int_{\partial B} \mathrm{d}S' \, \frac{\sqrt{\det \overline{\mu}} \, \hat{\mathbf{R}} \cdot \mathfrak{J}(\mathbf{r}')}{4\pi \left(\mathbf{R} \cdot \overline{\mu}^{-1} \cdot \mathbf{R}\right)^{1/2}} \right| &\leqslant \int_{\partial B} \mathrm{d}S' \, \frac{\sqrt{\det \overline{\mu}} \, \|\mathfrak{J}\|_\infty}{4\pi a \left(\hat{\mathbf{R}} \cdot \overline{\mu}^{-1} \cdot \hat{\mathbf{R}}\right)^{1/2}} \\
&\leqslant a \int_{4\pi} \mathrm{d}\Omega' \, \frac{\sqrt{\det \overline{\mu}} \, \|\mathfrak{J}\|_\infty}{4\pi \left(\hat{\mathbf{R}} \cdot \overline{\mu}^{-1} \cdot \hat{\mathbf{R}}\right)^{1/2}} \xrightarrow[a \to 0^+]{} 0
\end{aligned} \tag{5.88}$$

where $\int_{4\pi} \mathrm{d}\Omega'$ means integration over the complete solid angle subtended by $\mathbf{r}$. The last integral is finite and independent of a because $\hat{\mathbf{R}}$ depends on α' and β' only. Furthermore, on account of (E.76) we have

$$\begin{aligned}
\left| \int_{B(\mathbf{r},a)} \mathrm{d}V' \, \mathfrak{J}(\mathbf{r}') \cdot \frac{\sqrt{\det \overline{\mu}} \, \overline{\mu}^{-1} \cdot \mathbf{R}}{4\pi \left(\mathbf{R} \cdot \overline{\mu}^{-1} \cdot \mathbf{R}\right)^{3/2}} \right| &\leqslant \|\mathfrak{J}\|_\infty \int_{B(\mathbf{r},a)} \mathrm{d}V' \, \frac{\sqrt{\det \overline{\mu}} \, \|\overline{\mu}^{-1}\|}{4\pi R^2 \left(\hat{\mathbf{R}} \cdot \overline{\mu}^{-1} \cdot \hat{\mathbf{R}}\right)^{3/2}} \\
&= \|\mathfrak{J}\|_\infty \, a \int_{4\pi} \mathrm{d}\Omega' \, \frac{\sqrt{\det \overline{\mu}} \, \|\overline{\mu}^{-1}\|}{4\pi \left(\hat{\mathbf{R}} \cdot \overline{\mu}^{-1} \cdot \hat{\mathbf{R}}\right)^{3/2}} \xrightarrow[a \to 0^+]{} 0
\end{aligned} \tag{5.89}$$

where the result follows because the last integrand is finite and does not depend on a. This concludes the proof.

In light of (4.29) and (5.80) the magnetic induction in the anisotropic medium is given by

$$\mathfrak{B}(\mathbf{r}) = \nabla \times \int_{V_J} \mathrm{d}V' \, \frac{\sqrt{\det \overline{\mu}} \, \overline{\mu}^{-1} \cdot \mathfrak{J}(\mathbf{r}')}{4\pi \left(\mathbf{R} \cdot \overline{\mu}^{-1} \cdot \mathbf{R}\right)^{1/2}}$$

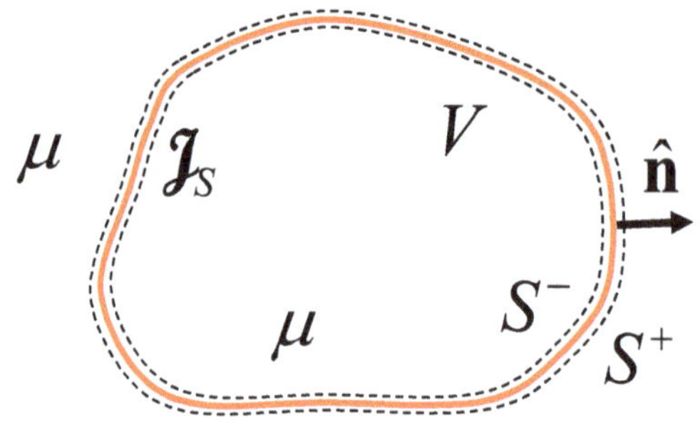

Figure 5.4 For the derivation of the magnetic vector potential generated by a stationary electric current sheet with density $\mathcal{J}_S(\mathbf{r})$ in a homogeneous isotropic magnetic medium. For visualization's sake the two sides of S (−−) are drawn slightly away from one another.

$$= \int_{V_J} dV' \, \nabla \frac{\sqrt{\det \overline{\mu}}}{4\pi \left(\mathbf{R} \cdot \overline{\mu}^{-1} \cdot \mathbf{R}\right)^{1/2}} \times \overline{\mu}^{-1} \cdot \mathcal{J}(\mathbf{r}')$$

$$= -\sqrt{\det \overline{\mu}} \int_{V_J} dV' \, \frac{\left(\overline{\mu}^{-1} \cdot \mathbf{R}\right) \times \left[\overline{\mu}^{-1} \cdot \mathcal{J}(\mathbf{r}')\right]}{4\pi \left(\mathbf{R} \cdot \overline{\mu}^{-1} \cdot \mathbf{R}\right)^{3/2}}$$

$$= \frac{\overline{\mu}}{\sqrt{\det \overline{\mu}}} \cdot \int_{V_J} dV' \, \frac{\mathcal{J}(\mathbf{r}') \times \mathbf{R}}{4\pi \left(\mathbf{R} \cdot \overline{\mu}^{-1} \cdot \mathbf{R}\right)^{3/2}}, \qquad \mathbf{r} \in \mathbb{R}^3 \tag{5.90}$$

where the interchange of integration and differentiation has been proved with (5.81) and subsequent arguments.

5.2 Vector potential due to surface currents

We suppose that a current sheet $\mathcal{J}_S(\mathbf{r})$ exists in a magnetic medium endowed with permeability μ and flows over a smooth closed surface $S := \partial V$ which is the boundary of a finite domain V. The relevant vector potential is the unique solution to the homogeneous problem (Figure 5.4)

$$\begin{cases} \nabla \times [\nabla \times \mathcal{A}(\mathbf{r})] = 0, & \mathbf{r} \in \mathbb{R}^3 \setminus S \\ \nabla \cdot \mathcal{A}(\mathbf{r}) = 0 & \mathbf{r} \in \mathbb{R}^3 \setminus S \\ \mathcal{A}(\mathbf{r})|_{S^+} = \mathcal{A}(\mathbf{r})|_{S^+}, & \mathbf{r} \in S \\ \mu \mathcal{J}_S(\mathbf{r}) = \hat{\mathbf{n}}(\mathbf{r}) \times [\nabla \times \mathcal{A}(\mathbf{r})|_{S^+} - \nabla \times \mathcal{A}(\mathbf{r})|_{S^-}], & \mathbf{r} \in S \\ \mathcal{A}(\mathbf{r}) = \mathcal{O}\left(\dfrac{1}{|\mathbf{r}|}\right), & |\mathbf{r}| \to +\infty \end{cases} \tag{5.91}$$

where the unit normal vector $\hat{\mathbf{n}}(\mathbf{r})$ on S points towards $\mathbb{R}^3 \setminus V$. The vector potential is continuous across S, whereas the tangential component of $\nabla \times \mathcal{A}$ suffers a jump owing to the presence of the current sheet.

Although in Section 4.7.1 we proved that uniqueness is achieved so long as on the domain boundary we assign $\hat{\mathbf{n}} \cdot \mathcal{A}$ and either $\hat{\mathbf{n}} \times \mathcal{A}$ or $\hat{\mathbf{n}} \times (\nabla \times \mathcal{A})$, clearly (5.91) includes constraints for the tangential components of both $\mathcal{A}$ and $\nabla \times \mathcal{A}$. This choice is explained by noticing that in actuality (5.91) is comprised of two vector Poisson equations, namely, one formulated in V and another one stated in the complementary domain. Proceeding as in Section 4.7.1 by defining separate difference potentials in V and $\mathbb{R}^3 \setminus \overline{V}$ it is possible to show that uniqueness requires specifying $\hat{\mathbf{n}} \times \mathcal{A}$ and $\hat{\mathbf{n}} \times (\nabla \times \mathcal{A})$ on ∂V simultaneously. Intuitively, from the standpoint of V we may enforce $\hat{\mathbf{n}} \cdot \mathcal{A}$ and, say, $\hat{\mathbf{n}} \times \mathcal{A}$, but these quantities are not available until we solve the problem outside V. Likewise, from the viewpoint of the complementary domain we may set $\hat{\mathbf{n}} \cdot \mathcal{A}$ and $\hat{\mathbf{n}} \times (\nabla \times \mathcal{A})$ which, however, are unknown unless we determine $\mathcal{A}$ in V. All in all, the matching conditions (4.48), (4.54) and (4.64) *couple* the two problems in V and $\mathbb{R}^3 \setminus \overline{V}$ and guarantee uniqueness.

That been said, we split the solution procedure into two parts, namely, finding the potential within V and in the complementary unbounded domain, so as to make use of the integral representations derived in Section 5.1.1. In order to apply (5.18) to $\mathbb{R}^3 \setminus \overline{V}$ we choose a ball $B(0, a)$ with radius a large enough for V to be enclosed by $B(0, a)$. In the limit as $a \to +\infty$ the contribution of the surface integrals over the sphere ∂B vanish in view of the asymptotic behavior (4.40) of $\mathcal{A}(\mathbf{r})$ and the static Green function. We obtain

$$\mathcal{A}(\mathbf{r}) = \int_S \mathrm{d}S'\, \hat{\mathbf{n}}(\mathbf{r}') \times [\nabla \times \mathcal{A}(\mathbf{r}')|_{S^+}] \frac{1}{4\pi R} + \int_S \mathrm{d}S'\, [\hat{\mathbf{n}}(\mathbf{r}') \times \mathcal{A}(\mathbf{r}')|_{S^+}] \times \nabla' \frac{1}{4\pi R} + \int_S \mathrm{d}S'\, \hat{\mathbf{n}}(\mathbf{r}') \cdot \mathcal{A}(\mathbf{r}')|_{S^+} \nabla' \frac{1}{4\pi R}, \qquad \mathbf{r} \in \mathbb{R}^3 \setminus \overline{V} \quad (5.92)$$

where $R = |\mathbf{r} - \mathbf{r}'|$, and we have indicated that within the integrals the potentials and the derivatives thereof are evaluated on the positive side of S.

For the vector potential in V we invoke (5.18) once again, viz.,

$$\mathcal{A}(\mathbf{r}) = -\int_S \mathrm{d}S'\, \hat{\mathbf{n}}(\mathbf{r}') \times [\nabla \times \mathcal{A}(\mathbf{r}')|_{S^-}] \frac{1}{4\pi R} - \int_S \mathrm{d}S'\, [\hat{\mathbf{n}}(\mathbf{r}') \times \mathcal{A}(\mathbf{r}')|_{S^-}] \times \nabla' \frac{1}{4\pi R} - \int_S \mathrm{d}S'\, \hat{\mathbf{n}}(\mathbf{r}') \cdot \mathcal{A}(\mathbf{r}')|_{S^-} \nabla' \frac{1}{4\pi R}, \qquad \mathbf{r} \in V \quad (5.93)$$

where the minus signs are due to the outward orientation of the unit normal on S. Within the integrals the potential and the derivatives thereof are evaluated on the negative side of S.

To proceed we recall that the integral representation (5.18) returns a null value when evaluated for observation points that fall outside the domain for which the formula was devised. Consequently, (5.92) yields a zero potential for $\mathbf{r} \in V$, whereas (5.93) returns zero in the complementary domain. Therefore, we may sum (5.92) and (5.93) side by side to arrive at

$$\begin{aligned}\mathcal{A}(\mathbf{r}) &= \int_S \mathrm{d}S'\, \hat{\mathbf{n}}(\mathbf{r}') \times [\nabla \times \mathcal{A}(\mathbf{r}')|_{S^+} - \nabla \times \mathcal{A}(\mathbf{r}')|_{S^-}] \frac{1}{4\pi R} \\ &+ \int_S \mathrm{d}S'\, [\hat{\mathbf{n}}(\mathbf{r}') \times \mathcal{A}(\mathbf{r}')|_{S^+} - \hat{\mathbf{n}}(\mathbf{r}') \times \mathcal{A}(\mathbf{r}')|_{S^-}] \times \nabla' \frac{1}{4\pi R} \\ &+ \int_S \mathrm{d}S'\, [\hat{\mathbf{n}}(\mathbf{r}') \cdot \mathcal{A}(\mathbf{r}')|_{S^+} - \hat{\mathbf{n}}(\mathbf{r}') \cdot \mathcal{A}(\mathbf{r}')|_{S^-}] \nabla' \frac{1}{4\pi R}, \qquad \mathbf{r} \in \mathbb{R}^3 \setminus S \quad (5.94)\end{aligned}$$

a single representation formula valid for every point away from S.

To finalize the result we recall that the tangential component of the curl of the potential suffers a jump proportional to the surface current density $\mathcal{J}_S(\mathbf{r})$, and that $\mathcal{A}(\mathbf{r})$ is continuous, whereby we get

$$\mathcal{A}(\mathbf{r}) = \mu \int_S \mathrm{d}S' \, \frac{\mathcal{J}_S(\mathbf{r}')}{4\pi R}, \qquad \mathbf{r} \in \mathbb{R}^3 \setminus S \tag{5.95}$$

as the desired solution to (5.91). Since the Cartesian components of the right member of (5.95) are in the form of single-layer potentials, the integral exists also for points $\mathbf{r} \in S$ (Section 2.10). It should be noted that failing to prescribe the continuity of $\hat{\mathbf{n}} \times \mathcal{A}$ across S would leave an unknown contribution in (5.94) and consequently formula (5.95) could not be obtained.

Example 5.1 (Vector potential and magnetic field of a rotating spherical layer of charge)

We suppose that a layer of electric charge with uniform density ϱ_{0S} is distributed over a rigid spherical surface S which is the boundary of a sphere with the center in the origin. If we further assume that the surface rotates with angular velocity ω around the z-axis of a system of Cartesian coordinates, then the motion gives rise to a convection current with surface density

$$\mathcal{J}_S(\mathbf{r}) := \varrho_{0S}\, \omega a \,\hat{\boldsymbol{\varphi}} \sin\vartheta, \qquad \mathbf{r} \in S \tag{5.96}$$

where ϑ and $\hat{\boldsymbol{\varphi}}$ are the polar angle and the azimuthal unit vector of a system of spherical coordinates. It is straightforward to check that $\mathcal{J}_S(\mathbf{r})$ is a solenoidal vector field on S, since the current is aligned with $\hat{\boldsymbol{\varphi}}$ but does not depend on the azimuthal angle φ. Therefore, $\mathcal{J}_S(\mathbf{r})$ produces a stationary magnetic field $\mathcal{H}(\mathbf{r})$ which we wish to compute by first finding the vector potential $\mathcal{A}(\mathbf{r})$ in the Coulomb gauge (see Section 5.1.1).

By virtue of definition (4.29), the Ampère law (4.9), and the matching conditions (4.54), (4.64) and (4.48) it follows that $\mathcal{A}(\mathbf{r})$ is the unique solution to the homogeneous boundary value problem

$$\begin{cases} \nabla \times [\nabla \times \mathcal{A}(\mathbf{r})] = 0, & \mathbf{r} \in \mathbb{R}^3 \setminus S \\ \nabla \cdot \mathcal{A}(\mathbf{r}) = 0 & \mathbf{r} \in \mathbb{R}^3 \setminus S \\ \mathcal{A}(\mathbf{r})|_{S^+} = \mathcal{A}(\mathbf{r})|_{S^+}, & \mathbf{r} \in S \\ \hat{\mathbf{n}}(\mathbf{r}) \times [\nabla \times \mathcal{A}(\mathbf{r})|_{S^+} - \nabla \times \mathcal{A}(\mathbf{r})|_{S^-}] = \mu_0 \varrho_{0S}\, \omega a \sin\vartheta \hat{\boldsymbol{\varphi}}, & \mathbf{r} \in S \\ \mathcal{A}(\mathbf{r}) = \mathcal{O}\left(\dfrac{1}{|\mathbf{r}|}\right), & |\mathbf{r}| \to +\infty \end{cases} \tag{5.97}$$

where we observe that the components $\mathcal{A}_r$ and $\mathcal{A}_\vartheta$ are not excited by $\mathcal{J}_S$ and thus are identically null. By applying (H.59), invoking (A.45) and performing a little algebra we can cast (5.97) into the alternative form

$$\begin{cases} \nabla^2 \mathcal{A}_\varphi(\mathbf{r}) - \dfrac{\mathcal{A}_\varphi(\mathbf{r})}{r^2 \sin^2\vartheta} = 0, & \mathbf{r} \in \mathbb{R}^3 \setminus S \\ \mathcal{A}_\varphi(\mathbf{r})|_{S^+} = \mathcal{A}_\varphi(\mathbf{r})|_{S^-}, & \mathbf{r} \in S \\ -\dfrac{1}{r}\dfrac{\partial}{\partial r}(r\mathcal{A}_\varphi)\Big|_{S^+} + \dfrac{1}{r}\dfrac{\partial}{\partial r}(r\mathcal{A}_\varphi)\Big|_{S^-} = \mu_0 \varrho_{0S}\, \omega a \sin\vartheta, & \mathbf{r} \in S \\ \mathcal{A}_\varphi(\mathbf{r}) = \mathcal{O}\left(\dfrac{1}{|\mathbf{r}|}\right), & |\mathbf{r}| \to +\infty \end{cases} \tag{5.98}$$

where $\mathcal{A}_\varphi$ depends on r and ϑ only. To solve the differential equation — which is neither of the Laplace type nor of the Helmholtz kind — we expand the Laplace operator in polar spherical coordinates according to (A.42) and then look for a solution in factorized form

$$\mathcal{A}_\varphi(r, \vartheta) := \Xi(r)\Theta(\vartheta) \tag{5.99}$$

in order to invoke the *separation argument* as in Section 3.5.1. After a few steps we arrive at the uncoupled equations

$$\frac{1}{\sin\vartheta}\frac{\partial}{\partial\vartheta}\left(\sin\vartheta\frac{\partial\Theta}{\partial\vartheta}\right) + n(n+1)\Theta(\vartheta) - \frac{\Theta(\vartheta)}{\sin^2\vartheta} = 0 \tag{5.100}$$

$$\frac{\partial}{\partial r}\left(r^2\frac{\partial\Xi}{\partial r}\right) - n(n+1)\Xi(r) = 0 \tag{5.101}$$

where $n(n+1)$ is the separation constant with n a whole number. The first equation is a special instance of the associated Legendre equation (3.79) with $m = 1$. The other one is the same as (3.80) with solutions given by (3.107). Since the vector potential is supposed to be regular on the z-axis and at infinity, the solution must be in form

$$\mathcal{A}_\varphi(r,\vartheta) = \begin{cases} \displaystyle\sum_{n=0}^{\infty} \frac{A_n}{r^{n+1}} P_n^1(\cos\vartheta), & r > a \\ \displaystyle\sum_{n=0}^{\infty} B_n r^n P_n^1(\cos\vartheta), & r < a \end{cases} \tag{5.102}$$

where the constants A_n and B_n are determined by fulfilling the boundary conditions on S. In this regard, from the second and the third line of (5.98) we have

$$\left(\frac{A_n}{a^{n+1}} - B_n a^n\right) P_n^1(\cos\vartheta) = 0 \tag{5.103}$$

$$\left[\frac{n}{a^{n+2}} A_n + (n+1)a^{n-1} B_n\right] P_n^1(\cos\vartheta) = 0 \tag{5.104}$$

for $n \neq 1$ and

$$\left(\frac{A_1}{a^2} - B_1 a\right) P_1^1(\cos\vartheta) = 0 \tag{5.105}$$

$$\left(\frac{A_1}{a^3} + 2B_1\right) P_1^1(\cos\vartheta) = -\mu_0 \varrho_{0S}\, \omega a P_1^1(\cos\vartheta) \tag{5.106}$$

for $n = 1$ in view of (H.121). Since these conditions must hold for any permissible value of ϑ, the coefficients of the Legendre functions on either side must be equal. As a result we find $A_n = 0 = B_n$ for $n \neq 1$ and

$$A_1 = -\frac{a^4}{3}\mu_0 \varrho_{0S}\,\omega, \qquad B_1 = -\frac{a}{3}\mu_0 \varrho_{0S}\,\omega \tag{5.107}$$

for $n = 1$, whereby the vector potential reads

$$\mathcal{A}(r,\vartheta) = \begin{cases} \dfrac{a^4}{3r^2}\mu_0 \varrho_{0S}\,\omega \hat{\boldsymbol{\varphi}} \sin\vartheta, & r > a \\ \dfrac{a}{3} r \mu_0 \varrho_{0S}\,\omega \hat{\boldsymbol{\varphi}} \sin\vartheta, & r < a. \end{cases} \tag{5.108}$$

Since $\mathcal{J}_S(\mathbf{r})$ is a surface current, we can also obtain the vector potential through the general integral solution (5.95), which becomes

$$\mathcal{A}(\mathbf{r}) = \mu_0 \varrho_{0S}\,\omega a \int_S \mathrm{d}S' \frac{\sin\vartheta'}{4\pi|\mathbf{r}-\mathbf{r}'|}(-\hat{\mathbf{x}}\sin\varphi' + \hat{\mathbf{y}}\cos\varphi') \tag{5.109}$$

in light of (5.96) and (A.13). To compute the integral in spherical coordinates (φ', ϑ') it is convenient to expand the kernel $1/|\mathbf{r} - \mathbf{r}'|$ in accordance with (3.235) and then invoke the orthogonality of Legendre and exponential functions over the intervals $[0, \pi]$ and $[0, 2\pi]$, respectively. By letting

$$r_< := \min\{r, a\}, \qquad r_> := \max\{r, a\} \tag{5.110}$$

for the x-component we have

$$\begin{aligned}
&\int_S \mathrm{d}S' \, \frac{-\sin\vartheta' \sin\varphi'}{4\pi|\mathbf{r} - \mathbf{r}'|} = \\
&= \sum_{n=0}^{+\infty} \sum_{m=-n}^{n} \int_S \mathrm{d}\Omega' \, a^2 P_1^1(\cos\vartheta') \frac{\mathrm{e}^{j\varphi'} - \mathrm{e}^{-j\varphi'}}{8\pi \mathrm{j}} \frac{r_<^n}{r_>^{n+1}} \frac{(n-m)!}{(n+m)!} P_n^m(\cos\vartheta') P_n^m(\cos\vartheta) \mathrm{e}^{\mathrm{j}m(\varphi'-\varphi)} \\
&= \sum_{n=0}^{+\infty} a^2 \int_0^\pi \mathrm{d}\vartheta' \sin\vartheta' P_1^1(\cos\vartheta') P_n^1(\cos\vartheta') \frac{(n-1)!}{(n+1)!} \frac{r_<^n}{r_>^{n+1}} \frac{1}{2} P_n^1(\cos\vartheta) \sin\varphi \\
&= \frac{a^2 r_<}{3 r_>^2} P_1^1(\cos\vartheta) \sin\varphi = -\frac{a^2 r_<}{3 r_>^2} \sin\vartheta \sin\varphi = \begin{cases} -\dfrac{a^3}{3r^2} \sin\vartheta \sin\varphi, & r > a \\ -\dfrac{r}{3} \sin\vartheta \sin\varphi, & r < a \end{cases}
\end{aligned} \tag{5.111}$$

on account of (H.129) and (H.128); similar calculations hold for the y-component. Putting the intermediate results together and using representation (A.13) backwards yield (5.108) again. The integral in (5.109) can be also computed by noticing that $\hat{\boldsymbol{\varphi}}' \sin\vartheta' = \hat{\mathbf{z}} \times \hat{\mathbf{r}}'$ and introducing another system of polar spherical coordinates (r', α', β') with polar axis aligned with $\hat{\mathbf{r}}$. Then, only the component of $\hat{\mathbf{r}}'$ parallel to $\hat{\mathbf{r}}$ contributes a non-zero value upon integration over $\beta' \in [0, 2\pi]$.

Since $\hat{\boldsymbol{\varphi}} \sin\vartheta = \hat{\mathbf{z}} \times \hat{\mathbf{r}}$, for $r \geqslant a$ it is possible to cast $\mathcal{A}(\mathbf{r})$ as in (4.87), namely,

$$\mathcal{A}(\mathbf{r}) = \mu_0 \underbrace{\frac{4\pi}{3} a^4 \varrho_{0S} \, \omega \, \hat{\mathbf{z}}}_{\text{dipole moment}} \times \frac{\hat{\mathbf{r}}}{4\pi r^2}, \qquad r \geqslant a \tag{5.112}$$

and interpret the potential as being produced by an equivalent magnetic dipole which is placed in $\mathbf{r} = 0$ (i.e., the center of the spherical surface S) and has moment

$$\mathbf{m}_{eq} = \frac{4\pi}{3} a^4 \varrho_{0S} \, \omega \hat{\mathbf{z}} = \frac{a^2}{3} Q \omega \hat{\mathbf{z}} \tag{5.113}$$

where $Q := 4\pi a^2 \varrho_{0S}$ is the total charge distributed on S.

Now, the magnetic field for $r \geqslant a$ can be computed either through (4.29) or, perhaps more simply in light of (4.86), by constructing the magnetic scalar potential (4.85) and taking the gradient thereof. The second approach gives

$$\begin{aligned}
\mathcal{H}(\mathbf{r}) &= -\nabla \left(\frac{\mathbf{m}_{eq} \cdot \hat{\mathbf{r}}}{4\pi r^2} \right) = \frac{m_{eq}}{4\pi r^3} \left(2\hat{\mathbf{r}} \cos\vartheta + \hat{\boldsymbol{\vartheta}} \sin\vartheta \right) = \frac{2\hat{\mathbf{r}}\hat{\mathbf{r}} - \hat{\boldsymbol{\vartheta}}\hat{\boldsymbol{\vartheta}}}{4\pi r^3} \cdot \mathbf{m}_{eq} \\
&= \frac{3\hat{\mathbf{r}}\hat{\mathbf{r}} - \overline{\mathfrak{I}}}{4\pi r^3} \cdot \mathbf{m}_{eq} = \frac{a^4}{3r^3} \varrho_{0S} \, \omega \left(2\hat{\mathbf{r}} \cos\vartheta + \hat{\boldsymbol{\vartheta}} \sin\vartheta \right).
\end{aligned} \tag{5.114}$$

For points inside the spherical surface S the magnetic field can be retrieved from (4.29), i.e.,

$$\mathcal{H}(\mathbf{r}) = \frac{1}{\mu_0} \nabla \times \mathcal{A}(\mathbf{r}) = \frac{2}{3} \varrho_{0S} \, \omega a \left(\hat{\mathbf{r}} \cos\vartheta - \hat{\boldsymbol{\vartheta}} \sin\vartheta \right) = \frac{2}{3} \varrho_{0S} \, \omega a \hat{\mathbf{z}} \tag{5.115}$$

by virtue of (A.37) and (A.16). From a practical viewpoint the rotating charge and the ensuing solenoidal current behave no differently than the permanently magnetized sphere which is examined in details in Example 5.4.

(End of Example 5.1)

5.3 Physical meaning of the vector potential

It is often stated that within the framework of classical electrodynamics the vector potential $\mathcal{A}(\mathbf{r})$ does not possess a physical meaning of its own, as does the electrostatic counterpart $\Phi(\mathbf{r})$ (Section 2.3), the reason being, perhaps, that $\mathcal{A}(\mathbf{r})$ is determined up to an arbitrary conservative vector field $\nabla\chi(\mathbf{r})$. Actually, we can offer two interpretations, namely, in terms of energy associated with steady current densities and in terms of *potential momentum* of a point charge [10].

In the first case we observe that, since currents are comprised of drifting electric charges after all, with the aid of (2.19) and (4.34) we may write down the total potential energy of a systems of charges $\varrho(\mathbf{r})$ and steady currents $\mathcal{J}(\mathbf{r})$ confined in a region of free space $V_J \subset \mathbb{R}^3$, namely,

$$\begin{aligned} W_e + W_h &:= \frac{1}{2}\int\limits_{\mathbb{R}^3} \mathrm{d}V\,[\mathcal{E}(\mathbf{r})\cdot\mathcal{D}(\mathbf{r}) + \mathcal{B}(\mathbf{r})\cdot\mathcal{H}(\mathbf{r})] \\ &= \frac{1}{2}\int\limits_{V_J} \mathrm{d}V\,[\Phi(\mathbf{r})\varrho(\mathbf{r}) + \mathcal{A}(\mathbf{r})\cdot\mathcal{J}(\mathbf{r})] \end{aligned} \tag{5.116}$$

having assumed, for the sake of argument, that the electrostatic potential vanishes at infinity ($\Phi_\infty = 0$) and the arbitrary conservative field $\nabla\chi(\mathbf{r})$ is null.

Evidently, both integrands in the formula above represent a volume density of potential energy $w_e(\mathbf{r}) + w_h(\mathbf{r})$. The first integrand involves the field entities and as such is valid in the whole space, whereas the second one yields $w_e(\mathbf{r}) + w_h(\mathbf{r})$ in terms of potentials and sources and hence it the applies only for points in the region occupied by $\varrho(\mathbf{r})$ and $\mathcal{J}(\mathbf{r})$. Since we can formally invert the equation $w_e = \Phi\varrho$ with respect to the potential (at least at points where $\varrho(\mathbf{r})$ is non-null), viz.,

$$\Phi(\mathbf{r}) = \frac{w_e(\mathbf{r})}{\varrho(\mathbf{r})}, \qquad \mathbf{r} \in V_J \tag{5.117}$$

we see that $\Phi(\mathbf{r})$ may be construed as a density of electrostatic energy per unit density of charge at the point $\mathbf{r} \in V_J$. In like manner, if we denote with $\hat{\mathbf{s}}(\mathbf{r})$ the local unit vector in the direction of $\mathcal{J}(\mathbf{r})$, we may write

$$\mathcal{A}(\mathbf{r})\cdot\hat{\mathbf{s}}(\mathbf{r}) = \frac{w_h(\mathbf{r})}{\hat{\mathbf{s}}(\mathbf{r})\cdot\mathcal{J}(\mathbf{r})} = \frac{w_h(\mathbf{r})}{\mathcal{J}(\mathbf{r})}, \qquad \mathbf{r} \in V_J \tag{5.118}$$

and interpret the component of the vector potential along $\hat{\mathbf{s}}(\mathbf{r})$ as a density of magnetic energy per unit density of current.

The second interpretation of $\mathcal{A}(\mathbf{r})$ is far more rigorous and satisfactory in that somehow it complements the meaning of the electric scalar potential. To establish the result we consider a point charge q located at $\mathbf{r} = \mathbf{r}_q$ and immersed in the stationary electric, $\mathcal{E}_J(\mathbf{r})$, and magnetic, $\mathcal{H}_J(\mathbf{r})$, fields produced by a steady current $\mathcal{J}(\mathbf{r}) \subset V_J$. This arrangement is pictorially shown in Figure 5.5. For the sake of argument, let charge and current exist in free space. The total electric field reads

$$\mathcal{E}(\mathbf{r}) := \mathcal{E}_q(\mathbf{r}, \mathbf{r}_q) + \mathcal{E}_J(\mathbf{r}) = -\nabla\frac{q}{4\pi\varepsilon_0|\mathbf{r} - \mathbf{r}_q|} + \mathcal{E}_J(\mathbf{r}) \tag{5.119}$$

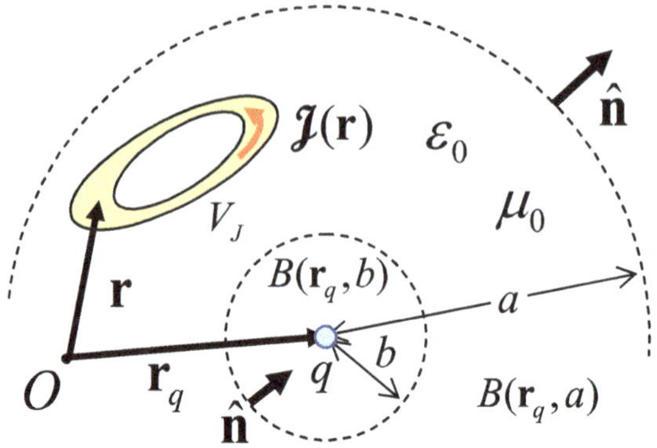

Figure 5.5 *A point charge (∘) at rest in a stationary field. The ball $B(\mathbf{r}_q, a)$ has been drawn (−−) only in part.*

for the charge generates its own electric field according to (2.15) and (2.29). Although the magnetic field of q is obviously null, as the charge is fixed in space, the volume density of electromagnetic momentum (1.332) is not and reads

$$\mathbf{g}_{em}(\mathbf{r}) = \underbrace{\frac{1}{c_0^2}\mathcal{E}_q(\mathbf{r}, \mathbf{r}_q) \times \mathcal{H}_J(\mathbf{r})}_{\text{interaction}} + \underbrace{\frac{1}{c_0^2}\mathcal{E}_J(\mathbf{r}) \times \mathcal{H}_J(\mathbf{r})}_{\text{self}} \tag{5.120}$$

with c_0 (the speed of light in free space) given by (1.53). The first summand in the right-hand member is the density of momentum available to the charge for motion and due to the interaction with the steady current $\mathcal{J}(\mathbf{r})$, whereas the second contribution is the density of self-momentum stored in the field $\mathcal{E}_J(\mathbf{r})$, $\mathcal{H}_J(\mathbf{r})$.

The total interaction momentum $\mathcal{G}_q(\mathbf{r}_q)$ of q follows by integrating the corresponding density over the whole space. To this purpose we begin with a finite region of space bounded by two concentric balls $B(\mathbf{r}_q, a)$ and $B(\mathbf{r}_q, b)$, i.e., $V := B(\mathbf{r}_q, a) \setminus B(\mathbf{r}_q, b)$, with $b < a$ (Figure 5.5). The smaller ball serves to exclude the location of the charge where the potential thereof is singular, and the radius a is chosen large enough to encompass V_J. Thus, we evaluate

$$\begin{aligned}
\mathcal{G}_q(\mathbf{r}_q; a, b) &:= \int_V \mathrm{d}V \frac{1}{c_0^2}\mathcal{E}_q(\mathbf{r}, \mathbf{r}_q) \times \mathcal{H}_J(\mathbf{r}) = -\int_V \mathrm{d}V \frac{1}{c_0^2}\nabla\Phi_q(\mathbf{r}, \mathbf{r}_q) \times \mathcal{H}_J(\mathbf{r}) \\
&= \int_V \mathrm{d}V \frac{1}{c_0^2}\Phi_q(\mathbf{r}, \mathbf{r}_q)\nabla \times \mathcal{H}_J(\mathbf{r}) - \int_V \mathrm{d}V \frac{1}{c_0^2}\nabla \times [\Phi_q(\mathbf{r}, \mathbf{r}_q)\mathcal{H}_J(\mathbf{r})] \\
&= \int_{V_J} \mathrm{d}V \frac{1}{c_0^2}\Phi_q(\mathbf{r}, \mathbf{r}_q)\mathcal{J}(\mathbf{r}) - \int_{\partial B(\mathbf{r}_q, a)} \mathrm{d}S \frac{1}{c_0^2}\hat{\mathbf{n}}(\mathbf{r}) \times \mathcal{H}_J(\mathbf{r})\Phi_q(\mathbf{r}, \mathbf{r}_q) \\
&\quad - \int_{\partial B(\mathbf{r}_q, b)} \mathrm{d}S \frac{1}{c_0^2}\hat{\mathbf{n}}(\mathbf{r}) \times \mathcal{H}_J(\mathbf{r})\Phi_q(\mathbf{r}, \mathbf{r}_q)
\end{aligned} \tag{5.121}$$

having invoked (4.9) and applied the integral theorem (H.91) with the unit normals pointing outwards V. The remaining volume integral is independent of a and b for the charge is located outside the current region V_J and the latter is bounded. We can show that the surface integrals vanish in the limit as $a \to +\infty$ and $b \to 0$, viz.,

$$\left| \int_{\partial B(\mathbf{r}_q,a)} \mathrm{d}S\, \hat{\mathbf{n}}(\mathbf{r}) \times \mathcal{H}_J(\mathbf{r})\Phi_q(\mathbf{r},\mathbf{r}_q) \right| \leqslant \int_{\partial B(\mathbf{r}_q,a)} \mathrm{d}S\, |\mathcal{H}_J(\mathbf{r})||\Phi_q(\mathbf{r},\mathbf{r}_q)|$$
$$\leqslant \int_{\partial B(\mathbf{r}_q,a)} \mathrm{d}S\, \frac{C_H}{a^2}\frac{|q|}{4\pi\varepsilon_0 a} = \frac{C_H|q|}{a\varepsilon_0} \xrightarrow[a\to+\infty]{} 0 \tag{5.122}$$

and

$$\left| \int_{\partial B(\mathbf{r}_q,b)} \mathrm{d}S\, \hat{\mathbf{n}}(\mathbf{r}) \times \mathcal{H}_J(\mathbf{r})\Phi_q(\mathbf{r},\mathbf{r}_q) \right| \leqslant \int_{\partial B(\mathbf{r}_q,b)} \mathrm{d}S\, |\mathcal{H}_J(\mathbf{r})||\Phi_b(\mathbf{r},\mathbf{r}_q)|$$
$$\leqslant \|\mathcal{H}_J\|_\infty \int_{\partial B(\mathbf{r}_q,b)} \mathrm{d}S\, \frac{|q|}{4\pi\varepsilon_0 b} = b\frac{\|\mathcal{H}_J\|_\infty\,|q|}{\varepsilon_0} \xrightarrow[b\to 0]{} 0 \tag{5.123}$$

by virtue of (2.29) and (4.37). Therefore, from (5.121) we have

$$\mathcal{G}_q(\mathbf{r}_q) = \lim_{\substack{a\to+\infty \\ b\to 0}} \mathcal{G}_q(\mathbf{r}_q;a,b) = \frac{1}{c_0^2}\int_{V_J} \mathrm{d}V\, \Phi_q(\mathbf{r},\mathbf{r}_q)\mathcal{J}(\mathbf{r}) = q\int_{V_J} \mathrm{d}V\, \frac{\mu_0\mathcal{J}(\mathbf{r})}{4\pi|\mathbf{r}_q-\mathbf{r}|} = q\mathcal{A}_J(\mathbf{r}_q) \tag{5.124}$$

on account of (1.115), (2.29) and the integral representation (5.29). From (5.124) we gather that $\mathcal{A}_J(\mathbf{r}_q)$ has the physical meaning of *potential momentum per unit charge* at the point $\mathbf{r}_q$ [10]. Such momentum — like the potential energy in the field — is available for the motion of the charge and may be converted into mechanical momentum (Example 5.2).

We emphasize that (5.124), being based on (5.29), holds under the assumption that $\nabla\cdot\mathcal{A}_J(\mathbf{r}) = 0$, i.e., the vector potential satisfies the Coulomb gauge (Section 5.1.1). However, if another gauge is chosen and the vector potential $\mathcal{A}_{J1}(\mathbf{r})$ satisfies (4.32), then the last step in (5.124) leads to

$$\mathcal{G}_q(\mathbf{r}_q) = q\mathcal{A}_{J1}(\mathbf{r}_q) + q\nabla\chi(\mathbf{r})|_{\mathbf{r}_q} \tag{5.125}$$

so that the potential momentum ultimately does not depend on the gauge.

Example 5.2 (Potential momentum of a charge outside an infinite solenoid)

The identification of the vector potential with the interaction momentum of a unitary electric charge fixed in space provides us with the tool for the analysis of the behavior of a point charge q placed *outside* the infinite solenoid investigated in Examples 4.2 and 4.3 (cf. [13, Section 17-4]).

Since the charge is fixed in space and the current I in the solenoid is steady, the total momentum of the charge is constant in accordance with (1.354). Suppose that at time $t = 0$ we start diminishing the current I so slowly that at any given time the configuration may be assumed as stationary and no radiation occurs. From (1.354), (5.124) and the computed vector potential (4.47) we get

$$\frac{\mathrm{d}}{\mathrm{d}t}\mathcal{G}_{mech}(t) = -\frac{\mathrm{d}}{\mathrm{d}t}q\mathcal{A}(\mathbf{r}_q,t) = -q\mu_0\frac{\mathcal{N}a^2}{2\tau}\frac{\mathrm{d}I}{\mathrm{d}t}\hat{\boldsymbol{\varphi}}, \qquad t \geqslant 0, \qquad \tau > a \tag{5.126}$$

having dropped the subscript J from the potential for it is inconsequential for the present discussion.

The equation above indicates that, although the charge was initially at rest, it acquires mechanical momentum in the azimuthal direction — and thus starts rotating — just because the magnetic vector potential at the location of the charge varies. The effect is all the more striking for we showed in Example 4.2 that the magnetic intensity $\mathcal{B}(\tau)$ is zero for points outside the solenoid, where the charge resides. While a charge 'feels' electric and magnetic fields in the way prescribed by the Lorentz force (1.4), here the motion does not seem due to the field directly, as it were. Rather, the *potential momentum* available for motion is converted into mechanical momentum [10, Section 6.2].

Then again, one can claim that the slowly changing magnetic intensity in the solenoid induces an electric field at the location of the charge in accordance with the Faraday law (1.20), and it is this electric field that sets the charge in motion (cf. [14, Example 7.8]). Indeed, by invoking the axial symmetry of the solenoid we can obtain the induced electric field through the global Faraday law (1.8) applied over a circle C which has radius $\tau > a$ and center the origin, and is oriented perpendicularly to the z-axis (Figure 4.4b). In symbols, this gives

$$\oint_{\partial C} \mathrm{d}s\, \hat{\mathbf{s}}(\mathbf{r}) \cdot \mathcal{E}(\mathbf{r}, t) = 2\pi\tau\mathcal{E}_{\varphi}(\tau, t) = -\mu_0 \mathcal{N}\frac{\mathrm{d}I}{\mathrm{d}t}\pi a^2, \qquad t \geqslant 0, \qquad \tau > a \tag{5.127}$$

whence it follows

$$\mathcal{E}_{\varphi}(\tau, t) = -\mu_0 \frac{\mathcal{N}a^2}{2\tau}\frac{\mathrm{d}I}{\mathrm{d}t}, \qquad t \geqslant 0, \qquad \tau > a \tag{5.128}$$

where we notice that $\mathcal{E}(\mathbf{r}, t)$ is oriented along $\hat{\boldsymbol{\varphi}}$ since $I(t)$ is decreasing. If the test charge is located somewhere on ∂C, according to (1.4) it experiences the force

$$\mathcal{F}_{em}(t) = -q\mu_0 \frac{\mathcal{N}a^2}{2\tau}\frac{\mathrm{d}I}{\mathrm{d}t}\hat{\boldsymbol{\varphi}}, \qquad t \geqslant 0, \qquad \tau > a \tag{5.129}$$

whereby we arrive again at the equation of motion (5.126).

We may compute the final velocity $\mathbf{v}_q(+\infty)$ acquired by the charge by integrating (5.126) for $t \in [0, +\infty)$, viz.,

$$m\mathbf{v}_q(+\infty) = \int_0^{+\infty} \mathrm{d}t\, \frac{\mathrm{d}}{\mathrm{d}t}\mathcal{G}_{mech}(t) = q\mu_0 \frac{\mathcal{N}a^2}{2\tau} I(0)\hat{\boldsymbol{\varphi}} = q\mathcal{A}(\mathbf{r}_q, 0) \tag{5.130}$$

where m is the mass of the charged particle, $I(0)$ is the initial value of the current flowing in the solenoid, and we have assumed that the current vanishes for $t \to +\infty$. This calculation shows that the potential momentum is entirely converted into mechanical momentum.

(End of Example 5.2)

The vector potential has physical meaning also in quantum electrodynamics. It can be shown that, as a charged particle q moves along a path γ through a region of space in which $\mathcal{A}(\mathbf{r})$ is non-zero, the *wave-function* $\psi(\mathbf{r}, t)$ of the particle [15, Chapter 2] acquires a phase shift given by [13, Section 15-5]

$$\Delta\phi = \frac{q}{\hbar}\int_{\gamma} \mathrm{d}s\, \hat{\mathbf{s}}(\mathbf{r}) \cdot \mathcal{A}(\mathbf{r}) \tag{5.131}$$

where

$$\hbar = \frac{h}{2\pi} \qquad \text{with} \qquad h = 6.62607015 \cdot 10^{-34}\,\mathrm{Js} \tag{5.132}$$

is the reduced Planck constant (which has the physical dimension of an angular momentum). Since the phase shift depends on the actual path followed by the particle, the latter behaves as a probe of sorts capable of detecting the magnetic vector potential. This phenomenon is known as the Aharonov-Bohm effect [16], [17, Section 2.2].

5.4 Geometrical meaning of the scalar potential

Despite being a many-valued scalar field, the magnetic potential $\Psi(\mathbf{r})$ defined in (4.68) has an interesting geometrical interpretation when it is generated by a thin current-carrying wire which forms a closed loop as in Figure 4.1a. We assume that the wire is immersed in an unbounded isotropic medium with permeability μ. More importantly, so long as the cross section of the wire is much smaller than the characteristic size or the maximum chord of the loop, we may think of the steady current density $\mathcal{J}(\mathbf{r}')$ as being a line current of intensity $I(\mathbf{r}')$ localized on the closed path $\gamma := \partial S$ where S is a quite arbitrary open smooth surface which has γ as its boundary, as is suggested in Figure 5.6. Further, since $\mathcal{J}(\mathbf{r}')$ is steady, then (4.2) demands that $I(\mathbf{r}')$ be equal to a constant value I_0.

On the grounds of the general solution (5.29) we argue that the magnetic vector potential generated by I_0 reads

$$\mathcal{A}(\mathbf{r}) = \mu \oint_\gamma \mathrm{d}s' \, \frac{I_0 \hat{\mathbf{s}}(\mathbf{r}')}{4\pi|\mathbf{r} - \mathbf{r}'|}, \qquad \mathbf{r} \notin \gamma \tag{5.133}$$

where $\hat{\mathbf{s}}(\mathbf{r}')$ denotes the unit vector tangential to γ (Figure 5.6). Actually, at the end of this section we will have the tools to put (5.133) on solid ground. Formula (5.133) is only valid for observation points $\mathbf{r}$ strictly *away* from the position of the loop because the line integral diverges for $\mathbf{r}' = \mathbf{r}$ even if it is interpreted in the Cauchy principal value sense. The occurrence of such infinity is both a consequence and a limitation of having modelled the source as an infinitely thin wire. Nonetheless, the stationary magnetic field $\mathcal{H}(\mathbf{r})$ generated by I_0 follows from (5.133) through (4.29) and the constitutive relationship (1.118), namely,

$$\mathcal{H}(\mathbf{r}) = \nabla \times \oint_\gamma \mathrm{d}s' \, \frac{I_0 \hat{\mathbf{s}}(\mathbf{r}')}{4\pi|\mathbf{r} - \mathbf{r}'|}, \qquad \mathbf{r} \notin \gamma \tag{5.134}$$

and we wish to show that

(1) $\mathcal{H}(\mathbf{r})$ can be derived alternatively from a magnetic scalar potential $\Psi(\mathbf{r})$;
(2) in particular, $\Psi(\mathbf{r})$ is proportional to the solid angle $\Omega(\mathbf{r})$ subtended by S and viewed from the observation point $\mathbf{r}$ (Figure 5.6) [18, pp. 36-42], [19, pp. 39-40].

To accomplish our task we first derive the integral representation of the solid angle $\Omega(\mathbf{r})$. We introduce the ball $B(\mathbf{r}, a)$ which is large enough to enclose the line γ and the chosen surface S. Then, we project γ onto the sphere ∂B by means of straight lines which emerge from $\mathbf{r}$, are tangential to γ and intersect $B(\mathbf{r}, a)$. This construction also yields a spherical sector $S_B \subset \partial B$ which constitutes the projection of S onto ∂B (Figure 5.6). The bundle of segments that connect γ to ∂S_B generates a surface S_L which is, in fact, part of a cone with apex in $\mathbf{r}$.

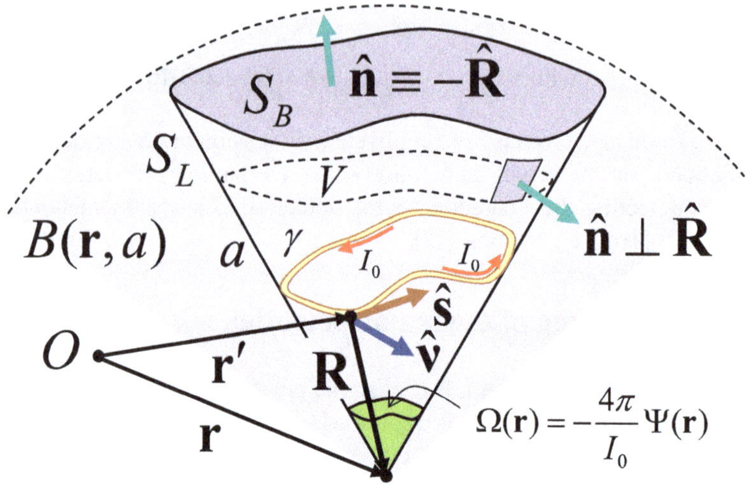

Figure 5.6 For illustrating the geometrical meaning of the magnetic scalar potential $\Psi(\mathbf{r})$ generated by a current loop γ. The ball $B(\mathbf{r}, a)$ has been outlined (−−) only in part.

We indicate with V the piecewise-smooth domain bounded by the surfaces S, S_B and S_L. We let $\mathbf{R} = \mathbf{r} - \mathbf{r}'$ and recall that $1/R = 1/|\mathbf{r} - \mathbf{r}'|$ is a harmonic function (i.e., solves the Laplace equation) for $\mathbf{r} \neq \mathbf{r}'$ [6, Section 8.7], [7, Section 2.2]. Then, we examine the vanishing volume integral

$$\begin{aligned} 0 &= \int_V dV' \, \nabla'^2 \frac{1}{|\mathbf{r} - \mathbf{r}'|} = \int_V dV' \, \nabla' \cdot \frac{\hat{\mathbf{R}}}{R^2} \\ &= \int_{S_B} dS' \, \frac{\hat{\mathbf{n}}(\mathbf{r}') \cdot \hat{\mathbf{R}}}{R^2} + \int_{S_L} dS' \, \underbrace{\frac{\hat{\mathbf{n}}(\mathbf{r}') \cdot \hat{\mathbf{R}}}{R^2}}_{=0} - \int_S dS' \, \frac{\hat{\mathbf{n}}(\mathbf{r}') \cdot \hat{\mathbf{R}}}{R^2} \\ &= -\frac{1}{a^2} \int_{S_B} dS' - \int_S dS' \, \frac{\hat{\mathbf{n}}(\mathbf{r}') \cdot \hat{\mathbf{R}}}{R^2} \end{aligned} \tag{5.135}$$

where we have applied the Gauss theorem (A.53) because $1/R$ is regular for $\mathbf{r} \neq \mathbf{r}'$. The contribution from S_L is identically null since thereon the unit normal $\hat{\mathbf{n}}(\mathbf{r}')$ is perpendicular to $\hat{\mathbf{R}}$ by construction. The leading minus sign of the contribution from S arises from the orientation of $\hat{\mathbf{n}}(\mathbf{r}')$, which, for $\mathbf{r}' \in S$, we have chosen as pointing inwards V and away from $\mathbf{r}$ for convenience (Figure 5.6). Lastly, the first integral in the rightmost member equals the negative of the ratio of the area of S_B to the square of the radius a, i.e., the solid angle $\Omega(\mathbf{r})$. Therefore, we find

$$\Omega(\mathbf{r}) := -\int_S dS' \, \frac{\hat{\mathbf{n}}(\mathbf{r}') \cdot \hat{\mathbf{R}}}{R^2} = \int_S dS' \, \hat{\mathbf{n}}(\mathbf{r}') \cdot \nabla \frac{1}{R}, \qquad \mathbf{r} \notin S \tag{5.136}$$

which is in agreement with the Gauss solid angle formula (2.248) in the special case where S is a closed surface and contains $\mathbf{r}$. Indeed, $\Omega(\mathbf{r})$ has the form of a double-layer potential [see (2.188)] with negative unitary density. What is more, a quick comparison of (5.136) and (3.10) suggests that $\Omega(\mathbf{r})$ is also proportional to the electrostatic scalar potential $\Phi(\mathbf{r})$ due to a uniform layer of electric dipoles distributed on S and aligned with $\hat{\mathbf{n}}(\mathbf{r}')$ (see Figure 3.2).

Secondly, we compute the gradient of $\Omega(\mathbf{r})$ by means of (5.136) by noticing that we are allowed to interchange derivatives and integral over S because the integrand is regular for $\mathbf{r} \notin S$. Thus, we compute

$$\nabla\left(\hat{\mathbf{n}}(\mathbf{r}')\cdot\nabla\frac{1}{R}\right) = \hat{\mathbf{n}}(\mathbf{r}')\cdot\nabla\nabla\frac{1}{R} = \nabla\times\left(\nabla\frac{1}{R}\times\hat{\mathbf{n}}(\mathbf{r}')\right) \tag{5.137}$$

where we have used (H.53) and (H.54) in succession while keeping in mind that $\hat{\mathbf{n}}(\mathbf{r}')$ is a constant vector with respect to $\mathbf{r}$, $\nabla(1/R)$ is a curl-free vector field, and $1/R$ is a harmonic function for $\mathbf{r} \neq \mathbf{r}'$. Then, we write

$$\begin{aligned}\nabla\Omega(\mathbf{r}) &= \int_S \mathrm{d}S'\,\nabla\left(\hat{\mathbf{n}}(\mathbf{r}')\cdot\nabla\frac{1}{R}\right) = \int_S \mathrm{d}S'\,\nabla\times\left(\nabla\frac{1}{R}\times\hat{\mathbf{n}}(\mathbf{r}')\right) = -\nabla\times\int_S \mathrm{d}S'\,\nabla'\frac{1}{R}\times\hat{\mathbf{n}}(\mathbf{r}')\\ &= -\nabla\times\int_S \mathrm{d}S'\,\nabla_s'\frac{1}{R}\times\hat{\mathbf{n}}(\mathbf{r}') = -\nabla\times\int_S \mathrm{d}S'\,\nabla_s'\times\frac{\hat{\mathbf{n}}(\mathbf{r}')}{R}\end{aligned} \tag{5.138}$$

where we have noticed that cross-multiplying with $\hat{\mathbf{n}}(\mathbf{r}')$ filters out the component of $\nabla'(1/R)$ perpendicular to S. The symbol ∇_s' [see (A.48)] denotes the surface gradient over S, and the last step follows from (H.80) since the surface curl of $\hat{\mathbf{n}}(\mathbf{r}')$ is identically null. We then apply the *surface curl theorem* (H.103) to the last integral, viz.,

$$\begin{aligned}\nabla\Omega(\mathbf{r}) &= -\nabla\times\oint_\gamma \mathrm{d}s'\,\hat{\boldsymbol{\nu}}(\mathbf{r}')\times\frac{\hat{\mathbf{n}}(\mathbf{r}')}{R} + \nabla\times\int_S \mathrm{d}S\; J(\mathbf{r}')\hat{\mathbf{n}}(\mathbf{r}')\times\frac{\hat{\mathbf{n}}(\mathbf{r}')}{R}\\ &= \nabla\times\oint_\gamma \mathrm{d}s'\,\frac{\hat{\mathbf{s}}(\mathbf{r}')}{|\mathbf{r}-\mathbf{r}'|}, \qquad \mathbf{r}\notin\gamma\end{aligned} \tag{5.139}$$

where $J(\mathbf{r}')$ indicates the first curvature of S given by (H.100) [8, Appendix 3]. The result follows because $\hat{\mathbf{n}} \times \hat{\boldsymbol{\nu}} = \hat{\mathbf{s}}$ on γ (Figure 5.6). All in all, we have turned the gradient of the solid angle into the curl of a suitable vector potential by means of purely mathematical considerations.

Finally, putting (5.139) to work in (5.134) yields the magnetic field in terms of the gradient of a scalar potential, namely,

$$\mathcal{H}(\mathbf{r}) = \frac{I_0}{4\pi}\nabla\times\oint_\gamma \mathrm{d}s'\,\frac{\hat{\mathbf{s}}(\mathbf{r}')}{|\mathbf{r}-\mathbf{r}'|} = \nabla\frac{I_0\Omega(\mathbf{r})}{4\pi} = -\nabla\Psi(\mathbf{r}), \qquad \mathbf{r}\notin\gamma \tag{5.140}$$

where the relevant magnetic scalar potential reads

$$\Psi(\mathbf{r}) = -\frac{I_0\Omega(\mathbf{r})}{4\pi} = \int_S \mathrm{d}S'\,\frac{\hat{\mathbf{R}}}{4\pi R^2}\cdot I_0\hat{\mathbf{n}}(\mathbf{r}'), \qquad \mathbf{r}\notin S \tag{5.141}$$

on account of (5.136). Additionally, on the grounds of (4.85) we claim that the vector field

$$\mathcal{M}_S(\mathbf{r}) := I_0\hat{\mathbf{n}}(\mathbf{r}), \qquad \mathbf{r}\in S \tag{5.142}$$

represents a uniform *surface density* of magnetic dipoles with moments perpendicular to S.

From a geometrical viewpoint, (5.141) tells us that the solid angle $\Omega(\mathbf{r})$ represents the magnetic scalar potential produced by a current of intensity $I_0 = -4\pi$ A flowing along γ. Not only do (5.140) and (5.141) prove our initial statements, they also show that a uniform current loop γ is equivalent to a fictitious distribution of magnetic dipoles over any smooth open surface S which has γ as boundary [20, Section 3.5]. Since in principle a dipole is made up of two infinitely close charges, we may also state that a uniform current loop γ is equivalent to a double layer of fictitious *magnetic* charges which have opposite sign and are distributed on either side of S (cf. Section 2.9).

At this stage we observe that we may also proceed backwards in our reasoning, namely, using arguments similar to those invoked in Section 3.1 for the electrostatic potential, we can first prove that the magnetostatic potential $\Psi(\mathbf{r})$ generated by the layer of magnetic dipoles with density (5.142) is indeed given by (5.141). Then, by applying (5.140) in reverse gear we express the magnetic field as the curl of a suitable vector potential. This finding provides the basis for claiming that the magnetic vector potential $\mathcal{A}(\mathbf{r})$ produced by a line current I_0 is indeed given by (5.133).

5.5 Multipole expansion of the vector potential

The behavior of the vector potential $\mathcal{A}(\mathbf{r})$ for points far away from the generating stationary currents was speculated in Section 4.2 by requiring that the magnetic energy be finite. We concluded that $\mathcal{A}(\mathbf{r})$ must decay at least as the inverse of the distance from the sources. This finding seems reasonable also because it agrees with the similar behavior of the electrostatic scalar potential $\Phi(\mathbf{r})$. However, since we now have the general expression (5.29) for the vector potential in a homogeneous isotropic unbounded medium we may examine the asymptotic behavior of $\mathcal{A}(\mathbf{r})$ rigorously.

As a matter fact, (5.29) is in the form of a convolution integral with a kernel which is the same three-dimensional static Green function derived in Section 2.6. Therefore, we may expand the function $1/|\mathbf{r}-\mathbf{r}'|$ in inverse powers of $r = |\mathbf{r}|$ as we did for the scalar potential; the resulting asymptotic expansion is still called the *multipole expansion* for the vector potential. We skip the details because the procedure is just based on (3.195), which when inserted into (5.29) yields [14, Section 5.4.3], [1, Section 4.5], [21, Section 5.7], [11, Chapter 11], [22]

$$\mathcal{A}(\mathbf{r}) = \frac{\mu}{4\pi r}\int_{V_J} dV'\mathcal{J}(\mathbf{r}') + \frac{\mu}{4\pi r^2}\int_{V_J} dV'\hat{\mathbf{r}}\cdot\mathbf{r}'\mathcal{J}(\mathbf{r}') \\ + \frac{1}{2}\frac{\mu}{4\pi r^3}\int_{V_J} dV'\hat{\mathbf{r}}\cdot(3\,\mathbf{r}'\mathbf{r}' - r'^2\overline{\mathbf{I}})\cdot\hat{\mathbf{r}}\mathcal{J}(\mathbf{r}') + o\left(\frac{1}{r^3}\right), \qquad r\to+\infty \qquad (5.143)$$

i.e., the first terms of the desired multipole expansion. In the present case, (3.195) and hence (5.143) are valid for points $\mathbf{r}\in\mathbb{R}^3\setminus B(0,a)$ where the radius a is large enough for the ball $B(0,a)$ to contain the region of the current V_J. Comparison of (5.143) with the corresponding formula (3.204) for the electrostatic potential suggests that the first term may represent the contribution of the *magnetic charge* in V_J and the second term may be interpreted as the potential of an equivalent magnetic dipole located at $\mathbf{r}$. Then again, we have repeatedly stated that isolated magnetic poles do not exist, and hence the first term in (5.143) ought to be zero. The second term possesses the right dependence on the distance r to be the contribution of a dipole, but needs some polishing to single out the magnetic dipole moment, as the integrand is a dyadic. The third term is a quadrupole moment of sorts but we will not attempt any further manipulation.

To confirm our expectations we first need to derive a few general properties of steady currents [2, 4]. Since we plan on using the results to evaluate the integrals of (5.143), we use $\mathbf{r}'$ as an independent variable and consider two differentiable functions $f(\mathbf{r}')$ and $g(\mathbf{r}')$. The following differential identity holds true

$$\nabla' \cdot [f(\mathbf{r}')g(\mathbf{r}')\mathfrak{J}(\mathbf{r}')] = f(\mathbf{r}')\mathfrak{J}(\mathbf{r}') \cdot \nabla' g(\mathbf{r}') + g(\mathbf{r}')\mathfrak{J}(\mathbf{r}') \cdot \nabla' f(\mathbf{r}') \tag{5.144}$$

for $\mathbf{r}' \in V_J$ on account of (H.51), (H.47) and $\nabla' \cdot \mathfrak{J}(\mathbf{r}') = 0$.

If we set $f(\mathbf{r}') = x'$ and $g(\mathbf{r}') = 1$, we find

$$\nabla' \cdot [x'\mathfrak{J}(\mathbf{r}')] = \nabla' x' \cdot \mathfrak{J}(\mathbf{r}') = \hat{\mathbf{x}} \cdot \mathfrak{J}(\mathbf{r}') = \mathfrak{J}_x(\mathbf{r}'), \qquad \mathbf{r}' \in V_J \tag{5.145}$$

which integrated over V_J gives

$$\int_{V_J} \mathrm{d}V' \, \mathfrak{J}_x(\mathbf{r}') = \int_{V_J} \mathrm{d}V' \, \nabla' \cdot [x'\mathfrak{J}(\mathbf{r}')] = \int_{\partial V_J} \mathrm{d}S' \, \hat{\mathbf{n}}(\mathbf{r}') \cdot \mathfrak{J}(\mathbf{r}')x' = 0 \tag{5.146}$$

because $\mathfrak{J}(\mathbf{r}')$ is confined in V_J by hypothesis. By repeating the calculation for $f(\mathbf{r}') = y'$ and $f(\mathbf{r}') = z'$ and putting the Cartesian components of the current density back together we conclude

$$\int_{V_J} \mathrm{d}V' \, \mathfrak{J}(\mathbf{r}') = 0 \tag{5.147}$$

i.e., the volume integral of a steady current is null.

Next, if we choose $f(\mathbf{r}') = x'^2$ and $g(\mathbf{r}') = 1/2$ we find

$$\nabla' \cdot \left[x'^2\mathfrak{J}(\mathbf{r}')\right]/2 = x'\nabla' x' \cdot \mathfrak{J}(\mathbf{r}') = x'\hat{\mathbf{x}} \cdot \mathfrak{J}(\mathbf{r}') = x'\mathfrak{J}_x(\mathbf{r}'), \qquad \mathbf{r}' \in V_J \tag{5.148}$$

which integrated over V_J gives

$$\int_{V_J} \mathrm{d}V' \, x'\mathfrak{J}_x(\mathbf{r}') = \frac{1}{2}\int_{V_J} \mathrm{d}V' \, \nabla' \cdot \left[x'^2\mathfrak{J}(\mathbf{r}')\right] = \frac{1}{2}\int_{\partial V_J} \mathrm{d}S' \, \hat{\mathbf{n}}(\mathbf{r}') \cdot \mathfrak{J}(\mathbf{r}')x' = 0 \tag{5.149}$$

again because $\mathfrak{J}(\mathbf{r}')$ is confined in V_J. Repeating for $f(\mathbf{r}') = y'^2$ and $f(\mathbf{r}') = z'^2$ and combining the results allows us to conclude

$$\int_{V_J} \mathrm{d}V' \, [x'\mathfrak{J}_x(\mathbf{r}') + y'\mathfrak{J}_y(\mathbf{r}') + z'\mathfrak{J}_z(\mathbf{r}')] = \int_{V_J} \mathrm{d}V' \, \mathbf{r}' \cdot \mathfrak{J}(\mathbf{r}') = 0. \tag{5.150}$$

Lastly, if we set $f(\mathbf{r}') = x'$ and $g(\mathbf{r}') = y'$ we have

$$\nabla' \cdot [x'y'\mathfrak{J}(\mathbf{r}')] = x'\mathfrak{J}_y(\mathbf{r}') + y'\mathfrak{J}_x(\mathbf{r}'), \qquad \mathbf{r}' \in V_J \tag{5.151}$$

which integrated over V_J yields

$$\int_{V_J} \mathrm{d}V' \, [x'\mathfrak{J}_y(\mathbf{r}') + y'\mathfrak{J}_x(\mathbf{r}')] = \int_{\partial V_J} \mathrm{d}S' \, \hat{\mathbf{n}}(\mathbf{r}') \cdot \mathfrak{J}(\mathbf{r}')x'y' = 0 \tag{5.152}$$

for the usual reason. Analogous expressions hold true for the other pairs of Cartesian components of $\mathbf{r}'$ and $\mathfrak{J}(\mathbf{r}')$.

In light of (5.147) it is apparent that the first term in the right-hand side of (5.143) is indeed zero. This finding also implies that $\mathcal{A}(\mathbf{r})$ falls off more rapidly than what we had initially surmised in (4.36), because the leading term in the multipole expansion decays as the inverse square of the distance. In this regard, we observe

$$\begin{aligned}\int_{V_J} dV'\,\hat{\mathbf{r}}\cdot\mathbf{r}'\mathcal{J}_x(\mathbf{r}') &= \frac{1}{r}\int_{V_J} dV'\,(xx'+yy'+zz')\mathcal{J}_x(\mathbf{r}')\\ &= \frac{1}{2r}\int_{V_J} dV'\,\left(zz'\mathcal{J}_x - zx'\mathcal{J}_z + yy'\mathcal{J}_x - yx'\mathcal{J}_y\right)\\ &= \frac{1}{2r}\int_{V_J} dV'\,[z\hat{\mathbf{y}}\cdot(\mathbf{r}'\times\mathcal{J}) - y\hat{\mathbf{z}}\cdot(\mathbf{r}'\times\mathcal{J})]\\ &= \frac{1}{2r}\int_{V_J} dV'\,(\mathbf{r}\times\hat{\mathbf{x}})\cdot(\mathbf{r}'\times\mathcal{J}) = \frac{1}{2}\int_{V_J} dV'\,[\mathbf{r}'\times\mathcal{J}(\mathbf{r}')]\times\hat{\mathbf{r}}\cdot\hat{\mathbf{x}}\end{aligned} \tag{5.153}$$

by virtue of (5.149), (5.152) and similar identities, and finally (H.13). Two other formulas follow by starting with J_y and J_z. By putting these results together we find

$$\int_{V_J} dV'\,\hat{\mathbf{r}}\cdot\mathbf{r}'\mathcal{J}(\mathbf{r}') = \frac{1}{2}\int_{V_J} dV'\,[\mathbf{r}'\times\mathcal{J}(\mathbf{r}')]\times\hat{\mathbf{r}} \tag{5.154}$$

and the latter allows us to write (5.143) as

$$\mathcal{A}(\mathbf{r}) = \mu\frac{1}{2}\int_{V_J} dV'\,[\mathbf{r}'\times\mathcal{J}(\mathbf{r}')]\times\frac{\hat{\mathbf{r}}}{4\pi r^2} + o\left(\frac{1}{r^2}\right), \qquad r\to+\infty. \tag{5.155}$$

Comparison with (4.87) prompts us to define the term

$$\mathbf{m} := \frac{1}{2}\int_{V_J} dV'\,\mathbf{r}'\times\mathcal{J}(\mathbf{r}') \tag{5.156}$$

as the equivalent magnetic dipole moment of the steady current $\mathcal{J}(\mathbf{r}')$. The magnetic dipole moment is always invariant under translations, say, $\mathbf{r}' = \mathbf{r}'' + \mathbf{u}$

$$\mathbf{m}_{O'} := \frac{1}{2}\int_{V_J} dV''\,(\mathbf{r}''+\mathbf{u})\times\mathcal{J}(\mathbf{r}''+\mathbf{u}) = \mathbf{m}_{O''} + \frac{\mathbf{u}}{2}\times\underbrace{\int_{V_J} dV''\,\mathcal{J}(\mathbf{r}''+\mathbf{u})}_{=0} \tag{5.157}$$

where $\mathbf{m}_{O'}$ and $\mathbf{m}_{O''}$ denote the moments with respect to the primed and double-primed coordinates. The last integral above is null by virtue of (5.147), in that $\mathcal{J}(\mathbf{r}''+\mathbf{u})$ is a steady current density.

Since (5.155) holds for distances $|\mathbf{r}|$ which are large as compared to the characteristic size of the volume V_J, we have just shown that any steady current behaves asymptotically as a magnetic dipole. Equivalently, the vector potential of a 'small' current density is well approximated by the vector potential produced by a magnetic dipole with moment given by (5.156).

The equivalence between steady currents and magnetic dipoles is ubiquitous in the stationary limit of the Maxwell equations. For instance, it also applies to the surface integral of $\hat{\mathbf{n}}(\mathbf{r}')\times[\nabla\times\mathcal{A}(\mathbf{r}')]$ over ∂V_1 in the representation (5.18) of the vector potential. With reference to Figure 5.1a, suppose that all the currents are confined in V_J and in V_1, so that we may let the surface S recede to

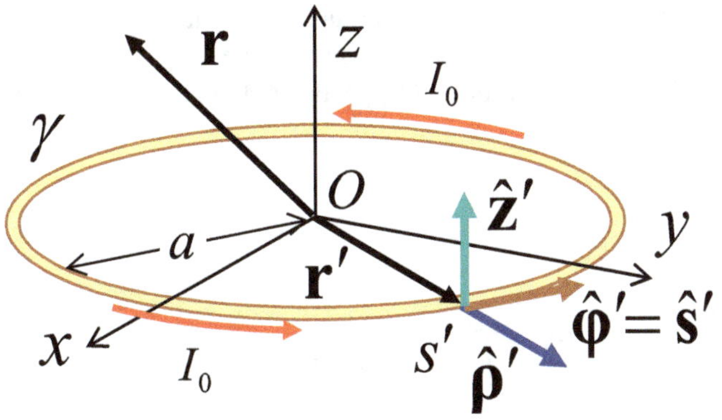

Figure 5.7 Geometrical setup for computing the equivalent magnetic dipole moment of a small circular loop of line current I_0.

infinity. We can write the contribution of $\hat{\mathbf{n}}(\mathbf{r}')\times[\nabla\times\mathcal{A}(\mathbf{r}')] = \hat{\mathbf{n}}(\mathbf{r}')\times\mathcal{B}(\mathbf{r}')$ as a multipole expansion in that the kernel is again $1/|\mathbf{r}-\mathbf{r}'|$. In symbols, we have

$$\int_{\partial V_1} dS' \, \frac{\hat{\mathbf{n}}(\mathbf{r}') \times \mathcal{B}(\mathbf{r}')}{4\pi R} = \frac{1}{4\pi r} \int_{\partial V_1} dS' \, \hat{\mathbf{n}}(\mathbf{r}') \times \mathcal{B}(\mathbf{r}') + \frac{1}{4\pi r^2} \frac{1}{2} \int_{\partial V_1} dS' \, \{\mathbf{r}' \times [\hat{\mathbf{n}}(\mathbf{r}') \times \mathcal{B}(\mathbf{r}')]\} \times \hat{\mathbf{r}} + o\left(\frac{1}{r^2}\right), \qquad r \to +\infty \tag{5.158}$$

whence it is apparent that $\hat{\mathbf{n}} \times \mathcal{B}/\mu$ plays the role of an equivalent surface density of steady current flowing on ∂V_1. Indeed, the first integral vanishes

$$\int_{\partial V_1} dS' \, \hat{\mathbf{n}}(\mathbf{r}') \times \mathcal{B}(\mathbf{r}') = \int_{V_1} dV' \, \nabla' \times \mathcal{B}(\mathbf{r}') = \mu_1 \int_{V_{J1}} dV' \, \mathcal{J}_1(\mathbf{r}) = 0 \tag{5.159}$$

because (4.9) holds in V_1 and the current $\mathcal{J}_1(\mathbf{r})$, $\mathbf{r} \in V_{J1} \subset V_1$ is steady. We have applied (H.91) on the grounds that $\hat{\mathbf{n}} \times \mathcal{B}$ is continuous across ∂V_{J1}. The second term of (5.158) is the magnetic dipole contribution.

Example 5.3 (Magnetic dipole moment of a small circular loop of uniform current)

The calculation of the equivalent dipole moment of a current through (5.156) may be accomplished analytically only for simple geometries. One of them is constituted by a current-carrying wire shaped in the form of a circular loop (Figure 5.7).

Since we are interested in the magnetic vector potential at distances which are large as compared to the radius a of the loop, we may also assume that the cross section of the wire is small with respect to a. As a result, we can think of the current density $\mathcal{J}(\mathbf{r}')$ as being a line current of intensity I concentrated on the circumference $\gamma := \{\mathbf{r}' \in \mathbb{R}^3 : |\mathbf{r}'| = a\}$ and tangent to γ in any point, whereby the magnetic dipole moment (5.156) passes over into a line integral along γ, viz.,

$$\mathbf{m} = \frac{1}{2} \oint_{\gamma} ds' \mathbf{r}' \times I(s') \hat{\mathbf{s}}' \tag{5.160}$$

where s' is the arc coordinate along γ and $\hat{\mathbf{s}}'$ the unit vector tangent to γ. Besides, since $\mathcal{J}(\mathbf{r}')$ is steady, then (4.2) implies that I is equal to a constant value, say, I_0. It is convenient to introduce a system of circular cylindrical coordinates (Appendix A.1) to represent the source point $\mathbf{r}'$, so we have

$$\mathbf{r}' = a\hat{\boldsymbol{\rho}}', \qquad \hat{\mathbf{s}}' = \hat{\boldsymbol{\varphi}}', \qquad \mathrm{d}s' = a\,\mathrm{d}\varphi' \tag{5.161}$$

and (5.160) gives

$$\mathbf{m} = \frac{1}{2}\int_0^{2\pi} \mathrm{d}\varphi' a^2 \hat{\boldsymbol{\rho}}' \times I_0 \hat{\boldsymbol{\varphi}}' = \frac{I_0 a^2}{2}\hat{\mathbf{z}} \int_0^{2\pi} \mathrm{d}\varphi' = I_0 \pi a^2 \hat{\mathbf{z}}. \tag{5.162}$$

In words, the magnetic dipole moment of a small circular loop of current is oriented perpendicularly to the plane of the loop and proportional to the current and the area of the circle bounded by the loop; this result was experimentally demonstrated by Ampère. Furthermore, formula (5.162) justifies the physical dimensions (Am^2) of $\mathbf{m}$ introduced in Section 4.5.

Finally, the vector potential produced by the current loop can be computed through (4.87). It should be kept in mind, though, that this is just the first term of the multipole expansion (5.155).

(End of Example 5.3)

5.6 Magnetization vector

In Section 4.5 we introduced the notion of static magnetic dipoles and derived the magnetic scalar and vector potentials on the grounds of the mathematical similarity between the electrostatic equations and the local form of the Ampère and magnetic Gauss laws. All in all, it really looks like as if a magnetic dipole were comprised of a pair of magnetic point charges.

Then again, in Section 5.5 we have shown that any finite-size stationary current density $\mathcal{J}(\mathbf{r})$ can be modelled to a first approximation as an equivalent magnetic dipole. So the implication is clear: magnetic dipoles do not exist, but rather they are just a useful mathematical expedient for describing current loops or, more generally, arbitrary current densities. However, do we really need magnetic dipoles at all?

For instance, in Section 5.1.2 we used one as a singular source to obtain the integral representation of the magnetic induction field. In truth, a consistent description of magnetism in free space requires only one magnetic entity, namely, the induction $\mathcal{B}(\mathbf{r})$. This is somehow attested by the simple, almost trivial, constitutive relationship (1.113), which renders the distinction between $\mathcal{B}(\mathbf{r})$ and $\mathcal{H}(\mathbf{r})$ in vacuum no more than a convention. Things are rather different, though, if we include magnetic media and permanent magnets in the picture, as it were. Therefore, we would like to substantiate the linear constitutive relationship (1.118) introduced in Section 1.6.

We start with the realistic idea of a material body as a collection of atoms or molecules held tightly together by mutual electric forces. The classical notion of electrons distributed and rotating in a 'cloud' around a nucleus will serve our purposes (see Figure 5.8a). We may think of the rotating electrons of an atom as a tiny electric current density or, in light of the multipole expansion derived in Section 5.5, as an elementary magnetic dipole; either way, each electron cloud produces a magnetic field [8, Section 6.6]. In this regard, it is worthwhile recalling that the positive orientation of the electric current is conventionally chosen as the direction of the positive charges. So, if the circular arrows in Figure 5.8a denote the direction of the net current flow, the electrons — which are negatively charged — actually rotate in the opposite direction.

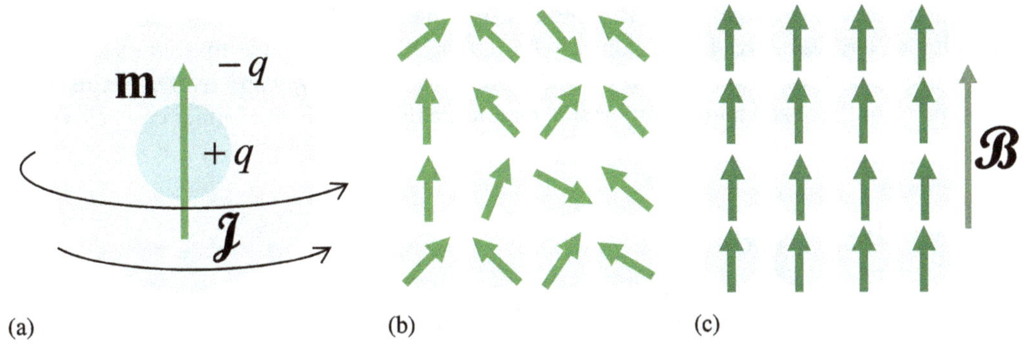

Figure 5.8 For defining the magnetization vector: (a) electron cloud and equivalent magnetic dipole; (b) the magnetic dipoles in a material medium are ordinarily oriented at random; (c) an external induction field causes the dipoles to orient coherently.

In ordinary conditions the net magnetic field generated by all the electrons in a material body is zero, essentially because the dipole moments are randomly oriented and the individual contributions do not add coherently, as illustrated in Figure 5.8b. On the other hand, if we subject the body to an external induction field generated by, e.g., a coil of wire carrying a stationary current, the dipoles tend to align with $\mathcal{B}(\mathbf{r})$ (Figure 5.8c) in order to minimize their potential energy (see Section 4.5). As a result, the combined effect of all the dipoles may give rise to a non-null macroscopic contribution which may enhance or oppose the external magnetic induction: we wish to quantify this statement.

Since a material body contains a huge number of magnetic dipoles, we may conveniently define a continuous volumetric distribution of dipole moments in a bounded region V_M. We introduce the **magnetization vector** $\mathcal{M}(\mathbf{r})$ (physical dimension: $\text{Am}^2/\text{m}^3 = \text{A/m}$) as

$$\mathcal{M}(\mathbf{r}) := \lim_{\Delta V \to 0} \frac{\Delta \mathbf{m}(\mathbf{r})}{\Delta V} = \frac{\mathrm{d}\mathbf{m}(\mathbf{r})}{\mathrm{d}V}, \qquad \mathbf{r} \in V_M \tag{5.163}$$

that is, a volume density of magnetic dipoles [23, Chapter 47], [24, Chapter 13], [25, Chapter 13], [11, Section 13.2]. A material body which acquires a non-zero magnetization vector when exposed to an external magnetic induction field is said to be *magnetized*.

We remark that $\mathcal{M}(\mathbf{r})$ is a secondary entity of quantity inasmuch as it depends on the applied induction field (cf. with the polarization vector). In most cases, then, the magnetization disappears as soon as the external induction field has been turned off. However, if the latter is strong enough, it may orient the magnetic dipoles coherently once and for all, so that the combined effect is maintained even after removing the external induction field [26]. Permanent magnets are precisely objects in which the magnetic dipoles are coherently aligned parallel to one another in the absence of an external magnetizing field $\mathcal{B}(\mathbf{r})$.

According to (4.88), which gives the magnetic vector potential of an elementary magnetic dipole in free space, we may surmise that a steady current $\mathcal{J}(\mathbf{r})$, $\mathbf{r} \in V_J$, $V_J \cap V_M = \emptyset$ in the presence of a magnetic body produces the vector potential

$$\mathcal{A}(\mathbf{r}) = \mu_0 \int_{V_J} \mathrm{d}V' \frac{\mathcal{J}(\mathbf{r}')}{4\pi R} + \mu_0 \int_{V_M} \mathrm{d}V' \, \mathcal{M}(\mathbf{r}') \times \nabla' \frac{1}{4\pi R}, \qquad \mathbf{r} \in \mathbb{R}^3 \tag{5.164}$$

in light of the integral representation (5.18). The effect of $\mathcal{J}(\mathbf{r})$ is augmented with a contribution that accounts for the state of magnetization of the body in V_M. Notice that both the current density and the magnetization vector exist in free space.

Under the assumption that $\mathcal{M}(\mathbf{r})$ is continuously differentiable for $\mathbf{r} \in V_M$, we show that the contribution of the magnetization vector may be interpreted as being due to volume and surface electric current densities flowing inside V_M and on the boundary ∂V_M, respectively. We distinguish the occurrence of the observation point outside and inside the body. In the former case, by virtue of (H.50) we write the integral over V_M as follows

$$\mu_0 \int_{V_M} \mathrm{d}V' \mathcal{M}(\mathbf{r}') \times \nabla' \frac{1}{4\pi R} = -\mu_0 \int_{V_M} \mathrm{d}V' \nabla' \times \frac{\mathcal{M}(\mathbf{r}')}{4\pi R} + \mu_0 \int_{V_M} \mathrm{d}V' \frac{\nabla' \times \mathcal{M}(\mathbf{r}')}{4\pi R}, \qquad \mathbf{r} \notin V_M \tag{5.165}$$

and we may safely apply the curl theorem (H.91) because $\mathcal{M}(\mathbf{r}')/R$ is of class $\mathcal{C}^1(V_M)^3 \cap \mathcal{C}(\overline{V}_M)^3$.

For observation points inside the body, we exclude the singularity of the kernel with a ball $B(\mathbf{r}, a)$ and split the integral into two parts, namely, over $V_a := V_M \setminus B(\mathbf{r}, a)$ and $B(\mathbf{r}, a)$

$$\begin{aligned}
\mu_0 \int_{V_M} \mathrm{d}V' \mathcal{M}(\mathbf{r}') \times \nabla' \frac{1}{4\pi R} &= \\
&= \mu_0 \int_{V_a} \mathrm{d}V' \mathcal{M}(\mathbf{r}') \times \nabla' \frac{1}{4\pi R} + \mu_0 \int_{B(\mathbf{r},a)} \mathrm{d}V' \mathcal{M}(\mathbf{r}') \times \nabla' \frac{1}{4\pi R} \\
&= -\mu_0 \int_{V_a} \mathrm{d}V' \nabla' \times \frac{\mathcal{M}(\mathbf{r}')}{4\pi R} + \mu_0 \int_{V_a} \mathrm{d}V' \frac{\nabla' \times \mathcal{M}(\mathbf{r}')}{4\pi R} + \mu_0 \int_{B(\mathbf{r},a)} \mathrm{d}V' \mathcal{M}(\mathbf{r}') \times \nabla' \frac{1}{4\pi R} \\
&= -\mu_0 \int_{\partial V_M} \mathrm{d}S' \frac{\hat{\mathbf{n}}(\mathbf{r}') \times \mathcal{M}(\mathbf{r}')}{4\pi R} - \mu_0 \int_{\partial B} \mathrm{d}S' \frac{\hat{\mathbf{n}}(\mathbf{r}') \times \mathcal{M}(\mathbf{r}')}{4\pi R} \\
&\quad + \mu_0 \int_{V_a} \mathrm{d}V' \frac{\nabla' \times \mathcal{M}(\mathbf{r}')}{4\pi R} + \mu_0 \int_{B(\mathbf{r},a)} \mathrm{d}V' \mathcal{M}(\mathbf{r}') \times \nabla' \frac{1}{4\pi R}
\end{aligned} \tag{5.166}$$

where we have applied (H.91) to the integral over V_a in that $\mathcal{M}(\mathbf{r}')/R$ is of class $\mathcal{C}^1(V_a)^3 \cap \mathcal{C}(\overline{V}_a)^3$. We examine the last four integrals in the limit as $a \to 0$. As regards the integrals over ∂B and $B(\mathbf{r}, a)$ we have

$$\left| \int_{\partial B} \mathrm{d}S' \frac{\hat{\mathbf{n}}(\mathbf{r}') \times \mathcal{M}(\mathbf{r}')}{4\pi R} \right| \leqslant \|\mathcal{M}\|_\infty \int_{\partial B} \mathrm{d}S' \frac{1}{4\pi a} = \|\mathcal{M}\|_\infty a \xrightarrow[a \to 0]{} 0 \tag{5.167}$$

$$\left| \int_{B(\mathbf{r},a)} \mathrm{d}V' \mathcal{M}(\mathbf{r}') \times \nabla' \frac{1}{4\pi R} \right| \leqslant \int_{B(\mathbf{r},a)} \mathrm{d}V' \frac{\|\mathcal{M}\|_\infty}{4\pi R^2} = \|\mathcal{M}\|_\infty a \xrightarrow[a \to 0]{} 0 \tag{5.168}$$

whereas the integral over ∂V_M is unaffected by the limiting process. Concerning the integral over V_a we pick a ball $B(\mathbf{r}, b)$ with radius b large enough so that $V_a \subset B(\mathbf{r}, b)$ and estimate

$$\left| \int_{V_a} \mathrm{d}V' \frac{\nabla' \times \mathcal{M}(\mathbf{r}')}{4\pi R} \right| \leqslant \int_{V_a} \mathrm{d}V' \frac{\|\nabla' \times \mathcal{M}\|_\infty}{4\pi R} \leqslant \int_{B(\mathbf{r},b)\setminus B(\mathbf{r},a)} \mathrm{d}V' \frac{\|\nabla' \times \mathcal{M}\|_\infty}{4\pi R}$$

$$= \frac{1}{2}(b^2 - a^2) \left\| \nabla' \times \mathcal{M} \right\|_\infty \leqslant \frac{1}{2} b^2 \left\| \nabla' \times \mathcal{M} \right\|_\infty \tag{5.169}$$

so this contribution is bounded for any value of $a < b$. With these intermediate results we have proved that the following identity holds

$$\mu_0 \int_{V_M} \mathrm{d}V'\, \mathcal{M}(\mathbf{r}') \times \nabla' \frac{1}{4\pi R} = \mu_0 \int_{V_M} \mathrm{d}V' \frac{\nabla' \times \mathcal{M}(\mathbf{r}')}{4\pi R} - \mu_0 \int_{\partial V_M} \mathrm{d}S' \frac{\hat{\mathbf{n}}(\mathbf{r}') \times \mathcal{M}(\mathbf{r}')}{4\pi R} \tag{5.170}$$

for observation points $\mathbf{r} \in \mathbb{R}^3$. By comparing this expression with the vector potential (5.18) generated by a volume current density in a homogeneous unbounded isotropic medium, we conclude that the effect of the magnetic body may be interpreted as a distribution of equivalent volume and surface electric currents with densities [1, Section 4.10] [10, Chapter E]

$$\mathcal{J}_{VM}(\mathbf{r}) := \nabla \times \mathcal{M}(\mathbf{r}), \qquad \mathbf{r} \in V_M \tag{5.171}$$

$$\mathcal{J}_{MS}(\mathbf{r}) := -\hat{\mathbf{n}}(\mathbf{r}) \times \mathcal{M}(\mathbf{r}), \qquad \mathbf{r} \in \partial V_M \tag{5.172}$$

with the unit normal pointing outward V_M. This result shows the intimate equivalence between magnetism and electricity. If the distribution of magnetic dipoles in the medium is uniform, the volume current density vanishes. Conversely, on the boundary of a magnetized body a surface electric current density always exists owing to the presence of unbalanced equivalent dipoles along the boundary.

Finally, we discuss the Ampère law in the presence of magnetic materials. The situation is quite the opposite of the electric case because $\mathcal{B}(\mathbf{r})$ is the fundamental entity and the sources (the currents) appear in the Ampère law rather than in the magnetic Gauss law. We model the magnetic body as a distribution of equivalent current densities existing in free space and given by (5.171) and (5.172).

We consider a finite stationary current $\mathcal{J}(\mathbf{r})$ flowing in a tube V_J and a magnetic material which occupies a bounded region V_M (Figure 5.9). The boundary ∂V_M is smooth and the unit normal thereon $\hat{\mathbf{n}}(\mathbf{r})$ is oriented outwards V_M. For the application of the Ampère law in global form we consider a smooth open surface S which intersects V_M and V_J along the open surfaces S_M and S_J, respectively. We denote with $\hat{\boldsymbol{\nu}}(\mathbf{r})$ the unit vector normal to S and with $\hat{\mathbf{s}}(\mathbf{r})$ the unit vector tangent to the boundary ∂S. Moreover, the unit vector $\hat{\mathbf{s}}(\mathbf{r})$ tangent to the boundary ∂S_M is oriented consistently with $\hat{\mathbf{s}}(\mathbf{r})$ on ∂S. Lastly, the unit vector $\hat{\boldsymbol{\tau}}$ tangent to ∂V_M and perpendicular to ∂S_M is defined as $\hat{\boldsymbol{\tau}}(\mathbf{r}) := \hat{\mathbf{n}}(\mathbf{r}) \times \hat{\mathbf{s}}(\mathbf{r})$ for $\mathbf{r} \in \partial S_M$.

The continuity of the tangential components of the magnetic field across the boundary of $\mathcal{J}(\mathbf{r})$ was already addressed in Section 1.2.2 and hence we shall take it for granted in the following derivation. Here we are interested in the jump conditions for the magnetic induction across ∂V_M, because a magnetization vector $\mathcal{M}(\mathbf{r})$ is induced in the body in response to the induction field $\mathcal{B}(\mathbf{r})$ generated by $\mathcal{J}(\mathbf{r})$.

In writing down the Ampère law (4.5) on S we have to include all the current densities that cross the surface S (see Figure 5.9), namely, the 'true' current $\mathcal{J}(\mathbf{r})$ but also the currents due to the magnetization and given by (5.171) and (5.172). Furthermore, we write $\mathcal{H}(\mathbf{r}) = \mathcal{B}(\mathbf{r})/\mu_0$ since on the one hand the distinction is purely formal in free space and, on the other, we want to emphasize that $\mathcal{B}(\mathbf{r})$ is the primary entity that is modified by the magnetic body. In symbols, we have

$$\frac{1}{\mu_0} \oint_{\partial S} \mathrm{d}s\, \hat{\mathbf{s}}(\mathbf{r}) \cdot \mathcal{B}(\mathbf{r}) = \int_{S_J} \mathrm{d}S\, \hat{\boldsymbol{\nu}}(\mathbf{r}) \cdot \mathcal{J}(\mathbf{r}) + \underbrace{\int_{S_M} \mathrm{d}S\, \hat{\boldsymbol{\nu}}(\mathbf{r}) \cdot \mathcal{J}_{VM}(\mathbf{r}) + \oint_{\partial S_M} \mathrm{d}s\, \hat{\boldsymbol{\tau}}(\mathbf{r}) \cdot \mathcal{J}_{SM}(\mathbf{r})}_{=I_{VM}+I_{SM}=0}$$

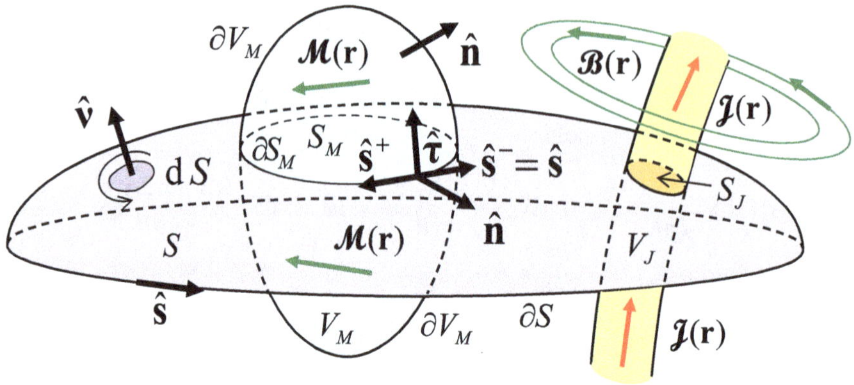

Figure 5.9 For determining the local Ampère law in the presence of a magnetized body; the current tube V_J has been drawn only partially.

$$= \int_{S_J} \mathrm{d}S\, \hat{\boldsymbol{\nu}}(\mathbf{r}) \cdot \mathcal{J}(\mathbf{r}) \tag{5.173}$$

where the surface integrals have been restricted to the respective supports of $\mathcal{J}(\mathbf{r})$ and $\mathcal{M}(\mathbf{r})$. The line integral along ∂S_M constitutes the 'flux' of the surface current density (5.172) and represents the current flowing over ∂V_M across ∂S_M and in the direction indicated by $\hat{\boldsymbol{\tau}}(\mathbf{r})$. Still, the total magnetization current crossing $\overline{S}_M := S_M \cup \partial S_M$ is null, and this follows from (5.171) and the Stokes theorem applied to the flux integral through S_M, which is then transformed into the negative of the integral of $\mathcal{J}_{SM}(\mathbf{r})$ over ∂S_M on account of (5.172).

We would like to obtain the local form of the Ampère law valid for all points in space, but we may not apply the Stokes theorem to the circulation of $\mathcal{B}(\mathbf{r})$ inasmuch as the latter may not be of class $\mathcal{C}^1(S)^3 \cap \mathcal{C}(\overline{S})^3$ due to the presence of the magnetized body. Thus, as usual, we state the global Ampère law separately over S_M, $S \setminus S_M$ and S. We indicate with ∂S_M^+ and ∂S_M^- the positive and the negative side of the loop ∂S_M with tangent unit vectors $\hat{\mathbf{s}}^+$ and $\hat{\mathbf{s}}^-$ defined so as to point clockwise and counterclockwise with respect to the normal $\hat{\boldsymbol{\nu}}(\mathbf{r})$ (Figure 5.9). Then, we have

$$\frac{1}{\mu_0} \oint_{\partial S_M} \mathrm{d}s\, \hat{\mathbf{s}}^-(\mathbf{r}) \cdot \mathcal{B}^-(\mathbf{r}) = \int_{S_M} \mathrm{d}S\, \hat{\boldsymbol{\nu}}(\mathbf{r}) \cdot \nabla \times \mathcal{M}(\mathbf{r}) \tag{5.174}$$

$$\frac{1}{\mu_0} \oint_{\partial S} \mathrm{d}s\, \hat{\mathbf{s}}(\mathbf{r}) \cdot \mathcal{B}(\mathbf{r}) + \frac{1}{\mu_0} \oint_{\partial S_M} \mathrm{d}s\, \hat{\mathbf{s}}^+(\mathbf{r}) \cdot \mathcal{B}^+(\mathbf{r}) = \int_{S_J} \mathrm{d}S\, \hat{\boldsymbol{\nu}}(\mathbf{r}) \cdot \mathcal{J}(\mathbf{r}) \tag{5.175}$$

and

$$\frac{1}{\mu_0} \oint_{\partial S} \mathrm{d}s\, \hat{\mathbf{s}}(\mathbf{r}) \cdot \mathcal{B}(\mathbf{r}) = \int_{S_J} \mathrm{d}S\, \hat{\boldsymbol{\nu}}(\mathbf{r}) \cdot \mathcal{J}(\mathbf{r}) + \int_{S_M} \mathrm{d}S\, \hat{\boldsymbol{\nu}}(\mathbf{r}) \cdot \nabla \times \mathcal{M}(\mathbf{r}) - \oint_{\partial S_M} \mathrm{d}s\, \hat{\boldsymbol{\tau}}(\mathbf{r}) \cdot \hat{\mathbf{n}}(\mathbf{r}) \times \mathcal{M}(\mathbf{r}) \tag{5.176}$$

where we have included the contribution of the surface magnetization current density on ∂V_M and hence through ∂S_M. The left member of (5.175) is justified by observing that $S \setminus S_M$ is multiply connected and the boundary thereof is comprised of two disjoint loops ∂S and ∂S_M.

By summing (5.174) and (5.175) side by side and combining the line integrals along ∂S_M we obtain

$$
\begin{aligned}
\frac{1}{\mu_0}\oint_{\partial S} \mathrm{d}s\,\hat{\mathbf{s}}(\mathbf{r})\cdot\mathcal{B}(\mathbf{r}) + \frac{1}{\mu_0}\oint_{\partial S_M} \mathrm{d}s\,\hat{\mathbf{s}}^{+}(\mathbf{r})\cdot\mathcal{B}^{+}(\mathbf{r}) + \frac{1}{\mu_0}\oint_{\partial S_M} \mathrm{d}s\,\hat{\mathbf{s}}^{-}(\mathbf{r})\cdot\mathcal{B}^{-}(\mathbf{r}) = \\
= \frac{1}{\mu_0}\oint_{\partial S} \mathrm{d}s\,\hat{\mathbf{s}}(\mathbf{r})\cdot\mathcal{B}(\mathbf{r}) + \frac{1}{\mu_0}\oint_{\partial S_M} \mathrm{d}s\,\hat{\mathbf{s}}(\mathbf{r})\cdot[\mathcal{B}^{-}(\mathbf{r}) - \mathcal{B}^{+}(\mathbf{r})] \\
= \int_{S_J} \mathrm{d}S\,\hat{\boldsymbol{\nu}}(\mathbf{r})\cdot\mathcal{J}(\mathbf{r}) + \int_{S_M} \mathrm{d}S\,\hat{\boldsymbol{\nu}}(\mathbf{r})\cdot\nabla\times\mathcal{M}(\mathbf{r})
\end{aligned}
\tag{5.177}
$$

and this relation should be identical with (5.176) for consistency. Evidently, this happens if we require

$$
\oint_{\partial S_M} \mathrm{d}s\,\hat{\mathbf{s}}(\mathbf{r})\cdot\left\{\frac{1}{\mu_0}[\mathcal{B}^{-}(\mathbf{r}) - \mathcal{B}^{+}(\mathbf{r})] - \mathcal{M}(\mathbf{r})\right\} = 0 \tag{5.178}
$$

and, if we assume that the vector fields $\mathcal{B}^{-}(\mathbf{r})$, $\mathcal{B}^{+}(\mathbf{r})$ and $\mathcal{M}(\mathbf{r})$ are continuous for $\mathbf{r}\in\partial S_M$, then the mean value theorem [6] applied to the integral above yields

$$
\hat{\mathbf{s}}(\mathbf{r}_0)\cdot\left\{\frac{1}{\mu_0}[\mathcal{B}^{-}(\mathbf{r}_0) - \mathcal{B}^{+}(\mathbf{r}_0)] - \mathcal{M}(\mathbf{r}_0)\right\} = 0 \tag{5.179}
$$

with $\mathbf{r}_0\in\partial S_M$ a suitable point. Since the shape and the position of S and hence of ∂S_M is arbitrary, we conclude

$$
\frac{1}{\mu_0}\hat{\mathbf{s}}(\mathbf{r})\cdot[\mathcal{B}^{+}(\mathbf{r}) - \mathcal{B}^{-}(\mathbf{r})] = -\hat{\mathbf{s}}(\mathbf{r})\cdot\mathcal{M}(\mathbf{r}), \qquad \mathbf{r}\in\partial V_M \tag{5.180}
$$

i.e., the tangential component of the magnetic induction suffers a jump due to the layer of unbalanced magnetization dipoles distributed over ∂V_M. We may rephrase the result as

$$
\mathcal{J}_{SM}(\mathbf{r}) := -\hat{\mathbf{n}}(\mathbf{r})\times\mathcal{M}(\mathbf{r}) = \frac{1}{\mu_0}\hat{\mathbf{n}}(\mathbf{r})\times[\mathcal{B}^{+}(\mathbf{r}) - \mathcal{B}^{-}(\mathbf{r})], \qquad \mathbf{r}\in\partial V_M \tag{5.181}
$$

in view of (5.172) and $\hat{\mathbf{s}}(\mathbf{r}) = \hat{\boldsymbol{\tau}}(\mathbf{r})\times\hat{\mathbf{n}}(\mathbf{r})$ for $\mathbf{r}\in\partial V_M$ (Figure 5.9). In light of (1.113) the matching condition (5.181) says that the tangential component of the magnetic field is discontinuous due to the presence of the surface magnetization current over ∂V_M.

Furthermore, in (5.174) and (5.175) we may apply the Stokes theorem independently on S_M and $S\setminus S_M$ where $\mathcal{B}(\mathbf{r})/\mu_0$ is continuously differentiable. The mean value theorem invoked on the resulting flux integrals and the arbitrariness of S lead to the following form for the local Ampère law in each point $\mathbf{r}\in\mathbb{R}^3$

$$
\frac{1}{\mu_0}\nabla\times\mathcal{B}(\mathbf{r}) = \begin{cases}\mathcal{J}(\mathbf{r}), & \mathbf{r}\in\mathbb{R}^3\setminus\overline{V}_M \\ \nabla\times\mathcal{M}(\mathbf{r}), & \mathbf{r}\in V_M\end{cases} \tag{5.182}
$$

subject to the matching condition (5.181). Clearly, we have not introduced the magnetic field, and the Ampère law is stated in free space where all the currents reside. Equation (5.182) is the Ampère law in the form of Maxwell and Boffi [2], and is the counterpart of (3.256) for electric entities.

To arrive at the more familiar form of the Ampère law we group the magnetic induction and the magnetization vector together and define the magnetic field as

$$\mathcal{H}(\mathbf{r}) := \begin{cases} \dfrac{1}{\mu_0}\mathcal{B}(\mathbf{r}), & \mathbf{r} \in \mathbb{R}^3 \setminus \overline{V}_M \\ \dfrac{1}{\mu_0}\mathcal{B}(\mathbf{r}) - \mathcal{M}(\mathbf{r}), & \mathbf{r} \in V_M \end{cases} \tag{5.183}$$

whereby (5.182) becomes

$$\nabla \times \mathcal{H}(\mathbf{r}) = \mathcal{J}(\mathbf{r}), \qquad \mathbf{r} \in \mathbb{R}^3 \tag{5.184}$$

with the boundary condition

$$\hat{\mathbf{n}}(\mathbf{r}) \times \underbrace{\left[\frac{1}{\mu_0}\mathcal{B}^-(\mathbf{r}) - \mathcal{M}(\mathbf{r})\right]}_{\mathcal{H}^-(\mathbf{r})} = \hat{\mathbf{n}}(\mathbf{r}) \times \underbrace{\frac{\mathcal{B}^+(\mathbf{r})}{\mu_0}}_{\mathcal{H}^+(\mathbf{r})}, \qquad \mathbf{r} \in \partial V_M \tag{5.185}$$

which is the same statement of the continuity of the tangential component of $\mathcal{H}$ discussed in Section 1.7.

We continue the analysis of the magnetization vector by showing that a magnetic body may also be considered as an equivalent distribution of fictitious magnetic charges. To this purpose we need to examine the magnetic scalar potential $\Psi(\mathbf{r})$ which is generated by $\mathcal{M}(\mathbf{r})$, $\mathbf{r} \in V_M$. Since the potential of a solitary magnetic dipole is given by (4.85), we speculate that in the presence of a magnetic medium we have

$$\Psi(\mathbf{r}) = \int_{V_M} \mathrm{d}V'\, \mathcal{M}(\mathbf{r}') \cdot \nabla' \frac{1}{4\pi R}, \qquad \mathbf{r} \in \mathbb{R}^3 \setminus \overline{V}_J \tag{5.186}$$

where we have excluded observation points within the spatial region occupied by the current that produces the primary magnetic induction. By comparing (5.186) to the analogous expression (3.239) for the electric potential $\Phi(\mathbf{r})$ we notice that a volume integral over the charge density is missing precisely because we dismiss the occurrence of *true* magnetic charges.

We can transform the right-hand side of (5.186) as we did for the electrostatic case in Section 3.7. Taking due care of the singular character of the integrand while applying the Gauss theorem we arrive at

$$\int_{V_M} \mathrm{d}V'\, \mathcal{M}(\mathbf{r}') \cdot \nabla' \frac{1}{4\pi R} = \frac{1}{\mu_0} \int_{\partial V_M} \mathrm{d}S'\, \frac{\mu_0 \hat{\mathbf{n}}(\mathbf{r}') \cdot \mathcal{M}(\mathbf{r}')}{4\pi R} - \frac{1}{\mu_0} \int_{V_M} \mathrm{d}V'\, \frac{\mu_0 \nabla' \cdot \mathcal{M}(\mathbf{r}')}{4\pi R} \tag{5.187}$$

for observation points $\mathbf{r} \in \mathbb{R}^3 \setminus \overline{V}_J$. Finally, if we compare (5.187) with the electric scalar potential (2.169) we are led to interpret $\Psi(\mathbf{r})$ as being generated by fictitious or equivalent magnetic charges (physical dimension: Wb) with densities

$$\varrho_{VM}(\mathbf{r}) := -\mu_0 \nabla \cdot \mathcal{M}(\mathbf{r}), \qquad \mathbf{r} \in V_M \tag{5.188}$$

$$\varrho_{MS}(\mathbf{r}) := \mu_0 \hat{\mathbf{n}}(\mathbf{r}) \cdot \mathcal{M}(\mathbf{r}), \qquad \mathbf{r} \in \partial V_M \tag{5.189}$$

which are distributed in the region V_M and on the boundary thereof [11, Section 13.4]. The unit normal $\hat{\mathbf{n}}(\mathbf{r})$ over ∂V_M is positively oriented toward $\mathbb{R}^3 \setminus V_M$, as in Figure 5.9. This result may be regarded as the magnetic version of the Poisson theorem obtained for the polarization vector in Section 3.7. The explicit appearance of the permeability in (5.188) and (5.189) is motivated by mere dimensional consistency and is a result of the different definitions adopted for $\mathcal{P}$ and $\mathcal{M}$ (also see the footnote on page 721). Using (5.188) and (5.189) as sources in the right-hand side of the global magnetic Gauss law (1.10) and carrying out manipulations perfectly similar to those effected in Section 3.7 provides an alternative way of introducing the magnetic field $\mathcal{H}(\mathbf{r})$ consistent with (5.183), as it should be. An interesting problem for which both polarization and magnetization vectors are used is addressed in Example 9.6.

In summary, we have shown that magnetism is the macroscopic manifestation of the combined effect of the tiny electronic currents existing in magnetic bodies. Additionally, we have explained that $\mathcal{B}(\mathbf{r})$ is the fundamental magnetic entity and that $\mathcal{H}(\mathbf{r})$ is only needed to describe the magnetization.

The actual relationship between the magnetization vector and the external induction field may be complicated, non-linear and even exhibit hysteresis (memory) [4], [26, Chapter 14]. For linear isotropic media it is possible to assume

$$\mathcal{M}(\mathbf{r}) := \frac{\chi_m(\mathbf{r})}{\mu_0}\mathcal{B}(\mathbf{r}), \qquad \mathbf{r} \in V_M \tag{5.190}$$

where $\chi_m(\mathbf{r})$ is called the *magnetic susceptibility*. Thereby, thanks to (5.183) we may write the constitutive relationship (1.118) as

$$\mathcal{B}(\mathbf{r}) = \mu_0[1 + \chi_m(\mathbf{r})]\mathcal{H}(\mathbf{r}) = \mu_0\mu_r(\mathbf{r})\mathcal{H}(\mathbf{r}) = \mu(\mathbf{r})\mathcal{H}(\mathbf{r}), \qquad \mathbf{r} \in V_M \tag{5.191}$$

with $\mu(\mathbf{r})$ the *magnetic permeability*. In the more general situation of anisotropic character of the magnetization we may introduce a dyadic magnetic susceptibility and define

$$\overline{\mu}(\mathbf{r}) = \mu_0\left[1 + \overline{\chi}_m(\mathbf{r})\right], \qquad \mathbf{r} \in V_M \tag{5.192}$$

as dyadic permeability. Magnetic media are customarily classified according to the range of values taken on by the susceptibility. In particular, we have [27, Chapter 21], [3, Chapter 6], [24, Chapter 13], [26, Section 14.1], [25, Chapter 13]

- **diamagnetic** media (e.g., water, hydrogen, bismuth and copper) if $\chi_m(\mathbf{r}) < 0$; atoms and molecules have no net permanent magnetic dipoles but, when exposed to an external magnetic induction $\mathcal{B}$, the electrons start rotating (precess) around $\mathcal{B}$ thus giving rise to small magnetic moments opposite to $\mathcal{B}$ [3, Section 6.2];
- **paramagnetic** media (e.g., aluminum, manganese and rare-earth salts) if $\chi_m(\mathbf{r}) > 0$; atoms and molecules do have net permanent magnetic moments which, due to thermal motion, are randomly oriented so that the magnetization vector is null; in the presence of an external magnetic induction $\mathcal{B}$ the magnetic dipoles tend to align with $\mathcal{B}$ — because this arrangement minimizes the magnetic energy (4.78) — thus causing the onset of a net magnetization [3, Section 6.3];
- **ferromagnetic** materials (e.g., iron, nickel and cobalt) if $\chi_m(\mathbf{r}) = \chi_m(\mathbf{r};\mathcal{B}) \gg 1$, and hence they can be regarded as the limiting case of paramagnetic media [13, Chapter 37], [21, Section 15.8], [24, Chapter 14], [3, Section 6.4], [27, Chapter 21]; in **anti-ferromagnetic** substances minimum energy is attained when adjacent dipole moments are antiparallel, whereby the net magnetization is null; in **ferrimagnetic** media (e.g., ferrites, magnetite) adjacent dipole moments are antiparallel and have slightly different magnitudes, so that a small net non-zero magnetization exists.

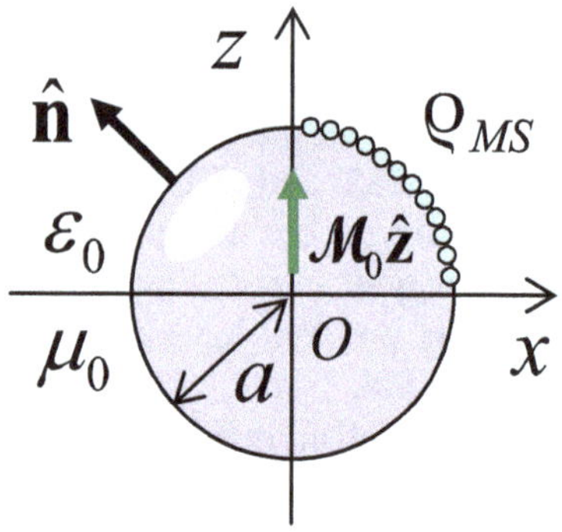

Figure 5.10 Permanently magnetized sphere with uniform magnetization $\mathcal{M}_0\hat{\mathbf{z}}$.

Example 5.4 (Magnetic field and induction produced by a permanently magnetized sphere)

We suppose that a permanent magnet of spherical shape is immersed in free space, occupies the ball $B(0, a)$ and has uniform magnetization

$$\mathcal{M}(\mathbf{r}) = \mathcal{M}_0\hat{\mathbf{z}}, \qquad \mathbf{r} \in B(0, a) \tag{5.193}$$

which is aligned with the polar axis of a system of spherical coordinates (r, ϑ, φ) (Figure 5.10). In order to determine the magnetic entities we notice that the Maxwell-Boffi point of view is appropriate for this problem inasmuch as the magnetization actually represents a source term rather than an unknown. Indeed, the magnetization can be equivalently modelled by means of fictitious magnetic charges in accordance with (5.188) and (5.189). In particular, as the magnetization is uniform, only a layer of magnetic charges with surface density

$$\varrho_{MS}(\mathbf{r}) = \mu_0\hat{\mathbf{n}}(\mathbf{r}) \cdot \mathcal{M}(\mathbf{r}) = \mu_0\mathcal{M}_0 \cos\vartheta, \qquad \mathbf{r} \in \partial B \tag{5.194}$$

need be introduced on the boundary of the magnet.

In view of (4.9) and (5.183) the magnetostatic entities due to $\varrho_{MS}(\mathbf{r})$ solve the homogeneous equations

$$\nabla \times \mathcal{H}(\mathbf{r}) = 0, \qquad \mathbf{r} \in \mathbb{R}^3 \tag{5.195}$$

$$\nabla \cdot \mathcal{B}(\mathbf{r}) = \begin{cases} \mu_0\nabla \cdot \mathcal{H}(\mathbf{r}) = 0, & \mathbf{r} \in \mathbb{R}^3 \setminus B[0, a] \\ \mu_0\nabla \cdot [\mathcal{H}(\mathbf{r}) + \mathcal{M}_0\hat{\mathbf{z}}] = \mu_0\nabla \cdot \mathcal{H}(\mathbf{r}) = 0, & \mathbf{r} \in B(0, a) \end{cases} \tag{5.196}$$

subject to suitable boundary conditions.

Since $\mathcal{H}(\mathbf{r})$ is curl-free everywhere, it can be derived from a single-valued scalar potential $\Psi(\mathbf{r})$ formally as in (4.68). Then, it follows from (5.196) that $\Psi(\mathbf{r})$ is the unique solution to the boundary value problem [cf. (3.1)]

$$\begin{cases} \nabla^2\Psi(\mathbf{r}) = 0, & \mathbf{r} \in \mathbb{R}^3 \setminus \partial B \\ \Psi^+(\mathbf{r}) = \Psi^-(\mathbf{r}), & \mathbf{r} \in \partial B \\ \left.\dfrac{\partial\Psi}{\partial\hat{\mathbf{n}}}\right|_{\mathbf{r}\in\partial B^-} - \left.\dfrac{\partial\Psi}{\partial\hat{\mathbf{n}}}\right|_{\mathbf{r}\in\partial B^+} = \mathcal{M}_0 \cos\vartheta, & \mathbf{r} \in \partial B \\ \Psi(\mathbf{r}) = \mathcal{O}\left(\dfrac{1}{|\mathbf{r}|}\right), & |\mathbf{r}| \to +\infty \end{cases} \tag{5.197}$$

where Ψ^+ and Ψ^- denote the potential on the positive and negative side of ∂B, with the unit normal $\hat{\mathbf{n}}$ on ∂B pointing into the unbounded background medium (Figure 5.10).

The continuity of $\Psi(\mathbf{r})$ across ∂B is a consequence of (5.185), and can be proved as we did in Section 2.4 for the electric scalar potential. The jump of the normal derivative of $\Psi(\mathbf{r})$ across ∂B results from (4.49) by using $\mathcal{B}(\mathbf{r}) = \mu_0[\mathcal{H}(\mathbf{r}) + \mathcal{M}_0\hat{\mathbf{z}}]$ on the negative side of ∂B. We claim that (5.197) admits only one solution inasmuch as it is of the same type of Laplace equations discussed in Sections 2.5.1 and 3.1 in relation to the electrostatic potential $\Phi(\mathbf{r})$.

Since the geometry and the magnetic charge distribution (5.194) are rotationally symmetric around the z-axis, we seek the solution in the form (Section 3.5.1)

$$\Psi(\mathbf{r}) = \begin{cases} \displaystyle\sum_{n=0}^{+\infty} \frac{A_n}{r^{n+1}} P_n(\cos\vartheta), & r > a \\ \displaystyle\sum_{n=0}^{+\infty} B_n r^n P_n(\cos\vartheta), & r < a \end{cases} \tag{5.198}$$

because the potential must be regular at the center of the magnet and vanish at infinity. Enforcing the matching conditions over ∂B provides the requirements

$$\sum_{n=0}^{+\infty} \left(\frac{A_n}{a^{n+1}} - B_n a^n \right) P_n(\cos\vartheta) = 0 \tag{5.199}$$

$$\sum_{n=0}^{+\infty} \left(\frac{n+1}{a^{n+2}} A_n + n a^{n-1} B_n \right) P_n(\cos\vartheta) = \mathcal{M}_0 P_1(\cos\vartheta) \tag{5.200}$$

which must be met for any value of the polar angle ϑ. Since the Legendre polynomials are orthogonal to each other over $\vartheta \in [0, \pi]$ (Appendix H.5) we obtain

$$\frac{A_n}{a^{n+1}} - B_n a^n = 0 \qquad\qquad \frac{n+1}{a^{n+2}} A_n + n a^{n-1} B_n = 0 \tag{5.201}$$

for $n \neq 1$ and

$$\frac{A_1}{a^2} - aB_1 = 0 \qquad\qquad 2\frac{A_1}{a^3} + B_1 = \mathcal{M}_0 \tag{5.202}$$

for $n = 1$. The first set of equations represents a homogeneous linear system which admits only the trivial solution $A_n = 0 = B_n$, $n \neq 1$. For $n = 1$ from the second set we have

$$A_1 = \frac{\mathcal{M}_0}{3} a^3 \qquad\qquad B_1 = \frac{\mathcal{M}_0}{3} \tag{5.203}$$

whereby the magnetostatic potential reads

$$\Psi(\mathbf{r}) = \begin{cases} \dfrac{a^3}{3r^2} \mathcal{M}_0 \cos\vartheta, & r > a \\ \dfrac{\mathcal{M}_0}{3} r \cos\vartheta, & r < a. \end{cases} \tag{5.204}$$

We may cast the expression of $\Psi(\mathbf{r})$ for $\mathbf{r} \in \mathbb{R}^3 \setminus B[0, a]$ as

$$\Psi(\mathbf{r}) = \underbrace{\frac{4}{3}\pi a^3 \mathcal{M}_0 \hat{\mathbf{z}}}_{\text{dipole moment}} \cdot \frac{\hat{\mathbf{r}}}{4\pi r^2} = \frac{m_{eq} \cos\vartheta}{4\pi r^2} \tag{5.205}$$

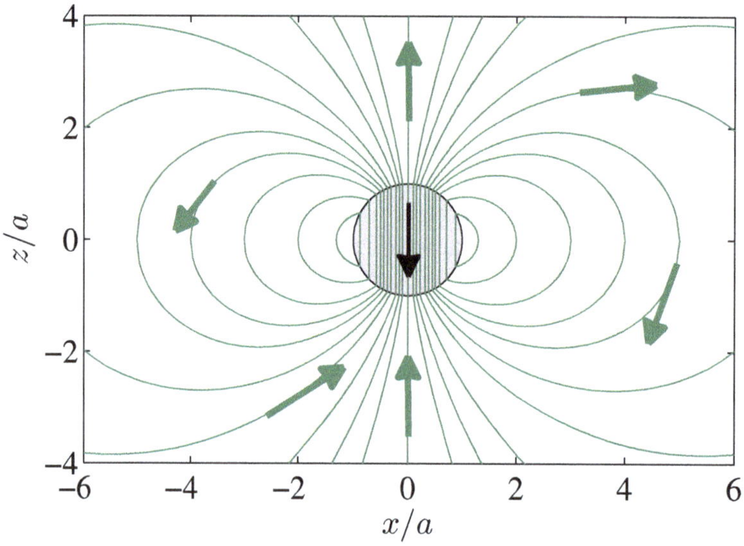

Figure 5.11 *Streamlines (−) of the magnetic field $\mathcal{H}(\mathbf{r})$ produced in the xOz plane by a permanently magnetized sphere (shaded region) with uniform magnetization $\mathcal{M}_0\hat{\mathbf{z}}$.*

which by comparison with (4.85) shows that the effect of the magnet is equivalent to that of an elementary magnetic dipole which has moment

$$\mathbf{m}_{eq} = \frac{4}{3}\pi a^3 \mathcal{M}_0 \hat{\mathbf{z}} \tag{5.206}$$

and is located in the origin (the center of the magnetized sphere). Then again, since $\mathcal{M}(\mathbf{r})$ physically represents the volume density of magnetic dipoles existing for $\mathbf{r} \in B(0, a)$, evidently we arrive at the same result by integrating $\mathcal{M}(\mathbf{r})$ over the spatial region occupied by the magnet.

Finally, the magnetic entities read [4, Section 5.10], [8], [11, Section 13.4.1]

$$\mathcal{H}(\mathbf{r}) = -\nabla\Psi(\mathbf{r}) = \begin{cases} \dfrac{a^3\mathcal{M}_0}{3r^3}(2\hat{\mathbf{r}}\cos\vartheta + \hat{\boldsymbol{\vartheta}}\sin\vartheta) = \dfrac{3\hat{\mathbf{r}}\hat{\mathbf{r}} - \overline{\mathcal{I}}}{4\pi r^3}\cdot\mathbf{m}_{eq}, & r > a \\ -\dfrac{1}{3}\mathcal{M}_0\hat{\mathbf{z}}, & r < a \end{cases} \tag{5.207}$$

$$\mathcal{B}(\mathbf{r}) = \begin{cases} \mu_0\dfrac{3\hat{\mathbf{r}}\hat{\mathbf{r}} - \overline{\mathcal{I}}}{4\pi r^3}\cdot\mathbf{m}_{eq}, & r > a \\ \mu_0\mathcal{H}(\mathbf{r}) + \mu_0\mathcal{M}(\mathbf{r}) = \dfrac{2}{3}\mu_0\mathcal{M}_0\hat{\mathbf{z}}, & r < a \end{cases} \tag{5.208}$$

and it is readily checked that the ϑ-component of $\mathcal{H}(\mathbf{r})$ (tangential to ∂B) and the radial component of $\mathcal{B}(\mathbf{r})$ (normal to ∂B) are continuous across the surface of the magnet, as is required by the boundary conditions (4.48) and (4.49). The streamlines of the magnetic field, some of which are shown in Figure 5.11, are seen to originate from the fictitious magnetic charges distributed on the surface of the magnet. By contrast, the lines of force of the magnetic intensity form closed loops which look formally the same as the streamlines of $\mathcal{H}(\mathbf{r})$ in Figure 5.11 except for $\mathcal{B}(\mathbf{r})$ having the same orientation as the magnetization inside the magnet.

We conclude by noticing that in view of (5.172) the magnetization is also equivalent to a surface current on ∂B with density

$$\mathcal{J}_S(\mathbf{r}) = -\hat{\mathbf{r}} \times \mathcal{M}_0\hat{\mathbf{z}} = \mathcal{M}_0\hat{\boldsymbol{\varphi}} \sin\vartheta, \qquad \mathbf{r} \in \partial B \tag{5.209}$$

and hence the magnetic entities can be equivalently derived from the vector potential $\mathcal{A}(\mathbf{r})$ generated by $\mathcal{J}_S(\mathbf{r})$ as was done in Example 5.1.

(End of Example 5.4)

5.7 Magnetic forces between steady currents

In this section we concern ourselves with the calculation of the magnetic forces due to densities of steady electric currents in an unbounded isotropic homogeneous medium endowed with permeability μ [11, Section 12.3]. The starting point is the magnetic part of the Lorentz force density (1.6) [10, pp. 106 and ff.], [28, Chapter 4]

$$\mathbf{f}_m(\mathbf{r}) := \mathcal{J}(\mathbf{r}) \times \mathcal{B}(\mathbf{r}), \qquad \mathbf{r} \in V_J \tag{5.210}$$

where V_J is the region supposedly occupied by $\mathcal{J}(\mathbf{r})$. Of course, the total magnetic force $\mathcal{F}_m$ acting on $\mathcal{J}(\mathbf{r})$ follows by integrating (5.210) over V_J.

We wish to show that the net force $\mathcal{F}_m$ experienced by $\mathcal{J}(\mathbf{r})$ and due to the magnetic induction $\mathcal{B}(\mathbf{r})$ produced by $\mathcal{J}(\mathbf{r})$ itself *vanishes*. For the proof we begin by recalling that $\mathcal{B}(\mathbf{r})$ is related to $\mathcal{J}(\mathbf{r})$ by means of the volume contribution of the integral representation (5.57). All in all, we have

$$\begin{aligned}
\mathcal{F}_m &= \int_{V_J} \mathrm{d}V\, \mathcal{J}(\mathbf{r}) \times \mathcal{B}(\mathbf{r}) = \mu \int_{V_J} \mathrm{d}V\, \mathcal{J}(\mathbf{r}) \times \int_{V_J} \mathrm{d}V'\, \mathcal{J}(\mathbf{r}') \times \nabla' \frac{1}{4\pi R} \\
&= \mu \int_{V_J} \mathrm{d}V\, \mathcal{J}(\mathbf{r}) \times \int_{V_J} \mathrm{d}V'\, \nabla \frac{1}{4\pi R} \times \mathcal{J}(\mathbf{r}') \\
&= \mu \int_{V_J} \mathrm{d}V \int_{V_J} \mathrm{d}V'\, \mathcal{J}(\mathbf{r}) \cdot \mathcal{J}(\mathbf{r}') \nabla \frac{1}{4\pi R} - \mu \int_{V_J} \mathrm{d}V \int_{V_J} \mathrm{d}V' \left[\mathcal{J}(\mathbf{r}) \cdot \nabla \frac{1}{4\pi R}\right] \mathcal{J}(\mathbf{r}') \\
&= -\mu \int_{V_J} \mathrm{d}V \int_{V_J} \mathrm{d}V'\, \frac{\mathcal{J}(\mathbf{r}) \cdot \mathcal{J}(\mathbf{r}')(\mathbf{r} - \mathbf{r}')}{4\pi|\mathbf{r} - \mathbf{r}'|^3} - \mu \int_{V_J} \mathrm{d}V'\, \mathcal{J}(\mathbf{r}') \int_{V_J} \mathrm{d}V\, \nabla \cdot \frac{\mathcal{J}(\mathbf{r})}{4\pi R}
\end{aligned} \tag{5.211}$$

where $R = |\mathbf{r} - \mathbf{r}'|$, and we have used the vector identity (H.14). In the last step above we have exploited the solenoidal nature of $\mathcal{J}(\mathbf{r})$ [see (4.11)] and also swapped the order of integration because both domain integrals exist. To apply the Gauss theorem (A.53) to the inner volume integral in the rightmost member we isolate the singular point $\mathbf{r} = \mathbf{r}'$ with a small ball $B(\mathbf{r}', a) \subset V_J$ and observe

$$\begin{aligned}
\int_{V_J} \mathrm{d}V\, \nabla \cdot \frac{\mathcal{J}(\mathbf{r})}{4\pi R} &= \int_{V_J \backslash B(\mathbf{r}',a)} \mathrm{d}V\, \nabla \cdot \frac{\mathcal{J}(\mathbf{r})}{4\pi R} + \int_{B(\mathbf{r}',a)} \mathrm{d}V\, \mathcal{J}(\mathbf{r}) \cdot \nabla \frac{1}{4\pi R} \\
&= \underbrace{\int_{\partial V_J} \mathrm{d}S\, \frac{\hat{\mathbf{n}}(\mathbf{r}) \cdot \mathcal{J}(\mathbf{r})}{4\pi R}}_{=0} - \int_{\partial B} \mathrm{d}S\, \frac{\hat{\mathbf{R}} \cdot \mathcal{J}(\mathbf{r})}{4\pi a} - \int_{B(\mathbf{r}',a)} \mathrm{d}V\, \frac{\mathcal{J}(\mathbf{r}) \cdot \hat{\mathbf{R}}}{4\pi R^2}
\end{aligned} \tag{5.212}$$

where the unit normal $\hat{\mathbf{n}}(\mathbf{r})$ on ∂V_J points outward V_J, and $\hat{\mathbf{n}}(\mathbf{r}) \equiv -\hat{\mathbf{R}}$ for $\mathbf{r} \in \partial B$. Evidently, the integral over ∂V_J vanishes because $\mathcal{J}(\mathbf{r})$ is confined to V_J by hypothesis and does not have a component perpendicular to ∂V_J. The other two contributions are finite for $a > 0$ and vanish in the limit as $a \to 0^+$, namely,

$$\left| \int_{\partial B} \mathrm{d}S \, \frac{\hat{\mathbf{R}} \cdot \mathfrak{J}(\mathbf{r})}{4\pi a} \right| \leqslant \int_{\partial B} \mathrm{d}S \, \frac{|\hat{\mathbf{R}} \cdot \mathfrak{J}(\mathbf{r})|}{4\pi a} \leqslant \|\mathfrak{J}\|_\infty \, a \xrightarrow[a\to 0]{} 0 \tag{5.213}$$

$$\left| \int_{B(\mathbf{r}',a)} \mathrm{d}V \, \frac{\mathfrak{J}(\mathbf{r}) \cdot \hat{\mathbf{R}}}{4\pi R^2} \right| \leqslant \int_{B(\mathbf{r}',a)} \mathrm{d}V \, \frac{|\mathfrak{J}(\mathbf{r}) \cdot \hat{\mathbf{R}}|}{4\pi R^2} \leqslant \|\mathfrak{J}\|_\infty \, a \xrightarrow[a\to 0]{} 0 \tag{5.214}$$

having invoked the Cauchy-Schwarz inequality (D.151) with $|\hat{\mathbf{R}}| = 1$ and integrated in polar spherical coordinates centered in $\mathbf{r}'$. Going back to (5.211) we obtain [2, Section 3.3.6], [10, Section 4.3]

$$\mathfrak{F}_m = -\mu \int_{V_J} \mathrm{d}V \int_{V_J} \mathrm{d}V' \, \mathfrak{J}(\mathbf{r}) \cdot \mathfrak{J}(\mathbf{r}') \frac{\mathbf{r} - \mathbf{r}'}{4\pi|\mathbf{r} - \mathbf{r}'|^3} \tag{5.215}$$

a formula which exhibits *odd symmetry* with respect to the interchange of source ($\mathbf{r}'$) and observation ($\mathbf{r}$) points. More specifically, if we swap the order of integration in (5.215) and then change the names of the dummy variables according to $\mathbf{r} = \boldsymbol{\xi}'$ and $\mathbf{r}' = \boldsymbol{\xi}$, we find

$$\begin{aligned} \mathfrak{F}_m &= \mu \int_{V_J} \mathrm{d}V' \int_{V_J} \mathrm{d}V \, \mathfrak{J}(\mathbf{r}') \cdot \mathfrak{J}(\mathbf{r}) \frac{\mathbf{r}' - \mathbf{r}}{4\pi|\mathbf{r}' - \mathbf{r}|^3} \\ &= \mu \int_{V_J} \mathrm{d}V \int_{V_J} \mathrm{d}V' \, \mathfrak{J}(\boldsymbol{\xi}) \cdot \mathfrak{J}(\boldsymbol{\xi}') \frac{\boldsymbol{\xi} - \boldsymbol{\xi}'}{4\pi|\boldsymbol{\xi} - \boldsymbol{\xi}'|^3} = -\mathfrak{F}_m \end{aligned} \tag{5.216}$$

where the last step follows because the expression is formally equal to the negative of the rightmost-hand side of (5.215). Now, the only way the same force can have two opposite orientations is for $\mathfrak{F}_m$ to be null, as anticipated.

We can extend (5.215) to an array of N current-carrying regions by defining

$$\mathfrak{J}(\mathbf{r}) := \sum_{n=1}^{N} \mathfrak{J}_n(\mathbf{r}), \qquad \mathbf{r} \in \bigcup_{n=1}^{N} V_{Jn} \tag{5.217}$$

which when inserted into (5.215) yields

$$\mathfrak{F}_m = -\mu \sum_{n=1}^{N} \sum_{l=1}^{N} \int_{V_{Jn}} \mathrm{d}V \int_{V_{Jl}} \mathrm{d}V' \, \mathfrak{J}_n(\mathbf{r}) \cdot \mathfrak{J}_l(\mathbf{r}') \frac{\mathbf{r} - \mathbf{r}'}{4\pi|\mathbf{r} - \mathbf{r}'|^3} \tag{5.218}$$

where each term of the double sum, viz.,

$$\mathfrak{F}_{nl} = -\mu \int_{V_{Jn}} \mathrm{d}V \int_{V_{Jl}} \mathrm{d}V' \, \mathfrak{J}_n(\mathbf{r}) \cdot \mathfrak{J}_l(\mathbf{r}') \frac{\mathbf{r} - \mathbf{r}'}{4\pi|\mathbf{r} - \mathbf{r}'|^3}, \qquad n, l \in \{1, \dots, N\} \tag{5.219}$$

is interpreted as the magnetic force exerted on $\mathfrak{J}_n(\mathbf{r})$ due to the magnetic induction generated by $\mathfrak{J}_l(\mathbf{r})$. In particular, the force exerted on $\mathfrak{J}_l(\mathbf{r})$ by $\mathfrak{J}_n(\mathbf{r})$ reads

$$\mathfrak{F}_{ln} = -\mu \int_{V_{Jl}} \mathrm{d}V \int_{V_{Jn}} \mathrm{d}V' \, \mathfrak{J}_l(\mathbf{r}) \cdot \mathfrak{J}_n(\mathbf{r}') \frac{\mathbf{r} - \mathbf{r}'}{4\pi|\mathbf{r} - \mathbf{r}'|^3}$$

$$= \mu \int_{V_{Jn}} \mathrm{d}V' \int_{V_{Jl}} \mathrm{d}V\, \mathcal{J}_n(\mathbf{r}') \cdot \mathcal{J}_l(\mathbf{r}) \frac{\mathbf{r}' - \mathbf{r}}{4\pi|\mathbf{r}' - \mathbf{r}|^3}$$

$$= \mu \int_{V_{Jn}} \mathrm{d}V \int_{V_{Jl}} \mathrm{d}V'\, \mathcal{J}_n(\boldsymbol{\xi}) \cdot \mathcal{J}_l(\boldsymbol{\xi}') \frac{\boldsymbol{\xi} - \boldsymbol{\xi}'}{4\pi|\boldsymbol{\xi} - \boldsymbol{\xi}'|^3} = -\mathfrak{F}_{nl}, \qquad n, l \in \{1, \dots, N\} \tag{5.220}$$

where we have inverted the order of integration and then changed the names of the dummy variables according to $\mathbf{r}' = \boldsymbol{\xi}$ and $\mathbf{r} = \boldsymbol{\xi}'$. Finally, comparison with (5.219) yields the very last step. In essence, this result is a statement of Newton's law of action and reaction [29, Chapter 10]. Since (5.220) holds in particular for $n = l$ we are led to conclude that $\mathfrak{F}_{nn} = 0$, which all in all proves that the net magnetic force (5.218) for a system of N current-carrying regions is zero as well.

When the stationary current in question depends only on two coordinates and is infinitely extended along the third spatial direction, say, the z-axis, the magnetic problem becomes inherently two-dimensional and (5.215) takes on a special form. While (5.210) still applies, we may not integrate $\mathbf{f}_m(\mathbf{r})$ over the unbounded cylindrical region $V_J := S_J \times \mathbb{R}$ because $\mathcal{J}(\mathbf{r})$ does not fall off as $|z| \to +\infty$, and hence the integral diverges. The best we can do is integrate $\mathbf{f}_m(\mathbf{r})$ over a cross section S_J perpendicular to $\hat{\mathbf{z}}$, and thus obtain a formula for the line density of force.

We define the transverse position vector $\boldsymbol{\tau} := \mathbf{r} - z\hat{\mathbf{z}}$ and assume a current density in the form $\mathcal{J}(\mathbf{r}) = \mathcal{J}_z(\boldsymbol{\tau})\hat{\mathbf{z}}$, valid for $\boldsymbol{\tau} \in S_J$. We point out that the shape of S_J need not be a circle nor does $\mathcal{J}_z(\boldsymbol{\tau})$ have to be uniform. By recalling the expression of the magnetic induction produced by a two-dimensional z-aligned current density, viz., [1, Chapter 4]

$$\mathcal{B}(\boldsymbol{\tau}) = \mu \int_{S_J} \mathrm{d}S'\, \mathcal{J}_z(\boldsymbol{\tau}')\hat{\mathbf{z}} \times \frac{\boldsymbol{\tau} - \boldsymbol{\tau}'}{2\pi|\boldsymbol{\tau} - \boldsymbol{\tau}'|^2}, \qquad \boldsymbol{\tau} \in \mathbb{R}^2 \tag{5.221}$$

and by integrating the Lorentz force density over S_J we find

$$\mathfrak{F}_m = \mu \int_{S_J} \mathrm{d}S\, \mathcal{J}_z(\boldsymbol{\tau})\hat{\mathbf{z}} \times \int_{S_J} \mathrm{d}S'\, \mathcal{J}_z(\boldsymbol{\tau}')\hat{\mathbf{z}} \times \frac{\boldsymbol{\tau} - \boldsymbol{\tau}'}{2\pi|\boldsymbol{\tau} - \boldsymbol{\tau}'|^2}$$

$$= -\mu \int_{S_J} \mathrm{d}S \int_{S_J} \mathrm{d}S'\, \mathcal{J}_z(\boldsymbol{\tau})\mathcal{J}_z(\boldsymbol{\tau}') \frac{\boldsymbol{\tau} - \boldsymbol{\tau}'}{2\pi|\boldsymbol{\tau} - \boldsymbol{\tau}'|^2} \tag{5.222}$$

on account of (H.14) and the fact that $\hat{\mathbf{z}} \cdot (\boldsymbol{\tau} - \boldsymbol{\tau}') = 0$. We should keep in mind that in (5.222) $\mathfrak{F}_m$ carries the physical dimensions of N/m and is directed perpendicularly to all the currents. Obviously, we can repeat the reasoning applied to (5.215) to conclude that the rightmost member of (5.222) is also null.

Quite often in practical applications one deals with more than one current density physically carried by comparatively long wires with finite cross sections, in which instance the two-dimensional approximation may turn out to be well-suited. If we define the current density as

$$\mathcal{J}(\mathbf{r}) = \mathcal{J}(\boldsymbol{\tau}) := \sum_{n=1}^{N} \mathcal{J}_{zn}(\boldsymbol{\tau})\hat{\mathbf{z}}, \qquad \boldsymbol{\tau} \in \bigcup_{n=1}^{N} S_{Jn} \tag{5.223}$$

we can immediately generalize (5.222) for the calculation of the force between a number N of different current densities all aligned with the z-axis, namely,

$$\mathfrak{F}_m = -\mu \sum_{n=1}^{N} \sum_{l=1}^{N} \int_{S_{Jn}} \mathrm{d}S \int_{S_{Jl}} \mathrm{d}S'\, \mathcal{J}_{zn}(\boldsymbol{\tau})\mathcal{J}_{zl}(\boldsymbol{\tau}') \frac{\boldsymbol{\tau} - \boldsymbol{\tau}'}{2\pi|\boldsymbol{\tau} - \boldsymbol{\tau}'|^2} \tag{5.224}$$

where with evident notation S_{Jn} indicates the cross section of the nth current.

For uniform current densities (5.224) becomes

$$\mathfrak{F}_m = -\mu \sum_{n=1}^{N} \sum_{l=1}^{N} \frac{I_n I_l}{A_n A_l} \int_{S_{Jn}} \mathrm{d}S \int_{S_{Jl}} \mathrm{d}S' \, \frac{\boldsymbol{\tau} - \boldsymbol{\tau}'}{2\pi |\boldsymbol{\tau} - \boldsymbol{\tau}'|^2} = [J]^T [F][J] \tag{5.225}$$

where A_n represents the area of S_{Jn}. The rightmost-hand side gives $\mathfrak{F}_m$ in symbolic matrix form where

$$[J] := \begin{pmatrix} I_1/A_1 \\ \vdots \\ I_N/A_N \end{pmatrix}, \qquad [F] := \begin{pmatrix} \mathbf{f}_{11} & \cdots & \mathbf{f}_{1N} \\ \vdots & \ddots & \vdots \\ \mathbf{f}_{N1} & \cdots & \mathbf{f}_{NN} \end{pmatrix} \tag{5.226}$$

and the vectorial entries

$$\mathbf{f}_{nl} = -\mu \int_{S_{Jn}} \mathrm{d}S \int_{S_{Jl}} \mathrm{d}S' \, \frac{\boldsymbol{\tau} - \boldsymbol{\tau}'}{2\pi |\boldsymbol{\tau} - \boldsymbol{\tau}'|^2}, \qquad n, l \in \{1, \ldots, N\} \tag{5.227}$$

represent a line density of force between two unitary current densities. The element $\mathbf{f}_{ln}$ follows from (5.227) by systematically swapping the indices n and l in the formula. Therefore, we have

$$\begin{aligned} \mathbf{f}_{ln} &= -\mu \int_{S_{Jl}} \mathrm{d}S \int_{S_{Jn}} \mathrm{d}S' \, \frac{\boldsymbol{\tau} - \boldsymbol{\tau}'}{2\pi |\boldsymbol{\tau} - \boldsymbol{\tau}'|^2} = \mu \int_{S_{Jn}} \mathrm{d}S' \int_{S_{Jl}} \mathrm{d}S \, \frac{\boldsymbol{\tau}' - \boldsymbol{\tau}}{2\pi |\boldsymbol{\tau}' - \boldsymbol{\tau}|^2} \\ &= \mu \int_{S_{Jn}} \mathrm{d}S \int_{S_{Jl}} \mathrm{d}S' \, \frac{\boldsymbol{\xi} - \boldsymbol{\xi}'}{2\pi |\boldsymbol{\xi} - \boldsymbol{\xi}'|^2} = -\mathbf{f}_{nl}, \qquad n, l \in \{1, \ldots, N\} \end{aligned} \tag{5.228}$$

where first we have inverted the order of integration and then changed the names of the dummy variables according to $\boldsymbol{\tau}' = \boldsymbol{\xi}$ and $\boldsymbol{\tau} = \boldsymbol{\xi}'$. Finally, comparison with (5.227) yields the very last step. Also this result is consistent with Newton's law of action and reaction [29, Chapter 10].

Since (5.228) holds in particular for $n = l$ we cannot help but conclude that $\mathbf{f}_{nn} = 0$. Hence, the symbolic matrix $[F]$ is anti-symmetric, and the vanishing of the diagonal entries implies that the self-force exerted by a straight uniform current on itself is null. For a system of N such currents, it follows that the total force is null, because

$$\mathfrak{F}_m = [J]^T [F][J] = [J]^T [F]^T [J] = -[J]^T [F][J] = -\mathfrak{F}_m \tag{5.229}$$

in view of the odd symmetry of $[F]$. Since only the uniformity of $\mathfrak{J}_{zn}$, $n = 1, \ldots, N$, has been invoked, the result holds for N uniform straight currents with arbitrary cross sections S_{Jn}.

Example 5.5 (Magnetic force between cylindrical uniform stationary currents)

When the sections S_{Jn} are circles $B_2(\boldsymbol{\tau}_n, a_n)$ we may compute integrals of the type in (5.227) by capitalizing on the findings of Example 4.1 where, with the aid of the Ampère law, we determined the magnetic field $\mathcal{H}(\mathbf{r})$ of an infinitely-long straight circular cylindrical current distributed symmetrically around and along the z-axis.

Indeed, for current densities defined by

$$\mathfrak{J}_n(\mathbf{r}) = \mathfrak{J}_n(\boldsymbol{\tau}) = \begin{cases} \mathfrak{J}_n \hat{\mathbf{z}} = \dfrac{I_n}{\pi a_n^2} \hat{\mathbf{z}}, & \boldsymbol{\tau} \in B_2(\boldsymbol{\tau}_n, a_n) \\ 0, & \boldsymbol{\tau} \notin B_2(\boldsymbol{\tau}_n, a_n) \end{cases} \qquad n = 1, 2, \ldots, N \tag{5.230}$$

from (4.16) we obtain

$$\mathcal{B}_n(\boldsymbol{\tau}) = \begin{cases} \dfrac{1}{2}\mu \mathcal{J}_n \hat{\mathbf{z}} \times (\boldsymbol{\tau} - \boldsymbol{\tau}_n), & |\boldsymbol{\tau} - \boldsymbol{\tau}_n| < a_n \\ \dfrac{1}{2}\mu \mathcal{J}_n a_n^2 \dfrac{\hat{\mathbf{z}} \times (\boldsymbol{\tau} - \boldsymbol{\tau}_n)}{|\boldsymbol{\tau} - \boldsymbol{\tau}_n|^2}, & |\boldsymbol{\tau} - \boldsymbol{\tau}_n| > a_n \end{cases} \tag{5.231}$$

having noticed that the magnetic induction — produced by $\mathcal{J}_n(\boldsymbol{\tau})$ — depends on the distance from the symmetry axis of $\mathcal{J}_n(\mathbf{r})$ and describes circles centered on points on the symmetry axis. Indeed, the vector $\hat{\mathbf{z}} \times (\boldsymbol{\tau} - \boldsymbol{\tau}_n)$ is tangential to the circle $B_2(\boldsymbol{\tau}_n, |\boldsymbol{\tau} - \boldsymbol{\tau}_n|)$ in the xOy plane. By comparing this result with the formal integral solution in two dimensions [2, 4]

$$\mathcal{B}_n(\boldsymbol{\tau}) = \mu \mathcal{J}_n \int\limits_{B_2(\boldsymbol{\tau}_n, a_n)} \mathrm{d}S' \, \frac{\hat{\mathbf{z}} \times (\boldsymbol{\tau} - \boldsymbol{\tau}')}{2\pi |\boldsymbol{\tau} - \boldsymbol{\tau}'|^2}, \qquad \boldsymbol{\tau} \in \mathbb{R}^2 \tag{5.232}$$

we arrive at the formula

$$\int\limits_{B_2(\boldsymbol{\tau}_n, a_n)} \mathrm{d}S' \, \frac{\boldsymbol{\tau} - \boldsymbol{\tau}'}{2\pi |\boldsymbol{\tau} - \boldsymbol{\tau}'|^2} = \begin{cases} \dfrac{1}{2}(\boldsymbol{\tau} - \boldsymbol{\tau}_n), & |\boldsymbol{\tau} - \boldsymbol{\tau}_n| < a_n \\ \dfrac{a_n^2}{2} \dfrac{\boldsymbol{\tau} - \boldsymbol{\tau}_n}{|\boldsymbol{\tau} - \boldsymbol{\tau}_n|^2}, & |\boldsymbol{\tau} - \boldsymbol{\tau}_n| > a_n. \end{cases} \tag{5.233}$$

We now have handy tools for evaluating the magnetic force between any two non-overlapping currents of the type (5.230). According to (5.225) and (5.227) the line density of force exerted by $\mathcal{J}_l$ on $\mathcal{J}_n$ reads

$$\begin{aligned} \mathcal{F}_{nl} &= -\mu \mathcal{J}_n \mathcal{J}_l \int\limits_{B_2(\boldsymbol{\tau}_n, a_n)} \mathrm{d}S \int\limits_{B_2(\boldsymbol{\tau}_l, a_l)} \mathrm{d}S' \, \frac{\boldsymbol{\tau} - \boldsymbol{\tau}'}{2\pi |\boldsymbol{\tau} - \boldsymbol{\tau}'|^2} = -\mu \mathcal{J}_n \mathcal{J}_l \int\limits_{B_2(\boldsymbol{\tau}_n, a_n)} \mathrm{d}S \, \frac{a_l^2}{2} \frac{\boldsymbol{\tau} - \boldsymbol{\tau}_l}{|\boldsymbol{\tau} - \boldsymbol{\tau}_l|^2} \\ &= \mu \mathcal{J}_n \mathcal{J}_l \pi a_l^2 \int\limits_{B_2(\boldsymbol{\tau}_n, a_n)} \mathrm{d}S \, \frac{\boldsymbol{\tau}_l - \boldsymbol{\tau}}{2\pi |\boldsymbol{\tau}_l - \boldsymbol{\tau}|^2} = \mu \mathcal{J}_n \mathcal{J}_l \pi a_l^2 \frac{a_n^2}{2} \frac{\boldsymbol{\tau}_l - \boldsymbol{\tau}_n}{|\boldsymbol{\tau}_l - \boldsymbol{\tau}_n|^2} \\ &= -I_n I_l \frac{\mu}{2\pi} \frac{\boldsymbol{\tau}_n - \boldsymbol{\tau}_l}{|\boldsymbol{\tau}_n - \boldsymbol{\tau}_l|^2}, \qquad |\boldsymbol{\tau}_n - \boldsymbol{\tau}_l| \geqslant a_l + a_n \end{aligned} \tag{5.234}$$

where we have made use of (5.233) twice. Since the vector $\boldsymbol{\tau}_n - \boldsymbol{\tau}_l$ provides the position of the axis of $\mathcal{J}_n$ with respect to the axis of $\mathcal{J}_l$, the force points towards $\mathcal{J}_l$ if $-I_n I_l < 0$ and towards $\mathcal{J}_n$ if $-I_n I_l > 0$. In other words, the force is attractive (resp. repulsive) if the currents flow in the same (resp. the opposite) direction.

We may deduce the formula for $\mathcal{F}_{ln}$, the force exerted by $\mathcal{J}_n$ on $\mathcal{J}_l$, by systematically swapping the indices in (5.234), whereby we conclude immediately that $\mathcal{F}_{ln} = -\mathcal{F}_{nl}$, as expected. More importantly, the self force should be null, in accordance with the general result (5.228). By definition, we have

$$\mathcal{F}_{nn} = -\mu \mathcal{J}_n^2 \int\limits_{B_2(\boldsymbol{\tau}_n, a_n)} \mathrm{d}S \int\limits_{B_2(\boldsymbol{\tau}_n, a_n)} \mathrm{d}S' \, \frac{\boldsymbol{\tau} - \boldsymbol{\tau}'}{2\pi |\boldsymbol{\tau} - \boldsymbol{\tau}'|^2} = -\frac{\mu}{2} \mathcal{J}_n^2 \int\limits_{B_2(\boldsymbol{\tau}_n, a_n)} \mathrm{d}S \, (\boldsymbol{\tau} - \boldsymbol{\tau}_n) = 0 \tag{5.235}$$

having made use of (5.233). The last step follows by integrating in local polar coordinates (ξ, α) centered on $\boldsymbol{\tau}_n$ and by noticing that $\boldsymbol{\xi} := \boldsymbol{\tau} - \boldsymbol{\tau}_n$ is a radial vector emerging from $\boldsymbol{\tau}_n$.

Since the force in (5.234) depends only on the currents I_n and I_l but not on the radii of S_{Jn} and S_{Jl} we can extend the formula to infinitely-long straight *line* currents by taking the limit for vanishing a_n and a_l. This modification requires we demand, e.g.,

$$I_n = \lim_{a_n \to 0^+} \int_{B_2(\tau_n, a_n)} dS\, \hat{\mathbf{z}} \cdot \mathfrak{J}(\boldsymbol{\tau}) = \lim_{a_n \to 0^+} \mathfrak{J}_n \pi a_n^2 \tag{5.236}$$

and similarly for I_l. In practice, the current density must grow infinitely large in order for the product $\mathfrak{J}_n \pi a_n^2$ to remain finite. In the notation of distributions (Appendix C) we would write

$$\mathfrak{J}_n(\boldsymbol{\tau}, z) = I_n \hat{\mathbf{z}}\, \delta^{(2)}(\boldsymbol{\tau} - \boldsymbol{\tau}_n)\,, \qquad n = 1, \ldots, N \tag{5.237}$$

where $\delta^{(2)}(\bullet)$ is the two-dimensional Dirac distribution. Hence, (5.234) is also valid for two line currents whose positions in the xOy plane are indicated by $\boldsymbol{\tau}_n$ and $\boldsymbol{\tau}_l$. Furthermore, from the viewpoint of two currents $\mathfrak{J}_n$ and $\mathfrak{J}_l$ of the type (5.230), (5.234) states that the magnetic force can be computed as if $\mathfrak{J}_n$ and $\mathfrak{J}_l$ *were* actually concentrated on the symmetry axes.

(End of Example 5.5)

References

[1] Stratton JA. Electromagnetic theory. London, UK: McGraw-Hill; 1941.

[2] Rothwell EJ, Cloud MJ. Electromagnetics. London, UK: CRC Press; 2001.

[3] Schwinger J, De Raad LL, Milton KA, et al. Classical electrodynamics. Perseus Books; 1998.

[4] Jackson JD. Classical Electrodynamics. 3rd ed. Chichester, UK: Wiley; 1999.

[5] Zhou P. Numerical Analysis of Electromagnetic Fields. Berlin Heidelberg: Springer-Verlag; 1993.

[6] Makarov B, Podkorytov A. Real Analysis: Measures, Integrals and Applications. Universitext. London, UK: Springer; 2013.

[7] Dautray R, Lions JL. Mathematical Analysis and Numerical Methods for Science and Technology - Physical Origins and Classical Methods. vol. 1. Berlin Heidelberg: Springer-Verlag; 1990.

[8] Van Bladel JG. Electromagnetic Fields. Piscataway, NJ: IEEE Press; 2007.

[9] Guru BS, Hiziroglu HR. Electromagnetic field theory fundamentals. 2nd ed. New York, NY: Cambridge University Press; 2004.

[10] Konopinski EJ. Electromagnetic Fields and Relativistic Particles. New York, NY: McGraw Hill; 1981.

[11] Zangwill A. Modern electrodynamics. Cambridge, UK: Cambrigde University Press; 2013.

[12] Lindell IV. Methods for electromagnetic field analysis. Piscataway, NJ: IEEE Press; 1992.

[13] Feynmann RP, Leighton R, Sands M. The Feynmann lectures on physics. vol. 2. Reading, MA: Addison-Wesley; 1964.

[14] Griffith DJ. Introduction to electrodynamics. 3rd ed. Upper Saddle River, NJ: Prentice Hall; 1999.

[15] Sherwin CW. Introduction to quantum mechanics. New York, NY: Henry Holt and Company; 1959.

[16] Aharonov Y, Bohm D. Significance of electromagnetic potentials in quantum theory. Physical Review. 1959;115:485–491.

[17] Thouless DJ. Topological quantum numbers in nonrelativistic physics. London, UK: World Scientific; 1998.
[18] Maxwell JC. A Treatise on Electricity and Magnetism. vol. 2. Oxford, UK: Clarendon Press; 1873.
[19] Tonnelat MA. The principles of electromagnetic theory and of relativity. Dordrecht-Holland, NL: D. Reidel Publishing Company; 1966.
[20] Bassanini P, Elcrat A. Mathematical Theory of Electromagnetism. Creative Commons; 2009.
[21] Helrich CS. The classical theory of fields - Electromagnetism. Berlin Heidelberg: Springer-Verlag; 2012.
[22] Raab RE, Lange OLD. Multipole Theory in Electromagnetism. Oxford, UK: Clarendon Press; 2005.
[23] Mason M, Weaver W. The electromagnetic field. New York, NY: Dover Publications, Inc.; 1929.
[24] Sommerfeld A. Electrodynamics. vol. 3 of Lectures on theoretical physics. New York, NY: Academic Press; 1952.
[25] Popović Z, Popović BD. Introductory Electromagnetics. Upper Saddle River, NJ: Prentice Hall; 2000.
[26] Shen LC, Kong JA. Applied electromagnetism. Monterey, CA: Brooks/Cole; 1983.
[27] Lorrain P, Corson DR, Lorrain F. Electromagnetic Fields and Waves. 3rd ed. New York, NY: W. H. Freeman and Company; 1988.
[28] Cottingham WN, Greenwood DA. Electricity and Magnetism. Cambridge, UK: Cambridge University Press; 1991.
[29] Feynmann RP, Leighton R, Sands M. The Feynmann lectures on physics. vol. 1. Reading, MA: Addison-Wesley; 1964.

Chapter 6

Properties of electromagnetic fields

6.1 Principle of superposition

Perhaps one of the most striking features of the electromagnetic theory, and hence of Maxwell equations, is that they are *linear* in the vast majority of practical situations. This is trivially true in free space, whereas in the presence of material media linearity of the equations is predicated on the type of constitutive relationships (1.110) and (1.111). As long as the functional form of (1.110) and (1.111), no matter how complicated, involves linear operators only, then Maxwell's equations are still linear. On the other hand, non-linear constitutive relationships typically arise when a material body is exposed to strong electric (as in the Kerr effect [1, Section V.4]) or magnetic fields or the medium exhibits hysteresis, as is the case for ferromagnetic materials [2].

So long as the Maxwell equations are linear, then the so-called **principle of superposition** holds [3, Chapter 23]. In words, this means that any linear combination of two or more solutions to (1.20), (1.23), (1.34) and (1.44) still represents a solution of the same equations subject to the same boundary conditions. As a matter of fact, we have already implicitly exploited this principle in many a situation, e.g., for the definition of time-harmonic fields in Section 1.5 and the construction of the Dirichlet Green function (3.37). What is more, all the integral representations for fields and potentials discussed in Sections 2.7, 3.2, 5.1.1 and 5.1.2 were derived by exploiting the linearity of Maxwell's equations and the principle of superposition. In this regard, we may even interpret an integral involving sources and a suitable Green function as the linear combination of *infinitely many* elementary solutions.

In order to prove the principle of superposition consider, for instance, the Ampère-Maxwell law (1.34) applied to two problems comprised of the same boundary conditions but different current densities $\mathcal{J}_1(\mathbf{r}, t)$ and $\mathcal{J}_2(\mathbf{r}, t)$, as is suggested in Figure 6.1. In symbols, we have

$$\nabla \times \mathcal{H}_1(\mathbf{r}, t) = \frac{\partial}{\partial t}\mathcal{D}_1(\mathbf{r}, t) + \mathcal{J}_1(\mathbf{r}, t), \qquad (\mathbf{r}, t) \in V \times \mathbb{R} \tag{6.1}$$

$$\nabla \times \mathcal{H}_2(\mathbf{r}, t) = \frac{\partial}{\partial t}\mathcal{D}_2(\mathbf{r}, t) + \mathcal{J}_2(\mathbf{r}, t), \qquad (\mathbf{r}, t) \in V \times \mathbb{R} \tag{6.2}$$

for points $\mathbf{r} \in V$. Thanks to the linearity of the differential operators we are allowed to sum these equations side by side to arrive at

$$\nabla \times [\mathcal{H}_1(\mathbf{r}, t) + \mathcal{H}_2(\mathbf{r}, t)] = \frac{\partial}{\partial t}[\mathcal{D}_1(\mathbf{r}, t) + \mathcal{D}_2(\mathbf{r}, t)] + \mathcal{J}_1(\mathbf{r}, t) + \mathcal{J}_2(\mathbf{r}, t) \tag{6.3}$$

which is the Ampère-Maxwell law involving the field entities $\mathcal{H}_1 + \mathcal{H}_2$ and $\mathcal{D}_1 + \mathcal{D}_2$ and the current density $\mathcal{J}_1 + \mathcal{J}_2$. While this step is permissible irrespective of the nature of the constitutive relationship (1.110) for the displacement vector, the following chain of equalities

$$\mathcal{D}_1 + \mathcal{D}_2 = f_D(\mathcal{E}_1, \mathcal{H}_1) + f_D(\mathcal{E}_2, \mathcal{H}_2) = f_D(\mathcal{E}_1 + \mathcal{E}_2, \mathcal{H}_1 + \mathcal{H}_2) \tag{6.4}$$

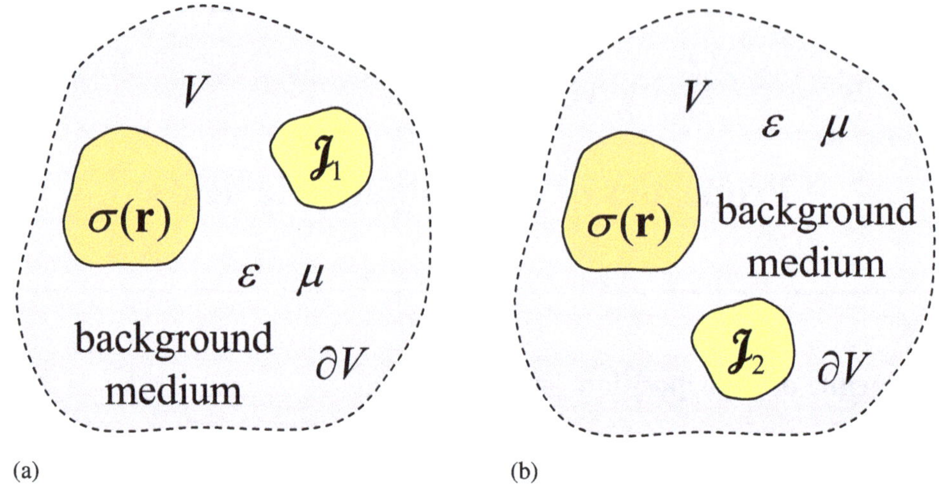

Figure 6.1 For illustrating the superposition principle: two electromagnetic problems involving different sources but the same boundary conditions.

holds only if f_D involves linear operations. If the latter condition is not true, then the solution to (6.3) is not equal to the sum of the separate solutions to (6.1) and (6.2).

Perfectly similar derivations and conclusions apply to the other three Maxwell equations and to the continuity equation (1.46).

From the previous discussion it may appear that the superposition principle has rather limited usefulness, in that the application thereof heavily relies on the availability of solutions. Besides, if we know how to solve a problem in closed form for a given distribution of sources, most likely we can do that for any source. Thus, why bother? Well, for at least a couple of good reasons having to do with the construction of approximate solutions to Maxwell's equations by means of numerical approaches (Chapter 14).

First of all, linearity allows us to *try* a representation of an unknown field in terms of elementary solutions (e.g., see discussion in Section 3.10) or even suitable arbitrary functions. Usually, the goal of this procedure is to cast the original differential problem into an algebraic system of equations [4–12] which is easier to solve with the methods of linear algebra [13–15]. Secondly, in a geometrical setup as that of Figure 6.1 we may want to separate the unknown electromagnetic field into two parts: 1) the field produced by the sources in the unbounded background medium *in the absence* of any material body; 2) the secondary field produced by the charges (either conduction or polarization or both) induced on the surface or in the bulk of the bodies and existing in the background medium. For instance, in electrostatics the mathematical tool for such separation is provided by the integral representation (2.160). The rationale for the splitting is that, most likely, we know how to determine the first field, and thus we just have to try an expansion in terms of elementary functions only for the induced sources (and secondary field) which are most certainly unknown.

It is worthwhile emphasizing that the principle of superposition does not apply to the Poynting theorem (1.292) in local form. Of course, we may write it down for the two problems of Figure 6.1, say,

$$\nabla \cdot \mathcal{S}_1(\mathbf{r}, t) + \frac{\mathrm{d}}{\mathrm{d}t}(w_{e1} + w_{h1}) + \sigma(\mathbf{r})|\mathcal{E}_1(\mathbf{r}, t)|^2 = -\mathcal{E}_1(\mathbf{r}, t) \cdot \mathcal{J}_1(\mathbf{r}, t) \tag{6.5}$$

$$\nabla \cdot \mathcal{S}_2(\mathbf{r},t) + \frac{\mathrm{d}}{\mathrm{d}t}(w_{e2} + w_{h2}) + \sigma(\mathbf{r})|\mathcal{E}_2(\mathbf{r},t)|^2 = -\mathcal{E}_2(\mathbf{r},t) \cdot \mathcal{J}_2(\mathbf{r},t) \tag{6.6}$$

and may even sum these equations side by side (linearity is not an issue)

$$\nabla \cdot [\mathcal{S}_1(\mathbf{r},t) + \mathcal{S}_2(\mathbf{r},t)] + \frac{\mathrm{d}}{\mathrm{d}t}(w_{e1} + w_{e2} + w_{h1} + w_{h2}) + \sigma(\mathbf{r})|\mathcal{E}_1(\mathbf{r},t)|^2 + \sigma(\mathbf{r})|\mathcal{E}_2(\mathbf{r},t)|^2 = -\mathcal{E}_1(\mathbf{r},t) \cdot \mathcal{J}_1(\mathbf{r},t) - \mathcal{E}_2(\mathbf{r},t) \cdot \mathcal{J}_2(\mathbf{r},t) \tag{6.7}$$

but regrettably the latter equation *does not* constitute a statement of the local Poynting theorem for the problem described by (6.3) and the three remaining pertinent Maxwell equations. This is equivalent to saying that the energy density associated with the sum of two electromagnetic fields, in general, is not equal to the sum of the energies associated with each one of the fields separately. What is more, $\mathcal{S}_1(\mathbf{r},t) + \mathcal{S}_2(\mathbf{r},t)$ *is not* the Poynting vector pertaining to the sum field.

As an exception to this rule, we mention the eigenfunction representation of the fields inside a classic hollow-pipe waveguide with conducting walls (Section 11.2.6) [16–19]. Since the eigenfunctions are *orthogonal* with respect to a suitable inner product defined on the cross section of the waveguide, one can combine the Poynting theorem *in global form* (1.276) or (1.314) and still obtain the correct power and energy balance relevant for the combination of eigenfunctions. This means that the power or energy associated with the sum of any two eigenfunctions equals the sum of the energies carried by each eigenfunction separately.

Example 6.1 (Electrostatic field of a semi-infinite straight line density of charge)

We wish to determine potential and electric field produced by a uniform density of charge ϱ_L placed along a semi-infinite straight line which, with no loss of generality, we may take as the negative z-axis of three superimposed systems of Cartesian, circular cylindrical and polar spherical coordinates (Figure 6.2a). While the potential and the field in question must be independent of the polar angle φ for mere reasons of symmetry, Φ and $\mathcal{E}$ must still depend on ρ and z or r and ϑ, and this behavior together with the singular nature of the source makes the direct solution of the relevant Poisson equation somewhat tricky.

Alternatively, we may resort to the principle of superposition backwards, as it were, and regard the original problem as stemming from the combination of two auxiliary distributions of charges, namely,

(1) an infinite line density of charge $\varrho_L/2$ placed along the z-axis (Figure 6.2b),
(2) an infinite piecewise-constant line density defined as (Figure 6.2c)

$$\varrho_L(z) := \frac{1}{2}\varrho_L U(-z) - \frac{1}{2}\varrho_L U(z) \tag{6.8}$$

where $U(\bullet)$ denote a one-dimensional step function.

We solve each auxiliary problem separately and then add the solutions. The inherent advantage of the configurations sketched in Figures 6.2b and 6.2c is that the charge density exhibits either even or odd mirror symmetry about the plane $z = 0$.

The problem of Figure 6.2b can be tackled with elementary means, because the electric field $\mathcal{E}_1$ depends only on ρ and is radially directed. Indeed, the φ-component must vanish inasmuch as (2.2) demands

$$0 = \oint_C \mathrm{d}s\, \hat{\mathbf{s}}(\mathbf{r}) \cdot \mathcal{E}_1(\mathbf{r}) = \int_0^{2\pi} \mathrm{d}\varphi\, a\mathcal{E}_{1\varphi}(a) = 2\pi a \mathcal{E}_{1\varphi}(a) \tag{6.9}$$

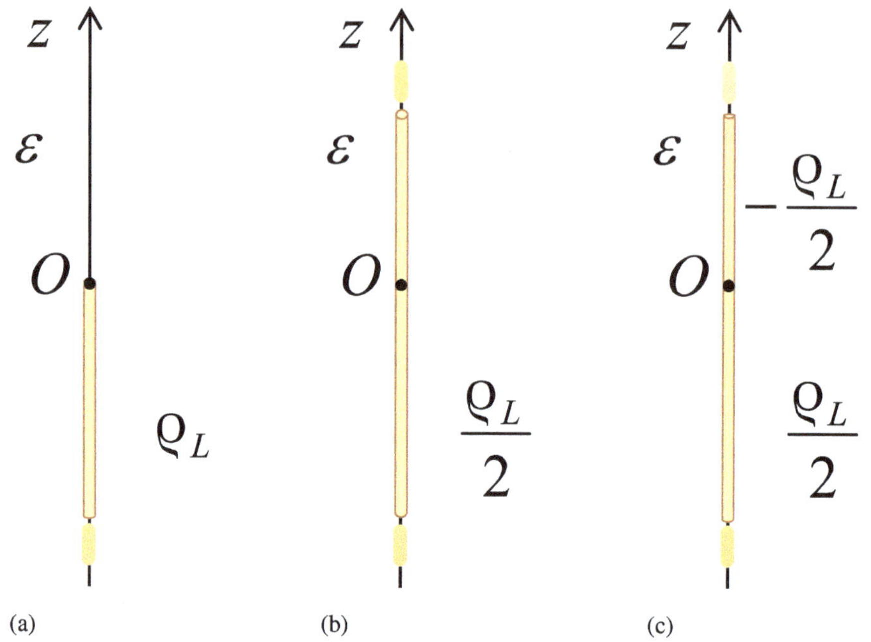

Figure 6.2 Superposition principle applied to an electrostatic problem: (a) original semi-infinite line charge distribution and (b), (c) auxiliary doubly-infinite charge distributions.

where C is the boundary of a circle of radius a with center on the z-axis and lying in a plane perpendicular to $\hat{\mathbf{z}}$. The z-component vanishes, too, owing to the even symmetry of the charge with respect to any plane $z = z_0$. To compute $\mathcal{E}_{1\rho}(\rho)$ we consider the cylinder

$$V := \{\mathbf{r} \in \mathbb{R}^3 : \rho \leqslant a, z \in [z_1, z_2], \varphi \in [0, 2\pi[\} \tag{6.10}$$

and apply the Gauss law (2.1), viz.,

$$\frac{1}{2}\varrho_L(z_2 - z_1) = \int_{\partial V} \mathrm{d}S\, \hat{\mathbf{n}}(\mathbf{r}) \cdot \varepsilon \boldsymbol{\mathcal{E}}_1(\rho) = (z_2 - z_1) \int_0^{2\pi} \mathrm{d}\varphi\, a\varepsilon \mathcal{E}_{1\rho}(a) = 2\pi a(z_2 - z_1)\varepsilon \mathcal{E}_{1\rho}(a) \tag{6.11}$$

whence we conclude

$$\boldsymbol{\mathcal{E}}_1(\rho) = \frac{\varrho_L \hat{\boldsymbol{\rho}}}{4\pi\varepsilon\rho} \tag{6.12}$$

on renaming the arbitrary radius a to ρ. The field falls off as the inverse of the distance from the charge, since the source is infinitely extended. The scalar potential $\Phi_1(\rho)$ can be recovered by integrating the electric field in accordance with (2.15), viz.,

$$\Phi_1(\rho) = \frac{\varrho_L}{4\pi\varepsilon} \log \frac{C_\Phi}{\rho} \tag{6.13}$$

where C_Φ is an arbitrary positive constant with the physical dimension of a length. Clearly, the condition $\rho = C_\Phi$ defines the circular cylinder over which the potential vanishes. Also, notice that $\Phi_1(\rho)$ is singular both at the location of the source and at infinity.

The problem of Figure 6.2c is better tackled by computing the scalar potential first. Since the charge density depends on z, we expect Φ_2 and $\mathcal{E}_2$ to be functions of ρ and z, and hence the electric field will also have a component parallel to the z-axis. For the scalar potential we rely on the general formula (2.169), which, since the charge density is concentrated on a line, passes over into an improper integral along the z-axis, namely,

$$\Phi_2(\rho, z) = \int_{\mathbb{R}} \mathrm{d}z' \, \frac{[U(-z') - U(z')]\varrho_L/2}{4\pi\varepsilon\sqrt{\rho^2 + (z - z')^2}} \tag{6.14}$$

although the integral diverges for $\rho = 0$ owing to the singularity in $z' = z$. To proceed we first integrate over the finite interval $[-h, h]$, $h > 0$, and only afterwards we take the limit as $h \to +\infty$, i.e.,

$$\begin{aligned}\Phi_2(\rho, z) &= \lim_{h\to+\infty}\left[\int_{-h}^{0} \mathrm{d}z' \, \frac{\varrho_L/2}{4\pi\varepsilon\sqrt{\rho^2 + (z - z')^2}} - \int_{0}^{h} \mathrm{d}z' \, \frac{\varrho_L/2}{4\pi\varepsilon\sqrt{\rho^2 + (z - z')^2}}\right] \\ &= \lim_{h\to+\infty} \frac{\varrho_L}{8\pi\varepsilon}\left\{\left[-\log\left(z - z' + \sqrt{\rho^2 + (z - z')^2}\right)\right]_{-h}^{0} + \left[\log\left(z - z' + \sqrt{\rho^2 + (z - z')^2}\right)\right]_{0}^{h}\right\} \\ &= \lim_{h\to+\infty} \frac{\varrho_L}{8\pi\varepsilon} \log \frac{\left(z + h + \sqrt{\rho^2 + (z + h)^2}\right)\left(z - h + \sqrt{\rho^2 + (z - h)^2}\right)}{\left(z + \sqrt{\rho^2 + z^2}\right)^2}\end{aligned} \tag{6.15}$$

where we have carried out the integrals with the help of formula 200.01 in [20]. To compute the limit — which, as it stands, amounts to an indeterminate form of the type $\infty \times (\infty - \infty)$ — we write the product in the numerator as

$$\begin{aligned}&(h + z)\left[\sqrt{1 + \frac{\rho^2}{(h + z)^2}} + 1\right](h - z)\left[\sqrt{1 + \frac{\rho^2}{(h - z)^2}} - 1\right] = \\ &= (h^2 - z^2)\left[\sqrt{1 + \frac{\rho^2}{(h + z)^2}} + 1\right]\left[\frac{\rho^2}{2(h - z)^2} - \frac{\rho^4}{8(h - z)^4} + o\left(\frac{\rho^4}{(h - z)^4}\right)\right] \\ &= \left[\sqrt{1 + \frac{\rho^2}{(h + z)^2}} + 1\right]\left[\frac{h + z}{h - z}\frac{\rho^2}{2} - \frac{h + z}{(h - z)^3}\frac{\rho^4}{8} + o\left(\frac{h + z}{(h - z)^3}\rho^4\right)\right]\end{aligned} \tag{6.16}$$

and this quantity tends to ρ^2 as h grows infinitely large. Therefore, the potential reads

$$\Phi_2(\rho, z) = \frac{\varrho_L}{8\pi\varepsilon} \log \frac{\rho^2}{\left(z + \sqrt{\rho^2 + z^2}\right)^2} = \frac{\varrho_L}{4\pi\varepsilon} \log \frac{\rho}{z + \sqrt{\rho^2 + z^2}} \tag{6.17}$$

though we may also express the result in polar spherical coordinates, viz.,

$$\Phi_2(\vartheta) = \frac{\varrho_L}{4\pi\varepsilon} \log \tan \frac{\vartheta}{2} \tag{6.18}$$

whence we see the potential is constant on one-sheeted cones which are symmetric around the z-axis and have opening ϑ and apex in the origin. We determine the electric field from either (6.17) or (6.18)

$$\mathcal{E}_2(\rho, z) = \frac{\varrho_L}{4\pi\varepsilon}\left(-\frac{\hat{\boldsymbol{\rho}}}{\rho} + \frac{\rho}{\rho^2 + z^2 + z\sqrt{\rho^2 + z^2}} + \frac{\hat{\mathbf{z}}}{\sqrt{\rho^2 + z^2}}\right) \tag{6.19}$$

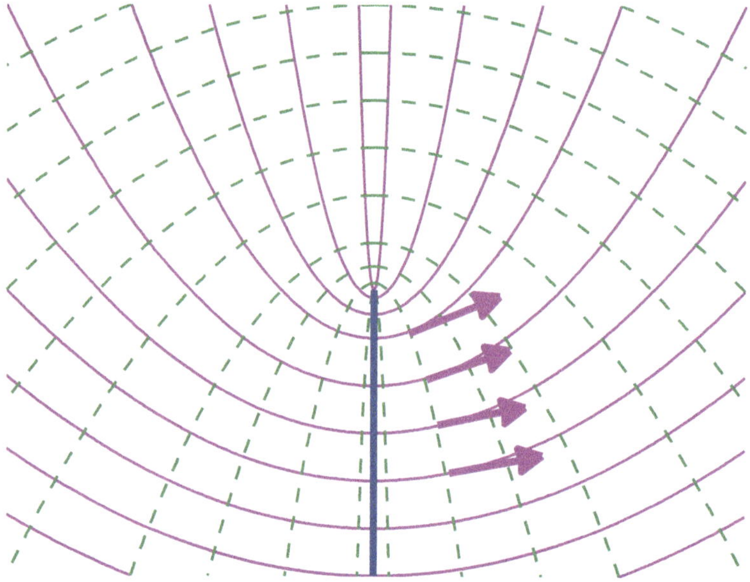

Figure 6.3 Lines of constant $\Phi(\mathbf{r})$ (−−) and streamlines of $\mathcal{E}(\mathbf{r})$ (−) in the xOz plane for the semi-infinite line charge density of Figure 6.2a.

$$\mathcal{E}_2(r,\vartheta) = -\frac{\varrho_L}{4\pi\varepsilon}\frac{\hat{\boldsymbol{\vartheta}}}{r\sin\vartheta} \tag{6.20}$$

on account of (2.15), (A.27) and (A.28). Not surprisingly, the electric field streamlines are half-circumferences which, if $\varrho_L > 0$, start on the lower density and end onto the upper one (Figure 6.2c). Incidentally, the image principle (Section 3.9) tells us that the solution to the problem of a semi-infinite line density $-\varrho_L/2$ along the positive z-axis in the presence of a conducting half-space in $z < 0$ is still given by (6.18) and (6.20) for $\vartheta \in [0, \pi/2]$.

Finally, by summing Φ_1 and Φ_2 as well as $\mathcal{E}_1$ and $\mathcal{E}_2$ we find the solution pertinent to the problem of Figure 6.2a, i.e.,

$$\Phi(\mathbf{r}) = \frac{\varrho_L}{4\pi\varepsilon}\log\frac{C_\Phi}{z+\sqrt{\rho^2+z^2}} = \frac{\varrho_L}{4\pi\varepsilon}\log\frac{C_\Phi}{r(1+\cos\vartheta)} \tag{6.21}$$

$$\mathcal{E}(\mathbf{r}) = \frac{\varrho_L}{4\pi\varepsilon\sqrt{\rho^2+z^2}}\left(\frac{\boldsymbol{\rho}}{z+\sqrt{\rho^2+z^2}}+\hat{\mathbf{z}}\right) = \frac{\varrho_L}{4\pi\varepsilon r}\left(\frac{\hat{\boldsymbol{\rho}}\sin\vartheta}{1+\cos\vartheta}+\hat{\mathbf{z}}\right). \tag{6.22}$$

Evidently, both $\Phi(\mathbf{r})$ and $\mathcal{E}(\mathbf{r})$ are singular on the negative z-axis, i.e., at the location of the charge. Besides, the equipotential surfaces and the electric-field streamlines constitute two families of paraboloids of revolution that are mutually orthogonal at any point in space. As a matter of fact, such paraboloids represent two sets of canonical surfaces of a system of *parabolic coordinates* in which the Laplace equation can be separated [21, pp. 34-36]. As an example, a few lines of constant Φ and some lines of force [22, Section 4.1] are shown in Figure 6.3 for points in the xOz plane.

(End of Example 6.1)

6.2 Well-posedness of the Maxwell equations

The following brief discussion concerns the ability of Maxwell's equations to describe the electromagnetic phenomena in a way which is useful for practical applications. In other words, given the full set of Maxwell's equations supplemented with suitable initial and boundary conditions, we wonder whether the problem can actually be solved or is an impossible one. Moreover, in case we have found a solution, we wonder whether the problem admits yet other solutions and, all in all, what the properties are of these solutions.

All these legitimate questions are summarized in the definition of 'well-posedness' proposed by J. Hadamard [23], [24, Section 2.7]. A mathematical or physical problem is said to be *well-posed*, if the following conditions are satisfied:

- (WP 1) The problem admits *at least* one solution (existence).
- (WP 2) The problem has *at most* one solution (uniqueness).
- (WP 3) The solution is *stable*, i.e., it depends continuously on the data supplied (initial and boundary conditions, sources).

By contrast, a problem which does not meet any or all of the conditions for well-posedness is said to be *ill-posed* [25]. When the problem is described by a linear set of equations, e.g., written symbolically in the form

$$\mathfrak{L}\{v\} = w \tag{6.23}$$

where $\mathfrak{L}\{\bullet\}$ denotes a suitable linear operator, conditions (WP 2) and (WP 3) are equivalent to

- (WP 2) The homogeneous problem has only the zero solution (uniqueness).
- (WP 3) If the data supplied (initial and boundary conditions, sources) approach zero, then the solution tends to zero, too.

In terms of properties of operators between vectors spaces (Appendix D.3) requirements (WP 1), (WP 2) and (WP 3) can be rephrased by stating that $\mathfrak{L}\{\bullet\}$ must be *surjective*, $\mathfrak{L}\{\bullet\}$ must be *injective*, and finally that $\mathfrak{L}\{\bullet\}$ must be *continuous* (or equivalently bounded, if $\mathfrak{L}\{\bullet\}$ is linear).

The claim of existence is generally more difficult to prove mathematically, though, if a set of equations is meant to describe a physical problem for which one already has experimental evidence, solutions are likely to exist. Uniqueness amounts to showing that, if two solutions are supposed to exist, then they must necessarily be the same solution (e.g., see Sections 2.5.1 and 2.5.2). Again, this property is crucial for a set of equations to be predictive and useful for design purposes. If the solution is not unique, then one cannot possibly know whether or not the realized system will behave as expected. It should be noticed that existence and uniqueness are strictly related to the initial and boundary conditions chosen for the problem. The third condition for well-posedness implies that small variations of the sources should cause small variations of the solution. In this regard, the 'distance' between different sources or solutions is usually measured by means of some norm in the appropriate function space.

A typical well-posed problem in electromagnetics is constituted by the *direct electromagnetic scattering*. Specifically, one wishes to compute the field produced by some sources in a background medium in the presence of a material body, as is suggested in Figure 6.4. It pays off to separate the problem into two parts, namely, calculation of the field generated by the sources without the body and calculation of the field due to the secondary sources induced on the surface of the body. The former is referred to as the *incident field*, the latter is called the *scattered field*. The problem is described by the Maxwell equations in local form with suitable boundary and initial conditions, in

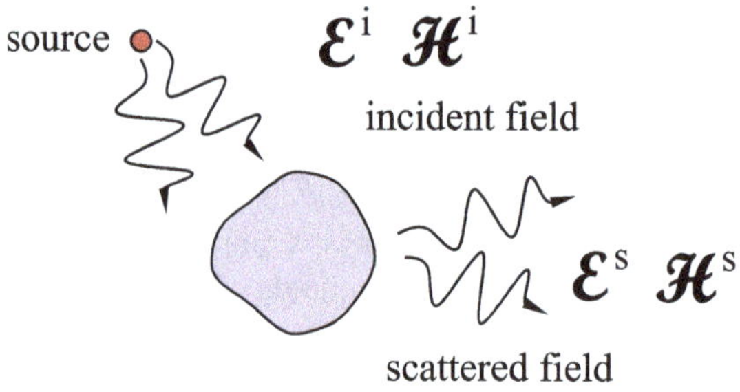

Figure 6.4 Example of well-posed problem: direct electromagnetic scattering.

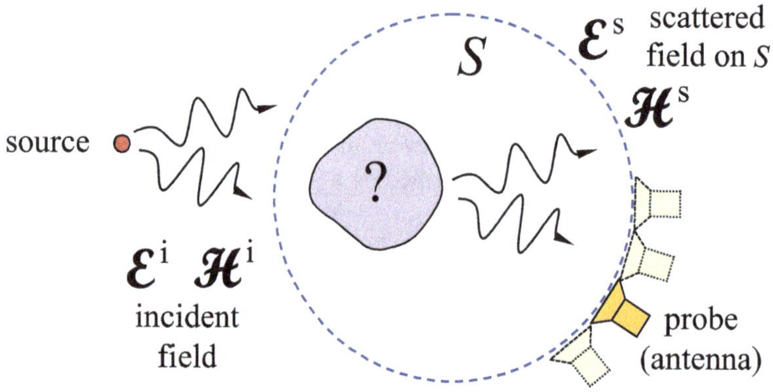

Figure 6.5 Example of ill-posed problem: inverse electromagnetic scattering.

which case one aims at computing the scattered field. Alternatively, one may formulate the problem as an integral equation (Chapter 13) by invoking one of the integral representations (equivalence principles) of Chapter 10. Then, the goal becomes the calculation of the secondary sources on the surface or in the bulk of the body.

Conversely, a notorious ill-posed problem is the so-called *inverse electromagnetic scattering* or *inverse profiling* in which one tries to extract information about a material body from the field it scatters back when illuminated by a known incident field [25,26], [27, Chapter 9], [28, Chapter 9]. A possible setup is sketched in Figure 6.5 where the scattered field is sampled over a sphere by means of an antenna which is placed in a set of discrete locations. Since the unknown variable may be the shape or the position of the object or even the material properties thereof (e.g., the permittivity), the problem as it stands is inherently ill-posed not for the lack of a solution, but rather because there may be infinitely many objects and positions thereof which give rise to the same scattered field.

6.3 Uniqueness in the time domain

In this Section we address the question of uniqueness of the solutions to the Maxwell equations when the sources exhibit a general time-dependence. Thus, the discussion extends and complements the analogous ones concerning static electric fields (Section 2.5.2) and stationary magnetic fields (Section 4.7.2).

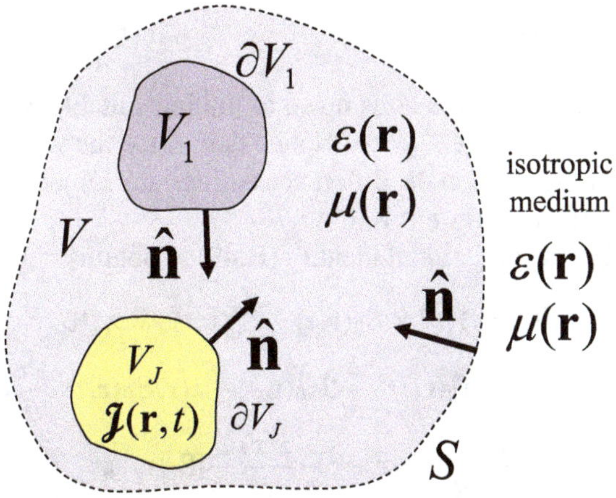

Figure 6.6 For proving uniqueness of the solutions to the Maxwell equations in the time domain in a bounded region of space.

6.3.1 Bounded regions

To begin with, we consider an unbounded homogeneous isotropic background medium endowed with linear constitutive parameters $\varepsilon(\mathbf{r})$ and $\mu(\mathbf{r})$. We exclude a connected bounded volume $V_1 \subset \mathbb{R}^3$ which may be occupied by a material medium yet unspecified. We denote with V the region of space surrounded by ∂V_1 and a 'large' smooth closed surface, so that $\partial V := S \cup \partial V_1$ is the boundary of V. We take the unit normal to ∂V positively oriented inwards V. Finally, time-dependent sources (charges and currents) are confined within a volume $V_J \subset V$. This geometry is sketched in Figure 6.6.

In our quest of the necessary conditions for the solutions to be unique, we suppose that the Maxwell equations are solved by two distinct electromagnetic fields — which we denote with subscripts 1 and 2 — for $(\mathbf{r}, t) \in V \times \mathbb{R}^+$, and try to determine whether said fields coincide, contrary to the hypothesis (proof by contradiction). Remarkably, all the manipulations involved in the proof are possible in view of the principle of superposition.

As is customary [29, 30], we write down the local form of the Faraday law and of the Ampère-Maxwell law for the two presumed solutions

$$\nabla \times \mathcal{E}_1(\mathbf{r}, t) + \mu(\mathbf{r})\frac{\partial}{\partial t}\mathcal{H}_1(\mathbf{r}, t) = 0, \qquad (\mathbf{r}, t) \in V \times \mathbb{R}^+ \tag{6.24}$$

$$\nabla \times \mathcal{H}_1(\mathbf{r}, t) - \varepsilon(\mathbf{r})\frac{\partial}{\partial t}\mathcal{E}_1(\mathbf{r}, t) = \mathcal{J}(\mathbf{r}, t), \qquad (\mathbf{r}, t) \in V \times \mathbb{R}^+ \tag{6.25}$$

and

$$\nabla \times \mathcal{E}_2(\mathbf{r}, t) + \mu(\mathbf{r})\frac{\partial}{\partial t}\mathcal{H}_2(\mathbf{r}, t) = 0, \qquad (\mathbf{r}, t) \in V \times \mathbb{R}^+ \tag{6.26}$$

$$\nabla \times \mathcal{H}_2(\mathbf{r}, t) - \varepsilon(\mathbf{r})\frac{\partial}{\partial t}\mathcal{E}_2(\mathbf{r}, t) = \mathcal{J}(\mathbf{r}, t), \qquad (\mathbf{r}, t) \in V \times \mathbb{R}^+ \tag{6.27}$$

with the values of $\mathcal{E}_1(\mathbf{r}, 0) = \mathcal{E}_2(\mathbf{r}, 0)$ and $\mathcal{H}_1(\mathbf{r}, 0) = \mathcal{H}_2(\mathbf{r}, 0)$ (initial conditions) assigned for $\mathbf{r} \in V$. For ease of manipulation we define the difference entities $\mathcal{E}_0 := \mathcal{E}_1 - \mathcal{E}_2$ and $\mathcal{H}_0 := \mathcal{H}_1 - \mathcal{H}_2$. Thanks to the principle of superposition, the equations above can be combined to show that $\mathcal{E}_0(\mathbf{r}, t)$ and $\mathcal{H}_0(\mathbf{r}, t)$ solve the Faraday law and the source-free Ampère-Maxwell law, namely,

$$\nabla \times \mathcal{E}_0(\mathbf{r}, t) + \mu(\mathbf{r})\frac{\partial}{\partial t}\mathcal{H}_0(\mathbf{r}, t) = 0, \qquad (\mathbf{r}, t) \in V \times \mathbb{R}^+ \tag{6.28}$$

$$\nabla \times \mathcal{H}_0(\mathbf{r}, t) - \varepsilon(\mathbf{r})\frac{\partial}{\partial t}\mathcal{E}_0(\mathbf{r}, t) = 0, \qquad (\mathbf{r}, t) \in V \times \mathbb{R}^+ \tag{6.29}$$

whence we see that proving uniqueness boils down to finding suitable conditions for the difference fields to be identically null for $(\mathbf{r}, t) \in V \times \mathbb{R}^+$. Notice that, since the values of the two solutions are the same for $t = 0$ by assumption, then the difference entities are subject to the homogeneous initial conditions $\mathcal{E}_0(\mathbf{r}, 0) = 0 = \mathcal{H}_0(\mathbf{r}, 0)$ for $\mathbf{r} \in V$.

To proceed, we employ the differential identity (H.49) to obtain

$$\begin{aligned}\nabla \cdot [\mathcal{E}_0(\mathbf{r}, t) \times \mathcal{H}_0(\mathbf{r}, t)] &= \mathcal{H}_0(\mathbf{r}, t) \cdot \nabla \times \mathcal{E}_0(\mathbf{r}, t) - \mathcal{E}_0(\mathbf{r}, t) \cdot \nabla \times \mathcal{H}_0(\mathbf{r}, t) \\ &= -\mu(\mathbf{r})\mathcal{H}_0(\mathbf{r}, t) \cdot \frac{\partial}{\partial t}\mathcal{H}_0(\mathbf{r}, t) - \varepsilon(\mathbf{r})\mathcal{E}_0(\mathbf{r}, t) \cdot \frac{\partial}{\partial t}\mathcal{E}_0(\mathbf{r}, t) \\ &= -\frac{\mu(\mathbf{r})}{2}\frac{\partial}{\partial t}|\mathcal{H}_0(\mathbf{r}, t)|^2 - \frac{\varepsilon(\mathbf{r})}{2}\frac{\partial}{\partial t}|\mathcal{E}_0(\mathbf{r}, t)|^2\end{aligned} \tag{6.30}$$

on account of (6.28) and (6.29). Next, we integrate the left-hand side and the rightmost-hand side of this equation over V to find

$$\int_V \mathrm{d}V\, \nabla \cdot [\mathcal{E}_0(\mathbf{r}, t) \times \mathcal{H}_0(\mathbf{r}, t)] = -\frac{1}{2}\int_V \mathrm{d}V \left[\mu(\mathbf{r})\frac{\partial}{\partial t}|\mathcal{H}_0(\mathbf{r}, t)|^2 + \varepsilon(\mathbf{r})\frac{\partial}{\partial t}|\mathcal{E}_0(\mathbf{r}, t)|^2\right] \tag{6.31}$$

and we would like to apply the Gauss theorem to the first integral. This is possible so long as the vector field $\mathcal{E}_0 \times \mathcal{H}_0$ is continuously differentiable separately in $V \setminus \overline{V}_J$ and V_J, and $\hat{\mathbf{n}} \cdot \mathcal{E}_0 \times \mathcal{H}_0$ is continuous across ∂V_J. This is certainly true, if the tangential components of all fields $\mathcal{E}_l$, $\mathcal{H}_l$, $l = 1, 2$, are continuous across ∂V_J. The condition on $\mathcal{E}_l$ is reasonable because ∂V_J is not a material interface, whereas the requirement on $\mathcal{H}_l$ was already invoked while deriving the local form of the Ampère-Maxwell law (1.34) on page 7 and ff. in Section 1.2.2.

Therefore, we are allowed to write

$$\int_{S \cup \partial V_1} \mathrm{d}S\, \hat{\mathbf{n}}(\mathbf{r}) \cdot [\mathcal{E}_0(\mathbf{r}, t) \times \mathcal{H}_0(\mathbf{r}, t)] = \frac{1}{2}\int_V \mathrm{d}V \left[\mu(\mathbf{r})\frac{\partial}{\partial t}|\mathcal{H}_0(\mathbf{r}, t)|^2 + \varepsilon(\mathbf{r})\frac{\partial}{\partial t}|\mathcal{E}_0(\mathbf{r}, t)|^2\right] \tag{6.32}$$

where the minus sign has disappeared in view of the orientation chosen for the unit normal on ∂V (Figure 6.6). Now, we see that the flux integral vanishes identically, if *either one* of the following conditions is true

$$\mathcal{E}_0(\mathbf{r}, t) \times \hat{\mathbf{n}}(\mathbf{r}) = 0, \qquad (\mathbf{r}, t) \in \partial V \times \mathbb{R}^+ \tag{6.33}$$

$$\hat{\mathbf{n}}(\mathbf{r}) \times \mathcal{H}_0(\mathbf{r}, t) = 0, \qquad (\mathbf{r}, t) \in \partial V \times \mathbb{R}^+ \tag{6.34}$$

that is, the tangential component of either $\mathcal{E}_0(\mathbf{r}, t)$ or $\mathcal{H}_0(\mathbf{r}, t)$ must vanish on the boundary of the region of interest for all times. As a matter of fact, we may also enforce one condition on S and the other on ∂V_1 or the other way round. Either one of these constraints is fulfilled so long as the tangential component of $\mathcal{E}(\mathbf{r}, t)$ or $\mathcal{H}(\mathbf{r}, t)$ is assigned on ∂V for all times.

If the flux integral on ∂V vanishes, then (6.32) implies

$$\frac{\mathrm{d}}{\mathrm{d}t}\int_V \mathrm{d}V \left[\mu(\mathbf{r})|\mathcal{H}_0(\mathbf{r}, t)|^2 + \varepsilon(\mathbf{r})|\mathcal{E}_0(\mathbf{r}, t)|^2\right] = 0, \qquad t \in \mathbb{R}^+ \tag{6.35}$$

where the interchange of time derivative and integration over space coordinates is permissible, because the integrands and the time derivative thereof are continuous in $(\mathbf{r}, t) \in V \times \mathbb{R}^+$. By virtue of (1.270), (6.35) means that energy stored in the difference field for $\mathbf{r} \in V$ must be constant in time. More precisely, since the entities $\mathcal{E}_0(\mathbf{r}, t)$ and $\mathcal{H}_0(\mathbf{r}, t)$ — and hence the energy — are null for $t = 0$, it follows that

$$\int_V \mathrm{d}V \left[\mu(\mathbf{r})|\mathcal{H}_0(\mathbf{r}, t)|^2 + \varepsilon(\mathbf{r})|\mathcal{E}_0(\mathbf{r}, t)|^2\right] = 0, \qquad t \in \mathbb{R}^+ \tag{6.36}$$

and, since for ordinary media it holds $\varepsilon(\mathbf{r}) > 0$ and $\mu(\mathbf{r}) > 0$, the integrand above is a positive function of time, and the last equality requires

$$\mathcal{E}_0(\mathbf{r}, t) = 0 = \mathcal{H}_0(\mathbf{r}, t), \qquad (\mathbf{r}, t) \in V \times \mathbb{R}^+ \tag{6.37}$$

that is, the difference entities are zero everywhere in V for any time. As a result, the two solutions $\mathcal{E}_l(\mathbf{r}, t)$, $\mathcal{H}_l(\mathbf{r}, t)$ coincide, contrary to our hypothesis, and the solution is thus unique.

We could have reached the same conclusion by applying the Poynting theorem (1.276) to the source-free difference field in the region V. In fact, the procedure we have outlined is sometimes called the *energy method.* To elucidate, we follow this approach to extend the proof of uniqueness to the case where the material filling the region V (see Figure 6.6) has conduction losses described by a conductivity $\sigma(\mathbf{r}) > 0$. By defining the difference fields as before and invoking (1.276) we get

$$\int_{S \cup \partial V_1} \mathrm{d}S\, \hat{\mathbf{n}}(\mathbf{r}) \cdot \mathcal{E}_0(\mathbf{r}, t) \times \mathcal{H}_0(\mathbf{r}, t) = \frac{\mathrm{d}}{\mathrm{d}t}[W_e(t) + W_h(t)] + \int_V \mathrm{d}V\, \sigma(\mathbf{r})|\mathcal{E}_0(\mathbf{r}, t)|^2 \tag{6.38}$$

again with the unit normal $\hat{\mathbf{n}}(\mathbf{r})$ oriented positively inwards V. If either (6.33) or (6.34) is enforced, the flux integral vanishes, and we are left with

$$\frac{\mathrm{d}}{\mathrm{d}t}[W_e(t) + W_h(t)] = -\int_V \mathrm{d}V\, \sigma(\mathbf{r})|\mathcal{E}_0(\mathbf{r}, t)|^2 \leqslant 0, \qquad t \in \mathbb{R}^+ \tag{6.39}$$

which in words states that the total energy $W(t) := W_e(t) + W_h(t)$ associated with the difference field in the region V is at most a non-increasing function of time. But then, $W(t)$ must be positive and hence, thanks to the mean value theorem (A.89) we have

$$0 \leqslant W(t) = W(0) + \left.\frac{\mathrm{d}W}{\mathrm{d}t}\right|_{t_0} t \leqslant W(0) = 0, \qquad t \in \mathbb{R}^+ \tag{6.40}$$

with $t_0 \in [0, t]$ a suitable instant of time. The last equality holds because the difference field is identically null at $t = 0$. As a result, $W(t) = 0$ for $t \in \mathbb{R}^+$, and uniqueness follows from (6.36) and (6.37).

In summary, the Maxwell equations admit a unique solution in a bounded region V, if the following conditions are *all* satisfied:

(1) the sources are assigned for $t \geqslant 0$;
(2) the fields are assigned for $\mathbf{r} \in V$ and $t = 0$ (initial conditions);
(3) either $\mathcal{E}(\mathbf{r}, t) \times \hat{\mathbf{n}}(\mathbf{r})$ or $\hat{\mathbf{n}}(\mathbf{r}) \times \mathcal{H}(\mathbf{r}, t)$ are assigned for $(\mathbf{r}, t) \in \partial V \times \mathbb{R}^+$ (boundary conditions).

Concerning the boundary conditions, we observe that (6.33) holds identically if the medium in the complementary region $\mathbb{R}^3 \setminus \overline{V}$ is a PEC, in which case $\mathcal{E}(\mathbf{r}, t) \times \hat{\mathbf{n}}(\mathbf{r}) = 0$ for $(\mathbf{r}, t) \in \partial V \times \mathbb{R}^+$. Alternatively, if the medium filling $\mathbb{R}^3 \setminus \overline{V}$ is penetrable, we may first invoke the jump conditions (1.142) and (1.144) with $\mathcal{J}_S(\mathbf{r}, t) = 0$ on $S \cup \partial V_1$, and then make S recede to infinity and ∂V_1 shrink to a point.

It is important to notice that the presence of conduction losses in V is not an essential ingredient for the uniqueness of the solution (also see Section 6.4.1). Then again, if the boundary ∂V in Figure 6.6 is flush with a lossy medium described by an *impedance relationship* in time domain [cf. (6.78) and (6.86) further on], one can similarly show that the energy of the difference field in V decreases, whereby (6.40) still applies.

On a side note, we mention that the proof outlined above can also be employed to show that the electromagnetic field remains zero inside a conducting shell V made of PEC. Accordingly, since $\mathcal{E}(\mathbf{r},t) \times \hat{\mathbf{n}}(\mathbf{r}) = 0$ on $S = \partial V$, if the field was initially zero, i.e., $\mathcal{E}(\mathbf{r},0) = 0 = \mathcal{H}(\mathbf{r},0)$ and no sources are located inside V, then from either (6.36) or (6.40) we conclude that $\mathcal{E}(\mathbf{r},t) = 0 = \mathcal{H}(\mathbf{r},t)$ for $(\mathbf{r},t) \in \mathbb{R}^3 \times \mathbb{R}^+$. Devices which exploit this property are known as *Faraday cages* [31]; the analogous result for electrostatic fields is discussed in Section 3.3.1.

At a first glance it appears that only the tangential components of the fields play a role in determining uniqueness, whereas the components perpendicular to ∂V may, in fact, be arbitrary (cf. [32]). However, if we set

$$\mathcal{E}(\mathbf{r},t) \times \hat{\mathbf{n}}(\mathbf{r}) = \mathbf{f}(\mathbf{r},t), \qquad (\mathbf{r},t) \in \partial V \times \mathbb{R}^+ \tag{6.41}$$

$$\hat{\mathbf{n}}(\mathbf{r}) \times \mathcal{H}(\mathbf{r},t) = \mathbf{g}(\mathbf{r},t), \qquad (\mathbf{r},t) \in \partial V \times \mathbb{R}^+ \tag{6.42}$$

then, by taking the surface divergence of both sides of these equations and using (A.60) we find

$$\nabla_S \cdot \mathbf{f}(\mathbf{r},t) = \nabla_S \cdot [\mathcal{E}(\mathbf{r},t) \times \hat{\mathbf{n}}(\mathbf{r})] = \hat{\mathbf{n}}(\mathbf{r}) \cdot \nabla \times \mathcal{E}(\mathbf{r},t) = -\frac{\partial}{\partial t}\hat{\mathbf{n}}(\mathbf{r}) \cdot \mathcal{B}(\mathbf{r},t) \tag{6.43}$$

$$\nabla_S \cdot \mathbf{g}(\mathbf{r},t) = \nabla_S \cdot [\hat{\mathbf{n}}(\mathbf{r}) \times \mathcal{H}(\mathbf{r},t)] = -\hat{\mathbf{n}}(\mathbf{r}) \cdot \nabla \times \mathcal{H}(\mathbf{r},t) = -\frac{\partial}{\partial t}\hat{\mathbf{n}}(\mathbf{r}) \cdot \mathcal{D}(\mathbf{r},t) \tag{6.44}$$

and this shows that the normal components of the flux densities are not independent! The last step follows by virtue of (1.20) and (1.34) extended to the boundary whereon no true currents are present (they are in $V_J \subset V$ by hypothesis).

It is also worthwhile mentioning that we cannot arbitrarily assign the tangential components of $\mathcal{E}(\mathbf{r},t)$ or $\mathcal{H}(\mathbf{r},t)$ on the boundary ∂V due to the presence of the electric source $\mathcal{J}(\mathbf{r},t)$ and the initial conditions within the region of concern. Indeed, if $\mathcal{J}(\mathbf{r},t)$ is 'turned on' at $t = 0$, then by integrating the Faraday law (1.20) and the Ampère-Maxwell law (1.34) over V and for $t \geqslant 0$ we obtain

$$\int_0^{+\infty} \mathrm{d}t \int_V \mathrm{d}V\, \nabla \times \mathcal{E}(\mathbf{r},t) = -\int_0^{+\infty} \mathrm{d}t \int_V \mathrm{d}V\, \mu(\mathbf{r})\frac{\partial \mathcal{H}}{\partial t} \tag{6.45}$$

$$\int_0^{+\infty} \mathrm{d}t \int_V \mathrm{d}V\, \nabla \times \mathcal{H}(\mathbf{r},t) = \int_0^{+\infty} \mathrm{d}t \int_V \mathrm{d}V\, \varepsilon(\mathbf{r})\frac{\partial \mathcal{E}}{\partial t} + \int_0^{+\infty} \mathrm{d}t \int_{V_J} \mathrm{d}V\, \mathcal{J}(\mathbf{r},t) \tag{6.46}$$

whence we derive the solvability conditions

$$\int_0^{+\infty} \mathrm{d}t \int_{S \cup \partial V_1} \mathrm{d}S\, \mathcal{E}(\mathbf{r},t) \times \hat{\mathbf{n}}(\mathbf{r}) = \int_0^{+\infty} \mathrm{d}t \int_{S \cup \partial V_1} \mathrm{d}S\, \mathbf{f}(\mathbf{r},t) = -\int_V \mathrm{d}V\, \mu(\mathbf{r})[\mathcal{H}(\mathbf{r},t)]_{t=0}^{+\infty} \tag{6.47}$$

$$\int_0^{+\infty} \mathrm{d}t \int_{S \cup \partial V_1} \mathrm{d}S\, \hat{\mathbf{n}}(\mathbf{r}) \times \mathcal{H}(\mathbf{r},t) = \int_0^{+\infty} \mathrm{d}t \int_{S \cup \partial V_1} \mathrm{d}S\, \mathbf{g}(\mathbf{r},t)$$

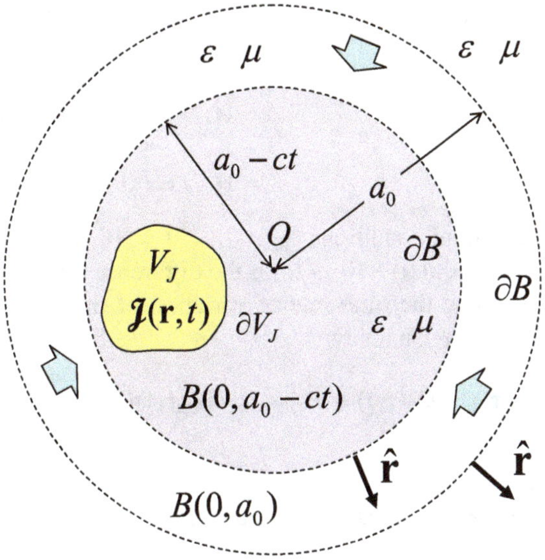

Figure 6.7 A shrinking ball (−−) for proving uniqueness of the solutions to the Maxwell equations in the time domain in an unbounded region.

$$= -\int_V dV\, \varepsilon(\mathbf{r})[\mathcal{E}(\mathbf{r},t)]_{t=0}^{+\infty} - \int_0^{+\infty} dt \int_{V_J} dV\, \mathcal{J}(\mathbf{r},t) \tag{6.48}$$

where we have invoked the curl theorem (H.91) with the unit normal positively oriented inwards V in line with Figure 6.6. The direct contribution of the fields may vanish at $t = 0$ for homogeneous initial conditions and also for infinitely long times if the source is 'switched off' at some point.

6.3.2 Unbounded regions

For the sake of simplicity we consider the fields generated by a finite current density $\mathcal{J}(\mathbf{r},t)$ in an unbounded homogeneous isotropic medium endowed with constitutive parameters ε and μ. This is not a serious limitation, since we already covered the question of uniqueness in the presence of penetrable and conducting bodies with smooth boundaries in Section 6.3.1, in which case we know that uniqueness is guaranteed if we also assign the tangential component of either $\mathcal{E}(\mathbf{r},t)$ or $\mathcal{H}(\mathbf{r},t)$ on the surfaces of the objects.

The proof based on (6.32) may be extended to the whole space with $V := \mathbb{R}^3$ and $V_1 = \emptyset$. Essentially, we only have to conclude, if it is possible, that the electromagnetic energy stored in the field remains finite and the flux integral over S vanishes, as we let the surface recede to infinity (see Figure 6.6). But these results were obtained in (1.283) and (1.284) on the grounds of the boundary conditions (1.260) and (1.261).

An alternative proof [33, Section 4.3] uses the energy method applied to a ball $B(0, a(t))$ of shrinking radius $a(t) := a_0 - ct$, $t \in [0, a_0/c]$, where c denotes the speed of light (1.209) in the underlying medium, and a_0 is the radius of the ball at time $t = 0$, as is shown in Figure 6.7. To begin with we presume that two solutions, indicated with subscripts 1 and 2, may exist in $B(0, a(t))$.

The difference entities $\mathcal{E}_0 := \mathcal{E}_1 - \mathcal{E}_2$ and $\mathcal{H}_0 := \mathcal{H}_1 - \mathcal{H}_2$ satisfy the source-free Faraday law and Ampère-Maxwell law

$$\nabla \times \mathcal{E}_0(\mathbf{r}, t) + \mu \frac{\partial}{\partial t} \mathcal{H}_0(\mathbf{r}, t) = 0, \qquad (\mathbf{r}, t) \in B(0, a(t)) \times [0, a_0/c] \tag{6.49}$$

$$\nabla \times \mathcal{H}_0(\mathbf{r}, t) - \varepsilon \frac{\partial}{\partial t} \mathcal{E}_0(\mathbf{r}, t) = 0, \qquad (\mathbf{r}, t) \in B(0, a(t)) \times [0, a_0/c] \tag{6.50}$$

subject to the homogeneous initial conditions $\mathcal{E}_0(\mathbf{r}, 0) = 0 = \mathcal{H}_0(\mathbf{r}, 0)$. We may conclude that the solution is unique for $(\mathbf{r}, t) \in B(0, a(t)) \times [0, a_0/c]$ if the difference field is identically zero.

To this purpose we examine the time rate of variation of the electromagnetic energy of the difference field in the ball $B(0, a(t))$, namely,

$$\begin{aligned}\frac{\mathrm{d}W}{\mathrm{d}t} &:= \frac{1}{2}\frac{\mathrm{d}}{\mathrm{d}t} \int\limits_{B(0,a(t))} \mathrm{d}V \, [\mathcal{E}_0(\mathbf{r}, t) \cdot \mathcal{D}_0(\mathbf{r}, t) + \mathcal{H}_0(\mathbf{r}, t) \cdot \mathcal{B}_0(\mathbf{r}, t)] \\ &= \frac{1}{2}\frac{\mathrm{d}}{\mathrm{d}t} \int\limits_0^{a_0 - ct} \mathrm{d}r \int\limits_{S(r)} \mathrm{d}S \, [\mathcal{E}_0(\mathbf{r}, t) \cdot \mathcal{D}_0(\mathbf{r}, t) + \mathcal{H}_0(\mathbf{r}, t) \cdot \mathcal{B}_0(\mathbf{r}, t)]\end{aligned} \tag{6.51}$$

where $S(r)$ is a sphere with center the origin and radius r with $0 \leqslant r \leqslant a(t)$. The stored energy changes because the electromagnetic field depends on time and the radius of the ball decreases. Having cast the volume integration into the cascade of a one-dimensional integral with varying limits and a surface integral over $S(r)$ which does not depend on time, we can obviate the usage of the Reynolds transport theorem [29, Appendix A.2], [34].

Indeed, under the assumption that the fields and their time derivatives are continuous for $(\mathbf{r}, t) \in B(0, a(t)) \times [0, a_0/c]$ we can compute the derivative by means of the Leibniz theorem for differentiation of an integral [35, Formula 3.3.7], [20, Formula 69.3]

$$\begin{aligned}\frac{\mathrm{d}W}{\mathrm{d}t} &= \frac{1}{2} \int\limits_0^{a_0 - ct} \mathrm{d}r \int\limits_{S(r)} \mathrm{d}S \, \frac{\partial}{\partial t} [\mathcal{E}_0(\mathbf{r}, t) \cdot \mathcal{D}_0(\mathbf{r}, t) + \mathcal{H}_0(\mathbf{r}, t) \cdot \mathcal{B}_0(\mathbf{r}, t)] \\ &\quad - \frac{c}{2} \int\limits_{\partial B} \mathrm{d}S \, [\mathcal{E}_0(\mathbf{r}, t) \cdot \mathcal{D}_0(\mathbf{r}, t) + \mathcal{H}_0(\mathbf{r}, t) \cdot \mathcal{B}_0(\mathbf{r}, t)] \\ &= \int\limits_{B(0,a(t))} \mathrm{d}V \left[\mathcal{E}_0(\mathbf{r}, t) \cdot \frac{\partial}{\partial t} \mathcal{D}_0(\mathbf{r}, t) + \mathcal{H}_0(\mathbf{r}, t) \cdot \frac{\partial}{\partial t} \mathcal{B}_0(\mathbf{r}, t) \right] \\ &\quad - \frac{c}{2} \int\limits_{\partial B} \mathrm{d}S \left[\varepsilon |\mathcal{E}_0(\mathbf{r}, t)|^2 + \mu |\mathcal{H}_0(\mathbf{r}, t)|^2 \right] \\ &= \int\limits_{B(0,a(t))} \mathrm{d}V \, \nabla \cdot [\mathcal{H}_0(\mathbf{r}, t) \times \mathcal{E}_0(\mathbf{r}, t)] - \frac{1}{2} \int\limits_{\partial B} \mathrm{d}S \left[\frac{1}{Z} |\mathcal{E}_0(\mathbf{r}, t)|^2 + Z |\mathcal{H}_0(\mathbf{r}, t)|^2 \right]\end{aligned} \tag{6.52}$$

on account of (6.49) and (6.50) and the differential identity (H.49). The scalar Z is the intrinsic impedance of the background medium defined in (1.358). Notice that in the first step of (6.52) the Leibniz rule has been used also to move the time derivative past the integration over $S(r)$ as well. We may apply the Gauss theorem to the integral over the ball, for the fields are of class $\mathcal{C}^1(B)^3 \cap \mathcal{C}(\overline{B})^3$

$$\int\limits_{B(0,a(t))} \mathrm{d}V \, \nabla \cdot [\mathcal{H}_0(\mathbf{r}, t) \times \mathcal{E}_0(\mathbf{r}, t)] = \int\limits_{\partial B} \mathrm{d}S \, \hat{\mathbf{r}} \cdot [\mathcal{H}_0(\mathbf{r}, t) \times \mathcal{E}_0(\mathbf{r}, t)] \leqslant \int\limits_{\partial B} \mathrm{d}S \, |\mathcal{H}_0(\mathbf{r}, t)| |\mathcal{E}_0(\mathbf{r}, t)| \tag{6.53}$$

having used the Cauchy-Schwarz inequality (D.151) on the integrand. By substituting this last estimate into (6.52) we find

$$\frac{\mathrm{d}W}{\mathrm{d}t} \leqslant \int_{\partial B} \mathrm{d}S\, |\mathcal{H}_0(\mathbf{r},t)||\mathcal{E}_0(\mathbf{r},t)| - \frac{1}{2Z}\int_{\partial B} \mathrm{d}S\, \left[|\mathcal{E}_0(\mathbf{r},t)|^2 + Z^2|\mathcal{H}_0(\mathbf{r},t)|^2\right]$$
$$= -\frac{1}{2Z}\int_{\partial B} \mathrm{d}S\, \left[|\mathcal{E}_0(\mathbf{r},t)| - Z|\mathcal{H}_0(\mathbf{r},t)|\right]^2 \leqslant 0 \tag{6.54}$$

that is, the electromagnetic energy $W(t)$ in the shrinking ball is a non-increasing function of time for $t \in [0, a_0/c]$.

Proceeding as in Section 6.3.1 we recognize that $W(t)$ is perforce a non-negative function of time and we apply the mean value theorem (A.89) to write

$$0 \leqslant W(t) = W(0) + \left.\frac{\mathrm{d}W}{\mathrm{d}t}\right|_{t=t_0} t \leqslant W(0) = 0, \qquad t \in [0, a_0/c] \tag{6.55}$$

with $t_0 \in [0, t] \subseteq [0, a_0/c]$ a suitable time. The last equality holds in view of the homogeneous initial conditions for the difference field. Then, we conclude

$$0 = W(t) := \int_{B(0,a(t))} \mathrm{d}V \frac{1}{2}\left[\varepsilon|\mathcal{E}_0(\mathbf{r},t)|^2 + \mu|\mathcal{H}_0(\mathbf{r},t)|^2\right], \qquad t \in [0, a_0/c] \tag{6.56}$$

which requires

$$\mathcal{E}_0(\mathbf{r},t) = 0 = \mathcal{H}_0(\mathbf{r},t), \qquad t \in [0, a_0/c] \tag{6.57}$$

whence the difference field is zero and, contrary to our hypothesis, the two solutions coincide for $(\mathbf{r}, t) \in B(0, a(t)) \times [0, a_0/c]$.

Finally, uniqueness in the whole space and for all times follows by taking the limit as $a_0 \to +\infty$, since $B(0, a(t)) \to \mathbb{R}^3$ and the time interval $[0, a_0/c] \to \mathbb{R}^+$. In so doing, we need not worry about the finiteness of the integrals in (6.54) and (6.56). In the former, since the integral is always a positive quantity, if it diverges then the estimate holds true *a fortiori*. In the latter, the energy is finite in light of the asymptotic behavior (1.260) and (1.261), as already argued at the beginning.

In summary, the solution to the Maxwell equations in the whole space for $t \geqslant 0$ is unique if the following conditions are met:

(1) the sources are assigned for $t \geqslant 0$;
(2) the fields are assigned for $\mathbf{r} \in \mathbb{R}^3$ and $t = 0$;
(3) the fields fall off as the inverse square of the distance, as formulated in (1.260) and (1.261).

A special instance of unbounded domain is represented by the inside of an infinite cylinder $V := S \times \mathbb{R}$, where S denotes a flat surface. While the shape of S may be quite arbitrary, for the moment we suppose that the boundary $\gamma := \partial S$ is smooth. Additionally, we indicate with $\varepsilon > 0$, $\mu > 0$ and $\sigma > 0$ the constitutive parameters of the lossy isotropic homogeneous medium that 'fills' the cylinder. This geometry covers the important practical case of uniform hollow-pipe waveguides with metallic or PEC walls (see Section 11.2 and Figure 11.1).

In order to discuss the uniqueness of the fields generated in V by a current $\mathcal{J}(\mathbf{r}, t)$ situated in $V_J \subset V$ and turned on at time $t = 0$, we introduce a system of Cartesian coordinates with the z-axis

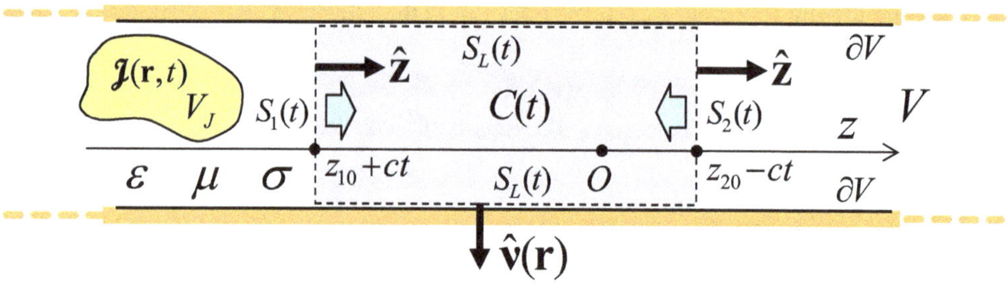

Figure 6.8 A shrinking cylinder (−−) for proving uniqueness of the solutions to the Maxwell equations in the time domain in an unbounded cylindrical region.

perpendicular to S. We modify the proof given above for the whole space by applying the energy method to a shrinking finite-sized cylinder, namely,

$$C(t) := S \times [z_1(t), z_2(t)], \qquad z_1(t) = z_{10} + ct, \qquad z_2(t) = z_{20} - ct \tag{6.58}$$

where z_{10} and z_{20}, $z_{10} < z_{20}$, are the z-coordinates of any two sections S_{10} and S_{20} of V, and c is the speed of light (1.209) in the underlying medium. The time variable ranges in the interval $[0, (z_{20} - z_{10})/(2c)]$, and for $t = (z_{20} - z_{10})/(2c)$ the moving bases $S_1(t)$ and $S_2(t)$ of $C(t)$ come to coincide with the section of V identified by $z = (z_{10} + z_{20})/2$. A longitudinal view of the setup described so far is sketched in Figure 6.8.

The difference entities $\mathcal{E}_0 := \mathcal{E}_1 - \mathcal{E}_2$ and $\mathcal{H}_0 := \mathcal{H}_1 - \mathcal{H}_2$ satisfy the source-free Faraday law and Ampère-Maxwell law

$$\nabla \times \mathcal{E}_0(\mathbf{r}, t) = -\mu \frac{\partial}{\partial t} \mathcal{H}_0(\mathbf{r}, t), \qquad (\mathbf{r}, t) \in C(t) \times \left[0, \frac{z_{20} - z_{10}}{2c}\right] \tag{6.59}$$

$$\nabla \times \mathcal{H}_0(\mathbf{r}, t) = \varepsilon \frac{\partial}{\partial t} \mathcal{E}_0(\mathbf{r}, t) + \sigma \mathcal{E}_0(\mathbf{r}, t), \qquad (\mathbf{r}, t) \in C(t) \times \left[0, \frac{z_{20} - z_{10}}{2c}\right] \tag{6.60}$$

subject to the homogeneous initial conditions $\mathcal{E}_0(\mathbf{r}, 0) = 0 = \mathcal{H}_0(\mathbf{r}, 0)$. The time rate of variation of the electromagnetic energy associated with the difference electromagnetic field in $C(t)$ is computed by starting with the definition

$$\begin{aligned} \frac{\mathrm{d}W}{\mathrm{d}t} &:= \frac{1}{2} \frac{\mathrm{d}}{\mathrm{d}t} \int_{C(t)} \mathrm{d}V \left[\mathcal{E}_0(\mathbf{r}, t) \cdot \mathcal{D}_0(\mathbf{r}, t) + \mathcal{H}_0(\mathbf{r}, t) \cdot \mathcal{B}_0(\mathbf{r}, t)\right] \\ &= \frac{1}{2} \frac{\mathrm{d}}{\mathrm{d}t} \int_{z_{10}+ct}^{z_{20}-ct} \mathrm{d}z \int_{S(z)} \mathrm{d}S \left[\mathcal{E}_0(\mathbf{r}, t) \cdot \mathcal{D}_0(\mathbf{r}, t) + \mathcal{H}_0(\mathbf{r}, t) \cdot \mathcal{B}_0(\mathbf{r}, t)\right] \end{aligned} \tag{6.61}$$

where $S(z)$ denotes a cross section of V located between the planes $z = z_1(t)$ and $z = z_1(t)$. Assuming that the fields and their temporal derivatives are continuous for $(\mathbf{r}, t) \in C(t) \times [0, (z_{20} - z_{10})/(2c)]$, we may invoke the Leibniz rule [35, Formula 3.3.7], [20, Formula 69.3] to obtain

$$\frac{\mathrm{d}W}{\mathrm{d}t} = \frac{1}{2} \int_{z_{10}+ct}^{z_{20}-ct} \mathrm{d}z \int_{S(z)} \mathrm{d}S \, \frac{\partial}{\partial t} \left[\mathcal{E}_0(\mathbf{r}, t) \cdot \mathcal{D}_0(\mathbf{r}, t) + \mathcal{H}_0(\mathbf{r}, t) \cdot \mathcal{B}_0(\mathbf{r}, t)\right]$$

$$
\begin{aligned}
&- \frac{c}{2} \sum_{l=1}^{2} \int_{S_l(t)} \mathrm{d}S \, [\mathcal{E}_0(\mathbf{r},t) \cdot \mathcal{D}_0(\mathbf{r},t) + \mathcal{H}_0(\mathbf{r},t) \cdot \mathcal{B}_0(\mathbf{r},t)] \\
&= \int_{C(t)} \mathrm{d}V \left[\mathcal{E}_0(\mathbf{r},t) \cdot \frac{\partial}{\partial t} \mathcal{D}_0(\mathbf{r},t) + \mathcal{H}_0(\mathbf{r},t) \cdot \frac{\partial}{\partial t} \mathcal{B}_0(\mathbf{r},t) \right] \\
&\quad - \frac{c}{2} \sum_{l=1}^{2} \int_{S_l(t)} \mathrm{d}S \, \left[\varepsilon |\mathcal{E}_0(\mathbf{r},t)|^2 + \mu |\mathcal{H}_0(\mathbf{r},t)|^2 \right] \\
&= \int_{C(t)} \mathrm{d}V \, \nabla \cdot [\mathcal{H}_0(\mathbf{r},t) \times \mathcal{E}_0(\mathbf{r},t)] - \int_{C(t)} \mathrm{d}V \, \sigma |\mathcal{E}_0(\mathbf{r},t)|^2 \\
&\quad - \frac{1}{2Z} \sum_{l=1}^{2} \int_{S_l(t)} \mathrm{d}S \, \left[|\mathcal{E}_0(\mathbf{r},t)|^2 + Z^2 |\mathcal{H}_0(\mathbf{r},t)|^2 \right]
\end{aligned} \tag{6.62}
$$

by virtue of (1.117), (1.118), (6.59), (6.60), the differential identity (H.49), and definition (1.358). We denote the lateral surface of $C(t)$ with $S_L(t)$ and apply the Gauss theorem (A.53) to the integral over $C(t)$

$$
\begin{aligned}
\int_{C(t)} \mathrm{d}V \, \nabla \cdot [\mathcal{H}_0(\mathbf{r},t) \times \mathcal{E}_0(\mathbf{r},t)] &= \int_{S_L(t)} \mathrm{d}S \, \hat{\boldsymbol{\nu}}(\mathbf{r}) \cdot [\mathcal{H}_0(\mathbf{r},t) \times \mathcal{E}_0(\mathbf{r},t)] \\
&+ \int_{S_2(t)} \mathrm{d}S \, \hat{\mathbf{z}} \cdot [\mathcal{H}_0(\mathbf{r},t) \times \mathcal{E}_0(\mathbf{r},t)] - \int_{S_1(t)} \mathrm{d}S \, \hat{\mathbf{z}} \cdot [\mathcal{H}_0(\mathbf{r},t) \times \mathcal{E}_0(\mathbf{r},t)]
\end{aligned} \tag{6.63}
$$

where $\hat{\boldsymbol{\nu}}(\mathbf{r})$ indicates the outwards unit normal on $S_L(t)$. The flux through $S_L(t)$ is null if either $\hat{\boldsymbol{\nu}} \times \mathcal{H}_0$ or $\mathcal{E}_0 \times \hat{\boldsymbol{\nu}}$ vanishes on the boundary of V at all times of interest. The fluxes through the bases $S_l(t)$, $l = 1, 2$, can be estimated as

$$
\int_{S_2(t)} \mathrm{d}S \, \hat{\mathbf{z}} \cdot [\mathcal{H}_0(\mathbf{r},t) \times \mathcal{E}_0(\mathbf{r},t)] \leqslant \int_{S_2(t)} \mathrm{d}S \, |\mathcal{H}_0(\mathbf{r},t)||\mathcal{E}_0(\mathbf{r},t)| \tag{6.64}
$$

$$
- \int_{S_1(t)} \mathrm{d}S \, \hat{\mathbf{z}} \cdot [\mathcal{H}_0(\mathbf{r},t) \times \mathcal{E}_0(\mathbf{r},t)] \leqslant \int_{S_1(t)} \mathrm{d}S \, |\mathcal{H}_0(\mathbf{r},t)||\mathcal{E}_0(\mathbf{r},t)| \tag{6.65}
$$

thanks to the Cauchy-Schwarz inequality (D.151). By using these findings back in (6.62) we get

$$
\begin{aligned}
\frac{\mathrm{d}W}{\mathrm{d}t} &\leqslant \sum_{l=1}^{2} \int_{S_l(t)} \mathrm{d}S \, |\mathcal{H}_0(\mathbf{r},t)||\mathcal{E}_0(\mathbf{r},t)| - \int_{C(t)} \mathrm{d}V \, \sigma |\mathcal{E}_0(\mathbf{r},t)|^2 \\
&\quad - \frac{1}{2Z} \sum_{l=1}^{2} \int_{S_l(t)} \mathrm{d}S \, \left[|\mathcal{E}_0(\mathbf{r},t)|^2 + Z^2 |\mathcal{H}_0(\mathbf{r},t)|^2 \right] \\
&= - \int_{C(t)} \mathrm{d}V \, \sigma |\mathcal{E}_0(\mathbf{r},t)|^2 - \frac{1}{2Z} \sum_{l=1}^{2} \int_{S_l(t)} \mathrm{d}S \, [|\mathcal{E}_0(\mathbf{r},t)| - Z|\mathcal{H}_0(\mathbf{r},t)|]^2 \leqslant 0
\end{aligned} \tag{6.66}
$$

which shows that the energy stored in the shrinking cylinder is a non-increasing function of time for $t \in [0, (z_{20} - z_{10})/(2c)]$. The subsequent conclusion that $W(t)$ remains zero in $C(t)$ at all times arises from the homogeneous initial conditions, (6.66) and an expression analogous to (6.40). Therefore, we have

$$0 = W(t) := \int_{C(t)} \mathrm{d}V \frac{1}{2} \left[\varepsilon |\mathcal{E}_0(\mathbf{r}, t)|^2 + \mu |\mathcal{H}_0(\mathbf{r}, t)|^2 \right], \qquad t \in \left[0, \frac{z_{20} - z_{10}}{2c}\right] \tag{6.67}$$

which necessarily requires

$$\mathcal{E}_0(\mathbf{r}, t) = 0 = \mathcal{H}_0(\mathbf{r}, t), \qquad t \in \left[0, \frac{z_{20} - z_{10}}{2c}\right] \tag{6.68}$$

whence we see that the difference field is zero and, contrary to our initial hypothesis, the two solutions coincide for $(\mathbf{r}, t) \in C(t) \times [0, (z_{20} - z_{10})/(2c)]$.

Uniqueness in $V := S \times \mathbb{R}$ at all times is proved by letting $z_{10} \to -\infty$ and $z_{20} \to +\infty$ since, as a result, $C(t) \to V$ and $[0, (z_{20} - z_{10})/(2c)] \to \mathbb{R}^+$. Also for this geometry the convergence of the integrals in (6.66) and (6.67) is not an issue. Indeed, the surface integrals over $S_l(t)$ in (6.66) should be bounded inasmuch as $S_l(t)$ is of finite extent and physical fields produced by the localized source $\mathcal{J}(\mathbf{r}, t)$ in $V_J \subset V$ cannot grow as $|z| \to +\infty$. Moreover, even if the domain integral over $C(t) \to V$ in (6.66) were divergent, then the last estimate would be all the more valid. As regards (6.67) we may claim that the energy remains finite because the fields are null for $|z| \to +\infty$ on the grounds that it takes an infinite amount of time for any disturbance produced by $\mathcal{J}(\mathbf{r}, t)$ to reach the infinitely remote end sides of V along the z-axis. If, on other hand, the initial configuration is supposed to be static or stationary, and the field obeys the special asymptotic boundary conditions [cf. (1.260) and (1.261)]

$$\mathcal{E}(\mathbf{r}, t) = \mathcal{O}\left(\frac{1}{|z|}\right), \qquad |z| \to +\infty, \qquad \mathbf{r} \in V := S \times \mathbb{R} \tag{6.69}$$

$$\mathcal{H}(\mathbf{r}, t) = \mathcal{O}\left(\frac{1}{|z|}\right), \qquad |z| \to +\infty, \qquad \mathbf{r} \in V := S \times \mathbb{R} \tag{6.70}$$

then convergence is guaranteed. For example, the electric energy in V reads

$$\begin{aligned} W_e(t) &:= \int_V \mathrm{d}V \frac{\varepsilon}{2} |\mathcal{E}_0(\mathbf{r}, t)|^2 \leqslant \int_{V_0} \mathrm{d}V \frac{\varepsilon}{2} |\mathcal{E}_0(\mathbf{r}, t)|^2 + 2 \int_S \mathrm{d}S \int_{z_0}^{+\infty} \mathrm{d}z \frac{\varepsilon}{2} \frac{b_E^2}{z^2} \\ &= \int_{V_0} \mathrm{d}V \frac{\varepsilon}{2} |\mathcal{E}_0(\mathbf{r}, t)|^2 + \varepsilon A_S \frac{b_E^2}{z_0} \end{aligned} \tag{6.71}$$

where $V_0 := S \times [-z_0, z_0]$, $z_0 > 0$, is a finite-sized cylinder, A_S is the area of S, and finally the coordinate z_0 and the constant $b_E > 0$ are such that

$$|\mathcal{E}(\mathbf{r}, t)| \leqslant \frac{b_E}{|z|} \qquad \text{for} \qquad |z| \geqslant z_0 \tag{6.72}$$

on account of (6.69). Analogous estimates can be obtained for the magnetic energy and the contribution of dissipated power in (6.66).

In summary, the solution to the Maxwell equations in an infinite cylindrical region of space with smooth boundary is unique for $t \geqslant 0$ if the following conditions are met:

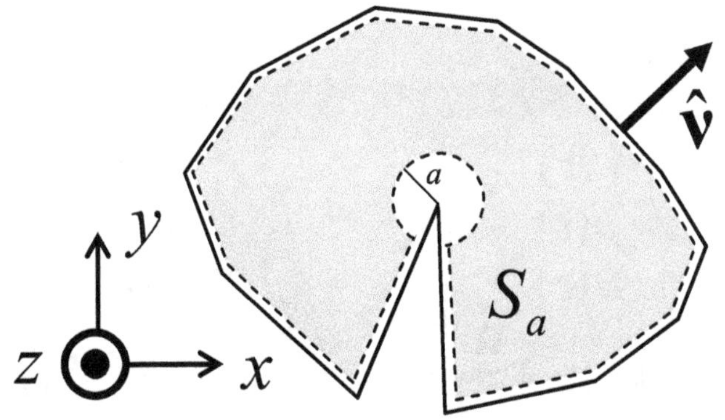

Figure 6.9 *Cross-sectional view of V and $C_a(t)$ for proving uniqueness of the solutions to the Maxwell equations in the time domain in an unbounded cylindrical region with piecewise-smooth boundary and one sharp edge.*

(1) the sources are assigned for $t \geqslant 0$;
(2) the fields are assigned for $\mathbf{r} \in V := S \times \mathbb{R}$ and $t = 0$ (initial conditions);
(3) either $\boldsymbol{\mathcal{E}}(\mathbf{r}, t) \times \hat{\boldsymbol{\nu}}(\mathbf{r})$ or $\hat{\boldsymbol{\nu}}(\mathbf{r}) \times \boldsymbol{\mathcal{H}}(\mathbf{r}, t)$ is assigned for $(\mathbf{r}, t) \in \partial V \times \mathbb{R}^+$ (boundary conditions);
(4) for $|z| \to +\infty$ the fields are either null or satisfy the asymptotic behavior (6.69) and (6.70).

Besides, it should be noted that the hypothesis of a finite non-null conductivity σ within V is not crucial for the proof, since the last part of (6.66) holds even for $\sigma = 0$.

We now elaborate a bit on the question of the smoothness of ∂S that was assumed at the beginning. Since in the course of the proof just outlined we invoked the Gauss theorem, and the latter is valid even if the region of concern has a piecewise-smooth boundary, one might expect that the uniqueness of solutions, as proved, still holds if S has corners or tips and hence, the boundary ∂V has reentrant wedges or sharp edges. However, while this conclusion is true for reentrant wedges, it is actually fallacious in the presence of sharp edges, because the electromagnetic field, in general, is singular at the location of sharp edges, as we argued in Section 3.10 [36]. Worse yet, if the field is singular, the validity of the Leibniz rule, which led us to (6.62), is not guaranteed. As a consequence, neither the Leibniz rule nor the Gauss theorem can be invoked straightaway, and this invalidates estimate (6.66).

Therefore, if V possesses a sharp edge, the shrinking cylinder $C(t)$ we used in (6.62) must be traded for a cylinder- or prism-like domain $C_a(t) := S_a \times [z_1(t), z_2(t)]$ (Figure 6.9) where the flat surface S_a — depending on a parameter $a > 0$ — is devised so as to exclude the edge of V prior to the application of the Leibniz rule and the Gauss theorem. Subsequently, in the limit as the modified boundary $\partial C_a(t)$ is made to approach $\partial C(t)$ one obtains an additional constraint — the so-called *edge condition* — that must be imposed on the Poynting vector (1.280) to conclude again that $W(t)$ is a non-increasing function of time, and this in turn ensures uniqueness [37, Section 9.2], [38, Section 4.7].

6.4 Uniqueness in the frequency domain

We continue our discussion of uniqueness by examining the case of time-harmonic fields and sources. One notable consequence of the periodic dependence on time is that the initial conditions are meaningless (see Section 1.5). More importantly, unlike the Maxwell equations in the time domain, the

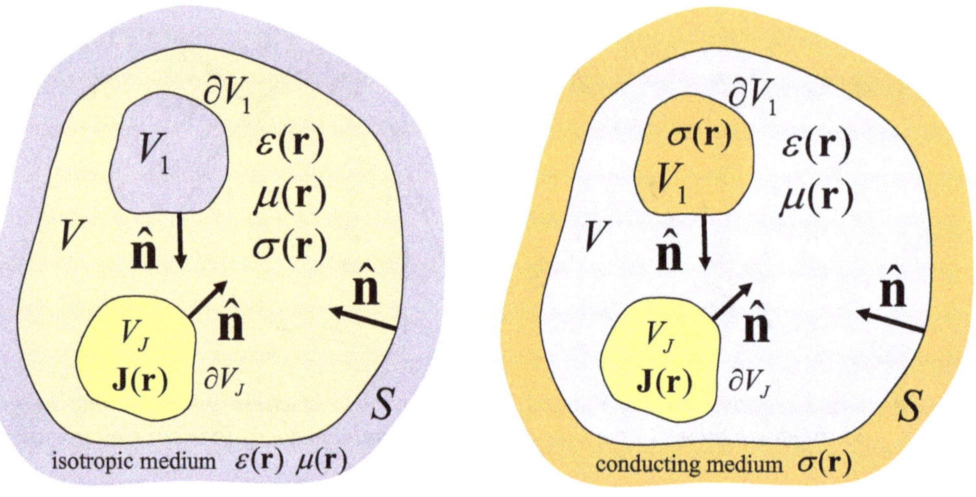

(a) *lossy medium* (b) *lossy boundary*

Figure 6.10 For proving uniqueness of the solutions to the time-harmonic Maxwell equations in a finite region of space in the presence of losses.

time-harmonic counterpart may not have a unique solution in a finite volume, even though suitable boundary conditions are assigned. In infinite domains other difficulties arise, because the fields cannot be considered null or static in character even outside a sufficiently large ball. In fact, having had an infinite amount of time to propagate even at a finite speed c or c_0, the fields exist everywhere in space at any given moment. Therefore, by considering the cases of bounded and unbounded regions of space separately, we discuss what additional constraints we need to enforce for the solution to be unique.

6.4.1 Bounded regions

We start with a bounded region of space V with smooth or piecewise smooth boundary ∂V which contains a time-harmonic volume current density $\mathbf{J}(\mathbf{r})$ confined in the volume $V_J \subset V$. The region V_1 is excluded, as it may host other unknown materials and sources as yet unspecified. The medium outside $V \cup V_1$ may be any, since the constitutive parameters do not enter the present analysis, but for the sake of simplicity we assume that $\varepsilon(\mathbf{r})$ and $\mu(\mathbf{r})$ are the same parameters relevant to the medium which fills V, as is suggested in Figure 6.10a. Additionally — and we shall see that this is a crucial point — we make the hypothesis that the medium in V is lossy, i.e., endowed with electric conductivity $\sigma(\mathbf{r}) \in \mathbb{R}^+$.

In principle, we may proceed as we did for the case of general time dependence, namely, start with the assumption that two field solutions exist, say, $\mathbf{E}_1(\mathbf{r})$, $\mathbf{H}_1(\mathbf{r})$ and $\mathbf{E}_2(\mathbf{r})$, $\mathbf{H}_2(\mathbf{r})$, and invoke the principle of superposition to derive an equation for the differences $\mathbf{E}_0(\mathbf{r}) := \mathbf{E}_1(\mathbf{r}) - \mathbf{E}_2(\mathbf{r})$ and $\mathbf{H}_0(\mathbf{r}) := \mathbf{H}_1(\mathbf{r}) - \mathbf{H}_2(\mathbf{r})$. More simply, since $\mathbf{E}_0(\mathbf{r})$ and $\mathbf{H}_0(\mathbf{r})$ must satisfy the source-free time-harmonic Maxwell equations for $\mathbf{r} \in V$, we may apply the complex Poynting theorem (1.314) to get

$$\frac{1}{2}\int_V \mathrm{d}V\, \sigma(\mathbf{r})|\mathbf{E}_0(\mathbf{r})|^2 - \mathrm{Re}\int_{S\cup\partial V_1} \mathrm{d}S\, \frac{1}{2}\hat{\mathbf{n}}(\mathbf{r}) \cdot \mathbf{E}_0(\mathbf{r}) \times \mathbf{H}_0^*(\mathbf{r}) = 0 \tag{6.73}$$

where the minus sign is a consequence of the orientation chosen for the unit normal on ∂V (see Figure 6.10a). The flux integral has the formal meaning of average complex radiated power *entering* the region V and, if *either one* of the following conditions is verified

$$\mathbf{E}_0(\mathbf{r}) \times \hat{\mathbf{n}}(\mathbf{r}) = 0, \qquad \mathbf{r} \in S \cup \partial V_1 \tag{6.74}$$

$$\hat{\mathbf{n}}(\mathbf{r}) \times \mathbf{H}_0(\mathbf{r}) = 0, \qquad \mathbf{r} \in S \cup \partial V_1 \tag{6.75}$$

it is identically null. In which case, then (6.73) leads to

$$\frac{1}{2} \int_V \mathrm{d}V \, \sigma(\mathbf{r}) |\mathbf{E}_0(\mathbf{r})|^2 = 0 \tag{6.76}$$

which we may interpret as a statement that the average power dissipated by the difference field in V due to conduction losses must, in fact, be zero. Furthermore, since $\sigma(\mathbf{r}) > 0$ for ordinary media, the left-hand side of (6.76) represents a weighted squared norm of $\mathbf{E}_0(\mathbf{r})$ in V. Therefore, (6.76) yields

$$\mathbf{E}_0(\mathbf{r}) = 0, \qquad \mathbf{r} \in V \tag{6.77}$$

by the very definitions (D.8) and (D.9) of norm. Then, the Faraday law (1.99) implies that $\mathbf{H}_0(\mathbf{r})$ is also null for $\mathbf{r} \in V$. Consequently, the two electromagnetic fields $\mathbf{E}_1(\mathbf{r})$, $\mathbf{H}_1(\mathbf{r})$ and $\mathbf{E}_2(\mathbf{r})$, $\mathbf{H}_2(\mathbf{r})$ coincide, contrary to our working hypothesis, and the solution is indeed unique [18, 31, 39, 40], [41, Chapter VI].

In conclusion, the time-harmonic Maxwell equations admit a unique solution in a bounded region V, if the following three conditions are *all* satisfied:

(1) the sources are specified;
(2) the medium filling V is lossy;
(3) either $\mathbf{E}(\mathbf{r}) \times \hat{\mathbf{n}}(\mathbf{r})$ or $\hat{\mathbf{n}}(\mathbf{r}) \times \mathbf{H}(\mathbf{r})$ are assigned on (different parts of) ∂V (boundary conditions).

Uniqueness of solutions is guaranteed also in the case where the medium filling V is lossless, provided the tangential electric and magnetic fields on the boundary $\partial V := S \cup \partial V_1$ are linked by an *approximate* boundary condition of the type

$$\mathbf{E}(\mathbf{r}) \times \hat{\mathbf{n}}(\mathbf{r}) = Z_S(\mathbf{r}) \hat{\mathbf{n}}(\mathbf{r}) \times [\mathbf{H}(\mathbf{r}) \times \hat{\mathbf{n}}(\mathbf{r})], \qquad \mathbf{r} \in \partial V \tag{6.78}$$

where $Z_S(\mathbf{r})$ is called the *surface impedance* of ∂V and has physical dimension of ohms (Ω). We refer to (6.78) as an *impedance relationship* or a *Leontovich boundary condition* [42–45]. Since the unit vector $\hat{\mathbf{n}}(\mathbf{r})$, perpendicular to ∂V, points inwards V, calculation of the average power radiated through ∂V with (1.303) indicates that a positive power efflux occurs from V into the complementary domain $\mathbb{R}^3 \setminus \overline{V}$ where the lossy medium resides. The real part of $Z_S(\mathbf{r})$, when it is non-null, accounts for the high, though finite conductivity of the material in $\mathbb{R}^3 \setminus \overline{V}$ flush with the boundary ∂V (Figure 6.10b). We shall derive an expression for $Z_S(\mathbf{r})$ in Section 7.1, page 426, when studying the propagation of electromagnetic plane waves in a lossy half-space.

To prove uniqueness, as usual we assume that two solutions exist and introduce the difference fields $\mathbf{E}_0(\mathbf{r}) := \mathbf{E}_1(\mathbf{r}) - \mathbf{E}_2(\mathbf{r})$ and $\mathbf{H}_0(\mathbf{r}) := \mathbf{H}_1(\mathbf{r}) - \mathbf{H}_2(\mathbf{r})$. The principle of superposition implies that $\mathbf{E}_0(\mathbf{r})$ and $\mathbf{H}_0(\mathbf{r})$ are related by (6.78) as well. If we apply the Poynting theorem (1.314) to the difference fields in the bounded region of Figure 6.10b, (6.73) reduces to

$$\mathrm{Re} \int_{S \cup \partial V_1} \mathrm{d}S \, \hat{\mathbf{n}}(\mathbf{r}) \cdot \mathbf{E}_0(\mathbf{r}) \times \mathbf{H}_0^*(\mathbf{r}) = 0 \tag{6.79}$$

because $\sigma(\mathbf{r}) = 0$ for $\mathbf{r} \in V$. From this condition we find

$$0 = \mathrm{Re} \int\limits_{S \cup \partial V_1} \mathrm{d}S\, \mathbf{H}_0^*(\mathbf{r}) \cdot \hat{\mathbf{n}}(\mathbf{r}) \times \mathbf{E}_0(\mathbf{r}) = -\mathrm{Re} \int\limits_{S \cup \partial V_1} \mathrm{d}S\, \mathbf{H}_0^*(\mathbf{r}) \cdot Z_S(\mathbf{r}) \hat{\mathbf{n}}(\mathbf{r}) \times [\mathbf{H}_0(\mathbf{r}) \times \hat{\mathbf{n}}(\mathbf{r})]$$

$$= - \int\limits_{S \cup \partial V_1} \mathrm{d}S\, \mathrm{Re}\{Z_S(\mathbf{r})\} |\mathbf{H}_0(\mathbf{r}) \times \hat{\mathbf{n}}(\mathbf{r})|^2 \tag{6.80}$$

in light of (6.78). Since the integrand in the rightmost-hand side is a positive quantity ($\mathrm{Re}\{Z_S\} > 0$), we obtain the following constraints

$$\hat{\mathbf{n}}(\mathbf{r}) \times \mathbf{H}_0(\mathbf{r}) = 0, \qquad \mathbf{E}_0(\mathbf{r}) \times \hat{\mathbf{n}}(\mathbf{r}) = 0, \qquad \mathbf{r} \in \partial V \tag{6.81}$$

that is, the tangential components of the difference fields must vanish on the boundary. Furthermore, by taking the surface divergence and using (A.60) we have

$$0 = \nabla_s \cdot [\hat{\mathbf{n}}(\mathbf{r}) \times \mathbf{H}_0(\mathbf{r})] = -\hat{\mathbf{n}}(\mathbf{r}) \cdot \nabla \times \mathbf{H}_0(\mathbf{r}) = -\mathrm{j}\,\omega \hat{\mathbf{n}}(\mathbf{r}) \cdot \mathbf{D}_0(\mathbf{r}), \qquad \mathbf{r} \in \partial V \tag{6.82}$$

$$0 = \nabla_s \cdot [\mathbf{E}_0(\mathbf{r}) \times \hat{\mathbf{n}}(\mathbf{r})] = \hat{\mathbf{n}}(\mathbf{r}) \cdot \nabla \times \mathbf{E}_0(\mathbf{r}) = -\mathrm{j}\,\omega \hat{\mathbf{n}}(\mathbf{r}) \cdot \mathbf{B}_0(\mathbf{r}), \qquad \mathbf{r} \in \partial V \tag{6.83}$$

whence we conclude that also the normal components of the difference flux densities vanish on the boundary.

To finalize the proof we need the *Stratton-Chu integral representation* of $\mathbf{E}_0(\mathbf{r})$ and $\mathbf{H}_0(\mathbf{r})$ for $\mathbf{r} \in V$, a result which we shall derive in Section 10.2. In words, the time-harmonic difference fields in a bounded source-free region are fully determined by the values of the tangential components of $\mathbf{E}_0(\mathbf{r})$ and $\mathbf{H}_0(\mathbf{r})$ and of the normal components of $\mathbf{D}_0(\mathbf{r})$ and $\mathbf{B}_0(\mathbf{r})$ on the boundary. Since all these quantities have been shown to be null, and the constitutive relations (1.117), (1.118) apply, with reference to the geometry of Figure 6.10b we have

$$\mathbf{E}_1(\mathbf{r}) - \mathbf{E}_2(\mathbf{r}) = \mathbf{E}_0(\mathbf{r}) = -\mathrm{j}\,\omega\mu \int\limits_{S \cup \partial V_1} \mathrm{d}S'\, \hat{\mathbf{n}}(\mathbf{r}') \times \mathbf{H}_0(\mathbf{r}') \frac{\mathrm{e}^{-\mathrm{j}\,k|\mathbf{r}-\mathbf{r}'|}}{4\pi|\mathbf{r}-\mathbf{r}'|}$$

$$+ \int\limits_{S \cup \partial V_1} \mathrm{d}S'\, \hat{\mathbf{n}}(\mathbf{r}') \cdot \mathbf{E}_0(\mathbf{r}') \nabla' \frac{\mathrm{e}^{-\mathrm{j}\,k|\mathbf{r}-\mathbf{r}'|}}{4\pi|\mathbf{r}-\mathbf{r}'|} + \int\limits_{S \cup \partial V_1} \mathrm{d}S'\, \nabla' \frac{\mathrm{e}^{-\mathrm{j}\,k|\mathbf{r}-\mathbf{r}'|}}{4\pi|\mathbf{r}-\mathbf{r}'|} \times [\mathbf{E}_0(\mathbf{r}') \times \hat{\mathbf{n}}(\mathbf{r}')] = 0 \tag{6.84}$$

$$\mathbf{H}_1(\mathbf{r}) - \mathbf{H}_2(\mathbf{r}) = \mathbf{H}_0(\mathbf{r}) = -\mathrm{j}\,\omega\varepsilon \int\limits_{S \cup \partial V_1} \mathrm{d}S'\, \mathbf{E}_0(\mathbf{r}') \times \hat{\mathbf{n}}(\mathbf{r}') \frac{\mathrm{e}^{-\mathrm{j}\,k|\mathbf{r}-\mathbf{r}'|}}{4\pi|\mathbf{r}-\mathbf{r}'|}$$

$$+ \int\limits_{S \cup \partial V_1} \mathrm{d}S'\, \hat{\mathbf{n}}(\mathbf{r}') \cdot \mathbf{H}_0(\mathbf{r}') \nabla' \frac{\mathrm{e}^{-\mathrm{j}\,k|\mathbf{r}-\mathbf{r}'|}}{4\pi|\mathbf{r}-\mathbf{r}'|} - \int\limits_{S \cup \partial V_1} \mathrm{d}S'\, \nabla' \frac{\mathrm{e}^{-\mathrm{j}\,k|\mathbf{r}-\mathbf{r}'|}}{4\pi|\mathbf{r}-\mathbf{r}'|} \times [\hat{\mathbf{n}}(\mathbf{r}') \times \mathbf{H}_0(\mathbf{r}')] = 0 \tag{6.85}$$

for all points $\mathbf{r} \in V$. Consequently, the two solutions $\mathbf{E}_l(\mathbf{r})$, $\mathbf{H}_l(\mathbf{r})$, $l = 1, 2$, coincide, contrary to our assumption, and the solution is unique [46, Section 3.5.3].

This conclusion may be extended to the case of a boundary ∂V endowed with a dyadic surface impedance $\overline{\mathbf{Z}}_S(\mathbf{r})$ and governed by the impedance relationship [42–45]

$$\hat{\mathbf{n}}(\mathbf{r}) \times [\mathbf{E}(\mathbf{r}) \times \hat{\mathbf{n}}(\mathbf{r})] = \overline{\mathbf{Z}}_S(\mathbf{r}) \cdot [\hat{\mathbf{n}}(\mathbf{r}) \times \mathbf{H}(\mathbf{r})], \qquad \mathbf{r} \in \partial V \tag{6.86}$$

with

$$\mathbf{H}^*(\mathbf{r}) \times \hat{\mathbf{n}}(\mathbf{r}) \cdot \frac{\overline{\mathbf{Z}}_S(\mathbf{r}) + \overline{\mathbf{Z}}_S^H(\mathbf{r})}{2} \cdot \mathbf{H}(\mathbf{r}) \times \hat{\mathbf{n}}(\mathbf{r}) > 0, \qquad \mathbf{r} \in \partial V \tag{6.87}$$

that is, the Hermitian part of $\overline{\mathbf{Z}}_S(\mathbf{r})$ [see (E.55)] must be a positive definite dyadic. With these hypotheses (6.80) passes over into

$$
\begin{aligned}
0 &= -\int\limits_{\partial S\cup V_1} \mathrm{d}S\,\mathrm{Re}\left\{\mathbf{H}_0^*(\mathbf{r})\times\hat{\mathbf{n}}(\mathbf{r})\cdot\overline{\mathbf{Z}}_S(\mathbf{r})\cdot\mathbf{H}_0(\mathbf{r})\times\hat{\mathbf{n}}(\mathbf{r})\right\} \\
&= -\frac{1}{2}\int\limits_{\partial S\cup V_1} \mathrm{d}S\,\mathbf{H}_0^*(\mathbf{r})\times\hat{\mathbf{n}}(\mathbf{r})\cdot\left[\overline{\mathbf{Z}}_S(\mathbf{r})+\overline{\mathbf{Z}}_S^H(\mathbf{r})\right]\cdot\mathbf{H}_0(\mathbf{r})\times\hat{\mathbf{n}}(\mathbf{r})
\end{aligned}
\tag{6.88}
$$

whereby $\mathbf{H}_0(\mathbf{r})\times\hat{\mathbf{n}}(\mathbf{r}) = 0$ for $\mathbf{r}\in\partial V$, because the integrand is a positive quantity. From this point on the proof proceeds as before.

Unlike the analogous result for fields with general time dependence, it is apparent from (6.76) and (6.80) that the uniqueness in the frequency domain is predicated on the presence of conduction losses, however tiny, in V or in $\mathbb{R}^3\setminus\overline{V}$. In fact, our proof breaks down if $\sigma(\mathbf{r}) = 0$ for $\mathbf{r}\in V$, and try as we may, we cannot use energy arguments to infer that there is only one possible solution. The root cause of this uncertainty, so to speak, is that the source-free time-harmonic Maxwell equations, subject to homogeneous boundary conditions, do admit *non-trivial solutions* for particular values of the angular frequency ω [18,30,39,40,47,48]. For instance, the system of equations

$$
\begin{cases}
\nabla\times\mathbf{E}_\nu(\mathbf{r}) + \mathrm{j}\,\omega_\nu\mu(\mathbf{r})\mathbf{H}_\nu(\mathbf{r}) = 0 & \mathbf{r}\in V\\
\nabla\times\mathbf{H}_\nu(\mathbf{r}) - \mathrm{j}\,\omega_\nu\varepsilon(\mathbf{r})\mathbf{E}_\nu(\mathbf{r}) = 0 & \mathbf{r}\in V\\
\nabla\cdot[\varepsilon(\mathbf{r})\mathbf{E}_\nu(\mathbf{r})] = 0 & \mathbf{r}\in V\\
\nabla\cdot[\mu(\mathbf{r})\mathbf{H}_\nu(\mathbf{r})] = 0 & \mathbf{r}\in V\\
\mathbf{E}_\nu(\mathbf{r})\times\hat{\mathbf{n}}(\mathbf{r}) = 0 & \mathbf{r}\in\partial V
\end{cases}
\qquad \nu := (m,n,p)\in\mathbb{N}^3
\tag{6.89}
$$

is solved by infinitely many real-valued vector fields called *eigenfunctions* for specific angular frequencies ω_ν which in turn are called *eigenvalues* (Appendix D.7). The set of equations supplemented with the dual constraint $\hat{\mathbf{n}}(\mathbf{r})\times\mathbf{H}_\nu(\mathbf{r}) = 0$ admits eigenfunctions as well, though the corresponding angular frequencies will be different in general. With these types of boundary conditions the volume V becomes a *cavity* with walls made of PEC or perfect magnetic conducting (PMC, see Section 6.6) medium, and the eigenvalues ω_ν represent the angular frequencies at which the cavity *resonates* (see Section 11.1).

To elaborate, suppose we are trying to solve the time-harmonic equations in a lossless bounded region V with source terms $\mathbf{J}(\mathbf{r})$, $\rho(\mathbf{r})$, $\mathbf{r}\in V_J\subset V$, and the angular frequency of interest happens to coincide with one of the eigenvalues, say, ω_η. Then, it can be shown that a solution $\mathbf{E}_1(\mathbf{r})$, $\mathbf{H}_1(\mathbf{r})$ *can* exist if, and only if,

$$
\left(\mathbf{E}_\eta,\mathbf{J}\right)_V := \int\limits_{V_J}\mathrm{d}V\,\mathbf{E}_\eta(\mathbf{r})\cdot\mathbf{J}(\mathbf{r}) = 0
\tag{6.90}
$$

that is, $\mathbf{J}(\mathbf{r})$ is *orthogonal* to the electric eigenfunction $\mathbf{E}_\eta(\mathbf{r})$ associated with ω_η. What is more, even though (6.90) holds and somehow we manage to establish that the field $\mathbf{E}_1(\mathbf{r})$, $\mathbf{H}_1(\mathbf{r})$ *is* a solution, then the family of fields

$$
\mathbf{E}_2(\mathbf{r}) = \mathbf{E}_1(\mathbf{r}) + c_\eta\mathbf{E}_\eta(\mathbf{r}),\qquad \mathbf{H}_2(\mathbf{r}) = \mathbf{H}_1(\mathbf{r}) + c_\eta\mathbf{H}_\eta(\mathbf{r}),\qquad \mathbf{r}\in V
\tag{6.91}
$$

with $c_\eta\in\mathbb{C}$ an arbitrary constant, solves the Maxwell equations, too, in light of (6.89) and the principle of superposition. For frequencies ω other than the eigenvalues ω_ν, $\nu\in\mathbb{N}^3$, the associated

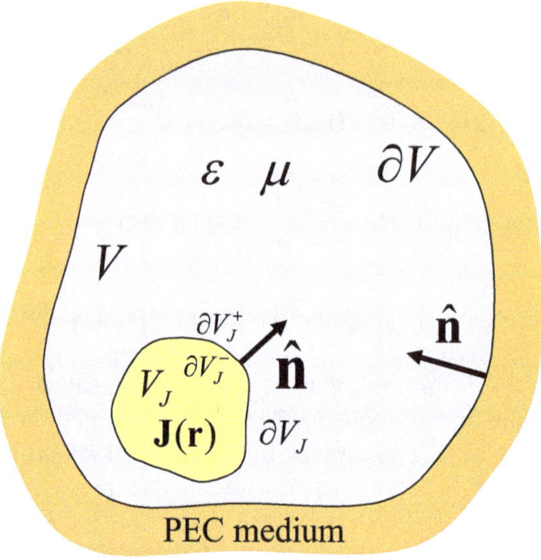

Figure 6.11 For studying the uniqueness of the solutions to the time-harmonic Maxwell equations in a finite lossless region of space with PEC boundary conditions.

homogeneous system of Maxwell equations has only the trivial solution $\mathbf{E}_0(\mathbf{r}) = 0 = \mathbf{H}_0(\mathbf{r})$, $\mathbf{r} \in V$, which is clearly orthogonal to *any* current $\mathbf{J}(\mathbf{r})$ in the sense of (6.90). As a result, the solution to the cavity problem is unique if ω does not coincide with an eigenfrequency. A practical situation is extensively investigated in Example 6.2 further on.

In order to prove these claims we consider the time-harmonic Maxwell equations in a lossless finite region of space V with smooth boundary and PEC boundary conditions (1.169), i.e., the cavity problem sketched in Figure 6.11. Since the derivation relies heavily on concepts from functional analysis, such as Sobolev spaces of vector functions (Appendix D.1) [49], the Riesz representation theorem of linear functionals (Appendix D.5) [50, Chapter 10], [51, Chapter 1], [52, Theorem 7.16], and the Fredholm theory for integral equations (Appendix D.8) [53], we content ourselves with the outline of the main steps for the sake of completeness.

The proof begins by first obtaining the wave equation (or vector Helmholtz equation) for the electric field [cf. (1.234)] and then by separating $\mathbf{E}(\mathbf{r})$ and $\mathbf{J}(\mathbf{r})$ into lamellar and solenoidal parts by virtue of the *Helmholtz decomposition* [29,30,38,40,54], [55, Section 1.9], which we shall prove in all generality in Section 8.1. As a result, we end up with two uncoupled problems, namely, for the lamellar component and for the solenoidal one: the analysis of the lamellar part is trivial, whereas the equation for the solenoidal part can be transformed into an integral equation for which the Riesz-Fredholm theory is applicable. In particular, since the key point consists of showing that a certain linear operator is *compact* (and this is not simple by all means) we shall take this notion for granted.

To study the solutions of the wave equation

$$\begin{cases} \nabla \times \nabla \times \mathbf{E} - k^2\mathbf{E}(\mathbf{r}) = -\mathrm{j}\,\omega\mu\mathbf{J}(\mathbf{r}), & \mathbf{r} \in V \\ \mathbf{E}(\mathbf{r}) \times \hat{\mathbf{n}}(\mathbf{r}) = 0, & \mathbf{r} \in \partial V \end{cases} \tag{6.92}$$

where k is a real wavenumber and $\mathbf{J}(\mathbf{r})$ is confined to a smaller region $V_J \subset V$, we invoke the Helmholtz decomposition in V. Specifically, we let

$$\mathbf{E}(\mathbf{r}) = \mathbf{E}_S(\mathbf{r}) + \nabla\Psi, \qquad \mathbf{J}(\mathbf{r}) = \mathbf{J}_S(\mathbf{r}) + \nabla\Upsilon, \qquad \mathbf{r} \in V \tag{6.93}$$

where $\mathbf{E}_S(\mathbf{r})$ and $\mathbf{J}_S(\mathbf{r})$ are the solenoidal (divergence-free) parts of field and source, respectively, and the gradients of the potentials $\Psi(\mathbf{r})$ and $\Upsilon(\mathbf{r})$ are clearly lamellar (curl-free) vector fields. Implicit in (6.93) is the assumption that V is a normal domain, so that harmonic fields are absent in the representation of $\mathbf{E}(\mathbf{r})$ and $\mathbf{J}(\mathbf{r})$. Since $\hat{\mathbf{n}}(\mathbf{r}) \cdot \mathbf{J}(\mathbf{r}) = 0$ on ∂V_J by hypothesis but $\hat{\mathbf{n}}(\mathbf{r}) \times \mathbf{J}(\mathbf{r})$ may be discontinuous across ∂V_J, we formally obtain the decomposition of $\mathbf{J}(\mathbf{r})$ for $\mathbf{r} \in V_J$ and extend $\mathbf{J}_S(\mathbf{r})$ and $\nabla\Upsilon$ to fields which vanish in $V \setminus \overline{V}_J$. In particular, it is possible to determine $\Upsilon(\mathbf{r})$ so that it is continuous across ∂V_J and vanish in $V \setminus \overline{V}_J$.

Inserting the Ansatz (6.93) into (6.92) leads to

$$\begin{cases} -k^2\nabla\Psi = -\mathrm{j}\,\omega\mu\nabla\Upsilon, & \mathbf{r} \in V \\ \Psi(\mathbf{r}) = 0 = \Upsilon(\mathbf{r}), & \mathbf{r} \in \partial V \end{cases} \qquad \text{(lamellar part)} \tag{6.94}$$

$$\begin{cases} \nabla \times \nabla \times \mathbf{E}_S - k^2\mathbf{E}_S(\mathbf{r}) = -\mathrm{j}\,\omega\mu\mathbf{J}_S(\mathbf{r}), & \mathbf{r} \in V \\ \mathbf{E}_S(\mathbf{r}) \times \hat{\mathbf{n}}(\mathbf{r}) = 0, & \mathbf{r} \in \partial V \end{cases} \qquad \text{(solenoidal part)} \tag{6.95}$$

where we have exploited the fact that $\Psi(\mathbf{r})$ can be determined so as to vanish on the boundary ∂V. Therefore, both $\Psi(\mathbf{r})$ and $\Upsilon(\mathbf{r})$ are elements of the Sobolev space $\mathcal{H}_0^1(V)$ given by (D.84), whereas $\mathbf{E}_S(\mathbf{r})$ belongs to the vector space $\mathcal{H}_{0S}(\mathrm{curl}, V)$ defined in (D.98) as the restriction of the Sobolev space $\mathcal{H}_0(\mathrm{curl}, V)$ to solenoidal fields in V. We mention that the fields $\mathbf{E}_S(\mathbf{r})$ and $\Psi(\mathbf{r})$ are proportional to the time-harmonic electrodynamic potentials $\mathbf{A}_E(\mathbf{r})$ and $\Phi_E(\mathbf{r})$ under the *Coulomb gauge* (Section 8.2) in which instance $\mathbf{A}_E(\mathbf{r})$ is indeed solenoidal. Moreover, (6.94) and (6.95) say that the lamellar (resp. solenoidal) part of the electric field depends solely on the lamellar (resp. solenoidal) part of the current density.

From (6.94) we obtain

$$\nabla\left[k^2\Psi(\mathbf{r}) - \mathrm{j}\,\omega\mu\Upsilon(\mathbf{r})\right] = 0, \qquad \mathbf{r} \in V \tag{6.96}$$

which is solved by

$$k^2\Psi(\mathbf{r}) - \mathrm{j}\,\omega\mu\Upsilon(\mathbf{r}) = C_L, \qquad \mathbf{r} \in V \tag{6.97}$$

where C_L is a constant to be determined. However, since both potentials vanish on ∂V by hypothesis, C_L must be equal to zero, and we find

$$\Psi(\mathbf{r}) = \mathrm{j}\,\frac{\omega\mu}{k^2}\Upsilon(\mathbf{r}) = -\frac{\Upsilon(\mathbf{r})}{\mathrm{j}\,\omega\varepsilon}, \qquad \mathbf{r} \in V, \qquad \Psi, \Upsilon \in \mathcal{H}_0^1(V) \tag{6.98}$$

which in view of the assumptions on $\Upsilon(\mathbf{r})$ implies that $\Psi(\mathbf{r})$ is null outside the source region or, equivalently, that $\mathbf{E}(\mathbf{r})$ is purely solenoidal in $V \setminus \overline{V}_J$. Indeed, the latter property follows by taking the divergence of the vector Helmholtz equations in (6.92) for $\mathbf{r} \notin \overline{V}_J$ and *a priori* from the Gauss law (1.100). More importantly, on account of (6.98) and decomposition (6.93) the lamellar part of $\mathbf{E}(\mathbf{r})$ is uniquely determined once $\Upsilon(\mathbf{r})$ is known.

For the solenoidal part of the solution we introduce a vector test function $\boldsymbol{\phi}(\mathbf{r}) \in \mathcal{H}_{0S}(\mathrm{curl}, V)$ and obtain the weak or distributional form of (6.95) by taking the inner product (D.68) in $\mathcal{L}^2(V)^3$. This yields

$$(\nabla \times \boldsymbol{\phi}, \nabla \times \mathbf{E}_S)_{\mathcal{L}^2(V)^3} - k^2\,(\boldsymbol{\phi}, \mathbf{E}_S)_{\mathcal{L}^2(V)^3} = -\mathrm{j}\,\omega\mu\,(\boldsymbol{\phi}, \mathbf{J}_S)_{\mathcal{L}^2(V)^3} \tag{6.99}$$

because

$$\int_V \mathrm{d}V\,\boldsymbol{\phi}^*(\mathbf{r}) \cdot \nabla \times \nabla \times \mathbf{E}_S = \int_V \mathrm{d}V\,\nabla \times \boldsymbol{\phi}^* \cdot \nabla \times \mathbf{E}_S - \int_V \mathrm{d}V\,\nabla \cdot [\boldsymbol{\phi}^*(\mathbf{r}) \times (\nabla \times \mathbf{E}_S)]$$

$$= \int_V dV \, \nabla \times \boldsymbol{\phi}^* \cdot \nabla \times \mathbf{E}_S + j\,\omega\mu \int_{V \setminus \overline{V}_J} dV \, \nabla \cdot [\boldsymbol{\phi}^*(\mathbf{r}) \times \mathbf{H}(\mathbf{r})] + j\,\omega\mu \int_{V_J} dV \, \nabla \cdot [\boldsymbol{\phi}^*(\mathbf{r}) \times \mathbf{H}(\mathbf{r})]$$

$$= \int_V dV \, \nabla \times \boldsymbol{\phi}^* \cdot \nabla \times \mathbf{E}_S - j\,\omega\mu \int_{\partial V} dS \, \hat{\mathbf{n}}(\mathbf{r}) \cdot \boldsymbol{\phi}^*(\mathbf{r}) \times \mathbf{H}(\mathbf{r})$$

$$- j\,\omega\mu \int_{\partial V_J} dS \, \hat{\mathbf{n}}(\mathbf{r}) \cdot [\boldsymbol{\phi}_+^*(\mathbf{r}) \times \mathbf{H}_+(\mathbf{r}) - \boldsymbol{\phi}_-^*(\mathbf{r}) \times \mathbf{H}_-(\mathbf{r})] = (\nabla \times \boldsymbol{\phi}, \nabla \times \mathbf{E}_S)_{\mathcal{L}^2(V)^3} \tag{6.100}$$

by virtue of identity (H.49), the Faraday law (1.99) and the Gauss theorem (A.53) applied separately in $V \setminus \overline{V}_J$ and V_J, in that the relevant vector fields may not be differentiable throughout V. The symbols $\boldsymbol{\phi}_-$, $\mathbf{H}_-$ (resp. $\boldsymbol{\phi}_+$, $\mathbf{H}_+$) denote fields evaluated on the negative (resp. positive) side of ∂V_J, as indicated by the local normal (Figure 6.11). The flux integrals are null in view of the vanishing of $\hat{\mathbf{n}} \times \boldsymbol{\phi}$ on ∂V, the continuity of $\hat{\mathbf{n}} \times \mathbf{H}$ across ∂V_J [cf. (1.27)] and the additional assumption that $\hat{\mathbf{n}} \times \boldsymbol{\phi}$ be continuous across ∂V_J.

In accordance with (6.99) $\mathbf{E}_S(\mathbf{r})$ does not have to be twice differentiable in V, whereas we simply require that $\mathbf{E}_S(\mathbf{r})$, $\nabla \times \mathbf{E}_S$ and $\mathbf{J}_S(\mathbf{r})$ be square-integrable. Therefore, we seek solutions in the Sobolev space $\mathcal{H}_{0S}(\text{curl}, V)$ and write the weak form as

$$(\boldsymbol{\phi}, \mathbf{E}_S)_\star - k^2 (\boldsymbol{\phi}, \mathbf{E}_S)_{\mathcal{L}^2(V)^3} = -j\,\omega\mu (\boldsymbol{\phi}, \mathbf{J}_S)_{\mathcal{L}^2(V)^3} \tag{6.101}$$

where $(\bullet, \bullet)_\star$ is the alternative inner product in $\mathcal{H}_{0S}(\text{curl}, V)$ defined in (D.99).

We notice that the inner product $(\boldsymbol{\phi}, \mathbf{E}_S)_{\mathcal{L}^2(V)}$ is obviously a linear functional, say,

$$\mathfrak{L}\{\bullet\} : \begin{cases} \mathcal{H}_{0S}(\text{curl}, V) \longrightarrow \mathbb{C} \\ \boldsymbol{\phi} \longmapsto (\boldsymbol{\phi}, \mathbf{E}_S)_{\mathcal{L}^2(V)^3} \end{cases} \tag{6.102}$$

which for a fixed $\mathbf{E}_S$ maps $\boldsymbol{\phi} \in \mathcal{H}_{0S}(\text{curl}, V)$ onto a complex number. This functional is also bounded because by virtue of the Cauchy-Schwarz inequality (D.145), definition (D.95) and inequality (D.104) we have

$$|(\boldsymbol{\phi}, \mathbf{E}_S)_{\mathcal{L}^2(V)^3}| \leqslant \|\boldsymbol{\phi}\|_{\mathcal{L}^2(V)^3} \|\mathbf{E}_S\|_{\mathcal{L}^2(V)^3} \leqslant \|\boldsymbol{\phi}\|_{\mathcal{H}_{0S}(\text{curl}, V)} \|\mathbf{E}_S\|_{\mathcal{L}^2(V)^3} \leqslant M \|\boldsymbol{\phi}\|_\star \|\mathbf{E}_S\|_{\mathcal{L}^2(V)^3} \tag{6.103}$$

whereby in view of definition (D.124) we find

$$\|\mathfrak{L}\{\bullet\}\| := \sup_{\boldsymbol{\phi} \in \mathcal{H}_{0S}(\text{curl}, V)} \frac{|(\boldsymbol{\phi}, \mathbf{E}_S)_{\mathcal{L}^2(V)^3}|}{\|\boldsymbol{\phi}\|_\star} \leqslant M \|\mathbf{E}_S\|_{\mathcal{L}^2(V)^3}, \qquad \boldsymbol{\phi}(\mathbf{r}) \neq 0 \tag{6.104}$$

and $\|\mathbf{E}_S\|_{\mathcal{L}^2(V)^3}$ is finite, since $\mathbf{E}_S(\mathbf{r})$ is square-integrable on V. Thus, according to the Riesz representation theorem of linear functionals (D.168), $\mathfrak{L}\{\boldsymbol{\phi}\}$ can be written as an inner product of the type (D.99), namely,

$$(\boldsymbol{\phi}, \mathbf{E}_S)_{\mathcal{L}^2(V)^3} = \mathfrak{L}\{\boldsymbol{\phi}\} = (\boldsymbol{\phi}, \mathbf{g}_E)_\star \tag{6.105}$$

where $\mathbf{g}_E(\mathbf{r}) \in \mathcal{H}_{0S}(\text{curl}, V)$ is a unique vector field that depends on $\mathbf{E}_S$. Therefore, we define the operator

$$\mathfrak{A}\{\bullet\} : \begin{cases} \mathcal{L}^2(V)^3 \longrightarrow \mathcal{H}_{0S}(\text{curl}, V) \\ \mathbf{E}_S \longmapsto \mathbf{g}_E \end{cases} \tag{6.106}$$

which can be shown to be linear and bounded. Indeed, for any two fields $\mathbf{E}_{S1}$, $\mathbf{E}_{S2}$ of $\mathcal{L}^2(V)^3$ and two complex constants α_1, α_2 we have

$$\begin{aligned}(\boldsymbol{\phi}, \mathfrak{A}\{\alpha_1\mathbf{E}_{S1} + \alpha_2\mathbf{E}_{S2}\})_\star &= \\ &= (\boldsymbol{\phi}, \alpha_1\mathbf{E}_{S1} + \alpha_2\mathbf{E}_{S2})_{\mathcal{L}^2(V)^3} = \alpha_1(\boldsymbol{\phi}, \mathbf{E}_{S1})_{\mathcal{L}^2(V)^3} + \alpha_2(\boldsymbol{\phi}, \mathbf{E}_{S2})_{\mathcal{L}^2(V)^3} \\ &= \alpha_1(\boldsymbol{\phi}, \mathfrak{A}\{\mathbf{E}_{S1}\})_\star + \alpha_2(\boldsymbol{\phi}, \mathfrak{A}\{\mathbf{E}_{S2}\})_\star = (\boldsymbol{\phi}, \alpha_1\mathfrak{A}\{\mathbf{E}_{S1}\} + \alpha_2\mathfrak{A}\{\mathbf{E}_{S2}\})_\star\end{aligned} \tag{6.107}$$

whence we find

$$(\boldsymbol{\phi}, \mathfrak{A}\{\alpha_1\mathbf{E}_{S1} + \alpha_2\mathbf{E}_{S2}\} - \alpha_1\mathfrak{A}\{\mathbf{E}_{S1}\} - \alpha_2\mathfrak{A}\{\mathbf{E}_{S2}\})_\star = 0 \tag{6.108}$$

thanks to definition (6.105) and the linearity of the inner products in $\mathcal{L}^2(V)^3$ and $\mathcal{H}_{0S}(\text{curl}, V)$. Since this last condition must hold true for any $\boldsymbol{\phi}(\mathbf{r}) \in \mathcal{H}_{0S}(\text{curl}, V)$ we let

$$\boldsymbol{\phi}(\mathbf{r}) = \mathfrak{A}\{\alpha_1\mathbf{E}_{S1} + \alpha_2\mathbf{E}_{S2}\} - \alpha_1\mathfrak{A}\{\mathbf{E}_{S1}\} - \alpha_2\mathfrak{A}\{\mathbf{E}_{S2}\}, \qquad \mathbf{r} \in V \tag{6.109}$$

whereby we get

$$\|\mathfrak{A}\{\alpha_1\mathbf{E}_{S1} + \alpha_2\mathbf{E}_{S2}\} - \alpha_1\mathfrak{A}\{\mathbf{E}_{S1}\} - \alpha_2\mathfrak{A}\{\mathbf{E}_{S2}\}\|_\star^2 = 0 \tag{6.110}$$

which, on account of the properties (D.8) and (D.9) of a norm, can only be true if the element vanishes and hence

$$\mathfrak{A}\{\alpha_1\mathbf{E}_{S1} + \alpha_2\mathbf{E}_{S2}\} = \alpha_1\mathfrak{A}\{\mathbf{E}_{S1}\} + \alpha_2\mathfrak{A}\{\mathbf{E}_{S2}\} \tag{6.111}$$

as required of a linear operator. As regards the boundedness of $\mathfrak{A}\{\bullet\}$ we have

$$\begin{aligned}\|\mathfrak{A}\{\mathbf{E}_S\}\|_\star^2 &= (\mathfrak{A}\{\mathbf{E}_S\}, \mathfrak{A}\{\mathbf{E}_S\})_\star = (\mathfrak{A}\{\mathbf{E}_S\}, \mathbf{g}_E)_\star = (\mathfrak{A}\{\mathbf{E}_S\}, \mathbf{E}_S)_{\mathcal{L}^2(V)^3} \\ &\leqslant \|\mathfrak{A}\{\mathbf{E}_S\}\|_{\mathcal{L}^2(V)^3}\,\|\mathbf{E}_S\|_{\mathcal{L}^2(V)^3} \leqslant \|\mathfrak{A}\{\mathbf{E}_S\}\|_{\mathcal{H}_{0S}(\text{curl},V)}\,\|\mathbf{E}_S\|_{\mathcal{L}^2(V)^3} \\ &\leqslant M\,\|\mathfrak{A}\{\mathbf{E}_S\}\|_\star\,\|\mathbf{E}_S\|_{\mathcal{L}^2(V)^3}\end{aligned} \tag{6.112}$$

where we have used (6.105) and (D.104). By dividing through by $\|\mathfrak{A}\{\mathbf{E}_S\}\|_\star\,\|\mathbf{E}_S\|_{\mathcal{L}^2(V)^3}$ and passing to the supremum we find

$$\|\mathfrak{A}\{\bullet\}\| := \sup_{\mathbf{E}_S \in \mathcal{L}^2(V)^2} \frac{\|\mathfrak{A}\{\mathbf{E}_S\}\|_\star}{\|\mathbf{E}_S\|_{\mathcal{L}^2(V)}} \leqslant M \qquad \mathbf{E}_S(\mathbf{r}) \neq 0 \tag{6.113}$$

in light of definition (D.124). Since the Sobolev space $\mathcal{H}_{0S}(\text{curl}, V)$ is embedded in $\mathcal{L}^2(V)^3$ according to (D.105), the *imbedding* operator (D.136) here reads

$$\mathfrak{J}\{\bullet\} : \begin{cases} \mathcal{H}_{0S}(\text{curl}, V) \longrightarrow \mathcal{L}^2(V)^3 \\ \mathbf{E}_S \longmapsto \mathbf{E}_S \end{cases} \tag{6.114}$$

it formally transforms an element $\mathbf{E}_S \in \mathcal{H}_{0S}(\text{curl}, V)$ into the same field though regarded as a member of $\mathcal{L}^2(V)^3$ and, since V is finite, $\mathfrak{J}\{\bullet\}$ can be shown to be compact (this is the essence of the *Sobolev embedding theorem* [56, Chapter IV], which we take for granted). Therefore, we can rewrite (6.105) as

$$(\boldsymbol{\phi}, \mathbf{g}_E)_\star = (\boldsymbol{\phi}, \mathfrak{A}\{\mathbf{E}_S\})_\star = \left(\boldsymbol{\phi}, \tilde{\mathfrak{A}}\{\mathbf{E}_S\}\right)_\star \tag{6.115}$$

where the operator

$$\tilde{\mathfrak{A}}\{\bullet\} := \mathfrak{A}\{\mathfrak{I}\{\bullet\}\} : \mathcal{H}_{0S}(\text{curl}, V) \longrightarrow \mathcal{L}^2(V)^3 \longrightarrow \mathcal{H}_{0S}(\text{curl}, V) \tag{6.116}$$

is compact because so is $\mathfrak{I}\{\bullet\}$, whereas $\mathfrak{A}\{\bullet\}$ is bounded, and the successive application of a compact and a bounded operator yields a compact operator [51, Proposition 4.2, p. 41].

By performing similar manipulations on the right-hand side of (6.95) we arrive at

$$(\boldsymbol{\phi}, \mathbf{E}_S)_\star - k^2 \left(\boldsymbol{\phi}, \tilde{\mathfrak{A}}\{\mathbf{E}_S\}\right)_\star = -\mathrm{j}\,\omega\mu\,(\boldsymbol{\phi}, \mathfrak{A}\{\mathbf{J}_S\})_\star \tag{6.117}$$

which we can cast as

$$\left(\boldsymbol{\phi}, \mathbf{E}_S - k^2\tilde{\mathfrak{A}}\{\mathbf{E}_S\} + \mathrm{j}\,\omega\mu\mathfrak{A}\{\mathbf{J}_S\}\right)_\star = 0 \tag{6.118}$$

thanks to the linearity of the inner product (D.99). Since (6.118) must hold for any test function $\boldsymbol{\phi} \in \mathcal{H}_{0S}(\text{curl}, V)$, we may choose

$$\boldsymbol{\phi}(\mathbf{r}) = \mathbf{E}_S(\mathbf{r}) - k^2\tilde{\mathfrak{A}}\{\mathbf{E}_S\} + \mathrm{j}\,\omega\mu\mathfrak{A}\{\mathbf{J}_S\}, \qquad \mathbf{r} \in V \tag{6.119}$$

whereby we obtain

$$\left\|\mathbf{E}_S(\mathbf{r}) - k^2\tilde{\mathfrak{A}}\{\mathbf{E}_S\} - \mathrm{j}\,\omega\mu\mathfrak{A}\{\mathbf{J}_S\}\right\|_\star^2 = 0 \tag{6.120}$$

which can only be true if the element vanishes and hence

$$\mathbf{E}_S(\mathbf{r}) - k^2\tilde{\mathfrak{A}}\{\mathbf{E}_S\} = -\mathrm{j}\,\omega\mu\mathfrak{A}\{\mathbf{J}_S\}, \qquad \mathbf{r} \in V, \qquad \mathbf{E}_S \in \mathcal{H}_{0S}(\text{curl}, V), \qquad \mathbf{J}_S \in \mathcal{L}^2(V)^3 \tag{6.121}$$

owing to the properties (D.8) and (D.9) of a norm. Expression (6.121) is a *Fredholm integral equation* of the second kind (see Section 13.1) where $\tilde{\mathfrak{A}}\{\bullet\}$ is compact, self-adjoint [see (D.179)] and positive definite. Indeed, we have

$$\left(\tilde{\mathfrak{A}}\{\boldsymbol{\phi}\}, \mathbf{E}_S\right)_\star = \left[\left(\mathbf{E}_S, \tilde{\mathfrak{A}}\{\boldsymbol{\phi}\}\right)_\star\right]^* = \left[(\mathbf{E}_S, \boldsymbol{\phi})_{\mathcal{L}^2(V)^3}\right]^* = (\boldsymbol{\phi}, \mathbf{E}_S)_{\mathcal{L}^2(V)^3} = \left(\boldsymbol{\phi}, \tilde{\mathfrak{A}}\{\mathbf{E}_S\}\right)_\star \tag{6.122}$$

and

$$\left(\mathbf{E}_S, \tilde{\mathfrak{A}}\{\mathbf{E}_S\}\right)_\star = (\mathbf{E}_S, \mathbf{E}_S)_{\mathcal{L}^2(V)^3} = \|\mathbf{E}_S\|^2_{\mathcal{L}^2(V)^3} > 0 \tag{6.123}$$

having repeatedly invoked definition (6.105). Further, we can put (6.121) in the standard form (D.245) by including the constant k^2 in the definition of the operator $\tilde{\mathfrak{A}}\{\bullet\}$.

From the *Fredholm alternative* (Appendix D.8) [57, Chapter 8], [52, Chapter 8], [58, Chapter 1], [53,59,60] we know that (6.121) has a unique solution for any source term $\mathfrak{A}\{\mathbf{J}_S\} \in \mathcal{H}_{0S}(\text{curl}, V)$, if the associated homogeneous equation

$$\mathbf{e}(\mathbf{r}) - k^2\tilde{\mathfrak{A}}\{\mathbf{e}\} = 0, \qquad \mathbf{r} \in V, \qquad \mathbf{e} \in \mathcal{H}_{0S}(\text{curl}, V) \tag{6.124}$$

admits only the trivial solution $\mathbf{e}(\mathbf{r}) = 0$. Or else, if for a given wavenumber (6.124) has $N < +\infty$ non-trivial solutions $\mathbf{e}_l(\mathbf{r})$, $l = 1, \ldots, N$, then (6.121) is solvable if, and only if, the right-hand side is orthogonal — with respect to the inner product (D.99) — to $\mathbf{e}_l(\mathbf{r})$, and in light of (6.105) this constraint translates into the requirement

$$(\mathbf{e}_l, \mathfrak{A}\{\mathbf{J}_S\})_\star = (\mathbf{e}_l, \mathbf{J}_S)_{\mathcal{L}^2(V)^3} = \int_{V_J} \mathrm{d}V\, \mathbf{e}_l^*(\mathbf{r}) \cdot \mathbf{J}_S(\mathbf{r}) = 0, \qquad l = 1, \ldots, N \tag{6.125}$$

in which case (6.121) and hence (6.95) have infinitely many solutions in the form

$$\mathbf{E}_S(\mathbf{r}) = \mathbf{E}_{S0}(\mathbf{r}) + \sum_{l=1}^{N} c_l \mathbf{e}_l(\mathbf{r}), \qquad \mathbf{r} \in V \tag{6.126}$$

where $\mathbf{E}_{S0}(\mathbf{r})$ is a particular solution and c_l are arbitrary constants. Now, combining the previous expression with the Helmholtz decomposition (6.93) yields the result anticipated by (6.91). Moreover, we can rephrase the orthogonality condition in terms of $\mathbf{J}(\mathbf{r})$ inasmuch as $\mathbf{e}_l(\mathbf{r})$ — being solenoidal — is orthogonal to the lamellar part of the current, namely,

$$\begin{aligned}\int_{V_J} \mathrm{d}V\, \mathbf{e}_l^*(\mathbf{r}) \cdot \mathbf{J}(\mathbf{r}) &= \int_{V_J} \mathrm{d}V\, \mathbf{e}_l^*(\mathbf{r}) \cdot [\mathbf{J}_S(\mathbf{r}) + \nabla\Upsilon] = 0 + \int_{V_J} \mathrm{d}V\, \nabla \cdot (\mathbf{e}_l^*\Upsilon) \\ &= \int_{\partial V_J} \mathrm{d}S\, \hat{\mathbf{n}}(\mathbf{r}) \cdot \mathbf{e}_l^*(\mathbf{r})\Upsilon(\mathbf{r}) = 0\end{aligned} \tag{6.127}$$

thanks to the Gauss theorem and the vanishing of $\Upsilon(\mathbf{r})$ on ∂V_J by construction. Last but not least, since $\mathbf{H}(\mathbf{r})$ is related to $\mathbf{E}(\mathbf{r})$ by the Faraday law (1.99), the magnetic field is uniquely determined when so is the electric field.

On a related score, it was also shown [61] that, when the medium in V is lossless and the angular frequency is an eigenvalue ω_ν, uniqueness is ensured by enforcing one of these two sets of boundary conditions

$$\mathbf{E}(\mathbf{r};\omega_\nu) \times \hat{\mathbf{n}}(\mathbf{r}) = 0, \qquad \left.\frac{\partial \mathbf{E}(\mathbf{r};\omega)}{\partial\omega}\right|_{\omega_\nu} \times \hat{\mathbf{n}}(\mathbf{r}) = 0, \qquad \mathbf{r} \in \partial V \tag{6.128}$$

$$\hat{\mathbf{n}}(\mathbf{r}) \times \mathbf{H}(\mathbf{r};\omega_\nu) = 0, \qquad \hat{\mathbf{n}}(\mathbf{r}) \times \left.\frac{\partial \mathbf{H}(\mathbf{r};\omega)}{\partial\omega}\right|_{\omega_\nu} = 0, \qquad \mathbf{r} \in \partial V \tag{6.129}$$

that is, also the derivative of the tangential fields with respect to the angular frequency must be specified. From a strictly mathematical viewpoint, this certainly settles the question of uniqueness, though it is not clear how to enforce the requirements on the derivatives in practice.

Then again, perhaps we have been too fussy about ensuring uniqueness at all costs. After all, we remarked in Section 1.5 that time-harmonic fields are just an abstraction, a convenient idealized representation of electromagnetic phenomena which, after a transient has finished, repeat themselves over and over again with some period $T = 2\pi/\omega$. So, keeping that in mind, if we tackle the problem of interest in the time domain — where it belongs — we achieve uniqueness as long as suitable initial and boundary conditions together with the sources are prescribed.

Alternatively, if we really wish to work in the frequency domain (there are compelling reasons for this choice, e.g., ordinarily it simplifies the process of solving the equations) we may assume a small non-zero conductivity $\sigma(\mathbf{r})$ for $\mathbf{r} \in V$ on the grounds that any material medium exhibit some losses. Still, if we do not fancy the idea of introducing artificial losses, after we have solved the 'lossy problem', we may take the limit as $\sigma(\mathbf{r}) \to 0$ in order to obtain the solution relevant to the lossless case. The latter approach is fruitful so long as we present the solution in a form that allows us to actually compute the limit, although, as pointed out by D. S. Jones in a different context [37, Section 9.8], a priori there is no guarantee that the limiting procedure should yield the desired solution.

Example 6.2 (Uniqueness in a circular cylindrical cavity with PEC walls)

To get a feeling of the role played by conduction losses in determining the uniqueness of solutions in a cavity with PEC boundaries and of the importance of condition (6.90), we consider a finite right

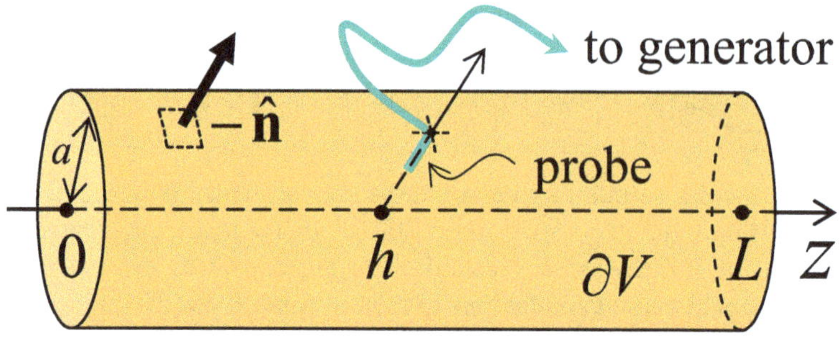

Figure 6.12 Circular cylindrical cavity with PEC walls.

circular cylinder with height L and radius a [62, Section 6.2]. We choose a system of cylindrical coordinates (τ, φ, z) in which the z-axis is aligned with the axis of the cylinder (Figure 6.12). The problem of finding both eigenfunctions and source-driven solutions within the cylinder may be tackled with the general approach for cavities presented in Section 11.1 [30]. However, thanks to the translational symmetry of the cylinder along its axis, it is also possible to consider this geometry as a length of hollow-pipe circular waveguide terminated with PEC walls at both ends. To this purpose, we recall that the time-harmonic electric and magnetic fields in a hollow-pipe waveguide admit a representation of the type (Section 11.2) [16–18, 63], [64, Chapter 13]

$$\mathbf{E}(\mathbf{r}) = \mathbf{E}_t(\mathbf{r}) + E_z(\mathbf{r})\hat{\mathbf{z}} = \sum_{m,n} \mathbf{e}_{mn}(\boldsymbol{\tau})V_{mn}(z) + E_z(\mathbf{r})\hat{\mathbf{z}} \tag{6.130}$$

$$\mathbf{H}(\mathbf{r}) = \mathbf{H}_t(\mathbf{r}) + H_z(\mathbf{r})\hat{\mathbf{z}} = \sum_{m,n} \mathbf{h}_{mn}(\boldsymbol{\tau})I_{mn}(z) + H_z(\mathbf{r})\hat{\mathbf{z}} \tag{6.131}$$

where

- the fields $\mathbf{E}_t(\mathbf{r})$ and $\mathbf{H}_t(\mathbf{r})$ are called the *transverse components*, in that they are perpendicular to the axis of the waveguide;
- the fields $E_z(\mathbf{r})\hat{\mathbf{z}}$ and $H_z(\mathbf{r})\hat{\mathbf{z}}$ are called the *longitudinal components*, in that they are parallel to the axis of the waveguide; $E_z(\mathbf{r})$ and $H_z(\mathbf{r})$ can be expressed in terms of the transverse components;
- m and n are the two *modal* indices;
- $\mathbf{e}_{mn}(\boldsymbol{\tau})$, $\mathbf{h}_{mn}(\boldsymbol{\tau})$ are called the electric and magnetic *transverse modal eigenfunctions*; they depend only on τ and φ, and are determined solely by the shape of the waveguide cross section;
- $V_{mn}(z)$, $I_{mn}(z)$ are called the *modal voltages* and *currents*; they depend only on z, but are determined by the boundary conditions at the ends of the waveguide and the properties of the medium that fills the waveguide.

Furthermore, the modal voltages and currents satisfy a system of ordinary first-order differential equations

$$-\frac{\mathrm{d}V_{mn}}{\mathrm{d}z} = \mathrm{j}\, k_{zmn} Z_{mn} I_{mn}(z) \tag{6.132}$$

$$-\frac{\mathrm{d}I_{mn}}{\mathrm{d}z} = \mathrm{j}\, \frac{k_{zmn}}{Z_{mn}} V_{mn}(z) \tag{6.133}$$

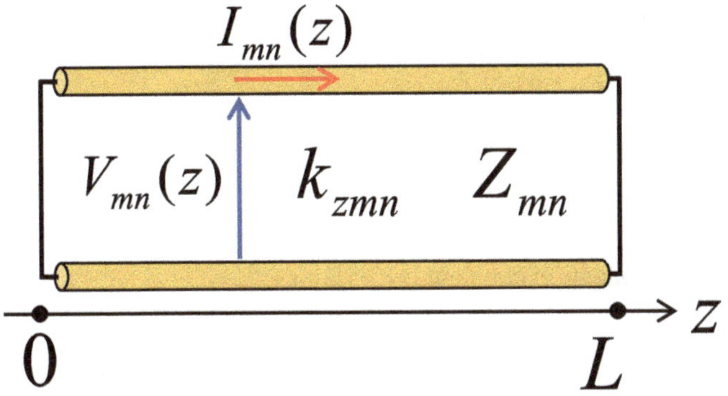

Figure 6.13 Transmission-line model of a source-free lossless cylindrical cavity.

that is, the typical time-harmonic *telegraph equations* of a uniform transmission line [2, 18, 19, 31, 62–73] with (longitudinal) propagation constant $k_{zmn}(\omega)$ and characteristic impedance $Z_{mn}(\omega)$. In particular, the propagation constant is related to the angular frequency through the relationship

$$k_{zmn} = \sqrt{\omega^2\left(\varepsilon - \mathrm{j}\frac{\sigma}{\omega}\right)\mu - k_{cmn}^2} = \sqrt{k^2 - k_{cmn}^2} \tag{6.134}$$

where ε, μ and σ are the constitutive parameters of the homogeneous, possibly lossy material that fills the cavity, k is the possibly complex wavenumber defined in (1.249), and k_{cmn} are the transverse propagation constants, which depend only on the shape of the waveguide cross section. We observe that the requirement $\mathbf{E}(\mathbf{r}) \times \hat{\mathbf{n}}(\mathbf{r}) = 0$ at $z = 0$ and $z = L$ translates into constraints for the modal voltages

$$V_{mn}(0) = 0 \qquad\qquad V_{mn}(L) = 0 \tag{6.135}$$

by virtue of (6.130), whereas for $\tau = a$ (i.e., on the lateral surface of the cylinder) the boundary condition is fulfilled by the electric transverse eigenfunctions $\mathbf{e}_{mn}(\boldsymbol{\tau})$ (Section 11.2).

We now examine three situations:

(i) the medium filling the cavity is lossless ($\sigma = 0$), and the cavity is source-free;
(ii) the medium filling the cavity is lossless ($\sigma = 0$), and the cavity is excited with a small probe (Figure 6.12);
(iii) the medium filling the cavity is lossy ($\sigma > 0$), and the cavity is again excited with the same probe.

(i) Source-free lossless cavity

We set $\sigma = 0$ in (6.134) and observe that, since the angular frequency ω affects only the modal voltages and currents through k_{zmn} and Z_{mn}, we may compute the eigenvalues ω_ν of the cylindrical cavity by solving the transmission-line equations (6.132) and (6.133) subject to (6.135) for each pair of indices (m, n). Thus, we have effectively turned the original cavity problem into the solution of infinitely many transmission-line circuits of the type sketched in Figure 6.13.

By eliminating the current $I_{mn}(z)$ from (6.132) with the aid of (6.133) we obtain one-dimensional homogeneous wave equations for the modal voltages

$$\frac{\mathrm{d}^2 V_{mn}}{\mathrm{d}z^2} + k_{zmn}^2 V_{mn}(z) = 0, \qquad z \in [0, L] \tag{6.136}$$

subject to the boundary condition (6.135). As is well known, the general solutions are

$$V_{mn}(z) = A_{nm}\sin(k_{zmn}z) + B_{nm}\cos(k_{zmn}z) \tag{6.137}$$

where A_{mn} and B_{mn} are suitable constants. The constraints (6.135) demand that

$$0 = V_{mn}(0) = B_{nm} \qquad 0 = V_{mn}(L) = A_{nm}\sin(k_{zmn}L) \tag{6.138}$$

and, if the longitudinal propagation constant is chosen as

$$k_{zmn} = k_{zmnp} = \frac{p\pi}{L}, \qquad p \in \mathbb{N}\setminus\{0\} \tag{6.139}$$

the second equation is solved for any value of A_{nm}. Evidently, the last condition determines the eigenvalues ω_ν on account of (6.134) and (1.248), namely,

$$\omega_\nu = \sqrt{\frac{k_{cmn}^2}{\varepsilon\mu} + \frac{1}{\varepsilon\mu}\left(\frac{p\pi}{L}\right)^2} = \omega_{mnp} \tag{6.140}$$

and this formula clarifies that ν is, in fact, a triple of indices. The admissible modal voltages read

$$V_{mnp}(z) = A_{mnp}\sin\left(\frac{p\pi}{L}z\right), \qquad p \in \mathbb{N}\setminus\{0\} \tag{6.141}$$

where the amplitudes A_{mnp} may be chosen real, positive and such that

$$\|V_{mnp}\|_2^2 = \int_0^L \mathrm{d}z\,|V_{mnp}(z)|^2 = \frac{L}{2}(A_{mnp})^2 = 1, \qquad p \in \mathbb{N}\setminus\{0\} \tag{6.142}$$

though this step is inessential for our discussion. The modal currents follow from (6.132) once $V_{mnp}(z)$ has been found. Now, modal voltages and currents constitute the eigenfunctions of the transmission-line circuit in Figure 6.13 but, in tandem with the transverse eigenfunctions in (6.130) and (6.131), they also define the eigenfunctions of the cylindrical cavity, that is, the solution to (6.89).

This analysis shows that (6.89) admits non-trivial solutions, if the medium filling the bounded region V is lossless.

(ii) Source-driven lossless cavity

We assume $\sigma = 0$ again and suppose that the cylindrical cavity is excited by means of a small probe inserted through a hole cut in the lateral surface a distance h from the bottom of the cylinder. Depending on position and shape of the probe, the electric current flowing thereon may couple to one or more modes in the cavity (Section 11.2.6) [16, 18, 40, 66]. For our purposes, though, it is sufficient and correct to model the effect of the probe as lumped ideal current generators of strength $I_{Gmn}(\omega)$ connected in parallel to the transmission line in the section $z = h$. The behavior of the resulting circuit, which is shown in Figure 6.14, is governed by the distributional inhomogeneous transmission-line equations

$$-\frac{\mathrm{d}V_{mn}}{\mathrm{d}z} = \mathrm{j}\,k_{zmn}Z_{mn}I_{mn}(z) \tag{6.143}$$

$$-\frac{\mathrm{d}I_{mn}}{\mathrm{d}z} = \mathrm{j}\,\frac{k_{zmn}}{Z_{mn}}V_{mn}(z) + I_{Gmn}\delta(z-h) \tag{6.144}$$

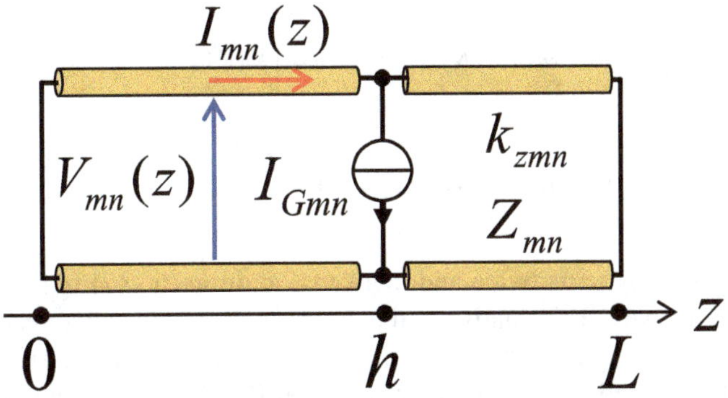

Figure 6.14 Transmission-line model of source-driven cylindrical cavity.

where the Dirac δ-distribution (Appendix C) signifies that the current generator is concentrated at $z = h$ and is responsible for a jump of the modal current, namely, $I_{mn}(h^-) = I_{mn}(h^+) + I_{Gmn}$, according to the Kirchhoff current law (KCL). By contrast, the voltage $V_{mn}(z)$ is continuous across $z = h$. (In Section 11.2.6 we shall employ the alternative, classic solution strategy which does not make use of distributions.)

The corresponding wave equation for the modal voltages reads

$$\frac{\mathrm{d}^2 V_{mn}}{\mathrm{d}z^2} + k_{zmn}^2 V_{mn}(z) = \mathrm{j}\, k_{zmn} Z_{mn} I_{Gmn} \delta(z - h) \tag{6.145}$$

and is subject to the same boundary conditions (6.135). Although we could solve (6.145) directly, motivated by (6.141) we try a representation of the solution in the form

$$V_{mn}(z) = \sum_{p=1}^{+\infty} A_{mnp} \sin(k_{zmnp} z), \qquad z \in [0, L] \tag{6.146}$$

with k_{zmnp} given by (6.139). This approach allows us to highlight the role played by the eigenfunctions and, more importantly, the absence of losses. Besides, in this way the boundary conditions in $z = 0$ and $z = L$ are automatically satisfied. Using the voltage eigenfunctions, which are continuous, is justified inasmuch as $V_{mn}(z)$ is itself continuous in $[0, L]$, as observed above, whereas (6.145) says that the first derivative of $V_{mn}(z)$ is finite but different on either side of $z = h$.

Since the set of eigenfunctions $\{\sin(k_{zmnp} z)\}_{p=1}^{\infty}$ is orthogonal for $z \in [0, L]$, the convergence of the series defined in (6.146) is guaranteed by the Bessel inequality (D.115) so long as we formally choose

$$(A_{mnp})^* = \frac{2}{L} \int_0^L \mathrm{d}z' \, [V_{mn}(z')]^* \sin(k_{zmnp} z') = \frac{\left(V_{mn}(z'), \sin(k_{zmnp} z')\right)_{[0,L]}}{\left\|\sin(k_{zmnp} z')\right\|_2^2} \tag{6.147}$$

as is prescribed by (D.113). To determine the unknown expansion coefficients A_{mnp} in practice we multiply (6.145) by $\sin(k_{zmnp} z)$ and integrate along z from 0 to L, viz.,

$$\mathrm{j}\, k_{zmn} Z_{mn} I_{Gmn} \underbrace{\int_0^L \mathrm{d}z \, \sin(k_{zmnp} z) \delta(z - h)}_{= \sin(k_{zmnp} h)} = \int_0^L \mathrm{d}z \, \sin(k_{zmnp} z) \left[\frac{\mathrm{d}^2 V_{mn}}{\mathrm{d}z^2} + k_{zmn}^2 V_{mn}(z) \right]$$

$$= \int_0^L \mathrm{d}z \left[\frac{\mathrm{d}^2}{\mathrm{d}z^2} \sin(k_{zmnp}z) + k_{zmn}^2 \sin(k_{zmnp}z) \right] V_{mn}(z)$$

$$= (k_{zmn}^2 - k_{zmnp}^2) \int_0^L \mathrm{d}z \, \sin(k_{zmnp}z) V_{mn}(z) = \frac{L}{2} \left(k_{zmn}^2 - k_{zmnp}^2 \right) A_{mnp} \tag{6.148}$$

where we have integrated by parts twice and invoked definition (6.147) in the last step. Parenthetically, the calculations above are alternative to the common approach which resorts to 1) substituting expansion (6.146) into (6.145), 2) carrying out the second-order derivative, 3) multiplying both sides by $\sin(k_{zmns})$, $s \in \mathbb{N} \setminus \{0\}$, and finally 4) integrating over $[0, L]$. Steps 2) and 4) require interchanging the order of differentiation and integration, respectively, with the summation over p, and these operations — which we have avoided — are not fully justified because the derivative of $V_{mn}(z)$ is discontinuous, though it is remarkable that in the end one obtains the same correct relationship (6.148).

By tracing our steps back through the chain of equalities (6.148) we arrive at

$$A_{mnp} \left(k_{zmn}^2 - k_{zmnp}^2 \right) = \frac{2}{L} \mathrm{j}\, k_{zmn} Z_{mn} I_{Gmn} \sin(k_{zmnp}h), \qquad p \in \mathbb{N} \setminus \{0\} \tag{6.149}$$

and, in principle, this algebraic equation can be used to find A_{mnp} and finalize the solution of (6.145). However, since the inversion of (6.149) is predicated on a division by the factor $k_{zmn}^2 - k_{zmnp}^2$, we need to make sure that the latter is non-zero. Therefore, we are presented with three possibilities.

(a) If the operating frequency of the probe is such that $k_{zmn}(\omega) \neq k_{zmnp}$ for all values of p, then

$$A_{mnp} = \frac{2}{L} \mathrm{j}\, k_{zmn} Z_{mn} I_{Gmn} \frac{\sin(k_{zmnp}h)}{k_{zmn}^2 - k_{zmnp}^2} \tag{6.150}$$

and we can write the modal voltages as

$$V_{mn}(z) = \frac{2}{L} \mathrm{j}\, k_{zmn} Z_{mn} I_{Gmn} \sum_{p=1}^{+\infty} \frac{\sin(k_{zmnp}h) \sin(k_{zmnp}z)}{k_{zmn}^2 - k_{zmnp}^2}, \qquad z \in [0, L] \tag{6.151}$$

so the solution *is* unique. This happens because the homogeneous equation associated with (6.145) has only the trivial solution $V_0(z) = 0$.

(b) If, on the contrary, the operating frequency of the probe is such that the propagation constant k_{zmn} coincides with the eigenvalue k_{zmnl}, then for $p = l$ (6.149) becomes

$$0 = \frac{2}{L} \mathrm{j}\, k_{zmnl} Z_{mn} I_{Gmn} \sin(k_{zmnl}h) \tag{6.152}$$

which can be fulfilled only if also the right-hand side vanishes, i.e., if the longitudinal position of the probe is adjusted so that $k_{zmnl}h = \pi$. This means that the current flowing on the probe is orthogonal to the electric eigenfunction associated with ω_{mnl} [see (6.90)]. Even so, the coefficient A_{mnl} remains undetermined, and the best we can do is write

$$V_{mn}(z) = \frac{2}{L} \mathrm{j}\, k_{zmnl} Z_{mn} I_{Gmn} \sum_{\substack{p=1 \\ p \neq l}}^{+\infty} \frac{\sin(k_{zmnp}h) \sin(k_{zmnl}z)}{k_{zmnl}^2 - k_{zmnp}^2} + A_{mnl} \sin(k_{zmnl}z) \tag{6.153}$$

whereby we see that a solution exists but is not unique, because the last term above comes with an arbitrary amplitude.

(c) If under the hypotheses of case (b) the right member of (6.152) does not vanish, the equation is impossible, and the cavity problem has no solution.

Incidentally, we notice that the summation appearing in (6.151), viz.,

$$G_{mn}(z, h; \omega) := \frac{2}{L} \mathrm{j}\, k_{zmn} Z_{mn} \sum_{p=1}^{+\infty} \frac{\sin(k_{zmnp} h) \sin(k_{zmnp} z)}{k_{zmn}^2 - k_{zmnp}^2} \tag{6.154}$$

is the *eigenfunction expansion* or *spectral representation* of the time-harmonic Green function relevant to the circuit of Figure 6.14. Physically, $G_{mn}(z, h; \omega)$ is the modal voltage produced at z (observation point) by a lumped unitary current generator at h (source point) and as such obeys the boundary conditions (6.135) [17].

(iii) Source-driven lossy cavity

Finally, we briefly examine the case in which the medium in the cylindrical cavity is indeed endowed with a non-zero conductivity σ. The transmission-line circuit of Figure 6.14 is still valid, but now k_{zmm} and Z_{mn} are complex quantities for any value of the angular frequency ω.

If we try again a solution in the form (6.146), by following the same steps as before, we arrive at relationship (6.149) for the expansion coefficients. Nevertheless, in the presence of losses (6.149) can always be inverted because for no real value of ω does the factor $k_{zmn}^2 - k_{zmnp}^2$ vanish, as the eigenvalues k_{zmnp} are real [see (6.139)]. The representation (6.151) is invariably valid, and there exists only one solution for any angular frequency, as predicted by the uniqueness theorem.

This analysis sheds light on the importance of losses in determining the time-harmonic solution univocally in a bounded domain.

(End of Example 6.2)

6.4.2 *Unbounded regions*

Most practical problems concerned with propagation and scattering of electromagnetic waves entail solving the full-fledged Maxwell equations in unbounded regions of space, as is suggested in Figure 6.4. Moreover, it is often convenient to make the assumption of time-harmonic regime and work in the frequency domain. Therefore, it is important to find the necessary conditions that can guarantee uniqueness of the time-harmonic solutions in infinitely extended spatial domains.

In fact, while discussing the average power flow $\langle P_F \rangle$ at infinity in Section 1.10.2, we already argued that the time-harmonic fields behave asymptotically as

$$\mathbf{E}(\mathbf{r}) = \mathcal{O}\left(\frac{1}{|\mathbf{r}|}\right), \qquad |\mathbf{r}| \to +\infty \tag{6.155}$$

$$\mathbf{H}(\mathbf{r}) = \mathcal{O}\left(\frac{1}{|\mathbf{r}|}\right), \qquad |\mathbf{r}| \to +\infty \tag{6.156}$$

uniformly with respect to $\mathbf{r}$. Nonetheless, we shall see that (6.155) and (6.156) alone do not ensure uniqueness if the isotropic background medium is lossless. Then, we begin by discussing the lossy case.

We consider a ball $B(0, a_1)$ with radius a_1 large enough to enclose the source region V_J and a bounded volume V_1 in which an unspecified material medium is contained, as is elucidated in Figure 6.15. We suppose that the surface-wise multiply-connected region $V_{a_1} := B(0, a_1) \setminus \overline{V}_1$ is filled with a possibly inhomogeneous material endowed with finite conductivity $\sigma(\mathbf{r})$, however small. In the limit as $a_1 \to +\infty$, V_{a_1} passes over into $\mathbb{R}^3 \setminus \overline{V}_1$, i.e., the unbounded region of space external to V_1.

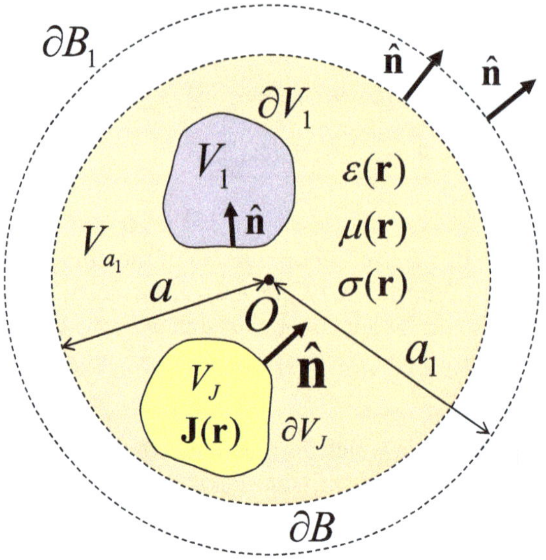

Figure 6.15 For proving uniqueness of the solutions to the time-harmonic Maxwell equations in a lossy unbounded region.

Now, if two solutions are possible, say, $\mathbf{E}_l(\mathbf{r})$, $\mathbf{H}_l(\mathbf{r})$, $l = 1, 2$, we may define the difference fields $\mathbf{E}_0(\mathbf{r}) := \mathbf{E}_1(\mathbf{r}) - \mathbf{E}_2(\mathbf{r})$ and $\mathbf{H}_0(\mathbf{r}) := \mathbf{H}_1(\mathbf{r}) - \mathbf{H}_2(\mathbf{r})$ which satisfy the source-free Maxwell equations for points $\mathbf{r} \in V_{a_1}$. An application of the Poynting theorem (1.314) to $\mathbf{E}_0(\mathbf{r})$ and $\mathbf{H}_0(\mathbf{r})$ in V_{a_1} yields

$$\frac{1}{2}\mathrm{Re}\int_{\partial B_1} \mathrm{d}S\ \hat{\mathbf{n}}(\mathbf{r}) \cdot \mathbf{E}_0(\mathbf{r}) \times \mathbf{H}_0^*(\mathbf{r}) + \frac{1}{2}\int_{V_{a_1}} \mathrm{d}V\, \sigma(\mathbf{r})|\mathbf{E}_0(\mathbf{r})|^2$$
$$= -\frac{1}{2}\mathrm{Re}\int_{\partial V_1} \mathrm{d}S\ \hat{\mathbf{n}}(\mathbf{r}) \cdot \mathbf{E}_0(\mathbf{r}) \times \mathbf{H}_0^*(\mathbf{r}) \quad (6.157)$$

where $\partial B_1 := \{\mathbf{r} \in \mathbb{R}^3 : |\mathbf{r}| = a_1\}$ and the unit normals are oriented outwards V_{a_1}. Because of the leading minus sign the term in the right-hand side represents the average flow of power that *leaves* V_1 and enters V_{a_1}. However, since V_1 may contain at most lossy media — but no sources — if a power efflux indeed takes place, then power must *enter* V_1 and, as a result, we infer that the right member is a non-positive quantity. Then, from (6.157) we get the estimate

$$\frac{1}{2}\mathrm{Re}\int_{\partial B_1} \mathrm{d}S\ \hat{\mathbf{n}}(\mathbf{r}) \cdot \mathbf{E}_0(\mathbf{r}) \times \mathbf{H}_0^*(\mathbf{r}) + \frac{1}{2}\int_{V_{a_1}} \mathrm{d}V\, \sigma(\mathbf{r})|\mathbf{E}_0(\mathbf{r})|^2 \leqslant 0 \quad (6.158)$$

which we wish to examine in the limit as $a_1 \to +\infty$. At a cursory glance it would appear that the asymptotic conditions (6.155) and (6.156) by themselves do not ensure the convergence of the integral over V_{a_1} in (6.158) for infinitely large values of a_1, essentially because the difference electric field does not decay rapidly enough.

To show that this expectation is flawed we consider a smaller ball $B(0,a)$ with radius $a < a_1$ (Figure 6.15) and apply the Poynting theorem (1.314) to the spherical shell $B(0,a_1) \setminus B(0,a)$, namely,

$$\underbrace{\frac{1}{2}\mathrm{Re}\int_{\partial B_1} \mathrm{d}S\, \hat{\mathbf{n}}(\mathbf{r}) \cdot \mathbf{E}_0(\mathbf{r}) \times \mathbf{H}_0^*(\mathbf{r})}_{=\langle P_F(a_1)\rangle} + \int_{B(0,a_1)\setminus B(0,a)} \mathrm{d}V\, \frac{1}{2}\sigma(\mathbf{r})|\mathbf{E}_0(\mathbf{r})|^2$$
$$= \underbrace{\frac{1}{2}\mathrm{Re}\int_{\partial B} \mathrm{d}S\, \hat{\mathbf{n}}(\mathbf{r}) \cdot \mathbf{E}_0(\mathbf{r}) \times \mathbf{H}_0^*(\mathbf{r})}_{=\langle P_F(a)\rangle} \tag{6.159}$$

where the unit normals on ∂B_1 and $\partial B := \{\mathbf{r} \in \mathbb{R}^3 : |\mathbf{r}| = a\}$ point inwards $\mathbb{R}^3 \setminus B(0,a_1)$ and $B(0,a_1) \setminus B(0,a)$, respectively. With these choices, the real parts of the flux integrals represent the average outward-flowing powers across ∂B_1 and ∂B, as indicated.

Since the medium is lossy by hypothesis and $\sigma(\mathbf{r}) > 0$, (6.159) implies

$$\langle P_F(a_1)\rangle < \langle P_F(a)\rangle \qquad \text{for} \qquad a_1 > a \tag{6.160}$$

whence we determine that the power flow is a strictly decreasing function of the radial distance $r > a_1 > a$, inasmuch as we can repeat the same reasoning for ever larger concentric balls. Besides, since there are no sources in $\mathbb{R}^3 \setminus B(0,a_1)$ — nor anywhere else for that matter — the power flow $\langle P_F(a_1)\rangle$ is non-negative and we get

$$0 \leqslant \lim_{a_1\to+\infty} \frac{1}{2}\mathrm{Re}\int_{\partial B_1} \mathrm{d}S\, \hat{\mathbf{n}}(\mathbf{r}) \cdot \mathbf{E}_0(\mathbf{r}) \times \mathbf{H}_0^*(\mathbf{r}) = \langle P_F(\infty)\rangle < \langle P_F(a)\rangle \tag{6.161}$$

since $\langle P_F(a_1)\rangle$ is dominated by $\langle P_F(a)\rangle$. We can arrive at the same conclusion also by invoking (6.155) and (6.156).

Next, we solve (6.159) formally with respect to the volume integral to obtain an estimate of the average power that is lost within $B(0,a_1) \setminus B(0,a)$, viz.,

$$\int_{B(0,a_1)\setminus B(0,a)} \mathrm{d}V\, \frac{1}{2}\sigma(\mathbf{r})|\mathbf{E}_0(\mathbf{r})|^2 = \langle P_F(a)\rangle - \langle P_F(a_1)\rangle < \langle P_F(a)\rangle = \frac{1}{2}\int_{\partial B} \mathrm{d}S\, \mathrm{Re}\{\hat{\mathbf{r}} \cdot \mathbf{E}_0(\mathbf{r}) \times \mathbf{H}_0^*(\mathbf{r})\}$$
$$\leqslant \frac{1}{2}\int_{\partial B} \mathrm{d}S\, |\mathbf{E}_0(\mathbf{r})||\mathbf{H}_0^*(\mathbf{r})| \leqslant \frac{1}{2}\int_{\partial B} \mathrm{d}S\, \frac{C_E C_H}{a^2} = 2\pi C_E C_H \tag{6.162}$$

where we have used the Cauchy-Schwarz inequality (D.151), and assumed that a is large enough that, given two suitable positive constants C_E, C_H, we have

$$|\mathbf{E}_0| \leqslant \frac{C_E}{a}, \qquad |\mathbf{H}_0| \leqslant \frac{C_H}{a} \tag{6.163}$$

on the grounds of (6.155) and (6.156). Thus, in the limit as $a_1 \to +\infty$ we find

$$\lim_{a_1\to+\infty} \int_{B(0,a_1)\setminus B(0,a)} \mathrm{d}V\, \frac{1}{2}\sigma(\mathbf{r})|\mathbf{E}_0(\mathbf{r})|^2 < 2\pi C_E C_H \tag{6.164}$$

because the power dissipated in the shell is bounded by a positive constant independent of a_1. Thanks to this intermediate result we can conclude that the integral over V_{a_1} in (6.158) remains *finite* in the limit as $a_1 \to +\infty$. We just have to split the integration into two parts, i.e., over the shell $B(0, a_1) \setminus B(0, a)$ and the domain $B(0, a) \setminus \overline{V}_1$.

Finally, from (6.158), (6.161) and (6.164) it follows that

$$\langle P_F(\infty)\rangle + \frac{1}{2}\int_{\mathbb{R}^3\setminus\overline{V}_1} \mathrm{d}V\,\sigma(\mathbf{r})|\mathbf{E}_0(\mathbf{r})|^2 \leqslant 0 \tag{6.165}$$

and this condition necessarily implies

$$\langle P_F(\infty)\rangle = 0 = \frac{1}{2}\int_{\mathbb{R}^3\setminus\overline{V}_1} \mathrm{d}V\,\sigma(\mathbf{r})|\mathbf{E}_0(\mathbf{r})|^2 \tag{6.166}$$

inasmuch as both $\langle P_F(\infty)\rangle$ and the average dissipated power are non-negative quantities. Therefore, we conclude that $\mathbf{E}_0(\mathbf{r}) = 0$ for $\mathbf{r} \in \mathbb{R}^3 \setminus \overline{V}_1$; that $\mathbf{H}_0(\mathbf{r})$ vanishes as well follows from the Faraday law (1.99). Finally, since $\mathbf{E}_1(\mathbf{r}) = \mathbf{E}_2(\mathbf{r})$ and $\mathbf{H}_1(\mathbf{r}) = \mathbf{H}_2(\mathbf{r})$ contrary to our working hypothesis, the solution is unique for $\mathbf{r} \in \mathbb{R}^3 \setminus \overline{V}_1$.

D. S. Jones provides an alternative proof [37, p. 566] based on the radiation conditions which we shall discuss further on below. Yet another approach frequently used to argue that the dissipated power remains finite and $\langle P_F(a_1)\rangle \to 0$ in the limit as $a_1 \to +\infty$ consists of invoking the notion that the solutions to Maxwell's equations in the frequency domain behave as *spherical waves* very far away from localized sources (Section 9.6) [18, 74]. The presence of losses in the background medium then causes said waves to be *exponentially damped* as $a_1 \to +\infty$. Clearly, this *additional* piece of information is more than sufficient to conclude that in (6.158) the flux integral vanishes at infinity and the dissipated power is finite. Still, to be fastidious, one should preliminarily show that all time-harmonic solutions to the Maxwell equations behave asymptotically as spherical waves. This statement can, in fact, be proved with the aid of the integral representations of the fields that we shall derive in Section 10.2.

One might object that the assumption of losses, albeit tiny, everywhere in space is unrealistic, though we may introduce small losses as a mathematical expedient for determining a solution and then take the limit as $\sigma(\mathbf{r}) \to 0$ (cf. [37, Section 9.8]). Regardless, unlike the case of a bounded region discussed in Section 6.4.1, when the background medium is lossless, the solution can be made unique by imposing suitable boundary conditions on the fields at infinity. This expedient works essentially because the source-driven Maxwell equations in an infinite domain admit *at most two* independent solutions. To see that this is true, for simplicity we take a look at the wave equations (1.238) and (1.239) for sources in a homogeneous unbounded medium, viz.,

$$(\nabla^2 + k^2)\mathbf{E}(\mathbf{r}) = \frac{1}{\varepsilon}\nabla\rho(\mathbf{r}) + \mathrm{j}\,\omega\mu\mathbf{J}(\mathbf{r}) \tag{6.167}$$

$$(\nabla^2 + k^2)\mathbf{H}(\mathbf{r}) = -\nabla \times \mathbf{J}(\mathbf{r}) \tag{6.168}$$

with the *real* wavenumber k shown explicitly [see (1.248)]. Evidently, these equations are unaffected by the substitution $k \leftrightarrow -k$, since the Helmholtz operator in the left-hand side depends on the square of the wavenumber. The practical consequence is that, if a solution has been found for a given value of k, replacing k with $-k$ in the formula provides the other one. These two possibilities represent wave-like solutions, granted, but how do we choose the 'right' one?

Well, it is an experimental fact that accelerated charges (and hence, time-varying currents) generate electromagnetic waves which *emanate* from the source rather than *enter* it (see Section 1.9). As nicely put by A. Sommerfeld [75, p. 189], sources must not be 'sinks' of electromagnetic energy.

Thus, between the only two possible independent solutions we need to pick up the one that constitutes an *outgoing* wave, i.e., a wave leaving the source. The other one is called an *ingoing* or incoming wave, and travels from infinity towards the source, an occurrence which is physically meaningless. The outgoing wave is associated with a positive wavenumber, say, $k > 0$, whereas the ingoing wave is characterized by a negative wavenumber $-k$. By requiring that the solution be an outgoing wave we achieve uniqueness of solutions in unbounded loss-free regions of space. We wish to specify this choice in a rigorous manner and prove that Maxwell's equations supplemented with this criterion admit only one solution.

As it turns out, the constraints (6.155) and (6.156) alone are not sufficient to discriminate between ingoing and outgoing waves. In fact, the necessary additional boundary conditions at infinity are *postulated* in the form

$$\lim_{r\to+\infty} r[\mathbf{E}(\mathbf{r}) - Z\mathbf{H}(\mathbf{r}) \times \hat{\mathbf{r}}] = 0 \tag{6.169}$$

$$\lim_{r\to+\infty} r[\hat{\mathbf{r}} \times \mathbf{E}(\mathbf{r}) - Z\mathbf{H}(\mathbf{r})] = 0 \tag{6.170}$$

or alternatively as

$$\mathbf{E}(\mathbf{r}) - Z\mathbf{H}(\mathbf{r}) \times \hat{\mathbf{r}} = \mathcal{O}\left(\frac{1}{|\mathbf{r}|^2}\right), \qquad |\mathbf{r}| \to +\infty \tag{6.171}$$

$$\hat{\mathbf{r}} \times \mathbf{E}(\mathbf{r}) - Z\mathbf{H}(\mathbf{r}) = \mathcal{O}\left(\frac{1}{|\mathbf{r}|^2}\right), \qquad |\mathbf{r}| \to +\infty \tag{6.172}$$

uniformly with respect to $\mathbf{r}$. They are referred to as the *Silver-Müller radiation conditions* (1949) [37, Section 9.1], [41, p. 13], [29]. We shall see in Section 10.2 that (6.171) and (6.172) are justified by (though not derived from) the integral representations of Stratton and Chu. The scalar Z is the intrinsic impedance of the background medium as defined in (1.358). The analogous constraints for scalar fields and the scalar Helmholtz equation (Section 8.5.4) are called the *Sommerfeld radiation conditions* [39,76], though this name is frequently associated also with (6.171) and (6.172).

In words, the Silver-Müller conditions require that the dominant contributions of $\mathbf{E}(\mathbf{r})$ and $\mathbf{H}(\mathbf{r})$ — which do obey (6.155) and (6.156) — cancel each other so as to leave a leading term which falls off as the inverse square of the distance from the sources; therefore, (6.171) and (6.172) imply (6.155) and (6.156). Besides, dot-multiplying (6.171) and (6.172) with $\hat{\mathbf{r}}$ yields

$$\hat{\mathbf{r}} \cdot \mathbf{E}(\mathbf{r}) = \mathcal{O}\left(\frac{1}{|\mathbf{r}|^2}\right), \qquad |\mathbf{r}| \to +\infty \tag{6.173}$$

$$\hat{\mathbf{r}} \cdot \mathbf{H}(\mathbf{r}) = \mathcal{O}\left(\frac{1}{|\mathbf{r}|^2}\right), \qquad |\mathbf{r}| \to +\infty \tag{6.174}$$

which, in combination with (6.155) and (6.156), mean that the radial components of the fields fall off more rapidly than the components perpendicular to $\hat{\mathbf{r}}$ (see Section 9.6).

We now proceed to show that the difference fields $\mathbf{E}_0(\mathbf{r})$ and $\mathbf{H}_0(\mathbf{r})$ in the region $\mathbb{R}^3 \setminus V_1$ are null. First of all from the Poynting theorem (1.314) applied to the source-free domain $V_{a_1} := B(0, a_1) \setminus V_1$ we derive

$$\frac{1}{2}\mathrm{Re} \int_{\partial B_1} dS\, \hat{\mathbf{r}} \cdot \mathbf{E}_0(\mathbf{r}) \times \mathbf{H}_0^*(\mathbf{r}) = -\frac{1}{2}\mathrm{Re} \int_{\partial V_1} dS\, \hat{\mathbf{n}}(\mathbf{r}) \cdot \mathbf{E}_0(\mathbf{r}) \times \mathbf{H}_0^*(\mathbf{r}) = 0 \tag{6.175}$$

thanks to (6.74) and (6.75), as the tangential components of $\mathbf{E}_l(\mathbf{r})$ and $\mathbf{H}_l(\mathbf{r})$ are assigned on ∂V_1. Next, we consider the following algebraic identity for $\mathbf{r} \in \mathbb{R}^3 \setminus V_1$

$$|\mathbf{E}_0(\mathbf{r}) - Z\mathbf{H}_0(\mathbf{r}) \times \hat{\mathbf{r}}|^2 + |\hat{\mathbf{r}} \times \mathbf{E}_0(\mathbf{r}) - Z\mathbf{H}_0(\mathbf{r})|^2 + |\hat{\mathbf{r}} \cdot \mathbf{E}_0(\mathbf{r})|^2 + Z^2|\hat{\mathbf{r}} \cdot \mathbf{H}_0(\mathbf{r})|^2$$

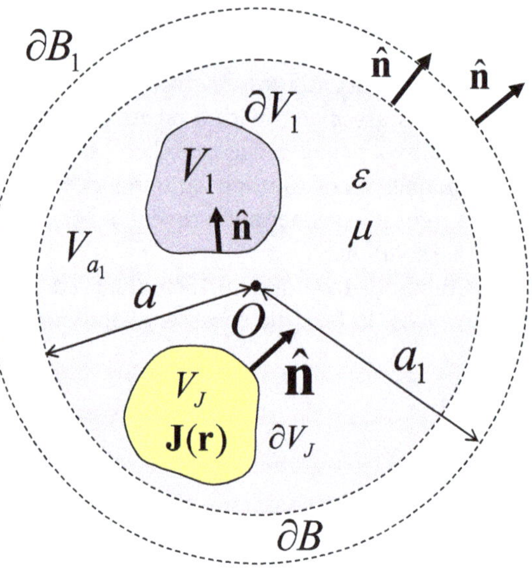

Figure 6.16 For proving uniqueness of solutions to the time-harmonic Maxwell equations in a lossless unbounded region.

$$= 2\left[|\mathbf{E}_0(\mathbf{r})|^2 + Z^2|\mathbf{H}_0(\mathbf{r})|^2\right] - 4Z\,\mathrm{Re}\{\hat{\mathbf{r}}\cdot\mathbf{E}_0(\mathbf{r})\times\mathbf{H}_0^*(\mathbf{r})\} \quad (6.176)$$

and by integrating over ∂B_1 we obtain the estimate

$$\int_{\partial B_1} \mathrm{d}S\,\left[|\mathbf{E}_0(\mathbf{r})|^2 + Z^2|\mathbf{H}_0(\mathbf{r})|^2\right] =$$
$$= \frac{1}{2}\int_{\partial B_1} \mathrm{d}S\,|\mathbf{E}_0(\mathbf{r}) - Z\mathbf{H}_0(\mathbf{r})\times\hat{\mathbf{r}}|^2 + \frac{1}{2}\int_{\partial B_1} \mathrm{d}S\,|\hat{\mathbf{r}}\times\mathbf{E}_0(\mathbf{r}) - Z\mathbf{H}_0(\mathbf{r})|^2$$
$$+ \frac{1}{2}\int_{\partial B_1} \mathrm{d}S\,\left[|\hat{\mathbf{r}}\cdot\mathbf{E}_0(\mathbf{r})|^2 + Z^2|\hat{\mathbf{r}}\cdot\mathbf{H}_0(\mathbf{r})|^2\right] \leqslant 2\pi\frac{C}{a_1^2} \quad (6.177)$$

in light of (6.175) and the Silver-Müller conditions, which apply to the difference fields as well. The positive constant C is chosen so that (6.171)-(6.174) hold true simultaneously for $r = a_1$.

We introduce the ball $B(0, a)$ with radius $a < a_1$ but still large enough to contain V_1 (Figure 6.16) and consider the volume integral

$$\int_{B(0,a_1)\setminus B(0,a)} \mathrm{d}V\,\left[|\mathbf{E}_0(\mathbf{r})|^2 + Z^2|\mathbf{H}_0(\mathbf{r})|^2\right] = \int_a^{a_1} \mathrm{d}r \int_{S(r)} \mathrm{d}S\,\left[|\mathbf{E}_0(\mathbf{r})|^2 + Z^2|\mathbf{H}_0(\mathbf{r})|^2\right] \leqslant \int_a^{a_1} \mathrm{d}r\,\frac{2\pi C}{r^2}$$
$$= 2\pi C\left(\frac{1}{a} - \frac{1}{a_1}\right) \leqslant 2\pi\frac{C}{a} \xrightarrow[a\to+\infty]{} 0 \quad (6.178)$$

where $S(r)$ is the sphere having radius $r \in [a, a_1]$ and center in the origin, and we have applied (6.177).

Finally, on account of (1.238) and (1.239) we observe that each Cartesian component of $\mathbf{E}_0(\mathbf{r})$ and $\mathbf{H}_0(\mathbf{r})$ satisfies a source-free scalar Helmholtz equation and meets the hypotheses of the Rellich theorem (which we shall prove in Section 8.5.2) [37, Section 9.1], [25, 77]. Accordingly, a suitable positive constant M exists such that

$$\int_{B(0,a_1)\setminus B(0,a)} \mathrm{d}V \left[|\mathbf{E}_0(\mathbf{r})|^2 + Z^2|\mathbf{H}_0(\mathbf{r})|^2\right] \geqslant Ma_1 > Ma \tag{6.179}$$

for sufficiently large a_1. This finding, though, is in contrast with (6.178), and we are led to conclude that the Cartesian components of $\mathbf{E}_0(\mathbf{r})$ and $\mathbf{H}_0(\mathbf{r})$ cannot be non-trivial solutions to the homogeneous scalar Helmholtz equation. Then, it must necessarily hold $\mathbf{E}_0(\mathbf{r}) = 0 = \mathbf{H}_0(\mathbf{r})$ for $\mathbf{r} \in \mathbb{R}^3 \setminus V_1$, whereby uniqueness follows. Parenthetically, we mention that the Silver-Müller conditions are overly demanding and that, as was proved by C. H. Wilcox [78], the milder requirements

$$\lim_{a_1\to+\infty} \int_{\partial B_1} \mathrm{d}S\ |\mathbf{E}_0(\mathbf{r}) - Z\mathbf{H}_0(\mathbf{r}) \times \hat{\mathbf{r}}|^2 = 0 \tag{6.180}$$

$$\lim_{a_1\to+\infty} \int_{\partial B_1} \mathrm{d}S\ |\hat{\mathbf{r}} \times \mathbf{E}_0(\mathbf{r}) - Z\mathbf{H}_0(\mathbf{r})|^2 = 0 \tag{6.181}$$

are sufficient to ensure uniqueness as well.

A key ingredient of the proof is constituted by (6.175) which in turn is based on the boundary conditions on ∂V_1. If the medium filling V_1 is a PEC, then $\mathbf{E}(\mathbf{r}) \times \hat{\mathbf{n}}(\mathbf{r}) = 0$ holds for $\mathbf{r} \in \partial V_1$. If the medium is penetrable with constitutive parameters $\varepsilon_1(\mathbf{r})$ and $\mu_1(\mathbf{r})$, then we invoke the jump conditions (1.142) and (1.144) with $\mathbf{J}_S(\mathbf{r}) = 0$ and apply the Poynting theorem within V_1 to show that

$$\int_{\partial V_1} \mathrm{d}S\ \hat{\mathbf{n}}(\mathbf{r}) \cdot \mathbf{E}(\mathbf{r}) \times \mathbf{H}^*(\mathbf{r}) = 0 \tag{6.182}$$

on either side of ∂V_1, whence (6.175) still holds. Notice that this conclusion is independent on the actual value of the constitutive parameters and hence, if we choose $\varepsilon_1(\mathbf{r}) = \varepsilon$ and $\mu_1(\mathbf{r}) = \mu$, we have the starting point for the extension of the proof to the whole space.

6.5 Magnetic charges and currents

We began Chapter 1 by reeling off the fundamental entities of the electromagnetic field and the sources thereof. Up to now we have shown that static electric charges produce electric fields, electric charges in uniform motion produce electric and magnetic fields, and finally electric charges in arbitrary (accelerated) motion may even generate electromagnetic waves. Furthermore, in Section 5.6 we showed that the magnetized state of a material body is actually related to unbalanced microscopic electronic currents which can be conveniently pictured and described as *equivalent* magnetic dipoles. However, unlike electric dipoles, which can indeed be made up of two electric charges, magnetic dipoles cannot be similarly realized by pairing two *magnetic charges*, for the simple reason that so far isolated magnetic monopoles have not been found yet. If they exist, then magnetic monopoles are certainly elusive and not as abundant in nature as electric charges.

To make things even worse, it would appear that magnetic monopoles have no use or place in the theory, since all the electromagnetic phenomena so nicely predicted by the Maxwell equations are accounted for by electric sources only. And yet, P. Dirac put forward quite a compelling argument in favor of the existence of magnetic charges [79]. He reasoned and showed that, in a quantum-mechanical description of the electromagnetic interactions, the occurrence of just one magnetic monopole with charge q_M would explain why the electric charge in turn is *quantized* (Example 6.3). This is as good a time as any to recall that the notion of electric charge 'continuously smeared' in space and described by a density function $\varrho(\mathbf{r}, t)$ is an abstraction, and that charge is indeed *discrete* (Figure 1.1). In 1976 it was also speculated that electric charges moving *faster* than light would behave as magnetic monopoles [80].

Even though the search for magnetic monopoles is still open, meanwhile we may introduce magnetic charges and currents in the theory by invoking simple arguments of symmetry — or lack thereof — in the classic Maxwell equations. In fact, it is precisely the absence of magnetic charges, as it were, that separates the Faraday law (1.20) and the magnetic Gauss law (1.23) on the one hand from the Ampère-Maxwell law (1.34) and the Gauss law (1.44) on the other, as was suggested in Figure 1.5. We may thus formally complement the list of sources of electromagnetic fields with the following two:

- $\varrho_M(\mathbf{r}, t)$: impressed magnetic charge density, which carries physical dimension of weber per cubic meter (Wb/m^3)
- $\mathcal{J}_M(\mathbf{r}, t)$: impressed volume magnetic current density with physical dimension of volt per square meter (V/m^2)

and for reasons of symmetry and dimensional consistency they must be related to the entities of intensity $\mathcal{E}$ and $\mathcal{B}$. Hence, the extended set of Maxwell's equations (sometimes called the *Dirac symmetrized Maxwell equations* [81, Section 1.4]) in local form reads

$$\nabla \times \mathcal{E}(\mathbf{r}, t) = -\frac{\partial}{\partial t}\mathcal{B}(\mathbf{r}, t) - \mathcal{J}_M(\mathbf{r}, t) \tag{6.183}$$

$$\nabla \times \mathcal{H}(\mathbf{r}, t) = \frac{\partial}{\partial t}\mathcal{D}(\mathbf{r}, t) + \mathcal{J}(\mathbf{r}, t) \tag{6.184}$$

$$\nabla \cdot \mathcal{D}(\mathbf{r}, t) = \varrho(\mathbf{r}, t) \tag{6.185}$$

$$\nabla \cdot \mathcal{B}(\mathbf{r}, t) = \varrho_M(\mathbf{r}, t) \tag{6.186}$$

for any $(\mathbf{r}, t) \in \mathbb{R}^3 \times \mathbb{R}$. Moreover, taking the divergence of (6.183) and combining the resulting equation with (6.186) yields the continuity equation (or conservation law) for magnetic charges

$$\nabla \cdot \mathcal{J}_M(\mathbf{r}, t) + \frac{\partial}{\partial t}\varrho_M(\mathbf{r}, t) = 0 \tag{6.187}$$

having swapped time and space derivative with the proviso that $\mathcal{B}(\mathbf{r}, t)$ is twice differentiable with respect to $\mathbf{r}$ and t. Perfectly analogous equations hold true for time-harmonic fields, and stationary magnetic currents may be envisioned as well.

With these additions, we may speculate that the magnetic lines of force may begin and end in points occupied by magnetic charges with opposite sign and that, conversely, the electric lines of force may form closed loops around $\mathcal{J}_M$.

Although magnetic charges and currents are an abstraction, they play an important role as *equivalent sources* in all the integral representations of potentials and fields. For instance, by comparing like contributions in the surface integrals in (3.21) and (5.58) we are prompted to interpret $\hat{\mathbf{n}}(\mathbf{r}) \cdot \mathcal{B}(\mathbf{r})$ as a layer of equivalent magnetic charges on $S \cup \partial V_1$, and $\mathcal{E}(\mathbf{r}) \times \hat{\mathbf{n}}(\mathbf{r})$ as an equivalent surface density of steady magnetic current flowing on $S \cup \partial V_1$. In Chapter 10 we shall extend these definitions to time-harmonic fields.

Example 6.3 (Momenta of a magnetic charge in the presence of an electric charge)

The electric (resp. magnetic) field produced by a stationary electric (resp. magnetic) charge in free space read

$$\mathcal{E}(\mathbf{r}') = \frac{q\hat{\mathbf{r}}'}{4\pi\varepsilon_0 r'^2}, \qquad \mathcal{H}(\mathbf{r}') = \frac{q_M}{4\pi\mu_0}\frac{\mathbf{r}' - \mathbf{r}}{|\mathbf{r}' - \mathbf{r}|^3}, \qquad \mathbf{r}' \in \mathbb{R}^3 \setminus \{0, \mathbf{r}\} \tag{6.188}$$

where the magnetic field is obtained by applying the duality principle (to be discussed in Section 6.7) to formula (2.14). A pair of electric and magnetic charges is known as a *Thomson's dipole*. Although linear and angular electromagnetic momentum are naturally associated with motion, still they can be computed for the system comprised of the stationary charges q and q_M, since the densities $\mathbf{g}_{em}(\mathbf{r}')$ and $\mathbf{l}_{em}(\mathbf{r}')$ are defined in terms of $\mathcal{E}(\mathbf{r}')$ and $\mathcal{H}(\mathbf{r}')$. From (1.332) and (1.380) we get

$$\mathbf{g}_{em}(\mathbf{r}') = \frac{qq_M}{(4\pi)^2}\frac{\mathbf{r}' \times (\mathbf{r}' - \mathbf{r})}{r'^2|\mathbf{r}' - \mathbf{r}|^3} = \frac{qq_M}{(4\pi)^2}\nabla'\frac{1}{r'} \times \nabla'\frac{1}{|\mathbf{r}' - \mathbf{r}|} = \frac{qq_M}{(4\pi)^2}\nabla' \times \left(\frac{1}{r'}\nabla'\frac{1}{R}\right) \tag{6.189}$$

and

$$\begin{aligned}
\mathbf{l}_{em}(\mathbf{r}') &= \mathbf{r}' \times \left[\frac{q\mu_0}{4\pi}\frac{\hat{\mathbf{r}}'}{r'^2} \times \mathcal{H}(\mathbf{r}')\right] = \frac{q}{4\pi r'}\hat{\mathbf{r}}' \times [\hat{\mathbf{r}}' \times \mathcal{B}(\mathbf{r}')] \\
&= \frac{q}{4\pi r'}\mathcal{B}(\mathbf{r}') \cdot (\hat{\mathbf{r}}'\hat{\mathbf{r}}' - \overline{\mathcal{I}}) = -\frac{q}{4\pi}\mathcal{B}(\mathbf{r}') \cdot \nabla'\hat{\mathbf{r}}' \\
&= -\frac{q}{4\pi}\nabla' \cdot [\mathcal{B}(\mathbf{r}')\hat{\mathbf{r}}'] = \frac{qq_M}{(4\pi)^2}\nabla' \cdot \left(\frac{\hat{\mathbf{R}}\hat{\mathbf{r}}'}{R^2}\right), \qquad \mathbf{r}' \notin \{0, \mathbf{r}\}
\end{aligned} \tag{6.190}$$

on account of identities (H.14), (H.50) and (H.71). We notice that the density of linear momentum, being in the form of a curl of some vector field, is *solenoidal* [cf. (A.39)]. The density of angular momentum can be written as the divergence of a dyadic field only for $\mathbf{r}' \neq \mathbf{r}$, that is, for points where the magnetic induction is solenoidal.

To compute the total momenta $\mathcal{G}_{em}$ and $\mathcal{L}_{em}$ we need to integrate the densities just found over all points in space. Since $\mathbf{g}_{em}(\mathbf{r}')$ and $\mathbf{l}_{em}(\mathbf{r}')$ are singular at the location of the charges and the integration domain is unbounded, we exclude the points $\mathbf{r}' = 0$ and $\mathbf{r}' = \mathbf{r}$ with two balls $B_1(0, a)$ and $B_2(\mathbf{r}, b)$, with $a + b < r$, and integrate over the surface-wise multiply connected region $V := B(\mathbf{r}, d) \setminus (B_1(0, a) \cup B_2(\mathbf{r}, b))$, with $d > r + a$ (Figure 6.17). In the limit as $a \to 0^+$, $b \to 0^+$ and $d \to +\infty$, V passes over into $\mathbb{R}^3$. To proceed we first derive an estimate of $1/R$, namely,

$$|r - r'| \leqslant |\mathbf{r} - \mathbf{r}'| = R \qquad \Longrightarrow \qquad \frac{1}{R} \leqslant \frac{1}{|r - r'|} \tag{6.191}$$

on the grounds of inequality (H.20).

For the linear momentum we have

$$\begin{aligned}
\mathcal{G}_{em} &= \lim_{\substack{a,b\to 0^+ \\ d\to+\infty}} \int_V \mathrm{d}V'\, \frac{qq_M}{(4\pi)^2}\nabla' \times \left(\frac{1}{r'}\nabla'\frac{1}{R}\right) = \frac{qq_M}{(4\pi)^2} \lim_{\substack{a,b\to 0^+ \\ d\to+\infty}} \int_{\partial V} \mathrm{d}S'\, \hat{\mathbf{n}}(\mathbf{r}') \times \frac{1}{r'}\nabla'\frac{1}{R} \\
&= \frac{qq_M}{(4\pi)^2} \lim_{a\to 0^+} \int_{\partial B_1} \mathrm{d}S'\, \hat{\mathbf{n}}(\mathbf{r}') \times \frac{1}{r'}\nabla'\frac{1}{R} + \frac{qq_M}{(4\pi)^2} \lim_{b\to 0^+} \int_{\partial B_2} \mathrm{d}S'\, \hat{\mathbf{n}}(\mathbf{r}') \times \frac{1}{r'}\nabla'\frac{1}{R} \\
&\quad + \frac{qq_M}{(4\pi)^2} \lim_{d\to+\infty} \int_{\partial B} \mathrm{d}S'\, \hat{\mathbf{n}}(\mathbf{r}') \times \frac{1}{r'}\nabla'\frac{1}{R}
\end{aligned} \tag{6.192}$$

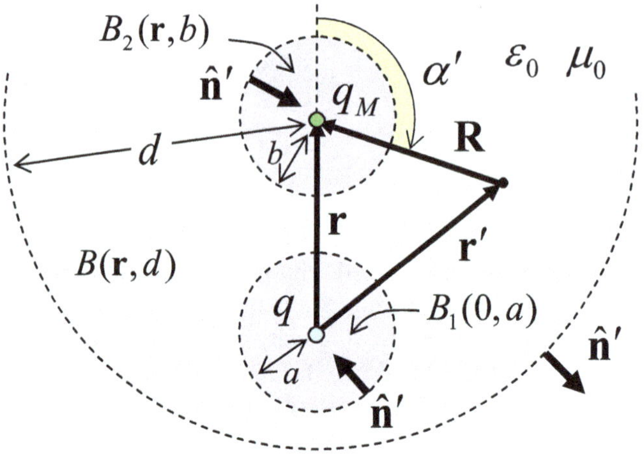

Figure 6.17 Geometry for the calculation of the linear and angular momentum due to a pair of electric and magnetic charges in free space. The large ball $B(\mathbf{r}, d)$ has been outlined (−−) only in part.

where we have invoked the curl theorem (H.91) with the unit normal pointing outward V, since the relevant vector field is regular in V. The three surface integrals are either estimated or evaluated as follows

$$\left|\int_{\partial B_1} \mathrm{d}S'\, \hat{\mathbf{n}}(\mathbf{r}') \times \frac{1}{r'}\nabla'\frac{1}{R}\right| \leqslant \int_{\partial B_1} \mathrm{d}S'\, \frac{1}{r'R^2} \leqslant \int_{\partial B_1} \mathrm{d}S'\, \frac{1}{a(r-a)^2} = \frac{4\pi a}{(r-a)^2} \xrightarrow[a\to 0^+]{} 0 \tag{6.193}$$

$$\int_{\partial B_2} \mathrm{d}S'\, \hat{\mathbf{n}}(\mathbf{r}') \times \frac{1}{r'}\nabla'\frac{1}{R} = \int_{\partial B_2} \mathrm{d}S'\, \hat{\mathbf{R}} \times \frac{1}{r'}\frac{\hat{\mathbf{R}}}{b^2} = 0 \tag{6.194}$$

$$\int_{\partial B} \mathrm{d}S'\, \hat{\mathbf{n}}(\mathbf{r}') \times \frac{1}{r'}\nabla'\frac{1}{R} = -\int_{\partial B} \mathrm{d}S'\, \hat{\mathbf{R}} \times \frac{1}{r'}\frac{\hat{\mathbf{R}}}{d^2} = 0 \tag{6.195}$$

where we have used (6.191) with $r' = a < r$. We conclude that the total momentum $\mathcal{G}_{em}$ vanishes. While this may seem a foregone result, given that the charges are fixed in space, surprisingly the total angular momentum is *not* null.

To prove this assertion we compute

$$\begin{aligned}\mathcal{L}_{em} &= \lim_{\substack{a,b\to 0^+\\ d\to+\infty}} \int_V \mathrm{d}V'\, \frac{qq_M}{(4\pi)^2}\nabla'\cdot\left(\frac{\hat{\mathbf{R}}\hat{\mathbf{r}}'}{R^2}\right) = \frac{qq_M}{(4\pi)^2} \lim_{\substack{a,b\to 0^+\\ d\to+\infty}} \int_{\partial V} \mathrm{d}S'\, \hat{\mathbf{n}}(\mathbf{r}')\cdot\frac{\hat{\mathbf{R}}\hat{\mathbf{r}}'}{R^2}\\ &= \frac{qq_M}{(4\pi)^2} \lim_{a\to 0^+} \int_{\partial B_1} \mathrm{d}S'\, \hat{\mathbf{n}}(\mathbf{r}')\cdot\frac{\hat{\mathbf{R}}\hat{\mathbf{r}}'}{R^2} + \frac{qq_M}{(4\pi)^2} \lim_{b\to 0^+} \int_{\partial B_2} \mathrm{d}S'\, \hat{\mathbf{n}}(\mathbf{r}')\cdot\frac{\hat{\mathbf{R}}\hat{\mathbf{r}}'}{R^2}\\ &\quad + \frac{qq_M}{(4\pi)^2} \lim_{d\to+\infty} \int_{\partial B} \mathrm{d}S'\, \hat{\mathbf{n}}(\mathbf{r}')\cdot\frac{\hat{\mathbf{R}}\hat{\mathbf{r}}'}{R^2}\end{aligned} \tag{6.196}$$

where use has been made of the Gauss theorem for dyadic fields (H.105) with the unit normal positively oriented outwards V. The surface integral over ∂B_1 is estimated as follows

$$\left| \int_{\partial B_1} \mathrm{d}S' \, \hat{\mathbf{n}}(\mathbf{r}') \cdot \frac{\hat{\mathbf{R}}\hat{\mathbf{r}}'}{R^2} \right| \leqslant \int_{\partial B_1} \mathrm{d}S' \frac{1}{R^2} \leqslant \int_{\partial B_1} \mathrm{d}S' \frac{1}{(r-a)^2} = \frac{4\pi a^2}{(r-a)^2} \xrightarrow[a \to 0^+]{} 0 \tag{6.197}$$

again on account of (6.191). For the integrals over ∂B_2 and ∂B we introduce a system of local polar spherical coordinates (R, α', β') with center on the magnetic charge and polar axis aligned with $\hat{\mathbf{r}}$ (Figure 6.17). Then, by observing upfront that only the component of $\mathbf{R}$ parallel to $\hat{\mathbf{r}}$ contributes to the result, we have

$$\begin{aligned} \int_{\partial B_2} \mathrm{d}S' \, \hat{\mathbf{n}}(\mathbf{r}') \cdot \frac{\hat{\mathbf{R}}\hat{\mathbf{r}}'}{R^2} &= \int_{\partial B_2} \mathrm{d}S' \frac{1}{R^2} \frac{\mathbf{r}-\mathbf{R}}{|\mathbf{r}-\mathbf{R}|} = \int_{\partial B_2} \mathrm{d}S' \frac{\hat{\mathbf{r}}(r + R\cos\alpha')}{R^2 \left(r^2 + R^2 + 2Rr\cos\alpha'\right)^{1/2}} \\ &= 2\pi\hat{\mathbf{r}} \int_0^\pi \mathrm{d}\alpha' \frac{(r + b\cos\alpha')\sin\alpha'}{\left(r^2 + b^2 + 2br\cos\alpha'\right)^{1/2}} = \frac{2\pi}{b}\hat{\mathbf{r}} \int_{-b}^{b} \mathrm{d}\xi \frac{r+\xi}{(r^2 + b^2 + 2r\xi)^{1/2}} \\ &= \frac{2\pi}{b}\hat{\mathbf{r}} \left[\frac{(r+b)^2}{r} - \frac{(r-b)|r-b|}{r} - \frac{(r+b)^3}{3r^2} + \frac{|r-b|^3}{3r^2} \right] \\ &= 4\pi\hat{\mathbf{r}} \left(1 - \frac{b^2}{3r^2}\right) \xrightarrow[b \to 0^+]{} 4\pi\hat{\mathbf{r}} \end{aligned} \tag{6.198}$$

$$\begin{aligned} \int_{\partial B} \mathrm{d}S' \, \hat{\mathbf{n}}(\mathbf{r}') \cdot \frac{\hat{\mathbf{R}}\hat{\mathbf{r}}'}{R^2} &= -\int_{\partial B} \mathrm{d}S' \frac{1}{R^2} \frac{\mathbf{r}-\mathbf{R}}{|\mathbf{r}-\mathbf{R}|} = -\frac{2\pi}{d}\hat{\mathbf{r}} \int_{-d}^{d} \mathrm{d}\xi \frac{r+\xi}{(r^2 + d^2 + 2r\xi)^{1/2}} \\ &= -\frac{2\pi}{d}\hat{\mathbf{r}} \left[\frac{(r+d)^2}{r} - \frac{(r-d)|r-d|}{r} - \frac{(r+d)^3}{3r^2} + \frac{|r-d|^3}{3r^2} \right] \\ &= -\frac{8\pi}{3d}\hat{\mathbf{r}} \xrightarrow[d \to +\infty]{} 0 \end{aligned} \tag{6.199}$$

where the intermediate results in the penultimate steps hold under the assumption that $b < r$ and $d > r$, respectively.

By inserting these findings back in (6.196) we arrive at the concise expression

$$\mathcal{L}_{em} = \frac{q q_M}{4\pi} \hat{\mathbf{r}} \tag{6.200}$$

for the angular momentum stored in the field. It turns out $\mathcal{L}_{em}$ depends only on the charges, not even the distance thereof. Now, if the angular momentum of q is quantized (i.e., discrete) then we have

$$\mathcal{L}_{em} = \frac{n}{2}\hbar = \frac{q q_M}{4\pi}, \qquad n \in \mathbb{N} \setminus \{0\} \tag{6.201}$$

where $\hbar$ is the reduced Planck constant given in (5.132). From this relation it follows that necessarily q can only come in discrete amounts. This is the semi-classical quantization argument first proposed by P. Dirac [82, Problem 8, p. 131], [83, Problem 8.12, p. 362], [79].

(End of Example 6.3)

Example 6.4 (Field of a truncated coaxial cable)

A coaxial cable is a guiding structure made of two aligned conducting cylinders one of which is larger and hollow so as to accommodate the other, smaller one. A dielectric medium is employed to keep the conductors in place and to prevent them from touching one another. As a waveguide, a coaxial cable can sustain infinitely many types of electromagnetic modes (Section 11.2) [16, 18, 19, 63]. However, as a classic transmission line, a coaxial cable is operated at frequencies low enough so that only the fundamental mode — called a transverse-electric-magnetic (TEM) mode — propagates [67, Section 9.4].

In a system of cylindrical coordinates (τ, φ, z) where the z-axis coincides with the axis of the cable, the electric and magnetic fields of the TEM mode in the time-harmonic regime read (see Example 11.1 further on)

$$\mathbf{E}_{TEM}(\mathbf{r}) = \frac{V_0 \hat{\boldsymbol{\tau}}}{\tau \log(a/b)} e^{-jkz}, \qquad b \leqslant \tau \leqslant a, \qquad z \geqslant 0 \tag{6.202}$$

$$\mathbf{H}_{TEM}(\mathbf{r}) = \frac{V_0 \hat{\boldsymbol{\varphi}}}{Z\tau \log(a/b)} e^{-jkz}, \qquad b \leqslant \tau \leqslant a, \qquad z \geqslant 0 \tag{6.203}$$

where V_0 is the modal voltage in $z = 0$ but also the strength of the ideal voltage generator that feeds the cable, $k = \omega\sqrt{\varepsilon\mu}$ is the relevant wavenumber, Z given by (1.358) is the intrinsic impedance of the dielectric medium filling the region between the conductors, and a and b $(a > b)$ are the radii of the two coaxial cylinders. The TEM mode is static in character, with the electric field perpendicular to the conductors and to the axis of the cable, and the magnetic field forming circular loops around the inner conductor.

If the cable is 'truncated' in the section $z = L$, as is pictured in Figure 6.18, to a first approximation the field in the cable may be written as

$$\mathbf{E}_{TEM}(\mathbf{r}) = \frac{V_0 \hat{\boldsymbol{\tau}}}{\tau \log(a/b)} \frac{\cos[k(z-L)]}{\cos(kL)}, \qquad b \leqslant \tau \leqslant a, \qquad 0 \leqslant z \leqslant L \tag{6.204}$$

$$\mathbf{H}_{TEM}(\mathbf{r}) = \frac{V_0 \hat{\boldsymbol{\varphi}}}{jZ\tau \log(a/b)} \frac{\sin[k(z-L)]}{\cos(kL)}, \qquad b \leqslant \tau \leqslant a, \qquad 0 \leqslant z \leqslant L \tag{6.205}$$

which is tantamount to assuming the boundary condition $\hat{\mathbf{z}} \times \mathbf{H}(\mathbf{r}) = 0$ for $z = L$, so that the TEM mode is totally reflected back. As a matter of fact, the situation is a bit more complicated than that: the TEM wave impinging on the discontinuity is partly reflected back towards the generator and partly radiated in the outer background medium, say, free space, in the form of outgoing electromagnetic waves.

Since the field of the TEM mode alone cannot satisfy the boundary conditions on the cross section of the inner conductor $(\tau < b,\ z = L)$, virtually all the remaining higher-order modes are excited with such amplitudes that the resulting electric field becomes perpendicular to the cross section. Likewise, on the exterior surface of the outer conductor $(\tau = a^+, 0 \leqslant z \leqslant L)$ a surface electric current $\mathbf{J}_S(\mathbf{r})$ is induced that produces a secondary electric field in order to make the tangential component of the resulting electric field equal to zero, as it should be.

All in all, both $\mathbf{J}_S(\mathbf{r})$ and the field in the aperture S_C of the coaxial cable are responsible for the generation of outgoing electromagnetic waves, though, in practice the dominant contribution comes exactly from the electric field of the TEM mode. In particular, we may think of the waves as being generated by an *equivalent* magnetic surface current density given by

$$\mathbf{J}_{MS}(\mathbf{r}) := \mathbf{E}(\mathbf{r}) \times \hat{\mathbf{z}} \approx -\frac{V_0 \hat{\boldsymbol{\varphi}}}{\tau \log(a/b)} \frac{1}{\cos(kL)}, \qquad b \leqslant \tau \leqslant a, \qquad z = L \tag{6.206}$$

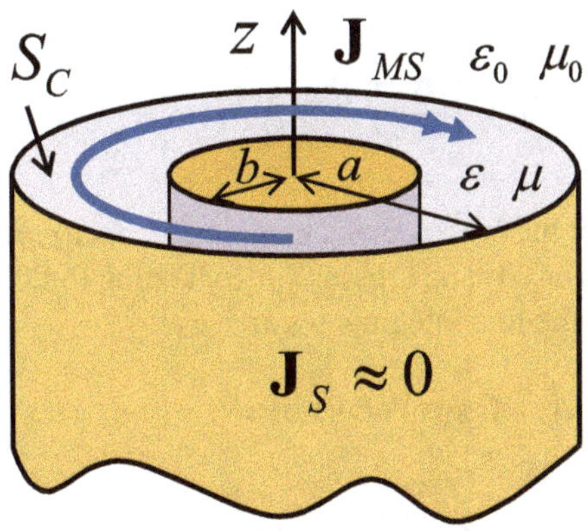

Figure 6.18 Equivalent magnetic surface current density $\mathbf{J}_{MS}(\mathbf{r})$ defined on the aperture S_C of a truncated coaxial cable immersed in free space.

which flows on the aperture S_C of the cable, as illustrated in Figure 6.18. In the limit as $\omega \to 0$ the TEM mode reduces to the electrostatic field between the conductors, $\mathbf{J}_{MS}(\mathbf{r})$ becomes a steady magnetic surface current density $\mathfrak{J}_{MS}(\mathbf{r})$, and the electrostatic field $\mathcal{E}(\mathbf{r})$ for observation points $\mathbf{r}$ outside the cable may be well approximated with the aid of the integral representation (3.21)

$$\mathcal{E}(\mathbf{r}) \approx \int_{S_C} \mathrm{d}S' \; \nabla' \frac{1}{4\pi R}\bigg|_{z'=L} \times \mathfrak{J}_{MS}(\mathbf{r}') \approx \frac{V_0}{4\pi \log(a/b)} \int_b^a \mathrm{d}\tau' \int_0^{2\pi} \mathrm{d}\varphi' \, \hat{\boldsymbol{\varphi}}' \times \frac{\mathbf{r} - \boldsymbol{\tau}' - L\hat{\mathbf{z}}}{|\mathbf{r} - \boldsymbol{\tau}' - L\hat{\mathbf{z}}|^3} \tag{6.207}$$

where we have neglected, among other contributions, the surface charge $\hat{\mathbf{z}} \cdot \mathcal{E}(\mathbf{r})\varepsilon_0$ on the cross section of the inner conductor.

(End of Example 6.4)

6.6 Boundary conditions with magnetic sources

Having extended the Maxwell equations in local form with magnetic charges and currents, it is necessary to revise the boundary conditions derived in Section 1.7 in order to account for the possible, though formal occurrence of surface densities $\varrho_{MS}(\mathbf{r})$ and $\mathfrak{J}_{MS}(\mathbf{r})$ at the interface between two material media. In particular, it is the jump relations derived from the Faraday law and the magnetic Gauss law that need updating, since magnetic sources do not affect the Ampère-Maxwell law and the electric Gauss law. If the underlying medium has piecewise continuous constitutive parameters, then (6.183)-(6.186) hold separately in each region where $\mathfrak{B}(\mathbf{r}, t)$ and $\mathfrak{D}(\mathbf{r}, t)$ are continuously differentiable. To obtain the relevant matching conditions we start with the global form of the equations.

With reference to the geometry of Figure 1.12 we notice that the global Faraday law (1.8) is certainly valid when stated over S_1, S_2 and S, namely,

$$\oint_{\partial S_1 \cup \gamma_0} \mathrm{d}s \, \hat{\mathbf{s}}_1(\mathbf{r}) \cdot \mathcal{E}_1(\mathbf{r}, t) = -\int_{S_1} \mathrm{d}S \, \hat{\boldsymbol{\nu}}_1(\mathbf{r}) \cdot \left[\frac{\partial \mathfrak{B}_1(\mathbf{r}, t)}{\partial t} + \mathfrak{J}_{M1}(\mathbf{r}, t)\right] \tag{6.208}$$

$$\oint_{\partial S_2 \cup \gamma_0} \mathrm{d}s \, \hat{\mathbf{s}}_2(\mathbf{r}) \cdot \mathcal{E}_2(\mathbf{r}, t) = -\int_{S_2} \mathrm{d}S \, \hat{\boldsymbol{\nu}}_2(\mathbf{r}) \cdot \left[\frac{\partial \mathfrak{B}_2(\mathbf{r}, t)}{\partial t} + \mathfrak{J}_{M2}(\mathbf{r}, t)\right] \tag{6.209}$$

and

$$\oint_{\partial S} \mathrm{d}s\, \hat{\mathbf{s}}(\mathbf{r}) \cdot \mathcal{E}(\mathbf{r}, t) = -\int_S \mathrm{d}S\, \hat{\boldsymbol{\nu}}(\mathbf{r}) \cdot \left[\frac{\partial \mathcal{B}(\mathbf{r}, t)}{\partial t} + \mathcal{J}_M(\mathbf{r}, t)\right] + \int_{\gamma_0} \mathrm{d}S\, \hat{\boldsymbol{\nu}}(\mathbf{r}) \cdot \mathcal{J}_{MS}(\mathbf{r}, t) \tag{6.210}$$

where we have also allowed for the presence of a magnetic *surface current density* $\mathcal{J}_{MS}(\mathbf{r}, t)$ (physical dimension: V/m) which flows on ∂V and crosses γ_0. By adding (6.208) and (6.209) side by side and performing a little algebra on the line integrals we arrive at

$$\begin{aligned}
&\oint_{\partial S_1 \cup \gamma_0} \mathrm{d}s\, \hat{\mathbf{s}}_1(\mathbf{r}) \cdot \mathcal{E}_1(\mathbf{r}, t) + \oint_{\partial S_2 \cup \gamma_0} \mathrm{d}s\, \hat{\mathbf{s}}_2(\mathbf{r}) \cdot \mathcal{E}_2(\mathbf{r}, t) = \\
&\quad = \oint_{\partial S} \mathrm{d}s\, \hat{\mathbf{s}}(\mathbf{r}) \cdot \mathcal{E}(\mathbf{r}, t) + \int_{\gamma_0} \mathrm{d}s\, \hat{\boldsymbol{\tau}}(\mathbf{r}) \cdot [\mathcal{E}_1(\mathbf{r}, t) - \mathcal{E}_2(\mathbf{r}, t)] \\
&\quad = -\int_S \mathrm{d}S\, \hat{\boldsymbol{\nu}}(\mathbf{r}) \cdot \left[\frac{\partial \mathcal{B}(\mathbf{r}, t)}{\partial t} + \mathcal{J}_M(\mathbf{r}, t)\right]
\end{aligned} \tag{6.211}$$

whence we obtain

$$\begin{aligned}
\oint_{\partial S} \mathrm{d}s\, \hat{\mathbf{s}}(\mathbf{r}) \cdot \mathcal{E}(\mathbf{r}, t) = -\int_S \mathrm{d}S\, \hat{\boldsymbol{\nu}}(\mathbf{r}) \cdot \left[\frac{\partial \mathcal{B}(\mathbf{r}, t)}{\partial t} + \mathcal{J}_M(\mathbf{r}, t)\right] \\
- \int_{\gamma_0} \mathrm{d}s\, \hat{\boldsymbol{\tau}}(\mathbf{r}) \cdot [\mathcal{E}_1(\mathbf{r}, t) - \mathcal{E}_2(\mathbf{r}, t)]
\end{aligned} \tag{6.212}$$

which should be identical with (6.210) for consistency. Evidently, the two expressions coincide if we *postulate* the condition

$$\int_{\gamma_0} \mathrm{d}s\, \{\hat{\boldsymbol{\nu}}(\mathbf{r}) \cdot \mathcal{J}_{MS}(\mathbf{r}, t) - \hat{\boldsymbol{\tau}}(\mathbf{r}) \cdot [\mathcal{E}_1(\mathbf{r}, t) - \mathcal{E}_2(\mathbf{r}, t)]\} = 0 \tag{6.213}$$

and, if we assume that the scalar fields $\hat{\boldsymbol{\nu}}(\mathbf{r}) \cdot \mathcal{J}_{MS}(\mathbf{r}, t)$ and $\hat{\boldsymbol{\tau}}(\mathbf{r}) \cdot \mathcal{E}_l(\mathbf{r}, t)$, $l = 1, 2$, are continuous functions of $\mathbf{r} \in \gamma_0$, the mean value theorem [84] applied to the last line integral provides

$$\hat{\boldsymbol{\tau}}(\mathbf{r}_0) \cdot [\mathcal{E}_1(\mathbf{r}_0, t) - \mathcal{E}_2(\mathbf{r}_0, t)] - \hat{\boldsymbol{\nu}}(\mathbf{r}_0) \cdot \mathcal{J}_{MS}(\mathbf{r}_0, t) = 0 \tag{6.214}$$

where $\mathbf{r}_0 \in \gamma_0$ is a suitable point. Since the surface S is arbitrary and can be chosen so as to intersect ∂V along any possible curve $\gamma_0 \subset \partial V$, on account of (1.133) it follows

$$[\mathcal{E}_1(\mathbf{r}, t) - \mathcal{E}_2(\mathbf{r}, t)] \times \hat{\mathbf{n}}(\mathbf{r}) = \mathcal{J}_{MS}(\mathbf{r}, t), \qquad \mathbf{r} \in \partial V \tag{6.215}$$

i.e., the (rotated) tangential component of the electric field is either continuous across the material interface ∂V — if $\mathcal{J}_{MS}(\mathbf{r}, t) = 0$ — or suffers a jump equal to the magnetic surface current density on ∂V; this relation extends (1.144). From (6.208) and (6.209) we recover the local form of the Faraday law in V_1 and V_2.

As regards the magnetic Gauss law we consider the geometry of Figure 1.13 and observe that (1.10) can be stated separately in W_1, W_2 and W, viz.,

$$\int_{\partial W_1 \cup S_0} \mathrm{d}S\, \hat{\boldsymbol{\nu}}_1(\mathbf{r}) \cdot \mathcal{B}_1(\mathbf{r}, t) = \int_{W_1} \mathrm{d}V\, \varrho_{M1}(\mathbf{r}, t) \tag{6.216}$$

$$\int_{\partial W_2 \cup S_0} \mathrm{d}S\, \hat{\boldsymbol{\nu}}_2(\mathbf{r}) \cdot \mathcal{B}_2(\mathbf{r},t) = \int_{W_2} \mathrm{d}V\, \varrho_{M2}(\mathbf{r},t) \tag{6.217}$$

$$\int_{\partial W} \mathrm{d}S\, \hat{\boldsymbol{\nu}}(\mathbf{r}) \cdot \mathcal{B}(\mathbf{r},t) = \int_{W} \mathrm{d}V\, \varrho_{M}(\mathbf{r},t) + \int_{S_0} \mathrm{d}S\, \varrho_{MS}(\mathbf{r},t) \tag{6.218}$$

where we have also allowed for the presence of an impressed magnetic surface charge density $\varrho_{MS}(\mathbf{r},t)$ (Wb/m^2) on S_0. Summing (6.216) and (6.217) side by side and combining the surface integrals leads to

$$\begin{aligned}\int_{W} \mathrm{d}V\, \varrho_{M}(\mathbf{r},t) &= \int_{\partial W_1 \cup S_0} \mathrm{d}S\, \hat{\boldsymbol{\nu}}_1(\mathbf{r}) \cdot \mathcal{B}_1(\mathbf{r},t) + \int_{\partial W_2 \cup S_0} \mathrm{d}S\, \hat{\boldsymbol{\nu}}_2(\mathbf{r}) \cdot \mathcal{B}_2(\mathbf{r},t) \\ &= \int_{\partial W} \mathrm{d}S\, \hat{\boldsymbol{\nu}}(\mathbf{r}) \cdot \mathcal{B}(\mathbf{r},t) + \int_{S_0} \mathrm{d}S\, \hat{\mathbf{n}}(\mathbf{r}) \cdot [\mathcal{B}_2(\mathbf{r},t) - \mathcal{B}_1(\mathbf{r},t)]\end{aligned} \tag{6.219}$$

whence we get

$$\int_{\partial W} \mathrm{d}S\, \hat{\boldsymbol{\nu}}(\mathbf{r}) \cdot \mathcal{B}(\mathbf{r},t) = \int_{W} \mathrm{d}V\, \varrho_{M}(\mathbf{r},t) - \int_{S_0} \mathrm{d}S\, \hat{\mathbf{n}}(\mathbf{r}) \cdot [\mathcal{B}_2(\mathbf{r},t) - \mathcal{B}_1(\mathbf{r},t)] \tag{6.220}$$

which must coincide with (6.218) for consistency. To make this happen we *postulate* the condition

$$\int_{S_0} \mathrm{d}S\, \{\hat{\mathbf{n}}(\mathbf{r}) \cdot [\mathcal{B}_2(\mathbf{r},t) - \mathcal{B}_1(\mathbf{r},t)] + \varrho_{MS}(\mathbf{r},t)\} = 0 \tag{6.221}$$

and, if we assume that the scalar fields $\varrho_{MS}(\mathbf{r},t)$ and $\hat{\mathbf{n}}(\mathbf{r}) \cdot \mathcal{B}_l(\mathbf{r},t)$, $l = 1, 2$, are continuous functions of $\mathbf{r} \in \partial V$, the mean value theorem [84] applied to the last integral provides

$$\hat{\mathbf{n}}(\mathbf{r}_0) \cdot [\mathcal{B}_1(\mathbf{r}_0,t) - \mathcal{B}_2(\mathbf{r}_0,t)] = \varrho_{MS}(\mathbf{r}_0,t) \tag{6.222}$$

with $\mathbf{r}_0 \in S_0$ a suitable point. But, since the domain W is arbitrary, it follows

$$\hat{\mathbf{n}}(\mathbf{r}) \cdot [\mathcal{B}_1(\mathbf{r},t) - \mathcal{B}_2(\mathbf{r},t)] = \varrho_{MS}(\mathbf{r},t), \qquad \mathbf{r} \in \partial V \tag{6.223}$$

i.e., the normal component of the displacement vector is either continuous across ∂V — if $\varrho_{MS}(\mathbf{r},t) = 0$ — or suffers a jump equal to the surface charge density on ∂V; this relation extends (1.157). From (6.216) and (6.217) the local form of the magnetic Gauss law is recovered separately for $\mathbf{r} \in V_1$ and $\mathbf{r} \in V_2$.

Since it is plausible that magnetic charges cannot exist inside an electric conductor or, in the limit of infinite conductivity, on the surface of a PEC, then the boundary conditions (1.168)-(1.171) need no modification. Conversely, for mere reasons of symmetry we may entertain the idea of magnetic conductors as carriers of free magnetic charges and also introduce the magnetic conductivity $\sigma_M(\mathbf{r})$. The latter parameter allows us to define magnetic conduction currents as

$$\mathcal{J}_{cM}(\mathbf{r}) := \sigma_M(\mathbf{r})\mathcal{H}(\mathbf{r},t) \tag{6.224}$$

which is the analogue of the Ohm law (1.120) for conductors. From (6.224) we infer that $\sigma_M(\mathbf{r})$ carries the physical dimension of Ω/m. In the limit as $\sigma_M(\mathbf{r}) \to +\infty$ we obtain the special case of a Perfect Magnetic Conductor (PMC). The magnetic field then vanishes identically

$$\mathcal{H}(\mathbf{r},t) = \lim_{\sigma_M \to +\infty} \frac{\mathcal{J}_{cM}(\mathbf{r},t)}{\sigma_M(\mathbf{r})} = 0 \tag{6.225}$$

if the magnetic conduction current is to remain finite. As a result, *inside* a PMC $\mathcal{B}(\mathbf{r}, t) = 0$ owing to (1.118) and no magnetic charge can exist in accordance with the Gauss magnetic law (6.186); indeed, the charge is 'pushed' to the surface of the PMC. The Ampère-Maxwell law yields $\partial\mathcal{D}/\partial t = 0$ whereby the displacement vector must be constant inside a PMC. Nevertheless, if we postulate the electromagnetic field was zero at some point in time, we are led to take $\mathcal{D}(\mathbf{r}, t) = 0$. Hence, since $\mathcal{E}(\mathbf{r}, t) = 0 = \mathcal{B}(\mathbf{r}, t)$ from the Faraday law we determine $\mathcal{J}_{cM}(\mathbf{r}, t) = 0$, namely, no magnetic conduction current is possible inside a PMC. Needless to say, electric charges cannot exist inside magnetic conductors.

We may now obtain the boundary conditions at the surface of a PMC. With reference to Figures 1.12 and 1.13, if the magnetic conductivity σ_{M1} in medium 1 vanishes and the medium in V_2 is a PMC, in light of the previous discussion the matching conditions (1.142), (6.215), (1.155) and (6.223) become

$$\hat{\mathbf{n}}(\mathbf{r}) \times \mathcal{H}_1(\mathbf{r}, t) = 0, \qquad \mathbf{r} \in \partial V \tag{6.226}$$

$$\mathcal{E}_1(\mathbf{r}, t) \times \hat{\mathbf{n}}(\mathbf{r}) = \mathcal{J}_{MS}(\mathbf{r}, t), \qquad \mathbf{r} \in \partial V \tag{6.227}$$

$$\hat{\mathbf{n}}(\mathbf{r}) \cdot \mathcal{D}_1(\mathbf{r}, t) = 0, \qquad \mathbf{r} \in \partial V \tag{6.228}$$

$$\hat{\mathbf{n}}(\mathbf{r}) \cdot \mathcal{B}_1(\mathbf{r}, t) = \varrho_{MS}(\mathbf{r}, t), \qquad \mathbf{r} \in \partial V \tag{6.229}$$

and

$$\nabla_s \cdot \mathcal{J}_{MS}(\mathbf{r}, t) + \frac{\partial}{\partial t}\varrho_{MS}(\mathbf{r}, t) = 0, \qquad \mathbf{r} \in \partial V \tag{6.230}$$

represents the magnetic counterpart of (1.172).

6.7 Duality transformations

Thanks to the introduction of magnetic charges the classic Maxwell equations take on, if possible, an even more symmetric and self-contained structure. Indeed, as is evident from (6.183)-(6.186), each electric entity now possesses a magnetic counterpart and vice-versa. This perfect correspondence could not be achieved without the presence of the magnetic sources, fictitious as they are. Actually, we can push this connection a little further.

For instance, let us consider the Ampère-Maxwell law (1.34). If we systematically swap electric and magnetic entities according to the following substitutions

$$\mathcal{H} \Longrightarrow \mathcal{E}', \qquad \mathcal{D} \Longrightarrow -\mathcal{B}', \qquad \mathcal{J} \Longrightarrow -\mathcal{J}'_M \tag{6.231}$$

we arrive at

$$\nabla \times \mathcal{E}'(\mathbf{r}, t) = -\frac{\partial}{\partial t}\mathcal{B}'(\mathbf{r}, t) - \mathcal{J}'_M(\mathbf{r}, t) \tag{6.232}$$

that is, the Faraday law for the primed entities $\mathcal{E}'$, $\mathcal{B}'$ and $\mathcal{J}'_M$. In words, with the substitution (6.231) the Ampère-Maxwell law is transformed into the Faraday law. A similar conclusion holds true also for the corresponding laws in global form.

Next, let us examine the Gauss law (1.44) in an isotropic medium endowed with permittivity ε. If we perform the substitutions

$$\mathcal{E} \Longrightarrow \mathcal{H}', \qquad \varepsilon \Longrightarrow -\mu', \qquad \varrho \Longrightarrow -\varrho'_M \tag{6.233}$$

Table 6.1 Duality transformations

Time domain		Frequency domain	
Primary entities	**Dual entities**	**Primary entities**	**Dual entities**
$\mathcal{E}$	$\mathcal{H}$	$\mathbf{E}$	$\mathbf{H}$
$\mathcal{H}$	$\mathcal{E}$	$\mathbf{H}$	$\mathbf{E}$
$\mathcal{D}$	$-\mathcal{B}$	$\mathbf{D}$	$-\mathbf{B}$
$\mathcal{B}$	$-\mathcal{D}$	$\mathbf{B}$	$-\mathbf{D}$
$\mathcal{P}$	$-\mu_0\mathcal{M}^\dagger$	$\mathbf{P}$	$-\mu_0\mathbf{M}^\dagger$
$\mu_0\mathcal{M}^\dagger$	$-\mathcal{P}$	$\mu_0\mathbf{M}^\dagger$	$-\mathbf{P}$
$\overline{\overline{\mathcal{T}}}_e$	$-\overline{\overline{\mathcal{T}}}_h$	$\overline{\overline{\mathbf{T}}}_e$	$-\overline{\overline{\mathbf{T}}}_h$
$\overline{\overline{\mathcal{T}}}_h$	$-\overline{\overline{\mathcal{T}}}_e$	$\overline{\overline{\mathbf{T}}}_h$	$-\overline{\overline{\mathbf{T}}}_e$
ϱ	$-\varrho_M$	ρ	$-\rho_M$
ϱ_M	$-\varrho$	ρ_M	$-\rho$
$\mathfrak{J}$	$-\mathfrak{J}_M$	$\mathbf{J}$	$-\mathbf{J}_M$
$\mathfrak{J}_M$	$-\mathfrak{J}$	$\mathbf{J}_M$	$-\mathbf{J}$
ε	$-\mu$	ε	$-\mu$
μ	$-\varepsilon$	μ	$-\varepsilon$
σ	$-\sigma_M$	σ	$-\sigma_M$
σ_M	$-\sigma$	σ_M	$-\sigma$
χ_e	χ_m	χ_e	χ_m
χ_m	χ_e	χ_m	χ_e
		$Z := (\mu/\varepsilon)^{1/2}$	$-Y := -1/Z$
		Y	$-Z$
		$k := \omega(\varepsilon\mu)^{1/2}$	k
W_e	$-W_h$	$\langle W_e \rangle$	$-\langle W_h \rangle$
W_h	$-W_e$	$\langle W_h \rangle$	$-\langle W_e \rangle$

(†) The explicit appearance of the permeability is due to the different definitions (3.257) and (5.183) adopted for $\mathcal{P}$ and $\mathcal{M}$, respectively.

we obtain

$$\nabla \cdot [\mu' \mathcal{H}'(\mathbf{r}, t)] = \varrho'_M(\mathbf{r}, t) \tag{6.234}$$

that is, the transformation (6.233) turns the Gauss law into the magnetic Gauss law for the primed quantities $\mu'\mathcal{H}'$ and ϱ'_M. Again, this result applies to the global form as well.

We could go on and repeat the procedure with the remaining Maxwell equations to find out that each one of them is exactly transformed into another, so that in the end we obtain the full set of equations satisfied by the primed quantities. Moreover, the transformation applies also to the continuity equations (1.46) and (6.187).

The property just described is referred to as the *duality principle* [46, Section 4.2], [73, Section 1.7], [85], and is, in effect, a way of associating primary electric or magnetic entities with their *dual* magnetic or electric counterpart. The principle works also in the frequency domain, and an extended list of transformations is laid out in Table 6.1, where we have dropped the prime, because its sole purpose was to make the transformation more evident. It is important to mention that the transformation of Table 6.1 is not the only possible one and that, accordingly, other consistent correspondences between primary and dual entities can be found [18, Table 3.2], [74, Table 7.2], [86, Section 1.2], [72, Section 1.11].

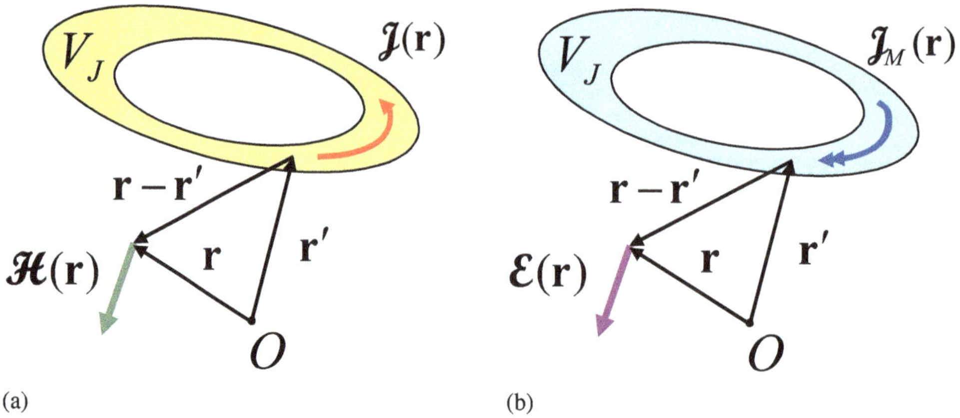

Figure 6.19 Dual problems: (a) magnetic field generated by a steady electric current $\mathcal{J}(\mathbf{r})$ and (b) electric field generated by a steady magnetic current $\mathcal{J}_M(\mathbf{r})$.

The duality transformations turn out quite handy especially in tandem with the superposition principle, and here is how it works. Suppose we wish to solve the Maxwell equations (6.183)-(6.186) with electric and magnetic sources. By invoking superposition first, we split up the task into two parts: 1) finding the fields produced by the electric sources alone first, 2) finding the fields generated by the magnetic sources alone. Then, summing the contributions of all sources will provide us with the final result. However, to determine the effect of the magnetic sources we do not have to start over. Rather, if we have found a formula that relates electric sources to fields (e.g., in the static limit the integral representations of Section 3.2) then we may apply duality to obtain the formula which relates magnetic sources to fields.

Example 6.5 (Electric field generated by a steady magnetic current density)

We consider a steady electric current density $\mathcal{J}(\mathbf{r})$ confined in a contour-wise multiply-connected region $V_J \subset \mathbb{R}^3$ in free space, as is shown in Figure 6.19a. The stationary magnetic field generated by $\mathcal{J}(\mathbf{r})$ can be computed by means of the integral representation (5.61) with $\mu = \mu_0$

$$\mathcal{H}(\mathbf{r}) = \frac{\mathcal{B}(\mathbf{r})}{\mu_0} = \int_{V_J} \mathrm{d}V' \, \mathcal{J}(\mathbf{r}') \times \nabla' \frac{1}{4\pi|\mathbf{r}-\mathbf{r}'|}, \qquad \mathbf{r} \in \mathbb{R}^3. \tag{6.235}$$

If we subject this formula to the duality transformation of Table 6.1, we obtain

$$\mathcal{E}(\mathbf{r}) = -\int_{V_J} \mathrm{d}V' \, \mathcal{J}_M(\mathbf{r}') \times \nabla' \frac{1}{4\pi|\mathbf{r}-\mathbf{r}'|}, \tag{6.236}$$

which represents the stationary electric field produced by a steady magnetic current $\mathcal{J}_M(\mathbf{r}')$ flowing in V_J in free space (Figure 6.19b).

From a different perspective, if the integral in (6.235) has been computed, then we may interpret the result also as the electric field due to $-\mathcal{J}_M(\mathbf{r}')$. If the shape or the position of $\mathcal{J}_M(\mathbf{r}')$ or both are different than those of $\mathcal{J}(\mathbf{r}')$, then the duality principle provides us at least with the necessary formula for computing the solution.

(End of Example 6.5)

6.8 Reciprocity theorems

Broadly speaking, the term *reciprocity* indicates a form of symmetry for electromagnetic fields produced by two set of sources. This means that, under quite general circumstances, we can swap the *role* of sources and observers without altering the nature of the electromagnetic entities detected by the observers. In free space or in an unbounded isotropic medium it seems obvious that the fields detected by an observer should depend only on his position relative to the source. Therefore, if we even swap the *physical position* of source and observer, the measured fields ought not to be affected. For instance, we already found this property to be true by examining the structure of the electrostatic Green function in Section 3.4.

The situation changes a bit if the medium is not homogeneous, say, if the observer makes his or her measurements in the presence of a material body. Of course, now there is a way to distinguish different points in space — relative to the body — and hence it is expected that the measured fields cannot depend just on the distance from the source. In this case, swapping the *position* of source and observer will alter the results of the experiment in general, save for special symmetric geometries, e.g., if the body of concern is a PEC sphere equally spaced from source and observer. By contrast, interchanging the *role* is precisely what reciprocity is about.

To elucidate, we consider the two antennas shown in Figure 6.20a. Antenna ① (the source) is radiating while antenna ② (the observer) is receiving and, since the system is linear, the current I_{21} flowing at the terminals of antenna ② is proportional to the voltage V_1 applied to the terminals of antenna ①. Next, suppose that antenna ② becomes the source whereas antenna ① detects the transmitted fields, as is suggested in Figure 6.20b. Again, the current I_{12} flowing at the terminals of antenna ① is proportional to the voltage V_2 of the generator that drives antenna ②. Now, reciprocity provides us with the following relation

$$\frac{I_{21}}{V_1} = \frac{I_{12}}{V_2} \tag{6.237}$$

which, if the generators have unitary strength, implies that the currents flowing at the antenna terminals are the *same* in both cases. This suggests that we can swap the role of source and observer and nothing changes. Notice that the conclusion is independent of the shape of the two antennas — they may be different — and holds true even in the presence of a material body. In fact, it can be shown that *geometrical symmetry* implies reciprocity, but the converse is not true.

In the following we wish to give these ideas a rigorous mathematical form and prove (6.237). Incidentally, some Authors refer to (6.237) as the reciprocity, though that result is just one of the many consequences of the reciprocity theorem. We shall address the time-harmonic regime first, because it is easier and more interesting for practical applications. Besides, we shall also give counterexamples, namely, situations in which reciprocity as described above is violated, though a more general result can be stated as well.

6.8.1 *Frequency domain*

To obtain the result in the time-harmonic regime we consider a finite region of space V bounded by a smooth surface $S := \partial V$. To gain more generality we also allow for the presence of a material body inside a smaller volume $V_1 \subset V$. Reciprocity is a relation between two *different* sets of sources and fields in the same domain, though with the same time dependence in the form $\exp(\mathrm{j}\,\omega t)$. Therefore, we define two scenarios — labelled with subscripts a and b — which in the context of reciprocity are referred to as *states*.

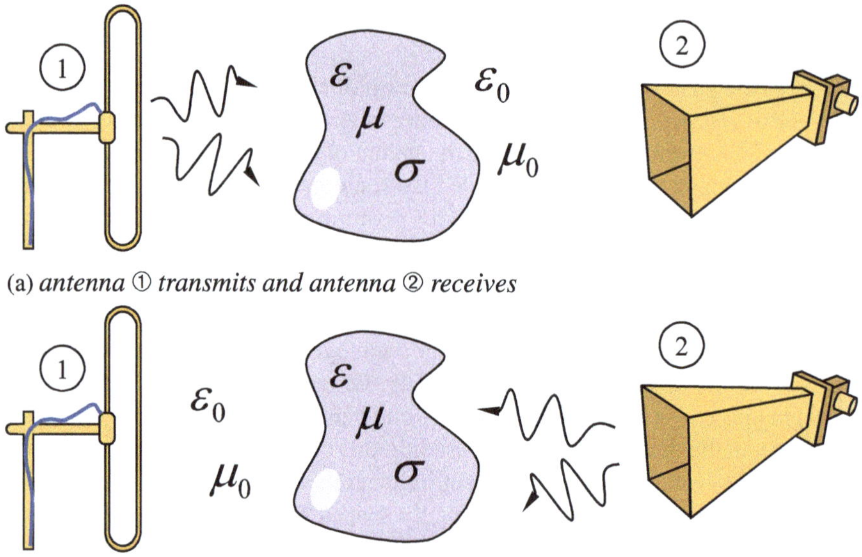

(a) *antenna ① transmits and antenna ② receives*

(b) *the role of transmitter and receiver is interchanged*

Figure 6.20 *For illustrating reciprocity in the presence of a material body.*

State (a) The underlying medium which fills $V \setminus V_1$ is endowed with constitutive parameters ε_a, μ_a and σ_a, whereas the material which occupies V_1 is anisotropic with dyadic permittivity $\overline{\boldsymbol{\varepsilon}}_a$ and permeability $\overline{\boldsymbol{\mu}}_a$. The sources are volume electric and magnetic current densities confined in a volume $V_{Ja} \subset V$. This scenario is exemplified in Figure 6.21a.

State (b) The underlying medium which fills $V \setminus V_1$ is endowed with constitutive parameters ε_b, μ_b and σ_b, whereas the material which occupies V_1 is anisotropic with dyadic permittivity $\overline{\boldsymbol{\varepsilon}}_b$ and permeability $\overline{\boldsymbol{\mu}}_b$. The sources are volume electric and magnetic current densities confined in a volume $V_{Jb} \subset V$. This scenario is pictured in Figure 6.21b.

We remark that the two sets of sources are *different* and, more importantly, may be located in different parts of V. By contrast, we require the material boundaries S and ∂V_1 to be the same in the two states, but for the moment we do not specify the relation — if any should exist — between the constitutive parameters pertinent to state (a) and state (b).

Evidently, sources and fields of each state so defined must be related by the Maxwell equations (6.183)-(6.186). However, for our purposes we just need the extended Faraday and Ampère-Maxwell laws, which we write down for convenience

$$\nabla \times \mathbf{E}_a(\mathbf{r}) = -\mathrm{j}\,\omega \mathbf{B}_a(\mathbf{r}) - \mathbf{J}_{Ma}(\mathbf{r}) \tag{6.238}$$

$$\nabla \times \mathbf{H}_a(\mathbf{r}) = \mathrm{j}\,\omega \mathbf{D}_a(\mathbf{r}) + \sigma_a \mathbf{E}_a(\mathbf{r}) + \mathbf{J}_a(\mathbf{r}) \tag{6.239}$$

$$\nabla \times \mathbf{E}_b(\mathbf{r}) = -\mathrm{j}\,\omega \mathbf{B}_b(\mathbf{r}) - \mathbf{J}_{Mb}(\mathbf{r}) \tag{6.240}$$

$$\nabla \times \mathbf{H}_b(\mathbf{r}) = \mathrm{j}\,\omega \mathbf{D}_b(\mathbf{r}) + \sigma_b \mathbf{E}_b(\mathbf{r}) + \mathbf{J}_b(\mathbf{r}) \tag{6.241}$$

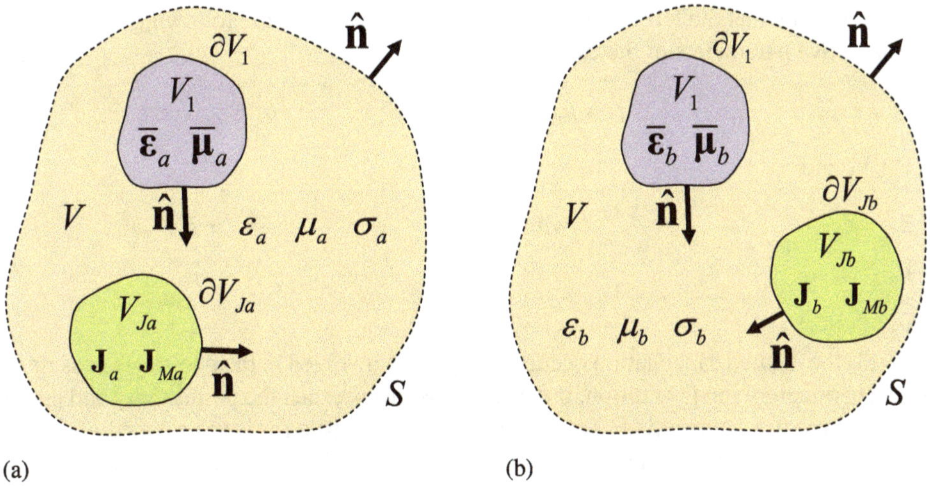

Figure 6.21 For deriving the reciprocity theorem: sources and matter for states (a) and (b).

for points $\mathbf{r} \in V$. Now, we dot-multiply the first equation with $\mathbf{H}_b(\mathbf{r})$ and the fourth equation with $\mathbf{E}_a(\mathbf{r})$, namely,

$$\mathbf{H}_b(\mathbf{r}) \cdot \nabla \times \mathbf{E}_a(\mathbf{r}) = -\mathrm{j}\,\omega \mathbf{H}_b(\mathbf{r}) \cdot \mathbf{B}_a(\mathbf{r}) - \mathbf{H}_b(\mathbf{r}) \cdot \mathbf{J}_{Ma}(\mathbf{r}) \tag{6.242}$$

$$\mathbf{E}_a(\mathbf{r}) \cdot \nabla \times \mathbf{H}_b(\mathbf{r}) = \mathrm{j}\,\omega \mathbf{E}_a(\mathbf{r}) \cdot \mathbf{D}_b(\mathbf{r}) + \sigma_b \mathbf{E}_a(\mathbf{r}) \cdot \mathbf{E}_b(\mathbf{r}) + \mathbf{E}_a(\mathbf{r}) \cdot \mathbf{J}_b(\mathbf{r}) \tag{6.243}$$

and subtract the resulting equations side by side to arrive at

$$\begin{aligned}\nabla \cdot [\mathbf{E}_a(\mathbf{r}) \times \mathbf{H}_b(\mathbf{r})] = &-\mathrm{j}\,\omega[\mathbf{H}_b(\mathbf{r}) \cdot \mathbf{B}_a(\mathbf{r}) + \mathbf{E}_a(\mathbf{r}) \cdot \mathbf{D}_b(\mathbf{r})] \\ &- \mathbf{H}_b(\mathbf{r}) \cdot \mathbf{J}_{Ma}(\mathbf{r}) - \mathbf{E}_a(\mathbf{r}) \cdot \sigma_b \mathbf{E}_b(\mathbf{r}) - \mathbf{E}_a(\mathbf{r}) \cdot \mathbf{J}_b(\mathbf{r})\end{aligned} \tag{6.244}$$

on account of the differential identity (H.49). Intuition suggests that by swapping the indices a and b we obtain the other equation which would follow by manipulating (6.239) and (6.240) in a similar consistent fashion. Irrespective of the strategy, we obtain

$$\begin{aligned}\nabla \cdot [\mathbf{E}_b(\mathbf{r}) \times \mathbf{H}_a(\mathbf{r})] = &-\mathrm{j}\,\omega[\mathbf{H}_a(\mathbf{r}) \cdot \mathbf{B}_b(\mathbf{r}) + \mathbf{E}_b(\mathbf{r}) \cdot \mathbf{D}_a(\mathbf{r})] \\ &- \mathbf{H}_a(\mathbf{r}) \cdot \mathbf{J}_{Mb}(\mathbf{r}) - \mathbf{E}_b(\mathbf{r}) \cdot \sigma_a \mathbf{E}_a(\mathbf{r}) - \mathbf{E}_b(\mathbf{r}) \cdot \mathbf{J}_a(\mathbf{r})\end{aligned} \tag{6.245}$$

which we wish to combine with (6.244). In so doing, we aim at getting rid of the terms that involve the constitutive parameters either explicitly or through the flux densities. To this purpose, we examine the expressions

$$\mathbf{E}_a(\mathbf{r}) \cdot \mathbf{D}_b(\mathbf{r}) - \mathbf{E}_b(\mathbf{r}) \cdot \mathbf{D}_a(\mathbf{r}) = \begin{cases}(\varepsilon_b - \varepsilon_a)\mathbf{E}_a(\mathbf{r}) \cdot \mathbf{E}_b(\mathbf{r}), & \mathbf{r} \in V \setminus V_1 \\ \mathbf{E}_a(\mathbf{r}) \cdot (\overline{\varepsilon}_b - \overline{\varepsilon}_a^T) \cdot \mathbf{E}_b(\mathbf{r}), & \mathbf{r} \in V_1\end{cases} \tag{6.246}$$

$$\mathbf{H}_a(\mathbf{r}) \cdot \mathbf{B}_b(\mathbf{r}) - \mathbf{H}_b(\mathbf{r}) \cdot \mathbf{B}_a(\mathbf{r}) = \begin{cases}(\mu_b - \mu_a)\mathbf{H}_a(\mathbf{r}) \cdot \mathbf{H}_b(\mathbf{r}), & \mathbf{r} \in V \setminus V_1 \\ \mathbf{H}_a(\mathbf{r}) \cdot (\overline{\mu}_b - \overline{\mu}_a^T) \cdot \mathbf{H}_b(\mathbf{r}), & \mathbf{r} \in V_1\end{cases} \tag{6.247}$$

$$\mathbf{E}_a(\mathbf{r}) \cdot \sigma_b \mathbf{E}_b(\mathbf{r}) - \mathbf{E}_b(\mathbf{r}) \cdot \sigma_a \mathbf{E}_a(\mathbf{r}) = (\sigma_b - \sigma_a)\mathbf{E}_b(\mathbf{r}) \cdot \mathbf{E}_a(\mathbf{r}), \qquad \mathbf{r} \in V \setminus V_1 \tag{6.248}$$

whence we conclude that the left-hand sides may vanish, as long as the following conditions are met by the constitutive parameters of the two states:

$$\varepsilon = \varepsilon_a = \varepsilon_b, \qquad \mu = \mu_a = \mu_b, \qquad \sigma = \sigma_a = \sigma_b \tag{6.249}$$

for $\mathbf{r} \in V \setminus V_1$ and

$$\overline{\varepsilon} = \overline{\varepsilon}_a = \overline{\varepsilon}_b \qquad \text{with} \qquad \overline{\varepsilon}^T = \overline{\varepsilon} \tag{6.250}$$

$$\overline{\mu} = \overline{\mu}_a = \overline{\mu}_b \qquad \text{with} \qquad \overline{\mu}^T = \overline{\mu} \tag{6.251}$$

for $\mathbf{r} \in V_1$.

In words, the desired cancellation occurs if the media involved in the two states possess the *same* constitutive parameters and in addition, if they are anisotropic, then the permittivity and permeability are symmetric dyadics (Appendix E.2). The latter is trivially true for isotropic media, since we may write $\overline{\varepsilon} = \varepsilon\overline{\mathfrak{I}}$ and so forth. Media characterized by symmetric dyadic constitutive parameters are called *reciprocal*.

Under these hypotheses, by subtracting (6.244) and (6.245) side by side we obtain the first important result

$$\begin{aligned}\nabla \cdot [\mathbf{E}_a(\mathbf{r}) \times \mathbf{H}_b(\mathbf{r}) - \mathbf{E}_b(\mathbf{r}) \times \mathbf{H}_a(\mathbf{r})] = {} & [\mathbf{E}_b(\mathbf{r}) \cdot \mathbf{J}_a(\mathbf{r}) - \mathbf{H}_b(\mathbf{r}) \cdot \mathbf{J}_{Ma}(\mathbf{r})] \\ & - [\mathbf{E}_a(\mathbf{r}) \cdot \mathbf{J}_b(\mathbf{r}) - \mathbf{H}_a(\mathbf{r}) \cdot \mathbf{J}_{Mb}(\mathbf{r})], \qquad \mathbf{r} \in V\end{aligned} \tag{6.252}$$

a relation which goes by the name of *Lorentz reciprocity theorem* or Lorentz lemma [18], [16, Section 1.10]; it is, in fact, the differential form of the *reciprocity theorem* in the time-harmonic regime. It is worthwhile noticing that each term is a dot or cross product of entities belonging to either state. In particular, the right-hand side involves dot products between the fields of one state and the sources of the other. The proof can be generalized to include magnetic losses or dyadic conductivities as well, and the result holds even if the media are inhomogeneous, so long as they are reciprocal, as established above.

If the media in the complementary domain $\mathbb{R}^3 \setminus \overline{V}$ are reciprocal too, then we can extend the range of validity of (6.252) to the whole space. However, to obtain the next result we integrate both sides of (6.252) for $\mathbf{r} \in V$ and apply the Gauss theorem. Even though any possible discontinuity in the properties of the materials inside V does not enter (6.252) explicitly, still the fields may be discontinuous across ∂V_1. So, in principle we should apply the divergence theorem separately in $V \setminus \overline{V}_1$ and V_1, as we did repeatedly in Section 1.7. Nevertheless, since the fields of both states satisfy the jump conditions (1.142) and (6.215) with $\mathfrak{J}_S(\mathbf{r}, t) = 0 = \mathfrak{J}_{MS}(\mathbf{r}, t)$, the extra contributions over ∂V_1 vanish identically. We do not concern ourselves with jumps across ∂V_{Ja} and ∂V_{Jb}, as these surfaces are not material interfaces by assumption. By contrast, since the application of the divergence theorem relies on the regularity of the fields throughout V, the presence of *singular sources* (such as the time-harmonic Hertzian dipole of Example 9.3) within V_{Ja} and V_{Jb} is forbidden in our derivation.

In the end, by taking the unit normal on S oriented positively outward V (Figure 6.21) we find

$$\begin{aligned}&\int_S \mathrm{d}S\, \hat{\mathbf{n}}(\mathbf{r}) \cdot [\mathbf{E}_a(\mathbf{r}) \times \mathbf{H}_b(\mathbf{r}) - \mathbf{E}_b(\mathbf{r}) \times \mathbf{H}_a(\mathbf{r})] \\ &= \int_{V_{Ja}} \mathrm{d}V\, [\mathbf{E}_b(\mathbf{r}) \cdot \mathbf{J}_a(\mathbf{r}) - \mathbf{H}_b(\mathbf{r}) \cdot \mathbf{J}_{Ma}(\mathbf{r})] - \int_{V_{Jb}} \mathrm{d}V\, [\mathbf{E}_a(\mathbf{r}) \cdot \mathbf{J}_b(\mathbf{r}) - \mathbf{H}_a(\mathbf{r}) \cdot \mathbf{J}_{Mb}(\mathbf{r})]\end{aligned} \tag{6.253}$$

which is known as the reciprocity theorem in global form [3,17,18,29,30,72,74,87,88]. The volume integrals in the right-hand side (carried out over the regions occupied by the sources of each state) were given the name of *reactions* by Rumsey (1954) [89], [90, Section 6.2.A]. In symbols, we have

$$\langle \mathbf{f}_b, \mathbf{g}_a \rangle := \int\limits_{V_{Ja}} \mathrm{d}V\, [\mathbf{E}_b(\mathbf{r}) \cdot \mathbf{J}_a(\mathbf{r}) - \mathbf{H}_b(\mathbf{r}) \cdot \mathbf{J}_{Ma}(\mathbf{r})], \qquad V_{Ja} \subset V \setminus V_1 \tag{6.254}$$

the reaction between the fields of state (b) and the sources of state (a), and

$$\langle \mathbf{f}_a, \mathbf{g}_b \rangle := \int\limits_{V_{Jb}} \mathrm{d}V\, [\mathbf{E}_a(\mathbf{r}) \cdot \mathbf{J}_b(\mathbf{r}) - \mathbf{H}_a(\mathbf{r}) \cdot \mathbf{J}_{Mb}(\mathbf{r})], \qquad V_{Jb} \subset V \setminus V_1 \tag{6.255}$$

the reaction between the fields of state (a) and the sources of state (b). With these definitions the reciprocity theorem can be written succinctly as

$$\int\limits_{S} \mathrm{d}S\, \hat{\mathbf{n}}(\mathbf{r}) \cdot [\mathbf{E}_a(\mathbf{r}) \times \mathbf{H}_b(\mathbf{r}) - \mathbf{E}_b(\mathbf{r}) \times \mathbf{H}_a(\mathbf{r})] = \langle \mathbf{f}_b, \mathbf{g}_a \rangle - \langle \mathbf{f}_a, \mathbf{g}_b \rangle\,. \tag{6.256}$$

which is a powerful result with far-reaching implications. For example, (6.253) allows proving (6.237), the symmetry properties of the Green functions, the orthogonality of the modal eigenfunctions in a classic hollow-pipe waveguide (Section 11.2), and it may even help formulate scattering and radiation problems in terms of integral equations (Chapter 13).

While a strong similarity between the Poynting theorem (1.314) and (6.253) is undeniable — indeed, all the integrals have the physical dimension of a *power* — yet the terms related by the global reciprocity theorem do not represent average powers for two reasons. For one thing, the average of the dot or cross product of two time-harmonic vector fields involves the complex conjugate of either fields (see Section 1.10.2), whereas in (6.253) no phasor enters with its complex conjugate.[1] Secondly, and perhaps more importantly, an average power involves the sources and the fields they generate, whereas the reactions combine sources of one state with the fields of the other state.

We now investigate under which circumstances the flux integral in (6.256) vanishes and the relationship reduces to

$$\langle \mathbf{f}_b, \mathbf{g}_a \rangle = \langle \mathbf{f}_a, \mathbf{g}_b \rangle \tag{6.257}$$

which is referred to as the *reaction theorem* [29] or the Carson theorem [91] or the *strong form* of the reciprocity theorem. While studying the integral representations of Sections 3.2 and 5.1.2 we found out that similar surface integrals invariably represented the contribution of sources outside the region of interest. We may wonder whether the same conclusion applies to (6.256).

Suppose that the sources pertinent to state (a) and state (b) are both located outside V, as is suggested in Figure 6.22. An application of (6.256) then yields

$$\int\limits_{S} \mathrm{d}S\, \hat{\mathbf{n}}(\mathbf{r}) \cdot [\mathbf{E}_a(\mathbf{r}) \times \mathbf{H}_b(\mathbf{r}) - \mathbf{E}_b(\mathbf{r}) \times \mathbf{H}_a(\mathbf{r})] = 0 \tag{6.258}$$

because the reactions are null by definition ($V_{Ja} \not\subset V$ and $V_{Jb} \not\subset V$) and (6.257) is trivially verified. In conclusion, the flux integral through S is null whenever the region bounded by S *does not contain* the sources of both states.

[1] Still, a formula which involves the conjugate entities and sources of state (b) can indeed be obtained [30, Section 8.6.2].

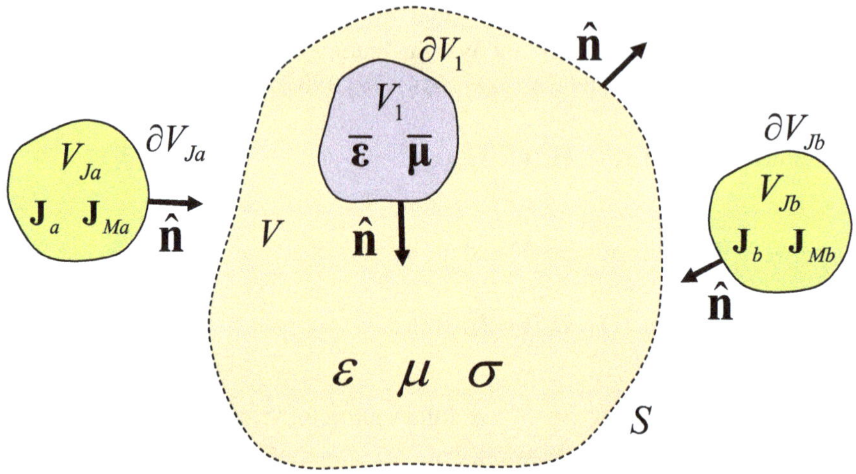

Figure 6.22 Special case of the reciprocity theorem: sources for states (a) and (b) located outside V.

Keeping this result in mind we go on to consider a ball $B(0,d)$ with radius d large enough so that $V \subset B(0,d)$ and apply the reciprocity theorem to the surface-wise multiply-connected region $B(0,d) \setminus V$, which, by construction, is devoid of sources, as they are all placed inside V (Figure 6.23). We assume for simplicity that the medium filling $\mathbb{R}^3 \setminus V$ is isotropic and thus reciprocal. Since the reactions are null in $B(0,d) \setminus V$, (6.256) provides

$$\int_{\partial B} dS\, \hat{\mathbf{n}}(\mathbf{r}) \cdot [\mathbf{E}_a(\mathbf{r}) \times \mathbf{H}_b(\mathbf{r}) - \mathbf{E}_b(\mathbf{r}) \times \mathbf{H}_a(\mathbf{r})]$$

$$= \int_S dS\, \hat{\mathbf{n}}(\mathbf{r}) \cdot [\mathbf{E}_a(\mathbf{r}) \times \mathbf{H}_b(\mathbf{r}) - \mathbf{E}_b(\mathbf{r}) \times \mathbf{H}_a(\mathbf{r})] \quad (6.259)$$

with the unit normals on ∂B and S oriented positively outwards $B(0,d)$ and V, respectively (Figure 6.23). We can show that the integral over the sphere ∂B vanishes in the limit as $d \to +\infty$, inasmuch as the fields of both states satisfy the Silver-Müller radiation conditions (6.171) and (6.172). In symbols, we have

$$\left| \int_{\partial B} dS\, \hat{\mathbf{r}} \cdot [\mathbf{E}_a(\mathbf{r}) \times \mathbf{H}_b(\mathbf{r}) - \mathbf{E}_b(\mathbf{r}) \times \mathbf{H}_a(\mathbf{r})] \right|$$

$$= \left| \int_{\partial B} dS\, \{[\hat{\mathbf{r}} \times \mathbf{E}_a(\mathbf{r}) - Z\mathbf{H}_a(\mathbf{r})] \cdot \mathbf{H}_b(\mathbf{r}) - [\hat{\mathbf{r}} \times \mathbf{E}_b(\mathbf{r}) - Z\mathbf{H}_b(\mathbf{r})] \cdot \mathbf{H}_a(\mathbf{r})\} \right|$$

$$\leqslant \int_{\partial B} dS\, [|\hat{\mathbf{r}} \times \mathbf{E}_a(\mathbf{r}) - Z\mathbf{H}_a(\mathbf{r})|\, |\mathbf{H}_b(\mathbf{r})| + |\hat{\mathbf{r}} \times \mathbf{E}_b(\mathbf{r}) - Z\mathbf{H}_b(\mathbf{r})|\, |\mathbf{H}_a(\mathbf{r})|]$$

$$\leqslant \int_{4\pi} d\Omega\, d^2 \left(\frac{M_a}{d^2} \frac{N_b}{d} + \frac{M_b}{d^2} \frac{N_a}{d} \right) = \frac{4\pi}{d}(M_a N_b + M_b N_a) \xrightarrow[d \to +\infty]{} 0 \quad (6.260)$$

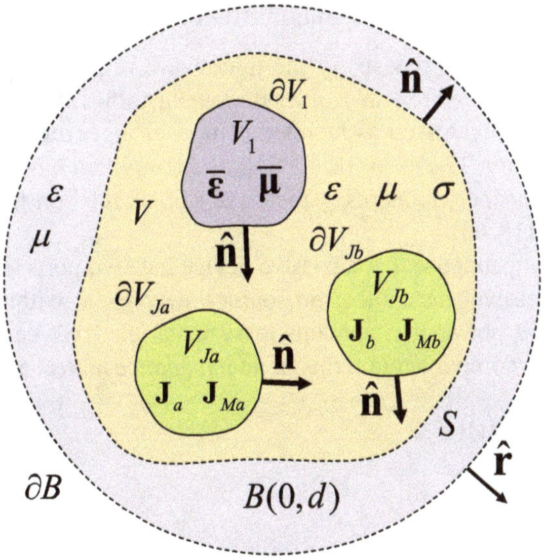

Figure 6.23 For deriving the reaction theorem: sources and matter for state (a) and state (b).

on account of (H.19), (D.152), (D.151), (6.156) and (6.172); the scalar Z is defined in (1.358) and is the intrinsic impedance of the medium that fills $\mathbb{R}^3 \setminus \overline{V}$. Combining this result with (6.259) we find

$$\int_S \mathrm{d}S\, \hat{\mathbf{n}}(\mathbf{r}) \cdot [\mathbf{E}_a(\mathbf{r}) \times \mathbf{H}_b(\mathbf{r}) - \mathbf{E}_b(\mathbf{r}) \times \mathbf{H}_a(\mathbf{r})] = 0 \tag{6.261}$$

that is, the flux integral through S is null whenever the region bounded by S *contains* the sources of both states! Now, this statement is more useful, because in tandem with (6.253) proves the reaction theorem (6.257) when the reactions are not zero.

Lastly, we consider the case of impedance or admittance boundary conditions for $\mathbf{r} \in S$, namely

$$\hat{\mathbf{n}}(\mathbf{r}) \times [\mathbf{E}(\mathbf{r}) \times \hat{\mathbf{n}}(\mathbf{r})] = \overline{\mathbf{Z}}_S(\mathbf{r}) \cdot [\mathbf{H}(\mathbf{r}) \times \hat{\mathbf{n}}(\mathbf{r})], \qquad \mathbf{r} \in S \tag{6.262}$$

$$\hat{\mathbf{n}}(\mathbf{r}) \times [\mathbf{H}(\mathbf{r}) \times \hat{\mathbf{n}}(\mathbf{r})] = \overline{\mathbf{Y}}_S(\mathbf{r}) \cdot [\hat{\mathbf{n}}(\mathbf{r}) \times \mathbf{E}(\mathbf{r})], \qquad \mathbf{r} \in S \tag{6.263}$$

where $\overline{\mathbf{Z}}_S(\mathbf{r})$ and $\overline{\mathbf{Y}}_S(\mathbf{r})$ are the surface impedance and admittance of the conducting medium flush with S (see Example 7.2).[2] It is easy to check that the flux integral through S vanishes as well and (6.257) applies, provided $\overline{\mathbf{Z}}_S$ and $\overline{\mathbf{Y}}_S$ are symmetric dyadics. The limiting cases $\overline{\mathbf{Z}}_S(\mathbf{r}) \to 0$ and $\overline{\mathbf{Y}}_S(\mathbf{r}) \to 0$ represent a PEC and a PMC boundary, respectively.

In summary, the reaction theorem (6.257) holds true if any one of these conditions is true

(i) all the sources are outside V;
(ii) all the sources are inside V;
(iii) the fields satisfy impedance or admittance boundary conditions on S, and the dyadics $\overline{\mathbf{Z}}_S(\mathbf{r})$, $\overline{\mathbf{Y}}_S(\mathbf{r})$ are symmetric.

We now discuss some important results which follow from the reciprocity and the reaction theorems.

[2]Notice that the right-hand sides of (6.78) and (6.262) are different owing to the opposite orientation chosen for the unit normal $\hat{\mathbf{n}}(\mathbf{r})$ for $\mathbf{r} \in S$.

Example 6.6 (Symmetry of impedance and admittance matrices)

Electromagnetic components and devices in the microwave regime may have two or more *ports* through which they are connected to waveguides or coaxial cables. Since the electromagnetic field within a waveguide can be expressed as a superposition of eigenfunctions weighted with modal voltages and currents (Section 11.2) [16–18, 63, 64], as exemplified by (6.130) and (6.131), we can describe the behavior of linear components by means of a linear relationship between modal voltages and currents at each port [18, 63].

For the sake argument, suppose that a passive device has two ports ① and ② connected to two metallic waveguides. 'Passive' means that no sources are present within the device, which, as a result, can at most *absorb* power if it contains lossy materials. We can write the equations that govern the behavior of the component in terms of an *impedance matrix* as

$$\begin{pmatrix}[V_1]\\ [V_2]\end{pmatrix} = \begin{pmatrix}[Z_{11}] & [Z_{12}]\\ [Z_{21}] & [Z_{22}]\end{pmatrix}\begin{pmatrix}[I_1]\\ [I_2]\end{pmatrix} \tag{6.264}$$

where

$$[V_l] := \begin{pmatrix}\vdots\\ V_{mn}^{(l)}\\ \vdots\end{pmatrix}, \qquad [I_l] := \begin{pmatrix}\vdots\\ I_{mn}^{(l)}\\ \vdots\end{pmatrix}, \qquad l \in \{1, 2\} \tag{6.265}$$

are infinite dimensional vectors of modal voltages and currents with (m, n) a pair of modal indices, and

$$[Z_{ls}] := \begin{pmatrix} & \vdots & \\ \cdots & Z_{mn,m'n'}^{(ls)} & \cdots\\ & \vdots & \end{pmatrix}, \qquad l, s \in \{1, 2\} \tag{6.266}$$

are infinite dimensional matrices. We wish to show that the impedance matrix in (6.264) is symmetric, if the component is comprised of reciprocal media in accordance with (6.249)-(6.251).

To this purpose we apply the reciprocity theorem (6.253) to a domain V which encompasses the component and protrudes a bit into the two waveguides, as is suggested in Figure 6.24. The boundary ∂V is comprised of three parts, namely, the surface S_1 which spans the cross section of waveguide ①, the surface S_2 which likewise spans the cross section of waveguide ②, and the surface S_w which is flush with the walls of the waveguides and of the device. We choose the unit normal $\hat{\mathbf{n}}(\mathbf{r})$ on ∂V positively oriented inwards V. In this way, the positive direction of propagation in each waveguide is the one for which the waves *enter* the device.

Both state (a) and state (b) are characterized by the presence of sources (which excite the modes) located *outside* V. Consequently, the reactions (6.254) and (6.255) are null by construction. Besides, if we assume that the walls of the waveguides and of the device are made of PEC, (6.253) reduces to

$$\int_{S_1} \mathrm{d}S\, \hat{\mathbf{n}}_1 \cdot [\mathbf{E}_a(\mathbf{r}) \times \mathbf{H}_b(\mathbf{r}) - \mathbf{E}_b(\mathbf{r}) \times \mathbf{H}_a(\mathbf{r})] + \int_{S_2} \mathrm{d}S\, \hat{\mathbf{n}}_2 \cdot [\mathbf{E}_a(\mathbf{r}) \times \mathbf{H}_b(\mathbf{r}) - \mathbf{E}_b(\mathbf{r}) \times \mathbf{H}_a(\mathbf{r})] = 0 \tag{6.267}$$

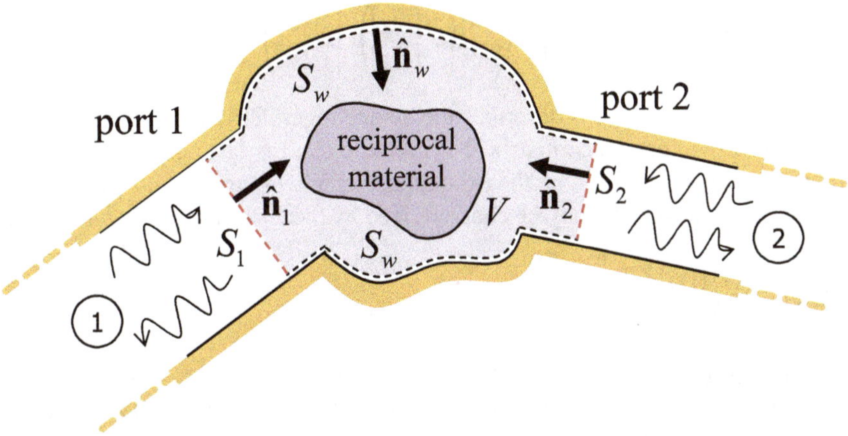

Figure 6.24 For applying the reciprocity theorem to a two-port device connected to two waveguides.

because the integral over S_w vanishes due to the boundary condition of the type (1.169). Only the transverse components of the fields in the waveguides enter in this expression — the longitudinal ones are filtered out, so to speak, by the dot-product with the unit normals $\hat{\mathbf{n}}_1$ and $\hat{\mathbf{n}}_2$. So, on account of the expansions (6.130) and (6.131), which apply to both state (a) and state (b), we have

$$\begin{aligned}
&\int_{S_1} dS\, \hat{\mathbf{n}}_1 \cdot [\mathbf{E}_a(\mathbf{r}) \times \mathbf{H}_b(\mathbf{r}) - \mathbf{E}_b(\mathbf{r}) \times \mathbf{H}_a(\mathbf{r})] = \\
&= \int_{S_1} dS\, \hat{\mathbf{n}}_1 \cdot \sum_{m,n} \mathbf{e}^{(1)}_{mn} V^{(1)}_{a,mn} \times \sum_{m',n'} \mathbf{h}^{(1)}_{m'n'} I^{(1)}_{b,m'n'} - \int_{S_1} dS\, \hat{\mathbf{n}}_1 \cdot \sum_{m,n} \mathbf{e}^{(1)}_{mn} V^{(1)}_{b,mn} \times \sum_{m',n'} \mathbf{h}^{(1)}_{m'n'} I^{(1)}_{a,m'n'} \\
&= \sum_{m,n} \sum_{m',n'} \left[V^{(1)}_{a,mn} I^{(1)}_{b,m'n'} - V^{(1)}_{b,mn} I^{(1)}_{a,m'n'} \right] \underbrace{\int_{S_1} dS\, \hat{\mathbf{n}}_1 \cdot \mathbf{e}^{(1)}_{mn} \times \mathbf{h}^{(1)}_{m'n'}}_{=\delta_{m,m'}\delta_{n,n'}} \\
&= \sum_{m,n} \left[V^{(1)}_{a,mn} I^{(1)}_{b,mn} - V^{(1)}_{b,mn} I^{(1)}_{a,mn} \right] = [I_{1b}]^T [V_{1a}] - [I_{1a}]^T [V_{1b}]
\end{aligned} \tag{6.268}$$

where we have taken into account the fact the eigenfunctions are orthogonal and normalized to unity (Section 11.2.5) [16, 18]. Similar manipulations of the integral over S_2 yield

$$\int_{S_2} dS\, \hat{\mathbf{n}}_2 \cdot [\mathbf{E}_a(\mathbf{r}) \times \mathbf{H}_b(\mathbf{r}) - \mathbf{E}_b(\mathbf{r}) \times \mathbf{H}_a(\mathbf{r})] = [I_{2b}]^T [V_{2a}] - [I_{2a}]^T [V_{2b}]. \tag{6.269}$$

Now, the reciprocity relation (6.267) implies

$$\begin{aligned}
0 &= [I_{1b}]^T [V_{1a}] - [I_{1a}]^T [V_{1b}] + [I_{2b}]^T [V_{2a}] - [I_{2a}]^T [V_{2b}] \\
&= \begin{pmatrix} [I_{1b}]^T & [I_{2b}]^T \end{pmatrix} \begin{pmatrix} [V_{1a}] \\ [V_{2a}] \end{pmatrix} - \begin{pmatrix} [I_{1a}]^T & [I_{2a}]^T \end{pmatrix} \begin{pmatrix} [V_{1b}] \\ [V_{2b}] \end{pmatrix}
\end{aligned} \tag{6.270}$$

and we may write

$$0 = \begin{pmatrix} [I_{1b}]^T & [I_{2b}]^T \end{pmatrix} \begin{pmatrix} [Z_{11}] & [Z_{12}] \\ [Z_{21}] & [Z_{22}] \end{pmatrix} \begin{pmatrix} [I_{1a}] \\ [I_{2a}] \end{pmatrix} - \begin{pmatrix} [I_{1a}]^T & [I_{2a}]^T \end{pmatrix} \begin{pmatrix} [Z_{11}] & [Z_{12}] \\ [Z_{21}] & [Z_{22}] \end{pmatrix} \begin{pmatrix} [I_{1b}] \\ [I_{2b}] \end{pmatrix}$$

$$= \left([I_{1b}]^T \quad [I_{2b}]^T\right)\left[\begin{pmatrix}[Z_{11}] & [Z_{12}]\\ [Z_{21}] & [Z_{22}]\end{pmatrix} - \begin{pmatrix}[Z_{11}] & [Z_{12}]\\ [Z_{21}] & [Z_{22}]\end{pmatrix}^T\right]\begin{pmatrix}[I_{1a}]\\ [I_{2a}]\end{pmatrix} \tag{6.271}$$

since the governing relation (6.264) describes the device regardless of the state of excitation of the waveguides. Only if the impedance matrix is symmetric can the last equation be satisfied, for the modal voltages and currents depend on the excitations, which, in turn, are completely arbitrary.

In principle, the same device may be characterized by means of an *admittance matrix*, viz.,

$$\begin{pmatrix}[I_1]\\ [I_2]\end{pmatrix} = \begin{pmatrix}[Y_{11}] & [Y_{12}]\\ [Y_{21}] & [Y_{22}]\end{pmatrix}\begin{pmatrix}[Y_1]\\ [Y_2]\end{pmatrix} \tag{6.272}$$

and with similar steps it can be proved that the matrix is symmetric. Alternatively, since the admittance matrix is the inverse of the impedance matrix in (6.264), symmetry follows directly.

The generalization of the proof to multi-port devices is obvious.

(End of Example 6.6)

Example 6.7 (Impressed electric currents on PEC surfaces do not radiate)

In Figure 6.4 we gave a schematic representation of a direct scattering problem. When the object illuminated by the incident field $\mathcal{E}^{\mathrm{i}}$, $\mathcal{H}^{\mathrm{i}}$ is a PEC body, on the boundary thereof an electric surface current density $\mathcal{J}_S$ is *induced* that generates a *secondary* (scattered) electromagnetic field $\mathcal{E}^{\mathrm{s}}$, $\mathcal{H}^{\mathrm{s}}$. The latter 1) cancels the incident electromagnetic field *inside* the body and 2) cancels the tangential component of $\mathcal{E}^{\mathrm{i}}$ on the *surface* of the body. This happens because the incident field is not zero within the body by definition nor does it satisfy the boundary condition (1.169) on the surface of the object. Besides, the scattered electromagnetic field $\mathcal{E}^{\mathrm{s}}$, $\mathcal{H}^{\mathrm{s}}$ is non-zero outside the body!

A totally different situation occurs if we place an *impressed* electric current in close proximity of a PEC body. In the limit as said current is made to flow on the surface of the PEC object, then it generates no field. Even though this configuration is not easy to realize (a surface current is an abstraction after all) still the effect has practical consequences and it explains why an antenna does not radiate efficiently or at all if it is mounted too close to a metallic interface.

One way to prove the statement above — in the time-harmonic regime — is to make a clever usage of the reciprocity theorem. We consider a PEC body which occupies a region of space V_1 immersed in an unbounded homogeneous isotropic medium with constitutive parameters ε and μ, and we define two states in $V := \mathbb{R}^3 \setminus \overline{V}_1$ as follows:

State (a) There exist no sources in V, but the electromagnetic field $\mathbf{E}_a(\mathbf{r})$, $\mathbf{H}_a(\mathbf{r})$ is *supposedly* produced by an *impressed* surface current density $\mathbf{J}_{Sa}(\mathbf{r}) := \hat{\mathbf{n}}(\mathbf{r}) \times \mathbf{H}_a(\mathbf{r})$ for $\mathbf{r} \in \partial V_1$ and, of course, satisfies the condition $\mathbf{E}_a(\mathbf{r}) \times \hat{\mathbf{n}}(\mathbf{r}) = 0$ for $\mathbf{r} \in \partial V_1$. We choose the unit normal $\hat{\mathbf{n}}(\mathbf{r})$ on ∂V_1 oriented towards V (Figure 6.25a).

State (b) The source of the field $\mathbf{E}_b(\mathbf{r})$, $\mathbf{H}_b(\mathbf{r})$ is a volume electric current density $\mathbf{J}_b(\mathbf{r})$ confined in the domain $V_{Jb} \subset V$. Also this field must perforce satisfy the condition $\mathbf{E}_b(\mathbf{r})\times\hat{\mathbf{n}}(\mathbf{r}) = 0$ for $\mathbf{r} \in \partial V_1$ (Figure 6.25b).

Moreover, since the sources are of finite extent, all fields satisfy the Silver-Müller radiation conditions (6.171) and (6.172). With these positions, the general form of the reciprocity theorem (6.253) reduces to

$$\int\limits_{\partial V_1} \mathrm{d}S\, \hat{\mathbf{n}}(\mathbf{r}) \cdot [\mathbf{E}_a(\mathbf{r}) \times \mathbf{H}_b(\mathbf{r}) - \mathbf{E}_b(\mathbf{r}) \times \mathbf{H}_a(\mathbf{r})] = \int\limits_{V_{Jb}} \mathrm{d}V\, \mathbf{E}_a(\mathbf{r}) \cdot \mathbf{J}_b(\mathbf{r}) \tag{6.273}$$

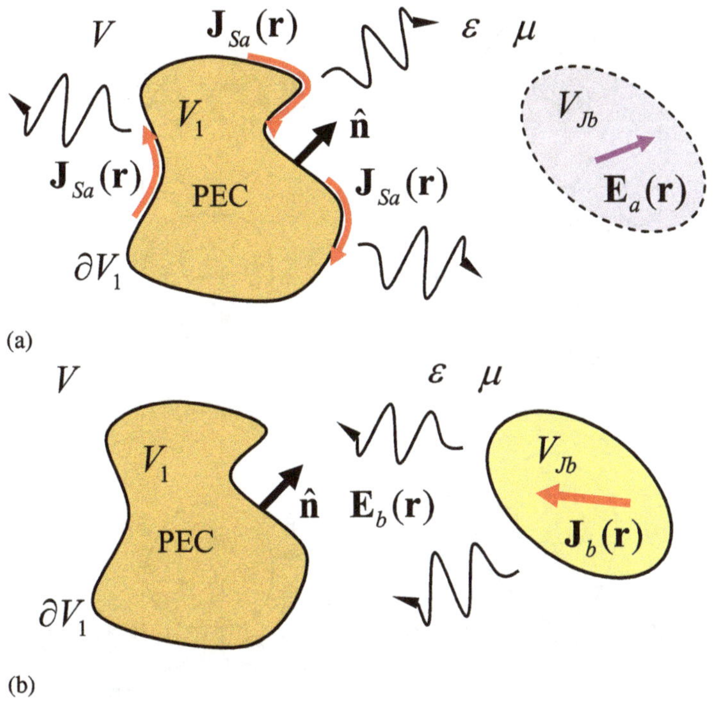

Figure 6.25 Applying reciprocity to show that an impressed surface electric current flush with a PEC boundary does not radiate: state (a) and state (b).

having taken due care of the orientation of the normal for the flux integral. We can convince ourselves that the latter is identically null by casting the integrand into the form

$$[\hat{\mathbf{n}}(\mathbf{r}) \times \mathbf{E}_a(\mathbf{r})] \cdot \mathbf{H}_b(\mathbf{r}) - [\hat{\mathbf{n}}(\mathbf{r}) \times \mathbf{E}_b(\mathbf{r})] \cdot \mathbf{H}_a(\mathbf{r}) = 0 \tag{6.274}$$

where the result follows because the tangential electric field pertinent to both states vanishes on the surface of the PEC object.

Therefore, we are left with a reaction integral which we may formally evaluate with the mean value theorem [84], as $\mathbf{E}_a(\mathbf{r})$ is differentiable in V and, in particular, in V_{Jb}, whereas $\mathbf{J}_b(\mathbf{r})$ is regular[3] for $\mathbf{r} \in V_{Jb}$. In symbols, we have

$$0 = \int_{V_{Jb}} \mathrm{d}V\, \mathbf{E}_a(\mathbf{r}) \cdot \mathbf{J}_b(\mathbf{r}) = \hat{V}\mathbf{E}_a(\mathbf{r}_0) \cdot \mathbf{J}_b(\mathbf{r}_0) \tag{6.275}$$

where $\mathbf{r}_0 \in V_{Jb}$ is a suitable point and $\hat{V}$ denotes the volume of V_{Jb}. Barring the very special case where $\mathbf{E}_a(\mathbf{r}_0)$ is perpendicular to $\mathbf{J}_b(\mathbf{r}_0)$, the last equation implies that the component of $\mathbf{E}_a(\mathbf{r}_0)$ aligned with $\mathbf{J}_b(\mathbf{r}_0)$ is null. Since the shape and the position of V_{Jb} are arbitrary and so is $\mathbf{J}_b(\mathbf{r})$, we can repeat the procedure *ad libitum* and conclude that $\mathbf{E}_a(\mathbf{r}) = 0$ for $\mathbf{r} \in V$, as we wished to prove. As there are no magnetic currents in state (a), the Faraday law (1.99) requires $\mathbf{H}_a(\mathbf{r}) = 0$ for $\mathbf{r} \in V$. Alternatively, we may reach the same conclusion by defining the state (b) with a magnetic current density $\mathbf{J}_{Mb}(\mathbf{r})$ for $\mathbf{r} \in V_{Jb}$.

[3] We excluded the occurrence of singular sources in the derivation of (6.253).

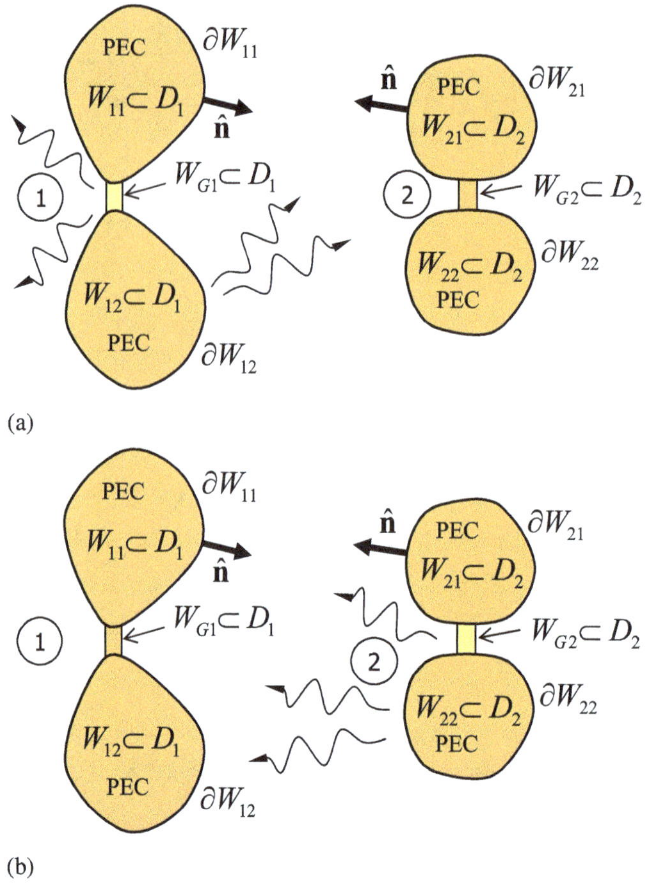

Figure 6.26 Application of reciprocity to two antennas in free space: states (a) and (b).

Incidentally, an application of the duality principle to the previous result allows stating that impressed magnetic currents placed on the boundary of a PMC body do not radiate either.

(End of Example 6.7)

Example 6.8 (Equivalence of mutual admittances of two antennas)

In this example we use the reciprocity theorem (6.256) to prove identity (6.237), which holds true for any two antennas as long as they are operated in the presence of reciprocal media. Since, in particular, the pair of antennas can be interpreted as a two-port system or device — which we can describe by means of an admittance matrix $[Y] \in \mathbb{C}^{2,2}$ — relationship (6.237) is a statement of the symmetry of $[Y]$.

Before we begin the derivation proper, we need to introduce a suitable model of an antenna *port*. For the sake of argument we focus on two metallic or, rather, PEC antennas, labelled ① and ②, which radiate in free space. We assume that each antenna occupies a volume $D_l := W_{l1} \cup W_{Gl} \cup W_{l2}$, $l \in \{1, 2\}$, as is suggested in Figures 6.26a and 6.26b. The volumes W_{l1} and W_{l2} are filled with PEC, whereas W_{Gl} — called the antenna *gap* — is a 'small' cylindrical region which represents the port.

Ordinarily, antennas are connected to generators through coaxial cables, and hence the sources are essentially equivalent magnetic surface current densities that flow on the cross section of the truncated cables (see Example 6.4). Here, we make the hypothesis that, when the antenna in question

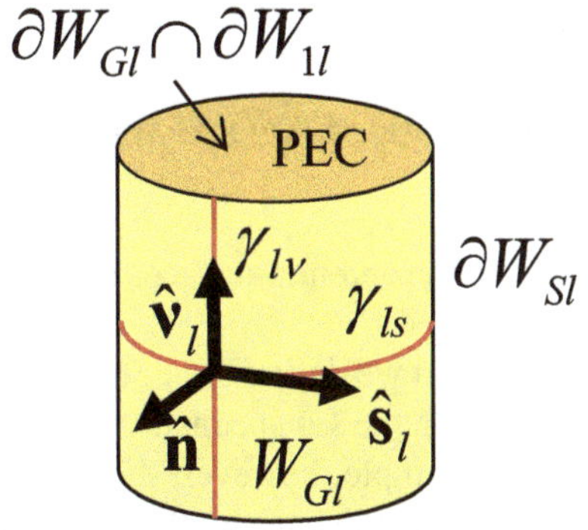

Figure 6.27 Close-up of the antenna gaps W_{Gl} (see Figure 6.26) and related geometrical quantities.

is excited, such equivalent sources are located *inside* the gap W_{Gl}. Thanks to the boundary condition (1.169) at the surface of a PEC body, the total tangential electric field (due to the impressed sources in the gaps and the secondary ones induced on the conductors) is zero for $\mathbf{r} \in \partial W_{l1} \cup \partial W_{l2}$. Within the gaps — that is, for observation points $\mathbf{r}$ *close* the equivalent sources — the field is static in nature [cf. (1.252)] and perpendicular to the surfaces $\partial W_{l1} \cap \partial W_{Gl}$ and $\partial W_{l2} \cap \partial W_{Gl}$. Conversely, when an antenna is not excited (i.e., the generator is turned off) the relevant port is *short-circuited*. This occurrence is described by assuming that the gap W_{Gl} is filled with PEC, as is the case for antenna ② in Figure 6.26a or antenna ① in Figure 6.26b. The latter hypothesis implies that the total tangential electric field over the surfaces $\partial W_{Sl} := \partial W_{Gl} \cap \partial D_l$ is zero.

The antenna gap, along with the properties just outlined, is a good, though approximate model of the region of space near the aperture of the truncated coaxial cable (Figure 6.18). This model may be used provided the gap is comparatively smaller than the antenna size, in order that the field be practically uniform in the source region.

We are now ready to apply (6.256) to the unbounded region $\mathbb{R}^3 \setminus (D_1 \cup D_2)$. The reactions are null by construction, as the sources are confined within W_{Gl}, and the integral over the sphere at infinity vanishes because all fields satisfy the Silver-Müller radiation conditions (6.171) and (6.172). We take the unit normal $\hat{\mathbf{n}}(\mathbf{r})$ on ∂D_l positively oriented towards the exterior of D_l. Furthermore, on the surfaces ∂W_{Sl} (see Figure 6.27) we define local sets of orthogonal coordinates (s_l, z_l) identified by the unit vectors $\hat{\mathbf{s}}_l(\mathbf{r})$ and $\hat{\mathbf{v}}_l$, so that

$$\hat{\mathbf{s}}_l(\mathbf{r}) := \hat{\mathbf{v}}_l \times \hat{\mathbf{n}}(\mathbf{r}), \qquad \mathbf{r} \in \partial W_{Sl}, \qquad l \in \{1, 2\} \tag{6.276}$$

with $\hat{\mathbf{v}}_l$ parallel to the axis of the cylinder W_{Gl} and tangent to the coordinate lines γ_{lv}. The vector $\hat{\mathbf{s}}_l(\mathbf{r})$ is tangent to the other coordinate lines γ_{ls}, which are circumferences.

We define the two states as follows:

State (a) Antenna ① in D_1 is excited, whereas antenna ② is switched off (Figure 6.26a). This entails turning off the source within W_{G2} and filling it with PEC. As a result, the tangential electric field $\mathbf{E}_a(\mathbf{r}) \times \hat{\mathbf{n}}(\mathbf{r}) = 0$ for points $\mathbf{r} \in \partial D_2$, i.e., on the metallic surfaces but also on the partial gap surface ∂W_{S2}.

State (b) The situation is reversed: antenna ② is excited and antenna ① is switched off (Figure 6.26b). Hence, $\mathbf{E}_b(\mathbf{r}) \times \hat{\mathbf{n}}(\mathbf{r}) = 0$ for points $\mathbf{r} \in \partial D_1$, that is, on the metallic surfaces but also on the partial gap surface W_{S1}.

With these positions, (6.256) reduces to

$$\int_{\partial W_{S1}} \mathrm{d}S\, \hat{\mathbf{n}}(\mathbf{r}) \cdot \mathbf{E}_a(\mathbf{r}) \times \mathbf{H}_b(\mathbf{r}) - \int_{\partial W_{S2}} \mathrm{d}S\, \hat{\mathbf{n}}(\mathbf{r}) \cdot \mathbf{E}_b(\mathbf{r}) \times \mathbf{H}_a(\mathbf{r}) = 0 \tag{6.277}$$

because the tangential electric fields on the antenna surfaces are zero everywhere except on ∂W_{Sl}, as is prescribed by the chosen states.

The attentive Reader might object that we derived (6.256) by demanding that the medium which fills the region of interest be endowed with the *same* constitutive parameters in state (a) and state (b), according to (6.249)-(6.251). In this example, however, the properties of the medium in the gaps W_{Gl} are chosen *differently* in either state. Is this wrong, then? No, it is not, and we are not violating the stipulations for which (6.256) is valid, inasmuch as the gaps W_{Gl} are *outside* the region of interest $\mathbb{R}^3 \setminus (D_1 \cup D_2)$, and hence we are allowed to alter the geometry or the properties of whatever medium we place in D_l if this suits our purposes, so long as we do not modify the shape of ∂D_l.

We now manipulate the flux integrals in order to exhibit voltages and currents explicitly. At the port of antenna ① we have

$$\begin{aligned}
\int_{\partial W_{S1}} \mathrm{d}S\, \hat{\mathbf{n}}(\mathbf{r}) \cdot \mathbf{E}_a(\mathbf{r}) \times \mathbf{H}_b(\mathbf{r}) &= \int_{\partial W_{S1}} \mathrm{d}S\, \hat{\mathbf{s}}_1(\mathbf{r}) \times \hat{\nu}_1 \cdot \mathbf{E}_a(\mathbf{r}) \times \mathbf{H}_b(\mathbf{r}) \\
&= \int_{\partial W_{S1}} \mathrm{d}S\, \hat{\mathbf{s}}_1(\mathbf{r}) \cdot \hat{\nu}_1 \times [\mathbf{E}_a(\mathbf{r}) \times \mathbf{H}_b(\mathbf{r})] \\
&= \int_{\partial W_{S1}} \mathrm{d}S\, \hat{\mathbf{s}}_1(\mathbf{r}) \cdot [\mathbf{E}_a(\mathbf{r})\hat{\nu}_1 \cdot \mathbf{H}_b(\mathbf{r}) - \mathbf{H}_b(\mathbf{r})\hat{\nu}_1 \cdot \mathbf{E}_a(\mathbf{r})] \\
&= -\int_{\partial W_{S1}} \mathrm{d}S\, \hat{\mathbf{s}}_1(\mathbf{r}) \cdot \mathbf{H}_b(\mathbf{r})\, \hat{\nu}_1 \cdot \mathbf{E}_a(\mathbf{r}) \\
&= \underbrace{-\int_{\gamma_{1\nu}} \mathrm{d}z_1\, \hat{\nu}_1 \cdot \mathbf{E}_a(z_1)}_{=V_1} \underbrace{\oint_{\gamma_{1s}} \mathrm{d}s_1\, \hat{\mathbf{s}}_1(\mathbf{r}) \cdot \mathbf{H}_b(\mathbf{r})}_{=I_{12}} = V_1 I_{12}
\end{aligned} \tag{6.278}$$

having used, seriatim, (6.276), (H.13), (H.14), and the fact that $\hat{\mathbf{s}}_1(\mathbf{r}) \cdot \mathbf{E}_a(\mathbf{r}) = 0$, as the field is static in the gap. Finally, the last step follows from the Ansatz (2.15) and the Ampère law (4.5) again on the grounds of the static character of $\mathbf{E}_a(\mathbf{r})$ and $\mathbf{H}_b(\mathbf{r})$ for $\mathbf{r} \in \partial W_{S1}$. In particular, $V_1 := \Phi_{11} - \Phi_{12}$ is the potential difference between the conductors in W_{11} and W_{12}, and I_{12} indicates the current flowing through port 1 (which is short circuited) when antenna ② is excited.

In like manner, at the port of antenna ② we have

$$\begin{aligned}
\int_{\partial W_{S2}} \mathrm{d}S\, \hat{\mathbf{n}}(\mathbf{r}) \cdot \mathbf{E}_b(\mathbf{r}) \times \mathbf{H}_a(\mathbf{r}) &= \int_{\partial W_{S2}} \mathrm{d}S\, \hat{\mathbf{s}}_2(\mathbf{r}) \times \hat{\nu}_2 \cdot \mathbf{E}_b(\mathbf{r}) \times \mathbf{H}_a(\mathbf{r}) \\
&= \int_{\partial W_{S2}} \mathrm{d}S\, \hat{\mathbf{s}}_2(\mathbf{r}) \cdot \hat{\nu}_2 \times [\mathbf{E}_b(\mathbf{r}) \times \mathbf{H}_a(\mathbf{r})] \\
&= \int_{\partial W_{S2}} \mathrm{d}S\, \hat{\mathbf{s}}_2(\mathbf{r}) \cdot [\mathbf{E}_b(\mathbf{r})\hat{\nu}_2 \cdot \mathbf{H}_a(\mathbf{r}) - \mathbf{H}_a(\mathbf{r})\hat{\nu}_2 \cdot \mathbf{E}_b(\mathbf{r})]
\end{aligned}$$

$$= - \int_{\partial W_{S2}} \mathrm{d}S\, \hat{\mathbf{s}}_2(\mathbf{r}) \cdot \mathbf{H}_a(\mathbf{r})\, \hat{\boldsymbol{\nu}}_2 \cdot \mathbf{E}_b(\mathbf{r})$$

$$= \underbrace{- \int_{\gamma_{2v}} \mathrm{d}z_2\, \hat{\boldsymbol{\nu}}_2 \cdot \mathbf{E}_b(z_2)}_{=V_2} \underbrace{\oint_{\gamma_{2s}} \mathrm{d}s_2\, \hat{\mathbf{s}}_2(\mathbf{r}) \cdot \mathbf{H}_a(\mathbf{r})}_{=I_{21}} = V_2 I_{21} \tag{6.279}$$

where the voltage $V_2 := \Phi_{21} - \Phi_{22}$ is the potential difference between the conductors in W_{21} and W_{22}, and the notation I_{21} signifies the current I_2 flowing through port 2 (which is short circuited) when antenna ① is excited.

Inserting these two results into (6.277) yields (6.237). The ratios

$$Y_{12} := \frac{I_{12}}{V_2}, \qquad\qquad Y_{21} := \frac{I_{21}}{V_1} \tag{6.280}$$

are the off-diagonal elements of the admittance matrix $[Y]$ and are referred to as the *mutual admittances* between the ports.

We conclude by mentioning that, provided the antennas operate in the presence of reciprocal media, (6.237) holds true irrespective of the model adopted for the antenna ports. For instance, in [92] the Author employed a more sophisticated model which relies on the representation of the magnetic current on the aperture of a truncated coaxial cable in terms of the electric and magnetic transverse eigenfunctions of the TEM mode (also see Example 6.4).

(End of Example 6.8)

From a mathematical viewpoint reciprocity and especially the reaction theorem (6.257) mean that the linear operators involved in the equations are *self-adjoint* (Appendix D.6). In fact, we may even interpret the reactions as inner products of the type (D.68) in a suitable vector space. In this regard, we first study a simpler one-dimensional problem to lay the groundwork for the result in the context of Maxwell's equations.

We consider the space $\mathfrak{C}^2(I)$ of twice-differentiable real scalar fields defined over the interval $I := [x_1, x_2] \subset \mathbb{R}$. A suitable inner product over I is

$$(f_a, f_b)_I := \int_I \mathrm{d}x\, f_a(x) f_b(x) \tag{6.281}$$

where $f_a(x)$ and $f_b(x)$ are any two fields in $\mathfrak{C}^2(I) \subset \mathcal{L}^2(I)$ [see definition (D.64)]. We say that a linear operator

$$\mathfrak{L}\{\bullet\} : \begin{cases} \mathfrak{C}^2(I) \longrightarrow \mathfrak{C}^2(I) \\ f(x) \longmapsto h(x) \end{cases} \tag{6.282}$$

is *self-adjoint* with respect to the chosen inner product if the following identity holds

$$(f_b, \mathfrak{L}\{f_a\})_I = (\mathfrak{L}\{f_b\}, f_a)_I \tag{6.283}$$

which in words means that the operator $\mathfrak{L}\{\bullet\}$ can be applied indifferently to either function involved in the inner product (Appendix D.3) [50, Section 10.4].

Now, suppose that $f_a(x)$ and $f_b(x)$ satisfy the harmonic equations

$$\mathfrak{L}\{f_a\} := \left(\frac{\mathrm{d}^2}{\mathrm{d}x^2} + k^2\right) f_a(x) = -h_a(x), \qquad x \in I \tag{6.284}$$

$$\mathfrak{L}\{f_b\} := \left(\frac{\mathrm{d}^2}{\mathrm{d}x^2} + k^2\right) f_b(x) = -h_b(x), \qquad x \in I \tag{6.285}$$

with $h_a(x)$ and $h_b(x)$ being two source terms in $\mathcal{C}^2(I)$, and $k \in \mathbb{R}$. The harmonic equation is the one-dimensional scalar counterpart of the wave equation (1.238), and may be employed to describe, among other wave phenomena, the small-amplitude oscillations of a string [93, Chapter 49]. We wish to ascertain whether or not $\mathfrak{L}\{\bullet\}$ thus defined is self-adjoint. As it turns out, the property depends on the boundary conditions imposed on the solutions for $x \in \{x_1, x_2\}$.

We have

$$\begin{aligned}
(f_b, \mathfrak{L}\{f_a\})_I &:= \int_I \mathrm{d}x\, f_b(x)\left(\frac{\mathrm{d}^2 f_a}{\mathrm{d}x^2} + k^2 f_a(x)\right) \\
&= \left[f_b(x)\frac{\mathrm{d}f_a}{\mathrm{d}x} - f_a(x)\frac{\mathrm{d}f_b}{\mathrm{d}x}\right]_{x_1}^{x_2} + \int_I \mathrm{d}x \left(\frac{\mathrm{d}^2 f_b}{\mathrm{d}x^2} + k^2 f_b(x)\right) f_a(x) \\
&= \left[f_b(x)\frac{\mathrm{d}f_a}{\mathrm{d}x} - f_a(x)\frac{\mathrm{d}f_b}{\mathrm{d}x}\right]_{x_1}^{x_2} + (\mathfrak{L}\{f_b\}, f_a)_I
\end{aligned} \tag{6.286}$$

having integrated by parts twice. Evidently, (6.283) is true for $\mathfrak{L}\{\bullet\}$ provided the finite increment in the right-hand side above vanishes. This is possible if we impose on $f_a(x)$ and $f_b(x)$ one of the following sets of boundary conditions

$$f(x_1) = 0 \qquad\qquad f(x_2) = 0 \tag{6.287}$$

$$\left.\frac{\mathrm{d}f}{\mathrm{d}x}\right|_{x_1} = 0 \qquad\qquad \left.\frac{\mathrm{d}f}{\mathrm{d}x}\right|_{x_2} = 0 \tag{6.288}$$

$$\left.\frac{\mathrm{d}f}{\mathrm{d}x}\right|_{x_1} = C_1 f(x_1) \qquad\qquad \left.\frac{\mathrm{d}f}{\mathrm{d}x}\right|_{x_2} = C_2 f(x_2) \tag{6.289}$$

where C_1 and C_2 are two real constants. These requirements are called *Dirichlet*, *Neumann* and *Robin* boundary conditions, respectively. When they are not met, the operator is not self-adjoint with respect to the inner product (6.281). This also hints that if we choose another inner product, then $\mathfrak{L}\{\bullet\}$ may become self-adjoint.

For instance, if we introduce the inner product (cf. [6, Section 1.7])

$$((f_b, \mathfrak{L}\{f_a\}))_I := \int_I \mathrm{d}x\, f_b(x)\left(\frac{\mathrm{d}^2 f_a}{\mathrm{d}x^2} + k^2 f_a(x)\right) - \left[\frac{\mathrm{d}f_a}{\mathrm{d}x} f_b(x)\right]_{x_1}^{x_2} \tag{6.290}$$

then (6.286) passes over into

$$((f_b, \mathfrak{L}\{f_a\}))_I = ((\mathfrak{L}\{f_b\}, f_a))_I \tag{6.291}$$

whereby the operator is self-adjoint regardless of the boundary conditions in x_1 and x_2. We may now extend these ideas to the time-harmonic Maxwell equations.

For the sake of simplicity we consider a bounded region $V \subset \mathbb{R}^3$ filled with a homogeneous anisotropic medium endowed with symmetric dyadic constitutive parameters $\overline{\varepsilon}$, $\overline{\mu}$ and $\sigma\overline{\mathcal{I}}$. Then, we already know that reciprocity holds and the reaction theorem (6.257) applies when either (6.262) or (6.263) are satisfied for $\mathbf{r} \in \partial V$. In particular, by introducing a function space whose elements are the abstract six-dimensional complex vectors

$$\mathbf{f}(\mathbf{r}) := \begin{bmatrix} \mathbf{E}(\mathbf{r}) \\ \mathbf{H}(\mathbf{r}) \end{bmatrix}, \qquad \mathbf{r} \in V \tag{6.292}$$

$$\mathbf{g}(\mathbf{r}) := \begin{bmatrix} \mathbf{J}(\mathbf{r}) \\ -\mathbf{J}_M(\mathbf{r}) \end{bmatrix}, \qquad \mathbf{r} \in V_J \subset V \tag{6.293}$$

the meaning of the reactions (6.254) and (6.255) as inner products over V becomes apparent. Moreover, direct inspection shows that

$$\langle \mathbf{f}_b, \mathbf{g}_a \rangle = \langle \mathbf{g}_a, \mathbf{f}_b \rangle \tag{6.294}$$

i.e., the reactions are symmetric inner products. If we write the local extended Faraday law and the local Ampère-Maxwell law in symbolic matrix form as

$$\mathfrak{L}\{\mathbf{f}\} := \begin{bmatrix} -\left(\mathrm{j}\,\omega\overline{\varepsilon} + \sigma\overline{\mathfrak{I}}\right)\cdot & \nabla \times \overline{\mathfrak{I}}\cdot \\ \nabla \times \overline{\mathfrak{I}}\cdot & \mathrm{j}\,\omega\overline{\mu}\cdot \end{bmatrix} \begin{bmatrix} \mathbf{E}(\mathbf{r}) \\ \mathbf{H}(\mathbf{r}) \end{bmatrix} = \begin{bmatrix} \mathbf{J}(\mathbf{r}) \\ -\mathbf{J}_M(\mathbf{r}) \end{bmatrix} \tag{6.295}$$

then (6.257) states that the operator $\mathfrak{L}\{\bullet\}$ is self-adjoint.[4] Indeed, we have

$$\langle \mathbf{f}_b, \mathfrak{L}\{\mathbf{f}_a\} \rangle = \langle \mathbf{f}_b, \mathbf{g}_a \rangle = \langle \mathbf{f}_a, \mathbf{g}_b \rangle = \langle \mathbf{g}_b, \mathbf{f}_a \rangle = \langle \mathfrak{L}\{\mathbf{f}_b\}, \mathbf{f}_a \rangle \tag{6.296}$$

whereby the assertion is proved.

When neither of the boundary conditions (6.262) and (6.263) is met, the flux integral in (6.253) does not vanish. As a result, the operator $\mathfrak{L}\{\bullet\}$ is not self-adjoint with respect to the inner product (6.254) or (6.255). However, we may define alternative inner products as [87]

$$((\mathbf{f}_b, \mathfrak{L}\{\mathbf{f}_a\}))_V := \langle \mathbf{f}_b, \mathfrak{L}\{\mathbf{f}_a\} \rangle + \int_{\partial V} \mathrm{d}S\, \mathbf{E}_b(\mathbf{r}) \cdot \mathbf{H}_a(\mathbf{r}) \times \hat{\mathbf{n}}(\mathbf{r}) \tag{6.297}$$

$$((\mathbf{f}_b, \mathfrak{L}\{\mathbf{f}_a\}))_V := \langle \mathbf{f}_b, \mathfrak{L}\{\mathbf{f}_a\} \rangle + \int_{\partial V} \mathrm{d}S\, \mathbf{H}_b(\mathbf{r}) \cdot \mathbf{E}_a(\mathbf{r}) \times \hat{\mathbf{n}}(\mathbf{r}) \tag{6.298}$$

and also

$$((\mathbf{f}_b, \mathfrak{L}\{\mathbf{f}_a\}))_V := \langle \mathbf{f}_b, \mathfrak{L}\{\mathbf{f}_a\} \rangle + \frac{1}{2}\int_{\partial V} \mathrm{d}S\, [\mathbf{E}_b(\mathbf{r}) \times \mathbf{H}_a(\mathbf{r}) + \mathbf{H}_b(\mathbf{r}) \times \mathbf{E}_a(\mathbf{r})] \cdot \hat{\mathbf{n}}(\mathbf{r}) \tag{6.299}$$

whereby the identity

$$((\mathbf{f}_b, \mathfrak{L}\{\mathbf{f}_a\}))_V = ((\mathfrak{L}\{\mathbf{f}_b\}, \mathbf{f}_a))_V \tag{6.300}$$

invariably follows from (6.253).

[4]Notice that $\nabla \times \overline{\mathfrak{I}} \cdot \mathbf{E}(\mathbf{r}) = \nabla \times \mathbf{E}(\mathbf{r})$ and so forth.

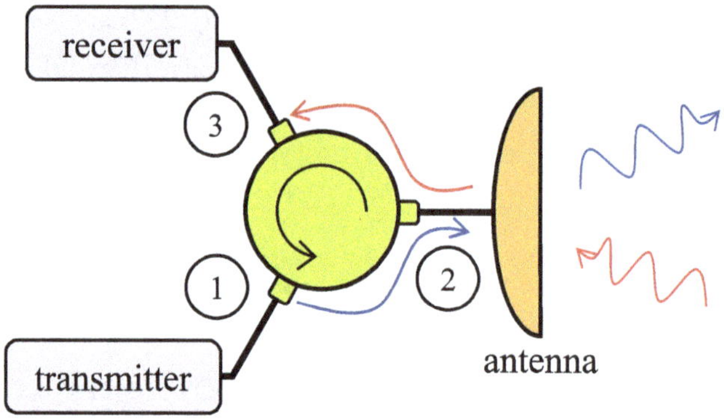

Figure 6.28 A circulator used as a duplexer to separate transmitted and received signals in a radar system.

6.8.2 Non-reciprocal media

The results of the previous sections were derived under the assumption that the constitutive parameters of the media contained in the volume V satisfy the symmetry conditions (6.249)-(6.251). Therefore, reciprocity — as embodied by the relations (6.252), (6.256) and (6.257) — is violated whenever symmetry in a wide sense is broken, an occurrence which sometimes may even be desirable.

A typical example is constituted by cold magnetized plasma which is characterized by a non-symmetric dyadic permittivity $\overline{\varepsilon}(\mathbf{r};\omega)$ [94], [95, Chapter 5], [86, Chapter 6], [73, Section 8.9]. Detailed calculations based on a classical model of electrons and ions (cf. Chapter 12) show that responsible for the breaking of the symmetry is the external magnetic field $\mathcal{B}_0(\mathbf{r})$ in which the plasma is immersed. From a physical and microscopic point of view, $\mathcal{B}_0(\mathbf{r})$ introduces a *preferred* axis around which the charged particles that form the plasma tend to rotate. Clearly, it is not possible to swap source and observer in a magnetized plasma, inasmuch as the fields detected by the observer depend on the way the particles rotate with respect to the source. Of course, removing the cause of asymmetry (i.e., setting $\mathcal{B}_0(\mathbf{r}) = 0$) makes the plasma isotropic [95, Section 5.7] and hence reciprocal in accordance with (6.249).

Magnetized ferrites are another example of non-reciprocal materials [40]. In this case, an external magnetic field $\mathcal{B}_0(\mathbf{r})$ causes the permeability dyadic $\overline{\mu}$ to become non-symmetric. Ferrites are used in electromagnetic components, such as isolators and circulators, which precisely exploit the different behavior of the electromagnetic field under an interchange of source and observer. An isolator lets the guided electromagnetic waves through only, say, from left to right, whereas it reflects the waves travelling from right to left. For this reason, the function of an isolator is to protect the generator from possibly unexpected waves reflected back by a mismatched load.

A ferrite circulator [96], [19, Section 9.6] works according to the same principle. In its simplest form a circulator has three ports, and the presence of a magnetized ferrite endows the device with a preferred direction of propagation of the energy, say, in a counterclockwise fashion from port ① to port ② to port ③. Circulators find application, e.g., as *duplexers* in radar systems where the same antenna is employed both in transmission and reception in order to save space and money. A practical setup is sketched in Figure 6.28: in transmission the signal from the generator at port ① travels towards the antenna at port ② but in reception the signal captured by the antenna travels towards the receiving circuit connected to port ③.

In the presence of non-reciprocal media it is still possible to obtain the relations (6.252), (6.256) and (6.257), so long as we define the two states pictured in Figure 6.21 in a suitable manner.

For the sake of simplicity, we suppose that the medium which fills $V \setminus V_1$ is isotropic, and hence reciprocal, with parameters $\varepsilon = \varepsilon_a = \varepsilon_b$ and $\mu = \mu_a = \mu_b$. For state (a) within V_1 we allow for non-reciprocal media characterized by non-symmetric dyadic permittivity $\overline{\varepsilon}_a(\mathbf{r})$ and permeability $\overline{\mu}_a(\mathbf{r})$. So, how should we choose, if possible, the parameters $\overline{\varepsilon}_b(\mathbf{r})$ and $\overline{\mu}_b(\mathbf{r})$ of the medium in V_1 for state (b) in order for the left-hand sides of (6.246) and (6.247) to vanish identically?

Well, the answer is already contained in (6.246) and (6.247), as it were. Apparently, if we let

$$\overline{\varepsilon}_b(\mathbf{r}) := [\overline{\varepsilon}_a(\mathbf{r})]^T, \qquad \mathbf{r} \in V_1 \tag{6.301}$$

$$\overline{\mu}_b(\mathbf{r}) := [\overline{\mu}_a(\mathbf{r})]^T, \qquad \mathbf{r} \in V_1 \tag{6.302}$$

then we achieve the desired cancellation of the terms which involve the material properties, and (6.252), (6.256) and (6.257) hold true again. However, we cannot help but emphasize a subtle, though substantial difference behind this result.

In the previous section, we concluded that reciprocity holds if both state (a) and state (b) involve the *same medium* which in turn must satisfy the symmetry relations (6.250) and (6.251). Here, on the contrary, for state (b) we deliberately choose a *different material* than that of state (a) so that the left-hand sides of (6.246) and (6.247) happen to be null. In essence, we are combining two problems which have in common only the *shape* of the region of interest V (this also concerns the shape of V_1). Indeed, the relevant Maxwell equations for the two states read

$$\nabla \times \mathbf{E}_a(\mathbf{r}) = -\mathrm{j}\,\omega\overline{\mu}_a(\mathbf{r}) \cdot \mathbf{H}_a(\mathbf{r}) - \mathbf{J}_{Ma}(\mathbf{r}) \tag{6.303}$$

$$\nabla \times \mathbf{H}_a(\mathbf{r}) = \mathrm{j}\,\omega\overline{\varepsilon}_a(\mathbf{r}) \cdot \mathbf{E}_a(\mathbf{r}) + \sigma_a\mathbf{E}_a(\mathbf{r}) + \mathbf{J}_a(\mathbf{r}) \tag{6.304}$$

$$\nabla \times \mathbf{E}_b(\mathbf{r}) = -\mathrm{j}\,\omega\overline{\mu}_a^T(\mathbf{r}) \cdot \mathbf{H}_b(\mathbf{r}) - \mathbf{J}_{Mb}(\mathbf{r}) \tag{6.305}$$

$$\nabla \times \mathbf{H}_b(\mathbf{r}) = \mathrm{j}\,\omega\overline{\varepsilon}_a^T(\mathbf{r}) \cdot \mathbf{E}_b(\mathbf{r}) + \sigma_b\mathbf{E}_b(\mathbf{r}) + \mathbf{J}_b(\mathbf{r}) \tag{6.306}$$

for $\mathbf{r} \in V$. The particular state (b) which meets the requirements (6.301) and (6.302) is called the *adjoint problem* (in the frequency domain).

6.8.3 Time domain

Theorems similar to (6.253) and (6.257) can be obtained for fields and sources with general time dependence [3,17,30] provided we define state (b) as a suitable *adjoint problem* (in time domain) of state (a) in order to accomplish the cancellation of terms which involve the constitutive parameters when we combine the relevant curl equations. Since the procedure also entails integration in time, the adjoint problem is obtained from the original one by means of *time and space inversions*. The latter transformations amount to systematically substituting the temporal and spatial derivatives in the Maxwell equations with the inverted counterparts, viz., $-\partial/\partial t$ and $-\nabla\times$ or $-\nabla\cdot$, respectively. The rationale for this choice will become clear soon.

To keep the derivation lucid we consider a finite domain V with smooth boundary $S := \partial V$ filled with an isotropic homogeneous medium endowed with constant constitutive parameters ε, μ and σ. The two states are described below:

State (a) For $(\mathbf{r}, t) \in V \times \{t \in \mathbb{R} : t \geqslant t_1\}$ the fields satisfy the curl equations

$$\nabla \times \mathcal{E}_a(\mathbf{r}, t) = -\mu\frac{\partial}{\partial t}\mathcal{H}_a(\mathbf{r}, t) - \mathcal{J}_{Ma}(\mathbf{r}, t) \tag{6.307}$$

$$\nabla \times \mathcal{H}_a(\mathbf{r}, t) = \varepsilon\frac{\partial}{\partial t}\mathcal{E}_a(\mathbf{r}, t) + \sigma\mathcal{E}_a(\mathbf{r}, t) + \mathcal{J}_a(\mathbf{r}, t) \tag{6.308}$$

subject to the initial conditions

$$\mathcal{E}_a(\mathbf{r}, t) = 0 = \mathcal{H}_a(\mathbf{r}, t), \qquad t \leqslant t_1 \tag{6.309}$$

with the sources confined in a volume $V_{Ja} \subset V$ and null for $t < t_1$.

State (b) For $(\mathbf{r}, t) \in V \times \{t \in \mathbb{R} : t \leqslant t_2\}$ the fields satisfy the time- and space-reversed Faraday and Ampère-Maxwell laws

$$-\nabla \times \mathcal{E}_b(\mathbf{r}, t) = \mu \frac{\partial}{\partial t} \mathcal{H}_b(\mathbf{r}, t) - \mathcal{J}_{Mb}(\mathbf{r}, t) \tag{6.310}$$

$$-\nabla \times \mathcal{H}_b(\mathbf{r}, t) = -\varepsilon \frac{\partial}{\partial t} \mathcal{E}_b(\mathbf{r}, t) + \sigma \mathcal{E}_b(\mathbf{r}, t) + \mathcal{J}_b(\mathbf{r}, t) \tag{6.311}$$

subject to the 'final' conditions

$$\mathcal{E}_b(\mathbf{r}, t) = 0 = \mathcal{H}_b(\mathbf{r}, t), \qquad t \geqslant t_2 \tag{6.312}$$

with the sources confined in a volume $V_{Jb} \subset V$ and null for $t > t_2$.

As pointed out by Felsen and Marcuvitz [17, Section 1.1b], the set of time- and space-reversed Maxwell equations admits *ingoing-wave* solutions, characterized by an *advanced* time $t_a := t + |\mathbf{r} - \mathbf{r}'|/c$ [cf. (1.251)], where c is the speed of light in the underlying medium.

If we choose $t_1 < t_2$, then we can combine the equations pertinent to the two states for times $t \in [t_1, t_2]$ during which interval all of them are valid. In particular, we dot-multiply (6.307) by $\mathcal{H}_b(\mathbf{r}, t)$ and (6.311) by $\mathcal{E}_a(\mathbf{r}, t)$, and sum the resulting expressions side by side to get

$$\begin{aligned} \nabla \cdot [\mathcal{E}_a(\mathbf{r}, t) \times \mathcal{H}_b(\mathbf{r}, t)] + \mu \mathcal{H}_b(\mathbf{r}, t) \cdot \frac{\partial \mathcal{H}_a}{\partial t} + \varepsilon \mathcal{E}_a(\mathbf{r}, t) \cdot \frac{\partial \mathcal{E}_b}{\partial t} \\ = \sigma \mathcal{E}_a(\mathbf{r}, t) \cdot \mathcal{E}_b(\mathbf{r}, t) + \mathcal{E}_a(\mathbf{r}, t) \cdot \mathcal{J}_b(\mathbf{r}, t) - \mathcal{H}_b(\mathbf{r}, t) \cdot \mathcal{J}_{Ma}(\mathbf{r}, t) \end{aligned} \tag{6.313}$$

for $\mathbf{r} \in V$ and $t \in [t_1, t_2]$ and by virtue of (H.49). Another equation follows if we dot-multiply (6.308) by $\mathcal{E}_b(\mathbf{r}, t)$ and (6.310) by $\mathcal{H}_a(\mathbf{r}, t)$, and then sum side by side[5]

$$\begin{aligned} \nabla \cdot [\mathcal{E}_b(\mathbf{r}, t) \times \mathcal{H}_a(\mathbf{r}, t)] + \varepsilon \mathcal{E}_b(\mathbf{r}, t) \cdot \frac{\partial \mathcal{E}_a}{\partial t} + \mu \mathcal{H}_a(\mathbf{r}, t) \cdot \frac{\partial \mathcal{H}_b}{\partial t} \\ = -\sigma \mathcal{E}_b(\mathbf{r}, t) \cdot \mathcal{E}_a(\mathbf{r}, t) + \mathcal{H}_a(\mathbf{r}, t) \cdot \mathcal{J}_{Mb}(\mathbf{r}, t) - \mathcal{E}_b(\mathbf{r}, t) \cdot \mathcal{J}_a(\mathbf{r}, t) \end{aligned} \tag{6.314}$$

again for $\mathbf{r} \in V$ and $t \in [t_1, t_2]$. Next, we sum (6.313) and (6.314) side by side

$$\begin{aligned} \nabla \cdot [\mathcal{E}_a(\mathbf{r}, t) \times \mathcal{H}_b(\mathbf{r}, t) + \mathcal{E}_b(\mathbf{r}, t) \times \mathcal{H}_a(\mathbf{r}, t)] + \frac{\partial}{\partial t}[\varepsilon \mathcal{E}_a(\mathbf{r}, t) \cdot \mathcal{E}_b(\mathbf{r}, t) + \mu \mathcal{H}_a(\mathbf{r}, t) \cdot \mathcal{H}_b(\mathbf{r}, t)] \\ = \mathcal{E}_a(\mathbf{r}, t) \cdot \mathcal{J}_b(\mathbf{r}, t) + \mathcal{H}_a(\mathbf{r}, t) \cdot \mathcal{J}_{Mb}(\mathbf{r}, t) - \mathcal{E}_b(\mathbf{r}, t) \cdot \mathcal{J}_a(\mathbf{r}, t) - \mathcal{H}_b(\mathbf{r}, t) \cdot \mathcal{J}_{Ma}(\mathbf{r}, t) \end{aligned} \tag{6.315}$$

where the appearance of the time derivative of the vector field in brackets is a consequence of the time-reversal in state (b). More importantly, integration of said contribution for $t \in [t_1, t_2]$ is trivial and yields

$$\int_{t_1}^{t_2} \mathrm{d}t \, \frac{\partial}{\partial t}[\varepsilon \mathcal{E}_a(\mathbf{r}, t) \cdot \mathcal{E}_b(\mathbf{r}, t) + \mu \mathcal{H}_a(\mathbf{r}, t) \cdot \mathcal{H}_b(\mathbf{r}, t)]$$

[5]Notice that simply swapping the subscripts 'a' and 'b' in (6.313) does not do the trick, inasmuch as (6.310) and (6.311) do not follow from (6.307) and (6.308) by means of that transformation.

$$= [\varepsilon \mathcal{E}_a(\mathbf{r},t)\cdot\mathcal{E}_b(\mathbf{r},t) + \mu\mathcal{H}_a(\mathbf{r},t)\cdot\mathcal{H}_b(\mathbf{r},t)]_{t=t_1}^{t_2} = 0 \quad (6.316)$$

because $\mathcal{E}_a(\mathbf{r},t_1) = 0 = \mathcal{H}_a(\mathbf{r},t_1)$ and $\mathcal{E}_b(\mathbf{r},t_2) = 0 = \mathcal{H}_b(\mathbf{r},t_2)$. Therefore, by integrating (6.315) over $V\times[t_1,t_2]$ we arrive at

$$\int_{t_1}^{t_2} \mathrm{d}t \int_S \mathrm{d}S\, \hat{\mathbf{n}}(\mathbf{r})\cdot[\mathcal{E}_a(\mathbf{r},t)\times\mathcal{H}_b(\mathbf{r},t) + \mathcal{E}_b(\mathbf{r},t)\times\mathcal{H}_a(\mathbf{r},t)]$$
$$= \int_{t_1}^{t_2} \mathrm{d}t \int_{V_{Jb}} \mathrm{d}V\, [\mathcal{E}_a(\mathbf{r},t)\cdot\mathcal{J}_b(\mathbf{r},t) + \mathcal{H}_a(\mathbf{r},t)\cdot\mathcal{J}_{Mb}(\mathbf{r},t)]$$
$$- \int_{t_1}^{t_2} \mathrm{d}t \int_{V_{Ja}} \mathrm{d}V\, [\mathcal{E}_b(\mathbf{r},t)\cdot\mathcal{J}_a(\mathbf{r},t) + \mathcal{H}_b(\mathbf{r},t)\cdot\mathcal{J}_{Ma}(\mathbf{r},t)] \quad (6.317)$$

where we have applied the Gauss theorem under the assumption that the fields are differentiable in V for $t\in[t_1,t_2]$. Equation (6.317) is the time-domain analogue of the reciprocity theorem (6.253).

If the flux integral happens to vanish, then (6.317) reduces to

$$\int_{t_1}^{t_2} \mathrm{d}t \int_{V_{Jb}} \mathrm{d}V\, [\mathcal{E}_a(\mathbf{r},t)\cdot\mathcal{J}_b(\mathbf{r},t) + \mathcal{H}_a(\mathbf{r},t)\cdot\mathcal{J}_{Mb}(\mathbf{r},t)]$$
$$= \int_{t_1}^{t_2} \mathrm{d}t \int_{V_{Ja}} \mathrm{d}V\, [\mathcal{E}_b(\mathbf{r},t)\cdot\mathcal{J}_a(\mathbf{r},t) + \mathcal{H}_b(\mathbf{r},t)\cdot\mathcal{J}_{Ma}(\mathbf{r},t)] \quad (6.318)$$

which is the counterpart of (6.257) in the time domain. For instance, the flux integral is null if the fields satisfy impedance boundary conditions for $\mathbf{r}\in S$

$$\hat{\mathbf{n}}(\mathbf{r})\times[\mathcal{E}_a(\mathbf{r},t)\times\hat{\mathbf{n}}(\mathbf{r})] = \overline{\mathcal{Z}}_S\cdot[\hat{\mathbf{n}}(\mathbf{r})\times\mathcal{H}_a(\mathbf{r},t)], \qquad \mathbf{r}\in S \quad (6.319)$$

$$\hat{\mathbf{n}}(\mathbf{r})\times[\mathcal{E}_b(\mathbf{r},t)\times\hat{\mathbf{n}}(\mathbf{r})] = -\overline{\mathcal{Z}}_S^{T}\cdot[\hat{\mathbf{n}}(\mathbf{r})\times\mathcal{H}_b(\mathbf{r},t)], \qquad \mathbf{r}\in S \quad (6.320)$$

where $\overline{\mathcal{Z}}_S$ is the dyadic surface impedance of the conducting medium flush with S. The special case of PEC is obtained with $\overline{\mathcal{Z}}_S = 0$. The impedance relationship for state (b) involves the transpose of $\overline{\mathcal{Z}}_S$, and the minus sign is a consequence of the spatial inversion of (6.308) and (6.311). Similar conclusions can be reached by assuming admittance boundary conditions on S.

6.9 Other symmetry relationships

Relations between the electromagnetic entities pertinent to two sets of solutions of Maxwell's equations may also be derived for electrostatic fields (Chapters 2 and 3) and magnetic fields produced by steady currents (Chapters 4 and 5).

6.9.1 *Electrostatic fields*

To gain some generality we examine fields and potentials inside a finite region of space V bounded by a smooth surface S and the smooth boundary ∂V_1 of an excluded volume V_1 which is filled with

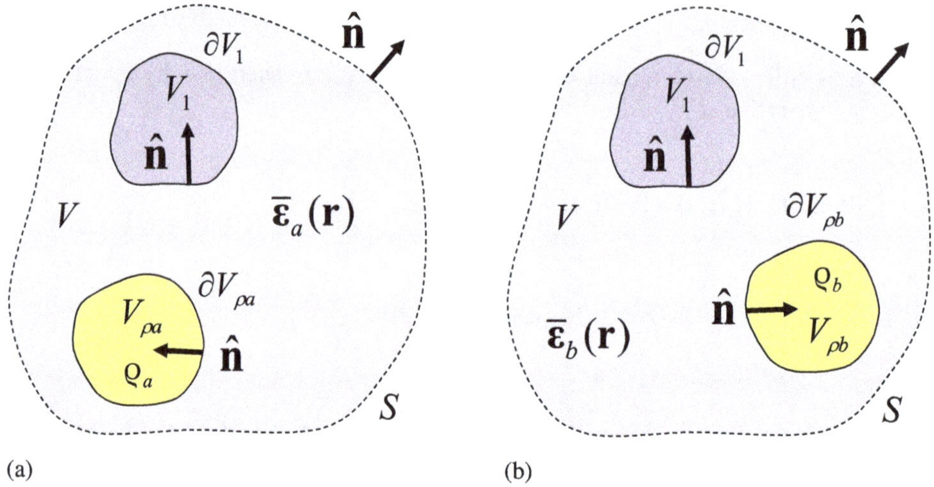

Figure 6.29 For deriving a symmetry relation for static fields: charges and matter for state (a) and state (b).

a medium yet unspecified. The unit normal $\hat{\mathbf{n}}(\mathbf{r})$ on $\partial V := S \cup \partial V_1$ points outwards V. Finally, we consider two different sets of charges and, as we did in Section 6.8, define two states:

State (a) The permittivity of the medium in V is $\overline{\varepsilon}_a(\mathbf{r})$, and the charge density $\varrho_a(\mathbf{r})$ is confined in a volume $V_{\varrho a} \subset V$ (Figure 6.29a).

State (b) The permittivity of the medium in V is $\overline{\varepsilon}_b(\mathbf{r})$, and the charge density $\varrho_b(\mathbf{r})$ is confined in a volume $V_{\varrho b} \subset V$ (Figure 6.29b).

Charges, fields and potentials of both states are related by the Gauss law (2.3) and definition (2.15)

$$\nabla \cdot [\overline{\varepsilon}_a(\mathbf{r}) \cdot \mathcal{E}_a(\mathbf{r})] = \varrho_a(\mathbf{r}) \tag{6.321}$$

$$\mathcal{E}_a(\mathbf{r}) = -\nabla \Phi_a(\mathbf{r}) \tag{6.322}$$

$$\nabla \cdot [\overline{\varepsilon}_b(\mathbf{r}) \cdot \mathcal{E}_b(\mathbf{r})] = \varrho_b(\mathbf{r}) \tag{6.323}$$

$$\mathcal{E}_b(\mathbf{r}) = -\nabla \Phi_b(\mathbf{r}) \tag{6.324}$$

for points $\mathbf{r} \in V$. For (6.321) and (6.323) to hold everywhere in V, the dyadic fields $\overline{\varepsilon}_a(\mathbf{r})$ and $\overline{\varepsilon}_b(\mathbf{r})$ must be continuous, otherwise we should write the Gauss law separately for each part of V in which the continuity of $\overline{\varepsilon}_a(\mathbf{r})$ and $\overline{\varepsilon}_b(\mathbf{r})$ may be assumed (see Sections 1.7 and 2.4).

We multiply (6.321) by $\Phi_b(\mathbf{r})$ and (6.323) by $\Phi_a(\mathbf{r})$

$$\Phi_b(\mathbf{r})\nabla \cdot [\overline{\varepsilon}_a(\mathbf{r}) \cdot \mathcal{E}_a(\mathbf{r})] = \Phi_b(\mathbf{r})\varrho_a(\mathbf{r}) \tag{6.325}$$

$$\Phi_a(\mathbf{r})\nabla \cdot [\overline{\varepsilon}_b(\mathbf{r}) \cdot \mathcal{E}_b(\mathbf{r})] = \Phi_a(\mathbf{r})\varrho_b(\mathbf{r}) \tag{6.326}$$

and subtract the resulting relations to obtain

$$\begin{aligned}\nabla \cdot [\Phi_b(\mathbf{r})\overline{\varepsilon}_a(\mathbf{r}) \cdot \mathcal{E}_a(\mathbf{r}) - \Phi_a(\mathbf{r})\overline{\varepsilon}_b \cdot \mathcal{E}_b(\mathbf{r})] &= \mathcal{E}_a(\mathbf{r}) \cdot \overline{\varepsilon}_b(\mathbf{r}) \cdot \mathcal{E}_b(\mathbf{r}) \\ &- \mathcal{E}_b(\mathbf{r}) \cdot \overline{\varepsilon}_a(\mathbf{r}) \cdot \mathcal{E}_a(\mathbf{r}) + \Phi_b(\mathbf{r})\varrho_a(\mathbf{r}) - \Phi_a(\mathbf{r})\varrho_b(\mathbf{r})\end{aligned} \tag{6.327}$$

by virtue of the differential identity (H.51), (6.322) and (6.324). The first two terms in the right-hand side read

$$\mathcal{E}_a(\mathbf{r}) \cdot \overline{\varepsilon}_b(\mathbf{r}) \cdot \mathcal{E}_b(\mathbf{r}) - \mathcal{E}_b(\mathbf{r}) \cdot \overline{\varepsilon}_a(\mathbf{r}) \cdot \mathcal{E}_a(\mathbf{r}) = \mathcal{E}_a(\mathbf{r}) \cdot \left\{\overline{\varepsilon}_b(\mathbf{r}) - [\overline{\varepsilon}_a(\mathbf{r})]^T\right\} \cdot \mathcal{E}_b(\mathbf{r}), \qquad \mathbf{r} \in V \tag{6.328}$$

whence we conclude that they cancel out if

$$\overline{\varepsilon}(\mathbf{r}) = \overline{\varepsilon}_a(\mathbf{r}) = \overline{\varepsilon}_b(\mathbf{r}) \qquad \text{with} \qquad [\overline{\varepsilon}(\mathbf{r})]^T = \overline{\varepsilon}(\mathbf{r}) \tag{6.329}$$

i.e., the dielectric is reciprocal, in line with the similar finding in (6.250) for the time-harmonic regime. With this assumption, for points $\mathbf{r} \in V$ we obtain

$$\nabla \cdot [\Phi_b(\mathbf{r})\overline{\varepsilon}(\mathbf{r}) \cdot \mathcal{E}_a(\mathbf{r}) - \Phi_a(\mathbf{r})\overline{\varepsilon}(\mathbf{r}) \cdot \mathcal{E}_b(\mathbf{r})] = \Phi_b(\mathbf{r})\varrho_a(\mathbf{r}) - \Phi_a(\mathbf{r})\varrho_b(\mathbf{r}) \tag{6.330}$$

a symmetry relation in differential form for fields, potentials and sources of the two states. To obtain the corresponding global form we integrate (6.330) over V and apply the Gauss theorem, viz.,

$$\int\limits_{S \cup \partial V_1} \mathrm{d}S\ \hat{\mathbf{n}}(\mathbf{r}) \cdot [\Phi_b(\mathbf{r})\overline{\varepsilon}(\mathbf{r}) \cdot \mathcal{E}_a(\mathbf{r}) - \Phi_a(\mathbf{r})\overline{\varepsilon}(\mathbf{r}) \cdot \mathcal{E}_b(\mathbf{r})] = \int\limits_{V_{\varrho a}} \mathrm{d}V\, \Phi_b(\mathbf{r})\varrho_a(\mathbf{r}) - \int\limits_{V_{\varrho b}} \mathrm{d}V\, \Phi_a(\mathbf{r})\varrho_b(\mathbf{r}) \tag{6.331}$$

under the further hypotheses that $\overline{\varepsilon}(\mathbf{r})$ is continuously differentiable in V and that $\hat{\mathbf{n}}(\mathbf{r}) \cdot \mathcal{D}_a(\mathbf{r})$, $\hat{\mathbf{n}}(\mathbf{r}) \cdot \mathcal{D}_b(\mathbf{r})$ are continuous across $\partial V_{\varrho a}$ and $\partial V_{\varrho b}$ [see the discussion in support of the derivation of (1.44)]. It is interesting to determine the conditions for which the flux integral in (6.331) vanishes.

To begin with, we place the charge densities $\varrho_a(\mathbf{r})$ and $\varrho_b(\mathbf{r})$ in $\mathbb{R}^3 \setminus \overline{V}$. Then, the integrals over $V_{\varrho a}$ and $V_{\varrho b}$ vanish by construction. From (6.331) we conclude that the flux integral is null if the sources are located *outside* the region of concern.

We suppose that the unbounded region $\mathbb{R}^3 \setminus \overline{V}$ is a dielectric medium endowed with permittivity $\overline{\varepsilon}(\mathbf{r})$ and that $\overline{\varepsilon}(\mathbf{r})$ is continuously differentiable for $\mathbf{r} \in \mathbb{R}^3 \setminus (S \cup \partial V_1)$. We consider a ball $B(0, a)$ with radius large enough so that $V \subset B(0, a)$ and apply (6.331) to the regions $B(0, a) \setminus (V \cup V_1)$ and V_1 with the sources within V. Then, since $B(0, a) \setminus (V \cup V_1)$ and V_1 are devoid of charges, we find

$$\int\limits_{\partial B} \mathrm{d}S\ \hat{\mathbf{n}}(\mathbf{r}) \cdot [\Phi_b(\mathbf{r})\overline{\varepsilon}(\mathbf{r}) \cdot \mathcal{E}_a(\mathbf{r}) - \Phi_a(\mathbf{r})\overline{\varepsilon}(\mathbf{r}) \cdot \mathcal{E}_b(\mathbf{r})] + \int\limits_{S} \mathrm{d}S\ \hat{\mathbf{n}}(\mathbf{r}) \cdot [\Phi_b(\mathbf{r})\overline{\varepsilon}(\mathbf{r}) \cdot \mathcal{E}_a(\mathbf{r}) - \Phi_a(\mathbf{r})\overline{\varepsilon}(\mathbf{r}) \cdot \mathcal{E}_b(\mathbf{r})] = 0 \tag{6.332}$$

$$\int\limits_{\partial V_1} \mathrm{d}S\ \hat{\mathbf{n}}(\mathbf{r}) \cdot [\Phi_b(\mathbf{r})\overline{\varepsilon}(\mathbf{r}) \cdot \mathcal{E}_a(\mathbf{r}) - \Phi_a(\mathbf{r})\overline{\varepsilon}(\mathbf{r}) \cdot \mathcal{E}_b(\mathbf{r})] = 0 \tag{6.333}$$

with the unit normal $\hat{\mathbf{n}}(\mathbf{r})$ oriented positively outwards $B(0, a) \setminus (V \cup V_1)$ and V_1. The surface integral over the sphere ∂B vanishes in the limit as $a \to +\infty$ in light of the asymptotic behavior (2.20) and (2.21). Hence, by inserting these results into (6.331) we conclude that

$$\int\limits_{V_{\varrho a}} \mathrm{d}V\, \Phi_b(\mathbf{r})\varrho_a(\mathbf{r}) = \int\limits_{V_{\varrho b}} \mathrm{d}V\, \Phi_a(\mathbf{r})\varrho_b(\mathbf{r}) \tag{6.334}$$

if all the sources are placed *inside* the region of concern. This identity is sometimes called the *Green reciprocation theorem* [30, Section 3.1], [29]. Since we may interpret the volume integrals as inner products over V, and the charge density is related to the potential through the Poisson equation (2.26), (6.334) states that the linear operator

$$\mathfrak{L}\{\bullet\} := \nabla \cdot [\overline{\varepsilon}(\mathbf{r}) \cdot \nabla\{\bullet\}] \tag{6.335}$$

is self-adjoint (Appendix D.6) provided $\overline{\varepsilon}(\mathbf{r})$ is a symmetric dyadic field in the region of interest.

It remains to examine the presence of a conductor which, for the sake of argument, we place in the excluded region V_1. The flux integral over ∂V_1 yields

$$\begin{aligned}
&\int_{\partial V_1} \mathrm{d}S\ \hat{\mathbf{n}}(\mathbf{r}) \cdot [\Phi_b(\mathbf{r})\overline{\varepsilon}(\mathbf{r}) \cdot \mathcal{E}_a(\mathbf{r}) - \Phi_a(\mathbf{r})\overline{\varepsilon}(\mathbf{r}) \cdot \mathcal{E}_b(\mathbf{r})] = \\
&\qquad = \Phi_{b1} \int_{\partial V_1} \mathrm{d}S\ \hat{\mathbf{n}}(\mathbf{r}) \cdot \mathcal{D}_a(\mathbf{r}) - \Phi_{a1} \int_{\partial V_1} \mathrm{d}S\ \hat{\mathbf{n}}(\mathbf{r}) \cdot \mathcal{D}_b(\mathbf{r}) \\
&\qquad = -\Phi_{b1} Q_{a1} + \Phi_{a1} Q_{b1}
\end{aligned} \tag{6.336}$$

on the grounds of the boundary conditions (1.182) and (2.57). The total charges induced or placed on the surface ∂V_1 as well as the potentials of the conductor may be different for the two states, whereby the flux integral over ∂V_1 is non-null in general. The symmetry relation (6.331) becomes

$$\begin{aligned}
&\int_{S} \mathrm{d}S\ \hat{\mathbf{n}}(\mathbf{r}) \cdot [\Phi_b(\mathbf{r})\overline{\varepsilon}(\mathbf{r}) \cdot \mathcal{E}_a(\mathbf{r}) - \Phi_a(\mathbf{r})\overline{\varepsilon}(\mathbf{r}) \cdot \mathcal{E}_b(\mathbf{r})] \\
&\qquad = \Phi_{b1} Q_{a1} + \int_{V_{\varrho a}} \mathrm{d}V\ \Phi_b(\mathbf{r})\varrho_a(\mathbf{r}) - \Phi_{a1} Q_{b1} - \int_{V_{\varrho b}} \mathrm{d}V\ \Phi_a(\mathbf{r})\varrho_b(\mathbf{r})
\end{aligned} \tag{6.337}$$

where the flux integral over S vanishes again if all the sources are inside V under the same hypotheses that led to (6.334).

Example 6.9 (Symmetry of the capacitance matrix)

The capacitance matrix $[C]$ with entries given by (2.86) can be proven to be symmetric by applying the symmetry relation (6.331) to the source-free region of space between a system of conductors. For the sake of argument we consider two conductors which occupy two finite domains V_1 and V_2 and are immersed in an isotropic dielectric medium with permittivity ε. The following proof is easily extended to the case of N conductors. We introduce the ball $B(0, a)$ with radius a large enough so that it encloses V_1 and V_2 and define the two states as follows.

State (a) Conductor ① is held at a potential Φ_{01}, and conductor ② is grounded (Figure 6.30a).
State (b) Conductor ① is grounded, and conductor ② is held at a potential Φ_{02} (Figure 6.30b).

Then, by using (6.331) in the surface-wise multiply connected domain $B(0, a) \setminus (V_1 \cup V_2)$ first and by taking the limit as $a \to +\infty$ we find

$$\begin{aligned}
0 &= -\int_{\partial V_1} \mathrm{d}S\ \Phi_{a1}(\mathbf{r})\hat{\mathbf{n}}(\mathbf{r}) \cdot \mathcal{D}_{b1}(\mathbf{r}) + \int_{\partial V_2} \mathrm{d}S\ \Phi_{b2}(\mathbf{r})\hat{\mathbf{n}}(\mathbf{r}) \cdot \mathcal{D}_{a2}(\mathbf{r}) \\
&= -\Phi_{01} \int_{\partial V_1} \mathrm{d}S\ \hat{\mathbf{n}}(\mathbf{r}) \cdot \mathcal{D}_{b1}(\mathbf{r}) + \Phi_{02} \int_{\partial V_2} \mathrm{d}S\ \hat{\mathbf{n}}(\mathbf{r}) \cdot \mathcal{D}_{a2}(\mathbf{r}) = \Phi_{01} Q_{b1} - \Phi_{02} Q_{a2}
\end{aligned} \tag{6.338}$$

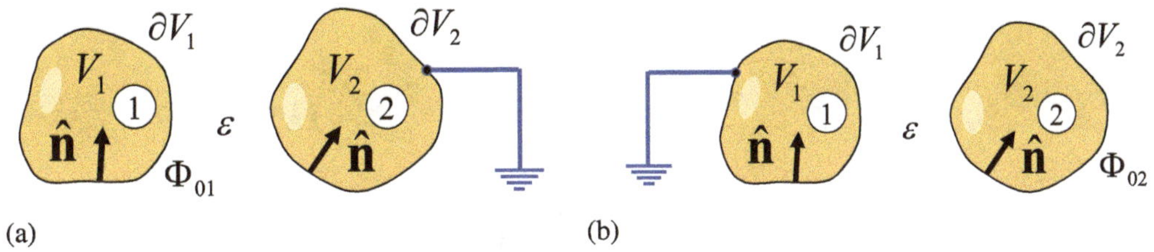

Figure 6.30 For proving the symmetry of the capacitance matrix [C]: a system of two conductors in state (a) and state (b).

because the surface integral over ∂B vanishes in light of the asymptotic behavior (2.20) and (2.21), and we have used the jump condition (1.182) with due regard to the orientation of the unit normal on ∂V_1 and ∂V_2. Since Q_{b1} (resp. Q_{a2}) is the charge induced on conductor ① (resp. ②) due to the potential Φ_{02} (resp. Φ_{01}) we have

$$Q_{b1} = C_{12}\Phi_{02} \qquad\qquad Q_{a2} = C_{21}\Phi_{01} \tag{6.339}$$

according to a special instance of (2.87). Now, inserting (6.339) into (6.338) yields

$$(C_{12} - C_{21})\Phi_{01}\Phi_{02} = 0 \tag{6.340}$$

whence it follows that C_{12} and C_{21} must coincide in view of the arbitrariness of the potentials Φ_{01} and Φ_{02}.

(End of Example 6.9)

We remark that in the derivation of (6.331) we have tacitly excluded the presence of singular sources such as point charges and elementary dipoles. We may fill this gap by expressing the charge density in terms of suitable Dirac delta distributions (Appendix C) and invoking the sifting property thereof to give a meaning to the volume integrals. Nonetheless, we follow the classic approach, which albeit a little lengthier, provides more insight into the role played by the various contributions.

For the sake of simplicity we consider a bounded volume V filled with the same homogeneous isotropic dielectric medium with permittivity ε for both state (a) and state (b). The sources are two point charges q_a and q_b located in $\mathbf{r}_a$ and $\mathbf{r}_b$ inside V. Since the potential and the field produced by q_a and q_b are singular at their respective locations, we exclude $\mathbf{r}_a$ and $\mathbf{r}_b$ by means of two balls $B(\mathbf{r}_a, a)$ and $B(\mathbf{r}_b, b)$, as is suggested in Figure 6.31. Then, (6.330) becomes

$$\nabla \cdot [\Phi_b(\mathbf{r})\varepsilon\boldsymbol{\mathcal{E}}_a(\mathbf{r}) - \Phi_a(\mathbf{r})\varepsilon\boldsymbol{\mathcal{E}}_b(\mathbf{r})] = 0 \tag{6.341}$$

for points $\mathbf{r} \in V \setminus (B(\mathbf{r}_a, a) \cup B(\mathbf{r}_b, b))$. We may go on to integrate and apply the Gauss theorem to find

$$\int_S \mathrm{d}S\, \hat{\mathbf{n}}(\mathbf{r}) \cdot [\Phi_b(\mathbf{r})\varepsilon\boldsymbol{\mathcal{E}}_a(\mathbf{r}) - \Phi_a(\mathbf{r})\varepsilon\boldsymbol{\mathcal{E}}_b(\mathbf{r})] + \int_{\partial B(\mathbf{r}_a,a)} \mathrm{d}S\, \hat{\mathbf{n}}(\mathbf{r}) \cdot [\Phi_b(\mathbf{r})\varepsilon\boldsymbol{\mathcal{E}}_a(\mathbf{r}) - \Phi_a(\mathbf{r})\varepsilon\boldsymbol{\mathcal{E}}_b(\mathbf{r})]$$
$$+ \int_{\partial B(\mathbf{r}_b,b)} \mathrm{d}S\, \hat{\mathbf{n}}(\mathbf{r}) \cdot [\Phi_b(\mathbf{r})\varepsilon\boldsymbol{\mathcal{E}}_a(\mathbf{r}) - \Phi_a(\mathbf{r})\varepsilon\boldsymbol{\mathcal{E}}_b(\mathbf{r})] = 0 \tag{6.342}$$

with the unit normal $\hat{\mathbf{n}}(\mathbf{r})$ pointing outwards $V \setminus B_a(\mathbf{r}_a, a) \setminus B_b(\mathbf{r}_b, b)$. We wish to evaluate the last two integrals in the limit as the radii a and b approach zero.

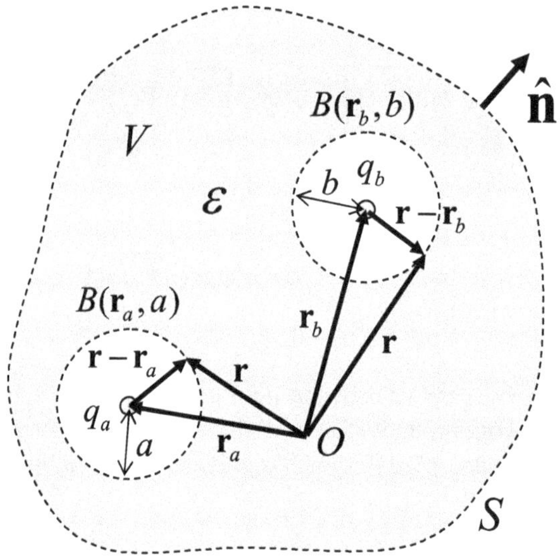

Figure 6.31 For deriving a symmetry relation for static fields produced by point charges.

To this purpose, we observe

$$\Phi_a(\mathbf{r}) = \frac{q_a}{4\pi\varepsilon|\mathbf{r}-\mathbf{r}_a|} + g_a(\mathbf{r}) = \frac{q_a}{4\pi\varepsilon a} + g_a(\mathbf{r}), \qquad |\mathbf{r}-\mathbf{r}_a| = a \tag{6.343}$$

$$\Phi_b(\mathbf{r}) = \frac{q_b}{4\pi\varepsilon|\mathbf{r}-\mathbf{r}_b|} + g_b(\mathbf{r}) = \frac{q_b}{4\pi\varepsilon b} + g_b(\mathbf{r}), \qquad |\mathbf{r}-\mathbf{r}_b| = b \tag{6.344}$$

$$\mathcal{E}_a(\mathbf{r}) = \frac{q_a(\mathbf{r}-\mathbf{r}_a)}{4\pi\varepsilon|\mathbf{r}-\mathbf{r}_a|^3} - \nabla g_a(\mathbf{r}) = -\frac{q_a\hat{\mathbf{n}}(\mathbf{r})}{4\pi\varepsilon a^2} - \nabla g_a(\mathbf{r}), \qquad |\mathbf{r}-\mathbf{r}_a| = a \tag{6.345}$$

$$\mathcal{E}_b(\mathbf{r}) = \frac{q_b(\mathbf{r}-\mathbf{r}_b)}{4\pi\varepsilon|\mathbf{r}-\mathbf{r}_b|^3} - \nabla g_b(\mathbf{r}) = -\frac{q_b\hat{\mathbf{n}}(\mathbf{r})}{4\pi\varepsilon b^2} - \nabla g_b(\mathbf{r}), \qquad |\mathbf{r}-\mathbf{r}_b| = b \tag{6.346}$$

in view of (2.123) and (2.131). The functions $g_a(\mathbf{r})$ and $g_b(\mathbf{r})$ are two harmonic scalar fields in $\mathbf{r} \in V$, account for the possible different properties of the medium outside V, and can be obtained from (2.160), though they are inconsequential for the derivation of the result.

With these positions we have

$$\begin{aligned}
&\int_{\partial B(\mathbf{r}_a,a)} \mathrm{d}S\ \hat{\mathbf{n}}(\mathbf{r}) \cdot [\Phi_b(\mathbf{r})\varepsilon\mathcal{E}_a(\mathbf{r}) - \Phi_a(\mathbf{r})\varepsilon\mathcal{E}_b(\mathbf{r})] = \\
&= -\int_{\partial B(\mathbf{r}_a,a)} \mathrm{d}S\ \left\{\Phi_b(\mathbf{r})\left[\frac{q_a}{4\pi a^2} + \varepsilon\hat{\mathbf{n}}(\mathbf{r})\cdot\nabla g_a(\mathbf{r})\right] + \left[\frac{q_a}{4\pi a} + \varepsilon g_a(\mathbf{r})\right]\hat{\mathbf{n}}(\mathbf{r})\cdot\mathcal{E}_b(\mathbf{r})\right\} \\
&= -\Phi_b(\mathbf{r}_0)\left[q_a + 4\pi a^2\varepsilon\hat{\mathbf{n}}(\mathbf{r}_0)\cdot\nabla g_a(\mathbf{r})|_{\mathbf{r}=\mathbf{r}_0}\right] \\
&\quad - \left[a q_a + 4\pi a^2\varepsilon g_a(\mathbf{r}_0)\right]\hat{\mathbf{n}}(\mathbf{r}_0)\cdot\mathcal{E}_b(\mathbf{r}_0) \xrightarrow[a\to 0]{} -\Phi_b(\mathbf{r}_a)q_a
\end{aligned} \tag{6.347}$$

having applied the mean value theorem [84] with $\mathbf{r}_0 = \mathbf{r}_a - a\hat{\mathbf{n}}(\mathbf{r}_0)$ a suitable point on the sphere $|\mathbf{r}| = a$, whereon the integrand is regular. In like manner, we find

$$\int_{\partial B(\mathbf{r}_b,b)} \mathrm{d}S\ \hat{\mathbf{n}}(\mathbf{r}) \cdot [\Phi_b(\mathbf{r})\varepsilon\mathcal{E}_a(\mathbf{r}) - \Phi_a(\mathbf{r})\varepsilon\mathcal{E}_b(\mathbf{r})]$$

$$= \left[bq_b + 4\pi b^2 \varepsilon g_b(\mathbf{r}_0)\right] \hat{\mathbf{n}}(\mathbf{r}_0) \cdot \mathcal{E}_a(\mathbf{r}_0) + \Phi_a(\mathbf{r}_0)\left[q_b + 4\pi b^2 \varepsilon \hat{\mathbf{n}}(\mathbf{r}_0) \cdot \nabla g_b(\mathbf{r})|_{\mathbf{r}=\mathbf{r}_0}\right] \xrightarrow[b\to 0]{} \Phi_a(\mathbf{r}_b)q_b \tag{6.348}$$

for the same reasons and with $\mathbf{r}_0 = \mathbf{r}_b - b\hat{\mathbf{n}}(\mathbf{r}_0)$ a point on the sphere $|\mathbf{r}| = b$. By inserting these results into (6.342) and rearranging the various contributions we arrive at

$$\int_S \mathrm{d}S\ \hat{\mathbf{n}}(\mathbf{r}) \cdot [\Phi_b(\mathbf{r})\varepsilon\mathcal{E}_a(\mathbf{r}) - \Phi_a(\mathbf{r})\varepsilon\mathcal{E}_b(\mathbf{r})] = \Phi_b(\mathbf{r}_a)q_a - \Phi_a(\mathbf{r}_b)q_b \tag{6.349}$$

which is the desired extension of (6.331) for point charges. The extension of the reciprocation theorem follows by showing that the flux through S vanishes when no sources are present outside V.

Finally, we consider two elementary electrostatic dipoles (see Example 2.5) with moments $\mathbf{p}_a$ and $\mathbf{p}_b$ located in $\mathbf{r}_a$ and $\mathbf{r}_b$ inside the region V filled with a homogeneous isotropic dielectric medium with permittivity ε for both state (a) and state (b). We exclude the dipoles with two balls $B(\mathbf{r}_a, a)$ and $B(\mathbf{r}_b, b)$, whereby (6.331) passes over into (6.342) once again. The geometry is the same as that of Figure 6.31 with the dipoles in lieu of the charges. To compute the integrals in the limit as a and b tend to zero, we observe

$$\Phi_a(\mathbf{r}) = \frac{\mathbf{p}_a \cdot (\mathbf{r} - \mathbf{r}_a)}{4\pi\varepsilon|\mathbf{r} - \mathbf{r}_a|^3} + g_a(\mathbf{r}) = -\frac{\mathbf{p}_a \cdot \hat{\mathbf{n}}(\mathbf{r})}{4\pi\varepsilon a^2} + g_a(\mathbf{r}), \qquad |\mathbf{r} - \mathbf{r}_a| = a \tag{6.350}$$

$$\Phi_b(\mathbf{r}) = \frac{\mathbf{p}_b \cdot (\mathbf{r} - \mathbf{r}_b)}{4\pi\varepsilon|\mathbf{r} - \mathbf{r}_b|^3} + g_b(\mathbf{r}) = -\frac{\mathbf{p}_b \cdot \hat{\mathbf{n}}(\mathbf{r})}{4\pi\varepsilon b^2} + g_b(\mathbf{r}), \qquad |\mathbf{r} - \mathbf{r}_b| = b \tag{6.351}$$

in light of (2.42). As before, the functions $g_a(\mathbf{r})$ and $g_b(\mathbf{r})$ are two harmonic scalar fields for $\mathbf{r} \in V$, which account for the possible different properties of the medium outside V. For the electric fields we observe

$$\begin{aligned}\mathcal{E}_a(\mathbf{r}) &= -\nabla \frac{\mathbf{p}_a \cdot (\mathbf{r} - \mathbf{r}_a)}{4\pi\varepsilon|\mathbf{r} - \mathbf{r}_a|^3} - \nabla g_a(\mathbf{r}) \\ &= \nabla \times \left(\mathbf{p}_a \times \frac{\mathbf{r} - \mathbf{r}_a}{4\pi\varepsilon|\mathbf{r} - \mathbf{r}_a|^3}\right) - \nabla g_a(\mathbf{r}), \qquad \mathbf{r} \in V \setminus B(\mathbf{r}_a, a)\end{aligned} \tag{6.352}$$

$$\begin{aligned}\mathcal{E}_b(\mathbf{r}) &= -\nabla \frac{\mathbf{p}_b \cdot (\mathbf{r} - \mathbf{r}_b)}{4\pi\varepsilon|\mathbf{r} - \mathbf{r}_b|^3} - \nabla g_b(\mathbf{r}) \\ &= \nabla \times \left(\mathbf{p}_b \times \frac{\mathbf{r} - \mathbf{r}_b}{4\pi\varepsilon|\mathbf{r} - \mathbf{r}_b|^3}\right) - \nabla g_b(\mathbf{r}), \qquad \mathbf{r} \in V \setminus B(\mathbf{r}_b, b)\end{aligned} \tag{6.353}$$

on account of the differential identities (H.53) and (H.54). Writing the singular part of electric field as the curl of a vector potential is a necessary step to ensure that the surface integrals remain finite in the limit for vanishing a and b. For $|\mathbf{r} - \mathbf{r}_a| = a$ this goal is accomplished as follows

$$\begin{aligned}\int_{\partial B(\mathbf{r}_a,a)} \mathrm{d}S\ \hat{\mathbf{n}}(\mathbf{r}) \cdot \Phi_b(\mathbf{r})\varepsilon\mathcal{E}_a(\mathbf{r}) &= \int_{\partial B(\mathbf{r}_a,a)} \mathrm{d}S\ \hat{\mathbf{n}}(\mathbf{r}) \cdot \Phi_b(\mathbf{r}) \left[\nabla \times \left(\mathbf{p}_a \times \frac{\mathbf{r} - \mathbf{r}_a}{4\pi|\mathbf{r} - \mathbf{r}_a|^3}\right) - \varepsilon\nabla g_a(\mathbf{r})\right] \\ &= \int_{\partial B(\mathbf{r}_a,a)} \mathrm{d}S\ \hat{\mathbf{n}}(\mathbf{r}) \cdot \left[\nabla \times \left(\Phi_b(\mathbf{r})\mathbf{p}_a \times \frac{\mathbf{r} - \mathbf{r}_a}{4\pi|\mathbf{r} - \mathbf{r}_a|^3}\right)\right.\end{aligned}$$

$$+ \mathcal{E}_b(\mathbf{r}) \times \left(\mathbf{p}_a \times \frac{\mathbf{r} - \mathbf{r}_a}{4\pi|\mathbf{r} - \mathbf{r}_a|^3} \right) - \Phi_b(\mathbf{r})\varepsilon\nabla g_a(\mathbf{r}) \Bigg] \tag{6.354}$$

thanks to the differential identity (H.52) and definition (2.15). The first integrand in the rightmost-hand side contributes naught, as can be shown by applying the Stokes theorem (A.55) over a closed surface (i.e., a sphere) for the relevant fields are regular thereon. Next, we have

$$\begin{aligned}
&\int_{\partial B(\mathbf{r}_a,a)} \mathrm{d}S\ \hat{\mathbf{n}}(\mathbf{r}) \cdot [\Phi_b(\mathbf{r})\varepsilon\mathcal{E}_a(\mathbf{r}) - \Phi_a(\mathbf{r})\varepsilon\mathcal{E}_b(\mathbf{r})] = \\
&= \int_{\partial B(\mathbf{r}_a,a)} \mathrm{d}S\ \hat{\mathbf{n}}(\mathbf{r}) \cdot \left\{ \mathcal{E}_b(\mathbf{r}) \times \left[\frac{\hat{\mathbf{n}}(\mathbf{r}) \times \mathbf{p}_a}{4\pi a^2} \right] - \Phi_b(\mathbf{r})\varepsilon\nabla g_a(\mathbf{r}) + \left[\frac{\mathbf{p}_a \cdot \hat{\mathbf{n}}(\mathbf{r})}{4\pi a^2} - \varepsilon g_a(\mathbf{r}) \right] \mathcal{E}_b(\mathbf{r}) \right\} \\
&= \int_{\partial B(\mathbf{r}_a,a)} \mathrm{d}S \left\{ \frac{\hat{\mathbf{n}}(\mathbf{r})}{4\pi a^2} \cdot [\hat{\mathbf{n}}(\mathbf{r})\mathcal{E}_b(\mathbf{r}) \cdot \mathbf{p}_a - \mathbf{p}_a\mathcal{E}_b(\mathbf{r}) \cdot \hat{\mathbf{n}}(\mathbf{r})] \right. \\
&\quad \left. - \Phi_b(\mathbf{r})\varepsilon\hat{\mathbf{n}}(\mathbf{r}) \cdot \nabla g_a(\mathbf{r}) + \left[\frac{\mathbf{p}_a \cdot \hat{\mathbf{n}}(\mathbf{r})}{4\pi a^2} - \varepsilon g_a(\mathbf{r}) \right] \hat{\mathbf{n}}(\mathbf{r}) \cdot \mathcal{E}_b(\mathbf{r}) \right\} \\
&= \int_{\partial B(\mathbf{r}_a,a)} \mathrm{d}S \left[\frac{\mathcal{E}_b(\mathbf{r}) \cdot \mathbf{p}_a}{4\pi a^2} - \Phi_b(\mathbf{r})\varepsilon\hat{\mathbf{n}}(\mathbf{r}) \cdot \nabla g_a(\mathbf{r}) - \varepsilon g_a(\mathbf{r})\hat{\mathbf{n}}(\mathbf{r}) \cdot \mathcal{E}_b(\mathbf{r}) \right] \\
&= \mathcal{E}_b(\mathbf{r}_0) \cdot \mathbf{p}_a - 4\pi a^2\Phi_b(\mathbf{r}_0)\varepsilon\hat{\mathbf{n}}(\mathbf{r}_0) \cdot \nabla g_a(\mathbf{r})|_{\mathbf{r}=\mathbf{r}_0} - 4\pi a^2\varepsilon g_a(\mathbf{r}_0)\hat{\mathbf{n}}(\mathbf{r}_0) \cdot \mathcal{E}_b(\mathbf{r}_0) \\
&\xrightarrow[a\to 0]{} \mathcal{E}_b(\mathbf{r}_a) \cdot \mathbf{p}_a
\end{aligned} \tag{6.355}$$

having used (H.14) and applied the mean value theorem [84] with $\mathbf{r}_0 = \mathbf{r}_a - a\hat{\mathbf{n}}(\mathbf{r}_0)$ a suitable point on the sphere $|\mathbf{r}| = a$, whereon the integrand is regular. By means of similar steps the integral over $|\mathbf{r}| = b$ yields

$$\int_{\partial B(\mathbf{r}_b,b)} \mathrm{d}S\ \hat{\mathbf{n}}(\mathbf{r}) \cdot [\Phi_b(\mathbf{r})\varepsilon\mathcal{E}_a(\mathbf{r}) - \Phi_a(\mathbf{r})\varepsilon\mathcal{E}_b(\mathbf{r})] \xrightarrow[b\to 0]{} -\mathcal{E}_a(\mathbf{r}_b) \cdot \mathbf{p}_b. \tag{6.356}$$

With these intermediate results (6.342) becomes

$$\begin{aligned}
\int_S \mathrm{d}S\ \hat{\mathbf{n}}(\mathbf{r}) \cdot [\Phi_b(\mathbf{r})\varepsilon\mathcal{E}_a(\mathbf{r}) - \Phi_a(\mathbf{r})\varepsilon\mathcal{E}_b(\mathbf{r})] &= \mathcal{E}_a(\mathbf{r}_b) \cdot \mathbf{p}_b - \mathcal{E}_b(\mathbf{r}_a) \cdot \mathbf{p}_a \\
&= \mathbf{p}_a \cdot \nabla\Phi_b(\mathbf{r})|_{\mathbf{r}=\mathbf{r}_a} - \mathbf{p}_b \cdot \nabla\Phi_a(\mathbf{r})|_{\mathbf{r}=\mathbf{r}_b}
\end{aligned} \tag{6.357}$$

in the limit for vanishing a and b. The extension of the reciprocation theorem follows by showing that the flux integral is null when no sources are located outside V.

By invoking linearity and superposition it is also easy to generalize these formulas to many point charges and many elementary dipoles and even combine them with (6.331) and (6.337) provided $\overline{\overline{\varepsilon}}(\mathbf{r}) = \varepsilon\overline{\overline{\mathcal{I}}}$.

6.9.2 Stationary fields and steady currents

The entities of interest are magnetic fields and vector potentials generated by two sets of steady currents in a magnetic medium. We consider a finite region of space V bounded by a smooth surface S and the smooth boundary ∂V_1 of an excluded volume V_1 occupied by matter unspecified for the time being. The geometrical setup is the same as that of Figure 6.29 but with the solenoidal current densities in place of charge densities. We define the following two states:

State (a) The permeability of the medium in V is $\overline{\mu}_a(\mathbf{r})$, and the current density $\mathcal{J}_a(\mathbf{r})$ is confined in a volume $V_{Ja} \subset V$.

State (b) The permeability of the medium in V is $\overline{\mu}_b(\mathbf{r})$, and the charge density $\mathcal{J}_b(\mathbf{r})$ is confined in a volume $V_{Jb} \subset V$.

The relevant currents, fields and vector potentials satisfy the Ampère law (4.9) and definition (4.29)

$$\nabla \times \mathcal{H}_a(\mathbf{r}) = \mathcal{J}_a(\mathbf{r}) \tag{6.358}$$

$$\mathcal{B}_a(\mathbf{r}) = \overline{\mu}_a(\mathbf{r}) \cdot \mathcal{H}_a(\mathbf{r}) = \nabla \times \mathcal{A}_a(\mathbf{r}) \tag{6.359}$$

$$\nabla \times \mathcal{H}_b(\mathbf{r}) = \mathcal{J}_b(\mathbf{r}) \tag{6.360}$$

$$\mathcal{B}_b(\mathbf{r}) = \overline{\mu}_b(\mathbf{r}) \cdot \mathcal{H}_b(\mathbf{r}) = \nabla \times \mathcal{A}_b(\mathbf{r}) \tag{6.361}$$

for points $\mathbf{r} \in V$. We recall that (6.359) and (6.361) are a consequence of the local magnetic Gauss law (1.23), which in turn holds everywhere in V provided the dyadic fields $\overline{\mu}_a(\mathbf{r})$ and $\overline{\mu}_b(\mathbf{r})$ are at least continuous for $\mathbf{r} \in V$. If the latter were not true, we should write (6.359) and (6.361) separately in each part of V where the continuity of $\overline{\mu}_a(\mathbf{r})$ and $\overline{\mu}_b(\mathbf{r})$ is granted (see Sections 1.7 and 4.3).

We proceed by dot-multiplying (6.358) by $\mathcal{A}_b(\mathbf{r})$ and (6.360) by $\mathcal{A}_a(\mathbf{r})$ and, by subtracting the resulting expressions, we find

$$\begin{aligned}\nabla \cdot [\mathcal{H}_a(\mathbf{r}) \times \mathcal{A}_b(\mathbf{r}) - \mathcal{H}_b(\mathbf{r}) \times \mathcal{A}_a(\mathbf{r})] &= \mathcal{H}_b(\mathbf{r}) \cdot \overline{\mu}_a(\mathbf{r}) \cdot \mathcal{H}_a(\mathbf{r}) \\ &\quad - \mathcal{H}_a(\mathbf{r}) \cdot \overline{\mu}_b(\mathbf{r}) \cdot \mathcal{H}_b(\mathbf{r}) + \mathcal{A}_b(\mathbf{r}) \cdot \mathcal{J}_a(\mathbf{r}) - \mathcal{A}_a(\mathbf{r}) \cdot \mathcal{J}_b(\mathbf{r})\end{aligned} \tag{6.362}$$

thanks to the differential identity (H.49), (6.359) and (6.361). The first two contributions in the right-hand side cancel out if

$$\overline{\mu}(\mathbf{r}) = \overline{\mu}_a(\mathbf{r}) = \overline{\mu}_b(\mathbf{r}) \qquad \text{with} \qquad [\overline{\mu}(\mathbf{r})]^T = \overline{\mu}(\mathbf{r}) \tag{6.363}$$

that is, the magnetic medium is reciprocal, in keeping with conclusion (6.251) previously obtained in the frequency domain. With this hypothesis for $\mathbf{r} \in V$ we arrive at

$$\nabla \cdot [\mathcal{H}_a(\mathbf{r}) \times \mathcal{A}_b(\mathbf{r}) - \mathcal{H}_b(\mathbf{r}) \times \mathcal{A}_a(\mathbf{r})] = \mathcal{A}_b(\mathbf{r}) \cdot \mathcal{J}_a(\mathbf{r}) - \mathcal{A}_a(\mathbf{r}) \cdot \mathcal{J}_b(\mathbf{r}) \tag{6.364}$$

a symmetry relation in local form for magnetic fields, vector potentials and currents of the two states. The global counterpart follows by integrating (6.364) over V

$$\begin{aligned}&\int\limits_{S \cup \partial V_1} \mathrm{d}S\, \hat{\mathbf{n}}(\mathbf{r}) \cdot [\mathcal{H}_a(\mathbf{r}) \times \mathcal{A}_b(\mathbf{r}) - \mathcal{H}_b(\mathbf{r}) \times \mathcal{A}_a(\mathbf{r})] \\ &\qquad = \int\limits_{V_{Ja}} \mathrm{d}V\, \mathcal{A}_b(\mathbf{r}) \cdot \mathcal{J}_a(\mathbf{r}) - \int\limits_{V_{Jb}} \mathrm{d}V\, \mathcal{A}_a(\mathbf{r}) \cdot \mathcal{J}_b(\mathbf{r})\end{aligned} \tag{6.365}$$

where we have applied the Gauss theorem under the assumption that $\hat{\mathbf{n}}(\mathbf{r}) \times \mathcal{H}_a(\mathbf{r})$ and $\hat{\mathbf{n}}(\mathbf{r}) \times \mathcal{H}_b(\mathbf{r})$ are continuous through ∂V_{Ja} and ∂V_{Jb} (see Section 1.2.2). Furthermore, singular sources such as magnetic dipoles are excluded for the fields and potentials to be continuously differentiable in V and especially in $V_{Ja} \cup V_{Jb}$. It is useful to discuss the conditions for which the flux integral vanishes.

If the currents are located in $\mathbb{R}^3 \setminus \overline{V}$, the volume integrals over V_{Ja} and V_{Jb} vanish by construction. From (6.365) we conclude that flux integral over $S \cup \partial V_1$ is null if the sources are placed *outside* the region of interest.

Next, we suppose that the unbounded region $\mathbb{R}^3 \setminus \overline{V}$ is filled with a magnetic medium whose permeability is $\overline{\mu}(\mathbf{r})$ and that the latter is continuously differentiable for $\mathbf{r} \in \mathbb{R}^3 \setminus (S \cup \partial V_1)$. We consider a ball $B(0, a)$ with radius a large enough for V to be contained in $B(0, a)$. We apply (6.365) to the regions $B(0, a) \setminus (V \cup V_1)$ and V_1 which are devoid of currents. Therefore, we have

$$\int_{\partial B} \mathrm{d}S\; \hat{\mathbf{n}}(\mathbf{r}) \cdot [\mathcal{H}_a(\mathbf{r}) \times \mathcal{A}_b(\mathbf{r}) - \mathcal{H}_b(\mathbf{r}) \times \mathcal{A}_a(\mathbf{r})] + \int_{S} \mathrm{d}S\; \hat{\mathbf{n}}(\mathbf{r}) \cdot [\mathcal{H}_a(\mathbf{r}) \times \mathcal{A}_b(\mathbf{r}) - \mathcal{H}_b(\mathbf{r}) \times \mathcal{A}_a(\mathbf{r})] = 0 \qquad (6.366)$$

and

$$\int_{\partial V_1} \mathrm{d}S\; \hat{\mathbf{n}}(\mathbf{r}) \cdot [\mathcal{H}_a(\mathbf{r}) \times \mathcal{A}_b(\mathbf{r}) - \mathcal{H}_b(\mathbf{r}) \times \mathcal{A}_a(\mathbf{r})] = 0 \qquad (6.367)$$

with the unit normal $\hat{\mathbf{n}}(\mathbf{r})$ oriented positively outwards $B(0, a) \setminus (V \cup V_1)$ and V_1. The flux across the sphere ∂B vanishes in the limit as $a \to +\infty$ thanks to the asymptotic behavior (4.36) and (4.37). Thus, by using these findings in (6.365) we obtain

$$\int_{V_{Ja}} \mathrm{d}V\, \mathcal{A}_b(\mathbf{r}) \cdot \mathcal{J}_a(\mathbf{r}) = \int_{V_{Jb}} \mathrm{d}V\, \mathcal{A}_a(\mathbf{r}) \cdot \mathcal{J}_b(\mathbf{r}) \qquad (6.368)$$

as long as all currents are confined within V. A steady current density is related to the vector potential through the vector Poisson equation (4.44) and, conversely, the vector potential is derived from the current density by means of the integral formula (5.80). Therefore, (6.368) states that the linear operators

$$\mathfrak{L}\{\bullet\} := \nabla \times \{[\overline{\mu}(\mathbf{r})]^{-1} \cdot \nabla \times \{\bullet\}\}, \qquad \mathbf{r} \in V_J \qquad (6.369)$$

$$\mathfrak{L}\{\bullet\} := \sqrt{\det \overline{\mu}} \int_{V_J} \mathrm{d}V' \frac{\overline{\mu}^{-1} \cdot \{\bullet\}}{4\pi \left(\overline{\mu}^{-1} : \mathbf{RR}\right)^{1/2}}, \qquad \mathbf{r} \in V_J \qquad (6.370)$$

are self-adjoint (Appendix D.6) if the dyadic field $\overline{\mu}(\mathbf{r})$ is symmetric in the domain of interest.

The extension of (6.365) to magnetic dipoles of moments $\mathbf{m}_a$ and $\mathbf{m}_b$ located at $\mathbf{r}_a$ and $\mathbf{r}_b$ can be done by writing the current densities as distributions (see Sections 4.5 and 5.6)

$$\mathcal{J}_a(\mathbf{r}) = \nabla \times [\mathbf{m}_a \delta^{(3)}(\mathbf{r} - \mathbf{r}_a)] \qquad \mathcal{J}_b(\mathbf{r}) = \nabla \times [\mathbf{m}_b \delta^{(3)}(\mathbf{r} - \mathbf{r}_b)] \qquad (6.371)$$

and invoking the sifting property of the Dirac delta distribution to give the volume integrals a meaning. Conversely, the classic approach consists of applying (6.365) to a source-free region $V \setminus (B(\mathbf{r}_a, a) \cup B(\mathbf{r}_b, b))$. We assume that the underlying magnetic medium is isotropic and homogeneous with permeability μ for both states. The geometry is the same as that of Figure 6.31 except for the presence of magnetic dipoles in lieu of point charges. Then, (6.365) yields

$$\int_{\partial V} \mathrm{d}S\; \hat{\mathbf{n}}(\mathbf{r}) \cdot [\mathcal{H}_a(\mathbf{r}) \times \mathcal{A}_b(\mathbf{r}) - \mathcal{H}_b(\mathbf{r}) \times \mathcal{A}_a(\mathbf{r})]$$

$$+ \int_{\partial B(\mathbf{r}_a,a)} \mathrm{d}S\ \hat{\mathbf{n}}(\mathbf{r}) \cdot [\mathcal{H}_a(\mathbf{r}) \times \mathcal{A}_b(\mathbf{r}) - \mathcal{H}_b(\mathbf{r}) \times \mathcal{A}_a(\mathbf{r})]$$

$$+ \int_{\partial B(\mathbf{r}_b,b)} \mathrm{d}S\ \hat{\mathbf{n}}(\mathbf{r}) \cdot [\mathcal{H}_a(\mathbf{r}) \times \mathcal{A}_b(\mathbf{r}) - \mathcal{H}_b(\mathbf{r}) \times \mathcal{A}_a(\mathbf{r})] = 0 \quad (6.372)$$

with the unit normal $\hat{\mathbf{n}}(\mathbf{r})$ pointing outwards $V \setminus (B_a(\mathbf{r}_a, a) \cup B_b(\mathbf{r}_b, b))$. We wish to evaluate the last two integrals in the limit as the radii a and b approach zero.

In accordance with (4.85) and (4.88) we write the magnetic scalar and vector potentials relevant to the two states as

$$\Psi_a(\mathbf{r}) = \frac{\mathbf{m}_a \cdot (\mathbf{r} - \mathbf{r}_a)}{4\pi|\mathbf{r} - \mathbf{r}_a|^3} + g_a(\mathbf{r}) = -\frac{\mathbf{m}_a \cdot \hat{\mathbf{n}}(\mathbf{r})}{4\pi a^2} + g_a(\mathbf{r}) \quad (6.373)$$

$$\mathcal{A}_a(\mathbf{r}) = \mu\mathbf{m}_a \times \frac{\mathbf{r} - \mathbf{r}_a}{4\pi|\mathbf{r} - \mathbf{r}_a|^3} + \mathbf{f}_a(\mathbf{r}) = -\mu\mathbf{m}_a \times \frac{\hat{\mathbf{n}}(\mathbf{r})}{4\pi a^2} + \mathbf{f}_a(\mathbf{r}) \quad (6.374)$$

for $|\mathbf{r} - \mathbf{r}_a| = a$ and

$$\Psi_b(\mathbf{r}) = \frac{\mathbf{m}_b \cdot (\mathbf{r} - \mathbf{r}_b)}{4\pi|\mathbf{r} - \mathbf{r}_b|^3} + g_b(\mathbf{r}) = -\frac{\mathbf{m}_b \cdot \hat{\mathbf{n}}(\mathbf{r})}{4\pi b^2} + g_b(\mathbf{r}) \quad (6.375)$$

$$\mathcal{A}_b(\mathbf{r}) = \mu\mathbf{m}_b \times \frac{\mathbf{r} - \mathbf{r}_b}{4\pi|\mathbf{r} - \mathbf{r}_b|^3} + \mathbf{f}_b(\mathbf{r}) = -\mu\mathbf{m}_b \times \frac{\hat{\mathbf{n}}(\mathbf{r})}{4\pi b^3} + \mathbf{f}_b(\mathbf{r}) \quad (6.376)$$

for $|\mathbf{r} - \mathbf{r}_b| = b$. The scalar fields $g_a(\mathbf{r})$ and $g_b(\mathbf{r})$ are harmonic for $\mathbf{r} \in V$, represent the effect of possibly different media outside V, and can be obtained from the magnetic counterpart of the integral representation (2.160). The solenoidal vector fields $\mathbf{f}_a(\mathbf{r})$ and $\mathbf{f}_b(\mathbf{r})$ — which also account for the presence of different media outside V — can be rigorously expressed with the aid of the integral representation (5.18). The combined usage of scalar and vector potential allows identifying and eliminating the terms that in the surface integrals do not converge in the limit as $a \to 0$ and $b \to 0$. For $\mathbf{r} \in \partial B(\mathbf{r}_a, a)$ we have

$$\int_{\partial B(\mathbf{r}_a,a)} \mathrm{d}S\ \hat{\mathbf{n}}(\mathbf{r}) \cdot [\mathcal{H}_a(\mathbf{r}) \times \mathcal{A}_b(\mathbf{r}) - \mathcal{H}_b(\mathbf{r}) \times \mathcal{A}_a(\mathbf{r})] =$$

$$= - \int_{\partial B(\mathbf{r}_a,a)} \mathrm{d}S\ \hat{\mathbf{n}}(\mathbf{r}) \cdot [\nabla\Psi_a(\mathbf{r}) \times \mathcal{A}_b(\mathbf{r}) + \mathcal{H}_b(\mathbf{r}) \times \mathcal{A}_a(\mathbf{r})]$$

$$= - \int_{\partial B(\mathbf{r}_a,a)} \mathrm{d}S\ \hat{\mathbf{n}}(\mathbf{r}) \cdot \{\nabla \times [\Psi_a(\mathbf{r})\mathcal{A}_b(\mathbf{r})] - \Psi_a(\mathbf{r})\mathcal{B}_b(\mathbf{r}) + \mathcal{H}_b(\mathbf{r}) \times \mathcal{A}_a(\mathbf{r})\}$$

$$= \int_{\partial B(\mathbf{r}_a,a)} \mathrm{d}S\ [\hat{\mathbf{n}}(\mathbf{r}) \cdot \mathcal{B}_b(\mathbf{r})\Psi_a(\mathbf{r}) - \hat{\mathbf{n}}(\mathbf{r}) \cdot \mathcal{H}_b(\mathbf{r}) \times \mathcal{A}_a(\mathbf{r})] \quad (6.377)$$

thanks to the differential identity (H.50) and definition (4.29). Since both $\Psi_a(\mathbf{r})$ and $\mathcal{A}_b(\mathbf{r})$ are continuously differentiable over the sphere $|\mathbf{r} - \mathbf{r}_a| = a$ we have applied the Stokes theorem (A.55) over a closed surface.

We continue by evaluating

$$\int_{\partial B(\mathbf{r}_a,a)} \mathrm{d}S\ [\hat{\mathbf{n}}(\mathbf{r}) \cdot \mathcal{B}_b(\mathbf{r})\Psi_a(\mathbf{r}) - \hat{\mathbf{n}}(\mathbf{r}) \cdot \mathcal{H}_b(\mathbf{r}) \times \mathcal{A}_a(\mathbf{r})] =$$

$$
\begin{aligned}
&= \int_{\partial B(\mathbf{r}_a,a)} \mathrm{d}S \left\{ \hat{\mathbf{n}}(\mathbf{r}) \cdot \mathcal{B}_b(\mathbf{r}) \left[g_a(\mathbf{r}) - \frac{\mathbf{m}_a \cdot \hat{\mathbf{n}}(\mathbf{r})}{4\pi a^2} \right] - \hat{\mathbf{n}}(\mathbf{r}) \cdot \mathcal{H}_b(\mathbf{r}) \times \left[\mathbf{f}_a(\mathbf{r}) - \mu \mathbf{m}_a \times \frac{\hat{\mathbf{n}}(\mathbf{r})}{4\pi a^2} \right] \right\} \\
&= \int_{\partial B(\mathbf{r}_a,a)} \mathrm{d}S \left[\hat{\mathbf{n}}(\mathbf{r}) \cdot \mathcal{B}_b(\mathbf{r}) g_a(\mathbf{r}) - \hat{\mathbf{n}}(\mathbf{r}) \cdot \mathcal{H}_b(\mathbf{r}) \times \mathbf{f}_a(\mathbf{r}) - \frac{\mathcal{B}_b(\mathbf{r}) \cdot \mathbf{m}_a}{4\pi a^2} \right] \\
&= 4\pi a^2 \hat{\mathbf{n}}(\mathbf{r}) \cdot \mathcal{B}_b(\mathbf{r}_0) g_a(\mathbf{r}_0) - 4\pi a^2 \hat{\mathbf{n}}(\mathbf{r}_0) \cdot \mathcal{H}_b(\mathbf{r}_0) \times \mathbf{f}_a(\mathbf{r}_0) - \mathcal{B}_b(\mathbf{r}_0) \cdot \mathbf{m}_a \\
&\xrightarrow[a \to 0]{} -\mathcal{B}_b(\mathbf{r}_a) \cdot \mathbf{m}_a
\end{aligned}
\tag{6.378}
$$

having used (H.14) and the mean value theorem [84] with $\mathbf{r}_0 = \mathbf{r}_a - a\hat{\mathbf{n}}(\mathbf{r}_0)$ a suitable point on the sphere $|\mathbf{r} - \mathbf{r}_a| = a$, whereon the integrand is regular. In like manner, the integral over the sphere $|\mathbf{r} - \mathbf{r}_b| = b$ becomes

$$
\int_{\partial B(\mathbf{r}_b,b)} \mathrm{d}S \, \hat{\mathbf{n}}(\mathbf{r}) \cdot [\mathcal{H}_a(\mathbf{r}) \times \mathcal{A}_b(\mathbf{r}) - \mathcal{H}_b(\mathbf{r}) \times \mathcal{A}_a(\mathbf{r})] \xrightarrow[b \to 0]{} \mathcal{B}_a(\mathbf{r}_b) \cdot \mathbf{m}_b
\tag{6.379}
$$

whereby (6.372) passes over into

$$
\begin{aligned}
&\int_{\partial V} \mathrm{d}S \, \hat{\mathbf{n}}(\mathbf{r}) \cdot [\mathcal{H}_a(\mathbf{r}) \times \mathcal{A}_b(\mathbf{r}) - \mathcal{H}_b(\mathbf{r}) \times \mathcal{A}_a(\mathbf{r})] = \\
&\quad = \mathbf{m}_a \cdot \mathcal{B}_b(\mathbf{r}_a) - \mathbf{m}_b \cdot \mathcal{B}_a(\mathbf{r}_b) \\
&\quad = \mathbf{m}_a \cdot \nabla \times \mathcal{A}_b(\mathbf{r})|_{\mathbf{r}=\mathbf{r}_a} - \mathbf{m}_b \cdot \nabla \times \mathcal{A}_a(\mathbf{r})|_{\mathbf{r}=\mathbf{r}_b} \\
&\quad = \mu \mathbf{m}_b \cdot \nabla \Psi_a(\mathbf{r})|_{\mathbf{r}=\mathbf{r}_b} - \mu \mathbf{m}_a \cdot \nabla \Psi_b(\mathbf{r})|_{\mathbf{r}=\mathbf{r}_a}
\end{aligned}
\tag{6.380}
$$

in the limit as $a \to 0$ and $b \to 0$.

References

[1] Coelho R. Physics of Dielectrics for the Engineer. Amsterdam, NL: Elsevier; 1979.

[2] Shen LC, Kong JA. Applied electromagnetism. Monterey, CA: Brooks/Cole; 1983.

[3] de Hoop AT. Handbook of radiation and scattering of waves. London, UK: Academic Press; 1995.

[4] Zienkiewicz OC. The Finite Element Method in Engineering Science. London, UK: McGraw-Hill; 1971.

[5] Peterson AF, Ray SL, Mittra R. Computational Methods for Electromagnetics. Piscataway, NJ: IEEE Press; 1998.

[6] Harrington RF. Field Computation by Moment Methods. New York, NY: MacMillan Publishing Co., Inc.; 1968.

[7] Jin JM. The Finite Element Method in Electromagnetics. New York, NY: John Wiley & Sons, Inc.; 1993.

[8] Sadiku MNO. Numerical techniques in electromagnetics. 2nd ed. Boca Raton, FL: CRC Press; 2001.

[9] R C Booton J. Computational Methods for Electromagnetics and Microwaves. New York, NY: John Wiley & Sons, Inc.; 1992.

[10] Bastos JPA, Sadowski N. Electromagnetic modeling by finite element methods. New York, NY: Marcel Dekker, Inc.; 2003.

[11] Gibson WC. The Method of Moments in electromagnetics. Boca Raton, FL: Chapman & Hall/CRC; 2008.

[12] Zhou P. Numerical Analysis of Electromagnetic Fields. Berlin Heidelberg: Springer-Verlag; 1993.

[13] Bau III D, Trefethen LN. Numerical linear algebra. Philadelphia, PA: Soci. Indus. Ap. Math.; 1997.

[14] Golub GH, van Loan CF. Matrix Computations. Baltimore, MD: Johns Hopkins University Press; 1996.

[15] Saad Y. Iterative methods for sparse linear systems. 2nd ed. Philadelfia, PA: Society for Industrial and Applied Mathematics; 2003.

[16] Collin RE. Field Theory of Guided Waves. Piscataway, NJ: IEEE press; 1991.

[17] Felsen LB, Marcuvitz N. Radiation and scattering of waves. Piscataway, NJ: IEEE Press; 2001.

[18] Harrington RF. Time-harmonic Electromagnetic Fields. London, UK: McGraw-Hill; 1961.

[19] Pozar D. Microwave Engineering. 4th ed. New York, NY: John Wiley & Sons, Inc.; 2012.

[20] Dwight HB. Tables of Integrals and Other Mathematical Data. 4th ed. New York, NY: MacMillan Publishing Co., Inc.; 1961.

[21] Moon P, Spencer DE. Field Theory Handbook: Including Coordinate Systems, Differential Equations and Their Solutions. 2nd ed. Berlin Heidelberg: Springer-Verlag; 1971.

[22] Holt CA. Introduction to Electromagnetic Fields and Waves. New York, NY: John Wiley & Sons, Inc.; 1963.

[23] Hadamard J. Sur les problèmes aux dérivées partielles et leur signification physique. Princeton University Bulletin. 1902;p. 49–52.

[24] Davis JL. Wave Propagation in Electromagnetic Media. New York, NY: Springer-Verlag; 1990.

[25] Colton DL, Kress R. Inverse Acoustic and Electromagnetic Scattering Theory. 3rd ed. New York, NY: Springer; 2013.

[26] Hopcraft KI, Smith PR. An Introduction to Electromagnetic Inverse Scattering. Dordrecht, NL: Springer Science+Business Media; 1992.

[27] Chew WC. Waves and Fields in Inhomogenous Media. New York, NY: Van Nostrand Teinhold; 1990.

[28] Jones DS. Methods in Electromagnetic Wave Propagation. 2nd ed. Piscataway, NJ: IEEE Press; 1994.

[29] Rothwell EJ, Cloud MJ. Electromagnetics. London, UK: CRC Press; 2001.

[30] Van Bladel JG. Electromagnetic Fields. Piscataway, NJ: IEEE Press; 2007.

[31] Kraus JD, Carver KR. Electromagnetics. 2nd ed. Tokyo, Japan: McGraw-Hill; 1973.

[32] Zhou XL. On independence, completeness of Maxwell's equations and uniqueness theorems in electromagnetics. Progress in Electromagnetic Research. 2006;64:117–134.

[33] Bassanini P, Elcrat A. Mathematical Theory of Electromagnetism. Creative Commons; 2009.

[34] Whitaker S. Introduction to fluid mechanics. Englewood Cliffs, NJ: Prentice-Hall; 1968.

[35] Abramowitz M, Stegun IA. Handbook of mathematical functions. New York, NY: Dover Publications, Inc.; 1965.

[36] Meixner J. The behavior of electromagnetic fields at edges. IEEE Transactions on Antennas Propagation. 1972 July;20(4):442–446.

[37] Jones DS. The Theory of Electromagnetism. Oxford, UK: Pergamon; 1964.

[38] Van Bladel JG. Singular Electromagnetic Fields and Sources. Piscataway, NJ: IEEE Press; 1991.

[39] Stratton JA. Electromagnetic theory. London, UK: McGraw-Hill; 1941.

[40] Collin RE. Foundations for Microwave Engineering. New York, NY: McGraw-Hill; 1992.
[41] Müller C. Foundations of the Mathematical Theory of Electromagnetic Waves. Berlin Heidelberg: Springer-Verlag; 1969.
[42] Leontovich MA. Investigation of propagation of radio waves, part II. USSR Academy of Sciences, Physical Series. 1944;.
[43] Senior TBA. Impedance boundary conditions for imperfectly conducting surfaces. Applied Scientific Research, Section B. 1960 December;8(1):418. Available from: http://dx.doi.org/10.1007/BF02920074.
[44] Mitzner KM. An integral equation approach to scattering from a body of finite conductivity. Radio Science. 1967;2:1459–1470.
[45] Hoppe DJ, Rahmat-Samii Y. Impedance Boundary Conditions in Electromagnetics. Washington, DC: Taylor & Francis; 1995.
[46] Lindell IV. Methods for electromagnetic field analysis. Piscataway, NJ: IEEE Press; 1992.
[47] Kurokawa K. The expansions of electromagnetic fields in cavities. IRE Trans Microwave Theory Tech. 1958;p. 178–187.
[48] Schelkunoff SA. On representation of electromagnetic fields in cavities in terms of natural modes of oscillation. Journal of Applied Physics. 1955;26:1231–1234.
[49] Adams RA, Fournier JJF. Sobolev spaces. 2nd ed. New York, NY: Academic Press; 2003.
[50] Muscat J. Functional Analysis. London, UK: Springer; 2014.
[51] Conway JB. A Course in Functional Analysis. 2nd ed. Graduate Texts in Mathematics. New York, NY: Springer-Verlag; 1990.
[52] Bowers A, Kalton NJ. An introductory course in functional analysis. Universitext. New York, NY: Springer Science+Business Media; 2014.
[53] Colton DL, Kress R. Integral Equation Methods in Scattering Theory. New York, NY: John Wiley & Sons, Inc.; 1983.
[54] Helrich CS. The classical theory of fields - Electromagnetism. Berlin Heidelberg: Springer-Verlag; 2012.
[55] Zangwill A. Modern electrodynamics. Cambridge, UK: Cambrigde University Press; 2013.
[56] Dautray R, Lions JL. Mathematical Analysis and Numerical Methods for Science and Technology - Functional and Variational Methods. vol. 2. Berlin Heidelberg: Springer-Verlag; 1990.
[57] Bassanini P, Elcrat AR. Theory and applications of partial differential equations. Mathematical Concepts And Methods In Science And Engineering. New York, NY: Plenum Press; 1997.
[58] Atkinson KE. The Numerical Solution of Integral Equations of the Second Kind. Cambridge, UK: Cambridge University Press; 1997.
[59] Kantorovich LV, Akilov GP. Functional analysis in normed spaces. Oxford, UK: Pergamon Press; 1964. Translated by D. E. Brown.
[60] Rudin W. Functional Analysis. New York, NY: McGraw-Hill; 1973.
[61] Chu Q, Liang C. The uniqueness theorem of electromagnetic fields in lossless regions. IEEE Transactions on Antennas and Propagation. 1993 Feb;41(2):245–246.
[62] Jin JM. Theory and Computation of Electromagnetic Fields. 2nd ed. Hoboken, NJ: IEEE Press; 2015.
[63] Marcuvitz N. Waveguide Handbook. 2nd ed. Electromagnetic Waves Series. London, UK: The Institution of Engineering and Technology; 1985.
[64] Milton KA, Schwinger J. Electromagnetic Radiation: Variational Methods, Waveguides and Accelerators. Berlin Heidelberg: Springer-Verlag; 2006.
[65] Hayt WH, Buck JA. Engineering Electromagnetics. 8th ed. New York, NY: McGraw-Hill; 2012. International edition.

[66] Guru BS, Hiziroglu HR. Electromagnetic field theory fundamentals. 2nd ed. New York, NY: Cambridge University Press; 2004.

[67] Orfanidis SJ. Electromagnetic Waves and Antennas. www.ece.rutgers.edu/~orfanidi/ewa; 2004.

[68] Adler RB, Chu LJ, Fano RM. Electromagnetic energy transmission and radiation. New York, NY: John Wiley & Sons, Inc.; 1960.

[69] Popović Z, Popović BD. Introductory Electromagnetics. Upper Saddle River, NJ: Prentice Hall; 2000.

[70] Schwab AJ. Field Theory Concepts. Berlin Heidelberg: Springer-Verlag; 1988.

[71] Schelkunoff SA. Electromagnetic Waves. Princeton, NJ: D. Van Nostrand Company; 1943.

[72] Weeks WL. Electromagnetic Theory for Engineering Applications. New York, NY: John Wiley & Sons, Inc.; 1964.

[73] Zhang K, Li D. Electromagnetic Theory for Microwaves and Optoelectronics. Berlin Heidelberg: Springer-Verlag; 1998.

[74] Balanis CA. Advanced Engineering Electromagnetics. New York, NY: John Wiley & Sons, Inc.; 2005.

[75] Sommerfeld A. Partial Differential Equations in Physics. vol. 1 of Lectures on theoretical physics. New York, NY: Academic Press; 1949.

[76] Sommerfeld A. Electrodynamics. vol. 3 of Lectures on theoretical physics. New York, NY: Academic Press; 1952.

[77] Rellich F. Über das asymptotische Verhalten von Lösungen von $\triangle u + \lambda u = 0$ in unendlichen Gebieten. Jber Deutsch Math Verein. 1943;53:57–65.

[78] Wilcox CH. An expansion theorem for electromagnetic fields. Comm Pure Appl Math. 1956;9:115–134.

[79] Jackson JD. Classical Electrodynamics. 3rd ed. Chichester, UK: Wiley; 1999.

[80] Recani E, Mignani R. Magnetic monopole and tachyons in special relativity. Physics Letters. 1976;62B:41–43.

[81] Thidé B. Electromagnetic Field Theory. New York, NY: Dover Publications, Inc.; 2011.

[82] Schwinger J, De Raad LL, Milton KA, et al. Classical electrodynamics. Perseus Books; 1998.

[83] Griffith DJ. Introduction to electrodynamics. 3rd ed. Upper Saddle River, NJ: Prentice Hall; 1999.

[84] Makarov B, Podkorytov A. Real Analysis: Measures, Integrals and Applications. Universitext. London, UK: Springer; 2013.

[85] Schelkunoff SA. Electromagnetic Fields. New York, NY: Blaisdell Pub. Co.; 1963.

[86] Papas CH. Theory of electromagnetic wave propagation. New York, NY: Dover Publications, Inc.; 1988.

[87] Chew WC. A New Look at Reciprocity and Energy Conservation Theorems in Electromagnetics. IEEE Transactions on Antennas and Propagation. 2008 April;56(4):970–975.

[88] Jordan EC, Balmain KG. Electromagnetic Waves and Radiating Systems. 2nd ed. Upper Saddle River, NJ: Prentice Hall; 1968.

[89] Rumsey VH. Reaction Concept in Electromagnetic Theory. Physics Review. 1954 June;94(6):1483–1491.

[90] Kong JA. Electromagnetic Wave Theory. 2nd ed. New York, NY: Wiley; 1990.

[91] Carson JR. Reciprocal theorems in radio communications. Proc IRE. 1929 June;17: 952–1491.

[92] Lancellotti V, Milanesio D, Maggiora R, et al. TOPICA: an accurate and efficient numerical tool for analysis and design of ICRF antennas. Nuclear Fusion. 2006;46(7):S476–S499.

[93] Feynmann RP, Leighton R, Sands M. The Feynmann lectures on physics. vol. 1. Reading, MA: Addison-Wesley; 1964.
[94] Stix TH. Plasma Waves. New York, NY: Springer-Verlag; 1992.
[95] Fitzpatrick R. Plasma Physics: An Introduction. Boca Raton, FL: CRC Press; 2014.
[96] Linkhart DK. Microwave Circulator Design. 2nd ed. Norwood, MA: Artech House Microwave Library; 2014.

Chapter 7
Electromagnetic waves

7.1 Time-domain uniform plane waves

We resume the discussion on electromagnetic waves started in Section 1.8 by looking for the simplest solutions to the D'Alembert equations (1.207) and (1.215).

We already know from the multipole expansion of the electrostatic scalar potential (3.204) and of the magnetic stationary vector potential (5.155) that the fields produced by localized distributions of fixed charges and steady currents must fall off as some power of the inverse of the distance from the sources. Besides, the same behavior also characterizes the fields and waves generated by a point charge in arbitrary motion, as prescribed by (1.252) and (1.253). Therefore, it stands to reason that the solutions to (1.207) and (1.215) follow the general spatial dependence with the distance away from the sources provided the latter are of finite extent. This statement can be proved by extending the multipole expansion of Section 3.6 to the three-dimensional Green function of Section 8.7 [1, Chapter 16], [2, Section 6.8].

Conversely, we may ask ourselves what kind of charge and current distributions are needed to generate fields and waves which *do not decay* with the distance. In the electrostatic regime, for instance, it can be shown with elementary means (i.e., the global Gauss law) that a planar infinite layer of charge with surface density $\varrho_S(\mathbf{r})$ generates a static *uniform* electric field which is everywhere perpendicular to the source [3, Section 5-6]. A similar conclusion can be reached for the stationary magnetic field produced by a planar infinite surface density of steady current $\mathfrak{J}_S(\mathbf{r})$, the required tool being the global Ampère law (4.5).

All in all, these observations suggest that a planar infinite layer of accelerated charges can generate an electromagnetic wave whose components do not decay with the distance from the source. On the other hand, the symmetry of the configuration demands that the solution depend only on said distance and, of course, on time.

In order to show that the D'Alembert equation admits wave-like solutions we choose a system of Cartesian coordinates and let $\Upsilon \in \{\mathcal{E}_x, \mathcal{E}_y, \mathcal{E}_z, \mathcal{H}_x, \mathcal{H}_y, \mathcal{H}_z\}$ denote any Cartesian component of the electromagnetic entities $\boldsymbol{\mathcal{E}}(\mathbf{r}, t)$ and $\boldsymbol{\mathcal{H}}(\mathbf{r}, t)$. In particular, we consider the source-free equation

$$\left(\frac{\partial^2}{\partial x^2} + \frac{\partial^2}{\partial y^2} + \frac{\partial^2}{\partial z^2} - \frac{1}{c^2}\frac{\partial^2}{\partial x^2}\right)\Upsilon(\mathbf{r}, t) = 0 \tag{7.1}$$

and seek the solutions — if any are possible — that depend on time and the auxiliary spatial coordinate

$$s := \mathbf{r} \cdot \hat{\mathbf{s}} = x s_x + y s_y + z s_z \qquad s \in \mathbb{R} \tag{7.2}$$

with $\hat{\mathbf{s}}$ a constant unit vector. Physically, s represents the distance with sign measured from the origin along the straight line defined by $\mathbf{r} = s\hat{\mathbf{s}}$.

Furthermore, we introduce the auxiliary variables (D'Alembert Ansatz)

$$\xi := s - ct, \qquad \eta := s + ct \tag{7.3}$$

whereby

$$\Upsilon(\mathbf{r}, t) = \Upsilon(s, t) = \Upsilon(\xi, \eta) \tag{7.4}$$

and express the differential operators in terms of ξ and η. To this purpose, we compute

$$\frac{\partial \Upsilon}{\partial x} = \frac{\partial \Upsilon}{\partial \xi} s_x + \frac{\partial \Upsilon}{\partial \eta} s_x \qquad \frac{\partial^2 \Upsilon}{\partial x^2} = \frac{\partial^2 \Upsilon}{\partial \xi^2} s_x^2 + 2\frac{\partial^2 \Upsilon}{\partial \eta \partial \xi} s_x^2 + \frac{\partial^2 \Upsilon}{\partial \eta^2} s_x^2 \tag{7.5}$$

$$\frac{\partial \Upsilon}{\partial y} = \frac{\partial \Upsilon}{\partial \xi} s_y + \frac{\partial \Upsilon}{\partial \eta} s_y \qquad \frac{\partial^2 \Upsilon}{\partial y^2} = \frac{\partial^2 \Upsilon}{\partial \xi^2} s_y^2 + 2\frac{\partial^2 \Upsilon}{\partial \eta \partial \xi} s_y^2 + \frac{\partial^2 \Upsilon}{\partial \eta^2} s_y^2 \tag{7.6}$$

$$\frac{\partial \Upsilon}{\partial z} = \frac{\partial \Upsilon}{\partial \xi} s_z + \frac{\partial \Upsilon}{\partial \eta} s_z \qquad \frac{\partial^2 \Upsilon}{\partial z^2} = \frac{\partial^2 \Upsilon}{\partial \xi^2} s_z^2 + 2\frac{\partial^2 \Upsilon}{\partial \eta \partial \xi} s_z^2 + \frac{\partial^2 \Upsilon}{\partial \eta^2} s_z^2 \tag{7.7}$$

$$\frac{\partial \Upsilon}{\partial t} = -\frac{\partial \Upsilon}{\partial \xi} c + \frac{\partial \Upsilon}{\partial \eta} c \qquad \frac{\partial^2 \Upsilon}{\partial t^2} = \frac{\partial^2 \Upsilon}{\partial \xi^2} c^2 - 2\frac{\partial^2 \Upsilon}{\partial \eta \partial \xi} c^2 + \frac{\partial^2 \Upsilon}{\partial \eta^2} c^2 \tag{7.8}$$

which, when substituted back into (7.1), yield the *normal form* of the wave equation [4, pp. 682–683]

$$0 = \Box^2 \Upsilon(\mathbf{r}, t) = 4\frac{\partial^2}{\partial \eta \partial \xi} \Upsilon(\xi, \eta) \tag{7.9}$$

having made use of the normalization condition $s_x^2 + s_y^2 + s_z^2 = 1$. This latter form implies that the source-free D'Alembert equation is satisfied by arbitrary well-behaved functions which depend on ξ and η separately, and hence the general solution may be written as

$$\Upsilon(\xi, \eta) = f(\xi) + g(\eta) \tag{7.10}$$

or, by reverting to the original variables, as

$$\Upsilon(\mathbf{r}, t) = f(\mathbf{r} \cdot \hat{\mathbf{s}} - ct) + g(\mathbf{r} \cdot \hat{\mathbf{s}} + ct) \tag{7.11}$$

with $f(\xi)$ and $g(\eta)$ being twice differentiable. The first contribution to $\Upsilon(\mathbf{r}, t)$ constitutes a *wave* that travels along the positive direction specified by $\hat{\mathbf{s}}$ whereas the second one is a wave that moves in the opposite direction.

To convince ourselves, we set $g(\eta) = 0$ and consider the pairs (s, t) for which ξ is equal to a constant, say, ξ_0. Suppose that an observer whose position is given by s_1 detects a field component $\Upsilon(\xi, \eta) = f(\xi_0)$ at time t_1. A second observer located farther along the way at $s_2 > s_1$ detects the same value of the field component at time $t_2 > t_1$. Owing to the definition of ξ (7.3) the pairs of spatial-temporal coordinates are not independent, but rather satisfy

$$\xi_0 = s_1 - ct_1 = s_2 - ct_2 \tag{7.12}$$

whence the quantity

$$t_2 - t_1 = \frac{1}{c}(s_2 - s_1) \tag{7.13}$$

is the time taken for the disturbance to travel from s_1 to s_2. Obviously, we may repeat this thought experiment for any value of ξ and, consequently, of $f(\xi)$, and the overall outcome may be pictured as

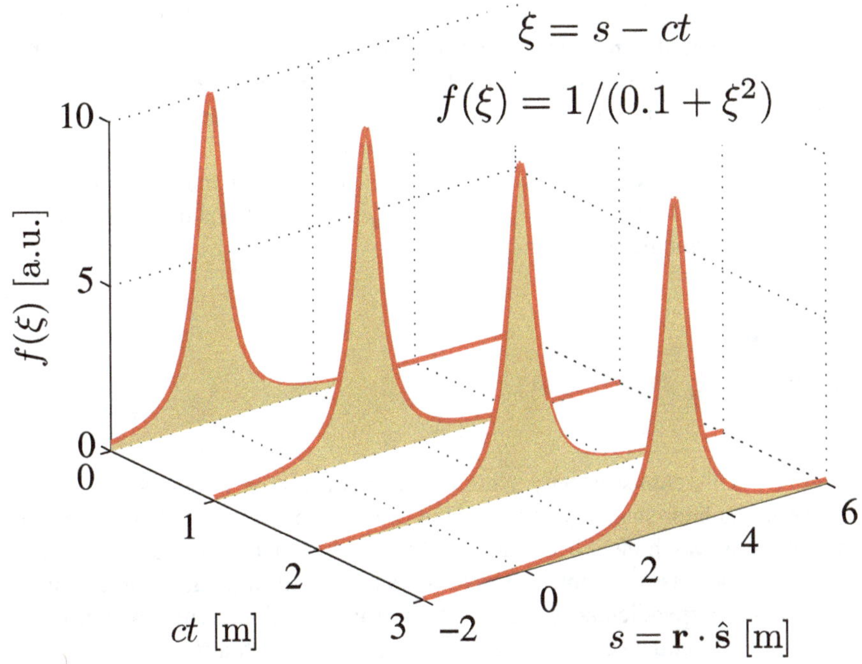

Figure 7.1 Snapshots of a plane wave propagating along $\hat{\mathbf{s}}$ *in a homogeneous isotropic medium.*

the *rigid translation* of the function $f(\mathbf{r} \cdot \hat{\mathbf{s}} - ct)$ in the positive direction set by $\hat{\mathbf{s}}$. This phenomenon is pictorially described in Figure 7.1 for the special choice $f(\xi) = 1/(0.1 + \xi^2)$.

What can be said about the velocity of propagation? Well, the latter must coincide with the velocity of an observer who wishes to 'chase after' the wave and detect always the same value of the field component $\Upsilon(\xi, \eta) = f(\xi)$, as if the field were frozen in time. From the observer's point of view then ξ must be constant in time

$$0 = \frac{\partial \xi}{\partial t} = \frac{\mathrm{d}\mathbf{r}}{\mathrm{d}t} \cdot \hat{\mathbf{s}} - c = \frac{\mathrm{d}s}{\mathrm{d}t} - c \tag{7.14}$$

which indicates c, the speed of light in the underlying medium, as the velocity of the wave. A similar reasoning carried out for $\Upsilon(\xi, \eta) = g(\eta)$ allows concluding that $g(\eta)$ represents a wave travelling in the direction $-\hat{\mathbf{s}}$ with velocity c.

Since the equations

$$\mathbf{r} \cdot \hat{\mathbf{s}} - ct = \xi \qquad\qquad \mathbf{r} \cdot \hat{\mathbf{s}} + ct = \eta \tag{7.15}$$

represent two families of planes perpendicular to the vector $\hat{\mathbf{s}}$, for fixed ξ and η the solutions $f(\xi)$ and $g(\eta)$ are constant over any plane orthogonal to the direction of propagation, hence the insightful name of *uniform plane waves* [5,6], [7, Section 16.3].

7.2 Time-harmonic plane waves

A special, though very useful, type of plane waves is represented by solutions which are periodic in time and oscillate with angular frequency ω [8, Chapters 7, 8], [2, Chapter 4], [5, Chapter 8], [9,

Chapter IV], [10, Chapter 11], [11, Chapter 4], [12, Chapter 6], [13, Chapter 27], [14, Chapter 2]. In which case, as we already know from Section 1.5, the Maxwell equations simplify in that the time-dependence may be taken in the form $\exp(\mathrm{j}\,\omega t)$ and implied in the calculations. The rules for going back and forth between the general form of the equations and the one valid for time-harmonic fields boil down to (1.88) and (1.94) with the exponent $n = 1, 2$.

A source of plane waves ought to be planar, infinitely extended and oscillating with frequency ω, which is why we obviate the issue by looking for general solutions of the *source-free* Maxwell equations in the time-harmonic regime, namely,

$$\nabla \times \mathbf{E}(\mathbf{r}) = -\mathrm{j}\,\omega \mathbf{B}(\mathbf{r}) \tag{7.16}$$

$$\nabla \times \mathbf{H}(\mathbf{r}) = \mathrm{j}\,\omega \mathbf{D}(\mathbf{r}) \tag{7.17}$$

$$\nabla \cdot \mathbf{D}(\mathbf{r}) = 0 \tag{7.18}$$

$$\nabla \cdot \mathbf{B}(\mathbf{r}) = 0 \tag{7.19}$$

for points $\mathbf{r} \in \mathbb{R}^3$. This choice is only a minor limitation for, if solutions do exist, we may try and combine them at a later stage, by invoking the principle of superposition, in order to construct more general solutions which meet boundary conditions on planar interfaces at finite locations (see Example 7.1). We shall also find that the field associated with a plane wave in some cases may be constant on planes perpendicular to the direction of propagation, hence the attribute *uniform* or *homogeneous*. As a consequence, the power carried along by such a wave turns out to be infinite, though the surface density (1.306) remains finite. This conclusion should come as no surprise, since the source is infinitely extended, too.

A plane wave characterized by a single frequency is termed *monochromatic*. Truly monochromatic waves in reality do not exist in that all practical sources of radiation emit waves which may at best be synthesized by a *bundle* or *packet* of plane waves oscillating at different frequencies. However, this is good enough a reason for studying the properties of such solutions.

7.2.1 Lossless isotropic medium

We assume that the underlying medium is isotropic, lossless and homogeneous, whereby the time-harmonic counterparts of the constitutive relationships (1.117) and (1.118) apply. To begin with we make the working hypothesis that the complex vector fields $\mathbf{E}(\mathbf{r})$ and $\mathbf{H}(\mathbf{r})$ may be written as

$$\mathbf{E}(\mathbf{r}) = \mathbf{E}_0 \mathrm{e}^{-\mathrm{j}\,\mathbf{k}\cdot\mathbf{r}} \tag{7.20}$$

$$\mathbf{H}(\mathbf{r}) = \mathbf{H}_0 \mathrm{e}^{-\mathrm{j}\,\mathbf{k}\cdot\mathbf{r}} \tag{7.21}$$

where $\mathbf{E}_0$, $\mathbf{H}_0$ and $\mathbf{k}$ are complex *constant* vectors yet to be specified. For reasons that will become clear in a moment $\mathbf{k}$ is called the *wavevector* (physical dimension: 1/m) and is usually divided into its real and imaginary parts as

$$\mathbf{k} = \mathbf{k}' - \mathrm{j}\,\mathbf{k}'' \tag{7.22}$$

where

- $\mathbf{k}' := \mathrm{Re}\{\mathbf{k}\}$ provides the direction of *propagation* of the wave,
- $\mathbf{k}'' := -\mathrm{Im}\{\mathbf{k}\}$ indicates the direction of *damping* or *attenuation*, having explicitly factored out a minus sign.[1]

[1] As a word of caution, we point out that some Authors (e.g., [15]) define the wavevector as $\mathbf{k} = \mathbf{k}' + \mathrm{j}\,\mathbf{k}''$ whereby $\mathbf{k}''$ points in the direction of *amplification* of the wave.

In fact, if we take a look at the electromagnetic field that in the time domain corresponds with the Ansatz (7.20), namely,

$$\begin{aligned}\mathcal{E}(\mathbf{r},t) &= \mathrm{Re}\left\{\mathbf{E}_0 e^{\mathrm{j}(\omega t-\mathbf{k}\cdot\mathbf{r})}\right\} = \mathrm{Re}\left\{(\mathbf{E}_0' + \mathrm{j}\,\mathbf{E}_0'')e^{\mathrm{j}(\omega t-\mathbf{k}'\cdot\mathbf{r})}\right\} e^{-\mathbf{k}''\cdot\mathbf{r}} \\ &= \mathbf{E}_0' e^{-\mathbf{k}''\cdot\mathbf{r}} \cos(\omega t - \mathbf{k}'\cdot\mathbf{r}) - \mathbf{E}_0'' e^{-\mathbf{k}''\cdot\mathbf{r}} \sin(\omega t - \mathbf{k}'\cdot\mathbf{r})\end{aligned} \tag{7.23}$$

we see that the surmised solution possesses the wave-like character expressed in general form by (7.11) provided we let

$$\mathbf{k}' = \frac{\omega}{c}\hat{\mathbf{s}} \qquad \mathbf{k}'' = 0 \tag{7.24}$$

but, more generally, the presence of the real exponential indicates that, as the wave travels, it is attenuated in the direction of $\mathbf{k}''$. The same conclusion is reached for the magnetic field $\mathcal{H}(\mathbf{r},t)$.

When considered in the space of points $\mathbf{r} \in \mathbb{R}^3$ the relation

$$\omega t - \mathbf{k}'\cdot\mathbf{r} = \phi_0 \qquad \phi_0 \in \mathbb{R} \tag{7.25}$$

describes a family of planes parameterized by the time $t \in \mathbb{R}$ and perpendicular to $\mathbf{k}'$. Such surfaces are called *planes of constant phase* in light of definitions (7.20) and (7.21). In like manner, the relation

$$\mathbf{k}''\cdot\mathbf{r} = A_0 \qquad A_0 \in \mathbb{R} \tag{7.26}$$

represents a family of planes orthogonal to $\mathbf{k}''$. Since the magnitude of the fields (7.20) and (7.21) is constant for points $\mathbf{r}$ prescribed by (7.26), the corresponding surfaces are termed *planes of constant amplitude*. Moreover, it is apparent from (7.23) that the planes of constant phase and constant amplitude coincide when $\mathbf{k}'' = 0$, in which instance the field (7.20) is also *periodic in space*. While the relation between the temporal period T and the angular frequency ω is already known and given by (1.87), to determine the shortest spatial period $\lambda > 0$ we fix the time t and consider two distinct planes of constant phase defined by (Figure 7.2)

$$\pi_l := \{\mathbf{r} \in \mathbb{R}^3 : \omega t - \mathbf{k}'\cdot\mathbf{r} = \phi_l\}, \qquad l \in \{1,2\} \tag{7.27}$$

then we pick up two points $\mathbf{r}_1 \in \pi_1$ and $\mathbf{r}_2 \in \pi_2$ such that

$$\mathbf{r}_2 = \mathbf{r}_1 + \lambda\hat{\mathbf{s}} \tag{7.28}$$

i.e., $\mathbf{r}_2$ is the projection of $\mathbf{r}_1$ onto π_2 or, equivalently, the other way around. Finally, we require that the electric field vectors in $\mathbf{r}_1$ and $\mathbf{r}_2$ be related as

$$\mathcal{E}(\mathbf{r}_1,t) = \mathcal{E}(\mathbf{r}_2,t) = \mathcal{E}(\mathbf{r}_1 + \lambda\hat{\mathbf{s}},t) \tag{7.29}$$

on account of the very definition of periodicity. Since the fundamental period of sine and cosine functions is 2π, in light of (7.23) with $\mathbf{k}'' = 0$ we see that (7.29) can be satisfied if we require

$$\phi_1 - \phi_2 = (\omega t - \mathbf{k}'\cdot\mathbf{r}_1) - (\omega t - \mathbf{k}'\cdot\mathbf{r}_2) = \mathbf{k}'\cdot(\mathbf{r}_2 - \mathbf{r}_1) = 2\pi \tag{7.30}$$

whence we find

$$\lambda = \frac{2\pi c}{\omega} = cT \tag{7.31}$$

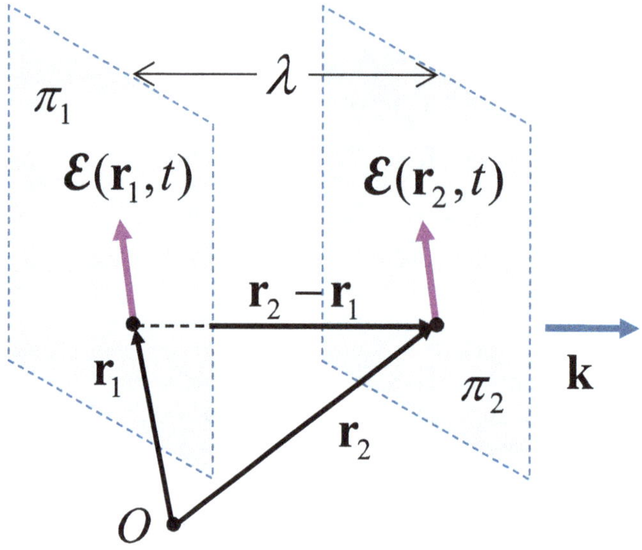

Figure 7.2 Two planes of constant phase for determining the spatial period of a time-harmonic plane wave in an isotropic lossless medium.

by virtue of (7.24), (7.28) and (1.87). The spatial period λ (physical dimension: m) is called the *wavelength* and, by construction, represents the distance between two neighboring planes of constant phase characterized by ϕ_0 and $\phi_0 - 2\pi$. Alternatively, thanks to (7.29) the wavelength can be construed as the shortest distance between two crests or troughs of the wave.

An ideal stationary observer 'sees' the planes of constant phase drift by with velocity $\mathbf{v}_p = v_p\hat{\mathbf{s}}$ in the direction indicated by $\mathbf{k}'$, which is why v_p is called the *phase velocity* of the wave. To find v_p we notice that, if the observer travelled at the same speed $\mathbf{v}_p$, then he would invariably detect the same value of the electromagnetic field. The latter appears to be constant in time if so are the arguments of sine and cosine in (7.23). Thus, by differentiating (7.25) with respect to time we get

$$0 = \frac{\partial \phi_0}{\partial t} = \omega - \mathbf{k}' \cdot \frac{\mathrm{d}\mathbf{r}}{\mathrm{d}t} = \omega - \mathbf{k}' \cdot \mathbf{v}_p = \omega - \frac{\omega}{c} v_p \tag{7.32}$$

owing to (7.24) and the fact that $\mathbf{v}_p$ is a vector parallel to $\mathbf{k}'$ by assumption. Solving for v_p yields

$$v_p = c = \frac{\lambda}{T} \tag{7.33}$$

also on account of (7.31). In words, plane waves in a homogeneous isotropic unbounded lossless medium travel at the speed of light (1.209). From the rightmost-hand side of (7.33) — which is the ratio of a length to a time — we also gather that the period T may be interpreted as the time it takes for the wave to sweep a distance equal to λ in the direction of propagation.

We now proceed to derive the conditions we require of $\mathbf{E}_0$, $\mathbf{H}_0$ and $\mathbf{k}$ in order for (7.20) and (7.21) to solve (7.16)-(7.19). To this purpose, we substitute (7.20) and (7.21) into (7.16)-(7.19) and obtain

$$\nabla \mathrm{e}^{-\mathrm{j}\mathbf{k}\cdot\mathbf{r}} \times \mathbf{E}_0 = -\mathrm{j}\,\omega\mu \mathbf{H}_0 \mathrm{e}^{-\mathrm{j}\mathbf{k}\cdot\mathbf{r}} \tag{7.34}$$

$$\nabla \mathrm{e}^{-\mathrm{j}\mathbf{k}\cdot\mathbf{r}} \times \mathbf{H}_0 = \mathrm{j}\,\omega\varepsilon \mathbf{E}_0 \mathrm{e}^{-\mathrm{j}\mathbf{k}\cdot\mathbf{r}} \tag{7.35}$$

$$\nabla e^{-j\mathbf{k}\cdot\mathbf{r}} \cdot \mathbf{E}_0 = 0 \tag{7.36}$$

$$\nabla e^{-j\mathbf{k}\cdot\mathbf{r}} \cdot \mathbf{H}_0 = 0 \tag{7.37}$$

on account of (H.52) for $\mathbf{E}_0$ and $\mathbf{H}_0$ are constant vectors. For the calculation of the gradient of the complex exponential, which contains the spatial dependence of the wave, we may utilize Cartesian coordinates and observe

$$\nabla e^{-j\mathbf{k}\cdot\mathbf{r}} = \nabla\left(e^{-jk_x x}e^{-jk_y y}e^{-jk_z z}\right) = -j\,k_x\hat{\mathbf{x}}e^{-j\mathbf{k}\cdot\mathbf{r}} - j\,k_y\hat{\mathbf{y}}e^{-j\mathbf{k}\cdot\mathbf{r}} - j\,k_z\hat{\mathbf{z}}e^{-j\mathbf{k}\cdot\mathbf{r}} = -j\,\mathbf{k}\,e^{-j\mathbf{k}\cdot\mathbf{r}} \tag{7.38}$$

in light of (H.47) and (A.26). The above calculation shows that the operator ∇ applied to $\exp(-j\,\mathbf{k}\cdot\mathbf{r})$ produces the very same exponential multiplied with $-j\,\mathbf{k}$. This property is analogous to the effect of $\partial/\partial t$ operating on $\exp(j\,\omega t)$ that was derived in (1.88). By using (7.38) we arrive at

$$\mathbf{k}\times\mathbf{E}_0 = \omega\mu\mathbf{H}_0 \tag{7.39}$$

$$\mathbf{k}\times\mathbf{H}_0 = -\omega\varepsilon\mathbf{E}_0 \tag{7.40}$$

$$\mathbf{k}\cdot\mathbf{E}_0 = 0 \tag{7.41}$$

$$\mathbf{k}\cdot\mathbf{H}_0 = 0 \tag{7.42}$$

after dividing through by $-j$ and implying the common multiplicative factor $\exp(-j\,\mathbf{k}\cdot\mathbf{r})$. The latter simplification is permissible for $\exp(-j\,\mathbf{k}\cdot\mathbf{r})$ never vanishes for $\mathbf{k}\cdot\mathbf{r}\in\mathbb{C}$, and consequently the algebraic relations (7.39)-(7.42) must hold true for any $\mathbf{r}\in\mathbb{R}^3$ and $t\in\mathbb{R}$.

At this stage we can make the following observations.

(1) Our working hypothesis, summarized by (7.20) and (7.21), has turned the original set of time-harmonic Maxwell's equations into an *algebraic* system of eight scalar equations for the six components of the vectors $\mathbf{E}_0$ and $\mathbf{H}_0$. Thus, it would appear that the system is truly over-determined. However, this conclusion is fallacious for (7.41) and (7.42) may be derived from the remaining equations by dot-multiplying through with $\mathbf{k}$. This property is consistent with the time-harmonic Gauss laws being dependent on the other two Maxwell's equations, as we argued at the end of Section 1.5.

(2) The algebraic conditions (7.39) and (7.40) signify that $\mathbf{E}_0$ and $\mathbf{H}_0$ are perpendicular to each other. The orthogonality, though, must be interpreted in a broader sense, as $\mathbf{E}_0$ and $\mathbf{H}_0$ may be complex vectors. The remaining relationships (7.41) and (7.42) demand that both $\mathbf{E}_0$ and $\mathbf{H}_0$ be perpendicular to $\mathbf{k}$. Once again the orthogonality must be understood in the sense of a vanishing dot-product when $\mathbf{k}$, $\mathbf{E}_0$ and $\mathbf{H}_0$ are complex vectors. In summary, among all the vectors $\mathbf{E}_0$, $\mathbf{H}_0$ and $\mathbf{k}$ only those which form an *orthogonal triplet* may represent a plane wave.

(3) Since we did not include sources in the Maxwell equations, the algebraic counterpart is a linear homogeneous system. As a consequence, (7.39) and (7.40) will admit non-trivial solutions if the 6×6 complex matrix of the system is singular [16–18], which may happen only for particular values of $\mathbf{k}$ and ω. If the angular frequency is fixed, then we are left with the task of finding the permissible values of the wavevector, which plays the role of a threefold eigenvalue. Alternatively, we may assign the wavevector and determine the eigenvalue ω.

To determine the solutions to (7.39) and (7.40) we obtain a reduced set of equations which involve either $\mathbf{E}_0$ or $\mathbf{H}_0$. For instance, by cross-multiplying (7.39) through with $\mathbf{k}$ we find

$$\begin{aligned} 0 &= \mathbf{k}\times(\mathbf{k}\times\mathbf{E}_0) + \omega\mu\mathbf{k}\times\mathbf{H}_0 = \mathbf{k}(\mathbf{k}\cdot\mathbf{E}_0) - \mathbf{E}_0(\mathbf{k}\cdot\mathbf{k}) + \omega^2\varepsilon\mu\mathbf{E}_0 \\ &= \left[\mathbf{k}\mathbf{k} - (\mathbf{k}\cdot\mathbf{k} - \omega^2\varepsilon\mu)\overline{\mathfrak{I}}\right]\cdot\mathbf{E}_0 \end{aligned} \tag{7.43}$$

having made use of (H.14) and (7.40). It is fruitful to introduce a system of polar spherical coordinates (k, α, β) in the space of wavevectors $\mathbf{k}$ and let

$$\mathbf{k} = k\hat{\mathbf{k}} \tag{7.44}$$

where we emphasize that, in general, k *is not* the magnitude of $\mathbf{k}$ if the latter is a complex vector. Indeed, we have

$$k^2 = \mathbf{k} \cdot \mathbf{k} \qquad |\mathbf{k}|^2 = \mathbf{k} \cdot \mathbf{k}^* \tag{7.45}$$

and only when $\mathbf{k}$ and $\hat{\mathbf{k}}$ are real does k coincide with the magnitude $|\mathbf{k}|$! Additionally, we express the identity dyadic as

$$\overline{\mathfrak{I}} = \hat{\mathbf{k}}\hat{\mathbf{k}} + \hat{\alpha}\hat{\alpha} + \hat{\beta}\hat{\beta} = \hat{\mathbf{k}}\hat{\mathbf{k}} + \overline{\mathfrak{I}}_{tk} \tag{7.46}$$

where $\hat{\alpha}$ and $\hat{\beta}$ constitute the remaining unit vectors, and $\overline{\mathfrak{I}}_{tk}$ is the *identity dyadic transverse to* k or $\mathbf{k}$. With these positions (7.43) reads

$$\left[\omega^2\varepsilon\mu\hat{\mathbf{k}}\hat{\mathbf{k}} - \left(\mathbf{k}\cdot\mathbf{k} - \omega^2\varepsilon\mu\right)\left(\hat{\alpha}\hat{\alpha} + \hat{\beta}\hat{\beta}\right)\right] \cdot \mathbf{E}_0 = 0 \tag{7.47}$$

which, since the elementary dyads are independent, can be further split as

$$\omega^2\varepsilon\mu\hat{\mathbf{k}}\hat{\mathbf{k}} \cdot \mathbf{E}_0 = 0 \tag{7.48}$$

$$\left(\mathbf{k}\cdot\mathbf{k} - \omega^2\varepsilon\mu\right)\left(\hat{\alpha}\hat{\alpha} + \hat{\beta}\hat{\beta}\right) \cdot \mathbf{E}_0 = 0 \tag{7.49}$$

whereby it becomes apparent that (7.41) is implied for $\omega \neq 0$. Condition (7.49) may be regarded as the algebraic counterpart of the wave equation (1.207) in case no sources are contemplated. Formally the same equations hold for the magnetic field vector $\mathbf{H}_0$, a conclusion which we may draw simply by invoking the principle of duality (Section 6.7).

Non-trivial solutions are possible if we choose $\mathbf{k}$ and ω so that

$$\mathbf{k}\cdot\mathbf{k} - \omega^2\varepsilon\mu = 0 \tag{7.50}$$

is satisfied. This constraint — which makes the dyadic in (7.49) singular and non-invertible (Appendix E) — is known as the *dispersion relationship* for plane waves (in a homogeneous isotropic lossless unbounded medium). When ω and the constitutive parameters of the material have been assigned, (7.50) yields the threefold eigenvalues $\mathbf{k}$. However, since the dispersion relation amounts to just one condition, whereas to specify $\mathbf{k}$ we need to provide, e.g., the Cartesian coordinates or magnitude, direction and orientation and so forth, we are left with two degrees of freedom. This indeterminacy is essentially due to the absence of boundary conditions imposed on the solutions we are seeking.

For real wavevectors, (7.50) lends itself to an interesting geometrical interpretation. By expressing $\mathbf{k}$ in Cartesian coordinates and recalling (1.209) we have

$$k_x^2 + k_y^2 + k_z^2 = \omega^2\varepsilon\mu = \frac{\omega^2}{c^2} \tag{7.51}$$

which clearly is the equation of the surface of a sphere with radius ω/c and centered in the origin of the space of points $(k_x, k_y, k_z) \in \mathbb{R}^3$. More generally, equations of the form $f(\mathbf{k}, \omega) = 0$ describe *dispersion surfaces* parameterized by ω. In particular, the dispersion relation (7.51) only prescribes the *magnitude* of $\mathbf{k} \in \mathbb{R}^3$ as

$$|\mathbf{k}| = \frac{\omega}{c} \tag{7.52}$$

while leaving direction and orientation (two degrees of freedom) undefined. In other words, plane-wave solutions in the form (7.20) and (7.21) can exist and propagate in *any direction* in a homogeneous isotropic lossless unbounded medium provided the real wavenumber has the right magnitude set by the frequency and the relevant speed of light.

From (7.49) we glean that once the wavevector has been selected in accordance with (7.50), we may still choose $\mathbf{E}_0$ in infinitely many ways, so long as (7.41) is satisfied. This means that for a given direction of propagation $\hat{\mathbf{k}}$ infinitely many plane waves exist. In fact, we need to specify just two linearly independent solutions, e.g.,

$$\mathbf{E}_0 = E_0\hat{\boldsymbol{\alpha}} \qquad\qquad \mathbf{E}_0 = E_0\hat{\boldsymbol{\beta}} \tag{7.53}$$

which we may adopt as an orthogonal basis to express all the other waves associated with the wavevector $\mathbf{k}$.

What about the magnetic field? Shall we start over or perhaps apply the duality principle? As a matter of fact, although (7.49) does not contain $\mathbf{H}_0$ by construction and a similar equation holds for $\mathbf{H}_0$, the latter must also satisfy (7.39) and (7.42). Therefore, once we have assigned $\mathbf{k}$ and $\mathbf{E}_0$, we are compelled to determine $\mathbf{H}_0$ as

$$\mathbf{H}_0 = \frac{\mathbf{k}}{\omega\mu} \times \mathbf{E}_0 \tag{7.54}$$

which makes (7.42) automatically satisfied.

Since the dispersion relationship (7.50) is invariant under the substitution $\mathbf{k} \to -\mathbf{k}$, we conclude that for any given $\mathbf{k}$, (7.49) is solved by two plane waves propagating in opposite directions. The waves associated with $\mathbf{k}$ and $-\mathbf{k}$ are termed *progressive* and *regressive*, respectively. In summary, we have proved that the Ansatz given by (7.20) and (7.21) is viable and leads to two distinct solutions

$$\mathbf{E}(\mathbf{r}) = \mathbf{E}_0 e^{-j\mathbf{k}\cdot\mathbf{r}} \qquad \mathbf{H}(\mathbf{r}) = \frac{\mathbf{k}}{\omega\mu} \times \mathbf{E}_0 e^{-j\mathbf{k}\cdot\mathbf{r}} \qquad \text{(progressive wave)} \tag{7.55}$$

$$\mathbf{E}(\mathbf{r}) = \mathbf{E}_0 e^{j\mathbf{k}\cdot\mathbf{r}} \qquad \mathbf{H}(\mathbf{r}) = -\frac{\mathbf{k}}{\omega\mu} \times \mathbf{E}_0 e^{j\mathbf{k}\cdot\mathbf{r}} \qquad \text{(regressive wave)} \tag{7.56}$$

under conditions (7.41), (7.42) and (7.50).

We still have to entertain the possibility of a complex wavevector and investigate the effect that this occurrence has on the solutions. To this purpose we insert the splitting (7.22) into the dispersion relation (7.50) which is valid irrespective of the nature of $\mathbf{k}$. In symbols, we have

$$\begin{aligned}\omega^2\varepsilon\mu &= (\mathbf{k}' - j\,\mathbf{k}'')\cdot(\mathbf{k}' - j\,\mathbf{k}'') = \mathbf{k}'\cdot\mathbf{k}' - \mathbf{k}''\cdot\mathbf{k}'' - 2j\,\mathbf{k}'\cdot\mathbf{k}'' \\ &= k'^2 - k''^2 - 2j\,\mathbf{k}'\cdot\mathbf{k}''\end{aligned} \tag{7.57}$$

where $k' > 0$ and $k'' \geqslant 0$ denote the *magnitude* of the *real* vectors $\mathbf{k}'$ and $\mathbf{k}''$. By separating real and imaginary parts of the dispersion relationship we obtain a system of two equations

$$k'^2 - k''^2 = \omega^2\varepsilon\mu \tag{7.58}$$

$$\mathbf{k}'\cdot\mathbf{k}'' = 0 \tag{7.59}$$

which we can satisfy in the following two ways.

(1) If $\mathbf{k}'' = 0$, then (7.59) is trivially true, $\mathbf{k}$ is a real vector, say,

$$\mathbf{k} = \mathbf{k}' = k'\hat{\mathbf{s}} \tag{7.60}$$

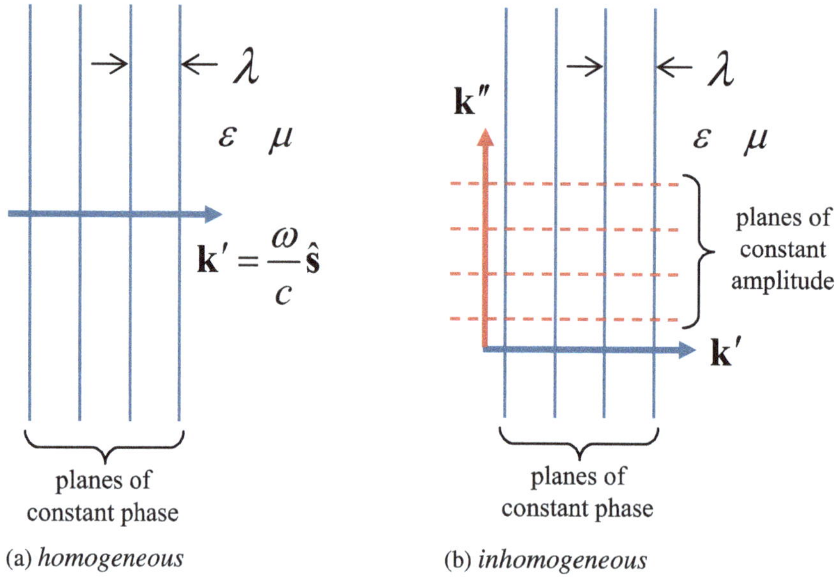

Figure 7.3 Plane waves in a homogeneous isotropic lossless unbounded medium.

and the magnitude of $\mathbf{k}'$ is derived from (7.58). This is the same result we have stipulated in (7.24). Solutions of this type — for which the planes of constant phase and constant amplitude coincide — are said *homogeneous* or *uniform* plane waves (Figure 7.3a).

(2) If $\mathbf{k}$ is a complex vector, then (7.59) requires that $\mathbf{k}'$ and $\mathbf{k}''$ be orthogonal in the ordinary real vector space, whereas (7.58) allows setting $k' > \omega(\varepsilon\mu)^{1/2}$ once $k'' > 0$ has been chosen. In particular, with reference to (7.44) and (1.209) we have

$$k = \sqrt{k'^2 - k''^2} = \frac{\omega}{c}, \qquad \hat{\mathbf{k}} = \frac{\mathbf{k}}{k} = \frac{c}{\omega}(\mathbf{k}' - \mathrm{j}\,\mathbf{k}'') \tag{7.61}$$

that is, the wavenumber k is real but the 'unit' vector $\hat{\mathbf{k}}$ is complex. In this regard, we notice that $\hat{\mathbf{k}} \cdot \hat{\mathbf{k}} = 1$ by construction, but as for the magnitude of $\hat{\mathbf{k}}$ we have

$$|\hat{\mathbf{k}}|^2 = \hat{\mathbf{k}} \cdot \hat{\mathbf{k}}^* = \frac{k'^2 + k''^2}{k'^2 - k''^2} = \frac{k'^2 + k''^2}{\omega^2\mu\varepsilon} = \frac{2c^2k''^2}{\omega^2} + 1 > 1 \tag{7.62}$$

i.e., $|\hat{\mathbf{k}}|$ exceeds unity. The locus of points $(k', k'') \in \mathbb{R}^2$ which solve (7.58) constitutes a family of rectangular hyperbolas with asymptotes given by $k'' = \pm k'$ [19, Section 8.4]. Solutions of this type — characterized by planes of constant phase and constant amplitude perpendicular to each other — are termed *inhomogeneous plane waves* (Figure 7.3b).

When $\mathbf{k}$ is real, (7.54) becomes

$$\mathbf{H}_0 = \frac{k'}{\omega\mu}\hat{\mathbf{s}} \times \mathbf{E}_0 = \sqrt{\frac{\varepsilon}{\mu}}\,\hat{\mathbf{s}} \times \mathbf{E}_0 = \frac{1}{Z}\hat{\mathbf{s}} \times \mathbf{E}_0 \tag{7.63}$$

in light of (7.58). The real quantity Z, defined (1.358), is the intrinsic impedance of the background medium. Inhomogeneous plane waves are damped in the positive direction prescribed by $\mathbf{k}''$. Such attenuation, though, is a *reactive* phenomenon and is not due to losses in the material. We shall

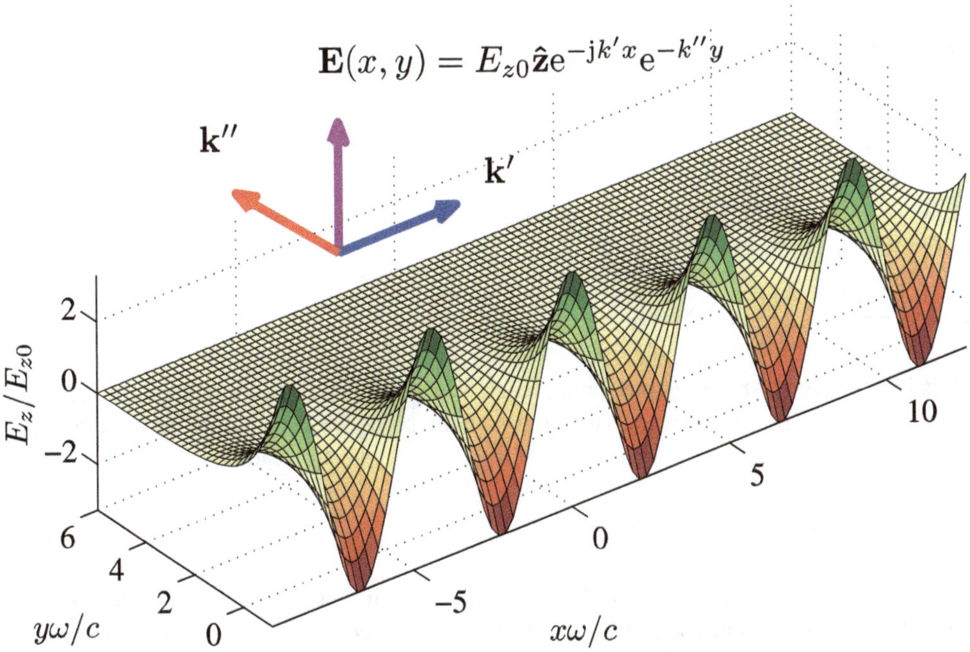

Figure 7.4 Snapshot for t = 0 of the normalized electric field of an x-propagating inhomogeneous plane wave in a lossless isotropic medium.

show that, as a result, there is no conversion of power into heat as the wave travels on. For example, Figure 7.4 shows the electric field $\mathbf{E}(x, y) = E_z(x, y)\hat{\mathbf{z}}$ for $k'' = \omega/c$ in (7.58) for points on any plane perpendicular to $\hat{\mathbf{z}}$ at time time $t = 0$.

We may compute the average power density radiated by the sources and carried along by a plane wave by first evaluating the complex Poynting vector (1.304). Since electric and magnetic fields are related by (7.39) and (7.40) we may obtain an expression which involves $\mathbf{E}_0$ only, viz.,

$$\begin{aligned}
\mathbf{S}(\mathbf{r}) &:= \frac{1}{2}\mathbf{E}(\mathbf{r}) \times \mathbf{H}^*(\mathbf{r}) = \frac{1}{2}\mathbf{E}_0 e^{-j\mathbf{k}\cdot\mathbf{r}} \times \left(\frac{\mathbf{k}^*}{\omega\mu} \times \mathbf{E}_0^* e^{j\mathbf{k}^*\cdot\mathbf{r}}\right) \\
&= \frac{e^{-2\mathbf{k}''\cdot\mathbf{r}}}{2\omega\mu}\left[|\mathbf{E}_0|^2\mathbf{k}^* - (\mathbf{E}_0 \cdot \mathbf{k}^*)\mathbf{E}_0^*\right] \\
&= \frac{e^{-2\mathbf{k}''\cdot\mathbf{r}}}{2\omega\mu}\left[|\mathbf{E}_0|^2\mathbf{k}^* + \underbrace{(\mathbf{k}^* \cdot \mathbf{E}_0^*)}_{=0}\mathbf{E}_0 - (\mathbf{k}^* \cdot \mathbf{E}_0)\mathbf{E}_0^*\right] \\
&= \frac{e^{-2\mathbf{k}''\cdot\mathbf{r}}}{2\omega\mu}\left[|\mathbf{E}_0|^2\mathbf{k}^* + \mathbf{k}^* \times (\mathbf{E}_0 \times \mathbf{E}_0^*)\right] \\
&= \frac{e^{-2\mathbf{k}''\cdot\mathbf{r}}}{2\omega\mu}\left(|\mathbf{E}_0|^2\mathbf{k}^* + j\,\mathbf{k}^* \times \mathrm{Im}\{\mathbf{E}_0 \times \mathbf{E}_0^*\}\right)
\end{aligned} \tag{7.64}$$

where we have recalled that $\mathbf{E}_0$ and $\mathbf{k}$ are orthogonal and noticed that $\mathbf{E}_0 \times \mathbf{E}_0^*$ is a purely imaginary vector on account of (H.10) and (B.6).

We examine the behavior of homogeneous and inhomogeneous plane waves separately. In the former case $\mathbf{k} = \mathbf{k}^* = \mathbf{k}'$, so we may use (H.14) and (7.41) to simplify the expression of the Poynting vector as

$$\mathbf{S} = \frac{|\mathbf{E}_0|^2}{2\omega\mu}\mathbf{k}' \tag{7.65}$$

whereby, according to (1.306), the *average surface power density* flowing through a plane perpendicular to the unit vector $\hat{\mathbf{n}}$ reads

$$p_R := \left\langle \frac{\mathrm{d}P_F}{\mathrm{d}S} \right\rangle = \frac{|\mathbf{E}_0|^2}{2\omega\mu}\mathbf{k}' \cdot \hat{\mathbf{n}} = \frac{|\mathbf{E}_0|^2}{2Z}\hat{\mathbf{s}} \cdot \hat{\mathbf{n}} \tag{7.66}$$

which is independent of $\mathbf{r}$ and, in particular, reaches its maximum when $\hat{\mathbf{n}} \equiv \hat{\mathbf{s}}$, i.e., across a plane of constant phase. The radiated power associated with the plane wave is infinite, as anticipated.

For inhomogeneous waves the real part of the Poynting vector reads

$$\mathrm{Re}\{\mathbf{S}\} = \frac{\mathrm{e}^{-2\mathbf{k}''\cdot\mathbf{r}}}{2\omega\mu}\left(|\mathbf{E}_0|^2\mathbf{k}' - \mathbf{k}'' \times \mathrm{Im}\{\mathbf{E}_0 \times \mathbf{E}_0^*\}\right) \tag{7.67}$$

and this indicates that — in view of (7.59) — the direction of power flow lies in planes orthogonal to $\mathbf{k}''$, though in general it is not perpendicular to the planes of constant phase owing to the presence of the second term within brackets above. In the special instance where $\mathbf{E}_0$ is real the *average surface power density* flowing through a plane perpendicular to the unit vector $\hat{\mathbf{n}}$ becomes

$$p_R(\mathbf{r}) = \left\langle \frac{\mathrm{d}P_F}{\mathrm{d}S} \right\rangle = \frac{|\mathbf{E}_0|^2}{2\omega\mu}\mathbf{k}' \cdot \hat{\mathbf{n}}\,\mathrm{e}^{-2\mathbf{k}''\cdot\mathbf{r}} \tag{7.68}$$

and in a given point $\mathbf{r}$, $p_R(\mathbf{r})$ is maximum across planes of constant phase. By applying the complex Poynting theorem (1.314) to a source-free bounded domain V we obtain

$$\langle P_F \rangle = \int_{\partial V} \mathrm{d}S\, \frac{|\mathbf{E}_0|^2}{2\omega\mu}\mathbf{k}' \cdot \hat{\mathbf{n}}(\mathbf{r})\,\mathrm{e}^{-2\mathbf{k}''\cdot\mathbf{r}} = 0 \tag{7.69}$$

since the background medium is lossless by hypothesis. Splitting the boundary ∂V into two adjoining non-overlapping parts, say, ∂V_1 and ∂V_2 with $\partial V_1 \cap \partial V_2 = \emptyset$, leads to

$$\int_{\partial V_1} \mathrm{d}S\, \frac{|\mathbf{E}_0|^2}{2\omega\mu}\mathbf{k}' \cdot \hat{\mathbf{n}}_1(\mathbf{r})\,\mathrm{e}^{-2\mathbf{k}''\cdot\mathbf{r}} = \int_{\partial V_2} \mathrm{d}S\, \frac{|\mathbf{E}_0|^2}{2\omega\mu}\mathbf{k}' \cdot \hat{\mathbf{n}}_2(\mathbf{r})\,\mathrm{e}^{-2\mathbf{k}''\cdot\mathbf{r}} \tag{7.70}$$

where we have chosen $\hat{\mathbf{n}}_1(\mathbf{r}) = -\hat{\mathbf{n}}(\mathbf{r})$ on ∂V_1 and $\hat{\mathbf{n}}_2(\mathbf{r}) = \hat{\mathbf{n}}(\mathbf{r})$ on ∂V_2. The left (right) member represents the average power that enters (leaves) V through ∂V_1 (∂V_2). Thus, the power is conserved as the plane wave travels along. A special and simple case obtains when we choose V as a cylinder with axis parallel to $\mathbf{k}'$ and any cross section. More precisely, (7.69) passes over into

$$\int_{S_1} \mathrm{d}S\, \frac{|\mathbf{E}_0|^2}{2\omega\mu}\mathbf{k}' \cdot \hat{\mathbf{n}}_1\,\mathrm{e}^{-2\mathbf{k}''\cdot\mathbf{r}} = \underbrace{\int_{S_L} \mathrm{d}S\, \frac{|\mathbf{E}_0|^2}{2\omega\mu}\mathbf{k}' \cdot \hat{\mathbf{n}}(\mathbf{r})\,\mathrm{e}^{-2\mathbf{k}''\cdot\mathbf{r}}}_{=0} + \int_{S_2} \mathrm{d}S\, \frac{|\mathbf{E}_0|^2}{2\omega\mu}\mathbf{k}' \cdot \hat{\mathbf{n}}_2\,\mathrm{e}^{-2\mathbf{k}''\cdot\mathbf{r}} \tag{7.71}$$

where we have separated the flux integral into the flow across the bases S_1 and S_2 of the cylinder and across its lateral surface S_L. Besides, $\hat{\mathbf{n}}_1 = -\hat{\mathbf{n}}(\mathbf{r})$ on S_1 and $\hat{\mathbf{n}}_2 = \hat{\mathbf{n}}(\mathbf{r})$ on S_2. Since by construction

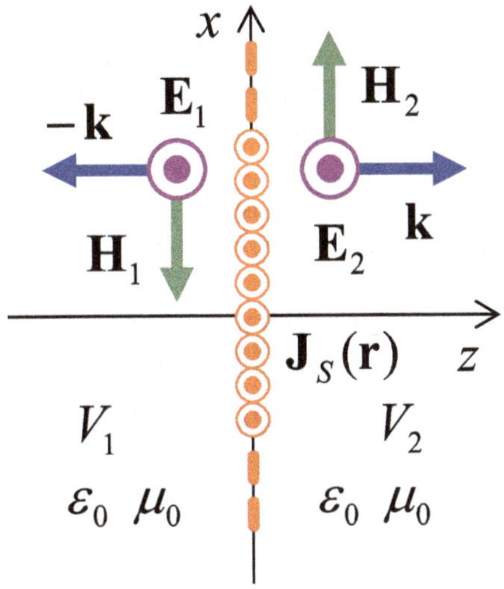

Figure 7.5 Uniform plane waves generated by an infinite electric current sheet $\mathbf{J}_S(\mathbf{r})$.

$\mathbf{k}' \cdot \hat{\mathbf{n}}(\mathbf{r}) = 0$ for $\mathbf{r} \in S_L$, no power enters or leaves the cylinder through the lateral surface. Therefore, (7.71) shows that an inhomogeneous plane wave does not lose power as it propagates.

Example 7.1 (Uniform plane waves generated by an infinite planar current sheet)

We wish to determine the solution to the time-harmonic Maxwell equations in free space when the source is an electric current *sheet* of infinite extent, i.e., a planar surface current whose phasor is given by (Figure 7.5)

$$\mathbf{J}_S(\mathbf{r}) = \mathbf{J}_S(x, y, z) = J_{0S}\,\hat{\mathbf{y}}, \qquad x, y \in \mathbb{R}, \qquad z = 0 \tag{7.72}$$

with J_{0S} a known complex constant. Since this ideal singular source divides the space into two complementary half-spaces, say, $V_1 := \{\mathbf{r} \in \mathbb{R}^3 : z < 0\}$ and $V_2 := \{\mathbf{r} \in \mathbb{R}^3 : z \geqslant 0\}$, and besides it is planar and independent of x and y, we expect the solution to be in the form

$$\mathbf{H}_1(z) = \mathbf{H}_{01} e^{\mathrm{j} k_0 z}, \qquad \mathbf{E}_1(z) = -\frac{k_0}{\omega\varepsilon_0}\hat{\mathbf{z}} \times \mathbf{H}_{01} e^{\mathrm{j} k_0 z}, \qquad \mathbf{r} \in V_1 \tag{7.73}$$

$$\mathbf{H}_2(z) = \mathbf{H}_{02} e^{-\mathrm{j} k_0 z}, \qquad \mathbf{E}_2(z) = \frac{k_0}{\omega\varepsilon_0}\hat{\mathbf{z}} \times \mathbf{H}_{02} e^{-\mathrm{j} k_0 z}, \qquad \mathbf{r} \in V_2 \tag{7.74}$$

that is, comprised of two uniform plane waves propagating away from $\mathbf{J}_S(\mathbf{r})$ in the negative (subscript '1') and the positive (subscript '2') z-direction. The electric fields in this Ansatz are derived from the general relationships (7.39) and (7.40). The other two waves that can be obtained from (7.73) and (7.74) by substituting k_0 with $-k_0$ are not physical, because they travel from infinity *towards* the current sheet and, by hypothesis, in $z = \pm\infty$ there are no sources which can sustain such waves.

At a first glance it would appear that we can compute the two constant *vectors* $\mathbf{H}_{01}$ and $\mathbf{H}_{02}$ by enforcing the matching conditions (1.196) and (1.197) at the location of the sheet, i.e., for $z = 0$. However, since (1.196) and (1.197) involve the components tangential to the plane $z = 0$, those constraints amount to *four* scalar relations, whereas $\mathbf{H}_{01}$ and $\mathbf{H}_{02}$ entail *six* scalar components. Therefore, we see that we can fix four components at best, unless we are given additional information.

To this purpose, we turn to the source $\mathbf{J}_S(\mathbf{r})$ and exploit the symmetry thereof to try and reduce the number of unknowns. More precisely, we can conclude that $H_{z1} = 0 = H_{z2}$ by reasoning as we did in Example 4.1 to rule out the occurrence of a radial component in the presence of a steady circular-cylindrical current density. In the present problem we should execute a one-hundred-eighty-degree rigid rotation of the sheet around the z-axis and also exploit the linearity of the Maxwell equations to run into a contradiction. As a matter of fact, the z-component of the magnetic field must be null on account of (7.42), if (7.73) and (7.74) are to represent plane waves travelling along the z-direction.

Next, we suppose for the moment that both the x- and y-component of the magnetic field are non-zero, and notice in particular that by virtue of (1.196) applied for $z = 0$, namely,

$$0 = \hat{\mathbf{x}} \cdot \mathbf{J}_S(\mathbf{r}) = \hat{\mathbf{x}} \cdot \hat{\mathbf{z}} \times [\mathbf{H}_2(0) - \mathbf{H}_1(0)] = -\hat{\mathbf{y}} \cdot [\mathbf{H}_2(0) - \mathbf{H}_1(0)] \tag{7.75}$$

we get

$$H_{01y} = H_{02y} \tag{7.76}$$

because the surface current (7.72) is directed along $\hat{\mathbf{y}}$. Further, the second parts of (7.73) and (7.74) become

$$\mathbf{E}_1(z) = -\frac{k_0}{\omega\varepsilon_0}(H_{01x}\hat{\mathbf{y}} - H_{01y}\hat{\mathbf{x}})\mathrm{e}^{\mathrm{j}k_0 z}, \qquad \mathbf{r} \in V_1 \tag{7.77}$$

$$\mathbf{E}_2(z) = \frac{k_0}{\omega\varepsilon_0}(H_{02x}\hat{\mathbf{y}} - H_{02y}\hat{\mathbf{x}})\mathrm{e}^{-\mathrm{j}k_0 z}, \qquad \mathbf{r} \in V_2 \tag{7.78}$$

whereby the difference of the x-components in $z = 0$ reads

$$\hat{\mathbf{x}} \cdot [\mathbf{E}_2(0) - \mathbf{E}_1(0)] = -\frac{k_0}{\omega\varepsilon_0}(H_{01y} + H_{02y}) = -\frac{2k_0}{\omega\varepsilon_0}H_{01y} \tag{7.79}$$

and this requires

$$H_{01y} = H_{02y} = 0 \tag{7.80}$$

on account of (1.197).

In summary, we are down to just one non-vanishing component for the magnetic field (H_x) and, consequently, only one for the electric field (E_y). By enforcing (1.196) and (1.197) for $z = 0$ we arrive at the algebraic linear system

$$J_{0S} = \hat{\mathbf{y}} \cdot \hat{\mathbf{z}} \times (\mathbf{H}_{02} - \mathbf{H}_{01}) = H_{02x} - H_{01x} \tag{7.81}$$

$$0 = (\mathbf{E}_{02} - \mathbf{E}_{01}) \times \hat{\mathbf{z}} \cdot \hat{\mathbf{x}} = \frac{k_0}{\omega\varepsilon_0}(H_{02x} + H_{01x}) \tag{7.82}$$

which is readily solved to give

$$H_{02x} = -H_{01x} = \frac{J_{0S}}{2}. \tag{7.83}$$

By substituting all of our findings back into (7.73) and (7.74) we get (Figure 7.5)

$$\mathbf{H}_1(z) = -\frac{J_{0S}}{2}\hat{\mathbf{x}}\,\mathrm{e}^{\mathrm{j}k_0 z}, \qquad \mathbf{E}_1(z) = \frac{1}{2}Z_0 J_{S0}\hat{\mathbf{y}}\,\mathrm{e}^{\mathrm{j}k_0 z}, \qquad \mathbf{r} \in V_1 \tag{7.84}$$

$$\mathbf{H}_2(z) = \frac{J_{0S}}{2}\hat{\mathbf{x}}\,\mathrm{e}^{-\mathrm{j}k_0 z}, \qquad \mathbf{E}_2(z) = \frac{1}{2}Z_0 J_{S0}\hat{\mathbf{y}}\,\mathrm{e}^{-\mathrm{j}k_0 z}, \qquad \mathbf{r} \in V_2 \tag{7.85}$$

where $Z_0 = (\mu_0/\varepsilon_0)^{1/2}$ is the intrinsic impedance of free space.

(End of Example 7.1)

7.2.2 *Lossy isotropic medium*

We extend the search for time-harmonic plane-wave solutions to the case of a homogeneous isotropic unbounded lossy space endowed with conductivity σ. Correspondingly, we modify the Ampère-Maxwell law (7.17) by including a conduction current $\mathbf{J}_c(\mathbf{r}) = \sigma\mathbf{E}(\mathbf{r})$ (Section 1.6)

$$\nabla \times \mathbf{H}(\mathbf{r}) = (\mathrm{j}\,\omega\varepsilon + \sigma)\mathbf{E}(\mathbf{r}) \tag{7.86}$$

and we assume the forms (7.20) and (7.21). By repeating the same manipulations as before we arrive at the reduced equation for $\mathbf{E}_0$

$$\left[\mathbf{k}\cdot\mathbf{k} - \omega^2\left(\varepsilon - \mathrm{j}\frac{\sigma}{\omega}\right)\mu\right]\left(\hat{\alpha}\hat{\alpha} + \hat{\beta}\hat{\beta}\right)\cdot\mathbf{E}_0 = 0 \tag{7.87}$$

which we may regard as the algebraic counterpart of the wave equation (1.216) in the absence of localized sources. The conditions for the existence of non-trivial solutions reads

$$\mathbf{k}\cdot\mathbf{k} - \omega^2\varepsilon\mu\left(1 - \mathrm{j}\frac{\sigma}{\omega\varepsilon}\right) = 0 \tag{7.88}$$

and constitutes the *dispersion relationship* for plane waves in a homogeneous isotropic lossy unbounded medium. Since (7.88) is not affected by the substitution $\mathbf{k} \to -\mathbf{k}$, *progressive* and *regressive* waves are possible as well.

By separating real and imaginary parts of the dispersion relation we obtain

$$k'^2 - k''^2 = \omega^2\varepsilon\mu \tag{7.89}$$

$$2\mathbf{k}'\cdot\mathbf{k}'' = \omega\mu\sigma \tag{7.90}$$

whereby we see that the wavevector is necessarily complex due to the presence of a non-vanishing conductivity. In fact, we have two possibilities:

(1) If we take $\mathbf{k}'' = a\mathbf{k}'$ with $a \in \mathbb{R}^+$, a little algebra allows showing that (7.89) and (7.90) are solved by

$$a = \sqrt{1 + \left(\frac{\omega\varepsilon}{\sigma}\right)^2} - \frac{\omega\varepsilon}{\sigma} > 0 \tag{7.91}$$

$$k'^2 = \frac{\omega^2}{2c^2}\left[\sqrt{1 + \left(\frac{\sigma}{\omega\varepsilon}\right)^2} + 1\right] > 0 \tag{7.92}$$

$$k''^2 = \frac{\omega^2}{2c^2}\left[\sqrt{1 + \left(\frac{\sigma}{\omega\varepsilon}\right)^2} - 1\right] \geqslant 0 \tag{7.93}$$

with c given by (1.209). We have discarded the inadmissible solution $a < 0$ and $k'^2 < 0$. Then, with reference to (7.44) we have

$$k = k' - \mathrm{j}\,k'' = k'(1 - \mathrm{j}\,a), \qquad \hat{\mathbf{k}} = \frac{\mathbf{k}}{k} = \frac{\mathbf{k}'(1 - \mathrm{j}\,a)}{k'(1 - \mathrm{j}\,a)} = \frac{\mathbf{k}'}{k'} \tag{7.94}$$

that is, the wavenumber is complex, but the unit vector $\hat{\mathbf{k}}$ is real. Solutions of this type — for which the planes of constant phase and constant amplitude are all parallel to one another — are also said *uniform* or *homogeneous plane waves* (Figure 7.6a). Indeed, (7.92) and (7.93) lead back to (7.24) as $\sigma \to 0^+$.

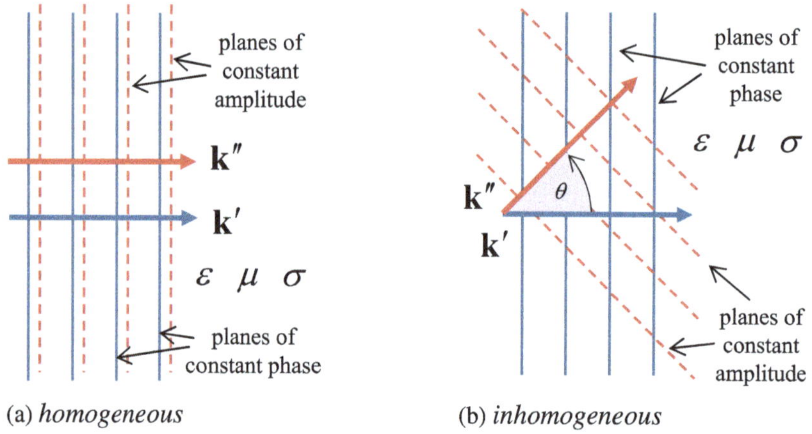

Figure 7.6 Plane waves in a homogeneous isotropic lossy unbounded medium.

(2) If $\mathbf{k}'$ and $\mathbf{k}''$ form an angle θ such that $0 < |\theta| < \pi/2$, solving (7.89) and (7.90) now yields

$$k'^2 = \frac{\omega^2}{2c^2}\left[\sqrt{1+\left(\frac{\sigma}{\omega\varepsilon\cos\theta}\right)^2}+1\right] > 0 \tag{7.95}$$

$$k''^2 = \frac{\omega^2}{2c^2}\left[\sqrt{1+\left(\frac{\sigma}{\omega\varepsilon\cos\theta}\right)^2}-1\right] \geqslant 0 \tag{7.96}$$

which may include (7.92) and (7.93) as a special case. As can be inferred from (7.90), right angles between $\mathbf{k}'$ and $\mathbf{k}''$ are precluded for $\sigma \neq 0$, whereas angles θ such that $\pi/2 < |\theta| \leqslant \pi$ result in $\mathbf{k}' \cdot \mathbf{k}'' = k'k'' \cos\theta < 0$, which is inadmissible because k', k'' and σ are all positive quantities. Solutions of this type — for which planes of constant phase and constant amplitude form an angle θ — are called *inhomogeneous* plane waves (Figure 7.6b).

Both types of waves are damped as they travel along $\mathbf{k}'$ owing to the *lossy nature* of the underlying medium, and hence power is converted into heat. The attenuation takes place along the direction of propagation for uniform waves, as can be seen in Figure 7.7. The electric field $\mathbf{E}(x) = E_z(x)\hat{\mathbf{z}}$ is relevant to a medium for which $\omega\varepsilon = 4\sigma$ in (7.91)-(7.93). By contrast, inhomogeneous waves are attenuated along $\mathbf{k}''$. As an example, plotted in Figure 7.8 is the electric field $\mathbf{E}(x, y) = E_z(x, y)\hat{\mathbf{z}}$ in a medium for which $\omega\varepsilon = 10\sigma$. It is assumed that $\mathbf{k}' = k'\hat{\mathbf{x}}$ and $\mathbf{k}'' = k''(\hat{\mathbf{x}}\cos\theta + \hat{\mathbf{y}}\sin\theta)$ with $\theta = \pi/3$ in (7.95) and (7.96). The surface plots of Figures 7.7 and 7.8 are valid for points on any plane perpendicular to $\hat{\mathbf{z}}$ at time $t = 0$.

For convenience, the four types of plane waves are summarized in Table 7.1 together with the properties of the medium and of the wavevector.

Two special cases are represented by the propagation of plane waves in a good, though lossy dielectric and a good conductor. In the former the magnitude of the conduction current $\mathbf{J}_c(\mathbf{r})$ is negligible with respect to the magnitude of the displacement current $\mathrm{j}\,\omega\mathbf{D}(\mathbf{r})$, which translates into the condition

$$\sigma \ll \omega\varepsilon \tag{7.97}$$

for a suitable combination of conductivity, angular frequency and permittivity. In a good conductor the converse is true, and the conduction current dominates over the displacement current. In either

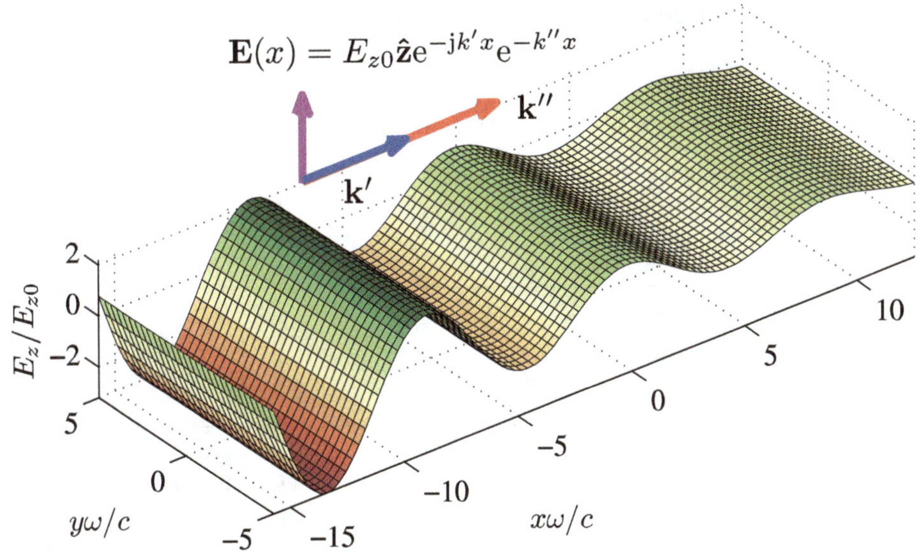

Figure 7.7 *Snapshot for t = 0 of the normalized electric field of a homogeneous plane wave in a lossy isotropic medium with* $\omega\varepsilon = 4\sigma$.

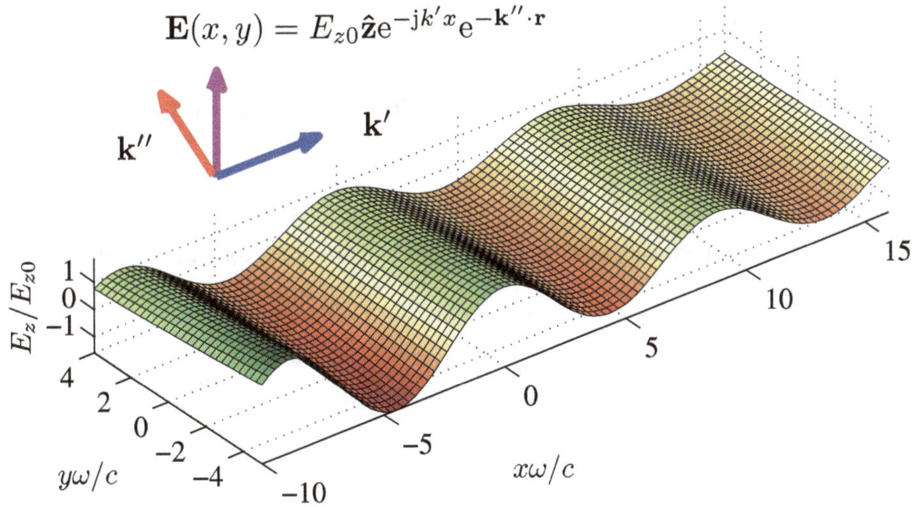

Figure 7.8 *Snapshot for t = 0 of the normalized electric field of an inhomogeneous plane wave in a lossy isotropic medium with* $\omega\varepsilon = 10\sigma$ *and* $\theta = \pi/3$.

limiting situation we can obtain an approximated solution of the dispersion relationship (7.88) for homogeneous plane waves.

By letting $\mathbf{k} \cdot \mathbf{k} = k^2$ with $k \in \mathbb{C}$ for a good dielectric we have

$$k = \omega\sqrt{\varepsilon\mu}\sqrt{1 - \mathrm{j}\frac{\sigma}{\omega\varepsilon}} \approx \omega\sqrt{\varepsilon\mu}\left(1 - \mathrm{j}\frac{\sigma}{2\omega\varepsilon}\right) = k' - \mathrm{j}k'' \tag{7.98}$$

Table 7.1 Summary of properties of time-harmonic plane waves in a homogeneous isotropic medium

Medium	**Wave type**	**Wavenumber**	**Wavevector**	**Field planes**[†]
Lossless ($\sigma = 0$)	Uniform	$k \in \mathbb{R}$	$\mathbf{k} = \mathbf{k}'$ $\mathbf{k}'' = 0$ $\hat{\mathbf{k}} \in \mathbb{R}^3$	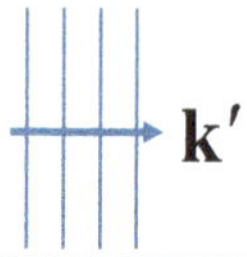
Lossless ($\sigma = 0$)	Inhomogeneous	$k \in \mathbb{R}$	$\mathbf{k}'' \neq 0$ $\mathbf{k}' \cdot \mathbf{k}'' = 0$ $\hat{\mathbf{k}} \in \mathbb{C}^3$	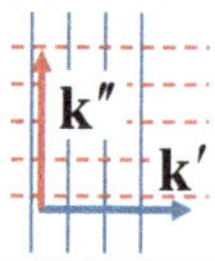
Lossy ($\sigma > 0$)	Uniform	$k \in \mathbb{C}$	$\mathbf{k}'' = a\mathbf{k}''$ $\mathbf{k} = (1 - \mathrm{j}\, a)\mathbf{k}'$ $\hat{\mathbf{k}} \in \mathbb{R}^3$	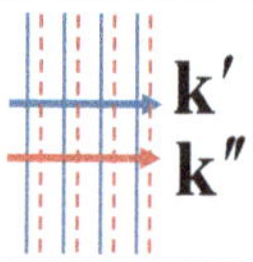
Lossy ($\sigma > 0$)	Inhomogeneous	$k \in \mathbb{C}$	$\mathbf{k} = \mathbf{k}' - \mathrm{j}\,\mathbf{k}''$ $0 < \lvert\theta\rvert < \pi/2$ $\hat{\mathbf{k}} \in \mathbb{C}^3$	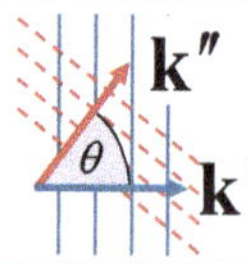

(†) Traces of planes of constant phase (–) and constant amplitude (– –).

having truncated the binomial expansion with respect to $\mathrm{j}\,\sigma/(\omega\varepsilon)$ [20, Formula 5.3] to the first two terms. The last step provides

$$k' \approx \frac{\omega}{c}, \qquad k'' \approx \frac{\sigma}{2} Z, \qquad \sigma \ll \omega\varepsilon \tag{7.99}$$

where Z is the intrinsic impedance introduced in (1.358). In the limit for small losses the real part of the complex wavenumber is independent of σ.

For a good conductor we have [21, Section 2.8]

$$k = \omega\sqrt{\varepsilon\mu}\sqrt{1 - \mathrm{j}\frac{\sigma}{\omega\varepsilon}} \approx \omega\sqrt{\varepsilon\mu}\sqrt{-\mathrm{j}}\sqrt{\frac{\sigma}{\omega\varepsilon}} = \sqrt{\frac{\omega\mu\sigma}{2}}(1 - \mathrm{j}) = k' - \mathrm{j}\,k'' \tag{7.100}$$

where we have picked up the square root of $-\mathrm{j}$ [20, Formula 58.3] that has positive real part and negative imaginary part so as to be consistent with (7.24). The other choice would yield an amplified wave as opposed to a damped one, an occurrence which is non-physical in a lossy medium. From the rightmost-hand side we get

$$k' = k'' \approx \sqrt{\frac{\omega\mu\sigma}{2}}, \qquad \sigma \gg \omega\varepsilon \tag{7.101}$$

and the quantity

$$\delta_{sd} := \frac{1}{k''} = \sqrt{\frac{2}{\omega\mu\sigma}} \tag{7.102}$$

is called the *penetration* or *skin depth* (physical dimension: m) [22, Appendix G.4], [23, Section 10-16] for it constitutes a measure of how deep or how far a plane wave can penetrate in the conductor

Table 7.2 Conductivity, permeability and skin depth of some good conductors

Material	σ [Ω^{-1}/m]	μ/μ_0	δ_{sd} [cm] @ 60 Hz	δ_{sd} [mm] @ 1 KHz	δ_{sd} [mm] @ 1 MHz	δ_{sd} [μm] @ 3 GHz
Aluminium	3.54×10^7	1.00	1.1	2.7	0.085	1.6
Chromium	3.8×10^7	1.00	1.0	2.6	0.081	1.5
Copper	5.80×10^7	1.00	0.85	2.1	0.066	1.2
Gold	4.50×10^7	1.00	0.97	2.38	0.075	1.4
Magnetic iron	1.0×10^7	2.0×10^2	0.14	0.35	0.011	0.20
Nickel	1.3×10^7	1.0×10^2	0.18	4.4	0.014	0.26
Silver	6.15×10^7	1.00	0.83	2.03	0.064	1.2
Tin	0.87×10^7	1.00	2.21	5.41	0.171	3.12
Zinc	1.86×10^7	1.00	1.51	3.70	0.117	2.14

before being so damped as to become negligible for all practical purposes. We will make this statement more rigorous in Example 7.2 and also in Example 9.7 further on. For reference, the conductivity, the permeability and the skin depth of various good conductors are listed in Table 7.2 (e.g., [24, Table 1.2], [25, Table 12.1]).

Example 7.2 (Uniform plane waves in a good conductor)

In a system of Cartesian coordinates we consider the infinite planar interface at $z = 0$ between the half-spaces $V_1 := \{\mathbf{r} \in \mathbb{R}^3 : z < 0\}$ and $V_2 := \{\mathbf{r} \in \mathbb{R}^3 : z \geqslant 0\}$. In order to elucidate the physical meaning of the skin depth defined in (7.102) we suppose that the medium in V_1 is free space whereas a good conductor endowed with constitutive parameters ε_2, μ_2 and σ_2 occupies V_2, as is sketched in Figure 7.9.

A uniform plane wave characterized by $\mathbf{E}_{01} = E_{x1}\hat{\mathbf{x}}$ and a real wavevector $\mathbf{k}_1 = k_{z1}\hat{\mathbf{z}} = \omega(\varepsilon_0\mu_0)^{1/2}\hat{\mathbf{z}} > 0$ solves the Maxwell equations in free space but cannot be a valid solution in the presence of the conducting half-space [26, Section 9.1.3], [27, Section 2.9]. In fact, on the one hand the electric field of the wave does not satisfy the boundary condition (1.144) for $z = 0$ and, on the other, the wavevector $\mathbf{k}_2 = k_{z2}\hat{\mathbf{z}}$ in the conductor should be complex in keeping with (7.88). We shall see in Section 7.4 that a combination of two uniform plane waves in V_1 and one uniform plane wave in the lossy medium that fills V_2 can solve the Maxwell equations subject to the time-harmonic jump conditions (1.196) and (1.197) for $z = 0$ and with $\mathbf{J}_S(\mathbf{r}) = 0$. For the time being we just pretend that such solution has been found and, as a result, that

$$\mathbf{E}_2(\mathbf{r}) = E_{x2}\hat{\mathbf{x}}\,\mathrm{e}^{-\mathrm{j}\mathbf{k}_2\cdot\mathbf{r}} = E_{x2}\hat{\mathbf{x}}\,\mathrm{e}^{-\mathrm{j}(k_2'-\mathrm{j}k_2'')z} = E_{x2}\hat{\mathbf{x}}\,\mathrm{e}^{-\mathrm{j}k_2'z}\mathrm{e}^{-k_2''z}, \qquad z \geqslant 0 \tag{7.103}$$

is the electric field of the uniform plane wave travelling in the conductor away from the interface and towards $z = +\infty$.

This wave is damped in the conductor due to the interaction of the field with the free electrons which ultimately transfer the acquired kinetic energy to the lattice, and the process results in an increase of the temperature of the medium. In particular, the magnitude of the electric field reads

$$|\mathbf{E}_2(\mathbf{r})| = |\mathbf{E}_2(z)| = |E_{x2}|\mathrm{e}^{-k_2''z}, \qquad z \geqslant 0 \tag{7.104}$$

and the function

$$f(z) = |E_{x2}|(1 - k_2''z), \qquad z \geqslant 0 \tag{7.105}$$

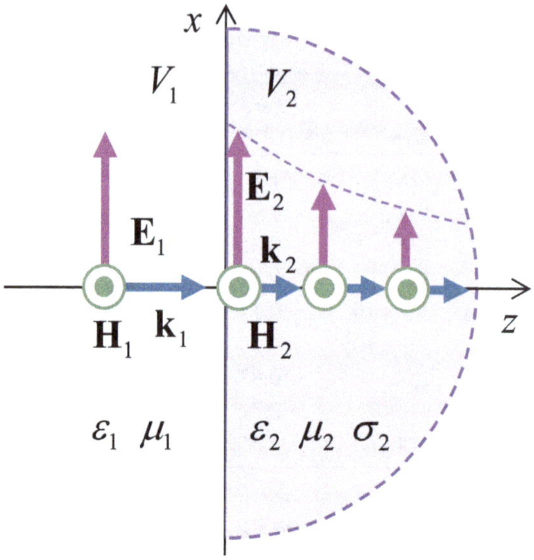

Figure 7.9 Damping of a uniform plane wave in a lossy half space.

represents the straight line that is tangent to the graph of $|\mathbf{E}_2(z)|$ in the point $(0, |E_{x2}|)$. We know from calculus that the (negative) slope of said tangent provides an indication of the decay rate of the function $|\mathbf{E}_2(z)|$. Since the intersection of the tangent with the horizontal axis occurs for

$$z = \frac{1}{k_2''} = \delta_{sd} = \sqrt{\frac{2}{\omega\mu_2\sigma_2}} \tag{7.106}$$

we see that the shorter the skin depth the faster the wave is attenuated in the conductor. Alternatively, since for $z = 0$ and $z = \delta_{sd}$ we have

$$|\mathbf{E}_2(0)| = |E_{x2}| \qquad |\mathbf{E}_2(\delta_{sd})| = |E_{x2}|\mathrm{e}^{-k_2''\delta_{sd}} = \frac{1}{\mathrm{e}}|\mathbf{E}_2(0)| \tag{7.107}$$

we may state that, after travelling a distance δ_{sd} within V_2, the amplitude of the wave has decreased by a factor $1/\mathrm{e} \approx 0.367$.

Notice that the skin depth depends on the frequency and so does the conductivity (Chapter 12). As can be gathered from Table 7.2 at relatively low frequencies a wave can penetrate for larger distances in the conductor, whereas at higher frequencies the conduction current

$$\mathbf{J}_{c2}(z) = \sigma_2\mathbf{E}_2 = \sigma_2 E_{x2}\hat{\mathbf{x}}\,\mathrm{e}^{-\mathrm{j}k_2'z}\mathrm{e}^{-k_2''z} \tag{7.108}$$

is essentially confined to a thin layer underneath the interface. Eventually, in the limit as $\sigma_2 \to +\infty$ the medium in the half-space V_2 becomes a PEC (see Section 1.6), the skin depth vanishes, and the wave cannot penetrate at all. This observation provides a basis for the special boundary condition (1.169).

The relation between $\mathbf{E}_2(\mathbf{r})$ and the magnetic field associated with the uniform plane wave in the conductor is provided by (7.55) with the wavevector $\mathbf{k}_2$ constrained by the dispersion relation

(7.88). We set $\hat{\mathbf{n}} = -\hat{\mathbf{z}}$, the unit vector perpendicular to the interface and positively oriented towards V_1 (outward the conductor) and we cast (7.55) as

$$\frac{\omega\mu_2}{k_2}\mathbf{H}_2(\mathbf{r}) = \hat{\mathbf{z}} \times \mathbf{E}_2(\mathbf{r}) = \mathbf{E}_2(\mathbf{r}) \times \hat{\mathbf{n}}, \qquad \mathbf{r} \in V_2 \tag{7.109}$$

$$\hat{\mathbf{n}} \times [\mathbf{E}_2(\mathbf{r}) \times \hat{\mathbf{n}}] = \frac{\omega\mu_2}{k_2}\hat{\mathbf{n}} \times \mathbf{H}_2(\mathbf{r}), \qquad \mathbf{r} \in V_2 \tag{7.110}$$

where by virtue of (7.100) the quantity

$$\frac{\omega\mu_2}{k_2} \approx \sqrt{\frac{\omega\mu_2}{2\sigma_2}}(1+\mathrm{j}) = R_S(1+\mathrm{j}) = Z_S, \qquad \sigma_2 \gg \omega\varepsilon_2 \tag{7.111}$$

is, strictly speaking, the intrinsic impedance of the lossy medium. However, (7.110) also holds for $z = 0$ and in light of the matching conditions (1.142) and (1.144) it constitutes a relation for the tangential components of $\mathbf{E}_1(\mathbf{r})$ and $\mathbf{H}_1(\mathbf{r})$ for $z = 0$. In other words, we have found

$$\hat{\mathbf{n}} \times [\mathbf{E}_1(\mathbf{r}) \times \hat{\mathbf{n}}] = Z_S \hat{\mathbf{n}} \times \mathbf{H}_1(\mathbf{r}), \qquad z = 0 \tag{7.112}$$

that is, the Leontovich boundary condition (6.78) [28,29]. Therefore, we see that Z_S given by (7.111) plays the role of a surface impedance. We emphasize that (7.112) is *exact* for a planar interface, though Z_S has been derived by means of the asymptotic calculation (7.100) of the wavenumber. When the boundary of the conducting body is curved but smooth and the radius of curvature is on average small with respect to the wavelength in the surrounding medium, (7.112) may be used as an approximate boundary condition to formulate scattering problems with integral equations (see Section 13.2.6).

(End of Example 7.2)

As regards the power flow associated with plane waves in a lossy medium we expect that, unlike the lossless case, energy is converted into heat and this phenomenon causes the amplitude of the wave to decrease as it travels along. The complex Poynting vector can be computed as in (7.64) so long as μ is real. The alternative expression involving the magnetic field reads

$$\begin{aligned}
\mathbf{S}(\mathbf{r}) &:= \frac{1}{2}\mathbf{E}(\mathbf{r}) \times \mathbf{H}^*(\mathbf{r}) = \frac{1}{2}\left(\frac{\mathbf{k} \times \mathbf{H}_0 e^{-\mathrm{j}\,\mathbf{k}\cdot\mathbf{r}}}{\mathrm{j}\,\sigma - \omega\varepsilon}\right) \times \mathbf{H}_0^* e^{\mathrm{j}\,\mathbf{k}^*\cdot\mathbf{r}} \\
&= \frac{e^{-2\mathbf{k}''\cdot\mathbf{r}}}{2(\mathrm{j}\,\sigma - \omega\varepsilon)}\left[\mathbf{H}_0(\mathbf{H}_0^* \cdot \mathbf{k}) - \mathbf{k}|\mathbf{H}_0|^2\right] \\
&= \frac{e^{-2\mathbf{k}''\cdot\mathbf{r}}}{2(\mathrm{j}\,\sigma - \omega\varepsilon)}\Big[(\mathbf{k} \cdot \mathbf{H}_0^*)\mathbf{H}_0 - \underbrace{(\mathbf{k} \cdot \mathbf{H}_0)}_{=0}\mathbf{H}_0^* - \mathbf{k}|\mathbf{H}_0|^2\Big] \\
&= \frac{e^{-2\mathbf{k}''\cdot\mathbf{r}}}{2(\mathrm{j}\,\sigma - \omega\varepsilon)}\left[\mathbf{k} \times (\mathbf{H}_0 \times \mathbf{H}_0^*) - \mathbf{k}|\mathbf{H}_0|^2\right] \\
&= \frac{e^{-2\mathbf{k}''\cdot\mathbf{r}}}{2(\omega\varepsilon - \mathrm{j}\,\sigma)}\left(\mathbf{k}|\mathbf{H}_0|^2 + \mathrm{j}\,\mathbf{k} \times \mathrm{Im}\{\mathbf{H}_0^* \times \mathbf{H}_0\}\right)
\end{aligned} \tag{7.113}$$

having used (H.14) and noticed that the complex vector $\mathbf{H}_0^* \times \mathbf{H}_0$ is purely imaginary.

From (7.64) we see that for uniform plane waves the real part of the Poynting vector is

$$\mathrm{Re}\{\mathbf{S}\} = \frac{e^{-2\mathbf{k}''\cdot\mathbf{r}}}{2\omega\mu}\left(\mathbf{k}'|\mathbf{E}_0|^2 - a\mathbf{k}' \times \mathrm{Im}\{\mathbf{E}_0 \times \mathbf{E}_0^*\}\right) \tag{7.114}$$

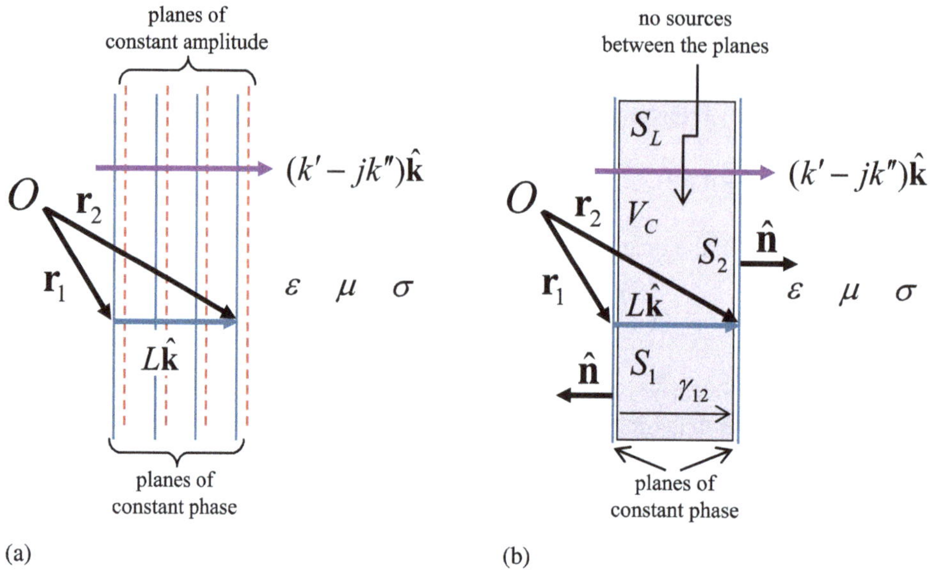

Figure 7.10 Uniform plane waves in a homogeneous isotropic lossy unbounded medium: (a) geometry for defining the attenuation along $\hat{\mathbf{k}}$ *and (b) geometry for determining the special power balance* (7.125).

with the proportionality factor a given by (7.91). This finding indicates that, when $\mathbf{E}_0$ is a complex vector, the power flow is directed at an angle with $\mathbf{k}'$, despite $\mathbf{k}'$ and $\mathbf{k}''$ being parallel and pointing in the same direction. Only when $\mathbf{E}_0$ is real does the power travel perpendicularly to the planes of constant phase and constant amplitude. Under this hypothesis the average *surface density of power* flowing through a plane perpendicular to the unit vector $\hat{\mathbf{n}}$ is given by

$$p_R(\mathbf{r}) := \left\langle \frac{\mathrm{d}P_F}{\mathrm{d}S} \right\rangle = \frac{|\mathbf{E}_0|^2}{2\omega\mu} k'\hat{\mathbf{k}} \cdot \hat{\mathbf{n}}\, \mathrm{e}^{-2k''\hat{\mathbf{k}}\cdot\mathbf{r}} \tag{7.115}$$

and, for a fixed $\mathbf{r}$, reaches its maximum across planes of constant phase or amplitude. However, due to the finite conductivity, the power flow is not constant along $\hat{\mathbf{k}}$. If we denote with $\mathbf{r}_2$ and $\mathbf{r}_1$ two points on two planes of constant phase (Figure 7.10a) so that

$$\mathbf{r}_2 - \mathbf{r}_1 = L\hat{\mathbf{k}} \tag{7.116}$$

where L is the distance travelled by the wave, we have

$$p_R(\mathbf{r}_2) - p_R(\mathbf{r}_1) = \frac{|\mathbf{E}_0|^2}{2\omega\mu} k' \left(\mathrm{e}^{-2k''\hat{\mathbf{k}}\cdot\mathbf{r}_2} - \mathrm{e}^{-2k''\hat{\mathbf{k}}\cdot\mathbf{r}_1} \right) = \frac{|\mathbf{E}_0|^2}{2\omega\mu} k' \left(\mathrm{e}^{-2k''L} - 1 \right) \mathrm{e}^{-2k''\hat{\mathbf{k}}\cdot\mathbf{r}_1} < 0 \tag{7.117}$$

because the real exponential is smaller than one for negative values of the argument. Furthermore, the ratio of power densities

$$\frac{p_R(\mathbf{r}_2)}{p_R(\mathbf{r}_1)} = \frac{\mathrm{e}^{-2k''\hat{\mathbf{k}}\cdot\mathbf{r}_2}}{\mathrm{e}^{-2k''\hat{\mathbf{k}}\cdot\mathbf{r}_1}} = \mathrm{e}^{-2k''L} < 1 \tag{7.118}$$

defines the *attenuation* undergone by the wave as it sweeps the distance L in the direction indicated by $\hat{\mathbf{k}}$. To show that the attenuation is indeed due to losses we resort to the Poynting theorem.

In particular, as the power lost between two planes of constant phase or amplitude is evidently infinite, we compute the *surface density* of dissipated power by applying (1.308) to a circular cylinder V_C which has the bases flush with the planes of concern, as is sketched in Figure 7.10b

$$\langle P_C \rangle := \frac{1}{2} \int_{V_C} \mathrm{d}V \, \sigma |\mathbf{E}(\mathbf{r})|^2 = \frac{\sigma}{2} \pi b^2 |\mathbf{E}_0|^2 \int_{\gamma_{12}} \mathrm{d}s \, \mathrm{e}^{-2\mathbf{k}'' \cdot \mathbf{r}} \tag{7.119}$$

where b is the radius of the cylinder and γ_{12} is the rectilinear path defined by the parametric representation

$$\mathbf{r} = \mathbf{r}_1 + \xi(\mathbf{r}_2 - \mathbf{r}_1) = \mathbf{r}_1 + \xi L \hat{\mathbf{k}}, \qquad \xi \in [0, 1]. \tag{7.120}$$

The *surface density of dissipated power* follows from

$$\begin{aligned} p_C &:= \lim_{b \to 0^+} \frac{\langle P_C \rangle}{\pi b^2} = \frac{\sigma}{2} |\mathbf{E}_0|^2 \int_{\gamma_{12}} \mathrm{d}s \, \mathrm{e}^{-2\mathbf{k}'' \cdot \mathbf{r}} = \frac{\sigma}{2} L |\mathbf{E}_0|^2 \mathrm{e}^{-2\mathbf{k}'' \cdot \mathbf{r}_1} \int_0^1 \mathrm{d}\xi \, \mathrm{e}^{-2k'' L \xi} \\ &= \frac{\sigma}{2} L |\mathbf{E}_0|^2 \mathrm{e}^{-2\mathbf{k}'' \cdot \mathbf{r}_1} \frac{1 - \mathrm{e}^{-2k'' L}}{2k'' L} = \frac{1}{2} |\mathbf{E}_0|^2 \frac{\sigma}{2k''} (\mathrm{e}^{-2\mathbf{k}'' \cdot \mathbf{r}_1} - \mathrm{e}^{-2\mathbf{k}'' \cdot \mathbf{r}_2}) > 0. \end{aligned} \tag{7.121}$$

Application of the Poynting theorem (1.314) to the same source-free cylinder considered above yields

$$\begin{aligned} 0 &= \frac{1}{2} \int_{V_C} \mathrm{d}V \, \sigma |\mathbf{E}(\mathbf{r})|^2 + \mathrm{Re} \int_{\partial V_C} \mathrm{d}S \, \hat{\mathbf{n}}(\mathbf{r}) \cdot \mathbf{S}(\mathbf{r}) \\ &= \frac{1}{2} \int_{V_C} \mathrm{d}V \, \sigma |\mathbf{E}(\mathbf{r})|^2 + \mathrm{Re} \int_{S_2} \mathrm{d}S \, \hat{\mathbf{n}}(\mathbf{r}) \cdot \mathbf{S}(\mathbf{r}) \\ &\quad + \mathrm{Re} \int_{S_1} \mathrm{d}S \, \hat{\mathbf{n}}(\mathbf{r}) \cdot \mathbf{S}(\mathbf{r}) + \mathrm{Re} \int_{S_L} \mathrm{d}S \, \hat{\mathbf{n}}(\mathbf{r}) \cdot \mathbf{S}(\mathbf{r}) \end{aligned} \tag{7.122}$$

where we have split the flux integral into three contributions, namely, the flow across the bases S_1 and S_2 of the cylinder and across its lateral surface S_L. Since $\hat{\mathbf{k}} \cdot \hat{\mathbf{n}} = 0$ over S_L, there is no power efflux through the lateral surface in accordance with (7.115). The integrals over S_1 and S_2 are evaluated as

$$\int_{S_1} \mathrm{d}S \, \hat{\mathbf{n}}(\mathbf{r}) \cdot \mathbf{S}(\mathbf{r}) = -\pi b^2 \frac{|\mathbf{E}_0|^2}{2\omega\mu} k' \mathrm{e}^{-2\mathbf{k}'' \cdot \mathbf{r}_1} = -\pi b^2 p_R(\mathbf{r}_1) \tag{7.123}$$

$$\int_{S_2} \mathrm{d}S \, \hat{\mathbf{n}}(\mathbf{r}) \cdot \mathbf{S}(\mathbf{r}) = \pi b^2 \frac{|\mathbf{E}_0|^2}{2\omega\mu} k' \mathrm{e}^{-2\mathbf{k}'' \cdot \mathbf{r}_2} = \pi b^2 p_R(\mathbf{r}_2) \tag{7.124}$$

because the integrands are constant on the circles S_1 and S_2, which are contained on planes of constant amplitude. We interpret $p_R(\mathbf{r}_1)$ and $p_R(\mathbf{r}_2)$ as the average surface densities of power radiated through the planes $\mathbf{k}'' \cdot (\mathbf{r} - \mathbf{r}_1) = 0$ and $\mathbf{k}'' \cdot (\mathbf{r} - \mathbf{r}_2) = 0$. By inserting these results into (7.122), dividing through by πb^2 and taking the limit for vanishing radius b we arrive at the special instance of power balance

$$p_R(\mathbf{r}_1) - p_R(\mathbf{r}_2) = p_C \tag{7.125}$$

on account of the first part of (7.121), which defines p_C. The left-hand side of this expression is the negative of the power difference obtained in (7.117), and this proves that the radiated power is dissipated due to conduction losses. In fact, the coincidence of (7.117) and (7.121) as well as the validity of (7.125) become evident once we notice that

$$\frac{k'}{\omega\mu} = \frac{\sigma}{2k''} \tag{7.126}$$

which is just a rearrangement of (7.90) for homogeneous plane waves.

Finally, by turning once again to the general expression of the Poynting vector (7.64) we deduce that the average surface density of power carried through a plane perpendicular to $\hat{\mathbf{n}}$ by an inhomogeneous wave is given by

$$p_R(\mathbf{r}) := \left\langle \frac{\mathrm{d}P_F}{\mathrm{d}S} \right\rangle = \frac{|\mathbf{E}_0|^2}{2\omega\mu} \mathbf{k}' \cdot \hat{\mathbf{n}}\, \mathrm{e}^{-2\mathbf{k}''\cdot\mathbf{r}} \tag{7.127}$$

provided $\mathbf{E}_0 \in \mathbb{R}^3$. Since $\mathbf{k}''$ forms an angle θ with $\mathbf{k}'$ (Figure 7.6b) it is possible to decompose $\mathbf{k}''$ into the sum of a vector parallel to $\mathbf{k}'$ plus a vector orthogonal to $\mathbf{k}'$, namely,

$$\mathbf{k}'' = \mathbf{k}''_{\parallel} + \mathbf{k}''_{\perp} = \frac{\mathbf{k}'' \cdot \mathbf{k}'}{k'^2}\mathbf{k}' + \frac{\mathbf{k}' \times (\mathbf{k}'' \times \mathbf{k}')}{k'^2} \tag{7.128}$$

by virtue of (H.14) with $\mathbf{a} = \mathbf{c} = \mathbf{k}'$ and $\mathbf{b} = \mathbf{k}''$. Then, we can cast $p_R(\mathbf{r})$ as

$$p_R(\mathbf{r}) = \frac{|\mathbf{E}_0|^2}{2\omega\mu} \mathbf{k}' \cdot \hat{\mathbf{n}}\, \mathrm{e}^{-2\mathbf{k}''_{\parallel}\cdot\mathbf{r}} \underbrace{\mathrm{e}^{-2\mathbf{k}''_{\perp}\cdot\mathbf{r}}}_{\substack{\text{constant} \\ \text{along } \mathbf{k}'}} \tag{7.129}$$

having separated the real exponential into two factors, as indicated, the second of which remains constant as the wave propagates along $\mathbf{k}'$. The first exponential causes the wave to decay along $\mathbf{k}'$ and is responsible for the power dissipation, as in (7.115) for the homogeneous waves. The second exponential plays the same role as the one in (7.68) for the inhomogeneous waves in a *lossless* medium, that is to say, it does not account for loss of power along $\mathbf{k}''_{\perp}$, perpendicularly to the direction of propagation.

7.3 Polarization of plane waves

With the term *polarization* of a wave or a field we indicate the way a time-harmonic electromagnetic entity oscillates in time in a given point $\mathbf{r}$ [23, Chapter 11], [30, Section 3.1.B], [10, Section 11.5]. Owing to the periodic character of the wave the tip of the vector associated with the entity (in the time domain) describes a closed curve γ in general. Accordingly, we say that the polarization is

- *elliptical*, if the curve γ is the boundary of an ellipsis (Figure 7.11a);
- *circular*, if the curve γ is the boundary of a circle (Figure 7.11b);
- *linear*, if the curve γ degenerates into part of a straight line (Figure 7.11c).

For a monochromatic plane wave in the form (7.20) and (7.21) we can ascertain the state of polarization by examining either one of the complex vectors $\mathbf{E}_0$ and $\mathbf{H}_0$, insofar as electric and magnetic fields are related by (7.55) or (7.56) [31, Section 1.4.2], [14, Section 2.2]. From (7.23) we see that the vectors $\mathbf{E}'_0$ and $\mathbf{E}''_0$ specify a plane orthogonal to the unit vector

$$\hat{\mathbf{n}} = \frac{\mathbf{E}'_0 \times \mathbf{E}''_0}{|\mathbf{E}'_0 \times \mathbf{E}''_0|} \tag{7.130}$$

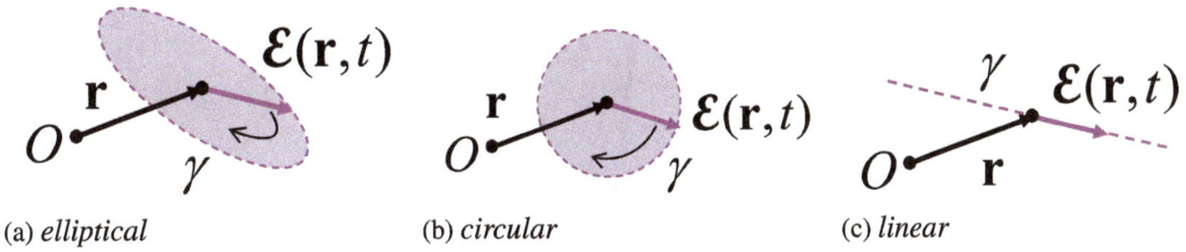

Figure 7.11 Polarization of a time-harmonic wave in a lossless medium.

which is well-defined, unless $\mathbf{E}_0'$ and $\mathbf{E}_0''$ are parallel to one another. The angle α between $\mathbf{E}_0'$ and $\mathbf{E}_0''$ is computed through

$$\cos\alpha = \frac{\mathbf{E}_0' \cdot \mathbf{E}_0''}{|\mathbf{E}_0'||\mathbf{E}_0''|} \tag{7.131}$$

and when $\alpha = \pi/2$, $\mathbf{E}_0'$ and $\mathbf{E}_0''$ represent the principal axes of an ellipsis [19, Section 8.4] with center given by $\mathbf{r}$. More generally, to determine the axes $\mathbf{C}_0'$ and $\mathbf{C}_0''$ we observe that they may be expressed as a linear combination of $\mathbf{E}_0'$ and $\mathbf{E}_0''$ and hence we may represent the same electric field in the alternative form (Figure 7.12)

$$\mathcal{E}(\mathbf{r},t) = \mathrm{Re}\left\{(\mathbf{C}_0' + \mathrm{j}\,\mathbf{C}_0'')\mathrm{e}^{\mathrm{j}\,\omega t + \mathrm{j}\beta - \mathrm{j}\,\mathbf{k}\cdot\mathbf{r}}\right\} \tag{7.132}$$

where β/ω is interpreted as the time taken for the electric field to rotate by an angle $\beta \in]0, \pi/2[$. Comparison of (7.23) and (7.132) yields

$$\mathbf{C}_0' + \mathrm{j}\,\mathbf{C}_0'' = (\mathbf{E}_0' + \mathrm{j}\,\mathbf{E}_0'')\mathrm{e}^{-\mathrm{j}\beta} \tag{7.133}$$

which we may further cast into

$$\mathbf{C}_0' = \mathbf{E}_0' \cos\beta + \mathbf{E}_0'' \sin\beta \tag{7.134}$$

$$\mathbf{C}_0'' = -\mathbf{E}_0' \sin\beta + \mathbf{E}_0'' \cos\beta \tag{7.135}$$

by separating real and imaginary parts. The angle β follows by requiring the axes of the ellipsis to be orthogonal, namely,

$$\begin{aligned} 0 = \mathbf{C}_0' \cdot \mathbf{C}_0'' &= (\mathbf{E}_0' \cos\beta + \mathbf{E}_0'' \sin\beta) \cdot (\mathbf{E}_0'' \cos\beta - \mathbf{E}_0' \sin\beta) \\ &= \mathbf{E}_0' \cdot \mathbf{E}_0''(\cos^2\beta - \sin^2\beta) + (|\mathbf{E}_0''|^2 - |\mathbf{E}_0'|^2)\sin\beta\,\cos\beta \\ &= \mathbf{E}_0' \cdot \mathbf{E}_0'' \cos(2\beta) + \frac{1}{2}(|\mathbf{E}_0''|^2 - |\mathbf{E}_0'|^2)\sin(2\beta) \end{aligned} \tag{7.136}$$

whence

$$\tan(2\beta) = \frac{2\mathbf{E}_0' \cdot \mathbf{E}_0''}{|\mathbf{E}_0'|^2 - |\mathbf{E}_0''|^2} \tag{7.137}$$

under the hypotheses that $|\mathbf{E}_0'| \neq |\mathbf{E}_0''|$ and $\mathbf{E}_0' \cdot \mathbf{E}_0'' \neq 0$. If the magnitudes of $\mathbf{E}_0'$ and $\mathbf{E}_0''$ are equal but the two vectors are not orthogonal, the angle $\beta = \pi/4$, and the polarization is still elliptical.

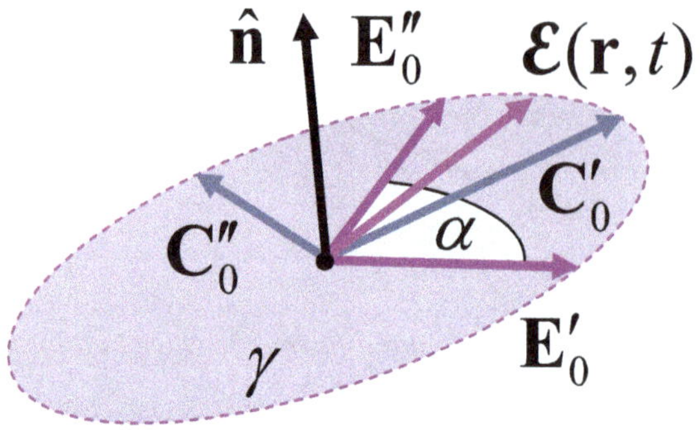

Figure 7.12 Geometry for determining the principal axes $\mathbf{C}_0'$ and $\mathbf{C}_0''$ of the polarization ellipsis.

Conversely, if $|\mathbf{E}_0'| = |\mathbf{E}_0''|$ and $\mathbf{E}_0' \cdot \mathbf{E}_0'' = 0$ the plane wave is circularly polarized. Indeed, application of (7.23) under the previous conditions yields

$$|\mathcal{E}(\mathbf{r}, t)| = \left|\mathbf{E}_0' \cos(\omega t - \mathbf{k}' \cdot \mathbf{r})e^{-\mathbf{k}'' \cdot \mathbf{r}} - \mathbf{E}_0'' \sin(\omega t - \mathbf{k}' \cdot \mathbf{r})e^{-\mathbf{k}'' \cdot \mathbf{r}}\right| = |\mathbf{E}_0'|e^{-\mathbf{k}'' \cdot \mathbf{r}} \tag{7.138}$$

that is, at a given point in space the magnitude of the electric field remains constant in time and the tip of the vector $\mathcal{E}(\mathbf{r}, t)$ describes a circumference of radius $|\mathbf{E}_0'|$. Nonetheless, we still have two options inasmuch as the electric field may rotate in a clockwise or counterclockwise fashion with respect to the positive direction indicated by the normal vector $\hat{\mathbf{n}}$.

To elaborate, we choose a *real* unit vector $\hat{\mathbf{e}}_1$ in the polarization plane and construct another vector $\hat{\mathbf{e}}_2$ orthogonal to $\hat{\mathbf{e}}_1$ as $\hat{\mathbf{e}}_2 = \hat{\mathbf{n}} \times \hat{\mathbf{e}}_1$. In this way, $\hat{\mathbf{e}}_1$, $\hat{\mathbf{e}}_2$ and $\hat{\mathbf{n}}$ form a right-handed orthogonal triplet. Next we define the following two *complex* unit vectors

$$\hat{\mathbf{e}}_{CW} = \frac{\hat{\mathbf{e}}_1 + \mathrm{j}\,\hat{\mathbf{e}}_2}{\sqrt{2}} \qquad \hat{\mathbf{e}}_{CCW} = \frac{\hat{\mathbf{e}}_1 - \mathrm{j}\,\hat{\mathbf{e}}_2}{\sqrt{2}} \tag{7.139}$$

where CW and CCW stand for clockwise and counterclockwise. We claim, for instance, that $\hat{\mathbf{e}}_{CW}$ represents a circularly polarized field which rotates in a clockwise fashion. Indeed, the real and imaginary parts of $\hat{\mathbf{e}}_{CW}$ are orthogonal and have the same magnitude. It remains to show that the tip of the vector in time domain moves along a circumference according to a left-handed screw with respect to $\hat{\mathbf{n}}$. To this purpose, we write

$$\mathbf{e}_{CW}(t) = \mathrm{Re}\left\{\hat{\mathbf{e}}_{CW} e^{\mathrm{j}\omega t}\right\} = \frac{1}{\sqrt{2}}\hat{\mathbf{e}}_1 \cos(\omega t) - \frac{1}{\sqrt{2}}\hat{\mathbf{e}}_2 \sin(\omega t) \tag{7.140}$$

and sample the field at times $t \in \{0, T/4, T/2, 3T/4\}$ to get (Figure 7.13)

$$\mathbf{e}_{CW}(0) = \frac{1}{\sqrt{2}}\hat{\mathbf{e}}_1 \qquad \mathbf{e}_{CW}\left(\frac{T}{4}\right) = -\frac{1}{\sqrt{2}}\hat{\mathbf{e}}_2 \tag{7.141}$$

$$\mathbf{e}_{CW}\left(\frac{T}{2}\right) = -\frac{1}{\sqrt{2}}\hat{\mathbf{e}}_1 \qquad \mathbf{e}_{CW}\left(\frac{3T}{4}\right) = \frac{1}{\sqrt{2}}\hat{\mathbf{e}}_2 \tag{7.142}$$

since $\omega T/4 = \pi/2$ and so forth. This analysis confirms that the polarization is clockwise circular. A similar investigation allows concluding that $\hat{\mathbf{e}}_{CCW}$ corresponds to a counterclockwise circular polarization.

Since the unit vectors $\hat{\mathbf{e}}_{CW}$ and $\hat{\mathbf{e}}_{CCW}$ are perpendicular in the sense that

$$\hat{\mathbf{e}}_{CW} \cdot \hat{\mathbf{e}}_{CCW}^* = 0 \tag{7.143}$$

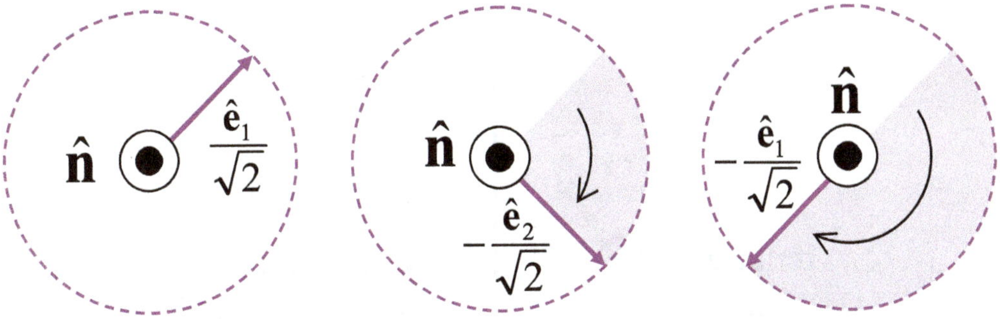

Figure 7.13 *Graphical representation of the circular clockwise polarization expressed in* (7.141) *and* (7.142).

we may use them as an orthonormal basis to represent any complex vector associated with a time-harmonic field. More specifically, we let

$$\mathbf{E}_0 = E_{CW}\hat{\mathbf{e}}_{CW} + E_{CCW}\hat{\mathbf{e}}_{CCW} \tag{7.144}$$

and the coefficients are computed through

$$E_{CW} = \hat{\mathbf{e}}^*_{CW} \cdot \mathbf{E}_0 \qquad\qquad E_{CCW} = \hat{\mathbf{e}}^*_{CCW} \cdot \mathbf{E}_0 \tag{7.145}$$

by virtue of (7.143). With this decomposition it is an easy matter to decide in which sense a circularly polarized field rotates: if $E_{CCW} = 0$ then the polarization is clockwise, whereas if $E_{CW} = 0$ the polarization is counterclockwise. Alternatively, we just have to analyze the field in time as we did above with $\mathbf{e}_{CW}(t)$. Notice that any field — regardless of the actual state of polarization — may always be written as the combination of two circularly polarized fields rotating in opposites senses. Therefore, if E_{CW} and E_{CCW} are both non-zero, we can only conclude that the polarization is not circular.

Lastly, with reference to (7.23) we see that linear polarization may occur if any one of the following conditions is true

$$\mathbf{E}''_0 = a\mathbf{E}'_0 \qquad\qquad \mathbf{E}''_0 = 0 \qquad\qquad \mathbf{E}'_0 = 0 \tag{7.146}$$

and the plane of polarization is undetermined. A linearly polarized uniform plane wave may be decomposed into two circular polarizations. For instance, if $\mathbf{E}''_0 = 0$ we may let

$$\hat{\mathbf{e}}_1 = \frac{\mathbf{E}'_0}{|\mathbf{E}'_0|} \qquad\qquad \hat{\mathbf{n}} = \hat{\mathbf{k}} \qquad\qquad \hat{\mathbf{e}}_2 = \hat{\mathbf{n}} \times \hat{\mathbf{e}}_1 \tag{7.147}$$

then apply (7.139), and compute the coefficients E_{CW} and E_{CCW} by means of (7.145).

7.4 Plane-wave propagation in layered isotropic media

In the previous section we showed that the source-free time-harmonic Maxwell equations admit the plane-wave solutions (7.20) and (7.21) in an unbounded isotropic homogeneous medium, provided the wavevector $\mathbf{k}$ and the wavenumber $k := \omega/c$ obey the dispersion relation (7.50). Here we wish to

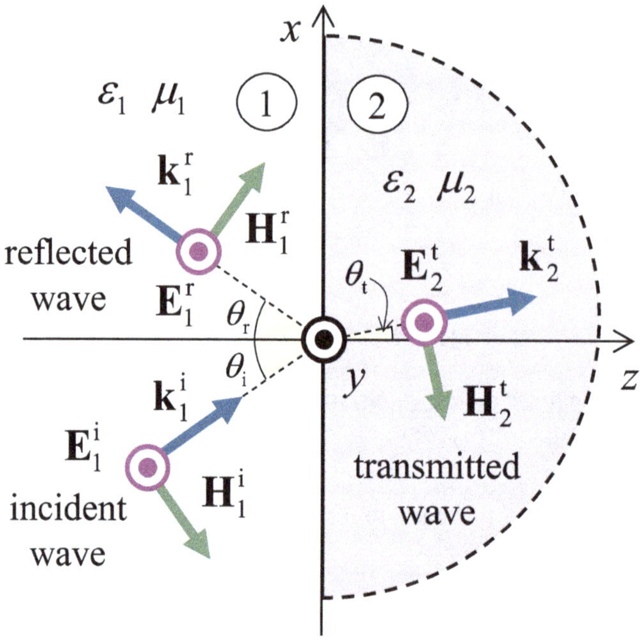

Figure 7.14 Reflection and transmission of uniform plane waves at a planar interface between two isotropic media: transverse-electric (TE) polarization.

develop mathematical tools for the calculation of time-harmonic plane-wave solutions within *layered* or *stratified isotropic* unbounded media. With the latter we mean a stack of finite-thickness adjoining planar slabs of penetrable dielectric or magnetic materials. The problem has practical interest for the analysis and design of, e.g., optical gratings, filters, mirrors at high frequencies, and planar optical waveguides [31–34].

We shall proceed incrementally by first considering the simple case of two half-spaces filled with different media, then a slab and eventually a multi-layered structure [10, Chapter 12], [25, Chapter 4], [35, Chapter 5], [27, Section 2.8], [31, Section 1.6], [30, Section 4.1.J], [36, Chapter X], [37].

7.4.1 Reflection and transmission at a planar interface

With reference to Figures 7.14 and 7.15 we choose a system of Cartesian coordinates with the z-axis perpendicular to the planar interface that separates two unbounded penetrable media indicated with labels 1 and 2. In particular, we place the origin O so that the interface coincides with the plane $z = 0$, medium 1 fills the half-space $z < 0$ and medium 2 occupies the half-space $z \geqslant 0$. Next, we suppose that a uniform plane wave, say,

$$\mathbf{E}_1^{\mathrm{i}}(\mathbf{r}) = \mathbf{E}_{10}^{\mathrm{i}} \mathrm{e}^{-\mathrm{j}\mathbf{k}_1^{\mathrm{i}}\cdot\mathbf{r}}, \qquad \mathbf{H}_1^{\mathrm{i}}(\mathbf{r}) = \mathbf{H}_{10}^{\mathrm{i}} \mathrm{e}^{-\mathrm{j}\mathbf{k}_1^{\mathrm{i}}\cdot\mathbf{r}}, \qquad z < 0 \tag{7.148}$$

is produced by an infinitely remote time-harmonic source located in medium 1 and 'propagates' towards the interface. We remark that, since in the time-harmonic regime uniform plane waves are periodic in time and space, there is no actual 'travelling' of a disturbance in the sense of Figure 1.15b, because the field has existed since an infinitely remote time in the whole space. The superscript 'i' stands for *incident wave*, and the reason for this attribute will be clarified in a moment. While the orientation of the wave fields $\mathbf{E}_1^{\mathrm{i}}$ and $\mathbf{H}_1^{\mathrm{i}}$ with respect to the interface can be arbitrary, we surely

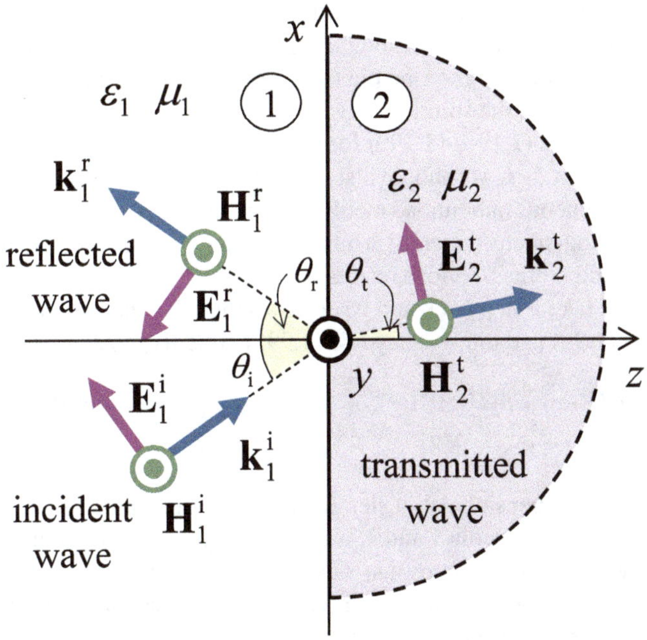

Figure 7.15 Reflection and transmission of uniform plane waves at a planar interface between two isotropic media: transverse-magnetic (TM) polarization.

cover all bases if we focus on just two mutually exclusive possibilities which we can combine to represent all plane waves in accordance with the principle of superposition (Section 6.1). These special choices are:

(1) the *transverse-electric* (TE) polarization in which the electric field is parallel to the interface and, in view of the adopted system of coordinates, is aligned with the y-axis; transverse-electric reminds us that **E** does not have a component perpendicular to the interface, i.e., aligned with the z-axis (Figure 7.14);

(2) the *transverse-magnetic* (TM) polarization in which the magnetic field is parallel to the interface and, in view of the adopted system of coordinates, is aligned with the y-axis; transverse-magnetic reminds us that **H** does not have a component perpendicular to the interface, i.e., aligned with the z-axis (Figure 7.15).

In both cases, once the orientation of the wavevector $\mathbf{k}_1^{\mathrm{i}}$ has been assigned subject to (7.41) or (7.42) and the dispersion relationship (7.50), the other field entity follows from (7.39) or (7.40). In symbols, this means

$$\mathbf{E}_1^{\mathrm{i}}(\mathbf{r}) = E_{10}^{\mathrm{i}}\hat{\mathbf{y}}\mathrm{e}^{-\mathrm{j}\mathbf{k}_1^{\mathrm{i}}\cdot\mathbf{r}} \qquad \mathbf{H}_1^{\mathrm{i}}(\mathbf{r}) = \frac{\mathbf{k}_1^{\mathrm{i}}}{\omega\mu_1}\times\hat{\mathbf{y}}E_{10}^{\mathrm{i}}\mathrm{e}^{-\mathrm{j}\mathbf{k}_1^{\mathrm{i}}\cdot\mathbf{r}} \qquad \text{(TE wave)} \tag{7.149}$$

$$\mathbf{H}_1^{\mathrm{i}}(\mathbf{r}) = H_{10}^{\mathrm{i}}\hat{\mathbf{y}}\mathrm{e}^{-\mathrm{j}\mathbf{k}_1^{\mathrm{i}}\cdot\mathbf{r}} \qquad \mathbf{E}_1^{\mathrm{i}}(\mathbf{r}) = \hat{\mathbf{y}}H_{10}^{\mathrm{i}}\times\frac{\mathbf{k}_1^{\mathrm{i}}}{\omega\varepsilon_1}\mathrm{e}^{-\mathrm{j}\mathbf{k}_1^{\mathrm{i}}\cdot\mathbf{r}} \qquad \text{(TM wave)} \tag{7.150}$$

where E_{10}^{i} and H_{10}^{i} are two *known*, possibly complex constants.

It only takes a moment's thought to realize that (7.149) and (7.150) cannot be the right solution for the problems sketched in Figure 7.14 and 7.15, respectively. Why? Well, first and foremost, the wavevector of a plane wave propagating in medium 2 ought to solve a dispersion relation such as (7.50) where the constitutive parameters, though, are ε_2 and μ_2, whereas $\mathbf{k}_1^{\mathrm{i}}$ is the valid wavevector for a wave in medium 1. Secondly, since the plane $z = 0$ separates two media endowed with different material properties, the fields of the solution — whatever strategy we adopt to arrive at it — ought to obey the matching conditions (1.196)-(1.199) for $z = 0$, whereas, in fact, the fields given by (7.149) and (7.150) fail to do so. But then, we should also remember that all the solutions obtained in Section 7.2 are valid for a homogeneous unbounded medium, whereas the problems of Figures 7.14 and 7.15 involve a piecewise-homogeneous material! Thus, we need to overhaul the solution procedure or at least devise a way to build on the previous results.

To this purpose, we take a step back and write down the relevant time-harmonic instance of the Ampère-Maxwell law (1.98) and the Faraday law (1.99) explicitly, viz.,

$$\nabla \times \mathbf{E}(\mathbf{r}) = -\mathrm{j}\,\omega[\mu_1 U(-z) + \mu_2 U(z)]\mathbf{H}(\mathbf{r}), \qquad z \neq 0 \tag{7.151}$$

$$\nabla \times \mathbf{H}(\mathbf{r}) = \mathrm{j}\,\omega[\varepsilon_1 U(-z) + \varepsilon_2 U(z)]\mathbf{E}(\mathbf{r}), \qquad z \neq 0 \tag{7.152}$$

where $U(\bullet)$ denotes a one-dimensional unit step function. These equations are valid everywhere but exactly at the interface between media 1 and 2, whereon we need to enforce the matching conditions (1.196) and (1.197) with no surface currents. Clearly, we may not assume a solution in the form (7.148) and work our way towards a dispersion relationship, as we did in Section 7.2, because the overall constitutive parameters are not constant.

As a matter of fact, it is more fruitful to cast (7.151) and (7.152) into two alternative, perfectly equivalent forms by means of elementary algebraic manipulations, namely,

$$\nabla \times \mathbf{E}(\mathbf{r}) = -\mathrm{j}\,\omega\mu_1\mathbf{H}(\mathbf{r}) \underbrace{-\,\mathrm{j}\,\omega(\mu_2 - \mu_1)U(z)\mathbf{H}(\mathbf{r}),}_{\text{magnetic source in } z>0} \qquad z \neq 0 \tag{7.153}$$

$$\nabla \times \mathbf{H}(\mathbf{r}) = \mathrm{j}\,\omega\varepsilon_1\mathbf{E}(\mathbf{r}) + \underbrace{\mathrm{j}\,\omega(\varepsilon_2 - \varepsilon_1)U(z)\mathbf{E}(\mathbf{r}),}_{\text{electric source in } z>0} \qquad z \neq 0 \tag{7.154}$$

and

$$\nabla \times \mathbf{E}(\mathbf{r}) = -\mathrm{j}\,\omega\mu_2\mathbf{H}(\mathbf{r}) \underbrace{-\,\mathrm{j}\,\omega(\mu_1 - \mu_2)U(-z)\mathbf{H}(\mathbf{r}),}_{\text{magnetic source in } z<0} \qquad z \neq 0 \tag{7.155}$$

$$\nabla \times \mathbf{H}(\mathbf{r}) = \mathrm{j}\,\omega\varepsilon_2\mathbf{E}(\mathbf{r}) + \underbrace{\mathrm{j}\,\omega(\varepsilon_1 - \varepsilon_2)U(-z)\mathbf{E}(\mathbf{r}),}_{\text{electric source in } z<0} \qquad z \neq 0 \tag{7.156}$$

which we interpret as follows.

In the set of equations (7.153) and (7.154) we favor the viewpoint of an ideal observer who sits somewhere in the half space $z < 0$. Accordingly, such observer formulates the problems of Figures 7.14 and 7.15 equivalently through (7.153) and (7.154) *as if* medium 1 filled the whole space and there existed infinitely extended volume current densities in the half space $z > 0$, as indicated. Physically, these currents — of electric and magnetic nature — arise from the difference between the properties of media 1 and 2. As a consequence, the observer in medium 1 detects the incident wave (7.148) but also a *reflected wave*, say,

$$\mathbf{E}_1^{\mathrm{r}}(\mathbf{r}) = \mathbf{E}_{10}^{\mathrm{r}}\mathrm{e}^{-\mathrm{j}\mathbf{k}_1^{\mathrm{r}}\cdot\mathbf{r}} \qquad \mathbf{H}_1^{\mathrm{r}}(\mathbf{r}) = \mathbf{H}_{10}^{\mathrm{r}}\mathrm{e}^{-\mathrm{j}\mathbf{k}_1^{\mathrm{r}}\cdot\mathbf{r}} \qquad z < 0 \tag{7.157}$$

which he ascribes to the equivalent sources in $z > 0$. The reflected wave is seen to emerge from the plane $z = 0$ and advance leftwards away from it.

By contrast, with the set of equations (7.155) and (7.156) we adopt the viewpoint of another ideal observer located in the half space $z > 0$. This observer formulates the problems through (7.155) and (7.156) *as if* medium 2 filled the whole space and infinitely extended volume current densities flowed in the half space $z < 0$. The origin of these currents is the same already stated above, and the observer — who is unaware of any infinitely far off sources in $z < 0$ — detects a *transmitted* or *refracted wave*

$$\mathbf{E}_2^{\mathrm{t}}(\mathbf{r}) = \mathbf{E}_{20}^{\mathrm{t}} \mathrm{e}^{-\mathrm{j}\mathbf{k}_2^{\mathrm{t}}\cdot\mathbf{r}} \qquad \mathbf{H}_2^{\mathrm{t}}(\mathbf{r}) = \mathbf{H}_{20}^{\mathrm{t}} \mathrm{e}^{-\mathrm{j}\mathbf{k}_2^{\mathrm{t}}\cdot\mathbf{r}} \qquad z > 0 \tag{7.158}$$

which she attributes to the equivalent sources in $z < 0$. The transmitted wave is seen to emerge from the plane $z = 0$ and advance rightwards away from it. It is important to notice that the wavevector ($\mathbf{k}_2^{\mathrm{t}}$) of the transmitted wave depends on the properties of medium 2.

With the aid of (7.153)-(7.156) we have showed that the correct solution to the problems of Figures 7.14 and 7.15 must perforce involve the original incident wave plus a reflected and a transmitted wave, the origin of which we have explained in terms of equivalent sources. In fact, the procedure is a special instance of the general *volume equivalence principle* which we shall address in Section 10.5. Truth be told, we have not solved the problems yet, rather we have only determined the ingredients, so to speak, that concur to the construction of the solution. In this regard, we remark that there is no incident wave coming in from infinitely remote sources in medium 2 for the simple reason that we have not entertained such possibility. We can deal with that occurrence independently, and then invoke the principle of superposition to write down a more general solution, if need be.

To proceed we first focus on the TE polarization of Figure 7.14 and write the fields of the reflected and transmitted waves, viz.,

$$\mathbf{E}_1^{\mathrm{r}}(\mathbf{r}) = E_{10}^{\mathrm{r}} \hat{\mathbf{y}} \mathrm{e}^{-\mathrm{j}\mathbf{k}_1^{\mathrm{r}}\cdot\mathbf{r}} \qquad \mathbf{H}_1^{\mathrm{r}}(\mathbf{r}) = \frac{\mathbf{k}_1^{\mathrm{r}}}{\omega\mu_1} \times \hat{\mathbf{y}} E_{10}^{\mathrm{r}} \mathrm{e}^{-\mathrm{j}\mathbf{k}_1^{\mathrm{r}}\cdot\mathbf{r}} \tag{7.159}$$

$$\mathbf{E}_2^{\mathrm{t}}(\mathbf{r}) = E_{10}^{\mathrm{t}} \hat{\mathbf{y}} \mathrm{e}^{-\mathrm{j}\mathbf{k}_2^{\mathrm{t}}\cdot\mathbf{r}} \qquad \mathbf{H}_2^{\mathrm{t}}(\mathbf{r}) = \frac{\mathbf{k}_2^{\mathrm{t}}}{\omega\mu_2} \times \hat{\mathbf{y}} E_{20}^{\mathrm{t}} \mathrm{e}^{-\mathrm{j}\mathbf{k}_2^{\mathrm{t}}\cdot\mathbf{r}} \tag{7.160}$$

where E_{10}^{r} and E_{10}^{t} are *unknown* constants. In order to determine E_{10}^{r} and E_{10}^{t} we demand that the total fields on either side of the interface meet the jump conditions (1.196) and (1.197). Is this requirement enough or rather too stringent? And how about the components of the fields perpendicular to the interface?

As it turns out, the field components tangential and perpendicular to a material interface are not independent, and hence if the tangential components satisfy the expected jump conditions so do the normal components automatically. Concerning the right number of constraints, let us state (1.196) and (1.197) explicitly for the problem of Figure 7.14, viz.,

$$[\mathbf{E}_1^{\mathrm{i}}(\mathbf{r}) + \mathbf{E}_1^{\mathrm{r}}(\mathbf{r})] \times \hat{\mathbf{z}} = \mathbf{E}_2^{\mathrm{t}}(\mathbf{r}) \times \hat{\mathbf{z}}, \qquad z = 0 \tag{7.161}$$

$$\hat{\mathbf{z}} \times [\mathbf{H}_1^{\mathrm{i}}(\mathbf{r}) + \mathbf{H}_1^{\mathrm{r}}(\mathbf{r})] = \hat{\mathbf{z}} \times \mathbf{H}_2^{\mathrm{t}}(\mathbf{r}), \qquad z = 0 \tag{7.162}$$

where there is no surface current density $\mathbf{J}_S(\mathbf{r})$ on the interface because neither material is a perfect conductor by hypothesis. Now, since (7.161) and (7.162) are two vector relations which amount to four scalar equations, it would appear that we have two conditions too many. However, at a closer look we find out that this is not the case.

Indeed, if we let $z = 0$, dot-multiply (7.161) with $\hat{\mathbf{x}}$ and (7.162) with $\hat{\mathbf{y}}$, on account of (7.149), (7.159) and (7.160) we get

$$E_{10}^{\mathrm{i}} \mathrm{e}^{-\mathrm{j}k_{x1}^{\mathrm{i}}x} + E_{10}^{\mathrm{r}} \mathrm{e}^{-\mathrm{j}k_{x1}^{\mathrm{r}}x} = E_{20}^{\mathrm{t}} \mathrm{e}^{-\mathrm{j}k_{x2}^{\mathrm{t}}x}, \qquad x \in \mathbb{R} \tag{7.163}$$

$$\frac{k_{z1}^{\mathrm{i}}}{\omega\mu_1} E_{10}^{\mathrm{i}} \mathrm{e}^{-\mathrm{j}k_{x1}^{\mathrm{i}}x} + \frac{k_{z1}^{\mathrm{r}}}{\omega\mu_1} E_{10}^{\mathrm{r}} \mathrm{e}^{-\mathrm{j}k_{x1}^{\mathrm{r}}x} = \frac{k_{z2}^{\mathrm{t}}}{\omega\mu_2} E_{20}^{\mathrm{t}} \mathrm{e}^{-\mathrm{j}k_{x2}^{\mathrm{t}}x}, \qquad x \in \mathbb{R} \tag{7.164}$$

whereas if we dot-multiply (7.161) with $\hat{\mathbf{y}}$ and (7.162) with $\hat{\mathbf{x}}$ we obtain two trivial identities. Thanks to our clever orientation of the Cartesian axes the three wavevectors $\mathbf{k}_1^{\rm i}$, $\mathbf{k}_1^{\rm r}$ and $\mathbf{k}_2^{\rm t}$ have no component along $\hat{\mathbf{y}}$. The set of equations (7.163) and (7.164) must be satisfied for any value of x, and this goal may be achieved if additionally we require

$$k_{x1}^{\rm r} = k_{x1}^{\rm i} \qquad\qquad k_{x2}^{\rm t} = k_{x1}^{\rm i} \tag{7.165}$$

which makes the three exponential factors all equal to each other. In words, (7.165) says that the component of the wavevector tangential to the interface is conserved. Equivalently, (7.165) constitutes a concise statement of the **Snell laws** (circa 1621) for the reflection and transmission of a light ray [3, Chapter 33], [31, Sections 1.5 and 3.2.2], namely,

- *law of reflection*: the reflected wave lies in the same plane as the incident wave, and the angle $\theta_{\rm r} \in [0, \pi/2]$ formed by the wavevector $\mathbf{k}_1^{\rm r}$ and the vector normal to the interface equals the angle $\theta_{\rm i} \in [0, \pi/2]$ formed by the wavevector $\mathbf{k}_1^{\rm i}$ and the same normal vector;
- *law of transmission*: the refracted wave lies in the same plane as the incident wave, and the angle $\theta_{\rm t} \in [0, \pi/2]$ formed by the wavevector $\mathbf{k}_2^{\rm t}$ and the vector normal to the interface is related to $\theta_{\rm i}$ by

$$\sqrt{\varepsilon_1\mu_1}\sin\theta_{\rm i} = \sqrt{\varepsilon_2\mu_2}\sin\theta_{\rm t}. \tag{7.166}$$

Since the complex exponential never vanishes for any value of the argument, (7.163) and (7.164) imply

$$E_{10}^{\rm i} + E_{10}^{\rm r} = E_{20}^{\rm t} \tag{7.167}$$

$$\frac{k_{z1}^{\rm i}}{\omega\mu_1}E_{10}^{\rm i} + \frac{k_{z1}^{\rm r}}{\omega\mu_1}E_{10}^{\rm r} = \frac{k_{z2}^{\rm t}}{\omega\mu_2}E_{20}^{\rm t} \tag{7.168}$$

on account of (7.165). We have ended up with a set of two algebraic linear equations in two unknowns. The solution reads

$$E_{10}^{\rm r} = \frac{\mu_1 k_{z2}^{\rm t} - \mu_2 k_{z1}^{\rm i}}{\mu_2 k_{z1}^{\rm r} - \mu_1 k_{z2}^{\rm t}}E_{10}^{\rm i} \qquad\qquad E_{20}^{\rm t} = \frac{\mu_2 k_{z1}^{\rm r} - \mu_2 k_{z1}^{\rm i}}{\mu_2 k_{z1}^{\rm r} - \mu_1 k_{z2}^{\rm t}}E_{10}^{\rm i} \tag{7.169}$$

which when substituted back into (7.159) and (7.160) yields all the relevant fields for the TE polarization.

The solution of the TM polarization of Figure 7.15 is only formally different, since in particular the wavevectors do not change. The reflected and transmitted waves read

$$\mathbf{H}_1^{\rm r}(\mathbf{r}) = H_{10}^{\rm r}\hat{\mathbf{y}}{\rm e}^{-{\rm j}\mathbf{k}_1^{\rm r}\cdot\mathbf{r}} \qquad\qquad \mathbf{E}_1^{\rm r}(\mathbf{r}) = \hat{\mathbf{y}}H_{10}^{\rm r} \times \frac{\mathbf{k}_1^{\rm r}}{\omega\varepsilon_1}{\rm e}^{-{\rm j}\mathbf{k}_1^{\rm r}\cdot\mathbf{r}} \tag{7.170}$$

$$\mathbf{H}_2^{\rm t}(\mathbf{r}) = H_{10}^{\rm t}\hat{\mathbf{y}}{\rm e}^{-{\rm j}\mathbf{k}_2^{\rm t}\cdot\mathbf{r}} \qquad\qquad \mathbf{E}_2^{\rm t}(\mathbf{r}) = \hat{\mathbf{y}}H_{20}^{\rm t} \times \frac{\mathbf{k}_2^{\rm t}}{\omega\varepsilon_2}{\rm e}^{-{\rm j}\mathbf{k}_2^{\rm t}\cdot\mathbf{r}} \tag{7.171}$$

where $H_{10}^{\rm r}$ and $H_{10}^{\rm t}$ are *unknown*, possibly complex constants which we determine by enforcing the matching conditions (7.161) and (7.162). By letting $z = 0$ and dot-multiplying (7.161) with $\hat{\mathbf{y}}$ and (7.162) with $\hat{\mathbf{x}}$, on account of (7.150), (7.170) and (7.171) we get

$$\frac{k_{z1}^{\rm i}}{\omega\varepsilon_1}H_{10}^{\rm i}{\rm e}^{-{\rm j}k_{x1}^{\rm i}x} + \frac{k_{z1}^{\rm r}}{\omega\varepsilon_1}H_{10}^{\rm r}{\rm e}^{-{\rm j}k_{x1}^{\rm r}x} = \frac{k_{z2}^{\rm t}}{\omega\varepsilon_2}H_{20}^{\rm t}{\rm e}^{-{\rm j}k_{x2}^{\rm t}x}, \qquad x \in \mathbb{R} \tag{7.172}$$

$$H^{\rm i}_{10} {\rm e}^{-{\rm j}k^{\rm i}_{x1}x} + H^{\rm r}_{10} {\rm e}^{-{\rm j}k^{\rm r}_{x1}x} = H^{\rm t}_{20} {\rm e}^{-{\rm j}k^{\rm t}_{x2}x}, \qquad x \in \mathbb{R} \tag{7.173}$$

which lead us to the linear algebraic system

$$\frac{k^{\rm i}_{z1}}{\omega\varepsilon_1} H^{\rm i}_{10} + \frac{k^{\rm r}_{z1}}{\omega\varepsilon_1} H^{\rm r}_{10} = \frac{k^{\rm t}_{z2}}{\omega\varepsilon_2} H^{\rm t}_{20} \tag{7.174}$$

$$H^{\rm i}_{10} + H^{\rm r}_{10} = H^{\rm t}_{20} \tag{7.175}$$

by virtue of (7.165). Solving for $H^{\rm r}_{10}$ and $H^{\rm t}_{10}$ yields

$$H^{\rm r}_{10} = \frac{\varepsilon_1 k^{\rm t}_{z2} - \varepsilon_2 k^{\rm i}_{z1}}{\varepsilon_2 k^{\rm r}_{z1} - \varepsilon_1 k^{\rm t}_{z2}} H^{\rm i}_{10} \qquad H^{\rm t}_{20} = \frac{\varepsilon_2 k^{\rm r}_{z1} - \varepsilon_2 k^{\rm i}_{z1}}{\varepsilon_2 k^{\rm r}_{z1} - \varepsilon_1 k^{\rm t}_{z2}} H^{\rm i}_{10} \tag{7.176}$$

which when inserted into (7.170) and (7.171) provides all the relevant fields for the TM polarization.

To finalize the calculation of the coefficients found in (7.169) and (7.176) we need to determine the z-components of the wavevectors. Since the uniform incident TE and TM waves are assigned in (7.149) and (7.150), we may take $k^{\rm i}_{x1} \in \mathbb{R}$ as known and then derive $k^{\rm i}_{z1}$ from the dispersion relation (7.51) with $k^{\rm i}_{y1} = 0$, i.e.,

$$k^{\rm i}_{z1} = +\sqrt{\omega^2\varepsilon_1\mu_1 - (k^{\rm i}_{x1})^2} \qquad k^{\rm i}_{x1} \leqslant \omega\sqrt{\varepsilon_1\mu_1} \tag{7.177}$$

where we take $k^{\rm i}_{z1}$ positive because the incident wave travels to the right towards the interface. For the reflected wave we recall (7.165) and resort to (7.51) again, viz.,

$$k^{\rm r}_{z1} = -\sqrt{\omega^2\varepsilon_1\mu_1 - (k^{\rm i}_{x1})^2} = -k^{\rm i}_{z1} \qquad k^{\rm i}_{x1} \leqslant \omega\sqrt{\varepsilon_1\mu_1} \tag{7.178}$$

where we must take $k^{\rm r}_{z1} < 0$ because the wave propagates leftwards away from the interface. Thirdly, applying the same procedure to the transmitted wave we have

$$k^{\rm t}_{z2} = \begin{cases} +\sqrt{\omega^2\varepsilon_2\mu_2 - (k^{\rm i}_{x1})^2} & k^{\rm i}_{x1} \leqslant \omega\sqrt{\varepsilon_2\mu_2} \\ -{\rm j}\sqrt{(k^{\rm i}_{x1})^2 - \omega^2\varepsilon_2\mu_2} & k^{\rm i}_{x1} > \omega\sqrt{\varepsilon_2\mu_2} \end{cases} \tag{7.179}$$

whereby we see that for certain combinations of $k^{\rm i}_{x1}$, angular frequency and constitutive parameters the transmitted wave may become inhomogeneous and damped along z (cf. Figure 7.4).

To elaborate, since in our setup $k^{\rm t}_{x2} = k^{\rm i}_{x1}$ is real and positive, we have

$$\mathbf{k}^{\rm t}_2 = k^{\rm i}_{x1}\hat{\mathbf{x}} - {\rm j}\sqrt{(k^{\rm i}_{x1})^2 - \omega^2\varepsilon_2\mu_2}\,\hat{\mathbf{z}} \tag{7.180}$$

so that the inhomogeneous wave propagates along the positive x-direction, right at the interface between the two media, and no power is carried away from the interface into medium 2. For this reason such phenomenon is referred to as *total internal reflection* (TIR) [38, Chapter 5], [30, Section 4.1.D] and it may only occur when

$$\omega\sqrt{\varepsilon_2\mu_2} \leqslant k^{\rm i}_{x1} \leqslant \omega\sqrt{\varepsilon_1\mu_1} \tag{7.181}$$

that is, if medium 1 is electromagnetically or optically 'denser' than medium 2. The situation is graphically illustrated in Figure 7.16a, which shows the cross sections for $k_y = 0$ of the dispersion relationship (7.51) for medium 1 (larger sphere) and medium 2 (smaller sphere). Since

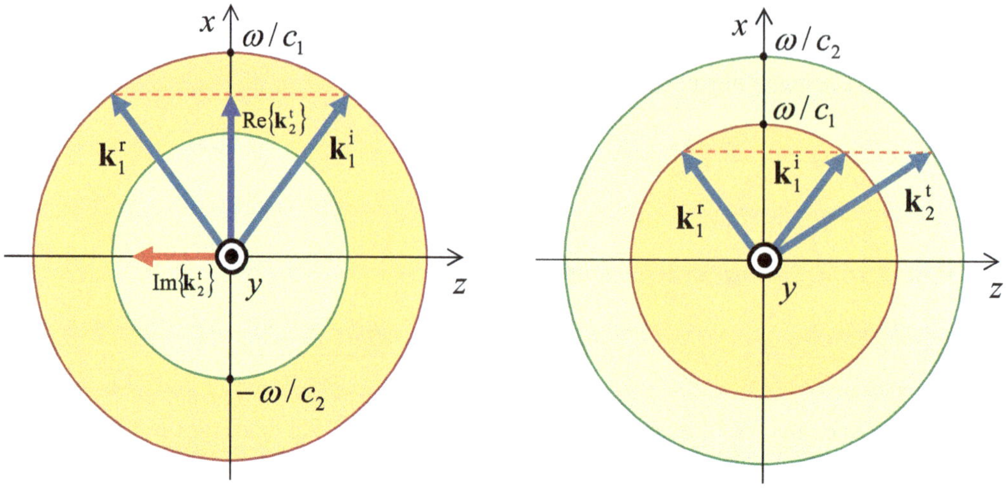

(a) $\varepsilon_1\mu_1 > \varepsilon_2\mu_2$ (b) $\varepsilon_1\mu_1 < \varepsilon_2\mu_2$

Figure 7.16 Graphical representation of dispersion relationships (7.51) *and wavevectors for the propagation on plane waves in the two isotropic media of Figures 7.14 and 7.15.*

$k^{\mathrm{i}}_{x1} > \omega\sqrt{\varepsilon_2\mu_2}$, then k^{t}_{z2} is imaginary. The transition from a transmitted uniform plane wave to an inhomogeneous one in medium 2 occurs when k^{t}_{z2} vanishes, i.e., if

$$k^{\mathrm{i}}_{x1} = \omega\sqrt{\varepsilon_2\mu_2} \tag{7.182}$$

which in terms of angles of incidence and transmission (Figures 7.14 and 7.15) means

$$\sin\theta_{\mathrm{i}} = \sin\theta_{\mathrm{c}} = \frac{k^{\mathrm{i}}_{x1}}{\omega\sqrt{\varepsilon_1\mu_1}} = \left(\frac{\varepsilon_2\mu_2}{\varepsilon_1\mu_1}\right)^{1/2} \qquad \sin\theta_{\mathrm{t}} = \frac{k^{\mathrm{t}}_{x2}}{\omega\sqrt{\varepsilon_2\mu_2}} = 1 \tag{7.183}$$

and θ_{c} is called *critical angle*. Notice that the critical angle is always real thanks to condition (7.181). If the incident plane wave impinges on the interface at an angle θ_{i} larger than θ_{c}, total internal reflection takes place in medium 1. On the contrary, when the constitutive parameters of media 1 and 2 are such that

$$k^{\mathrm{i}}_{x1} \leqslant \omega\sqrt{\varepsilon_1\mu_1} < \omega\sqrt{\varepsilon_2\mu_2} \tag{7.184}$$

k^{t}_{z2} is always real, and total internal reflection is precluded; this situation is exemplified in Figure 7.16b.

Total internal reflection is not restricted to just plane-wave propagation and in everyday life it may occur, e.g., at the interface between water ($\varepsilon_1 = 80\varepsilon_0$ at 20 °C) and air ($\varepsilon_2 = \varepsilon_0$) in a swimming-pool, if the narrow light beam of an underwater lamp impinges on the surface at an angle $\theta_{\mathrm{i}} > \theta_{\mathrm{c}} \approx 6.37°$. In modern telecommunications, total internal reflection makes it possible to guide electromagnetic waves in planar optical waveguides and, more generally, in optical fibers [32–34].

Having found the z-components of the wavevectors consistently, with a little algebra we can cast the solutions (7.169) and (7.176) as

$$E^{\mathrm{r}}_{10} = \frac{Z''_{\infty 2} - Z''_{\infty 1}}{Z''_{\infty 2} + Z''_{\infty 1}} E^{\mathrm{i}}_{10} \qquad E^{\mathrm{t}}_{10} = \frac{2Z''_{\infty 2}}{Z''_{\infty 2} + Z''_{\infty 1}} E^{\mathrm{i}}_{10} \qquad \text{(TE waves)} \tag{7.185}$$

$$H^{\mathrm{r}}_{10} = \frac{Y'_{\infty 2} - Y'_{\infty 1}}{Y'_{\infty 2} + Y'_{\infty 1}} H^{\mathrm{i}}_{10} \qquad H^{\mathrm{t}}_{10} = \frac{2Y'_{\infty 2}}{Y'_{\infty 2} + Y'_{\infty 1}} H^{\mathrm{i}}_{10} \qquad \text{(TM waves)} \tag{7.186}$$

where the quantities

$$Z''_{\infty 1} := \frac{\omega \mu_1}{k^{\mathrm{i}}_{z1}} \qquad Z''_{\infty 2} := \frac{\omega \mu_2}{k^{\mathrm{t}}_{z2}} \qquad \text{(TE waves)} \tag{7.187}$$

$$Y'_{\infty 1} := \frac{\omega \varepsilon_1}{k^{\mathrm{i}}_{z1}} \qquad Y'_{\infty 2} := \frac{\omega \varepsilon_2}{k^{\mathrm{t}}_{z2}} \qquad \text{(TM waves)} \tag{7.188}$$

are called *characteristic impedances* and *characteristic admittances* of media 1 and 2. It is easy to check that when $k^{\mathrm{i}}_{x1} = 0$ (i.e., for normal incidence)

$$Y'_{\infty 1} = 1/Z''_{\infty 1} \qquad Y'_{\infty 2} = 1/Z''_{\infty 2} \tag{7.189}$$

and, in particular, $Z''_{\infty 1}$ and $Z''_{\infty 2}$ degenerate into the intrinsic impedances of media 1 and 2 [see (1.358)]. In the presence of TIR in medium 1, $Z''_{\infty 2}$ and $Y'_{\infty 2}$ become purely imaginary, because so is k^{t}_{z2}. Besides, it is customary to define the following *reflection* and *transmission* coefficients

$$R_{TE} := \frac{Z''_{\infty 2} - Z''_{\infty 1}}{Z''_{\infty 2} + Z''_{\infty 1}} \qquad T_{TE} := \frac{2Z''_{\infty 2}}{Z''_{\infty 2} + Z''_{\infty 1}} \tag{7.190}$$

$$R_{TM} := \frac{Y'_{\infty 2} - Y'_{\infty 1}}{Y'_{\infty 2} + Y'_{\infty 1}} \qquad T_{TM} := \frac{2Y'_{\infty 2}}{Y'_{\infty 2} + Y'_{\infty 1}} \tag{7.191}$$

which obviously satisfy

$$1 + R_{TE} = T_{TE} \qquad 1 + R_{TM} = T_{TM} \tag{7.192}$$

in keeping with the enforced continuity conditions (7.161) and (7.162), respectively.

As an example, plots of the reflection coefficients versus $\sin\theta_{\mathrm{i}} = k^{\mathrm{i}}_{x1} c_1/\omega$ are drawn in Figure 7.17 for $\varepsilon_2 \in \{0.36\varepsilon_1, 2\varepsilon_1\}$ and the special case $\mu_1 = \mu_2 = \mu$. For ease of visualization the negative of R_{TM} is plotted. As can be seen, when $k^{\mathrm{i}}_{x1} = 0$, R_{TE} and $-R_{TM}$ coincide or, stated another way, R_{TE} and R_{TM} are the negative of each other. This result is more than reasonable, since for $\theta_{\mathrm{i}} = 0$ one obtains the TE polarization by executing a counterclockwise ninety-degree-rotation of the TM fields about the positive z-axis or, which is the same, the wavevector $\mathbf{k}^{\mathrm{i}}_1$. In fact, a plane wave propagating along the z-axis is simultaneously TE and TM or, for short, a transverse-electric-magnetic (TEM) wave with respect to z.

As the incident wave impinges against the interface at ever steeper angles, if $\varepsilon_2 > \varepsilon_1$, then R_{TE} remains negative and reaches -1 for $k^{\mathrm{i}}_{x1} = \omega/c_1$. By contrast, if $\varepsilon_2 < \varepsilon_1$, R_{TE} remains positive and tends to $+1$ for k^{i}_{x1} given by (7.182), since TIR occurs.

Regarding the TM polarization, as k^{i}_{x1} increases, if $\varepsilon_2 > \varepsilon_1$, $-R_{TM}$ increases, changes sign, and reaches $+1$ for for $k^{\mathrm{i}}_{x1} = \omega/c_1$, whereas if $\varepsilon_2 < \varepsilon_1$, $-R_{TM}$ decreases, changes sign, and tends to -1 again for k^{i}_{x1} given by (7.182). Therefore, irrespective of the ratio between ε_2 and ε_1, the

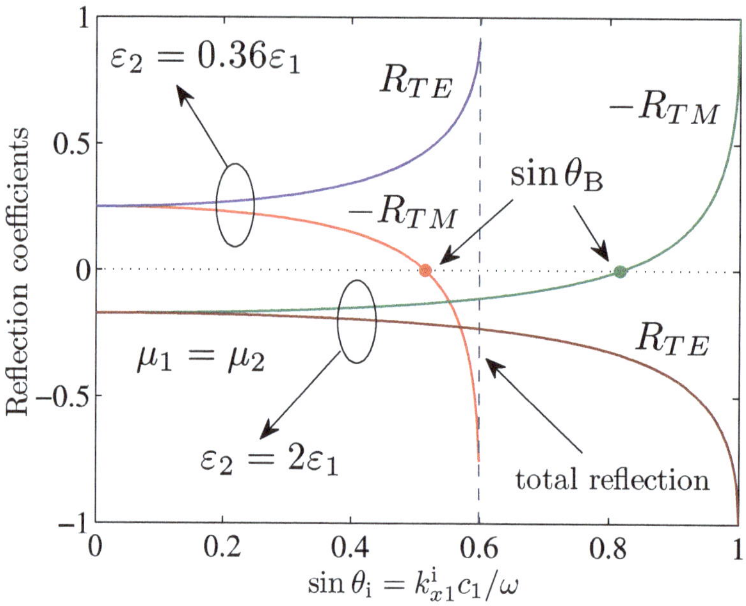

Figure 7.17 Reflection coefficients for the plane-wave problems of Figures 7.14 and 7.15.

TM-wave reflection coefficient vanishes for a certain angle of incidence. The specific value of k^i_{x1} which nullifies R_{TM} is found from [see (7.176)]

$$\varepsilon_1 k^t_{z2} - \varepsilon_2 k^i_{z1} = 0 \qquad \Longrightarrow \qquad k^i_{x1} = \omega\left(\frac{\mu\varepsilon_1\varepsilon_2}{\varepsilon_1 + \varepsilon_2}\right)^{1/2} \tag{7.193}$$

where the result is valid under the condition that $\varepsilon_1 \neq \varepsilon_2$. Rephrased in terms of angle of incidence and transmission, (7.193) means

$$\sin\theta_i = \sin\theta_B = \frac{k^i_{x1}}{\omega\sqrt{\varepsilon_1\mu}} = \left(\frac{\varepsilon_2}{\varepsilon_1 + \varepsilon_2}\right)^{1/2} < 1 \tag{7.194}$$

$$\sin\theta_t = \frac{k^t_{x2}}{\omega\sqrt{\varepsilon_2\mu}} = \left(\frac{\varepsilon_1}{\varepsilon_1 + \varepsilon_2}\right)^{1/2} < 1 \tag{7.195}$$

where θ_B is called the *Brewster angle* [39, Chapter 33], [10, Section 12.6], [30, Section 4.1.G]. For ordinary dielectric media endowed with positive permittivities, θ_B and the associated refraction angle θ_t are real. Moreover, from (7.191) we see that when $R_{TM} = 0$, then the characteristic admittances in medium 1 and 2 coincide, viz.,

$$Y'_{\infty 1} = Y'_{\infty 2} = \left(\frac{\varepsilon_1 + \varepsilon_2}{\mu}\right)^{1/2} \tag{7.196}$$

whereby T_{TM} equals 1.

The vanishing of R_{TM} implies that *no reflected wave* is generated if the TM incident plane wave impinges upon the interface at the Brewster angle! Therefore, for all practical purposes the incident TM wave behaves *as if* it does not encounter any interface, so to speak, and travels on unhindered, although it emerges in medium 2 at an angle θ_t that is different than θ_B. Further, going back to the equivalent equations (7.153) and (7.154) we notice that from the viewpoint of the observer in medium 1 there is no magnetic source in the half-space $z \geqslant 0$ — since $\mu_1 = \mu_2$ by hypothesis — and hence the observer ascribes the appearance of TE and TM reflected waves to the equivalent electric source only. Nevertheless, when the TM incident wave impinges at the Brewster angle, the infinitely many elementary contributions that make up the electric source generate fields which, by interfering destructively, produce no reflected TM wave altogether. The 'disappearance' of the TM reflected wave is not abrupt, though, because for incidence angles in a neighborhood of θ_B the relevant reflection coefficient is relatively small, as can be gathered from Figure 7.17.

We also notice that, under the conditions stated at the beginning of this discussion, there cannot exist a Brewster angle for the TE polarization, in that the relevant equation for k^{i}_{x1}, namely,

$$k^{\mathrm{i}}_{z1} - k^{\mathrm{t}}_{z2} = 0 \tag{7.197}$$

is impossible to satisfy, as is also clear geometrically from Figure 7.16b. Finally, perfectly dual considerations can be made by letting $\varepsilon_1 = \varepsilon_2 = \varepsilon$ and $\mu_1 \neq \mu_2$, in which case a Brewster angle is indeed found for the TE wave only.

We now have all the necessary pieces of information to write the solutions to the two problems of Figures 7.14 and 7.15 in the whole space. By combining (7.149), (7.159) and (7.160) in component form we find

$$E_y(x,z) = E^{\mathrm{i}}_{10}\left[\left(\mathrm{e}^{-\mathrm{j}k^{\mathrm{i}}_{z1}z} + R_{TE}\mathrm{e}^{\mathrm{j}k^{\mathrm{i}}_{z1}z}\right)U(-z) + T_{TE}\mathrm{e}^{-\mathrm{j}k^{\mathrm{t}}_{z2}z}U(z)\right]\mathrm{e}^{-\mathrm{j}k^{\mathrm{i}}_{x1}x} \tag{7.198}$$

$$H_x(x,z) = -E^{\mathrm{i}}_{10}\left[\frac{\mathrm{e}^{-\mathrm{j}k^{\mathrm{i}}_{z1}z} - R_{TE}\mathrm{e}^{\mathrm{j}k^{\mathrm{i}}_{z1}z}}{Z''_{\infty 1}}U(-z) - \frac{T_{TE}}{Z''_{\infty 2}}\mathrm{e}^{\mathrm{j}k^{\mathrm{t}}_{z2}z}U(z)\right]\mathrm{e}^{-\mathrm{j}k^{\mathrm{i}}_{x1}x} \tag{7.199}$$

$$H_z(x,z) = E^{\mathrm{i}}_{10}k^{\mathrm{i}}_{x1}\left[\frac{\mathrm{e}^{-\mathrm{j}k^{\mathrm{i}}_{z1}z} + R_{TE}\mathrm{e}^{\mathrm{j}k^{\mathrm{i}}_{z1}z}}{\omega\mu_1}U(-z) + \frac{T_{TE}}{\omega\mu_2}\mathrm{e}^{\mathrm{j}k^{\mathrm{t}}_{z2}z}U(z)\right]\mathrm{e}^{-\mathrm{j}k^{\mathrm{i}}_{x1}x} \tag{7.200}$$

for the TE polarization, and by combining (7.150), (7.170) and (7.171) we get

$$H_y(x,z) = H^{\mathrm{i}}_{10}\left[\left(\mathrm{e}^{-\mathrm{j}k^{\mathrm{i}}_{z1}z} + R_{TM}\mathrm{e}^{\mathrm{j}k^{\mathrm{i}}_{z1}z}\right)U(-z) + T_{TM}\mathrm{e}^{-\mathrm{j}k^{\mathrm{t}}_{z2}z}U(z)\right]\mathrm{e}^{-\mathrm{j}k^{\mathrm{i}}_{x1}x} \tag{7.201}$$

$$E_x(x,z) = H^{\mathrm{i}}_{10}\left[\frac{\mathrm{e}^{-\mathrm{j}k^{\mathrm{i}}_{z1}z} - R_{TM}\mathrm{e}^{\mathrm{j}k^{\mathrm{i}}_{z1}z}}{Y'_{\infty 1}}U(-z) + \frac{T_{TM}}{Y'_{\infty 2}}\mathrm{e}^{\mathrm{j}k^{\mathrm{t}}_{z2}z}U(z)\right]\mathrm{e}^{-\mathrm{j}k^{\mathrm{i}}_{x1}x} \tag{7.202}$$

$$E_z(x,z) = -H^{\mathrm{i}}_{10}k^{\mathrm{i}}_{x1}\left[\frac{\mathrm{e}^{-\mathrm{j}k^{\mathrm{i}}_{z1}z} + R_{TM}\mathrm{e}^{\mathrm{j}k^{\mathrm{i}}_{z1}z}}{\omega\varepsilon_1}U(-z) + \frac{T_{TM}}{\omega\varepsilon_2}\mathrm{e}^{\mathrm{j}k^{\mathrm{t}}_{z2}z}U(z)\right]\mathrm{e}^{-\mathrm{j}k^{\mathrm{i}}_{x1}x} \tag{7.203}$$

for the TM polarization, having explicitly used (7.165) and (7.178).

In preparation to the subject of Section 7.4.2 we wish to derive lumped-element circuit models of the problems of Figure 7.14 and 7.15. The present search for a circuit analogue partly justifies our choice of expressing the reflection coefficients as combinations of characteristic impedances and admittances in (7.190) and (7.191). Confessedly, a network equivalent is not really necessary for the relatively simple problem of reflection and transmission from an interface. However, this approach reveals its full potential when it is applied to far more complicated situations, such as the propagation in the presence of a layered medium, inasmuch as one can invoke well-established circuit concepts and techniques to solve a truly electromagnetic problem while, in fact, concealing the wave nature of the sought solution.

In short, the key step towards a network model of plane-wave reflection and transmission consists of casting the equations which relate tangential electric and magnetic fields on either side of the interface $z = 0$ into equivalent formats that we can construe, in a generalized way, as either the Kirchhoff voltage law (KVL) or the Kirchhoff current law (KCL). Indeed, by performing linear combinations of (7.167) and (7.168) on the one hand, and of (7.174) and (7.175) on the other we arrive at

$$2E^{\mathrm{i}}_{10} = Z''_{\infty 1}\frac{E^{\mathrm{t}}_{20}}{Z''_{\infty 2}} + E^{\mathrm{t}}_{20} \qquad \text{(KVL for TE waves)} \tag{7.204}$$

$$\frac{E^{\mathrm{i}}_{10} - E^{\mathrm{r}}_{10}}{Z''_{\infty 1}} = \frac{E^{\mathrm{t}}_{20}}{Z''_{\infty 2}} \qquad \text{(KCL for TE waves)} \tag{7.205}$$

and

$$\frac{H^{\mathrm{i}}_{10} - H^{\mathrm{r}}_{10}}{Y'_{\infty 1}} = \frac{H^{\mathrm{t}}_{20}}{Y'_{\infty 2}} \qquad \text{(KVL for TM waves)} \tag{7.206}$$

$$2H^{\mathrm{i}}_{10} = Y'_{\infty 1}\frac{H^{\mathrm{t}}_{20}}{Y'_{\infty 2}} + H^{\mathrm{t}}_{20} \qquad \text{(KCL for TM waves)} \tag{7.207}$$

where we have used (7.178) and definitions (7.187) and (7.188).

Next, we interpret the electric field coefficients as symbolic or *modal voltages* (physical dimension: V/m) and the magnetic field coefficients as symbolic or *modal currents* (physical dimension: A/m). The attribute modal is explained by observing that, since time-harmonic plane waves solve the source-free Maxwell equations, they represent, after all, the *eigenfunctions* or *modes* of a homogeneous unbounded medium (cf. Example 6.2 and Section 11.2). Proceeding with the circuit analogy we assign the various quantities in (7.204)-(7.207) the following meanings. For the TE polarization

- $2E^{\mathrm{i}}_{10}$ is the strength of a voltage generator with internal impedance $Z''_{\infty 1}$;
- $Z''_{\infty 2}$ represents the load connected in series to the generator;
- $E^{\mathrm{i}}_{10} - E^{\mathrm{r}}_{10}$ is the voltage drop across the internal impedance $Z''_{\infty 1}$;
- E^{t}_{20} is the voltage drop across the load;
- $E^{\mathrm{t}}_{20}/Z''_{\infty 1}$ is the current flowing through $Z''_{\infty 1}$ and the load;

whereas for the TM polarization

- $2H^{\mathrm{i}}_{10}$ is the strength of a current generator with internal admittance $Y'_{\infty 1}$;
- $Y'_{\infty 2}$ represents the load connected in parallel to the generator;
- $H^{\mathrm{i}}_{10} - H^{\mathrm{r}}_{10}$ is the current flowing through the internal admittance $Y'_{\infty 1}$;
- H^{t}_{20} is the current flowing through the load;
- $H^{\mathrm{t}}_{20}/Y'_{\infty 1}$ is the voltage drop across $Y'_{\infty 1}$ and the load.

The lumped-element circuits whose behavior is governed by (7.204)-(7.205) and (7.206)-(7.207) are sketched in Figures 7.18a and 7.18b; in many a respect, the two configurations are the dual of one another. We see that both $Z''_{\infty 2}$ and $Y'_{\infty 2}$ account for the effect of the medium which fills the half-space $z \geqslant 0$. So, the subscript '∞' has been included precisely as a reminder that the spatial region described by $Z''_{\infty 2}$ and $Y'_{\infty 2}$ is infinitely extended in particular along the positive z-direction. A similar consideration holds for $Z''_{\infty 1}$ and $Y'_{\infty 1}$, which — when the generators are turned off — represent the impedance and the admittance 'seen' in the section AA′ by 'looking' to the left in the setups of Figures 7.18a and 7.18b.

The distinctive advantage of the network equivalents lies in the possibility of analyzing the circuits rather than solving the original equations (7.167)-(7.168) and (7.174)-(7.175). For instance,

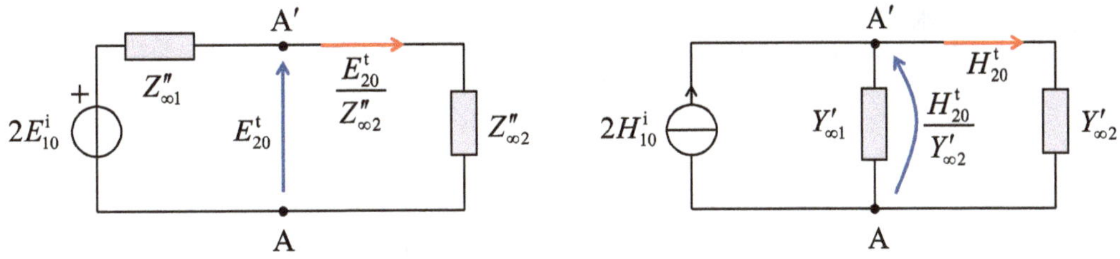

(a) *TE waves (Figure 7.14)* (b) *TM waves (Figure 7.15)*

Figure 7.18 Lumped-element circuit equivalents of plane-wave reflection and transmission at a planar interface.

by resorting to the formulas for voltage- and current-divider (which are known from network theory [40, 41]) we find immediately

$$E^{\rm t}_{20} = 2E^{\rm i}_{20}\frac{Z''_{\infty 2}}{Z''_{\infty 1} + Z''_{\infty 2}} \qquad H^{\rm t}_{20} = 2H^{\rm i}_{20}\frac{Y'_{\infty 2}}{Y'_{\infty 1} + Y'_{\infty 2}} \tag{7.208}$$

which coincide with the previous results in (7.185) and (7.186). In practice, when presented with an incident TE plane-wave in the half-space $z < 0$ we simply model the situation with a voltage generator plus its internal impedance in series or, if the wave is TM-polarized, with a current generator and its internal admittance in parallel. Likewise, the half-space $z > 0$, in which no sources exist, is readily modelled by means of just an impedance or an admittance, depending on the polarization. Incidentally, these recipes also tell us how to account for an incident wave that comes in from an infinitely remote source in medium 2: we just have to insert either a voltage or a current generator of strength $2E^{\rm i}_{20}$ or $2H^{\rm i}_{20}$ in the right parts of the circuits in Figures 7.18a and 7.18b.

The circuit analogy correctly applies even to the calculation of the average radiated power or, in this case, the average surface density thereof (Section 7.2). In particular, the power (density) absorbed by the load equals the power density carried over in medium 2 along the z-direction in the absence of TIR. For the TE polarization this means

$$\begin{aligned} p_R &= \frac{1}{2}\mathrm{Re}\{VI^*\} = \frac{1}{2}\mathrm{Re}\left\{E^{\rm t}_{20}\left(\frac{E^{\rm t}_{20}}{Z''_{\infty 2}}\right)^*\right\} = \frac{1}{2}\mathrm{Re}\left\{\frac{1}{Z''^*_{\infty 2}}\right\}|E^{\rm t}_{20}|^2 \\ &= \frac{1}{2}\frac{\mathrm{Re}\{Z''_{\infty 2}\}}{|Z''_{\infty 2}|^2}\frac{|Z''_{\infty 2}|^2}{|Z''_{\infty 1} + Z''_{\infty 2}|^2}4|E^{\rm i}_{20}|^2 = \frac{2\mathrm{Re}\{Z''_{\infty 2}\}}{|Z''_{\infty 1} + Z''_{\infty 2}|^2}|E^{\rm i}_{20}|^2 \end{aligned} \tag{7.209}$$

and one should check that p_R thus found indeed coincides with the result obtained through the complex Poynting vector in (7.66) with $\mathbf{E}_0 = E^{\rm t}_{20}\hat{\mathbf{y}}$, $\mathbf{k}' = \mathbf{k}^{\rm t}_2$ and $\hat{\mathbf{n}} = \hat{\mathbf{z}}$. The average power density in medium 2 clearly vanishes if TIR occurs, because $\mathrm{Re}\{Z''_{\infty 2}\} = 0$.

7.4.2 Network equivalent of a multi-layered medium

Next in line on the scale of complexity is the propagation of plane waves in the presence of a penetrable *slab* that is infinite in the x- and y-direction and extends between the planes $z = z_1$ to $z = z_2$ along z. To gain more generality, we suppose that the half-spaces to the left and to the right of the slab are filled with different isotropic homogeneous media which we label 1 and 3, whereas the slab material is denoted with 2. The electromagnetic problems for TE and TM polarizations are pictorially described in Figures 7.19 and 7.20.

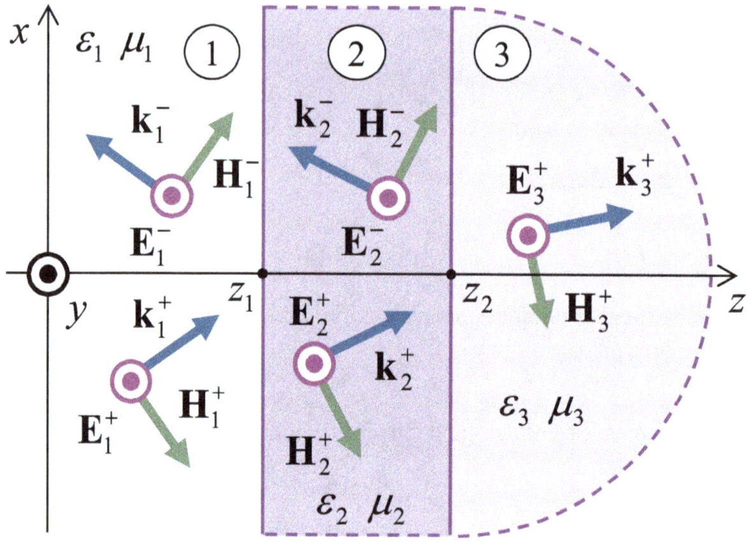

Figure 7.19 Reflection and transmission of uniform plane waves in the presence of a slab: transverse-electric (TE) polarization.

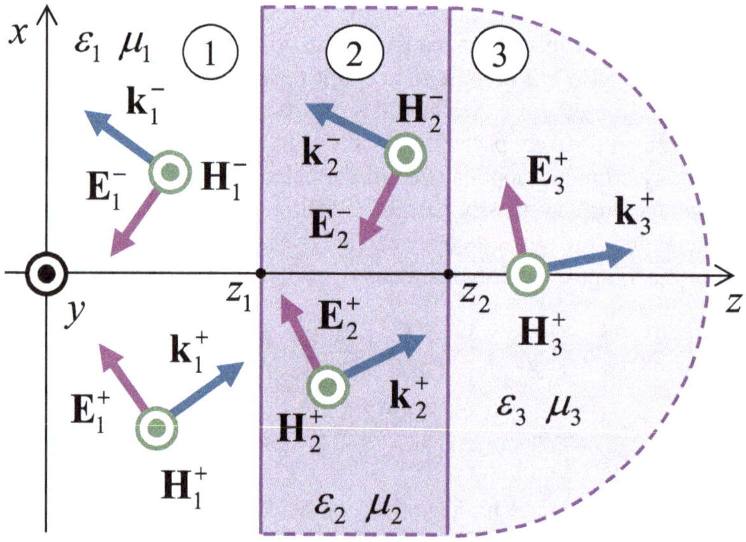

Figure 7.20 Reflection and transmission of uniform plane waves in the presence of a slab: transverse-magnetic (TM) polarization.

The incident plane waves — generated by some infinitely distant source located in medium 1 — are still given by (7.149) and (7.150) which we now know represent only a part of the total field for $z < z_1$. Indeed, from the viewpoint of an observer located somewhere in medium 1 a reflected wave is generated by equivalent sources that occupy the half-space $z > z_1$. Likewise, an observer sitting in medium 3 detects just one wave which he ascribes to equivalent sources in $z < z_2$, since he is not aware of any source in medium 1. Finally, an observer *inside* the slab may rearrange the Maxwell curl equations (1.99) and (1.98) for $z \in \mathbb{R} \setminus \{z_1, z_2\}$ as

$$\nabla \times \mathbf{E}(\mathbf{r}) = -\mathrm{j}\,\omega\mu_2\mathbf{H}(\mathbf{r}) \\ \underbrace{-\mathrm{j}\,\omega(\mu_1-\mu_2)U(z_1-z)\mathbf{H}(\mathbf{r})}_{\text{magnetic source in } z<z_1}\underbrace{-\mathrm{j}\,\omega(\mu_3-\mu_2)U(z-z_2)\mathbf{H}(\mathbf{r})}_{\text{magnetic source in } z>z_2} \tag{7.210}$$

$$\nabla \times \mathbf{H}(\mathbf{r}) = \mathrm{j}\,\omega\varepsilon_2\mathbf{E}(\mathbf{r}) \\ \underbrace{+\mathrm{j}\,\omega(\varepsilon_1-\varepsilon_2)U(z_1-z)\mathbf{E}(\mathbf{r})}_{\text{electric source in } z<z_1}\underbrace{+\mathrm{j}\,\omega(\varepsilon_3-\varepsilon_2)U(z-z_2)\mathbf{E}(\mathbf{r})}_{\text{electric source in } z>z_2} \tag{7.211}$$

and claim that the field she detects amounts, in fact, to the linear superposition of two plane waves produced by equivalent sources in $z < z_1$ and $z > z_2$, as indicated.

This preliminary qualitative analysis teaches us how to construct the complete solution. To be specific, we need a reflected wave in medium 1, a transmitted wave in medium 3 and a couple of waves travelling in opposite directions within the slab. Keeping that in mind, in principle we just have to write down all the waves formally and enforce the matching conditions (1.196) and (1.197) at the two material interfaces $z = z_1$ and $z = z_2$. Since each wave is determined up to an unknown multiplicative factor, by proceeding as we did in Section 7.4.1 we end up with two algebraic linear systems of rank four, one for each polarization. Admittedly, this approach becomes cumbersome and unpractical as soon as we consider more than one slab.

In fact, here is where a circuital view of the electromagnetic problems in Figures 7.19 and 7.20 comes in handy. Since we already obtained the pertinent circuits for a half-space containing the sources and, as a special case, a source-free half-space (Figures 7.18a and 7.18b) we may focus on deriving the network equivalent of the slab. Such model ought to take into account the finite thickness of the slab along z but also that waves can propagate from either interface towards the other. Therefore, unlike the model of a half-space — which, having only one interface, is described by a single-port lumped element — any physically sound equivalent circuit of the slab must be a two-port device. Depending upon what field entities we decide to sample at the ports, we obtain different sets of network parameters. We choose to keep track of the *total* electric and magnetic field components that are *tangential* to the interfaces $z = z_1$ and $z = z_2$. This is not a limitation because, once a set of network parameters has been found, we can invoke the well-known transformation rules for two-port devices to switch through alternative representations.

We start with the TE polarization and indicate the x-component of the wavevectors in all media just with k_x, because k_x is set by the incident wave in medium 1 and is unaffected by possible changes in the constitutive parameters [see (7.165)]. Having argued above that the total electromagnetic field within the slab is the sum of two plane waves, for $z \in [z_1, z_2]$ we make the Ansatz

$$\mathbf{E}_2(\mathbf{r}) = \hat{\mathbf{y}}\left[E_{20}^{+}\mathrm{e}^{-\mathrm{j}\,k_{z2}(z-z_2)} + E_{20}^{-}\mathrm{e}^{\mathrm{j}\,k_{z2}(z-z_2)}\right]\mathrm{e}^{-\mathrm{j}\,k_x x} \tag{7.212}$$

$$\mathbf{H}_2(\mathbf{r}) = -\hat{\mathbf{x}}\frac{1}{Z''_{\infty 2}}\left[E_{20}^{+}\mathrm{e}^{-\mathrm{j}\,k_{z2}(z-z_2)} - E_{20}^{-}\mathrm{e}^{\mathrm{j}\,k_{z2}(z-z_2)}\right]\mathrm{e}^{-\mathrm{j}\,k_x x} \\ + \hat{\mathbf{z}}\frac{k_x}{\omega\mu_2}\left[E_{20}^{+}\mathrm{e}^{-\mathrm{j}\,k_{z2}(z-z_2)} + E_{20}^{-}\mathrm{e}^{\mathrm{j}\,k_{z2}(z-z_2)}\right]\mathrm{e}^{-\mathrm{j}\,k_x x} \tag{7.213}$$

where k_{z2} is formally given by (7.179), $Z''_{\infty 2}$ is defined in the second part of (7.187), and the coefficients E_{20}^{+} and E_{20}^{-} are complex *unknown* constants. The expression of the total magnetic field within the slab can be constructed with the aid of (7.55) and (7.56) but, if in doubt, one can insert (7.212) into the local Faraday law (1.99) and perform a little algebra. Momentarily it will become apparent that we have taken the dependence on z in the form $z - z_2$ to facilitate the calculations.

Since we are interested in the field components *parallel* to the interfaces, and the x-dependence is already factored out and common to waves in media 1 and 3, we define two functions of z only, namely,

$$\hat{V}_2''(z) := E_{20}^{+}\mathrm{e}^{-\mathrm{j}k_{z2}(z-z_2)} + E_{20}^{-}\mathrm{e}^{\mathrm{j}k_{z2}(z-z_2)} \tag{7.214}$$

$$\hat{I}_2''(z) := \frac{E_{20}^{+}}{Z_{\infty 2}''}\mathrm{e}^{-\mathrm{j}k_{z2}(z-z_2)} - \frac{E_{20}^{-}}{Z_{\infty 2}''}\mathrm{e}^{\mathrm{j}k_{z2}(z-z_2)} \tag{7.215}$$

which we regard as the *symbolic* or *modal voltage* and *current* within the slab. It should be noted that these assumptions are consistent with the circuit viewpoint in (7.204) and (7.205), and the current $\hat{I}_2''(z)$ 'flows' in the *positive* z-direction. For our purposes it is expedient to re-write (7.214) and (7.215) by expanding the complex exponentials in terms of sines and cosines, namely,

$$\hat{V}_2''(z) = (E_{20}^{+} + E_{20}^{-})\cos[k_{z2}(z-z_2)] - \mathrm{j}(E_{20}^{+} - E_{20}^{-})\sin[k_{z2}(z-z_2)] \tag{7.216}$$

$$\hat{I}_2''(z) = \frac{E_{20}^{+} - E_{20}^{-}}{Z_{\infty 2}''}\cos[k_{z2}(z-z_2)] - \mathrm{j}\,\frac{E_{20}^{+} + E_{20}^{-}}{Z_{\infty 2}''}\sin[k_{z2}(z-z_2)] \tag{7.217}$$

which, when evaluated for $z = z_2$, provide

$$\hat{V}_2''(z_2) := V_2'' = E_{20}^{+} + E_{20}^{-} \qquad \hat{I}_2''(z_2) := I_2'' = \frac{1}{Z_{\infty 2}''}\left(E_{20}^{+} - E_{20}^{-}\right) \tag{7.218}$$

i.e., the modal voltage and current in the section $z = z_2$ or, with an eye towards the circuital description, at *port* 2. Thanks to these positions we arrive at

$$\hat{V}_2''(z) = V_2''\cos[k_{z2}(z-z_2)] - \mathrm{j}\,Z_{\infty 2}''I_2''\sin[k_{z2}(z-z_2)] \tag{7.219}$$

$$\hat{I}_2''(z) = -\mathrm{j}\,\frac{V_2''}{Z_{\infty 2}''}\sin[k_{z2}(z-z_2)] + I_2''\cos[k_{z2}(z-z_2)] \tag{7.220}$$

which, when evaluated now in $z = z_1$, yield

$$\hat{V}_2''(z_1) := V_1'' = V_2''\cos[k_{z2}(z_2-z_1)] + \mathrm{j}\,Z_{\infty 2}''I_2''\sin[k_{z2}(z_2-z_1)] \tag{7.221}$$

$$\hat{I}_2''(z_1) := I_1'' = \mathrm{j}\,\frac{V_2''}{Z_{\infty 2}''}\sin[k_{z2}(z_2-z_1)] + I_2''\cos[k_{z2}(z_2-z_1)] \tag{7.222}$$

that is, the modal voltage and current at port 1 as linear combinations of modal voltage and current at port 2. Therefore, (7.221) and (7.222) constitute the desired equivalent network description of the slab. We can write the governing equations succinctly as

$$[\Psi_1''] = [M''][\Psi_2''] \tag{7.223}$$

where the symbolic one-column vectors

$$[\Psi_1''] := \begin{pmatrix} V_1'' \\ I_1'' \end{pmatrix} \qquad [\Psi_2''] := \begin{pmatrix} V_2'' \\ I_2'' \end{pmatrix} \tag{7.224}$$

contain the *electrical state of the circuit* at ports 1 and 2, and

$$[M''] := \begin{pmatrix} \cos[k_{z2}(z_2-z_1)] & \mathrm{j}\,Z_{\infty 2}''\sin[k_{z2}(z_2-z_1)] \\ \dfrac{\mathrm{j}}{Z_{\infty 2}''}\sin[k_{z2}(z_2-z_1)] & \cos[k_{z2}(z_2-z_1)] \end{pmatrix} \tag{7.225}$$

is called the *chain matrix* or sometimes the *transmission matrix* of the two-port device equivalent to the slab for the TE polarization. The alternative designation, though, is likely to generate confusion, because the same name is also used for the matrix $[T]$ that relates incident and reflected fields — rather than the total ones — at the ports (e.g., [27, Chapter 4]).

The calculations for the TM case of Figure 7.20 differ only in the details. The total electromagnetic field within the slab reads

$$\mathbf{H}_2(\mathbf{r}) = \hat{\mathbf{y}}\left[H_{20}^{+}\mathrm{e}^{-\mathrm{j}k_{z2}(z-z_2)} + H_{20}^{-}\mathrm{e}^{\mathrm{j}k_{z2}(z-z_2)}\right]\mathrm{e}^{-\mathrm{j}k_x x} \tag{7.226}$$

$$\mathbf{E}_2(\mathbf{r}) = \hat{\mathbf{x}}\frac{1}{Y'_{\infty 2}}\left[H_{20}^{+}\mathrm{e}^{-\mathrm{j}k_{z2}(z-z_2)} - H_{20}^{-}\mathrm{e}^{\mathrm{j}k_{z2}(z-z_2)}\right]\mathrm{e}^{-\mathrm{j}k_x x} - \hat{\mathbf{z}}\frac{k_x}{\omega\varepsilon_2}\left[H_{20}^{+}\mathrm{e}^{-\mathrm{j}k_{z2}(z-z_2)} + H_{20}^{-}\mathrm{e}^{\mathrm{j}k_{z2}(z-z_2)}\right]\mathrm{e}^{-\mathrm{j}k_x x} \tag{7.227}$$

where $Y'_{\infty 2}$ is given by the second part of (7.188), and the coefficients H_{20}^{+} and H_{20}^{-} are complex *unknown* constants. The total electric field has been computed consistently by inserting (7.226) into the source-free instance of the local Ampère-Maxwell law (1.98). From the field components parallel to the interfaces we define two functions of z, viz.,

$$\hat{I}'_2(z) := H_{20}^{+}\mathrm{e}^{-\mathrm{j}k_{z2}(z-z_2)} + H_{20}^{-}\mathrm{e}^{\mathrm{j}k_{z2}(z-z_2)} \tag{7.228}$$

$$\hat{V}'_2(z) := \frac{H_{20}^{+}}{Y'_{\infty 2}}\mathrm{e}^{-\mathrm{j}k_{z2}(z-z_2)} - \frac{H_{20}^{-}}{Y'_{\infty 2}}\mathrm{e}^{\mathrm{j}k_{z2}(z-z_2)} \tag{7.229}$$

which we regard as *symbolic* or *modal current* and *voltage* for $z \in [z_1, z_2]$. These assumptions are compatible with the circuit viewpoint in (7.206) and (7.207), and $\hat{I}'_2(z)$ also 'flows' in the positive z-direction. By expanding the complex exponentials we arrive at

$$\hat{I}'_2(z) = (H_{20}^{+} + H_{20}^{-})\cos[k_{z2}(z-z_2)] - \mathrm{j}(H_{20}^{+} - H_{20}^{-})\sin[k_{z2}(z-z_2)] \tag{7.230}$$

$$\hat{V}'_2(z) = \frac{H_{20}^{+} - H_{20}^{-}}{Y'_{\infty 2}}\cos[k_{z2}(z-z_2)] - \mathrm{j}\,\frac{H_{20}^{+} + H_{20}^{-}}{Y'_{\infty 2}}\sin[k_{z2}(z-z_2)] \tag{7.231}$$

which, when evaluated in $z = z_2$, yield

$$\hat{I}'_2(z_2) := I'_2 = H_{20}^{+} + H_{20}^{-} \qquad \hat{V}'_2(z_2) := V'_2 = \frac{H_{20}^{+} - H_{20}^{-}}{Y'_{\infty 2}} \tag{7.232}$$

i.e., the modal current and voltage at *port* 2. With these definitions we find

$$\hat{I}'_2(z) = I'_2\cos[k_{z2}(z-z_2)] - \mathrm{j}\,Y'_{\infty 2}V'_2\sin[k_{z2}(z-z_2)] \tag{7.233}$$

$$\hat{V}'_2(z) = -\mathrm{j}\,\frac{I'_2}{Y'_{\infty 2}}\sin[k_{z2}(z-z_2)] + V'_2\cos[k_{z2}(z-z_2)] \tag{7.234}$$

which, when evaluated now in $z = z_1$, provide

$$\hat{I}'_2(z_1) := I'_1 = I'_2\cos[k_{z2}(z_2-z_1)] + \mathrm{j}\,Y'_{\infty 2}V'_2\sin[k_{z2}(z_2-z_1)] \tag{7.235}$$

$$\hat{V}'_2(z_1) := V'_1 = \mathrm{j}\,\frac{I'_2}{Y'_{\infty 2}}\sin[k_{z2}(z_2-z_1)] + V'_2\cos[k_{z2}(z_2-z_1)] \tag{7.236}$$

that is, the modal current and voltage at port 1 as linear combinations of modal current and voltage at port 2. Since the order of these governing equations is inconsequential, we can write (7.235) and (7.236) in compact form as

$$[\Psi'_1] = [M'][\Psi'_2] \tag{7.237}$$

where

$$[\Psi_1'] := \begin{pmatrix} V_1' \\ I_1' \end{pmatrix}, \qquad [\Psi_2'] := \begin{pmatrix} V_2' \\ I_2' \end{pmatrix} \tag{7.238}$$

and

$$[M'] := \begin{pmatrix} \cos[k_{z2}(z_2 - z_1)] & \dfrac{\mathrm{j}}{Y'_{\infty 2}} \sin[k_{z2}(z_2 - z_1)] \\ \mathrm{j}\, Y'_{\infty 2} \sin[k_{z2}(z_2 - z_1)] & \cos[k_{z2}(z_2 - z_1)] \end{pmatrix} \tag{7.239}$$

denotes the *chain matrix* of the slab for the TM polarization. Actually, the distinction is purely conceptual in that, if we agree to let $Z'_{\infty 2} := 1/Y'_{\infty 2}$, $[M']$ becomes formally identical with $[M'']$ in (7.225).

According to (7.223) and (7.237) the chain matrix provides the link between the *electrical state of the circuit* at port 2 and the state at port 1. However, this arrangement may sound odd because, when the source is located in medium 1, somehow we know or enforce the state at port 1 and wish to compute the state at port 2. In other words, at least for the setups of Figures 7.19 and 7.20 the chain matrix as it stands provides us with an *output-input* description of a linear system, rather than an *input-output* one, as it is more logical. In this regard, it is important to notice that, if in addition we contemplate a plane wave impinging on the slab from medium 3, such wave cannot but depend upon x through the multiplicative factor $\exp(-\mathrm{j}\, k_x x)$, since the latter was assumed to be common to all fields and then implied in the derivation.

By inspection of (7.225) and (7.239) we can ascertain that the determinant [16–18] of $[M'']$ and $[M']$ is equal to one. Still, this property can be proved *a priori* by applying the reciprocity theorem (6.256) in Section 6.8 to a cuboidal domain $V := \{\mathbf{r} \in \mathbb{R}^3 : x \in [x_1, x_2], y \in [y_1, y_2], z \in [z_1, z_2]\}$ entirely contained in the slab region. The boundary of V is comprised of six rectangles, say, $S_{x1}, S_{x2}, S_{y1}\ S_{y2}, S_{z1}$ and S_{z2}, that are perpendicular to the Cartesian axes. To be more specific, S_{x1} and S_{x2} lie on the planes $x = x_1$ and $x = x_2$, S_{y1} and S_{y2} on the planes $y = y_1$ and $y = y_2$, and finally S_{z1} and S_{z2} are contained in the planes $z = z_1$ and $z = z_2$. We assume that the sources are located at infinity in medium 1 for state (a) and at infinity in medium 3 for state (b). As a result, the reactions between fields and sources are null, whereby (6.256) reads

$$\begin{aligned} 0 = &\int_{S_{x2}} \mathrm{d}S\, \hat{\mathbf{x}} \cdot (\mathbf{E}_a \times \mathbf{H}_b - \mathbf{E}_b \times \mathbf{H}_a) - \int_{S_{x1}} \mathrm{d}S\, \hat{\mathbf{x}} \cdot (\mathbf{E}_a \times \mathbf{H}_b - \mathbf{E}_b \times \mathbf{H}_a) \\ &+ \int_{S_{y2}} \mathrm{d}S\, \hat{\mathbf{y}} \cdot (\mathbf{E}_a \times \mathbf{H}_b - \mathbf{E}_b \times \mathbf{H}_a) - \int_{S_{y1}} \mathrm{d}S\, \hat{\mathbf{y}} \cdot (\mathbf{E}_a \times \mathbf{H}_b - \mathbf{E}_b \times \mathbf{H}_a) \\ &+ \int_{S_{z2}} \mathrm{d}S\, \hat{\mathbf{z}} \cdot (\mathbf{E}_a \times \mathbf{H}_b - \mathbf{E}_b \times \mathbf{H}_a) - \int_{S_{z1}} \mathrm{d}S\, \hat{\mathbf{z}} \cdot (\mathbf{E}_a \times \mathbf{H}_b - \mathbf{E}_b \times \mathbf{H}_a) \end{aligned} \tag{7.240}$$

and we examine the surface integrals separately for, e.g., the TE polarization. From (7.212)-(7.215) — which hold true for both state (a) and state (b) — on S_{x1} we have

$$\begin{aligned} \hat{\mathbf{x}} \cdot [\mathbf{E}_a(\mathbf{r}) \times \mathbf{H}_b(\mathbf{r}) - \mathbf{E}_b(\mathbf{r}) \times \mathbf{H}_a(\mathbf{r})] &= E_{ya}(\mathbf{r}) H_{zb}(\mathbf{r}) - E_{yb}(\mathbf{r}) H_{za}(\mathbf{r}) \\ &= \frac{k_x}{\omega \mu_2} \hat{V}''_{2a}(z) \hat{V}''_{2b}(z) e^{-2\mathrm{j} k_x x_1} - \frac{k_x}{\omega \mu_2} \hat{V}''_{2b}(z) \hat{V}''_{2a}(z) e^{-2\mathrm{j} k_x x_1} = 0 \end{aligned} \tag{7.241}$$

and the same conclusion is reached for the integrand on S_{x2}. Even more simply, on S_{y1} and S_{y2} we have

$$\hat{\mathbf{y}} \cdot [\mathbf{E}_a(\mathbf{r}) \times \mathbf{H}_b(\mathbf{r}) - \mathbf{E}_b(\mathbf{r}) \times \mathbf{H}_a(\mathbf{r})] = 0 \tag{7.242}$$

because the electric field is parallel to $\hat{\mathbf{y}}$ for the TE polarization. Thirdly, on any plane perpendicular to the z-axis with $z \in [z_1, z_2]$ we have

$$\begin{aligned}\hat{\mathbf{z}} \cdot [\mathbf{E}_a(\mathbf{r}) \times \mathbf{H}_b(\mathbf{r}) - \mathbf{E}_b(\mathbf{r}) \times \mathbf{H}_a(\mathbf{r})] &= -E_{ya}(\mathbf{r})H_{xb}(\mathbf{r}) + E_{yb}(\mathbf{r})H_{xa}(\mathbf{r}) \\ &= \left[\hat{V}''_{2a}(z)\hat{I}''_{2b}(z) - \hat{V}''_{2b}(z)\hat{I}''_{2a}(z)\right] \mathrm{e}^{-2\mathrm{j}k_x x}\end{aligned} \tag{7.243}$$

whence we see that only the integrals over S_{z1} and S_{z2} may contribute. Thanks to these intermediate findings (7.240) simplifies to

$$\begin{aligned}0 = \left[\hat{V}''_{2a}(z_2)\hat{I}''_{2b}(z_2) - \hat{V}''_{2b}(z_2)\hat{I}''_{2a}(z_2)\right] \int_{S_{z2}} \mathrm{d}S\, \mathrm{e}^{-2\mathrm{j}k_x x} \\ - \left[\hat{V}''_{2a}(z_1)\hat{I}''_{2b}(z_1) - \hat{V}''_{2b}(z_1)\hat{I}''_{2a}(z_1)\right] \int_{S_{z1}} \mathrm{d}S\, \mathrm{e}^{-2\mathrm{j}k_x x}\end{aligned} \tag{7.244}$$

and, since the integrals of $\exp(2\mathrm{j}k_x x)$ over S_{z1} and S_{z2} always coincide and are non-zero in general — for $k_x \in \mathbb{R}_0^+$ they vanish if $k_x(x_2 - x_1) = \pi n$, $n \in \mathbb{N} \setminus \{0\}$ — this equation requires

$$V''_{a2}I''_{b2} - V''_{b2}I''_{a2} = V''_{a1}I''_{b1} - V''_{b1}I''_{a1} \tag{7.245}$$

where V''_{a2} means 'modal voltage at port 2 due to the source of state (a)' and so forth with evident notation. The algebraic relationship (7.245) is a statement of reciprocity for the slab in terms of equivalent network quantities.

Now we finalize the definitions of state (a) and state (b) by choosing the material properties of media 1 and 3. One of the simplest configurations consists of assuming that in state (a) the half-space $z > z_2$ is occupied by a PEC, whereas in state (b) the half-space $z < z_1$ in turn is filled with a PEC. In light of (7.212) and (7.214) these additional hypotheses imply

$$V''_{a2} = 0 \qquad\qquad V''_{b1} = 0 \tag{7.246}$$

because the electric-field component tangential to a PEC surface vanishes [see (1.169)]. Keeping this into account, from (7.223) and the inverse relationship thereof we also get

$$V''_{a1} = M''_{12}I''_{a2} \qquad\qquad V''_{b2} = -\frac{M''_{12}}{\det([M''])}I''_{b1} \tag{7.247}$$

where M''_{12} denotes the entry of $[M'']$ in the first row and second column. Inserting all this information back into (7.245) leads to

$$\frac{M''_{12}}{\det([M''])}I''_{b1}I''_{a2} = M''_{12}I''_{a2}I''_{b1} \tag{7.248}$$

and this equation can be true for arbitrary non-null values of the modal currents only if

$$\det([M'']) = 1 \tag{7.249}$$

as we wanted to prove. Alternatively, we may suppose that in state (a) the half-space $z > z_2$ is filled with a PMC (Section 6.6), whereas in state (b) the half-space $z < z_1$ in turn is occupied by a PMC. Then, from (7.213) and (7.215) we conclude that

$$I''_{a2} = 0 \qquad\qquad I''_{b1} = 0 \tag{7.250}$$

because the magnetic-field component tangential to a PMC surface is null [see (6.226)]. With this position, from (7.237) and the inverse relationship thereof we also find

$$I''_{a1} = M''_{21} V''_{a2} \qquad\qquad I''_{b2} = -\frac{M''_{21}}{\det([M''])} V''_{b1} \tag{7.251}$$

where M''_{21} denotes the entry of $[M'']$ in the second row and first column. Substituting all these particular values into (7.245) yields

$$-\frac{M''_{21}}{\det([M''])} V''_{a2} V''_{b1} = -M''_{21} V''_{a2} V''_{b1} \tag{7.252}$$

which again may hold true for arbitrary non-zero values of the modal voltages only if (7.249) is satisfied. It should be noted that the actual form of the entries of $[M'']$ has played no role in the proof. Moreover, with similar steps we can show that (7.249) must be true also for $[M']$.

The chain matrices depend on the constitutive parameters ε_2 and μ_2 through k_{z2} and the characteristic impedances $Z''_{\infty 2}$ and $Z'_{\infty 2}$. By construction the difference $z_2 - z_1$ equals the thickness of the slab along z. Indeed, if we let $z_2 - z_1 = 0$, both $[M'']$ and $[M']$ degenerate into the identity matrix of rank two. From a practical standpoint a slab of zero thickness is indistinguishable from a single interface between two media, in which instance (7.223) and (7.237) correctly state that voltages and currents are *continuous*.

Further, it can be shown that a piece of uniform classical transmission line [2, 5, 6, 8, 10, 14, 21, 23–25, 42–44] of length $z_2 - z_1$ is also described by a chain matrix which has the same structure as that of $[M'']$ and $[M']$. Therefore, pushing forward our circuital analogy, we may claim that the network equivalent of the slab is a length of transmission line with propagation constant k_{z2} and characteristic impedance $Z''_{\infty 2}$ or $Z'_{\infty 2}$. This is no more than an insightful interpretation, though, since at this stage $[M'']$ and $[M']$ already contain all the information we need to solve the propagation problems.

Having derived the chain matrix of a penetrable slab, we may assemble the electric circuits equivalent to the electromagnetic problems of Figures 7.19 and 7.20 by using the models for half-spaces that we obtained in the previous section. For the TE polarization, assuming an incident wave in the form

$$\mathbf{E}_1^+(\mathbf{r}) = E_{10}^+ \hat{\mathbf{y}} e^{-j\mathbf{k}_1^+ \cdot \mathbf{r}}, \qquad \mathbf{H}_1^+(\mathbf{r}) = \frac{\mathbf{k}_1^+}{\omega\mu_1} \times \hat{\mathbf{y}} E_{10}^+ e^{-j\mathbf{k}_1^+ \cdot \mathbf{r}}, \qquad z < z_1 \tag{7.253}$$

we model medium 1 and the impinging wave by means of a voltage generator of strength $2E_{10}^+$ and internal impedance $Z''_{\infty 1}$ connected to port 1 of the device equivalent to the slab, whereas we replace the source-free medium 3 with its characteristic impedance $Z''_{\infty 3}$ connected to port 2. The resulting circuit is sketched in Figure 7.21a. Similar considerations for the TM polarization lead us to the circuit of Figure 7.21b.

It is rather obvious that the slabs of Figures 7.19 and 7.20 'look the same' when viewed from the left or from the right along the z-direction. Consequently, if we subject the slabs to a one-hundred-eighty-degree rotation around the x-axis (or the y-axis for that matter) the electromagnetic problems of concern are not modified. From a circuit viewpoint this property means that the two-port device

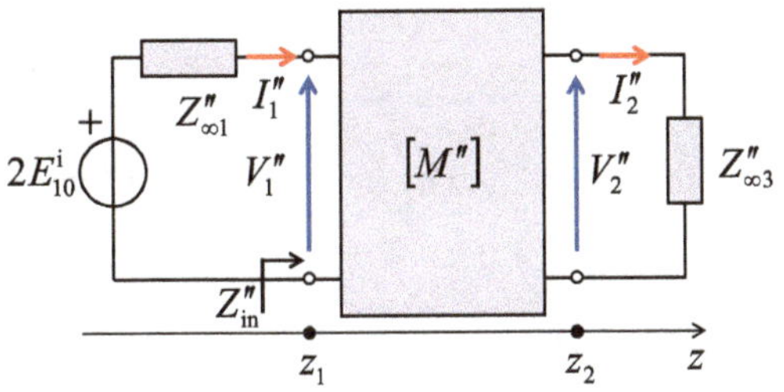

(a) *TE waves (Figure 7.19)*

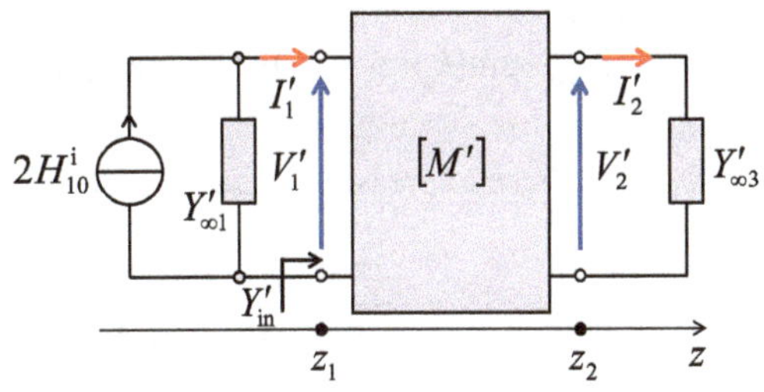

(b) *TM waves (Figure 7.20)*

Figure 7.21 Lumped-element circuit equivalents of plane-wave reflection and transmission in the presence of a slab.

equivalent to the slab is *symmetric*, that is, one can flip over the device to exchange port 1 and port 2 and connect it back reversed between sections $z = z_1$ and $z = z_2$ in Figures 7.21a and 7.21b without affecting the behavior of the network in the slightest. To find out the additional constraints that symmetry requires of the entries of $[M'']$ and $[M']$, we consider (7.223) and observe that swapping port 1 and port 2 leads to the set of governing equations

$$\begin{pmatrix} V''_2 \\ -I''_2 \end{pmatrix} = \begin{pmatrix} M''_{11} & M''_{12} \\ M''_{21} & M''_{22} \end{pmatrix} \begin{pmatrix} V''_1 \\ -I''_1 \end{pmatrix} \qquad \Longrightarrow \qquad \begin{pmatrix} V''_2 \\ I''_2 \end{pmatrix} = \begin{pmatrix} M''_{11} & -M''_{12} \\ -M''_{21} & M''_{22} \end{pmatrix} \begin{pmatrix} V''_1 \\ I''_1 \end{pmatrix} \tag{7.254}$$

where the minus signs result from the fact that I''_1 flows into port 1 and I''_2 flows out of port 2. Since the two-port device is still described by (7.223), the latter can be used to eliminate V''_1 and I''_1 from (7.254) and obtain the identity

$$\begin{pmatrix} V''_2 \\ I''_2 \end{pmatrix} = \begin{pmatrix} M''_{11} & -M''_{12} \\ -M''_{21} & M''_{22} \end{pmatrix} \begin{pmatrix} M''_{11} & M''_{12} \\ M''_{21} & M''_{22} \end{pmatrix} \begin{pmatrix} V''_2 \\ I''_2 \end{pmatrix} = \begin{pmatrix} M''^2_{11} - M''_{12}M''_{21} & (M''_{11} - M''_{22})M''_{12} \\ (M''_{22} - M''_{11})M''_{21} & M''^2_{22} - M''_{12}M''_{21} \end{pmatrix} \begin{pmatrix} V''_2 \\ I''_2 \end{pmatrix} \tag{7.255}$$

which for arbitrary non-zero values of V''_2 and I''_2 can only be true if we demand

$$M''_{11} = M''_{22} \qquad\qquad \det[M''] = 1 \tag{7.256}$$

whereby we conclude that *physical symmetry* implies *reciprocity*, but the converse is not true. Indeed, we see from (7.225) and (7.239) that (7.256) is satisfied. It is worthwhile remarking that in this case physical symmetry does not mean that a chain matrix is symmetric (cf. Example 6.6)!

We continue the discussion by solving the circuit of Figure 7.21a to determine the electrical states at ports 1 and 2. The solution strategy consists of three main steps, viz.,

(i) finding the input impedance $Z''_{\rm in}$ at port 1, that is, the impedance 'seen' by an ideal voltage generator connected directly to port 1;
(ii) computing V''_1 and I''_1,
(iii) inverting $[M'']$ and obtaining V''_2 and I''_2.

Indeed, since port 2 is closed on $Z''_{\infty 3}$ and the modal current I''_2 flows into $Z''_{\infty 3}$ as indicated, from the Ohm law we have

$$V''_2 = Z''_{\infty 3} I''_2 \tag{7.257}$$

which, once inserted into (7.223), provides us with

$$V''_1 = \left\{ Z''_{\infty 3} \cos[k_{z2}(z_2 - z_1)] + {\rm j}\, Z''_{\infty 2} \sin[k_{z2}(z_2 - z_1)] \right\} I''_2 \tag{7.258}$$

$$I_1'' = \left\{ \mathrm{j}\,\frac{Z_{\infty 3}''}{Z_{\infty 2}''}\sin[k_{z2}(z_2 - z_1)] + \cos[k_{z2}(z_2 - z_1)] \right\} I_2'' \tag{7.259}$$

and dividing these expressions side by side to eliminate the yet unknown common factor I_2'' yields

$$Z_{\mathrm{in}}'' := \frac{V_1''}{I_1''} = Z_{\infty 2}'' \frac{Z_{\infty 3}''\cos[k_{z2}(z_2 - z_1)] + \mathrm{j}\,Z_{\infty 2}''\sin[k_{z2}(z_2 - z_1)]}{Z_{\infty 2}''\cos[k_{z2}(z_2 - z_1)] + \mathrm{j}\,Z_{\infty 3}''\sin[k_{z2}(z_2 - z_1)]} \tag{7.260}$$

the desired input impedance at port 1. As a sanity check, we observe that, if the slab and the half-space $z > z_2$ were filled with the same material, then Z_{in}'' ought to reduce to $Z_{\infty 2}''$, i.e., the characteristic impedance of a half-space filled with medium 2. Letting $Z_{\infty 2}'' = Z_{\infty 3}''$ in (7.260) confirms that this expectation is indeed correct. From the generator's perspective, Z_{in}'' plays the role of a load connected in series to $Z_{\infty 1}''$ and perfectly equivalent to the combination of $[M'']$ and $Z_{\infty 3}''$. Therefore, by solving the circuit in the section $z = z_1$ we find the electrical state of the circuit at port 1, viz.,

$$I_1'' = \frac{2E_{10}^+}{Z_{\infty 1}'' + Z_{\mathrm{in}}''} \qquad V_1'' = 2E_{10}^+ \frac{Z_{\mathrm{in}}''}{Z_{\infty 1}'' + Z_{\mathrm{in}}''}. \tag{7.261}$$

which in tandem with the inverse of (7.223) give

$$V_2'' = 2E_{10}^+ \frac{Z_{\mathrm{in}}''\cos[k_{z2}(z_2 - z_1)] - \mathrm{j}\,Z_{\infty 2}''\sin[k_{z2}(z_2 - z_1)]}{Z_{\mathrm{in}}'' + Z_{\infty 1}''} \tag{7.262}$$

$$I_2'' = \frac{2E_{10}^+}{Z_{\infty 2}''} \frac{Z_{\infty 2}''\cos[k_{z2}(z_2 - z_1)] - \mathrm{j}\,Z_{\mathrm{in}}''\sin[k_{z2}(z_2 - z_1)]}{Z_{\infty 1}'' + Z_{\mathrm{in}}''} \tag{7.263}$$

although we may obtain either V_2'' or I_2'' from the other and the Ohm law (7.257). This completes the analysis of the equivalent circuit of Figure 7.21a.

Our ultimate goal, though, is the determination of the field reflected in medium 1, the field transmitted in medium 3, and the total field within the slab, whereas the circuital model is just a convenient tool. Recalling that Z_{in}'' accounts for the cascade of the slab and medium 3, with the help of definition (7.190) we write straightaway

$$R_{TE} := \frac{Z_{\mathrm{in}}'' - Z_{\infty 1}''}{Z_{\mathrm{in}}'' + Z_{\infty 1}''} \tag{7.264}$$

i.e., the reflection coefficient at the interface $z = z_1$. Thus, the reflected wave in medium 1 reads

$$\mathbf{E}_1^-(\mathbf{r}) = R_{TE}E_{10}^+\hat{\mathbf{y}}e^{-\mathrm{j}\mathbf{k}_1^-\cdot\mathbf{r}}, \qquad \mathbf{H}_1^-(\mathbf{r}) = R_{TE}\frac{\mathbf{k}_1^-}{\omega\mu_1} \times \hat{\mathbf{y}}E_{10}^+e^{-\mathrm{j}\mathbf{k}_1^-\cdot\mathbf{r}}, \qquad z < z_1 \tag{7.265}$$

in accordance with (7.159). Secondly, since voltages and currents are continuous across section $z = z_2$ and there exists only one wave in medium 3, we observe that

$$V_2'' = E_{30}^+ \qquad I_2'' = \frac{V_2''}{Z_{\infty 3}''} = \frac{E_{30}^+}{Z_{\infty 3}''} = -H_{30}^+ \tag{7.266}$$

whence, as suggested by (7.160), we can write

$$\mathbf{E}_3^+(\mathbf{r}) = V_2''\hat{\mathbf{y}}e^{-\mathrm{j}\mathbf{k}_3^+\cdot\mathbf{r}}, \qquad \mathbf{H}_3^+(\mathbf{r}) = V_2''\frac{\mathbf{k}_1^-}{\omega\mu_3} \times \hat{\mathbf{y}}e^{-\mathrm{j}\mathbf{k}_3^+\cdot\mathbf{r}}, \qquad z \geqslant z_2 \tag{7.267}$$

with V_2'' given explicitly by (7.262). Finally, in order to determine the total field within the slab we insert (7.262) and (7.263) into (7.219) and (7.220) and, after performing a little algebra, we get

$$\hat{V}_2''(z) = 2E_{10}^+ \frac{Z_{\text{in}}'' \cos[k_{z2}(z - z_1)] - \mathrm{j}\, Z_{\infty 2}'' \sin[k_{z2}(z - z_1)]}{Z_{\infty 1}'' + Z_{\text{in}}''} \tag{7.268}$$

$$\hat{I}_2''(z) = \frac{2E_{10}^+}{Z_{\infty 2}''} \frac{Z_{\infty 2}'' \cos[k_{z2}(z - z_1)] - \mathrm{j}\, Z_{\text{in}}'' \sin[k_{z2}(z - z_1)]}{Z_{\infty 1}'' + Z_{\text{in}}''} \tag{7.269}$$

whose striking similarity to (7.262) and (7.263) — they differ only in z_2 being traded for the z-coordinate — should not go unnoticed. In fact, this is a consequence of the chain-matrix representation of the slab. Knowing the modal voltage and current for $z \in [z_1, z_2]$ allows us to express the electromagnetic field within the slab as

$$\mathbf{E}_2(\mathbf{r}) = 2E_{10}^+ \frac{Z_{\text{in}}'' \cos[k_{z2}(z - z_1)] - \mathrm{j}\, Z_{\infty 2}'' \sin[k_{z2}(z - z_1)]}{Z_{\infty 1}'' + Z_{\text{in}}''} \mathrm{e}^{-\mathrm{j}\, k_x x} \hat{\mathbf{y}} \tag{7.270}$$

$$\begin{aligned} \mathbf{H}_2(\mathbf{r}) = &- \frac{2E_{10}^+}{Z_{\infty 2}''} \frac{Z_{\infty 2}'' \cos[k_{z2}(z - z_1)] - \mathrm{j}\, Z_{\text{in}}'' \sin[k_{z2}(z - z_1)]}{Z_{\infty 1}'' + Z_{\text{in}}''} \mathrm{e}^{-\mathrm{j}\, k_x x} \hat{\mathbf{x}} \\ &+ 2E_{10}^+ \frac{k_x}{\omega \mu_2} \frac{Z_{\text{in}}'' \cos[k_{z2}(z - z_1)] - \mathrm{j}\, Z_{\infty 2}'' \sin[k_{z2}(z - z_1)]}{Z_{\infty 1}'' + Z_{\text{in}}''} \mathrm{e}^{-\mathrm{j}\, k_x x} \hat{\mathbf{z}} \end{aligned} \tag{7.271}$$

by virtue of (7.212)-(7.215). To shed light on the physical meaning of this result we substitute the explicit expression of Z_{in}'' from (7.260) into (7.270) and we also expand sines and cosines in terms of complex exponentials.

After lengthy and not particularly instructive manipulations we arrive at an alternative formula for the electric field, viz.,

$$\mathbf{E}_2(\mathbf{r}) = T_{12}'' E_{10}^+ \frac{\mathrm{e}^{-\mathrm{j}\, k_{z2}(z - z_1)} + R_{32}'' \mathrm{e}^{\mathrm{j}\, k_{z2}(z - z_2)} \mathrm{e}^{-\mathrm{j}\, k_{z2}(z_2 - z_1)}}{1 - R_{12}'' R_{32}'' \mathrm{e}^{-\mathrm{j}\, 2k_{z2}(z_2 - z_1)}} \mathrm{e}^{\mathrm{j}\, k_x x} \hat{\mathbf{y}} \tag{7.272}$$

where

$$T_{12}'' := \frac{2Z_{\infty 2}''}{Z_{\infty 1}'' + Z_{\infty 2}''}, \qquad R_{12}'' := \frac{Z_{\infty 1}'' - Z_{\infty 2}''}{Z_{\infty 1}'' + Z_{\infty 2}''}, \qquad R_{32}'' := \frac{Z_{\infty 3}'' - Z_{\infty 2}''}{Z_{\infty 3}'' + Z_{\infty 2}''} \tag{7.273}$$

are suitable transmission and reflection coefficients at the material interfaces $z = z_1$ and $z = z_2$. More specifically,

- T_{12}'' gives the normalized amplitude of the electric field of a TE plane wave transmitted from medium 1 into the slab, *as if* the thickness of the slab were infinite or, equivalently, in the absence of the material interface at $z = z_2$,
- R_{12}'' gives the normalized amplitude of the electric field of a TE plane wave reflected inwards the slab by the interface with medium 1,
- R_{32}'' gives the normalized amplitude of the electric field of a TE plane wave reflected inwards the slab by the interface with medium 3.

All these definitions are consistent with (7.190). Moreover, R_{12}'' and R_{TE} in (7.190) are the negative of each other, the reason being that they provide the amplitude of waves reflected in opposite directions with respect to the planar interface between media 1 and 2. Last but not least, the electric field in (7.272) has precisely the structure we supposed in (7.212), with the unknown amplitudes E_{20}^+ and E_{20}^- explicitly computed.

Next, when k_{z2} is real so is $Z''_{\infty 2}$, and the magnitude of $R''_{12}R''_{32}\exp[-\mathrm{j}\,2k_{z2}(z_2 - z_1)]$ is smaller than one. Thus, we may invoke the Neumann series [20, Formula 9.04], [45, Formula 0.231], [46, Formula 3.1.10] for the denominator of (7.272) and obtain the representation

$$\mathbf{E}_2(\mathbf{r}) = T''_{12}E^+_{10}\left[\mathrm{e}^{-\mathrm{j}\,k_{z2}(z-z_1)} + R''_{32}\mathrm{e}^{\mathrm{j}\,k_{z2}(z-z_2)}\mathrm{e}^{-\mathrm{j}\,k_{z2}(z_2-z_1)}\right] \times \sum_{n=0}^{+\infty}\left(R''_{12}R''_{32}\right)^n \mathrm{e}^{-\mathrm{j}\,2k_{z2}(z_2-z_1)n}\mathrm{e}^{-\mathrm{j}\,k_x x}\hat{\mathbf{y}} \quad (7.274)$$

which manifestly exhibits the form of an *infinite superposition* of TE plane waves of ever decreasing amplitude that carom back and forth between the two interfaces. To elaborate, we write down the first terms of the series explicitly, viz.,

$$\begin{aligned}\mathbf{E}_2(\mathbf{r}) = {} & \underbrace{T''_{12}\mathrm{e}^{-\mathrm{j}\,k_{z2}(z-z_1)}\mathrm{e}^{-\mathrm{j}\,k_x x}E^+_{10}\hat{\mathbf{y}}}_{\text{transmission across } z=z_1} + \underbrace{T''_{12}R''_{32}\mathrm{e}^{\mathrm{j}\,k_{z2}(z-z_2)}\mathrm{e}^{-\mathrm{j}\,k_{z2}(z_2-z_1)}\mathrm{e}^{-\mathrm{j}\,k_x x}E^+_{10}\hat{\mathbf{y}}}_{\text{first reflection at } z=z_2} \\ & + \underbrace{T''_{12}R''_{12}R''_{32}\mathrm{e}^{-\mathrm{j}\,k_{z2}(z-z_1)}\mathrm{e}^{-\mathrm{j}\,2k_{z2}(z_2-z_1)}\mathrm{e}^{-\mathrm{j}\,k_x x}E^+_{10}\hat{\mathbf{y}}}_{\text{second reflection at } z=z_1} \\ & + \underbrace{T''_{12}R''_{12}(R''_{32})^2\mathrm{e}^{\mathrm{j}\,k_{z2}(z-z_2)}\mathrm{e}^{-\mathrm{j}\,3k_{z2}(z_2-z_1)}\mathrm{e}^{-\mathrm{j}\,k_x x}E^+_{10}\hat{\mathbf{y}}}_{\text{third reflection at } z=z_2} + \cdots \end{aligned} \quad (7.275)$$

which we interpret as follows (cf. with Figure 12.12 in Section 12.5)

- the first contribution to the field is the wave *transmitted* from medium 1 into the slab, *as if* the second material interface in $z = z_2$ were absent, since T''_{12} does not account for medium 3;
- the second wave 'travels' leftwards and is due to the *reflection* of the first at the material interface between the slab and medium 3;
- the third wave 'advances' rightwards and arises from the *reflection* of the second at the interface between the slab and medium 1;
- the fourth wave 'travels' again to the left and is caused by the *reflection* of the third at the interface between the slab and medium 3, and so on so forth.

After each reflection the amplitude of the wave that 'emerges' from either interface becomes smaller because $R''_{12} < 1$ and $R''_{32} < 1$. For obvious reasons this explanation is termed a *multiple-reflections expansion* of the field. Come to think of it, we could have adopted this point of view and started with the infinite sum (7.275) to find the field within the slab. Besides, since at each reflection also refracted waves are produced either in medium 1 or medium 3, we would end up with analogous expansions that, once summed, yield the reflected wave (7.265) and the refracted one (7.267).

We infer from (7.209) that the average power absorbed by the load at port 2 in the circuit of Figure 7.21a equals the surface density of average power (7.66) flowing through the interface $z = z_2$ into medium 3. Thus, to find the power that, after travelling in the slab, is transmitted into medium 3, on account of (7.257) we write

$$p_R = \frac{1}{2}\mathrm{Re}\left\{V''_2 I''^*_2\right\} = \frac{1}{2}\mathrm{Re}\left\{\frac{1}{Z''^*_{\infty 3}}\right\}|V''_2|^2 = \frac{1}{2}\mathrm{Re}\left\{\frac{1}{Z''^*_{\infty 3}}\right\}\frac{|T''_{12}|^2|T''_{23}|^2\left|\mathrm{e}^{-\mathrm{j}\,2k_{z2}(z_2-z_1)}\right|}{\left|1 - R''_{12}R''_{32}\mathrm{e}^{-\mathrm{j}\,2k_{z2}(z_2-z_1)}\right|^2}|E^+_{10}|^2 \quad (7.276)$$

where we have determined the modal voltage V''_2 from (7.272) evaluated in $z = z_2$, and

$$T''_{23} := 1 + R''_{32} = \frac{2Z''_{\infty 3}}{Z''_{\infty 2} + Z''_{\infty 3}} \quad (7.277)$$

is the transmission coefficient at interface $z = z_2$ from the slab into medium 3. Formula (7.262) may be used as well, but that expression does not serves our present purposes well because the dependence upon $k_{z2}(z_2 - z_1)$ is partly concealed in Z''_{in}.

First and foremost, if the incident plane wave (7.253) impinges against the slab at such an angle that

$$k_x \leqslant \omega\sqrt{\varepsilon_3\mu_3} \tag{7.278}$$

then k_{z3} and hence $Z''_{\infty3}$ are real [cf. with (7.179) and (7.187)], and (7.276) tells us that a certain amount of power, albeit small, is *invariably* transferred to medium 3! This conclusion may not come as a surprise at all when k_{z2} is real, too, that is, if

$$k_x \leqslant \omega\sqrt{\varepsilon_2\mu_2} \leqslant \omega\sqrt{\varepsilon_3\mu_3} \tag{7.279}$$

because then no TIR occurs at the interface $z = z_1$, in which case (7.276) yields

$$p_R = \frac{|E^+_{10}|^2}{2Z''_{\infty3}} \frac{(T''_{12})^2(T''_{23})^2}{1 - 2R''_{12}R''_{32}\cos[2k_{z2}(z_2 - z_1)] + (R''_{12}R''_{32})^2}. \tag{7.280}$$

However, even though the incident wave does undergo TIR at the interface $z = z_1$, the power transfer through the slab is *not completely eliminated*, as one may incorrectly reckon on the grounds of the previous result (7.209). Indeed, by assuming

$$\omega\sqrt{\varepsilon_2\mu_2} \leqslant k_x \leqslant \omega\sqrt{\varepsilon_3\mu_3} \tag{7.281}$$

and letting for simplicity

$$k_{z2} = -\mathrm{j}\,|k_{z2}| \in \mathbb{I} \tag{7.282}$$

we can cast (7.276) as (notice that T''_{12}, T''_{23}, R''_{12} and R''_{32} are all complex now)

$$p_R = \frac{|E^+_{10}|^2}{2Z''_{\infty3}} \frac{|T''_{12}|^2|T''_{23}|^2\mathrm{e}^{-2|k_{z2}|(z_2-z_1)}}{1 - 2\mathrm{Re}\left\{R''_{12}R''_{23}\right\}\mathrm{e}^{-2|k_{z2}|(z_2-z_1)} + \mathrm{e}^{-4|k_{z2}|(z_2-z_1)}} > 0 \tag{7.283}$$

where in the denominator we have also used the fact that $|R''_{12}| = 1 = |R''_{32}|$.

This phenomenon can be explained in two ways. From (7.260) we gather that the input impedance Z''_{in} retains a real part under hypothesis (7.281). Therefore, in the circuit of Figure 7.21a from the generator's viewpoint the equivalent load Z''_{in} is invariably lossy, and hence power is necessarily absorbed. In terms of electromagnetic fields we see from (7.212) and (7.213) that, when k_{z2} is imaginary, two *inhomogeneous* (evanescent) plane waves exist in the slab. One wave falls off exponentially away from the section $z = z_1$ and into medium 2; this is what we expect based on the knowledge gained in Section 7.4.1 on TIR at an interface between two half-spaces. The second wave *grows exponentially* towards the section $z = z_2$, and it is associated with the uniform wave transmitted into medium 3. The resulting *interference* of two inhomogeneous waves — which individually do not carry power along the z-direction — causes active power to flow into medium 3 nonetheless. The situation is sometimes called *frustrated total internal reflection* (FTIR) presumably because the second interface thwarts the 'shielding' effect of the first. A practical setup for the experimental demonstration of FTIR with light rays is sketched in [3, Section 33-6].

Does this puzzling and worrisome occurrence of an exponentially growing wave imply that the electric field within the slab may grow infinitely large, if so does the thickness $z_2 - z_1$? As a matter

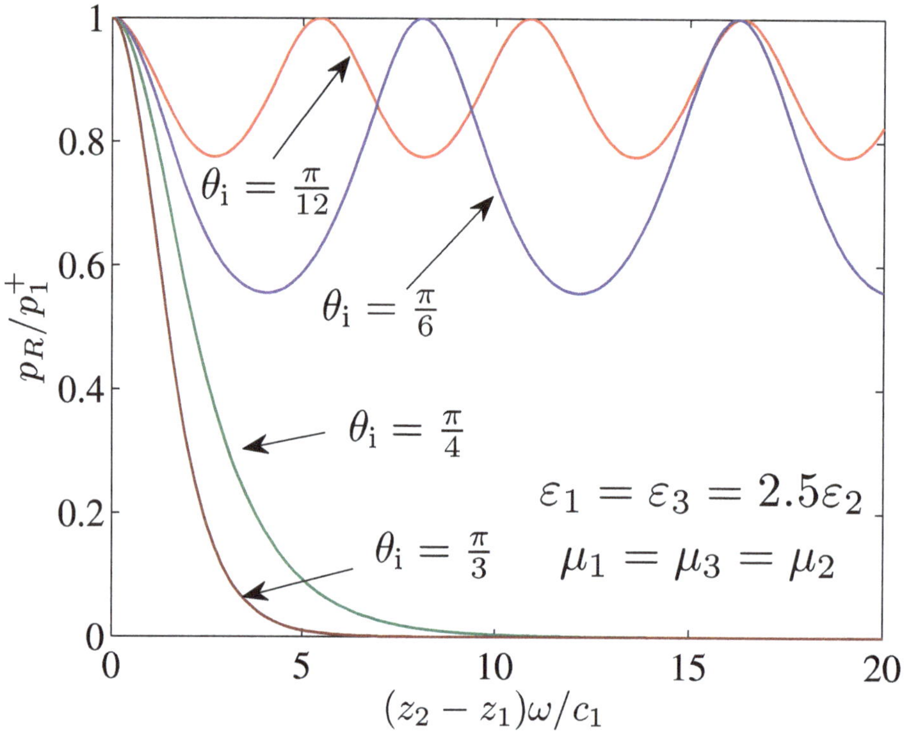

Figure 7.22 Example of transmitted power density for the plane-wave problem of Figure 7.19.

fact, no, it does not, because the second wave, after all, is generated by the reflection of the first at the interface $z = z_2$. Therefore, as we make the slab grow ever thicker, the second interface gets moved farther and farther away from the first, so that the amplitude of the evanescent reflected wave in section $z = z_2$ — where it is maximum — diminishes accordingly. In the limit as $z_2 - z_1 \to +\infty$, we see from (7.272) and (7.282) that the electric field in the slab — well, now a half-space — is comprised of just one inhomogeneous exponentially decaying wave. We reach the same conclusion with the aid of the multiple-reflections expansion (7.275) in which all the waves but the first disappear when k_{z2} is imaginary and $z_2 - z_1 \to +\infty$! Last but not least, under the same condition the power transferred beyond the interface $z = z_1$ drops to zero, as we gather from (7.283).

As an example, for the specific choice of constitutive parameters $\varepsilon_1 = \varepsilon_3 = 2.5\varepsilon_2$ and $\mu_1 = \mu_3 = \mu_2$ a few instances of the average transmitted power density computed by means of either (7.280) or (7.283) are drawn in Figure 7.22 as a function of the slab thickness $z_2 - z_1$. Each line represents the ratio p_R/p_1^+ where

$$p_1^+ := \frac{|E_{10}^+|^2}{2Z''_{\infty 1}} \tag{7.284}$$

is the surface density of power that is carried by the incident plane wave (7.253) across any plane perpendicular to the z-axis. The parameter of the lines is the incidence angle $\theta_i \in \{\pi/12, \pi/6, \pi/4, \pi/3\}$. From (7.183) we find that FTIR occurs for $\sin\theta_i = k_x c_1/\omega \geqslant \sqrt{2/5}$. For θ_i smaller than the critical angle θ_c the transmitted power exhibits a typical interference-like behavior, that is to say, as the slab grows relatively thicker, p_R oscillates periodically between maxima (equal to p_1^+) and minima whose value depend on the incidence angle. In practice, if the angular frequency and the constitutive parameters of medium 2 are assigned, one can adjust the thickness of the slab so as to maximize the power transfer into medium 3 for a specific angle of incidence. When the incident wave impinges upon the slab with an angle θ_i which exceeds θ_c, FTIR occurs and the transmitted power tends monotonically to zero with increasing slab thickness. Moreover, the larger θ_i is, the faster the power falls off with $z_2 - z_1$. As we gather from (7.283), the decay is not simply exponential owing to the combined effect of the two inhomogeneous plane waves that exist inside the slab.

We mention in passing that FTIR is *mathematically* analogous to the quantum-mechanical *tunnelling*, according to which a *wave-particle*, say, an electron, can indeed cross a barrier of potential energy [47]. In the framework of classical mechanics we expect the electron to act as a particle and as such to 'bounce' against the barrier and turn around, so long as it does not possess enough kinetic energy to 'jump' over the barrier, so to speak. By contrast, when we describe the electron by means of a complex wave-function $\psi(\mathbf{r}, t)$ [48, Chapter 2] — whose magnitude squared, $|\psi(\mathbf{r}, t)|^2$, represents the volume density of *probability* that the electron may be found in the point $\mathbf{r}$ at time t — calculations show that $|\psi(\mathbf{r}, t)|^2$ is not zero beyond the barrier [cf. with (7.283)], meaning that the electron actually passes *through* the obstacle. In particular, it turns out that inside the potential barrier the electron is described by two evanescent wave-functions, pretty much as the electromagnetic field within the slab of Figure 7.19 is the sum of two inhomogeneous plane waves.

Under condition (7.281) the chain matrix (7.225) may be written as

$$[M''] = \begin{pmatrix} \cosh[|k_{z2}|(z_2 - z_1)] & \mathrm{j}\,\dfrac{\omega\mu_2}{|k_{z2}|}\sinh[|k_{z2}|(z_2 - z_1)] \\ -\mathrm{j}\,\dfrac{|k_{z2}|}{\omega\mu_2}\sinh[|k_{z2}|(z_2 - z_1)] & \cosh[|k_{z2}|(z_2 - z_1)] \end{pmatrix} \tag{7.285}$$

having made use of (7.282). Although in (7.285) sines and cosines have been replaced by their hyperbolic counterparts, the properties of $[M'']$ remain valid. Still, the entries of $[M'']$ may grow quite large, if so does the quantity $|k_{z2}|(z_2 - z_1)$, i.e., if the slab happens to be relatively thick. While in principle there is nothing wrong with the chain matrix, one needs to exercise care when writing computer programs for the automated solution of equivalent circuits such as that of Figure 7.21a. Evidently, the very storing of $[M'']$ in a computer memory at run-time may cause numerical overflow in view of the exponential grow of hyperbolic sines and cosines. Confessedly, this difficulty is one glitch of the chain matrix representation. It is worth noting, though, that formulas (7.272) and (7.283) are numerically stable, even for large values of $|k_{z2}|(z_2 - z_1)$. Besides, when (7.281) holds true, the input impedance (7.260) — a key ingredient to the circuital approach — is better written as

$$Z''_{\mathrm{in}} = \mathrm{j}\,|Z''_{\infty 2}|\frac{Z''_{\infty 3} + \mathrm{j}\,|Z''_{\infty 2}|\tanh[|k_{z2}|(z_2 - z_1)]}{\mathrm{j}\,|Z''_{\infty 2}| + Z''_{\infty 3}\tanh[|k_{z2}|(z_2 - z_1)]} \tag{7.286}$$

because the hyperbolic tangent remains bounded for $|k_{z2}|(z_2 - z_1) \gg 1$ [20, Formula 654.3].

We conclude by remarking that all the considerations made for the TE polarization apply, *mutatis mutandis*, to the TM case of Figures 7.15 and 7.18b. The details are left to the Reader.

In order to solve problems of plane-wave propagation that involve more than one slab we can easily generalize the network approach developed so far. For the sake of argument, we suppose that a stack of $N - 1$ slabs — each one of them having different constitutive parameters and thicknesses — is 'sandwiched' between two half-spaces filled with medium 1 (for $z < z_1$) and medium $N + 1$ (for $z \geqslant z_N$). The nth slab, $n = 2, \ldots, N$, extends from z_{n-1} to z_n and is endowed with permittivity ε_n and permeability μ_n.

The equivalent circuits of the half-spaces are the same as those drawn in Figures 7.21a and 7.21b. The circuital model of the stack consists of a cascade of $N - 1$ devices (e.g., pieces of transmission lines) which in turn are governed by the equations

$$[\Psi''_{n-1}] = [M''_n][\Psi''_n], \qquad [\Psi'_{n-1}] = [M'_n][\Psi'_n], \qquad n = 2, \ldots, N \tag{7.287}$$

where $[M''_n]$ and $[M'_n]$ denote the chain matrices for TE and TM polarizations. Thanks to these very definitions, in any inner section $z = z_n$, $n = 2, \ldots, N-1$, which is the material interface between two adjacent slabs, the abstract vectors $[\Psi''_n]$ and $[\Psi'_n]$ play each a double role, namely,

- they contain the electrical state at the *output* port of the nth device,
- they also provide the electrical state at the *input* port of device number $n+1$.

Therefore, it is permissible and, in truth, quite tempting to combine the circuit equations (7.287) by eliminating the electrical states at the internal interfaces, viz.,

$$[\Psi''_1] = [M''_2]\cdots[M''_N][\Psi''_N] \qquad\qquad [\Psi'_1] = [M'_2]\cdots[M'_N][\Psi'_N] \tag{7.288}$$

whence we gather that the chain matrix of a cascade of devices can be obtained by *multiplying* the individual chain matrices in the order with which they appear in the circuit from left to right.

Once we have computed the total chain matrix of the stack for each polarization, the resulting equivalent network looks no different than the circuits in Figures 7.21a and 7.21b, and we can proceed with the solution as before. After the electrical state in section $z = z_N$ has been computed, we turn to the separate governing equations (7.287) and, proceeding backwards, we retrieve the electrical states in all the other interfaces. This is enough to obtain the electromagnetic field everywhere within the stack of slabs.

Regrettably, this neat procedure has a catch, that is, if one or more chain matrices take on the form (7.285) and the analogous one for the TM polarization, calculating the matrix products in (7.288) in a computer program may easily cause overflow, because the entries of $[M'']$ and $[M']$ may grow very large for relatively thick slabs. Of course, since k_x is assigned, one can ascertain beforehand in which slabs FTIR occurs by computing k_{zn} as in (7.179), but unfortunately this information *per se* does not solve the issue.

In fact, a better course of action consists of starting with the load in section $z = z_N$ and determine, e.g., for TE polarization, the input impedance in section $z = z_{N-1}$ by means of (7.260) or (7.285) depending on whether k_{zn} is real or imaginary. In like manner, since the input impedance thus found in turn constitutes a load for the device $N-1$, the input impedance in the preceding section $z = z_{N-2}$ can be computed, and so on so forth. These steps must be repeated until one gets the input impedance in section $z = z_1$, where it is directly connected to the generator.

This approach can be automated in a computer program and has the distinctive advantage of being stable, because neither (7.260) nor (7.285) involve functions which may diverge. Once the electrical state in section $z = z_1$ has been computed, we proceed forwards down the circuit and determine modal voltages and currents in the remaining sections. This can be accomplished by using the inverse relations of (7.287) for those slabs in which k_{zn} is real. On the contrary, if FTIR occurs, voltages, currents and fields are best obtained through formulas such as (7.272) and the like, because they involve at most real exponentials with negative arguments.

7.5 Time-domain uniform cylindrical waves

Plane waves truly are the simplest wave-like solution to Maxwell's equations. Next in line by way of complexity are circular *cylindrical* electromagnetic waves. While the amplitude of homogenous plane waves is constant on planes perpendicular to the direction of propagation, *uniform circular cylindrical waves* have constant amplitude on cylindrical surfaces which emanate from the source and are characterized by an ever-growing radius. Thus, to study these solutions it is convenient to introduce a system of cylindrical coordinates (ρ, φ, z) (Appendix A.1) in which the waves — if they are possible — propagate along the radial direction.

What type of sources may generate cylindrical waves? Well, if we require the field to be constant on the boundary of circular right cylinders whose axes coincide with the z-axis, then the time-varying current in question must be z-directed and invariant along φ and z. Since this property also implies that the source is infinitely extended along the z-direction, we see that cylindrical waves are still non-physical solutions to the Maxwell equations.

To show that cylindrical waves are indeed feasible we consider an ideal *line current source* placed along the z-axis in an unbounded homogeneous lossless medium endowed with permittivity ε and permeability μ. If the current is electric, then we may think of it as the limit for $a \to 0^+$ of an axis-symmetric cylindrical current density $\mathfrak{J}(\rho, t)$ with circular section of radius a (see Example 4.1). In order to compensate the effect of the ever-shrinking radius a, $\mathfrak{J}(\rho, t)$ must become infinite in such a way that

$$I(t) = \lim_{a\to 0^+} \int_{B_2(0,a)} \mathrm{d}S\, \hat{\mathbf{z}} \cdot \mathfrak{J}(\rho, t) = \lim_{a\to 0^+} 2\pi \int_0^a \mathrm{d}\rho\, \rho\, \mathfrak{J}(\rho, t) \tag{7.289}$$

i.e., the flux of $\mathfrak{J}(\rho, t)$ across circles perpendicular to the z-axis remains finite and independent of a; this condition may be realized by taking $\mathfrak{J}(\rho, t) = I(t)/(\pi a^2)$ for $\rho \leqslant a$. In like manner, if the current is magnetic (Section 6.5) we have

$$V(t) = \lim_{a\to 0^+} \int_{B_2(0,a)} \mathrm{d}S\, \hat{\mathbf{z}} \cdot \mathfrak{J}_M(\rho, t) = \lim_{a\to 0^+} 2\pi \int_0^a \mathrm{d}\rho\, \rho\, \mathfrak{J}_M(\rho, t) \tag{7.290}$$

by invoking the principle of duality as summarized in Table 6.1; in this case we may suppose $\mathfrak{J}_M(\rho, t) = V(t)/(\pi a^2)$ for $\rho \leqslant a$.

Since $\mathfrak{J}(\rho, t)$ and $\mathfrak{J}_M(\rho, t)$ point in the z-direction but do not depend on z, these currents are divergence-less. For cross sections of finite radius $a > 0$, the statement follows directly from (A.30). In the limit as $a \to 0^+$ we may consider a chunk of $\mathfrak{J}(\rho, t)$ bounded by two cross sections S_1 and S_2, and compute

$$\begin{aligned}
\lim_{a\to 0^+} \int_V \mathrm{d}V\, \nabla \cdot \mathfrak{J}(\mathbf{r}) &= \lim_{a\to 0^+} \int_{\partial V} \mathrm{d}S\, \hat{\mathbf{n}}(\mathbf{r}) \cdot \mathfrak{J}(\mathbf{r}) \\
&= \lim_{a\to 0^+} \int_{S_2} \mathrm{d}S\, \hat{\mathbf{z}} \cdot \mathfrak{J}(\mathbf{r}) - \lim_{a\to 0^+} \int_{S_1} \mathrm{d}S\, \hat{\mathbf{z}} \cdot \mathfrak{J}(\mathbf{r}) \\
&= I(t) - I(t) = 0
\end{aligned} \tag{7.291}$$

where we have used (7.289). As a result, from the conservation of charge (1.17) we find

$$0 = \lim_{a\to 0^+} \int_V \mathrm{d}V\, \nabla \cdot \mathfrak{J}(\mathbf{r}) + \frac{\mathrm{d}}{\mathrm{d}t} \lim_{a\to 0^+} \int_V \mathrm{d}V\, \varrho(\rho, t) = \int_{z_1}^{z_2} \mathrm{d}z\, \frac{\mathrm{d}}{\mathrm{d}t} \varrho_L(t) \tag{7.292}$$

where the coordinates z_1 and z_2 give the positions of S_1 and S_2 along the z-axis, and $\varrho_L(t)$ denotes the line density of charge associated with $I(t)$. It follows from the above that, as the charges move along the z-axis, the density thereof must remain constant in time. Similar remarks apply to $\mathcal{J}_M(\rho, t)$ and ϱ_{ML}.

Definitions (7.289) and (7.290) are consistent with (C.12) and are, in fact, the telltale of two-dimensional Dirac δ-distributions centered in $\boldsymbol{\rho} = 0$. Therefore, we could write the limiting densities as (Appendix C)

$$\mathcal{J}(\rho, t) := I(t)\delta^{(2)}(\boldsymbol{\rho})\hat{\mathbf{z}} \qquad \varrho(\rho) := \varrho_L \delta^{(2)}(\boldsymbol{\rho}) \tag{7.293}$$

$$\mathcal{J}_M(\rho, t) := V(t)\delta^{(2)}(\boldsymbol{\rho})\hat{\mathbf{z}} \qquad \varrho_M(\rho) := \varrho_{ML} \delta^{(2)}(\boldsymbol{\rho}) \tag{7.294}$$

and extend Maxwell's equations to the realm of generalized functions. Instead, for the moment we obviate the need for specifying the nature of the source by excluding the z-axis from the formulation.

By expressing the curl and the divergence operators in cylindrical coordinates in (1.13), (1.20), (1.23) and (1.44) with the aid of (A.30) and (A.33) we have

$$-\hat{\boldsymbol{\varphi}}\frac{\partial \mathcal{E}_z}{\partial \rho} + \hat{\mathbf{z}}\frac{1}{\rho}\frac{\partial}{\partial \rho}(\rho \mathcal{E}_\varphi) = -\mu \frac{\partial \mathcal{H}}{\partial t}, \qquad \rho > 0 \tag{7.295}$$

$$-\hat{\boldsymbol{\varphi}}\frac{\partial \mathcal{H}_z}{\partial \rho} + \hat{\mathbf{z}}\frac{1}{\rho}\frac{\partial}{\partial \rho}(\rho \mathcal{H}_\varphi) = -\varepsilon \frac{\partial \mathcal{E}}{\partial t}, \qquad \rho > 0 \tag{7.296}$$

$$\frac{1}{\rho}\frac{\partial}{\partial \rho}(\rho \mathcal{E}_\rho) = 0, \qquad \rho > 0 \tag{7.297}$$

$$\frac{1}{\rho}\frac{\partial}{\partial \rho}(\rho \mathcal{H}_\rho) = 0, \qquad \rho > 0 \tag{7.298}$$

having used the fact that the cylindrical components of $\mathcal{E}(\mathbf{r}, t)$ and $\mathcal{H}(\mathbf{r}, t)$ are functions of ρ and t only. Since the vector fields in the left-hand sides of (7.295) and (7.296) possess no radial components, together with (7.297) and (7.298), (7.295) and (7.296) require

$$\mathcal{E}_\rho(\rho, t) = \frac{C_{E\rho}}{\rho}, \qquad \mathcal{H}_\rho(\rho, t) = \frac{C_{H\rho}}{\rho}, \qquad \rho > 0 \tag{7.299}$$

where $C_{E\rho}$ and $C_{H\rho}$ are two constants.

If we suppose that a non-zero electric charge density ϱ_L was present on the z-axis prior to the onset of the current at time $t = 0$, then the electromagnetic field was static and non-zero for $t < 0$. In particular, by invoking azimuthal symmetry and the Gauss law (1.16) we have (cf. Example 6.1)

$$\mathcal{E}(\rho, t) = \frac{\varrho_L}{2\pi\varepsilon\rho}\hat{\boldsymbol{\rho}}, \qquad \mathcal{H}(\rho, t) = 0, \qquad t < 0 \tag{7.300}$$

whereby we are led to choose $C_{E\rho} = \varrho_L/(2\pi\varepsilon)$ and $C_{H\rho} = 0$. On the other hand, if we admit that positive and negative charges alike are present in equal amount on the z-axis so that $\varrho_L = 0$ for $t < 0$, then the field was null everywhere in space, whereby we must set $C_{E\rho} = C_{H\rho} = 0$. The current $I(t)$ — turned on at $t = 0$ — results from the collective drift of positive and negative charges, though the motion thereof will be such that ϱ_L remains zero at all times since, as we argued above, the charge density is a constant. Perfectly dual observations apply to the case of magnetic charges on the z-axis.

The radial components of the field — when they are not null — form the *static* part, whereas the remaining components (those perpendicular to $\hat{\boldsymbol{\rho}}$) constitute the *wave-like* part [cf. (1.252) and (1.253) for a point charge]. Therefore, we see that uniform cylindrical waves propagating along the radial direction cannot have a radial component. Since in our problem the static part is invariably given by (7.299), we let $C_{E\rho} = C_{H\rho} = 0$ and focus on the calculation of the wave-like part of the solution, which we indicate with $\mathcal{E}$ and $\mathcal{H}$ for simplicity.

In this regard, we can still satisfy the Faraday law and the Ampère-Maxwell law by assuming that $\mathcal{E}$ and $\mathcal{H}$ have only two non-zero components all in all.

(a) If we choose $\mathcal{H}_z = 0$, then (7.295) and (7.296) together demand

$$\mathcal{E}_\varphi(\rho, t) = \frac{C_{E\varphi}}{\rho}, \qquad \rho > 0 \tag{7.301}$$

where $C_{E\varphi}$ is a constant. Once again we need to set $C_{E\varphi} = 0$ inasmuch as, according to (7.300), the φ-component of the electric field was null for $t < 0$, even if ϱ_L was non-zero. Hence, we are left with $\mathcal{E}_z$ and $\mathcal{H}_\varphi$, whereby the solution is termed a *transverse-magnetic* (TM) wave with respect to z or, alternatively, an *E-polarized* wave. This type of wave is generated by the ideal line electric current (7.289).

(b) If we suppose $\mathcal{E}_z = 0$, then (7.295) and (7.296) yield

$$\mathcal{H}_\varphi(\rho, t) = \frac{C_{H\varphi}}{\rho}, \qquad \rho > 0 \tag{7.302}$$

where the constant $C_{H\varphi}$ is necessarily null as before, in view of the static solution dual to (7.300). We have two non-zero components, namely, $\mathcal{H}_z$ and $\mathcal{E}_\varphi$, and the solution is called a *transverse-electric* (TE) wave with respect to z or also an *H-polarized* wave. The latter is produced by the ideal line magnetic current (7.290).

It is apparent that both TM and TE solutions are transverse waves in that the electromagnetic field is orthogonal to $\hat{\boldsymbol{\rho}}$, which in turn is aligned with the direction of propagation. Besides, electric and magnetic fields are invariably perpendicular to each other.

To proceed we write the wave equations (1.204) and (1.213) in cylindrical coordinates with the help of (A.44), viz.,

$$\hat{\mathbf{z}}\frac{1}{\rho}\frac{\partial}{\partial\rho}\left(\rho\frac{\partial\mathcal{E}_z}{\partial\rho}\right) + \hat{\boldsymbol{\varphi}}\frac{\partial}{\partial\rho}\left[\frac{1}{\rho}\frac{\partial}{\partial\rho}(\rho\mathcal{E}_\varphi)\right] = \frac{1}{c^2}\left(\hat{\mathbf{z}}\frac{\partial^2\mathcal{E}_z}{\partial t^2} + \hat{\boldsymbol{\varphi}}\frac{\partial^2\mathcal{E}_\varphi}{\partial t^2}\right), \qquad \rho > 0 \tag{7.303}$$

$$\hat{\mathbf{z}}\frac{1}{\rho}\frac{\partial}{\partial\rho}\left(\rho\frac{\partial\mathcal{H}_z}{\partial\rho}\right) + \hat{\boldsymbol{\varphi}}\frac{\partial}{\partial\rho}\left[\frac{1}{\rho}\frac{\partial}{\partial\rho}(\rho\mathcal{H}_\varphi)\right] = \frac{1}{c^2}\left(\hat{\mathbf{z}}\frac{\partial^2\mathcal{H}_z}{\partial t^2} + \hat{\boldsymbol{\varphi}}\frac{\partial^2\mathcal{H}_\varphi}{\partial t^2}\right), \qquad \rho > 0 \tag{7.304}$$

having taken into account that $\mathcal{E}_\rho = 0 = \mathcal{H}_\rho$. Clearly, they separate into four scalar equations of which those involving the z-components are simpler. Therefore, we may consider

$$\mathcal{H}_z = 0 = \mathcal{E}_\varphi, \qquad \frac{1}{\rho}\frac{\partial}{\partial\rho}\left(\rho\frac{\partial\mathcal{E}_z}{\partial\rho}\right) - \frac{1}{c^2}\frac{\partial^2\mathcal{E}_z}{\partial t^2} = 0, \qquad \rho > 0 \tag{7.305}$$

for TM waves and

$$\mathcal{E}_z = 0 = \mathcal{H}_\varphi, \qquad \frac{1}{\rho}\frac{\partial}{\partial\rho}\left(\rho\frac{\partial\mathcal{H}_z}{\partial\rho}\right) - \frac{1}{c^2}\frac{\partial^2\mathcal{H}_z}{\partial t^2} = 0, \qquad \rho > 0 \tag{7.306}$$

for TE waves. The remaining non-zero components are not independent and may be computed with the aid of either (7.295) or (7.296) once $\mathcal{E}_z$ and $\mathcal{H}_z$ have been determined.

The solution of (7.305) and (7.306) is more easily carried out in the frequency domain or for periodic time dependence with single angular frequency ω. In the former case we introduce the Fourier transformation with respect to time [4, Section 4.8], [49] of the relevant electromagnetic entities, whereas in the latter we apply the transformation rules developed in Section 1.5 for the time-harmonic regime.

For the sake of argument, suppose we look for TM solutions that are null for $t < 0$ and decay fast enough to admit a temporal Fourier transform, viz.,

$$\mathfrak{F}\{\mathcal{E}_z(\rho, t)\} := \int_{\mathbb{R}} \mathrm{d}t\, \mathrm{e}^{-\mathrm{j}\omega t} \mathcal{E}_z(\rho, t) = E_z(\rho; \omega) \tag{7.307}$$

then, the electric field in time domain follows from the inversion formula, i.e.,

$$\mathfrak{F}^{-1}\{E_z(\rho; \omega)\} := \frac{1}{2\pi} \int_{\mathbb{R}} \mathrm{d}\omega\, \mathrm{e}^{\mathrm{j}\omega t} E_z(\rho; \omega) = \mathcal{E}_z(\rho, t) \tag{7.308}$$

if we know how to compute $E_z(\rho; \omega)$. To this purpose we insert (7.308) into (7.305)

$$\begin{aligned} 0 &= \frac{1}{\rho}\frac{\partial}{\partial \rho}\left[\rho \frac{\partial}{\partial \rho} \int_{\mathbb{R}} \mathrm{d}\omega\, \mathrm{e}^{\mathrm{j}\omega t} E_z(\rho; \omega)\right] - \frac{1}{c^2}\frac{\partial^2}{\partial t^2} \int_{\mathbb{R}} \mathrm{d}\omega\, \mathrm{e}^{\mathrm{j}\omega t} E_z(\rho; \omega) \\ &= \int_{\mathbb{R}} \mathrm{d}\omega\, \mathrm{e}^{\mathrm{j}\omega t} \frac{1}{\rho}\frac{\partial}{\partial \rho}\left[\rho \frac{\partial}{\partial \rho} E_z(\rho; \omega)\right] - \frac{1}{c^2} \int_{\mathbb{R}} \mathrm{d}\omega\, \frac{\partial^2 \mathrm{e}^{\mathrm{j}\omega t}}{\partial t^2} E_z(\rho; \omega) \\ &= \int_{\mathbb{R}} \mathrm{d}\omega\, \mathrm{e}^{\mathrm{j}\omega t} \left\{\frac{1}{\rho}\frac{\partial}{\partial \rho}\left[\rho \frac{\partial}{\partial \rho} E_z(\rho; \omega)\right] + \frac{\omega^2}{c^2} E_z(\rho; \omega)\right\} \end{aligned} \tag{7.309}$$

where we have inverted the order of integration and differentiation under the hypotheses that

- as a function of ω and $\rho > 0$, $E_z(\rho; \omega) \exp(\mathrm{j}\,\omega t)$ has continuous derivatives with respect to ω and ρ up to the second order for $(\omega, \rho) \in \mathbb{R} \times \mathbb{R}^+$,
- similarly, as a function of ω and t, $E_z(\rho; \omega) \exp(\mathrm{j}\,\omega t)$ has continuous derivatives with respect to ω and t up to the second order for $(\omega, t) \in \mathbb{R}^2$.

Since the factor $\exp(\mathrm{j}\,\omega t)$ never vanishes, (7.309) is satisfied if the function within braces is invariably null. With this observation and the definition of wavenumber (1.248), (7.309) requires

$$\frac{1}{\rho}\frac{\partial}{\partial \rho}\left[\rho \frac{\partial}{\partial \rho} E_z(\rho; \omega)\right] + k^2 E_z(\rho; \omega) = 0, \qquad \rho > 0 \tag{7.310}$$

which constitutes the frequency-domain counterpart of (7.305) and is a special instance of the general Helmholtz equation (1.238). We notice that, if we assume periodic time-dependence in the form

$$\mathcal{E}_z(\rho, t) = \mathrm{Re}\left\{E_z(\rho; \omega)\mathrm{e}^{\mathrm{j}\omega t}\right\}, \qquad \rho > 0 \tag{7.311}$$

we arrive again at (7.310). Regardless, given that the time variable does not appear explicitly any more, the Helmholtz equation is simpler to solve.

By performing the change of independent variable $\xi = k\rho$ (7.310) passes over into

$$\frac{1}{\xi}\frac{\partial}{\partial \xi}\left[\xi \frac{\partial}{\partial \xi} E_z(\xi; \omega)\right] + E_z(\xi; \omega) = 0, \qquad \xi > 0 \tag{7.312}$$

that is, the Bessel equation (H.136) with separation constant $\nu = 0$. Absent any other condition (7.312) is solved by any linear combination of Bessel functions $J_0(\xi)$, $Y_0(\xi)$, $H_0^{(1)}(\xi)$ and $H_0^{(2)}(\xi)$ [50,

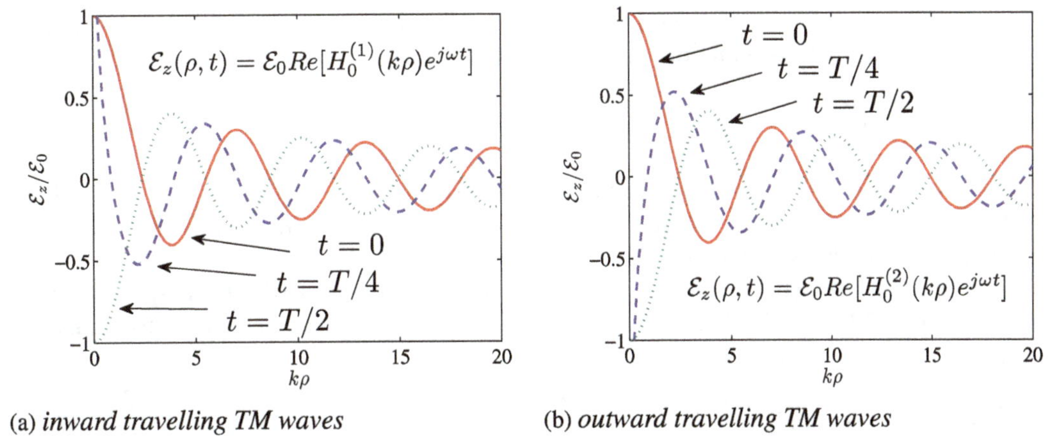

Figure 7.23 For choosing the right solution to (7.310).

Chapter IV]. In fact, each one of the latter does provide a wave-like solution, though only one is the right candidate for representing a wave which, originating on the z-axis, travels away *towards infinity*. To help us make the right decision we examine the asymptotic behavior of the Bessel functions.

From the asymptotic expansions (H.158) and (H.159) for large argument and Figure 3.7a we see the $J_0(\xi)$ and $Y_0(\xi)$ behave as trigonometric functions which are damped with the square root of the distance from the origin, and hence $J_0(\xi)$ and $Y_0(\xi)$ are better suited to express *standing-wave* solutions which occur in bounded regions. Since we are interested in travelling-wave solutions $J_0(\xi)$ and $Y_0(\xi)$ must be discarded.

By contrast, from (H.160) and (H.161) we see that $H_0^{(1)}(\xi)$ and $H_0^{(2)}(\xi)$ behave as complex exponentials for large values of their argument, and thus we may expect that they exhibit the right spatial dependence which results in a travelling wave in the time domain. If we provisionally invoke the time-harmonic regime for simplicity, we may quickly recover the corresponding electric field in the time domain from

$$\mathcal{E}_z(\rho,t) = \mathcal{E}_0 \mathrm{Re}\left\{H_0^{(1)}(k\rho)\mathrm{e}^{\mathrm{j}\omega t}\right\} = \mathcal{E}_0[J_0(k\rho)\cos(\omega t) - Y_0(k\rho)\sin(\omega t)] \tag{7.313}$$

$$\mathcal{E}_z(\rho,t) = \mathcal{E}_0 \mathrm{Re}\left\{H_0^{(2)}(k\rho)\mathrm{e}^{\mathrm{j}\omega t}\right\} = \mathcal{E}_0[J_0(k\rho)\cos(\omega t) + Y_0(k\rho)\sin(\omega t)] \tag{7.314}$$

with $\mathcal{E}_0$ denoting an arbitrary real constant. Snapshots of the resulting TM waves are plotted in Figure 7.23 for $t \in \{0, T/4, T/2\}$ with $T = 2\pi/\omega$. We see that the Hankel function of the first kind gives rise to a wave which, as the time goes by, travels towards the z-axis and enters the line source. Since this occurrence is utterly non-physical we are left with the Hankel function of the second kind, which indeed represents a wave emanating from the source and propagating towards infinity.

Accordingly, thanks to the linearity of the Helmholtz equation (7.310) we may write the general solution as

$$E_z(\rho;\omega) = \frac{1}{4\mathrm{j}} H_0^{(2)}(k\rho)E_0(\omega), \qquad \rho > 0 \tag{7.315}$$

where the dimensionless multiplicative constant $1/(4\,\mathrm{j})$ is included for convenience, and the factor $E_0(\omega)$ is determined by the spectral properties of the source, that is, the temporal Fourier transform of $I(t)$ in (7.289). We notice that $E_z(\rho;\omega)$ exhibits a logarithmic singularity [see (H.157)] on the

z-axis ($\rho = 0$) where the ideal line current is located. More importantly, since $E_0(\omega)$ accounts for the effect of sources with general time dependence, we identify the function

$$G(\rho;\omega) := \frac{1}{4\mathrm{j}} H_0^{(2)}(k\rho), \qquad \rho > 0 \tag{7.316}$$

as the z-component of the electric field of a TM cylindrical wave generated by a source with unitary $E_0(\omega)$. We call $G(\rho;\omega)$ the frequency-domain *two-dimensional Green function*, for it allows constructing every other solution to (7.310).

As regards the magnetic field component of the TM wave, we observe that the frequency-domain counterpart of (7.295) may be derived by substituting $\mathcal{E}_z$ and $\mathcal{H}_\varphi$ with the inverse Fourier transform thereof, and the result reads

$$-\frac{\mathrm{d}}{\mathrm{d}\rho} E_z(\rho;\omega) = -\mathrm{j}\,\omega\mu H_\varphi(\rho;\omega) \tag{7.317}$$

a relation we may use to express $H_\varphi(\rho;\omega)$ in terms of $E_z(\rho;\omega)$ given by (7.315). To be specific, by applying the recursion formula (H.167) with $\nu = 0$ to obtain the derivative of the Hankel function of zero order we have

$$H_\varphi(\rho;\omega) = -\frac{k}{\mathrm{j}\,\omega\mu} E_0(\omega) \frac{1}{4\mathrm{j}} H_1^{(2)}(k\rho) = \frac{1}{4Z} H_1^{(2)}(k\rho) E_0(\omega), \qquad \rho > 0 \tag{7.318}$$

having recalled definition (1.248) and introduced the intrinsic impedance (1.358) in the background medium.

To relate $E_0(\omega)$ to the source $I(t)$ we consider the frequency-domain counterpart of the global Ampère-Maxwell law (1.13)

$$\oint_{\partial S} \mathrm{d}s\, \hat{\mathbf{s}}(\mathbf{r}) \cdot \mathbf{H}(\mathbf{r};\omega) = \mathrm{j}\,\omega\varepsilon \int_S \mathrm{d}S\, \hat{\mathbf{n}}(\mathbf{r}) \cdot \mathbf{E}(\mathbf{r};\omega) + \int_S \mathrm{d}S\, \hat{\mathbf{n}}(\mathbf{r}) \cdot \mathbf{J}(\mathbf{r};\omega) \tag{7.319}$$

where for the problem of interest here we choose S as a circle of radius a centered in the origin and lying in the plane xOy. With $\hat{\mathbf{n}}(\mathbf{r}) = \hat{\mathbf{z}}$ and by virtue of (7.289) the flux of $\mathbf{J}(\mathbf{r};\omega)$ is just $\tilde{I}(\omega)$, the Fourier transform of $I(t)$. Only $E_z(\rho;\omega)$ and $H_\varphi(\rho;\omega)$ are non-zero for TM waves, and hence we have

$$2\pi a H_\varphi(a;\omega) = 2\pi\,\mathrm{j}\,\omega\varepsilon \int_0^a \mathrm{d}\rho\, \rho E_z(\rho;\omega) + \tilde{I}(\omega) \tag{7.320}$$

$$\frac{\pi a}{2Z} E_0(\omega) H_1^{(2)}(ka) = \omega\varepsilon \frac{\pi}{2} E_0(\omega) \int_0^a \mathrm{d}\rho\, \rho H_0^{(2)}(k\rho) + \tilde{I}(\omega) \tag{7.321}$$

in light of (7.315) and (7.318). To compute the integral we let $\xi = k\rho$ and derive the functional relation

$$\frac{\mathrm{d}}{\mathrm{d}\xi}\left[\xi H_1^{(2)}(\xi)\right] = \xi H_0^{(2)}(\xi) \tag{7.322}$$

from the recursion formula (H.166) with $\nu = 1$, whereby we have

$$\int_0^a \mathrm{d}\rho\, \rho H_0^{(2)}(k\rho) = \frac{1}{k^2} \lim_{b\to 0^+} \int_b^{ka} \mathrm{d}\xi\, \xi H_0^{(2)}(\xi) = \frac{1}{k^2} \lim_{b\to 0^+} \left[\xi H_1^{(2)}(\xi)\right]_b^{ka}$$

$$
\begin{aligned}
&= \frac{a}{k} H_1^{(2)}(ka) - \frac{1}{k^2} \lim_{b \to 0^+} b H_1^{(2)}(b) \\
&= \frac{a}{k} H_1^{(2)}(ka) - \frac{1}{k^2} \lim_{b \to 0^+} b \left(\frac{b}{2} + \mathrm{j} \frac{2}{\pi b} \right) \\
&= \frac{a}{k} H_1^{(2)}(ka) - \frac{2\mathrm{j}}{\pi k^2}
\end{aligned}
\tag{7.323}
$$

where the limiting process is necessary inasmuch as the Hankel function is singular in the origin. We have used the small-argument asymptotic expansions of $J_1(\xi)$ and $Y_1(\xi)$ on account of (H.152) and (H.153). Substituting this result into (7.321) and cancelling like terms with opposite signs yields

$$
E_0(\omega) = -\mathrm{j}\,\omega\mu\tilde{I}(\omega) \tag{7.324}
$$

a result which is independent of the arbitrary radius a. With this finding we can finalize the solution of (7.310), viz.,

$$
E_z(\rho;\omega) = -\frac{\omega\mu}{4} H_0^{(2)}(k\rho)\tilde{I}(\omega), \qquad \rho > 0 \tag{7.325}
$$

$$
H_\varphi(\rho;\omega) = \frac{k}{4\mathrm{j}} H_1^{(2)}(k\rho)\tilde{I}(\omega), \qquad \rho > 0. \tag{7.326}
$$

Having found an expression for $E_z(\rho;\omega)$ we compute the solution to (7.305) through (7.308), namely,

$$
\begin{aligned}
\mathcal{E}_z(\rho,t) &= \frac{1}{2\pi} \int_{\mathbb{R}} \mathrm{d}\omega\, \mathrm{e}^{\mathrm{j}\omega t} \frac{1}{4\mathrm{j}} H_0^{(2)}(k\rho) E_0(\omega) \\
&= \int_{\mathbb{R}} \mathrm{d}t'\, \mathcal{G}(\rho, t-t')\mathcal{E}_0(t') = \mathcal{G}(\rho,t) * \mathcal{E}_0(t) \\
&= -\mu \int_{\mathbb{R}} \mathrm{d}t'\, \mathcal{G}(\rho, t-t') \frac{\mathrm{d}I}{\mathrm{d}t'} = -\mathcal{G}(\rho,t) * \mu \frac{\mathrm{d}}{\mathrm{d}t} I(t)
\end{aligned}
\tag{7.327}
$$

where we have invoked the convolution theorem [4, Section 4.8] and indicated with $\mathcal{G}(\rho,t)$ and $\mathcal{E}_0(t)$ the inverse Fourier transforms of the Green function and $E_0(\omega)$, respectively. In particular, it can be shown that the two-dimensional Green function in time domain reads

$$
\mathcal{G}(\rho,t) = \frac{U(t-\rho/c)}{2\pi\sqrt{t^2 - (\rho/c)^2}} \tag{7.328}
$$

where $U(\bullet)$ denotes the unitary step function; the details of the derivation are postponed until Section 7.6.

The calculations required for TE waves and an ideal magnetic current $V(t)$ flowing along the z-axis are analogous and lead to a result for $\mathcal{H}_z(\rho,t)$ which is the dual of (7.327). In this regard, since $G(\rho;\omega)$ does not change, we may also interpret (7.316) as the z-component of the magnetic field of a TE wave produced by an ideal magnetic line current source with unitary spectral amplitude $H_0(\omega)$.

Example 7.3 (Propagation of a uniform cylindrical TM wave in free space)

To shed light on the generation and propagation of cylindrical waves we consider an ideal electric line current $I(t)$ aligned with the z-axis and defined by

$$
I(t) = -2\pi I_0 \left\{ \frac{t}{t_1}[U(t) - U(t-t_1)] + \frac{t_2 - t}{t_2 - t_1}[U(t-t_1) - U(t-t_2)] \right\} \tag{7.329}
$$

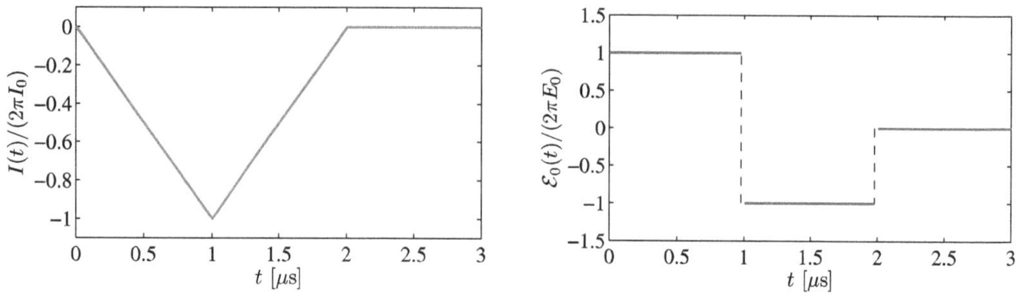

(a) $I(t)$ (b) $\mathcal{E}_0(t)$

Figure 7.24 Electric current and associated function $\mathcal{E}_0(t)$ for the generation of cylindrical TM waves.

i.e., a triangular pulse which extends from $t = 0$ up to $t_2 = 2\ \mu$s and, if $I_0 > 0$, reaches its minimum for $t_1 = 1\ \mu$s, as is shown in Figure 7.24a. The function $\mathcal{E}_0(t)$ which enters (7.327) reads (Figure 7.24b)

$$\begin{aligned}\mathcal{E}_0(t) &= -\mu\frac{\mathrm{d}I}{\mathrm{d}t} = 2\pi\mu I_0[U(t) - 2U(t-t_1) + U(t-t_2)] \\ &= 2\pi E_0[U(t) - 2U(t-t_1) + U(t-t_2)]\end{aligned} \tag{7.330}$$

whence by assuming that the line current exists in free space ($c = c_0$) the electric field of the TM wave is given by the convolution integral

$$\mathcal{E}_z(\rho,t) = E_0\int_{-\infty}^{+\infty}\mathrm{d}t'\,[U(t-t') - 2U(t-t'-t_1) + U(t-t'-t_2)]\frac{U(t'-\rho/c_0)}{\sqrt{t'^2-\rho^2/c_0^2}} \tag{7.331}$$

which we can evaluate analytically. To this purpose we observe that the combination of step functions amounts to a piecewise-constant function, and hence it is expedient to use the indefinite integral [20, Formula 260.01]

$$\int\frac{\mathrm{d}t'}{\sqrt{t'^2-\rho^2/c_0^2}} = \log\left[\frac{c_0}{\rho}\left(t' + \sqrt{t'^2-\frac{\rho^2}{c_0^2}}\right)\right], \qquad t'^2 \geqslant \frac{\rho^2}{c_0^2} \tag{7.332}$$

easily computed with the change of dummy variable $t' = (\rho/c_0)\cosh\xi$. Furthermore, the product of step functions reduces the integration range to finite intervals that depend on the values of ρ/c_0, t, t_1 and t_2. Therefore, we distinguish four cases:

(i) $t < \rho/c_0$

$$\mathcal{E}_z(\rho,t) = 0 \tag{7.333}$$

(ii) $t - t_1 < \rho/c_0 < t$

$$\mathcal{E}_z(\rho,t) = E_0\int_{\rho/c_0}^{t}\frac{\mathrm{d}t'}{\sqrt{t'^2-\rho^2/c_0^2}} = E_0\log\left[\frac{c_0}{\rho}\left(t + \sqrt{t^2-\frac{\rho^2}{c_0^2}}\right)\right] \tag{7.334}$$

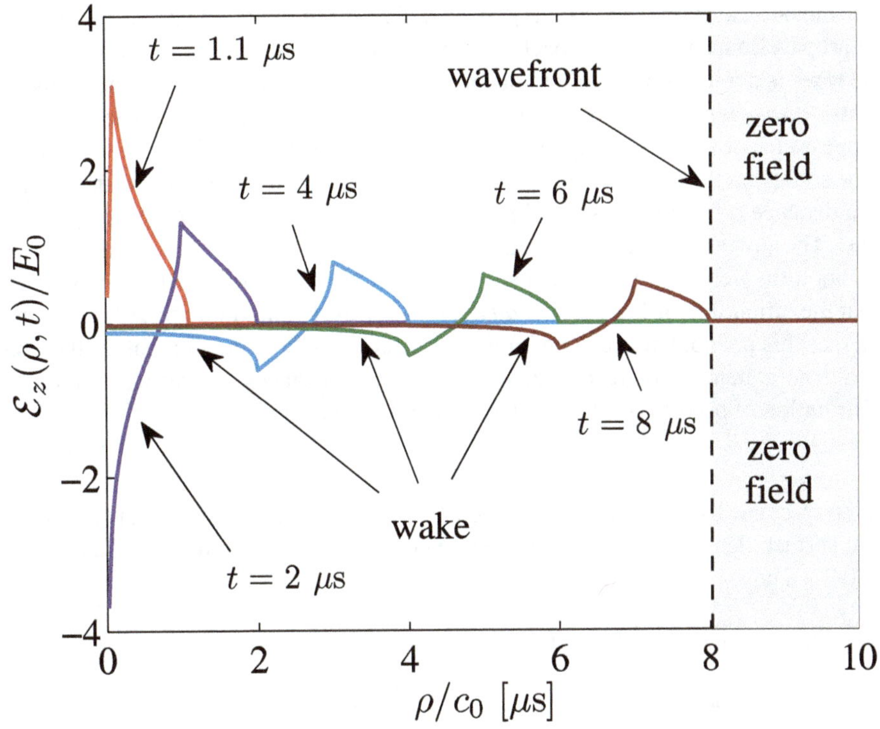

Figure 7.25 Snapshots of the electric field $\mathcal{E}_z(\rho, t)$ of a TM cylindrical wave as a function of the normalized distance ρ/c_0 from the source.

(iii) $t - t_2 < \rho/c_0 < t - t_1$

$$\begin{aligned}\mathcal{E}_z(\rho, t) &= E_0 \int_{t-t_1}^{t} \frac{\mathrm{d}t'}{\sqrt{t'^2 - \rho^2/c_0^2}} - E_0 \int_{\rho/c_0}^{t-t_1} \frac{\mathrm{d}t'}{\sqrt{t'^2 - \rho^2/c_0^2}} \\ &= E_0 \log \frac{\rho}{c_0} \frac{t + \sqrt{t^2 - \rho^2/c_0^2}}{\left(t - t_1 + \sqrt{(t-t_1)^2 - \rho^2/c_0^2}\right)^2}\end{aligned} \tag{7.335}$$

(iv) $\rho/c_0 < t - t_2$

$$\begin{aligned}\mathcal{E}_z(\rho, t) &= E_0 \int_{t-t_1}^{t} \frac{\mathrm{d}t'}{\sqrt{t'^2 - \rho^2/c_0^2}} - E_0 \int_{t-t_2}^{t-t_1} \frac{\mathrm{d}t'}{\sqrt{t'^2 - \rho^2/c_0^2}} \\ &= E_0 \log \frac{\left(t + \sqrt{t^2 - \rho^2/c_0^2}\right)\left(t - t_2 + \sqrt{(t-t_2)^2 - \rho^2/c_0^2}\right)}{\left(t - t_1 + \sqrt{(t-t_1)^2 - \rho^2/c_0^2}\right)^2}.\end{aligned} \tag{7.336}$$

The electric field component is plotted in Figure 7.25 for values of the normalized distance ρ/c_0 up to 10 μs and times $t \in \{1.1, 2, 4, 6, 8\}\,\mu$s. First of all, we notice that for any given time t there exists a *quiet zone* outside the cylinder of radius $\rho/c_0 = t$. This is the region of space that, due to the finite velocity of propagation, has not been reached yet by the disturbance (cf. Figure 1.15b). Secondly, we see that the wave is not just a more or less distorted replica of the function $\mathcal{E}_0(t)$ but it also comprises a tail of sorts or *wake* which trails behind even long after the current $I(t)$ has been turned off.

The appearance of the wake may be chalked up to the combined effect of the assumed *infinite* length of the source on the one hand and of the finite speed of light on the other. Indeed, an observer located at a distance ρ from the z-axis will keep receiving contributions from outlying parts of the line source. The greater the distance from the observer the longer the disturbances will take to reach him, but ultimately they will arrive and add up to yield a non-zero electric field. Finally, we observe that the amplitude of the wave is *damped* as it travels on, although the medium is lossless by hypothesis. This phenomenon can be intuitively understood by noticing that, as the time goes by, the electromagnetic field becomes non-zero on cylindrical surfaces of ever larger radius while the angular distribution of power (i.e., for $\varphi \in [0, 2\pi[$) remains constant.

(End of Example 7.3)

If we interpret the TM solution (7.325) and (7.326) as a monochromatic wave, we may compute the average surface density of radiated power. Starting with the definition of complex Poynting vector (1.304) we have

$$\begin{aligned}\mathbf{S}(\rho) &:= \frac{1}{2}\mathbf{E}(\rho) \times \mathbf{H}^*(\rho) = -\frac{1}{2}E_z(\rho)H_\varphi^*(\rho)\hat{\boldsymbol{\rho}} \\ &= -\frac{1}{2}\frac{\omega\mu k}{4\mathrm{j}}H_0^{(2)}(k\rho)\left[H_1^{(2)}(k\rho)\right]^*\left|\tilde{I}(\omega)\right|^2\hat{\boldsymbol{\rho}} \\ &= \frac{Zk^2}{8}\mathrm{j}\,H_0^{(2)}(k\rho)\left[H_1^{(2)}(k\rho)\right]^*\left|\tilde{I}(\omega)\right|^2\hat{\boldsymbol{\rho}}\end{aligned} \tag{7.337}$$

whence we see that the power efflux is radially directed and thus perpendicular to the wavefront. The surface power density through a cylinder of radius ρ coaxial with the z-axis reads

$$\begin{aligned}p_R = \left\langle\frac{\mathrm{d}P_F}{\mathrm{d}S}\right\rangle &:= \mathrm{Re}\{\mathbf{S}(\rho)\cdot\hat{\boldsymbol{\rho}}\} = \frac{Zk^2}{8}\left|\tilde{I}(\omega)\right|^2\mathrm{Re}\left\{\mathrm{j}\,H_0^{(2)}(k\rho)\left[H_1^{(2)}(k\rho)\right]^*\right\} \\ &= \frac{Zk^2}{8}\left|\tilde{I}(\omega)\right|^2\left[J_1(k\rho)Y_0(k\rho) - J_0(k\rho)Y_1(k\rho)\right] \\ &= \frac{Zk^2}{8}\left|\tilde{I}(\omega)\right|^2\frac{2}{\pi k\rho} = \frac{kZ}{4\pi\rho}\left|\tilde{I}(\omega)\right|^2\end{aligned} \tag{7.338}$$

where we have used the definition of Hankel functions (H.138) and (H.139), the fact that k is real, and finally the cross product of Bessel functions (H.168) with $\nu = 0$. The surface density of radiated power is constant along φ and z but falls off as the inverse of the distance from the source. Hence, the wave carries and infinite amount of power, though the latter is constant across any cylindrical surface of radius ρ and finite height.

7.6 The two-dimensional time-domain Green function

The two-dimensional Green function in time domain (7.328) is given formally by the following inverse Fourier integral

$$\mathcal{G}(\rho, t) = \frac{1}{2\pi}\int_{\mathbb{R}} \mathrm{d}\omega \frac{e^{\mathrm{j}\omega t}}{4\mathrm{j}} H_0^{(2)}(k\rho) \tag{7.339}$$

which is quite difficult to evaluate, not least in view of the logarithmic singularity of the Hankel function for $\rho = 0$ [see (H.157)] and the associated branch line (Appendix B.3). A powerful technique consists of writing the Hankel function as the unilateral Laplace transform of a function of time, whereby the Green function (7.328) is obtained by inspection [51, Section 5.2c].

The starting point is the wave equation (1.207) for the electric field, which we specialize to the case of cylindrical TM waves by recalling from Section 7.5 that only the z-component $\mathcal{E}_z$ is non-zero

$$\nabla_{\mathrm{t}}^2 \mathcal{E}_z(\boldsymbol{\rho}, t) - \frac{1}{c^2}\frac{\partial^2}{\partial t^2}\mathcal{E}_z(\boldsymbol{\rho}, t) = \mu \frac{\partial}{\partial t}\mathcal{J}_z(\boldsymbol{\rho}, t) \tag{7.340}$$

where

$$\nabla_{\mathrm{t}}^2 := \nabla_{\mathrm{t}} \cdot \nabla_{\mathrm{t}} = \nabla^2 - \frac{\partial^2}{\partial z^2} \tag{7.341}$$

is the *transverse Laplace operator* with respect to z. The TM waves which solve (7.340) are not necessarily uniform for we allow $\mathcal{J}_z$ to depend on ρ and φ. We turn (7.340) into an algebraic equation by expressing the unknown electric field and the source as the inverse Fourier transform with respect to time and space of suitable functions of ω and $\mathbf{k}_{\mathrm{t}} = k_x\hat{\mathbf{x}} + k_y\hat{\mathbf{y}}$, namely,

$$\mathcal{E}_z(\boldsymbol{\rho}, t) = \mathfrak{F}^{-1}\{\mathfrak{F}_2^{-1}\{\tilde{E}_z\}\} := \frac{1}{2\pi}\int_{\mathbb{R}} \mathrm{d}\omega\, \mathrm{e}^{\mathrm{j}\,\omega t}\frac{1}{(2\pi)^2}\iint_{\mathbb{R}^2} \mathrm{d}^2 k_t\, \mathrm{e}^{-\mathrm{j}\,\mathbf{k}_{\mathrm{t}}\cdot\boldsymbol{\rho}}\tilde{E}_z(\mathbf{k}_{\mathrm{t}}, \omega) \tag{7.342}$$

$$\mathcal{J}_z(\boldsymbol{\rho}, t) = \mathfrak{F}^{-1}\{\mathfrak{F}_2^{-1}\{\tilde{J}_z\}\} := \frac{1}{2\pi}\int_{\mathbb{R}} \mathrm{d}\omega\, \mathrm{e}^{\mathrm{j}\,\omega t}\frac{1}{(2\pi)^2}\iint_{\mathbb{R}^2} \mathrm{d}^2 k_t\, \mathrm{e}^{-\mathrm{j}\,\mathbf{k}_{\mathrm{t}}\cdot\boldsymbol{\rho}}\tilde{J}_z(\mathbf{k}_{\mathrm{t}}, \omega) \tag{7.343}$$

where $\mathbf{k}_{\mathrm{t}}$ (physical dimension: 1/m) is called the *spatial frequency*. By inserting the representations (7.342) and (7.343) into (7.340) we readily arrive at

$$\frac{1}{2\pi}\int_{\mathbb{R}} \mathrm{d}\omega\, \mathrm{e}^{\mathrm{j}\,\omega t}\frac{1}{(2\pi)^2}\iint_{\mathbb{R}^2} \mathrm{d}^2 k_t\, \mathrm{e}^{-\mathrm{j}\,\mathbf{k}_{\mathrm{t}}\cdot\boldsymbol{\rho}}\left[\left(-\mathbf{k}_{\mathrm{t}}\cdot\mathbf{k}_{\mathrm{t}} + \frac{\omega^2}{c^2}\right)\tilde{E}_z - \mathrm{j}\,\omega\mu\tilde{J}_z\right] = 0 \tag{7.344}$$

having swapped the order of integration and temporal and spatial derivatives under the following assumptions:

- the integrand $\tilde{E}_z(\boldsymbol{\rho};\omega)\exp(\mathrm{j}\,\omega t - \mathrm{j}\,\mathbf{k}_{\mathrm{t}}\cdot\boldsymbol{\rho})$ as a function of $\mathbf{k}_{\mathrm{t}}$, ω and t has continuous derivatives with respect to $\mathbf{k}_{\mathrm{t}}$, ω and t up to the second order for $(\mathbf{k}_{\mathrm{t}}, \omega, t) \in \mathbb{R}^2 \times \mathbb{R} \times \mathbb{R}^+$;
- the integrand $\tilde{E}_z(\boldsymbol{\rho};\omega)\exp(\mathrm{j}\,\omega t - \mathrm{j}\,\mathbf{k}_{\mathrm{t}}\cdot\boldsymbol{\rho})$ as a function of $\mathbf{k}_{\mathrm{t}}$, ω and $\boldsymbol{\rho}$ has continuous derivatives with respect to k_x, k_y, ω, ρ and φ up to the second order for $(\mathbf{k}_{\mathrm{t}}, \omega, \boldsymbol{\rho}) \in \mathbb{R}^2 \times \mathbb{R} \times \mathbb{R}^2$;
- the integrand $\tilde{J}_z(\boldsymbol{\rho};\omega)\exp(\mathrm{j}\,\omega t - \mathrm{j}\,\mathbf{k}_{\mathrm{t}}\cdot\boldsymbol{\rho})$ as function of $\mathbf{k}_{\mathrm{t}}$, ω and t is endowed with continuous derivatives with respect to $\mathbf{k}_{\mathrm{t}}$, ω and t up to the first order for $(\mathbf{k}_{\mathrm{t}}, \omega, t) \in \mathbb{R}^2 \times \mathbb{R} \times \mathbb{R}^+$.

Since the exponential function $\exp(\mathrm{j}\,\omega t - \mathrm{j}\,\mathbf{k}_{\mathrm{t}}\cdot\boldsymbol{\rho})$ never vanishes for any value of its argument, the equation above is satisfied only if we require

$$\left(\mathbf{k}_{\mathrm{t}}\cdot\mathbf{k}_{\mathrm{t}} - \frac{\omega^2}{c^2}\right)\tilde{E}_z(\mathbf{k}_{\mathrm{t}}, \omega) = -\mathrm{j}\,\omega\mu\tilde{J}_z(\mathbf{k}_{\mathrm{t}}, \omega) \tag{7.345}$$

which constitutes the algebraic (spectral) counterpart of the wave equation (7.340). The solution is now trivial, i.e.,

$$\tilde{E}_z(\mathbf{k}_{\mathrm{t}}, \omega) = -\frac{\mathrm{j}\,\omega\mu\tilde{J}_z(\mathbf{k}_{\mathrm{t}}, \omega)}{\mathbf{k}_{\mathrm{t}}\cdot\mathbf{k}_{\mathrm{t}} - \dfrac{\omega^2}{c^2}} = -\frac{\mathrm{j}\,\omega\mu\tilde{J}_z(\mathbf{k}_{\mathrm{t}}, \omega)}{k_x^2 + k_y^2 - k^2} = -\mathrm{j}\,\omega\mu\tilde{G}(\mathbf{k}_{\mathrm{t}}, \omega)\tilde{J}_z(\mathbf{k}_{\mathrm{t}}, \omega) \tag{7.346}$$

where with evident notation $\tilde{G}(\mathbf{k}_t, \omega)$ is the *spectral Green function* in the domain of the temporal and spatial frequencies.

We go back to the frequency domain by performing the inverse Fourier transformation with respect to space only, viz.,

$$E_z(\rho; \omega) = \frac{1}{(2\pi)^2} \iint_{\mathbb{R}^2} \mathrm{d}^2 k_t\, \mathrm{e}^{-\mathrm{j}\,\mathbf{k}_t\cdot\boldsymbol{\rho}} \frac{-\mathrm{j}\,\omega\mu \tilde{J}_z(\mathbf{k}_t, \omega)}{k_x^2 + k_y^2 - k^2} = -\mathrm{j}\,\omega\mu \int_{\mathbb{R}^2} \mathrm{d}S'\, G(\boldsymbol{\rho} - \boldsymbol{\rho}'; \omega) J_z(\boldsymbol{\rho}'; \omega) \qquad (7.347)$$

where we have applied the convolution theorem [4, Section 4.8] for the two-dimensional transformation, and in particular $G(\boldsymbol{\rho}; \omega)$ is the formal inverse Fourier transform of $\tilde{G}(\mathbf{k}_t, \omega)$.

We wish to show that $G(\boldsymbol{\rho}; \omega)$ is the same as the Green function (7.316). Since the latter is the cylindrical TM wave generated by an ideal electric line current in $\boldsymbol{\rho} = 0$ we insert the current density

$$J_z(\boldsymbol{\rho}'; \omega) := \frac{\tilde{I}(\omega)}{\pi a^2} U(a - \rho'), \qquad \rho' > 0 \qquad (7.348)$$

into (7.347) and examine the behavior of the electric field $E_z(\boldsymbol{\rho}; \omega)$ in the limit as $a \to 0^+$. For observation points $\boldsymbol{\rho}$ outside the source region ($\rho > a$) we have

$$\begin{aligned}
\left|E_z(\rho; \omega) + \mathrm{j}\,\omega\mu\tilde{I}(\omega) G(\rho; \omega)\right| &= \left|-\mathrm{j}\,\frac{\omega\mu}{\pi a^2}\tilde{I}(\omega) \int_{\rho' \leqslant a} \mathrm{d}S'\, \left[G(\boldsymbol{\rho} - \boldsymbol{\rho}'; \omega) - G(\rho; \omega)\right]\right| \\
&\leqslant \frac{\omega\mu}{\pi a^2}\left|\tilde{I}(\omega)\right| \int_{\rho' \leqslant a} \mathrm{d}S'\, \left|G(\boldsymbol{\rho} - \boldsymbol{\rho}'; \omega) - G(\rho; \omega)\right| \\
&\leqslant \frac{\omega\mu}{\pi a^2}\left|\tilde{I}(\omega)\right| \int_{\rho' \leqslant a} \mathrm{d}S'\, M|\boldsymbol{\rho}'| = \frac{\omega\mu}{\pi a^2}\left|\tilde{I}(\omega)\right| 2\pi M \frac{a^3}{3} \xrightarrow[a \to 0^+]{} 0
\end{aligned} \qquad (7.349)$$

where $M > 0$ is a suitable constant. We have been able to invoke estimate (A.90) for the scalar field $G(\boldsymbol{\rho} - \boldsymbol{\rho}'; \omega)$ because the latter is regular for $\boldsymbol{\rho} \neq \boldsymbol{\rho}'$. Since we have just proved the identity

$$E_z(\rho; \omega) = -\mathrm{j}\,\omega\mu\tilde{I}(\omega) G(\rho; \omega) \qquad (7.350)$$

comparison with (7.325) leads us to conclude that the Green function $G(\rho; \omega)$ is the same as in (7.316). As a byproduct we have also found a general solution to the Helmholtz equation (7.310) or (1.238) for $\mathbf{r} \in \mathbb{R}^3$ as a superposition or convolution integral (7.347) over the current density $J_z(\boldsymbol{\rho}; \omega)$. The role played by $G(\boldsymbol{\rho} - \boldsymbol{\rho}'; \omega)$ is analogous to that of the static Green functions examined in Chapters 2 and 4.

More importantly, $G(\rho; \omega)$ is the inverse Fourier transform of $\tilde{G}(\mathbf{k}_t, \omega)$ by definition, and thus we have proved the spectral representation

$$G(\rho; \omega) = \frac{1}{4\mathrm{j}} H_0^{(2)}(k\rho) = \frac{1}{(2\pi)^2} \int_{\mathbb{R}} \mathrm{d}k_x \int_{\mathbb{R}} \mathrm{d}k_y \frac{\mathrm{e}^{-\mathrm{j}\,k_x x - \mathrm{j}\,k_y y}}{k_x^2 + k_y^2 - k^2} \qquad (7.351)$$

which we can construe as the expansion of the Hankel function as a continuous superposition of plane waves. This result also follows from the solution of the distributional Helmholtz equation with a two-dimensional Dirac distribution $\delta^{(2)}(\boldsymbol{\rho})$ as a source term [52, Section 4.6].

As anticipated, we exploit (7.351) to obtain the Green function in time domain. The procedure is two-fold: first we carry out the integral with respect to k_x with the aid of the Cauchy theorem

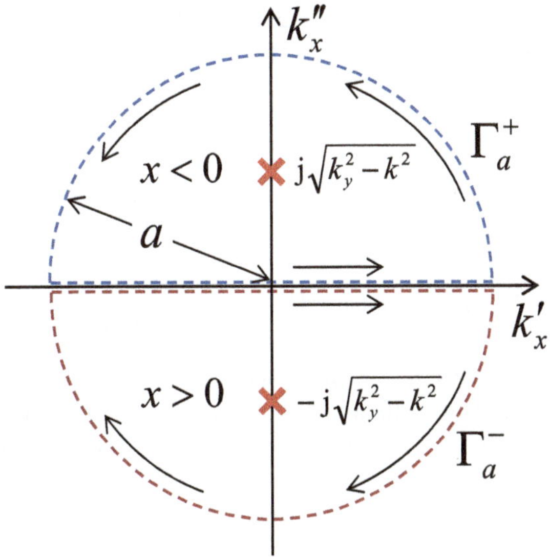

Figure 7.26 Inverse Fourier transformation of the two-dimensional spectral Green function: poles (×) and contours (−−) for integration in the complex plane k_x.

of residues (Appendix B) and secondly we manipulate the remaining integral to arrive at a Laplace transformation. We shall regard k_x and k_y as complex variables whenever this assumption serves our purposes.

For the integration along the real axis in the complex plane $k_x = k_x' + \mathrm{j}\, k_x''$ we consider the contours Γ_a^+ and Γ_a^- shown in Figure 7.26 and formally defined as

$$\Gamma_a^+ := \{k_x \in \mathbb{C} : k_x = \xi, \xi \in [-a, a] \cup k_x = a e^{\mathrm{j}\alpha}, \alpha \in [0, \pi]\} \tag{7.352}$$

$$\Gamma_a^- := \{k_x \in \mathbb{C} : k_x = \xi, \xi \in [-a, a] \cup k_x = a e^{-\mathrm{j}\alpha}, \alpha \in [0, \pi]\} \tag{7.353}$$

with $a > 0$. Both Γ_a^+ and Γ_a^- are comprised of part of the real axis and a half-circumference of radius a. With our definition, the orientation of Γ_a^+ is counterclockwise and that of Γ_a^- is clockwise. As a function of k_x with k_y fixed the integrand in (7.351) possesses singularities given by the roots of the algebraic equation

$$k_x^2 = k^2 - k_y^2 \tag{7.354}$$

which is trivially solved. Nonetheless, the location of the roots in the plane k_x is affected by the value we assign to k_y, though the final result — the Green function we are trying to compute — must not depend on this intermediate choice. Thus, for the moment we suppose $k_y \in \mathbb{R}$ and $|k_y| > k$ whereby the singularities in question become

$$k_{x1} = \mathrm{j}\, \sqrt{k_y^2 - k^2} \qquad\qquad k_{x2} = -\mathrm{j}\, \sqrt{k_y^2 - k^2} \tag{7.355}$$

i.e., two first-order poles located on the imaginary axis (Figure 7.26). For $k_y = k$ a single second-order pole appears in the origin, whereas for $|k_y| < k$ the poles are first-order but located on the real axis. The latter two occurrences just make the integration process more involved.

It is apparent that Γ_a^+ encircles the pole k_{x1} and Γ_a^- encircles the pole k_{x2}. Therefore, application of the Cauchy theorem (B.57) and calculation of the relevant residues in the poles through (B.46) yields

$$\oint_{\Gamma_a^+} \mathrm{d}k_x \frac{\mathrm{e}^{-\mathrm{j}k_x x}}{k_x^2 + k_y^2 - k^2} = 2\pi\mathrm{j} \lim_{k_x \to k_{x1}} \frac{(k_x - k_{x1})\mathrm{e}^{-\mathrm{j}k_x x}}{(k_x - k_{x1})(k_x - k_{x2})} = \pi \frac{\mathrm{e}^{\sqrt{k_y^2 - k^2}\, x}}{\sqrt{k_y^2 - k^2}} \tag{7.356}$$

$$\oint_{\Gamma_a^-} \mathrm{d}k_x \frac{\mathrm{e}^{-\mathrm{j}k_x x}}{k_x^2 + k_y^2 - k^2} = -2\pi\mathrm{j} \lim_{k_x \to k_{x2}} \frac{(k_x - k_{x2})\mathrm{e}^{-\mathrm{j}k_x x}}{(k_x - k_{x1})(k_x - k_{x2})} = \pi \frac{\mathrm{e}^{-\sqrt{k_y^2 - k^2}\, x}}{\sqrt{k_y^2 - k^2}} \tag{7.357}$$

with due regard to the different orientation of Γ_a^+ and Γ_a^-.

Next, we split the contour integrals into two contributions each and we investigate whether or not we may take the limit as $a \to +\infty$. This step is necessary since our goal is an integral over the whole real axis $k_x = k_x'$. From (7.356) and the definition of Γ_a^+ we get

$$\pi \frac{\mathrm{e}^{\sqrt{k_y^2 - k^2}\, x}}{\sqrt{k_y^2 - k^2}} = \int_{-a}^{a} \mathrm{d}\xi \frac{\mathrm{e}^{-\mathrm{j}\xi x}}{\xi^2 + k_y^2 - k^2} + \int_0^{\pi} \mathrm{d}\alpha \frac{\mathrm{j}\, a\mathrm{e}^{\mathrm{j}\alpha}\, \mathrm{e}^{-\mathrm{j}ax\cos\alpha}\, \mathrm{e}^{ax\sin\alpha}}{a^2\mathrm{e}^{2\mathrm{j}\alpha} + k_y^2 - k^2} \tag{7.358}$$

where the left member is independent of a. Since $a\sin\alpha > 0$, the second term in the right-hand side may remain finite as $a \to +\infty$ only if $x < 0$, owing to the presence of the real exponential function $\exp(ax\sin\alpha)$. Then, we observe

$$\left| \int_0^{\pi} \mathrm{d}\alpha \frac{\mathrm{j}\, a\mathrm{e}^{\mathrm{j}\alpha}\, \mathrm{e}^{-\mathrm{j}ax\cos\alpha}\, \mathrm{e}^{ax\sin\alpha}}{a^2\mathrm{e}^{2\mathrm{j}\alpha} + k_y^2 - k^2} \right| \leqslant \int_0^{\pi} \mathrm{d}\alpha \frac{a}{|a^2\mathrm{e}^{2\mathrm{j}\alpha} + k_y^2 - k^2|} \leqslant \int_0^{\pi} \mathrm{d}\alpha \frac{Ma}{a^2} = \frac{M\pi}{a} \xrightarrow[a \to +\infty]{} 0 \tag{7.359}$$

where $M > 1$ is a suitable constant.[2] Hence, in the limit as $a \to +\infty$ the contour integral along Γ_a^+ passes over into an improper integral along the real axis, and we conclude

$$\int_{-\infty}^{+\infty} \mathrm{d}k_x \frac{\mathrm{e}^{-\mathrm{j}k_x x}}{k_x^2 + k_y^2 - k^2} = \pi \frac{\mathrm{e}^{\sqrt{k_y^2 - k^2}\, x}}{\sqrt{k_y^2 - k^2}} \tag{7.360}$$

valid for $x < 0$. A perfectly similar analysis for the contour integral along Γ_a^- yields

$$\int_{-\infty}^{+\infty} \mathrm{d}k_x \frac{\mathrm{e}^{-\mathrm{j}k_x x}}{k_x^2 + k_y^2 - k^2} = \pi \frac{\mathrm{e}^{-\sqrt{k_y^2 - k^2}\, x}}{\sqrt{k_y^2 - k^2}} \tag{7.361}$$

which on the contrary holds true for $x > 0$. Inserting these two results into (7.351) provides us with the integral representation

$$\frac{1}{4\mathrm{j}} H_0^{(2)}(k\rho) = \frac{1}{4\pi} \int_{\mathbb{R}} \mathrm{d}k_y \frac{\mathrm{e}^{-\mathrm{j}k_y y - |x|\sqrt{k_y^2 - k^2}}}{\sqrt{k_y^2 - k^2}} \tag{7.362}$$

[2]This estimate holds for $M > 1$ and follows from the reverse triangle inequality (H.20) by noticing that

$$\frac{1}{|z - z_0|} \leqslant \frac{1}{|z| - |z_0|} \leqslant \frac{M}{|z|} \qquad \text{when} \qquad |z| \geqslant \frac{M}{M-1}|z_0|$$

where $z, z_0 \in \mathbb{C} \setminus \{0\}$ and $|z| > |z_0|$.

which despite the working hypothesis on k_y is indeed valid for $k_y \in \mathbb{C}$. We wish to cast this integral into an alternative form which we can construe as a Laplace transformation with respect to time.

As a function of the complex variable $k_y = k_y' + \mathrm{j}\, k_y''$ the integrand in (7.362) is many-valued and becomes infinite for $k_y = \pm k$, where it has two algebraic branch points of order one. Actually, we may slightly deform the integration path so as to steer clear of the singularities. The line γ comprised of pieces of the real axis and two small half-circumferences centered in $\pm k$ does the trick (Figure 7.27a).

The choice of the branch cuts is quite arbitrary (Appendix B.3). For instance, we might introduce a cut that connects the branch points directly along the real axis. This is not convenient, though, for the path γ would cross the branch line, and this occurrence in turn complicates the integration by far. A better solution is constituted by the lines

$$\gamma_b^+ := \{k_y \in \mathbb{C} : k_y = \xi, \xi \geqslant k\} \qquad \gamma_b^- := \{k_y \in \mathbb{C} : k_y = \xi, \xi \leqslant -k\} \tag{7.363}$$

where it is understood that γ runs parallel to γ_b^- but lies in the third quadrant and parallel to γ_b^+ but lies in the first quadrant, as is pictorially highlighted in Figure 7.27a. It is not difficult to see that

$$\mathrm{Im}\sqrt{k_y^2 - k^2} = 0 \tag{7.364}$$

for points $k_y \in \gamma_b^+ \cup \gamma_b^-$. Since the function $f(k_y) = (k_y^2 - k^2)^{1/2}$ is many-valued, namely, it takes on two opposite values for each k_y, we arbitrarily decide that the complex plane shown in Figure 7.27a is mapped to the values of $f(k_y)$ which have positive imaginary part, as indicated. The values of $f(k_y)$ with negative imaginary part are obtained from points k_y belonging to another plane superimposed to the previous one. The two planes — called *sheets* of the *Riemann surface* associated with the function — are conceptually distinct though 'glued' along the branch cuts. Since the imaginary part of $f(k_y)$ is null on $\gamma_b^+ \cup \gamma_b^-$, a path that crosses a branch line begins on one sheet and continues on the other.

We proceed by letting

$$|x| = \rho\cos\varphi, \qquad y = \rho\sin\varphi, \qquad \varphi \in \left[-\frac{\pi}{2}, \frac{\pi}{2}\right] \tag{7.365}$$

where the limitation imposed on the range of the polar angle φ is necessary for the consistency of the representation of $|x|$, which is always positive. Nevertheless, this restriction must be inconsequential for the final result which, as we know, does not depend on φ for symmetry reasons.

An important step of the second part of the procedure is represented by the change of dummy variable [51]

$$k_y = k\sin w \tag{7.366}$$

which for $w \in \mathbb{C}$ defines a single-valued analytic function from the complex plane w onto the complex plane k_y. We examine a few cases

$$k_y(w) = k\sin(\mathrm{j}\, w'') = \mathrm{j}\, k\sinh w'', \qquad w = \mathrm{j}\, w'' \tag{7.367}$$

$$k_y(w) = k\sin(w'), \qquad w = w' \in \left[-\frac{\pi}{2}, \frac{\pi}{2}\right] \tag{7.368}$$

$$k_y(w) = k\sin\left(\frac{\pi}{2} + \mathrm{j}\, w''\right) = k\cosh w'', \qquad w = \frac{\pi}{2} + \mathrm{j}\, w'' \tag{7.369}$$

$$k_y(w) = k\sin\left(-\frac{\pi}{2} + \mathrm{j}\, w''\right) = -k\cosh w'', \qquad w = -\frac{\pi}{2} + \mathrm{j}\, w'' \tag{7.370}$$

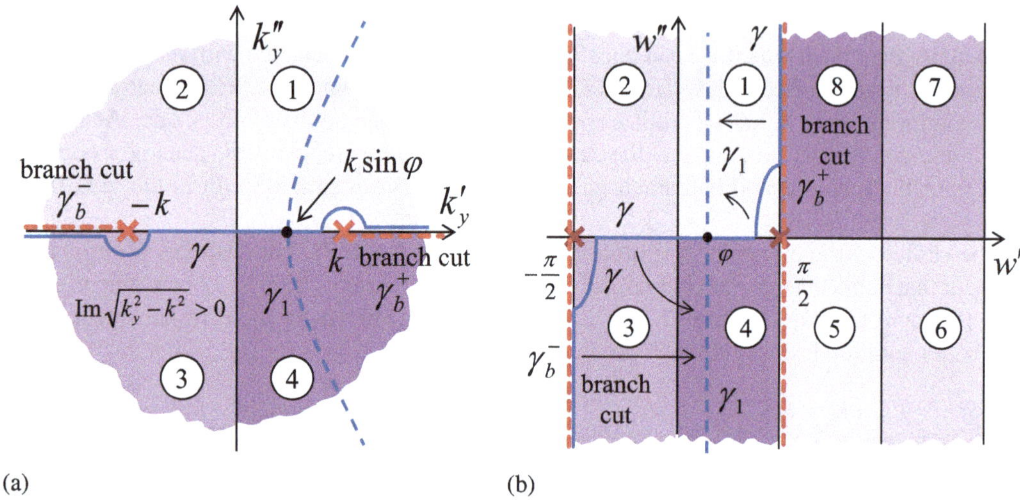

Figure 7.27 Inverse Fourier transformation of the two-dimensional spectral Green function: (a) singularities (×), branch cuts (••), old (−) and new (−−) integration paths in the complex plane k_y; (b) singularities (×), branch cuts (••), old (−) and new (−−) integration paths in the complex plane w.

which allow us to conclude that the four unbounded rectangular regions indicated with the labels ①, ②, ③ and ④ in Figure 7.27b are mapped to the four quadrants of the complex plane k_y. In particular, the path γ outlined in Figure 7.27b has the line γ in the complex plane k_y as image, and hence it represents the integration path in the plane w.

According to (7.366) and for w belonging to the strip $w' \in [-\pi/2, \pi/2]$ the function $f(k_y)$ becomes

$$f(k_y) = f(w) = \sqrt{k^2 \sin^2 w - k^2} = \mathrm{j}\, k \cos w \tag{7.371}$$

under the additional hypothesis that the four rectangular regions ①, ②, ③ and ④ are mapped to the values of $f(k_y)$ which have positive imaginary part. For the sake of completeness, we notice that the rectangular regions labelled with ⑤, ⑥, ⑦ and ⑧ are mapped to the values of $f(k_y) = f(w)$ with negative imaginary part.[3] Finally, the lines $w = (\pi/2)^+ + \mathrm{j}\, w''$ and $w = -(\pi/2)^- + \mathrm{j}\, w''$ are transformed into the branch cuts γ_b^+ and γ_b^- in the plane k_y. Branch points and associated cuts are drawn in Figure 7.27b as well, but we remark that these singularities belong to $f(w)$ and not to the function (7.366).

In light of (7.365) and (7.366) we have

$$-\mathrm{j}\, k_y y - |x| \sqrt{k_y^2 - k^2} = -\mathrm{j}\, k\rho \sin w \sin\varphi - \mathrm{j}\, k\rho \cos w \cos\varphi = -\mathrm{j}\, k\rho \cos(w - \varphi) \tag{7.372}$$

and with the differential given by $\mathrm{d}k_y = k \cos w \, \mathrm{d}w$ (7.362) passes over into (cf. [50, Chapter 19])

$$\frac{1}{4\mathrm{j}} H_0^{(2)}(k\rho) = \frac{1}{4\pi \mathrm{j}} \int_\gamma \mathrm{d}w \, \mathrm{e}^{-\mathrm{j}\, k\rho \cos(w-\varphi)} \tag{7.373}$$

[3]These regions come into play if instead of (7.365) we choose $\varphi \in [\pi/2, 3\pi/2]$, $|x| = -\rho\cos\varphi$, $y = \rho \sin\varphi$ and in lieu of (7.371) we employ $f(k_y) = f(w) = -\mathrm{j}\, k \cos w$. Regardless, in the end we arrive at the same formula (7.380).

having used (7.371). Since integrating along γ is cumbersome we may deform the path into the straight line γ_1 (Figure 7.27b) which runs parallel to the imaginary axis and encompasses the point $w = \varphi$. This deformation is permissible for the integrand in (7.373) is analytic in the four regions ①, ②, ③ and ④ and in the process we do not cross any branch cut. Of course, due to the correspondence between the complex planes w and k_y we could have deformed the path directly in (7.362), though it is not easy to figure out what the path should be [13]. In fact, the parametric representation of the line γ_1 in the plane k_y reads

$$\gamma_1 : \begin{cases} k_y' = k \sin\varphi \cosh w'' \\ k_y'' = k \cos\varphi \sinh w'' \end{cases} \qquad w'' \in \mathbb{R}, \qquad \varphi \in \left[-\frac{\pi}{2}, \frac{\pi}{2}\right] \tag{7.374}$$

which by eliminating w'' with a little algebra becomes

$$\gamma_1 : \frac{k_y'^2}{k^2 \sin^2\varphi} - \frac{k_y''^2}{k^2 \cos^2\varphi} = 1, \qquad \text{with} \qquad \begin{matrix} k_y' > 0, & \varphi \in [0, \pi/2] \\ k_y' < 0, & \varphi \in [-\pi/2, 0] \end{matrix} \tag{7.375}$$

that is, the equation of a branch of hyperbola [19, Section 8.4] which intersects the real axis for $k_y' = k \sin\varphi$ and has focus in $k_y' = k\,\mathrm{sign}(\varphi)$, where sign(•) is the *signum* function (Figure 7.27a). Finally, for $\varphi = 0$ the line γ_1 degenerates into the imaginary axis $k_y' = 0$.

Next, we make the change of variable

$$w = \varphi + \mathrm{j}\xi, \qquad \xi \in \mathbb{R} \tag{7.376}$$

whereby we obtain

$$\frac{1}{4\pi \mathrm{j}} \int_{\gamma_1} \mathrm{d}w\, \mathrm{e}^{-\mathrm{j}k\rho\cos(w-\varphi)} = \frac{1}{4\pi} \int_{-\infty}^{+\infty} \mathrm{d}\xi\, \mathrm{e}^{-\mathrm{j}k\rho\cosh\xi} = \frac{1}{2\pi} \int_{0}^{+\infty} \mathrm{d}\xi\, \mathrm{e}^{-\mathrm{j}k\rho\cosh\xi} \tag{7.377}$$

for the integrand is an even function of ξ. In order to write the last integral as a unilateral Laplace transformation with respect to time we let

$$k = \frac{\omega}{c} = -\frac{\mathrm{j}\,s}{c} \tag{7.378}$$

$$t = \frac{\rho}{c}\cosh\xi, \qquad \mathrm{d}t = \frac{\rho}{c}\sinh\xi\,\mathrm{d}\xi = \frac{\rho}{c}\sqrt{\cosh^2\xi - 1}\,\mathrm{d}\xi = \sqrt{t^2 - \frac{\rho^2}{c^2}}\,\mathrm{d}\xi, \qquad \xi \geqslant 0 \tag{7.379}$$

and from (7.362) and (7.377) we find

$$\frac{1}{4\mathrm{j}} H_0^{(2)}\left(-\mathrm{j}\,s\frac{\rho}{c}\right) = \int_{\rho/c}^{+\infty} \mathrm{d}t\, \frac{\mathrm{e}^{-st}}{2\pi\sqrt{t^2 - \rho^2/c^2}} \tag{7.380}$$

which is the desired result and completes the derivation. Our goal is the calculation of the integral in (7.339) which on account of the change of variable (7.378) we may cast as

$$\mathcal{G}(\rho, t) = \frac{1}{2\pi \mathrm{j}} \int_{-\mathrm{j}\infty}^{+\mathrm{j}\infty} \mathrm{d}s\, \frac{\mathrm{e}^{st}}{4\mathrm{j}} H_0^{(2)}\left(-\mathrm{j}\,s\frac{\rho}{c}\right) \tag{7.381}$$

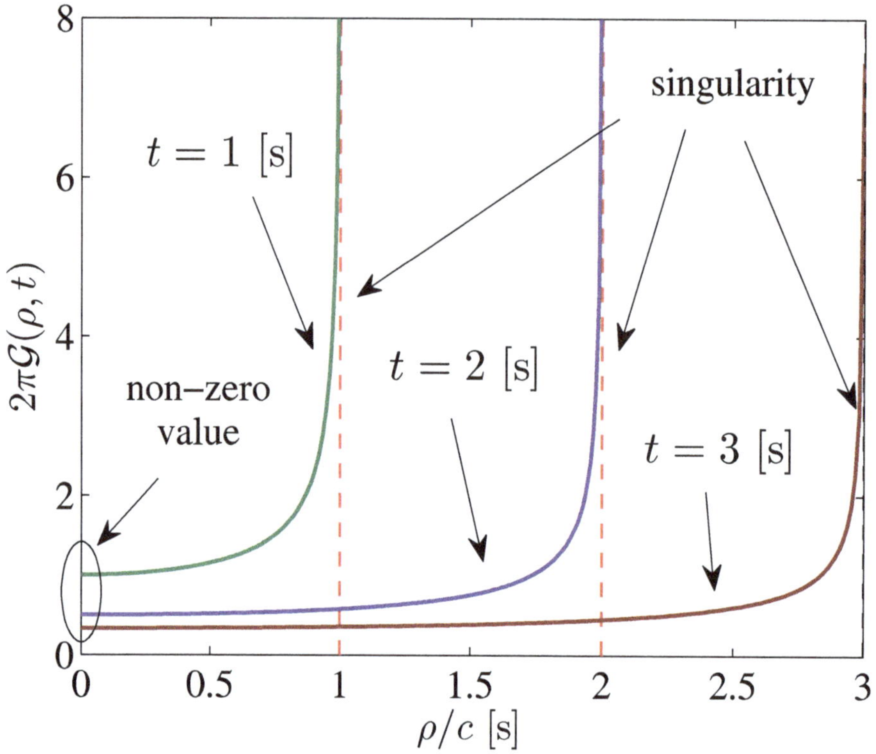

Figure 7.28 Three snapshots of the two-dimensional Green function (7.382).

that is, the inverse Laplace transformation associated with the direct one (7.380). Since according to the Lerch theorem [53], [54, Chapter 2] the inverse Laplace transform is unique, we conclude that

$$\mathcal{G}(\rho, t) = \frac{U(t - \rho/c)}{2\pi\sqrt{t^2 - \rho^2/c^2}} \tag{7.382}$$

is the two-dimensional Green function in the time domain. Some Authors (e.g., [15]) normalize the spectral Green function $G(\rho; \omega)$ differently, whereby the factor 2π in the denominator of (7.382) is absent.

As expected, the Green function depends only on the radial coordinate, and thus (7.382) is valid for $\varphi \in [0, 2\pi[$. The step function $U(t - \rho/c)$ signals that $\mathcal{G}(\rho, t) = 0$ for $\rho > ct$. The Green function is a special cylindrical wave which travels with finite speed given by c away from the z-axis where the impulsive singular source is located. Therefore, it takes a time $t = \rho/c$ for the disturbance to reach any point on the cylindrical surface of radius ρ. Figure 7.28 shows three instances of the function $\mathcal{G}(\rho, t)$ for increasingly larger values of time. The Green function is singular for $t = \rho/c$ and remains non-zero for $\rho = 0$ even if the impulsive source is active only at $t = 0$. This phenomenon may be explained by observing that the source — though concentrated on the z-axis — is infinitely extended along z. Thus, at any given time the disturbances generated by ever farther source points along z arrive in the origin (or any other point, for that matter) and add up coherently.

7.7 Time-domain transverse electric-magnetic spherical waves

In our quest for wave-like solutions thus far we have somewhat, though correctly, simplified our task by assuming that the electromagnetic entities of concern depended on just one spatial coordinate which turned out to be the direction of propagation of the wave. This hypothesis worked nicely and produced the desired result for both plane and cylindrical waves. Therefore, it seems logical to wonder whether or not wave-like solutions may exist that, while still depending on just the distance from a localized source, can propagate with uniform amplitude on ever growing *spheres*.

What kind of source would do the job? Well, based on similar arguments invoked for plane and cylindrical waves, one might surmise that in a system of polar spherical coordinates (r, ϑ, φ) a *spherically symmetric* distribution of current $\mathcal{J}(r,t)$ confined in a ball $B(0,a)$ might produce uniform spherical waves characterized only by the distance $r > a$ from the origin and obviously the time variable. Unfortunately, this speculation turns out to be faulty because such a source does not generate electromagnetic waves at all, as we shall argue extensively in Section 7.8 [55, Section 6.1].

The simplest spherical waves we can come up with must depend on the *distance* r from the source and the *elevation angle* ϑ, though we may still presume azimuthal symmetry. In theory, these waves are emitted by oscillating point charges, but the latter are no more practical than infinite planar current densities (for plane waves) or infinitely long line currents (for cylindrical waves). Nevertheless, we suppose that an ideal current *concentrated in the origin* of the system of coordinates possesses the necessary spatial dependence on r and θ so as to produce waves whose field distribution does not depend on φ and has no radial components, viz., $\mathcal{E}_r(r,t) = 0 = \mathcal{H}_r(r,t)$. Solutions of this type are called *transverse electric-magnetic* with respect to r ($\mathrm{TEM_r}$ for short) precisely because there is no field component in the direction of propagation $\hat{\mathbf{r}}$. Parenthetically, according to this definition, also the plane waves and the cylindrical waves discussed in the Sections 7.2 and 7.5 are TEM with respect to the directions of propagation thereof.

We obviate the need for expressing the source explicitly by excluding the troublesome point $\mathbf{r} = 0$, and for the problem of concern we write down the Maxwell equations in local form in spherical coordinates (Appendix A.1)

$$\frac{\hat{\mathbf{r}}}{r\sin\vartheta}\frac{\partial}{\partial\vartheta}(\mathcal{E}_\varphi \sin\vartheta) - \frac{\hat{\boldsymbol{\vartheta}}}{r}\frac{\partial}{\partial r}(r\mathcal{E}_\varphi) + \frac{\hat{\boldsymbol{\varphi}}}{r}\frac{\partial}{\partial r}(r\mathcal{E}_\vartheta) = -\mu\frac{\partial \mathcal{H}}{\partial t}, \qquad r > 0 \tag{7.383}$$

$$\frac{\hat{\mathbf{r}}}{r\sin\vartheta}\frac{\partial}{\partial\vartheta}(\mathcal{H}_\varphi \sin\vartheta) - \frac{\hat{\boldsymbol{\vartheta}}}{r}\frac{\partial}{\partial r}(r\mathcal{H}_\varphi) + \frac{\hat{\boldsymbol{\varphi}}}{r}\frac{\partial}{\partial r}(r\mathcal{H}_\vartheta) = \varepsilon\frac{\partial \mathcal{E}}{\partial t}, \qquad r > 0 \tag{7.384}$$

$$\frac{1}{r\sin\vartheta}\frac{\partial}{\partial\vartheta}(\mathcal{E}_\vartheta \sin\vartheta) = 0, \qquad r > 0 \tag{7.385}$$

$$\frac{1}{r\sin\vartheta}\frac{\partial}{\partial\vartheta}(\mathcal{H}_\vartheta \sin\vartheta) = 0, \qquad r > 0 \tag{7.386}$$

having used (A.34), (A.31) and also the fact that $\mathcal{E}$ and $\mathcal{H}$ are function of r, ϑ and t only. Since the vector fields in the right-hand sides of (7.383) and (7.384) do not have a radial component by hypothesis, on the whole these equations demand

$$\mathcal{E}_\varphi(r,\vartheta,t) = \frac{f_{E\varphi}(r,t)}{\sin\vartheta}, \qquad r > 0, \qquad 0 < \vartheta < \pi \tag{7.387}$$

$$\mathcal{H}_\varphi(r,\vartheta,t) = \frac{f_{H\varphi}(r,t)}{\sin\vartheta}, \qquad r > 0, \qquad 0 < \vartheta < \pi \tag{7.388}$$

$$\mathcal{E}_\vartheta(r,\vartheta,t) = \frac{f_{E\vartheta}(r,t)}{\sin\vartheta}, \qquad r > 0, \qquad 0 < \vartheta < \pi \tag{7.389}$$

$$\mathcal{H}_\vartheta(r,\vartheta,t) = \frac{f_{H\vartheta}(r,t)}{\sin\vartheta}, \qquad r>0, \qquad 0<\vartheta<\pi \tag{7.390}$$

where the functions $f_{E\varphi}(r,t)$, $f_{H\varphi}(r,t)$ and so forth are left unspecified for the moment. Regardless, the field components of a TEM$_\mathrm{r}$ spherical wave diverge on the z-axis.

As was the case for cylindrical waves, we may suppose that only two components at a time, namely, one magnetic and one electric, are actually non-zero. This gives rise to two orthogonal polarizations which, when combined by virtue of the principle of superposition (Section 6.1) provide the general TEM$_\mathrm{r}$ solution. Therefore, we can make either one of the following choices

$$\mathcal{E}_\varphi = 0 = \mathcal{H}_\vartheta \qquad \mathcal{H}_\varphi = 0 = \mathcal{E}_\vartheta \tag{7.391}$$

and go on to better specify the nature of the functions $f_{E\varphi}(r,t)$ and the like.

If we let $\mathcal{E}_\varphi = 0 = \mathcal{H}_\vartheta$, we obtain the wave equations for $\mathcal{E}_\vartheta$ and $\mathcal{H}_\varphi$. By taking the curl of (7.383) and using (7.384) to eliminate $\nabla\times\boldsymbol{\mathcal{H}}$ we find

$$\frac{\hat{\boldsymbol{\vartheta}}}{r}\frac{\partial^2}{\partial r^2}(r\mathcal{E}_\vartheta) - \hat{\boldsymbol{\vartheta}}\varepsilon\mu\frac{\partial^2\mathcal{E}_\vartheta}{\partial t^2} = 0, \qquad r>0, \qquad 0<\vartheta<\pi \tag{7.392}$$

having noticed that $\partial(\mathcal{H}_\varphi\sin\vartheta)/\partial\vartheta = 0$ on account of (7.388) even though $\mathcal{H}_\varphi$ is not null. Likewise, by taking the curl of (7.384) and invoking (7.383) to express $\nabla\times\boldsymbol{\mathcal{E}}$ we have

$$\frac{\hat{\boldsymbol{\varphi}}}{r}\frac{\partial^2}{\partial r^2}(r\mathcal{H}_\varphi) - \hat{\boldsymbol{\varphi}}\varepsilon\mu\frac{\partial^2\mathcal{H}_\varphi}{\partial t^2} = 0, \qquad r>0, \qquad 0<\vartheta<\pi \tag{7.393}$$

with $\partial(\mathcal{E}_\vartheta\sin\vartheta)/\partial\vartheta = 0$ by virtue of (7.389) even though $\mathcal{E}_\vartheta \neq 0$. By substituting the expected functional forms (7.389) and (7.388) we arrive at

$$\frac{\partial^2}{\partial r^2}(rf_{E\vartheta}) - \frac{1}{c^2}\frac{\partial^2}{\partial t^2}(rf_{E\vartheta}) = 0, \qquad r>0 \tag{7.394}$$

$$\frac{\partial^2}{\partial r^2}(rf_{H\varphi}) - \frac{1}{c^2}\frac{\partial^2}{\partial t^2}(rf_{H\varphi}) = 0, \qquad r>0 \tag{7.395}$$

namely, two wave equations for $f_{E\vartheta}$ and $f_{H\varphi}$. These equations can be put in normal form by introducing the auxiliary variables (D'Alembert Ansatz) [4, pp. 682–683]

$$\xi := t - r/c \qquad \eta := t + r/c \tag{7.396}$$

and noticing that

$$\frac{\partial\Upsilon}{\partial r} = -\frac{1}{c}\frac{\partial\Upsilon}{\partial\xi} + \frac{1}{c}\frac{\partial\Upsilon}{\partial\eta} \qquad \frac{\partial^2\Upsilon}{\partial r^2} = \frac{1}{c^2}\frac{\partial^2\Upsilon}{\partial^2\xi} - \frac{2}{c^2}\frac{\partial^2\Upsilon}{\partial\xi\partial\eta} + \frac{1}{c^2}\frac{\partial^2\Upsilon}{\partial^2\eta} \tag{7.397}$$

$$\frac{1}{c}\frac{\partial\Upsilon}{\partial t} = \frac{1}{c}\frac{\partial\Upsilon}{\partial\xi} + \frac{1}{c}\frac{\partial\Upsilon}{\partial\eta} \qquad \frac{1}{c^2}\frac{\partial^2\Upsilon}{\partial t^2} = \frac{1}{c^2}\frac{\partial^2\Upsilon}{\partial^2\xi} + \frac{2}{c^2}\frac{\partial^2\Upsilon}{\partial\xi\partial\eta} + \frac{1}{c^2}\frac{\partial^2\Upsilon}{\partial^2\eta} \tag{7.398}$$

where $\Upsilon(r,t)$ denotes either $rf_{E\vartheta}$ or $rf_{H\varphi}$. Substituting back into (7.394) or (7.395) yields

$$\frac{\partial^2\Upsilon}{\partial r^2} - \frac{1}{c^2}\frac{\partial^2\Upsilon}{\partial t^2} = -\frac{4}{c^2}\frac{\partial^2\Upsilon}{\partial\xi\partial\eta} = 0, \qquad r>0 \tag{7.399}$$

which is solved by a linear combination of functions which depend separately on ξ and η. Therefore, we can write the solution to (7.394) and (7.395) as

$$f_{E\vartheta}(r,t) = \frac{1}{r}E_0\left(t-\frac{r}{c}\right) + \frac{1}{r}F_0\left(t+\frac{r}{c}\right), \qquad r>0 \tag{7.400}$$

$$f_{H\varphi}(r,t) = \frac{1}{r}H_0\left(t - \frac{r}{c}\right) + \frac{1}{r}G_0\left(t + \frac{r}{c}\right), \qquad r > 0 \tag{7.401}$$

where E_0, F_0, H_0 and G_0 are functions at least continuously differentiable with respect to their argument. However, we must set $F_0 = 0 = G_0$ because the corresponding contributions in (7.400) and (7.401) represent non-physical *ingoing waves* which travel from infinity and *enter* the singular source in the origin. Besides, of the two remaining, non-null functions, either E_0 or H_0 can be specified arbitrarily whereas the other one must be chosen so as to satisfy the Faraday law (7.383). Indeed, inserting the expression of $f_{E\vartheta}$ and $f_{H\varphi}$ into (7.383) and letting $\xi = t - r/c$ yields

$$\frac{1}{r}\frac{\partial}{\partial r}E_0\left(t - \frac{r}{c}\right) = -\frac{\mu}{r}\frac{\partial}{\partial t}H_0\left(t - \frac{r}{c}\right) \quad \Longrightarrow \quad \frac{1}{r}\frac{\mathrm{d}}{\mathrm{d}\xi}\left[\frac{1}{c}E_0(\xi) - \mu H_0(\xi)\right] = 0 \tag{7.402}$$

and the last equation is certainly solved if we take $H_0 = E_0/(\mu c)$. Putting everything together we have found the solution

$$\mathcal{E}_\vartheta(r,\theta,t) = \frac{E_0(t - r/c)}{r\sin\vartheta}, \qquad r > 0, \qquad 0 < \vartheta < \pi \tag{7.403}$$

$$\mathcal{H}_\varphi(r,\theta,t) = \frac{E_0(t - r/c)}{\mu c r\sin\vartheta}, \qquad r > 0, \qquad 0 < \vartheta < \pi \tag{7.404}$$

with $\mathcal{E}_\varphi = 0 = \mathcal{H}_\vartheta$. Since $\mu c = Z$, the intrinsic impedance of the background medium (1.358), direct inspection shows that $\mathcal{E}_\vartheta$ and $\mathcal{H}_\varphi$ are related as

$$\mathcal{H}_\varphi\hat{\boldsymbol{\varphi}} = \frac{1}{Z}\hat{\mathbf{r}} \times \mathcal{E}_\vartheta\hat{\boldsymbol{\vartheta}} \quad \Longrightarrow \quad \mathcal{H} = \frac{1}{Z}\hat{\mathbf{r}} \times \mathcal{E} \tag{7.405}$$

a connection we refer to as an *impedance relationship*. As the wave advances towards infinity, the field components fall off as the inverse of the distance from the source concentrated in the origin. This attenuation — not associated with losses, which are absent by hypothesis — is *geometrical* in character and due to the fact the wave field must become non-zero on ever larger spheres.

For $\mathcal{E}_\vartheta = 0 = \mathcal{H}_\varphi$ we should consider the wave equations for $\mathcal{E}_\varphi$ and $\mathcal{H}_\vartheta$ and follow analogous steps to find the functions $f_{E\varphi}$ and $f_{H\vartheta}$. However, the principle of duality, here applicable, yields the answer right away

$$\mathcal{E}_\varphi(r,\theta,t) = -\frac{H_0(t - r/c)}{\varepsilon c r\sin\vartheta}, \qquad r > 0, \qquad 0 < \vartheta < \pi \tag{7.406}$$

$$\mathcal{H}_\vartheta(r,\theta,t) = \frac{H_0(t - r/c)}{r\sin\vartheta}, \qquad r > 0, \qquad 0 < \vartheta < \pi \tag{7.407}$$

with $\mathcal{E}_\vartheta = 0 = \mathcal{H}_\varphi$. In this case $\varepsilon c = 1/Z$ and we have

$$\mathcal{E}_\varphi\hat{\boldsymbol{\varphi}} = -Z\mathcal{H}_\vartheta\hat{\mathbf{r}} \times \hat{\boldsymbol{\vartheta}} \quad \Longrightarrow \quad \mathcal{E} = Z\mathcal{H} \times \hat{\mathbf{r}} \tag{7.408}$$

for the corresponding impedance relationship.

From Section 1.7 we know that at the interface between a PEC body and an isotropic dielectric medium the electric field is perpendicular to the boundary of the PEC object. Likewise, we found in Section 6.6 that the magnetic field is perpendicular to the boundary of a PMC body immersed in an isotropic magnetic medium. Therefore, we conclude that (7.403), (7.404) are also solution to (7.392) in the infinite region of space $V := \{\mathbf{r} \in \mathbb{R}^3 : r > 0, \vartheta \in [\vartheta_1, \vartheta_2], \varphi \in [0, 2\pi]\}$ limited by the boundary of two PEC cones with openings ϑ_1 and $\vartheta_2 > \vartheta_1$. For this reason and in view of the field

Figure 7.29 Practical source of spherical TEM_r waves: biconical antenna made of two truncated metallic conical surfaces.

singularity in $\vartheta \in \{0, \pi\}$ practical sources of TEM_r spherical waves are constituted by the so-called *biconical antennas* (Figure 7.29) [44, Section 11.1], [30, Section 5.6], [56, Section 9.2]. In light of the principle of duality, (7.406) and (7.407) solve the dual of (7.392) in the same region V but with the cones made of a PMC material.

To evaluate the surface power density associated with a spherical TEM_r wave we compute the Poynting vector (1.280). For the two polarizations discussed above we have

$$\mathcal{S}(\mathbf{r}, t) = \mathcal{E}_\vartheta \mathcal{H}_\varphi \hat{\mathbf{r}} = \frac{[E_0(t - r/c)]^2}{Z r^2 \sin^2 \vartheta} \hat{\mathbf{r}}, \qquad r > 0, \qquad 0 < \vartheta < \pi \tag{7.409}$$

$$\mathcal{S}(\mathbf{r}, t) = -\mathcal{E}_\varphi \mathcal{H}_\vartheta \hat{\mathbf{r}} = \frac{Z[H_0(t - r/c)]^2}{r^2 \sin^2 \vartheta} \hat{\mathbf{r}}, \qquad r > 0, \qquad 0 < \vartheta < \pi \tag{7.410}$$

whence we see that the power efflux is radially directed. Although the sources are localized in the origin, the power density *diverges* on the z-axis, and the radiated power that flows through a sphere of radius r and center the origin is *infinite*.

We can relate the functions $E_0(t - r/c)$ and $H_0(t - r/c)$ to the singular currents $\mathcal{J}(r, \vartheta, t)$ and $\mathcal{J}_M(r, \vartheta, t)$ that supposedly 'flow' in $\mathbf{r} = 0$. To this end, it pays to start by considering current distributions in a ball $B(0, a)$ and then to take the limit for vanishing radius a. For the $\mathcal{E}_\vartheta$, $\mathcal{H}_\varphi$ polarization we examine an instance of the global Ampère-Maxwell law (1.13), namely

$$\oint_{\gamma(\vartheta_0)} \mathrm{d}s\, \hat{\mathbf{s}}(\mathbf{r}) \cdot \mathcal{H}_\varphi(r, \vartheta_0, t) \hat{\boldsymbol{\varphi}} = \int_{S(\vartheta_0)} \mathrm{d}S\, \hat{\mathbf{n}}(\mathbf{r}) \cdot \mathcal{E}_\vartheta(r, \vartheta_0, t) \hat{\boldsymbol{\vartheta}} + \int_{S(\vartheta_0)} \mathrm{d}S\, \hat{\mathbf{n}}(\mathbf{r}) \cdot \mathcal{J}(r, \vartheta_0, t) \tag{7.411}$$

where

$$S(\vartheta_0) := \{\mathbf{r} \in \mathbb{R}^3 : r \in [0, a], \vartheta = \vartheta_0, \varphi \in [0, 2\pi]\} \tag{7.412}$$

is a finite one-sheeted cone symmetric around the z-axis, with opening ϑ_0 and apex in $\mathbf{r} = 0$, and

$$\gamma(\vartheta_0) : \begin{cases} x = a \sin \vartheta_0 \cos(2\pi - \varphi) \\ y = a \sin \vartheta_0 \sin(2\pi - \varphi) \\ z = a \cos \vartheta_0 \end{cases} \qquad \varphi \in [0, 2\pi] \tag{7.413}$$

is the boundary of $S(\vartheta_0)$. To compute the circulation of the magnetic field we notice that $\hat{\mathbf{s}}(\mathbf{r}) = -\hat{\boldsymbol{\varphi}}$ and $\mathrm{d}s = -a\sin\vartheta_0\,\mathrm{d}\varphi$. By inserting (7.403) and (7.404) into (7.411) we obtain

$$-\int_0^{2\pi} \mathrm{d}\varphi\, a\sin\vartheta_0\hat{\boldsymbol{\varphi}}\cdot\frac{E_0(t-a/c)}{\mu ca\sin\vartheta_0}\hat{\boldsymbol{\varphi}} = \varepsilon\int_0^a \mathrm{d}r\int_0^{2\pi}\mathrm{d}\varphi\, r\sin\vartheta_0\hat{\boldsymbol{\vartheta}}\cdot\frac{\partial}{\partial t}\frac{E_0(t-r/c)}{r\sin\vartheta_0}\hat{\boldsymbol{\vartheta}} + I(t) \qquad (7.414)$$

where $I(t)$ is the electric current that flows through $S(\vartheta_0)$. Indeed, since the other two contributions in this equation are independent of ϑ_0, so must be $I(t)$. To evaluate the integral in the right-hand side we make the change of dummy variable $\xi = t - r/c$. Since all integrands are independent of φ, this yields

$$-\frac{2\pi}{\mu c}E_0(t-a/c) = -2\pi\varepsilon c\int_t^{t-a/c}\mathrm{d}\xi\,\frac{\mathrm{d}}{\mathrm{d}\xi}E_0(\xi) + I(t) = 2\pi\varepsilon c[E_0(t) - E_0(t-a/c)] + I(t) \qquad (7.415)$$

whence eventually we get

$$E_0(t) = -\frac{Z}{2\pi}I(t) \qquad (7.416)$$

by virtue of (1.358). This expression involves neither the opening of the cone nor the radius a, and hence the result remains valid also in the limit as $a \to 0^+$, when the current $\mathfrak{J}(r,\vartheta,t)$ is confined to the origin. Consequently, (7.403) and (7.404) become

$$\mathcal{E}_\vartheta(r,\theta,t) = -Z\frac{I(t-r/c)}{2\pi r\sin\vartheta}, \qquad r>0, \qquad 0<\vartheta<\pi \qquad (7.417)$$

$$\mathcal{H}_\varphi(r,\theta,t) = -\frac{I(t-r/c)}{2\pi r\sin\vartheta}, \qquad r>0, \qquad 0<\vartheta<\pi \qquad (7.418)$$

and for the $\mathcal{E}_\varphi$, $\mathcal{H}_\vartheta$ polarization we write straightaway

$$\mathcal{E}_\varphi(r,\theta,t) = \frac{V(t-r/c)}{2\pi r\sin\vartheta}, \qquad r>0, \qquad 0<\vartheta<\pi \qquad (7.419)$$

$$\mathcal{H}_\vartheta(r,\theta,t) = -\frac{1}{Z}\frac{V(t-r/c)}{2\pi r\sin\vartheta}, \qquad r>0, \qquad 0<\vartheta<\pi \qquad (7.420)$$

thanks to the principle of duality.

Example 7.4 (Propagation of a TEM$_\mathrm{r}$ spherical wave in free space)

We consider the TEM wave generated by a singular source $\mathfrak{J}(r,\vartheta,t)$ in free space ($c = c_0$) with the current $I(t)$ [see (7.411)] given by

$$I(t) = I_0\left[U\left(t-\frac{r}{c_0}\right) - U\left(t-\frac{r}{c_0}-t_1\right)\right]\sin\left[\left(t-\frac{r}{c_0}\right)\frac{\pi}{t_1}\right] \qquad (7.421)$$

i.e., a sinusoidal pulse of duration $t_1 = 3$ s. The normalized electric field component $\mathcal{E}_\vartheta$ computed through (7.417) is shown in Figure 7.30 for $\vartheta \in [5\pi/180, 175\pi/180]$ and $\varphi = 0$, which is enough because the wave is rotationally symmetric around the z-axis. As the time goes by the amplitude of the wave, while depending on the inverse of $\sin\vartheta$, decreases with the inverse of the radial distance r.

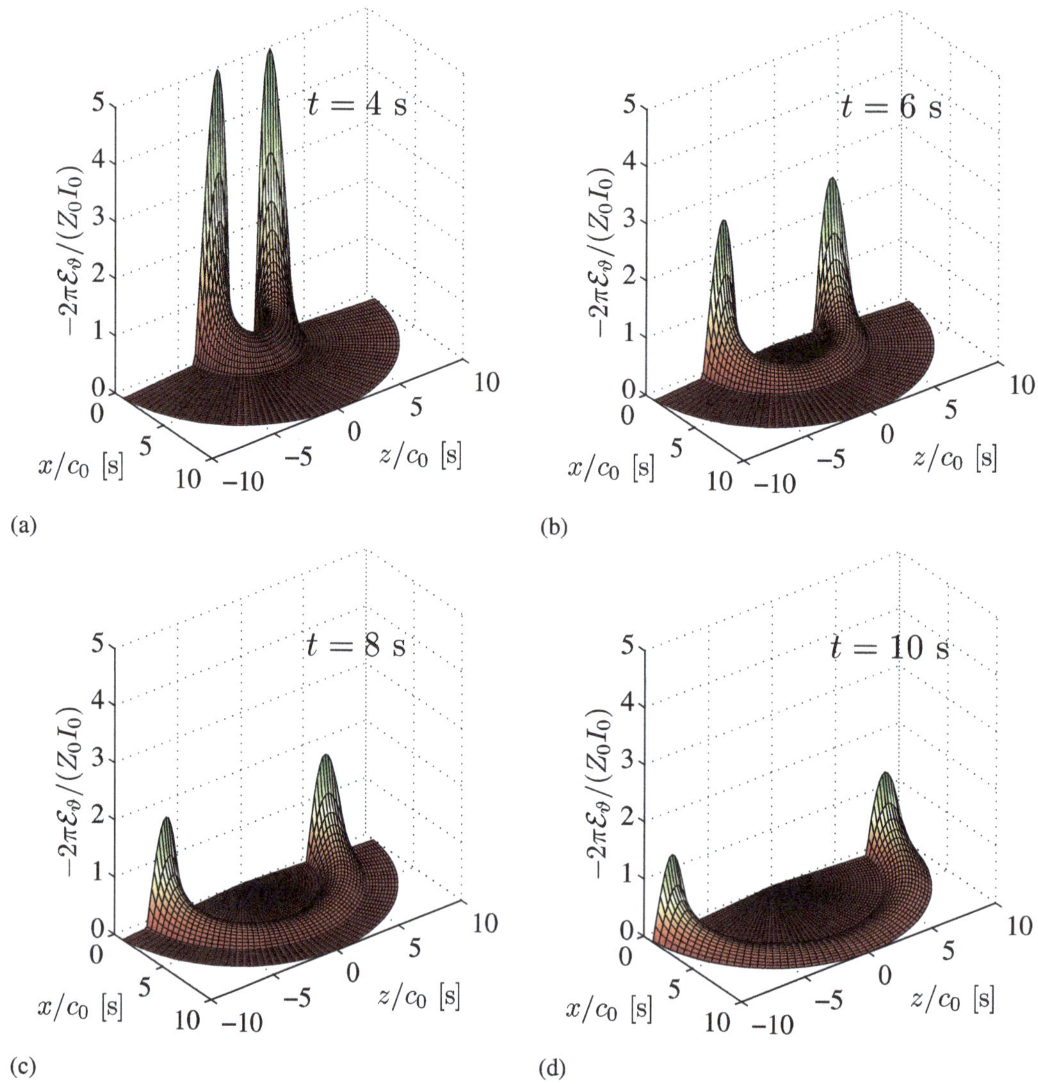

Figure 7.30 Four snapshots of the electric field $\mathcal{E}_\vartheta(r, \vartheta, t)$ of a TEM$_r$ spherical wave as a function of the normalized coordinates x/c_0 and z/c_0.

A quiet zone exists for $r/c_0 > t$, that is, outside a sphere of radius $c_0 t$ the field of the wave is zero because those points have yet to be reached by the disturbance. Unlike the two-dimensional case of cylindrical waves (Section 7.5) there is no wake trailing behind. Rather, a quiet zone exists also for $r/c_0 < t - t_1$, that is, for points inside a sphere of radius $c_0(t - t_1)$ after the disturbance has passed. The absence of a 'wake' (cf. Figure 7.25) is related to the finite extent of the source, no matter how singular that is.

(End of Example 7.4)

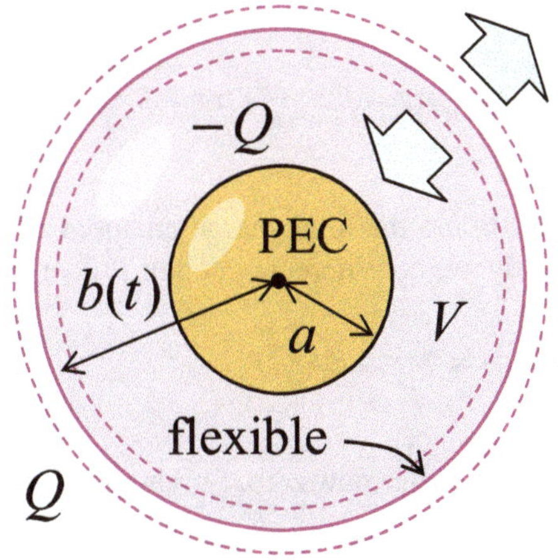

Figure 7.31 *An ideal non-radiating source made of two concentric charged spheres of which the larger external one oscillates back and forth along the radial direction as suggested by the arrows and the dashed lines.*

7.8 Non-radiating sources

In Section 7.7 we anticipated that highly symmetrical distributions of currents may not generate electromagnetic waves [55, Section 6.1], [15, 57]. In order to support this claim we consider a somewhat contrived ideal device comprised of two concentric spheres of radii a and b with $b > a > 0$ (Figure 7.31). The smaller, internal sphere may be solid and conducting inasmuch as we only require that it supports a charge $-Q$ and keeps it in place. The larger, external sphere is hollow and supposedly made of a flexible dielectric material which allows the radius b to be altered in time by applying some mechanical force.

A rubber balloon of sorts would do the trick, granted, but the real difficulty here is to inflate and deflate the balloon alternately in a perfectly symmetric way so that the outer surface indeed remains spherical all the time. To accommodate for the motion of the larger sphere we take the background medium as free space. We assume that a charge $Q > 0$ is distributed on the inside ($r = b^-$) of the extensible sphere. Obviously, if such a contraption can be realized, then spherical symmetry implies that the charge spreads uniformly with surface densities $\varrho_S = -Q/(4\pi a^2)$ and $\varrho_S = Q/(4\pi b^2)$.

As the external surface alternately increases and shrinks in time, it drags the charge Q along, and this motion gives rise to a current density of the form

$$\mathfrak{J}(\mathbf{r}, t) = \mathfrak{J}_r(r, t)\hat{\mathbf{r}} \tag{7.422}$$

since a radial displacement of charges is the only one taking place. We can be more specific on the actual expressions of $\mathfrak{J}_r(r, t)$ and $\varrho(r, t)$, but we postpone the details until further on in this section, since these quantities are not necessary for the calculation of the electromagnetic field. Indeed, as a consequence of the postulated spherical symmetry of the device the electric field must be radially directed, and thus a clever application of the Gauss law (1.16) enables us to compute $\mathcal{E}_r(r, t)$.

First of all, we observe that the electric field is zero for $r < a$, the inner sphere being solid and made of PEC [see discussion on p. 29]. Secondly, we pick up a ball $B(0, r)$ with radius $a < r < b(t)$

and consider

$$-Q = \int_{\partial B} \mathrm{d}S\, \hat{\mathbf{n}} \cdot \varepsilon_0 \mathcal{E}(r,t) = 4\pi r^2 \varepsilon_0 \mathcal{E}_r(r,t) \tag{7.423}$$

because $\mathcal{E}_r(r,t)$ is constant on ∂B and the total charge enclosed by ∂B is $-Q$, distributed on the surface of the inner sphere. Lastly, we consider a larger ball $B(0,r)$ with $r > b$ and examine

$$0 = Q - Q = \int_{\partial B} \mathrm{d}S\, \hat{\mathbf{n}} \cdot \varepsilon_0 \mathcal{E}(r,t) = 4\pi r^2 \varepsilon_0 \mathcal{E}_r(r,t) \tag{7.424}$$

because the total charge enclosed by ∂B is now zero. Solving for $\mathcal{E}_r(r,t)$ yields

$$\mathcal{E}(r,t) = \frac{Q}{4\pi\varepsilon_0 r^2}\{U[r - b(t)] - U(r-a)\}\hat{\mathbf{r}}, \qquad (\mathbf{r},t) \in \mathbb{R}^3 \times \mathbb{R}^+ \tag{7.425}$$

where $U(\bullet)$ is a one-dimensional step function. Clearly, the electric field is piecewise continuous and, in particular, vanishes outside the device while being static in character between the two spheres.

To determine the magnetic field we turn to the Faraday law (1.20), which requires computing the curl of $\mathcal{E}$. According to (A.37) we need to take the derivatives of $\mathcal{E}_r$ with respect to ϑ and φ. Since $\mathcal{E}_r$ is a constant — and hence continuous — function of ϑ and φ, the curl has still a meaning as an ordinary function. In particular, $\nabla \times \mathcal{E}$ vanishes everywhere, so that (1.20) gives

$$-\mu \frac{\partial}{\partial t}\mathcal{H}(r,t) = \nabla \times \mathcal{E}(r,t) = 0, \qquad (\mathbf{r},t) \in \mathbb{R}^3 \times \mathbb{R}^+ \tag{7.426}$$

whence we infer that $\mathcal{H}(r,t)$ must be a constant function of time. However, the magnetic field was certainly null before the charges on the balloon were set in motion [cf. (1.252) and (1.253)], and hence we conclude that the magnetic field remains null everywhere at all times.

All in all this is quite a surprising result: no radiation occurs in spite of the unquestionable radial acceleration of the charges spread on the extensible surface of the device! At a closer look, we may argue that no waves are generated because the motion of the outer layer of charges does not distort the electric field streamlines of the initial, static configuration [cf. Figures 1.15a and 1.15b]. In other words, at any given time the field remains electrostatic in character.

We now proceed by proving that in an isotropic homogeneous medium the most general time-varying electric current densities which do not produce electromagnetic waves take on the form [15, Section 2.10.9]

$$\mathcal{J}(\mathbf{r},t) = \varepsilon \frac{\partial}{\partial t}\nabla \Upsilon(\mathbf{r},t), \qquad (\mathbf{r},t) \in \mathbb{R}^3 \times \mathbb{R}^+ \tag{7.427}$$

where $\Upsilon(\mathbf{r},t)$ is a scalar field at least twice differentiable with respect to $\mathbf{r}$ and t. Clearly, $\Upsilon(\mathbf{r},t)$ is defined up to an inconsequential additive constant. More importantly, if the current density is confined to a finite spatial region $V \subset \mathbb{R}^3$, then $\Upsilon(\mathbf{r},t)$ must be independent of either time or position for $\mathbf{r} \notin V$. By inserting (7.427) into (1.20) and (1.34) we obtain

$$\nabla \times \mathcal{E} = -\mu \frac{\partial \mathcal{H}}{\partial t}, \qquad (\mathbf{r},t) \in \mathbb{R}^3 \times \mathbb{R}^+ \tag{7.428}$$

$$\nabla \times \mathcal{H} = \varepsilon \frac{\partial \mathcal{E}}{\partial t} + \varepsilon \frac{\partial}{\partial t}\nabla \Upsilon, \qquad (\mathbf{r},t) \in \mathbb{R}^3 \times \mathbb{R}^+ \tag{7.429}$$

which we may cast as

$$\nabla \times (\mathcal{E} + \nabla\Upsilon) = -\mu\frac{\partial\mathcal{H}}{\partial t}, \qquad (\mathbf{r}, t) \in \mathbb{R}^3 \times \mathbb{R}^+ \qquad (7.430)$$

$$\nabla \times \mathcal{H} = \varepsilon\frac{\partial}{\partial t}(\mathcal{E} + \nabla\Upsilon), \qquad (\mathbf{r}, t) \in \mathbb{R}^3 \times \mathbb{R}^+ \qquad (7.431)$$

thanks to (A.38). These equations can be construed as the Faraday law and the source-free Ampère-Maxwell law for the pair of electric and magnetic fields $\mathcal{E} + \nabla\Upsilon$ and $\mathcal{H}$. Since the sources are given — well, trivially so, as they are null — if we further provide initial conditions and require that the fields be regular at infinity, say,

$$\mathcal{E}(\mathbf{r}, t) + \nabla\Upsilon(\mathbf{r}, t) = \mathcal{O}\left(\frac{1}{|\mathbf{r}|^2}\right), \qquad |\mathbf{r}| \to +\infty \qquad (7.432)$$

$$\mathcal{H}(\mathbf{r}, t) = \mathcal{O}\left(\frac{1}{|\mathbf{r}|^2}\right), \qquad |\mathbf{r}| \to +\infty \qquad (7.433)$$

then the solution is unique by virtue of the arguments of Section 6.3.2. Direct inspection shows that (7.430) and (7.431) are solved by

$$\mathcal{E}(\mathbf{r}, t) = -\nabla\Upsilon(\mathbf{r}, t), \qquad (\mathbf{r}, t) \in \mathbb{R}^3 \times \mathbb{R}^+ \qquad (7.434)$$

$$\mathcal{H}(\mathbf{r}, t) = 0, \qquad (\mathbf{r}, t) \in \mathbb{R}^3 \times \mathbb{R}^+ \qquad (7.435)$$

and, as argued above, this is the only possible solution, so long as we also assume $\mathcal{H}(\mathbf{r}, 0) = 0$ as initial condition. The latter is physically plausible, because the field is at most electrostatic for $t \leqslant 0$, if the current $\mathcal{J}(\mathbf{r}, t)$ is turned on at $t = 0$. The electric field may or may not vanish outside the source region, depending on the value of $\nabla\Upsilon$ for $\mathbf{r} \notin V$ (see Example 7.5).

From (7.434) and the Gauss law (1.44) we also find

$$\varrho(\mathbf{r}, t) = -\varepsilon\nabla^2\Upsilon(\mathbf{r}, t), \qquad (\mathbf{r}, t) \in \mathbb{R}^3 \times \mathbb{R}^+ \qquad (7.436)$$

which gives the charge density associated with $\mathcal{J}(\mathbf{r}, t)$ in (7.427). Conversely, if $\varrho(\mathbf{r}, t)$ is known, (7.436) represents a Poisson equation that can be solved to determine $\Upsilon(\mathbf{r}, t)$. It is easy to check that assumptions (7.427) and (7.436) are consistent with the continuity equation (1.46). Then again, one could derive (7.436) from the conservation of charge and then verify that the Gauss law (1.44) indeed turns into a source-free equation for $\mathcal{E} + \nabla\Upsilon$.

Going back to the system of Figure 7.31, we observe that the densities of electric charge and current are distributions given by

$$\varrho(\mathbf{r}, t) = \frac{Q}{4\pi[b(t)]^2}\delta[b(t) - r] - \frac{Q}{4\pi a^2}\delta(a - r) \qquad (7.437)$$

$$\mathcal{J}(\mathbf{r}, t) = \frac{Q}{4\pi[b(t)]^2}\frac{\mathrm{d}b}{\mathrm{d}t}\delta[b(t) - r]\hat{\mathbf{r}} \qquad (7.438)$$

and, in particular, we have

$$\mathcal{J}(\mathbf{r}, t) = \varepsilon_0\frac{\partial}{\partial t}\nabla\frac{Q}{4\pi\varepsilon_0}\left\{\left[\frac{1}{b(t)} - \frac{1}{r}\right]U[b(t) - r] - \left(\frac{1}{a} - \frac{1}{r}\right)U(a - r)\right\} \qquad (7.439)$$

which constitutes a special case of (7.427) extended to distributions, in that the potential $\Upsilon(\mathbf{r}, t)$ is not twice-differentiable. To prove that (7.438) and (7.439) are equivalent we observe that the relevant potential is a continuous function of r for $r \geqslant 0$, and hence

$$\nabla\Upsilon = \hat{\mathbf{r}}\frac{\partial\Upsilon}{\partial r} = \frac{Q\hat{\mathbf{r}}}{4\pi\varepsilon_0 r^2}\{U[b(t) - r] - U(a - r)\} \tag{7.440}$$

whereas the subsequent differentiation with respect to time has a meaning in the sense of distributions in accordance with (C.40). Finally, by noticing that

$$\frac{1}{r^2}\delta[b(t) - r] = \frac{1}{[b(t)]^2}\delta[b(t) - r] \tag{7.441}$$

we recover (7.438).

Example 7.5 (Field generated by a pulsating spherical charge density)

We consider another thought experiment involving an aggregate of charges uniformly spread inside a sphere centered in the origin. The radius $b(t)$ of the sphere changes in time. By indicating with Q the total charge we may write the density as

$$\varrho(\mathbf{r}, t) = \frac{3Q}{4\pi[b(t)]^3}U[b(t) - r] \tag{7.442}$$

and from the continuity equation (1.46) extended to distributions we have

$$\begin{aligned}
-\frac{\partial}{\partial t}\varrho(\mathbf{r}, t) &= \frac{3Q}{4\pi[b(t)]^3}\frac{\mathrm{d}b}{\mathrm{d}t}\left\{\frac{3}{b(t)}U[b(t) - r] - \delta[b(t) - r]\right\} \\
&= \frac{3Q}{4\pi[b(t)]^3}\frac{\mathrm{d}b}{\mathrm{d}t}\frac{1}{r^2}\left\{\frac{3r^2}{b(t)}U[b(t) - r] - \frac{r^3}{b(t)}\delta[b(t) - r]\right\} \\
&= \frac{3Q}{4\pi[b(t)]^4}\frac{\mathrm{d}b}{\mathrm{d}t}\frac{1}{r^2}\frac{\partial}{\partial r}\left\{r^3 U[b(t) - r]\right\} = \frac{1}{r^2}\frac{\partial}{\partial r}\left[r^2\mathcal{J}_r(r, t)\right]
\end{aligned} \tag{7.443}$$

where we have noticed that

$$\delta[b(t) - r] = \frac{r}{b(t)}\delta[b(t) - r]. \tag{7.444}$$

By solving for $\mathcal{J}_r(r, t)$ we obtain

$$\begin{aligned}
\mathcal{J}(r, t) &= \frac{3Q}{4\pi[b(t)]^4}\mathbf{r}\frac{\mathrm{d}b}{\mathrm{d}t}U[b(t) - r] \\
&= \varepsilon_0\frac{\partial}{\partial t}\nabla\frac{Q}{4\pi\varepsilon_0}\left\{\frac{3[b(t)]^2 - r^2}{2[b(t)]^3}U[b(t) - r] + \frac{1}{r}U[r - b(t)]\right\}
\end{aligned} \tag{7.445}$$

whence the potential $\Upsilon(\mathbf{r}, t)$ can be retrieved. Actually, since $\Upsilon(\mathbf{r}, t)$ solves (7.436) it can also be computed by means of the integral formula (2.169), where the integral can be carried out by expanding the static Green function in a series of spherical harmonics as in (3.235).

The electric field is continuous, radially directed and reads

$$\mathcal{E}(r, t) = \frac{Q\hat{\mathbf{r}}}{4\pi\varepsilon_0}\left\{\frac{r}{[b(t)]^3}U[b(t) - r] + \frac{1}{r^2}U[r - b(t)]\right\} \tag{7.446}$$

which follows from (7.434) or also an application of the Gauss law (1.16) to a ball $B(0, r)$ by distinguishing the cases $r < b(t)$ and $r > b(t)$. As for the magnetic field, since $\mathcal{E}(r, t)$ is curl-free, the Faraday law (1.20) demands that $\mathcal{H}(r, t)$ be constant in time. However, the magnetic field was zero for $t < 0$ and thus if follows that $\mathcal{H}(r, t) = 0$.

Also in this case, despite the radial acceleration impressed to the charges inside the pulsating sphere, electromagnetic waves are not generated.

(End of Example 7.5)

Example 7.6 (Field generated by the sudden appearance of a point charge)

In a closed physical system electric charge is conserved, and stating that charges are not created or destroyed is equivalent to saying that the total charge in the universe is thus constant. But what if a point charge q were to materialize itself in no time? With this thought experiment we can show that the charge must indeed flow into the location, say $\mathbf{r} = 0$, and the accumulation process does not generate electromagnetic waves. Given the singular nature of sources and fields involved we need to resort to the Dirac δ-distributions to express charge and current densities.

Since the charge is concentrated in $\mathbf{r} = 0$ as of $t > 0$ we write $\varrho(\mathbf{r}, t)$ as (Appendix C)

$$\varrho(\mathbf{r}, t) = q\delta^{(3)}(\mathbf{r})\, U(t) = \frac{q}{4\pi r^2}\delta(r)U(t) = \frac{q}{4\pi r^2}\frac{\partial U(r)}{\partial r}U(t) \tag{7.447}$$

and if we interpret the continuity equation (1.46) in the sense of distributions we have

$$-\frac{\partial}{\partial t}\varrho(\mathbf{r}, t) = -\frac{q}{4\pi r^2}\frac{\partial U(r)}{\partial r}\delta(t) = \frac{1}{r^2}\frac{\partial}{\partial r}\left[r^2 \mathcal{J}_r(r, t)\right] = \nabla \cdot \mathcal{J}(r, t) \tag{7.448}$$

because we expect the current to be radially directed and to depend only on r on the grounds of the spherical symmetry of the charge density. The last equation is clearly satisfied if we take

$$\mathcal{J}(r, t) = -q\frac{U(r)}{4\pi r^2}\delta(t)\hat{\mathbf{r}} = \varepsilon_0 \frac{\partial}{\partial t}\nabla \frac{qU(r)U(t)}{4\pi\varepsilon_0 r} \tag{7.449}$$

which represents a current density permeating the whole space and existing for just $t = 0$. The charge does not just pop up in the origin, rather in no time it flows thereto isotropically from infinity. To put the current density in the form (7.427) we have made use of (C.40) upon letting $x = t$ and observed that

$$\nabla \frac{U(r)}{4\pi r} = \hat{\mathbf{r}}\frac{\partial}{\partial r}\frac{U(r)}{4\pi r} = \frac{\delta(r)}{4\pi r}\hat{\mathbf{r}} - \frac{U(r)}{4\pi r^2}\hat{\mathbf{r}} = \underbrace{\mathbf{r}\delta^{(3)}(\mathbf{r})}_{=0} - \frac{U(r)}{4\pi r^2}\hat{\mathbf{r}} = -\frac{U(r)}{4\pi r^2}\hat{\mathbf{r}} \tag{7.450}$$

on account of (C.31).

Concerning the electric field we may use (7.434) with the scalar potential $\Upsilon(\mathbf{r}, t)$ found in (7.449), viz.,

$$\mathcal{E}(\mathbf{r}, t) = -\nabla \frac{U(r)U(t)}{4\pi\varepsilon_0 r} = \frac{q}{4\pi\varepsilon_0 r^2}U(r)U(t)\hat{\mathbf{r}} \tag{7.451}$$

by virtue of (7.450). Alternatively, since for reasons of symmetry only $\mathcal{E}_r$ is non-null, the result follows from the global Gauss law (1.16) applied over the boundary of a ball $B(0, r)$, i.e.,

$$qU(t) = \int_{\partial B} \mathrm{d}S\, \hat{\mathbf{r}} \cdot \varepsilon_0 \mathcal{E}(\mathbf{r}, t) = 4\pi\varepsilon_0 r^2 \mathcal{E}_0(r, t) \tag{7.452}$$

whence we find again

$$\mathcal{E}(r,t) = \frac{q}{4\pi\varepsilon_0 r^2} U(r)U(t)\hat{\mathbf{r}} \tag{7.453}$$

that is, a static field for $t > 0$. Since the curl of $\mathcal{E}(r,t)$ has a meaning as an ordinary vector field and, in particular, vanishes everywhere, it follows from the Faraday law (1.20) that $\mathcal{H}(r,t)$ must be independent of time. However, since $\mathcal{H}(r,t)$ was null for $t < 0$, it remains null everywhere for any time.

All in all, these calculations prove that no waves are generated.

(End of Example 7.6)

We may also envision time-harmonic *non-radiating* electric current densities, and in a homogeneous isotropic medium they have the general expression [26, Section 7.12.2], [57]

$$\mathbf{J}(\mathbf{r}) = k^2\mathbf{f}(\mathbf{r}) - \nabla\times\nabla\times\mathbf{f}(\mathbf{r}), \qquad \mathbf{r}\in V \tag{7.454}$$

where k is the relevant wavenumber defined in (1.249), and $\mathbf{f}(\mathbf{r})$ denotes a complex vector field which *vanishes* outside the bounded region V and is at least of class $\mathcal{C}^3(\mathbb{R}^3)^3$.

The steps required for the proof are similar to those we followed in the time domain. In particular, we may insert (7.454) into the wave equations (1.234) and (1.236) to get

$$\nabla\times\nabla\times\mathbf{E}(\mathbf{r}) - k^2\mathbf{E}(\mathbf{r}) = -\mathrm{j}\,\omega\mu[k^2\mathbf{f}(\mathbf{r}) - \nabla\times\nabla\times\mathbf{f}(\mathbf{r})], \qquad \mathbf{r}\in\mathbb{R}^3 \tag{7.455}$$

$$\nabla\times\nabla\times\mathbf{H}(\mathbf{r}) - k^2\mathbf{H}(\mathbf{r}) = \nabla\times[k^2\mathbf{f}(\mathbf{r}) - \nabla\times\nabla\times\mathbf{f}(\mathbf{r})], \qquad \mathbf{r}\in\mathbb{R}^3 \tag{7.456}$$

whence

$$\nabla\times\nabla\times[\mathbf{E}(\mathbf{r}) - \mathrm{j}\,\omega\mu\mathbf{f}(\mathbf{r})] - k^2[\mathbf{E}(\mathbf{r}) - \mathrm{j}\,\omega\mu\mathbf{f}(\mathbf{r})] = 0, \qquad \mathbf{r}\in\mathbb{R}^3 \tag{7.457}$$

$$\nabla\times\nabla\times[\mathbf{H}(\mathbf{r}) + \nabla\times\mathbf{f}(\mathbf{r})] - k^2[\mathbf{H}(\mathbf{r}) + \nabla\times\mathbf{f}(\mathbf{r})] = 0, \qquad \mathbf{r}\in\mathbb{R}^3 \tag{7.458}$$

that is, the vector fields between brackets obey the source-free vector Helmholtz equation even in the region occupied by the current $\mathbf{J}(\mathbf{r})$. Hence, a possible solution for the electromagnetic field reads

$$\mathbf{E}(\mathbf{r}) = \mathrm{j}\,\omega\mu\,\mathbf{f}(\mathbf{r}), \qquad \mathbf{r}\in\mathbb{R}^3 \tag{7.459}$$

$$\mathbf{H}(\mathbf{r}) = -\nabla\times\mathbf{f}(\mathbf{r}), \qquad \mathbf{r}\in\mathbb{R}^3 \tag{7.460}$$

and since $\mathbf{E}(\mathbf{r})$ and $\mathbf{H}(\mathbf{r})$ trivially satisfy (6.155), (6.156) and the Silver-Müller conditions (6.171), (6.172), then the solution is *unique*. The electromagnetic field is null outside the region occupied by the sources. Besides, in the special instance where $\mathbf{f}(\mathbf{r})$ can be further written as the gradient of a scalar field $\upsilon(\mathbf{r})$, we have

$$\mathbf{J}(\mathbf{r}) = k^2\mathbf{f}(\mathbf{r}) = k^2\nabla\upsilon(\mathbf{r}), \qquad \mathbf{r}\in V \tag{7.461}$$

and the magnetic field vanishes everywhere.

References

[1] Jackson JD. Classical Electrodynamics. 3rd ed. Chichester, UK: Wiley; 1999.
[2] Harrington RF. Time-harmonic Electromagnetic Fields. London, UK: McGraw-Hill; 1961.

[3] Feynmann RP, Leighton R, Sands M. The Feynmann lectures on physics. vol. 2. Reading, MA: Addison-Wesley; 1964.
[4] Morse P, Feshbach H. Methods of Theoretical Physics. New York, NY: McGraw-Hill; 1953.
[5] Guru BS, Hiziroglu HR. Electromagnetic field theory fundamentals. 2nd ed. New York, NY: Cambridge University Press; 2004.
[6] Popović Z, Popović BD. Introductory Electromagnetics. Upper Saddle River, NJ: Prentice Hall; 2000.
[7] Zangwill A. Modern electrodynamics. Cambridge, UK: Cambrigde University Press; 2013.
[8] Adler RB, Chu LJ, Fano RM. Electromagnetic energy transmission and radiation. New York, NY: John Wiley & Sons, Inc.; 1960.
[9] Stratton JA. Electromagnetic theory. London, UK: McGraw-Hill; 1941.
[10] Hayt WH, Buck JA. Engineering Electromagnetics. 8th ed. New York, NY: McGraw-Hill; 2012. International edition.
[11] H P Neff J. Introductory Electromagnetics. New York, NY: John Wiley & Sons, Inc.; 1991.
[12] Cottingham WN, Greenwood DA. Electricity and Magnetism. Cambridge, UK: Cambridge University Press; 1991.
[13] de Hoop AT. Handbook of radiation and scattering of waves. London, UK: Academic Press; 1995.
[14] Zhang K, Li D. Electromagnetic Theory for Microwaves and Optoelectronics. Berlin Heidelberg: Springer-Verlag; 1998.
[15] Rothwell EJ, Cloud MJ. Electromagnetics. London, UK: CRC Press; 2001.
[16] Golub GH, van Loan CF. Matrix Computations. Baltimore, MD: Johns Hopkins University Press; 1996.
[17] Bau III D, Trefethen LN. Numerical linear algebra. Philadelphia, PA: Soci. Indus. Ap. Math.; 1997.
[18] Blyth TS, Robertson EF. Basic Linear Algebra. 2nd ed. Springer Undergraduate Mathematics Series. London, UK: Springer-Verlag; 2002.
[19] Coxeter HSM. Introduction to Geometry. 2nd ed. New York, NY: John Wiley & Sons; 1989.
[20] Dwight HB. Tables of Integrals and Other Mathematical Data. 4th ed. New York, NY: MacMillan Publishing Co., Inc.; 1961.
[21] Orfanidis SJ. Electromagnetic Waves and Antennas. www.ece.rutgers.edu/~orfanidi/ewa; 2004.
[22] Konopinski EJ. Electromagnetic Fields and Relativistic Particles. New York, NY: McGraw Hill; 1981.
[23] Kraus JD, Carver KR. Electromagnetics. 2nd ed. Tokyo, Japan: McGraw-Hill; 1973.
[24] Marcuvitz N. Waveguide Handbook. 2nd ed. Electromagnetic Waves Series. London, UK: The Institution of Engineering and Technology; 1985.
[25] Shen LC, Kong JA. Applied electromagnetism. Monterey, CA: Brooks/Cole; 1983.
[26] Van Bladel JG. Electromagnetic Fields. Piscataway, NJ: IEEE Press; 2007.
[27] Collin RE. Foundations for Microwave Engineering. New York, NY: McGraw-Hill; 1992.
[28] Leontovich MA. Investigation of propagation of radio waves, part II. USSR Academy of Sciences, Physical Series. 1944;.
[29] Senior TBA. Impedance boundary conditions for imperfectly conducting surfaces. Applied Scientific Research, Section B. 1960 December;8(1):418. Available from: http://dx.doi.org/10.1007/BF02920074.
[30] Kong JA. Electromagnetic Wave Theory. 2nd ed. New York, NY: Wiley; 1990.
[31] Born M, Wolf E. Principles of Optics. Oxford, UK: Pergamon Press; 1980.

[32] Rozzi T, Mongiardo M. Open electromagnetic waveguides. London, UK: Institution of Electrical Engineers; 1997.

[33] Marcuse D. Light Transission Optics. New York, NY: Van Nostrand Reinhold Company; 1972.

[34] Ghatak A, Lakshminarayanan V. Propagation Characteristics of Planar Waveguides. In: Calvo ML, Lakshminarayanan V, editors. Optical Waveguides. Optical Science and Engineering. Boca Raton, FL: CRC Press Taylor & Francis Group; 2007. p. 1–37.

[35] Balanis CA. Advanced Engineering Electromagnetics. New York, NY: John Wiley & Sons, Inc.; 2005.

[36] Slater JC, Frank NH. Electromagnetism. New York, NY: McGraw-Hill; 1947.

[37] Chew WC. Waves and Fields in Inhomogenous Media. New York, NY: Van Nostrand Teinhold; 1990.

[38] Dobbs R. Electromagnetic Waves. London, UK: Routledge & Kegan Paul; 1985.

[39] Feynmann RP, Leighton R, Sands M. The Feynmann lectures on physics. vol. 1. Reading, MA: Addison-Wesley; 1964.

[40] Brenner E, Javid M. Analysis of electric circuits. 2nd ed. McGraw-Hill electrical and electronic engineering series. London, UK: McGraw-Hill; 1967.

[41] Millmann J, Grabel A. Microelectronics. 2nd ed. New York, NY: McGraw-Hill; 1988.

[42] Pozar D. Microwave Engineering. 4th ed. New York, NY: John Wiley & Sons, Inc.; 2012.

[43] Schwab AJ. Field Theory Concepts. Berlin Heidelberg: Springer-Verlag; 1988.

[44] Schelkunoff SA. Electromagnetic Waves. Princeton, NJ: D. Van Nostrand Company; 1943.

[45] Gradshteyn IS, Ryzhik IM. Table of Integrals, Series, and Products. 5th ed. Amsterdam, NL: Academic Press, Inc.; 1994.

[46] Abramowitz M, Stegun IA. Handbook of mathematical functions. New York, NY: Dover Publications, Inc.; 1965.

[47] Razavy M. Quantum Theory of Tunneling. London, UK: World Scientific; 2003.

[48] Sherwin CW. Introduction to quantum mechanics. New York, NY: Henry Holt and Company; 1959.

[49] Stein EM, Weiss G. Introduction to Fourier Analysis on Euclidean Spaces. Princeton, NJ: Princeton University Press; 1971.

[50] Sommerfeld A. Partial Differential Equations in Physics. vol. 1 of Lectures on theoretical physics. New York, NY: Academic Press; 1949.

[51] Felsen LB, Marcuvitz N. Radiation and scattering of waves. Piscataway, NJ: IEEE Press; 2001.

[52] Dudley DG. Mathematical Foundations for Electromagnetic Theory. New York, NY: Wiley-Interscience; 1994.

[53] Lerch M. Sur un point de la théorie des fonctions génératrices d'Abel. Acta Mathematica. 1903;27:339.

[54] Cohen AM. Numerical Methods for Laplace Transform Inversion. vol. 5 of Inversion Formulae and Practical Results. New York, NY: Springer Science+Business Media; 2007.

[55] Lindell IV. Methods for electromagnetic field analysis. Piscataway, NJ: IEEE Press; 1992.

[56] Balanis CA. Antenna Theory: Analysis and Design. 2nd ed. New York, NY: John Wiley & Sons, Inc.; 1997.

[57] Devaney AJ, Wolf E. Radiating and nonradiating classical current distributions and the fields they generate. Physical Review D. 1973 August;8(4):1044–1047.

Chapter 8

Time-varying electromagnetic fields I

8.1 The Helmholtz decomposition

Maxwell's equations in local form are amenable to a few alternative representations, e.g., they can be stated in the language of differential forms [1, Chapter 4], [2–5] or in tensor notation in the four-dimensional space-time [6, Chapter 12], [7, Chapter 9], [1, Chapter 3], [8, Chapter 22], [9, Section 8.4], [10–13], and, of course, in the familiar vector-dyadic notation of Section 1.2.2. The latter form has the distinctive advantage of being *coordinate-free*, that is, (1.20), (1.34), (1.23), (1.44) and (1.46) express relationships between the electromagnetic entities without resorting to a particular system of coordinates which, once introduced, allow turning the Maxwell equations into scalar equations for the *components* of the entities.

In the spirit of the decomposition afforded by a system of coordinates, it is possible to separate a given vector field into parts or components, as it were, with different properties yet in quite an abstract fashion. The procedure goes by the name of *Helmholtz decomposition* and the result is provided by the *Helmholtz theorem* [14–18], [19, Section 2.6.1], [8, Section 1.9], [20, Appendix A], not to be confused with the Helmholtz *transport* theorem of Appendix A.6. We discuss the details for both unbounded and finite regions of space. Besides, since the formula only involves the spatial dependence of the field in question, we employ the notation for time-varying fields but we omit to show the time dependence explicitly, unless this turns out to be necessary, as in Section 8.2. Obviously, though, the decomposition applies to complex vector fields such as those introduced in the time-harmonic regime or the spectral domain (Chapter 11).

8.1.1 *Unbounded regions*

We begin by arguing that the basis for the Helmholtz decomposition is *geometrical* in character. Let $\mathcal{V}(\mathbf{r})$ be a generic vector field of class $\mathcal{C}^1(\mathbb{R}^3)^3$ and $\mathbf{u} \in \mathbb{R}^3$ a vector in the ordinary space. From the vector identity (H.14) applied with $\mathbf{a} = \mathbf{b} = \mathbf{u}$ and $\mathbf{c} = \mathcal{V}(\mathbf{r})$ we find

$$\mathcal{V}(\mathbf{r}) = \underbrace{\frac{\mathbf{u}[\mathbf{u} \cdot \mathcal{V}(\mathbf{r})]}{|\mathbf{u}|^2}}_{\text{longitudinal}} - \underbrace{\frac{\mathbf{u} \times [\mathbf{u} \times \mathcal{V}(\mathbf{r})]}{|\mathbf{u}|^2}}_{\text{transverse}} = \left[\frac{\mathbf{u}\mathbf{u}}{|\mathbf{u}|^2} - \frac{\mathbf{u} \times \left(\mathbf{u} \times \overline{\mathcal{I}}\right)}{|\mathbf{u}|^2}\right] \cdot \mathcal{V}(\mathbf{r}) \tag{8.1}$$

after reshuffling terms so as to solve for $\mathcal{V}(\mathbf{r})$. It is a simple matter to verify that this formula provides a means to express $\mathcal{V}(\mathbf{r})$ as the linear combination of two vectors, namely, one parallel to $\mathbf{u}$ (longitudinal part) and one perpendicular to $\mathbf{u}$ (transverse part). Indeed, the longitudinal part does not contribute when we cross-multiply both sides of (8.1) with $\mathbf{u}$, whereas the transverse part is filtered out when we dot-multiply by $\mathbf{u}$. Similarly, on the grounds of property (E.27) the quantity between brackets in the rightmost-hand side is an analogous splitting of the unitary dyadic $\overline{\mathcal{I}}$ into longitudinal and transverse dyadic parts.

Next, we invoke the differential identity (H.59) applied to $\mathcal{V}(\mathbf{r})$ to find

$$\mathcal{V}(\mathbf{r}) = \underbrace{\frac{\nabla[\nabla \cdot \mathcal{V}(\mathbf{r})]}{\nabla^2}}_{\text{lamellar}} - \underbrace{\frac{\nabla \times [\nabla \times \mathcal{V}(\mathbf{r})]}{\nabla^2}}_{\text{solenoidal}} = \mathcal{V}_L(\mathbf{r}) + \mathcal{V}_S(\mathbf{r})$$
$$= \left[\frac{\nabla\nabla}{\nabla^2} - \frac{\nabla \times \left(\nabla \times \overline{\mathcal{I}}\right)}{\nabla^2}\right] \cdot \mathcal{V}(\mathbf{r}), \qquad \mathbf{r} \in \mathbb{R}^3 \tag{8.2}$$

where we have *formally divided* through by $\nabla \cdot \nabla = \nabla^2$ [21, Chapter 1]. Such division is more than just symbolic, though, and it actually corresponds to the inversion of the Laplace operator and the solution of the Poisson equation with $\mathcal{V}(\mathbf{r})$ as unknown (cf. Section 4.2). Comparison with (8.1) suggests that, if we regard the del operator ∇ as an ordinary vector, (8.2) provides a decomposition of $\mathcal{V}(\mathbf{r})$ into a vector $\mathcal{V}_L(\mathbf{r})$ parallel to ∇ and a vector $\mathcal{V}_S(\mathbf{r})$ perpendicular to ∇. In particular,

- according to (A.38) $\mathcal{V}_L(\mathbf{r})$ is curl-free or conservative or lamellar, i.e.,

$$\nabla \times \mathcal{V}_L(\mathbf{r}) = 0, \qquad \mathbf{r} \in \mathbb{R}^3 \tag{8.3}$$

  and is called the *lamellar* or *longitudinal* part of $\mathcal{V}(\mathbf{r})$;
- in light of (A.39) $\mathcal{V}_S(\mathbf{r})$ is divergence-free or solenoidal, namely,

$$\nabla \cdot \mathcal{V}_S(\mathbf{r}) = 0, \qquad \mathbf{r} \in \mathbb{R}^3 \tag{8.4}$$

  and is called the *solenoidal* or *transverse* part of $\mathcal{V}(\mathbf{r})$.

Moreover, the dyadic between brackets in the rightmost-hand side of (8.2) constitutes a similar formal expansion of the identity dyadic. Notice that in writing (8.2) we do not resort to a system of coordinates, as anticipated. Some restrictions on the asymptotic behavior of $\mathcal{V}(\mathbf{r})$ for $|\mathbf{r}| \to +\infty$ are in order, but we will address them in a moment as they are not evident at this stage. The result (8.2) is rigorously proved if we can find a way to compute $\mathcal{V}_L(\mathbf{r})$ and $\mathcal{V}_S(\mathbf{r})$ in terms of the original field $\mathcal{V}(\mathbf{r})$. To this purpose from now on we indicate the position vector with $\mathbf{r}'$ and the corresponding differential operators with ∇'^2 and so forth.

To determine the lamellar part of $\mathcal{V}(\mathbf{r}')$ we observe that (8.3) is satisfied if we assume

$$\mathcal{V}_L(\mathbf{r}') = \nabla\Upsilon(\mathbf{r}'), \qquad \mathbf{r}' \in \mathbb{R}^3 \tag{8.5}$$

where $\Upsilon(\mathbf{r}')$ is a scalar field twice-differentiable for $\mathbf{r} \in \mathbb{R}^3$. This strategy is analogous to the approach followed in Section 2.2 for the laws of electrostatics. We go on to take the divergence of the Ansatz (8.2)

$$\nabla' \cdot \mathcal{V}(\mathbf{r}') = \nabla' \cdot \mathcal{V}_L(\mathbf{r}') + \nabla' \cdot \mathcal{V}_S(\mathbf{r}') = \nabla'^2\Upsilon(\mathbf{r}'), \qquad \mathbf{r}' \in \mathbb{R}^3 \tag{8.6}$$

on account of (8.4). Since the equation for $\Upsilon(\mathbf{r}')$ has the same structure as the Poisson equation (2.17) aside from immaterial constants, we recall from Section 2.5.1 that $\Upsilon(\mathbf{r}')$ is uniquely determined if we supplement suitable boundary conditions. For an unbounded region we demand that $\Upsilon(\mathbf{r}')$ vanish at infinity according to (2.20). In symbols, the problem

$$\begin{cases} \nabla'^2\Upsilon(\mathbf{r}') = \nabla' \cdot \mathcal{V}(\mathbf{r}'), & \mathbf{r}' \in \mathbb{R}^3 \\ \Upsilon(\mathbf{r}') = \mathcal{O}\left(\dfrac{1}{|\mathbf{r}'|}\right), & |\mathbf{r}'| \to +\infty \end{cases} \tag{8.7}$$

admits a unique solution. If the support of the field $\mathcal{V}(\mathbf{r}')$ is unbounded, i.e., $\mathcal{V}(\mathbf{r}')$ does not vanish outside a finite region of space, then we see that the condition

$$\mathcal{V}(\mathbf{r}') = \mathcal{O}\left(\frac{1}{|\mathbf{r}'|^2}\right), \qquad |\mathbf{r}'| \to +\infty \tag{8.8}$$

is induced by the asymptotic behavior of $\Upsilon(\mathbf{r}')$. To invert (8.7) we combine it with the special instance of the Poisson equation

$$\begin{cases} \nabla'^2 G(\mathbf{r},\mathbf{r}') = 0, & \mathbf{r}' \in \mathbb{R}^3 \setminus \{\mathbf{r}\} \\ G(\mathbf{r},\mathbf{r}') = \mathcal{O}\left(\frac{1}{|\mathbf{r}'|}\right), & |\mathbf{r}'| \to +\infty \end{cases} \tag{8.9}$$

which is solved by the three-dimensional static Green function (2.131). Multiplying the Poisson equation in (8.7) by $G(\mathbf{r},\mathbf{r}')$ and the one in (8.9) by $\Upsilon(\mathbf{r}')$ and then subtracting side by side the resulting expressions yields

$$\nabla' \cdot [G(\mathbf{r},\mathbf{r}')\nabla'\Upsilon(\mathbf{r}') - \Upsilon(\mathbf{r}')\nabla' G(\mathbf{r},\mathbf{r}')] = G(\mathbf{r},\mathbf{r}')\nabla' \cdot \mathcal{V}(\mathbf{r}'), \qquad \mathbf{r}' \in \mathbb{R}^3 \setminus \{\mathbf{r}\} \tag{8.10}$$

on account of the differential identity (H.51). Since the Green function is singular for $\mathbf{r}' = \mathbf{r}$ and the solution domain is unbounded we consider two concentric balls $B(\mathbf{r}, a)$ and $B(\mathbf{r}, b)$ with $a < b$ and integrate over the finite spherical shell $B(\mathbf{r}, b) \setminus B(\mathbf{r}, a)$ to obtain

$$\begin{aligned} \int\limits_{B(\mathbf{r},b)\setminus B(\mathbf{r},a)} \mathrm{d}V'\, G(\mathbf{r},\mathbf{r}')\nabla' \cdot \mathcal{V}(\mathbf{r}') = &\int\limits_{\partial B(\mathbf{r},b)} \mathrm{d}S'\, \hat{\mathbf{n}}(\mathbf{r}') \cdot [G(\mathbf{r},\mathbf{r}')\nabla'\Upsilon(\mathbf{r}') - \Upsilon(\mathbf{r}')\nabla' G(\mathbf{r},\mathbf{r}')] \\ &+ \int\limits_{\partial B(\mathbf{r},a)} \mathrm{d}S'\, \hat{\mathbf{n}}(\mathbf{r}') \cdot [G(\mathbf{r},\mathbf{r}')\nabla'\Upsilon(\mathbf{r}') - \Upsilon(\mathbf{r}')\nabla' G(\mathbf{r},\mathbf{r}')] \end{aligned} \tag{8.11}$$

where, thanks to the smoothness of the integrand for points $\mathbf{r}'$ within the shell, we have applied the Gauss theorem with the unit normal pointing outwards $B(\mathbf{r}, b) \setminus B(\mathbf{r}, a)$ on the boundary. In the limit as $a \to 0^+$ and $b \to +\infty$ the volume integral in the left-hand side becomes improper. However, we can prove it converges and at the same time find $\Upsilon(\mathbf{r})$ if we show that the flux integrals in the right-hand side are either finite or null.

Concerning the integral over $\partial B(\mathbf{r}, a)$ we observe that for $\mathbf{r}' \in \partial B(\mathbf{r}, a)$ we have

$$\mathbf{r}' = \mathbf{r} - a\hat{\mathbf{n}}(\mathbf{r}'), \qquad G(\mathbf{r},\mathbf{r}') = \frac{1}{4\pi a}, \qquad \hat{\mathbf{n}}(\mathbf{r}') \cdot \nabla' G(\mathbf{r},\mathbf{r}') = \frac{1}{4\pi a^2} \tag{8.12}$$

and by substituting into (8.11) we compute

$$\begin{aligned} &\int\limits_{\partial B(\mathbf{r},a)} \mathrm{d}S'\, \hat{\mathbf{n}}(\mathbf{r}') \cdot [G(\mathbf{r},\mathbf{r}')\nabla'\Upsilon(\mathbf{r}') - \Upsilon(\mathbf{r}')\nabla' G(\mathbf{r},\mathbf{r}')] = \\ &\quad = \frac{1}{4\pi a^2} \int\limits_{\partial B(\mathbf{r},a)} \mathrm{d}S'\, [a\hat{\mathbf{n}}(\mathbf{r}') \cdot \nabla'\Upsilon(\mathbf{r}') - \Upsilon(\mathbf{r}')] \\ &\quad = a\hat{\mathbf{n}}(\mathbf{r}'_0) \cdot \nabla'\Upsilon(\mathbf{r}')|_{\mathbf{r}'_0} - \Upsilon(\mathbf{r}'_0) \xrightarrow[a\to 0^+]{} -\Upsilon(\mathbf{r}) \end{aligned} \tag{8.13}$$

where, since the relevant integrand is regular on $\partial B(\mathbf{r}, a)$, the mean value theorem [22] has been invoked with $\mathbf{r}'_0 \in \partial B(\mathbf{r}, a)$ being a suitable point, and finally $\mathbf{r}'_0 \to \mathbf{r}$ as $a \to 0^+$.

As regards the integral over $\partial B(\mathbf{r}, b)$ we notice that for $\mathbf{r}' \in \partial B(\mathbf{r}, b)$ it holds

$$\mathbf{r}' = \mathbf{r} + b\hat{\mathbf{n}}(\mathbf{r}'), \qquad G(\mathbf{r}, \mathbf{r}') = \frac{1}{4\pi b}, \qquad \hat{\mathbf{n}}(\mathbf{r}') \cdot \nabla' G(\mathbf{r}, \mathbf{r}') = -\frac{1}{4\pi b^2} \tag{8.14}$$

and by substituting into (8.11) we estimate

$$\begin{aligned}
&\left| \int\limits_{\partial B(\mathbf{r},b)} \mathrm{d}S'\, \hat{\mathbf{n}}(\mathbf{r}') \cdot [G(\mathbf{r}, \mathbf{r}')\nabla'\Upsilon(\mathbf{r}') - \Upsilon(\mathbf{r}')\nabla' G(\mathbf{r}, \mathbf{r}')] \right| = \\
&\quad = \frac{1}{4\pi b^2} \left| \int\limits_{\partial B(\mathbf{r},b)} \mathrm{d}S'\, [b\hat{\mathbf{n}}(\mathbf{r}') \cdot \nabla'\Upsilon(\mathbf{r}') + \Upsilon(\mathbf{r}')] \right| \\
&\quad \leqslant \frac{1}{4\pi b^2} \int\limits_{\partial B(\mathbf{r},b)} \mathrm{d}S'\, [b|\hat{\mathbf{n}}(\mathbf{r}') \cdot \nabla'\Upsilon(\mathbf{r}')| + |\Upsilon(\mathbf{r}')|] \\
&\quad \leqslant \frac{1}{4\pi b^2} \int\limits_{\partial B(\mathbf{r},b)} \mathrm{d}S' \left[b\frac{C_\Upsilon}{(b-r)^2} + \frac{C_\Upsilon}{b-r} \right] = \frac{2b-r}{(b-r)^2} C_\Upsilon \xrightarrow[b\to+\infty]{} 0
\end{aligned} \tag{8.15}$$

with C_Υ a suitable positive constant involved in the asymptotic condition imposed on $\Upsilon(\mathbf{r}')$ in (8.7), since $r' = |\mathbf{R} - \mathbf{r}| \geqslant |b - r|$ in light of (H.20). Putting everything together we have proved that (8.11) passes over into

$$\Upsilon(\mathbf{r}) = -\int\limits_{\mathbb{R}^3} \mathrm{d}V' \frac{\nabla' \cdot \mathcal{V}(\mathbf{r}')}{4\pi|\mathbf{r} - \mathbf{r}'|}, \qquad \mathbf{r} \in \mathbb{R}^3 \tag{8.16}$$

in the limit as $a \to 0^+$ and $b \to +\infty$. The improper integral is finite because we have showed it is equal to the sum of two flux integrals which remain finite, though the result can also be proved as in (2.161).

To determine the solenoidal part of $\mathcal{V}(\mathbf{r})$ we notice that (8.4) is solved by

$$\mathcal{V}_S(\mathbf{r}') = \nabla' \times \Xi(\mathbf{r}'), \qquad \mathbf{r}' \in \mathbb{R}^3 \tag{8.17}$$

where $\Xi(\mathbf{r}')$ is a vector field twice-differentiable for $\mathbf{r} \in \mathbb{R}^3$. This approach is similar to the one we followed in Section 4.2 to handle the equations pertinent to stationary magnetic fields and steady currents. With this position we proceed to take the curl of the Ansatz (8.2)

$$\nabla' \times \mathcal{V}(\mathbf{r}') = \nabla' \times \mathcal{V}_L(\mathbf{r}') + \nabla' \times \mathcal{V}_S(\mathbf{r}') = \nabla' \times [\nabla' \times \Xi(\mathbf{r}')], \qquad \mathbf{r}' \in \mathbb{R}^3 \tag{8.18}$$

in view of (8.3). Apparently, we have an equation for $\Xi(\mathbf{r}')$ that is perfectly analogous to the vector Poisson equation for the magnetic potential $\mathcal{A}(\mathbf{r})$. We know from Section 4.7.1 that uniqueness of solutions is guaranteed if the boundary condition at infinity (4.36) is enforced but also the divergence of the field under investigation is assigned. Since $\Xi(\mathbf{r}')$ is nothing but an intermediate utilitarian quantity introduced to help us find the solenoidal part of $\mathcal{V}(\mathbf{r}')$ we are at liberty of choosing the divergence of $\Xi(\mathbf{r}')$ as we like and, why not, so as to simplify the task of solving the equations. The most convenient choice consists of demanding that $\Xi(\mathbf{r}')$ be solenoidal. In summary, the problem

$$\begin{cases} \nabla' \times [\nabla' \times \Xi(\mathbf{r}')] = \nabla' \times \mathcal{V}(\mathbf{r}'), & \mathbf{r}' \in \mathbb{R}^3 \\ \nabla' \cdot \Xi(\mathbf{r}') = 0 & \mathbf{r}' \in \mathbb{R}^3 \\ \Xi(\mathbf{r}') = \mathcal{O}\left(\dfrac{1}{|\mathbf{r}'|}\right), & |\mathbf{r}'| \to +\infty \end{cases} \tag{8.19}$$

admits a unique solution. Again, for a field $\mathcal{V}(\mathbf{r}')$ with unbounded support the asymptotic behavior of $\boldsymbol{\Xi}(\mathbf{r}')$ requires that $\mathcal{V}(\mathbf{r}')$ satisfy (8.8). Guided by the strategy employed in Section 5.1.1 we consider the auxiliary problem [cf. (5.5)]

$$\begin{cases} \nabla' \times \{\nabla' \times [\mathbf{u}G(\mathbf{r},\mathbf{r}')]\} = \nabla'\nabla' \cdot [\mathbf{u}G(\mathbf{r},\mathbf{r}')], & \mathbf{r}' \in \mathbb{R}^3 \setminus \{\mathbf{r}\} \\ \mathbf{u}G(\mathbf{r},\mathbf{r}') = \mathcal{O}\left(\dfrac{1}{|\mathbf{r}'|}\right), & |\mathbf{r}'| \to +\infty \end{cases} \tag{8.20}$$

where $G(\mathbf{r},\mathbf{r}')$ is the static Green function (2.131) and $\mathbf{u} \in \mathbb{R}^3$ is a really arbitrary constant vector which we will get rid of at the end. Actually, (8.20) is more of an identity, since the solution is already known. We dot-multiply (8.19) by $\mathbf{u}G(\mathbf{r},\mathbf{r}')$ and (8.20) by $\boldsymbol{\Xi}(\mathbf{r}')$, and we subtract the equations side by side to arrive at

$$\begin{aligned} \nabla' \cdot \{\boldsymbol{\Xi}(\mathbf{r}') \times [\nabla' G(\mathbf{r},\mathbf{r}') \times \mathbf{u}] + \boldsymbol{\Xi}(\mathbf{r}')\nabla' G(\mathbf{r},\mathbf{r}') \cdot \mathbf{u} - \mathbf{u}G(\mathbf{r},\mathbf{r}') \times [\nabla' \times \boldsymbol{\Xi}(\mathbf{r}')]\} \\ = \mathbf{u}G(\mathbf{r},\mathbf{r}') \cdot \nabla' \times \mathcal{V}(\mathbf{r}'), \qquad \mathbf{r} \in \mathbb{R}^3 \setminus \{\mathbf{r}\} \end{aligned} \tag{8.21}$$

having used (H.49)-(H.52) and the facts that $\boldsymbol{\Xi}(\mathbf{r}')$ is divergence-free by hypothesis and $\mathbf{u}$ is constant. For the same reasons stated above we consider two concentric balls $B(\mathbf{r},a)$ and $B(\mathbf{r},b)$ with $a < b$ and integrate over the finite spherical shell $B(\mathbf{r},b) \setminus B(\mathbf{r},a)$ to get

$$\begin{aligned} \int\limits_{B(\mathbf{r},b)\setminus B(\mathbf{r},a)} \mathrm{d}V'\, \mathbf{u} \cdot G(\mathbf{r},\mathbf{r}')\nabla' \times \mathcal{V}(\mathbf{r}') = \\ = \int\limits_{B(\mathbf{r},b)\setminus B(\mathbf{r},a)} \mathrm{d}V'\, \nabla' \cdot \{\boldsymbol{\Xi}(\mathbf{r}') \times [\nabla' G(\mathbf{r},\mathbf{r}') \times \mathbf{u}] + \boldsymbol{\Xi}(\mathbf{r}')\nabla' G(\mathbf{r},\mathbf{r}') \cdot \mathbf{u}\} \\ - \int\limits_{B(\mathbf{r},b)\setminus B(\mathbf{r},a)} \mathrm{d}V'\, \nabla' \cdot \{\mathbf{u}G(\mathbf{r},\mathbf{r}') \times [\nabla' \times \boldsymbol{\Xi}(\mathbf{r}')]\} \\ = \int\limits_{\partial B(\mathbf{r},a)} \mathrm{d}S'\, \hat{\mathbf{n}}(\mathbf{r}') \cdot \{\boldsymbol{\Xi}(\mathbf{r}') \times [\nabla' G(\mathbf{r},\mathbf{r}') \times \mathbf{u}] + \boldsymbol{\Xi}(\mathbf{r}')\nabla' G(\mathbf{r},\mathbf{r}') \cdot \mathbf{u}\} \\ - \int\limits_{\partial B(\mathbf{r},a)} \mathrm{d}S'\, \hat{\mathbf{n}}(\mathbf{r}') \cdot \mathbf{u}G(\mathbf{r},\mathbf{r}') \times [\nabla' \times \boldsymbol{\Xi}(\mathbf{r}')] \\ + \int\limits_{\partial B(\mathbf{r},b)} \mathrm{d}S'\, \hat{\mathbf{n}}(\mathbf{r}') \cdot \{\boldsymbol{\Xi}(\mathbf{r}') \times [\nabla' G(\mathbf{r},\mathbf{r}') \times \mathbf{u}] + \boldsymbol{\Xi}(\mathbf{r}')\nabla' G(\mathbf{r},\mathbf{r}') \cdot \mathbf{u}\} \\ - \int\limits_{\partial B(\mathbf{r},b)} \mathrm{d}S'\, \hat{\mathbf{n}}(\mathbf{r}') \cdot \mathbf{u}G(\mathbf{r},\mathbf{r}') \times [\nabla' \times \boldsymbol{\Xi}(\mathbf{r}')] \end{aligned} \tag{8.22}$$

where we have applied the Gauss theorem because the integrand is differentiable in $B(\mathbf{r},b) \setminus B(\mathbf{r},a)$. As before, in the limit as $a \to 0^+$ and $b \to +\infty$ the volume integral becomes improper but it converges since, as we shall show, the flux integrals in the right-hand side are finite.

For the two integrals over $\partial B(\mathbf{r},a)$ we have

$$\int\limits_{\partial B(\mathbf{r},a)} \mathrm{d}S'\, \hat{\mathbf{n}}(\mathbf{r}') \cdot \{\boldsymbol{\Xi}(\mathbf{r}') \times [\nabla' G(\mathbf{r},\mathbf{r}') \times \mathbf{u}] + \boldsymbol{\Xi}(\mathbf{r}')\nabla' G(\mathbf{r},\mathbf{r}') \cdot \mathbf{u}\} =$$

$$= \frac{1}{4\pi a^2} \int\limits_{\partial B(\mathbf{r},a)} \mathrm{d}S' \, \{\hat{\mathbf{n}}(\mathbf{r}') \cdot \{\Xi(\mathbf{r}') \times [\hat{\mathbf{n}}(\mathbf{r}') \times \mathbf{u}]\} + \hat{\mathbf{n}}(\mathbf{r}') \cdot \Xi(\mathbf{r}')\hat{\mathbf{n}}(\mathbf{r}') \cdot \mathbf{u}\}$$

$$= \frac{1}{4\pi a^2} \int\limits_{\partial B(\mathbf{r},a)} \mathrm{d}S' \, \Xi(\mathbf{r}') \cdot \mathbf{u} = \Xi(\mathbf{r} - a\hat{\mathbf{n}}(\mathbf{r}'_0)) \cdot \mathbf{u} \xrightarrow[a \to 0^+]{} \Xi(\mathbf{r}) \cdot \mathbf{u} \tag{8.23}$$

$$- \int\limits_{\partial B(\mathbf{r},a)} \mathrm{d}S' \, \hat{\mathbf{n}}(\mathbf{r}') \cdot \mathbf{u} G(\mathbf{r}, \mathbf{r}') \times [\nabla' \times \Xi(\mathbf{r}')] = -\frac{1}{4\pi a} \int\limits_{\partial B(\mathbf{r},a)} \mathrm{d}S' \, \hat{\mathbf{n}}(\mathbf{r}') \cdot \mathbf{u} \times [\nabla' \times \Xi(\mathbf{r}')]$$

$$= -a\hat{\mathbf{n}}(\mathbf{r}'_0) \cdot \mathbf{u} \times [\nabla' \times \Xi(\mathbf{r}')]|_{\mathbf{r}'_0} \xrightarrow[a \to 0^+]{} 0 \tag{8.24}$$

where we have employed the vector identity (H.14) and the mean value theorem [22] with $\mathbf{r}'_0 \in \partial B(\mathbf{r}, a)$ a suitable point.

For the two integrals over the larger sphere $\partial B(\mathbf{r}, b)$ we estimate

$$\left| \int\limits_{\partial B(\mathbf{r},b)} \mathrm{d}S' \, \hat{\mathbf{n}}(\mathbf{r}') \cdot \{\Xi(\mathbf{r}') \times [\nabla' G(\mathbf{r}, \mathbf{r}') \times \mathbf{u}] + \Xi(\mathbf{r}')\nabla' G(\mathbf{r}, \mathbf{r}') \cdot \mathbf{u}\} \right|$$

$$\leqslant \frac{1}{4\pi b^2} \int\limits_{\partial B(\mathbf{r},b)} \mathrm{d}S' \, \left| \hat{\mathbf{n}}(\mathbf{r}') \cdot \{\Xi(\mathbf{r}') \times [\hat{\mathbf{n}}(\mathbf{r}') \times \mathbf{u}] + \Xi(\mathbf{r}')\hat{\mathbf{n}}(\mathbf{r}') \cdot \mathbf{u}\} \right|$$

$$= \frac{1}{4\pi b^2} \int\limits_{\partial B(\mathbf{r},b)} \mathrm{d}S' \, \left| \Xi(\mathbf{r}') \cdot \mathbf{u} \right| \leqslant \frac{|\mathbf{u}|}{4\pi b^2} \int\limits_{\partial B(\mathbf{r},b)} \mathrm{d}S' \, |\Xi(\mathbf{r}')|$$

$$\leqslant \frac{|\mathbf{u}|}{4\pi b^2} \int\limits_{\partial B(\mathbf{r},b)} \mathrm{d}S' \, \frac{C_\Xi}{b - r} = \frac{C_\Xi}{b - r} |\mathbf{u}| \xrightarrow[b \to +\infty]{} 0 \tag{8.25}$$

$$\left| - \int\limits_{\partial B(\mathbf{r},a)} \mathrm{d}S' \, \hat{\mathbf{n}}(\mathbf{r}') \cdot \mathbf{u} G(\mathbf{r}, \mathbf{r}') \times [\nabla' \times \Xi(\mathbf{r}')] \right| =$$

$$= \frac{1}{4\pi b} \left| \int\limits_{\partial B(\mathbf{r},a)} \mathrm{d}S' \, \hat{\mathbf{n}}(\mathbf{r}') \cdot \mathbf{u} \times [\nabla' \times \Xi(\mathbf{r}')] \right| \leqslant \frac{|\mathbf{u}|}{4\pi b} \int\limits_{\partial B(\mathbf{r},a)} \mathrm{d}S' \, |\nabla' \times \Xi(\mathbf{r}')|$$

$$\leqslant \frac{|\mathbf{u}|}{4\pi b} \int\limits_{\partial B(\mathbf{r},a)} \mathrm{d}S' \, \frac{C_\Xi}{(b - r)^2} = \frac{bC_\Xi}{(b - r)^2} |\mathbf{u}| \xrightarrow[b \to +\infty]{} 0 \tag{8.26}$$

with C_Ξ a suitable positive constant involved in the asymptotic condition imposed on $\Xi(\mathbf{r}')$ in (8.19). By collecting these intermediate results we conclude that

$$\Xi(\mathbf{r}) = \int\limits_{\mathbb{R}^3} \mathrm{d}V' \, \frac{\nabla' \times \mathcal{V}(\mathbf{r}')}{4\pi|\mathbf{r} - \mathbf{r}'|}, \qquad \mathbf{r} \in \mathbb{R}^3 \tag{8.27}$$

in the limit as $a \to 0^+$ and $b \to +\infty$ and by virtue of the arbitrariness of $\mathbf{u}$. Again the integral converges for it is equal to the sum of two finite flux integrals.

Combining the expressions of $\Upsilon(\mathbf{r})$ and $\Xi(\mathbf{r})$ with (8.5) and (8.17) finally yields

$$\mathcal{V}(\mathbf{r}) = -\nabla \int_{\mathbb{R}^3} \mathrm{d}V' \frac{\nabla' \cdot \mathcal{V}(\mathbf{r}')}{4\pi|\mathbf{r}-\mathbf{r}'|} + \nabla \times \int_{\mathbb{R}^3} \mathrm{d}V' \frac{\nabla' \times \mathcal{V}(\mathbf{r}')}{4\pi|\mathbf{r}-\mathbf{r}'|}, \qquad \mathbf{r} \in \mathbb{R}^3 \tag{8.28}$$

which provides us with a practical recipe for the decomposition of $\mathcal{V}(\mathbf{r})$ into solenoidal and lamellar components. The result just proved is the *Helmholtz theorem* in an unbounded region of space [14–16, 20], [23, pp. 52-54] and applies to fields that meet the asymptotic condition (8.8), which is necessary to ensure the convergence of the improper integrals.

In words, a continuously differentiable vector field $\mathcal{V}(\mathbf{r})$ is *completely determined* in the whole space as long as both the divergence and the curl of $\mathcal{V}(\mathbf{r})$ are assigned. Of course, one still has to compute the volume integrals, possibly with the aid of numerical integration formulas, if need be. Nevertheless, the theoretical importance of the theorem is profound. For example, (8.28) allows understanding that the equations of electrostatics or those for stationary magnetic fields contain the right amount of information inasmuch as they specify both the curl and the divergence of the electromagnetic entity of interest.

The lamellar and solenoidal parts of $\mathcal{V}(\mathbf{r})$ are *orthogonal* with respect to the symmetric inner product (D.68) with $D = \mathbb{R}^3$, viz.,

$$(\mathcal{V}_L, \mathcal{V}_S)_{\mathbb{R}^3} := \int_{\mathbb{R}^3} \mathrm{d}V\, \mathcal{V}_L(\mathbf{r}) \cdot \mathcal{V}_S(\mathbf{r}) = 0 \tag{8.29}$$

where the improper integral converges in light of the asymptotic behavior of $\mathcal{V}_L(\mathbf{r})$ and $\mathcal{V}_S(\mathbf{r})$, which is the same as that of $\mathcal{V}(\mathbf{r})$ stated in (8.8). This also follows from (8.28) by expanding the kernel $1/|\mathbf{r}-\mathbf{r}'|$ as in (3.195) for $r \to +\infty$. For the proof of (8.29) we write $\mathcal{V}_L(\mathbf{r})$ and $\mathcal{V}_S(\mathbf{r})$ in terms of the auxiliary potentials $\Upsilon(\mathbf{r})$ and $\Xi(\mathbf{r})$ and integrate $\mathcal{V}_L(\mathbf{r}) \cdot \mathcal{V}_S(\mathbf{r})$ over the ball $B(0, b)$. In symbols, we have

$$\begin{aligned}\int_{B(0,b)} \mathrm{d}V\, \mathcal{V}_L(\mathbf{r}) \cdot \mathcal{V}_S(\mathbf{r}) &= \int_{B(0,b)} \mathrm{d}V\, \nabla\Upsilon(\mathbf{r}) \cdot \nabla \times \Xi(\mathbf{r}) = \int_{B(0,b)} \mathrm{d}V\, \nabla \cdot [\Xi(\mathbf{r}) \times \nabla\Upsilon(\mathbf{r})] \\ &= \int_{\partial B} \mathrm{d}S\, \hat{\mathbf{n}}(\mathbf{r}) \cdot [\Xi(\mathbf{r}) \times \nabla\Upsilon(\mathbf{r})]\end{aligned} \tag{8.30}$$

having applied identity (H.49) and the Gauss theorem. This is possible because according to (8.7) and (8.19) the potentials $\Upsilon(\mathbf{r})$ and $\Xi(\mathbf{r})$ are twice-differentiable in $\mathbb{R}^3$. Next, we need to show that the flux integral over ∂B vanishes as the radius b of the ball grows infinitely large. But the latter is true by virtue of the asymptotic behavior of $\Upsilon(\mathbf{r})$ and $\Xi(\mathbf{r})$, and this concludes the proof.

As a result, the Helmholtz decomposition (8.28) may also be regarded as the separation of the space of continuously differentiable vector fields $\mathcal{C}^1(\mathbb{R}^3)^3$ into two mutually *orthogonal subspaces* (see Appendix D.1), namely, the space of *lamellar fields* (i.e., those which are the gradient of a scalar field) and the space of *solenoidal fields* (i.e., those which are the curl of a vector field). We may wonder whether this splitting exhausts all the possibilities. Put another way, is there a vector field which is orthogonal to both $\mathcal{V}_L(\mathbf{r})$ and $\mathcal{V}_S(\mathbf{r})$ in the sense of (8.29)? In fact, we have proved (8.28) constructively by providing a way of computing $\mathcal{V}_L(\mathbf{r})$ and $\mathcal{V}_S(\mathbf{r})$, but what if assumption (8.2) does not cover all bases, so to speak? Perhaps we should have started with the more general Ansatz

$$\mathcal{V}(\mathbf{r}) = \mathcal{V}_L(\mathbf{r}) + \mathcal{V}_S(\mathbf{r}) + \mathcal{V}_H(\mathbf{r}), \qquad \mathbf{r} \in \mathbb{R}^3 \tag{8.31}$$

where $\mathcal{V}_H(\mathbf{r}) \in \mathcal{C}^1(\mathbb{R}^3)^3$ is a vector field regular at infinity such that

$$(\mathcal{V}_H, \mathcal{V}_L)_{\mathbb{R}^3} = 0 = (\mathcal{V}_H, \mathcal{V}_S)_{\mathbb{R}^3} \tag{8.32}$$

and $\mathcal{V}_L(\mathbf{r})$ and $\mathcal{V}_S(\mathbf{r})$ indicate again the lamellar and solenoidal parts of $\mathcal{V}(\mathbf{r}) \in \mathcal{C}^1(\mathbb{R}^3)^3$. We show by contradiction that, necessarily, $\mathcal{V}_H(\mathbf{r}) = 0$ for $\mathbf{r} \in \mathbb{R}^3$.

Suppose that $\mathcal{V}_H(\mathbf{r}) \neq 0$ and take the divergence and the curl of (8.31), namely,

$$\nabla \cdot \mathcal{V}(\mathbf{r}) = \nabla^2 \Upsilon(\mathbf{r}) + \nabla \cdot \mathcal{V}_H(\mathbf{r}) \tag{8.33}$$

$$\nabla \times \mathcal{V}(\mathbf{r}) = \nabla \times \nabla \times \Xi(\mathbf{r}) + \nabla \times \mathcal{V}_H(\mathbf{r}) \tag{8.34}$$

where we have also introduced the auxiliary potentials $\Upsilon(\mathbf{r})$ and $\Xi(\mathbf{r})$ on account of (8.5) and (8.17). Apparently, if $\mathcal{V}_L(\mathbf{r})$ and $\mathcal{V}_S(\mathbf{r})$ are to coincide with the fields found previously, then $\Upsilon(\mathbf{r})$ and $\Xi(\mathbf{r})$ must in turn satisfy (8.7) and (8.19). This is only possible if we require

$$\nabla \cdot \mathcal{V}_H(\mathbf{r}) = 0, \qquad \nabla \times \mathcal{V}_H(\mathbf{r}) = 0, \qquad \mathbf{r} \in \mathbb{R}^3 \tag{8.35}$$

that is, the vector field $\mathcal{V}_H(\mathbf{r})$ must be both solenoidal and lamellar. A field which is divergence-free and curl-free is said *harmonic*, and this motivates the subscript ‘H’ in the symbol chosen for the third contribution in (8.31). A conservative field in the whole space is the gradient of some single-valued scalar potential (cf. Sections 2.5.2 and 4.7.2). Thus, we may write

$$\mathcal{V}_H(\mathbf{r}) = \nabla f(\mathbf{r}) \tag{8.36}$$

where $f(\mathbf{r})$ must solve

$$\begin{cases} \nabla^2 f(\mathbf{r}) = 0, & \mathbf{r} \in \mathbb{R}^3 \\ f(\mathbf{r}) = \mathcal{O}\left(\dfrac{1}{|\mathbf{r}|}\right), & |\mathbf{r}| \to +\infty \end{cases} \tag{8.37}$$

since $\mathcal{V}_H(\mathbf{r})$ is also divergence-free. We consider a ball $B(0, a)$ and the identity

$$0 = \int_{B(0,a)} \mathrm{d}V\, f(\mathbf{r}) \nabla^2 f = \int_{\partial B} \mathrm{d}S\, f(\mathbf{r}) \hat{\mathbf{r}} \cdot \nabla f - \int_{B(0,a)} \mathrm{d}V\, |\nabla f|^2 \tag{8.38}$$

which follows from the application of (H.96). For the surface integral we estimate

$$\left| \int_{\partial B} \mathrm{d}S\, f(\mathbf{r}) \hat{\mathbf{r}} \cdot \nabla f \right| \leqslant \int_{\partial B} \mathrm{d}S\, |f(\mathbf{r})| |\nabla f| \leqslant \int_{\partial B} \mathrm{d}S\, \frac{C_f}{a} \frac{C_f}{a^2} = 4\pi \frac{C_f^2}{a} \tag{8.39}$$

with C_f a suitable positive constant. In the limit as $a \to +\infty$ from (8.38) we obtain

$$0 = \int_{\mathbb{R}^3} \mathrm{d}V\, |\nabla f|^2 = \|\nabla f\|_2^2 = \|\mathcal{V}_H\|_2^2 \tag{8.40}$$

which is verified only if $\nabla f = 0$, and thus $f(\mathbf{r})$ is a constant scalar field for $\mathbf{r} \in \mathbb{R}^3$. In particular, since $f(\mathbf{r})$ must vanish at infinity, we see that $f(\mathbf{r}) = 0$ everywhere. Regardless, $\mathcal{V}_H(\mathbf{r}) = 0$ for $\mathbf{r} \in \mathbb{R}^3$, contrary to our initial assumption. In this proof we have not used the orthogonality conditions (8.32), granted, but they are trivially satisfied because a null vector field is orthogonal to any other field. In conclusion, the decomposition given by (8.2) and (8.28) is the most general for vector fields in the whole space.

Example 8.1 (The Helmholtz decomposition of static and stationary fields in the whole space)

Static or stationary electric fields are solutions to (2.3) and (2.4) which, in tandem with the constitutive relation (1.117), precisely assign the divergence and the curl of the $\mathcal{E}(\mathbf{r})$ in the whole space. If the charge density is confined to a finite smooth volume V_ϱ, application of (8.28) yields

$$\mathcal{E}(\mathbf{r}) = -\nabla \int_{V_\varrho} dV' \frac{\varrho(\mathbf{r}')/\varepsilon}{4\pi|\mathbf{r}-\mathbf{r}'|}, \qquad \mathbf{r} \in \mathbb{R}^3 \tag{8.41}$$

which is the same result we may obtain by combining (2.15) and (2.169). In view of (2.4) and (8.27) the solenoidal part of the electric field is zero.

Stationary magnetic fields generated by solenoidal currents $\mathcal{J}(\mathbf{r})$ confined in a finite region V_J solve (4.9) and (4.10) complemented with the constitutive relation (1.118). By invoking the Helmholtz theorem (8.28) with $\mathcal{V}(\mathbf{r}) = \mathcal{B}(\mathbf{r})$ we find

$$\mathcal{B}(\mathbf{r}) = \nabla \times \int_{V_J} dV' \frac{\mu \mathcal{J}(\mathbf{r}')}{4\pi|\mathbf{r}-\mathbf{r}'|}, \qquad \mathbf{r} \in \mathbb{R}^3 \tag{8.42}$$

again the result arising from the combination of (4.29) and (5.29). The lamellar part of the magnetic induction is zero on account of (4.10) and (8.16).

The Helmholtz theorem applied to the magnetic vector potential $\mathcal{A}(\mathbf{r})$ (4.29) in the Coulomb gauge [see (5.1) or (8.106) further on] yields

$$\mathcal{A}(\mathbf{r}) = \nabla \times \int_{\mathbb{R}^3} dV' \frac{\nabla' \times \mathcal{A}(\mathbf{r}')}{4\pi|\mathbf{r}-\mathbf{r}'|}, \qquad \mathbf{r} \in \mathbb{R}^3 \tag{8.43}$$

because the solenoidal part of $\mathcal{A}(\mathbf{r})$ is zero. This representation does not look quite the same as (5.29) inasmuch as it does not involve the source $\mathcal{J}(\mathbf{r}')$ explicitly. However, if we swap integral and derivatives — which we proved to be feasible for the volume potential in Section 2.8 — we have

$$\begin{aligned}
\mathcal{A}(\mathbf{r}) &= \int_{\mathbb{R}^3} dV'\, \nabla \frac{1}{4\pi|\mathbf{r}-\mathbf{r}'|} \times [\nabla' \times \mathcal{A}(\mathbf{r}')] = -\int_{\mathbb{R}^3} dV'\, \nabla' \frac{1}{4\pi|\mathbf{r}-\mathbf{r}'|} \times [\nabla' \times \mathcal{A}(\mathbf{r}')] \\
&= \int_{\mathbb{R}^3} dV' \frac{\nabla' \times [\nabla' \times \mathcal{A}(\mathbf{r}')]}{4\pi|\mathbf{r}-\mathbf{r}'|} - \int_{\mathbb{R}^3} dV'\, \nabla' \times \frac{\nabla' \times \mathcal{A}(\mathbf{r}')}{4\pi|\mathbf{r}-\mathbf{r}'|} \\
&= \int_{V_J} dV' \frac{\mu \mathcal{J}(\mathbf{r}')}{4\pi|\mathbf{r}-\mathbf{r}'|} - \int_{\mathbb{R}^3} dV'\, \nabla' \times \frac{\nabla' \times \mathcal{A}(\mathbf{r}')}{4\pi|\mathbf{r}-\mathbf{r}'|}
\end{aligned} \tag{8.44}$$

by virtue of the differential identity (H.50). The last improper integral may be shown to be zero by means of a limiting process in which we first integrate over the spherical shell $B(\mathbf{r}, b) \setminus B(\mathbf{r}, a)$, $a < b$, so that the integral identity (H.91) may be applied. The resulting surface integrals over the spheres $\partial B(\mathbf{r}, a)$ and $\partial B(\mathbf{r}, b)$ vanish in the limit as $a \to 0^+$ and $b \to +\infty$ also because $\nabla' \times \mathcal{A}(\mathbf{r}')$ is regular in $\mathbf{r}' = \mathbf{r}$ and $\mathcal{A}(\mathbf{r}')$ satisfies the asymptotic condition (4.36). With these steps the equivalence of (5.29) and (8.43) is proved.

Since we have derived (8.28) under the assumption that $\mathcal{V}(\mathbf{r}) \in \mathcal{C}^1(\mathbb{R}^3)^3$, strictly speaking, the densities $\varrho(\mathbf{r})$ and $\mathcal{J}(\mathbf{r})$ have to be continuous. Nonetheless, (8.41), (8.42) and (8.43) were derived under milder requirements in Sections 3.2, 5.1.2 and 5.1.1.

(End of Example 8.1)

8.1.2 Bounded regions

The separation of a vector field $\mathcal{V}(\mathbf{r}) \in \mathcal{C}^1(V)^3$ into solenoidal and lamellar parts can also be accomplished in a finite region of space $V \subset \mathbb{R}^3$ with smooth or piecewise-smooth boundary ∂V [24]. In this regard, we may expect the decomposition to depend partly on the values the field and perhaps the derivatives thereof take on ∂V.

We begin by considering a simply connected domain V (see Figure 1.2b) and let [16, Appendix A], [20, Appendix A]

$$\begin{aligned}[1 - U(\mathbf{r})]\mathcal{V}(\mathbf{r}) &= -\nabla^2\mathbf{f}(\mathbf{r}) = \nabla \times \nabla \times \mathbf{f}(\mathbf{r}) - \nabla\nabla \cdot \mathbf{f}(\mathbf{r}) \\ &= \nabla \times \Xi(\mathbf{r}) + \nabla\Upsilon(\mathbf{r}) = \mathcal{V}_S(\mathbf{r}) + \mathcal{V}_L(\mathbf{r}), \qquad \mathbf{r} \in \mathbb{R}^3 \qquad (8.45)\end{aligned}$$

where $U(\mathbf{r})$ is the three-dimensional unit step function defined in (C.49), and $\mathbf{f}(\mathbf{r})$ denotes an auxiliary vector potential endowed with enough continuous derivatives. Since the function $1 - U(\mathbf{r})$ is equal to one in the region of interest and to zero outside V, multiplying $\mathcal{V}(\mathbf{r})$ by $1 - U(\mathbf{r})$ provides the restriction of the field $\mathcal{V}(\mathbf{r})$ to V, inasmuch as $\mathcal{V}(\mathbf{r})$ need not be zero outside V! By contrast, the vector field $-\nabla^2\mathbf{f}$ vanishes in $\mathbb{R}^3 \setminus \overline{V}$ by definition. The other potentials, $\Upsilon(\mathbf{r})$ and $\Xi(\mathbf{r})$, are analogous to those defined in (8.5) and (8.17), and will come in handy further on.

If we manage to find an expression for $\mathbf{f}(\mathbf{r})$, the rightmost-hand side of (8.45) will give the desired Helmholtz decomposition of $\mathcal{V}(\mathbf{r})$. Then again, $\mathbf{f}(\mathbf{r})$ is not univocally defined, because any vector field in the form

$$\mathbf{f}'(\mathbf{r}) = \mathbf{f}(\mathbf{r}) + \nabla\psi(\mathbf{r}), \qquad \mathbf{r} \in \mathbb{R}^3 \qquad (8.46)$$

where $\psi(\mathbf{r})$ is a harmonic potential (i.e., solves the Laplace equation) is also a solution to (8.45). We may exploit this degree of freedom to obtain as simple a solution as possible, and thus we require that $\mathbf{f}(\mathbf{r})$ obey the asymptotic behavior

$$\mathbf{f}(\mathbf{r}) = \mathcal{O}\left(\frac{1}{|\mathbf{r}|}\right), \qquad |\mathbf{r}| \to +\infty \qquad (8.47)$$

uniformly with respect to $\mathbf{r}$. According to (8.45) and (8.47) the Cartesian components of $\mathbf{f}(\mathbf{r})$ solve three uncoupled Poisson equations in $\mathbb{R}^3$ — whose 'source terms' are the scalar fields $\mathcal{V}_x(\mathbf{r})$, $\mathcal{V}_y(\mathbf{r})$ and $\mathcal{V}_z(\mathbf{r})$ restricted to V — and fall off as the inverse of the distance from the 'source'. Therefore, by virtue of the integral representations of Section 2.7 and, in particular, formula (2.169) applied to $f_x(\mathbf{r})$, $f_y(\mathbf{r})$ and $f_z(\mathbf{r})$ in succession we can write the solution straightaway as

$$\mathbf{f}(\mathbf{r}) = \int_V \mathrm{d}V' \, \frac{\mathcal{V}(\mathbf{r}')}{4\pi|\mathbf{r} - \mathbf{r}'|}, \qquad \mathbf{r} \in \mathbb{R}^3 \qquad (8.48)$$

because the 'vector density' $[1 - U(\mathbf{r})]\mathcal{V}(\mathbf{r})$ is confined to V by construction. Plugging (8.48) back into the first part of (8.45) yields

$$\mathcal{V}(\mathbf{r}) = \nabla \times \nabla \times \int_V \mathrm{d}V' \, \frac{\mathcal{V}(\mathbf{r}')}{4\pi|\mathbf{r} - \mathbf{r}'|} - \nabla\nabla \cdot \int_V \mathrm{d}V' \, \frac{\mathcal{V}(\mathbf{r}')}{4\pi|\mathbf{r} - \mathbf{r}'|}, \qquad \mathbf{r} \in V \qquad (8.49)$$

which is one possible form of the Helmholtz splitting for $\mathcal{V}(\mathbf{r})$ in V. Actually, (8.49) as it stands holds everywhere in space but, similarly to all the integral formulas we developed in Sections 2.7, 3.2 and 5.1, the combination of integrals and derivatives in the right-hand side returns zero when evaluated for observation points $\mathbf{r}$ in $\mathbb{R}^3 \setminus \overline{V}$ in light of the leftmost-hand side of (8.45). Indeed, invoking (H.59) yields

$$\begin{aligned}[1 - U(\mathbf{r})]\mathcal{V}(\mathbf{r}) &= -\nabla^2 \int_V \mathrm{d}V' \, \frac{\mathcal{V}(\mathbf{r}')}{4\pi|\mathbf{r} - \mathbf{r}'|} \\ &= - \int_V \mathrm{d}V' \, \mathcal{V}(\mathbf{r}')\nabla^2 \frac{1}{4\pi|\mathbf{r} - \mathbf{r}'|} = 0, \qquad \mathbf{r} \in \mathbb{R}^3 \setminus \overline{V} \end{aligned} \tag{8.50}$$

where we have swapped derivatives and integral on the grounds that $1/|\mathbf{r} - \mathbf{r}'|$ is regular when $\mathbf{r} \in \mathbb{R}^3 \setminus \overline{V}$ and $\mathbf{r}' \in V$. From a functional viewpoint we may interpret the dyadic integro-differential operator

$$\mathfrak{I}\{\bullet\} := \nabla \times \nabla \times \int_V \mathrm{d}V' \, \frac{\overline{\mathfrak{I}} \cdot \{\bullet\}}{4\pi|\mathbf{r} - \mathbf{r}'|} - \nabla\nabla \cdot \int_V \mathrm{d}V' \, \frac{\overline{\mathfrak{I}} \cdot \{\bullet\}}{4\pi|\mathbf{r} - \mathbf{r}'|}, \qquad \mathbf{r} \in V \tag{8.51}$$

as an identity (see Appendix D.3) which, for $\mathbf{r} \in V \subset \mathbb{R}^3$, is the explicit expression of the formal dyadic operator introduced in (8.2).

Compared to (8.28) decomposition (8.49) is unsatisfactory in that the integrals involve the field itself rather than curl and divergence thereof. We remedy to this glitch by 'moving' the inner curl operator and the divergence past the integrals and performing some algebra. Interchanging integration and differentiation is permitted on the grounds of (2.199), which was proved in Section 2.8, because the Cartesian components of $\mathbf{f}(\mathbf{r})$ are, in fact, static volume potentials. To be specific, expanding (2.199) in Cartesian coordinates and likewise turning to (A.32) and (A.29) for the curl and the divergence lead us to

$$\begin{aligned}\mathcal{V}(\mathbf{r}) &= \nabla \times \int_V \mathrm{d}V' \, \nabla \frac{1}{4\pi|\mathbf{r} - \mathbf{r}'|} \times \mathcal{V}(\mathbf{r}') - \nabla \int_V \mathrm{d}V' \, \nabla \frac{1}{4\pi|\mathbf{r} - \mathbf{r}'|} \cdot \mathcal{V}(\mathbf{r}') \\ &= -\nabla \times \int_V \mathrm{d}V' \, \nabla' \frac{1}{4\pi|\mathbf{r} - \mathbf{r}'|} \times \mathcal{V}(\mathbf{r}') + \nabla \int_V \mathrm{d}V' \, \nabla' \frac{1}{4\pi|\mathbf{r} - \mathbf{r}'|} \cdot \mathcal{V}(\mathbf{r}') \end{aligned} \tag{8.52}$$

and to proceed further we exclude the singular point $\mathbf{r} \in V$ with a ball $B(\mathbf{r}, a) \subset V$ and split the pertinent integration into two parts, viz., over $V_a := V \setminus B[\mathbf{r}, a]$ and over the ball (see Figure 2.8 for a similar setup). For the first integral this means

$$\begin{aligned}&- \int_V \mathrm{d}V' \, \nabla' \frac{1}{4\pi|\mathbf{r} - \mathbf{r}'|} \times \mathcal{V}(\mathbf{r}') = \\ &= \int_{V_a} \mathrm{d}V' \left[\frac{\nabla' \times \mathcal{V}}{4\pi|\mathbf{r} - \mathbf{r}'|} - \nabla' \times \frac{\mathcal{V}(\mathbf{r}')}{4\pi|\mathbf{r} - \mathbf{r}'|} \right] - \int_{B(\mathbf{r},a)} \mathrm{d}V' \, \nabla' \frac{1}{4\pi|\mathbf{r} - \mathbf{r}'|} \times \mathcal{V}(\mathbf{r}') \\ &= \int_{V_a} \mathrm{d}V' \, \frac{\nabla' \times \mathcal{V}}{4\pi|\mathbf{r} - \mathbf{r}'|} + \int_{\partial V} \mathrm{d}S' \, \frac{\hat{\mathbf{n}}(\mathbf{r}') \times \mathcal{V}(\mathbf{r}')}{4\pi|\mathbf{r} - \mathbf{r}'|} \\ &\quad + \int_{\partial B} \mathrm{d}S' \, \frac{\hat{\mathbf{R}} \times \mathcal{V}(\mathbf{r}')}{4\pi a} - \int_{B(\mathbf{r},a)} \mathrm{d}V' \, \frac{\hat{\mathbf{R}} \times \mathcal{V}(\mathbf{r}')}{4\pi R^2} \end{aligned} \tag{8.53}$$

having made use of (H.49) and the curl theorem (H.91) in V_a with the unit normal $\hat{\mathbf{n}}(\mathbf{r}')$ positively oriented inwards V_a. In the limit as $a \to 0^+$, the first integral in the rightmost member remains bounded, the second is unaffected being independent of a, the third and the last one vanish. More specifically, since $\mathcal{V}(\mathbf{r}) \in \mathcal{C}^1(V)^3$ we can make the following estimates

$$\left| \int_{V_a} dV' \, \frac{\nabla' \times \mathcal{V}}{4\pi|\mathbf{r}-\mathbf{r}'|} \right| \leqslant \int_{V_a} dV' \, \frac{\|\nabla' \times \mathcal{V}\|_\infty}{4\pi R} \leqslant \int_{B(\mathbf{r},b)\setminus B(\mathbf{r},a)} dV' \, \frac{\|\nabla' \times \mathcal{V}\|_\infty}{4\pi R} = \left\|\nabla' \times \mathcal{V}\right\|_\infty \int_a^b dR\, R$$
$$= \frac{1}{2}\left\|\nabla' \times \mathcal{V}\right\|_\infty (b^2 - a^2) \leqslant \frac{1}{2}\left\|\nabla' \times \mathcal{V}\right\|_\infty b^2 \tag{8.54}$$

$$\left| \int_{\partial B} dS' \, \frac{\hat{\mathbf{R}} \times \mathcal{V}(\mathbf{r}')}{4\pi a} \right| \leqslant \int_{\partial B} dS' \, \frac{\|\mathcal{V}\|_\infty}{4\pi a} = \|\mathcal{V}\|_\infty \, a \xrightarrow[a \to 0^+]{} 0 \tag{8.55}$$

$$\left| \int_{B(\mathbf{r},a)} dV' \, \frac{\hat{\mathbf{R}} \times \mathcal{V}(\mathbf{r}')}{4\pi R^2} \right| \leqslant \int_{B(\mathbf{r},a)} dV' \, \frac{\|\mathcal{V}\|_\infty}{4\pi R^2} = \|\mathcal{V}\|_\infty \, a \xrightarrow[a \to 0^+]{} 0 \tag{8.56}$$

where $B(\mathbf{r}, b)$ is a larger ball that contains V_a (see Figure 2.9 for a similar geometry). The steps required to handle the second integral in (8.52) differ only for the application of (H.51) and the Gauss theorem (H.90).

All in all, we have proved the alternative, classic form of the Helmholtz decomposition for bounded domains, namely,

$$\mathcal{V}(\mathbf{r}) = \nabla \times \int_V dV' \, \frac{\nabla' \times \mathcal{V}(\mathbf{r}')}{4\pi|\mathbf{r}-\mathbf{r}'|} + \nabla \times \int_{\partial V} dS' \, \frac{\hat{\mathbf{n}}(\mathbf{r}') \times \mathcal{V}(\mathbf{r}')}{4\pi|\mathbf{r}-\mathbf{r}'|}$$
$$- \nabla \int_V dV' \, \frac{\nabla' \cdot \mathcal{V}(\mathbf{r}')}{4\pi|\mathbf{r}-\mathbf{r}'|} - \nabla \int_{\partial V} dS' \, \frac{\hat{\mathbf{n}}(\mathbf{r}') \cdot \mathcal{V}(\mathbf{r}')}{4\pi|\mathbf{r}-\mathbf{r}'|}, \qquad \mathbf{r} \in V \tag{8.57}$$

which yields $\mathcal{V}(\mathbf{r})$ if the latter is known on the boundary ∂V and curl and divergence are assigned in V. This result is variably known as the *Clebsch-Helmholtz decomposition* of $\mathcal{V}(\mathbf{r})$, the *Poincaré identity*, or the *fundamental theorem of vector analysis* [25, Section 4.5]. Side by side comparison of (8.57) with the second part of (8.45) allows identifying the auxiliary potentials $\Xi(\mathbf{r})$ and $\Upsilon(\mathbf{r})$ as

$$\Xi(\mathbf{r}) = \int_V dV' \, \frac{\nabla' \times \mathcal{V}(\mathbf{r}')}{4\pi|\mathbf{r}-\mathbf{r}'|} + \int_{\partial V} dS' \, \frac{\hat{\mathbf{n}}(\mathbf{r}') \times \mathcal{V}(\mathbf{r}')}{4\pi|\mathbf{r}-\mathbf{r}'|}, \qquad \mathbf{r} \in V \tag{8.58}$$

$$\Upsilon(\mathbf{r}) = -\int_V dV' \, \frac{\nabla' \cdot \mathcal{V}(\mathbf{r}')}{4\pi|\mathbf{r}-\mathbf{r}'|} - \int_{\partial V} dS' \, \frac{\hat{\mathbf{n}}(\mathbf{r}') \cdot \mathcal{V}(\mathbf{r}')}{4\pi|\mathbf{r}-\mathbf{r}'|}, \qquad \mathbf{r} \in V \tag{8.59}$$

which unlike (8.27) and (8.16) contain boundary terms. We notice in passing that if $\mathcal{V}(\mathbf{r})$ is defined in $\mathbb{R}^3$ and additionally meets the asymptotic condition (8.8), then (8.57)-(8.59) may serve as the starting point to prove (8.28), the Helmholtz decomposition in the whole space. It suffices to choose $V := B(\mathbf{r}, b)$ and let the radius b grow infinitely large, in which case the surface integrals vanish.

Since in (8.57) the integrals over V are still volume potentials with vector densities, we are allowed to interchange integration and differentiation once more by invoking the special result

(2.199). Further manipulations are then performed by means of the same limiting process we devised above to arrive at (8.57). Thus, we leave the details to the Reader and quote the result, viz.,

$$\begin{aligned}\mathcal{V}(\mathbf{r}) = &\int_V \mathrm{d}V' \frac{\nabla' \times \nabla' \times \mathcal{V}(\mathbf{r}')}{4\pi|\mathbf{r}-\mathbf{r}'|} - \int_V \mathrm{d}V' \frac{\nabla'\nabla' \cdot \mathcal{V}(\mathbf{r}')}{4\pi|\mathbf{r}-\mathbf{r}'|} \\ &+ \int_{\partial V} \mathrm{d}S' \frac{\hat{\mathbf{n}}(\mathbf{r}') \times [\nabla' \times \mathcal{V}(\mathbf{r}')]}{4\pi|\mathbf{r}-\mathbf{r}'|} + \nabla \times \int_{\partial V} \mathrm{d}S' \frac{\hat{\mathbf{n}}(\mathbf{r}') \times \mathcal{V}(\mathbf{r}')}{4\pi|\mathbf{r}-\mathbf{r}'|} \\ &- \int_{\partial V} \mathrm{d}S' \frac{\hat{\mathbf{n}}(\mathbf{r}')\nabla' \cdot \mathcal{V}(\mathbf{r}')}{4\pi|\mathbf{r}-\mathbf{r}'|} - \nabla \int_{\partial V} \mathrm{d}S' \frac{\hat{\mathbf{n}}(\mathbf{r}') \cdot \mathcal{V}(\mathbf{r}')}{4\pi|\mathbf{r}-\mathbf{r}'|}, \qquad \mathbf{r} \in V \end{aligned} \tag{8.60}$$

which is yet another form of integral representation of $\mathcal{V}(\mathbf{r})$, though lamellar and solenoidal parts are no longer self-evident. Unlike (8.57), decomposition (8.60) requires extra information on the divergence of $\mathcal{V}(\mathbf{r})$ and the tangential curl thereof on the boundary ∂V.

The orthogonality of solenoidal and lamellar components of $\mathcal{V}(\mathbf{r})$ with respect to the inner product (D.68) with $D = V$ depends on the boundary conditions obeyed by $\mathcal{V}(\mathbf{r})$ on ∂V and, in general, may not occur. To be specific, by expressing $\mathcal{V}_S(\mathbf{r})$ and $\mathcal{V}_L(\mathbf{r})$ in terms of the auxiliary potentials (8.58) and (8.59) we have

$$\begin{aligned}(\mathcal{V}_L, \mathcal{V}_S)_V &:= \int_V \mathrm{d}V\, \mathcal{V}_L(\mathbf{r}) \cdot \mathcal{V}_S(\mathbf{r}) = \int_V \mathrm{d}V\, \nabla\Upsilon \cdot \nabla \times \Xi \\ &= \frac{1}{2}\int_V \mathrm{d}V\, \nabla \cdot [\Upsilon(\mathbf{r})\nabla \times \Xi] - \frac{1}{2}\int_V \mathrm{d}V\, \nabla \cdot [\nabla\Upsilon \times \Xi(\mathbf{r})] \\ &= -\frac{1}{2}\int_{\partial V} \mathrm{d}S\, \Upsilon(\mathbf{r})\hat{\mathbf{n}}(\mathbf{r}) \cdot \nabla \times \Xi + \frac{1}{2}\int_{\partial V} \mathrm{d}S\, \hat{\mathbf{n}}(\mathbf{r}) \cdot \nabla\Upsilon \times \Xi(\mathbf{r}) \\ &= \frac{1}{2}\int_{\partial V} \mathrm{d}S\, \Upsilon(\mathbf{r})\nabla_{\mathrm{s}} \cdot [\hat{\mathbf{n}}(\mathbf{r}) \times \Xi(\mathbf{r})] + \frac{1}{2}\int_{\partial V} \mathrm{d}S\, \hat{\mathbf{n}}(\mathbf{r}) \cdot \nabla\Upsilon \times \Xi(\mathbf{r})\end{aligned} \tag{8.61}$$

thanks to (H.49), (H.51), the Gauss theorem applied with the unit normal pointing inwards V, and finally (H.82). The conditions under which the surface integrals in the rightmost member vanish are quite apparent, but we should translate such requirements into constraints for $\mathcal{V}(\mathbf{r})$. In particular, we observe that

$$\hat{\mathbf{n}}(\mathbf{r}) \times \Xi(\mathbf{r}) = 0 \quad \Longrightarrow \quad \hat{\mathbf{n}}(\mathbf{r}) \cdot \nabla \times \Xi(\mathbf{r}) = -\nabla_{\mathrm{s}} \cdot [\hat{\mathbf{n}}(\mathbf{r}) \times \Xi(\mathbf{r})] = 0 \tag{8.62}$$

$$\Upsilon(\mathbf{r}) = 0 \quad \Longrightarrow \quad \hat{\mathbf{n}}(\mathbf{r}) \times \nabla\Upsilon(\mathbf{r}) = 0 \tag{8.63}$$

whence in light of (8.45) and (8.61) we conclude that if *either one* of the following conditions is met

$$\hat{\mathbf{n}}(\mathbf{r}) \cdot \mathcal{V}_S(\mathbf{r}) = 0, \qquad \mathbf{r} \in \partial V \tag{8.64}$$

$$\hat{\mathbf{n}}(\mathbf{r}) \times \mathcal{V}_L(\mathbf{r}) = 0, \qquad \mathbf{r} \in \partial V \tag{8.65}$$

then $\mathcal{V}_L(\mathbf{r})$ and $\mathcal{V}_S(\mathbf{r})$ are orthogonal in V. The combination of (8.63) and (8.65) produces the so-called *electric splitting* because $\Upsilon(\mathbf{r})$ and $\mathcal{V}_L(\mathbf{r})$ then obey the same matching conditions as electrostatic potential $\Phi(\mathbf{r})$ and field $\mathcal{E}(\mathbf{r})$ in a region V bounded by conducting walls. By contrast, the combination of (8.62) and (8.64) gives rise to the *magnetic splitting*, inasmuch as $\Xi(\mathbf{r})$ and $\mathcal{V}_S(\mathbf{r})$

satisfy the same matching conditions as stationary magnetic vector potential $\mathcal{A}(\mathbf{r})$ and field $\mathcal{H}(\mathbf{r})$ in a volume V flush with a PEC.

Nevertheless, owing to the way we arrived at $\Xi(\mathbf{r})$ and $\Upsilon(\mathbf{r})$, viz., *indirectly* after determining yet another vector potential, $\mathbf{f}(\mathbf{r})$, we have not subjected $\Xi(\mathbf{r})$ and $\Upsilon(\mathbf{r})$ to any boundary condition on ∂V. Therefore, even though (8.64) and (8.65) may also be derived from (8.61) rather than from the matching conditions of $\Xi(\mathbf{r})$ and $\Upsilon(\mathbf{r})$, we should check whether or not (8.62) and (8.63) are indeed satisfied and compatible with (8.58) and (8.59).

To this purpose, we observe that, for $\mathbf{r} \in V$, $\Xi(\mathbf{r})$ solves a vector Poisson equation such as (4.33) and is determined up to the gradient of a scalar field. Hence, we are at liberty to require that $\Xi(\mathbf{r})$ be *solenoidal* and hence we may resort to the integral representation (5.18) whereby for $\mathbf{r} \in V$ we may write

$$\begin{aligned}\Xi(\mathbf{r}) &= \int_V \mathrm{d}V' \frac{\nabla' \times \mathcal{V}(\mathbf{r}')}{4\pi|\mathbf{r}-\mathbf{r}'|} + \int_{\partial V} \mathrm{d}S'\, \hat{\mathbf{n}}(\mathbf{r}') \times [\nabla' \times \Xi(\mathbf{r}')] \frac{1}{4\pi R} \\ &\quad + \int_{\partial V} \mathrm{d}S'\, [\hat{\mathbf{n}}(\mathbf{r}') \times \Xi(\mathbf{r}')] \times \nabla' \frac{1}{4\pi R} + \int_{\partial V} \mathrm{d}S'\, \hat{\mathbf{n}}(\mathbf{r}') \cdot \Xi(\mathbf{r}') \nabla' \frac{1}{4\pi R} \\ &= \int_V \mathrm{d}V' \frac{\nabla' \times \mathcal{V}(\mathbf{r}')}{4\pi|\mathbf{r}-\mathbf{r}'|} + \int_{\partial V} \mathrm{d}S' \frac{\hat{\mathbf{n}}(\mathbf{r}') \times \mathcal{V}_S(\mathbf{r}')}{4\pi|\mathbf{r}-\mathbf{r}'|} \\ &\quad + \int_{\partial V} \mathrm{d}S'\, [\hat{\mathbf{n}}(\mathbf{r}') \times \Xi(\mathbf{r}')] \times \nabla' \frac{1}{4\pi R} + \int_{\partial V} \mathrm{d}S'\, \hat{\mathbf{n}}(\mathbf{r}') \cdot \Xi(\mathbf{r}') \nabla' \frac{1}{4\pi R}\end{aligned} \tag{8.66}$$

which is in agreement with (8.58) precisely if (8.62) and (8.65) hold true and in addition $\hat{\mathbf{n}}(\mathbf{r}) \cdot \Xi(\mathbf{r}) = 0$ for $\mathbf{r} \in \partial V$. Concerning this latter constraint, two remarks are in order. First, the vanishing of $\hat{\mathbf{n}} \cdot \Xi$ does not follow from our attempt to ensure the orthogonality of $\mathcal{V}_L(\mathbf{r})$ and $\mathcal{V}_S(\mathbf{r})$ but rather from the imposed solenoidal nature of $\Xi(\mathbf{r})$, in that

$$0 = \int_V \mathrm{d}V\, \nabla \cdot \Xi(\mathbf{r}) = -\int_{\partial V} \mathrm{d}S\, \hat{\mathbf{n}}(\mathbf{r}) \cdot \Xi(\mathbf{r}) \tag{8.67}$$

so $\hat{\mathbf{n}} \cdot \Xi = 0$ is the simplest way for the above equation to be fulfilled. Secondly, the contribution of $\hat{\mathbf{n}} \cdot \Xi$ — even though it were non-null — bears no consequence for $\mathcal{V}_S(\mathbf{r})$, because it is swept away when the curl of (8.66) is taken.

Next, we notice that, for $\mathbf{r} \in V$, $\Upsilon(\mathbf{r})$ solves a Poisson equation such as (2.17). Thus, making use of the integral representation (2.160) for $\mathbf{r} \in V$ we may write

$$\begin{aligned}\Upsilon(\mathbf{r}) &= -\int_V \mathrm{d}V' \frac{\nabla' \cdot \mathcal{V}(\mathbf{r}')}{4\pi|\mathbf{r}-\mathbf{r}'|} + \int_{\partial V} \mathrm{d}S' \left[\Upsilon(\mathbf{r}') \frac{\partial}{\partial \hat{\mathbf{n}}'} \frac{1}{4\pi R} - \frac{1}{4\pi R} \frac{\partial \Upsilon}{\partial \hat{\mathbf{n}}'}\right] \\ &= -\int_V \mathrm{d}V' \frac{\nabla' \cdot \mathcal{V}(\mathbf{r}')}{4\pi|\mathbf{r}-\mathbf{r}'|} + \int_{\partial V} \mathrm{d}S' \left[\Upsilon(\mathbf{r}') \frac{\partial}{\partial \hat{\mathbf{n}}'} \frac{1}{4\pi R} - \frac{\hat{\mathbf{n}}(\mathbf{r}') \cdot \mathcal{V}_L(\mathbf{r}')}{4\pi R}\right]\end{aligned} \tag{8.68}$$

and this alternative expression is consistent with (8.59) if (8.63) and (8.64) are true.

In summary, only one of (8.64) and (8.65) is required for the orthogonality of $\mathcal{V}_L$ and $\mathcal{V}_S$ but, in fact, they are both true because (8.62) and (8.63) are implied in representations (8.58) and (8.59). All in all, these checks unveil an internal consistency among the various actors involved in the construction of the Helmholtz decomposition of $\mathcal{V}(\mathbf{r})$ within the region V.

Example 8.2 (The Helmholtz decomposition of static electric fields in a bounded region)

Applying (8.57) to the static electric field $\mathcal{E}(\mathbf{r})$ produced by a charge density $\varrho(\mathbf{r})$ confined to a smaller domain $V_\varrho \subset V$ gives

$$\mathcal{E}(\mathbf{r}) = -\nabla \int_{V_\varrho} \mathrm{d}V' \frac{\varrho(\mathbf{r}')/\varepsilon}{4\pi|\mathbf{r}-\mathbf{r}'|} - \nabla \int_{\partial V} \mathrm{d}S' \frac{\hat{\mathbf{n}}(\mathbf{r}')\cdot\mathcal{E}(\mathbf{r}')}{4\pi|\mathbf{r}-\mathbf{r}'|} - \nabla\times \int_{\partial V} \mathrm{d}S' \frac{\mathcal{E}(\mathbf{r}')\times\hat{\mathbf{n}}(\mathbf{r}')}{4\pi|\mathbf{r}-\mathbf{r}'|}, \qquad \mathbf{r}\in V \quad (8.69)$$

where the fields $\varepsilon\hat{\mathbf{n}}\cdot\mathcal{E}$ and $\mathcal{E}\times\hat{\mathbf{n}}$ play the role of fictitious or equivalent or induced surface densities of electric charge and steady magnetic current on ∂V. This formula is consistent with (3.21) and, in fact, is well defined for observation points on the boundary since the surface integrals are single-layer potentials (Section 2.10).

Generally speaking, $\varepsilon\hat{\mathbf{n}}\cdot\mathcal{E}$ and $\mathcal{E}\times\hat{\mathbf{n}}$ account for the effect of media and sources in the complementary region $\mathbb{R}^3 \setminus \overline{V}$. If the boundary ∂V is flush with a conductor — it does not have to be a PEC in the static limit — the electrostatic field obeys (1.180) and (1.182). In which case, the last integral in (8.69) is absent and $\varrho_S := \varepsilon\hat{\mathbf{n}}\cdot\mathcal{E}$ represents the surface charge induced on the conductor. By constrast, if ∂V is flush with a magnetic conductor (not necessarily a PMC) no electric charge is induced on the boundary, and $\mathcal{J}_{MS} := \mathcal{E}\times\hat{\mathbf{n}}$ is an induced magnetic current which accounts for the very presence of the magnetic conductor in $\mathbb{R}^3 \setminus \overline{V}$.

In the right member of (8.69) the first two contributions together are the negative gradient of $\Upsilon(\mathbf{r})$, whereas the third one is the curl of $\Xi(\mathbf{r})$. Therefore, it appears that $\mathcal{E}(\mathbf{r})$ can be derived from a pair of scalar and vector potentials. Still, on account of (2.15) and the integral representation (2.160) of $\Phi(\mathbf{r})$, it is also possible to obtain $\mathcal{E}(\mathbf{r})$ from just one scalar potential! For the sake of consistency, the two approaches must be equivalent since the solution is unique. Indeed, by taking the gradient of the second term in the right-hand side of (2.160) for observation points $\mathbf{r}$ strictly away from the boundary we have

$$\begin{aligned}
-\nabla \int_{\partial V} \mathrm{d}S'\, \Phi(\mathbf{r}')\hat{\mathbf{n}}(\mathbf{r}')\cdot\nabla'\frac{1}{4\pi|\mathbf{r}-\mathbf{r}'|} &= -\int_{\partial V} \mathrm{d}S'\, \nabla\left[\Phi(\mathbf{r}')\hat{\mathbf{n}}(\mathbf{r}')\cdot\nabla'\frac{1}{4\pi R}\right] = \\
&= \int_{\partial V} \mathrm{d}S'\, \nabla\times\left[\Phi(\mathbf{r}')\hat{\mathbf{n}}(\mathbf{r}')\times\nabla'\frac{1}{4\pi R}\right] = \nabla\times\int_{\partial V} \mathrm{d}S'\, \Phi(\mathbf{r}')\hat{\mathbf{n}}(\mathbf{r}')\times\nabla'\frac{1}{4\pi R} \\
&= \underbrace{\nabla\times\int_{\partial V} \mathrm{d}S'\, \hat{\mathbf{n}}(\mathbf{r}')\times\nabla'\frac{\Phi(\mathbf{r}')}{4\pi R}}_{=0} - \nabla\times\int_{\partial V} \mathrm{d}S' \frac{\hat{\mathbf{n}}(\mathbf{r}')\times\nabla'\Phi(\mathbf{r}')}{4\pi R} \\
&= -\nabla\times\int_{\partial V} \mathrm{d}S' \frac{\mathcal{E}(\mathbf{r}')\times\hat{\mathbf{n}}(\mathbf{r}')}{4\pi|\mathbf{r}-\mathbf{r}'|}
\end{aligned} \quad (8.70)$$

whereby the equivalence of the two representations is established. In the proof we have interchanged derivatives and integral on the grounds that the integrand is regular for $\mathbf{r}' \neq \mathbf{r}$, and invoked formulas (H.53) and (H.54) to turn the gradient into a curl, as it were. Lastly, in the second-to-last step the first surface integral vanishes by virtue of the integral theorem (H.93) applied to a closed surface. From a physical viewpoint (8.70) indicates that a surface layer of electric dipoles with density $\boldsymbol{\tau}_S := \varepsilon\Phi\hat{\mathbf{n}}$ is equivalent to a stationary surface magnetic current density $\mathcal{J}_{MS} := -\nabla\Phi\times\hat{\mathbf{n}}$.

In spatial regions devoid of sources (here, this means $\mathbf{r} \notin V_\varrho \cup \partial V$) the static electric field is both solenoidal and lamellar, and this character ought to be embodied in (8.69). Taking the divergence first yields

$$\nabla \cdot \mathcal{E}(\mathbf{r}) = -\nabla^2 \int_{V_\varrho} \mathrm{d}V' \frac{\varrho(\mathbf{r}')/\varepsilon}{4\pi|\mathbf{r}-\mathbf{r}'|} - \nabla^2 \int_{\partial V} \mathrm{d}S' \frac{\hat{\mathbf{n}}(\mathbf{r}') \cdot \mathcal{E}(\mathbf{r}')}{4\pi|\mathbf{r}-\mathbf{r}'|}$$

$$= -\int_{V_\varrho} \mathrm{d}V' \frac{\varrho(\mathbf{r}')}{\varepsilon} \nabla'^2 \frac{1}{4\pi R} - \int_{\partial V} \mathrm{d}S' \, \hat{\mathbf{n}}(\mathbf{r}') \cdot \mathcal{E}(\mathbf{r}') \nabla'^2 \frac{1}{4\pi R} = 0 \tag{8.71}$$

where the interchange of integration and Laplace operator is allowed, since $1/R$ is regular for $\mathbf{r} \notin V_\varrho \cup \partial V$. The result follows because, under the stated hypotheses, $1/R$ is harmonic. For the same reasons, taking the curl next and using (H.59) gives

$$\nabla \times \mathcal{E}(\mathbf{r}) = -\nabla \times \nabla \times \int_{\partial V} \mathrm{d}S' \frac{\mathcal{E}(\mathbf{r}') \times \hat{\mathbf{n}}(\mathbf{r}')}{4\pi|\mathbf{r}-\mathbf{r}'|} = -\nabla\nabla \cdot \int_{\partial V} \mathrm{d}S' \frac{\mathcal{E}(\mathbf{r}') \times \hat{\mathbf{n}}(\mathbf{r}')}{4\pi|\mathbf{r}-\mathbf{r}'|}$$

$$= \nabla \int_{\partial V} \mathrm{d}S' \, \nabla' \frac{1}{4\pi R} \cdot \mathcal{E}(\mathbf{r}') \times \hat{\mathbf{n}}(\mathbf{r}') = \nabla \int_{\partial V} \mathrm{d}S' \, \nabla'_{\mathrm{s}} \frac{1}{4\pi R} \cdot \mathcal{E}(\mathbf{r}') \times \hat{\mathbf{n}}(\mathbf{r}')$$

$$= \nabla \int_{\partial V} \mathrm{d}S' \, \nabla'_{\mathrm{s}} \cdot \frac{\mathcal{E}(\mathbf{r}') \times \hat{\mathbf{n}}(\mathbf{r}')}{4\pi R} - \nabla \int_{\partial V} \mathrm{d}S' \frac{\nabla'_{\mathrm{s}} \cdot [\mathcal{E}(\mathbf{r}') \times \hat{\mathbf{n}}(\mathbf{r}')]}{4\pi R}$$

$$= \nabla \int_{\partial V} \mathrm{d}S' \, \hat{\mathbf{n}}(\mathbf{r}') \cdot \nabla' \times \frac{\mathcal{E}(\mathbf{r}')}{4\pi R} - \nabla \int_{\partial V} \mathrm{d}S' \frac{\hat{\mathbf{n}}(\mathbf{r}') \cdot \nabla' \times \mathcal{E}}{4\pi R} = 0 \tag{8.72}$$

where we have noticed that $\mathcal{E} \times \hat{\mathbf{n}}$ is tangential to ∂V and invoked identities (H.77) and (A.60) for the surface divergence. The last two integrals vanish by virtue of the Stokes theorem applied on a closed surface and because $\mathcal{E}(\mathbf{r})$ is lamellar on ∂V.

(End of Example 8.2)

Example 8.3 (The Helmholtz decomposition of stationary magnetic fields in a bounded region)

The stationary magnetic field $\mathcal{H}(\mathbf{r})$ produced by a steady current density $\mathcal{J}(\mathbf{r})$ confined to a smaller domain $V_J \subset V$ reads

$$\mathcal{H}(\mathbf{r}) = \nabla \times \int_{V_J} \mathrm{d}V' \frac{\mathcal{J}(\mathbf{r}')}{4\pi|\mathbf{r}-\mathbf{r}'|} + \nabla \times \int_{\partial V} \mathrm{d}S' \frac{\hat{\mathbf{n}}(\mathbf{r}') \times \mathcal{H}(\mathbf{r}')}{4\pi|\mathbf{r}-\mathbf{r}'|}$$

$$- \nabla \int_{\partial V} \mathrm{d}S' \frac{\hat{\mathbf{n}}(\mathbf{r}') \cdot \mathcal{H}(\mathbf{r}')}{4\pi|\mathbf{r}-\mathbf{r}'|}, \qquad \mathbf{r} \in V \tag{8.73}$$

where $\hat{\mathbf{n}} \times \mathcal{H}$ and $\mu\hat{\mathbf{n}} \cdot \mathcal{H}$ may be interpreted as fictitious surface densities of steady electric current and magnetic charge on ∂V. This representation is in line with (5.58) and also meaningful for points on the boundary.

On the whole, $\hat{\mathbf{n}} \times \mathcal{H}$ and $\mu\hat{\mathbf{n}} \cdot \mathcal{H}$ factor in the effect of sources and magnetic media located in the complementary region $\mathbb{R}^3 \setminus \overline{V}$. If the boundary ∂V is flush with a PEC, the stationary magnetic field obeys (4.50) and (4.51), whereby the last integral in (8.73) vanishes and $\mathcal{J}_S := \hat{\mathbf{n}} \times \mathcal{H}$ is a surface electric current induced on the conductor. Contrariwise, if ∂V is flush with a PMC, electric currents are absent, and $\varrho_{MS} := \mu\hat{\mathbf{n}} \cdot \mathcal{H}$ is an induced surface density of magnetic charge.

According to (8.73) the magnetic field can be derived from a pair of vector and scalar potentials, whereas on account of (4.29) and the integral representation (5.18) of $\mathcal{A}(\mathbf{r})$, it is also possible to obtain $\mathcal{H}(\mathbf{r}) = \mathcal{B}(\mathbf{r})/\mu$ from just one vector potential! Therefore, the curl of the last two contributions in formula (5.18) must pass over into the last term in the right member of (8.73). Specifically, for observation points $\mathbf{r}$ strictly away from the boundary we have

$$
\begin{aligned}
&\nabla \times \int_{\partial V} \mathrm{d}S'\,[\hat{\mathbf{n}}(\mathbf{r}') \times \mathcal{A}(\mathbf{r}')] \times \nabla' \frac{1}{4\pi|\mathbf{r}-\mathbf{r}'|} + \nabla \times \int_{\partial V} \mathrm{d}S'\,\hat{\mathbf{n}}(\mathbf{r}') \cdot \mathcal{A}(\mathbf{r}')\nabla' \frac{1}{4\pi|\mathbf{r}-\mathbf{r}'|} = \\
&= \int_{\partial V} \mathrm{d}S'\,\nabla \times \left\{[\hat{\mathbf{n}}(\mathbf{r}') \times \mathcal{A}(\mathbf{r}')] \times \nabla' \frac{1}{4\pi R}\right\} + \int_{\partial V} \mathrm{d}S'\,\hat{\mathbf{n}}(\mathbf{r}') \cdot \mathcal{A}(\mathbf{r}')\nabla \times \nabla' \frac{1}{4\pi R} \\
&= -\int_{\partial V} \mathrm{d}S'\,\nabla \left[\hat{\mathbf{n}}(\mathbf{r}') \times \mathcal{A}(\mathbf{r}') \cdot \nabla' \frac{1}{4\pi R}\right] = \nabla \int_{\partial V} \mathrm{d}S'\,\hat{\mathbf{n}}(\mathbf{r}') \cdot \nabla' \frac{1}{4\pi R} \times \mathcal{A}(\mathbf{r}') \\
&= \underbrace{\nabla \int_{\partial V} \mathrm{d}S'\,\hat{\mathbf{n}}(\mathbf{r}') \cdot \nabla' \times \frac{\mathcal{A}(\mathbf{r}')}{4\pi R}}_{=0} - \nabla \int_{\partial V} \mathrm{d}S'\,\frac{\hat{\mathbf{n}}(\mathbf{r}') \cdot \nabla' \times \mathcal{A}(\mathbf{r}')}{4\pi R} \\
&= -\mu\nabla \int_{\partial V} \mathrm{d}S'\,\frac{\hat{\mathbf{n}}(\mathbf{r}') \cdot \mathcal{H}(\mathbf{r}')}{4\pi|\mathbf{r}-\mathbf{r}'|}
\end{aligned}
\tag{8.74}
$$

which proves the equivalence of the two representations upon division by μ. In the procedure we have interchanged derivatives and integrals on the grounds that the integrand is regular for $\mathbf{r}' \neq \mathbf{r}$, and invoked formulas (H.54) and (H.53) to turn the curl into a gradient [cf. (4.86)]. In the second-to-last step the first flux integral vanishes by virtue of the Stokes theorem applied on a closed surface. Physically, (8.74) indicates that a layer of magnetic dipoles with density $\mathcal{M}_S := \hat{\mathbf{n}} \times \mathcal{A}/\mu$ is equivalent to a layer of magnetic charges with density $\varrho_{MS} := \hat{\mathbf{n}} \cdot \nabla \times \mathcal{A}$. This configuration is not the electromagnetic dual of the electric one which leads to (8.70), not least because the magnetic dipoles in question are tangential, rather than perpendicular, to ∂V.

The right-hand side of (8.73) ought to be solenoidal in regions devoid of magnetic charges and lamellar in current-free regions (here, for points $\mathbf{r} \notin V_J \cup \partial V$). Taking the divergence first yields

$$
\nabla \cdot \mathcal{H}(\mathbf{r}) = -\nabla^2 \int_{\partial V} \mathrm{d}S'\,\frac{\hat{\mathbf{n}}(\mathbf{r}') \cdot \mathcal{H}(\mathbf{r}')}{4\pi|\mathbf{r}-\mathbf{r}'|} = -\int_{\partial V} \mathrm{d}S'\,\hat{\mathbf{n}}(\mathbf{r}') \cdot \mathcal{H}(\mathbf{r}')\nabla'^2 \frac{1}{4\pi R} = 0
\tag{8.75}
$$

once again because $1/R$ is harmonic for $\mathbf{r} \notin V_J \cup \partial V$. Next, by taking the curl of (8.73) we obtain

$$
\begin{aligned}
\nabla \times \mathcal{H}(\mathbf{r}) &= \nabla \times \nabla \times \int_{V_J} \mathrm{d}V'\,\frac{\mathcal{J}(\mathbf{r}')}{4\pi|\mathbf{r}-\mathbf{r}'|} + \nabla \times \nabla \times \int_{\partial V} \mathrm{d}S'\,\frac{\hat{\mathbf{n}}(\mathbf{r}') \times \mathcal{H}(\mathbf{r}')}{4\pi|\mathbf{r}-\mathbf{r}'|} \\
&= \nabla\nabla \cdot \int_{V_J} \mathrm{d}V'\,\frac{\mathcal{J}(\mathbf{r}')}{4\pi|\mathbf{r}-\mathbf{r}'|} + \nabla\nabla \cdot \int_{\partial V} \mathrm{d}S'\,\frac{\hat{\mathbf{n}}(\mathbf{r}') \times \mathcal{H}(\mathbf{r}')}{4\pi|\mathbf{r}-\mathbf{r}'|}
\end{aligned}
\tag{8.76}
$$

where the last contribution can be proved to vanish as in (8.72) since $\hat{\mathbf{n}} \times \mathcal{H}$ is a tangential field. For the other one we have

$$
\nabla\nabla \cdot \int_{V_J} \mathrm{d}V'\,\frac{\mathcal{J}(\mathbf{r}')}{4\pi|\mathbf{r}-\mathbf{r}'|} = -\nabla \int_{V_J} \mathrm{d}V'\,\nabla' \frac{1}{4\pi R} \cdot \mathcal{J}(\mathbf{r}') = -\nabla \int_{V_J} \mathrm{d}V'\,\nabla' \cdot \frac{\mathcal{J}(\mathbf{r}')}{4\pi R}
$$

$$= -\nabla \int_{\partial V_J} dS' \frac{\hat{\mathbf{n}}(\mathbf{r}') \cdot \mathcal{J}(\mathbf{r}')}{4\pi R} = 0 \tag{8.77}$$

having used the solenoidal nature of $\mathcal{J}(\mathbf{r})$ and the fact the latter is confined to V_J by hypothesis.

Finally, since (8.60) involves the double curl of $\mathcal{V}(\mathbf{r})$, it is well-suited to obtain an expression for $\mathcal{A}(\mathbf{r})$, the stationary magnetic vector potential which solves (5.1). In symbols, we have

$$\begin{aligned}\mathcal{A}(\mathbf{r}) = &\int_{V_J} dV' \frac{\mu\mathcal{J}(\mathbf{r}')}{4\pi|\mathbf{r}-\mathbf{r}'|} + \int_{\partial V} dS' \frac{\hat{\mathbf{n}}(\mathbf{r}') \times [\nabla' \times \mathcal{A}(\mathbf{r}')]}{4\pi|\mathbf{r}-\mathbf{r}'|} \\ &+ \nabla \times \int_{\partial V} dS' \frac{\hat{\mathbf{n}}(\mathbf{r}') \times \mathcal{A}(\mathbf{r}')}{4\pi|\mathbf{r}-\mathbf{r}'|} - \nabla \int_{\partial V} dS' \frac{\hat{\mathbf{n}}(\mathbf{r}') \cdot \mathcal{A}(\mathbf{r}')}{4\pi|\mathbf{r}-\mathbf{r}'|}, \qquad \mathbf{r} \in V\end{aligned} \tag{8.78}$$

which is in agreement with (5.18).

(End of Example 8.3)

When the region V is multiply connected, decomposition (8.45) is not complete and must be extended by including a *harmonic* vector field $\mathcal{V}_H(\mathbf{r})$, viz.,

$$\mathcal{V}(\mathbf{r}) = \mathcal{V}_S(\mathbf{r}) + \mathcal{V}_L(\mathbf{r}) + \mathcal{V}_H(\mathbf{r}), \qquad \mathbf{r} \in V \tag{8.79}$$

where the three components in which $\mathcal{V}(\mathbf{r})$ is separated are orthogonal to one another with respect to the inner product (D.68) with $D = V$. Since this property was already proved in (8.61) for $\mathcal{V}_L(\mathbf{r})$ and $\mathcal{V}_S(\mathbf{r})$, we examine two more inner products, viz.,

$$\begin{aligned}(\mathcal{V}_H, \mathcal{V}_S)_V &= \int_V dV\, \mathcal{V}_H(\mathbf{r}) \cdot \nabla \times \Xi = \int_V dV\, \nabla \cdot (\Xi \times \mathcal{V}_H) \\ &= -\int_{\partial V} dS\, \hat{\mathbf{n}}(\mathbf{r}) \times \Xi(\mathbf{r}) \cdot \mathcal{V}_H(\mathbf{r}) = 0\end{aligned} \tag{8.80}$$

$$\begin{aligned}(\mathcal{V}_H, \mathcal{V}_L)_V &= \int_V dV\, \mathcal{V}_H(\mathbf{r}) \cdot \nabla \Upsilon = \int_V dV\, \nabla \cdot (\Upsilon \mathcal{V}_H) \\ &= -\int_{\partial V} dS\, \Upsilon(\mathbf{r}) \hat{\mathbf{n}}(\mathbf{r}) \cdot \mathcal{V}_H(\mathbf{r}) = 0\end{aligned} \tag{8.81}$$

where we have applied the Gauss theorem with unit normal oriented inwards V. The conclusions follow because $\hat{\mathbf{n}}(\mathbf{r}) \times \Xi(\mathbf{r})$ and $\Upsilon(\mathbf{r})$ vanish on ∂V according to (8.62) and (8.63).

If V is a surface-wise multiply-connected domain, that is, the boundary ∂V is comprised of N disjoint closed surfaces S_n, $n = 1, \dots, N$, we can find N single-valued harmonic scalar potentials $\Upsilon_n(\mathbf{r})$ which are the unique solutions to N Dirichlet problems (see discussion on page 95 and ff.), say,

$$\begin{cases}\nabla^2 \Upsilon_n(\mathbf{r}) = 0, & \mathbf{r} \in V \\ \Upsilon_n(\mathbf{r}) = \delta_{nm}, & \mathbf{r} \in S_m\end{cases} \qquad n, m \in \{1, \dots, N\} \tag{8.82}$$

where δ_{nm} is the Kronecker delta defined in (2.81). With these positions, the functions $\Upsilon_n(\mathbf{r})$ coincide with the *capacitary potentials* introduced in Section 2.5.1 and specifically in Example 2.6 for the spherical condenser. Since the capacitary potentials ultimately depend only on the shape of V, they can be computed once and for all.

The vector fields $\nabla\Upsilon_n$ are evidently harmonic in V, and hence it seems quite logical to express $\mathcal{V}_H(\mathbf{r})$ as a linear combination of $\nabla\Upsilon_n$, $n = 1, \dots, N$, and adjust the unknown coefficients so that (8.79) is satisfied. However, we must be careful here, because the N fields $\nabla\Upsilon_n$ are not all independent of each other! As it turns out, any one of them can be obtained from a linear combination of the remaining $N - 1$ ones. To prove this claim, we suppose that S_N is the largest boundary surface and, as such, encloses the others. (We may always number the surfaces so that the latter hypothesis is fulfilled.) We define the scalar potential

$$\Upsilon_H(\mathbf{r}) := \sum_{n=1}^{N} \upsilon_n \Upsilon_n(\mathbf{r}), \qquad \mathbf{r} \in V \tag{8.83}$$

where υ_n are arbitrary constant coefficients. Since $\Upsilon_H(\mathbf{r})$ clearly solves the Laplace equation in V by construction, we have

$$\begin{aligned} 0 &= -\int_V \mathrm{d}V\, \nabla^2\Upsilon_H(\mathbf{r}) = \int_{\partial V} \mathrm{d}S\, \hat{\mathbf{n}}(\mathbf{r}) \cdot \nabla\Upsilon_H(\mathbf{r}) \\ &= \sum_{n=1}^{N} \int_{S_n} \mathrm{d}S\, \hat{\mathbf{n}}(\mathbf{r}) \cdot \nabla\Upsilon_H(\mathbf{r}) = \sum_{n=1}^{N}\sum_{m=1}^{N} \upsilon_m \int_{S_n} \mathrm{d}S\, \hat{\mathbf{n}}(\mathbf{r}) \cdot \nabla\Upsilon_m(\mathbf{r}) \\ &= \upsilon_1 \sum_{n=1}^{N} \int_{S_n} \mathrm{d}S\, \hat{\mathbf{n}}(\mathbf{r}) \cdot \nabla\Upsilon_1(\mathbf{r}) + \dots + \upsilon_N \sum_{n=1}^{N} \int_{S_n} \mathrm{d}S\, \hat{\mathbf{n}}(\mathbf{r}) \cdot \nabla\Upsilon_N(\mathbf{r}) \end{aligned} \tag{8.84}$$

having applied the Gauss theorem with the unit normals on ∂V positively oriented inwards V. For the previous relationship to hold true in general, we must demand

$$\sum_{n=1}^{N} \int_{S_n} \mathrm{d}S\, \hat{\mathbf{n}}(\mathbf{r}) \cdot \nabla\Upsilon_m(\mathbf{r}) = \sum_{n=1}^{N} A_{nm} = 0, \qquad m = 1, \dots, N \tag{8.85}$$

because υ_1, υ_2 and so forth are arbitrary. Aside from a dimensional multiplicative factor, the terms of the sums in (8.85) are the entries of the $N \times N$ capacitance matrix $[C]$ in (2.86), though we prefer to adopt a different symbol in the interest of clarity. Now, (8.85) says that the rows of the matrix $[A]$ are linearly dependent, whereby the rank of $[A]$ is at best $N - 1$, and the matrix is singular [26].

With this intermediate finding we can show that the condition

$$\sum_{n=1}^{N} \upsilon_n \nabla\Upsilon_n(\mathbf{r}) = 0, \qquad \mathbf{r} \in V \tag{8.86}$$

may be satisfied without $\upsilon_1, \dots, \upsilon_N$ being all necessarily null. Indeed, taking the inner product with $\nabla\Upsilon_m(\mathbf{r})$ yields

$$\begin{aligned} 0 &= \left(\nabla\Upsilon_m, \sum_{n=1}^{N} \upsilon_n \nabla\Upsilon_n\right)_V = \sum_{n=1}^{N} \upsilon_n (\nabla\Upsilon_m, \nabla\Upsilon_n)_V = \sum_{n=1}^{N} \upsilon_n \int_V \mathrm{d}V\, \nabla \cdot (\Upsilon_m \nabla\Upsilon_n) \\ &= -\sum_{n=1}^{N} \upsilon_n \int_{\partial V} \mathrm{d}S\, \Upsilon_m(\mathbf{r})\hat{\mathbf{n}}(\mathbf{r}) \cdot \nabla\Upsilon_n = -\sum_{n=1}^{N} \upsilon_n \sum_{p=1}^{N} \int_{S_p} \mathrm{d}S\, \delta_{mp}\hat{\mathbf{n}}(\mathbf{r}) \cdot \nabla\Upsilon_n \end{aligned}$$

$$= -\sum_{n=1}^{N} v_n \int_{S_m} \mathrm{d}S\, \hat{\mathbf{n}}(\mathbf{r}) \cdot \nabla\Upsilon_n = -\sum_{n=1}^{N} A_{mn} v_n, \qquad m = 1, \ldots, N \tag{8.87}$$

or succinctly in matrix form

$$[A][v] = 0 \tag{8.88}$$

a homogeneous system which we know admits a non-trivial solution on account of (8.85) [26, Theorem 3.16]. It follows that the N vector fields $\nabla\Upsilon_n$ are not independent of each other, as we wanted to prove.

Thus, in a final step we write the harmonic part of $\mathcal{V}(\mathbf{r})$ as

$$\mathcal{V}_H(\mathbf{r}) = \sum_{n=1}^{N-1} v_n \nabla\Upsilon_n(\mathbf{r}), \qquad \mathbf{r} \in V \tag{8.89}$$

where we have arbitrarily excluded $\nabla\Upsilon_N$ from the representation. We assume that $\mathcal{V}(\mathbf{r})$ is real-valued for simplicity and observe that $\nabla\Upsilon_m$ — being harmonic in V — is orthogonal to $\mathcal{V}_L(\mathbf{r})$ and $\mathcal{V}_S(\mathbf{r})$ in accordance with (8.80) and (8.81). Then, from (8.79) and (8.89) with manipulations similar to those performed in (8.87) we get

$$\begin{aligned}(\nabla\Upsilon_m, \mathcal{V})_V = (\nabla\Upsilon_m, \mathcal{V}_H)_V &= -\sum_{n=1}^{N-1} v_n \int_{S_m} \mathrm{d}S\, \hat{\mathbf{n}}(\mathbf{r}) \cdot \nabla\Upsilon_n \\ &= -\sum_{n=1}^{N-1} A_{mn} v_n, \qquad m = 1, \ldots, N-1\end{aligned} \tag{8.90}$$

i.e., an algebraic linear system of rank $N - 1$ that can be solved to obtain $v_1, \ldots, v_{N-1}$, and this finalizes the splitting (8.79).

When V is contour-wise multiply connected (e.g., the inside of a torus) $\mathcal{V}_H(\mathbf{r})$ is a harmonic *Neumann vector field* and may be derived from a many-valued harmonic scalar potential $\Upsilon_H(\mathbf{r})$ which is determined only by the shape of V. We may take $\Upsilon_H(\mathbf{r})$ as one of the solutions to the Neumann problem [16, Section 4.11]

$$\begin{cases} \nabla^2\Upsilon_H(\mathbf{r}) = 0, & \mathbf{r} \in V \\ \hat{\mathbf{n}}(\mathbf{r}) \cdot \nabla\Upsilon_H(\mathbf{r}) = 0, & \mathbf{r} \in \partial V \\ \displaystyle\oint_{\gamma} \mathrm{d}s\, \hat{\mathbf{s}}(\mathbf{r}) \cdot \nabla\Upsilon_H(\mathbf{r}) = 1, & \gamma \subset V \end{cases} \tag{8.91}$$

where γ is a piecewise-smooth contour which is entirely contained in V and cannot be reduced to a point without crossing ∂V. Any such γ will do the trick, because if γ_1 and γ_2 are two contours which exhibit the stated property and can be deformed continuously into one another without crossing ∂V, then using the Stokes theorem gives

$$0 = \int_S \mathrm{d}S\, \hat{\mathbf{n}}(\mathbf{r}) \cdot \nabla \times \nabla\Upsilon_H = \oint_{\gamma_1} \mathrm{d}s\, \hat{\mathbf{s}}_1(\mathbf{r}) \cdot \nabla\Upsilon_H - \oint_{\gamma_2} \mathrm{d}s\, \hat{\mathbf{s}}_2(\mathbf{r}) \cdot \nabla\Upsilon_H \tag{8.92}$$

where $S \subset V$ is the open surface bounded by γ_1 and γ_2, and the latter are both oriented in a counterclockwise fashion with respect to the unit normal to S.

By assigning the value of the contour integral — which may be construed as a normalization condition of sorts — we determine the solution to (8.91) up to an additive constant. Indeed, if we only require that $\Upsilon_H(\mathbf{r})$ be harmonic with vanishing normal derivative on ∂V, and c_1, c_2 are two constants, then each member of the family of fields $c_1\Upsilon_H(\mathbf{r}) + c_2$ manifestly exhibits the same properties. Contrariwise, when the contour integral is included, and $\Upsilon_{H1}(\mathbf{r})$ and $\Upsilon_{H2}(\mathbf{r})$ are two solutions, the difference field $\Upsilon_{H0}(\mathbf{r}) = \Upsilon_{H1}(\mathbf{r}) - \Upsilon_{H2}(\mathbf{r})$ obeys the condition

$$\Upsilon_{H0}(\mathbf{r}_B) - \Upsilon_{H0}(\mathbf{r}_A) = \oint_\gamma \mathrm{d}s\, \hat{\mathbf{s}}(\mathbf{r}) \cdot \nabla\Upsilon_{H0}(\mathbf{r}) = 1 - 1 = 0 \tag{8.93}$$

where $\mathbf{r}_A$ and $\mathbf{r}_B \equiv \mathbf{r}_A$ indicate the endpoints of γ. From (8.93) we conclude that $\Upsilon_{H0}(\mathbf{r})$ is a *single-valued* potential. As a result, the Gauss theorem may be invoked in V without the intervention of branch surfaces (cf. Figure 4.11), and we have

$$\int_V \mathrm{d}V\, |\nabla\Upsilon_{H0}|^2 = \int_V \mathrm{d}V\, \nabla \cdot (\Upsilon_{H0}\nabla\Upsilon_{H0}) = -\int_{\partial V} \mathrm{d}S\; \Upsilon_{H0}(\mathbf{r})\hat{\mathbf{n}}(\mathbf{r}) \cdot \nabla\Upsilon_{H0} = 0 \tag{8.94}$$

which demands that $\Upsilon_{H0}(\mathbf{r})$ be constant in V. This in turn proves that $\Upsilon_{H1}(\mathbf{r})$ and $\Upsilon_{H2}(\mathbf{r})$ differ by a constant, though this is not an issue for the univocal derivation of $\mathcal{V}_H(\mathbf{r})$ from

$$\mathcal{V}_H(\mathbf{r}) = A\nabla\Upsilon_H(\mathbf{r}), \qquad \mathbf{r} \in V \tag{8.95}$$

in that the arbitrary additive constant is swept away when the gradient of $\Upsilon_H(\mathbf{r})$ is carried out. The multiplicative constant A is found from

$$(\nabla\Upsilon_H, \mathcal{V})_V = (\nabla\Upsilon_H, \mathcal{V}_H)_V = A\, \|\nabla\Upsilon_H\|_V^2 \qquad \Longrightarrow \qquad A = \frac{(\nabla\Upsilon_H, \mathcal{V})_V}{\|\nabla\Upsilon_H\|_V^2} \tag{8.96}$$

and this again finalizes the splitting (8.79).

8.2 Electrodynamic potentials and gauge transformations

Equipped with the results of Section 8.1 we wish to find, if possible, the solenoidal-lamellar decomposition of an electromagnetic field with a general time dependence. Since the four Maxwell equations in local form do relate the curl and the divergence of the electromagnetic entities to the sources (Figure 1.5) it would appear that we have the right ingredients for the application of the Helmholtz theorem (8.28). However, the task is a bit more complex than it was in the static or stationary case for we have to handle two fields, say, $\mathcal{E}(\mathbf{r}, t)$ and $\mathcal{B}(\mathbf{r}, t)$ and four coupled equations, whereby (8.28) cannot be immediately employed.

Therefore, we start over and try the following splitting

$$\mathcal{E}(\mathbf{r}, t) = \nabla\Upsilon_E(\mathbf{r}, t) + \nabla \times \Xi_E(\mathbf{r}, t), \qquad (\mathbf{r}, t) \in \mathbb{R}^3 \times \mathbb{R} \tag{8.97}$$

$$\mathcal{B}(\mathbf{r}, t) = \nabla\Upsilon_B(\mathbf{r}, t) + \nabla \times \Xi_B(\mathbf{r}, t), \qquad (\mathbf{r}, t) \in \mathbb{R}^3 \times \mathbb{R} \tag{8.98}$$

where $\Upsilon_E(\mathbf{r}, t)$, $\Xi_E(\mathbf{r}, t)$ and so forth are fields regular at infinity and twice differentiable with respect to the spatial coordinates. To gain more generality we also include magnetic sources in the picture, but we may work with only one type of charges and currents at a time on the grounds of linearity and the principle of superposition (Section 6.1).

Thus, we let $\mathcal{J}_M(\mathbf{r}, t) = 0 = \varrho_M(\mathbf{r}, t)$ and observe that in accordance with the Gauss magnetic law (1.23) the magnetic induction field $\mathcal{B}(\mathbf{r}, t)$ is solenoidal or, in other words, the lamellar part thereof must be identically null. As a consequence, the scalar potential $\Upsilon_B(\mathbf{r}, t)$ is not necessary and we simply let $\Upsilon_B(\mathbf{r}, t) = 0$. Actually, since $\Upsilon_B(\mathbf{r}, t)$ solves a problem of the type (8.37), it can be shown to be null as in (8.40).

We go on to substitute (8.97) and (8.98) into the Faraday law (1.20)

$$\nabla \times \nabla \Upsilon_E(\mathbf{r}, t) + \nabla \times \left[\nabla \times \Xi_E(\mathbf{r}, t) + \frac{\partial}{\partial t} \Xi_B(\mathbf{r}, t) \right] = 0 \tag{8.99}$$

where we have interchanged the order of temporal and spatial derivatives by virtue of the Schwarz theorem [27, pp. 235–236] since the vector field $\Xi_B(\mathbf{r}, t)$ is twice differentiable by hypothesis. Taking the curl nullifies the effect of the lamellar part of the electric field, and the equation is certainly satisfied if the quantity between brackets is curl-free as well, say,

$$\nabla \times \Xi_E(\mathbf{r}, t) + \frac{\partial}{\partial t} \Xi_B(\mathbf{r}, t) = \nabla f(\mathbf{r}, t) \tag{8.100}$$

where $f(\mathbf{r}, t)$ is an arbitrary twice differentiable scalar field. We use the last equation to eliminate $\nabla \times \Xi_E(\mathbf{r}, t)$ from (8.97) and (8.98), namely,

$$\mathcal{E}(\mathbf{r}, t) = -\frac{\partial}{\partial t} \Xi_B(\mathbf{r}, t) + \nabla[f(\mathbf{r}, t) + \Upsilon_E(\mathbf{r}, t)] \tag{8.101}$$

$$\mathcal{B}(\mathbf{r}, t) = \nabla \times \Xi_B(\mathbf{r}, t) \tag{8.102}$$

and finally, we let

$$f(\mathbf{r}, t) + \Upsilon_E(\mathbf{r}, t) = -\Phi_E \qquad \Xi_B(\mathbf{r}, t) = \mathcal{A}_E(\mathbf{r}, t) \tag{8.103}$$

to arrive at the familiar expressions [28, Chapter 1], [16, Chapter 7], [29, Chapter 31], [30, Section 1.7], [31, Section 1.3], [32, Section 2.3], [33, Chapter 37], [34, Section 13.1], [8, Section 15.3], [35, Section 3.3], [36, Chapter 16], [15,37,38]

$$\mathcal{E}(\mathbf{r}, t) = -\frac{\partial}{\partial t} \mathcal{A}_E(\mathbf{r}, t) - \nabla \Phi_E(\mathbf{r}, t), \qquad (\mathbf{r}, t) \in \mathbb{R}^3 \times \mathbb{R} \tag{8.104}$$

$$\mathcal{B}(\mathbf{r}, t) = \nabla \times \mathcal{A}_E(\mathbf{r}, t), \qquad (\mathbf{r}, t) \in \mathbb{R}^3 \times \mathbb{R} \tag{8.105}$$

where $\mathcal{A}_E(\mathbf{r}, t)$ is the *magnetic vector potential* (physical dimension: Vs/m or Wb/m or Tm) and $\Phi_E(\mathbf{r}, t)$ the *electric scalar potential* (physical dimension: V). The subscript 'E' appended to the potentials signals that they depend on electric sources only.

In summary, by starting with four potentials in (8.97) and (8.98) and closely examining the nature of the entities, we have found that $\mathcal{E}(\mathbf{r}, t)$ and $\mathcal{B}(\mathbf{r}, t)$ may be derived from just a vector potential and a scalar potential. Accordingly, this approach to the solution of the Maxwell equations is oft-times called the *mixed potential method.* In a three-dimensional space endowed with a system of coordinates, (8.104) and (8.105) state that the six scalar components of $\mathcal{E}(\mathbf{r}, t)$ and $\mathcal{B}(\mathbf{r}, t)$ are determined by means of four scalar quantities. Furthermore, we notice that (8.104) and (8.105) decouple and reduce to (2.15) and (4.29) for stationary fields.

We have to ascertain whether or not (8.104) and (8.105) constitute a Helmholtz decomposition of the electromagnetic field in the spirit of the splitting (8.2). The check is easily accomplished for the magnetic induction: the vector $\nabla \times \mathcal{A}_E(\mathbf{r}, t)$ is solenoidal in view of (A.39), and in any case we already know that $\mathcal{B}(\mathbf{r},t)$ is solenoidal in the absence of magnetic monopoles. Concerning the electric field, the vector $\nabla\Phi_E(\mathbf{r},t)$ is definitely lamellar on account of (A.38). However, the time derivative of $\mathcal{A}_E(\mathbf{r},t)$ may not be a solenoidal vector *per se*. Then again, the Helmholtz theorem (8.28) applied to $\mathcal{A}_E(\mathbf{r},t)$ tells us that the latter is completely specified provided we assign both the curl and the divergence of $\mathcal{A}_E(\mathbf{r},t)$.

A closer look at (8.104) and (8.105) reveals that, in effect, only $\nabla\times\mathcal{A}_E(\mathbf{r},t)$ has been prescribed so long as we regard $\mathcal{B}(\mathbf{r},t)$ as known. This implies on the one hand that at this stage $\mathcal{A}_E(\mathbf{r},t)$ is not unique and, on the other, that we are allowed to choose the divergence of $\mathcal{A}_E(\mathbf{r},t)$ as we please. We already encountered a similar degree of freedom while discussing the magnetic vector potential for stationary magnetic fields in Sections 4.2 and 4.7.1. The procedure of setting the divergence of $\mathcal{A}_E(\mathbf{r},t)$ is referred to as fixing the *gauge*. The special choice

$$\nabla\cdot\mathcal{A}_E(\mathbf{r},t)=0,\qquad (\mathbf{r},t)\in\mathbb{R}^3\times\mathbb{R} \tag{8.106}$$

with which (8.104) becomes a true Helmholtz decomposition, is called the *Coulomb* or *radiation* or *transverse gauge*. Since we may think of a divergence-free vector field as being perpendicular to the 'vector' ∇, i.e., transverse to ∇ [see (8.2)] this consideration may serve to justify the alternative name *transverse gauge* for the choice (8.106).

Depending on the gauge chosen, infinitely many pairs of potentials $\Phi_E(\mathbf{r},t)$ and $\mathcal{A}_E(\mathbf{r},t)$ are possible. Indeed, if somehow we have found the pair $\mathcal{A}_{E1}(\mathbf{r},t)$ and $\Phi_{E1}(\mathbf{r},t)$, we can determine another pair, say, $\mathcal{A}_{E2}(\mathbf{r},t)$ and $\Phi_{E2}(\mathbf{r},t)$, that produces the same electromagnetic entities $\mathcal{E}(\mathbf{r},t)$ and $\mathcal{B}(\mathbf{r},t)$. More specifically, from (8.98) we have

$$\mathcal{B}(\mathbf{r},t)=\nabla\times\mathcal{A}_{E2}(\mathbf{r},t)=\nabla\times[\mathcal{A}_{E1}(\mathbf{r},t)+\nabla\chi(\mathbf{r},t)] \tag{8.107}$$

where $\chi(\mathbf{r},t)$ is a scalar field twice differentiable for $(\mathbf{r},t)\in\mathbb{R}^3\times\mathbb{R}$. Therefore, if $\mathcal{A}_{E1}(\mathbf{r},t)$ and $\mathcal{A}_{E2}(\mathbf{r},t)$ are related by

$$\mathcal{A}_{E2}(\mathbf{r},t)=\mathcal{A}_{E1}(\mathbf{r},t)+\nabla\chi(\mathbf{r},t) \tag{8.108}$$

then the magnetic field is not altered. What about the electric field, though? Well, we need to modify $\Phi_{E1}(\mathbf{r},t)$ accordingly in order for $\mathcal{E}(\mathbf{r},t)$ to be independent of $\chi(\mathbf{r},t)$ as well. By inserting (8.108) into (8.97) we find

$$\begin{aligned}\mathcal{E}(\mathbf{r},t)&=-\frac{\partial}{\partial t}\mathcal{A}_{E2}(\mathbf{r},t)-\nabla\Phi_{E2}=-\frac{\partial}{\partial t}[\mathcal{A}_{E1}(\mathbf{r},t)+\nabla\chi(\mathbf{r},t)]-\nabla\Phi_{E2}(\mathbf{r},t)\\&=-\frac{\partial}{\partial t}\mathcal{A}_{E1}(\mathbf{r},t)-\nabla\left[\frac{\partial}{\partial t}\chi(\mathbf{r},t)+\Phi_{E2}(\mathbf{r},t)\right]\end{aligned} \tag{8.109}$$

which prompts us to set

$$\Phi_{E2}(\mathbf{r},t)=\Phi_{E1}(\mathbf{r},t)-\frac{\partial}{\partial t}\chi(\mathbf{r},t) \tag{8.110}$$

whereby also the electric field remains unchanged. Relationships (8.108) and (8.110) constitute a *gauge transformation* [29, Section 31.1] and can be employed to pass from the Coulomb gauge to any other useful pair in which the vector potential is not solenoidal. For instance, starting off with $\nabla\cdot\mathcal{A}_{E1}(\mathbf{r},t)=0$ we determine $\chi(\mathbf{r},t)$ by solving

$$\nabla\cdot\mathcal{A}_{E2}(\mathbf{r},t)=\nabla\cdot\mathcal{A}_{E1}(\mathbf{r},t)+\nabla^2\chi(\mathbf{r},t)=\nabla^2\chi(\mathbf{r},t),\qquad (\mathbf{r},t)\in\mathbb{R}^3\times\mathbb{R} \tag{8.111}$$

where the divergence of the new potential $\mathcal{A}_{E2}(\mathbf{r}, t)$ is assigned. Once $\chi(\mathbf{r}, t)$ is known, $\mathcal{A}_{E2}(\mathbf{r}, t)$ and $\Phi_{E2}(\mathbf{r}, t)$ follow from (8.108) and (8.110).

Of course, we still have to find a way to compute $\mathcal{A}_E(\mathbf{r}, t)$ and $\Phi_E(\mathbf{r}, t)$ explicitly. The process will provide us with a formal solution to the Maxwell equations in an unbounded homogeneous isotropic medium. Since the introduction of the potentials was motivated by the need for automatically satisfying the magnetic Gauss law and the Faraday law (the 'constraints' imposed on the electromagnetic entities of intensity) we have to use the constitutive relationships (1.117) and (1.118) to link the potentials to the entities of quantity $\mathcal{D}(\mathbf{r}, t)$ and $\mathcal{H}(\mathbf{r}, t)$ and hence to the sources by means of the Ampère-Maxwell law and the electric Gauss law.

By inserting (8.104) and (8.105) into (1.34) and (1.44) we readily obtain

$$\nabla \times [\nabla \times \mathcal{A}_E(\mathbf{r}, t)] + \frac{1}{c^2}\frac{\partial^2}{\partial t^2}\mathcal{A}_E(\mathbf{r}, t) + \frac{1}{c^2}\frac{\partial}{\partial t}\nabla\Phi_E(\mathbf{r}, t) = \mu\mathcal{J}(\mathbf{r}, t), \qquad (\mathbf{r}, t) \in \mathbb{R}^3 \times \mathbb{R} \qquad (8.112)$$

$$\nabla \cdot \frac{\partial}{\partial t}\mathcal{A}_E(\mathbf{r}, t) + \nabla^2\Phi_E(\mathbf{r}, t) = -\frac{1}{\varepsilon}\varrho(\mathbf{r}, t), \qquad (\mathbf{r}, t) \in \mathbb{R}^3 \times \mathbb{R} \qquad (8.113)$$

where the sources are confined to a volume $V_J = V_\varrho$. In principle, (8.112) and (8.113) may be solved for $\mathcal{A}_E(\mathbf{r}, t)$ and $\Phi_E(\mathbf{r}, t)$. Nonetheless, one of the difficulties posed by this second-order partial differential system is that the equations are *coupled.* We see then that the arbitrariness afforded by the gauge transformations can be exploited to simplify the task of solving (8.112) and (8.113).

For example, it only takes a few steps to show that in the Coulomb gauge (8.106) the equations decouple partially and take on the form

$$\nabla^2\Phi_E(\mathbf{r}, t) = -\frac{1}{\varepsilon}\varrho(\mathbf{r}, t), \qquad (\mathbf{r}, t) \in \mathbb{R}^3 \times \mathbb{R} \qquad (8.114)$$

$$\nabla^2\mathcal{A}_E(\mathbf{r}, t) - \frac{1}{c^2}\frac{\partial^2}{\partial t^2}\mathcal{A}_E(\mathbf{r}, t) = -\mu\mathcal{J}(\mathbf{r}, t) + \frac{1}{c^2}\frac{\partial}{\partial t}\nabla\Phi_E(\mathbf{r}, t), \qquad (\mathbf{r}, t) \in \mathbb{R}^3 \times \mathbb{R} \qquad (8.115)$$

on account of (H.59) and the Schwarz theorem since $\mathcal{A}_E(\mathbf{r}, t) \in \mathcal{C}^2(\mathbb{R}^3 \times \mathbb{R})^3$. In this way, we may solve the first equation for the scalar potential and then, using the result thus found, determine the vector potential from the second equation. Quite surprisingly, (8.114) is formally identical with the Poisson equation (2.190) which was derived for static electric fields, and this motivates the name Coulomb gauge. Conversely, (8.115) has indeed the structure of a wave equation for the vector potential, and this sounds more in line with the time-varying nature of the solutions we seek. The source term of $\mathcal{A}_E(\mathbf{r}, t)$ is an effective current density where the contribution depending on $\Phi_E(\mathbf{r}, t)$ is in the form of a non-radiating source (Section 7.8).

If we require $\Phi_E(\mathbf{r}, t)$ to be regular at infinity in accordance with (2.20), then we know from Section 2.160 that the unique solution to (8.114) reads

$$\Phi_E(\mathbf{r}, t) = \frac{1}{\varepsilon}\int_{V_\varrho} \mathrm{d}V' \,\frac{\varrho(\mathbf{r}, t)}{4\pi|\mathbf{r} - \mathbf{r}'|}, \qquad (\mathbf{r}, t) \in \mathbb{R}^3 \times \mathbb{R} \qquad (8.116)$$

where V_ϱ is the finite region occupied by the sources. The fundamental difference between (2.169) and (8.116) is that in the latter the charge density and hence the potential depend on position as well as the *present* time. This characteristic raises some issues which we will address in a moment. Meanwhile, we observe that $\Phi_E(\mathbf{r}, t)$ does not contribute to radiation in the sense of transport of energy, momentum and angular momentum away from the source (Section 1.9) because at large distances from V_ϱ the term $\nabla\Phi_E(\mathbf{r}, t)$ in (8.104) falls off as $1/r^2$ on account of the multipole expansion (3.204). Consequently, the radiation fields — which decay with the inverse of the distance from the source — are determined only by the vector potential $\mathcal{A}_E(\mathbf{r},t)$, and this motivates the alternative name radiation gauge reserved for (8.106).

We go on to examine (8.115) and observe that, since the left-hand side is a solenoidal vector field in light of (8.106) so must be the quantity in the right-hand side. This property follows immediately from (8.114) and the continuity equation (1.46). Further, the lamellar part of the right-hand side must vanish. To check whether this is the case we formally compute the lamellar part of the current density $\mathcal{J}(\mathbf{r},t)$ with the aid of the Helmholtz theorem in the whole space. From (8.28) we find

$$\begin{aligned}\mathcal{J}_L(\mathbf{r},t) &= -\nabla \int_{V_J} \mathrm{d}V' \, \frac{\nabla' \cdot \mathcal{J}(\mathbf{r}',t)}{4\pi|\mathbf{r}-\mathbf{r}'|} = \nabla \int_{V_\varrho} \mathrm{d}V' \, \frac{1}{4\pi|\mathbf{r}-\mathbf{r}'|}\frac{\partial}{\partial t}\varrho(\mathbf{r}',t) \\ &= \frac{\partial}{\partial t}\nabla \int_{V_\varrho} \mathrm{d}V' \, \frac{\varrho(\mathbf{r}',t)}{4\pi|\mathbf{r}-\mathbf{r}'|} = \varepsilon\frac{\partial}{\partial t}\nabla\Phi_E(\mathbf{r},t), \qquad (\mathbf{r},t) \in \mathbb{R}^3 \times \mathbb{R} \end{aligned} \tag{8.117}$$

having used the continuity equation (1.46), the Schwarz theorem, and (8.116). The interchange of spatial integration and time derivative is possible as the integrand and its derivative are continuous for $(\mathbf{r}',t) \in V_\varrho \times \mathbb{R}$ and observation points $\mathbf{r} \notin V_\varrho$. The result also holds for $\mathbf{r} \in V_\varrho$ since the improper integral converges (Section 2.7). Using the above expression for $\mathcal{J}_L(\mathbf{r},t)$ in (8.115) yields

$$\nabla^2 \mathcal{A}_E(\mathbf{r},t) - \frac{1}{c^2}\frac{\partial^2}{\partial t^2}\mathcal{A}_E(\mathbf{r},t) = -\mu\mathcal{J}_S(\mathbf{r},t) - \mu\mathcal{J}_L(\mathbf{r},t) + \varepsilon\mu\frac{\partial}{\partial t}\nabla\Phi_E(\mathbf{r},t) = -\mu\mathcal{J}_S(\mathbf{r},t) \tag{8.118}$$

that is, the right-hand side is indeed a solenoidal vector field, as speculated.

To summarize, so far we have learnt that in the Coulomb gauge:

(1) the scalar potential $\Phi_E(\mathbf{r},t)$ obeys the Poisson equation;
(2) the vector potential $\mathcal{A}_E(\mathbf{r},t)$ satisfies a typical D'Alembert equation and depends only upon the solenoidal part of the current density;
(3) owing to (8.104), (8.105) and (8.106) also the solenoidal parts of $\mathcal{E}(\mathbf{r},t)$ and $\mathcal{B}(\mathbf{r},t)$ result from just the solenoidal part of $\mathcal{J}(\mathbf{r},t)$.

The problem with (8.116) is that the speed of light c is not involved explicitly and any change in the distribution of charges within V_ϱ is reflected *instantaneously* in the potential in any point in space! It would thus appear that disturbances (electromagnetic waves) produced by accelerated charges can propagate with infinite velocity and hence *causality* is violated. Or is it? To resolve this conundrum we begin by observing that $\mathcal{A}_E(\mathbf{r},t)$ and $\Phi_E(\mathbf{r},t)$ are auxiliary quantities and that the fields $\mathcal{E}(\mathbf{r},t)$ and $\mathcal{B}(\mathbf{r},t)$ do obey wave equations, as we know from Section 1.8. Secondly, although the true source $\mathcal{J}(\mathbf{r},t)$ is localized, we see from (8.117) that $\mathcal{J}_L(\mathbf{r},t)$ extends throughout space and so does the solenoidal component $\mathcal{J}_S(\mathbf{r},t)$. This accounts for the instantaneous propagation of changes in the potential (8.116).

Although the Coulomb gauge allows decoupling the equations for the potentials, it does not represent the only choice to the effect. To elucidate we cast (8.112) into the form

$$\nabla\left[\nabla \cdot \mathcal{A}_E(\mathbf{r},t) + \frac{1}{c^2}\frac{\partial}{\partial t}\Phi_E(\mathbf{r},t)\right] - \nabla^2\mathcal{A}_E(\mathbf{r},t) + \frac{1}{c^2}\frac{\partial^2}{\partial t^2}\mathcal{A}_E(\mathbf{r},t) = \mu\mathcal{J}(\mathbf{r},t) \tag{8.119}$$

thanks to (H.59) and the Schwarz theorem. If we manage to nullify the quantity between brackets we obtain a typical wave equation for the vector potential. There are two ways to look at this goal. For instance, we could start with the pair $\mathcal{A}_{E1}(\mathbf{r},t)$ and $\Phi_{E1}(\mathbf{r},t)$ in the Coulomb gauge and apply transformations (8.108) and (8.110) to determine $\mathcal{A}_{E2}(\mathbf{r},t)$ and $\Phi_{E2}(\mathbf{r},t)$ endowed with the desired property. In symbols, this yields

$$\begin{aligned} 0 &= \nabla\cdot\mathcal{A}_{E2}(\mathbf{r},t) + \frac{1}{c^2}\frac{\partial}{\partial t}\Phi_{E2}(\mathbf{r},t) \\ &= \nabla\cdot[\mathcal{A}_{E1}(\mathbf{r},t) + \nabla\chi(\mathbf{r},t)] + \frac{1}{c^2}\frac{\partial}{\partial t}\Phi_{E1}(\mathbf{r},t) - \frac{1}{c^2}\frac{\partial^2}{\partial t^2}\chi(\mathbf{r},t) \\ &= \nabla^2\chi(\mathbf{r},t) - \frac{1}{c^2}\frac{\partial^2}{\partial t^2}\chi(\mathbf{r},t) + \frac{1}{c^2}\frac{\partial}{\partial t}\Phi_{E1}(\mathbf{r},t) \end{aligned} \tag{8.120}$$

which we interpret as a wave equation for the scalar field $\chi(\mathbf{r},t)$ to be used in (8.108) and (8.110). Alternatively, since (8.105) specifies only the curl of $\mathcal{A}_E(\mathbf{r},t)$ we may directly enforce the condition

$$\nabla\cdot\mathcal{A}_E(\mathbf{r},t) + \frac{1}{c^2}\frac{\partial}{\partial t}\Phi_E(\mathbf{r},t) = 0 \tag{8.121}$$

which goes by the name of *Lorenz gauge*, the Danish physicist who first proposed it in 1867 [15, 16, 38, 39]. Interestingly, (8.121) is in the form of a continuity equation, with Φ_E/c^2 being the 'source density' and $\mathcal{A}_E$ being the associated 'current density' [cf. (1.46)]. If we exploit the Lorenz condition to eliminate the vector potential from (8.113), in the end we arrive at

$$\nabla^2\Phi_E(\mathbf{r},t) - \frac{1}{c^2}\frac{\partial^2}{\partial t^2}\Phi_E(\mathbf{r},t) = -\frac{1}{\varepsilon}\varrho(\mathbf{r},t), \qquad (\mathbf{r},t)\in\mathbb{R}^3\times\mathbb{R} \tag{8.122}$$

$$\nabla^2\mathcal{A}_E(\mathbf{r},t) - \frac{1}{c^2}\frac{\partial^2}{\partial t^2}\mathcal{A}_E(\mathbf{r},t) = -\mu\mathcal{J}(\mathbf{r},t), \qquad (\mathbf{r},t)\in\mathbb{R}^3\times\mathbb{R} \tag{8.123}$$

which constitute a set of fully uncoupled wave equations for the potentials. The main advantage of (8.122) and (8.123) is that they are formally the same equation involving the D'Alembert operator, and hence we are done if we find a formula for the scalar potential. From a mathematical standpoint, under the Lorenz gauge the splitting (8.104) does not represent a Helmholtz decomposition of the electric field.

We conclude this section by listing the formulas pertinent to magnetic charges and currents which were nullified at the beginning of the discussion. In the presence of magnetic sources $\mathcal{J}_M(\mathbf{r},t)$ and $\varrho_M(\mathbf{r},t)$ with $\mathcal{J}(\mathbf{r},t) = 0 = \varrho(\mathbf{r},t)$ the displacement field $\mathcal{D}(\mathbf{r},t)$ is solenoidal whereas $\mathcal{H}(\mathbf{r},t)$ satisfies the source-free Ampère-Maxwell law (1.34). Thus, it is possible to let [35, pp. 46–47], [15, 28]

$$\mathcal{H}(\mathbf{r},t) = -\frac{\partial}{\partial t}\mathcal{A}_M(\mathbf{r},t) - \nabla\Phi_M(\mathbf{r},t), \qquad (\mathbf{r},t)\in\mathbb{R}^3\times\mathbb{R} \tag{8.124}$$

$$\mathcal{D}(\mathbf{r},t) = -\nabla\times\mathcal{A}_M(\mathbf{r},t), \qquad (\mathbf{r},t)\in\mathbb{R}^3\times\mathbb{R} \tag{8.125}$$

where $\mathcal{A}_M(\mathbf{r},t)$ is the *electric vector potential* (physical dimension: As/m or C/m) and $\Phi_M(\mathbf{r},t)$ the *magnetic scalar potential* (physical dimension: A). The subscript 'M' appended to the potentials means that they depend on magnetic sources only.

In a homogeneous lossless isotropic medium the magnetic potentials are solutions to the following fully uncoupled wave equations

$$\nabla^2\Phi_M(\mathbf{r},t) - \frac{1}{c^2}\frac{\partial^2}{\partial t^2}\Phi_M(\mathbf{r},t) = -\frac{1}{\mu}\varrho_M(\mathbf{r},t), \qquad (\mathbf{r},t)\in\mathbb{R}^3\times\mathbb{R} \tag{8.126}$$

$$\nabla^2\mathcal{A}_M(\mathbf{r},t) - \frac{1}{c^2}\frac{\partial^2}{\partial t^2}\mathcal{A}_M(\mathbf{r},t) = -\varepsilon\mathcal{J}_M(\mathbf{r},t), \qquad (\mathbf{r},t)\in\mathbb{R}^3\times\mathbb{R} \tag{8.127}$$

Table 8.1 Duality transformations for electrodynamic potentials

Time domain		**Frequency domain**	
Primary entities	**Dual entities**	**Primary entities**	**Dual entities**
$\mathcal{A}_E$	$\mathcal{A}_M$	$\mathbf{A}_E$	$\mathbf{A}_M$
$\mathcal{A}_M$	$\mathcal{A}_E$	$\mathbf{A}_M$	$\mathbf{A}_E$
Φ_E	Φ_M	Φ_E	Φ_M
Φ_M	Φ_E	Φ_M	Φ_E

provided we enforce the magnetic Lorenz gauge

$$\nabla \cdot \mathcal{A}_M(\mathbf{r}, t) + \frac{1}{c^2}\frac{\partial}{\partial t}\Phi_M(\mathbf{r}, t) = 0 \tag{8.128}$$

to resolve the ambiguity on the divergence of $\mathcal{A}_M(\mathbf{r}, t)$. In which instance (8.124) and (8.125) do not constitute a perfect Helmholtz splitting.

By invoking the principle of superposition (Section 6.1) we can write down the general decomposition of electromagnetic entities in the presence of both electric and magnetic sources, namely,

$$\mathcal{E}(\mathbf{r}, t) = -\nabla\Phi_E(\mathbf{r}, t) - \frac{\partial}{\partial t}\mathcal{A}_E(\mathbf{r}, t) - \frac{1}{\varepsilon}\nabla \times \mathcal{A}_M(\mathbf{r}, t), \qquad (\mathbf{r}, t) \in \mathbb{R}^3 \times \mathbb{R} \tag{8.129}$$

$$\mathcal{H}(\mathbf{r}, t) = \frac{1}{\mu}\nabla \times \mathcal{A}_E(\mathbf{r}, t) - \frac{\partial}{\partial t}\mathcal{A}_M(\mathbf{r}, t) - \nabla\Phi_M(\mathbf{r}, t), \qquad (\mathbf{r}, t) \in \mathbb{R}^3 \times \mathbb{R} \tag{8.130}$$

$$\mathcal{D}(\mathbf{r}, t) = -\varepsilon\nabla\Phi_E(\mathbf{r}, t) - \varepsilon\frac{\partial}{\partial t}\mathcal{A}_E(\mathbf{r}, t) - \nabla \times \mathcal{A}_M(\mathbf{r}, t), \qquad (\mathbf{r}, t) \in \mathbb{R}^3 \times \mathbb{R} \tag{8.131}$$

$$\mathcal{B}(\mathbf{r}, t) = \nabla \times \mathcal{A}_E(\mathbf{r}, t) - \mu\frac{\partial}{\partial t}\mathcal{A}_M(\mathbf{r}, t) - \mu\nabla\Phi_M(\mathbf{r}, t), \qquad (\mathbf{r}, t) \in \mathbb{R}^3 \times \mathbb{R} \tag{8.132}$$

on account of the constitutive relationships (1.117) and (1.118). Furthermore, it is straightforward to obtain the corresponding decompositions, wave equations and gauge conditions for time-harmonic fields and sources by applying the practical transformation rule $\partial/\partial t \to \mathrm{j}\,\omega$ developed in Section 1.5.

Finally, by inspection we may extend the duality transformations of Table 6.1 to the electromagnetic potentials in a consistent way. Apparently, (8.129) turns into (8.130) provided we make the substitutions

$$\mathcal{E} \Rightarrow \mathcal{H}', \qquad \varepsilon \Rightarrow -\mu, \qquad \mathcal{A}_E \Rightarrow \mathcal{A}'_M, \qquad \mathcal{A}_M \Rightarrow \mathcal{A}'_E, \qquad \Phi_E \Rightarrow \Phi'_M \tag{8.133}$$

and similar replacements hold for the other equations. The pairs of dual entities are summarized in Table 8.1 also for the time-harmonic regime.

In Section 8.4 we show how to recover a Helmholtz decomposition of the electromagnetic field, whereas we discuss the solution of (8.122), (8.123), (8.126) and (8.127) in Section 9.1.

8.3 Boundary conditions for the electrodynamic potentials

The electromagnetic potentials introduced in Section 8.2 can help us solve the Maxwell equations both in unbounded regions of space and finite ones. For the latter case, though, it is necessary to establish suitable matching conditions in order to 'stitch' solutions consistently across material interfaces (cf. Sections 2.4 and 4.3). In our quest we may build on the results of Section 1.7 by replacing the field entities with the representation thereof in terms of electromagnetic potentials.

We begin with the entities of intensity $\mathcal{E}(\mathbf{r},t)$ and $\mathcal{B}(\mathbf{r},t)$ generated by electric sources only and refer to the geometries of Figures 1.12 and 1.13.

By substituting (8.105) into (1.157) and applying (A.60) we find

$$0 = \nabla_s \cdot \{\hat{\mathbf{n}}(\mathbf{r}) \times [\mathcal{A}_{E1}(\mathbf{r},t) - \mathcal{A}_{E1}(\mathbf{r},t)]\}, \qquad \mathbf{r} \in \partial V \tag{8.134}$$

where $\nabla_s \cdot \{\bullet\}$ denotes the surface divergence on ∂V (cf. Appendix A.5). This conditions is essentially the same as (4.52), given that the time dependence does not appear explicitly in (1.157). Therefore, we reach the same conclusion, namely,

$$\hat{\mathbf{n}}(\mathbf{r}) \times \mathcal{A}_{E1}(\mathbf{r},t) = \hat{\mathbf{n}}(\mathbf{r}) \times \mathcal{A}_{E2}(\mathbf{r},t), \qquad \mathbf{r} \in \partial V \tag{8.135}$$

that is, the tangential component of magnetic vector potential due to electric sources is continuous across a material interface.

Secondly, by inserting (8.104) into (1.144) and taking an extra cross product with the unit normal $\hat{\mathbf{n}}(\mathbf{r})$ we get

$$\begin{aligned} 0 &= \hat{\mathbf{n}}(\mathbf{r}) \times [\mathcal{E}_1(\mathbf{r},t) \times \hat{\mathbf{n}}(\mathbf{r})] - \hat{\mathbf{n}}(\mathbf{r}) \times [\mathcal{E}_2(\mathbf{r},t) \times \hat{\mathbf{n}}(\mathbf{r})] \\ &= -\nabla_s \Phi_{E1}(\mathbf{r},t) - \frac{\partial \mathcal{A}_{E1t}}{\partial t} + \nabla_s \Phi_{E2}(\mathbf{r},t) + \frac{\partial \mathcal{A}_{E2t}}{\partial t} \\ &= -\nabla_s[\Phi_{E1}(\mathbf{r},t) - \Phi_{E2}(\mathbf{r},t)] - \frac{\partial}{\partial t}[\mathcal{A}_{E1t}(\mathbf{r},t) - \mathcal{A}_{E2t}(\mathbf{r},t)], \qquad \mathbf{r} \in \partial V \end{aligned} \tag{8.136}$$

where $\nabla_s\{\bullet\}$ indicates the surface gradient on ∂V (cf. Appendix A.5), we have used the continuity of $\Phi_{El}(\mathbf{r},t)$, $l = 1,2$, for points over ∂V, and $\mathcal{A}_{Elt}(\mathbf{r},t)$ denotes the component of $\mathcal{A}_{El}(\mathbf{r},t)$ that is tangential to the surface ∂V. Thanks to (8.135), (8.136) becomes formally identical with condition (2.50) which is valid in the static case. Hence, we conclude

$$\Phi_{E1}(\mathbf{r},t) - \Phi_{E2}(\mathbf{r},t) = M_E(t), \qquad \mathbf{r} \in \partial V \tag{8.137}$$

with $M_E(t)$ being a function of time. The electric scalar potential is continuous across ∂V, so long as we arbitrarily set $M_E(t) = 0$; indeed, in the static regime (8.137) reduces to (2.51).

Thirdly, we consider the normal component of $\mathcal{A}_E(\mathbf{r},t)$. As was the case for the steady-state analogue $\mathcal{A}(\mathbf{r})$ (Section 4.3) we can exploit the freedom afforded by the gauge transformations to *enforce* the continuity of $\hat{\mathbf{n}}(\mathbf{r}) \cdot \mathcal{A}_E(\mathbf{r},t)$ across a material interface. In the analysis of time-varying fields we have found it convenient to adopt the Lorenz gauge, inasmuch as this choice leads to two uncoupled equations for $\Phi_E(\mathbf{r},t)$ and $\mathcal{A}_E(\mathbf{r},t)$. However, since $\mathcal{A}_E(\mathbf{r},t)$ may not be continuously differentiable for $\mathbf{r} \in \mathbb{R}^3$ if the surface $\partial V \subset \mathbb{R}^3$ is a material boundary (see Figure 1.13) we need to enforce (8.121), (8.122) and (8.123) separately in V_1 and V_2. Hence, by integrating the gauge over W_1 and W_2 we find

$$\begin{aligned} 0 &= \int_{W_1} dV \left[\nabla \cdot \mathcal{A}_{E1}(\mathbf{r},t) + \frac{1}{c_1^2}\frac{\partial}{\partial t}\Phi_{E1}(\mathbf{r},t)\right] \\ &= \int_{\partial W_1 \cup S_0} dS\, \hat{\mathbf{v}}_1(\mathbf{r},t) \cdot \mathcal{A}_{E1}(\mathbf{r},t) + \int_{W_1} dV \frac{1}{c_1^2}\frac{\partial}{\partial t}\Phi_{E1}(\mathbf{r},t) \end{aligned} \tag{8.138}$$

$$0 = \int_{W_2} dV \left[\nabla \cdot \mathcal{A}_{E1}(\mathbf{r},t) + \frac{1}{c_2^2}\frac{\partial}{\partial t}\Phi_{E2}(\mathbf{r},t)\right]$$

$$= \int_{\partial W_2 \cup S_0} \mathrm{d}S\, \hat{\nu}_2(\mathbf{r},t) \cdot \mathcal{A}_{E2}(\mathbf{r},t) + \int_{W_2} \mathrm{d}V \frac{1}{c_2^2}\frac{\partial}{\partial t}\Phi_{E2}(\mathbf{r},t) \tag{8.139}$$

where c_1 and c_2 denote the speed of light in the media that fill V_1 and V_2, respectively, and we have applied the divergence theorem (A.53) in W_1 and W_2. By summing the above equations side by side and rearranging the surface integrals we get

$$\begin{aligned}
&\int_{\partial W_1 \cup S_0} \mathrm{d}S\, \hat{\nu}_1(\mathbf{r},t) \cdot \mathcal{A}_{E1}(\mathbf{r},t) + \int_{\partial W_2 \cup S_0} \mathrm{d}S\, \hat{\nu}_2(\mathbf{r},t) \cdot \mathcal{A}_{E2}(\mathbf{r},t) \\
&\quad + \int_{W_1} \mathrm{d}V \frac{1}{c_1^2}\frac{\partial}{\partial t}\Phi_{E1}(\mathbf{r},t) + \int_{W_2} \mathrm{d}V \frac{1}{c_2^2}\frac{\partial}{\partial t}\Phi_{E2}(\mathbf{r},t) \\
&\quad = \int_{\partial W} \mathrm{d}S\, \hat{\nu}(\mathbf{r},t) \cdot \mathcal{A}_{E}(\mathbf{r},t) + \int_{S_0} \mathrm{d}S\, \hat{\mathbf{n}}(\mathbf{r},t) \cdot [\mathcal{A}_{E2}(\mathbf{r},t) - \mathcal{A}_{E1}(\mathbf{r},t)] \\
&\quad + \int_{W} \mathrm{d}V \frac{1}{[c(\mathbf{r})]^2}\frac{\partial}{\partial t}\Phi_{E}(\mathbf{r},t) = 0
\end{aligned} \tag{8.140}$$

where $c(\mathbf{r})$ is a piecewise-constant function equal to the speed of light in V_1 and V_2 for $\mathbf{r} \in V_1$ and $\mathbf{r} \in V_2$. Now, we observe that *a priori* there is no reason why the global relation

$$\int_{\partial W} \mathrm{d}S\, \hat{\nu}(\mathbf{r}) \cdot \mathcal{A}_{E}(\mathbf{r},t) + \int_{W} \mathrm{d}V \frac{1}{[c(\mathbf{r})]^2}\frac{\partial}{\partial t}\Phi_{E}(\mathbf{r},t) = 0 \tag{8.141}$$

should hold because the Lorenz gauge is an arbitrary constraint in local form enforced separately in V_1 and V_2. Nonetheless, we may *require* that the condition

$$\hat{\mathbf{n}}(\mathbf{r}) \cdot \mathcal{A}_{E1}(\mathbf{r},t) = \hat{\mathbf{n}}(\mathbf{r}) \cdot \mathcal{A}_{E2}(\mathbf{r},t), \qquad \mathbf{r} \in \partial V \tag{8.142}$$

be true and thus achieve the vanishing of the flux integral over S_0. The reason is that transformations (8.108) and (8.110) do not alter the electromagnetic entities and, even though we enforce the Lorenz gauge separately in V_1 and V_2, this is not sufficient to specify the necessary auxiliary scalar fields $\chi_{E1}(\mathbf{r},t)$ and $\chi_{E2}(\mathbf{r},t)$. Indeed, if in accordance with (8.108) we let

$$\mathcal{A}_{El}(\mathbf{r},t) = \mathcal{A}'_{El}(\mathbf{r},t) + \nabla\chi_{El}(\mathbf{r},t), \qquad \mathbf{r} \in V_l, \qquad l = 1,2 \tag{8.143}$$

we may write the integral over S_0 as

$$\begin{aligned}
&\int_{S_0} \mathrm{d}S\, \hat{\mathbf{n}}(\mathbf{r},t) \cdot [\mathcal{A}_{E2}(\mathbf{r},t) - \mathcal{A}_{E1}(\mathbf{r},t)] \\
&\quad = \int_{S_0} \mathrm{d}S \left[\hat{\mathbf{n}}(\mathbf{r},t) \cdot \mathcal{A}'_{E2}(\mathbf{r},t) + \frac{\partial\chi_{E1}}{\partial\hat{\mathbf{n}}} - \hat{\mathbf{n}}(\mathbf{r},t) \cdot \mathcal{A}'_{E1}(\mathbf{r},t) - \frac{\partial\chi_{E2}}{\partial\hat{\mathbf{n}}}\right]
\end{aligned} \tag{8.144}$$

whereby we see that the normal derivatives of $\chi_{El}(\mathbf{r},t)$ can be chosen so as to nullify the flux integral over S_0.

The jump conditions of the entities of quantity $\mathcal{D}(\mathbf{r},t)$ and $\mathcal{H}(\mathbf{r},t)$ provide us with two additional requirements for the potentials. Substituting (8.104) into (1.155) yields

$$\begin{aligned}\varrho_S(\mathbf{r},t) &= -\hat{\mathbf{n}}(\mathbf{r})\cdot\varepsilon_1\left[\nabla\Phi_{E1}(\mathbf{r},t)+\frac{\partial\mathcal{A}_{E1}}{\partial t}\right]+\hat{\mathbf{n}}(\mathbf{r})\cdot\varepsilon_2\left[\nabla\Phi_{E2}(\mathbf{r},t)+\frac{\partial\mathcal{A}_{E2}}{\partial t}\right]\\ &= -\varepsilon_1\left[\frac{\partial}{\partial\hat{\mathbf{n}}}\Phi_{E1}(\mathbf{r},t)+\frac{\partial}{\partial t}\hat{\mathbf{n}}(\mathbf{r})\cdot\mathcal{A}_{E1}(\mathbf{r},t)\right]\\ &\quad+\varepsilon_2\left[\frac{\partial}{\partial\hat{\mathbf{n}}}\Phi_{E2}(\mathbf{r},t)+\frac{\partial}{\partial t}\hat{\mathbf{n}}(\mathbf{r})\cdot\mathcal{A}_{E2}(\mathbf{r},t)\right],\qquad \mathbf{r}\in\partial V\end{aligned}\tag{8.145}$$

which leads to

$$\varepsilon_2\frac{\partial}{\partial\hat{\mathbf{n}}}\Phi_{E2}(\mathbf{r},t)-\varepsilon_1\frac{\partial}{\partial\hat{\mathbf{n}}}\Phi_{E1}(\mathbf{r},t)+(\varepsilon_2-\varepsilon_1)\frac{\partial}{\partial t}\hat{\mathbf{n}}(\mathbf{r})\cdot\mathcal{A}_{E1}(\mathbf{r},t)=\varrho_S(\mathbf{r},t),\qquad \mathbf{r}\in\partial V\tag{8.146}$$

because (8.142) holds true at all times. We notice that (8.146) reduces to (2.48) in the static limit. Lastly, the combination of (8.105) and (1.142) gives

$$\frac{1}{\mu_1}\hat{\mathbf{n}}(\mathbf{r})\times[\nabla\times\mathcal{A}_{E1}(\mathbf{r},t)]-\frac{1}{\mu_2}\hat{\mathbf{n}}(\mathbf{r})\times[\nabla\times\mathcal{A}_{E2}(\mathbf{r},t)]=\mathcal{J}_S(\mathbf{r},t),\qquad \mathbf{r}\in\partial V.\tag{8.147}$$

To obtain the boundary conditions relevant to the electromagnetic potentials $\Phi_M(\mathbf{r},t)$ and $\mathcal{A}_M(\mathbf{r},t)$ defined in (8.124) and (8.125) and generated by magnetic sources $\varrho_M(\mathbf{r},t)$ and $\mathcal{J}_M(\mathbf{r},t)$ we apply the principle of duality. More specifically, we subject (8.135), (8.137), (8.142), (8.146) and (8.147) to the transformation rules listed in Tables 6.1 and 8.1. In symbols, we have

$$\hat{\mathbf{n}}(\mathbf{r})\times\mathcal{A}_{M1}(\mathbf{r},t)=\hat{\mathbf{n}}(\mathbf{r})\times\mathcal{A}_{M2}(\mathbf{r},t),\qquad \mathbf{r}\in\partial V\tag{8.148}$$

$$\Phi_{M1}(\mathbf{r},t)-\Phi_{M2}(\mathbf{r},t)=M_M(t),\qquad \mathbf{r}\in\partial V\tag{8.149}$$

$$\hat{\mathbf{n}}(\mathbf{r})\cdot\mathcal{A}_{M1}(\mathbf{r},t)=\hat{\mathbf{n}}(\mathbf{r})\cdot\mathcal{A}_{M2}(\mathbf{r},t),\qquad \mathbf{r}\in\partial V\tag{8.150}$$

where $M_M(t)$ may be set to zero, and also

$$\mu_2\frac{\partial}{\partial\hat{\mathbf{n}}}\Phi_{M2}(\mathbf{r},t)-\mu_1\frac{\partial}{\partial\hat{\mathbf{n}}}\Phi_{M1}(\mathbf{r},t)+(\mu_2-\mu_1)\frac{\partial}{\partial t}\hat{\mathbf{n}}(\mathbf{r})\cdot\mathcal{A}_{M1}(\mathbf{r},t)=\varrho_{MS}(\mathbf{r},t),\qquad \mathbf{r}\in\partial V\tag{8.151}$$

$$\frac{1}{\varepsilon_1}\hat{\mathbf{n}}(\mathbf{r})\times[\nabla\times\mathcal{A}_{M1}(\mathbf{r},t)]-\frac{1}{\varepsilon_2}\hat{\mathbf{n}}(\mathbf{r})\times[\nabla\times\mathcal{A}_{M2}(\mathbf{r},t)]=\mathcal{J}_{MS}(\mathbf{r},t),\qquad \mathbf{r}\in\partial V.\tag{8.152}$$

Finally, we specialize all the matching relations derived so far to the harmonic time-dependence (Section 1.5). The result follows immediately by replacing entities and sources with their phasors, as is prescribed by the rules of Table 1.1. We list the formulas hereinbelow

$$\hat{\mathbf{n}}(\mathbf{r})\times\mathbf{A}_{E1}(\mathbf{r})=\hat{\mathbf{n}}(\mathbf{r})\times\mathbf{A}_{E2}(\mathbf{r}),\qquad \mathbf{r}\in\partial V\tag{8.153}$$

$$\Phi_{E1}(\mathbf{r})-\Phi_{E2}(\mathbf{r})=M_E,\qquad \mathbf{r}\in\partial V\tag{8.154}$$

$$\hat{\mathbf{n}}(\mathbf{r})\cdot\mathbf{A}_{E1}(\mathbf{r})=\hat{\mathbf{n}}(\mathbf{r})\cdot\mathbf{A}_{E2}(\mathbf{r}),\qquad \mathbf{r}\in\partial V\tag{8.155}$$

$$\varepsilon_2\frac{\partial}{\partial\hat{\mathbf{n}}}\Phi_{E2}(\mathbf{r})-\varepsilon_1\frac{\partial}{\partial\hat{\mathbf{n}}}\Phi_{E1}(\mathbf{r})+\mathrm{j}\omega(\varepsilon_2-\varepsilon_1)\hat{\mathbf{n}}(\mathbf{r})\cdot\mathbf{A}_{E1}(\mathbf{r})=\rho_S(\mathbf{r}),\qquad \mathbf{r}\in\partial V\tag{8.156}$$

$$\frac{1}{\mu_1}\hat{\mathbf{n}}(\mathbf{r})\times[\nabla\times\mathbf{A}_{E1}(\mathbf{r})]-\frac{1}{\mu_2}\hat{\mathbf{n}}(\mathbf{r})\times[\nabla\times\mathbf{A}_{E2}(\mathbf{r})]=\mathbf{J}_S(\mathbf{r}),\qquad \mathbf{r}\in\partial V\tag{8.157}$$

$$\hat{\mathbf{n}}(\mathbf{r}) \times \mathbf{A}_{M1}(\mathbf{r}) = \hat{\mathbf{n}}(\mathbf{r}) \times \mathbf{A}_{M2}(\mathbf{r}), \qquad \mathbf{r} \in \partial V \tag{8.158}$$

$$\Phi_{M1}(\mathbf{r}) - \Phi_{M2}(\mathbf{r}) = M_M, \qquad \mathbf{r} \in \partial V \tag{8.159}$$

$$\hat{\mathbf{n}}(\mathbf{r}) \cdot \mathbf{A}_{M1}(\mathbf{r}) = \hat{\mathbf{n}}(\mathbf{r}) \cdot \mathbf{A}_{M2}(\mathbf{r}), \qquad \mathbf{r} \in \partial V \tag{8.160}$$

$$\mu_2 \frac{\partial}{\partial \hat{\mathbf{n}}} \Phi_{M2}(\mathbf{r}) - \mu_1 \frac{\partial}{\partial \hat{\mathbf{n}}} \Phi_{M1}(\mathbf{r}) + \mathrm{j}\,\omega(\mu_2 - \mu_1)\hat{\mathbf{n}}(\mathbf{r}) \cdot \mathbf{A}_{M1}(\mathbf{r}) = \rho_{MS}(\mathbf{r}), \qquad \mathbf{r} \in \partial V \tag{8.161}$$

$$\frac{1}{\varepsilon_1}\hat{\mathbf{n}}(\mathbf{r}) \times [\nabla \times \mathbf{A}_{M1}(\mathbf{r})] - \frac{1}{\varepsilon_2}\hat{\mathbf{n}}(\mathbf{r}) \times [\nabla \times \mathbf{A}_{M2}(\mathbf{r})] = \mathbf{J}_{MS}(\mathbf{r}), \qquad \mathbf{r} \in \partial V. \tag{8.162}$$

8.4 Hertzian potentials

In Section 8.2 we went to great lengths to derive the electromagnetic entities due to electric and magnetic sources from the four potentials $\mathcal{A}_E$, Φ_E, $\mathcal{A}_M$ and Φ_M. We concluded that, while the choice of the Lorenz gauges (8.121) and (8.128) has the undisputable merit of simplifying the governing equations for the potentials, still the resulting representations of $\mathcal{E}$ and $\mathcal{H}$ do not constitute a complete Helmholtz decomposition into solenoidal and lamellar parts that mimics the generic splitting (8.28).

It takes a moment's thought to realize that the scalar potentials Φ_E and Φ_M are not independent of $\mathcal{A}_E$ and $\mathcal{A}_M$, respectively, if conditions (8.121) and (8.128) must be obeyed. What is more, we can even solve the first-order partial differential equations (8.121) and (8.128) formally for the scalar potentials, namely,

$$\Phi_E(\mathbf{r}, t) = -c^2 \int_{-\infty}^{t} \mathrm{d}t'\, \nabla \cdot \mathcal{A}_E(\mathbf{r}, t') = -c^2 \nabla \cdot \int_{-\infty}^{t} \mathrm{d}t'\, \mathcal{A}_E(\mathbf{r}, t') \tag{8.163}$$

$$\Phi_M(\mathbf{r}, t) = -c^2 \int_{-\infty}^{t} \mathrm{d}t'\, \nabla \cdot \mathcal{A}_M(\mathbf{r}, t') = -c^2 \nabla \cdot \int_{-\infty}^{t} \mathrm{d}t'\, \mathcal{A}_M(\mathbf{r}, t') \tag{8.164}$$

where we have interchanged integration with respect to time and differentiation with respect to the space coordinates. This step — a special application of the Leibniz rule [40, Formula 3.3.7], [41, Formula 69.3], [42] for differentiation under the integral sign — is possible provided the six components of $\mathcal{A}_E(\mathbf{r}, t')$, $\mathcal{A}_M(\mathbf{r}, t')$ and the scalar functions $\nabla \cdot \mathcal{A}_E(\mathbf{r}, t')$, $\nabla \cdot \mathcal{A}_M(\mathbf{r}, t')$ are continuous for pairs $(\mathbf{r}, t')$ in the space-time domain of interest.

On the grounds of (8.163) and (8.164) we may define the auxiliary vector fields

$$\boldsymbol{\pi}_E(\mathbf{r}, t) := c^2 \int_{-\infty}^{t} \mathrm{d}t'\, \mathcal{A}_E(\mathbf{r}, t'), \qquad (\mathbf{r}, t) \in \mathbb{R}^3 \times \mathbb{R} \tag{8.165}$$

$$\boldsymbol{\pi}_M(\mathbf{r}, t) := c^2 \int_{-\infty}^{t} \mathrm{d}t'\, \mathcal{A}_M(\mathbf{r}, t'), \qquad (\mathbf{r}, t) \in \mathbb{R}^3 \times \mathbb{R} \tag{8.166}$$

which are known as the electric and magnetic *Hertzian potentials* [16, Section 7.3], [43, Section 6.3.3], [35, Section 8.2.1], [44, Section 1.7], [15,21], or *Hertz vectors* [28, Section 1.11], [45, Section 4.42], [8, Section 16.9], [46, Section 1.6], [47]. The physical dimensions carried by $\boldsymbol{\pi}_E$ and $\boldsymbol{\pi}_M$ are volt-meter (Vm) and ampere-meter (Am), respectively. The theoretical importance of the Ansatz (8.165), first proposed by H. Hertz in 1888, lies in the fact that, as a result, both Φ_E and $\mathcal{A}_E$ can be derived from a single vector field $\boldsymbol{\pi}_E$, i.e., three scalar fields. Of course, the same remark applies to (8.166) and the magnetic counterparts Φ_M and $\mathcal{A}_M$, all of which are needed whenever also magnetic sources are contemplated. More specifically, we have

$$\mathcal{A}_E(\mathbf{r}, t) = \frac{1}{c^2}\frac{\partial}{\partial t}\boldsymbol{\pi}_E(\mathbf{r}, t) \qquad \Phi_E(\mathbf{r}, t) = -\nabla \cdot \boldsymbol{\pi}_E(\mathbf{r}, t) \tag{8.167}$$

$$\mathcal{A}_M(\mathbf{r}, t) = \frac{1}{c^2}\frac{\partial}{\partial t}\boldsymbol{\pi}_M(\mathbf{r}, t) \qquad \Phi_M(\mathbf{r}, t) = -\nabla \cdot \boldsymbol{\pi}_M(\mathbf{r}, t) \tag{8.168}$$

which follow partly from the fundamental theorem of calculus [27, Theorem 6.21] applied to (8.165) and (8.166), and partly from the formal solutions (8.163) and (8.164).

In order to find the governing equation for the electric Hertz vector in a homogeneous lossless isotropic medium we may try eliminating $\mathcal{A}_E$ from (8.123) with the aid of the first part of (8.167). After a few manipulations this strategy yields

$$\frac{\partial}{\partial t}\left[\nabla^2\boldsymbol{\pi}_E(\mathbf{r}, t) - \frac{1}{c^2}\frac{\partial^2}{\partial t^2}\boldsymbol{\pi}_E(\mathbf{r}, t)\right] = -\frac{1}{\varepsilon}\mathcal{J}(\mathbf{r}, t), \qquad (\mathbf{r}, t) \in \mathbb{R}^3 \times \mathbb{R} \tag{8.169}$$

on account of the Schwarz theorem and (1.209). Now, were it not for the presence of the leading time derivative in the left-hand side, (8.169) would have the form a wave equation for $\boldsymbol{\pi}_E(\mathbf{r}, t)$. Then again, eliminating Φ_E from (8.122) thanks to the second part of (8.167) leads us to the condition

$$\nabla \cdot \left[\nabla^2\boldsymbol{\pi}_E(\mathbf{r}, t) - \frac{1}{c^2}\frac{\partial^2}{\partial t^2}\boldsymbol{\pi}_E(\mathbf{r}, t)\right] = \frac{1}{\varepsilon}\varrho(\mathbf{r}, t), \qquad (\mathbf{r}, t) \in \mathbb{R}^3 \times \mathbb{R} \tag{8.170}$$

which is not quite in the form of a wave equation either. Since the left members of (8.169) and (8.170) involve either the temporal derivative or the divergence of the same vectorial quantity (i.e., the time-varying field within brackets), we are prompted to let

$$\nabla^2\boldsymbol{\pi}_E(\mathbf{r}, t) - \frac{1}{c^2}\frac{\partial^2}{\partial t^2}\boldsymbol{\pi}_E(\mathbf{r}, t) = -\frac{1}{\varepsilon}\mathcal{P}(\mathbf{r}, t), \qquad (\mathbf{r}, t) \in \mathbb{R}^3 \times \mathbb{R} \tag{8.171}$$

where $\mathcal{P}(\mathbf{r}, t)$ is a *time-varying polarization vector* which has the same physical dimension (C/m^2) and meaning as the one we introduced in Section 3.7 for purely electrostatic fields.

Clearly, (8.171) does constitute an instance of the three-dimensional D'Alembert equation of Section 1.8. However, we need to verify whether defining $\mathcal{P}(\mathbf{r}, t)$ makes sense at all and, more importantly, what it represents from a physical viewpoint. To this purpose we observe that in light of (8.171) we may cast (8.169) and (8.170) alternatively as

$$\mathcal{J}(\mathbf{r}, t) = \frac{\partial}{\partial t}\mathcal{P}(\mathbf{r}, t), \qquad \varrho(\mathbf{r}, t) = -\nabla \cdot \mathcal{P}(\mathbf{r}, t) \tag{8.172}$$

whereby we see that

- $\mathcal{P}(\mathbf{r}, t)$ plays the role of volume density of time-varying *impressed electric dipole moments*,
- $\mathcal{J}(\mathbf{r}, t)$ can be regarded as an *impressed electric polarization current density*,
- $\varrho(\mathbf{r}, t)$ can be interpreted as an *impressed electric polarization charge density*.

For all these reasons, $\boldsymbol{\pi}_E(\mathbf{r}, t)$ is also called electric *polarization potential* [28, Section 1.11], [35, Section 8.2.1]. Although the conditions given in (8.172) are physically sound because they are compatible with the continuity equation (1.46) obeyed by $\mathcal{J}$ and ϱ, we should keep in mind that (8.172) is just an expedient which allows stating (8.171). In fact, the emphasis on the *impressed* character of $\mathcal{P}$, $\mathcal{J}$ and ϱ serves as a reminder that the latter are the *true* sources of the electromagnetic problem. Conversely, the effect of *bound* polarization charges possibly present in the underlying homogeneous isotropic lossless medium is already accounted for by the macroscopic parameter ε (see Sections 3.7 and 12.3.2).

We mention in passing that the constraints listed in (8.167) and (8.172) suffice to specify $\boldsymbol{\pi}_E$ and $\mathcal{P}$, respectively, only up to a solenoidal time-independent vector field. In other words, if we let

$$\boldsymbol{\pi}_E(\mathbf{r},t) = \boldsymbol{\pi}'_E(\mathbf{r},t) + \nabla \times \boldsymbol{\Xi}_E(\mathbf{r}) \tag{8.173}$$

$$\mathcal{P}(\mathbf{r},t) = \mathcal{P}'(\mathbf{r},t) - \varepsilon\nabla \times \Box^2\boldsymbol{\Xi}_E(\mathbf{r}) \tag{8.174}$$

with $\boldsymbol{\Xi}_E(\mathbf{r})$ being an arbitrary vector field function of position only, we see that $\boldsymbol{\pi}'_E(\mathbf{r},t)$ must obey the following D'Alembert equation

$$\nabla^2\boldsymbol{\pi}'_E(\mathbf{r},t) - \frac{1}{c^2}\frac{\partial^2}{\partial t^2}\boldsymbol{\pi}'_E(\mathbf{r},t) = -\frac{1}{\varepsilon}\mathcal{P}'(\mathbf{r},t), \qquad (\mathbf{r},t) \in \mathbb{R}^3 \times \mathbb{R} \tag{8.175}$$

which is formally the same as (8.171). Moreover, (8.167) holds true irrespective of whatever choice we make for $\boldsymbol{\Xi}_E(\mathbf{r})$. It is thus a matter of convenience to set $\boldsymbol{\Xi}_E(\mathbf{r}) = 0$, whereby we retrieve (8.171).

Perfectly dual equations and definitions hold true for the magnetic Hertzian potential $\boldsymbol{\pi}_M$ which is also known as the *Fitzgerald vector*. Indeed, with the help of (8.168) and the Schwarz theorem from (8.126) and (8.127) we find

$$\frac{\partial}{\partial t}\left[\nabla^2\boldsymbol{\pi}_M(\mathbf{r},t) - \frac{1}{c^2}\frac{\partial^2}{\partial t^2}\boldsymbol{\pi}_M(\mathbf{r},t)\right] = -\frac{1}{\mu}\mathcal{J}_M(\mathbf{r},t), \qquad (\mathbf{r},t) \in \mathbb{R}^3 \times \mathbb{R} \tag{8.176}$$

$$\nabla \cdot \left[\nabla^2\boldsymbol{\pi}_M(\mathbf{r},t) - \frac{1}{c^2}\frac{\partial^2}{\partial t^2}\boldsymbol{\pi}_M(\mathbf{r},t)\right] = \frac{1}{\mu}\varrho_M(\mathbf{r},t), \qquad (\mathbf{r},t) \in \mathbb{R}^3 \times \mathbb{R} \tag{8.177}$$

whereby we are led to require

$$\nabla^2\boldsymbol{\pi}_M(\mathbf{r},t) - \frac{1}{c^2}\frac{\partial^2}{\partial t^2}\boldsymbol{\pi}_M(\mathbf{r},t) = -\mathcal{M}(\mathbf{r},t), \qquad (\mathbf{r},t) \in \mathbb{R}^3 \times \mathbb{R} \tag{8.178}$$

with $\mathcal{M}(\mathbf{r},t)$ denoting a *time-varying magnetization vector* which has the same physical dimension (A/m) and meaning as the one introduced in Section 5.6 for stationary magnetic fields and steady currents. By inserting (8.178) into (8.176) and (8.177) we get[1]

$$\mathcal{J}_M(\mathbf{r},t) = \mu\frac{\partial}{\partial t}\mathcal{M}(\mathbf{r},t), \qquad \varrho_M(\mathbf{r},t) = -\mu\nabla \cdot \mathcal{M}(\mathbf{r},t) \tag{8.179}$$

whence we conclude that

- $\mathcal{M}(\mathbf{r},t)$ plays the role of volume density of time-varying *impressed magnetic dipole moments*,
- $\mathcal{J}_M(\mathbf{r},t)$ can be interpreted as an *impressed magnetization current density*,
- $\varrho_M(\mathbf{r},t)$ can be considered a density of *impressed magnetization charges*.

Besides, (8.179) is compatible with the conservation of magnetic charge (6.187). The effect of *bound* magnetization charges possibly occurring in the background homogeneous isotropic lossless medium is already included in the permeability μ (see Section 5.6).

[1]Since $\mathcal{H}(\mathbf{r},t)$ is a secondary entity but $\mathcal{E}(\mathbf{r},t)$ is a primary one, μ appears explicitly in combination with $\mathcal{M}(\mathbf{r},t)$ in (8.179) whereas in (8.172) the dimensional effect of ε is 'concealed' in $\mathcal{P}(\mathbf{r},t)$, so to speak.

In the context of the duality principle (Section 6.7) the Hertz vectors $\boldsymbol{\pi}_E$ and $\boldsymbol{\pi}_M$ are the dual of one another. This statement follows from the very definitions (8.165) and (8.166) because $\mathcal{A}_E$ and $\mathcal{A}_M$ in turn constitute a dual pair themselves (Section 8.2 and Table 8.1).

Lastly, to express the electromagnetic entities in terms of $\boldsymbol{\pi}_E$ and $\boldsymbol{\pi}_M$ we just make use of (8.167) and (8.168) in (8.129)-(8.132). In symbols, we have

$$\mathcal{E}(\mathbf{r},t) = \nabla\nabla\cdot\boldsymbol{\pi}_E(\mathbf{r},t) - \frac{1}{c^2}\frac{\partial^2}{\partial t^2}\boldsymbol{\pi}_E(\mathbf{r},t) - \mu\nabla\times\frac{\partial}{\partial t}\boldsymbol{\pi}_M(\mathbf{r},t) \tag{8.180}$$

$$\mathcal{H}(\mathbf{r},t) = \varepsilon\nabla\times\frac{\partial}{\partial t}\boldsymbol{\pi}_E(\mathbf{r},t) - \frac{1}{c^2}\frac{\partial^2}{\partial t^2}\boldsymbol{\pi}_M(\mathbf{r},t) + \nabla\nabla\cdot\boldsymbol{\pi}_M(\mathbf{r},t) \tag{8.181}$$

$$\mathcal{D}(\mathbf{r},t) = \varepsilon\nabla\nabla\cdot\boldsymbol{\pi}_E(\mathbf{r},t) - \frac{\varepsilon}{c^2}\frac{\partial^2}{\partial t^2}\boldsymbol{\pi}_E(\mathbf{r},t) - \frac{1}{c^2}\nabla\times\frac{\partial}{\partial t}\boldsymbol{\pi}_M(\mathbf{r},t) \tag{8.182}$$

$$\mathcal{B}(\mathbf{r},t) = \frac{1}{c^2}\nabla\times\frac{\partial}{\partial t}\boldsymbol{\pi}_E(\mathbf{r},t) - \frac{\mu}{c^2}\frac{\partial^2}{\partial t^2}\boldsymbol{\pi}_M(\mathbf{r},t) + \mu\nabla\nabla\cdot\boldsymbol{\pi}_M(\mathbf{r},t) \tag{8.183}$$

for $(\mathbf{r},t) \in \mathbb{R}^3 \times \mathbb{R}$. The expressions above can be further manipulated to arrive at true Helmholtz decompositions of the field entities. This step takes advantage of the differential identity (H.59) which allows writing

$$\begin{aligned}\nabla\nabla\cdot\boldsymbol{\pi}_E(\mathbf{r},t) - \frac{1}{c^2}\frac{\partial^2}{\partial t^2}\boldsymbol{\pi}_E(\mathbf{r},t) &= \nabla\times\nabla\times\boldsymbol{\pi}_E(\mathbf{r},t) + \nabla^2\boldsymbol{\pi}_E(\mathbf{r},t) - \frac{1}{c^2}\frac{\partial^2}{\partial t^2}\boldsymbol{\pi}_E(\mathbf{r},t)\\ &= \nabla\times\nabla\times\boldsymbol{\pi}_E(\mathbf{r},t) - \frac{1}{\varepsilon}\mathcal{P}(\mathbf{r},t)\end{aligned} \tag{8.184}$$

$$\begin{aligned}\nabla\nabla\cdot\boldsymbol{\pi}_M(\mathbf{r},t) - \frac{1}{c^2}\frac{\partial^2}{\partial t^2}\boldsymbol{\pi}_M(\mathbf{r},t) &= \nabla\times\nabla\times\boldsymbol{\pi}_M(\mathbf{r},t) + \nabla^2\boldsymbol{\pi}_M(\mathbf{r},t) - \frac{1}{c^2}\frac{\partial^2}{\partial t^2}\boldsymbol{\pi}_M(\mathbf{r},t)\\ &= \nabla\times\nabla\times\boldsymbol{\pi}_M(\mathbf{r},t) - \mathcal{M}(\mathbf{r},t)\end{aligned} \tag{8.185}$$

on account of the D'Alembert equations (8.171) and (8.178). Making use of these relationships back in (8.180)-(8.183) yields

$$\mathcal{E}(\mathbf{r},t) = \nabla\times\nabla\times\boldsymbol{\pi}_E(\mathbf{r},t) - \frac{1}{\varepsilon}\mathcal{P}(\mathbf{r},t) - \mu\nabla\times\frac{\partial}{\partial t}\boldsymbol{\pi}_M(\mathbf{r},t) \tag{8.186}$$

$$\mathcal{H}(\mathbf{r},t) = \varepsilon\nabla\times\frac{\partial}{\partial t}\boldsymbol{\pi}_E(\mathbf{r},t) - \mathcal{M}(\mathbf{r},t) + \nabla\times\nabla\times\boldsymbol{\pi}_M(\mathbf{r},t) \tag{8.187}$$

$$\mathcal{D}(\mathbf{r},t) = \varepsilon\nabla\times\nabla\times\boldsymbol{\pi}_E(\mathbf{r},t) - \mathcal{P}(\mathbf{r},t) - \frac{1}{c^2}\nabla\times\frac{\partial}{\partial t}\boldsymbol{\pi}_M(\mathbf{r},t) \tag{8.188}$$

$$\mathcal{B}(\mathbf{r},t) = \frac{1}{c^2}\nabla\times\frac{\partial}{\partial t}\boldsymbol{\pi}_E(\mathbf{r},t) - \mu\mathcal{M}(\mathbf{r},t) + \mu\nabla\times\nabla\times\boldsymbol{\pi}_M(\mathbf{r},t) \tag{8.189}$$

for $(\mathbf{r},t) \in \mathbb{R}^3 \times \mathbb{R}$. Clearly, all the contributions in the form of a curl or a double curl are solenoidal vector fields. What is more, (8.186)-(8.189) state that *all* the electromagnetic entities in a homogeneous isotropic medium are *divergence-free* in regions devoid of charges, in perfect agreement with the source-free instances of the Gauss laws (1.23) and (1.44). Then again, since the auxiliary fields $\mathcal{P}$ and $\mathcal{M}$ may not be solenoidal or lamellar per se, we may resort to the Helmholtz theorem in an unbounded space (8.28) to formally split $\mathcal{P}$ and $\mathcal{M}$ into their solenoidal and lamellar parts, as is suggested in (8.2). The entities of intensity may thus be written as

$$\mathcal{E}(\mathbf{r},t) = \underbrace{\nabla\times\nabla\times\boldsymbol{\pi}_E(\mathbf{r},t) - \frac{1}{\varepsilon}\mathcal{P}_S(\mathbf{r},t) - \mu\nabla\times\frac{\partial}{\partial t}\boldsymbol{\pi}_M(\mathbf{r},t)}_{\text{solenoidal part}} - \frac{1}{\varepsilon}\mathcal{P}_L(\mathbf{r},t) \tag{8.190}$$

$$\mathcal{B}(\mathbf{r},t) = \underbrace{\frac{1}{c^2}\nabla\times\frac{\partial}{\partial t}\boldsymbol{\pi}_E(\mathbf{r},t) - \mu\mathcal{M}_S(\mathbf{r},t) + \mu\nabla\times\nabla\times\boldsymbol{\pi}_M(\mathbf{r},t)}_{\text{solenoidal part}} - \mu\mathcal{M}_L(\mathbf{r},t) \tag{8.191}$$

and similar decompositions hold for $\mathcal{H}$ and $\mathcal{D}$. In conclusion, we have showed that a complete Helmholtz splitting of the electromagnetic entities can be achieved even when the Lorenz gauges (8.121) and (8.128) are imposed on the auxiliary potentials $\mathcal{A}_E$, Φ_E, $\mathcal{A}_M$ and Φ_M.

All the definitions and equations discussed so far have their counterparts for fields and sources in the time-harmonic regime (Section 1.5) [15,48]. In particular, the phasors of the Hertzian potentials are defined as

$$\boldsymbol{\Pi}_E(\mathbf{r}) := \frac{c^2}{\mathrm{j}\omega}\mathbf{A}_E(\mathbf{r}), \qquad \mathbf{r}\in\mathbb{R}^3 \tag{8.192}$$

$$\boldsymbol{\Pi}_M(\mathbf{r}) := \frac{c^2}{\mathrm{j}\omega}\mathbf{A}_M(\mathbf{r}), \qquad \mathbf{r}\in\mathbb{R}^3 \tag{8.193}$$

and the relevant expressions for the fields follow from (8.186)-(8.189) by invoking the transformation rules listed in Table 1.1. The wave equations (8.171) and (8.178) pass over into Helmholtz equations, the solution of which is discussed in Section 9.1.1.

8.5 The scalar Helmholtz equation

Second-order partial differential equations involving the Helmholtz operator (1.250) are routinely encountered in the mathematical formulation of problems that involve time-harmonic fields and potentials. In what follows we collect a few results which concern the scalar Helmholtz equation [49, Section 2.5], [16, Section 7.7], [43, Section 8.5] [50, Section II.7], [51]

$$\left(\nabla^2 + k^2\right)\Psi(\mathbf{r}) = -S(\mathbf{r}) \qquad \mathbf{r}\in V \tag{8.194}$$

namely,

(i) the fundamental solutions in polar spherical coordinates in the homogenous case, i.e., for $S(\mathbf{r}) = 0$;
(ii) an important theorem, due to F. Rellich, about the asymptotic behavior of the solutions $\Psi(\mathbf{r})$ outside a sphere;
(iii) an 'energy-balance' relationship analogous to the Poynting theorem in the frequency domain (1.314);
(iv) the question of the uniqueness of solutions in unbounded and bounded regions of space.

The rationale for this plan is that the solutions in spherical coordinates are used to prove the Rellich theorem, whereas the latter, in combination with the energy balance, is crucial to prove uniqueness in unbounded regions and, in fact, we already invoked the result in Section 6.4.2 for the time-harmonic Maxwell equations.

8.5.1 Polar spherical coordinates

With an eye to the steps taken in Section 3.5.1 for the Laplace equation we apply the method of the *separation of variables* [15,28,38,52,53], [54, Section 11.3], [46, Section 4.9] for the construction of solutions to the homogeneous Helmholtz equation

$$\nabla^2\Phi(\mathbf{r}) + k^2\Phi(\mathbf{r}) = 0, \qquad \mathbf{r}\in V\subseteq\mathbb{R}^3 \tag{8.195}$$

in regions of space which exhibit spherical symmetry. To this purpose we express the Helmholtz operator in a system of spherical coordinates (r, ϑ, φ) (see Figure A.2) and write the scalar potential $\Phi(\mathbf{r})$ as a series of known functions as we did in (3.69). The procedure for determining the set $\Psi_\nu(r, \vartheta, \varphi)$ differs from that for the Laplace operator only for the equations we obtain by applying the general principle of the separation of variables.

The starting point is the Helmholtz equation in spherical coordinates, viz.,

$$\frac{1}{r^2}\frac{\partial}{\partial r}\left(r^2\frac{\partial \Psi}{\partial r}\right) + \frac{1}{r^2 \sin\vartheta}\frac{\partial}{\partial \vartheta}\left(\sin\vartheta\frac{\partial \Psi}{\partial \vartheta}\right) + \frac{1}{r^2 \sin^2\vartheta}\frac{\partial^2 \Psi}{\partial \varphi^2} + k^2\Psi(r, \vartheta, \varphi) = 0 \tag{8.196}$$

to be solved in the region V. For instance, V may be the inside of a cone with vertex the origin or a wedge whose edge coincides with the z-axis. By inserting the Ansatz (3.71) into (8.196) and performing the same manipulations carried out in Section 3.5.1 we quickly arrive at the following three uncoupled equations [52]

$$\frac{\mathrm{d}^2 F_m}{\mathrm{d}\varphi^2} + m^2 F_m = 0 \tag{8.197}$$

$$\frac{1}{\sin\vartheta}\frac{\mathrm{d}}{\mathrm{d}\vartheta}\left(\sin\vartheta\frac{\mathrm{d}\Theta_{mn}}{\mathrm{d}\vartheta}\right) + \left[n(n+1) - \frac{m^2}{\sin^2\vartheta}\right]\Theta_{mn} = 0 \tag{8.198}$$

$$\frac{1}{r^2}\frac{\mathrm{d}}{\mathrm{d}r}\left(r^2\frac{\mathrm{d}R_n}{\mathrm{d}r}\right) + \left[k^2 - \frac{n(n+1)}{r^2}\right]R_n = 0 \tag{8.199}$$

where the separation constants m^2 and $n(n+1)$ are not necessarily whole numbers.

Any product of the type (3.71) formed with the solutions to (8.197)-(8.199) may be employed to expand general solutions to the Helmholtz equation in accordance with (3.69). As compared to the separation argument applied to the Laplace equation, only the equation for $R(r)$ is different, though (8.199) evidently reduces to (3.80) when $k = 0$. Therefore, in the interest of conciseness we refer to Section 3.5.1 for the details about the solution of (8.197) and (8.198) and focus instead on (8.199) which is new.

The resemblance between (8.199) and the Bessel equation (H.136) suggests that the solutions may be written in terms of Bessel functions of suitable order. In fact, (8.199) is sometimes called the *spherical Bessel equation* [40, Chapter 10]. To proceed we make a change of variables, viz., [15,38,53], [50, Section II.7]

$$\xi = kr, \qquad R_n(r) = \frac{B_n(\xi)}{\sqrt{\xi}} \tag{8.200}$$

and compute the required derivatives

$$r^2\frac{\mathrm{d}R_n}{\mathrm{d}r} = \frac{\xi^2}{k}\frac{\mathrm{d}R_n}{\mathrm{d}\xi} = \frac{\xi^{\frac{3}{2}}}{k}\frac{\mathrm{d}B_n}{\mathrm{d}\xi} - \frac{\xi^{\frac{1}{2}}}{2k}B_n(\xi) \tag{8.201}$$

$$\begin{aligned}\frac{1}{r^2}\frac{\mathrm{d}}{\mathrm{d}r}\left(r^2\frac{\mathrm{d}R_n}{\mathrm{d}r}\right) &= \frac{k^2}{\xi^2}\frac{\mathrm{d}}{\mathrm{d}\xi}\left(\xi^2\frac{\mathrm{d}R_n}{\mathrm{d}\xi}\right) = \frac{k^2}{\xi^{\frac{1}{2}}}\left[\frac{\mathrm{d}^2 B_n}{\mathrm{d}\xi^2} + \frac{1}{\xi}\frac{\mathrm{d}B_n}{\mathrm{d}\xi} - \frac{B_n(\xi)}{4\xi^2}\right] \\ &= \frac{k^2}{\xi^{\frac{1}{2}}}\left[\frac{1}{\xi}\frac{\mathrm{d}B_n}{\mathrm{d}\xi}\left(\xi\frac{\mathrm{d}B_n}{\mathrm{d}\xi}\right) - \frac{B_n(\xi)}{4\xi^2}\right].\end{aligned} \tag{8.202}$$

Substituting into (8.199) and rearranging yields

$$\frac{1}{\xi}\frac{\mathrm{d}B_n}{\mathrm{d}\xi}\left(\xi\frac{\mathrm{d}B_n}{\mathrm{d}\xi}\right)+\left[1-\frac{1}{\xi^2}\left(n+\frac{1}{2}\right)^2\right]B_n(\xi)=0 \tag{8.203}$$

which we recognize as the Bessel equation in standard form (H.136) where the index is $n+1/2$ [55]. Hence, the fundamental solutions to (8.203) — called *Bessel functions of fractional order* — are $J_{n+1/2}(\xi)$ and $Y_{n+1/2}(\xi)$. In light of (8.200) it is customary to define

$$j_n(\xi) := \sqrt{\frac{\pi}{2\xi}}J_{n+1/2}(\xi) \tag{8.204}$$

$$y_n(\xi) := \sqrt{\frac{\pi}{2\xi}}Y_{n+1/2}(\xi) \tag{8.205}$$

$$h_n^{(1)}(\xi) := \sqrt{\frac{\pi}{2\xi}}H_{n+1/2}^{(1)}(\xi) = j_n(\xi)+\mathrm{j}\,y_n(\xi) \tag{8.206}$$

$$h_n^{(2)}(\xi) := \sqrt{\frac{\pi}{2\xi}}H_{n+1/2}^{(2)}(\xi) = j_n(\xi)-\mathrm{j}\,y_n(\xi) \tag{8.207}$$

where the first two are known as *spherical Bessel functions* of the *first* and *second kind*, whereas the second two are called *spherical Hankel functions* of the *first* and *second kind.* Some Authors (e.g., [53], [56, pp. 51–52]) make different choices for the normalization constant in (8.204)-(8.207). It is remarkable that $j_n(\xi)$, $y_n(\xi)$ and combinations thereof admit closed-form expressions in terms of trigonometric functions [40, Formulas 10.1.11 and 10.1.12].

Clearly, any combination of the type

$$R_n(r)=\begin{cases} c_{n1}j_n(kr)+c_{n2}y_n(kr)\\ c_{n1}h_n^{(1)}(kr)+c_{n2}h_n^{(2)}(kr)\\ c_{n1}j_n(kr)+c_{n2}h_n^{(2)}(kr)\end{cases} \tag{8.208}$$

solves (8.199). As an example, if the region V is a spherical shell centered in the origin, then either $c_{n1}j_n(kr)+c_{n2}y_n(kr)$ or $c_{n1}h_n^{(1)}(kr)+c_{n2}h_n^{(2)}(kr)$ are well-suited. If V is the inside of a sphere, then only $j_n(kr)$ must be used, unless a singular point source is located in the origin, because $y_n(kr)$ exhibits an algebraic singularity for $r=0$. If wave-like solutions are sought outside a sphere, then $h_n^{(2)}(kr)$ must be chosen in that they correctly represent *outward-going* waves in the time-harmonic regime (cf. Figure 7.23). In particular, from (3.69) the special case for $n=0=m$

$$\Phi(\mathbf{r})=\Phi_{00}h_0^{(2)}(kr)=\mathrm{j}\,\frac{\mathrm{e}^{-\mathrm{j}kr}}{kr}\Phi_{00}, \qquad r>0 \tag{8.209}$$

provides the basis for the derivation of the three-dimensional Green function in the whole space (Section 8.7).

8.5.2 *The Rellich theorem*

Before we can prove uniqueness for (8.194) we need to derive a general result — first obtained by F. Rellich in 1943 [57, Section 9.1], [58, Section 2.5], [58, 59] — that concerns the asymptotic behavior of the solutions to the scalar Helmholtz equation. To be specific, if a *non-zero* single-valued twice-differentiable scalar field $\Upsilon(\mathbf{r})$ solves

$$(\nabla^2+k^2)\Upsilon(\mathbf{r})=0, \qquad \mathbf{r}\in\mathbb{R}^3 \tag{8.210}$$

with the real wavenumber $k > 0$, and if in addition there are *no sources* outside the ball $B[0, a]$, then it holds

$$\int_{B(0,a_1)\setminus B(0,a)} \mathrm{d}V\, |\Upsilon(\mathbf{r})|^2 \geqslant M a_1 \tag{8.211}$$

where $B(0, a_1)$ is a larger ball with radius $a_1 > a$, and M is a suitable real positive constant (cf. Figure 6.16). The integration is carried out over the spherical shell $B(0, a_1) \setminus B(0, a)$.

The proof of (8.211) is hinged on the representation of the scalar field $\Upsilon(\mathbf{r})$ as a superposition of fundamental solutions of the Helmholtz equation in polar spherical coordinates (r, ϑ, φ) (Section 8.5.1) for $r > a$, namely,

$$\Upsilon(\mathbf{r}) = \sum_{n=0}^{\infty} \sum_{m=-n}^{n} f_{mn}(r) Y_{mn}(\vartheta, \varphi) = \sum_{n=0}^{\infty} \sum_{m=-n}^{n} \left[c_{mn} h_n^{(1)}(kr) + d_{mn} h_n^{(2)}(kr) \right] Y_{mn}(\vartheta, \varphi) \tag{8.212}$$

where $h_n^{(1)}(kr)$ and $h_n^{(2)}(kr)$ are the spherical Hankel functions introduced in (8.206) and (8.207), and $Y_{mn}(\vartheta, \varphi)$ are the spherical harmonics defined in (3.102). The expansion 'coefficients' $f_{mn}(r)$ are related to $\Upsilon(\mathbf{r})$ and $Y_{mn}(\vartheta, \varphi)$ as follows

$$f_{m'n'}(r) = \int_{4\pi} \mathrm{d}\Omega\, Y^*_{m'n'}(\vartheta, \varphi) \Upsilon(\mathbf{r}), \qquad n' \in \mathbb{N}, \qquad m' \leqslant |n'| \tag{8.213}$$

thanks to the orthogonality relationship (3.103).

In particular, for large values of the radial coordinate, say, $r > a$, we have the asymptotic behavior

$$c_{mn} h_n^{(1)}(kr) + d_{mn} h_n^{(2)}(kr) = \mathrm{e}^{\mathrm{j}kr - \mathrm{j}(n+1)\frac{\pi}{2}} \frac{c_{mn}}{r} + \mathrm{e}^{-\mathrm{j}kr + \mathrm{j}(n+1)\frac{\pi}{2}} \frac{d_{mn}}{r} + \mathcal{O}\left(\frac{1}{r^2}\right) \tag{8.214}$$

and

$$\left| c_{mn} h_n^{(1)}(kr) + d_{mn} h_n^{(2)}(kr) \right|^2 = \frac{1}{r^2} \left[|c_{mn}|^2 + |d_{mn}|^2 + 2(-1)^{n+1} \mathrm{Re}\left\{ c^*_{mn} d_{mn} \mathrm{e}^{-2\mathrm{j}kr} \right\} + \mathcal{O}\left(\frac{1}{r}\right) \right] \tag{8.215}$$

in light of the asymptotic expansions (H.162) and (H.163). Then, by integrating $|\Upsilon(\mathbf{r})|^2$ over a sphere of radius $r > a$ and using the Bessel inequality (D.115) for the integral in (ϑ, φ) over the solid angle 4π, we find

$$\begin{aligned} r^2 \int_{4\pi} \mathrm{d}\Omega\, |\Upsilon(\mathbf{r})|^2 &\geqslant r^2 \sum_{n=0}^{\infty} \sum_{m=-n}^{n} |f_{mn}(r)|^2 \\ &= \sum_{n=0}^{\infty} \sum_{m=-n}^{n} \left[|c_{mn}|^2 + |d_{mn}|^2 + 2(-1)^{n+1} \mathrm{Re}\{ c^*_{mn} d_{mn} \mathrm{e}^{-2\mathrm{j}kr} \} \right] + \mathcal{O}\left(\frac{1}{r}\right) \end{aligned} \tag{8.216}$$

by virtue of (8.215). Finally, if we integrate $|\Upsilon(\mathbf{r})|^2$ over the spherical shell $B(0, a_1) \setminus B(0, a)$, on account of (8.216) we obtain

$$\int_{B(0,a_1)\setminus B(0,a)} \mathrm{d}V\, |\Upsilon(\mathbf{r})|^2 = \int_a^{a_1} \mathrm{d}r\, r^2 \int_{4\pi} \mathrm{d}\Omega\, |\Upsilon(\mathbf{r})|^2$$

$$\geqslant \int_a^{a_1} \mathrm{d}r \sum_{n=0}^{\infty} \sum_{m=-n}^{n} \left[|c_{mn}|^2 + |d_{mn}|^2 + 2(-1)^{n+1} \mathrm{Re}\{c_{mn}^* d_{mn} \mathrm{e}^{-2\mathrm{j}kr}\} \right] + \int_a^{a_1} \mathrm{d}r\, \mathcal{O}\left(\frac{1}{r}\right)$$
$$= \sum_{n=0}^{\infty} \sum_{m=-n}^{n} \left[(a_1 - a)\left(|c_{mn}|^2 + |d_{mn}|^2\right) - (-1)^n \mathrm{Re}\left\{ c_{mn}^* d_{mn} \frac{\mathrm{e}^{-2\mathrm{j}ka} - \mathrm{e}^{-2\mathrm{j}ka_1}}{\mathrm{j}k} \right\} \right] + \mathcal{O}(\log a_1) \tag{8.217}$$

where the logarithmic term is dominated by the first, because *at least one* of the expansion coefficients c_{mn}, d_{mn} must be non-null, otherwise $\Upsilon(\mathbf{r}) = 0$ contrary to our assumption, and this ends the proof of (8.211).

8.5.3 Conservation of 'energy'

Since the complex-valued scalar field $\Psi(\mathbf{r})$ may physically represent diverse quantities, e.g., the Cartesian components of the time-harmonic magnetic vector potential $\mathbf{A}_E(\mathbf{r};\omega)$, at this stage we do not have an understanding of what 'energy' means. Nevertheless, since it is reasonable to associate the idea of energy with the volume integrals of positive 'densities' such as $|\Psi(\mathbf{r})|^2$ and $|\nabla\Psi(\mathbf{r})|^2$, we can try to manipulate (8.194) to arrive at a balance equation. To gain a bit more generality we consider an excluded domain V_1 and a large surface S so that V is the surface-wise multiply-connected region of space bounded by S and ∂V_1 (cf. Figure 6.10a). Besides, the source $S(\mathbf{r})$ is confined to a region $V_S \subset V$. Lastly, we assume that the wavenumber k, given by (1.248), is possibly complex and written as $k' - \mathrm{j}k''$ with $k'' > 0$ in keeping with our choices in Chapter 7.

Now, if we multiply (8.194) by $\mathrm{j}k^*\Psi^*(\mathbf{r})$ and perform a little algebra we get

$$-\mathrm{j}k^*\Psi^*(\mathbf{r})S(\mathbf{r}) = \mathrm{j}k^*\Psi^*(\mathbf{r})\nabla^2\Psi(\mathbf{r}) + \mathrm{j}k^*\Psi^*(\mathbf{r})k^2\Psi(\mathbf{r})$$
$$= \nabla\cdot[\mathrm{j}k^*\Psi^*(\mathbf{r})\nabla\Psi(\mathbf{r})] - \mathrm{j}k^*|\nabla\Psi(\mathbf{r})|^2 + \mathrm{j}k|k|^2|\Psi(\mathbf{r})|^2 \tag{8.218}$$

whence by taking the real part of both sides and dividing through by two we find

$$\nabla\cdot\left[\mathrm{Re}\left\{\frac{1}{2}\mathrm{j}k^*\Psi^*(\mathbf{r})\nabla\Psi(\mathbf{r})\right\}\right] + \frac{k''}{2}\left[|k|^2|\Psi(\mathbf{r})|^2 + |\nabla\Psi(\mathbf{r})|^2\right]$$
$$= -\frac{1}{2}\mathrm{Re}\{\mathrm{j}k^*\Psi^*(\mathbf{r})S(\mathbf{r})\}, \qquad \mathbf{r}\in V \tag{8.219}$$

which we regard as the local form of the desired conservation theorem for 'energy', since it has the structure of a continuity equation with a source term [cf. (1.316)].

To interpret the result we let us be guided by the Poynting theorem in the frequency domain. Accordingly, the first term in (8.219) is the average volume density of 'power' flow at $\mathbf{r}$, and the vector field

$$\mathbf{S}_\Psi(\mathbf{r}) := \frac{1}{2}\mathrm{j}k^*\Psi^*(\mathbf{r})\nabla\Psi(\mathbf{r}) \tag{8.220}$$

may be considered the analogue of the complex Poynting vector (1.304). Of course, this definition is not unique, as any divergence-free vector field, say, $\nabla\times\mathbf{V}(\mathbf{r})$, can be added to the right-hand side of (8.220) without altering (8.219). The second term — always non-negative — constitutes the average volume density of 'power' dissipated into the background medium and, in particular, vanishes when k is real. Lastly, the right-hand side of (8.219) represents the average volume density of 'power' delivered by the source $S(\mathbf{r})$.

Integrating (8.219) over V and invoking the Gauss theorem (A.53) yields the global form of the balance equation, i.e.,

$$\frac{1}{2}\mathrm{Re}\int_{S\cup\partial V_1} dS\; \mathrm{j}k^*\Psi^*(\mathbf{r})\hat{\mathbf{n}}(\mathbf{r})\cdot\nabla\Psi(\mathbf{r}) + \frac{k''}{2}\int_V dV\,\left[|k|^2|\Psi(\mathbf{r})|^2 + |\nabla\Psi(\mathbf{r})|^2\right]$$
$$= -\frac{1}{2}\mathrm{Re}\int_{V_S} dV\; \mathrm{j}k^*\Psi^*(\mathbf{r})S(\mathbf{r}) \quad (8.221)$$

where the first term may represent the average flow of 'power' through the boundary $S \cup \partial V_1$. Using the Gauss theorem is possible (see discussion on page 10 and ff.) as long as the vector field $\Psi^*(\mathbf{r})\nabla\Psi(\mathbf{r})$ is continuously differentiable in $V \setminus \overline{V}_S$ and V_S, and $\Psi^*(\mathbf{r})\hat{\mathbf{n}}(\mathbf{r})\cdot\nabla\Psi(\mathbf{r})$ is continuous across ∂V_S.

8.5.4 *Uniqueness in unbounded regions*

We wish to determine the necessary conditions in order for (8.194) to have a unique solution in an unbounded region of space $\mathbb{R}^3 \setminus \overline{V}_1$ [60, Section 3.5.2], [51]. The excluded finite region V_1 has smooth boundary and may contain materials as yet unspecified, whereas the known source $S(\mathbf{r})$ is confined to a finite volume $V_S \subset \mathbb{R}^3 \setminus \overline{V}_1$. Thanks to (8.208) and (H.163) we do expect the existence of wave-like solutions which obey the asymptotic boundary condition

$$\Psi(\mathbf{r}) = \mathcal{O}\left(\frac{1}{|\mathbf{r}|}\right), \qquad |\mathbf{r}| \to +\infty \qquad (8.222)$$

uniformly with respect to $\mathbf{r}$. However, we shall find out further on that (8.222) by itself is not sufficient to ensure uniqueness in the absence of losses. Thus, we examine the lossy case first.

As is customary in the proofs by contradiction, we presume that two solutions $\Psi_1(\mathbf{r})$ and $\Psi_2(\mathbf{r})$ are actually possible and define the difference scalar field $\Psi_0(\mathbf{r}) := \Psi_1(\mathbf{r}) - \Psi_2(\mathbf{r})$ which obeys the homogeneous Helmholtz equation

$$(\nabla^2 + k^2)\Psi_0(\mathbf{r}) = 0, \qquad \mathbf{r} \in \mathbb{R}^3 \setminus \overline{V}_1 \qquad (8.223)$$

by virtue of the linearity of the operator and the principle of superposition (Section 6.1). We introduce two concentric balls $B(0,a)$ and $B(0,a_1)$ with $a < a_1$ and a large enough to contain V_S and V_1 (see Figure 6.16 for the same geometrical setup).

We begin by applying the balance equation (8.221) to the region $V_{a_1} := B(0,a_1) \setminus \overline{V}_1$, namely,

$$\frac{1}{2}\mathrm{Re}\int_{|\mathbf{r}|=a_1} dS\; \mathrm{j}k^*\Psi_0^*(\mathbf{r})\frac{\partial\Psi_0}{\partial r} + \frac{k''}{2}\int_{V_{a_1}} dV\,\left[|k|^2|\Psi_0(\mathbf{r})|^2 + |\nabla\Psi_0(\mathbf{r})|^2\right]$$
$$= -\frac{1}{2}\mathrm{Re}\int_{\partial V_1} dS\; \mathrm{j}k^*\Psi_0^*(\mathbf{r})\frac{\partial\Psi_0}{\partial\hat{\mathbf{n}}} \leqslant 0 \quad (8.224)$$

with the unit normal on $\partial B_1 := \{\mathbf{r} \in \mathbb{R}^3 : |\mathbf{r}| = a_1\}$ and ∂V_1 being oriented outwards V_{a_1}. The minus sign in the right member signals that 'power' *leaves* V_1 and *enters* V_{a_1} on average. Still, since V_1 is devoid of sources, if an energy flux across ∂V_1 occurs at all, then 'power' must *enter* V_1 and, consequently, we deduce that the right-hand side is a non-positive quantity, as indicated.

In the limit as $a_1 \to +\infty$, V_{a_1} passes over into $\mathbb{R}^3\setminus\overline{V}_1$, i.e., the unbounded region of space external to V_1. Nonetheless, as was the case for (6.158) it appears that (8.222) cannot ensure the convergence of the integral over V_{a_1} as the radius a_1 grows infinitely large. Then, we proceed by applying the conservation law (8.221) again, this time for points inside the spherical shell $B(0, a_1) \setminus B(0, a)$, viz.,

$$\frac{1}{2}\mathrm{Re} \int_{|\mathbf{r}|=a_1} \mathrm{d}S \; \mathrm{j} k^* \Psi_0^*(\mathbf{r}) \frac{\partial \Psi_0}{\partial r} + \int_{B(0,a_1)\setminus B(0,a)} \mathrm{d}V \frac{k''}{2} \left[|k|^2 |\Psi_0(\mathbf{r})|^2 + |\nabla \Psi_0(\mathbf{r})|^2 \right] = \frac{1}{2}\mathrm{Re} \int_{|\mathbf{r}|=a} \mathrm{d}S \; \mathrm{j} k^* \Psi_0^*(\mathbf{r}) \frac{\partial \Psi_0}{\partial r} \quad (8.225)$$

where the unit normals on ∂B_1 and $\partial B := \{\mathbf{r} \in \mathbb{R}^3 : |\mathbf{r}| = a\}$ point inwards $\mathbb{R}^3 \setminus B(0, a_1)$ and $B(0, a_1) \setminus B(0, a)$, respectively. With these choices, the flux integrals represent the average outward-flowing 'powers' through ∂B_1 and ∂B. Since the underlying medium is lossy and $k'' > 0$ by hypothesis, (8.225) implies

$$\frac{1}{2}\mathrm{Re} \int_{|\mathbf{r}|=a_1} \mathrm{d}S \; \mathrm{j} k^* \Psi_0^*(\mathbf{r}) \frac{\partial \Psi_0}{\partial r} < \frac{1}{2}\mathrm{Re} \int_{|\mathbf{r}|=a} \mathrm{d}S \; \mathrm{j} k^* \Psi_0^*(\mathbf{r}) \frac{\partial \Psi_0}{\partial r}, \qquad a_1 > a \quad (8.226)$$

whence we infer that the average 'power' is a strictly decreasing function of the radial coordinate $r > a_1 > a$, because we can repeat the reasoning *ad libitum* for concentric balls of increasingly larger radii. Moreover, since there are no sources in $\mathbb{R}^3 \setminus B[0, a_1]$ the 'power' flow across ∂B_1 is non-negative and we obtain

$$0 \leqslant \lim_{a_1 \to +\infty} \frac{1}{2}\mathrm{Re} \int_{|\mathbf{r}|=a_1} \mathrm{d}S \; \mathrm{j} k^* \Psi_0^*(\mathbf{r}) \frac{\partial \Psi_0}{\partial r} = C_\infty < \frac{1}{2}\mathrm{Re} \int_{|\mathbf{r}|=a} \mathrm{d}S \; \mathrm{j} k^* \Psi_0^*(\mathbf{r}) \frac{\partial \Psi_0}{\partial r} \quad (8.227)$$

where C_∞ is a finite positive constant denoting the 'power' that reaches infinity.

Next, we solve (8.225) formally with respect to the average 'power' dissipated within the shell, that is,

$$\begin{aligned}
&\int_{B(0,a_1)\setminus B(0,a)} \mathrm{d}V \frac{k''}{2} \left[|k|^2 |\Psi_0(\mathbf{r})|^2 + |\nabla \Psi_0(\mathbf{r})|^2 \right] = \\
&\quad = \frac{1}{2}\mathrm{Re} \int_{|\mathbf{r}|=a} \mathrm{d}S \; \mathrm{j} k^* \Psi_0^*(\mathbf{r}) \frac{\partial \Psi_0}{\partial r} - \frac{1}{2}\mathrm{Re} \int_{|\mathbf{r}|=a_1} \mathrm{d}S \; \mathrm{j} k^* \Psi_0^*(\mathbf{r}) \frac{\partial \Psi_0}{\partial r} \\
&\quad \leqslant \frac{1}{2} \int_{|\mathbf{r}|=a} \mathrm{d}S \; \mathrm{Re} \left\{ \mathrm{j} k^* \Psi_0^*(\mathbf{r}) \frac{\partial \Psi_0}{\partial r} \right\} \leqslant \frac{1}{2} \int_{|\mathbf{r}|=a} \mathrm{d}S \; |k| |\Psi_0(\mathbf{r})| \left| \frac{\partial \Psi_0}{\partial r} \right| \\
&\quad \leqslant \frac{1}{2} \int_{|\mathbf{r}|=a} \mathrm{d}S \; \frac{C_\Psi^2}{a^2} = 2\pi C_\Psi^2
\end{aligned} \quad (8.228)$$

where we have supposed that a is so large that, given a suitable positive constant C_Ψ, we have

$$|k| |\Psi_0(\mathbf{r})| \leqslant \frac{C_\Psi}{a}, \qquad \left| \frac{\partial \Psi_0}{\partial r} \right| \leqslant \frac{C_\Psi}{a} \quad (8.229)$$

in light of (8.222). Notice that there is no reason to support a faster decay of the radial derivative of $\Psi_0(\mathbf{r})$. At any rate, in the limit as $a_1 \to +\infty$ we find

$$\lim_{a_1 \to +\infty} \int_{B(0,a_1)\setminus B(0,a)} \mathrm{d}V \frac{k''}{2} \left[|k|^2 |\Psi_0(\mathbf{r})|^2 + |\nabla \Psi_0(\mathbf{r})|^2 \right] < 2\pi C_\Psi^2 \quad (8.230)$$

because the average 'power' lost in the shell is bounded by a positive constant independent of a_1. With this result we conclude that the integral over V_{a_1} in (8.224) remains *finite* as a_1 grows infinitely large. It is sufficient to separate the integral into two parts, i.e., over the shell $B(0, a_1) \setminus B(0, a)$ and the domain $B(0, a) \setminus \overline{V}_1$.

Finally, from (8.224), (8.227) and (8.230) we obtain

$$C_\infty + \frac{k''}{2} \int\limits_{\mathbb{R}^3 \setminus \overline{V}_1} \mathrm{d}V \left[|k|^2 |\Psi_0(\mathbf{r})|^2 + |\nabla \Psi_0(\mathbf{r})|^2 \right] \leqslant 0 \tag{8.231}$$

and this equation necessarily requires

$$C_\infty = 0 = \frac{k''}{2} \int\limits_{\mathbb{R}^3 \setminus \overline{V}_1} \mathrm{d}V \left[|k|^2 |\Psi_0(\mathbf{r})|^2 + |\nabla \Psi_0(\mathbf{r})|^2 \right] \tag{8.232}$$

because both C_∞ and the average lost 'power' are non-negative quantities. As a result, we conclude that $\Psi_0(\mathbf{r}) = 0$ for $\mathbf{r} \in \mathbb{R}^3 \setminus \overline{V}_1$. And this shows that, contrary to our hypothesis, the solution is indeed unique if the wavenumber k is complex.

The previous proof apparently breaks down when the underlying medium is lossless and k is real, though the solution to (8.194) can be made unique by postulating an additional suitable boundary condition at infinity. Evidently, if a solution has been found for a given $k > 0$, the scalar field obtained by replacing k with $-k < 0$ still solves (8.194), which, in effect, is not altered by the substitution $k \leftrightarrow -k$. We associate *outward-going* wave-like solutions with $k > 0$ and *inward-going* ones with a negative wavenumber. Since (8.222) cannot help us discriminate between these two types of waves, we shall determine the extra requirement constructively in the course of the proof by contradiction.

First of all, with reference to the same geometrical setup used above for the lossy case (cf. Figure 6.16) we apply the balance equation (8.221) to the difference scalar field $\Psi_0(\mathbf{r})$ for points $\mathbf{r} \in V_{a_1}$, viz.,

$$\operatorname{Re} \int\limits_{|\mathbf{r}|=a_1} \mathrm{d}S \; \mathrm{j} k \Psi_0^*(\mathbf{r}) \frac{\partial \Psi_0}{\partial r} = -\operatorname{Re} \int\limits_{\partial V_1} \mathrm{d}S \; \mathrm{j} k \Psi_0^*(\mathbf{r}) \frac{\partial \Psi_0}{\partial \hat{\mathbf{n}}} \leqslant 0 \tag{8.233}$$

with the unit normals on ∂B_1 and ∂V_1 positively oriented outwards V_{a_1}. As already observed, the flux integral over ∂V_1 — if it does not vanish — cannot but represent and average efflux of 'power' into V_1, and hence the right member is a non-positive quantity. More precisely, we may distinguish three cases depending upon the boundary conditions imposed for $\mathbf{r} \in \partial V_1$.

(1) If $\Psi(\mathbf{r})$ is assigned on ∂V_1 (Dirichlet boundary condition), then $\Psi_0(\mathbf{r}) = 0$, and the relevant integral vanishes.

(2) If $\hat{\mathbf{n}}(\mathbf{r}) \cdot \nabla \Psi(\mathbf{r})$ is prescribed on ∂V_1 (Neumann boundary condition), then $\hat{\mathbf{n}}(\mathbf{r}) \cdot \nabla \Psi_0(\mathbf{r}) = 0$, and again the flux integral vanishes.

(3) If a combination of $\Psi(\mathbf{r})$ and $\hat{\mathbf{n}}(\mathbf{r}) \cdot \nabla \Psi(\mathbf{r})$ is given on ∂V_1, say,

$$\alpha \Psi(\mathbf{r}) + \beta \hat{\mathbf{n}}(\mathbf{r}) \cdot \nabla \Psi(\mathbf{r}) = f(\mathbf{r}), \qquad \mathbf{r} \in \partial V_1, \qquad \alpha, \beta \in \mathbb{C} \setminus \{0\} \tag{8.234}$$

then $\alpha \Psi_0(\mathbf{r}) + \beta \hat{\mathbf{n}}(\mathbf{r}) \cdot \nabla \Psi_0(\mathbf{r}) = 0$ on ∂V_1, and we have

$$-\operatorname{Re} \int\limits_{\partial V_1} \mathrm{d}S \; \mathrm{j} k \Psi_0^*(\mathbf{r}) \frac{\partial \Psi_0}{\partial \hat{\mathbf{n}}} = -\operatorname{Im} \left\{ \frac{\alpha}{\beta} \right\} k \int\limits_{\partial V_1} \mathrm{d}S \; |\Psi_0(\mathbf{r})|^2 \leqslant 0 \tag{8.235}$$

because the complex numbers α and β ought to be chosen so that the second term above represents the flux of 'power' entering V_1 supposedly owing to losses on the boundary ∂V_1. Indeed, (8.234) can be regarded as the scalar counterpart of the impedance relationship (6.78) and is referred to as a *Robin boundary condition* [16].

Now, if either $\Psi_0(\mathbf{r}) = 0$ or $\hat{\mathbf{n}}(\mathbf{r}) \cdot \nabla\Psi_0(\mathbf{r}) = 0$ on ∂V_1, then (8.233) implies the vanishing of the flux integral over the sphere ∂B_1. Further, in the event Robin boundary conditions are envisioned, then (8.233) implies that the flux integral over ∂B_1 is negative, as if, on average, 'power' were *entering* the region V_{a_1} flowing through ∂B_1. However, such an occurrence is absurd, because there are no sources in $\mathbb{R}^3 \setminus B[0, a_1]$. Hence, we are led to conclude that

$$\operatorname{Re} \int\limits_{|\mathbf{r}|=a_1} \mathrm{d}S \; \mathrm{j}k\Psi_0^*(\mathbf{r})\frac{\partial \Psi_0}{\partial r} = 0 \tag{8.236}$$

for all types of boundary conditions on ∂V_1.

Secondly, we consider the following identity for points $\mathbf{r} \in V_{a_1}$

$$\left|\frac{\partial \Psi_0}{\partial r} + \mathrm{j}k\Psi_0(\mathbf{r})\right|^2 = \left|\frac{\partial \Psi_0}{\partial r}\right|^2 + k^2|\Psi_0(\mathbf{r})|^2 - 2\operatorname{Re}\left\{\mathrm{j}k\Psi_0^*(\mathbf{r})\frac{\partial \Psi_0}{\partial r}\right\} \tag{8.237}$$

and integrate over ∂B_1 to get

$$\int\limits_{|\mathbf{r}|=a_1} \mathrm{d}S \left|\frac{\partial \Psi_0}{\partial r} + \mathrm{j}k\Psi_0(\mathbf{r})\right|^2 = \int\limits_{|\mathbf{r}|=a_1} \mathrm{d}S \left[\left|\frac{\partial \Psi_0}{\partial r}\right|^2 + k^2|\Psi_0(\mathbf{r})|^2\right] - 2\operatorname{Re}\int\limits_{|\mathbf{r}|=a_1} \mathrm{d}S \; \mathrm{j}k\Psi_0^*(\mathbf{r})\frac{\partial \Psi_0}{\partial r} \tag{8.238}$$

where the last integral is null by virtue of (8.236). Besides, the application of the boundary condition (8.222) — which holds also for $\Psi_0(\mathbf{r})$ — allows us to conclude that the remaining integral in the right member remains finite and is possibly non-null for $a_1 \to +\infty$. But then, said integral *must vanish* if we require the left-hand side to tend to zero as $a_1 \to +\infty$. This occurs if we *postulate* the asymptotic behavior

$$\frac{\partial \Psi}{\partial r} + \mathrm{j}k\Psi(\mathbf{r}) = \mathcal{O}\left(\frac{1}{|\mathbf{r}|^2}\right), \qquad |\mathbf{r}| \to +\infty \tag{8.239}$$

uniformly with respect to $\mathbf{r}$. This additional requirement — which applies to $\Psi_0(\mathbf{r})$ as well — is known as the *Sommerfeld radiation condition* (1912) [15,16,28,38,61,62] and similarly to the Silver-Müller conditions (6.171) and (6.172) it serves to pick up the wave-like solution that travels away from the source. For instance, it is easy to check that the spherical Hankel function of the second kind and zeroth order, $h_0^{(2)}(kr)$, obeys (8.239). In truth, the Sommerfeld condition is unnecessarily stringent as, in fact, all we need is

$$\lim_{a_1 \to +\infty} \int\limits_{|\mathbf{r}|=a_1} \mathrm{d}S \left|\frac{\partial \Psi_0}{\partial r} + \mathrm{j}k\Psi_0(\mathbf{r})\right|^2 = 0 \tag{8.240}$$

as was pointed out by C. H. Wilcox [63]. At any rate, on account of (8.239) we have the estimate

$$\int\limits_{|\mathbf{r}|=a_1} \mathrm{d}S \left[\left|\frac{\partial \Psi_0}{\partial r}\right|^2 + k^2|\Psi_0(\mathbf{r})|^2\right] = \int\limits_{|\mathbf{r}|=a_1} \mathrm{d}S \left|\frac{\partial \Psi_0}{\partial r} + \mathrm{j}k\Psi_0(\mathbf{r})\right|^2 \leqslant \int\limits_{|\mathbf{r}|=a_1} \mathrm{d}S \, \frac{C_\Psi}{a_1^4} = \frac{4\pi C_\Psi}{a_1^2} \tag{8.241}$$

where C_Ψ is a suitable positive constant, and a_1 is supposedly large enough for (8.239) to hold true. The last estimate also implies that the first integral is a non-negative decreasing function of the radial distance.

Thirdly, we use the smaller ball $B(0, a)$, $a < a_1$, and consider the volume integral

$$\int\limits_{B(0,a_1)\setminus B(0,a)} \mathrm{d}V\, k^2|\Psi_0(\mathbf{r})|^2 \leqslant \int\limits_{B(0,a_1)\setminus B(0,a)} \mathrm{d}V \left[\left|\frac{\partial \Psi_0}{\partial r}\right|^2 + k^2|\Psi_0(\mathbf{r})|^2\right]$$

$$= \int_a^{a_1} \mathrm{d}r \int\limits_{S(r)} \mathrm{d}S \left[\left|\frac{\partial \Psi_0}{\partial r}\right|^2 + k^2|\Psi_0(\mathbf{r})|^2\right]$$

$$\leqslant \int_a^{a_1} \mathrm{d}r \frac{4\pi C_\Psi}{r^2} = 4\pi C_\Psi\left(\frac{1}{a} - \frac{1}{a_1}\right) \leqslant \frac{4\pi C_\Psi}{a} \xrightarrow[a\to+\infty]{} 0 \tag{8.242}$$

where $S(r)$ indicates the sphere which has radius $r \in [a, a_1]$ and is centered in the origin.

Finally, we observe that $\Psi_0(\mathbf{r})$ satisfies the source-free Helmholtz equation (8.223) and meets the hypotheses of the Rellich theorem (Section 8.5.2) [57, Section 9.1], [58, 59]. Consequently, we know from (8.211) that a positive constant M exists such that

$$\int\limits_{B(0,a_1)\setminus B(0,a)} \mathrm{d}V\, k^2|\Psi_0(\mathbf{r})|^2 \geqslant M a_1 > M a \tag{8.243}$$

for sufficiently large values of a_1. However, since this conclusion clashes with the opposite one found in (8.242) we are led to infer that $\Psi_0(\mathbf{r})$ cannot be a non-trivial solution to (8.223). Hence, $\Psi_0(\mathbf{r}) = 0$ must perforce hold for $\mathbf{r} \in \mathbb{R}^3 \setminus \overline{V}_1$, whereby uniqueness of solutions follows.

We may still wonder whether a twice-differentiable solution to (8.223) is possible that vanishes only outside the ball $B(0, a)$ where the expansion in spherical Hankel functions and spherical harmonics was used for the proof of the Rellich theorem. However, inside the same ball we can employ for $\Psi_0(\mathbf{r})$ the expansion in spherical Bessel functions of the first kind, since there is no singularity for $\mathbf{r} = 0$. Then, requiring that $\Psi_0(\mathbf{r})$ and $\partial\Psi_0/\partial r$ be continuous for $r = a$ (the radius a being large but otherwise arbitrary) forces all the coefficients of the expansion for $r < a$ to be necessarily null, whereby it follows that the solution vanishes everywhere in $\mathbb{R}^3 \setminus \overline{V}_1$, as stated.

In the absence of the excluded region V_1 the integral over ∂V_1 is identically null and (8.236) stems directly from the conservation law (8.221) applied for $\mathbf{r} \in B(0, a_1)$. In this way we obtain the proof of uniqueness in the whole space.

Not surprisingly, from a mathematical viewpoint we achieve uniqueness even if we posit the *absorbing condition*

$$\frac{\partial \Psi}{\partial r} - \mathrm{j}k\Psi(\mathbf{r}) = \mathcal{O}\left(\frac{1}{|\mathbf{r}|^2}\right), \qquad |\mathbf{r}| \to +\infty \tag{8.244}$$

which allows selecting the non-physical wave-like solution that travels from infinity into the source. The spherical Hankel function of the first kind and zeroth order, $h_0^{(1)}(kr)$, satisfies (8.244).

8.5.5 Uniqueness in bounded regions

The uniqueness of solutions to the Helmholtz equation (8.194) in a finite spatial region $V \subset \mathbb{R}^3$ with smooth boundary is *always* guaranteed provided the typical boundary conditions are prescribed

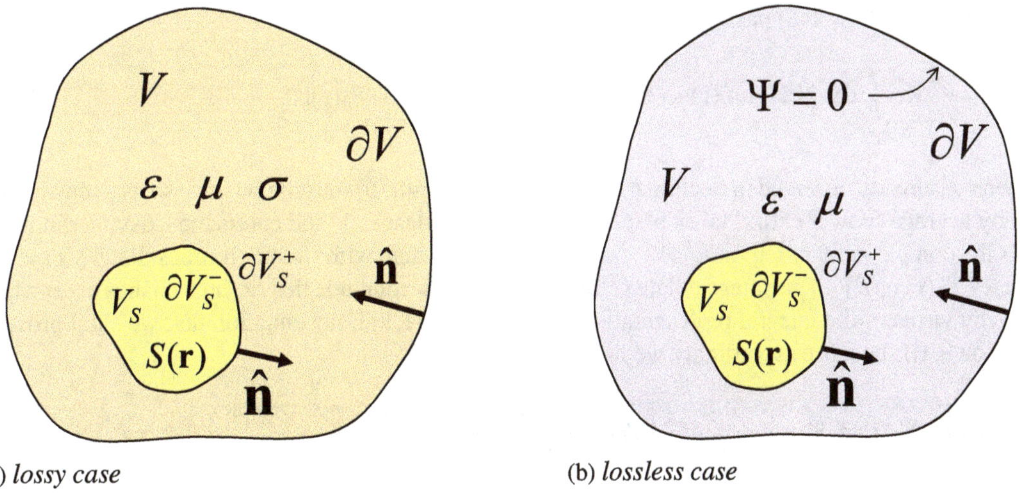

Figure 8.1 For studying the uniqueness of the solutions to the Helmholtz equation (8.194) *in a finite region of space.*

on ∂V and either the medium which fills V is lossy or the boundary ∂V is flush with a dissipative medium. This finding is analogous to the conclusion drawn in Section 6.4.1 for the time-harmonic Maxwell equations. By contrast, if no conduction losses are present anywhere in the closed domain $\overline{V}$ the situation is quite different. As it turns out, (8.194) may have, in fact, either a unique solution, or infinitely many solutions, or even no solution at all, depending on the specific value of the real wavenumber k and the chosen source $S(\mathbf{r})$.

We begin by examining the proof in the lossy case, which is way simpler, and suppose that the medium in V is endowed with constitutive parameters ε, μ and σ (Figure 8.1a) whereby $k = k' - \mathrm{j}k''$, $k'' > 0$, is the relevant complex wavenumber (1.249). If two solutions $\Psi_1(\mathbf{r})$ and $\Psi_2(\mathbf{r})$ are actually possible, we define the difference scalar field $\Psi_0(\mathbf{r}) := \Psi_1(\mathbf{r}) - \Psi_2(\mathbf{r})$ which obeys the homogeneous instance of the Helmholtz equation (8.194). Then, we can use the zero-energy argument by applying the 'energy balance' (8.221) to $\Psi_0(\mathbf{r})$ in the region V to arrive at the condition

$$-\frac{1}{2}\mathrm{Re}\int_{\partial V} \mathrm{d}S\ \mathrm{j}k^*\Psi_0^*(\mathbf{r})\hat{\mathbf{n}}(\mathbf{r})\cdot\nabla\Psi_0(\mathbf{r}) + \frac{k''}{2}\int_V \mathrm{d}V\ \left[|k|^2|\Psi_0(\mathbf{r})|^2 + |\nabla\Psi_0(\mathbf{r})|^2\right] = 0 \tag{8.245}$$

where the leading minus in the 'power' flux integral is due the unit normal $\hat{\mathbf{n}}(\mathbf{r})$ being positively oriented inwards V (Figure 8.1a). Now, we see that if the flux integral across ∂V is null, then (8.245) implies that the remaining volume integral — the 'energy' of the field $\Psi_0(\mathbf{r})$ — must vanish, too, and this happens only if $\Psi_0(\mathbf{r}) = 0 = \nabla\Psi_0(\mathbf{r})$ for $\mathbf{r} \in V$, whereby $\Psi_1(\mathbf{r}) = \Psi_2(\mathbf{r})$ and uniqueness is proved. In particular, we notice that the vanishing of the flux integral occurs if

(1) either $\Psi(\mathbf{r})$ is assigned on ∂V (Dirichlet boundary condition), and hence $\Psi_0(\mathbf{r}) = 0$
(2) or $\hat{\mathbf{n}}(\mathbf{r})\cdot\nabla\Psi(\mathbf{r})$ is prescribed on ∂V (Neumann boundary condition), and hence $\hat{\mathbf{n}}(\mathbf{r})\cdot\nabla\Psi_0(\mathbf{r}) = 0$.

On the other hand, if $\sigma = 0$ in V but the surmised solutions $\Psi_1(\mathbf{r})$ and $\Psi_2(\mathbf{r})$ obey the same Robin condition (8.235) on S [16], then we have

$$\alpha\Psi_0(\mathbf{r}) - \beta\hat{\mathbf{n}}(\mathbf{r})\cdot\nabla\Psi_0(\mathbf{r}) = 0, \qquad \alpha,\beta \in \mathbb{C}, \qquad \mathbf{r} \in S \tag{8.246}$$

and since $k'' = 0$ (8.245) passes over into

$$0 = -\frac{1}{2}\mathrm{Re}\int_{\partial V} \mathrm{d}S\ \mathrm{j}k\Psi_0^*(\mathbf{r})\hat{\mathbf{n}}(\mathbf{r})\cdot\nabla\Psi_0(\mathbf{r}) = \frac{k}{2}\int_{\partial V} \mathrm{d}S\ \mathrm{Im}\left\{\frac{\alpha}{\beta}\right\}|\Psi_0(\mathbf{r})|^2 \tag{8.247}$$

where, as already observed in Section 8.5.4, the quantity $\mathrm{Im}\{\alpha/\beta\}$ must be positive so as to ensure that if any average 'power efflux' takes place at all, the power leaves V and enters the lossy medium that fills the complementary region $\mathbb{R}^3 \setminus V$. Thus, (8.247) in tandem with (8.246) implies that $\Psi_0(\mathbf{r}) = 0 = \hat{\mathbf{n}}(\mathbf{r})\cdot\nabla\Psi_0(\mathbf{r})$ on ∂V. This intermediate result is sufficient to conclude that $\Psi_0(\mathbf{r})$ vanishes everywhere in V by virtue of the integral representation (9.10) for the Helmholtz equation that we shall prove in Section 9.1.1, because accordingly we have

$$\Psi_1(\mathbf{r}) - \Psi_2(\mathbf{r}) = \Psi_0(\mathbf{r}) = \int_{\partial V} \mathrm{d}S' \left[\Psi_0(\mathbf{r}')\frac{\partial}{\partial\hat{\mathbf{n}}'}\frac{\mathrm{e}^{-\mathrm{j}k|\mathbf{r}-\mathbf{r}'|}}{4\pi|\mathbf{r}-\mathbf{r}'|} - \frac{\mathrm{e}^{-\mathrm{j}k|\mathbf{r}-\mathbf{r}'|}}{4\pi|\mathbf{r}-\mathbf{r}'|}\frac{\partial\Psi_0}{\partial\hat{\mathbf{n}}'}\right] = 0, \quad \mathbf{r}\in V \tag{8.248}$$

whereby $\Psi_1(\mathbf{r}) = \Psi_2(\mathbf{r})$, and uniqueness follows again.

Next, we turn our attention to the more complicated lossless case and, for the sake of argument, we assume that $\Psi(\mathbf{r}) = 0$ on ∂V (Figure 8.1b). The Helmholtz equation in a lossless finite spatial region $V \subset \mathbb{R}^3$ with smooth boundary and Dirichlet boundary conditions is sometimes called an *interior homogeneous Dirichlet problem* with spectral parameter k^2. The question of uniqueness of solutions is addressed by transforming the problem into an integral equation for which the Riesz-Fredholm theory can be applied [64, Chapter 1], [65, Chapter 10], [66]. Since the key point of the proof consists of showing that a certain linear operator is *compact* (and this requires other notions from functional analysis) we content ourselves with the outline of the main steps.

To study the solutions to (8.194) where k is a real wavenumber and $S(\mathbf{r})$ is confined to a smaller region $V_S \subset V$ (Figure 8.1b) we introduce a test function $\phi(\mathbf{r})$ in the *Sobolev space* $\mathcal{H}_0^1(V)$ defined in (D.84) in Appendix D.1 [67], and obtain the weak or distributional form by taking the inner product (D.67) in $\mathcal{L}^2(V)$ of both sides of (8.194). This gives

$$\left(\phi, \nabla^2\Psi + k^2\Psi\right)_{\mathcal{L}^2(V)} = -\left(\nabla\phi, \nabla\Psi\right)_{\mathcal{L}^2(V)^3} + k^2\left(\phi, \Psi\right)_{\mathcal{L}^2(V)} = -\left(\phi, S\right)_{\mathcal{L}^2(V)} \tag{8.249}$$

because

$$\begin{aligned}
\int_V \mathrm{d}V\,\phi^*(\mathbf{r})\nabla^2\Psi &= \int_V \mathrm{d}V\,\nabla\cdot[\phi^*(\mathbf{r})\nabla\Psi] - \int_V \mathrm{d}V\,\nabla\phi^*\cdot\nabla\Psi \\
&= \int_{V\setminus\overline{V}_S} \mathrm{d}V\,\nabla\cdot[\phi^*(\mathbf{r})\nabla\Psi] + \int_{V_S} \mathrm{d}V\,\nabla\cdot[\phi^*(\mathbf{r})\nabla\Psi] - \int_V \mathrm{d}V\,\nabla\phi^*\cdot\nabla\Psi \\
&= -\int_{\partial V} \mathrm{d}S\,\phi^*(\mathbf{r})\hat{\mathbf{n}}(\mathbf{r})\cdot\nabla\Psi - \int_{\partial V_S} \mathrm{d}S\,\left[\phi_+^*(\mathbf{r})\hat{\mathbf{n}}(\mathbf{r})\cdot\nabla\Psi_+ - \phi_-^*(\mathbf{r})\hat{\mathbf{n}}(\mathbf{r})\cdot\nabla\Psi_-\right] \\
&\quad - \int_V \mathrm{d}V\,\nabla\phi^*\cdot\nabla\Psi = -\left(\nabla\phi, \nabla\Psi\right)_{\mathcal{L}^2(V)^3}
\end{aligned} \tag{8.250}$$

on account of identity (H.51) and the Gauss theorem (A.53) applied separately in $V \setminus \overline{V}_S$ and V_S, in that $\phi^*(\mathbf{r})\nabla\Psi$ may not be differentiable throughout V. The symbols ϕ_-, Ψ_- (resp. ϕ_+, Ψ_+) denote fields evaluated on the negative (resp. positive) side of ∂V_S, as indicated by the local normal

(Figure 8.1b). The flux integrals are null in view of the vanishing of ϕ on ∂V, and the additional assumption that $\hat{\mathbf{n}} \cdot \nabla\Psi$ and ϕ be continuous across ∂V_S.

In accordance with (8.249) $\Psi(\mathbf{r})$ is not required to be twice-differentiable in V, whereas we only need $\Psi(\mathbf{r})$, $\nabla\Psi$ and $S(\mathbf{r})$ to be square-integrable. Thus, we seek the solutions to (8.249) in the Sobolev space $\mathcal{H}_0^1(V)$ and write the weak form as

$$-(\phi, \Psi)_\star + k^2 (\phi, \Psi)_{\mathcal{L}^2(V)} = -(\phi, S)_{\mathcal{L}^2(V)} \tag{8.251}$$

where $(\bullet, \bullet)_\star$ is the alternative inner product (D.85) defined in $\mathcal{H}_0^1(V)$.

We notice that from the standpoint of $\phi(\mathbf{r})$ the inner product $(\phi, \Psi)_{\mathcal{L}^2(V)}$ is obviously a linear functional (Appendix D.3), say,

$$\mathfrak{L}\{\bullet\} : \begin{cases} \mathcal{H}_0^1(V) \longrightarrow \mathbb{C} \\ \phi \longmapsto (\phi, \Psi)_{\mathcal{L}^2(V)} \end{cases} \tag{8.252}$$

which for a fixed Ψ maps $\phi(\mathbf{r}) \in \mathcal{H}_0^1(V)$ onto a complex number. This functional is also bounded because by virtue of the Cauchy-Schwarz inequality (D.145) and the Poincaré inequality (D.89) [25, p. 415] we have

$$|(\phi, \Psi)_{\mathcal{L}^2(V)}| \leqslant \|\phi\|_{\mathcal{L}^2(V)} \|\Psi\|_{\mathcal{L}^2(V)} \leqslant M \|\phi\|_\star \|\Psi\|_{\mathcal{L}^2(V)} \tag{8.253}$$

whereby in view of definition (D.124) we find

$$\|\mathfrak{L}\{\bullet\}\| := \sup_{\phi \in \mathcal{H}_0^1(V)} \frac{|(\phi, \Psi)_{\mathcal{L}^2(V)}|}{\|\phi\|_\star} \leqslant M \|\Psi\|_{\mathcal{L}^2(V)}, \qquad \phi(\mathbf{r}) \neq 0 \tag{8.254}$$

and $\|\Psi\|_{\mathcal{L}^2(V)}$ is finite, since $\Psi(\mathbf{r})$ is square-integrable on V. Thus, according to the Riesz representation theorem of linear functionals (Appendix D.5) [65, Chapter 10], [68, Theorem 7.16], $\mathfrak{L}\{\phi\}$ can be written as an inner product of the type (D.85)

$$(\phi, \Psi)_{\mathcal{L}^2(V)} = \mathfrak{L}\{\phi\} = (\phi, g_\Psi)_\star \tag{8.255}$$

where $g_\Psi(\mathbf{r}) \in \mathcal{H}_0^1(V)$ is a unique function that depends on Ψ. Therefore, we define the operator

$$\mathfrak{A}\{\bullet\} : \begin{cases} \mathcal{L}^2(V) \longrightarrow \mathcal{H}_0^1(V) \\ \Psi \longmapsto g_\Psi \end{cases} \tag{8.256}$$

which we can prove to be linear and bounded (Appendix D.3). Indeed, for any two functions Ψ_1, Ψ_2 of $\mathcal{L}^2(V)$ and two complex constants α_1, α_2 we have

$$\begin{aligned}(\phi, \mathfrak{A}\{\alpha_1\Psi_1 + \alpha_2\Psi_2\})_\star &= (\phi, \alpha_1\Psi_1 + \alpha_2\Psi_2)_{\mathcal{L}^2(V)} = \alpha_1 (\phi, \Psi_1)_{\mathcal{L}^2(V)} + \alpha_2 (\phi, \Psi_2)_{\mathcal{L}^2(V)} \\ &= \alpha_1 (\phi, \mathfrak{A}\{\Psi_1\})_\star + \alpha_2 (\phi, \mathfrak{A}\{\Psi_2\})_\star = (\phi, \alpha_1\mathfrak{A}\{\Psi_1\} + \alpha_2\mathfrak{A}\{\Psi_2\})_\star\end{aligned} \tag{8.257}$$

whence we find

$$(\phi, \mathfrak{A}\{\alpha_1\Psi_1 + \alpha_2\Psi_2\} - \alpha_1\mathfrak{A}\{\Psi_1\} - \alpha_2\mathfrak{A}\{\Psi_2\})_\star = 0 \tag{8.258}$$

thanks to definition (8.255) and the linearity of the inner products in $\mathcal{L}^2(V)$ and $\mathcal{H}_0^1(V)$. Since this last condition must hold true for any $\phi(\mathbf{r}) \in \mathcal{H}_0^1(V)$ we let

$$\phi(\mathbf{r}) = \mathfrak{A}\{\alpha_1\Psi_1 + \alpha_2\Psi_2\} - \alpha_1\mathfrak{A}\{\Psi_1\} - \alpha_2\mathfrak{A}\{\Psi_2\}, \qquad \mathbf{r} \in V \tag{8.259}$$

whereby we get

$$\|\mathfrak{A}\{\alpha_1\Psi_1 + \alpha_2\Psi_2\} - \alpha_1\mathfrak{A}\{\Psi_1\} - \alpha_2\mathfrak{A}\{\Psi_2\}\|_\star^2 = 0 \tag{8.260}$$

which, on account of properties (D.8) and (D.9) of a norm, can only be true if the element vanishes and hence

$$\mathfrak{A}\{\alpha_1\Psi_1 + \alpha_2\Psi_2\} = \alpha_1\mathfrak{A}\{\Psi_1\} + \alpha_2\mathfrak{A}\{\Psi_2\} \tag{8.261}$$

as required of a linear operator. As for the boundedness of $\mathfrak{A}\{\bullet\}$ we consider

$$\begin{aligned}\|\mathfrak{A}\{\Psi\}\|_\star^2 &= (\mathfrak{A}\{\Psi\}, \mathfrak{A}\{\Psi\})_\star = (\mathfrak{A}\{\Psi\}, g_\Psi)_\star = (\mathfrak{A}\{\Psi\}, \Psi)_{\mathcal{L}^2(V)} \\ &\leqslant \|\mathfrak{A}\{\Psi\}\|_{\mathcal{L}^2(V)} \|\Psi\|_{\mathcal{L}^2(V)} \leqslant M\,\|\mathfrak{A}\{\Psi\}\|_\star \|\Psi\|_{\mathcal{L}^2(V)}\end{aligned} \tag{8.262}$$

where we have used (8.255), the Cauchy-Schwarz inequality (D.145), and the Poincaré inequality (D.89). Dividing through by $\|\mathfrak{A}\{\Psi\}\|_\star \|\Psi\|_{\mathcal{L}^2(V)}$ and passing to the supremum yields

$$\|\mathfrak{A}\{\bullet\}\| := \sup_{\Psi\in\mathcal{L}^2(V)} \frac{\|\mathfrak{A}\{\Psi\}\|_\star}{\|\Psi\|_{\mathcal{L}^2(V)}} \leqslant M, \qquad \Psi(\mathbf{r}) \neq 0 \tag{8.263}$$

in light of (D.124). Since the space $\mathcal{H}_0^1(V)$ is embedded in $\mathcal{L}^2(V)$ according to the general definition (D.34) and property (D.92), the *imbedding* operator (D.136) here reads

$$\mathfrak{I}\{\bullet\} : \begin{cases} \mathcal{H}_0^1(V) \longrightarrow \mathcal{L}^2(V) \\ \Psi \longmapsto \Psi \end{cases} \tag{8.264}$$

it formally transforms an element $\Psi \in \mathcal{H}_0^1(V)$ into the same element $\Psi \in \mathcal{L}^2(V)$ and, since the domain V is finite, $\mathfrak{I}\{\bullet\}$ can be shown to be compact (this is a special case of the *Rellich-Kondrachov embedding theorem*, which we take for granted [25, Proposition 3.13, p. 417]). Therefore, we can rewrite (8.255) as

$$(\phi, g_\Psi)_\star = (\phi, \mathfrak{A}\{\Psi\})_\star = \left(\phi, \tilde{\mathfrak{A}}\{\Psi\}\right)_\star \tag{8.265}$$

where the operator

$$\tilde{\mathfrak{A}}\{\bullet\} := \mathfrak{A}\{\mathfrak{I}\{\bullet\}\} : \mathcal{H}_0^1(V) \longrightarrow \mathcal{L}^2(V) \longrightarrow \mathcal{H}_0^1(V) \tag{8.266}$$

is compact because so is $\mathfrak{I}\{\bullet\}$, whereas $\mathfrak{A}\{\bullet\}$ is bounded, and the successive application of a compact and a bounded operator yields a compact operator [64, Proposition 4.2, p. 41].

By performing similar manipulations on the right-hand side of (8.249) we arrive at

$$-(\phi, \Psi)_\star + k^2\left(\phi, \tilde{\mathfrak{A}}\{\Psi\}\right)_\star = -(\phi, \mathfrak{A}\{S\})_\star \tag{8.267}$$

which can be cast as

$$\left(\phi, \Psi - k^2\tilde{\mathfrak{A}}\{\Psi\} - \mathfrak{A}\{S\}\right)_\star = 0 \tag{8.268}$$

thanks to the linearity of the inner product (D.85). Since (8.268) must hold for any choice of $\phi(\mathbf{r}) \in \mathcal{H}_0^1(V)$, in particular we may choose

$$\phi(\mathbf{r}) = \Psi(\mathbf{r}) - k^2\tilde{\mathfrak{A}}\{\Psi\} - \mathfrak{A}\{S\}, \qquad \mathbf{r} \in V \tag{8.269}$$

whereby we get

$$\left\|\Psi(\mathbf{r}) - k^2\tilde{\mathfrak{A}}\{\Psi\} - \mathfrak{A}\{S\}\right\|_\star^2 = 0 \tag{8.270}$$

which can only be true if the element vanishes and hence

$$\Psi(\mathbf{r}) - k^2\tilde{\mathfrak{A}}\{\Psi\} = \mathfrak{A}\{S\}, \qquad \mathbf{r} \in V, \qquad \Psi \in \mathcal{H}_0^1(V), \qquad S \in \mathcal{L}^2(V) \tag{8.271}$$

owing to properties (D.8) and (D.9) of a norm. Expression (8.271) is a *Fredholm integral equation* of the second kind (see Section 13.1) where $\tilde{\mathfrak{A}}\{\bullet\}$ is a compact, self-adjoint (see Appendix D.6) and even positive definite operator because

$$\left(\tilde{\mathfrak{A}}\{\phi\}, \Psi\right)_\star = \left[\left(\Psi, \tilde{\mathfrak{A}}\{\phi\}\right)_\star\right]^* = \left[(\Psi, \phi)_{\mathcal{L}^2(V)}\right]^* = (\phi, \Psi)_{\mathcal{L}^2(V)} = \left(\phi, \tilde{\mathfrak{A}}\{\Psi\}\right)_\star \tag{8.272}$$

and

$$\left(\Psi, \tilde{\mathfrak{A}}\{\Psi\}\right)_\star = (\Psi, \Psi)_{\mathcal{L}^2(V)} = \|\Psi\|_{\mathcal{L}^2(V)}^2 > 0 \tag{8.273}$$

having repeatedly invoked definition (8.255). Moreover, (8.271) can be put in the standard form (D.245) by including the constant k^2 in the definition of the operator $\tilde{\mathfrak{A}}\{\bullet\}$.

From the *Fredholm alternative* (Appendix D.8) [25, Chapter 8], [68, Chapter 8], [69, Chapter 1], [66,70,71] we know that (8.271) has a unique solution for any source term $\mathfrak{A}\{S\} \in \mathcal{H}_0^1(V)$, if the associated homogeneous equation

$$\Upsilon(\mathbf{r}) - k^2\tilde{\mathfrak{A}}\{\Upsilon\} = 0, \qquad \mathbf{r} \in V, \qquad \Upsilon \in \mathcal{H}_0^1(V) \tag{8.274}$$

admits only the trivial solution $\Upsilon(\mathbf{r}) = 0$. Or else, if for a given wavenumber $k \in \mathbb{R}$ (8.274) has $N < +\infty$ non-trivial solutions $\Upsilon_l(\mathbf{r})$, $l = 1, \ldots, N$, then (8.271) is solvable if, and only if, the right-hand side is orthogonal — with respect to the inner product (D.85) — to each and every $\Upsilon_l(\mathbf{r})$. In light of (8.255) this constraint translates into the requirement

$$(\Upsilon_l, \mathfrak{A}\{S\})_\star = (\Upsilon_l, S)_{\mathcal{L}^2(V)} = \int\limits_{V_S} \mathrm{d}V\, \Upsilon_l^*(\mathbf{r}) S(\mathbf{r}) = 0, \qquad l = 1, \ldots, N \tag{8.275}$$

in which case (8.271) and hence (8.195) — subject to $\Psi(\mathbf{r}) = 0$ on ∂V — have infinitely many solutions in the form

$$\Psi(\mathbf{r}) = \Psi_0(\mathbf{r}) + \sum_{l=1}^{N} c_l \Upsilon_n(\mathbf{r}), \qquad \mathbf{r} \in V \tag{8.276}$$

where $\Psi_0(\mathbf{r})$ is a particular solution and c_l are arbitrary constants (cf. Example 6.2).

Since $\tilde{\mathfrak{A}}\{\bullet\}$ is compact, self-adjoint and positive definite, there exists a set of real eigenvalues ν_n associated with orthogonal eigenvectors $\Upsilon_n(\mathbf{r})$ in the space $\mathcal{H}_0^1(V)$ endowed with the inner product (D.85). The eigenvalues are positive and tend to zero as $n \to +\infty$ (Appendix D.7) and the set of eigenvectors $\Upsilon_n(\mathbf{r})$ is complete with respect to the inner product (D.85). Therefore, from

$$\tilde{\mathfrak{A}}\{\Upsilon_n\} = \nu_n \Upsilon_n(\mathbf{r}), \qquad \mathbf{r} \in V, \qquad \Upsilon_n \in \mathcal{H}_0^1(V) \tag{8.277}$$

we obtain

$$\Upsilon_n(\mathbf{r}) - \frac{1}{\nu_n}\tilde{\mathfrak{A}}\{\Upsilon_n\} = 0, \qquad \mathbf{r} \in V, \qquad \Upsilon_n \in \mathcal{H}_0^1(V) \tag{8.278}$$

which by virtue of our previous derivations is equivalent to

$$\nabla^2 \Upsilon_n + \frac{1}{\nu_n} \Upsilon_n(\mathbf{r}) = 0, \qquad \mathbf{r} \in V, \qquad \Upsilon_n \in \mathcal{H}_0^1(V) \tag{8.279}$$

and this shows that the negative of the Laplace operator $-\nabla^2\{\bullet\}$ in V with homogeneous Dirichlet boundary conditions on ∂V has a set of positive eigenvalues $k_n^2 = 1/\nu_n$ which grow infinitely large and are associated with a complete set of eigenfunctions which vanish on ∂V. An alternative, constructive proof of this same conclusion can be found in [25, pp. 174-176]. It is important to notice that the eigenfunctions are also orthogonal with respect to the inner product in $\mathcal{L}^2(V)$. Indeed, we have

$$0 = (\Upsilon_m, \Upsilon_n)_\star = \frac{1}{\nu_n} \left(\Upsilon_m, \tilde{\mathfrak{A}}\{\Upsilon_n\}\right)_\star = k_n^2 (\Upsilon_m, \Upsilon_n)_{\mathcal{L}^2(V)} \tag{8.280}$$

on account of definitions (8.255) and (8.265).

8.6 Uniqueness of solutions to the D'Alembert equation

We devote this section to determining the necessary requirements that ensure a unique solution to the D'Alembert equation

$$\nabla^2 \Upsilon(\mathbf{r}, t) - \frac{1}{c^2} \frac{\partial^2}{\partial t^2} \Upsilon(\mathbf{r}, t) = -s(\mathbf{r}, t), \qquad \mathbf{r} \in V \qquad t \geqslant 0 \tag{8.281}$$

both in bounded domains ($V \subset \mathbb{R}^3$) and in the whole space ($V \equiv \mathbb{R}^3$). The generic scalar function $\Upsilon(\mathbf{r}, t)$ may represent, e.g., the electric potential $\Phi_E(\mathbf{r}, t)$ or one of the Cartesian components of $\mathcal{A}_M(\mathbf{r}, t)$, in accordance with (8.122) and (8.127), respectively. Further, we suppose that the source $s(\mathbf{r}, t)$ is confined to a finite region $V_S \subset V$ and is turned on at time $t = 0$.

We first derive an 'energy-balance' equation of sorts that will come in handy in our quest. Conceptually, we may attach the meaning of volume energy density to non-negative scalar fields such as $|\nabla \Upsilon(\mathbf{r}, t)|^2$ and $|\partial \Upsilon / \partial t|^2$, though the physical interpretation may vary depending on the quantity that $\Upsilon(\mathbf{r}, t)$ represents. To obtain the conservation law we begin by multiplying (8.281) by $\partial \Upsilon / \partial t$, viz.,

$$\begin{aligned} -s(\mathbf{r}, t) \frac{\partial \Upsilon}{\partial t} &= \frac{\partial \Upsilon}{\partial t} \nabla^2 \Upsilon(\mathbf{r}, t) - \frac{1}{c^2} \frac{\partial \Upsilon}{\partial t} \frac{\partial^2 \Upsilon}{\partial t^2} \\ &= \nabla \cdot \left[\frac{\partial \Upsilon}{\partial t} \nabla \Upsilon(\mathbf{r}, t) \right] - \nabla \frac{\partial \Upsilon}{\partial t} \cdot \nabla \Upsilon(\mathbf{r}, t) - \frac{1}{2c^2} \frac{\partial}{\partial t} \left| \frac{\partial \Upsilon}{\partial t} \right|^2 \\ &= \nabla \cdot \left[\frac{\partial \Upsilon}{\partial t} \nabla \Upsilon(\mathbf{r}, t) \right] - \frac{1}{2} \frac{\partial}{\partial t} \left[|\nabla \Upsilon(\mathbf{r}, t)|^2 + \frac{1}{c^2} \left| \frac{\partial \Upsilon}{\partial t} \right|^2 \right] \end{aligned} \tag{8.282}$$

where we have used (H.51) and the Schwarz theorem [27, pp. 235–236] to interchange spatial and temporal derivatives of $\Upsilon(\mathbf{r}, t)$. Since (8.282) has the structure of a continuity equation with a source term [cf. (1.316) and (8.219)], this relationship constitutes the desired 'energy balance' in local form.

Integrating over $V \subset \mathbb{R}^3$ yields the corresponding global form

$$\begin{aligned} \int_{\partial V} \mathrm{d}S \, \frac{\partial \Upsilon}{\partial t} \hat{\mathbf{n}}(\mathbf{r}) \cdot \nabla \Upsilon(\mathbf{r}, t) - \frac{1}{2} \frac{\mathrm{d}}{\mathrm{d}t} \int_V \mathrm{d}V \left[|\nabla \Upsilon(\mathbf{r}, t)|^2 + \frac{1}{c^2} \left| \frac{\partial \Upsilon}{\partial t} \right|^2 \right] \\ = - \int_{V_S} \mathrm{d}V \, s(\mathbf{r}, t) \frac{\partial \Upsilon}{\partial t} \qquad t \geqslant 0 \end{aligned} \tag{8.283}$$

having applied the Gauss theorem (A.53) with the unit normal $\hat{\mathbf{n}}(\mathbf{r})$ on ∂V positively oriented outward V. This step is permitted (see discussion on pages 10 and ff.) so long as the vector field $\partial\Upsilon/\partial t\,\nabla\Upsilon$ is continuously differentiable in $V \setminus \overline{V}_S$ and V_S, and the normal component $\partial\Upsilon/\partial t\,\hat{\mathbf{n}}\cdot\nabla\Upsilon$ is continuous across ∂V_S.

8.6.1 Bounded regions

If (8.281) were indeed solved in $V \subset \mathbb{R}^3$ by two distinct functions, say, $\Upsilon_1(\mathbf{r}, t)$ and $\Upsilon_2(\mathbf{r}, t)$, then the difference scalar field $\Upsilon_0 = \Upsilon_1 - \Upsilon_2$ would be a non-trivial solution to the homogeneous D'Alembert equation

$$\nabla^2\Upsilon_0(\mathbf{r}, t) - \frac{1}{c^2}\frac{\partial^2}{\partial t^2}\Upsilon_0(\mathbf{r}, t) = 0, \qquad \mathbf{r} \in V \qquad t \geqslant 0 \tag{8.284}$$

and besides, Υ_0 should obey the special instance of the conservation law (8.283), namely,

$$-\int_{\partial V} \mathrm{d}S\,\frac{\partial\Upsilon_0}{\partial t}\hat{\mathbf{n}}(\mathbf{r})\cdot\nabla\Upsilon_0(\mathbf{r}, t) = \frac{1}{2}\frac{\mathrm{d}}{\mathrm{d}t}\int_V \mathrm{d}V\left[|\nabla\Upsilon_0(\mathbf{r}, t)|^2 + \frac{1}{c^2}\left|\frac{\partial\Upsilon_0}{\partial t}\right|^2\right] \tag{8.285}$$

where we have taken the unit normal $\hat{\mathbf{n}}(\mathbf{r})$ on ∂V pointing inward V so as to conform with the viewpoint of an observer sitting inside V. This choice is also in line with the proof of uniqueness of the Poisson equation (Section 2.5.1) which, after all, represents the static limit of (8.281).

Next, if we suppose that the flux integral in the left-hand side of (8.285) somehow vanishes at all times, then the 'energy' must satisfy the constraint

$$W(t) := \frac{1}{2}\int_V \mathrm{d}V\left[|\nabla\Upsilon_0(\mathbf{r}, t)|^2 + \frac{1}{c^2}\left|\frac{\partial\Upsilon_0}{\partial t}\right|^2\right] = C_W, \qquad t \geqslant 0 \tag{8.286}$$

where $C_W \geqslant 0$ is a constant yet to be determined. In this regard, if the functions $\Upsilon_1(\mathbf{r}, t)$ and $\Upsilon_2(\mathbf{r}, t)$ as well as the temporal derivatives thereof are specified at $t = 0$ for $\mathbf{r} \in V$ (initial conditions) it follows that

$$\Upsilon_0(\mathbf{r}, 0) = 0 = \left.\frac{\partial\Upsilon_0}{\partial t}\right|_{t=0}, \qquad \mathbf{r} \in V \tag{8.287}$$

whence we infer that the constant C_W must be zero. As a result, it follows from (8.286) that the 'energy' vanishes at all times $t \geqslant 0$.

Moreover, since the integrand in (8.286) is a non-negative function of position and time, the integral over V vanishes only if

$$\nabla\Upsilon_0(\mathbf{r}, t) = 0 = \frac{\partial\Upsilon_0}{\partial t}, \qquad \mathbf{r} \in V \qquad t \geqslant 0 \tag{8.288}$$

and these conditions together imply

$$\Upsilon_0(\mathbf{r}, t) = C_0, \qquad \mathbf{r} \in V \qquad t \geqslant 0 \tag{8.289}$$

where $C_0 \in \mathbb{R}$ is yet another constant. However, we see that necessarily $C_0 = 0$ in light of the first part of (8.287). Hence, $\Upsilon_0(\mathbf{r}, t)$ is null everywhere in V at all times, whereby we conclude that $\Upsilon_1(\mathbf{r}, t) = \Upsilon_2(\mathbf{r}, t)$ and the solution is unique for $\mathbf{r} \in V$ and $t \geqslant 0$ contrary to our initial hypothesis.

Nonetheless, we still need to ascertain what circumstances determine the vanishing of the flux integral in (8.285). Clearly, this requirement is fulfilled if either one of the following boundary conditions is true

$$\begin{aligned} &\Upsilon_0(\mathbf{r},t)=0, && \mathbf{r}\in\partial V && t\geqslant 0 \\ &\hat{\mathbf{n}}(\mathbf{r})\cdot\nabla\Upsilon_0(\mathbf{r},t)=0, && \mathbf{r}\in\partial V && t\geqslant 0 \end{aligned} \tag{8.290}$$

which in turn are satisfied if we complement (8.281) by assigning either the scalar field $\Upsilon(\mathbf{r},t)$ or the normal derivative thereof on ∂V.

Additionally, given two real non-null constants or functions α and β with $\alpha\beta<0$, we can specify a mixed boundary condition of the type

$$\alpha\frac{\partial\Upsilon_0}{\partial t}+\beta\hat{\mathbf{n}}(\mathbf{r})\cdot\nabla\Upsilon_0(\mathbf{r},t)=0, \qquad \mathbf{r}\in\partial V \qquad t\geqslant 0 \tag{8.291}$$

which guarantees uniqueness of solutions to (8.281) as well, but the statement does not follow immediately from (8.285) inasmuch as the flux integral does not vanish. Instead, by using (8.291) to eliminate, e.g., the normal derivative from the flux integral we get

$$\frac{1}{2}\frac{\mathrm{d}}{\mathrm{d}t}\int_V \mathrm{d}V\left[|\nabla\Upsilon_0(\mathbf{r},t)|^2+\frac{1}{c^2}\left|\frac{\partial\Upsilon_0}{\partial t}\right|^2\right]=\int_{\partial V}\mathrm{d}S\,\frac{\alpha}{\beta}\left|\frac{\partial\Upsilon_0}{\partial t}\right|^2\leqslant 0 \tag{8.292}$$

which tells us that the time rate of variation of the energy-like integral is non-positive because $\alpha\beta<0$. Since the 'energy' $W(t)$ defined in (8.286) is a non-negative function of time, by applying the mean value theorem (A.89) we have

$$0\leqslant W(t)=W(0)+\left.\frac{\mathrm{d}W}{\mathrm{d}t}\right|_{t=t_0}t\leqslant W(0)=0, \qquad t\geqslant 0 \tag{8.293}$$

with $t_0\in[0,t]$ a suitable time. The last equality holds because the difference field $\Upsilon_0(\mathbf{r},t)$ and the time derivative thereof are null for $t=0$ in view of the initial conditions (8.287). Consequently, once again $W(t)=0$ at all times of interest, and uniqueness follows by means of the same arguments invoked above.

In summary, the D'Alembert equation (8.281) has a unique solution in a bounded region of space V if *all* of the following requirements are met:

(1) the source $s(\mathbf{r},t)$ is assigned for $t\geqslant 0$;
(2) the scalar field $\Upsilon(\mathbf{r},t)$ *and* the time derivative thereof are assigned for $\mathbf{r}\in V$ and $t=0$ (initial conditions);
(3) either $\Upsilon(\mathbf{r},t)$ or $\hat{\mathbf{n}}(\mathbf{r})\cdot\nabla\Upsilon(\mathbf{r},t)$ or $\alpha\partial\Upsilon/\partial t+\beta\hat{\mathbf{n}}(\mathbf{r})\cdot\nabla\Upsilon(\mathbf{r},t)$, $\alpha\beta<0$, is assigned for $(\mathbf{r},t)\in\partial V\times\mathbb{R}^+$ (boundary conditions).

On a related score, we observe that the problem

$$\begin{cases}\nabla^2\Upsilon(\mathbf{r},t)-\dfrac{1}{c^2}\dfrac{\partial^2}{\partial t^2}\Upsilon(\mathbf{r},t)=-s(\mathbf{r},t), & (\mathbf{r},t)\in V\times\mathbb{R}^+\\ \Upsilon(\mathbf{r},0)=0=\left.\dfrac{\partial\Upsilon}{\partial t}\right|_{t=0}, & \mathbf{r}\in V\\ \hat{\mathbf{n}}(\mathbf{r})\cdot\nabla\Upsilon(\mathbf{r},t)=g(\mathbf{r},t), & (\mathbf{r},t)\in\partial V\times\mathbb{R}^+\end{cases} \tag{8.294}$$

in principle should have a unique solution. However, *none* might exist if the function $g(\mathbf{r}, t)$ is specified arbitrarily, in that it is not independent of $s(\mathbf{r}, t)$. To find out the extra constraint we integrate the D'Alembert equation over V and for all times $t \geqslant 0$, viz.,

$$\int_0^{+\infty} \mathrm{d}t \int_{\partial V} \mathrm{d}S\, \hat{\mathbf{n}}(\mathbf{r}) \cdot \nabla \Upsilon(\mathbf{r}, t) + \frac{1}{c^2} \int_0^{+\infty} \mathrm{d}t \int_V \mathrm{d}V \frac{\partial^2 \Upsilon}{\partial t^2} = \int_0^{+\infty} \mathrm{d}t \int_{V_S} \mathrm{d}V\, s(\mathbf{r}, t) \tag{8.295}$$

where we have applied the Gauss theorem with the unit normal positively oriented inwards V. By using the initial and boundary conditions assumed in (8.294) we obtain

$$\int_0^{+\infty} \mathrm{d}t \int_{\partial V} \mathrm{d}S\, g(\mathbf{r}, t) + \frac{1}{c^2} \int_V \mathrm{d}V \left. \frac{\partial \Upsilon}{\partial t} \right|_{t \to +\infty} = \int_0^{+\infty} \mathrm{d}t \int_{V_S} \mathrm{d}V\, s(\mathbf{r}, t) \tag{8.296}$$

where the second integral in the left-hand side may vanish if the source is active only for a finite time interval. This limitation extends the analogous one (2.71) for the Poisson equation with Neumann boundary conditions (2.69).

8.6.2 *Unbounded regions*

We continue with the discussion of the necessary conditions that ensure uniqueness of solutions to (8.281) in the whole space, i.e., when $V \equiv \mathbb{R}^3$. For this task we specialize the procedure outlined in Section 6.3.2 for the full-fledged Maxwell equations [72, Section 4.3]. Accordingly, we introduce a ball $B(0, a(t))$ of decreasing radius $a(t) := a_0 - ct$, $t \in [0, a_0/c]$, as in Figure 6.7. Then, we consider the time rate of variation of the 'energy' associated with the difference scalar field $\Upsilon_0 = \Upsilon_1 - \Upsilon_2$ for points in $B(0, a(t))$, namely,

$$\begin{aligned} \frac{\mathrm{d}W}{\mathrm{d}t} &:= \frac{1}{2} \frac{\mathrm{d}}{\mathrm{d}t} \int_{B(0,a(t))} \mathrm{d}V \left[|\nabla \Upsilon_0(\mathbf{r}, t)|^2 + \frac{1}{c^2} \left| \frac{\partial \Upsilon_0}{\partial t} \right|^2 \right] \\ &= \frac{1}{2} \frac{\mathrm{d}}{\mathrm{d}t} \int_0^{a_0 - ct} \mathrm{d}r \int_{S(r)} \mathrm{d}S \left[|\nabla \Upsilon_0(\mathbf{r}, t)|^2 + \frac{1}{c^2} \left| \frac{\partial \Upsilon_0}{\partial t} \right|^2 \right] \end{aligned} \tag{8.297}$$

where $S(r)$ denotes the sphere with center the origin and radius $r \in [0, a_0 - ct]$. The 'energy' changes inasmuch as 1) the scalar field $\Upsilon_0(\mathbf{r}, t)$ depends on time and 2) the radius of the ball becomes ever smaller.

To compute the temporal derivative formally, we regard the inner integral in the second part of (8.297) as a function of the radial coordinate r and the time t, and invoke the Leibniz rule for the differentiation of the outer integral [40, Formula 3.3.7], [41, Formula 69.3], [42]. Specifically, under the assumption that the integrand and its time derivative are continuous for $(\mathbf{r}, t) \in B(0, a(t)) \times [0, a_0/c]$ we have

$$\begin{aligned} \frac{\mathrm{d}W}{\mathrm{d}t} &= \frac{1}{2} \int_0^{a_0 - ct} \mathrm{d}r \int_{S(r)} \mathrm{d}S \frac{\partial}{\partial t} \left[|\nabla \Upsilon_0(\mathbf{r}, t)|^2 + \frac{1}{c^2} \left| \frac{\partial \Upsilon_0}{\partial t} \right|^2 \right] - \frac{c}{2} \int_{\partial B} \mathrm{d}S \left[|\nabla \Upsilon_0(\mathbf{r}, t)|^2 + \frac{1}{c^2} \left| \frac{\partial \Upsilon_0}{\partial t} \right|^2 \right] \\ &= \int_{B(0,a(t))} \mathrm{d}V \frac{1}{2} \frac{\partial}{\partial t} \left[|\nabla \Upsilon_0(\mathbf{r}, t)|^2 + \frac{1}{c^2} \left| \frac{\partial \Upsilon_0}{\partial t} \right|^2 \right] - \frac{c}{2} \int_{\partial B} \mathrm{d}S \left[|\nabla \Upsilon_0(\mathbf{r}, t)|^2 + \frac{1}{c^2} \left| \frac{\partial \Upsilon_0}{\partial t} \right|^2 \right] \end{aligned}$$

$$= \int_{B(0,a(t))} \mathrm{d}V\, \nabla \cdot \left[\frac{\partial \Upsilon_0}{\partial t} \nabla \Upsilon_0(\mathbf{r}, t) \right] - \frac{c}{2} \int_{\partial B} \mathrm{d}S \left[|\nabla \Upsilon_0(\mathbf{r}, t)|^2 + \frac{1}{c^2} \left| \frac{\partial \Upsilon_0}{\partial t} \right|^2 \right] \tag{8.298}$$

where we have invoked the source-free instance of the local conservation law (8.282) for the function $\Upsilon_0(\mathbf{r}, t)$.

Next, under the assumption that the vector field $\partial \Upsilon_0/\partial t \nabla \Upsilon_0$ is of class $\mathcal{C}^1(B)^3 \cap \mathcal{C}(\overline{B})^3$ for any relevant value of time, we apply the Gauss theorem (A.53) over the ball to get

$$\begin{aligned} \frac{\mathrm{d}W}{\mathrm{d}t} &= \int_{\partial B} \mathrm{d}S\, \frac{\partial \Upsilon_0}{\partial t} \hat{\mathbf{r}} \cdot \nabla \Upsilon_0(\mathbf{r}, t) - \frac{c}{2} \int_{\partial B} \mathrm{d}S \left[|\nabla \Upsilon_0(\mathbf{r}, t)|^2 + \frac{1}{c^2} \left| \frac{\partial \Upsilon_0}{\partial t} \right|^2 \right] \\ &\leqslant \int_{\partial B} \mathrm{d}S \left| \frac{\partial \Upsilon_0}{\partial t} \right| |\nabla \Upsilon_0(\mathbf{r}, t)| - \frac{c}{2} \int_{\partial B} \mathrm{d}S \left[|\nabla \Upsilon_0(\mathbf{r}, t)|^2 + \frac{1}{c^2} \left| \frac{\partial \Upsilon_0}{\partial t} \right|^2 \right] \\ &= -\frac{c}{2} \int_{\partial B} \mathrm{d}S \left[|\nabla \Upsilon_0(\mathbf{r}, t)| - \frac{1}{c} \left| \frac{\partial \Upsilon_0}{\partial t} \right| \right]^2 \leqslant 0 \end{aligned} \tag{8.299}$$

also on account of the Cauchy-Schwarz inequality (D.151). This intermediate result implies that the time rate of variation of the 'energy' $W(t)$ within the ball is a non-increasing function of time. Lastly, by proceeding as in (8.293) we obtain

$$0 \leqslant W(t) = W(0) + \left. \frac{\mathrm{d}W}{\mathrm{d}t} \right|_{t=t_0} t \leqslant W(0) = 0, \qquad t \in [0, a_0/c] \tag{8.300}$$

with $t_0 \in [0, t] \subseteq [0, a_0/c]$ a suitable time. Since the energy must necessarily vanish in the ball for $t \in [0, a_0/c]$, then it follows that $\Upsilon_0(\mathbf{r}, t) = 0$ on account of the homogeneous initial conditions (8.287). This is sufficient to conclude that the solution to (8.281) is unique for $(\mathbf{r}, t) \in B(0, a(t)) \times [0, a_0/c]$.

Uniqueness in the whole space for all times of concern is proved by taking the limit as $a_0 \to +\infty$. In the process, if the last surface integral in (8.299) diverges, then the estimate of $\mathrm{d}W/\mathrm{d}t$ is all the more true. On the other hand, we should check whether the 'energy' integral in (8.297) remains finite. In fact, convergence is easily ensured by noticing that, since (8.281) is a wave equation and any disturbance produced by the source propagates with finite speed c, then there exists a distance d and a ball $B(0, d)$ outside which the function $\Upsilon(\mathbf{r}, t)$ is independent of time and reduces to the solution of the static limit of (8.281) (cf. Figure 1.15b). Therefore, it is reasonable to assume the asymptotic condition

$$\Upsilon(\mathbf{r}, t) = \Upsilon(\mathbf{r}) = \mathcal{O}\left(\frac{1}{|\mathbf{r}|}\right), \qquad |\mathbf{r}| \to +\infty \tag{8.301}$$

as we did for the electrostatic scalar potential in Chapter 2. Thanks to this requirement, by considering an even larger ball $B(0, d_1)$, $d_1 > d$, we obtain the estimate

$$\int_{B(0,d_1)\backslash B(0,d)} \mathrm{d}V\, |\nabla \Upsilon_0(\mathbf{r})|^2 \leqslant \int_{B(0,d_1)\backslash B(0,d)} \mathrm{d}V\, \frac{C_\Upsilon}{r^4} = 4\pi C_\Upsilon \left(\frac{1}{d} - \frac{1}{d_1} \right) \xrightarrow[d_1 \to +\infty]{} 4\pi \frac{C_\Upsilon}{d} \tag{8.302}$$

where C_Υ is a suitable positive constant. With this finding we can conclude that the 'energy' integral is finite for $a_0 \to +\infty$.

Alternatively, we can guarantee finiteness of $W(t)$ by assuming that

$$\Upsilon(\mathbf{r},t) = \mathcal{O}\left(\frac{1}{|\mathbf{r}|}\right), \qquad |\mathbf{r}| \to +\infty \tag{8.303}$$

where $\Upsilon(\mathbf{r},t)$ can still depend on time asymptotically, and additionally by postulating the *radiation condition* [10]

$$\hat{\mathbf{r}} \cdot \nabla\Upsilon(\mathbf{r},t) + \frac{1}{c}\frac{\partial \Upsilon}{\partial t} = \mathcal{O}\left(\frac{1}{|\mathbf{r}|^2}\right), \qquad |\mathbf{r}| \to +\infty \tag{8.304}$$

uniformly with respect to position and time. We shall support this latter requirement in Section 9.1.2 with the aid of the integral representation of $\Upsilon(\mathbf{r},t)$. Meanwhile we notice that (8.304) is the time-domain counterpart of the Sommerfeld condition (8.239) and similarly allows picking up the wave-like solutions to (8.281) that travel *away* from the source rather than into it. In view of the linearity of the D'Alembert operator (8.304) applies also to the difference field $\Upsilon_0(\mathbf{r},t)$, and hence we may consider the identity

$$\frac{1}{2}\left|\hat{\mathbf{r}} \cdot \nabla\Upsilon_0(\mathbf{r},t) + \frac{1}{c}\frac{\partial \Upsilon_0}{\partial t}\right|^2 = \frac{1}{2}|\hat{\mathbf{r}} \cdot \nabla\Upsilon_0(\mathbf{r},t)|^2 + \frac{1}{2c^2}\left|\frac{\partial \Upsilon_0}{\partial t}\right|^2 + \hat{\mathbf{r}} \cdot \nabla\Upsilon_0(\mathbf{r},t)\frac{1}{c}\frac{\partial \Upsilon_0}{\partial t} \tag{8.305}$$

which holds in particular for points in $\mathbb{R}^3 \setminus B(0,d)$. By integrating both sides over the sphere ∂B, we get

$$\begin{aligned}\frac{1}{2}\int\limits_{|\mathbf{r}|=d} \mathrm{d}S \left[|\hat{\mathbf{r}} \cdot \nabla\Upsilon_0(\mathbf{r},t)|^2 + \frac{1}{c^2}\left|\frac{\partial \Upsilon_0}{\partial t}\right|^2\right] + \int\limits_{|\mathbf{r}|=d} \mathrm{d}S\, \hat{\mathbf{r}} \cdot \nabla\Upsilon_0(\mathbf{r},t)\frac{1}{c}\frac{\partial \Upsilon_0}{\partial t} \\ = \frac{1}{2}\int\limits_{|\mathbf{r}|=d} \mathrm{d}S \left|\hat{\mathbf{r}} \cdot \nabla\Upsilon_0(\mathbf{r},t) + \frac{1}{c}\frac{\partial \Upsilon_0}{\partial t}\right|^2\end{aligned} \tag{8.306}$$

where on account of the balance equation (8.283) we recognize the flux integral in the left member as the instantaneous 'power' that possibly crosses ∂B in either direction. However, since we are dealing with the homogeneous problem (8.284), no sources are located within $B(0,d)$ or in the complementary unbounded domain $\mathbb{R}^3 \setminus B[0,d]$, and hence the flux integral over ∂B must vanish.

With this new piece of information and (8.304) we derive from (8.306) the estimate

$$\frac{1}{2}\int\limits_{|\mathbf{r}|=d} \mathrm{d}S \left[|\hat{\mathbf{r}} \cdot \nabla\Upsilon_0(\mathbf{r},t)|^2 + \frac{1}{c^2}\left|\frac{\partial \Upsilon_0}{\partial t}\right|^2\right] \leqslant \frac{1}{2}\int\limits_{|\mathbf{r}|=d} \mathrm{d}S\, \frac{M}{r^4} = 2\pi\frac{M}{d^2} \tag{8.307}$$

where M is a suitable positive constant. From (8.307) we conclude that the integral is a positive function that decreases with the square of the distance from the origin. Next, we observe that in polar spherical coordinates the *transverse gradient*

$$\nabla_t\Upsilon_0(\mathbf{r},t) = \nabla\Upsilon_0(\mathbf{r},t) - \hat{\mathbf{r}}\hat{\mathbf{r}} \cdot \nabla\Upsilon_0(\mathbf{r},t) \tag{8.308}$$

satisfies the asymptotic condition

$$\nabla_t\Upsilon_0(\mathbf{r},t) = \mathcal{O}\left(\frac{1}{|\mathbf{r}|^2}\right), \qquad |\mathbf{r}| \to +\infty \tag{8.309}$$

on account of (A.28) and (8.303). Finally, by considering the 'energy' stored in the spherical shell $B(0, d_1) \setminus B(0, d)$, $d_1 > d$, we have

$$\int\limits_{B(0,d_1)\setminus B(0,d)} \mathrm{d}V \, \frac{1}{2}\left[|\nabla\Upsilon_0(\mathbf{r},t)|^2 + \frac{1}{c^2}\left|\frac{\partial\Upsilon_0}{\partial t}\right|^2\right] =$$

$$= \frac{1}{2}\int\limits_d^{d_1} \mathrm{d}r \int\limits_{S(r)} \mathrm{d}S \left[|\hat{\mathbf{r}}\cdot\nabla\Upsilon_0(\mathbf{r},t)|^2 + |\nabla_t\Upsilon_0(\mathbf{r},t)|^2 + \frac{1}{c^2}\left|\frac{\partial\Upsilon_0}{\partial t}\right|^2\right]$$

$$\leqslant 2\pi \int\limits_d^{d_1} \mathrm{d}r \, \frac{C_\Upsilon}{r^2} = 2\pi C_\Upsilon\left(\frac{1}{d} - \frac{1}{d_1}\right) \xrightarrow[d_1\to+\infty]{} 2\pi\frac{C_\Upsilon}{d} \tag{8.310}$$

where $S(r)$ is the sphere of radius $r \in [d, d_1]$ and center the origin, and we have combined (8.307) and (8.309) by introducing a suitable positive constant C_Υ. This last estimate is sufficient to conclude that the 'energy' integral in (8.297) remains finite for $a_0 \to +\infty$.

In summary, the solution to the D'Alembert equation (8.281) in the *whole space* for $t \geqslant 0$ is unique if the following conditions are *all* satisfied:

(1) the source $s(\mathbf{r}, t)$ is assigned for $t \geqslant 0$;
(2) the scalar field $\Upsilon(\mathbf{r}, t)$ *and* the time derivative thereof are assigned for $\mathbf{r} \in V$ and $t = 0$ (initial conditions);
(3) either the asymptotic boundary condition (8.301) or the pair (8.303), (8.304) is prescribed.

8.7 The three-dimensional time-dependent Green function

In this Section we consider the problem of finding a fundamental solution to the Helmholtz and D'Alembert equations in an unbounded homogeneous isotropic medium. In electromagnetics such solutions are known as *Green functions* [19, Chapter 8], [30, Chapter 2], [17,21,60,73] and are used to construct explicit expressions for fields and potentials generated by arbitrary non-singular sources. This was the guiding principle behind the integral representations developed in Sections 2.7 and 5.1 for static and stationary fields, respectively.

In order to extend the idea to time-varying fields we consider the generic scalar equation

$$\mathfrak{L}\{\Upsilon(\mathbf{r},t)\} = -s(\mathbf{r},t), \qquad (\mathbf{r},t) \in \mathbb{R}^3 \times \mathbb{R}^+ \tag{8.311}$$

where $\mathfrak{L}\{\bullet\}$ denotes a linear *differential* operator involving spatial and temporal derivatives, $\Upsilon(\mathbf{r}, t)$ is a scalar field and $s(\mathbf{r}, t)$ is a known source with support the finite region V_S and null for $t < 0$. Linearity is a key ingredient as it allows using the principle of superposition and hence combine the effect of multiple sources. For an arbitrary forcing term $s(\mathbf{r}, t)$ an integral of the form

$$\Upsilon(\mathbf{r},t) = \int\limits_{\mathbb{R}^3} \mathrm{d}V' \int\limits_{\mathbb{R}} \mathrm{d}t' \mathcal{G}(\mathbf{r},\mathbf{r}',t,t')s(\mathbf{r}',t') = \int\limits_{V_S} \mathrm{d}V' \int\limits_0^t \mathrm{d}t' \mathcal{G}(\mathbf{r},\mathbf{r}',t,t')s(\mathbf{r}',t') \tag{8.312}$$

may be interpreted as an *integral transformation* which associates the source with the solution $\Upsilon(\mathbf{r}, t)$, but also as the *inverse* of $\mathfrak{L}\{\bullet\}$. The kernel of the transformation $\mathcal{G}(\mathbf{r}, \mathbf{r}', t, t')$ is the scalar Green function of the operator $\mathfrak{L}\{\bullet\}$. The integration with respect to time effectively ranges from 0 to t because the source is 'switched on' at $t' = 0$ and $\mathcal{G}(\mathbf{r}, \mathbf{r}', t, t') = 0$ for $t' > t$. The latter requirement is dictated by the principle of causality, which forbids the response of a physical system (at time t) to be affected by the future state of the source (at time $t' > t$). Stated another way, the *cause* may not precede the *effect*. From a physical viewpoint (8.312) may be seen as the *linear superposition* of the effects of infinitely many elementary sources each one of them located at a point $\mathbf{r}' \in V_S$ and active at the time t'.

To allow for singular sources localized at one or more points and for solutions $\Upsilon(\mathbf{r}, t)$ that are not differentiable, it is better to write (8.311) in a weak form or distributional sense (Appendix C). Let $\phi(\mathbf{r}, t) \in \mathcal{C}_0^\infty(\mathbb{R}^4)$ indicate a real-valued test function and define the symmetric inner product

$$(\Upsilon, \phi)_{\mathbb{R}^4} := \int_{\mathbb{R}^3} \mathrm{d}V \int_{\mathbb{R}} \mathrm{d}t\, \Upsilon(\mathbf{r}, t)\phi(\mathbf{r}, t) = (\phi, \Upsilon)_{\mathbb{R}^4} . \tag{8.313}$$

Next, we recall that the *adjoint operator* $\mathfrak{L}^\dagger\{\bullet\}$ is defined as [65, Section 10.4]

$$(\phi, \mathfrak{L}\{\Upsilon\})_{\mathbb{R}^4} = \left(\mathfrak{L}^\dagger\{\phi\}, \Upsilon\right)_{\mathbb{R}^4} \tag{8.314}$$

and that the operator $\mathfrak{L}\{\bullet\}$ is said to be be *self-adjoint* if $\mathfrak{L}^\dagger\{\bullet\} \equiv \mathfrak{L}\{\bullet\}$ (Appendix D.6). Thanks to these positions we may replace (8.311) with the following weak-form equation

$$\left(\Upsilon, \mathfrak{L}^\dagger\{\phi\}\right)_{\mathbb{R}^4} = -(s, \phi)_{\mathbb{R}^4} \tag{8.315}$$

$$\int_{\mathbb{R}^3} \mathrm{d}V \int_{\mathbb{R}} \mathrm{d}t\, \Upsilon(\mathbf{r}, t)\mathfrak{L}^\dagger\{\phi(\mathbf{r}, t)\} = -\int_{\mathbb{R}^3} \mathrm{d}V \int_{\mathbb{R}} \mathrm{d}t\, s(\mathbf{r}, t)\phi(\mathbf{r}, t) \tag{8.316}$$

which is far more general because $\phi(\mathbf{r}, t)$ can stand infinitely many derivatives. In practice, we employ the right-hand side of (8.314) when $\Upsilon(\mathbf{r}, t)$ cannot 'absorb' the derivatives implied in $\mathfrak{L}\{\bullet\}$.

Then, to find the governing equation for $\mathcal{G}(\mathbf{r}, \mathbf{r}', t, t')$ we insert the formal solution (8.312) into (8.316)

$$\int_{\mathbb{R}^3} \mathrm{d}V \int_{\mathbb{R}} \mathrm{d}t \left[\int_{V_S} \mathrm{d}V' \int_0^{+\infty} \mathrm{d}t' \mathcal{G}(\mathbf{r}, \mathbf{r}', t, t')s(\mathbf{r}', t')\right] \mathfrak{L}^\dagger\{\phi(\mathbf{r}, t)\}$$
$$= -\int_{V_S} \mathrm{d}V' \int_0^{+\infty} \mathrm{d}t'\, s(\mathbf{r}', t')\phi(\mathbf{r}', t') \tag{8.317}$$

whereby we obtain

$$\int_{V_S} \mathrm{d}V' \int_0^{+\infty} \mathrm{d}t' s(\mathbf{r}', t') \left[\int_{\mathbb{R}^3} \mathrm{d}V \int_{\mathbb{R}} \mathrm{d}t\, \mathcal{G}(\mathbf{r}, \mathbf{r}', t, t')\mathfrak{L}^\dagger\{\phi(\mathbf{r}, t)\} + \phi(\mathbf{r}', t')\right] = 0 \tag{8.318}$$

by swapping the order of integration. Since the last equation must hold true for any source $s(\mathbf{r}', t')$ we demand that

$$\int_{\mathbb{R}^3} \mathrm{d}V \int_{\mathbb{R}} \mathrm{d}t\, \mathcal{G}(\mathbf{r}, \mathbf{r}', t, t')\mathfrak{L}^\dagger\{\phi(\mathbf{r}, t)\} = -\phi(\mathbf{r}', t'), \qquad (\mathbf{r}', t') \in \mathbb{R}^3 \times \mathbb{R}^+ \tag{8.319}$$

or, equivalently, in short-hand notation (Appendix C) [15, Section 5.2.1]

$$\mathfrak{L}\{\mathcal{G}(\mathbf{r},\mathbf{r}',t,t')\} = -\delta^{(3)}(\mathbf{r}-\mathbf{r}')\,\delta(t-t') \tag{8.320}$$

by virtue of (8.314) applied formally and (C.4), (C.19). Equation (8.320) shows that $\mathcal{G}(\mathbf{r},\mathbf{r}',t,t')$ is best interpreted as a distribution and that the corresponding source term is a singular source localized both in space and time.

8.7.1 Frequency domain

In our quest for the fundamental solutions to the wave equations for the potentials (8.122), (8.123), (8.126) and (8.127) we first consider time-harmonic fields or, more generally, a representation in terms of temporal Fourier transforms [23, Section 4.8], [74] because this strategy allows us to single out the time dependence and focus on the spatial derivatives only.

We indicate with $\Upsilon(\mathbf{r},t)$ either the scalar potentials $\Phi_E(\mathbf{r},t)$ and $\Phi_M(\mathbf{r},t)$ or any one of the Cartesian components of $\mathcal{A}_E(\mathbf{r},t)$ and $\mathcal{A}_M(\mathbf{r},t)$. The scalar field $\Upsilon(\mathbf{r},t)$ satisfies a D'Alembert equation of the type

$$\nabla^2\Upsilon(\mathbf{r},t) - \frac{1}{c^2}\frac{\partial^2}{\partial t^2}\Upsilon(\mathbf{r},t) = -s(\mathbf{r},t), \qquad (\mathbf{r},t)\in\mathbb{R}^3\times\mathbb{R} \tag{8.321}$$

where $s(\mathbf{r},t)$ is proportional to the charge density or any Cartesian component of $\mathcal{J}(\mathbf{r},t)$ or $\mathcal{J}_M(\mathbf{r},t)$. We assume that $\Upsilon(\mathbf{r},t)$ and $s(\mathbf{r},t)$ can be obtained from the inverse temporal Fourier transformation of suitable complex scalar fields $\Psi(\mathbf{r};\omega)$ and $S(\mathbf{r};\omega)$, viz.,

$$\Upsilon(\mathbf{r},t) = \mathfrak{F}^{-1}\{\Psi(\mathbf{r})\} := \frac{1}{2\pi}\int_{\mathbb{R}} \mathrm{d}\omega\, e^{\mathrm{j}\omega t}\Psi(\mathbf{r};\omega) \tag{8.322}$$

$$s(\mathbf{r},t) = \mathfrak{F}^{-1}\{S(\mathbf{r})\} := \frac{1}{2\pi}\int_{\mathbb{R}} \mathrm{d}\omega\, e^{\mathrm{j}\omega t}S(\mathbf{r};\omega) \tag{8.323}$$

whereby (8.321) becomes

$$\frac{1}{2\pi}\int_{\mathbb{R}} \mathrm{d}\omega\, e^{\mathrm{j}\omega t}\left[\nabla^2\Psi(\mathbf{r};\omega) + \frac{\omega^2}{c^2}\Psi(\mathbf{r};\omega) + S(\mathbf{r};\omega)\right] = 0 \tag{8.324}$$

having swapped differential and integral operators under the hypotheses that

- as a function of $\mathbf{r}$ and ω the integrand $\exp(\mathrm{j}\,\omega t)\Psi(\mathbf{r};\omega)$ and its derivatives up to the second order are continuous for $(\mathbf{r},\omega)\in\mathbb{R}^3\times\mathbb{R}$;
- as a function of t and ω the integrand $\exp(\mathrm{j}\,\omega t)\Psi(\mathbf{r};\omega)$ and its derivatives up to the second order are continuous for $(t,\omega)\in\mathbb{R}\times\mathbb{R}$.

Since the last equation must be true at all times t, the scalar field $\Psi(\mathbf{r};\omega)$ must in turn solve the following Helmholtz equation [49, Section 2.5], [16, Section 7.7], [43, Section 8.5], [51]

$$\nabla^2\Psi(\mathbf{r};\omega) + \frac{\omega^2}{c^2}\Psi(\mathbf{r};\omega) = \nabla^2\Psi(\mathbf{r};\omega) + k^2\Psi(\mathbf{r};\omega) = -S(\mathbf{r};\omega), \qquad \mathbf{r}\in\mathbb{R}^3 \tag{8.325}$$

for the exponential function never vanishes. Owing to the linearity of the Helmholtz operator we may try seeking a solution in the form

$$\Psi(\mathbf{r};\omega) = \int_{\mathbb{R}^3} \mathrm{d}V'\, G(\mathbf{r},\mathbf{r}';\omega)S(\mathbf{r}';\omega) \tag{8.326}$$

where $G(\mathbf{r}, \mathbf{r}'; \omega)$ is the Green function in the frequency domain. To gain more generality, with the aid of the definition of weak Laplace operator (C.48) we interpret (8.325) in a weak sense, i.e.,

$$\int_{\mathbb{R}^3} \mathrm{d}V\, \Psi(\mathbf{r}) \left[\nabla^2 \phi(\mathbf{r}) + k^2 \phi(\mathbf{r})\right] = -\int_{\mathbb{R}^3} \mathrm{d}V\, \phi(\mathbf{r}) S(\mathbf{r}; \omega) \tag{8.327}$$

where $\phi(\mathbf{r}) \in \mathcal{C}_0^\infty(\mathbb{R}^3)$ is a test function (Appendix C). This step is necessary to allow for solutions $\Psi(\mathbf{r}; \omega)$ which are not twice differentiable. Inserting (8.326) into (8.327) yields

$$\int_{\mathbb{R}^3} \mathrm{d}V'\, S(\mathbf{r}'; \omega) \left\{ \int_{\mathbb{R}^3} \mathrm{d}V\, G(\mathbf{r}, \mathbf{r}'; \omega) \left[\nabla^2 \phi(\mathbf{r}) + k^2 \phi(\mathbf{r})\right] \right\} = -\int_{\mathbb{R}^3} \mathrm{d}V'\, \phi(\mathbf{r}') S(\mathbf{r}'; \omega) \tag{8.328}$$

where we have swapped the order of integration with respect to $\mathbf{r}$ and $\mathbf{r}'$ and changed the name of the dummy variable of integration in the right-hand side. This equation is true if the scalar field between braces solves

$$\int_{\mathbb{R}^3} \mathrm{d}V\, G(\mathbf{r}, \mathbf{r}'; \omega) \left[\nabla^2 \phi(\mathbf{r}) + k^2 \phi(\mathbf{r})\right] = -\phi(\mathbf{r}') \tag{8.329}$$

or with equivalent compact notation

$$\nabla^2 G(\mathbf{r}, \mathbf{r}'; \omega) + k^2 G(\mathbf{r}, \mathbf{r}'; \omega) = -\delta^{(3)}(\mathbf{r} - \mathbf{r}'), \qquad \mathbf{r}, \mathbf{r}' \in \mathbb{R}^3 \tag{8.330}$$

that is, the application of the weak Helmholtz operator to the Green function produces the three-dimensional Dirac delta distribution.

We are now ready to compute the Green function $G(\mathbf{r}, \mathbf{r}' = 0; \omega) = G(\mathbf{r})$ which is the unique solution to the problem [60, Chapter 5], [75, Section 26.2], [21]

$$\begin{cases} \nabla^2 G(\mathbf{r}) + k^2 G(\mathbf{r}) = 0, & \mathbf{r} \in \mathbb{R}^3 \setminus \{0\} \\ \hat{\mathbf{r}} \cdot \nabla G(\mathbf{r}) + \mathrm{j}\, k G(\mathbf{r}) = \mathcal{O}\left(\dfrac{1}{|\mathbf{r}|^2}\right), & |\mathbf{r}| \to +\infty \\ \displaystyle\int_{\partial B} \mathrm{d}S\, \hat{\mathbf{r}} \cdot \nabla G(\mathbf{r}) + \int_{B(0,a)} \mathrm{d}V\, k^2 G(\mathbf{r}) = -1 & \end{cases} \tag{8.331}$$

where $B(0, a)$, $a > 0$, is a 'small' ball which surrounds and isolates the point $\mathbf{r} = 0$, i.e., the location of the singular point source. The normalization condition — which is added to ensure uniqueness — is motivated by the analogous one we employed in (2.125) for the static Green function. Indeed, we cannot just invoke the boundary conditions of the proof described in Section 8.5.4, because the domain of interest is 'punctured' or, equivalently, the excluded region V_1 has shrunk to a point. Comparing (8.331) with the time-harmonic counterpart of (8.122) helps us understand that the normalization condition amounts to assuming the presence of a unitary point source in $\mathbf{r} = 0$. Then again, even though we interpret $G(\mathbf{r})$ as a special scalar potential solution, we cannot just invoke the global Gauss law (1.16) on $\nabla G(\mathbf{r})$ to normalize the Green function — as we did in the static case — because $\nabla G(\mathbf{r})$ constitutes only a part of the electric field in accordance with representation (8.104).

Spherical symmetry prompts us to introduce a system of polar coordinates (r, ϑ, φ) and to look for solutions that are independent of the angles ϑ and φ. Therefore, it is expedient to cast the Helmholtz equation in the form

$$\frac{1}{r^2} \frac{\mathrm{d}}{\mathrm{d}r} \left(r^2 \frac{\mathrm{d}G}{\mathrm{d}r} \right) + k^2 G(r) = \frac{1}{r} \frac{\mathrm{d}^2}{\mathrm{d}r^2} [r G(r)] + k^2 G(r) = 0, \qquad r > 0 \tag{8.332}$$

with a few manipulations. Since the latter (upon multiplication by $r \neq 0$) reduces to a special instance of the harmonic equation (8.197) for the unknown function $rG(r)$, we may write the general solution as

$$G(r) = c_1 \frac{\mathrm{e}^{-\mathrm{j}kr}}{r} + c_2 \frac{\mathrm{e}^{\mathrm{j}kr}}{r}, \qquad r > 0 \tag{8.333}$$

with c_1 and c_2 two constants. Clearly, the Helmholtz equation has infinitely many solutions, unless we enforce the Sommerfeld radiation condition (8.239) *and* the normalization relationship to fix c_1 and c_2.

It is a simple matter to check that the second term in the sum above does not satisfy (8.239) and hence must be discarded by setting $c_2 = 0$. The value of the remaining constant is dictated by the chosen normalization condition, which provides us with an algebraic equation for c_1, namely,

$$\begin{aligned} -1 &= \int_{\partial B} \mathrm{d}S\, \hat{\mathbf{r}} \cdot \nabla G(r) + \int_{B(0,a)} \mathrm{d}V\, k^2 G(r) \\ &= -4\pi a^2 c_1 \frac{\mathrm{e}^{-\mathrm{j}ka}}{a}\left(\mathrm{j}k + \frac{1}{a}\right) + 4\pi k^2 c_1 \left(\mathrm{j}a\frac{\mathrm{e}^{-\mathrm{j}ka}}{k} + \frac{\mathrm{e}^{-\mathrm{j}ka} - 1}{k^2}\right) = -4\pi c_1 \end{aligned} \tag{8.334}$$

whence we find $c_1 = 1/(4\pi)$. In this way, we arrive at

$$G(\mathbf{r}) = \frac{\mathrm{e}^{-\mathrm{j}kr}}{4\pi r} = \frac{\mathrm{e}^{-\mathrm{j}k|\mathbf{r}|}}{4\pi|\mathbf{r}|}, \qquad \mathbf{r} \in \mathbb{R}^3 \setminus \{0\} \tag{8.335}$$

which consistently reduces to the static solution (2.130) in the limit as $k \to 0^+$.

Given the relative simplicity of the Helmholtz equation it is instructive to determine the Green function through a *spectral approach*, that is, by turning the differential equation into an algebraic one with the aid of the three-dimensional spatial Fourier transformation [15]. To this purpose we return to (8.194) and suppose that the potential $\Psi(\mathbf{r})$ and the source $S(\mathbf{r})$ may be expressed as

$$\Psi(\mathbf{r}) = \mathfrak{F}_3^{-1}\{\tilde{\Psi}\} := \frac{1}{(2\pi)^3} \iiint_{\mathbb{R}^3} \mathrm{d}^3k\, \mathrm{e}^{-\mathrm{j}\mathbf{k}\cdot\mathbf{r}} \tilde{\Psi}(\mathbf{k}) \tag{8.336}$$

$$S(\mathbf{r}) = \mathfrak{F}_3^{-1}\{\tilde{S}\} := \frac{1}{(2\pi)^3} \iiint_{\mathbb{R}^3} \mathrm{d}^3k\, \mathrm{e}^{-\mathrm{j}\mathbf{k}\cdot\mathbf{r}} \tilde{S}(\mathbf{k}) \tag{8.337}$$

where $\tilde{\Psi}(\mathbf{k})$ and $\tilde{S}(\mathbf{k})$ are complex functions of the *spatial frequency* $\mathbf{k} = k_x\hat{\mathbf{x}} + k_y\hat{\mathbf{y}} + k_z\hat{\mathbf{z}}$ (physical dimension: 1/m). It is customary to use the symbol $\mathbf{k}$ for the spatial frequency and the wavevector as well (see Section 7.2) and, in effect, these quantities are only conceptually different. Thus, to avoid confusion we provisionally replace k (wavenumber) with ω/c in (8.194) and utilize the symbol k as a shorthand notation for the quantity $(\mathbf{k} \cdot \mathbf{k})^{1/2}$. Parenthetically, we remark that k is not necessarily the magnitude of $\mathbf{k}$ as the latter may complex if needed.

By inserting (8.336) and (8.337) into (8.194) we find

$$\frac{1}{(2\pi)^3} \iiint_{\mathbb{R}^3} \mathrm{d}^3k\, \mathrm{e}^{-\mathrm{j}\mathbf{k}\cdot\mathbf{r}} \left[\left(-\mathbf{k}\cdot\mathbf{k} + \frac{\omega^2}{c^2}\right)\tilde{\Psi}(\mathbf{k}) + \tilde{S}(\mathbf{k})\right] = 0 \tag{8.338}$$

having swapped the order of integration and spatial differentiation under the assumption that the integrand $\tilde{\Psi}(\mathbf{k})\exp(-\mathrm{j}\,\mathbf{k}\cdot\mathbf{r})$ as a function of $\mathbf{r}$ and $\mathbf{k}$ has continuous derivatives with respect to x, y, z,

k_x, k_y and k_z up to the second order for $(\mathbf{r}, \mathbf{k}) \in \mathbb{R}^3 \times \mathbb{R}^3$. Since the exponential function $\exp(-\mathrm{j}\,\mathbf{k}\cdot\mathbf{r})$ never vanishes for any value of the argument, the last equation is satisfied if we set

$$\left(-\mathbf{k}\cdot\mathbf{k} + \frac{\omega^2}{c^2}\right)\tilde{\Psi}(\mathbf{k}) = -\tilde{S}(\mathbf{k}) \tag{8.339}$$

and this algebraic relation represents the spectral equivalent to (8.194). The solution is trivial and reads

$$\tilde{\Psi}(\mathbf{k}) = \frac{\tilde{S}(\mathbf{k})}{\mathbf{k}\cdot\mathbf{k} - \omega^2/c^2} = \frac{\tilde{S}(\mathbf{k})}{k^2 - \omega^2/c^2} = \tilde{G}(\mathbf{k})\tilde{S}(\mathbf{k}) \tag{8.340}$$

where $\tilde{G}(\mathbf{k})$ is the *spectral* scalar Green function for the Helmholtz equation. Evidently, $\tilde{G}(\mathbf{k})$ represents the spectral scalar potential due to a source with unitary Fourier transform. In the space of the vectors $\mathbf{k} \in \mathbb{R}^3$ the Green function is *singular* for points over the sphere of radius $|\mathbf{k}| = \omega/c$, that is, when the magnitude of the spectral variable coincides with the wavenumber of the background isotropic medium. As a matter of fact, under such circumstances (8.339) may not be inverted.

We return to the spatial domain by inserting (8.340) into (8.336)

$$\Psi(\mathbf{r}) = \frac{1}{(2\pi)^3}\iiint_{\mathbb{R}^3} \mathrm{d}^3k\, \mathrm{e}^{-\mathrm{j}\mathbf{k}\cdot\mathbf{r}} \frac{\tilde{S}(\mathbf{k})}{k^2 - \omega^2/c^2} = \int_{\mathbb{R}^3} \mathrm{d}V'\, G(\mathbf{r} - \mathbf{r}')S(\mathbf{r}') \tag{8.341}$$

where we have applied the convolution theorem [23, Section 4.8] for the three-dimensional Fourier transformation and, in particular, $G(\mathbf{r})$ is the inverse Fourier transform of $\tilde{G}(\mathbf{k})$. To carry out the inverse Fourier transformation of $\tilde{G}(\mathbf{k})$ explicitly it is convenient to introduce a local system of polar spherical coordinates (k, α, β) in which the polar axis is aligned with the position vector $\mathbf{r}$ so that $\mathbf{k}\cdot\mathbf{r} = kr\cos\alpha$. In symbols, we have

$$\begin{aligned} G(\mathbf{r}) &= \frac{1}{(2\pi)^3}\iiint_{\mathbb{R}^3} \mathrm{d}^3k\, \frac{\mathrm{e}^{-\mathrm{j}\mathbf{k}\cdot\mathbf{r}}}{k^2 - \omega^2/c^2} = \frac{1}{(2\pi)^3}\int_0^{+\infty} \mathrm{d}k \int_0^{\pi} \mathrm{d}\alpha \int_0^{2\pi} \mathrm{d}\beta\, \frac{k^2 \mathrm{e}^{-\mathrm{j}kr\cos\alpha}}{k^2 - \omega^2/c^2}\sin\alpha \\ &= \frac{\mathrm{j}}{(2\pi)^2 r}\int_0^{+\infty} \mathrm{d}k\, \frac{k\mathrm{e}^{-\mathrm{j}kr}}{k^2 - \omega^2/c^2} - \frac{\mathrm{j}}{(2\pi)^2 r}\int_0^{+\infty} \mathrm{d}k\, \frac{k\mathrm{e}^{\mathrm{j}kr}}{k^2 - \omega^2/c^2} \end{aligned} \tag{8.342}$$

and in the last two integrals we perform the changes of dummy variable $k = \zeta$ and $k = -\zeta$, respectively, to arrive at

$$G(\mathbf{r}) = \frac{\mathrm{j}}{4\pi^2 r}\int_{\mathbb{R}} \mathrm{d}\zeta\, \frac{\zeta \mathrm{e}^{-\mathrm{j}\zeta r}}{\zeta^2 - \omega^2/c^2} \tag{8.343}$$

an improper integral we may compute with the aid of the Cauchy theorem (B.57) in the complex plane of the variable $\zeta = \zeta' + \mathrm{j}\,\zeta'' \in \mathbb{C}$. Meanwhile, since the integration variable is now called ζ in (8.343), we may go back to using the symbol k to denote the wavenumber as in (1.248).

We observe that the integrand has two poles on the real axis for $\zeta_1 = k = \omega/c$ and $\zeta_2 = -\zeta_1$, as is shown in Figure 8.2a. Thus, to give the integral a meaning we may integrate along an arbitrary path γ which avoids the singularities so long as γ begins in $\zeta = -\infty - \mathrm{j}\,\zeta''$ and ends in $\zeta = +\infty + \mathrm{j}\,\zeta''$ with $\zeta'' > 0$. This alteration is possible by virtue of property (B.32) because the integrand does not exhibit other singularities for $|\zeta| < +\infty$. However, it is not immediately apparent whether we

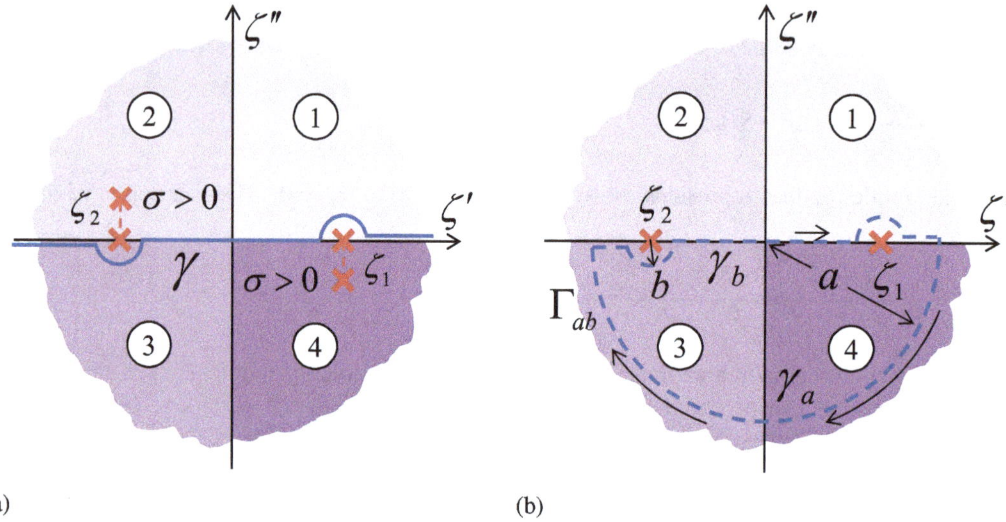

Figure 8.2 Inverse Fourier transformation of the three-dimensional spectral Green function: (a) poles (×), migration thereof in the presence of small losses, and path of integration (−) in the complex plane ζ; (b) poles (×) and integration contour $\Gamma_{ab} := \gamma_a \cup \gamma_b$ (−−) in the complex plane ζ.

must 'walk' around the poles by 'trespassing' in the upper or the lower imaginary half-plane. Such ambiguity is a consequence of the theoretical existence (discussed above) of outgoing and ingoing wave-like solutions and ultimately means that the solution is not unique.

In the spatial domain the way out consisted of enforcing the Sommerfeld radiation condition (8.239). In the spectral domain we resort to the conceptual trick of temporarily assuming that the underlying medium possesses a small non-null conductivity $\sigma \ll \omega\varepsilon$ [12, Section 1.5]. As a result, the wavenumber becomes complex and exhibits a small negative imaginary part [see (7.98)]

$$k = \frac{\omega}{c}\left(1 - \mathrm{j}\frac{\sigma}{2\omega\varepsilon}\right), \qquad \sigma \ll \omega\varepsilon \tag{8.344}$$

and this extension causes the poles ζ_1 and ζ_2 to 'drift' away from the real axis. More specifically, $\zeta_1 = k$ migrates into the lower imaginary plane and $\zeta_2 = -k$ into the upper imaginary plane (Figure 8.2a). Hence, when the medium is even a wee lossy the ambiguity in (8.343) disappears, and we may actually integrate along the real axis. This thought experiment also helps us understand that, in the limit as $\sigma \to 0^+$, we have to choose a path γ that partly runs in the lower (resp. upper) imaginary half-plane so as to steer clear of the pole ζ_2 (resp. ζ_1).

Therefore, in (8.343) we can trade the ambiguous integral along the real axis for a *perfectly equivalent* one along a curve, viz.,

$$G(\mathbf{r}) = \frac{\mathrm{j}}{4\pi^2 r}\int_\gamma \mathrm{d}\zeta\,\frac{\zeta e^{-\mathrm{j}\zeta r}}{\zeta^2 - k^2} \tag{8.345}$$

where $k = \omega/c \in \mathbb{R}^+$, and γ is the jagged continuous line which skirts the poles along two half-circumferences centered in ζ_1 and ζ_2, as is shown in Figure 8.2a. To apply the Cauchy theorem of residues (B.57), though, we need a convenient loop of which γ is only a part. Since for $r > 0$ the integrand decays exponentially as $\mathrm{Im}\{\zeta\} = \zeta'' \to -\infty$, we consider the contour $\Gamma_{ab} := \gamma_a \cup \gamma_b$ (the dashed line in Figure 8.2b) where

$$\gamma_a := \{\zeta \in \mathbb{C} : \zeta = a\exp(-\mathrm{j}\,\alpha), \alpha \in [0, \pi]\} \tag{8.346}$$

$$\begin{aligned}\gamma_b := \{&\zeta \in \mathbb{C} : \zeta = \xi, |\xi| < k - b, k + b < |\xi| < a \\ &\cup \zeta = \zeta_1 - b\exp(-\mathrm{j}\,\alpha) \cup \zeta = \zeta_2 - b\exp(\mathrm{j}\,\alpha), \alpha \in [0, \pi]\}\end{aligned} \tag{8.347}$$

with $a > b > 0$ and $b < k$. Now, thanks to (B.57) the integral along Γ_{ab} is readily evaluated as

$$\frac{\mathrm{j}}{4\pi^2 r}\oint_{\Gamma_{ab}} \mathrm{d}\zeta\, \frac{\zeta \mathrm{e}^{-\mathrm{j}\zeta r}}{\zeta^2 - k^2} = -2\pi\mathrm{j}\lim_{\zeta\to\zeta_1} \frac{(\zeta - \zeta_1)\zeta \mathrm{e}^{-\mathrm{j}\zeta r}}{(\zeta - \zeta_1)(\zeta - \zeta_2)}\frac{\mathrm{j}}{4\pi^2 r} = \frac{\mathrm{e}^{-\mathrm{j}kr}}{4\pi r} \tag{8.348}$$

because Γ_{ab} encircles only the pole ζ_1 in a clockwise fashion. We use (8.348), split the contour integral into two contributions, namely,

$$\frac{\mathrm{e}^{-\mathrm{j}kr}}{4\pi r} = \frac{\mathrm{j}}{4\pi^2 r}\int_{\gamma_a} \mathrm{d}\zeta\, \frac{\zeta \mathrm{e}^{-\mathrm{j}\zeta r}}{\zeta^2 - k^2} + \frac{\mathrm{j}}{4\pi^2 r}\int_{\gamma_b} \mathrm{d}\zeta\, \frac{\zeta \mathrm{e}^{-\mathrm{j}\zeta r}}{\zeta^2 - k^2} \tag{8.349}$$

and observe that, in the limit as $a \to +\infty$, the line γ_b reduces to the path γ (Figure 8.2a) and the corresponding integral passes over into the right-hand side of (8.345). Thus, if we can show that the remaining integral along γ_a vanishes as $a \to +\infty$, we have the desired result. We consider the estimate

$$\begin{aligned}\left|\int_{\gamma_a} \mathrm{d}\zeta\, \frac{\zeta \mathrm{e}^{-\mathrm{j}\zeta r}}{\zeta^2 - k^2}\right| &= \left|\int_0^\pi \mathrm{d}\alpha\,(-\mathrm{j}\,a^2)\mathrm{e}^{-2\mathrm{j}\,\alpha}\frac{\mathrm{e}^{-\mathrm{j}\,ar\cos\alpha}\mathrm{e}^{-ar\sin\alpha}}{a^2\mathrm{e}^{-2\mathrm{j}\,\alpha} - k^2}\right| \\ &\leqslant \int_0^\pi \mathrm{d}\alpha\, \frac{a^2\mathrm{e}^{-ar\sin\alpha}}{|a^2\mathrm{e}^{-2\mathrm{j}\,\alpha} - k^2|} \leqslant M\int_0^\pi \mathrm{d}\alpha\, \mathrm{e}^{-ar\sin\alpha} \leqslant 2M\int_0^{\pi/2} \mathrm{d}\alpha\, \mathrm{e}^{-ar\sin\alpha} \\ &\leqslant 2M\int_0^{\pi/2} \mathrm{d}\alpha\, \mathrm{e}^{-2ar\alpha/\pi} = M\frac{\pi}{ar}(1 - \mathrm{e}^{-ra}) \xrightarrow[a\to+\infty]{} 0\end{aligned} \tag{8.350}$$

where $M > 1$ is a suitable constant (see note on page 474) and we have employed the Jordan inequality (B.59).

In conclusion, we have found

$$G(\mathbf{r}) = G(r) = \frac{\mathrm{e}^{-\mathrm{j}kr}}{4\pi r}, \qquad \mathbf{r} \in \mathbb{R}^3 \setminus \{0\} \tag{8.351}$$

which coincides with the previous expression (8.335). If we had integrated along the alternative contour $\Gamma_{ab} := \gamma_a \cup \gamma_b'$ with γ_b' being the specular reflection of γ_b around the real axis ζ', the result would have been determined by the residue in the pole ζ_2. It is easy to check that this procedure would yield the Green function corresponding to an ingoing wave, which we discarded on the grounds of the physical behavior of sources and the Sommerfeld radiation condition.

We now show that (8.351) is the solution to the distributional Helmholtz equation (8.330) with $\mathbf{r}' = 0$. In light of (C.19) and (C.48) we see that (8.330) is a shorthand notation for

$$\int_{\mathbb{R}^3} \mathrm{d}V\, G(\mathbf{r})\left[\nabla^2\phi(\mathbf{r}) + k^2\phi(\mathbf{r})\right] = -\phi(0) \tag{8.352}$$

where $\phi(\mathbf{r}) \in \mathcal{C}_0^\infty(\mathbb{R}^3)$ is a test function with bounded support (Appendix C). Thus, we need to prove that the weak Helmholtz operator in the left-hand side returns the negative of the test function sampled in the origin. To this purpose we consider two concentric balls $B(0, a)$ and $B(0, b)$ with $a < b$ where the radius b is large enough for the support of $\phi(\mathbf{r})$ to be entirely contained in $B(0, b)$. Integration over the spherical shell $V := B(0, b) \setminus B(0, a)$ yields

$$\int_V \mathrm{d}V \left[\nabla^2\phi(\mathbf{r}) + k^2\phi(\mathbf{r})\right] G(\mathbf{r}) =$$

$$= \int_V \mathrm{d}V \left\{\nabla \cdot [G(\mathbf{r})\nabla\phi(\mathbf{r})] - \nabla G \cdot \nabla\phi(\mathbf{r}) + k^2 G(\mathbf{r})\phi(\mathbf{r})\right\}$$

$$= \int_V \mathrm{d}V\, \nabla \cdot [G(\mathbf{r})\nabla\phi(\mathbf{r}) - \phi(\mathbf{r})\nabla G(\mathbf{r})] + \int_V \mathrm{d}V\, \phi(\mathbf{r}) \underbrace{\left[\nabla^2 G(\mathbf{r}) + k^2 G(\mathbf{r})\right]}_{=0}$$

$$= \int_{|\mathbf{r}|=a} \mathrm{d}S\, \hat{\mathbf{n}}(\mathbf{r}) \cdot [G(\mathbf{r})\nabla\phi(\mathbf{r}) - \phi(\mathbf{r})\nabla G(\mathbf{r})] + \int_{|\mathbf{r}|=b} \mathrm{d}S\, \hat{\mathbf{n}}(\mathbf{r}) \cdot [G(\mathbf{r})\nabla\phi(\mathbf{r}) - \phi(\mathbf{r})\nabla G(\mathbf{r})] \quad (8.353)$$

where we have invoked the Gauss theorem (A.53) because the integrands are regular and differentiable in V. The flux integral over the sphere $\partial B_b := \{\mathbf{r} \in \mathbb{R}^3 : |\mathbf{r}| = b\}$ vanishes inasmuch as the test function is null for points thereon by assumption. For the remaining flux integral we observe

$$G(\mathbf{r}) = \frac{\mathrm{e}^{-\mathrm{j}ka}}{4\pi a}, \qquad \hat{\mathbf{n}}(\mathbf{r}) \cdot \nabla G(\mathbf{r}) = \frac{1 + \mathrm{j}ka}{4\pi a^2}\mathrm{e}^{-\mathrm{j}ka}, \qquad |\mathbf{r}| = a \quad (8.354)$$

and evaluate

$$\int_{|\mathbf{r}|=a} \mathrm{d}S\, \hat{\mathbf{n}}(\mathbf{r}) \cdot [G(\mathbf{r})\nabla\phi(\mathbf{r}) - \phi(\mathbf{r})\nabla G(\mathbf{r})] =$$

$$= \int_{|\mathbf{r}|=a} \mathrm{d}S \left[\frac{\hat{\mathbf{n}}(\mathbf{r}) \cdot \nabla\phi(\mathbf{r})}{4\pi a} - \phi(\mathbf{r})\frac{1 + \mathrm{j}ka}{4\pi a^2}\right] \mathrm{e}^{-\mathrm{j}ka}$$

$$= \mathrm{e}^{-\mathrm{j}ka}[a\hat{\mathbf{n}}(\mathbf{r}_0) \cdot \nabla\phi(\mathbf{r})|_{\mathbf{r}_0} - \phi(\mathbf{r}_0)(1 + \mathrm{j}ka)] \xrightarrow[a\to 0^+]{} -\phi(0) \quad (8.355)$$

where we have applied the mean value theorem [22] with $\mathbf{r}_0 = -a\hat{\mathbf{n}}(\mathbf{r}_0)$ a suitable point. Since $V \to \mathbb{R}^3$ in the limit as $a \to 0^+$ and $b \to +\infty$, this ends the proof.

The functional form of (8.335) and (8.351) indicates that the Green function depends only on the *distance* from the singular source. Thereby, we infer that the expression

$$G(\mathbf{r}, \mathbf{r}') = \frac{\mathrm{e}^{-\mathrm{j}k|\mathbf{r}-\mathbf{r}'|}}{4\pi|\mathbf{r} - \mathbf{r}'|} = \frac{\mathrm{e}^{-\mathrm{j}kR}}{4\pi R} = G(R), \qquad \mathbf{r}, \mathbf{r}' \in \mathbb{R}^3 \quad (8.356)$$

gives the Green function in the case where the point source is located in the point $\mathbf{r}'$.

8.7.2 Time domain

Since we already have at our disposal the Green function $G(\mathbf{r}, \mathbf{r}'; \omega)$ in (8.356), the quickest way to obtain the time-domain counterpart $\mathcal{G}(\mathbf{r}, \mathbf{r}', t, t')$ consists of deriving the generic potential $\Upsilon(\mathbf{r}, t)$ which solves (8.321) by carrying out the inverse Fourier transformation of $\Psi(\mathbf{r}; \omega)$ prescribed by

(8.322). Based on past developments, we can expect the resulting expression to be an integral which involves the source and the desired Green function.

Specifically, by inserting the integral representation (8.326) into (8.322), using (8.356) and recalling the definition of wavenumber (1.248) we get

$$
\begin{aligned}
\Upsilon(\mathbf{r},t) &= \frac{1}{2\pi}\int_{\mathbb{R}} \mathrm{d}\omega\, \mathrm{e}^{\mathrm{j}\,\omega t}\Psi(\mathbf{r};\omega) = \frac{1}{2\pi}\int_{\mathbb{R}} \mathrm{d}\omega\, \mathrm{e}^{\mathrm{j}\,\omega t}\int_{\mathbb{R}^3} \mathrm{d}V' \,\frac{\mathrm{e}^{-\mathrm{j}\,k|\mathbf{r}-\mathbf{r}'|}}{4\pi|\mathbf{r}-\mathbf{r}'|} S(\mathbf{r}';\omega) \\
&= \frac{1}{2\pi}\int_{\mathbb{R}} \mathrm{d}\omega \int_{\mathbb{R}^3} \mathrm{d}V' \,\exp\left[\mathrm{j}\,\omega\left(t-\frac{|\mathbf{r}-\mathbf{r}'|}{c}\right)\right]\frac{S(\mathbf{r}';\omega)}{4\pi|\mathbf{r}-\mathbf{r}'|} \\
&= \int_{\mathbb{R}^3} \mathrm{d}V' \,\frac{1}{4\pi|\mathbf{r}-\mathbf{r}'|}\underbrace{\frac{1}{2\pi}\int_{\mathbb{R}} \mathrm{d}\omega\, \exp\left[\mathrm{j}\,\omega\left(t-\frac{|\mathbf{r}-\mathbf{r}'|}{c}\right)\right] S(\mathbf{r}';\omega)}_{=\mathfrak{F}^{-1}\{S(\mathbf{r}';\omega)\}} \\
&= \int_{V_S} \mathrm{d}V' \,\frac{s(\mathbf{r}',t-|\mathbf{r}-\mathbf{r}'|/c)}{4\pi|\mathbf{r}-\mathbf{r}'|}
\end{aligned} \tag{8.357}
$$

where $V_S \subset \mathbb{R}^3$ is the region that contains the finite-sized source $s(\mathbf{r},t)$. Although the Green function (8.356) is evidently singular for $\mathbf{r}' = \mathbf{r}$, in the third line above we have swapped the order of integration because the singularity in $R = |\mathbf{r}' - \mathbf{r}|$ disappears when the volume integral is expanded in a local system of polar spherical coordinates (R,α,β) with center in $\mathbf{r}$. Next, we have recognized the resulting integral over ω as an inverse Fourier transform on account of (8.323). Further, we notice that the potential at time t depends on the value of the source at the earlier (retarded) time $t_r = t - |\mathbf{r}-\mathbf{r}'|/c$, though we consign the discussion of this topic to Section 9.1.2.

As far as the solution of the D'Alembert equation (8.321) is concerned, we could just stop at (8.357), in that the latter provides a means for computing $\Upsilon(\mathbf{r},t)$ in a homogeneous unbounded isotropic medium. Remarkably, when the source is constant in time, so is the potential, and (8.357) correctly reduces to the static solution (2.169) of the Poisson equation (2.17) aside from an inconsequential multiplicative factor. On the other hand, for the moment our goal is to 'extract' the time-domain Green function $\mathcal{G}(\mathbf{r},\mathbf{r}',t,t')$ from (8.357) by comparing that expression with the formal solution (8.312). However, the procedure is not straightforward because in (8.357) the integration over t' has been performed, so to speak. This is a minor hurdle which we circumvent by considering the symbolic integral (Appendix C)

$$
s\left(\mathbf{r}',t-\frac{|\mathbf{r}-\mathbf{r}'|}{c}\right) = \int_{\mathbb{R}} \mathrm{d}t'\,\delta\left(t-t'-\frac{|\mathbf{r}-\mathbf{r}'|}{c}\right)s(\mathbf{r}',t') \tag{8.358}
$$

which we regard as the *convolution product* (with respect to time) between the Dirac delta distribution $\delta(t-|\mathbf{r}-\mathbf{r}'|/c)$ and the *distribution* $s(\mathbf{r}',t)$ [15, 17, 76]. By substituting (8.358) into (8.357) — now interpreted in a distributional sense — we obtain

$$
\begin{aligned}
\Upsilon(\mathbf{r},t) &= \int_{V_S} \mathrm{d}V' \,\frac{1}{4\pi|\mathbf{r}-\mathbf{r}'|}\int_{\mathbb{R}} \mathrm{d}t'\,\delta\left(t-t'-\frac{|\mathbf{r}-\mathbf{r}'|}{c}\right)s(\mathbf{r}',t') \\
&= \int_{V_S} \mathrm{d}V' \int_{\mathbb{R}} \mathrm{d}t' \,\frac{\delta(t-t'-|\mathbf{r}-\mathbf{r}'|/c)}{4\pi|\mathbf{r}-\mathbf{r}'|} s(\mathbf{r}',t')
\end{aligned} \tag{8.359}
$$

whence we infer that the time-domain Green function reads

$$\mathcal{G}(\mathbf{r}, \mathbf{r}', t, t') = \frac{\delta(t - t' - |\mathbf{r} - \mathbf{r}'|/c)}{4\pi|\mathbf{r} - \mathbf{r}'|} \tag{8.360}$$

and is, more precisely, a distribution [12, Section 1.5], [16]. In (8.359) we may restrict the integration over t' from 0 up to $t_r = t - |\mathbf{r} - \mathbf{r}'|/c$, inasmuch as the source is turned on at $t' = 0$ and, owing to the finiteness of the speed of light c, an observer standing in a point $\mathbf{r}$ at time t can detect only the waves created at points $\mathbf{r}'$ such that $|\mathbf{r} - \mathbf{r}'| \leqslant c(t - t')$. Since $\mathcal{G}(\mathbf{r}, \mathbf{r}', t, t')$ depends on the retarded time t_r it is also called the *retarded Green function* of the D'Alembert equation (8.321).

As is typical of the Green functions in unbounded isotropic media, $\mathcal{G}(\mathbf{r}, \mathbf{r}', t, t')$ depends on the distance from the source point $|\mathbf{r} - \mathbf{r}'|$ but also on the time interval $t - t' \geqslant 0$, rather than separately on each one of the four variables $\mathbf{r}$, $\mathbf{r}'$, t and t'. Besides, $\mathcal{G}(\mathbf{r}, \mathbf{r}', t, t')$ exhibits the spatial and temporal dependence characteristic of a spherical wave [e.g., see (7.400)] emanating from the source point $\mathbf{r}'$. In fact, $\mathcal{G}(\mathbf{r}, \mathbf{r}', t, t')$ vanishes everywhere in space and time except at the location of the wavefront. The latter is the locus of points for which the argument of the delta distribution vanishes, viz.,

$$|\mathbf{r} - \mathbf{r}'| = c(t - t'), \qquad t' \leqslant t \tag{8.361}$$

and this represents an ever expanding spherical surface of radius $c(t - t')$ and center $\mathbf{r}'$.

In principle, another approach to the derivation of $\mathcal{G}(\mathbf{r}, \mathbf{r}', t, t')$ consists of directly computing the inverse Fourier transform of $G(\mathbf{r}, \mathbf{r}'; \omega)$ in $t - t'$

$$\mathcal{G}(\mathbf{r}, \mathbf{r}', t, t') := \frac{1}{2\pi} \int_{\mathbb{R}} \mathrm{d}\omega\, \mathrm{e}^{\mathrm{j}\,\omega(t-t')} \frac{\mathrm{e}^{-\mathrm{j}\,\omega|\mathbf{r}-\mathbf{r}'|/c}}{4\pi|\mathbf{r} - \mathbf{r}'|} \tag{8.362}$$

but the trouble is that this improper integral, as it stands, does not exist in the usual sense. The root cause of the problem is that $G(\mathbf{r}, \mathbf{r}'; \omega)$ does not decay for $\omega \to \pm\infty$ and hence, strictly speaking, $G(\mathbf{r}, \mathbf{r}'; \omega)$ does not qualify as the Fourier transform of an ordinary function of time [74], [62, Chapter 1], [23, Section 4.8].

Therefore, to give the integral in (8.362) a meaning, we proceed by examining the ordinary real-valued function

$$f(\mathbf{r}, t) := \frac{1}{2\pi} \int_{-\pi/t_0}^{\pi/t_0} \mathrm{d}\omega\, \exp\left[\mathrm{j}\,\omega\left(t - \frac{r}{c}\right)\right] = \frac{\sin\left[\frac{\pi}{t_0}\left(t - \frac{r}{c}\right)\right]}{\pi(t - r/c)} \tag{8.363}$$

where $t_0 > 0$ is an arbitrary time which we regard as a parameter in the definition of $f(\mathbf{r}, t)$. From the explicit expression in the rightmost member of (8.363) we gather that, as t_0 becomes ever smaller, $f(\mathbf{r}, t)$ grows increasingly narrower and peaked around $t = r/c$ where it reaches a local and absolute maximum equal to $1/t_0$. In the limit as $t_0 \to 0^+$, $f(\mathbf{r}, t)$ vanishes everywhere but for $t = r/c$ where it becomes infinitely large.

Under the same circumstances the defining integral of $f(\mathbf{r}, t)$ takes on a form analogous to the one of interest in (8.362), and thus it does not exist. However, by first integrating $f(\mathbf{r}, t)$ for positive values of t_0 with respect to time from $-\infty$ up to $+\infty$, we find

$$\int_{\mathbb{R}} \mathrm{d}t\, f(\mathbf{r}, t) = \int_{\mathbb{R}} \mathrm{d}t\, \frac{\sin\left[\frac{\pi}{t_0}\left(t - \frac{r}{c}\right)\right]}{\pi(t - r/c)} = \frac{2}{\pi} \int_0^{+\infty} \mathrm{d}\xi\, \frac{\sin\left(\frac{\pi}{t_0}\xi\right)}{\xi} = 1 \tag{8.364}$$

where we have changed the dummy variable according to $t = \xi + r/c$ and made use of Formula 4.3.142 in [40]. Since the result of the integration is a constant independent of t_0, we may take the limit for infinitesimal values of t_0 to obtain

$$\lim_{t_0 \to 0^+} \int_{\mathbb{R}} \mathrm{d}t\, f(\mathbf{r}, t) = \lim_{t_0 \to 0^+} \int_{\mathbb{R}} \mathrm{d}t \frac{1}{2\pi} \int_{-\pi/t_0}^{\pi/t_0} \mathrm{d}\omega \, \exp\left[\mathrm{j}\,\omega\left(t - \frac{r}{c}\right)\right] = 1 \tag{8.365}$$

whereby we realize that the inner integral behaves as the one-dimensional Dirac delta distribution of argument $t - r/c$. Formally, we can write (see Appendix C)

$$\lim_{t_0 \to 0^+} \int_{\mathbb{R}} \mathrm{d}t \frac{1}{2\pi} \int_{-\pi/t_0}^{\pi/t_0} \mathrm{d}\omega \, \exp\left[\mathrm{j}\,\omega\left(t - \frac{r}{c}\right)\right] = \int_{\mathbb{R}} \mathrm{d}t \, \delta\left(t - \frac{r}{c}\right) = 1 \tag{8.366}$$

and also

$$\frac{1}{2\pi} \int_{\mathbb{R}} \mathrm{d}\omega \, \exp\left[\mathrm{j}\,\omega\left(t - \frac{r}{c}\right)\right] = \delta\left(t - \frac{r}{c}\right) \tag{8.367}$$

which gives a meaning to the inverse Fourier transformation of the complex exponential. By interpreting the integral in (8.362) symbolically on account of the previous result, we obtain (8.360), as expected.

By now we know that $\mathcal{G}(\mathbf{r}, \mathbf{r}', t, t')$ exists only as a distribution. This feature should be contrasted with the nature of the time-harmonic counterpart $G(\mathbf{r}, \mathbf{r}'; \omega)$ which, at least so long as the observation point $\mathbf{r}$ and the source point $\mathbf{r}'$ do not coincide, can be treated as an ordinary function that solves the homogeneous Helmholtz equation (see Section 8.7.1). Thus, in order to attempt the direct inversion of the distributional D'Alembert equation

$$\nabla^2 \mathcal{G}(\mathbf{r}, t) - \frac{1}{c^2} \frac{\partial^2 \mathcal{G}}{\partial t^2} = -\delta^{(3)}(\mathbf{r})\, \delta(t) \tag{8.368}$$

we recall from (C.40) that $\delta(t)$ equals the temporal derivative of the one-dimensional unit step distribution $U(t)$. By introducing the auxiliary distribution $\Gamma(\mathbf{r}, t)$ such that

$$\mathcal{G}(\mathbf{r}, t) := \frac{\partial}{\partial t} \Gamma(\mathbf{r}, t) \tag{8.369}$$

we can transform (8.368) into

$$\frac{\partial}{\partial t} \left[\nabla^2 \Gamma(\mathbf{r}, t) - \frac{1}{c^2} \frac{\partial^2 \Gamma}{\partial t^2} \right] = -\delta^{(3)}(\mathbf{r}) \frac{\mathrm{d}}{\mathrm{d}t} U(t) \tag{8.370}$$

which in turn is solved if we can determine $\Gamma(\mathbf{r}, t)$ so that

$$\nabla^2 \Gamma(\mathbf{r}, t) - \frac{1}{c^2} \frac{\partial^2 \Gamma}{\partial t^2} = -\delta^{(3)}(\mathbf{r})\, U(t) \tag{8.371}$$

where now the source term is localized (singular) in space but not in time.

Since (8.371) still entails distributions, unless we steer clear of the origin $\mathbf{r} = 0$, as we did for the Helmholtz equation (8.330), to proceed we consider the following homogeneous problem for the *function* $\Gamma(\mathbf{r}, t)$

$$\begin{cases} \nabla^2\Gamma(\mathbf{r}, t) - \dfrac{1}{c^2}\dfrac{\partial^2\Gamma}{\partial t^2} = 0, & \mathbf{r} \in \mathbb{R}^3 \setminus \{0\}, \quad t \in \mathbb{R} \\ \hat{\mathbf{r}} \cdot \nabla\Gamma(\mathbf{r}, t) + \dfrac{1}{c}\dfrac{\partial\Gamma}{\partial t} = \mathcal{O}\left(\dfrac{1}{|\mathbf{r}|^2}\right), & |\mathbf{r}| \to +\infty \\ \displaystyle\int_{\partial B} \mathrm{d}S\, \hat{\mathbf{r}} \cdot \nabla\Gamma(\mathbf{r}, t) - \int_{B(0,a)} \mathrm{d}V\, \frac{1}{c^2}\frac{\partial^2\Gamma}{\partial t^2} = -U(t), & t \in \mathbb{R} \end{cases} \tag{8.372}$$

with $B(0, a)$, $a > 0$, denoting a ball which surrounds and isolates the point $\mathbf{r} = 0$ where the singular source is placed. The second requirement — which we shall better motivate in Section 9.1.2 — is the time-domain *radiation condition* (8.304), follows by taking the inverse Fourier transform of (8.239), and serves to pick up the physical *outward-travelling* waves. Finally, the last constraint is justified by the analogous one used in (8.331) but it differs from it in that both sides depend on time.

We expect $\Gamma(\mathbf{r}, t)$ to depend on the radial distance only, because the source is spherically symmetric. Hence, of the full Laplace operator cast in spherical coordinates (A.45) we retain only the derivative with respect to r and write the relevant D'Alembert equation as

$$\frac{1}{r}\frac{\partial^2}{\partial r^2}[r\Gamma(r, t)] - \frac{1}{c^2}\frac{\partial^2\Gamma}{\partial t^2} = 0, \qquad r > 0, \qquad t \in \mathbb{R} \tag{8.373}$$

which is evidently solved by

$$\Gamma(r, t) = \frac{1}{r}f\left(t - \frac{r}{c}\right) + \frac{1}{r}g\left(t + \frac{r}{c}\right), \qquad r > 0, \qquad t \in \mathbb{R} \tag{8.374}$$

where $f(\bullet)$ and $g(\bullet)$ are *arbitrary* functions at least twice-differentiable with respect to their arguments. We notice that both contributions in the right-hand side of (8.374) represent spherical waves with the expected spatial decay at infinity (see Section 7.7). However, the second term must be discarded, in that it is an *inward-going* wave, travelling from infinity into the source, which is a physical nonsense. So, we set $g(\bullet) = 0$ and go on to show that the first term does obey the radiation condition in (8.372). To this purpose, we observe

$$\begin{aligned} \hat{\mathbf{r}} \cdot \nabla\Gamma(r, t) + \frac{1}{c}\frac{\partial\Gamma}{\partial t} &= \frac{\partial}{\partial r}\left[\frac{1}{r}f\left(t - \frac{r}{c}\right)\right] + \frac{1}{cr}\frac{\partial}{\partial t}f\left(t - \frac{r}{c}\right) \\ &= -\frac{1}{r^2}f\left(t - \frac{r}{c}\right) - \frac{1}{cr}f'\left(t - \frac{r}{c}\right) + \frac{1}{cr}f'\left(t - \frac{r}{c}\right) \\ &= -\frac{1}{r^2}f\left(t - \frac{r}{c}\right) \end{aligned} \tag{8.375}$$

where the prime in $f'(\bullet)$ means 'derivative with respect to the argument'. The expression in the rightmost-hand side is consistent with the second constraint in (8.372) for any function $f(\bullet)$ that is bounded at infinity.

Lastly, we should be able to determine $f(\bullet)$ uniquely by imposing the third condition in (8.372). Indeed, since

$$\frac{\partial\Gamma}{\partial r} = -\frac{1}{r^2}f\left(t - \frac{r}{c}\right) - \frac{1}{cr}f'\left(t - \frac{r}{c}\right) \qquad \frac{1}{c^2}\frac{\partial^2\Gamma}{\partial t^2} = \frac{1}{rc^2}f''\left(t - \frac{r}{c}\right) \tag{8.376}$$

we have

$$-U(t) = \int_{\partial B} \mathrm{d}S\, \frac{\partial \Gamma}{\partial r} - \int_{B(0,a)} \mathrm{d}V\, \frac{1}{c^2}\frac{\partial^2 \Gamma}{\partial t^2}$$
$$= -4\pi f\left(t-\frac{a}{c}\right) - \frac{4\pi a}{c} f'\left(t-\frac{a}{c}\right) - 4\pi \int_0^a \mathrm{d}r\, \frac{r}{c^2} f''\left(t-\frac{r}{c}\right)$$
$$= -4\pi f\left(t-\frac{a}{c}\right) - \frac{4\pi a}{c} f'\left(t-\frac{a}{c}\right) - 4\pi \int_{t-a/c}^{t} \mathrm{d}\xi\, (t-\xi) f''(\xi)$$
$$= -4\pi f\left(t-\frac{a}{c}\right) - \frac{4\pi a}{c} f'\left(t-\frac{a}{c}\right) - 4\pi \left[(t-\xi) f'(\xi) + f(\xi)\right]_{\xi=t-a/c}^{t}$$
$$= -4\pi f(t) \tag{8.377}$$

where we have integrated in standard polar spherical coordinates. With this result we find

$$f\left(t-\frac{r}{c}\right)\bigg|_{r=0} = f(t) = \frac{U(t)}{4\pi} \tag{8.378}$$

which, when substituted back into (8.374) gives

$$\Gamma(r,t) = \frac{1}{4\pi r} U\left(t-\frac{r}{c}\right), \qquad r>0, \qquad t\in\mathbb{R} \tag{8.379}$$

where the right-hand side is a *function* discontinuous for $t = r/c > 0$ and singular for $r = 0$. When interpreted as a *distribution*, $\Gamma(r,t)$ can be differentiated with respect to time in accordance with (C.40), and by virtue of definition (8.369) we obtain

$$\mathcal{G}(\mathbf{r},t) = \frac{\delta(t-r/c)}{4\pi r} \tag{8.380}$$

which is the expected Green function solution to (8.368).

Checking *a posteriori* that $\mathcal{G}(\mathbf{r},t)$ in (8.380) does solve (8.368) is anything but trivial. To elucidate, we pick up a radially symmetric test function $\phi(r,t) \in \mathcal{C}_0^\infty(\mathbb{R}^2)$ and state (8.368) in a weak sense as (Appendix C)

$$\int_{\mathbb{R}^3} \mathrm{d}V \int_{\mathbb{R}} \mathrm{d}t \left(\nabla^2\phi - \frac{1}{c^2}\frac{\partial^2\phi}{\partial t^2}\right)\frac{\delta(t-r/c)}{4\pi r} = -\int_{\mathbb{R}^3} \mathrm{d}V \int_{\mathbb{R}} \mathrm{d}t\, \phi(r,t)\delta^{(3)}(\mathbf{r})\,\delta(t) \tag{8.381}$$

in keeping with definitions (C.48) and (C.37) for $m = 2$. At this stage the integrals are symbolic owing to the presence of the δ's. In particular, according to (C.19) and (C.4) — with x playing the role of t — the right member is just an extended notation for $-\phi(0,0)$, i.e., the test function evaluated in the origin at time $t = 0$. In the left-hand side it is more convenient to perform the symbolic integration over time first and then to tackle the volume integral in spherical coordinates, namely,

$$\int_{\mathbb{R}^3} \mathrm{d}V \int_{\mathbb{R}} \mathrm{d}t \left(\nabla^2\phi - \frac{1}{c^2}\frac{\partial^2\phi}{\partial t^2}\right)\frac{\delta(t-r/c)}{4\pi r} = \frac{1}{4\pi}\int_{\mathbb{R}^3} \mathrm{d}V\, \frac{1}{r}\left[\frac{1}{r}\frac{\partial^2}{\partial r^2}(r\phi) - \frac{1}{c^2}\frac{\partial^2\phi}{\partial t^2}\right]\bigg|_{t=r/c}$$
$$= \int_0^{+\infty} \mathrm{d}r \left[\frac{\partial^2}{\partial r^2}(r\phi) - \frac{1}{c^2}\frac{\partial^2}{\partial t^2}(r\phi)\right]\bigg|_{t=r/c} \tag{8.382}$$

again by virtue of (C.4). The spatial integrals have a meaning in the ordinary sense, inasmuch as the integrand is an infinitely differentiable *function* which depends on the radial distance r directly but also through the time variable, which the first integration has set equal to r/c, as indicated. For this reason in (8.382) we are not permitted to integrate the first term with respect to r straightaway because it involves the *partial* derivative $\partial/\partial r$ rather than the total one $\mathrm{d}/\mathrm{d}r$.

Thus, to proceed we observe that for a function $f(r,t)$ we have

$$\frac{\mathrm{d}}{\mathrm{d}r} f(r,t)\bigg|_{t=r/c} = \left[\frac{\partial}{\partial r} f(r,t) + \frac{\partial f}{\partial t}\frac{\partial t}{\partial r}\right]\bigg|_{t=r/c} = \frac{\partial}{\partial r} f\left(r, \frac{r}{c}\right) + \frac{1}{c}\frac{\partial f}{\partial t}\bigg|_{t=r/c} \tag{8.383}$$

where the symbol $\partial/\partial r$ means 'derivative with respect to the first argument'. Secondly, we write the second-order differential operator in (8.382) as the cascade of two first-order operators, viz.,

$$\left[\frac{\partial^2}{\partial r^2}(r\phi) - \frac{1}{c^2}\frac{\partial^2}{\partial t^2}(r\phi)\right]\bigg|_{t=r/c} = \left(\frac{\partial}{\partial r} + \frac{1}{c}\frac{\partial}{\partial t}\right)\left[\frac{\partial}{\partial r}(r\phi) - \frac{1}{c}\frac{\partial}{\partial t}(r\phi)\right]\bigg|_{t=r/c} \tag{8.384}$$

which can be seen as a formal application of the well-known algebraic identity $a^2-b^2 = (a+b)(a-b)$. More rigorously, the statement holds because the second-order mixed derivatives cancel out by virtue of the Schwarz theorem [27, pp. 235–236].

Now, in light of (8.383) we recognize the outer differential operator in the right member of (8.384) as the first-order total derivative with respect to r of a function of r and r/c. With this finding we manage to cast the last integrand in (8.382) as

$$\left[\frac{\partial^2}{\partial r^2}(r\phi) - \frac{1}{c^2}\frac{\partial^2}{\partial t^2}(r\phi)\right]\bigg|_{t=r/c} = \frac{\mathrm{d}}{\mathrm{d}r}\left[\frac{\partial}{\partial r}(r\phi) - \frac{1}{c}\frac{\partial}{\partial t}(r\phi)\right]\bigg|_{t=r/c} \tag{8.385}$$

which we are now allowed to integrate once with respect to r. In symbols, we get

$$\begin{aligned}\int_0^{+\infty} \mathrm{d}r \left[\frac{\partial^2}{\partial r^2}(r\phi) - \frac{1}{c^2}\frac{\partial^2}{\partial t^2}(r\phi)\right]\bigg|_{t=r/c} &= \int_0^{+\infty} \mathrm{d}r \frac{\mathrm{d}}{\mathrm{d}r}\left[\frac{\partial}{\partial r}(r\phi) - \frac{1}{c}\frac{\partial}{\partial t}(r\phi)\right]\bigg|_{t=r/c} \\ &= \left[\frac{\partial}{\partial r}(r\phi)\bigg|_{t=r/c} - \frac{1}{c}\frac{\partial}{\partial t}(r\phi)\bigg|_{t=r/c}\right]_{r=0}^{+\infty} \\ &= \left[\phi\left(r, \frac{r}{c}\right) + r\frac{\partial \phi}{\partial r}\bigg|_{t=r/c} - \frac{r}{c}\frac{\partial \phi}{\partial t}\bigg|_{t=r/c}\right]_{r=0}^{+\infty} = -\phi(0,0)\end{aligned} \tag{8.386}$$

having used the fact that $\phi(r,t)$ and its derivatives vanish at infinity because they all have bounded support as is prescribed for a test function (Appendix C). Since the result of the integration is a short-hand notation for the right-hand side of (8.381) we have proved that (8.380) solves (8.368).

By comparing (8.368) with (8.122) we notice that $\mathcal{G}(\mathbf{r},t)$ is proportional to the time-varying electric potential due to a point charge that is 'turned on and off' in $\mathbf{r} = 0$ in an infinitely short time lapse. Likewise, comparison of (8.371) and (8.122) suggests that $\Gamma(\mathbf{r},t)$ can be construed as the time-varying electric potential $\Phi_E(\mathbf{r},t)$ produced by a unitary point charge 'suddenly turned on' in the origin at $t = 0$ [see (8.104)]. More precisely, if we let

$$\varrho(\mathbf{r},t) = q\delta^{(3)}(\mathbf{r})\, U(t) \tag{8.387}$$

then

$$\Phi_E(\mathbf{r},t) = \frac{q}{\varepsilon}\Gamma(\mathbf{r},t) \tag{8.388}$$

and further, we have

$$\mathcal{G}(\mathbf{r},t) = \frac{\varepsilon}{q}\frac{\partial}{\partial t}\Phi_E(\mathbf{r},t) \tag{8.389}$$

on account of (8.369). This means we may also interpret the time-domain Green function as the temporal derivative of the electric potential produced by a unitary point charge switched on in the origin at $t = 0$. In particular, from (8.379) and (8.388) we gather that the relevant time-varying electric potential reads

$$\Phi_E(\mathbf{r},t) = \frac{q}{4\pi r\varepsilon}U\left(t-\frac{r}{c}\right), \qquad r>0, \qquad t\in\mathbb{R} \tag{8.390}$$

and, by the looks of it, there is no denying that $\Phi_E(\mathbf{r},t)$ behaves as a spherical *wave* propagating away from $\mathbf{r} = 0$. Nonetheless, we showed in Example 7.6 that the sudden appearance of a charge q in the origin generates the electrostatic field (7.6), a magnetic field which is *null* at all times, and last but not least, *no electromagnetic waves* at all. So, how do we reconcile these *seemingly* contradictory results?

Well, we need to remember from Section 8.2 that for a general time-dependence the electromagnetic field produced by electric sources is derived from a scalar potential $\Phi_E(\mathbf{r},t)$ *and* a magnetic vector potential $\mathcal{A}_E(\mathbf{r},t)$. As a consequence, if we also compute the latter and then turn to (8.104) and (8.105) to compute $\mathcal{E}(\mathbf{r},t)$ and $\mathcal{B}(\mathbf{r},t)$, we ought to obtain the same results laid out in Example 7.6.

To be specific, spherical symmetry suggests that $\mathcal{A}_E(\mathbf{r},t)$ depends only on the radial distance r. Besides, since the current density (7.449) associated with $\varrho(\mathbf{r},t)$ in (8.387) is radially directed, so is the magnetic vector potential $\mathcal{A}_E(r,t)$ on account of (8.123) written in polar spherical coordinates. Thus, noting that $\Phi_E(\mathbf{r},t)$ is available, we may determine the radial component $\hat{\mathbf{r}}\cdot\mathcal{A}_E(r,t)$ through the Lorenz gauge (8.121), instead of attempting the solution of the D'Alembert equation (8.123), a task which is far more complicated.

Due to the travelling discontinuity at $r = ct$, taking the required time derivative of $\Phi_E(r,t)$ in (8.121) produces a Dirac delta distribution, namely,

$$\frac{1}{r^2}\frac{\partial}{\partial r}\left[r^2\hat{\mathbf{r}}\cdot\mathcal{A}_E\right] = -\frac{q}{4\pi\varepsilon r c^2}\delta\left(t-\frac{r}{c}\right) \tag{8.391}$$

so the Lorenz gauge too must be interpreted in a distributional sense. For the solution, we resort to integrating both sides over a ball $B(0,a)$, keeping in mind that the integral in the right member is purely symbolic. By using standard spherical coordinates and exploiting the independence of φ and ϑ, we readily arrive at

$$\int_0^a \mathrm{d}r\,\frac{\partial}{\partial r}\left[r^2\hat{\mathbf{r}}\cdot\mathcal{A}_E\right] = -\frac{q}{4\pi\varepsilon c^2}\int_0^a \mathrm{d}r\, r\delta\left(t-\frac{r}{c}\right) \tag{8.392}$$

and by letting $r = c\tau$ in the symbolic integral we find

$$a^2\hat{\mathbf{r}}\cdot\mathcal{A}_E(a,t) = -\frac{q}{4\pi\varepsilon}\int_0^{a/c}\mathrm{d}\tau\,\tau\delta(t-\tau) = \begin{cases} -\dfrac{qt}{4\pi\varepsilon}, & t\in[0,a/c] \\ 0, & t\notin[0,a/c] \end{cases} \tag{8.393}$$

on account of (C.4). We just have to solve for $\hat{\mathbf{r}}\cdot\mathcal{A}_E(a,t)$ and, in order to highlight the dependence on the radial coordinate, rename the utilitarian variable a to r, viz.,

$$\mathcal{A}_E(r,t) = -\frac{qt}{4\pi\varepsilon r^2}\left[U(t) - U\left(t-\frac{r}{c}\right)\right]\hat{\mathbf{r}}, \qquad t\in\mathbb{R} \tag{8.394}$$

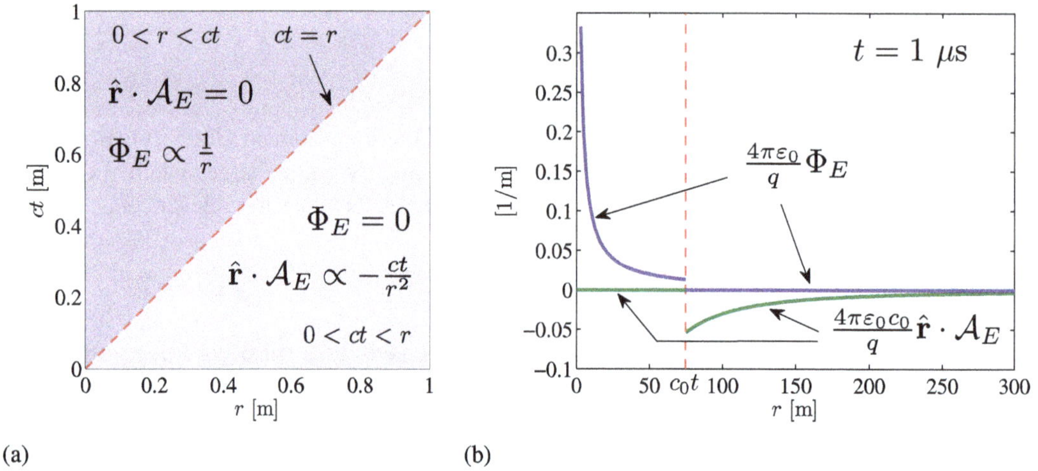

Figure 8.3 Properties of potentials (8.390) *and* (8.394)*: (a) regions of definition in the first quadrant of the plane rOt; (b) normalized plots (for a charge in free space) as a function of the radial coordinate r and for fixed t* > 0.

where we have also used two unit step functions to express the two cases in (8.393) in a compact form.

Before getting around to computing the electromagnetic field, we observe that the potentials $\Phi_E(r,t)$ and $\mathcal{A}_E(r,t)$ are singular for $r = 0 = t$ and, in particular, they do not admit limit as $r \to 0^+$ and $t \to 0^+$, because the result depends on the 'direction' one takes to approach the troublesome point $(r,t) = (0,0)$. This we can partly understand with the aid of Figure 8.3a which shows the two unbounded regions in which the straight line $t = r/c$ divides the first quadrant of the plane rOt. Also indicated in Figure 8.3a are the values assumed by the potentials in either region. In particular, $\Phi_E(r,t)$ and $\mathcal{A}_E(r,t)$ are discontinuous across the line $t = r/c$.

At time $t = 0^+ < r/c$, right after the appearance of the charge in the origin, $\Phi_E(r,t)$ vanishes everywhere but for $r = 0$ where it is infinitely large. By contrast, at time $t = 0^+ < r/c$, $\mathcal{A}_E(r,t)$ is everywhere null. Then, as time goes by, $\Phi_E(r,t)$ becomes non-zero within a sphere of ever growing radius $r = ct$, whereas $\mathcal{A}_E(r,t)$ becomes non-null outside the very same sphere. This behavior is illustrated in Figure 8.3b for a charge in free space at time $t = 1\ \mu$s. The vertical dashed line indicates the position of $r = c_0 t$, at which point (actually, the points over a sphere in the three-dimensional space) both $\Phi_E(r,t)$ and $\mathcal{A}_E(r,t)$ suffer a jump. In the limit as $t \to +\infty$, $\Phi_E(r,t)$ degenerates into the known electrostatic potential of a point charge, and $\mathcal{A}_E(r,t)$ vanishes everywhere.

Finally, we determine the radial component of the electric field $\mathcal{E}(r,t)$ through (8.104) interpreted in a distributional sense, i.e.,

$$\begin{aligned}
\mathcal{E}_r(r,t) &= -\frac{\partial}{\partial t}\hat{\mathbf{r}}\cdot\mathcal{A}_E(r,t) - \frac{\partial}{\partial r}\Phi_E(r,t)\\
&= \frac{q}{4\pi\varepsilon}\left\{\frac{1}{r^2}\left[U(t) - U\left(t-\frac{r}{c}\right) - t\delta\left(t-\frac{r}{c}\right)\right] - \left[-\frac{1}{r^2}U\left(t-\frac{r}{c}\right) - \frac{1}{cr}\delta\left(t-\frac{r}{c}\right)\right]\right\}\\
&= \frac{q}{4\pi\varepsilon r^2}\left[U(t) - t\delta\left(t-\frac{r}{c}\right) + \frac{r}{c}\delta\left(t-\frac{r}{c}\right)\right] = \frac{qU(t)}{4\pi\varepsilon r^2}
\end{aligned} \tag{8.395}$$

where the two terms proportional to the δ's are always null except when $t = r/c$, and hence they cancel one another. The calculations for the magnetic field $\mathfrak{B}(r,t)$ through (8.105) are trivial provided the curl is expressed in spherical coordinates with the help of (A.34). Since $\mathcal{A}_E(r,t)$ is radially directed and independent of ϑ and φ it follows immediately that $\mathfrak{B}(r,t) = 0$ at all times.

These findings are in perfect agreement with those of Example 7.6, so there is no contradiction. Despite the potentials (8.390) and (8.394) being wave-like solutions, the field generated by (8.387) is *electrostatic* in character. In this regard, it is important to realize that $\Phi_E(\mathbf{r},t)$ is *not* the auxiliary potential $\Upsilon(\mathbf{r},t)$ introduced in (7.427) to give the general form of the current density $\mathfrak{J}(\mathbf{r},t)$ pertinent to a non-radiating source!

References

[1] Misner CW, Thorne KS, Wheeler JA. Gravitation. San Francisco, CA: W. H. Freeman and Company; 1970.

[2] Warnick KF, Russer P. Differential forms and electromagnetic theory. Progress in Electromagnetic Research. 2014 July;148:83–112.

[3] Lindell IV. Differential Forms and Electromagnetic Materials. In: Capolino F, editor. Theory and Phenomema of Metamaterials. Boca Raton, FL: CRC Pres; 2009. p. 4.1–4.16.

[4] Lindell IV. Differential Forms in Electromagnetics. New York, NY: IEEE Press; 2004.

[5] Deschamps AG. Electromagnetics and differential forms. Proceedings of the IEEE. 1981;69(6):676–696.

[6] Griffith DJ. Introduction to electrodynamics. 3rd ed. Upper Saddle River, NJ: Prentice Hall; 1999.

[7] Tonnelat MA. The principles of electromagnetic theory and of relativity. Dordrecht-Holland, NL: D. Reidel Publishing Company; 1966.

[8] Zangwill A. Modern electrodynamics. Cambridge, UK: Cambrigde University Press; 2013.

[9] Kong JA. Electromagnetic Wave Theory. 2nd ed. New York, NY: Wiley; 1990.

[10] Pauli W. Theory of relativity. New York, NY: Dover Publications, Inc.; 1981.

[11] Van Bladel JG. Relativity and Engineering. Berlin Heidelberg : Springer Berlin Heidelberg; 1984.

[12] Milton KA, Schwinger J. Electromagnetic Radiation: Variational Methods, Waveguides and Accelerators. Berlin Heidelberg: Springer-Verlag; 2006.

[13] Post EJ. Formal structure of electromagnetics. Amsterdam, NL: North-Holland Publishing Company; 1962.

[14] Collin RE. Foundations for Microwave Engineering. New York, NY: McGraw-Hill; 1992.

[15] Rothwell EJ, Cloud MJ. Electromagnetics. London, UK: CRC Press; 2001.

[16] Van Bladel JG. Electromagnetic Fields. Piscataway, NJ: IEEE Press; 2007.

[17] Van Bladel JG. Singular Electromagnetic Fields and Sources. Piscataway, NJ: IEEE Press; 1991.

[18] Phillips HB. Vector Analysis. New York, NY: John Wiley & Sons; 1933.

[19] Helrich CS. The classical theory of fields - Electromagnetism. Berlin Heidelberg: Springer-Verlag; 2012.

[20] Oughstun KE. Electromagnetic and Optical Pulse Propagation. vol. 1. New York, NY: Springer Science+Business Media; 2006.

[21] Felsen LB, Marcuvitz N. Radiation and scattering of waves. Piscataway, NJ: IEEE Press; 2001.

[22] Makarov B, Podkorytov A. Real Analysis: Measures, Integrals and Applications. Universitext. London, UK: Springer; 2013.
[23] Morse P, Feshbach H. Methods of Theoretical Physics. New York, NY: McGraw-Hill; 1953.
[24] Bladel JV. A Discussion of Helmholtz' Theorem. Electromagnetics. 1993;13(1):95–110. Available from: https://doi.org/10.1080/02726349308908332.
[25] Bassanini P, Elcrat AR. Theory and applications of partial differential equations. Mathematical Concepts And Methods In Science And Engineering. New York, NY: Plenum Press; 1997.
[26] Blyth TS, Robertson EF. Basic Linear Algebra. 2nd ed. Springer Undergraduate Mathematics Series. London, UK: Springer-Verlag; 2002.
[27] Rudin W. Principles of mathematical analysis. 3rd ed. New York, NY: McGraw-Hill; 1976.
[28] Stratton JA. Electromagnetic theory. London, UK: McGraw-Hill; 1941.
[29] Schwinger J, De Raad LL, Milton KA, et al. Classical electrodynamics. Perseus Books; 1998.
[30] Collin RE. Field Theory of Guided Waves. Piscataway, NJ: IEEE press; 1991.
[31] Davis JL. Wave Propagation in Electromagnetic Media. New York, NY: Springer-Verlag; 1990.
[32] Dobbs R. Electromagnetic Waves. London, UK: Routledge & Kegan Paul; 1985.
[33] Lorrain P, Corson DR, Lorrain F. Electromagnetic Fields and Waves. 3rd ed. New York, NY: W. H. Freeman and Company; 1988.
[34] Orfanidis SJ. Electromagnetic Waves and Antennas. www.ece.rutgers.edu/~orfanidi/ewa; 2004.
[35] Thidé B. Electromagnetic Field Theory. New York, NY: Dover Publications, Inc.; 2011.
[36] Cottingham WN, Greenwood DA. Electricity and Magnetism. Cambridge, UK: Cambridge University Press; 1991.
[37] Konopinski EJ. Electromagnetic Fields and Relativistic Particles. New York, NY: McGraw Hill; 1981.
[38] Jackson JD. Classical Electrodynamics. 3rd ed. Chichester, UK: Wiley; 1999.
[39] Lorenz L. On the identity of the vibrations of light with electrical currents. Philos Mag. 1867;34:287–301.
[40] Abramowitz M, Stegun IA. Handbook of mathematical functions. New York, NY: Dover Publications, Inc.; 1965.
[41] Dwight HB. Tables of Integrals and Other Mathematical Data. 4th ed. New York, NY: MacMillan Publishing Co., Inc.; 1961.
[42] Flanders H. Differentiation under the integral sign. American Mathematical Monthly. 1973 June-July;80(6):615–627.
[43] Schwab AJ. Field Theory Concepts. Berlin Heidelberg: Springer-Verlag; 1988.
[44] Weeks WL. Electromagnetic Theory for Engineering Applications. New York, NY: John Wiley & Sons, Inc.; 1964.
[45] Biggs HF. The Electromagnetic Field. London, UK: Oxford University Press; 1934.
[46] Zhang K, Li D. Electromagnetic Theory for Microwaves and Optoelectronics. Berlin Heidelberg: Springer-Verlag; 1998.
[47] Essex EA. Hertz vector potentials of electromagnetic theory. American Journal of Physics. 1977;45(11):1099–1101. Available from: https://doi.org/10.1119/1.10955.
[48] Sein JJ. Solutions to Time-Harmonic Maxwell Equations with a Hertz Vector. Am J Phys. 1989;57:834.
[49] Anderson N. The Electromagnetic Field. New York, NY: Springer Science+Business Media; 1968.

[50] Dautray R, Lions JL. Mathematical Analysis and Numerical Methods for Science and Technology - Physical Origins and Classical Methods. vol. 1. Berlin Heidelberg: Springer-Verlag; 1990.
[51] Chew WC. Waves and Fields in Inhomogenous Media. New York, NY: Van Nostrand Teinhold; 1990.
[52] Moon P, Spencer DE. Field Theory Handbook: Including Coordinate Systems, Differential Equations and Their Solutions. 2nd ed. Berlin Heidelberg: Springer-Verlag; 1971.
[53] Harrington RF. Time-harmonic Electromagnetic Fields. London, UK: McGraw-Hill; 1961.
[54] Shen LC, Kong JA. Applied electromagnetism. Monterey, CA: Brooks/Cole; 1983.
[55] Watson GN. A Treatise on the Theory of Bessel Functions. 2nd ed. Cambridge Mathematical Library. Cambridge, UK: Cambridge University Press; 1995.
[56] Schelkunoff SA. Electromagnetic Waves. Princeton, NJ: D. Van Nostrand Company; 1943.
[57] Jones DS. The Theory of Electromagnetism. Oxford, UK: Pergamon; 1964.
[58] Colton DL, Kress R. Inverse Acoustic and Electromagnetic Scattering Theory. 3rd ed. New York, NY: Springer; 2013.
[59] Rellich F. Über das asymptotische Verhalten von Lösungen von $\Delta u + \lambda u = 0$ in unendlichen Gebieten. Jber Deutsch Math Verein. 1943;53:57–65.
[60] Lindell IV. Methods for electromagnetic field analysis. Piscataway, NJ: IEEE Press; 1992.
[61] Sommerfeld A. Electrodynamics. vol. 3 of Lectures on theoretical physics. New York, NY: Academic Press; 1952.
[62] Sommerfeld A. Partial Differential Equations in Physics. vol. 1 of Lectures on theoretical physics. New York, NY: Academic Press; 1949.
[63] Wilcox CH. An expansion theorem for electromagnetic fields. Comm Pure Appl Math. 1956;9:115–134.
[64] Conway JB. A Course in Functional Analysis. 2nd ed. Graduate Texts in Mathematics. New York, NY: Springer-Verlag; 1990.
[65] Muscat J. Functional Analysis. London, UK: Springer; 2014.
[66] Colton DL, Kress R. Integral Equation Methods in Scattering Theory. New York, NY: John Wiley & Sons, Inc.; 1983.
[67] Adams RA, Fournier JJF. Sobolev spaces. 2nd ed. New York, NY: Academic Press; 2003.
[68] Bowers A, Kalton NJ. An introductory course in functional analysis. Universitext. New York, NY: Springer Science+Business Media; 2014.
[69] Atkinson KE. The Numerical Solution of Integral Equations of the Second Kind. Cambridge, UK: Cambridge University Press; 1997.
[70] Kantorovich LV, Akilov GP. Functional analysis in normed spaces. Oxford, UK: Pergamon Press; 1964. Translated by D. E. Brown.
[71] Rudin W. Functional Analysis. New York, NY: McGraw-Hill; 1973.
[72] Bassanini P, Elcrat A. Mathematical Theory of Electromagnetism. Creative Commons; 2009.
[73] Tai CT. Dyadic Green's functions in electromagnetic theory. 2nd ed. Piscataway, NJ: IEEE Press; 1994.
[74] Stein EM, Weiss G. Introduction to Fourier Analysis on Euclidean Spaces. Princeton, NJ: Princeton University Press; 1971.
[75] de Hoop AT. Handbook of radiation and scattering of waves. London, UK: Academic Press; 1995.
[76] Jones DS. The Theory of Generalised Functions. Cambridge, UK: Cambridge University Press; 2009.

Chapter 9

Time-varying electromagnetic fields II

9.1 Integral representations of the potentials

The three-dimensional Green functions help us solve the D'Alembert equations (8.122), (8.123), (8.126) and (8.127) for the electrodynamic potentials. Actually, the extensive analysis carried out in Section 8.7 partially answers the question in the whole space inasmuch as the combination of (8.356) and (8.326) yields the solution in the frequency domain, whereas (8.357) solves the original D'Alembert equation.

9.1.1 Frequency domain

We wish to generalize the result by determining an integral representation of the potentials in the frequency domain in a bounded region of space. The procedure will put the surmised expression (8.326) on more solid ground.

To this purpose we consider two problems, namely, the auxiliary one

$$\begin{cases} \nabla'^2 G(\mathbf{r},\mathbf{r}') + k^2 G(\mathbf{r},\mathbf{r}') = 0, & \mathbf{r}' \in \mathbb{R}^3 \setminus \{\mathbf{r}\}, \quad \mathbf{r} \in V \setminus \overline{V}_S \\ \dfrac{\partial G}{\partial r'} + \mathrm{j}\, k G(\mathbf{r},\mathbf{r}') = \mathcal{O}\left(\dfrac{1}{|\mathbf{r}'|^2}\right), & |\mathbf{r}'| \to +\infty \\ \displaystyle\int_{\partial B} \mathrm{d}S'\, \hat{\mathbf{r}}' \cdot \nabla' G(\mathbf{r},\mathbf{r}') + \int_{B(\mathbf{r},a)} \mathrm{d}V'\, k^2 G(\mathbf{r},\mathbf{r}') = -1 & \end{cases} \tag{9.1}$$

and the one of interest

$$\begin{cases} \nabla'^2 \Psi(\mathbf{r}') + k^2 \Psi(\mathbf{r}') = -S(\mathbf{r}'), & \mathbf{r}' \in V \\ \Psi(\mathbf{r}') = f(\mathbf{r}') \quad \text{or} \quad \hat{\mathbf{n}}(\mathbf{r}') \cdot \nabla' \Psi = g(\mathbf{r}'), & \mathbf{r}' \in \partial V \end{cases} \tag{9.2}$$

where $V \subset \mathbb{R}^3$ is a finite region of space with smooth boundary ∂V, f and g are regular functions on ∂V, and $V_S \subset V$ is the volume which contains the source $S(\mathbf{r}')$. We already know that the first problem is solved by the Green function (8.356).

We multiply (9.1) by $\Psi(\mathbf{r}')$ and (9.2) by $G(\mathbf{r},\mathbf{r}') = G(R)$ to get

$$\begin{aligned} 0 &= \Psi(\mathbf{r})\nabla'^2 G(R) + k^2 \Psi(\mathbf{r}') G(R) \\ &= \nabla' \cdot [\Psi(\mathbf{r}')\nabla' G(R)] - \nabla'\Psi(\mathbf{r}') \cdot \nabla' G(R) + k^2 \Psi(\mathbf{r}') G(R) \end{aligned} \tag{9.3}$$

and

$$-G(R)S(\mathbf{r}') = G(R)\nabla'^2 \Psi(\mathbf{r}') + k^2 G(R)\Psi(\mathbf{r}')$$

$$= \nabla' \cdot [G(R)\nabla'\Psi(\mathbf{r}')] - \nabla' G(R) \cdot \nabla'\Psi(\mathbf{r}') + k^2 G(R)\Psi(\mathbf{r}') \tag{9.4}$$

which hold simultaneously for $\mathbf{r}' \in V \setminus \{\mathbf{r}\}$. However, since the Green function is singular for $\mathbf{r}' = \mathbf{r}$ we exclude that point with a ball $B(\mathbf{r}, a)$, where the radius a is chosen so that $B[\mathbf{r}, a] \subset V \setminus V_S$ (cf. Figure 2.8). We subtract the equations side by side and integrate over $V_a := V \setminus B(\mathbf{r}, a)$ with respect to $\mathbf{r}'$ to find

$$\int_{V_a} \mathrm{d}V' \, \nabla' \cdot [G(R)\nabla'\Psi(\mathbf{r}') - \Psi(\mathbf{r}')\nabla' G(R)] = -\int_{V_S} \mathrm{d}V' \, G(R) S(\mathbf{r}') \tag{9.5}$$

having restricted the integration to V_S in the right member. We may invoke the Gauss theorem because $G(R)\nabla'\Psi(\mathbf{r}') - \Psi(\mathbf{r}')\nabla' G(R)$ is a vector field of class $\mathcal{C}^1(V_a)^3 \cap \mathcal{C}(\overline{V}_a)^3$ if $S(\mathbf{r}')$ is continuous throughout V. On the other hand, if $S(\mathbf{r}')$ is discontinuous across ∂V_S we may apply the Gauss theorem separately in $V_a \setminus \overline{V}_S$ and V_S and impose suitable matching conditions for $\Psi(\mathbf{r}')$ and $\hat{\mathbf{n}}(\mathbf{r}') \cdot \nabla'\Psi(\mathbf{r}')$ on ∂V_S. By proceeding in either way, we obtain

$$\int_{\partial V} \mathrm{d}S' \, \hat{\mathbf{n}}(\mathbf{r}') \cdot [G(R)\nabla'\Psi(\mathbf{r}') - \Psi(\mathbf{r}')\nabla' G(R)] + \int_{\partial B} \mathrm{d}S' \, G(R)\hat{\mathbf{n}}(\mathbf{r}') \cdot \nabla'\Psi(\mathbf{r}')$$
$$- \int_{\partial B} \mathrm{d}S' \, \Psi(\mathbf{r}')\hat{\mathbf{n}}(\mathbf{r}') \cdot \nabla' G(R) = \int_{V_S} \mathrm{d}V' \, G(R) S(\mathbf{r}') \tag{9.6}$$

having taken the unit normal $\hat{\mathbf{n}}(\mathbf{r}')$ on ∂V and ∂B positively oriented inwards V_a.

To establish a relation which is independent of a we take the limit for vanishing radius. The procedure does not affect the integral over the sources or the contribution from the flux through ∂V. Concerning the flux through ∂B, for $\mathbf{r}' \in \partial B$ we observe that

$$\mathbf{r}' = \mathbf{r} + a\hat{\mathbf{n}}(\mathbf{r}'), \qquad G(R) = \frac{\mathrm{e}^{-\mathrm{j}ka}}{4\pi a}, \qquad \hat{\mathbf{n}}(\mathbf{r}') \cdot \nabla' G(R) = -(1 + \mathrm{j}\,ka)\frac{\mathrm{e}^{-\mathrm{j}ka}}{4\pi a^2} \tag{9.7}$$

on account of (8.356). Then, it follows

$$\int_{\partial B} \mathrm{d}S' \, G(R)\hat{\mathbf{n}}(\mathbf{r}') \cdot \nabla'\Psi(\mathbf{r}') = \frac{\mathrm{e}^{-\mathrm{j}ka}}{4\pi a} \int_{\partial B} \mathrm{d}S' \, \hat{\mathbf{n}}(\mathbf{r}') \cdot \nabla'\Psi(\mathbf{r}')$$
$$= a\mathrm{e}^{-\mathrm{j}ka} \hat{\mathbf{n}}(\mathbf{r}'_0) \cdot \nabla'\Psi(\mathbf{r}')|_{\mathbf{r}'_0} \xrightarrow[a \to 0^+]{} 0 \tag{9.8}$$

$$- \int_{\partial B} \mathrm{d}S' \, \Psi(\mathbf{r}')\hat{\mathbf{n}}(\mathbf{r}') \cdot \nabla' G(R) = (1 + \mathrm{j}\,ka)\frac{\mathrm{e}^{-\mathrm{j}ka}}{4\pi a^2} \int_{\partial B} \mathrm{d}S' \, \Psi(\mathbf{r}')$$
$$= (1 + \mathrm{j}\,ka)\mathrm{e}^{-\mathrm{j}ka}\Psi(\mathbf{r}'_0) \xrightarrow[a \to 0^+]{} \Psi(\mathbf{r}) \tag{9.9}$$

where we have invoked the mean value theorem since the integrands are regular for $\mathbf{r}' \in \partial B$, $\mathbf{r}'_0$ denotes a suitable point, and $\mathbf{r}'_0 \to \mathbf{r}$ as the radius vanishes. Collecting these intermediate results yields

$$\Psi(\mathbf{r}) = \int_{V_S} \mathrm{d}V' \, \frac{\mathrm{e}^{-\mathrm{j}kR}}{4\pi R} S(\mathbf{r}') + \int_{\partial V} \mathrm{d}S' \, \Psi(\mathbf{r}') \frac{\partial}{\partial \hat{\mathbf{n}}'} \frac{\mathrm{e}^{-\mathrm{j}kR}}{4\pi R} - \int_{\partial V} \mathrm{d}S' \, \frac{\mathrm{e}^{-\mathrm{j}kR}}{4\pi R} \frac{\partial \Psi}{\partial \hat{\mathbf{n}}'}$$

$$= \int_{V_S} \mathrm{d}V' \frac{\mathrm{e}^{-\mathrm{j}kR}}{4\pi R} S(\mathbf{r}') + \int_{\partial V} \mathrm{d}S' \, \Psi(\mathbf{r}')\hat{\mathbf{n}}(\mathbf{r}') \cdot \hat{\mathbf{R}} \left(\mathrm{j}k + \frac{1}{R} \right) \frac{\mathrm{e}^{-\mathrm{j}kR}}{4\pi R} - \int_{\partial V} \mathrm{d}S' \, \frac{\mathrm{e}^{-\mathrm{j}kR}}{4\pi R} \frac{\partial \Psi}{\partial \hat{\mathbf{n}}'}$$

$$= \int_{V_S} \mathrm{d}V' \frac{\mathrm{e}^{-\mathrm{j}kR}}{4\pi R} S(\mathbf{r}') + \int_{\partial V} \mathrm{d}S' \, \Psi(\mathbf{r}')\hat{\mathbf{n}}(\mathbf{r}') \cdot \hat{\mathbf{R}} \frac{\mathrm{e}^{-\mathrm{j}kR}}{4\pi R^2}$$

$$+ \int_{\partial V} \mathrm{d}S' \, \hat{\mathbf{n}}(\mathbf{r}') \cdot \left[\mathrm{j}k\hat{\mathbf{R}}\Psi(\mathbf{r}') - \nabla'\Psi \right] \frac{\mathrm{e}^{-\mathrm{j}kR}}{4\pi R} \tag{9.10}$$

where $\hat{\mathbf{n}}' = \hat{\mathbf{n}}(\mathbf{r}')$ and $R = |\mathbf{r} - \mathbf{r}'|$. This is the desired integral representation of $\Psi(\mathbf{r})$ in a bounded region [1,2], [3, Section 2.6], [4, Theorem 2.1 p. 16]. It is valid for $\mathbf{r} \notin V \setminus \overline{V_S}$ for the moment, but we can extend the formula for observation points within the source region.

If we choose $\mathbf{r} \in V_S \subset V$ and a ball $B(\mathbf{r}, a) \subset V_S$ and perform the same steps previously outlined we arrive at an expression similar to (9.6) in which the volume integral is over $V_S \setminus B(\mathbf{r}, a)$. If the source $S(\mathbf{r}')$ is at least continuous for $\mathbf{r}' \in V_S$, the integrals over ∂B yield the same result as before in the limit as $a \to 0^+$ because $\Psi(\mathbf{r}')$, $G(\mathbf{r}, \mathbf{r}')$ and the derivative thereof are regular for $\mathbf{r}' \in V_S \setminus B(\mathbf{r}, a)$. Concerning the integral over $V_S \setminus B(\mathbf{r}, a)$ we consider a larger ball $B(\mathbf{r}, b)$ with radius b large enough so that $B(\mathbf{r}, a) \subset V_S \subset B(\mathbf{r}, b)$ (cf. Figure 2.9) and estimate

$$\left| \int_{V_S \setminus B(\mathbf{r},a)} \mathrm{d}V' \frac{\mathrm{e}^{-\mathrm{j}kR}}{4\pi R} S(\mathbf{r}') \right| \leqslant \int_{V_S \setminus B(\mathbf{r},a)} \mathrm{d}V' \frac{|S(\mathbf{r}')|}{4\pi R} \leqslant \int_{V_S \setminus B(\mathbf{r},a)} \mathrm{d}V' \frac{\|S\|_\infty}{4\pi R} \leqslant \int_{B(\mathbf{r},b) \setminus B(\mathbf{r},a)} \mathrm{d}V' \frac{\|S\|_\infty}{4\pi R}$$

$$= \|S\|_\infty \int_a^b \mathrm{d}R \, R = \frac{1}{2} \|S\|_\infty (b^2 - a^2) \leqslant \frac{b^2}{2} \|S\|_\infty \tag{9.11}$$

and this shows that the integral remains bounded for values of a all the way down to zero.

The strong similarity existing between (9.10) and the analogous formula (2.160) for the electrostatic potential cannot go unnoticed. In fact, the only difference is constituted by the Green function and, by setting $k = 0$ in (9.10) we recover (2.160) apart from immaterial constants. Furthermore, following the same analysis as in Section 2.7 we may show that for points $\mathbf{r} \in \mathbb{R}^3 \setminus \overline{V}$ the representation (9.10) returns a null value, whereas for $\mathbf{r} \in \partial V$ the surface integrals must be interpreted as Cauchy principal values and the right-hand side is equal to half the value of the potential on ∂V, as in (2.167). The details are left to the Reader.

In words, (9.10) states that the scalar field $\Psi(\mathbf{r})$ — solution to the Helmholtz equation (9.2) for $\mathbf{r} \in V$ — can be written as a combination of three contributions, namely,

- the *volume* potential

$$\mathcal{V}_k(\mathbf{r}) := \int_{V_S} \mathrm{d}V' \frac{\mathrm{e}^{-\mathrm{j}k|\mathbf{r}-\mathbf{r}'|}}{4\pi|\mathbf{r} - \mathbf{r}'|} S(\mathbf{r}') = G(\mathbf{r}) * S(\mathbf{r}) \tag{9.12}$$

- the *single-layer* potential

$$\mathcal{S}_k(\mathbf{r}) := -\int_S \mathrm{d}S' \frac{\mathrm{e}^{-\mathrm{j}k|\mathbf{r}-\mathbf{r}'|}}{4\pi|\mathbf{r} - \mathbf{r}'|} \frac{\partial \Psi}{\partial \hat{\mathbf{n}}'} = G(\mathbf{r}) * [-\hat{\mathbf{n}}(\mathbf{r}) \cdot \nabla\Psi(\mathbf{r})] \tag{9.13}$$

- the *double-layer* potential

$$\mathcal{W}_k(\mathbf{r}) := \int_S \mathrm{d}S'\, \Psi(\mathbf{r}') \frac{\partial}{\partial \hat{\mathbf{n}}'} \frac{\mathrm{e}^{-\mathrm{j}k|\mathbf{r}-\mathbf{r}'|}}{4\pi|\mathbf{r}-\mathbf{r}'|} = \nabla G(\mathbf{r}) \overset{*}{,} [\hat{\mathbf{n}}(\mathbf{r})\Psi(\mathbf{r})]. \tag{9.14}$$

It is relatively straightforward to show that these potentials exhibit the same smoothness as their static counterparts (Sections 2.8, 2.9 and 2.10) because the singularity of the kernels for $\mathbf{r}' \to \mathbf{r}$ is the same as that of the three-dimensional static Green function (2.131).

With the aid of (9.10) we may write the electromagnetic potentials as

$$\begin{aligned}
\Phi_E(\mathbf{r}) &= \frac{1}{\varepsilon}\int_{V_\rho} \mathrm{d}V' \frac{\mathrm{e}^{-\mathrm{j}kR}}{4\pi R}\rho(\mathbf{r}') + \int_{\partial V} \mathrm{d}S'\, \Phi_E(\mathbf{r}') \frac{\partial}{\partial \hat{\mathbf{n}}'} \frac{\mathrm{e}^{-\mathrm{j}kR}}{4\pi R} - \int_{\partial V} \mathrm{d}S' \frac{\mathrm{e}^{-\mathrm{j}kR}}{4\pi R} \frac{\partial \Phi_E}{\partial \hat{\mathbf{n}}'} \\
&= \frac{1}{\varepsilon}\int_{V_\rho} \mathrm{d}V' \frac{\mathrm{e}^{-\mathrm{j}kR}}{4\pi R}\rho(\mathbf{r}') + \int_{\partial V} \mathrm{d}S'\, \Phi_E(\mathbf{r}')\hat{\mathbf{n}}(\mathbf{r}') \cdot \hat{\mathbf{R}} \frac{\mathrm{e}^{-\mathrm{j}kR}}{4\pi R^2} \\
&\quad + \int_{\partial V} \mathrm{d}S'\, \hat{\mathbf{n}}(\mathbf{r}') \cdot \left[\mathrm{j}\, k\hat{\mathbf{R}}\Phi_E(\mathbf{r}') - \nabla'\Phi_E\right] \frac{\mathrm{e}^{-\mathrm{j}kR}}{4\pi R}
\end{aligned} \tag{9.15}$$

$$\begin{aligned}
\mathbf{A}_E(\mathbf{r}) &= \mu \int_{V_J} \mathrm{d}V' \frac{\mathrm{e}^{-\mathrm{j}kR}}{4\pi R}\mathbf{J}(\mathbf{r}') + \int_{\partial V} \mathrm{d}S'\, \mathbf{A}_E(\mathbf{r}') \frac{\partial}{\partial \hat{\mathbf{n}}'} \frac{\mathrm{e}^{-\mathrm{j}kR}}{4\pi R} - \int_{\partial V} \mathrm{d}S' \frac{\mathrm{e}^{-\mathrm{j}kR}}{4\pi R} \frac{\partial \mathbf{A}_E}{\partial \hat{\mathbf{n}}'} \\
&= \mu \int_{V_J} \mathrm{d}V' \frac{\mathrm{e}^{-\mathrm{j}kR}}{4\pi R}\mathbf{J}(\mathbf{r}') + \int_{\partial V} \mathrm{d}S'\, \mathbf{A}_E(\mathbf{r}')\hat{\mathbf{n}}(\mathbf{r}') \cdot \hat{\mathbf{R}} \frac{\mathrm{e}^{-\mathrm{j}kR}}{4\pi R^2} \\
&\quad + \int_{\partial V} \mathrm{d}S'\, \hat{\mathbf{n}}(\mathbf{r}') \cdot \left[\mathrm{j}\, k\hat{\mathbf{R}}\mathbf{A}_E(\mathbf{r}') - \nabla'\mathbf{A}_E\right] \frac{\mathrm{e}^{-\mathrm{j}kR}}{4\pi R}
\end{aligned} \tag{9.16}$$

for points $\mathbf{r} \in V$; a similar pair of formulas holds for $\Phi_M(\mathbf{r})$ and $\mathbf{A}_M(\mathbf{r})$. As was the case for the integral representations of $\Phi(\mathbf{r})$ and $\mathcal{A}(\mathbf{r})$ in the static and stationary regimes, (9.15), (9.16) and the like constitute just the formal solution to the Helmholtz equation inasmuch as they involve both the potential and the normal derivative thereof for $\mathbf{r}' \in \partial V$. The latter two quantities are not independent and specifying them arbitrarily may lead to non-physical results.

Conversely, if the region of concern is unbounded we do obtain the integral solution (8.326). Essentially, this happens because the potential and its normal derivative are related explicitly by the Sommerfeld radiation condition (8.239). We need to show that the flux integrals over ∂V reduce to zero as we let the boundary recede to infinity. The task is relatively easier if we choose $V \equiv B(0, b)$ with the radius b large enough for (8.222) and (8.239) to hold for $\Psi(\mathbf{r}')$. We first derive an estimate involving $\hat{\mathbf{r}}' \cdot \hat{\mathbf{R}}$ by considering

$$\begin{aligned}
r^2 &= \mathbf{r}\cdot\mathbf{r} = (\mathbf{r}' + \mathbf{R}) \cdot (\mathbf{r}' + \mathbf{R}) = r'^2 + R^2 + 2Rr'\hat{\mathbf{r}}' \cdot \hat{\mathbf{R}} \\
&= (r' - R)^2 + 2Rr'(\hat{\mathbf{r}}' \cdot \hat{\mathbf{R}} + 1) \geqslant 2Rr'(\hat{\mathbf{r}}' \cdot \hat{\mathbf{R}} + 1)
\end{aligned} \tag{9.17}$$

whence we find

$$\hat{\mathbf{r}}' \cdot \hat{\mathbf{R}} + 1 \leqslant \frac{r^2}{2Rr'} \tag{9.18}$$

because $2Rr'$ is always a non-negative quantity. Then, we observe that

$$\hat{\mathbf{n}}(\mathbf{r}') = -\hat{\mathbf{r}}', \qquad -\hat{\mathbf{n}}(\mathbf{r}') \cdot \nabla'\Psi(\mathbf{r}') = \frac{\partial \Psi}{\partial r'}, \qquad \mathbf{r}' \in \partial B \tag{9.19}$$

and also

$$\begin{aligned}\hat{\mathbf{n}}(\mathbf{r}') \cdot \left[\mathrm{j}\,k\hat{\mathbf{R}}\Psi(\mathbf{r}') - \nabla'\Psi\right] &= -\mathrm{j}\,k\hat{\mathbf{r}}' \cdot \hat{\mathbf{R}}\Psi(\mathbf{r}') + \frac{\partial \Psi}{\partial r'} \\ &= \mathrm{j}\,k\Psi(\mathbf{r}') + \frac{\partial \Psi}{\partial r'} - \mathrm{j}\,k\left(\hat{\mathbf{r}}' \cdot \hat{\mathbf{R}} + 1\right)\Psi(\mathbf{r}'), \qquad \mathbf{r}' \in \partial B\end{aligned} \tag{9.20}$$

whereby we can estimate

$$\left| \int_{\partial B} \mathrm{d}S'\, \Psi(\mathbf{r}')\hat{\mathbf{n}}(\mathbf{r}') \cdot \hat{\mathbf{R}} \frac{\mathrm{e}^{-\mathrm{j}kR}}{4\pi R^2} \right| \leqslant \int_{\partial B} \mathrm{d}S'\, \frac{|\Psi(\mathbf{r}')|}{4\pi R^2} \leqslant \int_{\partial B} \mathrm{d}S'\, \frac{C_{\Psi 1}}{4\pi b(b-r)^2} = \frac{bC_{\Psi 1}}{(b-r)^2} \xrightarrow[b\to+\infty]{} 0 \tag{9.21}$$

$$\begin{aligned}&\left| \int_{\partial B} \mathrm{d}S'\, \hat{\mathbf{n}}(\mathbf{r}') \cdot \left[\mathrm{j}\,k\hat{\mathbf{R}}\Psi(\mathbf{r}') - \nabla'\Psi\right] \frac{\mathrm{e}^{-\mathrm{j}kR}}{4\pi R} \right| \\ &\leqslant \int_{\partial B} \mathrm{d}S'\, \frac{1}{4\pi R} \left|\hat{\mathbf{n}}(\mathbf{r}') \cdot \left[\mathrm{j}\,k\hat{\mathbf{R}}\Psi(\mathbf{r}') - \nabla'\Psi\right]\right| \\ &\leqslant \int_{\partial B} \mathrm{d}S'\, \frac{1}{4\pi R} \left[\left|\mathrm{j}\,k\Psi(\mathbf{r}') + \frac{\partial \Psi}{\partial r'}\right| + k\left|\hat{\mathbf{r}}' \cdot \hat{\mathbf{R}} + 1\right| |\Psi(\mathbf{r}')|\right] \\ &\leqslant \int_{\partial B} \mathrm{d}S'\, \frac{1}{4\pi(b-r)} \left[\frac{C_{\Psi 2}}{b^2} + k\frac{r^2}{2b(b-r)}\frac{C_{\Psi 1}}{b}\right] \\ &= \frac{1}{b-r}\left[C_{\Psi 2} + k\frac{C_{\Psi 1}r^2}{2(b-r)}\right] \xrightarrow[b\to+\infty]{} 0\end{aligned} \tag{9.22}$$

where use has been made of estimates (6.191) and (9.18), and $C_{\Psi 1} > 0$ and $C_{\Psi 2} > 0$ are two suitable constants involved in the asymptotic conditions (8.222) and (8.239).

With this result — applied in particular to the Cartesian components of $\mathcal{A}_E(\mathbf{r})$ in (9.16) — we have proved that the formulas

$$\Phi_E(\mathbf{r}) = \frac{1}{\varepsilon}\int_{V_\rho} \mathrm{d}V'\, \frac{\mathrm{e}^{-\mathrm{j}kR}}{4\pi R}\rho(\mathbf{r}'), \qquad \mathbf{r} \in \mathbb{R}^3 \tag{9.23}$$

$$\mathbf{A}_E(\mathbf{r}) = \mu \int_{V_J} \mathrm{d}V'\, \frac{\mathrm{e}^{-\mathrm{j}kR}}{4\pi R}\mathbf{J}(\mathbf{r}'), \qquad \mathbf{r} \in \mathbb{R}^3 \tag{9.24}$$

give the solution to the Helmholtz equations for the electrodynamic potentials in the frequency domain.

We recall that the wave equations (8.122) and (8.123) were derived under the Lorenz gauge (8.121) which in the frequency domain reads

$$\nabla \cdot \mathbf{A}_E(\mathbf{r};\omega) + \mathrm{j}\,\frac{\omega}{c^2}\Phi_E(\mathbf{r};\omega) = 0 \tag{9.25}$$

on account of inversion formulas such as (8.322). Therefore, it is expedient to verify that the electrodynamic potentials (9.23) and (9.24) do fulfill (9.25). Once again, the crucial step is the interchange of derivatives and integral, which may be proved as was done for the static volume potential on page 123, viz.,

$$\nabla \cdot \mu \int\limits_{V_J} \mathrm{d}V' \frac{\mathrm{e}^{-\mathrm{j}kR}}{4\pi R}\mathbf{J}(\mathbf{r}') = \mu \int\limits_{V_J} \mathrm{d}V' \, \nabla \frac{\mathrm{e}^{-\mathrm{j}kR}}{4\pi R} \cdot \mathbf{J}(\mathbf{r}'), \qquad \mathbf{r} \in \mathbb{R}^3 \tag{9.26}$$

as long as $\mathbf{J}(\mathbf{r})$ is bounded in V_J. For observation points $\mathbf{r} \in \mathbb{R}^3 \setminus \overline{V}_J$ we compute the divergence of $\mathbf{A}_E(\mathbf{r})$ as follows

$$\begin{aligned}
\mu \int\limits_{V_J} \mathrm{d}V' \, \nabla \frac{\mathrm{e}^{-\mathrm{j}kR}}{4\pi R} \cdot \mathbf{J}(\mathbf{r}') &= -\mu \int\limits_{V_J} \mathrm{d}V' \, \mathbf{J}(\mathbf{r}') \cdot \nabla' \frac{\mathrm{e}^{-\mathrm{j}kR}}{4\pi R} \\
&= \mu \int\limits_{V_J} \mathrm{d}V' \left\{ \frac{\mathrm{e}^{-\mathrm{j}kR}}{4\pi R} \nabla' \cdot \mathbf{J}(\mathbf{r}') - \nabla' \cdot \left[\mathbf{J}(\mathbf{r}') \frac{\mathrm{e}^{-\mathrm{j}kR}}{4\pi R} \right] \right\} \\
&= -\mathrm{j}\omega\mu \int\limits_{V_\rho} \mathrm{d}V' \frac{\mathrm{e}^{-\mathrm{j}kR}}{4\pi R} \rho(\mathbf{r}') - \mu \int\limits_{\partial V_J} \mathrm{d}S' \, \underbrace{\hat{\mathbf{n}}(\mathbf{r}') \cdot \mathbf{J}(\mathbf{r}')}_{=0} \frac{\mathrm{e}^{-\mathrm{j}kR}}{4\pi R} = -\mathrm{j}\frac{\omega}{c^2}\Phi_E(\mathbf{r})
\end{aligned} \tag{9.27}$$

where we have invoked (9.23) in the last step. The Gauss theorem has been applied in that the vector field $\mathbf{J}(\mathbf{r}')\exp(-\mathrm{j}kR)/R$ is of class $\mathcal{C}^1(V_J)^3 \cup \mathcal{C}(\overline{V}_J)^3$ for $\mathbf{r} \notin \overline{V}_J$.

Since the Green function is singular for $\mathbf{r} \in V_J$ we exclude the observation point with a ball $B(\mathbf{r}, a) \subset V_J$ and split the integral into two parts, namely, over $V_a := V_J \setminus B(\mathbf{r}, a)$ and over the ball (see Figure 2.8)

$$\begin{aligned}
\mu \int\limits_{V_J} \mathrm{d}V' \, \mathbf{J}(\mathbf{r}') \cdot \nabla \frac{\mathrm{e}^{-\mathrm{j}kR}}{4\pi R} &= -\mu \int\limits_{V_a} \mathrm{d}V' \, \mathbf{J}(\mathbf{r}') \cdot \nabla' \frac{\mathrm{e}^{-\mathrm{j}kR}}{4\pi R} - \mu \int\limits_{B(\mathbf{r},a)} \mathrm{d}V' \, \mathbf{J}(\mathbf{r}') \cdot \nabla' \frac{\mathrm{e}^{-\mathrm{j}kR}}{4\pi R} \\
&= \mu \int\limits_{V_a} \mathrm{d}V' \left\{ \frac{\mathrm{e}^{-\mathrm{j}kR}}{4\pi R} \nabla' \cdot \mathbf{J}(\mathbf{r}') - \nabla' \cdot \left[\mathbf{J}(\mathbf{r}') \frac{\mathrm{e}^{-\mathrm{j}kR}}{4\pi R} \right] \right\} - \mu \int\limits_{B(\mathbf{r},a)} \mathrm{d}V' \, \mathbf{J}(\mathbf{r}') \cdot \nabla' \frac{\mathrm{e}^{-\mathrm{j}kR}}{4\pi R} \\
&= -\mathrm{j}\omega\mu \int\limits_{V_a} \mathrm{d}V' \frac{\mathrm{e}^{-\mathrm{j}kR}}{4\pi R} \rho(\mathbf{r}') - \mu \int\limits_{\partial V_J} \mathrm{d}S' \, \underbrace{\hat{\mathbf{n}}(\mathbf{r}') \cdot \mathbf{J}(\mathbf{r}')}_{=0} \frac{\mathrm{e}^{-\mathrm{j}kR}}{4\pi R} \\
&\quad - \mu \int\limits_{\partial B} \mathrm{d}S' \, \hat{\mathbf{n}}(\mathbf{r}') \cdot \mathbf{J}(\mathbf{r}') \frac{\mathrm{e}^{-\mathrm{j}kR}}{4\pi R} - \mu \int\limits_{B(\mathbf{r},a)} \mathrm{d}V' \, \mathbf{J}(\mathbf{r}') \cdot \nabla' \frac{\mathrm{e}^{-\mathrm{j}kR}}{4\pi R} \\
&= -\mathrm{j}\frac{\omega}{c^2}\Phi_E(\mathbf{r}) - \mu \int\limits_{\partial B} \mathrm{d}S' \, \hat{\mathbf{n}}(\mathbf{r}') \cdot \mathbf{J}(\mathbf{r}') \frac{\mathrm{e}^{-\mathrm{j}kR}}{4\pi R} \\
&\quad - \mu \int\limits_{B(\mathbf{r},a)} \mathrm{d}V' \, \mathbf{J}(\mathbf{r}') \cdot \hat{\mathbf{R}} \left(\frac{\mathrm{j}k}{4\pi R} + \frac{1}{4\pi R^2} \right) \mathrm{e}^{-\mathrm{j}kR}
\end{aligned} \tag{9.28}$$

where we have applied the Gauss theorem because the integrand is continuously differentiable in V_a. Finally, we need to show that the last two integrals vanish in the limit as $a \to 0^+$. We estimate

$$\left| \int\limits_{\partial B} \mathrm{d}S' \, \hat{\mathbf{n}}(\mathbf{r}') \cdot \mathbf{J}(\mathbf{r}') \frac{\mathrm{e}^{-\mathrm{j}kR}}{4\pi R} \right| \leqslant \int\limits_{\partial B} \mathrm{d}S' \frac{\|\mathbf{J}\|_\infty}{4\pi a} = a \, \|\mathbf{J}\|_\infty \xrightarrow[a \to 0^+]{} 0 \tag{9.29}$$

$$\left| \int_{B(\mathbf{r},a)} \mathrm{d}V'\, \mathbf{J}(\mathbf{r}') \cdot \hat{\mathbf{R}} \left(\frac{\mathrm{j}k}{4\pi R} + \frac{1}{4\pi R^2} \right) \mathrm{e}^{-\mathrm{j}kR} \right| \leqslant \int_{B(\mathbf{r},a)} \mathrm{d}V'\, \|\mathbf{J}\|_\infty \left(\frac{k}{4\pi R} + \frac{1}{4\pi R^2} \right)$$
$$= \|\mathbf{J}\|_\infty \int_0^a \mathrm{d}R\,(kR+1) = \|\mathbf{J}\|_\infty \left(\frac{ka^2}{2} + a \right) \xrightarrow[a\to 0^+]{} 0 \tag{9.30}$$

and this proves that the Lorenz gauge is satisfied by (9.23) and (9.24). The verification that even the integral representations (9.15) and (9.16) satisfy the Lorenz gauge requires similar manipulations of the surface integrals (cf. Section 5.1.1).

Formula (9.16) has the advantage of treating the Cartesian components of the vector potential $\mathbf{A}_E(\mathbf{r})$ on a par with the scalar potential $\Phi_E(\mathbf{r})$. However, the effect of possible sources located externally to V enters (9.16) through a surface integral over ∂V that involves both $\mathbf{A}_E(\mathbf{r})$ and the normal derivative thereof. This feature makes it difficult to combine integral formulas pertinent to different regions of space — possibly filled with different media — because the matching conditions we derived at the end of Section 8.3 do not involve $\hat{\mathbf{n}} \cdot \nabla \mathbf{A}_E$. Therefore, similarly to what we did for the steady-state magnetic vector potential in Section 5.1.1 we wish to transform the surface integral into one which contains $\hat{\mathbf{n}}(\mathbf{r}') \times \mathbf{A}_E(\mathbf{r}')$, $\hat{\mathbf{n}}(\mathbf{r}') \cdot \mathbf{A}_E(\mathbf{r}')$ and $\hat{\mathbf{n}}(\mathbf{r}') \times [\nabla' \times \mathbf{A}_E(\mathbf{r}')]$ explicitly.

To lighten the notation we drop the explicit dependence on the source point $\mathbf{r}'$, write $\hat{\mathbf{n}}'$ in lieu of $\hat{\mathbf{n}}(\mathbf{r}')$ and use the symbol $G(R)$, with $R = |\mathbf{r} - \mathbf{r}'|$, to denote the time-harmonic Green function (8.356). Guided by the manipulations performed in (5.43) we write (cf. [2, Section 8.15])

$$\begin{aligned}
&\mathbf{A}_E \hat{\mathbf{n}}' \cdot \nabla' G - G(R) \hat{\mathbf{n}}' \cdot \nabla' \mathbf{A}_E = \\
&\quad = 2\mathbf{A}_E \hat{\mathbf{n}}' \cdot \nabla' G - G(R) \hat{\mathbf{n}}' \cdot \nabla' \mathbf{A}_E - \mathbf{A}_E \hat{\mathbf{n}}' \cdot \nabla' G \\
&\quad = 2\mathbf{A}_E \hat{\mathbf{n}}' \cdot \nabla' G - \hat{\mathbf{n}}' \cdot \nabla' [G(R)\mathbf{A}_E] \\
&\quad = 2\mathbf{A}_E \hat{\mathbf{n}}' \cdot \nabla' G + \hat{\mathbf{n}}' \times \{\nabla' \times [G(R)\mathbf{A}_E]\} \\
&\qquad - (\hat{\mathbf{n}}' \times \nabla') \times [G(R)\mathbf{A}_E] - \hat{\mathbf{n}}' \nabla' \cdot [G(R)\mathbf{A}_E]
\end{aligned} \tag{9.31}$$

where ∇' means differentiation with respect to $\mathbf{r}'$ and the last step follows from the differential identity (H.57). We proceed by transforming $\hat{\mathbf{n}}' \times [\nabla' \times (G\mathbf{A}_E)]$ and $\hat{\mathbf{n}}' \nabla' \cdot (G\mathbf{A}_E)$ with the aid of (H.50) and (H.51), namely,

$$\begin{aligned}
&\mathbf{A}_E \hat{\mathbf{n}}' \cdot \nabla' G - G(R) \hat{\mathbf{n}}' \cdot \nabla' \mathbf{A}_E = \\
&\quad = 2\mathbf{A}_E \hat{\mathbf{n}}' \cdot \nabla' G + G(R) \hat{\mathbf{n}}' \times (\nabla' \times \mathbf{A}_E) + \hat{\mathbf{n}}' \times (\nabla' G \times \mathbf{A}_E) \\
&\qquad - (\hat{\mathbf{n}}' \times \nabla') \times [G(R)\mathbf{A}_E] - \hat{\mathbf{n}}' \nabla' G \cdot \mathbf{A}_E - G(R) \hat{\mathbf{n}}' \nabla' \cdot \mathbf{A}_E \\
&\quad = 2\mathbf{A}_E \hat{\mathbf{n}}' \cdot \nabla' G + G(R) \hat{\mathbf{n}}' \times (\nabla' \times \mathbf{A}_E) + \hat{\mathbf{n}}' \cdot \mathbf{A}_E \nabla' G - \mathbf{A}_E \hat{\mathbf{n}}' \cdot \nabla' G \\
&\qquad - (\hat{\mathbf{n}}' \times \nabla') \times [G(R)\mathbf{A}_E] - \hat{\mathbf{n}}' \nabla' G \cdot \mathbf{A}_E - G(R) \hat{\mathbf{n}}' \nabla' \cdot \mathbf{A}_E \\
&\quad = \underbrace{\mathbf{A}_E \hat{\mathbf{n}}' \cdot \nabla' G}_{①} + G(R) \hat{\mathbf{n}}' \times (\nabla' \times \mathbf{A}_E) + \hat{\mathbf{n}}' \cdot \mathbf{A}_E \nabla' G \\
&\qquad - (\hat{\mathbf{n}}' \times \nabla') \times [G(R)\mathbf{A}_E] - \underbrace{\hat{\mathbf{n}}' \nabla' G \cdot \mathbf{A}_E}_{②} - G(R) \hat{\mathbf{n}}' \nabla' \cdot \mathbf{A}_E \\
&\quad = \nabla' G \times (\mathbf{A}_E \times \hat{\mathbf{n}}') + G(R) \hat{\mathbf{n}}' \times (\nabla' \times \mathbf{A}_E) + \hat{\mathbf{n}}' \cdot \mathbf{A}_E \nabla' G \\
&\qquad - (\hat{\mathbf{n}}' \times \nabla') \times [G(R)\mathbf{A}_E] - G(R) \hat{\mathbf{n}}' \nabla' \cdot \mathbf{A}_E
\end{aligned} \tag{9.32}$$

where we have also employed (H.14) backwards to combine the terms labelled with ① and ②. This result differs from the one obtained in the stationary regime obviously for the Green function but also, more importantly, for the presence of the divergence of $\mathbf{A}_E$ which is *absent* in (5.43) on account of the Coulomb gauge.

To finalize our derivation we simply insert (9.32) into (9.16), viz.,

$$\mathbf{A}_E(\mathbf{r}) = \mu \int_{V_J} dV' \frac{e^{-jkR}}{4\pi R}\mathbf{J}(\mathbf{r}')$$
$$+ \int_{\partial V} dS' \, [\hat{\mathbf{n}}(\mathbf{r}') \times \mathbf{A}_E(\mathbf{r}')] \times \nabla' \frac{e^{-jkR}}{4\pi R} + \int_{\partial V} dS' \, \hat{\mathbf{n}}(\mathbf{r}') \cdot \mathbf{A}_E(\mathbf{r}')\nabla' \frac{e^{-jkR}}{4\pi R}$$
$$+ \int_{\partial V} dS' \frac{e^{-jkR}}{4\pi R}\hat{\mathbf{n}}(\mathbf{r}') \times [\nabla' \times \mathbf{A}_E(\mathbf{r}')] - \int_{\partial V} dS' \frac{e^{-jkR}}{4\pi R}\hat{\mathbf{n}}(\mathbf{r}')\nabla' \cdot \mathbf{A}_E(\mathbf{r}') \quad (9.33)$$

having noticed that the integral of $(\hat{\mathbf{n}}' \times \nabla') \times [G(R)\mathbf{A}_E]$ over a closed surface vanishes in light of (H.94) [5, Eq. (A1.144)], [6, Eq. (C.26)]. The last integral in the right member of (9.33) may be rephrased in terms of the scalar potential $\Phi_E(\mathbf{r})$, because the divergence of $\mathbf{A}_E(\mathbf{r})$ is determined through the Lorenz gauge (9.25). Furthermore, since this very contribution becomes proportional to $j\omega$, then (9.33) reduces to (5.18) in the limit as $\omega \to 0$.

Unlike formula (9.16), which is well-defined for observation points on the boundary ∂V, (9.33) seems to run into trouble for $\mathbf{r} \in \partial V$ because of the two integrals involving the gradient of the Green function. These two contributions may be cast into an alternative format, namely,

$$\int_{\partial V} dS' \, [\hat{\mathbf{n}}(\mathbf{r}') \times \mathbf{A}_E(\mathbf{r}')] \times \nabla' G + \int_{\partial V} dS' \, \hat{\mathbf{n}}(\mathbf{r}') \cdot \mathbf{A}_E(\mathbf{r}')\nabla' G =$$
$$= \int_{\partial V} dS' \, \mathbf{A}_E(\mathbf{r}')\hat{\mathbf{n}}(\mathbf{r}') \cdot \nabla' G + \int_{\partial V} dS' \, \mathbf{A}_E(\mathbf{r}') \times [\nabla' G \times \hat{\mathbf{n}}(\mathbf{r}')]$$
$$= \int_{\partial V} dS' \, \mathbf{A}_E(\mathbf{r}')\hat{\mathbf{n}}(\mathbf{r}') \cdot \nabla' G + \int_{\partial V} dS' \, \mathbf{A}_E(\mathbf{r}') \times \{\nabla'_s \times [G(R)\hat{\mathbf{n}}(\mathbf{r}')]\}$$
$$= \int_{\partial V} dS' \, \mathbf{A}_E(\mathbf{r}')\hat{\mathbf{n}}(\mathbf{r}') \cdot \nabla' G + \int_{\partial V} dS' \, [\mathbf{A}_E(\mathbf{r}') - \mathbf{A}_E(\mathbf{r})] \times \{\nabla'_s \times [G(R)\hat{\mathbf{n}}(\mathbf{r}')]\}$$
$$= \int_{\partial V} dS' \, \mathbf{A}_E(\mathbf{r}')\hat{\mathbf{n}}(\mathbf{r}') \cdot \nabla' G + \int_{\partial V} dS' \, [\mathbf{A}_E(\mathbf{r}') - \mathbf{A}_E(\mathbf{r})] \times [\nabla' G \times \hat{\mathbf{n}}(\mathbf{r}')] \quad (9.34)$$

where we have used (H.14), (H.81) and the integral theorem (H.103), which here returns zero over a closed surface. Now, the first integral is in the form of a double-layer potential, and hence it is bounded. This term contributes half the value of the potential at $\mathbf{r}$ when the observation point tends to ∂V [cf. (2.244)]. The second term can be shown to be bounded by following the same procedure outlined in (2.256) and (2.257) provided

$$|\mathbf{A}_E(\mathbf{r}') - \mathbf{A}_E(\mathbf{r})| \leqslant M|\mathbf{r}' - \mathbf{r}|^{\nu}, \qquad \mathbf{r}, \mathbf{r}' \in \partial V, \qquad \nu \in]0, 1[\quad (9.35)$$

that is, the potential is Hölder continuous on ∂V. The latter is true because $\mathbf{A}_E(\mathbf{r})$ is twice-differentiable for points away from the sources. On the other hand, the second contribution can be reduced to integrals of the type (G.1), and the proof of Appendix G guarantees that this term is even Hölder continuous.

We conclude by noticing that the Cartesian components of the time-harmonic Hertz vectors introduced in Section 8.4 solve six equations of the type (9.2). Thus, in a homogeneous isotropic unbounded medium $\mathbf{\Pi}_E$ and $\mathbf{\Pi}_M$ are given by

$$\mathbf{\Pi}_E(\mathbf{r}) = \frac{1}{\varepsilon}\int_{V_J} \mathrm{d}V' \, \frac{\mathrm{e}^{-\mathrm{j}k|\mathbf{r}-\mathbf{r}'|}}{4\pi|\mathbf{r}-\mathbf{r}'|}\mathbf{P}(\mathbf{r}'), \qquad \mathbf{r} \in \mathbb{R}^3 \tag{9.36}$$

$$\mathbf{\Pi}_M(\mathbf{r}) = \int_{V_{J_M}} \mathrm{d}V' \, \frac{\mathrm{e}^{-\mathrm{j}k|\mathbf{r}-\mathbf{r}'|}}{4\pi|\mathbf{r}-\mathbf{r}'|}\mathbf{M}(\mathbf{r}'), \qquad \mathbf{r} \in \mathbb{R}^3 \tag{9.37}$$

on the grounds of the integral representation (9.10). The integrals are extended to the regions occupied by $\mathbf{P}$ and $\mathbf{M}$, which are the same as those pertinent to $\mathbf{J}$ and $\mathbf{J}_M$ in accordance with (8.172) and (8.179).

9.1.2 *Time domain*

Since our goal was the determination of the electrodynamic potentials in the time domain, the result follows through the application of (8.322) to $\Phi_E(\mathbf{r};\omega)$ and the Cartesian components of $\mathbf{A}_E(\mathbf{r};\omega)$. On account of (9.23) and (9.24) we have

$$\begin{aligned}
\Phi_E(\mathbf{r},t) &= \frac{1}{2\pi}\int_{\mathbb{R}} \mathrm{d}\omega \, \mathrm{e}^{\mathrm{j}\omega t}\frac{1}{\varepsilon}\int_{V_\rho} \mathrm{d}V' \, \frac{\mathrm{e}^{-\mathrm{j}kR}}{4\pi R}\rho(\mathbf{r}';\omega) \\
&= \frac{1}{\varepsilon}\int_{V_\rho} \mathrm{d}V' \, \frac{1}{4\pi|\mathbf{r}-\mathbf{r}'|}\underbrace{\frac{1}{2\pi}\int_{\mathbb{R}} \mathrm{d}\omega \, \exp\left[\mathrm{j}\,\omega\left(t-\frac{|\mathbf{r}-\mathbf{r}'|}{c}\right)\right]\rho(\mathbf{r}';\omega)}_{=\mathfrak{F}^{-1}\{\rho\}} \\
&= \frac{1}{\varepsilon}\int_{V_\rho} \mathrm{d}V' \, \frac{\varrho(\mathbf{r}',t-|\mathbf{r}-\mathbf{r}'|/c)}{4\pi|\mathbf{r}-\mathbf{r}'|} = \int_{V_\rho\cap B(\mathbf{r},ct)} \mathrm{d}V' \, \frac{\varrho(\mathbf{r}',t-|\mathbf{r}-\mathbf{r}'|/c)}{4\pi\varepsilon|\mathbf{r}-\mathbf{r}'|}
\end{aligned} \tag{9.38}$$

$$\begin{aligned}
\mathcal{A}_E(\mathbf{r},t) &= \frac{1}{2\pi}\int_{\mathbb{R}} \mathrm{d}\omega \, \mathrm{e}^{\mathrm{j}\omega t}\mu\int_{V_J} \mathrm{d}V' \, \frac{\mathrm{e}^{-\mathrm{j}kR}}{4\pi R}\mathbf{J}(\mathbf{r}';\omega) \\
&= \mu\int_{V_J} \mathrm{d}V' \, \frac{1}{4\pi|\mathbf{r}-\mathbf{r}'|}\underbrace{\frac{1}{2\pi}\int_{\mathbb{R}} \mathrm{d}\omega \, \exp\left[\mathrm{j}\,\omega\left(t-\frac{|\mathbf{r}-\mathbf{r}'|}{c}\right)\right]\mathbf{J}(\mathbf{r}';\omega)}_{=\mathfrak{F}^{-1}\{\mathbf{J}\}} \\
&= \mu\int_{V_J} \mathrm{d}V' \, \frac{\mathfrak{J}(\mathbf{r}',t-|\mathbf{r}-\mathbf{r}'|/c)}{4\pi|\mathbf{r}-\mathbf{r}'|} = \int_{V_J\cap B(\mathbf{r},ct)} \mathrm{d}V' \, \frac{\mu\mathfrak{J}(\mathbf{r}',t-|\mathbf{r}-\mathbf{r}'|/c)}{4\pi|\mathbf{r}-\mathbf{r}'|}
\end{aligned} \tag{9.39}$$

where we have recognized the integrals over $\mathbb{R}$ as inverse Fourier transformations with the time variable given by the quantity

$$t_r = t - \frac{|\mathbf{r}-\mathbf{r}'|}{c} = t - \frac{R}{c} \tag{9.40}$$

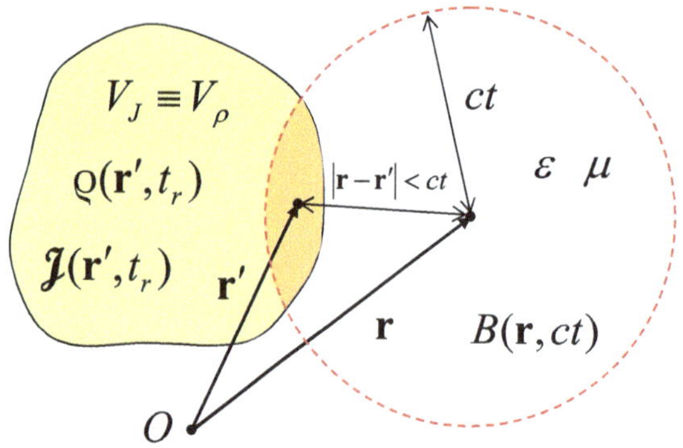

Figure 9.1 *For illustrating which part of the sources contributes to the retarded potentials in the observation point* $\mathbf{r}$ *for* $t > 0$.

i.e., the retarded time we introduced in (1.251) for a single charged particle. Expressions (9.38) and (9.39) yield the so-called *retarded potentials* [7, Section 2.5], [8, Section 8.1], [9, Chapter 4], [10, Chapter IV], [11, Section 1.5], [12, Section XIII.1] [2, 5, 13, 14]. Indeed, they can be interpreted by saying that the potentials 'detected' by an observer at the location $\mathbf{r}$ and at time t are due to the superposition of the elementary contributions generated by infinitesimal amounts of charge $\varrho(\mathbf{r}', t_r)\mathrm{d}V'$ when the latter occupied the *retarded position* $\mathbf{r}'(t_r)$ at the time t_r, as given by (9.40).

A few remarks will serve to explain the restricted domains of integration in the rightmost-hand sides of (9.38) and (9.39). If the charges are set in motion at time $t = 0$, then for negative times $\mathcal{J}(\mathbf{r}, t)$ was zero and $\varrho(\mathbf{r}, t)$ was at most constant. Likewise, if the configuration was stationary up to $t = 0$, $\varrho(\mathbf{r}, t)$ was constant and $\mathcal{J}(\mathbf{r}, t)$ solenoidal. Then, so long as we integrated over the whole source region, for negative times (9.38) and (9.39) would correctly return the static or stationary potentials found in Sections 2.7 and 5.1.1. By contrast, if we concern ourselves with wave-like solutions only, we may assume the charge density was zero for $t < 0$, whereby static or stationary solutions are not included in (9.38) and (9.39). More importantly, for a given point $\mathbf{r}$ and a time $t > 0$ only the charges that are accelerated for $t_r > 0$ *can* contribute to the potentials. By virtue of (9.40) this condition translates into

$$|\mathbf{r} - \mathbf{r}'| < ct \tag{9.41}$$

that is, only the charges which find themselves inside a ball of radius ct and center $\mathbf{r}$ may influence the potentials in $\mathbf{r}$, as is pictorially suggested in Figure 9.1 [15].

On a related score, it would appear that the electrodynamic potentials in the Lorenz gauge (8.121) might be systematically obtained from the static and stationary counterparts (2.169) and (5.29) by including the dependence on the retarded time of charge and current density. Make no mistake, though, because this viewpoint is deceitful and may serve at most to remember the formulas, but it certainly does not apply to the relevant expressions of the time-dependent electric and magnetic fields, that is, the general solution to the Maxwell equations in the whole space does not follow from (2.214) and (5.60) by simply allowing for the retarded time in the integration! The reason is that in computing the spatial derivatives of the potentials as indicated by (8.104) and (8.105) we need to account also for the dependence of $\varrho(\mathbf{r}', t_r)$ and $\mathcal{J}(\mathbf{r}', t_r)$ on the observation point $\mathbf{r}$ through the retarded time (9.40) [14, 16]. This is a significant detail, as it ultimately makes electromagnetic radiation possible (cf. Section 1.9).

Without the retarded-time dependence the gradient of Φ_E would give rise to an electrostatic field which, as we know, falls off as the inverse square of the distance from the source (Section 3.2). On the contrary, the presence of the retarded time produces a new contribution which decays only as $1/r$. In symbols, we have

$$\begin{aligned}\nabla\Phi_E &= \frac{1}{\varepsilon}\nabla\int_{V_J} \mathrm{d}V'\,\frac{\varrho(\mathbf{r}',t_r)}{4\pi R} = \frac{1}{\varepsilon}\int_{V_J}\mathrm{d}V'\,\nabla\frac{\varrho(\mathbf{r}',t_r)}{4\pi R}\\ &= \frac{1}{\varepsilon}\int_{V_J}\mathrm{d}V'\,\varrho(\mathbf{r}',t_r)\nabla\frac{1}{4\pi R} + \frac{1}{\varepsilon}\int_{V_J}\mathrm{d}V'\,\left.\frac{\partial\varrho}{\partial t}\right|_{t=t_r}\frac{\nabla t_r}{4\pi R}\\ &= -\frac{1}{\varepsilon}\int_{V_J}\mathrm{d}V'\,\varrho(\mathbf{r}',t_r)\frac{\mathbf{r}-\mathbf{r}'}{4\pi|\mathbf{r}-\mathbf{r}'|^3} + \frac{1}{\varepsilon}\int_{V_J}\mathrm{d}V'\,\frac{\nabla t_r}{4\pi|\mathbf{r}-\mathbf{r}'|}\frac{\partial\varrho}{\partial t_r}\end{aligned} \tag{9.42}$$

where we have moved the gradient inside the domain integral by virtue of an extension of theorem (2.199) proved in Section 2.8. In particular, from (9.40) we have

$$\nabla t_r = -\frac{\nabla R}{c} = -\frac{\mathbf{r}-\mathbf{r}'}{c|\mathbf{r}-\mathbf{r}'|} = -\frac{\hat{\mathbf{R}}}{c} \tag{9.43}$$

whereby we see that the second term in (9.42) behaves asymptotically as $1/r$. Similar steps and conclusions hold for the curl of $\mathcal{A}_E$.

Although we have determined the retarded potentials by resorting to the Fourier transformation with respect to time, the inhomogeneous D'Alembert equation (8.321) can be inverted directly in the temporal domain in a bounded smooth volume V containing a smaller region V_S where the sources reside. The procedure we describe is essentially based on changing the time variable and casting the equation into a form amenable to the application of the Gauss theorem [17]. Alternative strategies are the Kirchhoff method of integration described by J. A. Stratton [2, Section 8.1] and the Poisson method outlined in [10, Section IV.56], whereas yet another approach, which makes use of δ-distributions, can be found in [6, Section 13.2].

We denote with $\mathbf{r}$ and $\mathbf{r}'$ the observation point and the independent space variable, respectively, and also use the retarded time (9.40) as the independent time variable, whereby (8.321) becomes

$$\nabla'^2_{t_r}\Upsilon(\mathbf{r}',t_r) - \frac{1}{c^2}\frac{\partial^2}{\partial t_r^2}\Upsilon(\mathbf{r}',t_r) = -s(R,\alpha,\beta,t_r), \qquad (\mathbf{r}',t_r)\in\mathbb{R}^3\times\mathbb{R} \tag{9.44}$$

with (R,α,β) being a system of local spherical coordinates centered in $\mathbf{r}$. In particular, since $\mathbf{R} = \mathbf{r}-\mathbf{r}'$ is a position vector positively oriented from $\mathbf{r}'$ to $\mathbf{r}$, the pertinent radial unit vector is $-\hat{\mathbf{R}}$ (Figure 9.2). Further, since $\Upsilon(\mathbf{r}',t_r)$ depends on R directly and through the retarded time (9.40), the symbol $\nabla'^2_{t_r}$ means that the derivative with respect to R is carried out while keeping t_r constant.

With these positions we have $\Upsilon(\mathbf{r}',t_r) = \Upsilon(R,\alpha,\beta,t-R/c)$ and

$$\frac{\partial\Upsilon}{\partial t} = \frac{\partial\Upsilon}{\partial t_r}\frac{\partial t_r}{\partial t} = \frac{\partial\Upsilon}{\partial t_r} \tag{9.45}$$

$$\frac{\mathrm{d}\Upsilon}{\mathrm{d}R} = \frac{\partial\Upsilon}{\partial R} + \frac{\partial\Upsilon}{\partial t_r}\frac{\partial t_r}{\partial R} = \frac{\partial\Upsilon}{\partial R} - \frac{1}{c}\frac{\partial\Upsilon}{\partial t_r} \tag{9.46}$$

where

- $\mathrm{d}\Upsilon/\mathrm{d}R$ represents the derivative with respect to R with t held constant;
- $\partial\Upsilon/\partial R$ represents the derivative with respect to R with t_r held constant.

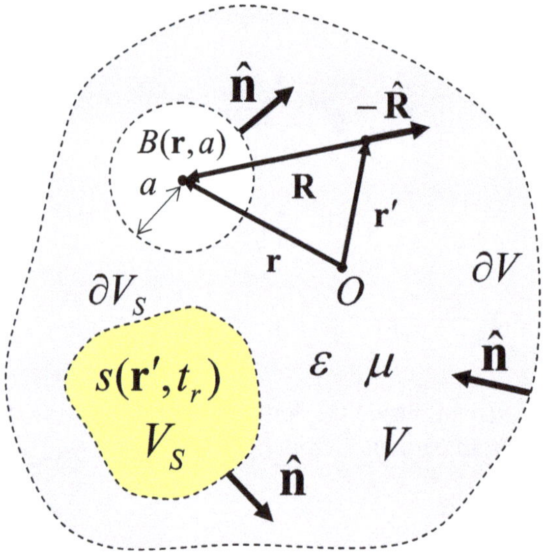

Figure 9.2 Geometry for the direct solution of the D'Alembert equation in the time domain.

As a consequence, the gradient of $\Upsilon(\mathbf{r}', t_r)$ and the divergence of a vector field $\boldsymbol{\Xi}(\mathbf{r}', t_r)$ can be computed in spherical coordinates (R, α, β) as follows

$$\begin{aligned}\nabla'\Upsilon(\mathbf{r}', t_r) &= -\hat{\mathbf{R}}\frac{\mathrm{d}\Upsilon}{\mathrm{d}R} + \frac{\hat{\boldsymbol{\alpha}}}{R\sin\alpha}\frac{\partial}{\partial\alpha}(\Upsilon\sin\alpha) + \frac{\hat{\boldsymbol{\beta}}}{R\sin\alpha}\frac{\partial\Upsilon}{\partial\beta} \\ &= -\hat{\mathbf{R}}\frac{\partial\Upsilon}{\partial R} + \frac{\hat{\boldsymbol{\alpha}}}{R\sin\alpha}\frac{\partial}{\partial\alpha}(\Upsilon\sin\alpha) + \frac{\hat{\boldsymbol{\beta}}}{R\sin\alpha}\frac{\partial\Upsilon}{\partial\beta} + \frac{\hat{\mathbf{R}}}{c}\frac{\partial\Upsilon}{\partial t_r} \\ &= \nabla'_{t_r}\Upsilon(\mathbf{r}', t_r) + \frac{\hat{\mathbf{R}}}{c}\frac{\partial\Upsilon}{\partial t_r}\end{aligned} \tag{9.47}$$

$$\begin{aligned}\nabla'\cdot\boldsymbol{\Xi}(\mathbf{r}', t_r) &= -\frac{1}{R^2}\frac{\mathrm{d}}{\mathrm{d}R}(R^2\Xi_R) + \frac{1}{R\sin\alpha}\frac{\partial}{\partial\alpha}(\Xi_\alpha\sin\alpha) + \frac{1}{R\sin\alpha}\frac{\partial\Xi_\beta}{\partial\beta} \\ &= -\frac{1}{R^2}\frac{\partial}{\partial R}(R^2\Xi_R) + \frac{1}{R\sin\alpha}\frac{\partial}{\partial\alpha}(\Xi_\alpha\sin\alpha) + \frac{1}{R\sin\alpha}\frac{\partial\Xi_\beta}{\partial\beta} + \frac{1}{c}\frac{\partial\Xi_R}{\partial t_r} \\ &= \nabla'_{t_r}\cdot\boldsymbol{\Xi}(\mathbf{r}', t_r) + \frac{1}{c}\frac{\partial\Xi_R}{\partial t_r}\end{aligned} \tag{9.48}$$

where $\Xi_R = \hat{\mathbf{R}}\cdot\boldsymbol{\Xi}$, and ∇' and ∇'_{t_r} mean that the derivative with respect to R is carried out as t or t_r is held constant. An application of (9.48) to the auxiliary field $\boldsymbol{\Xi}(\mathbf{r}', t_r) = \nabla'_{t_r}\Upsilon/R$ yields

$$\begin{aligned}\nabla'\cdot\left(\frac{\nabla'_{t_r}\Upsilon}{R}\right) &= \nabla'_{t_r}\cdot\left(\frac{\nabla'_{t_r}\Upsilon}{R}\right) + \frac{1}{Rc}\frac{\partial}{\partial t_r}\hat{\mathbf{R}}\cdot\nabla'_{t_r}\Upsilon \\ &= \frac{1}{R}\nabla'^2_{t_r}\Upsilon + \nabla'_{t_r}\frac{1}{R}\cdot\nabla'_{t_r}\Upsilon - \frac{1}{Rc}\frac{\partial^2\Upsilon}{\partial t_r\partial R} \\ &= \frac{1}{R}\nabla'^2_{t_r}\Upsilon - \frac{1}{R^2}\frac{\partial\Upsilon}{\partial R} - \frac{1}{Rc}\frac{\partial^2\Upsilon}{\partial t_r\partial R}, \qquad R > 0\end{aligned} \tag{9.49}$$

having noticed that $\hat{\mathbf{R}} \cdot \nabla'_{t_r} \Upsilon = -\partial\Upsilon/\partial R$. For the subsequent integration we need derivatives with respect to R with t held constant, thus we write

$$\begin{aligned}
\nabla' \cdot \left(\frac{\nabla'_{t_r} \Upsilon}{R} \right) &= \frac{1}{R} \nabla'^2_{t_r} \Upsilon - \frac{1}{R^2} \frac{\mathrm{d}\Upsilon}{\mathrm{d}R} - \frac{1}{R^2 c} \frac{\partial \Upsilon}{\partial t_r} - \frac{1}{Rc} \frac{\partial}{\partial t_r} \frac{\mathrm{d}\Upsilon}{\mathrm{d}R} - \frac{1}{Rc^2} \frac{\partial^2 \Upsilon}{\partial t_r^2} \\
&= \frac{1}{R} \left(\nabla'^2_{t_r} \Upsilon - \frac{1}{c^2} \frac{\partial^2 \Upsilon}{\partial t_r^2} \right) - \frac{1}{R^2} \frac{\mathrm{d}}{\mathrm{d}R} \left(\Upsilon + \frac{R}{c} \frac{\partial \Upsilon}{\partial t_r} \right) \\
&= -\frac{1}{R} s(\mathbf{r}', t_r) + \nabla' \cdot \left[\frac{\hat{\mathbf{R}}}{R^2} \left(\Upsilon + \frac{R}{c} \frac{\partial \Upsilon}{\partial t_r} \right) \right], \qquad R > 0
\end{aligned} \tag{9.50}$$

on account of the Schwarz theorem [18, pp. 235–236], (9.44), (9.46) and (9.48).

Since (9.50) is not defined for $R = 0$ we isolate the point $\mathbf{r}$ with a ball $B(\mathbf{r}, a)$ and integrate over the region $V_a := V \setminus B(\mathbf{r}, a)$ (Figure 9.2). To begin with, we suppose that $\mathbf{r} \notin \overline{V}_S$ and choose a small enough so that $B(\mathbf{r}, a) \cap V_S = \emptyset$. In symbols, we have

$$\begin{aligned}
\int_{V_S} \mathrm{d}V' \frac{s(\mathbf{r}', t_r)}{R} &= -\int_{V_a} \mathrm{d}V' \, \nabla' \cdot \left[\frac{\nabla'_{t_r} \Upsilon}{R} - \frac{\hat{\mathbf{R}}}{R^2} \left(\Upsilon + \frac{R}{c} \frac{\partial \Upsilon}{\partial t_r} \right) \right] \\
&= \int_{\partial B} \mathrm{d}S' \, \hat{\mathbf{n}}(\mathbf{r}') \cdot \left[\frac{\nabla'_{t_r} \Upsilon}{R} - \frac{\hat{\mathbf{R}}}{R^2} \left(\Upsilon + \frac{R}{c} \frac{\partial \Upsilon}{\partial t_r} \right) \right] \\
&\quad + \int_{\partial V} \mathrm{d}S' \, \hat{\mathbf{n}}(\mathbf{r}') \cdot \left[\frac{\nabla'_{t_r} \Upsilon}{R} - \frac{\hat{\mathbf{R}}}{R^2} \left(\Upsilon + \frac{R}{c} \frac{\partial \Upsilon}{\partial t_r} \right) \right]
\end{aligned} \tag{9.51}$$

where we have used the Gauss theorem with the unit normals on ∂V and ∂B pointing inwards V_a. The theorem may be applied so long as the vector field between brackets is continuously differentiable in V_a. Since $1/R$ and $1/R^2$ are regular functions for $\mathbf{r}' \in V_a$, then it is necessary that $\nabla'_{t_r} \Upsilon(\mathbf{r}', t_r) \in \mathcal{C}^1(V_a)^3 \cap \mathcal{C}(\overline{V}_a)^3$ and $\partial\Upsilon(\mathbf{r}', t_r)/\partial t_r \in \mathcal{C}^1(V_a) \cap \mathcal{C}(\overline{V}_a)$. The result also holds if $s(\mathbf{r}', t_r)$ is discontinuous across ∂V_S, the boundary of the source region, in which case we apply the theorem separately to $V_a \setminus \overline{V}_S$ and V_S and demand that $\Upsilon(\mathbf{r}', t_r)$, $\hat{\mathbf{n}}(\mathbf{r}') \cdot \nabla'_{t_r} \Upsilon(\mathbf{r}', t_r)$ and $\partial\Upsilon/\partial t_r$ be continuous across ∂V_S (cf. Section 1.2.2).

As regards the contribution of the flux integral over ∂B in the limit as $a \to 0^+$ (Figure 9.2) we observe that, for $\mathbf{r}' \in \partial B$, we have $\hat{\mathbf{n}}(\mathbf{r}') = -\hat{\mathbf{R}}$, $R = a$, $\mathbf{r}' = \mathbf{r} - a\hat{\mathbf{R}}$, $t_r = t - a/c$ and

$$\begin{aligned}
&\int_{\partial B} \mathrm{d}S' \, \hat{\mathbf{n}}(\mathbf{r}') \cdot \left[\frac{\nabla'_{t_r} \Upsilon}{R} - \frac{\hat{\mathbf{R}}}{R^2} \left(\Upsilon + \frac{R}{c} \frac{\partial \Upsilon}{\partial t_r} \right) \right] = \\
&= \frac{1}{a^2} \int_{\partial B} \mathrm{d}S' \left[-a\hat{\mathbf{R}} \cdot \nabla'_{t_r} \Upsilon + \Upsilon(\mathbf{r}', t_r) + \frac{a}{c} \frac{\partial \Upsilon}{\partial t_r} \right] \\
&= 4\pi \left[a \left. \frac{\partial \Upsilon}{\partial R} \right|_{\mathbf{r}' = \mathbf{r}'_0} + \Upsilon(\mathbf{r}'_0, t_r) + \frac{a}{c} \frac{\partial}{\partial t_r} \Upsilon(\mathbf{r}'_0, t_r) \right] \xrightarrow[a \to 0^+]{} 4\pi \Upsilon(\mathbf{r}, t)
\end{aligned} \tag{9.52}$$

having used the mean value theorem with $\mathbf{r}'_0 \in \partial B$ a suitable point.

Thus, by taking the limit for vanishing a in (9.51) we arrive at the desired integral representation of $\Upsilon(\mathbf{r}, t)$, viz.,

$$\Upsilon(\mathbf{r}, t) = \int_{V_S} \mathrm{d}V' \frac{s(\mathbf{r}', t - |\mathbf{r} - \mathbf{r}'|/c)}{4\pi|\mathbf{r} - \mathbf{r}'|} + \int_{\partial V} \mathrm{d}S' \frac{\hat{\mathbf{n}}(\mathbf{r}') \cdot (\mathbf{r} - \mathbf{r}')}{4\pi|\mathbf{r} - \mathbf{r}'|^3} \Upsilon(\mathbf{r}', t_r)$$

$$+\int_{\partial V} \mathrm{d}S' \frac{\hat{\mathbf{n}}(\mathbf{r}')}{4\pi|\mathbf{r}-\mathbf{r}'|} \cdot \left(\frac{\mathbf{r}-\mathbf{r}'}{c|\mathbf{r}-\mathbf{r}'|}\frac{\partial \Upsilon}{\partial t_r} - \nabla'_{t_r}\Upsilon\right) \tag{9.53}$$

for points $\mathbf{r} \in V \setminus \overline{V}_S$ [2, p. 427], [12, p. 171]. It is important to notice that all the time-varying quantities in the right member are evaluated at the *retarded time* (9.40). Incidentally, (9.53) coincides with the representation that can be obtained by inserting (9.10) into the inverse Fourier transformation (8.322).

The extension of the formula for observation points in the source region may be done by isolating $\mathbf{r}$ with a ball $B(\mathbf{r}, a)$ and showing that the volume integral over $V_S \setminus B(\mathbf{r}, a)$ remains finite (cf. Figure 2.9). This also follows from the analysis of Section 2.7 for the singularity of the kernel in the volume integral in (9.53) is the same as that of the static Green function (2.131). Likewise, if we choose the observation point $\mathbf{r}$ outside V, we conclude that the right-hand side of (9.53) vanishes. However, this does not necessarily mean that the true potential is zero in $\mathbf{r} \in \mathbb{R}^3 \setminus \overline{V}$ at any given time. Finally, for $\mathbf{r} \in \partial V$ we can prove that that the right-hand side — where the surface integrals have meaning as Cauchy principal value — yields half the value of the potential on ∂V.

The surface integrals in (9.53) constitute the contribution to the potential due to fields and waves generated by sources existing *outside* V, and in fact we can prove that it vanishes when all the sources are localized in $V_S \subset V$ (cf. the static and stationary cases). If we consider a ball $B(0, b)$ with radius $b > 0$ large enough so that $V \subset B(0, b)$ (Figure 9.3) and apply (9.53) to the region $B(0, b) \setminus \overline{V}$ for observation points $\mathbf{r} \in V$ we find

$$0 = \int_{\partial V} \mathrm{d}S' \frac{\hat{\mathbf{n}}(\mathbf{r}') \cdot \hat{\mathbf{R}}}{4\pi R^2} \Upsilon(\mathbf{r}', t_r) + \int_{\partial V} \mathrm{d}S' \frac{\hat{\mathbf{n}}(\mathbf{r}')}{4\pi R} \cdot \left(\frac{\hat{\mathbf{R}}}{c}\frac{\partial \Upsilon}{\partial t_r} - \nabla'_{t_r}\Upsilon\right)$$

$$+\int_{\partial B} \mathrm{d}S' \frac{\hat{\mathbf{n}}(\mathbf{r}') \cdot \hat{\mathbf{R}}}{4\pi R^2} \Upsilon(\mathbf{r}', t_r) + \int_{\partial B} \mathrm{d}S' \frac{\hat{\mathbf{n}}(\mathbf{r}')}{4\pi R} \cdot \left(\frac{\hat{\mathbf{R}}}{c}\frac{\partial \Upsilon}{\partial t_r} - \nabla'_{t_r}\Upsilon\right) \tag{9.54}$$

where the unit normal $\hat{\mathbf{n}}(\mathbf{r}')$ points outwards $B(0, b) \setminus V$. Since the integrals over ∂V are the same as in (9.53) under the assumption that the sources are confined in V_S, we need to show that the integrals over ∂B vanish in the limit as $b \to +\infty$. We observe

$$\hat{\mathbf{n}}(\mathbf{r}') = \hat{\mathbf{r}}', \qquad \hat{\mathbf{n}}(\mathbf{r}') \cdot \nabla'_{t_r}\Upsilon = \hat{\mathbf{r}}' \cdot \nabla'_{t_r}\Upsilon = \frac{\partial \Upsilon}{\partial r'}, \qquad \mathbf{r}' \in \partial B \tag{9.55}$$

and also

$$\hat{\mathbf{n}}(\mathbf{r}') \cdot \left(\frac{\hat{\mathbf{R}}}{c}\frac{\partial \Upsilon}{\partial t_r} - \nabla'_{t_r}\Upsilon\right) = \frac{\hat{\mathbf{r}}' \cdot \hat{\mathbf{R}}}{c}\frac{\partial \Upsilon}{\partial t_r} - \frac{\partial \Upsilon}{\partial r'}$$

$$= \left(\hat{\mathbf{r}}' \cdot \hat{\mathbf{R}} + 1\right)\frac{1}{c}\frac{\partial \Upsilon}{\partial t_r} - \left(\frac{1}{c}\frac{\partial \Upsilon}{\partial t_r} + \frac{\partial \Upsilon}{\partial r'}\right), \qquad \mathbf{r}' \in \partial B \tag{9.56}$$

whereby we can estimate

$$\left|\int_{\partial B} \mathrm{d}S' \frac{\hat{\mathbf{n}}(\mathbf{r}') \cdot \hat{\mathbf{R}}}{4\pi R^2} \Upsilon(\mathbf{r}', t_r)\right| \leqslant \int_{\partial B} \mathrm{d}S' \frac{|\Upsilon(\mathbf{r}', t_r)|}{4\pi R^2} \leqslant \int_{\partial B} \mathrm{d}S' \frac{|\Upsilon(\mathbf{r}', t_r)|}{4\pi (b-r)^2} \tag{9.57}$$

$$\left|\int_{\partial B} \mathrm{d}S' \frac{\hat{\mathbf{n}}(\mathbf{r}')}{4\pi R} \cdot \left(\frac{\hat{\mathbf{R}}}{c}\frac{\partial \Upsilon}{\partial t_r} - \nabla'_{t_r}\Upsilon\right)\right| \leqslant \int_{\partial B} \mathrm{d}S' \frac{1}{4\pi R}\left|\hat{\mathbf{n}}(\mathbf{r}') \cdot \left(\frac{\hat{\mathbf{R}}}{c}\frac{\partial \Upsilon}{\partial t_r} - \nabla'_{t_r}\Upsilon\right)\right|$$

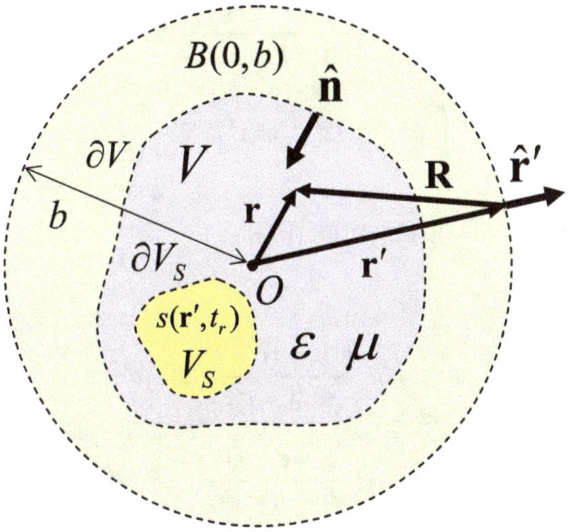

Figure 9.3 For showing that the surface integrals over ∂V in the representation (9.53) *vanish when all sources are located inside the region V.*

$$\leqslant \int_{\partial B} \mathrm{d}S' \frac{1}{4\pi R}\left[\frac{1}{c}\left|\hat{\mathbf{r}}' \cdot \hat{\mathbf{R}} + 1\right|\left|\frac{\partial \Upsilon}{\partial t_r}\right| + \left|\frac{1}{c}\frac{\partial \Upsilon}{\partial t_r} + \frac{\partial \Upsilon}{\partial r'}\right|\right]$$

$$\leqslant \int_{\partial B} \mathrm{d}S' \frac{1}{4\pi(b-r)}\left[\frac{1}{c}\frac{r^2}{2b(b-r)}\left|\frac{\partial \Upsilon}{\partial t_r}\right| + \left|\frac{1}{c}\frac{\partial \Upsilon}{\partial t_r} + \frac{\partial \Upsilon}{\partial r'}\right|\right] \tag{9.58}$$

on account of (9.18) and (6.191). The integrals in the rightmost-hand sides tend to zero if we speculate that Υ is asymptotically static and

$$\Upsilon(\mathbf{r}, t) = \Upsilon(\mathbf{r}) = \mathcal{O}\left(\frac{1}{|\mathbf{r}|}\right), \qquad |\mathbf{r}| \to +\infty \tag{9.59}$$

or if we *postulate* the asymptotic behavior

$$\Upsilon(\mathbf{r}, t) = \mathcal{O}\left(\frac{1}{|\mathbf{r}|}\right), \qquad |\mathbf{r}| \to +\infty \tag{9.60}$$

$$\frac{\partial}{\partial r}\Upsilon(\mathbf{r}, t) + \frac{1}{c}\frac{\partial}{\partial t}\Upsilon(\mathbf{r}, t) = \mathcal{O}\left(\frac{1}{|\mathbf{r}|^2}\right), \qquad |\mathbf{r}| \to +\infty \tag{9.61}$$

uniformly with respect to space and time [17]. The second constraint is the same as (8.304), viz., the time-domain counterpart of the *Sommerfeld radiation condition* (8.239) and allows selecting the outgoing wave-like solution. In conclusion, the surface integral over ∂V in (9.53) vanishes, if no sources are present outside the region V.

Thanks to (9.53) applied to $\Phi_E(\mathbf{r}, t)$ and the Cartesian components of $\mathcal{A}_E(\mathbf{r}, t)$ we arrive at the integral representations for the electrodynamic potentials due to electric sources, viz.,

$$\Phi_E(\mathbf{r}, t) = \frac{1}{\varepsilon}\int_{V_\varrho} \mathrm{d}V' \frac{\varrho(\mathbf{r}', t_r)}{4\pi|\mathbf{r}-\mathbf{r}'|} + \int_{\partial V} \mathrm{d}S' \frac{\hat{\mathbf{n}}(\mathbf{r}') \cdot (\mathbf{r}-\mathbf{r}')}{4\pi|\mathbf{r}-\mathbf{r}'|^3}\Phi_E(\mathbf{r}', t_r)$$

$$+\int_{\partial V} \mathrm{d}S' \, \frac{\hat{\mathbf{n}}(\mathbf{r}')}{4\pi|\mathbf{r}-\mathbf{r}'|} \cdot \left(\frac{\mathbf{r}-\mathbf{r}'}{c|\mathbf{r}-\mathbf{r}'|} \frac{\partial \Phi_E}{\partial t_r} - \nabla'_{t_r} \Phi_E \right)$$
$$= \frac{1}{\varepsilon} \int_{V_\varrho} \mathrm{d}V' \, \frac{\varrho(\mathbf{r}', t_r)}{4\pi R} + \int_{\partial V} \mathrm{d}S' \, \Phi_E(\mathbf{r}', t_r) \hat{\mathbf{n}}(\mathbf{r}') \cdot \nabla' \frac{1}{4\pi R}$$
$$- \int_{\partial V} \mathrm{d}S' \, \frac{1}{4\pi R} \hat{\mathbf{n}}(\mathbf{r}') \cdot \nabla' \Phi_E(\mathbf{r}', t_r) \tag{9.62}$$

$$\mathcal{A}_E(\mathbf{r}, t) = \mu \int_{V_J} \mathrm{d}V' \, \frac{\mathcal{J}(\mathbf{r}', t_r)}{4\pi|\mathbf{r}-\mathbf{r}'|} + \int_{\partial V} \mathrm{d}S' \, \frac{\hat{\mathbf{n}}(\mathbf{r}') \cdot (\mathbf{r}-\mathbf{r}')}{4\pi|\mathbf{r}-\mathbf{r}'|^3} \mathcal{A}_E(\mathbf{r}', t_r)$$
$$+ \int_{\partial V} \mathrm{d}S' \, \frac{\hat{\mathbf{n}}(\mathbf{r}')}{4\pi|\mathbf{r}-\mathbf{r}'|} \cdot \left(\frac{\mathbf{r}-\mathbf{r}'}{c|\mathbf{r}-\mathbf{r}'|} \frac{\partial \mathcal{A}_E}{\partial t_r} - \nabla'_{t_r} \mathcal{A}_E \right)$$
$$= \mu \int_{V_J} \mathrm{d}V' \, \frac{\mathcal{J}(\mathbf{r}', t_r)}{4\pi R} + \int_{\partial V} \mathrm{d}S' \, \mathcal{A}_E(\mathbf{r}', t_r) \hat{\mathbf{n}}(\mathbf{r}') \cdot \nabla' \frac{1}{4\pi R}$$
$$- \int_{\partial V} \mathrm{d}S' \, \frac{1}{4\pi R} \hat{\mathbf{n}}(\mathbf{r}') \cdot \nabla' \mathcal{A}_E(\mathbf{r}', t_r) \tag{9.63}$$

for points $\mathbf{r} \in V$ and t_r given by (9.40). The alternative expressions for the boundary terms are a consequence of (9.47) and (9.43). As regards $\Phi_E(\mathbf{r}, t)$, the surface integrals can be construed as the potentials due to layers of fictitious time-varying electric dipoles and charges whose surface densities are evaluated at the retarded time. The integrands of the surface terms of $\mathcal{A}_E(\mathbf{r}, t)$ can be transformed as in (9.32) whereby it becomes easier to identify fictitious layers of magnetic dipoles, charges and electric currents on ∂V.

The verification that $\Phi_E(\mathbf{r}, t)$ and $\mathcal{A}_E(\mathbf{r}, t)$ as given by (9.62) and (9.63) are related by the Lorenz gauge (8.121) can be done by carefully moving space and time derivatives inside the integrals [2, Section 8.2]. Alternatively, for the proof in the whole space we may examine the scalar field [10, p. 285]

$$f(\mathbf{r}, t) := \nabla \cdot \mathcal{A}_E(\mathbf{r}, t) + \frac{1}{c^2} \frac{\partial \Phi_E}{\partial t}, \qquad (\mathbf{r}, t) \in \mathbb{R}^3 \times \mathbb{R}^+ \tag{9.64}$$

which is regular at infinity because so are $\mathcal{A}_E$ and Φ_E given by (9.38) and (9.39). At time $t = 0$ $f(\mathbf{r}, t)$ vanishes along with its temporal derivative, and these initial conditions follow easily from (9.38) and (9.39) if the configuration was either static or stationary for $t < 0$. Further, $f(\mathbf{r}, t)$ obeys the homogeneous D'Alembert equation, since

$$\Box^2 f(\mathbf{r}, t) = \Box^2 \nabla \cdot \mathcal{A}_E(\mathbf{r}, t) + \frac{1}{c^2} \Box^2 \frac{\partial \Phi_E}{\partial t} = \nabla \cdot \Box^2 \mathcal{A}_E(\mathbf{r}, t) + \frac{1}{c^2} \frac{\partial}{\partial t} \Box^2 \Phi_E(\mathbf{r}, t)$$
$$= -\mu \nabla \cdot \mathcal{J} - \frac{1}{\varepsilon c^2} \frac{\partial \varrho}{\partial t} = -\mu \left(\nabla \cdot \mathcal{J} + \frac{\partial \varrho}{\partial t} \right) = 0 \tag{9.65}$$

in light of (8.122) and (8.123), the Schwarz theorem and the conservation of charge (1.46). Evidently, (9.65) is solved by $f(\mathbf{r}, t) = 0$, and since initial and boundary conditions at infinity are trivially fulfilled, this is also the only possible solution (see Section 8.6). Therefore, $\mathcal{A}_E$ and Φ_E satisfy the Lorenz gauge owing to the definition of $f(\mathbf{r}, t)$.

The relevant formulas for the potentials $\Phi_M(\mathbf{r},t)$ and $\mathcal{A}_M(\mathbf{r},t)$ due to magnetic charges and currents are obtained from (9.62) and (9.63) by invoking the duality principle.

Lastly, we observe that in order for either (8.180)-(8.183) or (8.186)-(8.189) to be useful one has to solve the wave equations for the Hertz vectors first. Since (8.171) and (8.178) each amount to three uncoupled D'Alembert equations for the Cartesian components of $\boldsymbol{\pi}_E$ and $\boldsymbol{\pi}_M$, in an unbounded isotropic lossless homogeneous medium we find

$$\boldsymbol{\pi}_E(\mathbf{r},t) = \frac{1}{\varepsilon}\int_{V_J} \mathrm{d}V' \, \frac{\mathcal{P}(\mathbf{r}', t - |\mathbf{r}-\mathbf{r}'|/c)}{4\pi|\mathbf{r}-\mathbf{r}'|}, \qquad (\mathbf{r},t) \in \mathbb{R}^3 \times \mathbb{R} \tag{9.66}$$

$$\boldsymbol{\pi}_M(\mathbf{r},t) = \int_{V_{J_M}} \mathrm{d}V' \, \frac{\mathcal{M}(\mathbf{r}', t - |\mathbf{r}-\mathbf{r}'|/c)}{4\pi|\mathbf{r}-\mathbf{r}'|}, \qquad (\mathbf{r},t) \in \mathbb{R}^3 \times \mathbb{R} \tag{9.67}$$

by comparison with (8.321) and by virtue of the integral representation (9.53). A similar remark holds true for the electromagnetic entities $\mathcal{E}(\mathbf{r},t)$ and $\mathcal{H}(\mathbf{r},t)$ which, since they satisfy the D'Alembert equations (1.207) and (1.215), can be obtained by applying formula (9.53) to the six Cartesian components of $\mathcal{E}(\mathbf{r},t)$ and $\mathcal{H}(\mathbf{r},t)$.

9.2 Potentials and fields of a point charge in uniform motion

The calculation of the electromagnetic field generated by a point charge q in uniform motion with velocity $v = |\mathbf{v}|$ in an unbounded isotropic medium is a classic problem [13, Chapter 14], [8, Section 10.6], [19, Section 3.9], [20, Sections 20.2, 23.2.4], [21, Section 8.3], [10, Section IV.59], [22, Section 5.4e], [17]. In particular, the relevant potentials $\Phi_E(\mathbf{r},t)$ and $\mathcal{A}_E(\mathbf{r},t)$ can even be computed by solving the wave equations (8.122) and (8.123) with elementary means instead of turning to the general integral formulas (9.62) and (9.63).

The configuration is ideal inasmuch as it is assumed that the particle has been and will be in uniform rectilinear motion forever. Without loss of generality we may conveniently suppose that the charge trajectory $\mathbf{w}(t)$ — a straight line — coincides with the z-axis of a system of Cartesian coordinates and hence let

$$\mathbf{w}(t) = \mathbf{v}t = vt\hat{\mathbf{z}}, \qquad t \in \mathbb{R} \tag{9.68}$$

under the additional hypothesis that the charge crosses the origin O at time $t = 0$. We distinguish three possibilities, namely, v smaller than, equal to or larger than the speed of light c in the background medium.

9.2.1 *Velocity smaller than c*

The electromagnetic field produced by a charge in uniform motion with $v < c$ was first observed experimentally by O. Heaviside in 1888. Since at any given time t the charge is located at $\mathbf{r} = \mathbf{w}(t)$, we can formally write the volume densities of charge and current in distributional form as

$$\varrho(\mathbf{r},t) = q\delta^{(3)}(\mathbf{r}-\mathbf{w}(t)) = q\delta(x)\delta(y)\delta(z-vt) \tag{9.69}$$

$$\mathcal{J}(\mathbf{r},t) = \mathbf{v}\varrho(\mathbf{r},t) = vq\delta(x)\delta(y)\delta(z-vt)\hat{\mathbf{z}} \tag{9.70}$$

on account of representation (C.25) for the three-dimensional Dirac distribution. Still, we can obviate the need for working with distributions by excluding the singular point $\mathbf{r} = \mathbf{w}(t)$ and by stating the D'Alembert equations for the potentials in an unbounded punctured domain as

$$\varepsilon\nabla^2\Phi_E(\mathbf{r},t) - \frac{\varepsilon}{c^2}\frac{\partial^2\Phi_E}{\partial t^2} = 0, \qquad \mathbf{r} \in \mathbb{R}^3 \setminus \{\mathbf{w}(t)\} \tag{9.71}$$

$$\nabla^2 \mathcal{A}_{Ez}(\mathbf{r},t) - \frac{1}{c^2}\frac{\partial^2 \mathcal{A}_{Ez}}{\partial t^2} = 0, \qquad \mathbf{r} \in \mathbb{R}^3 \setminus \{\mathbf{w}(t)\} \tag{9.72}$$

where we have noticed that $\mathcal{A}_{Ex} = \mathcal{A}_{Ey} = 0$ because the current density is aligned with the z-axis. These equations are subject to the asymptotic behavior (8.301) and a suitable normalization condition to ensure uniqueness.

In view of (9.69) and (9.70) we expect that the dependence of the potentials on z and t occurs through the function

$$Z(z,t) := z - vt \tag{9.73}$$

which can be regarded as an auxiliary spatial coordinate. Accordingly, we have

$$\frac{\partial\Phi_E}{\partial z} = \frac{\partial\Phi_E}{\partial Z}, \qquad \frac{\partial\Phi_E}{\partial t} = -v\frac{\partial\Phi_E}{\partial Z} \tag{9.74}$$

$$\frac{\partial\mathcal{A}_{Ez}}{\partial z} = \frac{\partial\mathcal{A}_{Ez}}{\partial Z}, \qquad \frac{\partial\mathcal{A}_{Ez}}{\partial t} = -v\frac{\partial\mathcal{A}_{Ez}}{\partial Z} \tag{9.75}$$

$$\frac{\partial^2\Phi_E}{\partial z^2} = \frac{\partial^2\Phi_E}{\partial Z^2}, \qquad \frac{1}{c^2}\frac{\partial^2\Phi_E}{\partial t^2} = \varsigma^2\frac{\partial^2\Phi_E}{\partial Z^2} \tag{9.76}$$

where the dimensionless factor

$$\varsigma := v/c \in [0, 1[\tag{9.77}$$

is called the *speed parameter* in the context of the special theory of relativity. In mathematical physics a fairly common symbol for the speed parameter is β [21, Section 4.1.1], [19, Section 1.5], [23, Chapter 8], though in this book we oft-times use β to denote an angle, whereas the letter ς seems a nice mnemonic for the word 'speed'. Furthermore, $\boldsymbol{\mathcal{A}}_E$ and Φ_E are related to each other through the Lorenz gauge (8.121), which in light of (9.74) and (9.75) requires

$$\frac{\partial}{\partial Z}\left(\mathcal{A}_{Ez} - \frac{v}{c^2}\Phi_E\right) = 0 \qquad \Longrightarrow \qquad \mathcal{A}_{Ez}(\mathbf{r},t) = \frac{v}{c^2}\Phi_E(\mathbf{r},t) \tag{9.78}$$

where the arbitrary constant brought in by the integration over Z must be set to zero if both potentials are to vanish at infinity in accordance with the asymptotic condition (8.301). As a result of the simple linear relation that exists between $\mathcal{A}_{Ez}$ and Φ_E, we can just focus on the calculation of the scalar potential.

Thanks to identities (9.76) the wave equation (9.71) becomes

$$\varepsilon\frac{\partial^2\Phi_E}{\partial x^2} + \varepsilon\frac{\partial^2\Phi_E}{\partial y^2} + \varepsilon\left(1-\varsigma^2\right)\frac{\partial^2\Phi_E}{\partial Z^2} = 0, \qquad (x,y,Z) \in \mathbb{R}^3 \setminus \{0\} \tag{9.79}$$

which, provided $v < c$ (i.e., $\varsigma < 1$), can be construed as a three-dimensional Poisson equation for a charge *at rest* in a *fictitious* anisotropic medium endowed with dyadic permittivity

$$\overline{\boldsymbol{\varepsilon}} = \varepsilon\hat{\mathbf{x}}\hat{\mathbf{x}} + \varepsilon\hat{\mathbf{y}}\hat{\mathbf{y}} + \varepsilon\left(1-\varsigma^2\right)\hat{\mathbf{z}}\hat{\mathbf{z}} = \varepsilon\overline{\mathbf{J}} - \varepsilon\varsigma^2\hat{\mathbf{z}}\hat{\mathbf{z}} \tag{9.80}$$

and this problem was addressed and solved in Section 2.6.2. Consequently, the electric potential is readily obtained by means of the relevant static Green function (2.148) in combination with definition (2.133), namely,

$$\Phi_E(\mathbf{r}, t) = \frac{q / \sqrt{1-\varsigma^2}}{4\pi\varepsilon \left[x^2 + y^2 + \dfrac{(z - vt)^2}{1 - \varsigma^2} \right]^{1/2}} \tag{9.81}$$

whereas the magnetic potential reads

$$\mathcal{A}_E(\mathbf{r}, t) = \frac{\mu q v \hat{\mathbf{z}} / \sqrt{1-\varsigma^2}}{4\pi \left[x^2 + y^2 + \dfrac{(z - vt)^2}{1 - \varsigma^2} \right]^{1/2}} \tag{9.82}$$

by virtue of (9.78) and definition (1.209). Next, the electric field follows through Ansatz (8.104) which, thanks to (9.78), (9.74) and (9.75), can be rephrased in terms of derivatives of the scalar potential only, namely,

$$\begin{aligned} \mathcal{E}(\mathbf{r}, t) &= -\frac{\partial \Phi_E}{\partial x}\hat{\mathbf{x}} - \frac{\partial \Phi_E}{\partial y}\hat{\mathbf{y}} - \left(1 - \varsigma^2\right)\frac{\partial \Phi_E}{\partial Z}\hat{\mathbf{z}} = \left(\frac{\mathbf{v}\mathbf{v}}{c^2} - \overline{\mathbf{J}}\right) \cdot \nabla \Phi_E \\ &= \frac{q}{4\pi\varepsilon\sqrt{1-\varsigma^2}} \frac{x\hat{\mathbf{x}} + y\hat{\mathbf{y}} + (z - vt)\hat{\mathbf{z}}}{\left[x^2 + y^2 + \dfrac{(z - vt)^2}{1 - \varsigma^2} \right]^{3/2}} \end{aligned} \tag{9.83}$$

and the magnetic induction is obtained through (8.105) applied to (9.82)

$$\mathcal{B}(\mathbf{r}, t) = \frac{\partial \mathcal{A}_{Ez}}{\partial y}\hat{\mathbf{x}} - \frac{\partial \mathcal{A}_{Ez}}{\partial x}\hat{\mathbf{y}} = \frac{\mu q v}{4\pi\sqrt{1-\varsigma^2}} \frac{x\hat{\mathbf{y}} - y\hat{\mathbf{x}}}{\left[x^2 + y^2 + \dfrac{(z - vt)^2}{1 - \varsigma^2} \right]^{3/2}} = \frac{1}{c^2}\mathbf{v} \times \mathcal{E}(\mathbf{r}, t) \tag{9.84}$$

and this clearly shows how magnetism is a relativistic phenomenon that arises from the very motion of the charge with respect to the observer. It is important to notice that the solution strategy of Section 2.6.2 breaks down if the dyadic $\overline{\boldsymbol{\varepsilon}}$ is not invertible — which in the present case happens for $\varsigma = 1$ — and hence formulas (9.81)-(9.84) are valid strictly for $\varsigma \in [0, 1[$.

From (9.81) and (9.82) we gather that the equipotential surfaces are ellipsoids or oblate spheroids [24, Fig. 1.07] centered on the particle and flattened along the direction of motion. The presence of the term $z - vt$ in the right-hand sides of (9.81) and (9.82) indicates that the equipotential surfaces are carried along by the charge without modification as time goes by. From (9.83) we see that the electric field streamlines are straight and originate from (resp. terminate onto) the particle if $q > 0$ (resp. $q < 0$) as is the case for a charge at rest. However, owing to the finite non-zero velocity the lines of force are not distributed isotropically around the particle but rather are more concentrated on either side of the plane $z = vt$, which is perpendicular to the direction of motion. This 'distortion' of the field pattern in comparison to the spherically symmetric static case (Figure 1.15a) can be appreciated with the aid of Figure 9.4 and is better quantified by the ratio

$$\frac{\mathcal{E}_{\perp}}{\mathcal{E}_{\parallel}} = \frac{1}{(1 - \varsigma^2)^{3/2}} \tag{9.85}$$

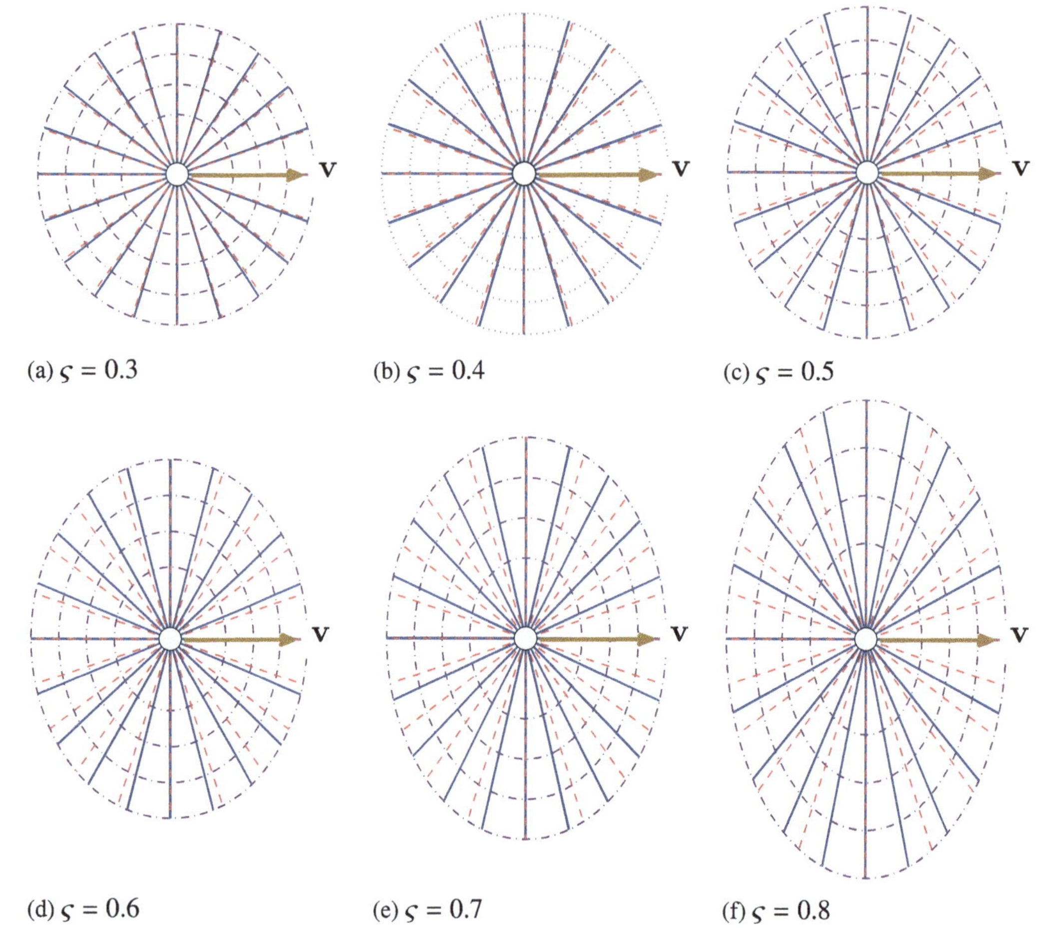

Figure 9.4 Lines of constant $\Phi_E(\mathbf{r}, t)$ $(-\cdot)$ and streamlines of $\mathcal{E}(\mathbf{r}, t)$ $(-)$ in the ZOx plane for a charge in uniform motion ($\mathbf{v} = v\hat{\mathbf{z}}$) for a few values of the speed parameter ς in (9.77); the streamlines of $\mathcal{E}(\mathbf{r})$ $(--)$ for the charge at rest ($\varsigma = 0$) are superimposed to highlight the distortion caused by the motion.

between the field in the plane $z = vt$ (perpendicular to the velocity) and the field on the z-axis (parallel to the velocity) on any sphere centered on the actual position of the charge. For $\varsigma = 0$ the ratio above reduces to unity, and this occurrence signals that the field is not distorted.

Both the electric field and the magnetic induction fall off as the inverse square of the distance from the charge. While this asymptotic behavior is in keeping with the static character of $\mathcal{E}(\mathbf{r}, t)$, we notice that $\mathcal{B}(\mathbf{r}, t)$ does not decrease as the inverse cube of the distance as would be expected of a stationary magnetic field (Section 5.5). This should not come as a surprise because, after all, the moving charge does not give rise to a steady current, as is confirmed by (9.70). But then, no waves are generated and no radiation takes place through a sphere centered on the *instantaneous* position of the particle. Indeed, from definition (1.281) the surface density of power flow reads

$$\frac{\mathrm{d}P_F}{\mathrm{d}S} := \hat{\mathbf{n}} \cdot \mathcal{E}(\mathbf{r}, t) \times \mathcal{H}(\mathbf{r}, t) = \hat{\mathbf{n}} \times \mathcal{E}(\mathbf{r}, t) \cdot \mathcal{H}(\mathbf{r}, t) = 0 \tag{9.86}$$

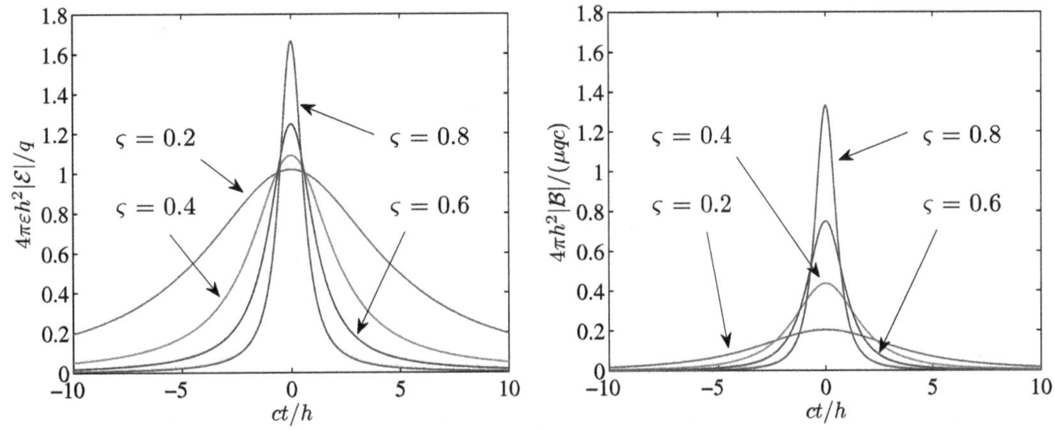

(a) *normalized* $|\mathcal{E}(h, 0, 0, t)|$ (b) *normalized* $|\mathcal{B}(h, 0, 0, t)|$

Figure 9.5 Time variation of electric and induction fields of a charge in uniform motion along the z-axis for a few values of the speed parameter ς *in* (9.77).

as the electric field and the unit vector $\hat{\mathbf{n}}$ are everywhere aligned on such sphere.

To study the time variation of the electromagnetic field we choose the location of the observer, e.g., as $\mathbf{r} = h\hat{\mathbf{x}}$ and with a little algebra obtain from (9.83) and (9.84) the normalized functions

$$\frac{4\pi\varepsilon h^2}{q}|\mathcal{E}(\mathbf{r}, t)| = \frac{\left(1-\varsigma^2\right)\left[1+(\varsigma ct/h)^2\right]^{1/2}}{[1-\varsigma^2+(\varsigma ct/h)^2]^{3/2}} \tag{9.87}$$

$$\frac{4\pi h^2}{\mu qc}|\mathcal{B}(\mathbf{r}, t)| = \frac{\varsigma\left(1-\varsigma^2\right)}{[1-\varsigma^2+(\varsigma ct/h)^2]^{3/2}} \tag{9.88}$$

which are plotted in Figures 9.5a and 9.5b versus ct/h and for a few values of ς. The field strength is maximum for $t = 0$, when the charge is located closest to the observer. For increasingly larger values of the particle velocity the electromagnetic field grows ever more intense in a small time interval and negligible elsewhere, so that for comparatively large ς's the observer registers a short pulse.

Unlike the general formulas (1.252) and (1.253), which give the electromagnetic field of a charge in arbitrary motion in terms of the *retarded* time t_r, (9.83) and (9.84) provide the same result as a function of the *present* time t. Therefore, it is instructive to show that (9.83) is precisely the Coulomb field of formula (1.252), wherein the acceleration field is absent as the velocity is constant, and also that (9.84) is equivalent to (1.253). To this purpose we refer to the geometrical construction of Figure 9.6 and notice that the separation between actual and retarded positions of the charge reads

$$v(t-t_r) = \varsigma c(t-t_r) = \varsigma R \tag{9.89}$$

on account of definition (1.251) where R is the distance between observer and retarded location. We shall also need the projection of $\mathbf{R}$ along the particle trajectory, viz.,

$$(\mathbf{R}\cdot\hat{\mathbf{z}})^2 = R^2 - x^2 - y^2 \tag{9.90}$$

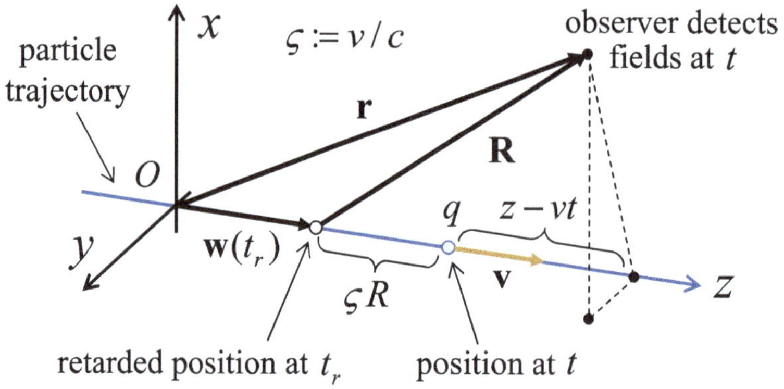

Figure 9.6 Relative positions of a point charge in uniform motion along the z-axis and an observer at rest (cf. Figure 1.16).

thanks to an obvious application of the Pythagorean theorem. For the vector in the numerator of (9.83) — the position of the observer with respect to the present position of the charge — we have

$$x\hat{\mathbf{x}} + y\hat{\mathbf{y}} + (z - vt)\hat{\mathbf{z}} = \mathbf{r} - \mathbf{w}(t_r) - v(t - t_r)\hat{\mathbf{z}} = \mathbf{R} - \varsigma R\hat{\mathbf{z}} = \frac{R}{c}\mathbf{u} \tag{9.91}$$

where the velocity-like vector $\mathbf{u}$ is defined as in (1.254). We then transform the denominator of the expression in (9.83) as follows

$$\begin{aligned}
&\frac{1}{\sqrt{1-\varsigma^2}}\left[x^2 + y^2 + \frac{(z-vt)^2}{1-\varsigma^2}\right]^{-3/2} = \\
&= \frac{1-\varsigma^2}{[(1-\varsigma^2)(x^2+y^2) + (z-vt)^2]^{3/2}} = \frac{1-\varsigma^2}{[x^2+y^2+(z-vt)^2 - \varsigma^2(x^2+y^2)]^{3/2}} \\
&= \frac{1-\varsigma^2}{\left[|\mathbf{R} - \varsigma R\hat{\mathbf{z}}|^2 - \varsigma^2(x^2+y^2)\right]^{3/2}} = \frac{1-\varsigma^2}{[R^2 + \varsigma^2 R^2 - 2\varsigma R\mathbf{R}\cdot\hat{\mathbf{z}} - \varsigma^2(x^2+y^2)]^{3/2}} \\
&= \frac{1-\varsigma^2}{[R^2 + \varsigma^2(\mathbf{R}\cdot\hat{\mathbf{z}})^2 - 2\varsigma R\mathbf{R}\cdot\hat{\mathbf{z}}]^{3/2}} = \frac{1-\varsigma^2}{(R - \varsigma\mathbf{R}\cdot\hat{\mathbf{z}})^3} = \frac{c(c^2 - v^2)}{(cR - \mathbf{R}\cdot\mathbf{v})^3} = \frac{c(c^2-v^2)}{(\mathbf{R}\cdot\mathbf{u})^3}
\end{aligned} \tag{9.92}$$

by virtue of (9.90) and (9.91). Lastly, using (9.91) and (9.92) in (9.83) yields the sought formula

$$\mathcal{E}(\mathbf{r}, t) = \frac{q}{4\pi\varepsilon}\frac{c\mathbf{R} - vR\hat{\mathbf{z}}}{(\mathbf{R}\cdot\mathbf{u})^3}(c^2 - v^2) = \frac{q}{4\pi\varepsilon}\frac{R\mathbf{u}(c^2 - v^2)}{(\mathbf{R}\cdot\mathbf{u})^3} \tag{9.93}$$

where $\mathbf{R}$ and $\mathbf{u}$ are vectors evaluated at the retarded time t_r. Finally, the magnetic induction (9.84) is easily cast into the alternative format (1.253), namely,

$$\mathcal{B}(\mathbf{r}, t) = \frac{\mathbf{u} + \mathbf{v}}{c^2} \times \mathcal{E}(\mathbf{r}, t) = \frac{c\hat{\mathbf{R}} - \mathbf{v} + \mathbf{v}}{c^2} \times \mathcal{E}(\mathbf{r}, t) = \frac{\hat{\mathbf{R}}}{c} \times \mathcal{E}(\mathbf{r}, t) \tag{9.94}$$

having noticed that the electric field is aligned with $\mathbf{u}$.

9.2.2 Velocity equal to c

If the particle velocity *equals* the speed of light, the procedure that led us to (9.81) needs to be revised. To this purpose, we temporarily consider the distributional wave equation

$$\nabla^2\Phi_E(\mathbf{r},t) - \frac{1}{c^2}\frac{\partial^2\Phi_E}{\partial t^2} = -\frac{q}{\varepsilon}\delta(x)\delta(y)\delta(z-ct) \tag{9.95}$$

and let

$$Z(z,t) := z - ct \tag{9.96}$$

whereby on account of (9.76) we get

$$\nabla_t^2\Phi_E(x,y,Z) = -\frac{q}{\varepsilon}\delta(x)\delta(y)\delta(Z) \tag{9.97}$$

where $\nabla_t^2\{\bullet\}$ is the transverse Laplace operator defined in (7.341). After the change of variable the remaining differential operator does not depend on Z explicitly and the source term is proportional with $\delta(Z)$. As a consequence, the solution must be a distribution in the form

$$\Phi_E(x,y,Z) = \Upsilon(x,y)\delta(Z) \tag{9.98}$$

where the auxiliary potential $\Upsilon(x,y)$, regarded as a distribution, satisfies

$$\nabla_t^2\Upsilon(x,y) = -\frac{q}{\varepsilon}\delta(x)\delta(y), \qquad (x,y)\in\mathbb{R}^2 \tag{9.99}$$

or, equivalently, as an ordinary function

$$\nabla_t^2\Upsilon(x,y) = 0, \qquad (x,y)\in\mathbb{R}^2\setminus\{0\} \tag{9.100}$$

which is a two-dimensional Poisson equation on an unbounded contour-wise multiply connected domain. Clearly, in view of definition (9.98) both (9.99) and (9.100) are only valid in the plane $Z = 0$.

In light of the axial symmetry of the problem the transverse Laplace operator is best written in polar coordinates (ρ,φ) — see (A.44) — and the trial solution reads

$$\Upsilon(x,y) = \Upsilon(\rho) = A\log\frac{1}{\rho}, \qquad \rho > 0 \tag{9.101}$$

with the real constant A set by a suitable normalization condition. In this regard, we apply the electric Gauss law (2.1) to the circular cylinder $V := B_2(0,a)\times[z_1,z_2]$ where $B_2(0,a)$ is a circle in the z-plane and the coordinates z_1 and z_2 must satisfy the condition $z_1 < ct < z_2$ in order for V to contain the charge. The bases of the cylinder, $B_2(z_1,a)$ and $B_2(z_2,a)$, contribute naught to the flux of $\mathfrak{D}$ through the boundary ∂V because the electric field is non-null and perpendicular to $\hat{\mathbf{z}}$ only on the plane $Z = 0$, namely,

$$\mathcal{E}(\rho,Z) = -\nabla_t\Phi_E(\rho,Z) = -\delta(Z)\nabla_t\Upsilon(\rho) \tag{9.102}$$

which follows by letting $\mathbf{v} = c\hat{\mathbf{z}}$ in the first part of (9.83). Therefore, keeping (C.3) into account for the formal integration along z, (2.1) provides

$$q = -\varepsilon\int_{z_1}^{z_2} \mathrm{d}z\,\delta(Z)\int_0^{2\pi}\mathrm{d}\varphi\, a\hat{\rho}\cdot\nabla_t\Upsilon = \varepsilon\int_0^{2\pi}\mathrm{d}\varphi\, a\hat{\rho}\cdot\hat{\rho}\frac{A}{a} = 2\pi\varepsilon A \qquad \Longrightarrow \qquad A = \frac{q}{2\pi\varepsilon} \tag{9.103}$$

whence finally we get

$$\Upsilon(\rho) = \frac{q}{2\pi\varepsilon} \log \frac{1}{\rho} \tag{9.104}$$

and

$$\Phi_E(\mathbf{r}, t) = \frac{q}{2\pi\varepsilon} \delta(z - ct) \log \frac{1}{\rho} \tag{9.105}$$

and also

$$\mathcal{A}_{Ez}(\mathbf{r}, t) = \frac{q}{2\pi\varepsilon c} \delta(z - ct) \log \frac{1}{\rho} = \frac{q\mu c}{2\pi} \delta(z - ct) \log \frac{1}{\rho} \tag{9.106}$$

by virtue of (9.78).

The electric field reads

$$\mathcal{E}(\mathbf{r}, t) = -\nabla_t \Phi_E = \frac{q}{2\pi\varepsilon} \frac{\hat{\boldsymbol{\rho}}}{\rho} \delta(z - ct) \tag{9.107}$$

and is radially directed and non-zero only on planes $z = ct$. The magnetic induction is computed as

$$\mathcal{B}(\mathbf{r}, t) = \frac{\partial \mathcal{A}_{Ez}}{\partial y} \hat{\mathbf{x}} - \frac{\partial \mathcal{A}_{Ez}}{\partial x} \hat{\mathbf{y}} = \frac{\mu q c}{2\pi} \frac{\hat{\mathbf{z}} \times \hat{\boldsymbol{\rho}}}{\rho} \delta(z - ct) = \frac{\mu q c}{2\pi} \frac{\hat{\boldsymbol{\varphi}}}{\rho} \delta(z - ct) \tag{9.108}$$

and the lines of force are circles with center on the z-axis. The presence of the Dirac delta distribution in (9.105)-(9.108) indicates that potentials and fields are non-null only on the plane $z = ct$ and are carried along by the charge. Besides, $\mathcal{E}$ and $\mathcal{B}$ are perpendicular to the direction of motion.

9.2.3 *Velocity larger than c*

This situation makes physical sense inasmuch as the conditions

$$\frac{1}{\sqrt{\varepsilon\mu}} = c < v < c_0 = \frac{1}{\sqrt{\varepsilon_0\mu_0}} \tag{9.109}$$

can be met in practice by charged particles moving through a material medium, and hence the velocity remains smaller than the speed of light in vacuum. If the particle velocity *exceeds* the speed of light in the background medium, it is convenient to introduce a time-like variable

$$T(z, t) := t - z/v \tag{9.110}$$

whereby

$$\frac{\partial \Phi_E}{\partial z} = -\frac{1}{v} \frac{\partial \Phi_E}{\partial T}, \qquad \frac{\partial \Phi_E}{\partial t} = \frac{\partial \Phi_E}{\partial T} \tag{9.111}$$

$$\frac{\partial^2 \Phi_E}{\partial z^2} = \frac{1}{v^2} \frac{\partial^2 \Phi_E}{\partial T^2}, \qquad \frac{\partial^2 \Phi_E}{\partial t^2} = \frac{\partial^2 \Phi_E}{\partial T^2} \tag{9.112}$$

and (9.71) becomes

$$\frac{\partial^2 \Phi_E}{\partial x^2} + \frac{\partial^2 \Phi_E}{\partial y^2} - \frac{\varsigma^2 - 1}{v^2} \frac{\partial^2 \Phi_E}{\partial T^2} = 0, \qquad (x, y, T) \in \mathbb{R}^3 \setminus \{0\} \tag{9.113}$$

where ς is given by (9.77). This is a two-dimensional D'Alembert equation for cylindrical waves in a *fictitious* medium where the speed of light is $V := v/(\varsigma^2 - 1)^{1/2}$.

The solution can be accomplished by resorting to the strategy devised in Section 7.6 for the calculation of the two-dimensional Green function $\mathcal{G}(\rho, t)$. Taking into account the different multiplicative constants the scalar potential reads

$$\Phi_E(x, y, T) = \frac{q}{2\pi v\varepsilon} \frac{1}{\sqrt{T^2 - \rho^2/V^2}}, \qquad T > \rho/V \tag{9.114}$$

and by reverting to the physical variables z and t we get

$$\Phi_E(\mathbf{r}, t) = \frac{q}{2\pi\varepsilon} \frac{U\left(vt - z - \rho\sqrt{\varsigma^2 - 1}\right)}{\left[(vt - z)^2 - (\varsigma^2 - 1)\rho^2\right]^{1/2}} \tag{9.115}$$

$$\mathcal{A}_E(\mathbf{r}, t) = \frac{\mu q \mathbf{v}}{2\pi} \frac{U\left(vt - z - \rho\sqrt{\varsigma^2 - 1}\right)}{\left[(vt - z)^2 - (\varsigma^2 - 1)\rho^2\right]^{1/2}} \tag{9.116}$$

where $U(\bullet)$ is the three-dimensional step function. The spatial region defined by

$$C := \left\{\mathbf{r} \in \mathbb{R}^3, t \in \mathbb{R} : vt - z \geqslant \rho\sqrt{\varsigma^2 - 1}\right\} \tag{9.117}$$

represents a family of one-sided cones (Figure 9.7) with opening angle ψ such that

$$\sin\psi = 1/\varsigma \tag{9.118}$$

and apex coinciding with the instantaneous position of the charge.

The electrodynamic potentials are non-zero for points inside the cones and become infinite on the boundary thereof. Thus, the fields for points strictly inside C are ordinary functions of the spatial and temporal coordinates. However, if the values on the boundary and outside the cone are also of concern, then we need to interpret the fields in the sense of distributions. In particular, the presence of the step function gives rise to a surface Dirac delta upon differentiation.

For the calculation of the spatial derivatives of the step function it is convenient to introduce a local system of polar spherical coordinates (R, α, β) with center on the charge and polar axis aligned with the trajectory thereof (Figure 9.7). It should be noticed that

$$\mathbf{R} := \mathbf{r} - \mathbf{w}(t) = \mathbf{r} - \mathbf{v}t \tag{9.119}$$

is the position of the observation point relative to the *present* location of the charge. Then, we have

$$vt - z = -R\cos\alpha \qquad \rho = R\sin\alpha \tag{9.120}$$

and

$$vt - z - \rho\sqrt{\varsigma^2 - 1} = -R\varsigma\sin(\alpha + \psi) = R\varsigma\sin(\alpha + \psi - \pi) \tag{9.121}$$

$$(vt - z)^2 - \rho^2(\varsigma^2 - 1) = R^2(1 - \varsigma^2\sin^2\alpha) \tag{9.122}$$

also on account of (9.118). With these positions the potentials can be written as

$$\Phi_E(\mathbf{r}, t) = \frac{q}{2\pi\varepsilon} \frac{U(R\varsigma\sin(\alpha + \psi - \pi))}{R\sqrt{1 - \varsigma^2\sin^2\alpha}} \tag{9.123}$$

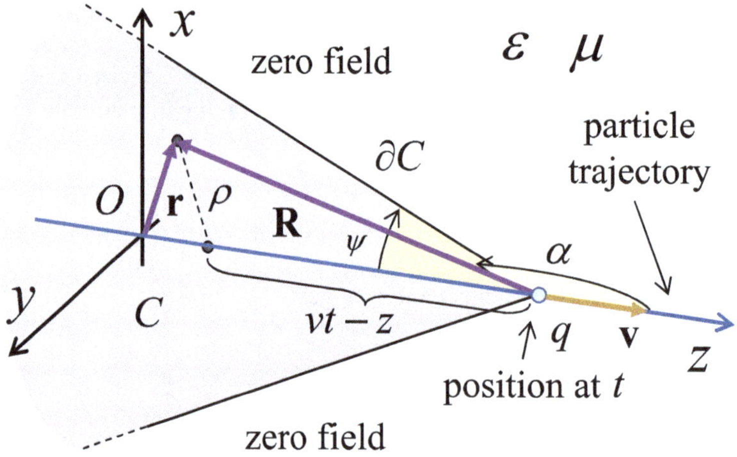

Figure 9.7 Conical shock wave produced by a point charge in uniform motion with velocity larger than the speed of light in the background medium.

$$\mathcal{A}_E(\mathbf{r},t) = \frac{\mu q \mathbf{v}}{2\pi} \frac{U\,(R\varsigma \sin(\alpha + \psi - \pi))}{R\sqrt{1-\varsigma^2 \sin^2 \alpha}} \tag{9.124}$$

wherefrom the singular behavior on the boundary of the cones (9.117) becomes even more apparent.

Since the argument of $U(\bullet)$ is a weird function of R and α, we resort to definition (C.43) for the calculation of the gradient. Moreover, since $U(\bullet)$ is non-zero inside unbounded conical regions, we introduce the finite domain $V_a := C \cap B(\mathbf{w}(t), a)$ which tends to C in the limit as a grows infinitely large. Then, we consider

$$\begin{aligned}
\nabla U &:= -\int_{\mathbb{R}^3} \mathrm{d}V\, U\,(R\varsigma \sin(\alpha + \psi - \pi))\, \nabla\phi(R,\alpha,\beta) \\
&= -\lim_{a\to+\infty} \int_{V_a} \mathrm{d}V\, \nabla\phi(R,\alpha,\beta) = -\lim_{a\to+\infty} \int_{\partial V_a} \mathrm{d}S\, \hat{\mathbf{n}}\, \phi(R,\alpha,\beta) \\
&= \lim_{a\to+\infty} \int_{\partial C\cap B(\mathbf{w},a)} \mathrm{d}S\, \hat{\alpha}\, \phi(R,\pi-\psi,\beta) - \underbrace{\lim_{a\to+\infty} \int_{\partial B(\mathbf{w},a)\cap C} \mathrm{d}S\, \hat{\mathbf{R}}\, \phi(R,\alpha,\beta)}_{=0} \\
&= \int_{\partial C} \mathrm{d}S\, \hat{\alpha}\, \phi(R,\pi-\psi,\beta) = \hat{\alpha} \frac{\delta(\alpha+\psi-\pi)}{R}
\end{aligned} \tag{9.125}$$

where $\phi(R,\alpha,\beta) \in \mathcal{C}_0^\infty(\mathbb{R}^3)$ is a three-dimensional test function. The Dirac delta in the above, being localized on the boundary of a cone, is of the surface type and 'travels' with the charge because of (9.120) [cf. (C.51) in Example C.1].

As for the temporal derivative of $U(\bullet)$ the format displayed in (9.123) is better suited because the time variable appears explicitly. Indeed, in this respect $U(\bullet)$ is nothing more than a one-dimensional function of t and hence by virtue of (C.40) we have

$$\frac{\partial U}{\partial t} = v\delta\left(vt - z - \rho\sqrt{\varsigma^2 - 1}\right) = v\delta\,(R\varsigma \sin(\alpha + \psi - \pi)) = v\frac{\delta(\alpha+\psi-\pi)}{\varsigma R} \tag{9.126}$$

with few manipulations based on formula (C.8) for the change of variable.

Now, keeping into account that $\Phi_E(\mathbf{r},t)$ amounts to the product of a scalar field $f(\mathbf{r},t)$ and the step function $U(\bullet)$, electric and magnetic field interpreted as distributions read

$$\begin{aligned}
\mathcal{E}(\mathbf{r},t) &= -\left[\nabla f(\mathbf{r},t) + \frac{\mathbf{v}}{c^2}\frac{\partial f}{\partial t}\right]U(\mathbf{r},t) - f(\mathbf{r},t)\left[\nabla U(\mathbf{r},t) + \frac{\mathbf{v}}{c^2}\frac{\partial U}{\partial t}\right] \\
&= -\frac{q}{2\pi\varepsilon}\frac{(\varsigma^2-1)(\mathbf{r}-\mathbf{v}t)}{\left[(vt-z)^2-(\varsigma^2-1)\rho^2\right]^{3/2}}U\left(vt-z-\rho\sqrt{\varsigma^2-1}\right) \\
&\quad -\frac{q}{2\pi\varepsilon}\frac{\hat{\boldsymbol{\alpha}}\sin\psi+\hat{\mathbf{z}}}{R\sqrt{1-\varsigma^2\sin^2\alpha}}\frac{\delta(\alpha+\psi-\pi)}{R\sin\psi} \\
&= -\frac{q}{2\pi\varepsilon\sqrt{\varsigma^2-1}}\frac{\mathbf{r}-\mathbf{v}t}{\left[\dfrac{(vt-z)^2}{\varsigma^2-1}-(x^2+y^2)\right]^{3/2}}U\left(vt-z-\rho\sqrt{\varsigma^2-1}\right) \\
&\quad +\frac{q}{2\pi\varepsilon}\frac{\hat{\mathbf{R}}\sqrt{\varsigma^2-1}}{R^2\sqrt{1-\varsigma^2\sin^2\alpha}}\delta(\alpha+\psi-\pi)
\end{aligned} \tag{9.127}$$

$$\begin{aligned}
\mathcal{B}(\mathbf{r},t) &= \nabla\times\mathcal{A}_E(\mathbf{r},t) = \nabla\mathcal{A}_{Ez}(\mathbf{r},t)\times\hat{\mathbf{z}} = \nabla\Phi_E(\mathbf{r},t)\times\frac{\mathbf{v}}{c^2} = \frac{1}{c^2}\mathbf{v}\times\mathcal{E}(\mathbf{r},t) \\
&= -\frac{q\mu}{2\pi\sqrt{\varsigma^2-1}}\frac{\mathbf{v}\times\mathbf{r}}{\left[\dfrac{(vt-z)^2}{\varsigma^2-1}-(x^2+y^2)\right]^{3/2}}U\left(vt-z-\rho\sqrt{\varsigma^2-1}\right) \\
&\quad +\frac{q\mu}{2\pi}\frac{\mathbf{v}\times\hat{\mathbf{R}}\sqrt{\varsigma^2-1}}{R^2\sqrt{1-\varsigma^2\sin^2\alpha}}\delta(\alpha+\psi-\pi)
\end{aligned} \tag{9.128}$$

also in view of (9.78). The fields are non-zero only inside the cones C, and a stationary observer detects a spiky (theoretically infinite) electromagnetic field which is tangential to the conical surfaces ∂C. This phenomenon — which was first observed experimentally by P. A. Cherenkov in 1934 [21, p. 142], [20, Section 23.7], [23, Section 5.1], [25] — is the electromagnetic analogue of the supersonic boom produced by aircraft flying at velocities larger than the speed of sound in the atmosphere. The conical shock-wave is also akin to the wedge-like wake produced by a speedboat on the water surface. Unlike the result for $v < c$, which is given in (9.83) and (9.84), the fields for $v > c$ come naturally separated into two parts. Intuitively, the term proportional to the Dirac delta accounts for half of the infinitely many streamlines, which have become tightly packed on the boundary of the cone C.

9.3 Electrodynamic potentials due to surface sources

Integral representation (9.10) — which constitutes the formal solution to the scalar Helmholtz equation (9.2) — indicates that the time-harmonic scalar potential $\Psi(\mathbf{r})$ in a given bounded region of space V can be written as a suitable combination of volume, single- and double-layer potentials. Having assumed that the generic source $S(\mathbf{r}')$ is confined to a volume V_S, the volume potential indeed represents the contribution of $S(\mathbf{r}')$, whereas the surface potentials may be interpreted as being due to surface densities of sources spread on the boundary ∂V. Furthermore, this was the viewpoint adopted in Section 3.1 to determine the electrostatic potential generated by a surface density of charge $\varrho_S(\mathbf{r}')$ and in Section 5.2 to find an expression for the magnetic vector potential due to a steady surface current density $\mathcal{J}_S(\mathbf{r}')$.

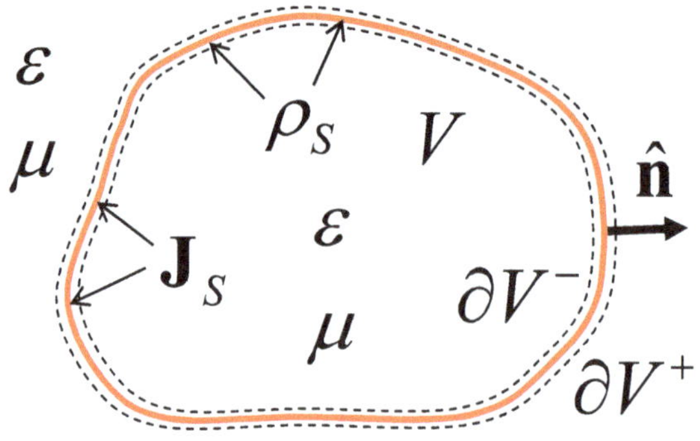

Figure 9.8 *For the derivation of the electrodynamic potentials generated by time-harmonic electric surface sources with density $\rho_S(\mathbf{r})$ and $\mathbf{J}_S(\mathbf{r})$ in a homogeneous isotropic medium. For visualization's sake the two sides of ∂V (−−) are drawn slightly away from one another.*

We wish to extend these arguments to the time-harmonic potentials $\Phi_E(\mathbf{r})$ and $\mathbf{A}_E(\mathbf{r})$ produced by surface densities $\rho_S(\mathbf{r}')$ and $\mathbf{J}_S(\mathbf{r}')$ which are confined on a smooth surface ∂V (Figure 9.8). This problem finds application in the construction of surface integral equations for the analysis of wave scattering and radiation in the presence of PEC objects (Section 13.2) or penetrable bodies comprised of homogeneous isotropic media (Section 13.3).

We assume that $\Phi_E(\mathbf{r})$ and $\mathbf{A}_E(\mathbf{r})$ are related by the Lorenz gauge (9.25) and that charges and currents are immersed in a penetrable isotropic medium endowed with constitutive parameters ε and μ. One way to arrive at the relevant formulas consists of interpreting the charge and current densities $\rho(\mathbf{r}')$ and $\mathbf{J}(\mathbf{r}')$ as distributions, i.e.,

$$\rho(\mathbf{r}') := \rho_S(\mathbf{r}')\delta_S(\mathbf{r}' - \mathbf{r}'_S) \qquad \mathbf{J}(\mathbf{r}') := \mathbf{J}_S(\mathbf{r}')\delta_S(\mathbf{r}' - \mathbf{r}'_S) \tag{9.129}$$

and then formally evaluating the integrals in (9.23) and (9.24) as convolution products with the surface δ-distribution [cf. (C.32)]. Here we follow the alternative approach based on two homogeneous boundary value problems for $\Phi_E(\mathbf{r})$ and $\mathbf{A}_E(\mathbf{r})$.

The scalar potential $\Phi_E(\mathbf{r})$ is the unique solution to the following Helmholtz equation

$$\begin{cases} \nabla^2\Phi_E(\mathbf{r}) + k^2\Phi_E(\mathbf{r}) = 0, & \mathbf{r} \in \mathbb{R}^3 \setminus \partial V \\ \Phi_E(\mathbf{r})|_{\mathbf{r}\in\partial V^-} = \Phi_E(\mathbf{r})|_{\mathbf{r}\in\partial V^+}, & \mathbf{r} \in \partial V \\ \rho_S(\mathbf{r}) = \varepsilon \left.\dfrac{\partial\Phi_E}{\partial\hat{\mathbf{n}}}\right|_{\mathbf{r}\in\partial V^-} - \varepsilon \left.\dfrac{\partial\Phi_E}{\partial\hat{\mathbf{n}}}\right|_{\mathbf{r}\in\partial V^+}, & \mathbf{r} \in \partial V \\ \mathrm{j}\,k\Phi_E(\mathbf{r}) + \dfrac{\partial\Phi_E}{\partial r} = \mathcal{O}\left(\dfrac{1}{|\mathbf{r}|^2}\right), & |\mathbf{r}| \to +\infty \end{cases} \tag{9.130}$$

where the unit vector $\hat{\mathbf{n}}(\mathbf{r})$ perpendicular to ∂V points towards the outside of V. The normal component of $\nabla\Phi_E(\mathbf{r})$ suffers a jump across the surface ∂V owing to the presence of the charges. The discontinuity is given by the matching condition (8.156) which in turn is a consequence of (1.198). By contrast, the potential *per se* is continuous through S in accordance with (8.154) with $M_E = 0$.

Regarding the question of uniqueness we notice that (9.130) is comprised of two problems, namely, one for the potential in V and one for the same quantity in the unbounded complementary domain. If we suppose that by happenstance two solutions $\Phi_{E1}(\mathbf{r})$ and $\Phi_{E2}(\mathbf{r})$ exist and apply the 'conservation of energy' (8.221) to the difference potential $\Phi_{E0} := \Phi_{E1} - \Phi_{E2}$ both in V and in $B(0, a_1) \setminus \overline{V}$ we obtain

$$\mathrm{Re} \int_{\partial V} \mathrm{d}S \; \mathrm{j} k \Phi_{E0}^*(\mathbf{r})|_{\mathbf{r} \in \partial V^-} \left. \frac{\partial \Phi_{E0}}{\partial \hat{\mathbf{n}}} \right|_{\mathbf{r} \in \partial V^-} = 0 \tag{9.131}$$

$$\mathrm{Re} \int_{|\mathbf{r}|=a_1} \mathrm{d}S \; \mathrm{j} k \Phi_{E0}^*(\mathbf{r}) \frac{\partial \Phi_{E0}}{\partial r} + \mathrm{Re} \int_{\partial V} \mathrm{d}S \; \mathrm{j} k \Phi_{E0}^*(\mathbf{r})|_{\mathbf{r} \in \partial V^+} \left. \frac{\partial \Phi_{E0}}{\partial \hat{\mathbf{n}}} \right|_{\mathbf{r} \in \partial V^+} = 0 \tag{9.132}$$

whence we gather that requiring Φ_{E0} and the normal derivative thereof to be continuous across ∂V is sufficient to accomplish the vanishing of the integral over the boundary of $B(0, a_1)$ as in (8.236). With this condition in hand the proof of uniqueness can continue as in Section 8.5.4 with the help of the Rellich theorem. The requirements on Φ_{E0} and $\hat{\mathbf{n}} \cdot \nabla \Phi_{E0}$ are met if we enforce the matching conditions (8.154) (with $M_E = 0$) and (8.156) precisely as stated in (9.130).

In order to solve (9.130) we exploit the integral representations derived in Section 9.1.1. More specifically, by separating the problem into two parts we write the formal solutions for $\mathbf{r} \in V$ and in the unbounded region $\mathbb{R}^3 \setminus \overline{V}$ with the help of (9.15). Since there are no volume sources within V (Figure 9.8), (9.15) yields

$$\Phi_E(\mathbf{r}) = -\int_{\partial V^-} \mathrm{d}S' \, \Phi_E(\mathbf{r}')|_{\partial V^-} \frac{\partial}{\partial \hat{\mathbf{n}}'} \frac{\mathrm{e}^{-\mathrm{j}kR}}{4\pi R} + \frac{1}{\varepsilon} \int_{\partial V^-} \mathrm{d}S' \, \frac{\mathrm{e}^{-\mathrm{j}kR}}{4\pi R} \, \varepsilon \left. \frac{\partial \Phi_E}{\partial \hat{\mathbf{n}}'} \right|_{\partial V^-}, \qquad \mathbf{r} \in V \tag{9.133}$$

where we have taken into account the fact that $\hat{\mathbf{n}}(\mathbf{r}')$ points outwards V. Likewise, since no volume sources exist outside V, we get

$$\Phi_E(\mathbf{r}) = \int_{\partial V^+} \mathrm{d}S' \, \Phi_E(\mathbf{r}')|_{\partial V^+} \frac{\partial}{\partial \hat{\mathbf{n}}'} \frac{\mathrm{e}^{-\mathrm{j}kR}}{4\pi R} - \frac{1}{\varepsilon} \int_{\partial V^+} \mathrm{d}S' \, \frac{\mathrm{e}^{-\mathrm{j}kR}}{4\pi R} \, \varepsilon \left. \frac{\partial \Phi_E}{\partial \hat{\mathbf{n}}'} \right|_{\partial V^+}, \qquad \mathbf{r} \in \mathbb{R}^3 \setminus \overline{V} \tag{9.134}$$

because the contribution of the surface integral at infinity vanishes thanks to the Sommerfeld radiation condition included in (9.130). Next, we recall that the right member of (9.133) returns a null value when evaluated for points $\mathbf{r}$ outside V. Similarly, the right member of (9.134) vanishes for $\mathbf{r} \in V$. Thus, by summing (9.133) and (9.134) side by side we obtain

$$\begin{aligned} \Phi_E(\mathbf{r}) = & \int_{\partial V} \mathrm{d}S' \, [\Phi_E(\mathbf{r}')|_{\partial V^+} - \Phi_E(\mathbf{r}')|_{\partial V^-}] \frac{\partial}{\partial \hat{\mathbf{n}}'} \frac{\mathrm{e}^{-\mathrm{j}kR}}{4\pi R} \\ & + \frac{1}{\varepsilon} \int_{\partial V} \mathrm{d}S' \, \frac{\mathrm{e}^{-\mathrm{j}kR}}{4\pi R} \left[\varepsilon \left. \frac{\partial \Phi_E}{\partial \hat{\mathbf{n}}'} \right|_{\partial V^-} - \varepsilon \left. \frac{\partial \Phi_E}{\partial \hat{\mathbf{n}}'} \right|_{\partial V^+} \right], \qquad \mathbf{r} \in \mathbb{R}^3 \setminus \partial V \end{aligned} \tag{9.135}$$

that is, an integral representation valid everywhere except on ∂V. The final desired solution to (9.130) follows by enforcing the boundary conditions across ∂V, namely,

$$\Phi_E(\mathbf{r}) = \frac{1}{\varepsilon} \int_{\partial V} \mathrm{d}S' \rho_S(\mathbf{r}') \frac{\mathrm{e}^{-\mathrm{j}kR}}{4\pi R}, \qquad \mathbf{r} \in \mathbb{R}^3 \tag{9.136}$$

which, since the right member is in the form of a single-layer potential (Section 2.10), is valid also for $\mathbf{r} \in \partial V$. Compared with (9.23), formula (9.136) conforms to the intuitive idea that, since the charge is confined to ∂V, the domain integral must be traded for one over a surface.

As an example, suppose that for some reason the current density $\mathbf{J}(\mathbf{r})$ flowing in a finite region of space $V_J \equiv V_\rho$ in a lossless background medium happens to have non-zero normal component on the boundary ∂V_J. Then, a layer of charge builds up on ∂V_J with density

$$\rho_S(\mathbf{r}) = \frac{\hat{\mathbf{n}}(\mathbf{r}) \cdot \mathbf{J}(\mathbf{r}) - \nabla_s \cdot \mathbf{J}_S}{\mathrm{j}\omega}, \qquad \mathbf{r} \in \partial V_J \tag{9.137}$$

in accordance with the surface continuity equation (1.200). Consequently, solution (9.23) must be augmented with a contribution in the form (9.136).

We continue with the determination of the vector potential $\mathbf{A}_E(\mathbf{r})$ which is the unique solution to the following Helmholtz equation

$$\begin{cases} \nabla^2 \mathbf{A}_E(\mathbf{r}) + k^2 \mathbf{A}_E(\mathbf{r}) = 0, & \mathbf{r} \in \mathbb{R}^3 \setminus \partial V \\ \mathbf{A}_E(\mathbf{r})|_{\mathbf{r}\in\partial V^-} = \mathbf{A}_E(\mathbf{r})|_{\mathbf{r}\in\partial V^+}, & \mathbf{r} \in \partial V \\ \mu \mathbf{J}_S(\mathbf{r}) = \hat{\mathbf{n}}(\mathbf{r}) \times \left[\nabla \times \mathbf{A}_E(\mathbf{r})|_{\mathbf{r}\in\partial V^+} - \nabla \times \mathbf{A}_E(\mathbf{r})|_{\mathbf{r}\in\partial V^-} \right], & \mathbf{r} \in \partial V \\ \mathbf{A}_E(\mathbf{r}) = \mathcal{O}\left(\dfrac{1}{|\mathbf{r}|}\right), & |\mathbf{r}| \to +\infty \\ \mathrm{j}k\hat{\mathbf{r}} \cdot \mathbf{A}_E(\mathbf{r}) + \nabla \cdot \mathbf{A}_E = \mathcal{O}\left(\dfrac{1}{|\mathbf{r}|^2}\right), & |\mathbf{r}| \to +\infty \\ \mathrm{j}k\hat{\mathbf{r}} \times \mathbf{A}_E(\mathbf{r}) + \nabla \times \mathbf{A}_E = \mathcal{O}\left(\dfrac{1}{|\mathbf{r}|^2}\right), & |\mathbf{r}| \to +\infty \end{cases} \tag{9.138}$$

where the matching conditions across ∂V are derived from (8.153), (8.155) and (8.157). The requirements at infinity, which are not obvious, can be explained by transforming the integrands of the four surface integrals of representation (9.33). For points on the boundary of a ball $B(\mathbf{r}, b)$ we have

$$\hat{\mathbf{n}}(\mathbf{r}') = -\hat{\mathbf{R}}, \qquad G(\mathbf{r}, \mathbf{r}') = \frac{\mathrm{e}^{-\mathrm{j}kb}}{4\pi b}, \qquad \nabla' G(\mathbf{r}, \mathbf{r}') = \left(\mathrm{j}k + \frac{1}{b}\right) \frac{\mathrm{e}^{-\mathrm{j}kb}}{4\pi b} \hat{\mathbf{R}} \tag{9.139}$$

and then

$$\begin{aligned}
&[\hat{\mathbf{n}}(\mathbf{r}') \times \mathbf{A}_E(\mathbf{r}')] \times \nabla' G + \hat{\mathbf{n}}(\mathbf{r}') \cdot \mathbf{A}_E(\mathbf{r}') \nabla' G \\
&\qquad + G(\mathbf{r}, \mathbf{r}') \hat{\mathbf{n}}(\mathbf{r}') \times (\nabla' \times \mathbf{A}_E) - G(\mathbf{r}, \mathbf{r}') \hat{\mathbf{n}}(\mathbf{r}') \nabla' \cdot \mathbf{A}_E = \\
&= \mathbf{A}_E(\mathbf{r}') \nabla' G \cdot \hat{\mathbf{n}}(\mathbf{r}') - \hat{\mathbf{n}}(\mathbf{r}') \mathbf{A}_E(\mathbf{r}') \cdot \nabla' G + \hat{\mathbf{n}}(\mathbf{r}') \cdot \mathbf{A}_E(\mathbf{r}') \nabla' G \\
&\qquad + G(\mathbf{r}, \mathbf{r}') \hat{\mathbf{n}}(\mathbf{r}') \times (\nabla' \times \mathbf{A}_E) - G(\mathbf{r}, \mathbf{r}') \hat{\mathbf{n}}(\mathbf{r}') \nabla' \cdot \mathbf{A}_E \\
&= \mathbf{A}_E(\mathbf{r}') \nabla' G \cdot \hat{\mathbf{n}}(\mathbf{r}') + \mathbf{A}_E(\mathbf{r}') \times \underbrace{[\nabla' G \times \hat{\mathbf{n}}(\mathbf{r}')]}_{=0} \\
&\qquad + G(\mathbf{r}, \mathbf{r}') \hat{\mathbf{n}}(\mathbf{r}') \times (\nabla' \times \mathbf{A}_E) - G(\mathbf{r}, \mathbf{r}') \hat{\mathbf{n}}(\mathbf{r}') \nabla' \cdot \mathbf{A}_E \\
&= -\mathbf{A}_E(\mathbf{r}') \left(\mathrm{j}k + \frac{1}{b}\right) G(\mathbf{r}, \mathbf{r}') - G(\mathbf{r}, \mathbf{r}') \hat{\mathbf{R}} \times (\nabla' \times \mathbf{A}_E) + G(\mathbf{r}, \mathbf{r}') \hat{\mathbf{R}} \nabla' \cdot \mathbf{A}_E \\
&= \left[-\mathrm{j}k \mathbf{A}_E(\mathbf{r}') - \frac{\mathbf{A}_E(\mathbf{r}')}{b} - \hat{\mathbf{R}} \times (\nabla' \times \mathbf{A}_E) + \hat{\mathbf{R}} \nabla' \cdot \mathbf{A}_E \right] G(\mathbf{r}, \mathbf{r}') \\
&= -\hat{\mathbf{R}} \times \left[\mathrm{j}k \mathbf{A}_E(\mathbf{r}') \times \hat{\mathbf{R}} + \nabla' \times \mathbf{A}_E(\mathbf{r}') \right] \frac{\mathrm{e}^{-\mathrm{j}kb}}{4\pi b}
\end{aligned}$$

$$-\hat{\mathbf{R}}\left[\mathrm{j}k\hat{\mathbf{R}}\cdot\mathbf{A}_E(\mathbf{r}')-\nabla'\cdot\mathbf{A}_E\right]\frac{\mathrm{e}^{-\mathrm{j}kb}}{4\pi b}-\mathbf{A}_E(\mathbf{r}')\frac{\mathrm{e}^{-\mathrm{j}kb}}{4\pi b^2} \tag{9.140}$$

where we have used (H.14) and the decomposition of $\mathbf{A}_E$ into parts parallel and perpendicular to $\hat{\mathbf{R}}$ [cf. (8.1)]. The three summands in the last row are in a form that can account, with few more steps, for the asymptotic conditions included in (9.138) upon noticing that $\hat{\mathbf{R}}$ points towards $\mathbf{r}$.

As regards uniqueness, we suppose that two solutions $\mathbf{A}_{El}(\mathbf{r})$, $l = 1, 2$, exist and define the difference potential $\mathbf{A}_{E0} := \mathbf{A}_{E1} - \mathbf{A}_{E2}$ both in V and in $B(0, a_1) \setminus \overline{V}$. From the Helmholtz equation in (9.138) and the differential identity (H.59) we get

$$\begin{aligned}0 &= \mathrm{j}k\mathbf{A}^*_{E0}\cdot\left(\nabla\nabla\cdot\mathbf{A}_{E0}-\nabla\times\nabla\times\mathbf{A}_{E0}+k^2\mathbf{A}_{E0}\right)\\&= \mathrm{j}k\nabla\cdot\left(\mathbf{A}^*_{E0}\nabla\cdot\mathbf{A}_{E0}\right)-\mathrm{j}k|\nabla\cdot\mathbf{A}_{E0}|^2\\&\quad+\mathrm{j}k\nabla\cdot\left(\mathbf{A}^*_{E0}\times\nabla\times\mathbf{A}_{E0}\right)+\mathrm{j}k|\nabla\times\mathbf{A}_{E0}|^2+\mathrm{j}k^3|\mathbf{A}_{E0}|^2\end{aligned} \tag{9.141}$$

by virtue of (H.49) and (H.51). Integrating the real part of the above identity over V and $B(0, a_1) \setminus \overline{V}$ and applying the Gauss theorem yields the conditions

$$\mathrm{Re}\int_{\partial V}\mathrm{d}S\ \mathrm{j}k\hat{\mathbf{n}}(\mathbf{r})\cdot\left[\mathbf{A}^*_{E0}(\mathbf{r})\nabla\cdot\mathbf{A}_{E0}+\mathbf{A}^*_{E0}(\mathbf{r})\times\nabla\times\mathbf{A}_{E0}\right]\Big|_{\partial V^-}=0 \tag{9.142}$$

and

$$\begin{aligned}&\mathrm{Re}\int_{|\mathbf{r}|=a_1}\mathrm{d}S\ \mathrm{j}k\hat{\mathbf{r}}\cdot\left[\mathbf{A}^*_{E0}(\mathbf{r})\nabla\cdot\mathbf{A}_{E0}+\mathbf{A}^*_{E0}(\mathbf{r})\times\nabla\times\mathbf{A}_{E0}\right]\\&\qquad=\mathrm{Re}\int_{\partial V}\mathrm{d}S\ \mathrm{j}k\hat{\mathbf{n}}(\mathbf{r})\cdot\left[\mathbf{A}^*_{E0}(\mathbf{r})\nabla\cdot\mathbf{A}_{E0}+\mathbf{A}^*_{E0}(\mathbf{r})\times\nabla\times\mathbf{A}_{E0}\right]\Big|_{\partial V^+}\end{aligned} \tag{9.143}$$

whence we gather that enforcing the continuity of $\mathbf{A}_{E0}$, $\hat{\mathbf{n}} \times (\nabla \times \mathbf{A}_{E0})$ and $\nabla \cdot \mathbf{A}_{E0}$ across ∂V is sufficient to obtain the vanishing of the integral over the boundary of $B(0, a_1)$. The requirements on $\mathbf{A}_{E0}$ and $\hat{\mathbf{n}} \times (\nabla \times \mathbf{A}_{E0})$ are met if we enforce the matching conditions (8.153), (8.155) and (8.157) precisely as stated in (9.138). Likewise, the constraint on the divergence is fulfilled because $\mathbf{A}_{El}$ and Φ_{El} are related to one another by the Lorenz gauge (9.25) and Φ_{El} is continuous across ∂V.

Next, we form the identity

$$\begin{aligned}&|\mathrm{j}k\hat{\mathbf{r}}\cdot\mathbf{A}_{E0}+\nabla\cdot\mathbf{A}_{E0}|^2+|\mathrm{j}k\hat{\mathbf{r}}\times\mathbf{A}_{E0}+\nabla\times\mathbf{A}_{E0}|^2\\&\quad=k^2|\mathbf{A}_{E0}|^2+|\nabla\cdot\mathbf{A}_{E0}|^2+|\nabla\times\mathbf{A}_{E0}|^2-2\mathrm{Re}\left\{\mathrm{j}k\hat{\mathbf{r}}\cdot\left[\mathbf{A}^*_{E0}\nabla\cdot\mathbf{A}_{E0}+\mathbf{A}^*_{E0}\times\nabla\times\mathbf{A}_{E0}\right]\right\}\end{aligned} \tag{9.144}$$

which by integrating over the boundary of $B(0, a_1)$ provides the estimate

$$\begin{aligned}&\int_{|\mathbf{r}|=a_1}\mathrm{d}S\ \left[k^2|\mathbf{A}_{E0}|^2+|\nabla\cdot\mathbf{A}_{E0}|^2+|\nabla\times\mathbf{A}_{E0}|^2\right]\\&\quad=\int_{|\mathbf{r}|=a_1}\mathrm{d}S\ |\mathrm{j}k\hat{\mathbf{r}}\cdot\mathbf{A}_{E0}+\nabla\cdot\mathbf{A}_{E0}|^2+\int_{|\mathbf{r}|=a_1}\mathrm{d}S\ |\mathrm{j}k\hat{\mathbf{r}}\times\mathbf{A}_{E0}+\nabla\times\mathbf{A}_{E0}|^2\leqslant\frac{4\pi C_A}{a_1^2}\end{aligned} \tag{9.145}$$

by virtue of (9.142) and (9.143). The positive constant C_A and the radius a_1 are chosen so that the asymptotic boundary conditions in (9.138) hold simultaneously. This calculation indicates that the quantity in the leftmost member of (9.145) is a decreasing function of the distance from the origin. Therefore, integrating over the spherical shell $B(0,a_1) \setminus B(0,a)$ with the radius $a < a_1$ but still large enough for V to be contained in $B(0,a)$ gives

$$\begin{aligned}\int\limits_{B(0,a_1)\setminus B(0,a)} \mathrm{d}V\, k^2|\mathbf{A}_{E0}|^2 &\leqslant \int\limits_{B(0,a_1)\setminus B(0,a)} \mathrm{d}V \left[k^2|\mathbf{A}_{E0}|^2 + |\nabla\cdot\mathbf{A}_{E0}|^2 + |\nabla\times\mathbf{A}_{E0}|^2\right] \\ &= \int\limits_a^{a_1} \mathrm{d}r \int\limits_{S(r)} \mathrm{d}S \left[k^2|\mathbf{A}_{E0}|^2 + |\nabla\cdot\mathbf{A}_{E0}|^2 + |\nabla\times\mathbf{A}_{E0}|^2\right] \\ &\leqslant \int\limits_a^{a_1} \mathrm{d}r \frac{4\pi C_A}{r^2} = 4\pi C_A\left(\frac{1}{a}-\frac{1}{a_1}\right) \leqslant \frac{4\pi C_A}{a} \xrightarrow[a\to+\infty]{} 0 \end{aligned} \tag{9.146}$$

where $S(r)$ is the sphere having radius $r \in [a, a_1]$ and center in the origin, and we have applied (9.145).

The final step consists of noticing that the Cartesian components of $\mathbf{A}_{E0}(\mathbf{r})$ satisfy three Helmholtz equations and meet the hypothesis of the Rellich theorem (Section 8.5.2). Hence, the integral of $|\mathbf{A}_{E0}|^2$ diverges in accordance with (8.211). Since this behavior clashes with the opposite one in (9.146) we conclude that $\mathbf{A}_{E0}$ must be the trivial null solution, and uniqueness for (9.138) is achieved (see discussion on page 536).

The formal solutions to (9.138) in V and the complementary domain are provided by (9.16) but also by the alternative expression (9.33) which is better-suited to enforce the jump conditions through ∂V. Since there are no volume sources by hypothesis (Figure 9.8), we get

$$\begin{aligned}\mathbf{A}_E(\mathbf{r}) = &-\int\limits_{\partial V^-} \mathrm{d}S'\,[\hat{\mathbf{n}}(\mathbf{r}')\times\mathbf{A}_E(\mathbf{r}')|_{\partial V^-}]\times\nabla'\frac{\mathrm{e}^{-\mathrm{j}kR}}{4\pi R} \\ &-\int\limits_{\partial V^-} \mathrm{d}S'\,\hat{\mathbf{n}}(\mathbf{r}')\cdot\mathbf{A}_E(\mathbf{r}')|_{\partial V^-}\nabla'\frac{\mathrm{e}^{-\mathrm{j}kR}}{4\pi R} - \int\limits_{\partial V^-} \mathrm{d}S'\,\frac{\mathrm{e}^{-\mathrm{j}kR}}{4\pi R}\hat{\mathbf{n}}(\mathbf{r}')\times[\nabla'\times\mathbf{A}_E(\mathbf{r}')]|_{\partial V^-} \\ &+\int\limits_{\partial V^-} \mathrm{d}S'\,\frac{\mathrm{e}^{-\mathrm{j}kR}}{4\pi R}\hat{\mathbf{n}}(\mathbf{r}')[\nabla'\cdot\mathbf{A}_E(\mathbf{r}')]|_{\partial V^-}, \qquad \mathbf{r}\in V\end{aligned} \tag{9.147}$$

and

$$\begin{aligned}\mathbf{A}_E(\mathbf{r}) = &\int\limits_{\partial V^+} \mathrm{d}S'\,[\hat{\mathbf{n}}(\mathbf{r}')\times\mathbf{A}_E(\mathbf{r}')|_{\partial V^+}]\times\nabla'\frac{\mathrm{e}^{-\mathrm{j}kR}}{4\pi R} \\ &+\int\limits_{\partial V^+} \mathrm{d}S'\,\hat{\mathbf{n}}(\mathbf{r}')\cdot\mathbf{A}_E(\mathbf{r}')|_{\partial V^+}\nabla'\frac{\mathrm{e}^{-\mathrm{j}kR}}{4\pi R} + \int\limits_{\partial V^+} \mathrm{d}S'\,\frac{\mathrm{e}^{-\mathrm{j}kR}}{4\pi R}\hat{\mathbf{n}}(\mathbf{r}')\times[\nabla'\times\mathbf{A}_E(\mathbf{r}')]|_{\partial V^+} \\ &-\int\limits_{\partial V^+} \mathrm{d}S'\,\frac{\mathrm{e}^{-\mathrm{j}kR}}{4\pi R}\hat{\mathbf{n}}(\mathbf{r}')[\nabla'\cdot\mathbf{A}_E(\mathbf{r}')]|_{\partial V^+}, \qquad \mathbf{r}\in\mathbb{R}^3\setminus\overline{V}\end{aligned} \tag{9.148}$$

where the right member of (9.147) returns a null field for $\mathbf{r}\in\mathbb{R}^3\setminus\overline{V}$ and the right-hand side of (9.148) vanishes for points inside V. Thus, we can obtain a single integral representation valid for $\mathbf{r}\in\mathbb{R}^3\setminus\partial V$ by summing these equations side by side. To speed up the derivation we observe beforehand that in the process the integrals involving $\hat{\mathbf{n}}(\mathbf{r}')\times\mathbf{A}_E(\mathbf{r}')$ and $\hat{\mathbf{n}}(\mathbf{r}')\cdot\mathbf{A}_E(\mathbf{r}')$ cancel out thanks to the continuity of $\mathbf{A}_E(\mathbf{r}')$ through ∂V. Thereby, we are left with

$$\mathbf{A}_E(\mathbf{r}) = \int_{\partial V} \mathrm{d}S' \,\frac{\mathrm{e}^{-\mathrm{j}kR}}{4\pi R}\{\hat{\mathbf{n}}(\mathbf{r}')\times[\nabla'\times\mathbf{A}_E(\mathbf{r}')]|_{\partial V^+} - \hat{\mathbf{n}}(\mathbf{r}')\times[\nabla'\times\mathbf{A}_E(\mathbf{r}')]|_{\partial V^-}\}$$
$$- \int_{\partial V} \mathrm{d}S' \,\frac{\mathrm{e}^{-\mathrm{j}kR}}{4\pi R}\hat{\mathbf{n}}(\mathbf{r}')\{[\nabla'\cdot\mathbf{A}_E(\mathbf{r}')]|_{\partial V^+} - [\nabla'\cdot\mathbf{A}_E(\mathbf{r}')]|_{\partial V^-}\}, \quad \mathbf{r}\in\mathbb{R}^3\setminus\partial V \quad (9.149)$$

which is also valid for $\mathbf{r} \in \partial V$ since the right-hand side is comprised of single-layer potentials (Section 2.10). The combination of the curls of $\mathbf{A}_E(\mathbf{r}')$ produces the surface current density $\mathbf{J}_S(\mathbf{r}')$, whereas for the difference of the divergences we have

$$[\nabla'\cdot\mathbf{A}_E(\mathbf{r}')]|_{\partial V^+} - [\nabla'\cdot\mathbf{A}_E(\mathbf{r}')]|_{\partial V^-} = \frac{\mathrm{j}\omega}{c^2}[\Phi_E(\mathbf{r}')|_{\partial V^-} - \Phi_E(\mathbf{r}')|_{\partial V^+}] = 0, \qquad \mathbf{r}'\in\partial V \qquad (9.150)$$

in light of the Lorenz gauge (9.25) and the continuity of Φ_E included in (9.130). In essence, even though (9.130) and (9.138) can be solved independently, the two problems are *coupled* because of the Lorenz gauges enforced in V and the complementary domain. Lastly, we obtain the final formula

$$\mathbf{A}_E(\mathbf{r}) = \mu\int_{\partial V} \mathrm{d}S' \,\frac{\mathrm{e}^{-\mathrm{j}kR}}{4\pi R}\mathbf{J}_S(\mathbf{r}'), \qquad \mathbf{r}\in\mathbb{R}^3 \qquad (9.151)$$

on account of the jump condition for the curl of $\mathbf{A}_E$.

Both (9.136) and (9.151) remain valid if the surface sources are confined to just a smaller open part S_0 of ∂V. This occurrence naturally presents itself in the study of the wave scattering from infinitely thin PEC objects (Section 13.2.1). Finally, the principle of duality (Tables 6.1 and 8.1) immediately provides the analogous formulas for $\Phi_M(\mathbf{r})$ and $\mathbf{A}_M(\mathbf{r})$ generated by surface magnetic sources $\rho_{MS}(\mathbf{r}')$ and $\mathbf{J}_{MS}(\mathbf{r}')$. Fictive as they are, $\rho_{MS}(\mathbf{r}')$ and $\mathbf{J}_{MS}(\mathbf{r}')$ come in handy in the study of electromagnetic radiation from apertures, as in Example 6.4.

9.4 Time-harmonic dyadic Green functions

Although we introduced $\Phi_E(\mathbf{r},t)$, $\mathcal{A}_E(\mathbf{r},t)$, $\Phi_M(\mathbf{r},t)$ and $\mathcal{A}_M(\mathbf{r},t)$ in an attempt to separate the entities of intensity $\mathcal{E}(\mathbf{r},t)$ and $\mathcal{B}(\mathbf{r},t)$ into solenoidal and lamellar parts — an endeavor which, truth be told, was only partially successful with the Lorenz gauge — we may also regard the electrodynamic potentials as convenient auxiliary fields which may help us find the solution to the Maxwell equations. Indeed, on the one hand, this was precisely the approach followed for static (Section 2.2) and stationary fields (Section 4.2), on the other, (8.104) and (8.105) in tandem with (9.62) and (9.63) formally provide the solution to (1.20), (1.23), (1.34) and (1.44) in a bounded region of space $V\subset\mathbb{R}^3$ when only electric sources are envisioned. Furthermore, the duality transformations of Tables 6.1 and 8.1 allow obtaining the retarded potentials due to magnetic sources.

In this section we focus on solving the time-harmonic Maxwell equations and consider sources confined in a volume V_S in an otherwise unbounded region of space endowed with constitutive parameters ε and μ. This assumption may seem a strong limitation, but in practice the result is relevant for most scattering and radiation problems, e.g., antennas for telecommunications. In particular, the calculations will lead us to the notion of *dyadic Green functions* in a homogeneous unbounded medium. Such dyadic fields — which may be regarded as fundamental solutions to the vector Helmholtz equations (1.234) and (1.236) — play the same role as the static scalar Green function (2.131) in the construction of integral representations for electromagnetic fields that are also known as equivalence principles (Chapter 10).

9.4.1 Observation points away from the sources

First of all, we show that $\mathbf{E}(\mathbf{r})$ and $\mathbf{H}(\mathbf{r})$ may actually be derived from just the vector potentials $\mathbf{A}_E(\mathbf{r})$ and $\mathbf{A}_M(\mathbf{r})$. Turning (8.129) and (8.130) into the frequency domain provides the decompositions

$$\mathbf{E}(\mathbf{r}) = -\mathrm{j}\,\omega\mathbf{A}_E(\mathbf{r}) - \nabla\Phi_E(\mathbf{r}) - \frac{1}{\varepsilon}\nabla\times\mathbf{A}_M(\mathbf{r}), \qquad \mathbf{r}\in\mathbb{R}^3 \tag{9.152}$$

$$\mathbf{H}(\mathbf{r}) = \frac{1}{\mu}\nabla\times\mathbf{A}_E(\mathbf{r}) - \mathrm{j}\,\omega\mathbf{A}_M(\mathbf{r}) - \nabla\Phi_M(\mathbf{r}), \qquad \mathbf{r}\in\mathbb{R}^3 \tag{9.153}$$

and by using the Lorenz gauge (9.25) and the dual thereof to eliminate the scalar potentials we find

$$\begin{aligned}\mathbf{E}(\mathbf{r}) &= -\mathrm{j}\,\omega\mathbf{A}_E(\mathbf{r}) + c^2\frac{\nabla\nabla}{\mathrm{j}\,\omega}\cdot\mathbf{A}_E(\mathbf{r}) - \frac{1}{\varepsilon}\nabla\times\mathbf{A}_M(\mathbf{r})\\ &= -\mathrm{j}\,\omega\left(\overline{\mathfrak{I}} + \frac{\nabla\nabla}{k^2}\right)\cdot\mathbf{A}_E(\mathbf{r}) - \frac{1}{\varepsilon}\nabla\times\mathbf{A}_M(\mathbf{r}), \qquad \mathbf{r}\in\mathbb{R}^3\end{aligned} \tag{9.154}$$

$$\begin{aligned}\mathbf{H}(\mathbf{r}) &= \frac{1}{\mu}\nabla\times\mathbf{A}_E(\mathbf{r}) - \mathrm{j}\,\omega\mathbf{A}_M(\mathbf{r}) + c^2\frac{\nabla\nabla}{\mathrm{j}\,\omega}\cdot\mathbf{A}_M(\mathbf{r})\\ &= \frac{1}{\mu}\nabla\times\mathbf{A}_E(\mathbf{r}) - \mathrm{j}\,\omega\left(\overline{\mathfrak{I}} + \frac{\nabla\nabla}{k^2}\right)\cdot\mathbf{A}_M(\mathbf{r}), \qquad \mathbf{r}\in\mathbb{R}^3\end{aligned} \tag{9.155}$$

where $\mathbf{A}_E(\mathbf{r})$ is given by (9.24) and $\mathbf{A}_M(\mathbf{r})$ by the dual expression involving $\mathbf{J}_M(\mathbf{r})$. Then, substituting for the explicit expressions of $\mathbf{A}_E(\mathbf{r})$ and $\mathbf{A}_M(\mathbf{r})$ yields

$$\mathbf{E}(\mathbf{r}) = -\mathrm{j}\,\omega\mu\left(\overline{\mathfrak{I}} + \frac{\nabla\nabla}{k^2}\right)\cdot\int_{V_S}\mathrm{d}V'\,\mathbf{J}(\mathbf{r}')\frac{\mathrm{e}^{-\mathrm{j}kR}}{4\pi R} - \nabla\times\int_{V_S}\mathrm{d}V'\,\mathbf{J}_M(\mathbf{r}')\frac{\mathrm{e}^{-\mathrm{j}kR}}{4\pi R} \tag{9.156}$$

$$\mathbf{H}(\mathbf{r}) = \nabla\times\int_{V_S}\mathrm{d}V'\,\mathbf{J}(\mathbf{r}')\frac{\mathrm{e}^{-\mathrm{j}kR}}{4\pi R} - \mathrm{j}\,\omega\varepsilon\left(\overline{\mathfrak{I}} + \frac{\nabla\nabla}{k^2}\right)\cdot\int_{V_S}\mathrm{d}V'\,\mathbf{J}_M(\mathbf{r}')\frac{\mathrm{e}^{-\mathrm{j}kR}}{4\pi R} \tag{9.157}$$

for points $\mathbf{r}\in\mathbb{R}^3$, since the vector potential is everywhere bounded [cf. (9.11)].

These formulas give the solution to the time-harmonic Maxwell equations in the whole space when the background medium is isotropic, homogeneous and lossless. As a matter of fact, (9.156) and (9.157) can be extended to the lossy case by simply introducing the complex wavenumber (1.249), even though we derived (9.24) and its dual without including a conduction current from the beginning. It should be noted that (9.156) and (9.157) hold only for angular frequencies $\omega \neq 0$, whereas no such restriction exists for (9.152) and (9.153). The explanation lies in the very usage of the Lorenz gauges to determine the scalar potentials, a procedure which clearly breaks down for $\omega = 0$.

Next, we wonder whether or not we may swap the order of integration and derivatives in (9.156) and (9.157). As was the case in the static and stationary regimes, the trouble is that the kernel of the integral — the scalar Green function (8.356) for time-harmonic fields — is singular for observations points $\mathbf{r}$ within the source region V_S. Nevertheless, continuity and first-order differentiability of the vector potentials are not an issue as these can be proved for the Cartesian components of $\mathbf{A}_E(\mathbf{r})$ and $\mathbf{A}_M(\mathbf{r})$ with the same approaches used in Section 2.8 for the static volume potential (2.187).

Let us start with the electric field generated by electric currents. Assuming that $\mathbf{r} \notin \overline{V}_S$ for the time being, we may write

$$\mathbf{E}(\mathbf{r}) = -\mathrm{j}\,\omega\mu \int_{V_S} \mathrm{d}V' \left(\overline{\mathfrak{I}} + \frac{\nabla\nabla}{k^2}\right) \cdot \left[\mathbf{J}(\mathbf{r}')\frac{\mathrm{e}^{-\mathrm{j}\,kR}}{4\pi R}\right] = -\mathrm{j}\,\omega\mu \int_{V_S} \mathrm{d}V' \left[\left(\overline{\mathfrak{I}} + \frac{\nabla\nabla}{k^2}\right)\frac{\mathrm{e}^{-\mathrm{j}\,kR}}{4\pi R}\right] \cdot \mathbf{J}(\mathbf{r}') \quad (9.158)$$

where it is customary to define the dyadic field

$$\overline{\mathbf{G}}_{EJ}(\mathbf{r},\mathbf{r}') := \left(\overline{\mathfrak{I}} + \frac{\nabla\nabla}{k^2}\right)\frac{\mathrm{e}^{-\mathrm{j}\,k|\mathbf{r}-\mathbf{r}'|}}{4\pi|\mathbf{r}-\mathbf{r}'|}, \qquad \mathbf{r} \notin \overline{V}_S, \qquad \mathbf{r}' \in V_S \quad (9.159)$$

which is called the *electric dyadic Green function* for electric sources [23, Section 5.2.A], [5,22,26–28] and carries the physical dimension of the inverse of a length. Some Authors (e.g. [29, Section 3.11]) define $\overline{\mathbf{G}}_{EJ}$ as the negative of the right-hand side of (9.159).

In like manner for the magnetic field generated by $\mathbf{J}(\mathbf{r}')$ we have

$$\mathbf{H}(\mathbf{r}) = \int_{V_S} \mathrm{d}V'\, \nabla \times \left[\mathbf{J}(\mathbf{r}')\frac{\mathrm{e}^{-\mathrm{j}\,kR}}{4\pi R}\right] = \int_{V_S} \mathrm{d}V'\, \nabla\frac{\mathrm{e}^{-\mathrm{j}\,kR}}{4\pi R} \times \mathbf{J}(\mathbf{r}') = \int_{V_S} \mathrm{d}V' \left(\nabla\frac{\mathrm{e}^{-\mathrm{j}\,kR}}{4\pi R} \times \overline{\mathfrak{I}}\right) \cdot \mathbf{J}(\mathbf{r}') \quad (9.160)$$

by virtue of (H.37). Notice, though, that the derivatives of the vector potential may be taken inside the integral even for observation points in the source region, and hence (9.160) is valid for $\mathbf{r} \in \mathbb{R}^3$. We may define the entity

$$\overline{\mathbf{G}}_{HJ}(\mathbf{r},\mathbf{r}') := \begin{cases} \nabla\dfrac{\mathrm{e}^{-\mathrm{j}\,k|\mathbf{r}-\mathbf{r}'|}}{4\pi|\mathbf{r}-\mathbf{r}'|} \times \overline{\mathfrak{I}}, & \mathbf{r} \in \mathbb{R}^3, \\ \nabla \times \overline{\mathbf{G}}_{EJ}, & \mathbf{r} \notin \overline{V}_S, \end{cases} \qquad \mathbf{r}' \in V_S \quad (9.161)$$

which is termed the *magnetic dyadic Green function* for electric sources [22, 26, 29] and has the physical dimension of the inverse of an area. The curl of a dyadic field is given by (E.92) or (E.93) in Cartesian coordinates.

Finally, similar steps carried out on the electric and magnetic fields generated by a magnetic current density $\mathbf{J}_M(\mathbf{r}')$ allow us to define

$$\overline{\mathbf{G}}_{HM}(\mathbf{r},\mathbf{r}') := \left(\overline{\mathfrak{I}} + \frac{\nabla\nabla}{k^2}\right)\frac{\mathrm{e}^{-\mathrm{j}\,k|\mathbf{r}-\mathbf{r}'|}}{4\pi|\mathbf{r}-\mathbf{r}'|}, \qquad \mathbf{r} \notin \overline{V}_S, \qquad \mathbf{r}' \in V_S \quad (9.162)$$

and

$$\overline{\mathbf{G}}_{EM}(\mathbf{r},\mathbf{r}') := \begin{cases} -\nabla\dfrac{\mathrm{e}^{-\mathrm{j}\,k|\mathbf{r}-\mathbf{r}'|}}{4\pi|\mathbf{r}-\mathbf{r}'|} \times \overline{\mathfrak{I}}, & \mathbf{r} \in \mathbb{R}^3, \\ -\nabla \times \overline{\mathbf{G}}_{HM}, & \mathbf{r} \notin \overline{V}_S, \end{cases} \qquad \mathbf{r}' \in V_S \quad (9.163)$$

which we refer to as the *magnetic* and *electric dyadic Green functions* for magnetic sources.

With these positions we may write (9.156) and (9.157) succinctly as

$$\mathbf{E}(\mathbf{r}) = -\mathrm{j}\,\omega\mu\,\overline{\mathbf{G}}_{EJ}(\mathbf{r}) \stackrel{*}{\cdot} \mathbf{J}(\mathbf{r}) + \overline{\mathbf{G}}_{EM}(\mathbf{r}) \stackrel{*}{\cdot} \mathbf{J}_M(\mathbf{r}), \qquad \mathbf{r} \notin \overline{V}_S \quad (9.164)$$

$$\mathbf{H}(\mathbf{r}) = \overline{\mathbf{G}}_{HJ}(\mathbf{r}) \stackrel{*}{\cdot} \mathbf{J}(\mathbf{r}) - \mathrm{j}\,\omega\varepsilon\,\overline{\mathbf{G}}_{HM}(\mathbf{r}) \stackrel{*}{\cdot} \mathbf{J}_M(\mathbf{r}), \qquad \mathbf{r} \notin \overline{V}_S \quad (9.165)$$

where the star-dot symbol ‘ $\stackrel{*}{\cdot}$ ’ indicates three-dimensional convolution and dot product between the operands.

9.4.2 *Observation points in the source region*

We would like to extend these ideas and the dyadic Green functions $\overline{\mathbf{G}}_{EJ}(\mathbf{r},\mathbf{r}')$, $\overline{\mathbf{G}}_{HM}(\mathbf{r},\mathbf{r}')$ to the case where the observation point $\mathbf{r}$ lies within the source region V_S (Figure 9.9). To this purpose, it is expedient to establish a result that concerns the application of the dyadic operator $\nabla\nabla\{\bullet\}$ to a radial function $\psi(R)$, namely,

$$\nabla\nabla\psi(R) = \nabla\left(\frac{\mathrm{d}\psi}{\mathrm{d}R}\hat{\mathbf{R}}\right) = \left(\frac{\mathrm{d}^2\psi}{\mathrm{d}R^2} - \frac{1}{R}\frac{\mathrm{d}\psi}{\mathrm{d}R}\right)\hat{\mathbf{R}}\hat{\mathbf{R}} + \frac{1}{R}\frac{\mathrm{d}\psi}{\mathrm{d}R}\overline{\mathcal{I}} \tag{9.166}$$

which is obtained by virtue of (A.28) and (H.52). We will also need the following estimates of the derivatives of the scalar Green function (8.356)

$$\left|\frac{\mathrm{d}}{\mathrm{d}R}G(\mathbf{r},\mathbf{r}')\right| = \left|-\left(\frac{1}{R}+\mathrm{j}k\right)\frac{\mathrm{e}^{-\mathrm{j}kR}}{4\pi R}\right| \leqslant \frac{c_2}{4\pi R^2} + \frac{c_1}{4\pi R} \tag{9.167}$$

$$\left|\frac{\mathrm{d}^2}{\mathrm{d}R^2}G(\mathbf{r},\mathbf{r}')\right| = \left|\left(\frac{2}{R^2}+\frac{2\mathrm{j}k}{R}-k^2\right)\frac{\mathrm{e}^{-\mathrm{j}kR}}{4\pi R}\right| \leqslant \frac{d_3}{4\pi R^3} + \frac{d_2}{4\pi R^2} + \frac{d_1}{4\pi R} \tag{9.168}$$

where c_1, c_2, d_1 etc. are suitable positive constants. For comparatively small values of R the singular behavior of the derivatives is dominated by the highest powers of $1/R$.

We examine the electric field produced by $\mathbf{J}(\mathbf{r}')$ — which is provided by (9.156) with $\mathbf{J}_M(\mathbf{r}') = 0$ — and notice that it is possible to move one derivative onto the kernel by virtue of essentially the same arguments invoked in Section 2.8 for the static volume potential $\mathcal{V}(\mathbf{r})$. In order to handle the subsequent derivative we exclude the observation point $\mathbf{r}$ by means of a convex domain $V_E \subset V_S$ [see definition (D.6)] with smooth or piecewise-smooth boundary ∂V_E, as is suggested in Figure 9.9, and split the integration into two parts, viz.,

$$\begin{aligned}\mathbf{E}(\mathbf{r}) &= -\mathrm{j}\omega\mu\int_{V_S}\mathrm{d}V'\,\frac{\mathrm{e}^{-\mathrm{j}kR}}{4\pi R}\mathbf{J}(\mathbf{r}') + \frac{\nabla}{\mathrm{j}\omega\varepsilon}\int_{V_S}\mathrm{d}V'\,\nabla\frac{\mathrm{e}^{-\mathrm{j}kR}}{4\pi R}\cdot\mathbf{J}(\mathbf{r}')\\ &= -\mathrm{j}\omega\mu\int_{V_S}\mathrm{d}V'\,\frac{\mathrm{e}^{-\mathrm{j}kR}}{4\pi R}\mathbf{J}(\mathbf{r}') + \frac{1}{\mathrm{j}\omega\varepsilon}\int_{V_S\setminus V_E}\mathrm{d}V'\,\nabla\nabla\frac{\mathrm{e}^{-\mathrm{j}kR}}{4\pi R}\cdot\mathbf{J}(\mathbf{r}')\\ &\quad + \frac{1}{\mathrm{j}\omega\varepsilon}\nabla\int_{V_E}\mathrm{d}V'\,\nabla\frac{\mathrm{e}^{-\mathrm{j}kR}}{4\pi R}\cdot\mathbf{J}(\mathbf{r}'), \qquad \mathbf{r}\in V_S\end{aligned} \tag{9.169}$$

where the last integral converges even for $\mathbf{r} \in \overline{V}_S$, since the singular behavior of the integrand is at most of the type $1/R^2$ [cf. (2.200)]. By contrast, according to (9.166)-(9.168) the dyadic differential operator $\nabla\nabla\{\bullet\}$ applied to the scalar Green function produces terms proportional to $1/R^3$ which cannot be integrated in V_E but remain finite in $V_S \setminus V_E$ nonetheless.

For ease of subsequent manipulation we introduce the scalar field

$$\mathcal{U}(\mathbf{r}) := \int_{V_E}\mathrm{d}V'\,\nabla\frac{\mathrm{e}^{-\mathrm{j}k|\mathbf{r}-\mathbf{r}'|}}{4\pi|\mathbf{r}-\mathbf{r}'|}\cdot\mathbf{J}(\mathbf{r}'), \qquad \mathbf{r}\in V_E \tag{9.170}$$

i.e., the integral over the excluded region V_E in (9.169), and go on to prove that $\nabla\mathcal{U}(\mathbf{r})$ may be cast into the alternative, fully equivalent form [29, 30]

$$\nabla\mathcal{U}(\mathbf{r}) := \nabla\int_{V_E}\mathrm{d}V'\,\nabla\frac{\mathrm{e}^{-\mathrm{j}kR}}{4\pi R}\cdot\mathbf{J}(\mathbf{r}')$$

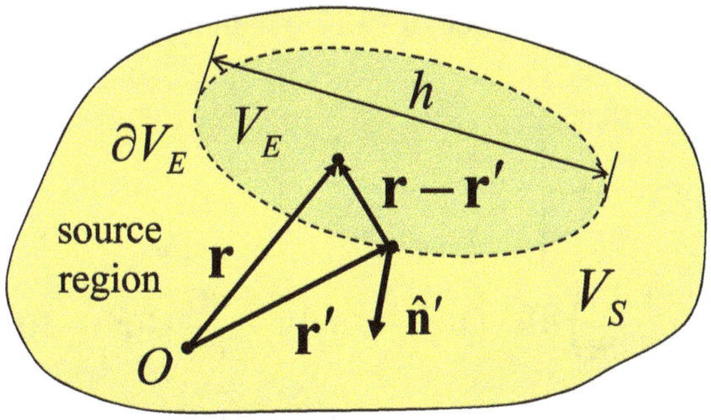

Figure 9.9 Close-up of the source region V_S and an excluded domain $V_E \subset V_S$ for determining the electric dyadic Green function in a point $\mathbf{r} \in V_S$.

$$= \int_{V_E} \mathrm{d}V' \left[\nabla\nabla \frac{\mathrm{e}^{-\mathrm{j}kR}}{4\pi R} \cdot \mathbf{J}(\mathbf{r}') - \nabla\nabla \frac{1}{4\pi R} \cdot \mathbf{J}(\mathbf{r}) \right] - \int_{\partial V_E} \mathrm{d}S' \, \hat{\mathbf{n}}(\mathbf{r}') \nabla \frac{1}{4\pi R} \cdot \mathbf{J}(\mathbf{r})$$
$$= \boldsymbol{\Xi}(\mathbf{r}) \tag{9.171}$$

under the additional hypothesis that $\mathbf{J}(\mathbf{r})$ is *Hölder continuous* in V_S and hence in V_E, namely,

$$|\mathbf{J}(\mathbf{r}_1) - \mathbf{J}(\mathbf{r}_2)| < M|\mathbf{r}_1 - \mathbf{r}_2|^\nu, \qquad \nu \in]0,1], \qquad \mathbf{r}_1, \mathbf{r}_2 \in V_S \tag{9.172}$$

where $M > 0$ is a suitable constant [31, p. 127]. A field satisfying (9.172) is also continuous, whereas the converse is generally false.

First of all, we observe that both integrals in the second part of (9.171) exist. As a matter of fact, the surface integral poses no problem at all because $\mathbf{r} \in V_E$ by construction while $\mathbf{r}' \in \partial V_E$ (Figure 9.9) and the kernel remains finite. For the volume integral we consider the identity

$$\int_{V_E} \mathrm{d}V' \left[\nabla\nabla \frac{\mathrm{e}^{-\mathrm{j}kR}}{4\pi R} \cdot \mathbf{J}(\mathbf{r}') - \nabla\nabla \frac{1}{4\pi R} \cdot \mathbf{J}(\mathbf{r}) \right]$$
$$= \int_{V_E} \mathrm{d}V' \, \nabla\nabla \frac{\mathrm{e}^{-\mathrm{j}kR}}{4\pi R} \cdot [\mathbf{J}(\mathbf{r}') - \mathbf{J}(\mathbf{r})] + \int_{V_E} \mathrm{d}V' \, \nabla\nabla \frac{\mathrm{e}^{-\mathrm{j}kR} - 1}{4\pi R} \cdot \mathbf{J}(\mathbf{r}) \tag{9.173}$$

where the scalar field

$$\tilde{G}(R) := \frac{\mathrm{e}^{-\mathrm{j}kR} - 1}{4\pi R} = \frac{\mathrm{e}^{-\mathrm{j}k|\mathbf{r}-\mathbf{r}'|} - 1}{4\pi |\mathbf{r} - \mathbf{r}'|}, \qquad \mathbf{r}, \mathbf{r}' \in \mathbb{R}^3 \tag{9.174}$$

is oft-times called the *regularized Green function* in that, unlike $G(R)$ in (8.356), $\tilde{G}(R)$ is bounded for $R \geqslant 0$ along with its derivatives. Indeed, we have

$$\lim_{R \to 0^+} \frac{\mathrm{d}^n}{\mathrm{d}R^n} \frac{\mathrm{e}^{-\mathrm{j}kR} - 1}{4\pi R} = \frac{(-\mathrm{j}k)^{n+1}}{4\pi(n+1)}, \qquad n \in \mathbb{N} \tag{9.175}$$

a result which can be proved by using the asymptotic expansion of the exponential function for small values of the argument $-\mathrm{j}kR$ [32, Formula 550.2]. Nevertheless, an application of (9.166) shows that the dyadic field $\nabla\nabla\tilde{G}$ is not bounded due to the appearance of dyads proportional to $1/R$, which can be integrated, though.

The first integral in the right-hand side of (9.173) is finite because it is dominated by a convergent integral, viz.,

$$\begin{aligned}
&\left|\int_{V_E} dV' \nabla\nabla \frac{e^{-jkR}}{4\pi R} \cdot [\mathbf{J}(\mathbf{r}') - \mathbf{J}(\mathbf{r})]\right| \\
&\quad \leqslant \int_{V_E} dV' \left| \left(\frac{d^2G}{dR^2} - \frac{1}{R}\frac{dG}{dR} \right) \hat{\mathbf{R}}\hat{\mathbf{R}} \cdot [\mathbf{J}(\mathbf{r}') - \mathbf{J}(\mathbf{r})] + \frac{1}{R}\frac{dG}{dR} [\mathbf{J}(\mathbf{r}') - \mathbf{J}(\mathbf{r})] \right| \\
&\quad \leqslant \int_{V_E} dV' \frac{d_3 + 2c_2}{4\pi R^3} M R^\nu \leqslant (d_3 + 2c_2) M \int_{B(\mathbf{r},b)} dV' \frac{R^{\nu-3}}{4\pi} = (d_3 + 2c_2) M \frac{b^\nu}{\nu}
\end{aligned} \tag{9.176}$$

by virtue of (9.166)-(9.168), the Cauchy-Schwarz inequality (D.151), and (9.172). The radius b is chosen large enough for V_E to be entirely contained in the ball $B(\mathbf{r}, b)$. Apparently, simple continuity of $\mathbf{J}(\mathbf{r})$ — which obtains for $\nu = 0$ in (9.172) — is not sufficient to guarantee the finiteness of the integral, whereby we see that the Hölder condition (9.172) is an important ingredient for somehow bringing second-order derivatives inside the volume integral over V_E.

The second integral in the right member of (9.173) is likewise dominated

$$\begin{aligned}
&\left|\int_{V_E} dV' \nabla\nabla \tilde{G}(R) \cdot \mathbf{J}(\mathbf{r})\right| \leqslant \int_{V_E} dV' \left| \left(\frac{d^2\tilde{G}}{dR^2} - \frac{1}{R}\frac{d\tilde{G}}{dR} \right) \hat{\mathbf{R}}\hat{\mathbf{R}} \cdot \mathbf{J}(\mathbf{r}) + \frac{1}{R}\frac{d\tilde{G}}{dR} \mathbf{J}(\mathbf{r}) \right| \\
&\quad \leqslant \int_{V_E} dV' \left(g_2 + 2\frac{g_1}{R} \right) \frac{|\mathbf{J}(\mathbf{r})|}{4\pi} \leqslant \int_{B(\mathbf{r},b)} dV' \left(g_2 + 2\frac{g_1}{R} \right) \frac{\|\mathbf{J}\|_\infty}{4\pi} = \left(\frac{b^3}{3} g_2 + b^2 g_1 \right) \|\mathbf{J}\|_\infty
\end{aligned} \tag{9.177}$$

where g_1 and g_2 denote the maximum values attained by $4\pi|\tilde{G}'(R)|$ and $4\pi|\tilde{G}''(R)|$ for $\mathbf{r}, \mathbf{r}' \in V_E$.

Secondly, we introduce the auxiliary scalar field $\Upsilon(\mathbf{r}; a)$

$$\Upsilon(\mathbf{r}; a) := \int_{V_E} dV' f\left(\frac{R}{a}\right) \nabla \frac{e^{-jkR}}{4\pi R} \cdot \mathbf{J}(\mathbf{r}'), \qquad a > 0, \qquad \mathbf{r} \in \mathbb{R}^3 \tag{9.178}$$

where $f(R/a)$ is the same smooth three-dimensional step function defined in (2.201) on page 123, and the parameter $a > 0$ is chosen so that $B(\mathbf{r}, 2a) \subset V_E$ when $\mathbf{r} \in V_E$. We observe that, for $a \to 0^+$, $\Upsilon(\mathbf{r}; a)$ tends to $\mathfrak{U}(\mathbf{r})$ in (9.170), thanks to the chosen form of $f(R/a)$. Actually, $\mathfrak{U}(\mathbf{r})$ and $\Upsilon(\mathbf{r}; a)$ even coincide when $V_E \cap B(\mathbf{r}, 2a) = \emptyset$. Otherwise, since $\mathbf{J}(\mathbf{r}')$ is bounded and $0 \leqslant 1 - f(R/a) \leqslant 1$ by construction, for $\mathbf{r} \in V_E$ we have

$$\begin{aligned}
|\mathfrak{U}(\mathbf{r}) - \Upsilon(\mathbf{r}; a)| &\leqslant \int_{B(\mathbf{r},2a)} dV' \left[1 - f\left(\frac{R}{a}\right) \right] \left| \nabla \frac{e^{-jkR}}{4\pi R} \right| |\mathbf{J}(\mathbf{r}')| \leqslant \int_{B(\mathbf{r},2a)} dV' \left(\frac{1}{R^2} + \frac{|k|}{R} \right) \frac{\|\mathbf{J}\|_\infty}{4\pi} \\
&= \|\mathbf{J}\|_\infty \int_0^{2a} d\tau' \, (1 + |k|\tau') = 2a(1 + |k|a) \, \|\mathbf{J}\|_\infty \xrightarrow[a \to 0^+]{} 0
\end{aligned} \tag{9.179}$$

where we have used the Cauchy-Schwarz inequality (D.151), computed the gradient with the help of (9.167), and integrated in local polar spherical coordinates (τ', α', β') centered on $\mathbf{r}$. The convergence is uniform in V_E because the leftmost member of (9.179) is dominated by a vanishing constant which does not depend on the observation point $\mathbf{r}$.

Since $f(R/a) = 0$ for $R \leqslant a$, the integrand in (9.178) is continuously differentiable for $(\mathbf{r}, \mathbf{r}') \in \mathbb{R}^3 \times V_E$, and we are allowed to compute $\nabla\Upsilon(\mathbf{r}; a)$ by moving the derivatives onto the kernel. We consider the following manipulations

$$
\begin{aligned}
\nabla\Upsilon(\mathbf{r}; a) &= \int_{V_E} \mathrm{d}V' \, \nabla\left[f\left(\frac{R}{a}\right)\nabla\frac{\mathrm{e}^{-\mathrm{j}kR}}{4\pi R}\right] \cdot \mathbf{J}(\mathbf{r}') \\
&= \int_{V_E} \mathrm{d}V' \, \nabla\left[f\left(\frac{R}{a}\right)\nabla\frac{\mathrm{e}^{-\mathrm{j}kR}}{4\pi R}\right] \cdot [\mathbf{J}(\mathbf{r}') - \mathbf{J}(\mathbf{r})] + \int_{V_E} \mathrm{d}V' \, \nabla\left[f\left(\frac{R}{a}\right)\nabla\frac{\mathrm{e}^{-\mathrm{j}kR} - 1}{4\pi R}\right] \cdot \mathbf{J}(\mathbf{r}) \\
&\quad - \int_{V_E} \mathrm{d}V' \, \nabla'\left[f\left(\frac{R}{a}\right)\nabla\frac{1}{4\pi R}\right] \cdot \mathbf{J}(\mathbf{r}) \\
&= \int_{V_E} \mathrm{d}V' \, \nabla\left[f\left(\frac{R}{a}\right)\nabla\frac{\mathrm{e}^{-\mathrm{j}kR}}{4\pi R}\right] \cdot [\mathbf{J}(\mathbf{r}') - \mathbf{J}(\mathbf{r})] + \int_{V_E} \mathrm{d}V' \, \nabla\left[f\left(\frac{R}{a}\right)\nabla\frac{\mathrm{e}^{-\mathrm{j}kR} - 1}{4\pi R}\right] \cdot \mathbf{J}(\mathbf{r}) \\
&\quad - \int_{\partial V_E} \mathrm{d}V' \, \hat{\mathbf{n}}(\mathbf{r}')\nabla\frac{1}{4\pi R} \cdot \mathbf{J}(\mathbf{r})
\end{aligned}
\tag{9.180}
$$

where in the last step we have applied the integral theorem (H.104) for the integrand is regular and differentiable throughout V_E and we have simplified the result by noticing that $f(R/a) = 1$ for $\mathbf{r}' \in \partial V_E$, as we have assumed $B(\mathbf{r}, 2a) \subset V_E$.

Next, we need to check that for $a \to 0^+$, $\nabla\Upsilon(\mathbf{r}; a)$ tends to the vector field in the second part of (9.171), i.e., the expected result which we have indicated with $\Xi(\mathbf{r})$ for convenience. If we subtract (9.171) and (9.180) side by side, we obtain

$$
\begin{aligned}
\Xi(\mathbf{r}) - \nabla\Upsilon(\mathbf{r}; a) &= \int_{V_E} \mathrm{d}V' \left[\nabla\nabla\frac{\mathrm{e}^{-\mathrm{j}kR}}{4\pi R} \cdot \mathbf{J}(\mathbf{r}') - \nabla\nabla\frac{1}{4\pi R} \cdot \mathbf{J}(\mathbf{r})\right] \\
&\quad - \int_{V_E} \mathrm{d}V' \left\{\nabla\left[f\left(\frac{R}{a}\right)\nabla\frac{\mathrm{e}^{-\mathrm{j}kR}}{4\pi R}\right] \cdot [\mathbf{J}(\mathbf{r}') - \mathbf{J}(\mathbf{r})] + \nabla\left[f\left(\frac{R}{a}\right)\nabla\frac{\mathrm{e}^{-\mathrm{j}kR} - 1}{4\pi R}\right] \cdot \mathbf{J}(\mathbf{r})\right\} \\
&= \int_{B(\mathbf{r},2a)} \mathrm{d}V' \, \nabla\left\{\left[1 - f\left(\frac{R}{a}\right)\right]\nabla\frac{\mathrm{e}^{-\mathrm{j}kR}}{4\pi R}\right\} \cdot [\mathbf{J}(\mathbf{r}') - \mathbf{J}(\mathbf{r})] \\
&\quad + \int_{B(\mathbf{r},2a)} \mathrm{d}V' \, \nabla\left\{\left[1 - f\left(\frac{R}{a}\right)\right]\nabla\frac{\mathrm{e}^{-\mathrm{j}kR} - 1}{4\pi R}\right\} \cdot \mathbf{J}(\mathbf{r})
\end{aligned}
\tag{9.181}
$$

where we have applied the splitting (9.173) again and also noticed that the function $1 - f(R/a)$ vanishes outside the ball $B(\mathbf{r}, 2a) \subset V_E$. Now, it is just a matter of showing that the last two integrals above reduce to zero when a becomes vanishingly small. For the first one we have

$$
\left|\int_{B(\mathbf{r},2a)} \mathrm{d}V' \, \nabla\left\{\left[1 - f\left(\frac{R}{a}\right)\right]\nabla\frac{\mathrm{e}^{-\mathrm{j}kR}}{4\pi R}\right\} \cdot [\mathbf{J}(\mathbf{r}') - \mathbf{J}(\mathbf{r})]\right|
$$

$$
\begin{aligned}
&\leqslant \int\limits_{B(\mathbf{r},2a)} \mathrm{d}V' \left| \left\{ \left[1 - f\left(\frac{R}{a}\right) \right] \nabla\nabla \frac{\mathrm{e}^{-\mathrm{j}kR}}{4\pi R} - \frac{f'(R/a)}{a} \hat{\mathbf{R}} \nabla \frac{\mathrm{e}^{-\mathrm{j}kR}}{4\pi R} \right\} \cdot [\mathbf{J}(\mathbf{r}') - \mathbf{J}(\mathbf{r})] \right| \\
&\leqslant \int\limits_{B(\mathbf{r},2a)} \mathrm{d}V' \left(\frac{d_3 + 2c_2}{4\pi R^3} M R^{\nu} + \frac{\|f'\|_\infty}{a} \frac{c_2}{4\pi R^2} M R^{\nu} \right) \\
&= M \left[(d_3 + 2c_2) \frac{(2a)^{\nu}}{\nu} + c_2 \frac{\|f'\|_\infty}{a} \frac{(2a)^{\nu+1}}{\nu+1} \right] \xrightarrow[a\to 0^+]{} 0
\end{aligned}
\tag{9.182}
$$

on account of (9.166)-(9.168), the Cauchy-Schwarz inequality (D.151), and (9.172). The symbol $f'(\bullet)$ signifies differentiation of $f(\bullet)$ with respect to the argument. Concerning the second integral we estimate

$$
\begin{aligned}
&\left| \int\limits_{B(\mathbf{r},2a)} \mathrm{d}V' \, \nabla \left\{ \left[1 - f\left(\frac{R}{a}\right) \right] \nabla \tilde{G}(R) \right\} \cdot \mathbf{J}(\mathbf{r}) \right| \\
&\quad \leqslant \int\limits_{B(\mathbf{r},2a)} \mathrm{d}V' \left| \left[1 - f\left(\frac{R}{a}\right) \right] \left[\left(\frac{\mathrm{d}^2\tilde{G}}{\mathrm{d}R^2} - \frac{1}{R}\frac{\mathrm{d}\tilde{G}}{\mathrm{d}R} \right) \hat{\mathbf{R}}\hat{\mathbf{R}} \cdot \mathbf{J}(\mathbf{r}) + \frac{1}{R}\frac{\mathrm{d}\tilde{G}}{\mathrm{d}R} \mathbf{J}(\mathbf{r}) \right] \right. \\
&\qquad \left. - \frac{f'(R/a)}{a} \hat{\mathbf{R}} \frac{\mathrm{d}\tilde{G}}{\mathrm{d}R} \hat{\mathbf{R}} \cdot \mathbf{J}(\mathbf{r}) \right| \\
&\quad \leqslant \int\limits_{B(\mathbf{r},2a)} \mathrm{d}V' \left[\left(g_2 + 2\frac{g_1}{R} \right) \frac{\|\mathbf{J}\|_\infty}{4\pi} + \frac{\|f'\|_\infty}{a} g_1 \frac{\|\mathbf{J}\|_\infty}{4\pi} \right] \\
&\quad = \left[\frac{(2a)^3}{3} g_2 + (2a)^2 g_1 + \frac{\|f'\|_\infty}{a} \frac{(2a)^3}{3} g_1 \right] \|\mathbf{J}\|_\infty \xrightarrow[a\to 0^+]{} 0
\end{aligned}
\tag{9.183}
$$

where again g_1 and g_2 denote the maximum values attained by $4\pi|\tilde{G}'(R)|$ and $4\pi|\tilde{G}''(R)|$ for $\mathbf{r}, \mathbf{r}' \in V_E$. The convergence is uniform for $\mathbf{r} \in V_E$ in that the quantities in the leftmost members of (9.182) and (9.183) are dominated by vanishing constants which are independent of $\mathbf{r}$.

Finally, we are left with the task of showing that the vector field $\nabla\Upsilon(\mathbf{r}; a)$ indeed approaches $\nabla\mathfrak{U}(\mathbf{r})$ with $\mathfrak{U}(\mathbf{r})$ given by (9.170) as $a \to 0^+$. Then, we pick up two points $\mathbf{r}_1$ and $\mathbf{r}_2$ in V_E and consider the following function of the real dimensionless parameter $\tau \in [0, 1]$

$$
\Psi((1-\tau)\mathbf{r}_1 + \tau\mathbf{r}_2) := \mathfrak{U}(\mathbf{r}_1) + \int\limits_0^\tau \mathrm{d}\tau' \, (\mathbf{r}_2 - \mathbf{r}_1) \cdot \boldsymbol{\Xi}((1-\tau')\mathbf{r}_1 + \tau'\mathbf{r}_2)
\tag{9.184}
$$

which arises from the line integral of the field $\boldsymbol{\Xi}(\mathbf{r})$ from $\mathbf{r}_1$ to $\mathbf{r}$ along the straight segment that connects $\mathbf{r}_1$ to $\mathbf{r}_2$. Definition (9.184) makes sense because by means of (9.182) and (9.183) we have proved that $\boldsymbol{\Xi}(\mathbf{r})$ is the uniform limit of the continuous field $\nabla\Upsilon(\mathbf{r}; a)$, and hence $\boldsymbol{\Xi}(\mathbf{r})$ is continuous [18, Chapter 7] and as such it can be integrated. More importantly, the function $\Psi((1-\tau)\mathbf{r}_1 + \tau\mathbf{r}_2)$ is constructed so as to coincide with the scalar field $\mathfrak{U}(\mathbf{r})$ when $\tau = 0$ or, equivalently, $\mathbf{r} = \mathbf{r}_1$. In like manner, the auxiliary scalar field $\Upsilon(\mathbf{r}; a)$ can be expressed as

$$
\Upsilon((1-\tau)\mathbf{r}_1 + \tau\mathbf{r}_2; a) = \Upsilon(\mathbf{r}_1; a) + \int\limits_0^\tau \mathrm{d}\tau' \, (\mathbf{r}_2 - \mathbf{r}_1) \cdot \nabla\Upsilon((1-\tau')\mathbf{r}_1 + \tau'\mathbf{r}_2; a)
\tag{9.185}
$$

when restricted to points $\mathbf{r}(\tau) := (1-\tau)\mathbf{r}_1 + \tau\mathbf{r}_2$. By subtracting (9.184) and (9.185) side by side and taking the absolute values we get

$$
\begin{aligned}
&|\Psi(\mathbf{r}(\tau)) - \Upsilon(\mathbf{r}(\tau);a)| = \\
&= \left| \mathcal{U}(\mathbf{r}_1) - \Upsilon(\mathbf{r}_1;a) + \int_0^\tau \mathrm{d}\tau' \, (\mathbf{r}_2 - \mathbf{r}_1) \cdot [\Xi(\mathbf{r}(\tau')) - \nabla\Upsilon(\mathbf{r}(\tau');a)] \right| \\
&\leqslant |\mathcal{U}(\mathbf{r}_1) - \Upsilon(\mathbf{r}_1;a)| + \int_0^1 \mathrm{d}\tau' \, |\mathbf{r}_2 - \mathbf{r}_1| \left| \Xi(\mathbf{r}(\tau')) - \nabla\Upsilon(\mathbf{r}(\tau');a) \right| \\
&\leqslant 2a(1+|k|a)\,\|\mathbf{J}\|_\infty + |\mathbf{r}_2 - \mathbf{r}_1| M \left(\frac{d_3 + 2c_2}{\nu} + 2c_2 \frac{\|f'\|_\infty}{\nu+1} \right)(2a)^\nu \\
&\quad + |\mathbf{r}_2 - \mathbf{r}_1| \left(\frac{2}{3} a g_2 + g_1 + \frac{2}{3} g_1 \left\| f' \right\|_\infty \right)(2a)^2 \, \|\mathbf{J}\|_\infty \xrightarrow[a \to 0^+]{} 0
\end{aligned}
\tag{9.186}
$$

on account of (9.179) applied with $\mathbf{r} = \mathbf{r}_1$ and estimates (9.182) and (9.183).

This intermediate result says that the functions $\Upsilon(\mathbf{r};a)$ tend to $\Psi(\mathbf{r})$ for points on the straight segment $\mathbf{r}(\tau) \in V_E$. But then, since $\mathbf{r}_1$ and $\mathbf{r}_2$ can be arbitrarily chosen in V_E, we conclude that $\Upsilon(\mathbf{r};a)$ approaches $\Psi(\mathbf{r})$ everywhere in V_E. Since $\Upsilon(\mathbf{r};a)$ tends also to the function $\mathcal{U}(\mathbf{r})$ in view of (9.179), and this limit is unique, we obtain that $\Psi(\mathbf{r})$ and $\mathcal{U}(\mathbf{r})$ coincide for $\mathbf{r} \in V_E$. By taking the derivative with respect to τ of both sides of (9.184) we obtain

$$
\frac{\mathrm{d}}{\mathrm{d}\tau}\Psi((1-\tau)\mathbf{r}_1 + \tau\mathbf{r}_2) = (\mathbf{r}_2 - \mathbf{r}_1) \cdot \nabla\Psi(\mathbf{r}(\tau)) = (\mathbf{r}_2 - \mathbf{r}_1) \cdot \Xi(\mathbf{r}(\tau))
\tag{9.187}
$$

whence we conclude

$$
\begin{aligned}
\nabla\Psi(\mathbf{r}) = \Xi(\mathbf{r}) &= \int_{V_E} \mathrm{d}V' \left[\nabla\nabla \frac{\mathrm{e}^{-\mathrm{j}kR}}{4\pi R} \cdot \mathbf{J}(\mathbf{r}') - \nabla\nabla \frac{1}{4\pi R} \cdot \mathbf{J}(\mathbf{r}) \right] \\
&\quad - \int_{\partial V_E} \mathrm{d}S' \, \hat{\mathbf{n}}(\mathbf{r}') \nabla \frac{1}{4\pi R} \cdot \mathbf{J}(\mathbf{r}), \qquad \mathbf{r} \in V_E
\end{aligned}
\tag{9.188}
$$

in view of the arbitrariness of $\mathbf{r}_1$ and $\mathbf{r}_2$ already argued above. Therefore, as we have shown that $\Psi(\mathbf{r})$ in (9.184) and $\mathcal{U}(\mathbf{r})$ in (9.170) are, in fact, the same potential, we obtain that (9.188) proves (9.171) and also that

$$
\nabla \left[\lim_{a \to 0^+} \Upsilon(\mathbf{r};a) \right] = \nabla\mathcal{U}(\mathbf{r}) = \nabla\Psi(\mathbf{r}) = \Xi(\mathbf{r}) = \lim_{a \to 0^+} \nabla\Upsilon(\mathbf{r};a), \qquad \mathbf{r} \in V_E
\tag{9.189}
$$

in view of (9.179), (9.182) and (9.183).

Now, by inserting (9.171) into (9.169) we may write the electric field generated by $\mathbf{J}(\mathbf{r}')$ in the source region as follows [28–30]

$$
\begin{aligned}
\mathbf{E}(\mathbf{r}) = &-\mathrm{j}\omega\mu \int_{V_S} \mathrm{d}V' \, \frac{\mathrm{e}^{-\mathrm{j}kR}}{4\pi R} \mathbf{J}(\mathbf{r}') + \frac{1}{\mathrm{j}\omega\varepsilon} \int_{V_S \setminus V_E} \mathrm{d}V' \, \nabla\nabla \frac{\mathrm{e}^{-\mathrm{j}kR}}{4\pi R} \cdot \mathbf{J}(\mathbf{r}') \\
&+ \frac{1}{\mathrm{j}\omega\varepsilon} \int_{V_E} \mathrm{d}V' \left[\nabla\nabla \frac{\mathrm{e}^{-\mathrm{j}kR}}{4\pi R} \cdot \mathbf{J}(\mathbf{r}') - \nabla\nabla \frac{1}{4\pi R} \cdot \mathbf{J}(\mathbf{r}) \right]
\end{aligned}
$$

$$- \frac{1}{\mathrm{j}\omega\varepsilon} \int_{\partial V_E} \mathrm{d}S'\, \hat{\mathbf{n}}(\mathbf{r}')\nabla \frac{1}{4\pi R} \cdot \mathbf{J}(\mathbf{r}), \qquad \mathbf{r} \in V_S \quad (9.190)$$

where the frequency-independent field

$$\overline{\mathbf{L}}(\mathbf{r}) := \int_{\partial V_E} \mathrm{d}S'\, \hat{\mathbf{n}}(\mathbf{r}')\nabla \frac{1}{4\pi R} = - \int_{\partial V_E} \mathrm{d}S'\, \frac{\hat{\mathbf{n}}(\mathbf{r}')\hat{\mathbf{R}}}{4\pi R^2}, \qquad \mathbf{r} \in V_S \quad (9.191)$$

vanishes for $\mathbf{r} \notin V_S$ by construction and is called the *depolarization* or *depolarizing* or *source dyadic* [1,2,28,29,33]. It is a simple matter to check that the integral over ∂V_E is left unaltered by a change of scale in the form $\mathbf{r}' = \alpha\tilde{\mathbf{r}}'$ and $\mathbf{r} = \alpha\tilde{\mathbf{r}}$, $\alpha > 0$, because the unit vectors $\hat{\mathbf{n}}'$ and $\hat{\mathbf{R}}$ are not affected by the process, whereas the scaling of $\mathrm{d}S'$ compensates that of R^2. Therefore, the depolarization dyadic depends on the *shape* but not on the *size* of the excluded region V_E. Although the dyad $\hat{\mathbf{n}}(\mathbf{r}')\hat{\mathbf{R}}$ is asymmetric in general — except at points $\mathbf{r}' \in \partial V_E$ where $\hat{\mathbf{n}}(\mathbf{r}')$ and $\hat{\mathbf{R}}$ are parallel, that is — $\overline{\mathbf{L}}(\mathbf{r})$ is, in fact, symmetric. This can be proved by starting from the identity

$$\begin{aligned} \hat{\mathbf{n}}(\mathbf{r}')\mathbf{J}(\mathbf{r}) \cdot \nabla \frac{1}{4\pi R} - \mathbf{J}(\mathbf{r}) \cdot \hat{\mathbf{n}}(\mathbf{r}')\nabla \frac{1}{4\pi R} &= - \left[\nabla' \frac{1}{4\pi R} \times \hat{\mathbf{n}}(\mathbf{r}') \right] \times \mathbf{J}(\mathbf{r}) \\ = - \left[\nabla'_{\mathrm{s}} \frac{1}{4\pi R} \times \hat{\mathbf{n}}(\mathbf{r}') \right] \times \mathbf{J}(\mathbf{r}) &= - \left[\nabla'_{\mathrm{s}} \times \frac{\hat{\mathbf{n}}(\mathbf{r}')}{4\pi R} \right] \times \mathbf{J}(\mathbf{r}) \end{aligned} \quad (9.192)$$

which is based on (H.15), (H.80) and (H.81). Integrating both sides over ∂V_E yields

$$\begin{aligned} \int_{\partial V_E} \mathrm{d}S'\, \hat{\mathbf{n}}(\mathbf{r}')\nabla \frac{1}{4\pi R} \cdot \mathbf{J}(\mathbf{r}) - \int_{\partial V_E} \mathrm{d}S'\, \nabla \frac{1}{4\pi R} \hat{\mathbf{n}}(\mathbf{r}') \cdot \mathbf{J}(\mathbf{r}) \\ = - \int_{\partial V_E} \mathrm{d}S' \left[\nabla'_{\mathrm{s}} \times \frac{\hat{\mathbf{n}}(\mathbf{r}')}{4\pi R} \right] \times \mathbf{J}(\mathbf{r}) = 0 \end{aligned} \quad (9.193)$$

by virtue of the integral theorem (H.103) because ∂V_E is a closed surface and $\hat{\mathbf{n}}(\mathbf{r}') \times \hat{\mathbf{n}}(\mathbf{r}') = 0$. From this equation we obtain immediately

$$\overline{\mathbf{L}}(\mathbf{r}) = - \int_{\partial V_E} \mathrm{d}S'\, \frac{\hat{\mathbf{n}}(\mathbf{r}')\hat{\mathbf{R}}}{4\pi R^2} = - \int_{\partial V_E} \mathrm{d}S'\, \frac{\hat{\mathbf{R}}\,\hat{\mathbf{n}}(\mathbf{r}')}{4\pi R^2} = \left[\overline{\mathbf{L}}(\mathbf{r}) \right]^T, \qquad \mathbf{r} \in V_S \quad (9.194)$$

since $\mathbf{J}(\mathbf{r})$ may be an arbitrary Hölder continuous vector field. Further, the trace of $\overline{\mathbf{L}}(\mathbf{r})$ is a constant, viz.,

$$\mathrm{Tr}\left\{ \overline{\mathbf{L}}(\mathbf{r}) \right\} = \overline{\mathbf{L}}(\mathbf{r}) : \overline{\mathbf{I}} = - \int_{\partial V_E} \mathrm{d}S'\, \frac{\hat{\mathbf{n}}(\mathbf{r}')\hat{\mathbf{R}}}{4\pi R^2} : \overline{\mathbf{I}} = \int_{\partial V_E} \mathrm{d}S'\, \frac{\hat{\mathbf{n}}(\mathbf{r}') \cdot (\mathbf{r}' - \mathbf{r})}{4\pi |\mathbf{r}' - \mathbf{r}|^3} = 1 \quad (9.195)$$

on account of the Gauss solid angle formula (2.248).

Example 9.1 (Depolarization dyadic for a sphere)

As an example, we compute $\overline{\mathbf{L}}(\mathbf{r})$ when $V_E \equiv B(\mathbf{r}, a)$, in which instance $\hat{\mathbf{n}}(\mathbf{r}') = -\hat{\mathbf{R}}$ and $R = a$ (see Figure 2.8 on page 111). In a local system of spherical coordinates (R, α', β') centered in $\mathbf{r}$ and with the polar axis parallel to $\hat{\mathbf{z}}$ we have to integrate the symmetric dyadic field [34, 35]

$$\begin{aligned}\hat{\mathbf{n}}(\mathbf{r}')\hat{\mathbf{n}}(\mathbf{r}') &= \cos^2\beta' \sin^2\alpha' \hat{\mathbf{x}}\hat{\mathbf{x}} + \frac{1}{2}\sin(2\beta')\sin^2\alpha' \hat{\mathbf{x}}\hat{\mathbf{y}} + \frac{1}{2}\cos\beta' \sin(2\alpha')\hat{\mathbf{x}}\hat{\mathbf{z}} \\ &+ \frac{1}{2}\sin(2\beta')\sin^2\alpha' \hat{\mathbf{y}}\hat{\mathbf{x}} + \sin^2\beta' \sin^2\alpha' \hat{\mathbf{y}}\hat{\mathbf{y}} + \frac{1}{2}\sin\beta' \sin(2\alpha')\hat{\mathbf{y}}\hat{\mathbf{z}} \\ &+ \frac{1}{2}\cos\beta' \sin(2\alpha')\hat{\mathbf{z}}\hat{\mathbf{x}} + \frac{1}{2}\sin\beta' \sin(2\alpha')\hat{\mathbf{z}}\hat{\mathbf{y}} + \cos^2\alpha' \hat{\mathbf{z}}\hat{\mathbf{z}}\end{aligned} \tag{9.196}$$

and from (9.191) we have

$$\overline{\mathbf{L}}(\mathbf{r}) = \frac{1}{4\pi}\int_0^{2\pi} \mathrm{d}\beta' \int_0^{\pi} \mathrm{d}\alpha' \, \sin\alpha' \, \hat{\mathbf{n}}(\mathbf{r}')\hat{\mathbf{n}}(\mathbf{r}') = \frac{1}{3}\overline{\mathbf{\mathcal{I}}}, \qquad \mathbf{r} \in V_S \tag{9.197}$$

as only the diagonal terms of $\hat{\mathbf{n}}(\mathbf{r}')\hat{\mathbf{n}}(\mathbf{r}')$ contribute non-null values owing to symmetry and periodicity of the trigonometric functions. Interestingly, the same result is obtained when V_E is a cube centered in $\mathbf{r}$ [29, 34, 36].

(End of Example 9.1)

Example 9.2 (Depolarization dyadic for a regular tetrahedron)

When the excluded region $V_E \subset V_S$ in Figure 9.9 has the shape of a regular tetrahedron whose center P coincides with the observation point $\mathbf{r}$, application of definition (9.191) yields

$$\begin{aligned}\overline{\mathbf{L}}(\mathbf{r}) &= -\int_{\partial V_E} \mathrm{d}S' \, \hat{\mathbf{n}}(\mathbf{r}')\nabla' \frac{1}{4\pi R} = -\sum_{l=1}^{4} \hat{\mathbf{n}}_l \int_{T_l} \mathrm{d}S' \, \nabla' \frac{1}{4\pi R_l} \\ &= -\sum_{l=1}^{4} \hat{\mathbf{n}}_l \int_{T_l} \mathrm{d}S' \left(\hat{\mathbf{n}}_l \hat{\mathbf{n}}_l \cdot \nabla' \frac{1}{4\pi R_l} + \nabla'_{\mathrm{s}} \frac{1}{4\pi R_l} \right) \\ &= -\sum_{l=1}^{4} \hat{\mathbf{n}}_l \hat{\mathbf{n}}_l \int_{T_l} \mathrm{d}S' \, \frac{\hat{\mathbf{n}}_l \cdot \hat{\mathbf{R}}}{4\pi R_l^2} - \sum_{l=1}^{4} \hat{\mathbf{n}}_l \oint_{\partial T_l} \mathrm{d}s' \, \frac{\hat{\boldsymbol{\nu}}_l(\mathbf{r}')}{4\pi R_l}\end{aligned} \tag{9.198}$$

where $\hat{\mathbf{n}}_l$ are the constant unit vectors perpendicular to the triangular facets T_l, $R_l = |\mathbf{r} - \mathbf{r}'_l|$ is the usual distance R though evaluated with the source point $\mathbf{r}' \equiv \mathbf{r}'_l$ on the triangle T_l, and $\nabla'_{\mathrm{s}}\{\bullet\}$ indicates the surface gradient computed over T_l with respect to the primed coordinates. Since the scalar field $1/R_l$ is regular, in the last step we have invoked the surface gradient theorem (H.102) — with curvature $J(\mathbf{r}') = 0$, in that the surface T_l is flat — and $\hat{\boldsymbol{\nu}}_l(\mathbf{r}')$ denotes the unit vector perpendicular to the boundary ∂T_l such that $\hat{\mathbf{n}}_l \cdot \hat{\boldsymbol{\nu}}_l(\mathbf{r}') = 0$.

For symmetry reasons the four flux integrals over the equilateral triangles T_l are identical to each other and, according to formula (5.136), equal to the negative of the solid angle subtended by T_l with respect to $\mathbf{r}$. As a result, we have

$$-\int_{T_l} \mathrm{d}S' \, \frac{\hat{\mathbf{n}}_l \cdot \hat{\mathbf{R}}}{R_l^2} = \pi, \qquad l = 1, \ldots, 4 \tag{9.199}$$

since the sum of the four contributions must needs give 4π.

The four integrals over ∂T_l can be split up into integrals along the three edges γ_{li}, $i = 1, 2, 3$, of T_l,[1] viz.,

$$\oint_{\partial T_l} \mathrm{d}s' \, \frac{\hat{\boldsymbol{\nu}}_l(\mathbf{r}')}{4\pi R_l} = \sum_{i=1}^{3} \hat{\boldsymbol{\nu}}_{li} \int_{\gamma_{li}} \mathrm{d}s' \, \frac{1}{4\pi R_{li}} = 0, \qquad l = 1, \ldots, 4 \tag{9.200}$$

where $\hat{\boldsymbol{\nu}}_{li}$ are the constant unit vectors perpendicular to the edges γ_{li}, and $R_{li} = |\mathbf{r} - \mathbf{r}'_{li}|$ denotes the distance R though evaluated with the source point $\mathbf{r}' \equiv \mathbf{r}'_{li}$ on the edge γ_{li}. The result follows inasmuch as the $3 \times 4 = 12$ integrals along γ_{il} are all identical for symmetry reasons (we do not even have to compute them) and also

$$\sum_{i=1}^{3} \hat{\boldsymbol{\nu}}_{li} = 0 \qquad l = 1, \ldots, 4 \tag{9.201}$$

because the three unit vectors $\hat{\boldsymbol{\nu}}_{li}$, $i = 1, 2, 3$, are coplanar and form angles of $2\pi/3$ radians between one another.

By putting these intermediate findings together we obtain the depolarization dyadic

$$\overline{\mathbf{L}}(\mathbf{r}) = \frac{1}{4} \sum_{l=1}^{4} \hat{\mathbf{n}}_l \hat{\mathbf{n}}_l, \qquad \mathbf{r} \in V_S \tag{9.202}$$

which is evidently symmetric and has unitary trace, as expected. A more general expression for the case where the observation point is not the center of a regular tetrahedron or the latter is not symmetric about its center can be obtained with the procedure outlined in Section 14.9.3.

(End of Example 9.2)

The excluded region V_E need not be small, though formula (9.190) certainly simplifies if V_E is reduced to the point $\mathbf{r}$. In the process, it is important that the limits be carried out in such a way that the volume of V_E decreases while the shape thereof retains its form. Stated another way, V_E must contract to zero through a *similarity transformation* of center $\mathbf{r}$, and to indicate this procedure we adopt the symbol $\lim_{|V_E| \to 0^+}$. As the characteristic size h of V_E is made to vanish, the contribution of V_E tends to zero on account of identity (9.173) and estimates (9.176), (9.177) for $b \to 0^+$. The integrals over $V_S \setminus V_E$ and ∂V_E remain finite, but the results depend on the *shape* of V_E — which is arbitrary — and hence, in principle, infinitely many different splittings are possible for a given V_S. Still, the combination of the two parts must invariably yield the same result, as expected, since the electric field is unique (Section 6.4.2). In particular, only when V_E is a sphere does the contribution of $V_S \setminus V_E$ converge to the Cauchy principal value of the integral for $h \to 0^+$ [37]. (To get a feeling for the problem, consider the well-known one-dimensional case

$$\int_{-1}^{1} \frac{\mathrm{d}\xi}{\xi} = \lim_{a \to 0^+} \left(\int_{-1}^{\alpha a} \frac{\mathrm{d}\xi}{\xi} + \int_{\beta a}^{1} \frac{\mathrm{d}\xi}{\xi} \right) = \log \frac{|\alpha|}{|\beta|} \tag{9.203}$$

where clearly the result depends upon the ratio $|\alpha|/|\beta|$. When $\alpha = \beta$, the excluded interval $[\alpha a, \beta a]$ is symmetric around the singular point $\xi = 0$ and one obtains zero, the Cauchy principal value of the

[1] Hopefully, there should be no mistaking the discrete index i for the imaginary unit (Appendix B) since throughout this text we always employ the symbol j for that task.

otherwise undefined integral.) Therefore, for $\mathbf{r} \in V_S$ we write

$$\begin{aligned}\mathbf{E}(\mathbf{r}) &= -\mathrm{j}\,\omega\mu \int_{V_S} \mathrm{d}V' \frac{\mathrm{e}^{-\mathrm{j}kR}}{4\pi R}\mathbf{J}(\mathbf{r}') + \lim_{|V_E|\to 0^+} \frac{1}{\mathrm{j}\,\omega\varepsilon} \int_{V_S\setminus V_E} \mathrm{d}V'\, \nabla\nabla \frac{\mathrm{e}^{-\mathrm{j}kR}}{4\pi R} \cdot \mathbf{J}(\mathbf{r}') \\ &\quad - \lim_{|V_E|\to 0^+} \frac{1}{\mathrm{j}\,\omega\varepsilon} \int_{\partial V_E} \mathrm{d}S'\, \hat{\mathbf{n}}(\mathbf{r}')\nabla \frac{1}{4\pi R} \cdot \mathbf{J}(\mathbf{r}) \\ &= -\mathrm{j}\,\omega\mu \lim_{|V_E|\to 0^+} \int_{V_S\setminus V_E} \mathrm{d}V' \left[\left(\overline{\mathfrak{I}} + \frac{\nabla\nabla}{k^2}\right)\frac{\mathrm{e}^{-\mathrm{j}kR}}{4\pi R}\right] \cdot \mathbf{J}(\mathbf{r}') - \frac{\overline{\mathbf{L}}(\mathbf{r}) \cdot \mathbf{J}(\mathbf{r})}{\mathrm{j}\,\omega\varepsilon} \\ &= -\mathrm{j}\,\omega\mu\, \mathrm{PV} \int_{V_S} \mathrm{d}V' \left[\left(\overline{\mathfrak{I}} + \frac{\nabla\nabla}{k^2}\right)\frac{\mathrm{e}^{-\mathrm{j}kR}}{4\pi R}\right] \cdot \mathbf{J}(\mathbf{r}') - \frac{\mathbf{J}(\mathbf{r})}{3\,\mathrm{j}\,\omega\varepsilon} \end{aligned} \tag{9.204}$$

where the last equality applies if $V_E \equiv B(\mathbf{r}, h/2)$ in Figure 9.9, in which case $\overline{\mathbf{L}}(\mathbf{r})$ is still given by (9.197).

Formulas (9.204) and, for a finite V_E, (9.190) tell us how to extend (9.158) for observation points $\mathbf{r} \in \mathbb{R}^3$. However, what can be said about the electric dyadic Green function for electric sources (9.159)? Through the combination with the depolarization dyadic $\overline{\mathbf{L}}(\mathbf{r})$ the electric current density $\mathbf{J}(\mathbf{r})$ appears algebraically in the expression of the electric field. In order to write such contribution as an integral over V_S instead, we have to interpret (9.204) in the realm of distributions where the following expression

$$\int_{V_S} \mathrm{d}V'\, \mathbf{J}(\mathbf{r}')\delta^{(3)}(\mathbf{r} - \mathbf{r}') = \begin{cases} \mathbf{J}(\mathbf{r}), & \mathbf{r} \in V_S \\ 0, & \mathbf{r} \in \mathbb{R}^3 \setminus \overline{V}_S \end{cases} \tag{9.205}$$

has a meaning as *convolution product* between the three-dimensional Dirac δ-distribution and $\mathbf{J}(\mathbf{r})$, and it is referred to as the *sifting property* of $\delta^{(3)}(\mathbf{r})$ (Appendix C). Therefore, (9.159) may be extended in a distributional sense as

$$\overline{\mathbf{G}}_{EJ}(\mathbf{r}, \mathbf{r}') := \mathrm{PV}\left(\overline{\mathfrak{I}} + \frac{\nabla\nabla}{k^2}\right)\frac{\mathrm{e}^{-\mathrm{j}k|\mathbf{r}-\mathbf{r}'|}}{4\pi|\mathbf{r} - \mathbf{r}'|} - \frac{\overline{\mathbf{L}}(\mathbf{r}')}{k^2}\delta^{(3)}(\mathbf{r} - \mathbf{r}'), \qquad \mathbf{r}, \mathbf{r}' \in V_S \tag{9.206}$$

for observation points in the source region. Evidently, this formula also extends the magnetic Green function (9.161) by virtue of duality.

With the aid of definitions (8.356), (9.159) and (9.161) we may compute the electric and magnetic dyadic Green functions $\overline{\mathbf{G}}_{EJ}(\mathbf{r}, \mathbf{r}')$, $\overline{\mathbf{G}}_{HJ}(\mathbf{r}, \mathbf{r}')$ explicitly. Since the scalar Green function (8.356) depends on the distance $R = |\mathbf{r} - \mathbf{r}'|$ it is expedient to introduce a local spherical system of coordinates (R, α, β) with polar axis aligned with $\hat{\mathbf{z}}$ and center the source point $\mathbf{r}'$. In this way the identity dyadic $\overline{\mathfrak{I}}$ may be written as $\hat{\mathbf{R}}\hat{\mathbf{R}} + \hat{\alpha}\hat{\alpha} + \hat{\beta}\hat{\beta}$ and we can also introduce the dyadic

$$\overline{\mathfrak{I}}_{\mathrm{t}R} = \overline{\mathfrak{I}} - \hat{\mathbf{R}}\hat{\mathbf{R}} = \hat{\alpha}\hat{\alpha} + \hat{\beta}\hat{\beta} \tag{9.207}$$

which is called the *identity dyadic transverse to R* or $\hat{\mathbf{R}}$. Evidently, dot-multiplying a vector field $\mathcal{V}(\mathbf{r})$ with $\overline{\mathfrak{I}}_{\mathrm{t}R}$ extracts the components of $\mathcal{V}(\mathbf{r})$ perpendicular, that is, transverse, to $\hat{\mathbf{R}}$, hence the name of $\overline{\mathfrak{I}}_{\mathrm{t}R}$.

From (9.166)-(9.168) we find

$$\overline{\mathbf{G}}_{EJ}(\mathbf{r}, \mathbf{r}') = \left[\left(\frac{3}{k^2R^2} + \frac{3\mathrm{j}}{kR} - 1\right)\hat{\mathbf{R}}\hat{\mathbf{R}} + \left(1 - \frac{1}{k^2R^2} - \frac{\mathrm{j}}{kR}\right)\overline{\mathfrak{I}}\right]\frac{\mathrm{e}^{-\mathrm{j}kR}}{4\pi R}$$

$$= \left[\left(\frac{2}{k^2R^2} + \frac{2\mathrm{j}}{kR}\right)\hat{\mathbf{R}}\hat{\mathbf{R}} + \left(1 - \frac{1}{k^2R^2} - \frac{\mathrm{j}}{kR}\right)\overline{\mathbf{J}}_{tR}\right]\frac{\mathrm{e}^{-\mathrm{j}kR}}{4\pi R} \tag{9.208}$$

$$\overline{\mathbf{G}}_{HJ}(\mathbf{r},\mathbf{r}') = -\left(\mathrm{j}k + \frac{1}{R}\right)\frac{\mathrm{e}^{-\mathrm{j}kR}}{4\pi R}\hat{\mathbf{R}} \times \overline{\mathbf{J}} = -\left(\mathrm{j}k + \frac{1}{R}\right)\frac{\mathrm{e}^{-\mathrm{j}kR}}{4\pi R}(\hat{\beta}\hat{\alpha} - \hat{\alpha}\hat{\beta}) \tag{9.209}$$

for observation points $\mathbf{r} \in \mathbb{R}^3 \setminus \{\mathbf{r}'\}$ (see Figure 9.9).

In the chosen system of coordinates the electric dyadic Green function is diagonal and thus symmetric, whereas the magnetic dyadic Green function is anti-symmetric. Besides, the non-zero entries essentially consist of negative powers of R, the distance from the source point. This is expected and consistent with the singular source of $\overline{\mathbf{G}}_{EJ}(\mathbf{r},\mathbf{r}')$ and $\overline{\mathbf{G}}_{HJ}(\mathbf{r},\mathbf{r}')$ being concentrated in a point and thus having finite spatial extent.

9.4.3 Governing equation of $\overline{\mathbf{G}}_{EJ}(\mathbf{r},\mathbf{r}')$

Our discussion so far has been centered on the *construction* of the dyadic Green functions $\overline{\mathbf{G}}_{EJ}(\mathbf{r},\mathbf{r}')$ and the like by starting from the known representation of electric and magnetic fields (8.129)-(8.132) in terms of scalar and vector potentials in the Lorenz gauge. It may be argued, though, that the dyadic Green functions *per se* are fundamental solutions to specific second-order partial differential equations, similarly to what we previously found for $G(\mathbf{r},\mathbf{r}';\omega)$ in Section 8.7.1. Therefore, to determine the governing equations, we carefully extend the procedure outlined in Section 8.7, while focusing on $\overline{\mathbf{G}}_{EJ}(\mathbf{r},\mathbf{r}')$ for the time being.

The time-harmonic electric field $\mathbf{E}(\mathbf{r})$ solves the vector Helmholtz equation (1.234) which for convenience we rewrite here in operator form [28, Chapter 3]

$$\mathfrak{L}\{\mathbf{E}(\mathbf{r})\} = -\mathrm{j}\,\omega\mu\mathbf{J}(\mathbf{r}), \qquad \mathbf{r} \in \mathbb{R}^3 \tag{9.210}$$

where

$$\mathfrak{L}\{\bullet\} := \nabla \times \nabla \times \overline{\mathbf{J}} \cdot \{\bullet\} - k^2\overline{\mathbf{J}} \cdot \{\bullet\} \tag{9.211}$$

is a *dyadic* second-order differential operator which transforms electric fields into volume electric current densities aside from an inconsequential multiplicative factor. Besides, k is a possibly complex wavenumber, and $\mathbf{J}(\mathbf{r})$ is a known source whose support is the finite region of space $V_S \subset \mathbb{R}^3$. The dyadic nature of $\mathfrak{L}\{\bullet\}$ is consistent with the fact that the electric field and the current density are not parallel to one another in general. More importantly, it should be noted that, since the scalar product between two dyadics is defined as well (Appendix E.2), $\mathfrak{L}\{\bullet\}$ can operate on vector and dyadic fields alike. In the latter case then, $\mathfrak{L}\{\bullet\}$ transforms a dyadic into another dyadic, and in fact the governing equation for $\overline{\mathbf{G}}_{EJ}(\mathbf{r},\mathbf{r}')$ must precisely be of this kind.

Next, in order to broaden the set of allowable sources (that is, to deal even with the singular ones) we wish to formulate (9.210) in a weak form or distributional sense (Appendix C). To this purpose we define the *symmetric* inner product

$$(\boldsymbol{\phi},\mathbf{E})_{\mathbb{R}^3} := \int_{\mathbb{R}^3} \mathrm{d}V\,\boldsymbol{\phi}(\mathbf{r}) \cdot \mathbf{E}(\mathbf{r}) = (\mathbf{E},\boldsymbol{\phi})_{\mathbb{R}^3} \tag{9.212}$$

where $\boldsymbol{\phi}(\mathbf{r}) \in \mathcal{C}_0^\infty(\mathbb{R}^3)^3$ denotes a real-valued vector test function with bounded support, so that the convergence of the improper integral is always guaranteed at least for electric fields which behave reasonably well in the region where $\boldsymbol{\phi}(\mathbf{r})$ is non-null.

We observe that the dyadic operator (9.211) satisfies the identity

$$(\boldsymbol{\phi}, \mathfrak{L}\{\mathbf{E}\})_{\mathbb{R}^3} = (\mathfrak{L}\{\boldsymbol{\phi}\}, \mathbf{E})_{\mathbb{R}^3} \tag{9.213}$$

i.e., $\mathfrak{L}\{\bullet\}$ is *self-adjoint* (Appendix D.6) [38, Section 10.4] with respect to the inner product (9.212). This statement is easily proved by applying the integral theorem (H.99) [5, Formula (A1.33)] with $\mathcal{A}(\mathbf{r}) = \mathbf{E}(\mathbf{r})$, $\mathcal{B}(\mathbf{r}) = \boldsymbol{\phi}(\mathbf{r})$ and the region V coinciding with the ball $B(0, a)$, viz.,

$$\begin{aligned}\int_{B(0,a)} \mathrm{d}V\, [\boldsymbol{\phi}(\mathbf{r}) \cdot \nabla \times \nabla \times \mathbf{E}(\mathbf{r}) - \mathbf{E}(\mathbf{r}) \cdot \nabla \times \nabla \times \boldsymbol{\phi}(\mathbf{r})] \\ = \int_{\partial B} \mathrm{d}S\, [\hat{\mathbf{n}}(\mathbf{r}) \times \mathbf{E}(\mathbf{r}) \cdot \nabla \times \boldsymbol{\phi}(\mathbf{r}) - \hat{\mathbf{n}}(\mathbf{r}) \times \boldsymbol{\phi}(\mathbf{r}) \cdot \nabla \times \mathbf{E}(\mathbf{r})]\end{aligned} \tag{9.214}$$

provided the electric field is at least twice differentiable. The right-hand side tends to zero as $a \to +\infty$ because $\boldsymbol{\phi}(\mathbf{r})$ has bounded support by hypothesis and hence the integrand is null for suitably large values of a. For the same reason also the domain integral in the left member remains finite for $a \to +\infty$. As a result, if we add and subtract the quantity $k^2\mathbf{E}(\mathbf{r}) \cdot \boldsymbol{\phi}(\mathbf{r})$ in the integrand of the left member and we rearrange the terms, after letting a grow infinitely large, we arrive at the integral identity

$$\int_{\mathbb{R}^3} \mathrm{d}V\, \boldsymbol{\phi}(\mathbf{r}) \cdot \left[\nabla \times \nabla \times \mathbf{E}(\mathbf{r}) - k^2\mathbf{E}(\mathbf{r})\right] = \int_{\mathbb{R}^3} \mathrm{d}V\, \left[\nabla \times \nabla \times \boldsymbol{\phi}(\mathbf{r}) - k^2\boldsymbol{\phi}(\mathbf{r})\right] \cdot \mathbf{E}(\mathbf{r}) \tag{9.215}$$

which is just the explicit expression of (9.213), as we wanted to prove.

The desired weak form of (9.210) now follows by taking the inner product of both sides with a test function $\boldsymbol{\phi}(\mathbf{r})$ and by invoking (9.213), viz.,

$$(\boldsymbol{\phi}, \mathfrak{L}\{\mathbf{E}\})_{\mathbb{R}^3} = (\mathfrak{L}\{\boldsymbol{\phi}\}, \mathbf{E})_{\mathbb{R}^3} = -\mathrm{j}\,\omega\mu\, (\boldsymbol{\phi}, \mathbf{J})_{\mathbb{R}^3} \tag{9.216}$$

where the second part of the equation must be used in lieu of (9.210) whenever the electric field cannot 'stand' the second-order derivatives implied in the operator $\mathfrak{L}\{\bullet\}$. Stated another way, the right member of (9.213) constitutes the weak form of the differential operator $\mathfrak{L}\{\bullet\}$.

Thanks to the principle of superposition (Section 6.1) we are entitled to look for solutions to (9.210) or (9.216) in the form

$$\mathbf{E}(\mathbf{r}) = -\mathrm{j}\,\omega\mu \int_{V_S} \mathrm{d}V'\, \overline{\mathbf{G}}_{EJ}(\mathbf{r}, \mathbf{r}') \cdot \mathbf{J}(\mathbf{r}') \tag{9.217}$$

where the right-hand side can be regarded as the weighted combination of the electric fields contributed by infinitely many elementary time-harmonic sources located at points $\mathbf{r}' \in V_S$. Further, if the Green function is a distribution and contains three-dimensional Dirac δ's as in (9.206), then it is necessary to interpret the integral symbolically.

To proceed with the derivation we insert the Ansatz (9.217) into (9.216)

$$-\mathrm{j}\,\omega\mu \int_{\mathbb{R}^3} \mathrm{d}V\, \mathfrak{L}\{\boldsymbol{\phi}\} \cdot \int_{V_S} \mathrm{d}V'\, \overline{\mathbf{G}}_{EJ}(\mathbf{r}, \mathbf{r}') \cdot \mathbf{J}(\mathbf{r}') = -\mathrm{j}\,\omega\mu \int_{V_S} \mathrm{d}V'\, \boldsymbol{\phi}(\mathbf{r}') \cdot \mathbf{J}(\mathbf{r}') \tag{9.218}$$

whereby we obtain

$$\int_{V_S} \mathrm{d}V' \left[\int_{\mathbb{R}^3} \mathrm{d}V \, \mathfrak{L}\{\boldsymbol{\phi}\} \cdot \overline{\mathbf{G}}_{EJ}(\mathbf{r},\mathbf{r}') - \boldsymbol{\phi}(\mathbf{r}') \right] \cdot \mathbf{J}(\mathbf{r}') = 0 \tag{9.219}$$

by *formally* swapping the order of integration. Since this last equation must be true for any current density $\mathbf{J}(\mathbf{r}')$ and arbitrary-shaped supports V_S, we require

$$\int_{\mathbb{R}^3} \mathrm{d}V \, \mathfrak{L}\{\boldsymbol{\phi}\} \cdot \overline{\mathbf{G}}_{EJ}(\mathbf{r},\mathbf{r}') = \boldsymbol{\phi}(\mathbf{r}'), \qquad \mathbf{r}' \in \mathbb{R}^3 \tag{9.220}$$

where the left-hand side can be construed as the weak form of $\mathfrak{L}\{\bullet\}$ applied to the dyadic field $\overline{\mathbf{G}}_{EJ}(\mathbf{r},\mathbf{r}')$ with respect to $\mathbf{r}$. Indeed, this follows by expanding $\overline{\mathbf{G}}_{EJ}(\mathbf{r},\mathbf{r}')$ as the sum of three dyads, viz.,

$$\overline{\mathbf{G}}_{EJ}(\mathbf{r},\mathbf{r}') = \mathbf{G}'_{EJx}(\mathbf{r},\mathbf{r}')\hat{\mathbf{x}} + \mathbf{G}'_{EJy}(\mathbf{r},\mathbf{r}')\hat{\mathbf{y}} + \mathbf{G}'_{EJz}(\mathbf{r},\mathbf{r}')\hat{\mathbf{z}} \tag{9.221}$$

and by invoking (9.212) three times, since the unit vectors $\hat{\mathbf{x}}$, $\hat{\mathbf{y}}$ and $\hat{\mathbf{z}}$ are constant. Moreover, the right member of (9.220) may be written symbolically as

$$\boldsymbol{\phi}(\mathbf{r}') = \int_{\mathbb{R}^3} \mathrm{d}V \, \boldsymbol{\phi}(\mathbf{r}) \cdot \overline{\mathfrak{I}} \, \delta^{(3)}(\mathbf{r}-\mathbf{r}') \tag{9.222}$$

by applying (C.19) three times on the Cartesian components of $\boldsymbol{\phi}(\mathbf{r}')$ and recalling the fundamental property (E.27) of the identity dyadic. All in all, (9.220) can be stated in equivalent short-hand notation as [22, Chapter 1], [28, Chapter 5], [29]

$$\nabla \times \nabla \times \overline{\mathbf{G}}_{EJ}(\mathbf{r},\mathbf{r}') - k^2 \overline{\mathbf{G}}_{EJ}(\mathbf{r},\mathbf{r}') = \delta^{(3)}(\mathbf{r}-\mathbf{r}') \, \overline{\mathfrak{I}} \tag{9.223}$$

which is the desired governing equation for $\overline{\mathbf{G}}_{EJ}(\mathbf{r},\mathbf{r}')$. Parenthetically, the principle of duality tells us that (9.223) is also the pertinent equation of the magnetic Green function $\overline{\mathbf{G}}_{HM}(\mathbf{r},\mathbf{r}')$ (9.162).

Showing that $\overline{\mathbf{G}}_{EJ}(\mathbf{r},\mathbf{r}')$ solves (9.223) for observation points $\mathbf{r} \neq \mathbf{r}'$ is relatively simple because, under the stated hypothesis, (9.223) reduces to a homogeneous wave equation for the dyadic field (9.159). Since the derivatives exist in an ordinary sense, we have

$$\begin{aligned} \nabla \times \nabla \times \overline{\mathbf{G}}_{EJ}(\mathbf{r},\mathbf{r}') &= \nabla \times \nabla \times \left(\frac{\mathrm{e}^{-\mathrm{j}kR}}{4\pi R} \overline{\mathfrak{I}} \right) = \nabla\nabla \frac{\mathrm{e}^{-\mathrm{j}kR}}{4\pi R} - \overline{\mathfrak{I}} \nabla^2 \frac{\mathrm{e}^{-\mathrm{j}kR}}{4\pi R} \\ &= \nabla\nabla \frac{\mathrm{e}^{-\mathrm{j}kR}}{4\pi R} + \overline{\mathfrak{I}} k^2 \frac{\mathrm{e}^{-\mathrm{j}kR}}{4\pi R} = k^2 \overline{\mathbf{G}}_{EJ}(\mathbf{r},\mathbf{r}') \end{aligned} \tag{9.224}$$

where we have used in succession (H.64), (H.68), and the fact that the time-harmonic scalar Green function (8.356) is the solution to the Helmholtz equation (8.330) for $\mathbf{r} \neq \mathbf{r}'$. The last step is a consequence of the definition of $\overline{\mathbf{G}}_{EJ}(\mathbf{r},\mathbf{r}')$, whereby the statement follows.

By contrast, the calculations are way more involved when we include the singular point $\mathbf{r}'$, in that we must interpret $\overline{\mathbf{G}}_{EJ}(\mathbf{r},\mathbf{r}')$ as the distribution given by (9.206) and employ the weak formulation (9.220). In particular, the left member of (9.220) becomes

$$\mathrm{PV} \int_{\mathbb{R}^3} \mathrm{d}V \, \mathfrak{L}\{\boldsymbol{\phi}\} \cdot \left(\overline{\mathfrak{I}} + \frac{\nabla\nabla}{k^2} \right) \frac{\mathrm{e}^{-\mathrm{j}kR}}{4\pi R} - \frac{1}{k^2} \int_{\mathbb{R}^3} \mathrm{d}V \, \mathfrak{L}\{\boldsymbol{\phi}\} \cdot \overline{\mathbf{L}}(\mathbf{r}) \, \delta^{(3)}(\mathbf{r}-\mathbf{r}')$$

$$= \mathrm{PV}\int_{\mathbb{R}^3} \mathrm{d}V\,\mathfrak{L}\{\boldsymbol{\phi}\}\cdot\left(\overline{\mathfrak{I}}+\frac{\nabla\nabla}{k^2}\right)\frac{\mathrm{e}^{-\mathrm{j}kR}}{4\pi R} - \mathfrak{L}\{\boldsymbol{\phi}\}|_{\mathbf{r}=\mathbf{r}'}\cdot\frac{\overline{\mathbf{L}}(\mathbf{r}')}{k^2} \qquad (9.225)$$

and we need to prove that the last expression above is identical with $\boldsymbol{\phi}(\mathbf{r}')$. To use the depolarization dyadic (9.197) which is relevant to an excluded region V_E of spherical shape, here we isolate the *source point* $\mathbf{r}'$ by means of a ball $B(\mathbf{r}',a)$ and also integrate over the spherical shell $V := B(\mathbf{r}',b)\setminus B(\mathbf{r}',a)$ with $b > a$. In the limit as a vanishes and b grows infinitely large the integral over V reduces to the one of interest in (9.225).

We observe that for $\mathbf{r}\in V$ the dyadic Green function $\overline{\mathbf{G}}_{EJ}(\mathbf{r},\mathbf{r}')$ is given by (9.159) and is regular. As a result, since $\mathfrak{L}\{\boldsymbol{\phi}\}$ involves the double curl of $\boldsymbol{\phi}(\mathbf{r})$ we may apply the dyadic integral theorem (H.109) [5, Formula (A4.75)] with $\mathcal{B}(\mathbf{r}) = \boldsymbol{\phi}(\mathbf{r})$ and $\overline{\mathcal{A}}(\mathbf{r}) = \overline{\mathbf{G}}_{EJ}(\mathbf{r},\mathbf{r}')$. In symbols, this step yields

$$\begin{aligned}&\int_V \mathrm{d}V\left[(\nabla\times\nabla\times\boldsymbol{\phi})\cdot\overline{\mathbf{G}}_{EJ}(\mathbf{r},\mathbf{r}') - \boldsymbol{\phi}(\mathbf{r})\cdot\nabla\times\nabla\times\overline{\mathbf{G}}_{EJ}\right]\\ &= \int_{|\mathbf{R}|=b}\mathrm{d}S\,\hat{\mathbf{R}}\cdot\left[\phi(\mathbf{r})\times(\nabla\times\overline{\mathbf{G}}_{EJ}) + (\nabla\times\boldsymbol{\phi})\times\overline{\mathbf{G}}_{EJ}(\mathbf{r},\mathbf{r}')\right]\\ &\qquad - \int_{|\mathbf{R}|=a}\mathrm{d}S\,\hat{\mathbf{R}}\cdot\left[\phi(\mathbf{r})\times(\nabla\times\overline{\mathbf{G}}_{EJ}) + (\nabla\times\boldsymbol{\phi})\times\overline{\mathbf{G}}_{EJ}(\mathbf{r},\mathbf{r}')\right] \qquad (9.226)\end{aligned}$$

where we emphasize that $\overline{\mathbf{G}}_{EJ}(\mathbf{r},\mathbf{r}')$ denotes the dyadic field (9.159). The integral in the left-hand side can be simplified by eliminating the double curl of $\overline{\mathbf{G}}_{EJ}(\mathbf{r},\mathbf{r}')$ with the aid of (9.224), which holds in V. The flux integral over the sphere of radius b vanishes if we choose the ball $B(\mathbf{r}',b)$ large enough to contain the bounded support of the test function $\boldsymbol{\phi}(\mathbf{r})$. Thus, we are left with

$$\begin{aligned}&\int_V \mathrm{d}V\left[\nabla\times\nabla\times\boldsymbol{\phi} - k^2\boldsymbol{\phi}(\mathbf{r})\right]\cdot\overline{\mathbf{G}}_{EJ}(\mathbf{r},\mathbf{r}')\\ &\qquad = -\int_{|\mathbf{R}|=a}\mathrm{d}S\,\hat{\mathbf{R}}\cdot\left[\boldsymbol{\phi}(\mathbf{r})\times(\nabla\times\overline{\mathbf{G}}_{EJ}) + (\nabla\times\boldsymbol{\phi})\times\overline{\mathbf{G}}_{EJ}(\mathbf{r},\mathbf{r}')\right] \qquad (9.227)\end{aligned}$$

where, by indicating with $G(R)$ the scalar Green function (8.356), we notice that

$$\hat{\mathbf{R}}\cdot\boldsymbol{\phi}(\mathbf{r})\times(\nabla\times\overline{\mathbf{G}}_{EJ}) = \hat{\mathbf{R}}\times\boldsymbol{\phi}(\mathbf{r})\cdot\nabla\times[G(R)\overline{\mathfrak{I}}] = \left[\hat{\mathbf{R}}\times\boldsymbol{\phi}(\mathbf{r})\right]\times\nabla G \qquad (9.228)$$

on account of (H.66) and (H.39), and also

$$\begin{aligned}\hat{\mathbf{R}}\cdot(\nabla\times\boldsymbol{\phi})\times\overline{\mathbf{G}}_{EJ}(\mathbf{r},\mathbf{r}') &= \hat{\mathbf{R}}\cdot(\nabla\times\boldsymbol{\phi})\times\left[\overline{\mathfrak{I}}G(R) + \frac{1}{k^2}\nabla\nabla G\right]\\ &= \hat{\mathbf{R}}\times(\nabla\times\boldsymbol{\phi})\,G(R) + \hat{\mathbf{R}}\cdot(\nabla\times\boldsymbol{\phi})\times\frac{1}{k^2}\nabla\nabla G\\ &= \hat{\mathbf{R}}\times(\nabla\times\boldsymbol{\phi})\,G(R) + \hat{\mathbf{R}}\cdot(\nabla\times\nabla\times\boldsymbol{\phi})\frac{\nabla G}{k^2} - \hat{\mathbf{R}}\cdot\nabla\times\left[(\nabla\times\boldsymbol{\phi})\frac{\nabla G}{k^2}\right] \qquad (9.229)\end{aligned}$$

by virtue of (H.39) again and (H.76) used with $\mathcal{A} = \nabla\times\boldsymbol{\phi}$ and $\mathcal{B} = \nabla G$. The surface integral of the last term above vanishes by virtue of the dyadic extension of the Stokes theorem (H.107) [5, Formula (A4.70)], because the sphere is a closed surface and the boundary thereof reduces to a point.

Alternatively, we may split the sphere into two hemispheres and apply (H.107) twice, in which case the line integrals along the boundaries of the hemispheres cancel each other out.

Thanks to these intermediate findings, from (9.225) and (9.227) we get

$$\mathrm{PV}\int_{\mathbb{R}^3} \mathrm{d}V\, \mathfrak{L}\{\boldsymbol{\phi}\}\cdot\left(\overline{\mathbf{I}}+\frac{\nabla\nabla}{k^2}\right)\frac{\mathrm{e}^{-\mathrm{j}kR}}{4\pi R} = -\lim_{a\to 0}\int_{|\mathbf{R}|=a} \mathrm{d}S\,\left[\hat{\mathbf{R}}\times\boldsymbol{\phi}(\mathbf{r})\right]\times\nabla G$$
$$-\lim_{a\to 0}\int_{|\mathbf{R}|=a} \mathrm{d}S\,\hat{\mathbf{R}}\times(\nabla\times\boldsymbol{\phi})\,G(R) - \frac{1}{k^2}\lim_{a\to 0}\int_{|\mathbf{R}|=a} \mathrm{d}S\,\hat{\mathbf{R}}\cdot(\nabla\times\nabla\times\boldsymbol{\phi})\nabla G \quad (9.230)$$

and we examine the three surface integrals individually. In this regard, it is convenient to introduce a local system of polar spherical coordinates (R,α,β) centered on $\mathbf{r}'$.

By starting with (A.28), (9.167) and (H.14) we manipulate the first integral as follows

$$\begin{aligned}
-\int_{|\mathbf{R}|=a} \mathrm{d}S\,\left[\hat{\mathbf{R}}\times\boldsymbol{\phi}(\mathbf{r})\right]\times\nabla G &= -\int_{4\pi}\mathrm{d}\Omega\, a^2\left.\frac{\mathrm{d}G}{\mathrm{d}R}\right|_{R=a}\left[\hat{\mathbf{R}}\times\boldsymbol{\phi}(\mathbf{r})\right]\times\hat{\mathbf{R}}\\
&= (1+\mathrm{j}\,ka)\frac{\mathrm{e}^{-\mathrm{j}\,ka}}{4\pi}\int_{4\pi}\mathrm{d}\Omega\,\boldsymbol{\phi}(\mathbf{r}) - (1+\mathrm{j}\,ka)\frac{\mathrm{e}^{-\mathrm{j}\,ka}}{4\pi}\int_{4\pi}\mathrm{d}\Omega\,\hat{\mathbf{R}}\hat{\mathbf{R}}\cdot\boldsymbol{\phi}(\mathbf{r})\\
&= (1+\mathrm{j}\,ka)\frac{\mathrm{e}^{-\mathrm{j}\,ka}}{4\pi}\int_{4\pi}\mathrm{d}\Omega\,[\boldsymbol{\phi}(\mathbf{r})-\boldsymbol{\phi}(\mathbf{r}')] + (1+\mathrm{j}\,ka)\mathrm{e}^{-\mathrm{j}\,ka}\boldsymbol{\phi}(\mathbf{r}')\\
&\quad - (1+\mathrm{j}\,ka)\frac{\mathrm{e}^{-\mathrm{j}\,ka}}{4\pi}\int_{4\pi}\mathrm{d}\Omega\,\hat{\mathbf{R}}\hat{\mathbf{R}}\cdot[\boldsymbol{\phi}(\mathbf{r})-\boldsymbol{\phi}(\mathbf{r}')]\\
&\quad - (1+\mathrm{j}\,ka)\mathrm{e}^{-\mathrm{j}\,ka}\boldsymbol{\phi}(\mathbf{r}')\cdot\frac{1}{4\pi}\int_{4\pi}\mathrm{d}\Omega\,\hat{\mathbf{R}}\hat{\mathbf{R}}
\end{aligned} \quad (9.231)$$

where $\int_{4\pi}\mathrm{d}\Omega$ signifies integration over the complete solid angle seen from $\mathbf{r}'$. Since $\boldsymbol{\phi}(\mathbf{r})$ is regular for $\mathbf{r}=\mathbf{r}'$, the first and third contributions in the rightmost member vanish in the limit as $a\to 0^+$, and hence we have

$$\begin{aligned}
-\lim_{a\to 0}\int_{|\mathbf{R}|=a} \mathrm{d}S\,\left[\hat{\mathbf{R}}\times\boldsymbol{\phi}(\mathbf{r})\right]\times\nabla G &= \boldsymbol{\phi}(\mathbf{r}')-\boldsymbol{\phi}(\mathbf{r}')\cdot\frac{1}{4\pi}\int_{4\pi}\mathrm{d}\Omega\,\hat{\mathbf{R}}\hat{\mathbf{R}}\\
&= \boldsymbol{\phi}(\mathbf{r}')-\boldsymbol{\phi}(\mathbf{r}')\cdot\overline{\mathbf{L}}(\mathbf{r}')
\end{aligned} \quad (9.232)$$

having recognized the integral of $\hat{\mathbf{R}}\hat{\mathbf{R}}/(4\pi)$ as the depolarization dyadic given by (9.197).

The second surface integral in (9.230) vanishes for $a\to 0^+$ inasmuch as it is dominated by a constant proportional to a, namely,

$$\left|\int_{|\mathbf{R}|=a}\mathrm{d}S\,\hat{\mathbf{R}}\times(\nabla\times\boldsymbol{\phi})\,G(R)\right| \leqslant \frac{a}{4\pi}\int_{4\pi}\mathrm{d}\Omega\,|\nabla\times\boldsymbol{\phi}| \leqslant a\,\|\nabla\times\boldsymbol{\phi}\|_\infty \quad (9.233)$$

as the curl of $\boldsymbol{\phi}(\mathbf{r})$ is regular.

With the aid of (A.28) and (9.167) the third surface integral in (9.230) can be transformed as follows

$$
\begin{aligned}
-\frac{1}{k^2}\int\limits_{|\mathbf{R}|=a} \mathrm{d}S\,\hat{\mathbf{R}}\cdot(\nabla\times\nabla\times\boldsymbol{\phi})\nabla G &= -\frac{1}{k^2}\int\limits_{4\pi}\mathrm{d}\Omega\; a^2\frac{\mathrm{d}G}{\mathrm{d}R}\bigg|_{R=a}\hat{\mathbf{R}}\hat{\mathbf{R}}\cdot(\nabla\times\nabla\times\boldsymbol{\phi})\\
&= (1+\mathrm{j}\,ka)\frac{\mathrm{e}^{-\mathrm{j}\,ka}}{4\pi k^2}\int\limits_{4\pi}\mathrm{d}\Omega\,\hat{\mathbf{R}}\hat{\mathbf{R}}\cdot(\nabla\times\nabla\times\boldsymbol{\phi})\\
&= (1+\mathrm{j}\,ka)\frac{\mathrm{e}^{-\mathrm{j}\,ka}}{4\pi k^2}\int\limits_{4\pi}\mathrm{d}\Omega\,\hat{\mathbf{R}}\hat{\mathbf{R}}\cdot(\nabla\times\nabla\times\boldsymbol{\phi}-\nabla\times\nabla\times\boldsymbol{\phi}|_{\mathbf{r}=\mathbf{r}'})\\
&\quad+(1+\mathrm{j}\,ka)\mathrm{e}^{-\mathrm{j}\,ka}\nabla\times\nabla\times\boldsymbol{\phi}|_{\mathbf{r}=\mathbf{r}'}\cdot\frac{1}{4\pi k^2}\int\limits_{4\pi}\mathrm{d}\Omega\,\hat{\mathbf{R}}\hat{\mathbf{R}}
\end{aligned}
\tag{9.234}
$$

where the first term in the rightmost-hand side vanishes for $a\to 0^+$ since the double curl of $\boldsymbol{\phi}(\mathbf{r})$ is regular for $\mathbf{r}=\mathbf{r}'$. Thus, we find

$$
-\frac{1}{k^2}\lim_{a\to 0^+}\int\limits_{|\mathbf{R}|=a}\mathrm{d}S\,\hat{\mathbf{R}}\cdot(\nabla\times\nabla\times\boldsymbol{\phi})\nabla G=\frac{1}{k^2}\nabla\times\nabla\times\boldsymbol{\phi}|_{\mathbf{r}=\mathbf{r}'}\cdot\overline{\mathbf{L}}(\mathbf{r}')
\tag{9.235}
$$

again by comparison with (9.197).

Lastly, by using (9.232), (9.233) and (9.235) in (9.230) we arrive at

$$
\begin{aligned}
\mathrm{PV}\int\limits_{\mathbb{R}^3}\mathrm{d}V\,\mathfrak{L}\{\boldsymbol{\phi}\}\cdot\left(\overline{\mathbf{I}}+\frac{\nabla\nabla}{k^2}\right)\frac{\mathrm{e}^{-\mathrm{j}\,kR}}{4\pi R} &= \boldsymbol{\phi}(\mathbf{r}')-\boldsymbol{\phi}(\mathbf{r}')\cdot\overline{\mathbf{L}}(\mathbf{r}')+\frac{1}{k^2}\nabla\times\nabla\times\boldsymbol{\phi}|_{\mathbf{r}=\mathbf{r}'}\cdot\overline{\mathbf{L}}(\mathbf{r}')\\
&= \boldsymbol{\phi}(\mathbf{r}')+\mathfrak{L}\{\boldsymbol{\phi}\}|_{\mathbf{r}=\mathbf{r}'}\cdot\frac{\overline{\mathbf{L}}(\mathbf{r}')}{k^2}
\end{aligned}
\tag{9.236}
$$

on account of definition (9.211). Since the contribution of the depolarization dyadic $\overline{\mathbf{L}}(\mathbf{r}')$ is compensated by the equal and opposite one in (9.225), it remains proved that the distributional Green dyadic (9.206) solves (9.220) or (9.223).

9.4.4 *Governing equation of* $\overline{\mathbf{G}}_{HJ}(\mathbf{r},\mathbf{r}')$

We continue the discussion with the determination of the differential equation satisfied by the magnetic dyadic Green function $\overline{\mathbf{G}}_{HJ}(\mathbf{r},\mathbf{r}')$ expressed by the first part of (9.161). Since much of the legwork has already been done while dealing with $\overline{\mathbf{G}}_{EJ}(\mathbf{r},\mathbf{r}')$, we can proceed more expeditiously. The time-harmonic magnetic field $\mathbf{H}(\mathbf{r})$ solves the vector Helmholtz equation (1.236) which, written in operator form, reads

$$
\mathfrak{L}\{\mathbf{H}(\mathbf{r})\}=\nabla\times\mathbf{J}(\mathbf{r}),\qquad \mathbf{r}\in\mathbb{R}^3
\tag{9.237}
$$

with the dyadic operator $\mathfrak{L}\{\bullet\}$ still given by (9.211). In order to state (9.237) in weak form we adopt the same inner product (9.212) and, for a continuous vector field $\mathbf{f}(\mathbf{r})$, observe that

$$
(\boldsymbol{\phi},\nabla\times\mathbf{f})_{\mathbb{R}^3}=(\nabla\times\boldsymbol{\phi},\mathbf{f})_{\mathbb{R}^3}
\tag{9.238}
$$

i.e., the curl operator is *self-adjoint* (Appendix D.6) [38, Section 10.4] with respect to (9.212) so long as $\hat{\mathbf{n}}(\mathbf{r})\times\mathbf{f}(\mathbf{r})$ is continuous across the boundary ∂V of a domain $V\subset\mathbb{R}^3$. This can be proved with the aid of the differential identity (H.49) and the Gauss theorem (A.53) applied to $B(0,a)\setminus\overline{V}$ and V, namely,

$$\int\limits_{B(0,a)\setminus\overline{V}} \mathrm{d}V\,[\mathbf{f}(\mathbf{r})\cdot\nabla\times\boldsymbol{\phi}-\boldsymbol{\phi}(\mathbf{r})\cdot\nabla\times\mathbf{f}] = \int\limits_{B(0,a)\setminus\overline{V}} \mathrm{d}V\,\nabla\cdot(\boldsymbol{\phi}\times\mathbf{f})$$

$$= \int\limits_{\partial B} \mathrm{d}S\,\hat{\mathbf{n}}(\mathbf{r})\cdot\boldsymbol{\phi}(\mathbf{r})\times\mathbf{f}(\mathbf{r}) - \int\limits_{\partial V} \mathrm{d}S\,\hat{\mathbf{n}}(\mathbf{r})\cdot\boldsymbol{\phi}(\mathbf{r})\times\mathbf{f}^{+}(\mathbf{r}) \tag{9.239}$$

and

$$\int\limits_{V} \mathrm{d}V\,[\mathbf{f}(\mathbf{r})\cdot\nabla\times\boldsymbol{\phi}-\boldsymbol{\phi}(\mathbf{r})\cdot\nabla\times\mathbf{f}] = \int\limits_{V} \mathrm{d}V\,\nabla\cdot(\boldsymbol{\phi}\times\mathbf{f}) = \int\limits_{\partial V} \mathrm{d}S\,\hat{\mathbf{n}}(\mathbf{r})\cdot\boldsymbol{\phi}(\mathbf{r})\times\mathbf{f}^{-}(\mathbf{r}) \tag{9.240}$$

with the unit vector perpendicular to ∂V pointing outwards V. By summing these identities side by side and using $\hat{\mathbf{n}}(\mathbf{r})\times\mathbf{f}^{+}(\mathbf{r}) = \hat{\mathbf{n}}(\mathbf{r})\times\mathbf{f}^{-}(\mathbf{r})$ for $\mathbf{r}\in\partial V$ we get

$$\int\limits_{B(0,a)} \mathrm{d}V\,[\mathbf{f}(\mathbf{r})\cdot\nabla\times\boldsymbol{\phi}-\boldsymbol{\phi}(\mathbf{r})\cdot\nabla\times\mathbf{f}] = \int\limits_{B(0,a)} \mathrm{d}V\,\nabla\cdot(\boldsymbol{\phi}\times\mathbf{f}) = \int\limits_{\partial B} \mathrm{d}S\,\hat{\mathbf{n}}(\mathbf{r})\cdot\boldsymbol{\phi}(\mathbf{r})\times\mathbf{f}(\mathbf{r}) \tag{9.241}$$

where the flux integral vanishes if the ball $B(0,a)$ entirely contains the bounded support of $\boldsymbol{\phi}(\mathbf{r})$. By taking the limit for infinitely large radius a we obtain

$$\int\limits_{\mathbb{R}^3} \mathrm{d}V\,\mathbf{f}(\mathbf{r})\cdot\nabla\times\boldsymbol{\phi} = \int\limits_{\mathbb{R}^3} \mathrm{d}V\,\boldsymbol{\phi}(\mathbf{r})\cdot\nabla\times\mathbf{f} \tag{9.242}$$

which is the explicit format of (9.238). As a matter of fact, the latter is also valid for the combination of a vector and a dyadic field or distribution, say,

$$\left(\boldsymbol{\phi},\nabla\times\overline{\mathbf{G}}\right)_{\mathbb{R}^3} = \left(\nabla\times\boldsymbol{\phi},\overline{\mathbf{G}}\right)_{\mathbb{R}^3} \tag{9.243}$$

which is proved in a similar way by starting with the differential identity (H.74) and invoking the divergence theorem for dyadics (H.105). Alternatively, (9.243) follows by applying (9.238) three times after expanding $\overline{\mathbf{G}}(\mathbf{r})$ as in (9.221).

In light of (9.212) and (9.238) the weak form of (9.237) reads

$$(\boldsymbol{\phi},\mathfrak{L}\{\mathbf{H}\})_{\mathbb{R}^3} = (\mathfrak{L}\{\boldsymbol{\phi}\},\mathbf{H})_{\mathbb{R}^3} = (\nabla\times\boldsymbol{\phi},\mathbf{J})_{\mathbb{R}^3} = (\boldsymbol{\phi},\nabla\times\mathbf{J})_{\mathbb{R}^3} \tag{9.244}$$

where the central part of this equation becomes mandatory if the magnetic field cannot 'stand' the derivatives implied in $\mathfrak{L}\{\bullet\}$ and the current density happens to be discontinuous or contains Dirac δ's.

By assuming solutions to (9.237) in the form

$$\mathbf{H}(\mathbf{r}) = \int\limits_{V_S} \mathrm{d}V'\,\overline{\mathbf{G}}_{HJ}(\mathbf{r},\mathbf{r}')\cdot\mathbf{J}(\mathbf{r}') \tag{9.245}$$

and by substituting into (9.244) we arrive at

$$\int\limits_{\mathbb{R}^3} \mathrm{d}V\,\mathfrak{L}\{\boldsymbol{\phi}\}\cdot\overline{\mathbf{G}}_{HJ}(\mathbf{r},\mathbf{r}') = \nabla\times\boldsymbol{\phi}|_{\mathbf{r}=\mathbf{r}'}, \qquad \mathbf{r}'\in\mathbb{R}^3 \tag{9.246}$$

after a few manipulations. Moreover, by invoking the sifting property (C.19) on the Cartesian components of $\nabla \times \boldsymbol{\phi}|_{\mathbf{r}'}$ we can write the right-hand side alternatively as

$$\nabla \times \boldsymbol{\phi}|_{\mathbf{r}=\mathbf{r}'} = \left(\nabla \times \boldsymbol{\phi}, \delta^{(3)}(\mathbf{r}-\mathbf{r}')\,\overline{\mathfrak{I}}\right)_{\mathbb{R}^3} = \left(\boldsymbol{\phi}, \nabla \times \left[\delta^{(3)}(\mathbf{r}-\mathbf{r}')\,\overline{\mathfrak{I}}\right]\right)_{\mathbb{R}^3} \tag{9.247}$$

on account of (9.212) and (9.243) applied formally. Therefore, (9.246) may be stated in short-hand notation as [22, Chapter 1], [29]

$$\nabla \times \nabla \times \overline{\mathbf{G}}_{HJ} - k^2\overline{\mathbf{G}}_{JH}(\mathbf{r},\mathbf{r}') = \nabla \times [\delta^{(3)}(\mathbf{r}-\mathbf{r}')\,\overline{\mathfrak{I}}] = \nabla\delta^{(3)}(\mathbf{r}-\mathbf{r}') \times \overline{\mathfrak{I}} \tag{9.248}$$

where the last step is a consequence of (H.66) used in a formal fashion. It is worthwhile noticing that by virtue of (H.63) this equation implies

$$\nabla \cdot \overline{\mathbf{G}}_{HJ}(\mathbf{r},\mathbf{r}') = 0 \tag{9.249}$$

i.e., $\overline{\mathbf{G}}_{HJ}(\mathbf{r},\mathbf{r}')$ is a *solenoidal* dyadic field or distribution (cf. Section 8.1), which is consistent with our definition (9.161). Owing to duality, (9.248) also represents the governing equation of $\overline{\mathbf{G}}_{EM}(\mathbf{r},\mathbf{r}')$ in (9.163).

We begin by showing that $\overline{\mathbf{G}}_{HJ}(\mathbf{r},\mathbf{r}')$ expressed in (9.161) solves (9.248) for $\mathbf{r} \neq \mathbf{r}'$, in which case the right-hand side is null and the derivatives exists in the ordinary sense. More precisely, we have

$$\begin{aligned}\nabla \times \nabla \times \overline{\mathbf{G}}_{HJ} - k^2\overline{\mathbf{G}}_{HJ}(\mathbf{r},\mathbf{r}') &= \nabla \times \nabla \times \nabla \times G(R)\overline{\mathfrak{I}} - k^2\nabla \times G(R)\overline{\mathfrak{I}} \\ &= \nabla \times \left[\nabla \times \nabla \times G(R)\overline{\mathfrak{I}} - k^2 G(R)\overline{\mathfrak{I}}\right] \\ &= \nabla \times \left[\nabla\nabla G - \overline{\mathfrak{I}}\nabla^2 G - k^2 G(R)\overline{\mathfrak{I}}\right] = 0\end{aligned} \tag{9.250}$$

where we have invoked (H.68), (H.64), and the fact that the time-harmonic scalar Green function (8.356) satisfies the homogeneous instance of the Helmholtz equation (8.330) for $\mathbf{r} \neq \mathbf{r}'$.

When we include the source point $\mathbf{r}'$, though, we must interpret $\overline{\mathbf{G}}_{HJ}(\mathbf{r},\mathbf{r}')$ as a distribution still given by (9.160) and resort to the weak formulation (9.246) which reads

$$\int_{\mathbb{R}^3} \mathrm{d}V\, \mathfrak{L}\{\boldsymbol{\phi}\} \cdot \nabla \times \frac{\mathrm{e}^{-\mathrm{j}kR}}{4\pi R}\overline{\mathfrak{I}} = \nabla \times \boldsymbol{\phi}|_{\mathbf{r}'} \tag{9.251}$$

and to prove that this relation is indeed true we follow the same approach used for $\overline{\mathbf{G}}_{EJ}(\mathbf{r},\mathbf{r}')$. Specifically, we consider the integral over the spherical shell $V := B(\mathbf{r}',b) \setminus B(\mathbf{r}',a)$ with $b > a$ and, thanks to the dyadic integral theorem (H.109) [5, Formula (A4.75)] with $\mathcal{B}(\mathbf{r}) = \boldsymbol{\phi}(\mathbf{r})$ and $\overline{\mathcal{A}}(\mathbf{r}) = \nabla \times G(R)\overline{\mathfrak{I}}$, we arrive at

$$\begin{aligned}&\int_V \mathrm{d}V \left[\nabla \times \nabla \times \boldsymbol{\phi} - k^2\boldsymbol{\phi}(\mathbf{r})\right] \cdot \nabla \times G(R)\overline{\mathfrak{I}} \\ &\qquad = -\int_{|\mathbf{R}|=a} \mathrm{d}S\, \hat{\mathbf{R}} \cdot \left\{\boldsymbol{\phi}(\mathbf{r}) \times \left[\nabla \times \nabla \times G(R)\overline{\mathfrak{I}}\right] + (\nabla \times \boldsymbol{\phi}) \times \left[\nabla \times G(R)\overline{\mathfrak{I}}\right]\right\}\end{aligned} \tag{9.252}$$

where we notice that

$$\hat{\mathbf{R}} \cdot (\nabla \times \boldsymbol{\phi}) \times \left[\nabla \times G(R)\overline{\mathfrak{I}}\right] = \left[\hat{\mathbf{R}} \times (\nabla \times \boldsymbol{\phi})\right] \times \nabla G \tag{9.253}$$

by virtue of (H.23), (H.66) and (H.39), and also that

$$\begin{aligned}\hat{\mathbf{R}} \cdot \boldsymbol{\phi}(\mathbf{r}) \times \left[\nabla \times \nabla \times G(R)\overline{\mathcal{I}}\right] &= \hat{\mathbf{R}} \cdot \boldsymbol{\phi}(\mathbf{r}) \times \left[\nabla\nabla G - \overline{\mathcal{I}}\nabla^2 G\right] \\ &= \hat{\mathbf{R}} \cdot \boldsymbol{\phi}(\mathbf{r}) \times \left[\nabla\nabla G + k^2 G(R)\overline{\mathcal{I}}\right] \\ &= \hat{\mathbf{R}} \cdot (\nabla \times \boldsymbol{\phi})\nabla G - \hat{\mathbf{R}} \cdot \nabla \times [\boldsymbol{\phi}(\mathbf{r})\nabla G] + k^2 \hat{\mathbf{R}} \times \boldsymbol{\phi}(\mathbf{r})\, G(R)\end{aligned} \tag{9.254}$$

having made use of (H.68), (H.76), (H.39) and the fact that $G(R)$ is a solution to (8.330) for $\mathbf{r} \neq \mathbf{r}'$. The surface integral of the second contribution in the last row vanishes by virtue of the dyadic version of the Stokes theorem (H.107) [5, Formula (A4.70)] applied over the sphere, which is a closed surface. In the limit as $a \to 0^+$ and $b \to +\infty$ we find

$$\begin{aligned}\int_{\mathbb{R}^3} \mathrm{d}V\, \mathfrak{L}\{\boldsymbol{\phi}\} \cdot \nabla \times \frac{\mathrm{e}^{-\mathrm{j}kR}}{4\pi R}\overline{\mathcal{I}} &= -\lim_{a\to 0^+} \int_{|\mathbf{R}|=a} \mathrm{d}S\, \hat{\mathbf{R}} \cdot (\nabla \times \boldsymbol{\phi})\nabla G \\ &\quad - k^2 \lim_{a\to 0^+} \int_{|\mathbf{R}|=a} \mathrm{d}S\, \hat{\mathbf{R}} \times \boldsymbol{\phi}(\mathbf{r})\, G(R) - \lim_{a\to 0^+} \int_{|\mathbf{R}|=a} \mathrm{d}S \left[\hat{\mathbf{R}} \times (\nabla \times \boldsymbol{\phi})\right] \times \nabla G\end{aligned} \tag{9.255}$$

and we examine the three surface integrals separately, though we can omit the details, since the calculations are quite similar to those carried out for $\overline{\mathbf{G}}_{EJ}(\mathbf{r}, \mathbf{r}')$. In practice, we find

$$\begin{aligned}-\lim_{a\to 0^+} \int_{|\mathbf{R}|=a} \mathrm{d}S\, \hat{\mathbf{R}} \cdot (\nabla \times \boldsymbol{\phi})\nabla G &= \lim_{a\to 0^+} (1 + \mathrm{j}\,ka)\mathrm{e}^{-\mathrm{j}\,ka} \nabla \times \boldsymbol{\phi}|_{\mathbf{r}'} \cdot \frac{1}{4\pi} \int_{4\pi} \mathrm{d}\Omega\, \hat{\mathbf{R}}\hat{\mathbf{R}} \\ &= \nabla \times \boldsymbol{\phi}|_{\mathbf{r}'} \cdot \overline{\mathbf{L}}(\mathbf{r}')\end{aligned} \tag{9.256}$$

$$-k^2 \lim_{a\to 0^+} \int_{|\mathbf{R}|=a} \mathrm{d}S\, \hat{\mathbf{R}} \times \boldsymbol{\phi}(\mathbf{r})\, G(R) = 0 \tag{9.257}$$

$$\begin{aligned}&-\lim_{a\to 0^+} \int_{|\mathbf{R}|=a} \mathrm{d}S \left[\hat{\mathbf{R}} \times (\nabla \times \boldsymbol{\phi})\right] \times \nabla G = \\ &= \lim_{a\to 0^+} \left[(1 + \mathrm{j}\,ka)\mathrm{e}^{-\mathrm{j}\,ka} \nabla \times \boldsymbol{\phi}|_{\mathbf{r}'} - (1 + \mathrm{j}\,ka)\mathrm{e}^{-\mathrm{j}\,ka} \nabla \times \boldsymbol{\phi}|_{\mathbf{r}'} \cdot \frac{1}{4\pi} \int_{4\pi} \mathrm{d}\Omega\, \hat{\mathbf{R}}\hat{\mathbf{R}}\right] \\ &= \nabla \times \boldsymbol{\phi}|_{\mathbf{r}'} - \nabla \times \boldsymbol{\phi}|_{\mathbf{r}'} \cdot \overline{\mathbf{L}}(\mathbf{r}')\end{aligned} \tag{9.258}$$

by comparison with (9.232), (9.233) and (9.235). Now, inserting (9.256)-(9.258) into (9.255) yields (9.251), as expected.

9.5 Regions of a localized time-harmonic source

Thanks to formulas (9.158) and (9.160) the problem of computing the electromagnetic field produced by a localized current density $\mathbf{J}(\mathbf{r}')$, $\mathbf{r}' \in V_S$, for observation points away from the source in a homogeneous unbounded medium is reduced to the calculation of suitable domain integrals. The direct

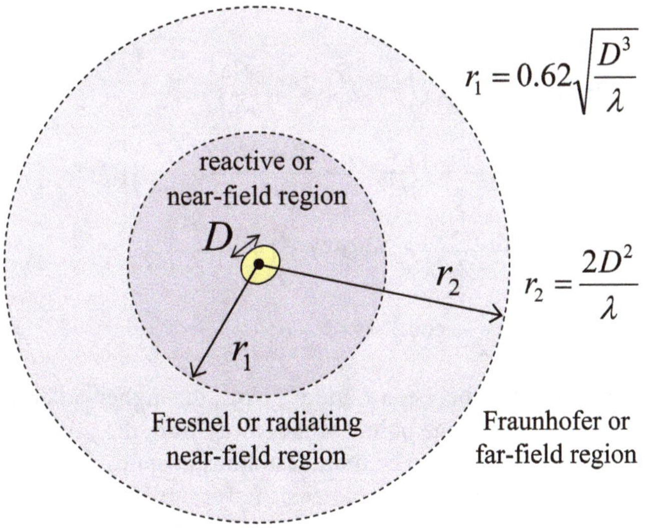

Figure 9.10 Classification of the spatial regions around a localized time-harmonic electromagnetic source according to the distance therefrom.

numerical calculation of said integrals is always possible but in order to gain insight into the properties of the field it is convenient to approximate the relevant dyadic Green functions according to the relative distance between V_S and the observation point $\mathbf{r}$ (Section 9.6).

The explicit expressions (9.159) and (9.161) tell us that $\overline{\mathbf{G}}_{EJ}(\mathbf{r}, \mathbf{r}')$ and $\overline{\mathbf{G}}_{HJ}(\mathbf{r}, \mathbf{r}')$ are given by a combination of terms proportional to negative powers of $R = |\mathbf{r} - \mathbf{r}'|$ as well as the *phase factor* $\exp(-\mathrm{j}kR)$. Thus, on the whole we may expect that higher-order powers of $1/R$ become predominant for small values of R, i.e., 'close' to the source, whereas lower-order powers remain significant and dominate 'far' away from the current. This suggests that we may identify *spherical regions* of space around V_S — loosely speaking, close, intermediate and far — in which the Green functions and hence formulas (9.158) and (9.160) can be conveniently approximated so as to facilitate the calculation of electromagnetic field. For the subsequent discussion we choose the origin O of a system of polar spherical coordinates (r, ϑ, φ) in the vicinity of V_S or, even better, inside V_S without loss of generality. In the technical literature [8, Section 8.2], [5, Section 7.7], [39] the three zones around V_S (Figure 9.10) are termed the reactive or *near-field* or Rayleigh region ($r < r_1$), the *radiating near-field* or Fresnel region ($r_1 < r < r_2$) and the *far-field* or Fraunhofer region ($r > r_2$).

Of course, there is no unique way of defining the radii r_1 and r_2 of the two separation surfaces. The most common approach consists of putting an upper bound to the error committed in the approximation of the phase factor. If we indicate with $\Delta\phi$ the difference between that exact argument of $\exp(-\mathrm{j}kR)$ and the chosen approximation, we require that

$$|\Delta\phi| \leqslant \frac{\pi}{8} \tag{9.259}$$

be true for all observations points in a given region. The maximum phase error $\pi/8$ is adopted because in turn the approximated evaluation of (9.158) and (9.160) produces results affected by a 'distortion' which is deemed acceptable as compared to the 'exact' solution.

To determine the values of the radii r_1 and r_2 on the grounds of the criterion (9.259) we need the asymptotic expansion of R in terms of inverse powers of r for relatively large values of r. This goal is accomplished by factoring out r first and then by applying the binomial expansion [32, Formula 5.3] up to the third order, namely,

$$\begin{aligned}
R = |\mathbf{r} - \mathbf{r}'| &= \left(r^2 - 2\mathbf{r}\cdot\mathbf{r}' + r'^2\right)^{\frac{1}{2}} = r\left(1 - 2\frac{\hat{\mathbf{r}}}{r}\cdot\mathbf{r}' + \frac{r'^2}{r^2}\right)^{\frac{1}{2}} \\
&= r\left[1 - \frac{\hat{\mathbf{r}}}{r}\cdot\mathbf{r}' + \frac{r'^2}{2r^2} - \frac{(\hat{\mathbf{r}}\cdot\mathbf{r}')^2}{2r^2} + \frac{\hat{\mathbf{r}}\cdot\mathbf{r}'}{2r^3}r'^2 + \frac{(\hat{\mathbf{r}}\cdot\mathbf{r}')^3}{2r^3} + o\left(\frac{r'^3}{r^3}\right)\right] \\
&= r - \hat{\mathbf{r}}\cdot\mathbf{r}' + \frac{r'^2 - (\hat{\mathbf{r}}\cdot\mathbf{r}')^2}{2r} + \frac{\hat{\mathbf{r}}\cdot\mathbf{r}'}{2r^2}[r'^2 - (\hat{\mathbf{r}}\cdot\mathbf{r}')^2] + o\left(\frac{r'^3}{r^2}\right) \\
&= r - r'\cos\psi + \frac{r'^2}{2r}\sin^2\psi + \frac{r'^3}{2r^2}\cos\psi\sin^2\psi + o\left(\frac{r'^3}{r^2}\right), \qquad r' \ll r
\end{aligned} \tag{9.260}$$

where $\psi \in [0, \pi]$ denotes the angle between $\mathbf{r}$ and $\mathbf{r}'$. Now, the higher-order terms of this representation grow ever smaller as we examine points farther away from the origin (i.e., the source in our geometrical setup). Evidently, we have to be more fastidious in approximating R while determining the inner boundary (r_1) of the Fresnel region, whereas fewer terms in the asymptotic expansion are required to find the outer boundary (r_2), which is also the beginning of the far-field region.

So, if we retain powers of $1/r$ up to the first order, we may take the phase error as

$$\Delta\phi \approx k\frac{r'^3}{2r^2}\cos\psi\sin^2\psi \tag{9.261}$$

by neglecting higher-order terms. Such error depends on the angle ψ and the source point r', but the range of the latter is bounded by the characteristic size D of V_S. In effect, we may suppose that $r' \lesseqgtr D/2$, if the origin O is set, give or take, in the centroid of V_S. The maximum value of the ψ-dependent part of $\Delta\phi$ can be found by nullifying the derivative with respect to ψ, viz.,

$$(-\sin^2\psi + 2\cos^2\psi)\sin\psi = 0, \qquad \psi \in [0, \pi] \tag{9.262}$$

which is solved formally by

$$\cos\psi_1 = \pm\frac{1}{\sqrt{3}}, \qquad \sin\psi_2 = 0. \tag{9.263}$$

Since the phase error vanishes for $\psi = \psi_2$, then we conclude that the maximum obtains for $\psi = \psi_1$ and hence we may require

$$|\Delta\phi| \leqslant |\Delta\phi|_{\max} \approx \frac{k}{2r^2}\left(\frac{D}{2}\right)^3\frac{2}{3\sqrt{3}} \leqslant \frac{\pi}{8} \tag{9.264}$$

which, when solved for r, provides

$$r \geqslant \sqrt{\frac{2}{3\sqrt{3}}}\sqrt{\frac{D^3}{\lambda}} = 0.62\sqrt{\frac{D^3}{\lambda}} = r_1 \tag{9.265}$$

as the inner radius of the Fresnel region.

The coarser approximation $R \approx r - \hat{\mathbf{r}}\cdot\mathbf{r}'$ has an interesting geometrical interpretation [39]. As Figure 9.11 suggests, it amounts to assuming that the vector $\mathbf{r}$ is nearly parallel to $\mathbf{R} = \mathbf{r} - \mathbf{r}'$. Therefore, we retrieve the magnitude of $\mathbf{R}$ by subtracting the projection of $\mathbf{r}'$ onto $\mathbf{r}$ (i.e., $\mathbf{r}'\cdot\hat{\mathbf{r}}$) from r. Besides, if we represent R by retaining zero-order terms only, the phase error reads

$$\Delta\phi \approx k\frac{r'^2}{2r}\sin^2\psi \tag{9.266}$$

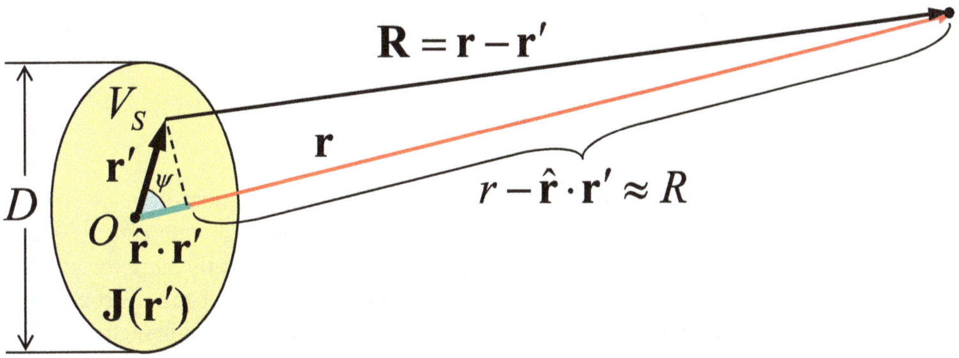

Figure 9.11 *Geometrical meaning of the far-field approximation.*

while neglecting all higher-order contributions. Evidently, this error is maximum for $\psi = \pi/2$ and, if we stick to the hypothesis that the origin O is the same as the centroid of V_S, for $r' \lessapprox D/2$, viz.

$$|\Delta\phi| \leqslant |\Delta\phi|_{\max} \approx \frac{k}{2r}\left(\frac{D}{2}\right)^2 \leqslant \frac{\pi}{8} \tag{9.267}$$

which yields

$$r \geqslant \frac{2D^2}{\lambda} = r_2 \tag{9.268}$$

for the minimum radius of the far-field zone.

In computing r_1 and r_2 above we have tacitly used the identity $k = 2\pi/\lambda$ which may appear to be based on (1.248) and (7.31). This latter one, though, was derived while looking for the spatial period of periodic plane waves. A time-harmonic current density of finite extent may generate electromagnetic waves which, though periodic in time, have arbitrary aperiodic spatial dependence as prescribed by (9.23) and (9.24). We can, however, introduce the wavelength through (7.31), though λ does not have the physical interpretation of the spatial period of the wave. Alternatively, since the phase factor is indeed a periodic function of R, we may define λ by looking for the minimum distance between two spheres around V_S such that the phase factor takes on the same value. In symbols, this reads

$$k(R + \lambda) = kR + 2\pi \tag{9.269}$$

which yields

$$k = \frac{2\pi}{\lambda} \tag{9.270}$$

in agreement with the previous assumption.

In summary, we have shown that the radii of the boundaries of the regions around the source are related to the wavelength λ in the background homogeneous lossless medium and, as is logical, to the characteristic size of V_S. Thus, although the classification is generally valid, the extent of the intermediate regions, that is to say, the position of the separation boundaries is not absolute as it may vary with the dimension of the source for a given frequency f and may change with f for a given source (recall that $\lambda = c/f$ with $c = 1/(\varepsilon\mu)^{1/2}$). It is worthwhile mentioning that the separation spheres of radii r_1 and r_2 are not physical interfaces between two media endowed with different constitutive parameters (cf. Section 1.7), therefore the fields as well as the potentials do not suffer any jump or discontinuity.

We go on to examine the properties of the fields in the three zones.

Reactive or near-field

This is the region of space closest to the source — hence the name near-field — and bounded by the sphere of radius r_1. Although the charges confined in V_S are in a state of accelerated periodic motion, the fields generated by the source happen to be predominantly *static* or *stationary* in character. This can be ascertained by taking a look at the explicit form of the dyadic Green functions (9.159) and (9.161) which in turn enter the calculation of the electromagnetic field as given by (9.158) and (9.160). For observation points very close to V_S the spatial variations of $\overline{\mathbf{G}}_{EJ}(\mathbf{r},\mathbf{r}')$ and $\overline{\mathbf{G}}_{HJ}(\mathbf{r},\mathbf{r}')$ are essentially determined by terms proportional to $1/|\mathbf{r}-\mathbf{r}'|^3$ and $1/|\mathbf{r}-\mathbf{r}'|^2$. Furthermore, the phase factor $\exp(-\mathrm{j}kR)$ can be safely approximated by 1.

In fact, direct comparison shows that the dominant term of $\overline{\mathbf{G}}_{HJ}(\mathbf{r},\mathbf{r}')$ is approximately the same as the kernel of the volume integral in the representation of the stationary magnetic field (5.58). The dominant terms of $\overline{\mathbf{G}}_{EJ}(\mathbf{r},\mathbf{r}')$ suggest that the electric field is essentially static as if it were due to distributions of fixed charges and static dipoles. Correspondingly, the average electromagnetic energy $\langle W_e\rangle + \langle W_h\rangle$ (Section 1.10.2) remains stored in the field around the source. There may be a periodic transformation of electric energy into magnetic one and viceversa but, barring any loss mechanism, the energy stays essentially constant. This property justifies the name reactive for the near-field zone.

Fresnel or radiating near field

This the intermediate region bounded by the spheres of radii r_1 and r_2. However, depending on the particular combination of wavelength and characteristic size, the Fresnel zone may not exist, if $r_1 > r_2$. This may happen if D is very small with respect to the wavelength.

The radiation fields [see (1.252) and (1.253)] predominate but the angular distribution of radiated power density (1.306) still depends upon the distance r. In other words, in the expression of $\langle \mathrm{d}P_F/dS\rangle$, which is a function of (r,ϑ,φ), it is not possible to separate the dependence on (ϑ,φ) from that on the radial distance from the source.

Fraunhofer or far-field

In this zone the contributions proportional to $1/R^3$ and $1/R^2$ are utterly negligible, the radiation fields dominate, and the angular distribution of power is independent of the distance r. This means that in the expression of $\langle \mathrm{d}P_F/dS\rangle$ we are able to separate the dependence on (ϑ,φ) from that on r. In fact, we shall show in Section 9.6 that the dominant field components can be written as

$$F(\mathbf{r};\omega) \approx g(\theta,\phi)\frac{\mathrm{e}^{-\mathrm{j}kr}}{4\pi r}, \qquad r \geqslant \frac{2D^2}{\lambda} \tag{9.271}$$

and also that the fields are essentially transverse (perpendicular) to $\hat{\mathbf{r}}$. That is, the radial components of the fields fall off faster than the others as can be surmised from (9.159) and (9.161). This property is in agreement with the asymptotic behavior (6.173) and (6.174) implied by the Silver-Müller conditions (6.171) and (6.172).

Example 9.3 (Field generated by an electric Hertzian dipole)

The Hertzian dipole is an elementary singular current concentrated in a point [1,27,29,39,40], [41, Chapter 7], [23, Section 5.3.A], [12, Sections XII.3-XII.4] and hence it is impracticable as a real device for the generation of electromagnetic waves. Then again, the relevant fields can be computed

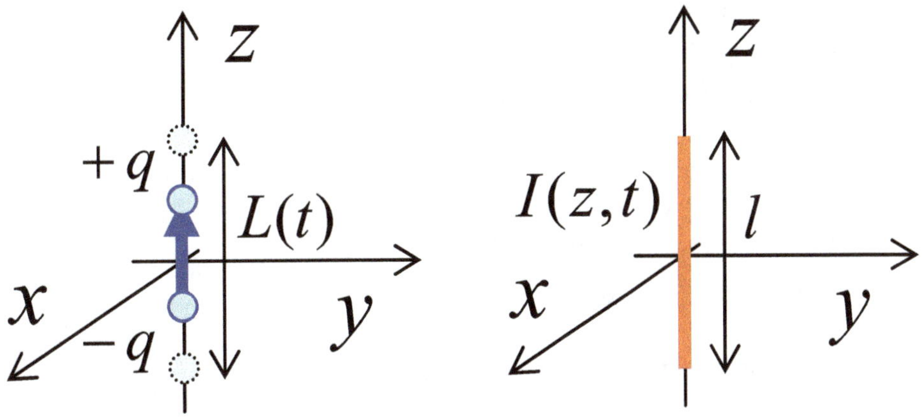

(a) *two oscillating point charges* (b) *line current*

Figure 9.12 Alternative ways of defining an electric Hertzian dipole.

in closed form and this allows us to take a look at the regions of the source explicitly so as to provide support to the previous discussion. However, for a Hertzian dipole it is not possible to determine the radii r_1 and r_2 with (9.265) and (9.268) because the characteristic size of the source is theoretically zero. Thus, we shall use the wavelength λ in the background medium as a reference.

Our first task is to determine an expression for the current density $\mathbf{J}(\mathbf{r};\omega)$ of the dipole. The name itself hints at some connection with the corresponding electrostatic dipole we discussed in Example 2.5. In fact, we can picture the Hertzian dipole as a pair of oscillating charges with opposite sign and placed an infinitesimal distance apart, but we could also start from a distribution of charge with finite non-zero spatial extent and then take the limit for vanishing size. Since this is precisely the idea underlying the definition of Dirac delta distribution that is outlined in Appendix C, we skip this step and suppose instead that two point charges $q > 0$ and $-q$ are located along the z-axis at $z = \pm L(t)/2$, as is sketched in Figure 9.12a.

To express the corresponding charge density mathematically we resort to the three-dimensional Dirac delta and write

$$\varrho(\mathbf{r},t) = q\delta^{(3)}\left(\mathbf{r} - \frac{L}{2}\hat{\mathbf{z}}\right) - q\delta^{(3)}\left(\mathbf{r} + \frac{L}{2}\hat{\mathbf{z}}\right) \tag{9.272}$$

where the distance $L(t) = l\cos(\omega t)$, $l > 0$, changes periodically in time with angular frequency ω. Since the charges are bound to move on a straight line parallel to $\hat{\mathbf{z}}$, we conclude that the current density $\mathfrak{J}(\mathbf{r},t) = \mathfrak{J}_z(\mathbf{r},t)\hat{\mathbf{z}}$. For this reason the continuity equation (1.46) — extended to distributions — reads

$$\frac{\partial\varrho}{\partial t} = -\frac{\partial\mathfrak{J}_z}{\partial z} \tag{9.273}$$

and can be solved to find $\mathfrak{J}_z(\mathbf{r},t)$, inasmuch as $\varrho(\mathbf{r},t)$ is given. For the calculation of the time derivative of $\varrho(\mathbf{r},t)$ we recall the Cartesian representation (C.25) of the three-dimensional delta. In symbols, we have

$$\begin{aligned}\frac{\partial\varrho}{\partial t} &= q\delta(x)\delta(y)\frac{\partial}{\partial t}\delta\left(z - \frac{L}{2}\right) - q\delta(x)\delta(y)\frac{\partial}{\partial t}\delta\left(z + \frac{L}{2}\right)\\ &= \omega q\frac{l}{2}\left[\delta(x)\delta(y)\delta'\left(z - \frac{L}{2}\right) + \delta(x)\delta(y)\delta'\left(z + \frac{L}{2}\right)\right]\sin(\omega t)\end{aligned}$$

$$= \omega q \frac{l}{2}\left[\delta(x)\delta(y)\frac{\partial}{\partial z}\delta\left(z-\frac{L}{2}\right)+\delta(x)\delta(y)\frac{\partial}{\partial z}\delta\left(z+\frac{L}{2}\right)\right]\sin(\omega t)$$

$$= \omega q \frac{l}{2}\frac{\partial}{\partial z}\left[\delta^{(3)}\left(\mathbf{r}-\frac{L}{2}\hat{\mathbf{z}}\right)+\delta^{(3)}\left(\mathbf{r}+\frac{L}{2}\hat{\mathbf{z}}\right)\right]\sin(\omega t) = -\frac{\partial \mathfrak{J}_z}{\partial z} \tag{9.274}$$

where the notation $\delta'(\bullet)$ signifies derivative with respect to the argument, and the differentiation of the δ's has a meaning in the weak sense defined in (C.36).

The continuity equation is certainly satisfied if we take

$$\mathfrak{J}_z(\mathbf{r},t) = -\omega q \frac{l}{2}\left[\delta^{(3)}\left(\mathbf{r}-\frac{L}{2}\hat{\mathbf{z}}\right)+\delta^{(3)}\left(\mathbf{r}+\frac{L}{2}\hat{\mathbf{z}}\right)\right]\sin(\omega t) \tag{9.275}$$

and in the limit as the maximum distance $l \to 0^+$ we get

$$\mathfrak{J}_z(\mathbf{r},t) = -\omega p\delta^{(3)}(\mathbf{r})\sin(\omega t) = \omega p\delta^{(3)}(\mathbf{r})\cos\left(\omega t+\frac{\pi}{2}\right) \tag{9.276}$$

where p is the same as the moment of an electrostatic dipole introduced in Example 2.5. In the limiting process we demand that q grow infinitely large in order for the product ql to remain finite and approach p. At this stage the dependence on time is factored out and hence we may pass over to the complex (phasor) notation for time-harmonic fields by letting (Section 1.5)

$$\mathfrak{J}_z(\mathbf{r},t) = \mathrm{Re}\{J_z(\mathbf{r};\omega)e^{\mathrm{j}\omega t}\} \tag{9.277}$$

which by comparison with (9.276) yields

$$J_z(\mathbf{r};\omega) = \mathrm{j}\,\omega p\delta^{(3)}(\mathbf{r}) \tag{9.278}$$

and finally

$$\mathbf{J}(\mathbf{r};\omega) = \mathrm{j}\,\omega \mathbf{p}\delta^{(3)}(\mathbf{r}) \tag{9.279}$$

with $\mathbf{p} = p\hat{\mathbf{z}}$. The complex vector $\mathrm{j}\,\omega\mathbf{p}$ may be regarded as the *moment* (physical dimension: Am) of the electric Hertzian dipole.

Alternatively, we can start with a periodic z-aligned line current (Figure 9.12b)

$$I(z,t) = |I_0|\left[U\left(z+\frac{l}{2}\right)-U\left(z-\frac{l}{2}\right)\right]\cos\left(\omega t+\frac{\pi}{2}\right)$$

$$= |I_0|l\left\{\frac{1}{l}\left[U\left(z+\frac{l}{2}\right)-U\left(z-\frac{l}{2}\right)\right]\right\}\cos\left(\omega t+\frac{\pi}{2}\right) \tag{9.280}$$

where $U(\bullet)$ denotes the one-dimensional step function. The chosen time-dependence for $I(z,t)$ is motivated by (9.276). The function between braces is non-zero only for $|z| \leqslant l/2$ and becomes steeper and steeper around $z = 0$ as the length l decreases. Still, the integral along z is independent of l, namely,

$$\int_{\mathbb{R}} \mathrm{d}z\,\frac{1}{l}\left[U\left(z+\frac{l}{2}\right)-U\left(z-\frac{l}{2}\right)\right] = 1 \tag{9.281}$$

hence we may write

$$\lim_{l\to 0^+}\int_{\mathbb{R}} \mathrm{d}z\,\frac{1}{l}\left[U\left(z+\frac{l}{2}\right)-U\left(z-\frac{l}{2}\right)\right] = \int_{\mathbb{R}} \mathrm{d}z\,\delta(z) = 1 \tag{9.282}$$

similarly to (C.3). In shorthand notation this means

$$I(z,t) = |I_0| l\delta(z)\cos\left(\omega t + \frac{\pi}{2}\right) \tag{9.283}$$

provided we let $|I_0|$ approach infinity in such a way that the product $|I_0|l$ remains finite. Therefore, the notation $|I_0|l$ means the limiting value of the product as $l \to 0^+$. (Notice that we employ the same symbol $I(z,t)$ to indicate both the function and the distribution.) Since the current is confined to the z-axis, we may write the current density of the Hertzian dipole as

$$\mathcal{J}(\mathbf{r},t) = |I_0| l\delta(x)\delta(y)\delta(z)\cos\left(\omega t + \frac{\pi}{2}\right)\hat{\mathbf{z}} = |I_0| l\delta^{(3)}(\mathbf{r})\cos\left(\omega t + \frac{\pi}{2}\right)\hat{\mathbf{z}} \tag{9.284}$$

and in complex notation (see Table 1.1)

$$\mathbf{J}(\mathbf{r};\omega) = j|I_0| l\delta^{(3)}(\mathbf{r})\,\hat{\mathbf{z}} = I_0 l\delta^{(3)}(\mathbf{r})\,\hat{\mathbf{z}} \tag{9.285}$$

where $I_0 \in \mathbb{C}$. Now, comparing (9.285) with (9.279) yields

$$I_0 l\hat{\mathbf{z}} = \mathrm{j}\,\omega\mathbf{p} \qquad \Longrightarrow \qquad \mathbf{p} = p\hat{\mathbf{z}} = \frac{I_0 l}{\mathrm{j}\,\omega}\hat{\mathbf{z}} \tag{9.286}$$

as an alternative representation for the static dipole moment [1]. We remark that $I_0 l$ indicates the limiting value of the product as $l \to 0^+$. However, in the following it will be convenient to assume that l is very small but not zero and that $|I_0|$ is large though not infinite.

Having found an expression for the electric current density we go on to compute the electric and magnetic fields produced by a Hertzian dipole located in the origin of a space filled with a homogeneous lossless isotropic medium. To this purpose we observe that the magnetic vector potential $\mathbf{A}_E(\mathbf{r};\omega)$ solves the Helmholtz equation

$$\nabla^2\mathbf{A}_E(\mathbf{r};\omega) + k^2\mathbf{A}_E(\mathbf{r};\omega) = -\mathrm{j}\,\omega\mu\mathbf{p}\delta^{(3)}(\mathbf{r}), \qquad \mathbf{r} \in \mathbb{R}^3 \tag{9.287}$$

that is, the time-harmonic counterpart of (8.123) though extended to distributions on account of (9.279). Evidently, each Cartesian component of $\mathbf{A}_E(\mathbf{r};\omega)$ satisfies an equation of the type (8.330). In Section 8.7.1 we proved that (8.330) is solved by the Green function (8.356), thus straightforward comparison for $\mathbf{r}' = 0$ along with the linearity of the Helmholtz operator provides

$$\mathbf{A}_E(\mathbf{r};\omega) = \mathrm{j}\,\omega\mu\frac{\mathrm{e}^{-\mathrm{j}kr}}{4\pi r}\mathbf{p} \tag{9.288}$$

as the solution to (9.287). Another way to obtain $\mathbf{A}_E(\mathbf{r};\omega)$ is to interpret the solution (9.24) as the *convolution product* between two distributions, namely, the scalar Green function (8.335) and the current (9.279). Since the delta distribution plays the role of *identity* or *neutral* element of the convolution product, we find (9.288) right away.

The magnetic vector potential is a function of the distance r from the dipole and is everywhere aligned with the moment $\mathrm{j}\,\omega\mathbf{p}$. Spherical symmetry is consistent with the dipole being a point current localized in the origin. Moreover, according to (9.288) we can interpret the three-dimensional Green function (8.335) as the component of the magnetic vector potential generated by and parallel to a Hertzian dipole of moment $\mathrm{j}\,\omega p$ numerically equal to $1/\mu$.

The calculation of $\mathbf{E}(\mathbf{r};\omega)$ and $\mathbf{H}(\mathbf{r};\omega)$ is just an application of (9.154) and (9.155) without the contribution of the potentials due to magnetic sources and $\mathbf{A}_E(\mathbf{r};\omega)$ given by (9.288), viz.,

$$\mathbf{E}(\mathbf{r};\omega) = -\mathrm{j}\,\omega\mu\left(\overline{\mathbf{J}} + \frac{\nabla\nabla}{k^2}\right)\frac{\mathrm{e}^{-\mathrm{j}kr}}{4\pi r}\cdot\mathrm{j}\,\omega\mathbf{p} = \omega^2\mu\overline{\mathbf{G}}_{EJ}(\mathbf{r},0;\omega)\cdot\mathbf{p} \tag{9.289}$$

$$\mathbf{H}(\mathbf{r};\omega) = \mathrm{j}\,\omega\nabla\times\frac{\mathrm{e}^{-\mathrm{j}kr}}{4\pi r}\mathbf{p} = \mathrm{j}\,\omega\nabla\frac{\mathrm{e}^{-\mathrm{j}kr}}{4\pi r}\times\mathbf{p} = \mathrm{j}\,\omega\overline{\mathbf{G}}_{HJ}(\mathbf{r},0;\omega)\cdot\mathbf{p} \tag{9.290}$$

where we have taken into account that **p** is a constant vector. Alternatively, (9.289) and (9.290) follow from (9.158) and (9.160) by interpreting the integral as the convolution product of the relevant dyadic Green function and the current. Since in view of (E.11) we can write (9.289) as

$$\mathbf{E}(\mathbf{r};\omega) = \omega^2\mu\mathbf{G}'_{EJx}(\mathbf{r},0;\omega)p_x + \omega^2\mu\mathbf{G}'_{EJy}(\mathbf{r},0;\omega)p_y + \omega^2\mu\mathbf{G}'_{EJz}(\mathbf{r},0;\omega)p_z \tag{9.291}$$

we see that in a system of Cartesian coordinates the vector $\mathbf{G}'_{EJx}(\mathbf{r},0;\omega)$ represents the electric field produced by a Hertzian dipole with moment $\mathrm{j}\,\omega p_x$ numerically equal to $\mathrm{j}/(\omega\mu)$. Similar considerations hold for the vectors $\mathbf{G}'_{EJy}(\mathbf{r},0;\omega)$ and $\mathbf{G}'_{EJz}(\mathbf{r},0;\omega)$.

Finally, by carrying out the required derivatives of the scalar Green function and assuming that the dipole is aligned with the z-axis as in Figure 9.12b we arrive at

$$\begin{aligned}\mathbf{E}(\mathbf{r};\omega) = \frac{I_0 l}{4\pi}Zk^2\left[\frac{\mathrm{j}}{kr} + \frac{1}{(kr)^2} - \frac{\mathrm{j}}{(kr)^3}\right]\mathrm{e}^{-\mathrm{j}kr}\sin\vartheta\hat{\boldsymbol{\vartheta}} \\ + \frac{I_0 l}{4\pi}Zk^2\left[\frac{2}{(kr)^2} - \frac{2\mathrm{j}}{(kr)^3}\right]\mathrm{e}^{-\mathrm{j}kr}\cos\vartheta\hat{\mathbf{r}}\end{aligned} \tag{9.292}$$

$$\mathbf{H}(\mathbf{r};\omega) = \frac{I_0 l}{4\pi}k^2\left[\frac{\mathrm{j}}{kr} + \frac{1}{(kr)^2}\right]\mathrm{e}^{-\mathrm{j}kr}\sin\vartheta\hat{\boldsymbol{\varphi}} \tag{9.293}$$

where Z is the intrinsic impedance (1.358) of the background medium, and we have used (9.286) to eliminate **p**.

Now, we see that for $r < \lambda$ the spatial variation of the field is dominated by terms proportional to $1/r^2$ and $1/r^3$ and the phase factor $\mathrm{e}^{-\mathrm{j}kr}$ is practically unitary, which is typical of the near-field and the Fresnel zones, although the latter are not neatly separated. The general form (9.292) and (9.293) can be approximated as

$$\mathbf{E}(\mathbf{r};\omega) \approx -\mathrm{j}Z\frac{I_0 l\mathrm{e}^{-\mathrm{j}kr}}{4\pi kr^3}(2\hat{\mathbf{r}}\cos\vartheta + \hat{\boldsymbol{\vartheta}}\sin\vartheta) \tag{9.294}$$

$$\mathbf{H}(\mathbf{r};\omega) \approx \frac{I_0 l\mathrm{e}^{-\mathrm{j}kr}}{4\pi r^2}\sin\vartheta\hat{\boldsymbol{\varphi}} \tag{9.295}$$

which clearly highlight the static and stationary character of the field. Actually, there is a striking similarity between (9.294) and the electric field (2.43) of a truly static dipole, provided we notice that

$$-\mathrm{j}\frac{ZI_0 l}{k} = \frac{p}{\varepsilon} \tag{9.296}$$

by virtue of (1.248), (9.286) and the definition of intrinsic impedance (1.358). Thanks to this substitution we see that (9.294) reduces exactly to (2.43) in the limit as the oscillation frequency ω vanishes. Further, the static magnetic field of a stationary line current can be obtained from formula (5.134). When a current I_0 has length l, is placed around the origin and is aligned with the z-axis, the resulting magnetic field is the same as that obtained from (9.295) when $k \to 0$.

The far-field zone may be identified for observation points $r \gg \lambda$ in which case the dominant terms read

$$\mathbf{E}(\mathbf{r};\omega) \approx \mathrm{j}kZ\frac{\mathrm{e}^{-\mathrm{j}kr}}{4\pi r}I_0 l\sin\vartheta\hat{\boldsymbol{\vartheta}} = -\mathrm{j}\omega\mu\frac{\mathrm{e}^{-\mathrm{j}kr}}{4\pi r}(\hat{\boldsymbol{\vartheta}}\hat{\boldsymbol{\vartheta}} + \hat{\boldsymbol{\varphi}}\hat{\boldsymbol{\varphi}})\cdot\mathrm{j}\omega\mathbf{p} \tag{9.297}$$

$$\mathbf{H}(\mathbf{r};\omega) \approx \mathrm{j}k\frac{\mathrm{e}^{-\mathrm{j}kr}}{4\pi r}I_0 l\sin\vartheta\hat{\boldsymbol{\varphi}} = -\mathrm{j}k\frac{\mathrm{e}^{-\mathrm{j}kr}}{4\pi r}(\hat{\boldsymbol{\varphi}}\hat{\boldsymbol{\vartheta}} - \hat{\boldsymbol{\vartheta}}\hat{\boldsymbol{\varphi}})\cdot\mathrm{j}\omega\mathbf{p} \tag{9.298}$$

where the expressions in the rightmost-hand side are valid for a Hertzian dipole with moment $\mathbf{p}$ arbitrarily oriented.

It is instructive to compute the complex Poynting vector (1.304) and the power flow (1.303) for a z-directed Hertzian dipole. From (9.292) and (9.293) we have

$$\begin{aligned}\mathbf{S}(\mathbf{r};\omega) &= \frac{1}{2}\left(E_\vartheta H_\varphi^* \hat{\mathbf{r}} - E_r H_\varphi^* \hat{\boldsymbol{\vartheta}}\right) \\ &= \frac{Z|I_0|^2 l^2}{2(4\pi)^2}\left[\sin^2\vartheta\left(\frac{k^2}{r^2} - \frac{\mathrm{j}}{kr^5}\right)\hat{\mathbf{r}} + \sin\vartheta\cos\vartheta\left(\frac{2\mathrm{j}}{kr^5} + \frac{2\mathrm{j}k}{r^3}\right)\hat{\boldsymbol{\vartheta}}\right]\end{aligned} \tag{9.299}$$

whence we observe that the Poynting vector possesses a non-zero *imaginary* component along $\hat{\boldsymbol{\vartheta}}$ which is significant close to the dipole but falls off very rapidly away from the source. The only real component of $\mathbf{S}(\mathbf{r};\omega)$ — the one which can contribute to the average power efflux $\langle P_F \rangle$ — is radially directed. (We remind that both Z and k are *real* because we have assumed that the homogeneous background medium is lossless.) Indeed, by invoking (1.306) we can write the average surface density of power flowing through the boundary of a domain V which contains the dipole as

$$\left\langle \frac{\mathrm{d}P_F}{\mathrm{d}S} \right\rangle = \frac{Z|I_0|^2 l^2}{2(4\pi)^2}\frac{k^2}{r^2}\sin^2\vartheta\,\hat{\mathbf{n}}\cdot\hat{\mathbf{r}} \tag{9.300}$$

where $\hat{\mathbf{n}}$ is the outward unit normal over ∂V. We emphasize that (9.300) is *exact* and valid regardless of the location of ∂V with respect to the dipole. But then, in light of the dependence on the inverse square of the distance from the dipole we may conclude that, in effect, only the radiation fields contribute to the power flow towards infinity.

To check this out, we place the boundary of V in the Fraunhofer region of the dipole, so that we are entitled to apply (9.297) and (9.298) to compute the Poynting vector. The resulting expression is approximated, granted, but when we use it in tandem with (1.306) to get the average surface density of power flow, we arrive again at (9.300)! This analysis confirms what we anticipated in Section 1.10.2, namely, that the radiated power is solely due to the acceleration fields, which represent the only significant contribution in the far-field zone. If we choose $V \equiv B(0,a)$ for simplicity to compute $\langle P_F \rangle$ — because the average radiated power is independent of the specific choice for V, and this property in turn is a consequence of (1.316) which shows that $\mathrm{Re}\{\mathbf{S}\}$ is solenoidal in source-free lossless regions — we have $\hat{\mathbf{n}}\cdot\hat{\mathbf{r}} = 1$ and

$$\begin{aligned}\langle P_F \rangle &= \mathrm{Re}\int_{\partial B} \mathrm{d}S\,\hat{\mathbf{r}}\cdot\mathbf{S}(\mathbf{r};\omega) = \frac{Z}{2}k^2|I_0|^2 l^2 \int_{\partial B}\mathrm{d}S\,\frac{\sin^2\vartheta}{(4\pi a)^2} \\ &= Z|I_0|^2\frac{\pi l^2}{4\lambda^2}\int_0^\pi \mathrm{d}\vartheta\,\sin^3\vartheta = \frac{\pi}{3}Z\frac{l^2}{\lambda^2}|I_0|^2\end{aligned} \tag{9.301}$$

on account of (9.270).

Then again, according to the Poynting theorem for time-harmonic fields (1.314), since the background medium is lossless, we should arrive at the same result by computing the average power $\langle P_S \rangle$ delivered by the source. In the present case, by virtue of (9.279) and (9.286) we have

$$\langle P_F \rangle = \langle P_S \rangle := -\frac{1}{2}\mathrm{Re}\int_{V_S}\mathrm{d}V\,\mathbf{E}^*(\mathbf{r})\cdot\mathbf{J}(\mathbf{r}) = -\frac{1}{2}\mathrm{Re}\int_{\mathbb{R}^3}\mathrm{d}V\,\mathbf{E}^*(\mathbf{r})\cdot\mathrm{j}\,\omega\,\mathbf{p}\,\delta^{(3)}(\mathbf{r})$$

$$= -\frac{1}{2}\mathrm{Re}\{\mathrm{j}\,\omega\mathbf{p}\cdot\mathbf{E}^*(0)\} = -\frac{1}{2}\mathrm{Re}\{I_0 l E_z^*(0)\} \tag{9.302}$$

having interpreted the volume integral in a distributional sense in light of (C.19). Admittedly, one may argue that (1.314) was derived for functions while excluding the occurrence of singular sources, a case in point being precisely the Hertzian dipole. And hence, why should we trust (9.302)?

Well, we can convince ourselves that (9.302) is valid by re-deriving it on physical grounds. We observe that the radiated power $\langle P_F\rangle$ must be equal to the average of the time rate of the work done by the *mechanical forces* [42, Chapters 13-14] which keep the charges moving back and forth along the z-axis (Figure 9.12a). Since at any time such forces must oppose the electromagnetic ones given by (1.4), for a maximum finite distance l of the two charges the instantaneous power delivered by the source reads [see (1.7)]

$$P_S(t;l) = \underbrace{\mathbf{v}_+(t)\cdot\left[-q\mathcal{E}\left(0,0,\frac{l}{2}\cos(\omega t),t\right)\right]}_{\text{time rate of work done on } q>0} + \underbrace{\mathbf{v}_-(t)\cdot\left[q\mathcal{E}\left(0,0,-\frac{l}{2}\cos(\omega t),t\right)\right]}_{\text{time rate of work done on } -q} \tag{9.303}$$

where $\mathbf{v}_+(t)$ and $\mathbf{v}_-(t) = -\mathbf{v}_+(t)$ denote the velocity of the positive and the negative charge. The electric field in question is the one generated by the charges and evaluated — in Cartesian coordinates — at the respective positions thereof. The velocities are the time derivative of the charge positions, i.e.,

$$\mathbf{v}_+(t) = \frac{\mathrm{d}}{\mathrm{d}t}\frac{l}{2}\cos(\omega t)\hat{\mathbf{z}} = -\frac{l}{2}\omega\sin(\omega t)\hat{\mathbf{z}} = -\mathbf{v}_-(t) \tag{9.304}$$

whereby in the limit as $l \to 0^+$ (9.303) yields

$$P_S(t) = \omega\mathbf{p}\sin(\omega t)\cdot\mathcal{E}(0,t) \tag{9.305}$$

thanks to the same hypotheses that underlie (9.276). Now, we know from previous calculations that $\mathcal{E}(0,t)$ is time-harmonic, so it can be written as

$$\mathcal{E}(0,t) = \mathrm{Re}\{\mathbf{E}(0)\mathrm{e}^{\mathrm{j}\,\omega t}\} \tag{9.306}$$

whence (9.305) turns out to be the dot-product of two time-harmonic fields. Thus, the average over a period is given in terms of phasors by (1.301), viz.,

$$\langle P_S\rangle = \frac{1}{2}\mathrm{Re}\{-\mathrm{j}\,\omega\mathbf{p}\cdot\mathbf{E}^*(0)\} = -\frac{1}{2}\mathrm{Re}\{I_0 l E_z^*(0)\} \tag{9.307}$$

having observed, with the aid of Table 1.1, that the phasor of $\omega\mathbf{p}\sin(\omega t)$ is $-\mathrm{j}\,\omega\mathbf{p}$. This expression coincides with the rightmost-hand side of (9.302), and this answers our question.

Nonetheless, at first blush we may be taken aback by formula (9.302) because the electric field (9.292) is evidently singular at the position of the dipole. However, detailed calculations show that the quantity in the rightmost member is indeed finite. From (9.292) it follows that the z-component of the electric field reads

$$\begin{aligned} E_z(\mathbf{r}) &= \frac{I_0 l}{4\pi}Zk^2\left\{\left[\frac{2}{(kr)^2} - \frac{2\mathrm{j}}{(kr)^3}\right]\cos^2\vartheta - \left[\frac{\mathrm{j}}{kr} + \frac{1}{(kr)^2} - \frac{\mathrm{j}}{(kr)^3}\right]\sin^2\vartheta\right\}\mathrm{e}^{-\mathrm{j}kr} \\ &= \frac{I_0 l}{4\pi}Zk^2\left\{2(kr-\mathrm{j})\cos^2\vartheta - \left[\mathrm{j}(kr)^2 + kr - \mathrm{j}\right]\sin^2\vartheta\right\}\frac{\mathrm{e}^{-\mathrm{j}kr}}{(kr)^3} \end{aligned} \tag{9.308}$$

and we consider the asymptotic expansion (truncated Taylor series) of $\mathrm{e}^{-\mathrm{j}kr}$ around the origin [43, Formula 4.2.1]

$$\mathrm{e}^{-\mathrm{j}kr} = 1 - \mathrm{j}\,kr - \frac{(kr)^2}{2} - \mathrm{j}\frac{(kr)^3}{6} + o\left((kr)^3\right), \qquad r \ll \lambda \tag{9.309}$$

since likewise (9.308) involves a rational function of powers of kr up to the third order. By inserting (9.309) into (9.308) and performing a little algebra we find

$$E_z(\mathbf{r}) = -\frac{I_0 l}{4\pi} Zk^2 \left\{ \frac{2}{3} + \mathrm{j}\left[\frac{1}{kr} + \frac{2}{(kr)^3}\right]\cos^2\vartheta + \frac{\mathrm{j}}{2}\left[\frac{1}{kr} - \frac{2}{(kr)^3}\right]\sin^2\vartheta + o(1) \right\}, \qquad r \ll \lambda \tag{9.310}$$

where $o(1)$ represents a complex function which for $r \to 0^+$ falls off at least as fast as kr. Since this last expression is undefined for $r = 0$, to compute $\langle P_S \rangle$ we use a limiting procedure, namely,

$$\langle P_S \rangle = -\lim_{r\to 0^+} \frac{1}{2}\mathrm{Re}\{I_0 l E_z^*(\mathbf{r})\} = \lim_{r\to 0^+} \frac{1}{2}\mathrm{Re}\left\{ \frac{|I_0|^2 l^2}{4\pi} Zk^2 \frac{2}{3} + o(1) \right\} = \frac{\pi}{3} Z \frac{l^2}{\lambda^2} |I_0|^2 \tag{9.311}$$

which coincides with (9.301) as expected. The result follows inasmuch as the troublesome singular terms of $E_z(\mathbf{r})$ are filtered out as we take the real part of $I_0 l E_z^*(\mathbf{r})$.

If we regard the Hertzian dipole as single-port lumped electrical device characterized by a resistance R_{rad} (physical dimension: Ω), then analogy with circuit theory [44, 45] prompts us to write the average radiated power as

$$\langle P_F \rangle = \frac{1}{2} R_{\mathrm{rad}} |I_0|^2 \tag{9.312}$$

whence we find

$$R_{\mathrm{rad}} = \frac{2\pi}{3} Z \left(\frac{l}{\lambda}\right)^2 \tag{9.313}$$

by comparison with (9.301) or (9.311). The quantity R_{rad} is called *radiation resistance* of the device and for the case unver investigation is a measure of the ability of the dipole to transfer power towards infinity. From the viewpoint of a generator connected to the dipole — if only this arrangement could be realized in practice — the device would absorb power which somehow would be dissipated. However, we have not included losses anywhere in the analysis, and hence R_{rad} just accounts for the average power 'lost' to radiation, so to speak. Nonetheless, the Hertzian dipole is not very good at radiating, because in theory it is vanishingly short as compared to the wavelength λ and, as a result, the radiation resistance is zero. The radiated power $\langle P_F \rangle$ is not zero, though, as we have supposed that $I_0 l$ — that is, the limiting value as $l \to 0^+$ — is finite and non-null.

(End of Example 9.3)

9.6 Fields in the Fraunhofer region of a time-harmonic source

We wish to derive explicit, though approximate expressions of the fields that are valid in the far-field zone of a localized source (Figure 9.10) [46, Section 26.11]. These results will provide theoretical evidence for (9.271) but also have practical relevance, e.g., in the analysis of antenna and radar systems [39,47]. For this task we may follow two strategies, namely, we either approximate $\overline{\mathbf{G}}_{EJ}(\mathbf{r},\mathbf{r}')$ and $\overline{\mathbf{G}}_{HJ}(\mathbf{r},\mathbf{r}')$ and determine the fields directly or we find an approximation to the electromagnetic potentials (9.23) and (9.24), whence the fields can be computed. The former approach seems natural and straightforward, since the dyadic Green functions have been introduced, but the latter procedure has the merit of reminding that $\overline{\mathbf{G}}_{EJ}(\mathbf{r},\mathbf{r}')$ and $\overline{\mathbf{G}}_{HJ}(\mathbf{r},\mathbf{r}')$ are not essential.

We begin with the approximation of $\overline{\mathbf{G}}_{EJ}(\mathbf{r},\mathbf{r}')$ and $\overline{\mathbf{G}}_{HJ}(\mathbf{r},\mathbf{r}')$ for observation points satisfying condition (9.268). In the Fraunhofer region we substitute R for $r-\hat{\mathbf{r}}\cdot\mathbf{r}'$ in the phase factor $\exp(-\mathrm{j}\,kR)$ so as to limit the phase error to at most $\pi/8$. However, we need not be so fussy about the calculation of $1/R$ because the fields are less sensitive to this term. In fact, we may replace R with r, while neglecting the terms proportional to $1/R^2$ and $1/R^3$. Finally, as is suggested in Figure 9.11, in the far-field zone the radial vector $\hat{\mathbf{R}}$ is nearly parallel to $\hat{\mathbf{r}}$ and the identity dyadic transverse to $\hat{\mathbf{R}}$ (9.207) passes over into

$$\overline{\mathfrak{I}} - \hat{\mathbf{r}}\hat{\mathbf{r}} = \hat{\boldsymbol{\vartheta}}\hat{\boldsymbol{\vartheta}} + \hat{\boldsymbol{\varphi}}\hat{\boldsymbol{\varphi}} = \overline{\mathfrak{I}}_{\mathrm{tr}} \tag{9.314}$$

that is, the identity dyadic transverse to $\hat{\mathbf{r}}$.

Armed with these assumptions we can simplify (9.208) and (9.209) as follows

$$\overline{\mathbf{G}}_{EJ}(\mathbf{r},\mathbf{r}';\omega) \approx \frac{\mathrm{e}^{-\mathrm{j}\,kr}}{4\pi r}\mathrm{e}^{\mathrm{j}\,k\hat{\mathbf{r}}\cdot\mathbf{r}'}(\hat{\boldsymbol{\vartheta}}\hat{\boldsymbol{\vartheta}} + \hat{\boldsymbol{\varphi}}\hat{\boldsymbol{\varphi}}) \tag{9.315}$$

$$\overline{\mathbf{G}}_{HJ}(\mathbf{r},\mathbf{r}';\omega) \approx -\mathrm{j}\,k\frac{\mathrm{e}^{-\mathrm{j}\,kr}}{4\pi r}\mathrm{e}^{\mathrm{j}\,k\hat{\mathbf{r}}\cdot\mathbf{r}'}\hat{\mathbf{r}}\times\overline{\mathfrak{I}}_{tr} = -\mathrm{j}\,k\frac{\mathrm{e}^{-\mathrm{j}\,kr}}{4\pi r}\mathrm{e}^{\mathrm{j}\,k\hat{\mathbf{r}}\cdot\mathbf{r}'}(\hat{\boldsymbol{\varphi}}\hat{\boldsymbol{\vartheta}} - \hat{\boldsymbol{\vartheta}}\hat{\boldsymbol{\varphi}}) \tag{9.316}$$

for $r \geqslant 2D^2/\lambda$. One of the advantages of these expressions is that the dependence on r and $\mathbf{r}'$ is fully separated, whereby the calculation of the integrals simplifies because r is a constant. The same remark holds for the dyadics $\overline{\mathfrak{I}}_{tr}$ and $\hat{\mathbf{r}}\times\overline{\mathfrak{I}}_{tr}$.

Substituting (9.315) and (9.316) into (9.158) and (9.160) yields

$$\mathbf{E}(\mathbf{r};\omega) \approx -\mathrm{j}\,\omega\mu\,\underbrace{\frac{\mathrm{e}^{-\mathrm{j}\,kr}}{4\pi r}}_{\substack{\text{spherical}\\ \text{wave}}}(\hat{\boldsymbol{\vartheta}}\hat{\boldsymbol{\vartheta}} + \hat{\boldsymbol{\varphi}}\hat{\boldsymbol{\varphi}})\cdot\underbrace{\int_{V_J}\mathrm{d}V'\,\mathrm{e}^{\mathrm{j}\,k\hat{\mathbf{r}}\cdot\mathbf{r}'}\mathbf{J}(\mathbf{r}';\omega)}_{\text{3-D spatial Fourier transform}} \tag{9.317}$$

$$\mathbf{H}(\mathbf{r};\omega) \approx -\mathrm{j}\,k\,\underbrace{\frac{\mathrm{e}^{-\mathrm{j}\,kr}}{4\pi r}}_{\substack{\text{spherical}\\ \text{wave}}}(\hat{\boldsymbol{\varphi}}\hat{\boldsymbol{\vartheta}} - \hat{\boldsymbol{\vartheta}}\hat{\boldsymbol{\varphi}})\cdot\underbrace{\int_{V_J}\mathrm{d}V'\,\mathrm{e}^{\mathrm{j}\,k\hat{\mathbf{r}}\cdot\mathbf{r}'}\mathbf{J}(\mathbf{r}';\omega)}_{\text{3-D spatial Fourier transform}} \tag{9.318}$$

which is the desired result [1,39], [16, Chapters 31-32], [48, Section 9.1], [23, Section 5.2.B]. The corresponding expressions for the fields generated by a time-harmonic magnetic current density $\mathbf{J}_M(\mathbf{r}';\omega)$ follow immediately from (9.317) and (9.318) by invoking the duality transformations of Table 6.1.

We now obtain (9.317) and (9.318) by first approximating the magnetic vector potential (9.24) and the derivatives thereof, and then by computing $\mathbf{E}$ and $\mathbf{H}$ through (9.156) and (9.157) [6, Section 13.7].

We begin with $\mathbf{A}_E(\mathbf{r};\omega)$ and observe

$$\mathbf{A}_E(\mathbf{r};\omega) \approx \mu\frac{\mathrm{e}^{-\mathrm{j}\,kr}}{4\pi r}\underbrace{\int_{V_J}\mathrm{d}V'\,\mathrm{e}^{\mathrm{j}\,k\hat{\mathbf{r}}\cdot\mathbf{r}'}\mathbf{J}(\mathbf{r}';\omega)}_{\text{function of }\vartheta,\,\varphi} = \mu G(r)\mathfrak{F}_3\{\mathbf{J}\}_{k\hat{\mathbf{r}}} \tag{9.319}$$

on account of the asymptotic expansion (9.260) stopped to the zeroth order for the calculation of the phase factor. We may be less precise for the denominator of the kernel and let $R \approx r$. The integral of the current — which does not depend on the distance from the source — may be construed as a three-dimensional spatial Fourier transform with spectral variable $k\hat{\mathbf{r}}$.

We continue with the calculation of the divergence of the vector potential by using polar spherical coordinates, viz.,

$$\begin{aligned}
\nabla \cdot \mathbf{A}_E(\mathbf{r};\omega) &\approx \mu \nabla \cdot \left[\frac{\mathrm{e}^{-\mathrm{j}kr}}{4\pi r} \int_{V_J} \mathrm{d}V' \, \mathrm{e}^{\mathrm{j}k\hat{\mathbf{r}}\cdot\mathbf{r}'} \mathbf{J}(\mathbf{r}';\omega) \right] \\
&= \mu \nabla \frac{\mathrm{e}^{-\mathrm{j}kr}}{4\pi r} \cdot \int_{V_J} \mathrm{d}V' \, \mathrm{e}^{\mathrm{j}k\hat{\mathbf{r}}\cdot\mathbf{r}'} \mathbf{J}(\mathbf{r}';\omega) + \underbrace{\mu \frac{\mathrm{e}^{-\mathrm{j}kr}}{4\pi r} \nabla \cdot \int_{V_J} \mathrm{d}V' \, \mathrm{e}^{\mathrm{j}k\hat{\mathbf{r}}\cdot\mathbf{r}'} \mathbf{J}(\mathbf{r}';\omega)}_{\mathcal{O}(1/r^2)} \\
&\approx -\mathrm{j}k\mu \frac{\mathrm{e}^{-\mathrm{j}kr}}{4\pi r} \hat{\mathbf{r}} \cdot \int_{V_J} \mathrm{d}V' \, \mathrm{e}^{\mathrm{j}k\hat{\mathbf{r}}\cdot\mathbf{r}'} \mathbf{J}(\mathbf{r}';\omega)
\end{aligned} \tag{9.320}$$

by virtue of the differential identity (H.51). In the last step we have neglected all the terms which are proportional to $1/r^2$.

Thirdly, we use the previous result to find $\nabla\nabla \cdot \mathbf{A}_E$, namely,

$$\begin{aligned}
\nabla\nabla \cdot \mathbf{A}_E(\mathbf{r};\omega) &\approx -\mathrm{j}k\mu \nabla \left[\frac{\mathrm{e}^{-\mathrm{j}kr}}{4\pi r} \hat{\mathbf{r}} \cdot \int_{V_J} \mathrm{d}V' \, \mathrm{e}^{\mathrm{j}k\hat{\mathbf{r}}\cdot\mathbf{r}'} \mathbf{J}(\mathbf{r}';\omega) \right] \\
&= -\mathrm{j}k\mu \nabla \frac{\mathrm{e}^{-\mathrm{j}kr}}{4\pi r} \left[\hat{\mathbf{r}} \cdot \int_{V_J} \mathrm{d}V' \, \mathrm{e}^{\mathrm{j}k\hat{\mathbf{r}}\cdot\mathbf{r}'} \mathbf{J}(\mathbf{r}';\omega) \right] \underbrace{- \mathrm{j}k\mu \frac{\mathrm{e}^{-\mathrm{j}kr}}{4\pi r} \nabla \left[\hat{\mathbf{r}} \cdot \int_{V_J} \mathrm{d}V' \, \mathrm{e}^{\mathrm{j}k\hat{\mathbf{r}}\cdot\mathbf{r}'} \mathbf{J}(\mathbf{r}';\omega) \right]}_{\mathcal{O}(1/r^2)} \\
&= -\mu k^2 \frac{\mathrm{e}^{-\mathrm{j}kr}}{4\pi r} \hat{\mathbf{r}}\hat{\mathbf{r}} \cdot \int_{V_J} \mathrm{d}V' \, \mathrm{e}^{\mathrm{j}k\hat{\mathbf{r}}\cdot\mathbf{r}'} \mathbf{J}(\mathbf{r}';\omega)
\end{aligned} \tag{9.321}$$

where we have made use of identity (H.47). In the last passage we have discarded terms proportional to $1/r^2$.

For the calculation of $\nabla \times \mathbf{A}_E$ we have

$$\begin{aligned}
\nabla \times \mathbf{A}_E(\mathbf{r};\omega) &\approx \nabla \times \left[\mu \frac{\mathrm{e}^{-\mathrm{j}kr}}{4\pi r} \int_{V_J} \mathrm{d}V' \, \mathrm{e}^{\mathrm{j}k\hat{\mathbf{r}}\cdot\mathbf{r}'} \mathbf{J}(\mathbf{r}';\omega) \right] \\
&= \mu \nabla \frac{\mathrm{e}^{-\mathrm{j}kr}}{4\pi r} \times \int_{V_J} \mathrm{d}V' \, \mathrm{e}^{\mathrm{j}k\hat{\mathbf{r}}\cdot\mathbf{r}'} \mathbf{J}(\mathbf{r}';\omega) + \underbrace{\mu \frac{\mathrm{e}^{-\mathrm{j}kr}}{4\pi r} \nabla \times \int_{V_J} \mathrm{d}V' \, \mathrm{e}^{\mathrm{j}k\hat{\mathbf{r}}\cdot\mathbf{r}'} \mathbf{J}(\mathbf{r}';\omega)}_{\mathcal{O}(1/r^2)} \\
&\approx -\mathrm{j}\mu k \frac{\mathrm{e}^{-\mathrm{j}kr}}{4\pi r} \hat{\mathbf{r}} \times \int_{V_J} \mathrm{d}V' \, \mathrm{e}^{\mathrm{j}k\hat{\mathbf{r}}\cdot\mathbf{r}'} \mathbf{J}(\mathbf{r}';\omega)
\end{aligned} \tag{9.322}$$

on account of identity (H.50). As before, we have retained only terms proportional to $1/r$.

With these approximations and (9.156) and (9.157) we can eventually determine electric and magnetic fields as

$$\mathbf{E}(\mathbf{r};\omega) \approx -\mathrm{j}\,\omega\mu\frac{\mathrm{e}^{-\mathrm{j}kr}}{4\pi r}\left(\overline{\mathfrak{I}} - \hat{\mathbf{r}}\hat{\mathbf{r}}\right)\cdot\int\limits_{V_J} \mathrm{d}V'\,\mathrm{e}^{\mathrm{j}k\hat{\mathbf{r}}\cdot\mathbf{r}'}\mathbf{J}(\mathbf{r}';\omega) \tag{9.323}$$

$$\mathbf{H}(\mathbf{r};\omega) \approx -\mathrm{j}\,k\frac{\mathrm{e}^{-\mathrm{j}kr}}{4\pi r}\hat{\mathbf{r}}\times\int\limits_{V_J} \mathrm{d}V'\,\mathrm{e}^{\mathrm{j}k\hat{\mathbf{r}}\cdot\mathbf{r}'}\mathbf{J}(\mathbf{r}';\omega) \tag{9.324}$$

which can be cast into the same form as (9.317) and (9.318) by means of few more manipulations. With these expressions it is evident that both $\mathbf{E}(\mathbf{r};\omega)$ and $\mathbf{H}(\mathbf{r};\omega)$ in the Fraunhofer zone depend explicitly on the vector potential $\mathbf{A}_E(\mathbf{r};\omega)$ written in the form (9.319).

Furthermore, the far-field approximation of $\mathbf{E}(\mathbf{r};\omega)$ and $\mathbf{H}(\mathbf{r};\omega)$ is also valid for finite sources localized on surfaces. In fact, the asymptotic expansions carried out in (9.319), (9.320), (9.321) and (9.322) basically affect the scalar Green function and as such can be repeated by starting with the vector potential given in (9.151). As a result we obtain

$$\mathbf{E}(\mathbf{r};\omega) \approx -\mathrm{j}\,\omega\mu\underbrace{\frac{\mathrm{e}^{-\mathrm{j}kr}}{4\pi r}}_{\substack{\text{spherical}\\ \text{wave}}}(\hat{\boldsymbol{\vartheta}}\hat{\boldsymbol{\vartheta}} + \hat{\boldsymbol{\varphi}}\hat{\boldsymbol{\varphi}})\cdot\underbrace{\int\limits_{S_J}\mathrm{d}S'\,\mathrm{e}^{\mathrm{j}k\hat{\mathbf{r}}\cdot\mathbf{r}'}\mathbf{J}_S(\mathbf{r}';\omega)}_{\text{2-D spatial Fourier transform}} \tag{9.325}$$

$$\mathbf{H}(\mathbf{r};\omega) \approx -\mathrm{j}\,k\underbrace{\frac{\mathrm{e}^{-\mathrm{j}kr}}{4\pi r}}_{\substack{\text{spherical}\\ \text{wave}}}(\hat{\boldsymbol{\varphi}}\hat{\boldsymbol{\vartheta}} - \hat{\boldsymbol{\vartheta}}\hat{\boldsymbol{\varphi}})\cdot\underbrace{\int\limits_{S_J}\mathrm{d}S'\,\mathrm{e}^{\mathrm{j}k\hat{\mathbf{r}}\cdot\mathbf{r}'}\mathbf{J}_S(\mathbf{r}';\omega)}_{\text{2-D spatial Fourier transform}} \tag{9.326}$$

where the surface integrals can be interpreted as generalized Fourier transformations over S_J.

As anticipated by (9.271), we clearly see that in (9.317), (9.318), (9.325) and (9.326) the dependence upon the radial coordinate r is separated from the angular variation with ϑ and φ, which is also implied in the integral over the current. Besides, the fields are transverse to $\hat{\mathbf{r}}$, that is, the radial components $E_r = \hat{\mathbf{r}}\cdot\mathbf{E}$, $H_r = \hat{\mathbf{r}}\cdot\mathbf{H}$ are negligible to this level of approximation. The phase factor combined with $1/r$ represents an *outgoing spherical wave* emanating from the origin O and propagating with velocity c. In fact, we recognize this term as the physical fundamental solution of the scalar Helmholtz equation (8.194) or, equivalently, the scalar Green function (8.335). Both the electric and the magnetic field involve the same integral of the electric current density $\mathbf{J}(\mathbf{r}';\omega)$. Finally, in the Fraunhofer region $\mathbf{E}$ and $\mathbf{H}$ are related as

$$\mathbf{H}(\mathbf{r};\omega) \approx \frac{1}{Z}\hat{\mathbf{r}}\times\mathbf{E}(\mathbf{r};\omega), \qquad r \geqslant \frac{2D^2}{\lambda} \tag{9.327}$$

a formula which we refer to as an *impedance relationship* [cf. the analogous ones valid for plane waves (7.63) and TEM spherical waves (7.405)]. The following vector fields

$$\mathbf{P}_{EJ}(\vartheta,\varphi;\omega) := \begin{cases} \overline{\mathfrak{I}}_{tr}\cdot\mathfrak{F}_3\{\mathbf{J}\}_{k\hat{\mathbf{r}}}, & \text{volume current} \\ \overline{\mathfrak{I}}_{tr}\cdot\mathfrak{F}_2\{\mathbf{J}_S\}_{k\hat{\mathbf{r}}}, & \text{surface current} \end{cases} \tag{9.328}$$

$$\mathbf{P}_{HJ}(\vartheta,\varphi;\omega) := \begin{cases} Z\,\hat{\mathbf{r}}\times\overline{\mathfrak{I}}_{tr}\cdot\mathfrak{F}_3\{\mathbf{J}\}_{k\hat{\mathbf{r}}}, & \text{volume current} \\ Z\,\hat{\mathbf{r}}\times\overline{\mathfrak{I}}_{tr}\cdot\mathfrak{F}_2\{\mathbf{J}_S\}_{k\hat{\mathbf{r}}}, & \text{surface current} \end{cases} \tag{9.329}$$

are called the *Schelkunoff radiation vectors* (physical dimensions: Am and Vm) and satisfy the relationship

$$\mathbf{P}_{HJ}(\vartheta,\varphi;\omega) = Z\hat{\mathbf{r}} \times \mathbf{P}_{EJ}(\vartheta,\varphi;\omega) \tag{9.330}$$

similarly to (9.327). Alternatively, $\mathbf{P}_{EJ}$ and $\mathbf{P}_{HJ}$ are also known as *generalized electric and magnetic moments*, respectively, in that they play a role analogous to the dipole moment of a Hertzian dipole in the radiation field (9.297) and (9.298).

When the radiating source is a single-port antenna the current *density* $\mathbf{J}(\mathbf{r})$ or $\mathbf{J}_S(\mathbf{r})$ is proportional to the input current $I(\omega)$ at the port in light of the linearity of the Maxwell equations (see Figure 6.26, Examples 6.8 and 9.4, and Section 10.6). Thus, if we divide the generalized electric moment by $I(\omega)$ we obtain the quantity [6, 39, 47]

$$\begin{aligned}\mathbf{h}_{\text{eff}}(\vartheta,\varphi;\omega) &:= \frac{\mathbf{P}_{EJ}(\vartheta,\varphi;\omega)}{I(\omega)}\\ &= (\hat{\boldsymbol{\vartheta}}\hat{\boldsymbol{\vartheta}} + \hat{\boldsymbol{\varphi}}\hat{\boldsymbol{\varphi}})\cdot\begin{cases}\dfrac{1}{I(\omega)}\displaystyle\int_{V_J} \mathrm{d}V'\, \mathrm{e}^{\mathrm{j}k\hat{\mathbf{r}}\cdot\mathbf{r}'}\,\mathbf{J}(\mathbf{r}';\omega), & \text{volume current}\\[2ex] \dfrac{1}{I(\omega)}\displaystyle\int_{S_J} \mathrm{d}S'\, \mathrm{e}^{\mathrm{j}k\hat{\mathbf{r}}\cdot\mathbf{r}'}\,\mathbf{J}_S(\mathbf{r}';\omega), & \text{surface current}\end{cases}\end{aligned} \tag{9.331}$$

which carries the physical dimension of a length and is called the *effective height* or *length* of the antenna, regardless of the antenna shape, though for linear antennas $\mathbf{h}_{\text{eff}}$ can, in fact, be related to the geometrical length of the device. The effective height is a vector field defined in the Fraunhofer region of the antenna (Figure 9.10) and as such depends only the polar angles ϑ and φ.

It is worthwhile mentioning that (9.327) is only approximate because the waves are not exactly spherical for a source which is not concentrated in a point and in our representation of the field components we have only retained the terms which fall off with the distance from the source. Nonetheless, it is important to keep in mind that the electric and the magnetic field are indeed *orthogonal* to one another at each point in space away from the source in a homogeneous isotropic medium [16].

In light of (9.327) and (9.330) the complex Poynting vector (1.304) in the far-field zone reads

$$\mathbf{S}(\mathbf{r};\omega) \approx \frac{1}{2Z}\mathbf{E}(\mathbf{r};\omega)\times[\hat{\mathbf{r}}\times\mathbf{E}^*(\mathbf{r};\omega)] = \frac{Z}{8}\frac{|\mathbf{P}_{EJ}(\vartheta,\varphi;\omega)|^2}{(\lambda r)^2}\hat{\mathbf{r}} \tag{9.332}$$

$$\mathbf{S}(\mathbf{r};\omega) \approx \frac{Z}{2}[\mathbf{H}(\mathbf{r};\omega)\times\hat{\mathbf{r}}]\times\mathbf{H}^*(\mathbf{r};\omega) = \frac{1}{8Z}\frac{|\mathbf{P}_{HJ}(\vartheta,\varphi;\omega)|^2}{(\lambda r)^2}\hat{\mathbf{r}} \tag{9.333}$$

which suggest that, if any power at all gets transferred from the source towards infinity by means of electromagnetic waves, then the flow occurs predominantly along the radial direction. While obtaining the Poynting theorem in the frequency domain (Section 1.10.2) we argued that the average power flow $\langle P_F\rangle$ through the boundary ∂V of a region of space V which contains the sources ($V_S \subseteq V$) is due only to the acceleration or *radiation field* whereas the velocity or *Coulomb field* contributes naught. As a consequence, even though (9.317), (9.318), (9.325) and (9.326) are valid only in the Fraunhofer region of $\mathbf{J}(\mathbf{r}')$, when such formulas are used to compute $\langle P_F\rangle$ with (1.303) we obtain the radiated power with no approximation at all. This statement has been verified in Example 9.3 for an electric Hertzian dipole. In symbols, (1.303) yields

$$\langle P_F\rangle = \int_{\partial V} \mathrm{d}S\,\hat{\mathbf{n}}(\mathbf{r})\cdot\frac{Z}{8}\frac{|\mathbf{P}_{EJ}(\vartheta,\varphi)|^2}{(\lambda r)^2}\hat{\mathbf{r}} = \frac{Z}{8\lambda^2}\int_{4\pi}\mathrm{d}\Omega\,|\mathbf{P}_{EJ}(\vartheta,\varphi)|^2 \tag{9.334}$$

where we have noticed that the area element $\mathrm{d}S$ on ∂V and the differential solid angle $\mathrm{d}\Omega$ subtended by $\mathbf{r}$ are related by (cf. Figure 2.19 and Section 5.4)

$$\mathrm{d}\Omega = \frac{\hat{\mathbf{n}}(\mathbf{r}) \cdot \hat{\mathbf{r}}}{r^2}\, \mathrm{d}S \tag{9.335}$$

and also introduced the symbol $\int_{4\pi} \mathrm{d}\Omega$ to signify integration over the whole solid angle. The rightmost-hand side of (9.334) clearly shows that $\langle P_F \rangle$ does not depend on the size and shape of ∂V. This conclusion was also reached with the application of the complex Poynting theorem in (1.317) to the source-free region $B(0, b) \setminus V$.

Since in the far-field region the angular dependence of $\mathbf{E}$ and $\mathbf{H}$ is the same for any distance $r \geqslant 2D^2/\lambda$, it is possible to draw a surface plot of $|\mathbf{E}|$ or $|\mathbf{H}|$ as a function of (ϑ, φ) while keeping r fixed. Even better, we may normalize the magnitude of the field to its maximum value, i.e., we consider the function

$$F(\vartheta, \varphi; \omega) := \frac{|\mathbf{E}(\mathbf{r}; \omega)|}{\max\limits_{\vartheta,\varphi} |\mathbf{E}(\mathbf{r}; \omega)|} = \frac{|\mathbf{H}(\mathbf{r}; \omega)|}{\max\limits_{\vartheta,\varphi} |\mathbf{H}(\mathbf{r}; \omega)|} \tag{9.336}$$

which is referred to as the *radiation pattern* of the source [27, 39, 49]. Thanks to the impedance relationship (9.327) the radiation pattern may be defined equivalently by means of the magnetic field. A three-dimensional plot of $F(\vartheta, \varphi)$ — sometimes called a *radiation solid* — may be obtained by drawing points at a distance $F(\vartheta, \varphi)$ from the origin for each pair of angles (ϑ, φ). In Cartesian coordinates this means

$$x = F(\vartheta, \varphi) \cos\varphi \sin\vartheta \qquad y = F(\vartheta, \varphi) \sin\varphi \sin\vartheta \qquad z = F(\vartheta, \varphi) \cos\vartheta \tag{9.337}$$

and these equations define a surface in a three-dimensional space.

Example 9.4 (Radiation field of a dipole antenna)

By and large a *dipole antenna* is a radiating electromagnetic device fashioned out of two straight lengths of conducting material or, quite simply, two wires connected to a generator of power by means of a transmission line, as is sketched in Figure 9.13 [6, 27, 39, 50, 51]. To understand the rationale behind this configuration we take a step back and consider a classic transmission line made of two parallel metallic wires [27,45,52–54].

Ordinarily, a transmission line is employed as a means to transport electromagnetic energy from a generator to a device (i.e., a load) ideally with negligible losses along the way. The free electrons in the conductors are set in motion by the generator and give rise to two time-harmonic currents densities $\mathbf{J}_1(\mathbf{r}'; \omega)$ and $\mathbf{J}_2(\mathbf{r}'; \omega)$ that flow in opposite directions on the parallel wires, as is suggested in Figure 9.14. The periodic motion of the electrons occurs predominantly along the axis of the wires which, as a result, we may take as infinitely thin also because this aspect is not important for the present discussion. In order to express the current densities mathematically we introduce a local system of Cartesian coordinates (x', y', z') with the z-axis aligned with the wires and the origin at the beginning of the line. Furthermore, we suppose that the wires lie in the xOz plane and write the currents as distributions, viz.,

$$\mathbf{J}_1(\mathbf{r}'; \omega) = I(z')\delta(x' - x_1)\delta(y')\hat{\mathbf{z}}, \qquad z' \in [0, L] \tag{9.338}$$

$$\mathbf{J}_2(\mathbf{r}'; \omega) = -I(z')\delta(x' - x_2)\delta(y')\hat{\mathbf{z}}, \qquad z' \in [0, L] \tag{9.339}$$

where the precise form of the line current $I(z')$ ultimately depends on the impedance $Z_L \in \mathbb{C}$ of the electromagnetic device (i.e., the load) connected to the end of the line ($z' = L$). In particular, we say

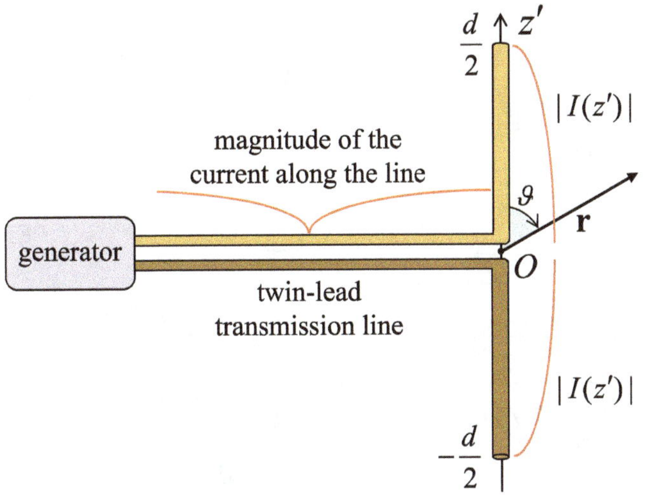

Figure 9.13 For the calculation of the radiation field of a dipole antenna.

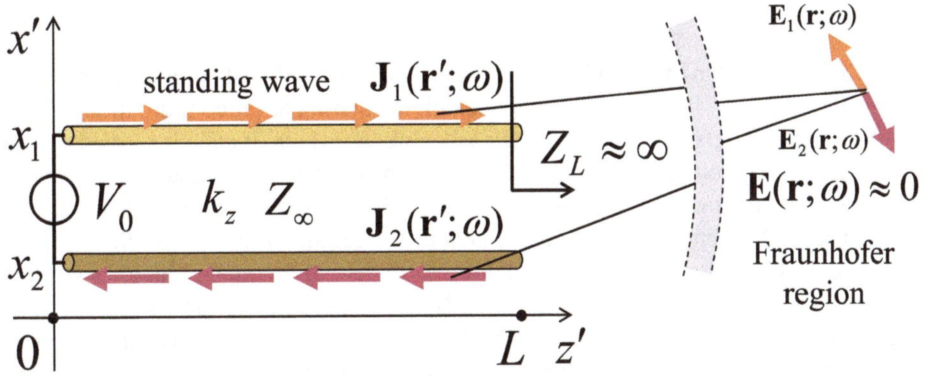

Figure 9.14 For illustrating the genesis of a dipole antenna: open-ended transmission line.

the line is left 'open' if it is not connected to a device. This situation can be modelled by means of an ideal load described by an impedance Z_L which is imaginary and infinite to a good approximation. In practice, though, at the end of the line the two conductors form a capacitor of sorts characterized by a tiny capacitance, which in turn accounts for the very large $|Z_L|$.

For the sake of completeness we write down the relevant line current $I(z')$ when $Z_L = \mathrm{j}\,\infty$, i.e.,

$$I(z') = \mathrm{j}\,\frac{V_0}{Z_\infty}\frac{\sin[k_z(L-z')]}{\cos(k_z L)}, \qquad z' \in [0, L] \tag{9.340}$$

where $V_0 \in \mathbb{C}$ is the phasor of the ideal voltage generator in $z' = 0$, and $k_z \in \mathbb{R}$ and $Z_\infty \in \mathbb{R}$ denote the propagation constant and the characteristic impedance of the transmission line (also see Example 6.2). Since $\mathrm{Re}\{Z_L\}$ is negligible in practice or outright null in theory, essentially no energy is extracted from the generator and no power is radiated. This is confirmed by noticing that $I(L) = 0$, i.e., no current flows through the 'load'.

From the viewpoint of the generation of electromagnetic waves, an open-ended transmission line is not an efficient radiating device because the electric fields $\mathbf{E}_1(\mathbf{r};\omega)$ and $\mathbf{E}_2(\mathbf{r};\omega)$ produced by

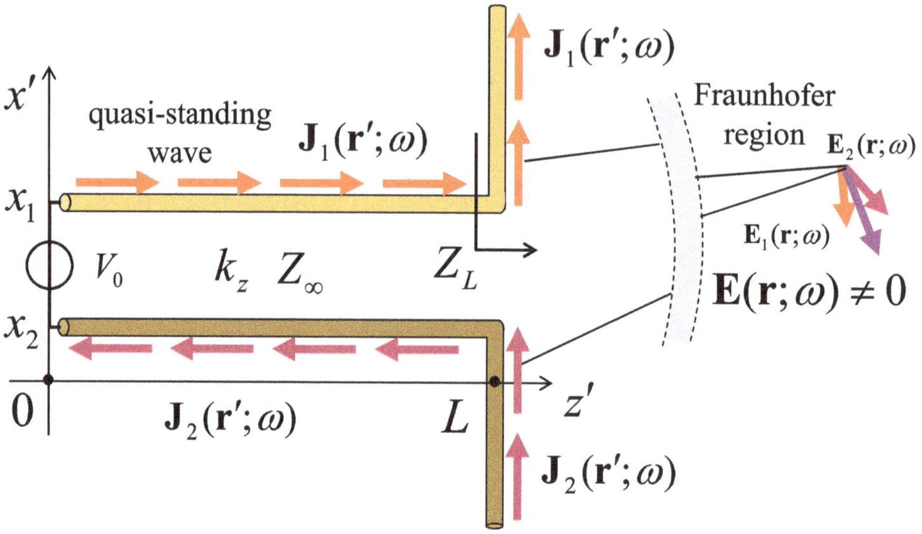

Figure 9.15 For illustrating the genesis of a dipole antenna: open and bent transmission line.

$\mathbf{J}_1(\mathbf{r}'; \omega)$ and $\mathbf{J}_2(\mathbf{r}'; \omega)$ in the Fraunhofer region cancel each other almost completely, as is pictorially described in Figure 9.14. This is a consequence of the symmetrical configuration of the line as well as of $\mathbf{J}_1(\mathbf{r}'; \omega)$ and $\mathbf{J}_2(\mathbf{r}'; \omega)$ flowing in opposite directions, and it can be verified by evaluating $\mathbf{E}_1(\mathbf{r}; \omega)$ and $\mathbf{E}_2(\mathbf{r}; \omega)$ through (9.317). The cancellation is not perfect, though, partly because the 'open' amounts to a small capacitance as already argued, and partly because the two wires are not superimposed but rather separated a distance which in our geometry reads $x_2 - x_1$.

The previous observations suggest that in order to achieve substantial radiation we should break the symmetry, as it were, by making the currents $\mathbf{J}_1(\mathbf{r}'; \omega)$ and $\mathbf{J}_2(\mathbf{r}'; \omega)$ flow in the same direction so as to generate fields that add up coherently in the Fraunhofer region. This can be accomplished by starting with a longer transmission line and 'bending' the final part of the wires at right angles with respect to the z-axis, as indicated in Figure 9.15. We may think of this configuration as being comprised of the original transmission line (of length L) terminated onto an electromagnetic device with finite impedance Z_L. Transmission line theory now yields

$$I(z') = \frac{V_0}{Z_\infty} \frac{(1 - \Gamma_L)\cos[k_z(L - z')] + \mathrm{j}(1 + \Gamma_L)\sin[k_z(L - z')]}{(1 + \Gamma_L)\cos(k_z L) + \mathrm{j}(1 - \Gamma_L)\sin(k_z L)}, \qquad z' \in [0, L] \tag{9.341}$$

where the quantity

$$\Gamma_L = \frac{Z_L - Z_\infty}{Z_L + Z_\infty}, \qquad |\Gamma_L| \leqslant 1 \tag{9.342}$$

is called the *reflection coefficient* of the load [45, 52]. A little algebra gives

$$I(L) = \frac{V_0}{Z_L \cos(k_z L) + \mathrm{j}\, Z_\infty \sin(k_z L)} \tag{9.343}$$

for the non-zero current entering the load. Moreover, a detailed analysis [6, 39, 47, 55] shows that $\mathrm{Re}\{Z_L\}$ is non-null, and thus the generator delivers energy to the load. Where does the energy go, since the wires are lossless? Well, the energy is carried to infinity by the electromagnetic waves generated by the currents on the bent part of the transmission line. Intuitively, by bending the wires we have forced $\mathbf{J}_1(\mathbf{r}';\omega)$ and $\mathbf{J}_2(\mathbf{r}';\omega)$ to flow in the same direction in the final stretch of the line (Figure 9.15). A 'bent' transmission line is a better radiating device because the electric fields $\mathbf{E}_1(\mathbf{r};\omega)$ and $\mathbf{E}_2(\mathbf{r};\omega)$ generated by the 'final parts' of $\mathbf{J}_1(\mathbf{r}';\omega)$ and $\mathbf{J}_2(\mathbf{r}';\omega)$ are almost parallel in the far zone.

Having offered an explanation for the shape of the dipole antenna of Figure 9.13 we continue with calculation of the radiation fields in a homogeneous isotropic *lossless* medium. To this purpose, we place the origin of the system of Cartesian coordinates (x', y', z') between the two arms or poles of the antenna which we take as aligned with the z-axis for simplicity. We need not worry about the currents flowing along the transmission line because, as we have argued, they contribute negligibly to the radiation field.

The 'exact' distribution of the current density along the dipole can be computed by means of numerical techniques such as the Method of Moments (Chapter 14) [39, 56, 57]. However, we may suppose that the current flows chiefly along the z-axis and actually write

$$\mathbf{J}(\mathbf{r}';\omega) = I(z')\delta(x')\delta(y')\hat{\mathbf{z}}, \qquad z' \in \left[-\frac{d}{2}, \frac{d}{2}\right] \tag{9.344}$$

if the cross section of the wires is very small as compared to the length d of the dipole antenna. The electric current $I(z')$ depends on the angular frequency but it must invariably vanish at the tips of the dipole, because the background medium is lossless by hypothesis and hence no conduction current can exist anywhere but in the region occupied by the dipole (cf. Section 1.7) and the transmission line, of course.

Classical analysis shows that we may assume [6, 39, 47, 55], [53, Section 14-10]

$$I(z') = I_0 \cos\left(\frac{\pi}{d}z'\right), \qquad z' \in \left[-\frac{d}{2}, \frac{d}{2}\right] \tag{9.345}$$

to a good approximation when the frequency is such that the length d of the dipole is about half the wavelength λ in the surrounding medium. This is an optimal situation because it corresponds to a zero of $\mathrm{Im}\{Z_L\}$, the impedance of the load. In which case the device 'absorbs' power that is radiated while the reactive energy stored in the field around the dipole is kept to a minimum, ideally to zero: We say that the antenna *resonates*.

With these positions we can apply (9.317) and (9.318) interpreted as distributions. Since the integrals along x' and y' are trivial in view of the sifting property (C.4), we are left with the integral along z'. In symbols, we have

$$\begin{aligned}\int_{V_J} \mathrm{d}V'\, \mathrm{e}^{\mathrm{j}k\hat{\mathbf{r}}\cdot\mathbf{r}'} I(z')\delta(x')\delta(y')\hat{\mathbf{z}} &= \hat{\mathbf{z}} \int_{-d/2}^{d/2} \mathrm{d}z' \mathrm{e}^{\mathrm{j}k\hat{\mathbf{r}}\cdot z'\hat{\mathbf{z}}} I(z') = I_0\hat{\mathbf{z}} \int_{-d/2}^{d/2} \mathrm{d}z' \mathrm{e}^{\mathrm{j}k\hat{\mathbf{r}}\cdot z'\hat{\mathbf{z}}} \cos\left(\frac{\pi}{d}z'\right) \\ &= \frac{\lambda}{\pi} I_0 \frac{\cos\left(\dfrac{kd}{2}\cos\vartheta\right)}{1-\left(\dfrac{kd}{\pi}\cos\vartheta\right)^2}\hat{\mathbf{z}} \approx \frac{\lambda}{\pi} I_0 \frac{\cos\left(\dfrac{\pi}{2}\cos\vartheta\right)}{\sin^2\vartheta}\hat{\mathbf{z}}\end{aligned} \tag{9.346}$$

on account of the hypothesis $d \approx \lambda/2$ and (9.270). The resulting function of ϑ exhibits a removable singularity for $\vartheta \in \{0, \pi\}$. Substituting into (9.317) and (9.318) yields

$$\mathbf{E}(\mathbf{r};\omega) \approx \mathrm{j}\,\omega\mu \frac{\mathrm{e}^{-\mathrm{j}kr}}{4\pi r} \frac{\lambda}{\pi} I_0 \frac{\cos\left(\dfrac{\pi}{2}\cos\vartheta\right)}{\sin\vartheta}\hat{\boldsymbol{\vartheta}}, \qquad r \geqslant \frac{2d^2}{\lambda} \tag{9.347}$$

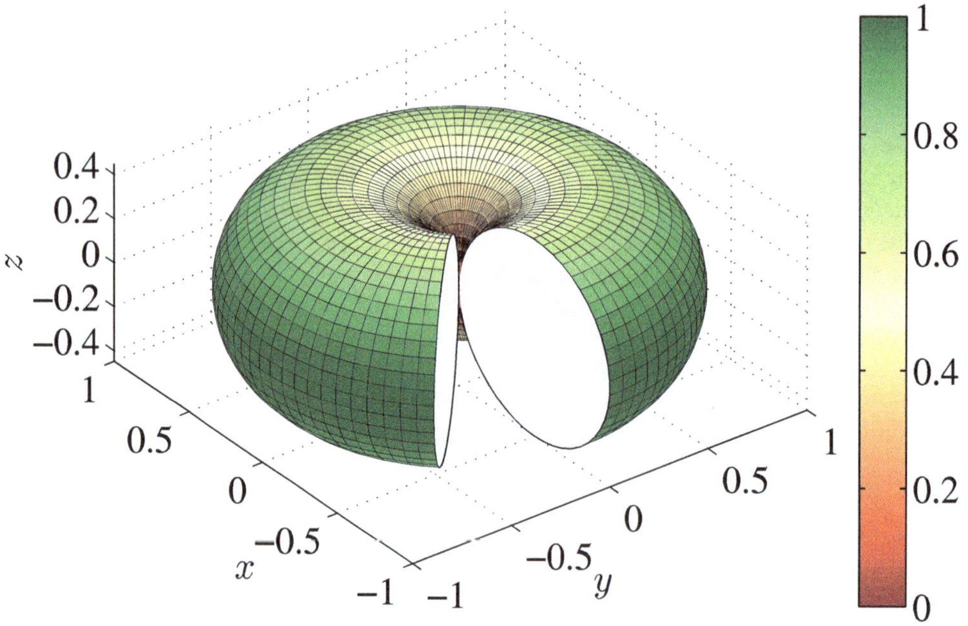

Figure 9.16 Radiation solid of the dipole antenna of Figure 9.13 for d = $\lambda/2$; a slice of the graph has been cut away for the sake of visualization.

$$\mathbf{H}(\mathbf{r};\omega) \approx \mathrm{j}\,k\frac{\mathrm{e}^{-\mathrm{j}kr}}{4\pi r}\frac{\lambda}{\pi}I_0\frac{\cos\left(\dfrac{\pi}{2}\cos\vartheta\right)}{\sin\vartheta}\hat{\boldsymbol{\varphi}}, \qquad r \geqslant \frac{2d^2}{\lambda} \tag{9.348}$$

having noticed that $\hat{\boldsymbol{\vartheta}} \cdot \hat{\mathbf{z}} = \sin\vartheta$ and $\hat{\boldsymbol{\varphi}} \cdot \hat{\mathbf{z}} = 0$. Electric and magnetic field are independent of φ for the current density is cylindrically symmetric around the z-axis. From definition (9.336) we see that the radiation pattern of the dipole is the function

$$F(\vartheta,\varphi) = \frac{\left|\cos\left(\dfrac{\pi}{2}\cos\vartheta\right)\right|}{|\sin\vartheta|} \tag{9.349}$$

which vanishes for observation points along the dipole axis ($\vartheta \in \{0, \pi\}$) and reaches its maximum for $\vartheta = \pi/2$, i.e., in the plane xOy, which is perpendicular to dipole axis. The resulting radiation solid for the dipole antenna is a torus and is plotted in Figure 9.16.

It is instructive to compute the complex Poynting vector and the average radiated power. From (1.304), (9.347) and (9.348) we find

$$\mathbf{S}(\mathbf{r};\omega) \approx \frac{Z|I_0|^2}{8\pi^2 r^2}\frac{\left[\cos\left(\dfrac{\pi}{2}\cos\vartheta\right)\right]^2}{\sin^2\vartheta}\hat{\mathbf{r}} \tag{9.350}$$

which is only approximate since we have used the far-field expressions for **E** and **H**. Nonetheless, numerical integration[2] of $\mathrm{Re}\{\mathbf{S}(\mathbf{r};\omega)\cdot\hat{\mathbf{r}}\}$ over the ball $B(0,r)$ yields the average radiated power $\langle P_F\rangle$

[2]The integral may also be expressed formally as a combination of integral sines $\mathrm{Si}(\xi)$ and integral cosines $\mathrm{Ci}(\xi)$, [43, Section 5.2], [39, Section 4.6].

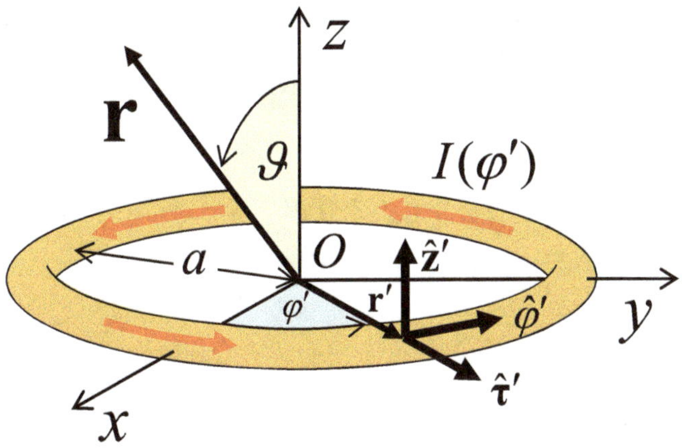

Figure 9.17 For the calculation of the radiation field of a loop antenna.

whereby the radiation resistance R_{rad} can be computed according to (9.312). In free space, with $Z = Z_0 \approx 377\ \Omega$, we find $R_{\text{rad}} \approx 73.1\ \Omega$. The result is approximated essentially because we have assumed the cosine-like current (9.345) instead of the true unknown $I(z')$ over the arms of the dipole antenna.

(End of Example 9.4)

Example 9.5 (Radiation field of a loop antenna)

A loop antenna is a radiating element comprised of a thin metallic wire in the shape of a circle and connected to a generator via a transmission line, as is pictured in Figure 9.17 [39, Chapter 5], [46, Section 26.10], [23, Section 5.3.B]. As was the case for the dipole antenna of Example 9.4, the periodic motion of the free electrons in the conductor occurs preferably parallel to the wire. Consequently, for the analysis we may suppose that the cross section of the wire is infinitely thin.

Owing to the cylindrical symmetry of the loop we choose three systems of Cartesian, cylindrical and polar spherical coordinates with the origin in the center of the antenna and the z-axis perpendicular to the plane that contains the loop. (We adopt the symbols τ' and $\hat{\boldsymbol{\tau}}'$ to indicate the radial distance from the z-axis and the polar vector.) Since the current is then concentrated on a circle of radius a, we express the current density along the loop in distributional form as

$$\mathbf{J}(\mathbf{r}'; \omega) = I(\varphi')\delta(\tau' - a)\delta(z')\hat{\boldsymbol{\varphi}}', \qquad \varphi' \in [0, 2\pi] \tag{9.351}$$

where the line current $I(\varphi')$ is a periodic function of the polar angle φ'.

To begin with, we examine the properties of a loop whose size is ‘small’ as compared to the wavelength λ in the background medium which we take homogeneous, isotropic and *lossless*. Since the length of the loop is $2\pi a$, the latter conditions translates into $ka \ll 1$ in light of (9.270). Accordingly, we may assume that the variation of the current $I(\varphi')$ along the loop is negligible and let

$$I(\varphi') = I_0, \qquad \varphi' \in [0, 2\pi], \qquad ka \ll 1 \tag{9.352}$$

where $I_0 \in \mathbb{R}$ is a constant.

With these positions we can compute the field in the Fraunhofer region with (9.317) and (9.318) interpreted as distributions. The domain integral is best carried out in cylindrical coordinates (τ', φ', z') because the integration along τ' and z' is trivial thanks to the sifting property (C.4). This gives

$$\int_{V_J} \mathrm{d}V' \, \mathrm{e}^{\mathrm{j}\,k\hat{\mathbf{r}}\cdot\mathbf{r}'} I_0 \delta(\tau' - a)\delta(z')\hat{\boldsymbol{\varphi}}' = I_0 a \int_0^{2\pi} \mathrm{d}\varphi' \mathrm{e}^{\mathrm{j}\,ka\hat{\mathbf{r}}\cdot\hat{\boldsymbol{\tau}}'} \hat{\boldsymbol{\varphi}}' \tag{9.353}$$

where

$$k\hat{\mathbf{r}} \cdot \mathbf{r}' = ka\hat{\mathbf{r}} \cdot \hat{\boldsymbol{\tau}}' = \sin\vartheta \cos(\varphi - \varphi') \tag{9.354}$$

$$\begin{aligned} \mathrm{e}^{\mathrm{j}\,ka\hat{\mathbf{r}}\cdot\hat{\boldsymbol{\tau}}'} &= 1 + \mathrm{j}\,ka\hat{\mathbf{r}} \cdot \hat{\boldsymbol{\tau}}' + o(ka) \\ &= 1 + \mathrm{j}\,ka \sin\vartheta \cos(\varphi - \varphi') + o(ka), \qquad ka \ll 1 \end{aligned} \tag{9.355}$$

by virtue of our assumptions. Thanks to the latter asymptotic expansion the integral can be carried out analytically as

$$\begin{aligned} I_0 a \int_0^{2\pi} \mathrm{d}\varphi' \mathrm{e}^{\mathrm{j}\,ka\hat{\mathbf{r}}\cdot\hat{\boldsymbol{\tau}}'} \hat{\boldsymbol{\varphi}}' &\approx I_0 a \int_0^{2\pi} \mathrm{d}\varphi' [1 + \mathrm{j}\,ka \sin\vartheta \cos(\varphi - \varphi')](-\hat{\mathbf{x}} \sin\varphi' + \hat{\mathbf{y}} \cos\varphi') \\ &= \frac{I_0 a^2 \mathrm{j}\,k}{2} \sin\vartheta \int_0^{2\pi} \mathrm{d}\varphi' \, \{-\hat{\mathbf{x}}[\sin\varphi - \sin(\varphi - 2\varphi')] + \hat{\mathbf{y}}[\cos\varphi + \cos(\varphi - 2\varphi')]\} \\ &= \mathrm{j}\,k^2 \pi a^2 \sin\vartheta \hat{\boldsymbol{\varphi}} = \mathrm{j}\,kAI_0 \sin\vartheta \hat{\boldsymbol{\varphi}} \end{aligned} \tag{9.356}$$

having used formulas 401.05 and 401.06 in [32]. By substituting into (9.317) and (9.318) we get

$$\mathbf{E}(\mathbf{r}; \omega) \approx Zk^2 I_0 A \frac{\mathrm{e}^{-\mathrm{j}\,kr}}{4\pi r} \sin\vartheta \hat{\boldsymbol{\varphi}}, \qquad r \geqslant 2\frac{(2a)^2}{\lambda} \tag{9.357}$$

$$\mathbf{H}(\mathbf{r}; \omega) \approx -k^2 I_0 A \frac{\mathrm{e}^{-\mathrm{j}\,kr}}{4\pi r} \sin\vartheta \hat{\boldsymbol{\vartheta}}, \qquad r \geqslant 2\frac{(2a)^2}{\lambda}. \tag{9.358}$$

The magnitude of the field is independent of φ, as expected, for the loop is symmetric around the z-axis. We also notice that the variation with ϑ is the same as the one we found for the electric Hertzian dipole in Example 9.3, though the polarization of electric and magnetic fields are interchanged in the two cases. Since I_0 is a constant we recognize the product $I_0 A$ as the magnetic moment of a small loop of stationary current examined in Example 5.3. In fact, by virtue of (1.248), (1.209) and (1.358) we may write

$$-k^2 I_0 A \sin\vartheta = k\frac{\omega\mu}{Z} I_0 A \hat{\mathbf{z}} \cdot \hat{\boldsymbol{\vartheta}} = -\frac{\mathrm{j}\,k}{Z} \mathrm{j}\,\omega \mathbf{p}_M \cdot \hat{\boldsymbol{\vartheta}} \tag{9.359}$$

where $\mathrm{j}\,\omega\mathbf{p}_M$ is the moment of an equivalent magnetic dipole located in the origin and oriented perpendicularly to the loop. This is consistent with the expression of $\mathbf{H}(\mathbf{r}; \omega)$ obtained by applying the duality principle to (9.317) in light of (9.286).

The complex Poynting vector in the Fraunhofer region and the average radiated power read

$$\mathbf{S}(\mathbf{r}; \omega) \approx \frac{Zk^4 A^2}{2(4\pi r)^2} |I_0|^2 \sin^2\vartheta \hat{\mathbf{r}} \qquad \langle P_F \rangle \approx Z\frac{k^4 A^2}{12\pi} |I_0|^2 \tag{9.360}$$

where $\langle P_F \rangle$ is approximate due to the assumption $ka \ll 1$. Comparison with formula (9.312) indicates that the radiation resistance of a small loop antenna is proportional with $(ka)^4$ and thus it is very small.

We now repeat the analysis for a loop of finite size and a constant current $I(z') = I_0$. Although with this latter hypothesis we can still compute the radiation fields analytically, we remark that the results are accurate so long as $a \leqslant \lambda/10$ [39]. For larger radii, the assumption of constant current is no longer valid, and one should represent $I(\tau', \varphi', z')$ as a Fourier series in the angle φ'. That been said, we evaluate

$$I_0 a \int_0^{2\pi} \mathrm{d}\varphi' \mathrm{e}^{\mathrm{j}ka\hat{\mathbf{r}}\cdot\hat{\mathbf{r}}'} \hat{\boldsymbol{\varphi}}' = I_0 a \int_0^{2\pi} \mathrm{d}\varphi' \mathrm{e}^{\mathrm{j}ka \sin\vartheta \cos(\varphi-\varphi')} (-\hat{\mathbf{x}} \sin\varphi' + \hat{\mathbf{y}} \cos\varphi')$$
$$= I_0 a \int_{\varphi-2\pi}^{\varphi} \mathrm{d}\xi \, \mathrm{e}^{\mathrm{j}ka \sin\vartheta \cos\xi} [-\hat{\mathbf{x}} \sin(\varphi-\xi) + \hat{\mathbf{y}} \cos(\varphi-\xi)] \tag{9.361}$$

having performed the change of dummy variable $\varphi - \varphi' = \xi$. Since the integrand is periodic with period 2π, we may integrate between any two convenient endpoints so long as we encompass a full period. In particular, if we temporarily choose the interval $[-\pi, \pi]$, we have

$$I_0 a \int_0^{2\pi} \mathrm{d}\varphi' \mathrm{e}^{\mathrm{j}ka\hat{\mathbf{r}}\cdot\hat{\mathbf{r}}'} \hat{\boldsymbol{\varphi}}' =$$
$$= I_0 a \int_{-\pi}^{\pi} \mathrm{d}\xi \, \mathrm{e}^{\mathrm{j}ka \sin\vartheta \cos\xi} [-\hat{\mathbf{x}}(\sin\varphi \cos\xi - \cos\varphi \sin\xi) + \hat{\mathbf{y}}(\cos\varphi \cos\xi + \sin\varphi \sin\xi)]$$
$$= I_0 a \int_{-\pi}^{\pi} \mathrm{d}\xi \, \mathrm{e}^{\mathrm{j}ka \sin\vartheta \cos\xi} (-\hat{\mathbf{x}} \sin\varphi + \hat{\mathbf{y}} \cos\varphi) \cos\xi \tag{9.362}$$

because $\cos\varphi \sin\xi$ and $\sin\varphi \sin\xi$ are odd functions of ξ and the argument of the exponential is an even function of ξ. Lastly, we change the integration interval again to $[0, 2\pi]$ and evaluate

$$I_0 a \int_0^{2\pi} \mathrm{d}\varphi' \mathrm{e}^{\mathrm{j}ka\hat{\mathbf{r}}\cdot\hat{\mathbf{r}}'} \hat{\boldsymbol{\varphi}}' = I_0 a (-\hat{\mathbf{x}} \sin\varphi + \hat{\mathbf{y}} \cos\varphi) \int_0^{2\pi} \mathrm{d}\xi \, \mathrm{e}^{\mathrm{j}ka \sin\vartheta \cos\xi} \cos\xi$$
$$= I_0 a 2\pi \mathrm{j} (-\hat{\mathbf{x}} \sin\varphi + \hat{\mathbf{y}} \cos\varphi) J_1(ka \sin\vartheta) = 2\pi \mathrm{j} I_0 a J_1(ka \sin\vartheta) \hat{\boldsymbol{\varphi}} \tag{9.363}$$

by virtue of the integral representation of the Bessel functions (H.175) for $n = 1$. Inserting this expression into (9.317) and (9.318) we arrive at

$$\mathbf{E}(\mathbf{r};\omega) \approx kZ \frac{\mathrm{e}^{-\mathrm{j}kr}}{2r} I_0 a J_1(ka \sin\vartheta) \hat{\boldsymbol{\varphi}}, \qquad r \geqslant 2\frac{(2a)^2}{\lambda} \tag{9.364}$$

$$\mathbf{H}(\mathbf{r};\omega) \approx -k \frac{\mathrm{e}^{-\mathrm{j}kr}}{2r} I_0 a J_1(ka \sin\vartheta) \hat{\boldsymbol{\vartheta}}, \qquad r \geqslant 2\frac{(2a)^2}{\lambda}. \tag{9.365}$$

The calculation of the average radiated power gives rise to an integral of Bessel functions over a sphere or radius r that can be evaluated with the aid of the Mellin transform and expressed in terms of generalized hypergeometric functions [58, Chapter 5].

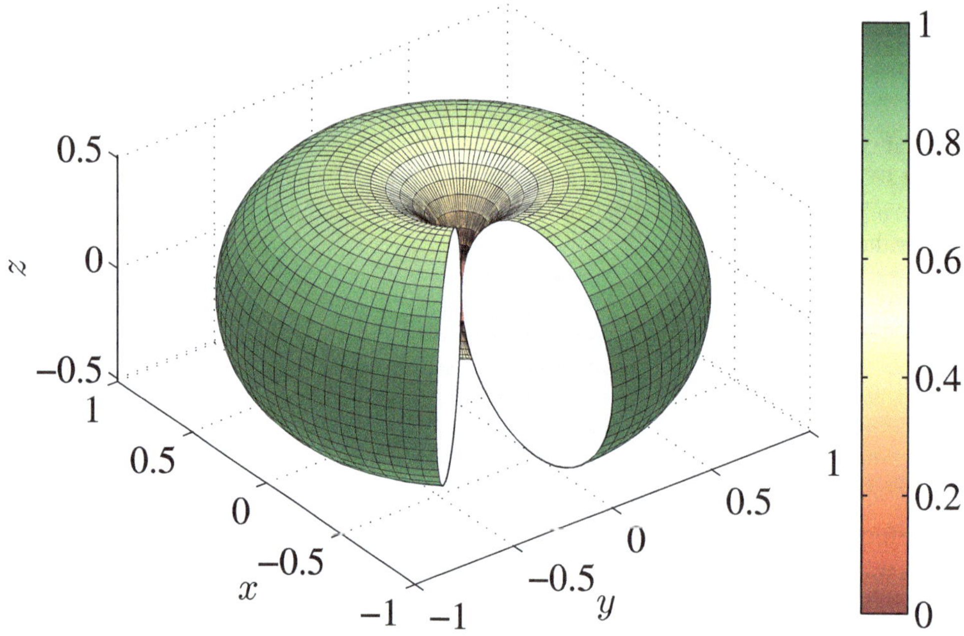

Figure 9.18 Radiation solid of the loop antenna of Figure 9.17 for $a = \lambda/10$; a slice of the graph has been cut away for the sake of visualization.

The radiation pattern of the loop antenna with uniform current distribution is the function

$$F(\vartheta, \varphi; \omega) = \frac{|J_1(ka \sin \vartheta)|}{\max\limits_{\vartheta} |J_1(ka \sin \vartheta)|} \tag{9.366}$$

which vanishes for observation points in a direction perpendicular to the plane of the loop ($\vartheta \in [0, \pi]$). The maximum may occur for the values of ϑ which solve

$$kaJ_0(ka \sin \vartheta) \cos \vartheta = 0, \qquad \vartheta \in [0, \pi] \tag{9.367}$$

and hence in particular for $\vartheta = \pi/2$; yet, the Bessel function $J_0(ka \sin \vartheta)$ does not vanish for $a \leqslant \lambda/10$. The radiation solid is shown in Figure 9.18 for $a = \lambda/10$.

(End of Example 9.5)

9.7 Symmetry properties of dyadic Green functions

In electromagnetics the Green functions (either scalar or dyadic) represent special solutions to the Maxwell equations and can be interpreted as the potentials or the fields produced by *singular sources* [29]. Depending on the properties of the medium where the singular source exists, the Green functions may satisfy certain symmetry relationships. For instance, the scalar static Green function (2.131) obeys (3.46), and the same holds true for the three-dimensional time-harmonic Green function (8.356), as can be seen by inspection. Thus, it is fruitful to find out what symmetry properties hold for the dyadic Green functions $\overline{\mathbf{G}}_{EJ}$, $\overline{\mathbf{G}}_{HM}$, $\overline{\mathbf{G}}_{HJ}$ and $\overline{\mathbf{G}}_{EM}$ introduced in Section 9.4. The results will come in handy in Section 10.1 to derive general integral representations for time-varying fields in the frequency domain.

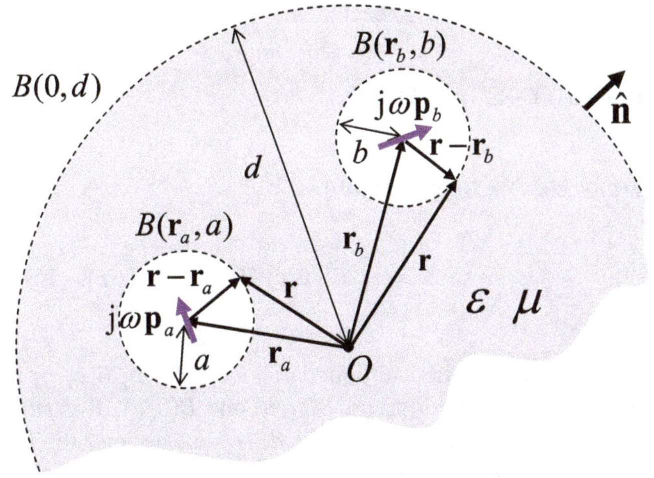

Figure 9.19 For deriving a special instance of the reciprocity theorem in the presence of Hertzian dipoles. The large ball $B(0, d)$ has been outlined (−−) only in part.

To draw our conclusions we need to apply the reciprocity theorem in its general form (6.253). However, care must be exercised because we obtained the result by explicitly forbidding the occurrence of singular sources, such as Hertzian dipoles, in order to deal with ordinary functions and invoke the Gauss theorem. Then again, to arrive at a symmetry relation which involves the dyadic Green functions we must necessarily handle the fields produced by singular sources, which are best described in terms of three-dimensional delta distributions (Appendix C).

Indeed, we could just interpret (6.253) in a distributional sense and evaluate the reactions 'trivially' as convolution products with $\delta^{(3)}(\mathbf{r})$ [1]. This strategy has the advantage of unveiling the desired symmetry properties with very little effort. We prefer to describe the classic approach which, by contrast, does not require using delta distributions and capitalizes on the knowledge of the fields produced by a Hertzian dipole (Example 9.3). Either way we will arrive at a reciprocity relationship which is sometimes named after Rayleigh and Carson.

To make a statement on the electric dyadic Green function $\overline{\mathbf{G}}_{EJ}(\mathbf{r}, \mathbf{r}')$ defined in (9.159) we consider electric sources only. Then, we apply (6.253) to the unbounded region of space complementary to two balls $B(\mathbf{r}_a, a)$ and $B(\mathbf{r}_b, b)$ with $B(\mathbf{r}_a, a) \cap B(\mathbf{r}_b, b) = \emptyset$, where $\mathbf{r}_a$ and $\mathbf{r}_b$ are the position vectors of two electric Hertzian dipoles in a homogeneous and isotropic medium (cf. Section 6.9). The two states for the reciprocity theorem are sketched in Figure 9.19 and described hereinbelow.

State (a) The Hertzian dipole characterized by a moment $\mathrm{j}\,\omega\mathbf{p}_a$, $\mathbf{p}_a = p_a\hat{\mathbf{z}}$, located at $\mathbf{r} = \mathbf{r}_a$ is turned on, whereas the Hertzian dipole with moment $\mathrm{j}\,\omega\mathbf{p}_b$, $\mathbf{p}_b = p_b\hat{\mathbf{z}}$, in $\mathbf{r} = \mathbf{r}_b$ is switched off. Notice that the field produced by $\mathrm{j}\,\omega\mathbf{p}_a$ is singular for $\mathbf{r} = \mathbf{r}_a$ but regular at the location of the other dipole.

State (b) The situation is just reversed: $\mathrm{j}\,\omega\mathbf{p}_a$ is off, whereas $\mathrm{j}\,\omega\mathbf{p}_b$ generates a field which is regular for $\mathbf{r} = \mathbf{r}_a$.

In our setup the reactions (6.254) and (6.255) are identically null because we have placed the sources in $B(\mathbf{r}_a, a)$ and $B(\mathbf{r}_b, b)$, i.e., outside the region of interest $\mathbb{R}^3 \setminus (B[\mathbf{r}_a, a] \cup B[\mathbf{r}_b, b])$, which is unbounded. Thus, we start by applying (6.253) to the surface-wise multiply connected domain $B(0, d) \setminus (B[\mathbf{r}_a, a] \cup B[\mathbf{r}_b, b])$ where $B(0, d)$ is a third ball with radius d large enough to contain both

$B(\mathbf{r}_a, a)$ and $B(\mathbf{r}_b, b)$. Consequently, (6.253) becomes

$$\int_{\partial B_a} \mathrm{d}S\ \hat{\mathbf{n}}(\mathbf{r}) \cdot [\mathbf{E}_a(\mathbf{r}) \times \mathbf{H}_b(\mathbf{r}) - \mathbf{E}_b(\mathbf{r}) \times \mathbf{H}_a(\mathbf{r})] + \int_{\partial B_b} \mathrm{d}S\ \hat{\mathbf{n}}(\mathbf{r}) \cdot [\mathbf{E}_a(\mathbf{r}) \times \mathbf{H}_b(\mathbf{r}) - \mathbf{E}_b(\mathbf{r}) \times \mathbf{H}_a(\mathbf{r})] = \int_{\partial B} \mathrm{d}S\ \hat{\mathbf{n}}(\mathbf{r}) \cdot [\mathbf{E}_a(\mathbf{r}) \times \mathbf{H}_b(\mathbf{r}) - \mathbf{E}_b(\mathbf{r}) \times \mathbf{H}_a(\mathbf{r})] \quad (9.368)$$

where ∂B_a, ∂B_b and ∂B denote the spherical boundaries of $B(\mathbf{r}_a, a)$, $B(\mathbf{r}_b, b)$ and $B(0, d)$. Besides, the unit normals are oriented outwards $B(\mathbf{r}_a, a)$, $B(\mathbf{r}_b, b)$ and $B(0, d)$. It is straightforward to show that the flux integral over ∂B vanishes in the limit as $d \to +\infty$, because the fields produced by the dipoles obey the Silver-Müller conditions (6.171) and (6.172). In fact, this was the argument used in (6.260) to obtain the reaction theorem (6.257).

In conclusion, the reciprocity theorem applied to $\mathbb{R}^3 \setminus (B[\mathbf{r}_a, a] \cup B[\mathbf{r}_b, b])$ implies

$$\int_{\partial B_a} \mathrm{d}S\ \hat{\mathbf{n}}(\mathbf{r}) \cdot [\mathbf{E}_a(\mathbf{r}) \times \mathbf{H}_b(\mathbf{r}) - \mathbf{E}_b(\mathbf{r}) \times \mathbf{H}_a(\mathbf{r})] = \int_{\partial B_b} \mathrm{d}S\ \hat{\mathbf{n}}(\mathbf{r}) \cdot [\mathbf{E}_b(\mathbf{r}) \times \mathbf{H}_a(\mathbf{r}) - \mathbf{E}_a(\mathbf{r}) \times \mathbf{H}_b(\mathbf{r})] \quad (9.369)$$

where we have changed sign to the integral over the sphere ∂B_b. In this way the flux integral in the right-hand side is formally obtained from the expression in the left member by swapping subscripts throughout. Since this practical recipe applies to the result of the integration as well, it is sufficient to examine the integral over ∂B_a in more details.

The fields generated by $\mathrm{j}\,\omega\mathbf{p}_a$ and $\mathrm{j}\,\omega\mathbf{p}_b$ can be written with the aid of the results of Example 9.3 and, more precisely, (9.289) and (9.290). Those formulas tell us that the field depends on the distance (r) from the location of the dipole, and thus here we have to substitute r for $R_a = |\mathbf{R}_a| = |\mathbf{r} - \mathbf{r}_a|$ and $R_b = |\mathbf{R}_b| = |\mathbf{r} - \mathbf{r}_b|$. In symbols, we have

$$\mathbf{E}_a(\mathbf{r}; \omega) = \omega^2 \mu \overline{\mathbf{G}}_{EJ}(\mathbf{r}, \mathbf{r}_a; \omega) \cdot \mathbf{p}_a = \omega^2 \mu \left(\overline{\mathbf{J}} + \frac{\nabla\nabla}{k^2} \right) \frac{\mathrm{e}^{-\mathrm{j}\,kR_a}}{4\pi R_a} \cdot \mathbf{p}_a \quad (9.370)$$

$$\mathbf{H}_a(\mathbf{r}; \omega) = \mathrm{j}\,\omega \overline{\mathbf{G}}_{HJ}(\mathbf{r}, \mathbf{r}_a; \omega) \cdot \mathbf{p}_a = \mathrm{j}\,\omega \nabla \frac{\mathrm{e}^{-\mathrm{j}\,kR_a}}{4\pi R_a} \times \mathbf{p}_a \quad (9.371)$$

and an obvious interchange of the subscripts 'a' and 'b' yields the analogous formulas that hold for the electromagnetic field entities produced by $\mathrm{j}\,\omega\mathbf{p}_b$.

The difficulty with the integrals in (9.369) lies in the highly singular character of the two terms $\nabla\nabla[\exp(-\mathrm{j}\,kR_a)/R_a]$ and $\nabla\nabla[\exp(-\mathrm{j}\,kR_b)/R_b]$ which, though regular on ∂B_a and ∂B_b, are not integrable in the limit as the radii a and b vanish (see Section 9.4). Therefore, we perform a little algebra to cast the integrands into an alternative, less singular format. In light of (9.370) for points $\mathbf{r} \in \mathbb{R}^3 \setminus (B[\mathbf{r}_a, a] \cup B[\mathbf{r}_b, b])$ we have

$$\mathbf{E}_a(\mathbf{r}) \times \mathbf{H}_b(\mathbf{r}) = -\mathbf{H}_b(\mathbf{r}) \times \omega^2 \mu \left(\overline{\mathbf{J}} + \frac{\nabla\nabla}{k^2} \right) \frac{\mathrm{e}^{-\mathrm{j}\,kR_a}}{4\pi R_a} \cdot \mathbf{p}_a$$

$$= -\omega^2\mu\mathbf{H}_b(\mathbf{r}) \times \mathbf{p}_a \frac{\mathrm{e}^{-\mathrm{j}kR_a}}{4\pi R_a} - \frac{1}{\varepsilon}\mathbf{H}_b(\mathbf{r}) \times \nabla \underbrace{\left(\nabla \frac{\mathrm{e}^{-\mathrm{j}kR_a}}{4\pi R_a} \cdot \mathbf{p}_a\right)}_{\text{scalar field}}$$

$$= \omega^2\mu\mathbf{p}_a \times \mathbf{H}_b(\mathbf{r})\frac{\mathrm{e}^{-\mathrm{j}kR_a}}{4\pi R_a} + \frac{1}{\varepsilon}\nabla \times \left[\mathbf{H}_b(\mathbf{r})\nabla \frac{\mathrm{e}^{-\mathrm{j}kR_a}}{4\pi R_a} \cdot \mathbf{p}_a\right] - \frac{1}{\varepsilon}\nabla \frac{\mathrm{e}^{-\mathrm{j}kR_a}}{4\pi R_a} \cdot \mathbf{p}_a \nabla \times \mathbf{H}_b(\mathbf{r})$$

$$= \omega^2\mu\mathbf{p}_a \times \mathbf{H}_b(\mathbf{r})\frac{\mathrm{e}^{-\mathrm{j}kR_a}}{4\pi R_a} + \frac{1}{\varepsilon}\nabla \times \left[\mathbf{H}_b(\mathbf{r})\nabla \frac{\mathrm{e}^{-\mathrm{j}kR_a}}{4\pi R_a} \cdot \mathbf{p}_a\right] - \nabla \frac{\mathrm{e}^{-\mathrm{j}kR_a}}{4\pi R_a} \cdot \mathbf{p}_a\,\mathrm{j}\,\omega\mathbf{E}_b(\mathbf{r}) \tag{9.372}$$

because $\mathbf{p}_a$ is a constant vector and the source-free local Ampère-Maxwell law (1.98) holds for the fields of state (b) when $\mathbf{r} \neq \mathbf{r}_b$. Thanks to the differential identity (H.50) we have effectively gotten rid of the troublesome dyadic field $\nabla\nabla[\exp(-\mathrm{j}\,kR_a)/R_a]$. In like manner, from (9.371) we have

$$-\mathbf{E}_b(\mathbf{r}) \times \mathbf{H}_a(\mathbf{r}) = -\mathbf{E}_b(\mathbf{r}) \times \left(\nabla \frac{\mathrm{e}^{-\mathrm{j}kR_a}}{4\pi R_a} \times \mathrm{j}\,\omega\mathbf{p}_a\right) = \mathrm{j}\,\omega\mathbf{p}_a\mathbf{E}_b(\mathbf{r}) \cdot \nabla \frac{\mathrm{e}^{-\mathrm{j}kR_a}}{4\pi R_a} - \mathrm{j}\,\omega\mathbf{E}_b(\mathbf{r}) \cdot \mathbf{p}_a \nabla \frac{\mathrm{e}^{-\mathrm{j}kR_a}}{4\pi R_a} \tag{9.373}$$

by virtue of (H.14). Finally, by combining (9.372) and (9.373) we find

$$\begin{aligned}
&\mathbf{E}_a(\mathbf{r}) \times \mathbf{H}_b(\mathbf{r}) - \mathbf{E}_b(\mathbf{r}) \times \mathbf{H}_a(\mathbf{r}) = \\
&= \omega^2\mu\mathbf{p}_a \times \mathbf{H}_b(\mathbf{r})\frac{\mathrm{e}^{-\mathrm{j}kR_a}}{4\pi R_a} + \frac{1}{\varepsilon}\nabla \times \left[\mathbf{H}_b(\mathbf{r})\nabla \frac{\mathrm{e}^{-\mathrm{j}kR_a}}{4\pi R_a} \cdot \mathbf{p}_a\right] \\
&\quad - \nabla \frac{\mathrm{e}^{-\mathrm{j}kR_a}}{4\pi R_a} \cdot \mathbf{p}_a\,\mathrm{j}\,\omega\mathbf{E}_b(\mathbf{r}) + \mathrm{j}\,\omega\mathbf{p}_a\mathbf{E}_b(\mathbf{r}) \cdot \nabla \frac{\mathrm{e}^{-\mathrm{j}kR_a}}{4\pi R_a} - \mathrm{j}\,\omega\mathbf{E}_b(\mathbf{r}) \cdot \mathbf{p}_a \nabla \frac{\mathrm{e}^{-\mathrm{j}kR_a}}{4\pi R_a} \\
&= \omega^2\mu\mathbf{p}_a \times \mathbf{H}_b(\mathbf{r})\frac{\mathrm{e}^{-\mathrm{j}kR_a}}{4\pi R_a} + \frac{1}{\varepsilon}\nabla \times \left[\mathbf{H}_b(\mathbf{r})\nabla \frac{\mathrm{e}^{-\mathrm{j}kR_a}}{4\pi R_a} \cdot \mathbf{p}_a\right] \\
&\quad + \nabla \frac{\mathrm{e}^{-\mathrm{j}kR_a}}{4\pi R_a} \times [\mathrm{j}\,\omega\mathbf{p}_a \times \mathbf{E}_b(\mathbf{r})] - \mathrm{j}\,\omega\mathbf{E}_b(\mathbf{r}) \cdot \mathbf{p}_a \nabla \frac{\mathrm{e}^{-\mathrm{j}kR_a}}{4\pi R_a}
\end{aligned} \tag{9.374}$$

having used (H.14) backwards. By letting

$$\hat{\mathbf{n}}(\mathbf{r}) = \hat{\mathbf{R}}_a = \frac{\mathbf{r} - \mathbf{r}_a}{|\mathbf{r} - \mathbf{r}_a|} = \frac{\mathbf{R}_a}{a}, \qquad \mathbf{r} \in \partial B_a \tag{9.375}$$

and noticing that

$$\nabla \frac{\mathrm{e}^{-\mathrm{j}kR_a}}{4\pi R_a} = -\left(\mathrm{j}\,k + \frac{1}{R_a}\right)\frac{\mathrm{e}^{-\mathrm{j}kR_a}}{4\pi R_a}\hat{\mathbf{R}}_a \tag{9.376}$$

we can cast the integral over ∂B_a as

$$\int_{\partial B_a} \mathrm{d}S\; \hat{\mathbf{n}}(\mathbf{r}) \cdot [\mathbf{E}_a(\mathbf{r}) \times \mathbf{H}_b(\mathbf{r}) - \mathbf{E}_b(\mathbf{r}) \times \mathbf{H}_a(\mathbf{r})] =$$

$$= \int_{\partial B_a} dS\ \omega^2\mu\hat{\mathbf{R}}_a \cdot \mathbf{p}_a \times \mathbf{H}_b(\mathbf{r})\frac{e^{-jkR_a}}{4\pi R_a} + \frac{1}{\varepsilon}\int_{\partial B_a} dS\ \hat{\mathbf{R}}_a \cdot \nabla \times \left[\mathbf{H}_b(\mathbf{r})\nabla\frac{e^{-jkR_a}}{4\pi R_a} \cdot \mathbf{p}_a\right]$$
$$+ \int_{\partial B_a} dS\ \hat{\mathbf{R}}_a \times \nabla\frac{e^{-jkR_a}}{4\pi R_a} \cdot [j\omega\mathbf{p}_a \times \mathbf{E}_b(\mathbf{r})] - \int_{\partial B_a} dS\ j\omega\mathbf{E}_b(\mathbf{r}) \cdot \mathbf{p}_a\hat{\mathbf{R}}_a \cdot \nabla\frac{e^{-jkR_a}}{4\pi R_a}$$
$$= \int_{\partial B_a} dS\ \omega^2\mu\hat{\mathbf{R}}_a \cdot \mathbf{p}_a \times \mathbf{H}_b(\mathbf{r})\frac{e^{-jkR_a}}{4\pi R_a} - \int_{\partial B_a} dS\ j\omega\mathbf{E}_b(\mathbf{r}) \cdot \mathbf{p}_a\hat{\mathbf{R}}_a \cdot \nabla\frac{e^{-jkR_a}}{4\pi R_a} \tag{9.377}$$

thanks to the Stokes theorem (A.55) applied over a closed surface and the fact that $\nabla(e^{-jkR_a}/R_a)$ is a vector field parallel to $\hat{\mathbf{R}}_a$. At this stage we may safely evaluate the remaining integrals in (9.377) in the limit as $a \to 0^+$. In particular, the magnetic field $\mathbf{H}_b(\mathbf{r})$ is regular and hence bounded for points $\mathbf{r} \in \partial B_a$. Therefore, we estimate

$$\left|\int_{\partial B_a} dS\ \omega^2\mu\hat{\mathbf{R}}_a \cdot \mathbf{p}_a \times \mathbf{H}_b(\mathbf{r})\frac{e^{-jkR_a}}{4\pi R_a}\right| \leqslant c_1 \int_{\partial B_a} dS\ \frac{1}{4\pi a} = c_1 a \tag{9.378}$$

with c_1 being a suitable positive constant. We compute the second integral with the aid of the mean value theorem [59]

$$-\int_{\partial B_a} dS\ j\omega\mathbf{E}_b(\mathbf{r}) \cdot \mathbf{p}_a\hat{\mathbf{R}}_a \cdot \nabla\frac{e^{-jkR_a}}{4\pi R_a} = \int_{\partial B_a} dS\ j\omega\mathbf{E}_b(\mathbf{r}) \cdot \mathbf{p}_a\left(jk + \frac{1}{a}\right)\frac{e^{-jka}}{4\pi a}$$
$$= j\omega\mathbf{E}_b(\mathbf{r}_0) \cdot \mathbf{p}_a\,(jka + 1)\,e^{-jka} \tag{9.379}$$

where $\mathbf{r}_0$ is a suitable point over ∂B_a. With these results in the limit as $a \to 0^+$ (9.377) yields

$$\lim_{a\to 0^+} \int_{\partial B_a} dS\ \hat{\mathbf{n}}(\mathbf{r}) \cdot [\mathbf{E}_a(\mathbf{r}) \times \mathbf{H}_b(\mathbf{r}) - \mathbf{E}_b(\mathbf{r}) \times \mathbf{H}_a(\mathbf{r})] = -j\omega\mathbf{E}_b(\mathbf{r}_a) \cdot \mathbf{p}_a \tag{9.380}$$

because $\mathbf{r}_0 \to \mathbf{r}_a$ as the ball $B(\mathbf{r}_a, a)$ shrinks to a point.

Since perfectly analogous steps or just the interchange of subscripts as argued before yield

$$\lim_{b\to 0^+} \int_{\partial B_b} dS\ \hat{\mathbf{n}}(\mathbf{r}) \cdot [\mathbf{E}_b(\mathbf{r}) \times \mathbf{H}_a(\mathbf{r}) - \mathbf{E}_a(\mathbf{r}) \times \mathbf{H}_b(\mathbf{r})] = -j\omega\mathbf{E}_a(\mathbf{r}_b) \cdot \mathbf{p}_b \tag{9.381}$$

in light of (9.369) we have just proved

$$\mathbf{p}_a \cdot \mathbf{E}_b(\mathbf{r}_a) = \mathbf{p}_b \cdot \mathbf{E}_a(\mathbf{r}_b) \tag{9.382}$$

a relationship which goes by the name of *Rayleigh-Carson theorem* [1].

Now we just need to apply (9.382) with (9.370) and the analogous representation of the electric field $\mathbf{E}_b(\mathbf{r})$ produced by $j\omega\mathbf{p}_b$. In symbols, we have

$$\mathbf{p}_a \cdot \omega^2\mu\overline{\mathbf{G}}_{EJ}(\mathbf{r}_a, \mathbf{r}_b; \omega) \cdot \mathbf{p}_b = \mathbf{p}_b \cdot \omega^2\mu\overline{\mathbf{G}}_{EJ}(\mathbf{r}_b, \mathbf{r}_a; \omega) \cdot \mathbf{p}_a$$
$$= \mathbf{p}_a \cdot \omega^2\mu\left[\overline{\mathbf{G}}_{EJ}(\mathbf{r}_b, \mathbf{r}_a; \omega)\right]^T \cdot \mathbf{p}_b \tag{9.383}$$

and, since the dipole moments as well as the position thereof are arbitrary, the previous equation is satisfied so long as

$$\overline{\mathbf{G}}_{EJ}(\mathbf{r}_a, \mathbf{r}_b; \omega) = \left[\overline{\mathbf{G}}_{EJ}(\mathbf{r}_b, \mathbf{r}_a; \omega)\right]^T, \qquad \mathbf{r}_a, \mathbf{r}_b \in \mathbb{R}^3, \qquad \mathbf{r}_a \neq \mathbf{r}_b \tag{9.384}$$

which is the sought symmetry relation for the electric dyadic Green function due to electric sources [1].

Since the magnetic dyadic Green function due to magnetic sources (9.162) is formally the same as $\overline{\mathbf{G}}_{EJ}(\mathbf{r}, \mathbf{r}'; \omega)$ we may conclude right away that

$$\overline{\mathbf{G}}_{HM}(\mathbf{r}_a, \mathbf{r}_b; \omega) = \left[\overline{\mathbf{G}}_{HM}(\mathbf{r}_b, \mathbf{r}_a; \omega)\right]^T, \qquad \mathbf{r}_a, \mathbf{r}_b \in \mathbb{R}^3, \qquad \mathbf{r}_a \neq \mathbf{r}_b \tag{9.385}$$

is the corresponding symmetry property. The actual proof requires the fields produced by two *magnetic Hertzian dipoles.*

To derive the mixed-type relation which involves $\overline{\mathbf{G}}_{EM}$ and $\overline{\mathbf{G}}_{HJ}$ we need to use an electric Hertzian dipole and a magnetic one. So, now is a good time to introduce the moment of a time-harmonic magnetic Hertzian dipole and its relation to the stationary magnetic dipole $\mathbf{m}$ presented in Sections 4.5 and 5.5.

In view of the defining equations (4.78) and (4.79) we see that the negative of the static magnetic moment $(-\mathbf{m})$ is not the dual of the electric static moment $\mathbf{p}$ for the simple reason that, according to Table 6.1, $\mathcal{B}(\mathbf{r})$ is not the dual entity of $\mathcal{E}(\mathbf{r})$. But then, in a linear isotropic medium endowed with magnetic permittivity μ, $\mathcal{B}(\mathbf{r})$ and $\mathcal{H}(\mathbf{r})$ are related by the constitutive relationship (1.118). Therefore, we may rewrite (4.78) and (4.79) as

$$W_h(\mathbf{r}) := -\mu\mathbf{m} \cdot \mathcal{H}(\mathbf{r}) \tag{9.386}$$

$$\mathcal{N}_h(\mathbf{r}) := \mu\mathbf{m} \times \mathcal{H}(\mathbf{r}) \tag{9.387}$$

and comparison with (4.76) and (4.77) leads us to assume

$$\mathbf{p}_M := \mu\mathbf{m} \tag{9.388}$$

as the definition of the static magnetic dipole that is consistent with the duality principle. In particular, $-\mathbf{p}_M$ is the dual of $\mathbf{p}$ and vice-versa. The physical dimension of $\mathbf{p}_M$ is weber meter (Wb m) in keeping with the alternative picture of the dipole as being formed by a pair of magnetic charges rather than being an infinitely small loop of electric current (cf. Example 5.3). If we go on to apply the duality transformations to the electric current density of (9.279) we find

$$\mathbf{J}_M(\mathbf{r}) = \mathrm{j}\,\omega\mathbf{p}_M\delta^{(3)}(\mathbf{r}) = \mathrm{j}\,\omega\mu\mathbf{m}\,\delta^{(3)}(\mathbf{r}) \tag{9.389}$$

whereby $\mathrm{j}\,\omega\mu\mathbf{m}$ (physical dimension: Vm) constitutes the expression of the moment of a time-harmonic magnetic Hertzian dipole. The transformations rules are listed in Table 9.1 for ease of reference.

We are now ready to apply the reciprocity theorem in the form (9.369) with the following two states.

State (a) An electric Hertzian dipole characterized by a moment $\mathrm{j}\,\omega\mathbf{p}_a$ and located at $\mathbf{r} = \mathbf{r}_a$ is turned on, whereas a magnetic Hertzian dipole with moment $\mathrm{j}\,\omega\mu\mathbf{m}_b$ in $\mathbf{r} = \mathbf{r}_b$ is switched off.

State (b) The situations is reversed: $\mathrm{j}\,\omega\mathbf{p}_a$ is off, whereas $\mathrm{j}\,\omega\mu\mathbf{m}_b$ generates a field which is regular for $\mathbf{r} = \mathbf{r}_a$.

Table 9.1 Duality transformations for dipole moments

Time domain†		Frequency domain†	
Primary entities	**Dual entities**	**Primary entities**	**Dual entities**
$\mathbf{p}$	$-\mu\mathbf{m}$	$\mathbf{p}$	$-\mu\mathbf{m}$
$\mu\mathbf{m}$	$-\mathbf{p}$	$\mu\mathbf{m}$	$-\mathbf{p}$
$\dfrac{\partial\mathbf{p}}{\partial t}$	$-\mu\dfrac{\partial\mathbf{m}}{\partial t}$	$\mathrm{j}\,\omega\mathbf{p}$	$-\mathrm{j}\,\omega\mu\mathbf{m}$
$\mu\dfrac{\partial\mathbf{m}}{\partial t}$	$-\dfrac{\partial\mathbf{p}}{\partial t}$	$\mathrm{j}\,\omega\mu\mathbf{m}$	$-\mathrm{j}\,\omega\mathbf{p}$

(†) Static, stationary, time-varying and time-harmonic vectors are all indicated with boldface Roman letters.

The fields due to $\mathrm{j}\,\omega\mu\mathbf{m}_b$ read

$$\mathbf{E}_b(\mathbf{r};\omega) = -\mathrm{j}\,\omega\mathbf{G}_{EM}(\mathbf{r},\mathbf{r}_b;\omega)\cdot\mu\mathbf{m}_b = -\mathrm{j}\,\omega\nabla\frac{\mathrm{e}^{-\mathrm{j}\,kR_a}}{4\pi R_a}\times\mu\mathbf{m}_b \tag{9.390}$$

$$\mathbf{H}_b(\mathbf{r};\omega) = \omega^2\varepsilon\overline{\mathbf{G}}_{HM}(\mathbf{r},\mathbf{r}_b;\omega)\cdot\mu\mathbf{m}_b = \omega^2\varepsilon\left(\overline{\mathbf{J}}+\frac{\nabla\nabla}{k^2}\right)\frac{\mathrm{e}^{-\mathrm{j}\,kR_b}}{4\pi R_b}\cdot\mu\mathbf{m}_b \tag{9.391}$$

formulas which we obtain by applying the duality principle to (9.370) and (9.371).

We do not have to repeat the calculation of the flux integral in the left-hand side of (9.369) because in deriving (9.380) we have not explicitly employed the specific form of the fields produced by the source of state (b), which consequently may be any. On the contrary, the flux integral in the right-hand side does depend on the choice of a magnetic dipole as a source, and thus the result is not given by (9.381). We may either embark on a detailed analysis as before or, somewhat more quickly, invoke the duality principle on (9.381). In so doing, we observe that the left-hand side is transformed into itself, and hence we find

$$\lim_{b\to 0^+}\int_{\partial B_b} \mathrm{d}S\ \hat{\mathbf{n}}(\mathbf{r})\cdot[\mathbf{E}_b(\mathbf{r})\times\mathbf{H}_a(\mathbf{r})-\mathbf{E}_a(\mathbf{r})\times\mathbf{H}_b(\mathbf{r})] = \mathrm{j}\,\omega\mathbf{H}_a(\mathbf{r}_b)\cdot\mu\mathbf{m}_b \tag{9.392}$$

in light of our previous discussion about the dual of $\mathbf{p}$. By enforcing (9.369) with (9.380) and (9.392) we arrive at

$$-\mathbf{p}_a\cdot\mathbf{E}_b(\mathbf{r}_a) = \mu\mathbf{m}_b\cdot\mathbf{H}_a(\mathbf{r}_b) \tag{9.393}$$

which is yet another instance of the Rayleigh-Carson theorem [1]. Now, since $\mathbf{E}_b(\mathbf{r}_a)$ is provided by (9.390) and $\mathbf{H}_a(\mathbf{r}_b)$ is the same as in (9.371), (9.393) becomes

$$\begin{aligned}-\mathbf{p}_a\cdot\overline{\mathbf{G}}_{EM}(\mathbf{r}_a,\mathbf{r}_b;\omega)\cdot\mathrm{j}\,\omega\mu\mathbf{m}_b &= \mu\mathbf{m}_b\cdot\overline{\mathbf{G}}_{HJ}(\mathbf{r}_b,\mathbf{r}_a;\omega)\cdot\mathrm{j}\,\omega\mathbf{p}_a\\ &= \mathrm{j}\,\omega\mathbf{p}_a\cdot\left[\overline{\mathbf{G}}_{HJ}(\mathbf{r}_b,\mathbf{r}_a;\omega)\right]^T\cdot\mu\mathbf{m}_b\end{aligned} \tag{9.394}$$

and, as the dipole moments and the positions thereof are arbitrary, we obtain

$$\overline{\mathbf{G}}_{EM}(\mathbf{r}_a,\mathbf{r}_b;\omega) = -\left[\overline{\mathbf{G}}_{HJ}(\mathbf{r}_b,\mathbf{r}_a;\omega)\right]^T,\qquad \mathbf{r}_a,\mathbf{r}_b\in\mathbb{R}^3,\qquad \mathbf{r}_a\neq\mathbf{r}_b \tag{9.395}$$

as the mixed-type symmetry relation involving electric and magnetic dyadic Green functions [1].

We conclude by mentioning that the reciprocity condition (9.369) can be derived also for fields and sources in a finite region V with arbitrary shape, so long as the electromagnetic field on ∂V satisfies impedance or admittance relationships in the form (6.262) or (6.263), in which case the contribution of the flux integral over ∂V vanishes. As a result, the symmetry properties (9.384), (9.385) and (9.395) apply to the relevant time-harmonic dyadic Green functions, even though they may not be available in closed form.

9.8 Quasi-static electromagnetic fields

The Maxwell equations provide the basis for the analysis and prediction of electromagnetic phenomena at *any* timescale. The importance of the full-fledged theoretical framework afforded by (1.20), (1.23), (1.34), (1.44) and (1.46) lies in the intimate relation between time-varying electric and magnetic fields. Indeed, the displacement current may be regarded as a source of magnetic fields — and especially so in regions of space devoid of electric currents — whereas a time-dependent magnetic induction field effectively plays the role of a source for the electric field. If either one of these two ingredients is absent, overlooked or neglected, we fail to have generation and propagation of electromagnetic waves (Section 1.8 and Chapter 7). Then again, in many practical situations the time rate of variation of either $\mathfrak{D}(\mathbf{r}, t)$ or $\mathfrak{B}(\mathbf{r}, t)$ happens to be *small* as compared to the time taken by the light to propagate across the spatial region of interest. When this is case, we may neglect either the displacement current or the time derivative of the magnetic induction field. The resulting sets of Maxwell equations are termed *electro-quasi-static* (EQS) and *magneto-quasi-static* (MQS), respectively [60, Section 6.2], [20, Chapter 14]. We wish to put this approximation on solid ground.

With reference to the wave scattering problem sketched in Figure 6.4 we suppose that the sources generate electromagnetic waves whose typical duration is T. The latter may represent the period of a time-harmonic wave (Section 1.5) as well as the temporal extension of a time-limited pulse. Secondly, we assume that the characteristic size of the object (e.g., its maximum chord) is L. We observe that the time taken by the light to travel the distance L is

$$T_{light} := \frac{L}{c_0} \tag{9.396}$$

whereas the distance covered by the light during the time interval T reads

$$L_{light} := c_0 T \tag{9.397}$$

with c_0 being the speed of light in free space given by (1.53). We are interested in the solution of the problem in the special instance where

$$\alpha := \frac{L}{L_{light}} \equiv \frac{T_{light}}{T} \ll 1 \tag{9.398}$$

that is, when either the object is very small compared to L_{light} or the characteristic time of the electromagnetic wave is very long compared to T_{light}. The meaning of these hypotheses (which are equivalent) is clear:

(1) for a given instant of time t the variation of the field with the space coordinates is negligible over the spatial region occupied by the object;

(2) in any given point $\mathbf{r}$ of the object the electromagnetic field changes so slowly in time as to appear nearly constant.

Consequently, can we somehow 'simplify' the Maxwell equations when (9.398) holds true and, as a result, α is a small parameter? To answer this question first of all we need to exhibit α explicitly in the sources and the electromagnetic entities [61, Chapter 15], [62, Chapter 6]. This goal makes sense because the pertinent electromagnetic field depends on L and T, which are input data to the problem.

Now, if $\alpha = 1$ — that is, $L \equiv L_{light}$ and $T_{light} = T$ — and $\varrho(\mathbf{r}, t)$ denotes the charge density responsible for generating the electromagnetic field in the presence of the obstacle, we may certainly write

$$\varrho(\mathbf{r}, t) \equiv \varrho(\mathbf{r}, \alpha t) = \varrho(\mathbf{r}, \tau) \tag{9.399}$$

with $\tau := \alpha t$ indicating a new time variable. The current density $\mathfrak{J}(\mathbf{r}, t)$ must satisfy

$$\nabla \cdot \mathfrak{J}(\mathbf{r}, t) = -\frac{\partial}{\partial t}\varrho(\mathbf{r}, \tau) = -\alpha\frac{\partial}{\partial \tau}\varrho(\mathbf{r}, \tau) \tag{9.400}$$

on account of the charge conservation (1.46). This equation states that $\mathfrak{J}$ depends *separately* on α and τ inasmuch as the parameter α appears explicitly in the right-hand side. Therefore, we write $\mathfrak{J}(\mathbf{r}, t) \equiv \mathfrak{J}(\mathbf{r}, \tau, \alpha)$. By inserting the current density in the Ampère-Maxwell law (1.34) and noticing that

$$\frac{\partial}{\partial t}\mathfrak{D}(\mathbf{r}, \alpha t) = \alpha\frac{\partial}{\partial \tau}\mathfrak{D}(\mathbf{r}, \tau) \tag{9.401}$$

we conclude that also the magnetic field $\mathcal{H}(\mathbf{r}, t)$ must depend separately on α and τ. It follows that the same is true for $\mathcal{E}(\mathbf{r}, t)$ by virtue of the Faraday law (1.20) and a suitable constitutive relationship for $\mathcal{B}$ and $\mathcal{H}$. Then, the relevant constitutive relationship which links $\mathfrak{D}$ to $\mathcal{E}$ indicates that $\mathfrak{D}(\mathbf{r}, \alpha t) \equiv \mathfrak{D}(\mathbf{r}, \alpha, \tau)$. Lastly, the Gauss law (1.44) requires we reconsider our initial assumption on the functional dependence of ϱ and that, more precisely, we write $\varrho(\mathbf{r}, \tau, \alpha)$.

All in all we have managed to cast the Maxwell equations (1.20), (1.23), (1.34), (1.44) and (1.46) into the alternative form

$$\nabla \times \mathcal{E}(\mathbf{r}, \tau, \alpha) = -\alpha\frac{\partial}{\partial \tau}\mathcal{B}(\mathbf{r}, \tau, \alpha) \tag{9.402}$$

$$\nabla \times \mathcal{H}(\mathbf{r}, \tau, \alpha) = \alpha\frac{\partial}{\partial \tau}\mathfrak{D}(\mathbf{r}, \tau, \alpha) + \mathfrak{J}(\mathbf{r}, \tau, \alpha) \tag{9.403}$$

$$\nabla \cdot \mathfrak{D}(\mathbf{r}, \tau, \alpha) = \varrho(\mathbf{r}, \tau, \alpha) \tag{9.404}$$

$$\nabla \cdot \mathcal{B}(\mathbf{r}, \tau, \alpha) = 0 \tag{9.405}$$

$$\nabla \cdot \mathfrak{J}(\mathbf{r}, \tau, \alpha) = -\alpha\frac{\partial}{\partial \tau}\varrho(\mathbf{r}, \tau, \alpha) \tag{9.406}$$

for any $(\mathbf{r}, \tau) \in \mathbb{R}^3 \times \mathbb{R}$ and $\alpha = 1$. We shall concede that this set of equations does not bring in any advantage to the solution of the original problem of Figure 6.4. However, since we have made the dependence on L or T explicit through α, we are now in a better position to wonder what happens to the equations and the electromagnetic field when we reduce the size L of the object for a given source or we extend the duration T of the source in the presence of an object with fixed size. Either way, the scaling parameter α becomes smaller than unity.

Since sources and fields are function of the parameter α as well, we may try asymptotic expansions in positive powers of α, namely,

$$\varrho(\mathbf{r}, \tau, \alpha) = \varrho^{(0)}(\mathbf{r}, \tau) + \alpha\varrho^{(1)}(\mathbf{r}, \tau) + \alpha^2\varrho^{(2)}(\mathbf{r}, \tau) + \ldots \tag{9.407}$$

$$\mathcal{J}(\mathbf{r}, \tau, \alpha) = \mathcal{J}^{(0)}(\mathbf{r}, \tau) + \alpha \mathcal{J}^{(1)}(\mathbf{r}, \tau) + \alpha^2 \mathcal{J}^{(2)}(\mathbf{r}, \tau) + \dots \tag{9.408}$$

$$\mathcal{E}(\mathbf{r}, \tau, \alpha) = \mathcal{E}^{(0)}(\mathbf{r}, \tau) + \alpha \mathcal{E}^{(1)}(\mathbf{r}, \tau) + \alpha^2 \mathcal{E}^{(2)}(\mathbf{r}, \tau) + \dots \tag{9.409}$$

$$\mathcal{H}(\mathbf{r}, \tau, \alpha) = \mathcal{H}^{(0)}(\mathbf{r}, \tau) + \alpha \mathcal{H}^{(1)}(\mathbf{r}, \tau) + \alpha^2 \mathcal{H}^{(2)}(\mathbf{r}, \tau) + \dots \tag{9.410}$$

$$\mathcal{D}(\mathbf{r}, \tau, \alpha) = \mathcal{D}^{(0)}(\mathbf{r}, \tau) + \alpha \mathcal{D}^{(1)}(\mathbf{r}, \tau) + \alpha^2 \mathcal{D}^{(2)}(\mathbf{r}, \tau) + \dots \tag{9.411}$$

$$\mathcal{B}(\mathbf{r}, \tau, \alpha) = \mathcal{B}^{(0)}(\mathbf{r}, \tau) + \alpha \mathcal{B}^{(1)}(\mathbf{r}, \tau) + \alpha^2 \mathcal{B}^{(2)}(\mathbf{r}, \tau) + \dots \tag{9.412}$$

for $\alpha \to 0^+$. Next, we insert (9.407)-(9.412) into (9.402)-(9.406) and, by invoking the linearity of the differential operators and interchanging the order or differentiation and summation, we group terms which involve like powers $\alpha^0, \alpha^1, \alpha^2, \dots$ For instance, (9.402) and (9.404) pass over into

$$\nabla \times \mathcal{E}^{(0)} + \alpha \left(\nabla \times \mathcal{E}^{(1)} + \frac{\partial \mathcal{B}^{(0)}}{\partial \tau} \right) + \alpha^2 \left(\nabla \times \mathcal{E}^{(2)} + \frac{\partial \mathcal{B}^{(1)}}{\partial \tau} \right) + \dots = 0 \tag{9.413}$$

$$\nabla \cdot \mathcal{D}^{(0)} - \varrho^{(0)} + \alpha \left(\nabla \cdot \mathcal{D}^{(1)} - \varrho^{(1)} \right) + \alpha \left(\nabla \cdot \mathcal{D}^{(2)} - \varrho^{(2)} \right) + \dots = 0 \tag{9.414}$$

and similar expressions follow from (9.403), (9.405) and (9.406). In order for all these equations to be true for any value of $\alpha \in \mathbb{R}^+$ we demand that the coefficients of like powers of α vanish simultaneously. Such procedure generates an infinite sequence of differential systems, namely,

$$\nabla \times \mathcal{E}^{(0)}(\mathbf{r}, \tau) = 0 \tag{9.415}$$

$$\nabla \times \mathcal{H}^{(0)}(\mathbf{r}, \tau) = \mathcal{J}^{(0)}(\mathbf{r}, \tau) \tag{9.416}$$

$$\nabla \cdot \mathcal{D}^{(0)}(\mathbf{r}, \tau) = \varrho^{(0)}(\mathbf{r}, \tau) \qquad \text{(zeroth order)} \tag{9.417}$$

$$\nabla \cdot \mathcal{B}^{(0)}(\mathbf{r}, \tau) = 0 \tag{9.418}$$

$$\nabla \cdot \mathcal{J}^{(0)}(\mathbf{r}, \tau) = 0 \tag{9.419}$$

$$\nabla \times \mathcal{E}^{(1)}(\mathbf{r}, \tau) = -\frac{\partial}{\partial \tau} \mathcal{B}^{(0)}(\mathbf{r}, \tau) \tag{9.420}$$

$$\nabla \times \mathcal{H}^{(1)}(\mathbf{r}, \tau) = \frac{\partial}{\partial \tau} \mathcal{D}^{(0)}(\mathbf{r}, \tau) + \mathcal{J}^{(1)}(\mathbf{r}, \tau) \tag{9.421}$$

$$\nabla \cdot \mathcal{D}^{(1)}(\mathbf{r}, \tau) = \varrho^{(1)}(\mathbf{r}, \tau) \qquad \text{(first order)} \tag{9.422}$$

$$\nabla \cdot \mathcal{B}^{(1)}(\mathbf{r}, \tau) = 0 \tag{9.423}$$

$$\nabla \cdot \mathcal{J}^{(1)}(\mathbf{r}, \tau) = -\frac{\partial}{\partial \tau} \varrho^{(0)}(\mathbf{r}, \tau) \tag{9.424}$$

and more generally, for $n \in \mathbb{N}$

$$\nabla \times \mathcal{E}^{(n)}(\mathbf{r}, \tau) = -\frac{\partial}{\partial \tau} \mathcal{B}^{(n-1)}(\mathbf{r}, \tau) \tag{9.425}$$

$$\nabla \times \mathcal{H}^{(n)}(\mathbf{r}, \tau) = \frac{\partial}{\partial \tau} \mathcal{D}^{(n-1)}(\mathbf{r}, \tau) + \mathcal{J}^{(n)}(\mathbf{r}, \tau) \tag{9.426}$$

$$\nabla \cdot \mathcal{D}^{(n)}(\mathbf{r}, \tau) = \varrho^{(n)}(\mathbf{r}, \tau) \qquad (n\text{th order}) \tag{9.427}$$

$$\nabla \cdot \mathcal{B}^{(n)}(\mathbf{r}, \tau) = 0 \tag{9.428}$$

$$\nabla \cdot \mathcal{J}^{(n)}(\mathbf{r}, \tau) = -\frac{\partial}{\partial \tau} \varrho^{(n-1)}(\mathbf{r}, \tau) \tag{9.429}$$

for any $(\mathbf{r}, \tau) \in \mathbb{R}^3 \times \mathbb{R}$, with the understanding that $\mathcal{B}^{(-1)}$, $\mathcal{D}^{(-1)}$ and $\varrho^{(-1)}$ are null by convention.

We observe that (9.415)-(9.419) are *formally* identical with the Maxwell equations in the static and stationary regimes. However, the electromagnetic entities involved *do* depend on time. What (9.415)-(9.419) say is that the zeroth-order fields are determined at a given time only by the value of the sources in the same instant and not by the previous values taken on by the fields. Besides, we may solve (9.415)-(9.419) for the zeroth-order fields $\mathcal{E}^{(0)}$, $\mathcal{D}^{(0)}$, $\mathcal{H}^{(0)}$ and $\mathcal{B}^{(0)}$ independently of the remaining systems.

Conversely, we see from (9.420)-(9.424) — which are formally identical with the general Maxwell equations for time-varying fields — that we need to know $\mathcal{D}^{(0)}$ and $\mathcal{B}^{(0)}$ in order to determine the first-order fields $\mathcal{E}^{(1)}$, $\mathcal{D}^{(1)}$, $\mathcal{H}^{(1)}$ and $\mathcal{B}^{(1)}$. In this respect, (9.420)-(9.424) are coupled to (9.415)-(9.419). In practice, we may regard the zeroth-order displacement current $\partial\mathcal{D}^{(0)}/\partial\tau$ as a source term for the first-order magnetic field $\mathcal{H}^{(1)}$. In like manner, the time variation of the zeroth-order magnetic induction plays the role of a source term for the first-order electric field $\mathcal{H}^{(1)}$. The procedure may be iterated *ad libitum* and, in effect, (9.425)-(9.429) for $n \geqslant 1$ state that, once we know the same entities and sources at the order $n-1$, we can determine the fields at the order n by solving equations valid for static or stationary fields! In some situations, though, we hardly need extend the calculation beyond the first order, provided (9.398) applies.

9.8.1 Electro-quasi-static regime

In the so-called electro-quasi-static (EQS) regime we make the approximation

$$\mathcal{E} \approx \mathcal{E}^{(0)}, \qquad \mathcal{D} \approx \mathcal{D}^{(0)}, \qquad \varrho \approx \varrho^{(0)} \tag{9.430}$$

and to the zeroth-order neglect the magnetic entities. Therefore, the relevant equations are (9.415) and (9.417). The magnetic entities are first-order in α and follow from (9.421) and (9.423) complemented with the continuity equation (9.424).

We summarize the EQS equations by reverting to the usual time variable t and dropping the superscripts (0) and (1), viz.,

$$\nabla \cdot \mathcal{D}(\mathbf{r}, t) = \varrho(\mathbf{r}, t) \tag{9.431}$$

$$\nabla \times \mathcal{E}(\mathbf{r}, t) = 0 \tag{9.432}$$

$$\nabla \times \mathcal{H}(\mathbf{r}, t) = \frac{\partial}{\partial t}\mathcal{D}(\mathbf{r}, t) + \mathcal{J}(\mathbf{r}, t) \tag{9.433}$$

$$\nabla \cdot \mathcal{B}(\mathbf{r}, t) = 0 \tag{9.434}$$

$$\nabla \cdot \mathcal{J}(\mathbf{r}, t) = -\frac{\partial}{\partial t}\varrho(\mathbf{r}, t) \tag{9.435}$$

supplemented with suitable constitutive relationships and boundary conditions. If a system is EQS in nature, then in the limit as the characteristic time (resp. the size) grows very large (resp. small) no stationary current can flow, which is why the magnetic entities are null to the zeroth order.

Since in the EQS regime the electric field is curl-free to the zeroth order, we may introduce an electric potential $\Phi(\mathbf{r}, t)$ as in (2.15) and, thanks to (9.431) and (1.117), we obtain a Poisson equation for $\Phi(\mathbf{r}, t)$. The main difference with respect to (2.26) lies in the fact that charge density and potential are also *slowly varying* functions of time. Nonetheless, we may apply all the results of Chapter 2 and in particular the integral representation (2.169) in an unbounded homogeneous isotropic medium. Once the electric entities have been found, we compute the magnetic entities by introducing a magnetic vector potential $\mathcal{A}(\mathbf{r}, t)$ and regarding the right-hand side of (9.433) as a known *total* electric current density.

9.8.2 Magneto-quasi-static regime

In the magneto-quasi-static (MQS) regime we make use of the approximations

$$\mathcal{H} \approx \mathcal{H}^{(0)}, \qquad \mathcal{B} \approx \mathcal{B}^{(0)}, \qquad \mathcal{J} \approx \mathcal{J}^{(0)} \tag{9.436}$$

and to the zeroth-order we discard the electric entities. The pertinent equations are (9.416) and (9.418) supplemented with condition (9.419) for solvability. The electric entities are first-order in α and are derived from (9.420), (9.422). In summary, the MQS equations read

$$\nabla \times \mathcal{H}(\mathbf{r}, t) = \mathcal{J}(\mathbf{r}, t) \tag{9.437}$$

$$\nabla \cdot \mathcal{B}(\mathbf{r}, t) = 0 \tag{9.438}$$

$$\nabla \cdot \mathcal{J}(\mathbf{r}, t) = 0 \tag{9.439}$$

$$\nabla \times \mathcal{E}(\mathbf{r}, t) = -\frac{\partial}{\partial t}\mathcal{B}(\mathbf{r}, t) \tag{9.440}$$

$$\nabla \cdot \mathcal{D}(\mathbf{r}, t) = \varrho(\mathbf{r}, t) \tag{9.441}$$

plus constitutive relations and boundary conditions. A system is MQS in nature if a *steady current* can still flow in the limit as the characteristic time (resp. the size) grows very large (resp. small) which is why the magnetic entities dominate to the zeroth order, whereas the electric ones are first order.

In the MQS regime we may define a vector potential $\mathcal{A}(\mathbf{r}, t)$ as in (4.29) and, by virtue of (9.437) and (1.118), write the Poisson vector equation for $\mathcal{A}(\mathbf{r}, t)$. The solution is still given by (5.29) in an unbounded homogeneous isotropic medium. Alternatively, since in spatial regions devoid of currents the magnetic field is curl-free, we may introduce a magnetic scalar potential $\Psi(\mathbf{r}, t)$ as in (4.68). Then, by invoking (9.438) and the constitutive relationship (1.118) we can write the Laplace equation for $\Psi(\mathbf{r}, t)$.

Once the magnetic entities have been determined, we compute the electric entities with the help of the remaining equations. Since according to (9.440) the electric field is not curl-free, we proceed by invoking the principle of superposition (Section 6.1). In practice, we notice that the electric field is due to two 'sources', namely, the charge density $\varrho(\mathbf{r}, t)$ and a magnetic current density $\partial\mathcal{B}/\partial t$. In symbols, we have $\mathcal{E} := \mathcal{E}_1 + \mathcal{E}_2$ and the uncoupled sets of equations

$$\nabla \times \mathcal{E}_1(\mathbf{r}, t) = 0 \qquad \nabla \cdot \mathcal{D}_1(\mathbf{r}, t) = \varrho(\mathbf{r}, t) \tag{9.442}$$

$$\nabla \times \mathcal{E}_2(\mathbf{r}, t) = -\frac{\partial}{\partial t}\mathcal{B}(\mathbf{r}, t) \qquad \nabla \cdot \mathcal{D}_2(\mathbf{r}, t) = 0 \tag{9.443}$$

plus suitable boundary conditions and constitutive relationships. The first set is solved by introducing a scalar electric potential $\Phi_1(\mathbf{r}, t)$ as in Chapter 2. For the second system we may resort to a *vector* electric potential $\mathcal{F}_2(\mathbf{r}, t)$ such that

$$\mathcal{D}_2(\mathbf{r}, t) = -\nabla \times \mathcal{F}_2(\mathbf{r}, t), \qquad \mathbf{r} \in \mathbb{R}^3 \tag{9.444}$$

and mirror the steps outlined in Chapter 4 for stationary magnetic fields.

Example 9.6 (Dielectric sphere slowly rotating in a uniform electrostatic field)

The problem of a rotating dielectric sphere exposed to a uniform electrostatic field $\mathcal{E}^{\mathrm{i}}(\mathbf{r}) = \mathcal{E}_0\hat{\mathbf{x}}$ may be treated in an elementary way, so long as the velocities involved are small compared to the speed of light. Or else, one has to resort to the full-fledged Maxwell equations suitably formulated in rotating systems of coordinates [19, 63].

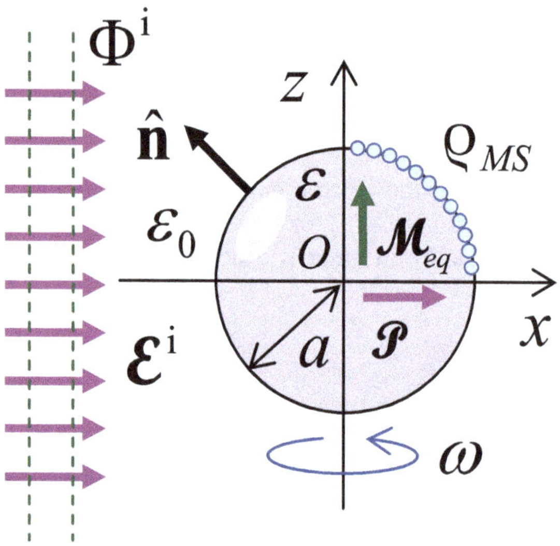

Figure 9.20 Rotating dielectric sphere immersed in an impressed uniform electrostatic field $\mathcal{E}^i$. The dashed lines are the traces of the planes (perpendicular to $\mathcal{E}^i$) whereon the electrostatic potential Φ^{i} is constant.

We suppose that the sphere in question, endowed with permittivity ε, is immersed in free space, occupies the ball $B(0, a)$, and rotates with uniform angular velocity $\omega \ll c_0/a$ around the z-axis, as is sketched in Figure 9.20. Although the electric field is static in the laboratory frame, the polarization vector $\mathcal{P}$ — which to an observer on the sphere appears to be *rotating* with velocity $-\omega$ — gives rise to an equivalent or effective magnetization $\mathcal{M}_{eq}(\mathbf{r})$ within the sphere. The electro-quasi-static regime is invoked first to derive the polarization vector $\mathcal{P}(\mathbf{r})$, then the magneto-quasi-static approximation is applied to find the magnetic scalar potential $\Psi(\mathbf{r})$.

To begin with, we observe that the polarization vector reads

$$\mathcal{P}(\mathbf{r}) = 3\varepsilon\frac{\varepsilon - \varepsilon_0}{\varepsilon + 2\varepsilon_0}\mathcal{E}_0\hat{\mathbf{x}}, \qquad r \leqslant a \tag{9.445}$$

thanks to the results of Example 3.6. Since $\mathcal{P}(\mathbf{r})$ is actually uniform, the volume density of polarization charges (3.246) is null within sphere. However, on the surface thereof we have

$$\begin{aligned}\varrho_{SP}(\mathbf{r}) &:= \hat{\mathbf{n}}(\mathbf{r}) \cdot \mathcal{P}(\mathbf{r}) = 3\varepsilon\frac{\varepsilon - \varepsilon_0}{\varepsilon + 2\varepsilon_0}\mathcal{E}_0\hat{\mathbf{r}} \cdot \hat{\mathbf{x}} \\ &= 3\varepsilon\frac{\varepsilon - \varepsilon_0}{\varepsilon + 2\varepsilon_0}\mathcal{E}_0 \sin\vartheta\cos\varphi, \qquad r = a\end{aligned} \tag{9.446}$$

and, since the sphere rotates around $\hat{\mathbf{z}}$, these charges are apparently dragged along the azimuthal direction with velocity $\mathbf{v}(t)$. The latter is by definition the time rate of change of the position vector of a point on the surface. Therefore, we have

$$\mathbf{v}(t) = \frac{\mathrm{d}\mathbf{r}}{\mathrm{d}t} = r\frac{\mathrm{d}\hat{\mathbf{r}}}{\mathrm{d}t} = r\frac{\mathrm{d}\varphi}{\mathrm{d}t}(-\sin\varphi\hat{\mathbf{x}} + \cos\varphi\hat{\mathbf{y}})\sin\vartheta = r\omega\sin\vartheta\hat{\boldsymbol{\varphi}}, \qquad r = a \tag{9.447}$$

on account of (A.11) and (A.13). Different points of the sphere move at different velocities, depending on the distance from the z-axis. The apparent motion of the polarization charges constitutes an electric surface current, i.e.,

$$\mathcal{J}_{SP}(\mathbf{r}, t) = \varrho_{SP}\mathbf{v}(t) = \hat{\mathbf{n}}(\mathbf{r}) \cdot \mathcal{P}(\mathbf{r})\mathbf{v}(t) = -\hat{\mathbf{n}}(\mathbf{r}) \times [\mathcal{P}(\mathbf{r}) \times \mathbf{v}(t)], \qquad r = a \tag{9.448}$$

where we have made use of the expansion of the triple cross product (H.14) and noticed that the velocity is perpendicular to the unit normal vector on ∂B. Finally, by comparing (9.448) with (5.172) we obtain

$$\mathcal{M}_{eq}(\mathbf{r}) = \mathcal{P}(\mathbf{r}) \times \mathbf{v}(t), \qquad r \leqslant a \tag{9.449}$$

which is the desired equivalent magnetization vector. Inserting the explicit expressions of $\mathcal{P}(\mathbf{r})$ and $\mathbf{v}(t)$ in the above expression provides

$$\begin{aligned}\mathcal{M}_{eq}(\mathbf{r}) &= 3\varepsilon\frac{\varepsilon - \varepsilon_0}{\varepsilon + 2\varepsilon_0}\mathcal{E}_0\hat{\mathbf{x}} \times \omega r \sin\vartheta\hat{\boldsymbol{\varphi}} = 3\omega\varepsilon\frac{\varepsilon - \varepsilon_0}{\varepsilon + 2\varepsilon_0}\mathcal{E}_0\hat{\mathbf{z}} r \sin\vartheta\cos\varphi \\ &= 3\omega\varepsilon\frac{\varepsilon - \varepsilon_0}{\varepsilon + 2\varepsilon_0}\mathcal{E}_0 x\hat{\mathbf{z}}\end{aligned} \tag{9.450}$$

where we have invoked the transformation rule (A.13). The weak time dependence of $\mathcal{M}_{eq}(\mathbf{r})$ results from the velocity constantly changing direction. What is more, the polarization charges on ∂B are, in fact, accelerated toward the z-axis and radiation should occur, as we know from (1.252) and (1.253) or more generally from (9.136) and (9.151). However, in the MQS regime radiation is neglected because so is the displacement current.

Next, we recall that according to (5.188) and (5.189) the state of magnetization may be interpreted equivalently as a distribution of *magnetic charges* within the sphere and on the surface thereof. For the problem of concern, in the bulk of the sphere ($r < a$) we have

$$\varrho_M(\mathbf{r}) := -\mu_0\nabla \cdot \mathcal{M}_{eq}(\mathbf{r}) = -3\omega\varepsilon\frac{\varepsilon - \varepsilon_0}{\varepsilon + 2\varepsilon_0}\mathcal{E}_0\frac{\partial x}{\partial z} = 0 \tag{9.451}$$

and on the surface ($r = a$) we have

$$\begin{aligned}\varrho_{MS}(\mathbf{r}) &:= \mu_0\hat{\mathbf{n}}(\mathbf{r}) \cdot \mathcal{M}_{eq}(\mathbf{r}) = 3\omega\varepsilon\mu_0\frac{\varepsilon - \varepsilon_0}{\varepsilon + 2\varepsilon_0}\mathcal{E}_0 a\cos\vartheta\sin\vartheta\cos\varphi \\ &= -\omega\varepsilon\mu_0\frac{\varepsilon - \varepsilon_0}{\varepsilon + 2\varepsilon_0}aP_2^1(\vartheta)\frac{\mathrm{e}^{\mathrm{j}\varphi} + \mathrm{e}^{-\mathrm{j}\varphi}}{2}\mathcal{E}_0\end{aligned} \tag{9.452}$$

where $P_2^1(\vartheta)$ denotes the associated Legendre function of order $n = 2$ and $m = 1$ as defined in (H.122).

From this point on the calculation of the magnetic field boils down to solving a boundary value problem for the magnetic scalar potential (4.68) produced by ϱ_{MS}. Notice that here $\Psi(\mathbf{r})$ is *single-valued* because there are no macroscopic electric current densities anywhere (cf. Figure 4.7). In symbols, $\Psi(\mathbf{r})$ is the unique solution to the boundary value problem [cf. (3.1)]

$$\begin{cases}\nabla^2\Psi(\mathbf{r}) = 0, & \mathbf{r} \in \mathbb{R}^3 \setminus \partial B \\ \Psi^+(\mathbf{r}) = \Psi^-(\mathbf{r}), & \mathbf{r} \in \partial B \\ \varrho_{MS}(\mathbf{r}) = \mu_0\left.\dfrac{\partial\Psi}{\partial\hat{\mathbf{n}}}\right|_{\mathbf{r}\in\partial B^-} - \mu_0\left.\dfrac{\partial\Psi}{\partial\hat{\mathbf{n}}}\right|_{\mathbf{r}\in\partial B^+}, & \mathbf{r} \in \partial B \\ \Psi(\mathbf{r}) = \mathcal{O}\left(\dfrac{1}{|\mathbf{r}|}\right), & |\mathbf{r}| \to +\infty\end{cases} \tag{9.453}$$

where Ψ^+ and Ψ^- denote the potential on the positive and negative side of ∂B, with the unit normal $\hat{\mathbf{n}}$ on ∂B pointing into the unbounded background medium. We claim uniqueness of solutions on the grounds that (9.453) is of the same type of Laplace equations discussed in Sections 2.5.1 and 3.1 with reference to the electrostatic potential $\Phi(\mathbf{r})$.

While the geometry is rotationally symmetric around the z-axis, yet the distribution of equivalent magnetic charge is not, whereby we are prompted to seek the solution in the form (Section 3.5.1)

$$\Psi(\mathbf{r}) = \begin{cases} \sum_{n=0}^{+\infty} \sum_{m=-n}^{+n} A_{nm} \left(\frac{a}{r}\right)^{n+1} P_n^m(\vartheta) e^{jm\varphi}, & r > a \\ \sum_{n=0}^{+\infty} \sum_{m=-n}^{+n} B_{nm} \left(\frac{r}{a}\right)^{n} P_n^m(\vartheta) e^{jm\varphi}, & r < a \end{cases} \tag{9.454}$$

because the potential must be regular at the center of the sphere as well as vanish at infinity. In order to ensure the continuity of $\Psi(\mathbf{r})$ across the boundary of the sphere we must choose

$$A_{nm} = B_{nm} \qquad n \geqslant 0 \qquad m \in [-n, n] \tag{9.455}$$

which, when used in tandem with the jump of the normal derivatives, provides us with the condition

$$\frac{1}{a} \sum_{n=0}^{+\infty} \sum_{m=-n}^{+n} A_{nm}(2n+1) P_n^m(\vartheta) e^{jm\varphi} = -\omega\varepsilon \frac{\varepsilon - \varepsilon_0}{\varepsilon + 2\varepsilon_0} a P_2^1(\vartheta) \frac{e^{j\varphi} + e^{-j\varphi}}{2} \mathcal{E}_0 \tag{9.456}$$

that has to hold true for any value of ϑ and φ. By pairing like terms proportional to the associated Legendre functions and the complex exponentials we get

$$\frac{1}{a} A_{nm}(2n+1) P_n^m(\vartheta) e^{jm\varphi} = 0 \qquad n \neq 2 \qquad m \neq \pm 1 \qquad m \in [-n, n] \tag{9.457}$$

and in particular

$$\frac{5}{a} A_{21} P_2^1(\vartheta) e^{j\varphi} = -\frac{1}{2} \omega\varepsilon \frac{\varepsilon - \varepsilon_0}{\varepsilon + 2\varepsilon_0} a P_2^1(\vartheta) e^{j\varphi} \mathcal{E}_0 \qquad n = 2 \qquad m = 1 \tag{9.458}$$

$$-\frac{5}{3!a} A_{2,-1} P_2^1(\vartheta) e^{-j\varphi} = -\frac{1}{2} \omega\varepsilon \frac{\varepsilon - \varepsilon_0}{\varepsilon + 2\varepsilon_0} a P_2^1(\vartheta) e^{-j\varphi} \mathcal{E}_0 \qquad n = 2 \qquad m = -1 \tag{9.459}$$

where we have traded $P_2^{-1}(\vartheta)$ for $P_2^1(\vartheta)$ with the aid of the functional relationship (H.129). Thus, we find

$$A_{nm} = 0 \qquad n \neq 2 \qquad m \neq \pm 1 \qquad m \in [-n, n] \tag{9.460}$$

and that the only two non-zero coefficients for the representation of $\Psi(\mathbf{r})$ read

$$A_{21} = -\frac{a^2}{10} \omega\varepsilon \frac{\varepsilon - \varepsilon_0}{\varepsilon + 2\varepsilon_0} \mathcal{E}_0 \qquad A_{2,-1} = \frac{3a^2}{5} \omega\varepsilon \frac{\varepsilon - \varepsilon_0}{\varepsilon + 2\varepsilon_0} \mathcal{E}_0 \tag{9.461}$$

whereby the scalar potential becomes

$$\begin{aligned} \Psi(\mathbf{r}) &= -\frac{\omega a^2}{10} \left(\frac{a}{r}\right)^3 \varepsilon \frac{\varepsilon - \varepsilon_0}{\varepsilon + 2\varepsilon_0} P_2^1(\vartheta) e^{j\phi} \mathcal{E}_0 + \frac{3\omega a^2}{5} \left(\frac{a}{r}\right)^3 \varepsilon \frac{\varepsilon - \varepsilon_0}{\varepsilon + 2\varepsilon_0} \left(-\frac{1}{3!}\right) P_2^1(\vartheta) e^{-j\phi} \mathcal{E}_0 \\ &= \frac{3}{5} \frac{a^5}{r^3} \omega\varepsilon \frac{\varepsilon - \varepsilon_0}{\varepsilon + 2\varepsilon_0} \mathcal{E}_0 \cos\vartheta \sin\vartheta \cos\varphi \end{aligned}$$

$$= \frac{3}{5}\frac{a^5}{r^5}\omega\varepsilon\frac{\varepsilon-\varepsilon_0}{\varepsilon+2\varepsilon_0}\mathcal{E}_0 xz, \qquad r > a \tag{9.462}$$

$$\begin{aligned}\Psi(\mathbf{r}) &= -\frac{\omega a^2}{10}\left(\frac{r}{a}\right)^2 \varepsilon\frac{\varepsilon-\varepsilon_0}{\varepsilon+2\varepsilon_0}P_2^1(\vartheta)\mathrm{e}^{\mathrm{j}\phi}\mathcal{E}_0 + \frac{3\omega a^2}{5}\left(\frac{r}{a}\right)^2 \varepsilon\frac{\varepsilon-\varepsilon_0}{\varepsilon+2\varepsilon_0}\left(-\frac{1}{3!}\right)P_2^1(\vartheta)\mathrm{e}^{-\mathrm{j}\phi}\mathcal{E}_0 \\ &= \frac{3}{5}\omega\varepsilon\frac{\varepsilon-\varepsilon_0}{\varepsilon+2\varepsilon_0}r^2\mathcal{E}_0\cos\vartheta\sin\vartheta\cos\varphi \\ &= \frac{3}{5}\omega\varepsilon\frac{\varepsilon-\varepsilon_0}{\varepsilon+2\varepsilon_0}\mathcal{E}_0 xz, \qquad r < a\end{aligned} \tag{9.463}$$

by virtue of the transformation rules (A.17).

A few equipotential lines relevant to a sphere of permittivity $\varepsilon = 8\varepsilon_0$ are plotted in Figure 9.21 for $\varphi \in \{0, \pi/8, \pi/4, 3\pi/8\}$ as a function of z/a and the normalized distance $r\sin\vartheta/a$ from the z-axis. In light of the even-odd symmetry of the potential we consider only the values $\vartheta \in [0, \pi/2]$ for the polar angle. The lines represent the loci of the points for which the dimensionless function

$$\frac{\Psi(\mathbf{r})}{\omega\varepsilon a^2\mathcal{E}_0} = \frac{3}{5}\left(\frac{a}{r_>}\right)^5 \frac{\varepsilon-\varepsilon_0}{\varepsilon+2\varepsilon_0}\frac{xz}{a^2}, \qquad r_> = \max\{r, a\} \tag{9.464}$$

is a constant and equal to the values partly indicated in the plots. The specific values of angular velocity, radius of the sphere, and impressed electric field will affect the actual values of the magnetic potential but not the distribution of equipotential lines given in Figure 9.21.

Outside the sphere the magnetic scalar potential falls off as the inverse of the cubed distance from the origin and this is, in fact, the only non-vanishing term in the spherical-harmonics representation of $\Psi(\mathbf{r})$. Therefore, guided by the multipole expansion of $\Phi(\mathbf{r})$ developed in Section 3.6, we interpret $\Psi(\mathbf{r})$ as being generated by a *magnetic quadrupole*, i.e., a set of four fictitious magnetic charges (cf. Figure 3.9). Comparison with (3.199) prompts us to compute the magnetic quadrupole moment (physical dimension: Wb m^2) as

$$\begin{aligned}\overline{\mathfrak{Q}}_M &:= \int_{\partial B} \mathrm{d}S\,(3\mathbf{r}\mathbf{r} - r^2\overline{\mathfrak{I}})\varrho_{MS}(\mathbf{r}) \\ &= 3\omega\varepsilon\mu_0\frac{\varepsilon-\varepsilon_0}{\varepsilon+2\varepsilon_0}\mathcal{E}_0 a^5 \int_0^{2\pi}\mathrm{d}\varphi\int_0^{\pi}\mathrm{d}\vartheta\,\sin^2\vartheta\cos\vartheta\cos\varphi(3\hat{\mathbf{r}}\hat{\mathbf{r}} - \overline{\mathfrak{I}})\end{aligned} \tag{9.465}$$

where the traceless symmetric dyadic in the integrand reads

$$\begin{aligned}3\hat{\mathbf{r}}\hat{\mathbf{r}} - \overline{\mathfrak{I}} &= (3\sin^2\vartheta\cos^2\varphi - 1)\hat{\mathbf{x}}\hat{\mathbf{x}} + 3\sin^2\vartheta\cos\varphi\sin\varphi\hat{\mathbf{x}}\hat{\mathbf{y}} + 3\sin\vartheta\cos\vartheta\cos\varphi\hat{\mathbf{x}}\hat{\mathbf{z}} \\ &\quad + 3\sin^2\vartheta\cos\varphi\sin\varphi\hat{\mathbf{y}}\hat{\mathbf{x}} + (3\sin^2\vartheta\sin^2\varphi - 1)\hat{\mathbf{y}}\hat{\mathbf{y}} + 3\sin\vartheta\cos\vartheta\sin\varphi\hat{\mathbf{y}}\hat{\mathbf{z}} \\ &\quad + 3\sin\vartheta\cos\vartheta\cos\varphi\hat{\mathbf{z}}\hat{\mathbf{x}} + 3\sin\vartheta\cos\vartheta\sin\varphi\hat{\mathbf{z}}\hat{\mathbf{y}} + (3\cos^2\vartheta - 1)\hat{\mathbf{z}}\hat{\mathbf{z}}\end{aligned} \tag{9.466}$$

on account of the transformation formula (A.11). It is easy to ascertain that most terms integrate to zero except those proportional to the dyads $\hat{\mathbf{x}}\hat{\mathbf{z}}$ and $\hat{\mathbf{z}}\hat{\mathbf{x}}$. At last, we find

$$\overline{\mathfrak{Q}}_M = 9\omega\varepsilon\mu_0\frac{\varepsilon-\varepsilon_0}{\varepsilon+2\varepsilon_0}\mathcal{E}_0 a^5(\hat{\mathbf{x}}\hat{\mathbf{z}} + \hat{\mathbf{z}}\hat{\mathbf{x}})\int_0^{2\pi}\mathrm{d}\varphi\,\cos^2\varphi\int_0^{\pi}\mathrm{d}\vartheta\,\sin^3\vartheta\cos^2\vartheta$$

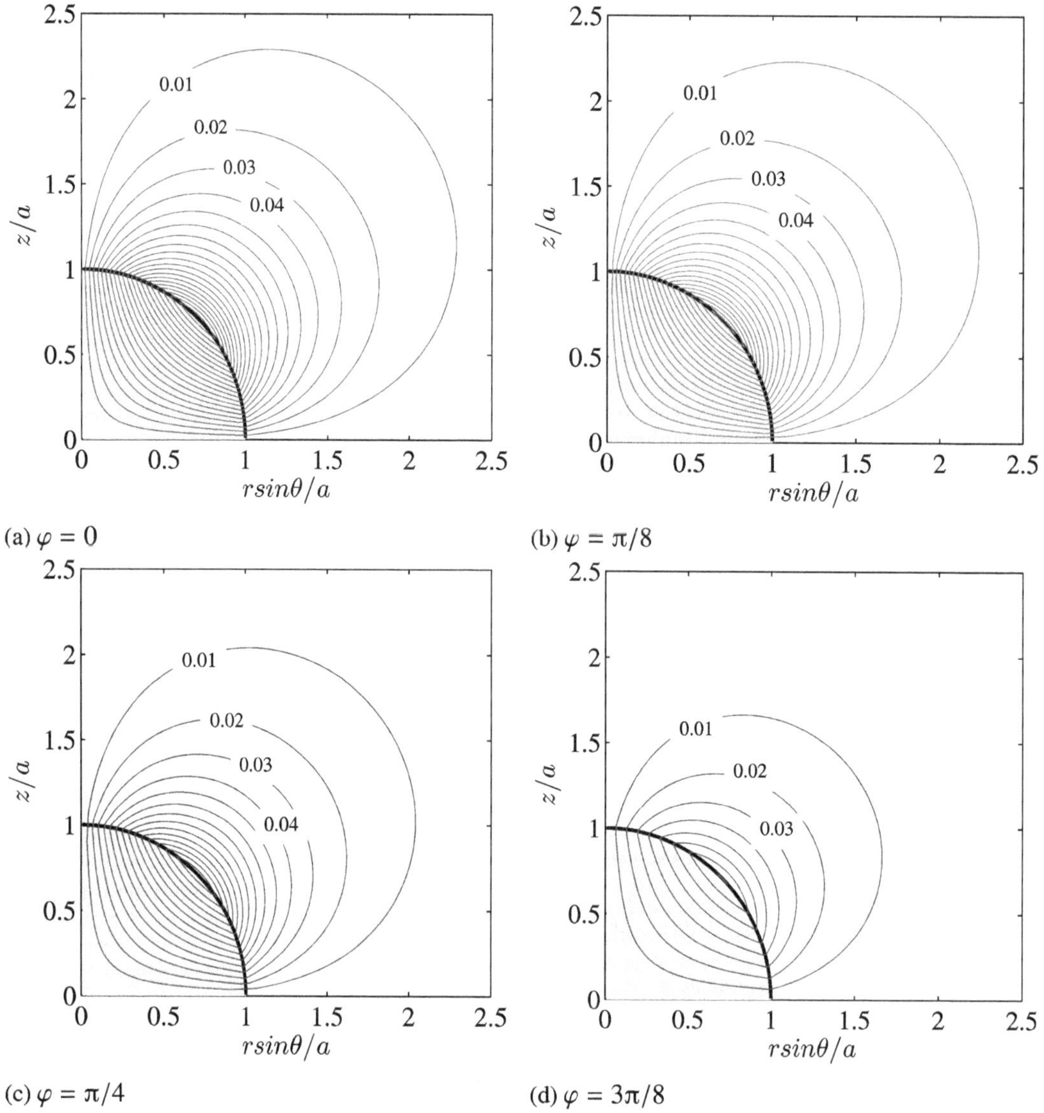

Figure 9.21 Equipotential lines of the normalized magnetic potential generated by a dielectric sphere ($\varepsilon = 8\varepsilon_0$) rotating in a uniform electrostatic field (Figure 9.20).

$$= 4\pi\mu_0 \frac{3}{5}\omega\varepsilon \frac{\varepsilon - \varepsilon_0}{\varepsilon + 2\varepsilon_0} \mathcal{E}_0 a^5 (\hat{\mathbf{x}}\hat{\mathbf{z}} + \hat{\mathbf{z}}\hat{\mathbf{x}}) \tag{9.467}$$

whereby the magnetic scalar potential is computed as

$$\Psi(\mathbf{r}) = \frac{1}{2}\frac{\hat{\mathbf{r}} \cdot \overline{\mathfrak{Q}}_M \cdot \hat{\mathbf{r}}}{4\pi\mu_0 r^3} = \frac{1}{2}\frac{\overline{\mathfrak{Q}}_M : \hat{\mathbf{r}}\hat{\mathbf{r}}}{4\pi\mu_0 r^3}, \qquad r \geqslant a \tag{9.468}$$

on the grounds of (3.204) and the principle of duality. In this regard, $\overline{\mathfrak{Q}}_M$ is the dual of $-\overline{\mathfrak{Q}}$ and viceversa. The result coincides with the one we obtained by solving the boundary value problem (9.453).

The second approach, though, is based on our knowing *a posteriori* that only the quadrupole term survives in the multipole expansion of $\Psi(\mathbf{r})$. Besides, the quadrupole interpretation holds only for the region outside the sphere. To push the equivalence even further we mention that (9.467) is produced by four magnetic point charges positioned at the corners of a square lying in the xOz plane and symmetrically placed around the origin. Specifying the edges with the vectors $\mathbf{u}_1 = b\hat{\mathbf{z}}$ and $\mathbf{u}_2 = -b\hat{\mathbf{x}}$ (see Figure 3.9) and using result (3.208) obtained in Example 3.5 for four electric charges allows us to write

$$\overline{\mathfrak{Q}}_M = 3b^2 q_M(\hat{\mathbf{x}}\hat{\mathbf{z}} + \hat{\mathbf{z}}\hat{\mathbf{x}}) \tag{9.469}$$

where q_M tends to infinity as $b \to 0$ so as to retrieve (9.467).

(End of Example 9.6)

A special formulation for MQS fields may be obtained when $\mathcal{J}$ takes the form of a conduction current $\mathcal{J}_c$ as given by (1.120) or (1.128) [54, Section 9.10], [15]. By starting with the curl of (9.437) we get

$$\nabla \times \nabla \times \mathcal{H}(\mathbf{r}, t) = \nabla \times \mathcal{J}(\mathbf{r}, t) = \sigma \nabla \times \mathcal{E}(\mathbf{r}, t), \qquad \mathbf{r} \in V \tag{9.470}$$

having assumed that the medium filling the domain V is an isotropic homogeneous conductor. Since the magnetic field is solenoidal and the electric field obeys the Faraday law (9.440), on account of (H.59) we find

$$\nabla^2 \mathcal{H}(\mathbf{r}, t) - \sigma\mu \frac{\partial}{\partial t} \mathcal{H}(\mathbf{r}, t) = 0, \qquad \mathbf{r} \in V \tag{9.471}$$

which has the typical form of a *diffusion equation* [8, Section G.4], [64, Section 3.4], [65, Chapter 2]. Comparison with the wave equation (1.217) for $\mathcal{H}$ in a lossy medium shows that — aside from the presence or absence of the source term $\mathcal{J}$ — in the MQS approximation the second-order time derivative of the magnetic field has been discarded. Indeed, by virtue of definition (1.209) we can retrieve (9.471) from (1.217) by letting the speed of light c in the medium grow infinitely large while keeping both σ and μ finite. Thus, here lies the well-known problem with diffusion-like equations: they imply that a 'disturbance' in the field propagates with infinite velocity.

Interestingly, in the MQS regime also the conduction current and the electric field satisfy the very same diffusion equation. Taking the curl of (1.120) and using (9.440) gives

$$\nabla \times \mathcal{J}(\mathbf{r}, t) = -\sigma\mu \frac{\partial}{\partial t} \mathcal{H}(\mathbf{r}, t) \tag{9.472}$$

but since $\mathcal{J}(\mathbf{r}, t)$ is solenoidal in consideration of (9.439), by taking the curl of $\nabla \times \mathcal{J}$ and invoking (H.59) we arrive at

$$\nabla^2 \mathcal{J}(\mathbf{r}, t) - \sigma\mu \frac{\partial}{\partial t} \mathcal{J}(\mathbf{r}, t) = 0, \qquad \mathbf{r} \in V \tag{9.473}$$

in view of (9.437). Lastly, the diffusion equation for $\mathcal{E}$ reads

$$\nabla^2 \mathcal{E}(\mathbf{r}, t) - \sigma\mu \frac{\partial}{\partial t} \mathcal{E}(\mathbf{r}, t) = 0, \qquad \mathbf{r} \in V \tag{9.474}$$

and it follows trivially from (9.473) through the Ohm law (1.120). Alternatively, also (9.474) may be derived from a wave equation, namely, (1.216), upon letting $c \to +\infty$.

Example 9.7 (Skin effect and skin depth in a conducting half space)

Although the MQS diffusion equations (9.471), (9.473) and (9.474) embody the unrealistic assumption of an infinitely large speed of light, they correctly provide the quantitative description of the *skin effect* [54, Section 9.10], [51, Chapter 20], [60, Section 6.2.4], [40, Section 20.B], [20, Section 14.10], [6]. This phenomenon (also known as the Kelvin effect) consists of the confinement of conduction currents into a thin layer right below the boundary of a conducting body. The decay of fields and current away from the surface of the conductor is *exponential*, and the decay constant (a length) is the so-called *skin depth*. In Example 7.2 we already examined the skin effect from the viewpoint of a uniform time-harmonic plane wave propagating in a homogeneous lossy half space. Here we wish to obtain the relevant result with the aid of (9.474).

With reference to Figure 7.9 again we look for time-harmonic quasi-static solutions in region V_2 in the form (Section 1.5)

$$\mathcal{E}(\mathbf{r}, t) = \text{Re}\{E_x(z)\hat{\mathbf{x}}e^{j\omega t}\}, \qquad z \geqslant 0 \tag{9.475}$$

which, when inserted into (9.474), allows writing

$$\frac{d^2 E_x}{dz^2} - j\omega\sigma_2\mu_2 E_x(z) = 0, \qquad z \geqslant 0 \tag{9.476}$$

thanks to the transformation rules for phasors that are listed in Table 1.1. If we posit a wave-like spatial dependence in the form

$$E_x(z) = E_{x0}e^{-jk_2 z} \tag{9.477}$$

then (9.476) yields the condition

$$\left(k_2^2 + j\omega\sigma_2\mu_2\right) E_{x0}e^{-jk_2 z} = 0, \qquad z \geqslant 0 \tag{9.478}$$

which can only be satisfied if

$$k_2^2 = -j\omega\sigma_2\mu_2 \tag{9.479}$$

because the complex exponential never vanishes and we must have $E_{x0} \neq 0$ in order to obtain non-trivial solutions. Finally, by choosing the square root of k_2^2 that has a negative imaginary part [32, Formula 58.3] we get

$$k_2 = \sqrt{\frac{\omega\sigma_2\mu_2}{2}}(1 - j) = \frac{1 - j}{\delta_{sd}} \tag{9.480}$$

where δ_{sd} is the *skin depth* previously defined in (7.102). The square root of k_2^2 with positive imaginary part must be discarded inasmuch as it gives rise to a solution $E_x(z)$ which grows exponentially as it travels in the conductor away from the interface $z = 0$. With this result the electric field and the conduction current in the region V_2 read

$$E_x(z) = E_{x0}e^{-jz/\delta_{sd}}e^{-z/\delta_{sd}}, \qquad J_x(z) = \sigma_2 E_{x0}e^{-jz/\delta_{sd}}e^{-z/\delta_{sd}} \tag{9.481}$$

whereas the magnetic field may be computed through the time-harmonic counterpart of (9.440). In symbols, we have

$$-j\omega\mu_2\mathbf{H}(z) = \nabla E_x(z) \times \hat{\mathbf{x}} = -jk_2 E_x(z)\hat{\mathbf{y}} \tag{9.482}$$

whence we see that $\mathbf{H}(z)$ is aligned with the y-axis. In particular

$$H_y(z) = \frac{1-\mathrm{j}}{\omega\mu_2\delta_{sd}}E_x(z) = E_{x0}\sqrt{\frac{\sigma_2}{\omega\mu_2}}\mathrm{e}^{-\mathrm{j}\pi/4}\mathrm{e}^{-\mathrm{j}z/\delta_{sd}}\mathrm{e}^{-z/\delta_{sd}} \tag{9.483}$$

which shows that the magnetic field lags behind $E_x(z)$ by a time equal to $T/8$ with $T := 2\pi/\omega$ being the period of the time-harmonic wave-form.

We observe that the complex wavenumber (7.100) is the same as (9.480). The former was obtained for a good conductor, which in the present problem means $\sigma_2 \gg \omega\varepsilon_2$, whereas the latter has been derived for slowly varying electromagnetic fields without any evident hypothesis concerning the conductivity σ_2. However, from the condition which qualifies a good conductor in V_2 we get

$$\omega\sigma_2\mu_2 \gg \omega^2\varepsilon_2\mu_2 = \frac{\omega^2}{c_2^2} \tag{9.484}$$

which we further rewrite as

$$\frac{\omega}{c_2} \ll \frac{\sqrt{2}}{\delta_{sd}} \tag{9.485}$$

on account of definition (7.102). Since the diffusion equations (9.471), (9.473) and (9.474) follow from the wave equations (Section 1.8) in the limit as the speed of light tends to infinity, when the MQS approximation for time-harmonic fields is valid, the ratio of the angular frequency to the velocity c_2 is certainly small enough for (9.485) to be true. But then, we have derived (9.485) under the hypothesis that the medium in V_2 is a good conductor, so the two requirements are equivalent.

(End of Example 9.7)

Example 9.8 (Skin effect and skin depth in a circular cylindrical conductor)

Another geometry amenable to a closed-form solution of (9.473) for the study of the skin effect is represented by an infinitely long cylindrical conductor with circular cross section. The conductor has radius a, conductivity σ, permeability μ_0, and is aligned with the z-axis. By assuming that sources and fields are time-harmonic we seek solutions for the phasor $\mathbf{J}(\mathbf{r})$ in the form

$$\mathbf{J}(\mathbf{r}) = J_z(\tau)\hat{\mathbf{z}}, \qquad \tau \leqslant a \tag{9.486}$$

where τ denotes the radial coordinate, i.e., the distance from the z-axis. Then, with the aid of (A.41) we write the diffusion equation (9.473) in cylindrical coordinates as

$$\frac{1}{\tau}\frac{\partial}{\partial\tau}\left(\tau\frac{\partial J_z}{\partial\tau}\right) - \underbrace{\mathrm{j}\,\omega\sigma\mu_0}_{-k^2}\,J_z(\tau) = 0 \tag{9.487}$$

which can be turned into the standard Bessel equation (H.136) of order $\nu = 0$ [43,66] with the change of independent variable $\xi = k\tau$, where

$$k := \sqrt{-\mathrm{j}\,\omega\sigma\mu_0} = \sqrt{\frac{\omega\sigma\mu_0}{2}}(1-\mathrm{j}) = \frac{1-\mathrm{j}}{\delta_{sd}} \tag{9.488}$$

constitutes the radial wavenumber.

The general non-singular solution to (9.487) reads

$$J_z(\tau) = AJ_0(k\tau), \qquad \tau \leqslant a \tag{9.489}$$

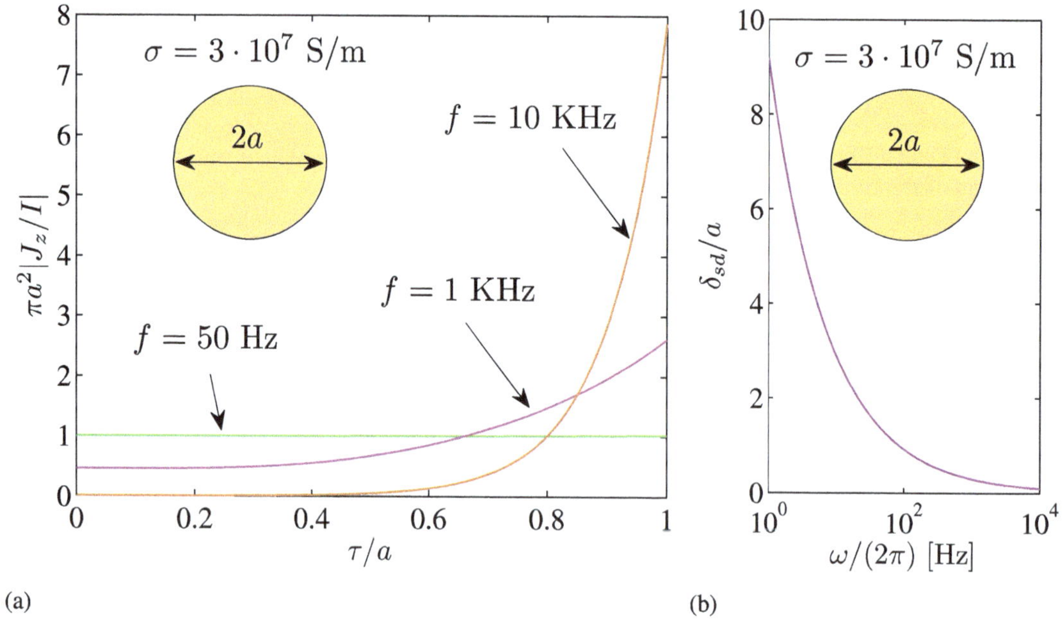

Figure 9.22 *Skin effect in a wire of circular cross section: (a) normalized current density versus normalized radial coordinate for various frequencies; (b) skin depth versus frequency.*

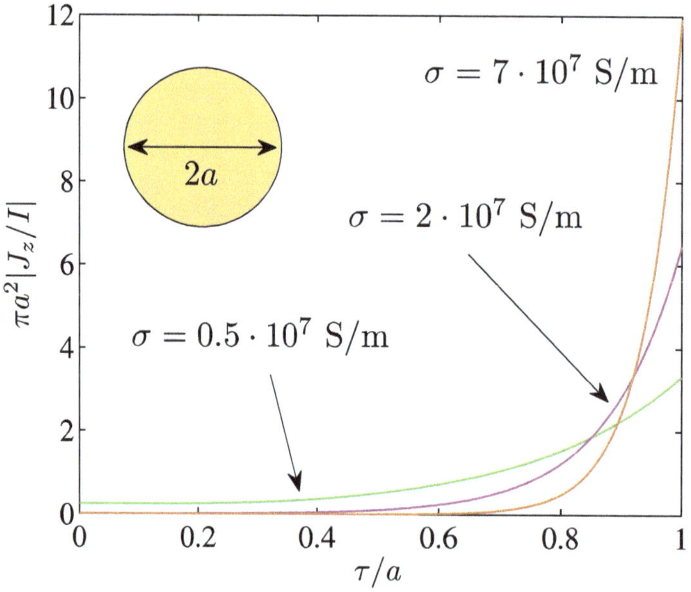

Figure 9.23 *Skin effect in a wire of circular cross section: normalized current density versus normalized radial coordinate for various conductivities at f = 10 KHz.*

where $A \in \mathbb{C}$ is an arbitrary constant and $J_0(\bullet)$ denotes the Bessel function of the first kind and order zero. If the current I which flows in the conductor is known, we can set the constant A by requiring that

$$I = \int_{\tau \leqslant a} \mathrm{d}S\, \hat{\mathbf{z}} \cdot \mathbf{J}(\tau) = 2\pi A \int_0^a \mathrm{d}\tau\, \tau J_0(k\tau) = 2\pi \frac{A}{k^2} \int_0^{ka} \mathrm{d}\xi\, \xi J_0(\xi) = 2\pi \frac{A}{k^2} ka J_1(ka) \tag{9.490}$$

having made use of the indefinite integral (H.172) with $n = 0$. Solving for A and substituting the result back into (9.489) yields

$$J_z(\tau) = \frac{Ik}{2\pi a} \frac{J_0(k\tau)}{J_1(ka)} = \underbrace{\frac{I}{\pi a^2}}_{\substack{\text{uniform} \\ \text{current density}}} \frac{ka}{2} \frac{J_0(k\tau)}{J_1(ka)}, \qquad \tau \leqslant a \tag{9.491}$$

a formula that shows how the uniform distribution of steady current is modified by the skin effect as the frequency (hidden in the complex wavenumber k) is increased. Indeed, the correction factor tends to 1 in the limit as $k \to 0$, as can be ascertained with the aid of the asymptotic expansion (H.152).

The current density is 'pushed' ever more towards the surface of the wire as the angular frequency ω becomes larger because the skin depth (7.102) diminishes accordingly. The phenomenon is illustrated quantitatively in Figure 9.22a for $\sigma = 3 \cdot 10^7$ S/m and various frequencies, and the relevant skin depth is plotted in Figure 9.22b. For larger values of the conductivity the decay of the current away from the boundary of the wire becomes steeper. This is shown in Figure 9.23 at $f = 10$ KHz for various conductivities. From Figures 9.22a and 9.23 we gather that at $f = 10$ KHz about 60 per cent of the conductor's section is devoid of current for all practical purposes.

(End of Example 9.8)

When a conducting body is either exposed to a time-varying induction field $\mathcal{B}(\mathbf{r}, t)$ or forced to move through a region of space where a static magnetic field $\mathcal{B}_0$ exists, electric currents appear within the metal, because the free electrons in the conductor can swirl around the streamlines of the impressed magnetic field (Figure 9.24). The motion of the electrons can be chalked up either to the electric field associated with $\mathcal{B}(\mathbf{r}, t)$ in accordance with the Faraday law (1.20) or, in case of slow motion of the conductor, to the electromotive force $\mathcal{E}(t)$ generated by a time-varying magnetic flux $\Psi_B(t)$, as prescribed by the kinematic form of the Faraday law (1.56) or the flux rule (1.61).

If the conductor of concern is not part of a closed electric circuit (cf. Examples 1.1 and 1.2) the induced currents tend to flow in loops that lie in planes perpendicular to the generating magnetic field. For this reason and also for the resemblance they bear to whirlpools created by an oar on the water surface, such currents are termed *eddy* and are sometimes named after the French physicist Léon Foucalt [64, Section 16-3], [20, Section 14.12], [53, Section 8-12]. In his experiment (1855) Foucalt noticed that a larger force was necessary in order to rotate a copper disk partly immersed in the static magnetic field produced by a permanent magnet. This happens because, according to the Lenz rule [13,55], an induced current generates in turn a magnetic field that hinders the variation of the one which created the current in the first place, in an attempt to restore the situation where no current was present. Alternatively, one may notice that also the eddy currents experience the action of the Lorentz force (1.4) which invariably tries to thwart the current flow.

In electrical machinery and components, such as transformers, eddy currents are unwanted because they give rise to additional ohmic losses, and these in turn negatively affect efficiency and

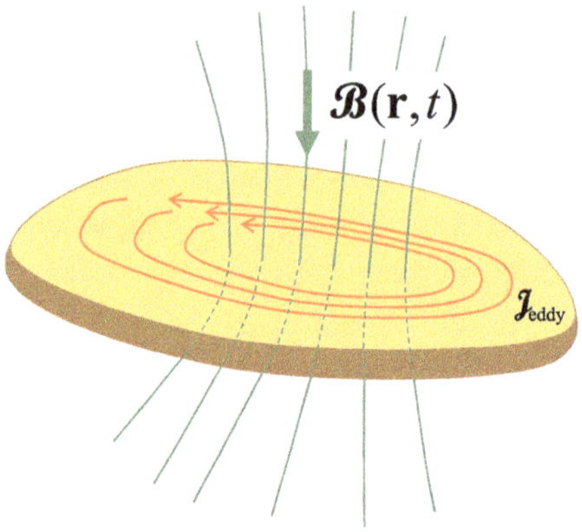

Figure 9.24 Generation of eddy currents in a conducting body exposed to a time-varying magnetic induction field.

performance, not least because the temperature of the conducting parts increases. Which is why eddy currents are used in induction heating, e.g., to heat up and melt metals with magnetic inductions the order of 100 μT and frequencies ranging from 50 Hz up to 1 MHz. Further, since eddy currents tend to counter the motion of the body where they occur, they find application in specially devised braking systems.

When condition (9.485) is met, we may study eddy currents with the magneto-quasi-static approximation and the trio of diffusion equations (9.471), (9.473) and (9.474), possibly turned into the frequency domain [67,68], [69, Section 4.4].

Example 9.9 (Induction heating of a circular cylindrical metallic bar)

An infinitely long metallic bar with finite conductivity σ and circular cross section of radius a is immersed in a uniform slowly-varying time-harmonic magnetic field aligned with the bar. In order to study the induced eddy currents we introduce a system of circular cylindrical coordinates (τ, φ, z) with the z-axis coinciding with the axis of the bar, whereby the phasor of the impressed magnetic field is $\mathbf{H}^{\mathrm{i}} = H_0\hat{\mathbf{z}}$.

The onset of eddy currents is governed by the subset of Maxwell equations

$$\nabla \times \mathbf{H}(\tau) = \mathbf{J}(\tau) = \sigma \mathbf{E}(\tau) \tag{9.492}$$

$$\nabla \times \mathbf{E}(\tau) = -\mathrm{j}\,\omega\mu_0 \mathbf{H}(\tau) - \mathrm{j}\,\omega\mu_0 \mathbf{H}^{\mathrm{i}} \tag{9.493}$$

for $\tau \leqslant a$ and

$$\nabla \times \mathbf{H}(\tau) = 0 \tag{9.494}$$

$$\nabla \times \mathbf{E}(\tau) = -\mathrm{j}\,\omega\mu_0 \mathbf{H}(\tau) - \mathrm{j}\,\omega\mu_0 \mathbf{H}^{\mathrm{i}} \tag{9.495}$$

for $\tau > a$, plus boundary conditions in $\tau \in \{0, a, +\infty\}$. In light of the axial and longitudinal symmetry of the problem all fields and currents are independent of φ and z, and the only non-zero components are E_φ, H_z and J_φ. Under the MQS hypothesis we neglect the impressed electric field associated with $\mathbf{H}^{\mathrm{i}}$, whereby E_φ represents the secondary field produced by $\mathbf{H}^{\mathrm{i}}$, and H_z is the secondary magnetic field in turn produced by $J_\varphi = \sigma E_\varphi$.

From (9.494) and (A.33) we gather that $H_z(\tau)$ must be constant outside the bar. If we also require $H_z(\tau)$ to vanish at infinity, then we must take $H_z(\tau) = 0$ for $\tau > a$. In fact, the situation is similar to that of Example 4.2 where we found a null magnetic field outside an infinitely long solenoid. The *total* magnetic field for $\tau > a$ is not zero, though, and equals H_0, the impressed part.

Taking the curl of (9.493) and using (9.492) to eliminate $\mathbf{H}(\tau)$ yields the following diffusion equation for E_φ inside the bar

$$\frac{1}{\tau}\frac{\mathrm{d}}{\mathrm{d}\tau}\left(\tau\frac{\mathrm{d}E_\varphi}{\mathrm{d}\tau}\right) - \mathrm{j}\,\omega\mu\sigma E_\varphi - \frac{E_\varphi}{\tau^2} = 0, \qquad \tau \leqslant a \tag{9.496}$$

by virtue of (A.44) and the fact that $\mathbf{H}^{\mathrm{i}}$ *per se* is approximately curl-free by assumption. By letting $k^2 = -\mathrm{j}\,\omega\mu_0\sigma$ and then $\xi = k\tau$ (9.496) is readily turned into the Bessel equation (H.136) of order $\nu = 1$, and hence the solution reads

$$E_\varphi(\tau) = AJ_1(k\tau) + BY_1(k\tau), \qquad \tau \leqslant a \tag{9.497}$$

where A and B are constants to be determined. However, we must exclude the contribution proportional to $Y_1(k\tau)$ by setting $B = 0$ inasmuch as the electric field is finite in $\tau = 0$.

The total magnetic field within the bar amounts to the sum of H_0 and H_z and can be retrieved from the Faraday law (9.493), viz.,

$$-\mathrm{j}\,\omega\mu_0(H_z + H_0) = \hat{\mathbf{z}}\cdot\nabla\times\mathbf{E} = \frac{1}{\tau}\frac{\mathrm{d}}{\mathrm{d}\tau}(\tau E_\varphi) = AkJ_0(k\tau), \qquad \tau \leqslant a \tag{9.498}$$

on account of (9.497), (A.33) and the recursion formula (H.166) with $\nu = 1$. Now, the z-component of the magnetic field — parallel to the boundary of the bar — must be continuous in $\tau = a$, and since the magnetic field for $\tau > a$ is just $H_0\hat{\mathbf{z}}$, as argued above, we require

$$H_z(a) + H_0 = -A\frac{kJ_0(ka)}{\mathrm{j}\,\omega\mu_0} = H_0 \tag{9.499}$$

whence we find

$$A = -\frac{\mathrm{j}\,\omega\mu_0}{kJ_0(ka)}H_0. \tag{9.500}$$

Inserting this expression back into (9.497) yields the electric field and the eddy current density, viz.,

$$E_\varphi(\tau) = -\frac{\mathrm{j}\,\omega\mu_0}{k}\frac{J_1(k\tau)}{J_0(ka)}H_0 = \frac{1-\mathrm{j}}{\sigma\delta_{sd}}\frac{J_1(k\tau)}{J_0(ka)}H_0, \qquad \tau \leqslant a \tag{9.501}$$

$$J_\varphi(\tau) = k\frac{J_1(k\tau)}{J_0(ka)}H_0 = \frac{1-\mathrm{j}}{\delta_{sd}}\frac{J_1(k\tau)}{J_0(ka)}H_0, \qquad \tau \leqslant a \tag{9.502}$$

with δ_{sd} indicating the skin depth previously defined in (7.102).

For the sake of completeness we also derive the electric field outside the bar. By taking the curl of (9.495) and recalling that the total magnetic field is lamellar for $\tau > a$ we obtain

$$\frac{1}{\tau}\frac{\mathrm{d}}{\mathrm{d}\tau}\left(\tau\frac{\mathrm{d}E_\varphi}{\mathrm{d}\tau}\right) - \frac{E_\varphi}{\tau^2} = 0, \qquad \tau > a \tag{9.503}$$

which has the solution

$$E_\varphi = C\tau + \frac{D}{\tau} = \frac{1-\mathrm{j}}{\sigma\delta_{sd}}\frac{J_1(ka)}{J_0(ka)}\frac{a}{\tau}H_0, \qquad \tau > a \tag{9.504}$$

where the constant C must be taken equal to zero because the electric field is regular at infinity, and the constant D is set by the matching condition (1.197) for $\tau = a$.

Despite the approximations made to arrive at (9.502), the current density correctly reduces to zero if the bar is made of a lossless dielectric, in which case $k = 0$. Perhaps it is not obvious that the current vanishes also for a PEC bar. To check this out we turn to the asymptotic expansion of the Bessel functions for large values of the argument, since $1/\delta_{sd}$ tends to infinity as $\sigma \to +\infty$. With the aid of (H.158) we find

$$\begin{aligned}\lim_{\sigma\to+\infty} k\frac{J_1(k\tau)}{J_0(ka)} &= \lim_{1/\delta_{sd}\to+\infty} \frac{1-\mathrm{j}}{\delta_{sd}}\sqrt{\frac{a}{\tau}}\frac{\sin(k\tau-\pi/4)}{\cos(ka-\pi/4)} \\ &= -(1+\mathrm{j})\sqrt{\frac{a}{\tau}}\lim_{1/\delta_{sd}\to+\infty}\frac{1-\mathrm{j}\,\mathrm{e}^{-2(1+\mathrm{j})\tau/\delta_{sd}}}{1+\mathrm{j}\,\mathrm{e}^{-2(1+\mathrm{j})a/\delta_{sd}}}\frac{1/\delta_{sd}}{\mathrm{e}^{(1+\mathrm{j})(a-\tau)/\delta_{sd}}} = 0. \end{aligned} \tag{9.505}$$

The average power dissipated by the eddy currents per unit length (that is, along the z-direction) can be computed with the aid of definition (1.308) applied to a portion of the bar bounded by the planes $z = z_1$ and $z = z_2$, i.e.,

$$\langle P_C\rangle = \frac{1}{2}\int_{z_1}^{z_2}\mathrm{d}z\int_0^{2\pi}\mathrm{d}\varphi\int_0^a \mathrm{d}\tau\,\tau\sigma|E_\varphi(\tau)|^2 = \int_{z_1}^{z_2}\mathrm{d}z\,p_C \tag{9.506}$$

whereby we have

$$\begin{aligned} p_C = \pi\sigma\int_0^a \mathrm{d}\tau\,\tau|E_\varphi(\tau)|^2 &= \frac{\pi\omega\mu_0}{|J_0(ka)|^2}|H_0|^2\int_0^a \mathrm{d}\tau\,\tau J_1(k\tau)J_1(k^*\tau) \\ &= \frac{\pi\omega\mu_0}{|J_0(ka)|^2}a\frac{k^*J_1(ka)J_0(k^*a)-kJ_1(k^*a)J_0(ka)}{k^2-(k^*)^2}|H_0|^2 \\ &= \frac{\pi a\omega\mu_0}{|J_0(ka)|^2}\frac{-2\,\mathrm{j}\,\mathrm{Im}\{kJ_1(k^*a)J_0(ka)\}}{-2\,\mathrm{j}\,\omega\mu_0\sigma}|H_0|^2 \\ &= \frac{\pi a}{\sigma\delta_{sd}}\frac{\mathrm{Im}\{(1-\mathrm{j})J_1(k^*a)J_0(ka)\}}{|J_0(ka)|^2}|H_0|^2 \end{aligned} \tag{9.507}$$

thanks to the integral formula (H.174).

(End of Example 9.9)

References

[1] Rothwell EJ, Cloud MJ. Electromagnetics. London, UK: CRC Press; 2001.

[2] Stratton JA. Electromagnetic theory. London, UK: McGraw-Hill; 1941.

[3] Hopcraft KI, Smith PR. An Introduction to Electromagnetic Inverse Scattering. Dordrecht, NL: Springer Science+Business Media; 1992.

[4] Colton DL, Kress R. Inverse Acoustic and Electromagnetic Scattering Theory. 3rd ed. New York, NY: Springer; 2013.

[5] Van Bladel JG. Electromagnetic Fields. Piscataway, NJ: IEEE Press; 2007.

[6] Orfanidis SJ. Electromagnetic Waves and Antennas. www.ece.rutgers.edu/~orfanidi/ewa; 2004.

[7] Anderson N. The Electromagnetic Field. New York, NY: Springer Science+Business Media; 1968.
[8] Konopinski EJ. Electromagnetic Fields and Relativistic Particles. New York, NY: McGraw Hill; 1981.
[9] Biggs HF. The Electromagnetic Field. London, UK: Oxford University Press; 1934.
[10] Mason M, Weaver W. The electromagnetic field. New York, NY: Dover Publications, Inc.; 1929.
[11] Milton KA, Schwinger J. Electromagnetic Radiation: Variational Methods, Waveguides and Accelerators. Berlin Heidelberg: Springer-Verlag; 2006.
[12] Slater JC, Frank NH. Electromagnetism. New York, NY: McGraw-Hill; 1947.
[13] Jackson JD. Classical Electrodynamics. 3rd ed. Chichester, UK: Wiley; 1999.
[14] Griffith DJ. Introduction to electrodynamics. 3rd ed. Upper Saddle River, NJ: Prentice Hall; 1999.
[15] Bassanini P, Elcrat A. Mathematical Theory of Electromagnetism. Creative Commons; 2009.
[16] Schwinger J, De Raad LL, Milton KA, et al. Classical electrodynamics. Perseus Books; 1998.
[17] Pauli W. Theory of relativity. New York, NY: Dover Publications, Inc.; 1981.
[18] Rudin W. Principles of mathematical analysis. 3rd ed. New York, NY: McGraw-Hill; 1976.
[19] Van Bladel JG. Relativity and Engineering. Berlin Heidelberg : Springer Berlin Heidelberg; 1984.
[20] Zangwill A. Modern electrodynamics. Cambridge, UK: Cambrigde University Press; 2013.
[21] Thidé B. Electromagnetic Field Theory. New York, NY: Dover Publications, Inc.; 2011.
[22] Felsen LB, Marcuvitz N. Radiation and scattering of waves. Piscataway, NJ: IEEE Press; 2001.
[23] Kong JA. Electromagnetic Wave Theory. 2nd ed. New York, NY: Wiley; 1990.
[24] Moon P, Spencer DE. Field Theory Handbook: Including Coordinate Systems, Differential Equations and Their Solutions. 2nd ed. Berlin Heidelberg: Springer-Verlag; 1971.
[25] Hosono T. Foundation of Electromagnetic Waves. Tokyo, Japan: Shokodo; 1973.
[26] Tai CT. Dyadic Green's functions in electromagnetic theory. 2nd ed. Piscataway, NJ: IEEE Press; 1994.
[27] Harrington RF. Time-harmonic Electromagnetic Fields. London, UK: McGraw-Hill; 1961.
[28] Lindell IV. Methods for electromagnetic field analysis. Piscataway, NJ: IEEE Press; 1992.
[29] Van Bladel JG. Singular Electromagnetic Fields and Sources. Piscataway, NJ: IEEE Press; 1991.
[30] Lee SW, Boersma J, Law CL, et al. Singularity in Green's function and its numerical evaluation. IEEE Transactions on Antennas and Propagation. 1980 May;28(3):311–317.
[31] Bassanini P, Elcrat AR. Theory and applications of partial differential equations. Mathematical Concepts And Methods In Science And Engineering. New York, NY: Plenum Press; 1997.
[32] Dwight HB. Tables of Integrals and Other Mathematical Data. 4th ed. New York, NY: MacMillan Publishing Co., Inc.; 1961.
[33] Faryad M, Lakhtakia A. Infinite-Space Dyadic Green Functions in Electromagnetism. Electromagnetics and Metamaterials. San Rafael, CA: Morgan & Claypool Publishers; 2018. Available from: http://dx.doi.org/10.1088/978-1-6817-4557-2.
[34] Yaghjian AD. Electric dyadic Green's functions in the source region. Proceedings of the IEEE. 1980 Feb;68(2):248–263.
[35] Wilcox CH. Debye potentials. J Math Mech. 1957 March;6.

[36] Avelin J, Arslan AN, *et al* JB. Electric fields in the source region: the depolarization dyadic for a cubic cavity. Electrical Engineering. 1998 November;81(4):199–202.
[37] Bladel JV. Some remarks on Green's dyadic for infinite space. IRE Transactions on Antennas and Propagation. 1961 Nov;9:563–566.
[38] Muscat J. Functional Analysis. London, UK: Springer; 2014.
[39] Balanis CA. Antenna Theory: Analysis and Design. 2nd ed. New York, NY: John Wiley & Sons, Inc.; 1997.
[40] Sommerfeld A. Electrodynamics. vol. 3 of Lectures on theoretical physics. New York, NY: Academic Press; 1952.
[41] Dobbs R. Electromagnetic Waves. London, UK: Routledge & Kegan Paul; 1985.
[42] Feynmann RP, Leighton R, Sands M. The Feynmann lectures on physics. vol. 1. Reading, MA: Addison-Wesley; 1964.
[43] Abramowitz M, Stegun IA. Handbook of mathematical functions. New York, NY: Dover Publications, Inc.; 1965.
[44] Collin RE. Foundations for Microwave Engineering. New York, NY: McGraw-Hill; 1992.
[45] Pozar D. Microwave Engineering. 4th ed. New York, NY: John Wiley & Sons, Inc.; 2012.
[46] de Hoop AT. Handbook of radiation and scattering of waves. London, UK: Academic Press; 1995.
[47] King RWP. The Theory of Linear Antennas. Cambridge, MA: Harvard University Press; 1956.
[48] Landau LD, Lifshitz EM. The classical theory of fields. 3rd ed. Oxford, UK: Pergamon Press; 1971.
[49] Visser HJ. Array and Phased Array Antenna Basics. New York, NY: Wiley-Interscience; 2005.
[50] Adler RB, Chu LJ, Fano RM. Electromagnetic energy transmission and radiation. New York, NY: John Wiley & Sons, Inc.; 1960.
[51] Popović Z, Popović BD. Introductory Electromagnetics. Upper Saddle River, NJ: Prentice Hall; 2000.
[52] Hayt WH, Buck JA. Engineering Electromagnetics. 8th ed. New York, NY: McGraw-Hill; 2012. International edition.
[53] Kraus JD, Carver KR. Electromagnetics. 2nd ed. Tokyo, Japan: McGraw-Hill; 1973.
[54] Guru BS, Hiziroglu HR. Electromagnetic field theory fundamentals. 2nd ed. New York, NY: Cambridge University Press; 2004.
[55] Holt CA. Introduction to Electromagnetic Fields and Waves. New York, NY: John Wiley & Sons, Inc.; 1963.
[56] Peterson AF, Ray SL, Mittra R. Computational Methods for Electromagnetics. Piscataway, NJ: IEEE Press; 1998.
[57] Harrington RF. Field Computation by Moment Methods. New York, NY: MacMillan Publishing Co., Inc.; 1968.
[58] Fikioris G. Mellin-Transform Method for Integral Evaluation. San Rafael, CA: Morgan & Claypool Publishers; 2007.
[59] Makarov B, Podkorytov A. Real Analysis: Measures, Integrals and Applications. Universitext. London, UK: Springer; 2013.
[60] Schwab AJ. Field Theory Concepts. Berlin Heidelberg: Springer-Verlag; 1988.
[61] Shen LC, Kong JA. Applied electromagnetism. Monterey, CA: Brooks/Cole; 1983.
[62] Fano RM, Chu LJ, Adler RB. Electromagnetic Fields, Energy, and Forces. New York, NY: Wiley; 1960.
[63] Penfield P, Haus H. Electrodynamics of Moving Media. Cambridge, MA: MIT Press; 1967.

[64] Feynmann RP, Leighton R, Sands M. The Feynmann lectures on physics. vol. 2. Reading, MA: Addison-Wesley; 1964.
[65] Morse P, Feshbach H. Methods of Theoretical Physics. New York, NY: McGraw-Hill; 1953.
[66] Watson GN. A Treatise on the Theory of Bessel Functions. 2nd ed. Cambridge Mathematical Library. Cambridge, UK: Cambridge University Press; 1995.
[67] Stoll RL. The analysis of eddy currents. London, UK: Oxford University Press; 1974.
[68] Kalaichelvan S. Boundary element methods for eddy current problems. In: Brebbia CA, editor. Topics in boundary elements research. vol. 6. Berlin: Springer-Verlag; 1989. p. 78–117.
[69] Binns KJ, Lawrenson PJ, Trowbridge CW. The analytical and numerical solution of electric and magnetic fields. Chichester: John Wiley & Sons, Inc.; 1992.

Index

www.ingramcontent.com/pod-product-compliance
Lightning Source LLC
LaVergne TN
LVHW062127200926
848484LV00007B/1778

9781839535642